P9-EDG-896

ELECTRICAL MOTOR CONTROLS

for Integrated Systems

Third Edition

S MEADOWS

Gary J. Rockis
Glen A. Mazur

AMERICAN TECHNICAL PUBLISHERS, INC.
HOMEWOOD, ILLINOIS 60430-4600

American Technical Publishers, Inc. Editorial Staff

Editor in Chief:
Jonathan F. Gosse

Production Manager:
Peter A. Zurlis

Technical Editors:
Peter A. Zurlis
James T. Gresens

Copy Editor:
Richard S. Stein

Cover Design:
Carl R. Hansen

Illustration/Layout:
Thomas E. Zabinski
Maria R. Aviles
Jennifer M. Hines
Gianna C. Butterfield
Sarah E. Kaducak

CD-ROM Development:
Carl R. Hansen
Gianna C. Butterfield
Sarah E. Kaducak

3 4 5 6 7 8 9 – 05 – 9 8 7 6 5 4 3 2

Printed in the United States of America

ISBN 0-8269-1207-9

Acknowledgments

The authors and publisher are grateful for the technical information and assistance provided by the following companies and organizations.

- ABB Power T&D Company Inc.
- Advanced Assembly Automation, Inc.
- AEMC Instruments
- Atlas Technologies Inc.
- Bacharach, Inc
- Baldor Electric Co.
- Banner Engineering Corp.
- Bergey Windpower Co., Inc.
- Boeing Commercial Airplane Group
- Bunn-O-Matic
- Carlo Gavazzi Inc. Electromatic Business Unit
- Cincinnati Milacron
- Cummins Power Generation
- Cutler-Hammer
- DoALL Company
- Eagle Signal Industrial Controls
- Electrical Apparatus Service Association, Inc.
- FANUC Robotics North America
- Fluke Corporation
- The Foxboro Company
- Furnas Electric Co.
- GE Motors & Industrial Systems
- General Electric Company
- Giddings & Lewis, Inc.
- Grayhill Inc.
- Greenlee Textron, Inc.
- Guardian Electric Mfg. Co.
- Handheld Products
- Heidelberg Harris, Inc.
- Honeywell
- Ideal Industries, Inc.
- Ircon, Inc.
- Justrite Manufacturing Company
- Klein Tools, Inc.
- Leeson Electric Corporation
- March Manufacturing, Inc.
- Milwaukee Electric Tool Corporation
- Namco Controls Corporation
- North American Industries, Inc.
- Omron Electronics
- Pandjiris, Inc.
- Panduit Corp.
- Products Unlimited
- Ridge Tool Company
- Rockwell Automation, Allen-Bradley Company, Inc.
- Rofin Sinar
- Ruud Lighting, Inc.
- Siemens Corporation
- The Sinco Group, Inc.
- Sonin Incorporated
- SPM Instrument, Inc.
- Sprecher + Schuh
- Square D Company
- SSAC Inc.
- The Stanley Works
- UE Systems, Inc.

Contents

Introduction

Electrical Motor Controls for Integrated Systems, Third Edition, is the industry-leading reference that covers electrical, motor, and mechanical devices and their use in industrial control circuits. This text provides the architecture for acquiring the knowledge and skills required in an advanced manufacturing environment. This environment integrates mechanical, electrical, and fluid power systems which has expanded the skill set needs of today's worker.

The text begins with basic electrical and motor theory, builds on circuit fundamentals, and reinforces comprehension through examples of industrial applications. Special emphasis is placed on the development of troubleshooting skills throughout the text. This text is a practical resource for technicians working in electrical, maintenance, manufacturing, industrial, boiler, and HVAC operations who have had some background in electrical theory.

This new edition is the product of an extensive revision and reorganization effort and features detailed illustrations, descriptive photographs, and informative factoids that supplement the concise text.

Expanded content in this edition includes the following:

- Electrical test tools and electrical test instruments
- Electrical safety
- Motor control wiring methods
- System integration
- Solid-state motor starters
- Programmable logic controllers
- Dynamic braking for electric motor drives
- Troubleshooting drive and motor circuits
- Preventive maintenance strategies and management systems

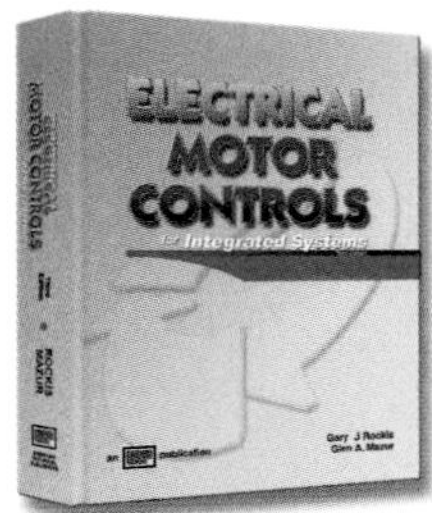

In addition to expanded coverage of new technology, the text was reorganized to present electrical, motor, and motor circuit concepts for greater clarity and efficiency. For example, solenoid, magnetism, DC generators, and DC motor material is now presented in Chapter 6. Chapter 7 now includes AC generators, transformers, and AC motors. Manual and magnetic contactor and motor starter material has been combined and included in Chapter 8.

Each chapter includes review questions to reinforce key concepts. Answers to the odd-numbered review questions are located in the back of this textbook. Answers to the even-numbered review questions are located in the Answer Key. The Appendix contains many useful tables, charts, and formulas. The Glossary defines over 500 technical terms related to electrical theory, motors, motor control circuits, and control devices.

The CD-ROM in the back of the book includes Chapter Quick Quizzes™, an Illustrated Glossary, and other related motor control material. Information about using the *Electrical Motor Controls* CD-ROM is included on the last page of the text.

The Publisher

Features

Chapter introductions provide an overview of key content found in the chapter.

Technical factoids provide informative facts related to the topic.

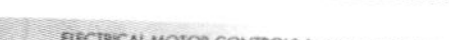

Industrial application photos supplement text and illustrations.

Detailed illustrations depict control and power circuits commonly found in industry.

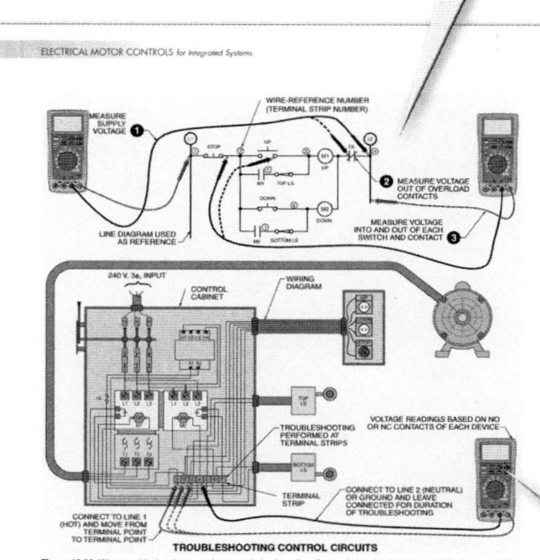

Industrial test instruments are shown in common troubleshooting applications.

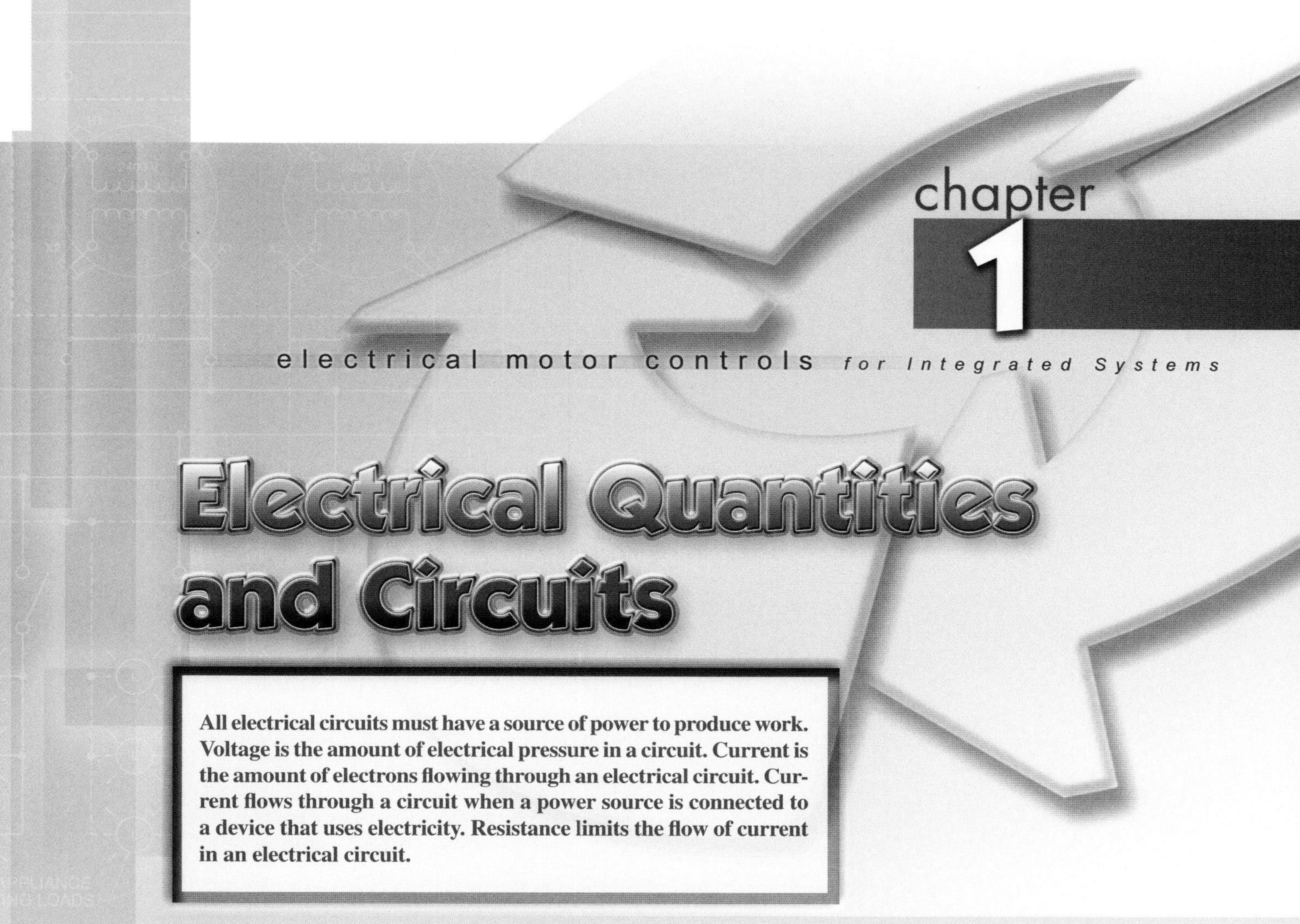

Electrical Quantities and Circuits

All electrical circuits must have a source of power to produce work. Voltage is the amount of electrical pressure in a circuit. Current is the amount of electrons flowing through an electrical circuit. Current flows through a circuit when a power source is connected to a device that uses electricity. Resistance limits the flow of current in an electrical circuit.

ENERGY

Energy is used to produce electricity. *Energy* is the capacity to do work. The two forms of energy are potential energy and kinetic energy. *Potential energy* is stored energy a body has due to its position, chemical state, or condition. For example, water behind a dam has potential energy because of its position. A battery has potential energy based on its chemical state. A compressed spring has potential energy because of its physical condition.

Kinetic energy is the energy of motion. Examples of kinetic energy include falling water, a rotating motor, or a released spring. Kinetic energy is released potential energy. Energy released when water falls through a dam is used to generate electricity. Energy released when a motor is connected to a battery is used to produce a rotating mechanical force. Energy released by a compressed spring is used to apply a braking force on a motor shaft.

The sources of energy used to produce electricity are coal, nuclear power, natural gas, and oil. Wind, solar power, and water also provide energy. These energy sources are used to produce work when converted to electricity, steam, heat, and mechanical force. Some energy sources, such as coal, oil, and natural gas, are consumed in use. Energy sources such as wind, solar power, and water are not consumed in use. **See Figure 1-1.**

Coal is used to produce approximately 50% of the electricity produced, nuclear power approximately 20%, natural gas approximately 18%, and oil approximately 3%. Wind, solar power, and water account for approximately 9% of the electricity produced. Wind and solar power are growing as sources of electricity.

Electricity is converted into motion, light, heat, sound, and visual outputs. Approximately 62% of all electricity is converted into rotary motion by motors. Three-phase motors use the largest amount of electricity in commercial and industrial applications. Three-phase motors are used because they are the most energy-efficient motors. **See Figure 1-2.**

ENERGY SOURCES			
Source	**Percent***	**Potential Energy → Electrical Energy**	**Description**
Coal	50%		FOSSIL FUEL → HEAT FROM COMBUSTION → STEAM → MECHANICAL FORCE DRIVES GENERATOR
Nuclear Power	20%		HEAT FROM NUCLEAR FISSION → STEAM → MECHANICAL FORCE DRIVES GENERATOR
Natural Gas	18%	 DANGER Natural Gas Pipeline	FOSSIL FUEL → HEAT FROM COMBUSTION → STEAM → MECHANICAL FORCE DRIVES GENERATOR
Oil	3%		FOSSIL FUEL → HEAT FROM COMBUSTION → STEAM → MECHANICAL FORCE DRIVES GENERATOR
Water (Hydro-electric) **Other (Wind/ Solar/ Water)**	9%		WIND: BLOWING WIND → MECHANICAL FORCE DRIVES GENERATOR SOLAR: LIGHT → ELECTRICAL ENERGY PRODUCED FALLING WATER → MECHANICAL FORCE DRIVES GENERATOR

* of total power produced

Figure 1-1. The forms of energy used to produce electricity include coal, nuclear power, natural gas, and oil.

Approximately 20% of all electricity is converted into light by lamps. The most common lamp used in residential lighting is the incandescent lamp. The most common lamps used in commercial and industrial lighting are fluorescent lamps for office installations and high-intensity discharge (HID) lamps for warehouse and factory installations. HID lamps include low-pressure sodium, mercury-vapor, metal-halide, and high-pressure sodium lamps. HID lamps are also the most common lamps used for exterior lighting applications.

Approximately 18% of all electricity is used to produce heat, linear motion, audible signals, and visual outputs. When the total number of individual electrical loads is considered, this group is the largest group of electricity-using components because it includes a large number of loads that consume very little power compared to motors.

Technical Fact

Coal, oil, and natural gas are called fossil fuels because they developed from the fossilized remains of prehistoric plants and animals.

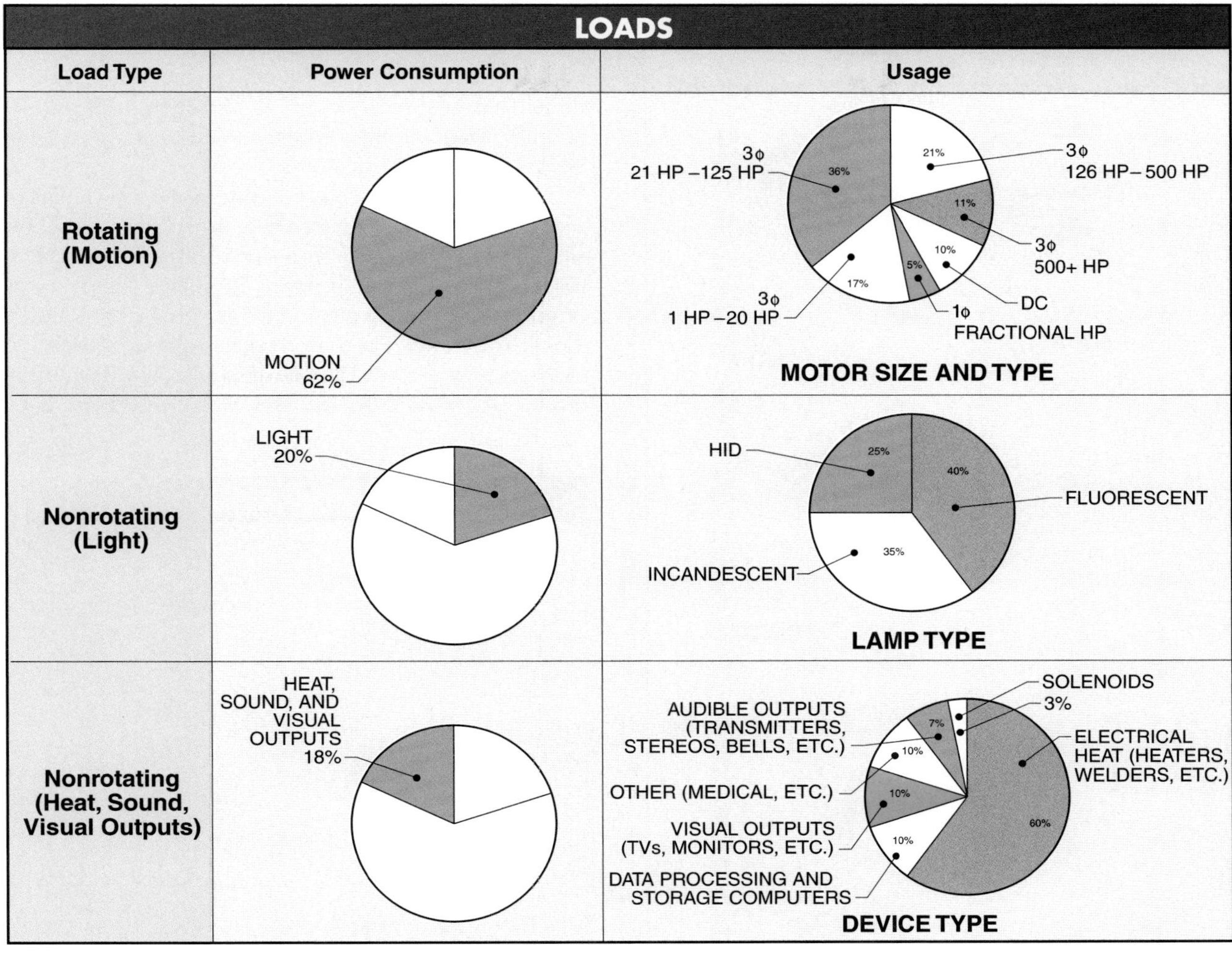

Figure 1-2. Electrical energy is used to produce motion, light, heat, sound, and visual outputs.

Production of Electricity

Electricity is produced by converting potential energy directly or indirectly into electricity. For example, solar cells convert solar energy directly into electricity. The majority of all electricity is produced indirectly by converting potential energy into electricity using a generator. A *generator (alternator)* is a device that converts mechanical energy into electrical energy by means of electromagnetic induction.

A generator produces electricity when magnetic lines of force are cut by a rotating wire coil (rotor). The magnetic lines of force are produced by the magnetic field present between the north and south poles of a permanent magnet or electromagnet (stator).

As the rotor rotates through the magnetic field, electric current flow is produced through the wire coil(s) of the rotor. **See Figure 1-3.** Electric current from the wire coil is conducted to the load through slip rings. The voltage produced by a generator depends on the strength of the magnetic field and the rotational speed of the rotor. The stronger the magnetic lines of force and the faster the rotational speed, the higher the voltage produced.

The output of a generator may be connected directly to the load, as in a portable generator located on a construction site; connected to transformers; or connected to a rectifier, as in an automobile alternator. In large generator applications, the generator output is connected to transformers. A *transformer* is an electric device that uses electromagnetism to change voltage from one level to another or to isolate one voltage from another. Transformers normally step up voltage so power can be transmitted at a lower current level. As AC voltage is increased, current is reduced for any fixed amount of power. Alternating current allows efficient transmission of electrical power between power stations and end users.

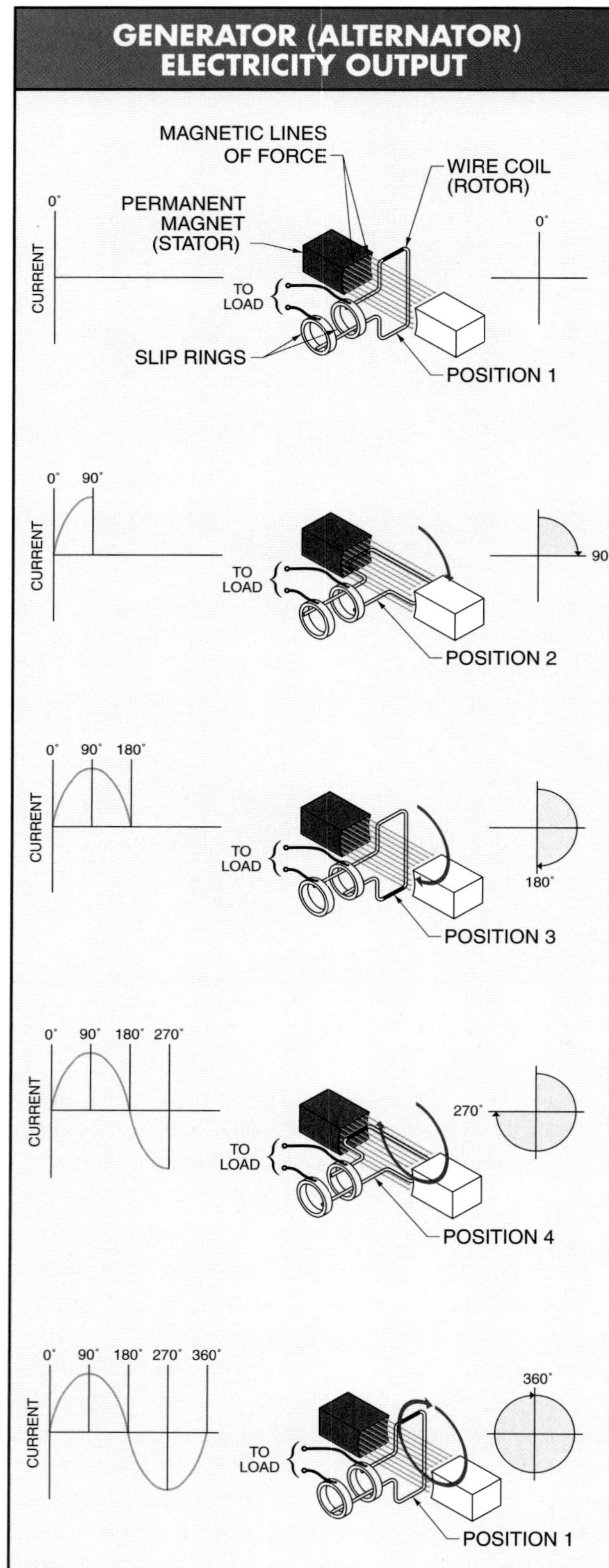

Figure 1-3. A generator produces electricity when magnetic flux lines are cut by a rotating wire coil (rotor).

Reduced current allows small conductors to be used to conduct electricity. The conductor current rating depends on the wire size, insulation used, conductor temperature rating, and wire type (copper or aluminum). The allowable amount of current a wire may safely carry is listed in National Electrical Code® (NEC®) tables.

An AC generator that has only one rotating coil produces a single-phase output. Single-phase generators are used for small power demands, but are not practical or economical for producing large amounts of power. To produce large amounts of power, three single-phase coils are coupled to produce three-phase power. The three separate coils are spaced 120 electrical degrees apart. The individual AC voltage outputs are phase 1 (A), phase 2 (B), and phase 3 (C). **See Figure 1-4.**

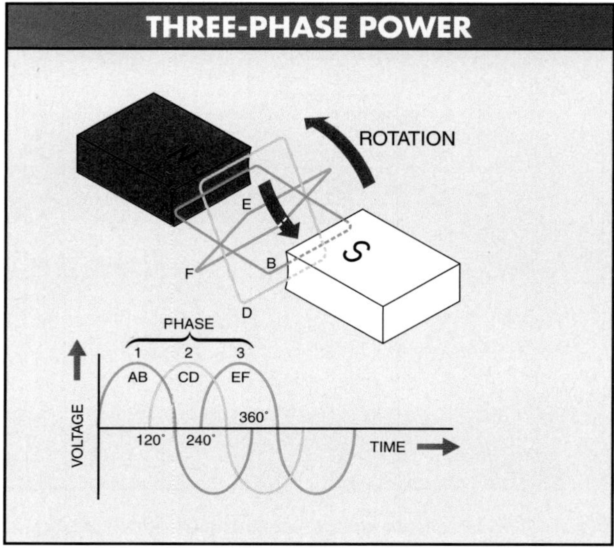

Figure 1-4. Three-phase output is generated by three separate coils spaced 120 electrical degrees apart in a generator.

Electrical Abbreviations/Prefixes

Electrical abbreviations are used to simplify the expression of common electrical terms and quantities. An *abbreviation* is a letter or combination of letters that represents a word. The exact abbreviation used normally depends on the use of the electrical unit. For example, voltage may be abbreviated using a capital letter E or V. A capital letter V is used to indicate voltage quantity because voltage is measured in volts. These abbreviations are often interchanged and both can be used to represent voltage. **See Figure 1-5.**

Prefixes are used to avoid long expressions of units that are smaller or larger than the base unit. A *base unit* is a number that does not include a metric prefix. To convert between different units, the decimal point is moved to the left or right, depending on the unit. The decimal point is moved to the left and a prefix is added to convert a large base value to a simpler term. For example, 1000 V can be written as 1 kV. The decimal point is moved to the right and a prefix is added to convert a small base value to a simpler term. For example, .001 V can be written as 1 mV. **See Appendix.**

COMMON ELECTRICAL QUANTITIES		
Equation Variable	**Name**	**Unit of Measure and Abbreviation**
E	voltage	volt—V
I	current	ampere—A
R	resistance	ohm—Ω
P	power	watt—W
P	power (apparent)	volt-amp—VA
C	capacitance	farad—F
L	inductance	henry—H
Z	impedance	ohm—Ω
G	conductance	siemens—S
f	frequency	hertz—Hz
T	period	second—s

Figure 1-5. Abbreviations are used to simplify the expression of common electrical terms and quantities.

VOLTAGE

All electrical circuits must have a source of power to produce work. The source of power used depends on the application and the amount of power required. All sources of power produce a set voltage level or voltage range.

Voltage (E) is the amount of electrical pressure in a circuit. Voltage is measured in volts (V). Voltage is also known as electromotive force (EMF) or potential difference. Voltage is produced when electrons are freed from atoms. An *atom* is the smallest particle that an element can be reduced to and still keep the properties of that element. The three principal parts of an atom are the electron, neutron, and proton. An *electron* is a negatively charged particle that orbits the nucleus of an atom. A *neutron* is a particle contained in the nucleus of an atom that has no electrical charge. A *proton* is a particle contained in the nucleus of an atom that has a positive electrical charge.

The electrons whirl around the nucleus, completing billions of trips around the nucleus each millionth of a second. The speed of the electrons causes the atoms to behave as if they were solid. In some materials, such as insulators (glass, rubber, etc.), electrons are not easily freed. In other materials, such as conductors (copper, aluminum, etc.), electrons are easily freed.

Alternating current produced at generating plants allows efficient transmission of electrical power between power stations and end users.

Voltage may be produced when electrons are freed from atoms by electromagnetism (generator), heat (thermocouple), light (photocell), chemical reaction (battery), pressure (piezoelectricity in strain gauge), and friction (static electricity). **See Figure 1-6.**

Voltage is either direct (DC) or alternating (AC). *DC voltage* is voltage that flows in one direction only. *AC voltage* is voltage that reverses its direction of flow at regular intervals. DC voltage is used in almost all portable equipment (automobiles, golf carts, flashlights, cameras, etc.). AC voltage is used in residential, commercial, and industrial lighting and power distribution systems.

DC Voltage

All DC voltage sources have a positive and a negative terminal. The positive and negative terminals establish polarity in a circuit. *Polarity* is the positive (+) or negative (–) state of an object. All points in a DC circuit have polarity.

The most common power sources that directly produce DC voltage are batteries and photocells. In addition to obtaining DC voltage directly from batteries and photocells, DC voltage is also obtained from a rectified AC voltage supply. **See Figure 1-7.** DC voltage is obtained any time an AC voltage is passed through a rectifier. A *rectifier* is a device that converts AC voltage to DC voltage by allowing the voltage and current to flow in only one direction. DC voltage obtained from a rectified AC voltage supply varies from almost pure DC voltage to half-wave DC voltage. Common DC voltage levels include 1.5 V, 6 V, 9 V, 12 V, 24 V, 36 V, and 125 V.

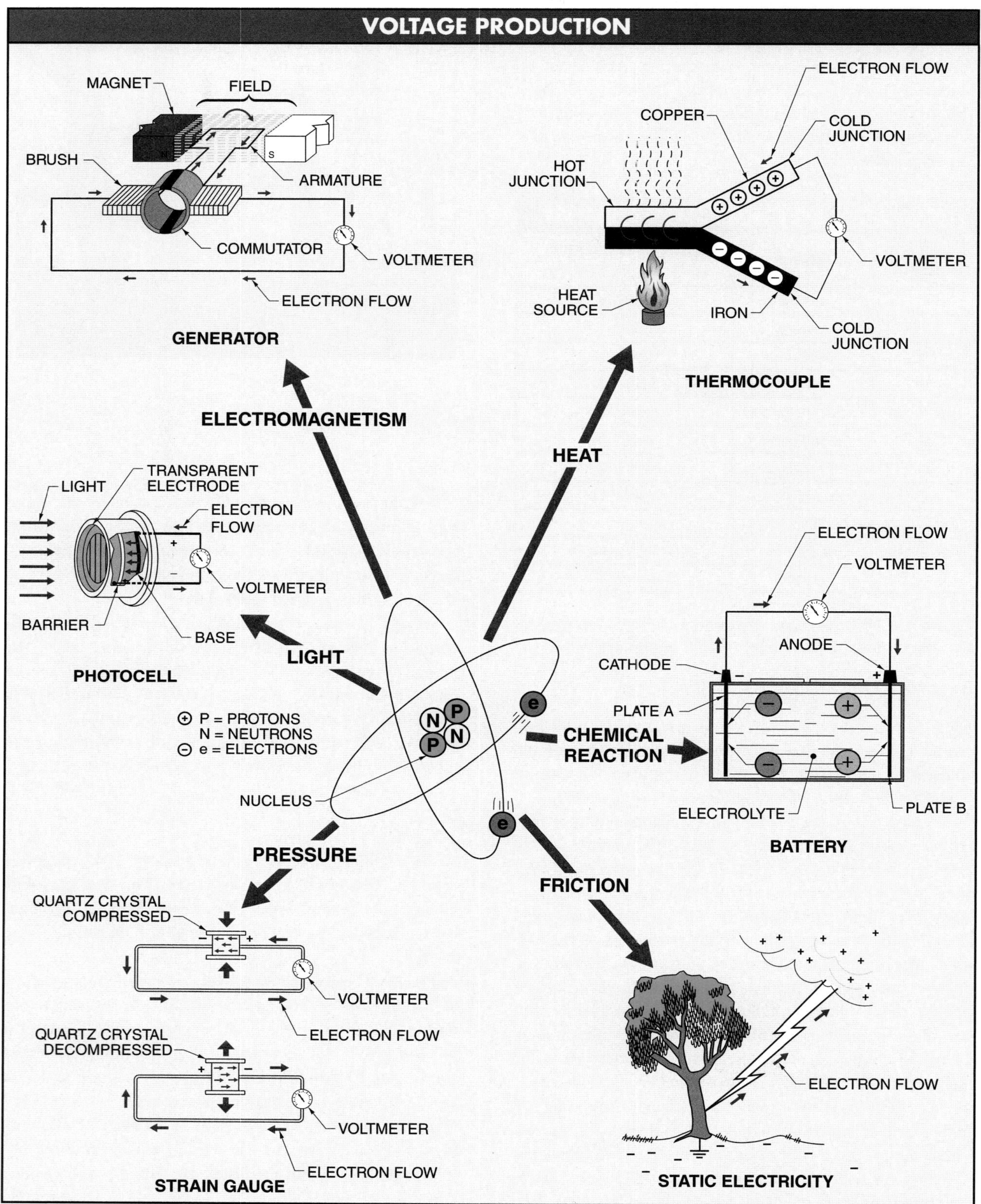

Figure 1-6. Voltage is produced by magnetism, heat, light, chemical action, pressure, and friction.

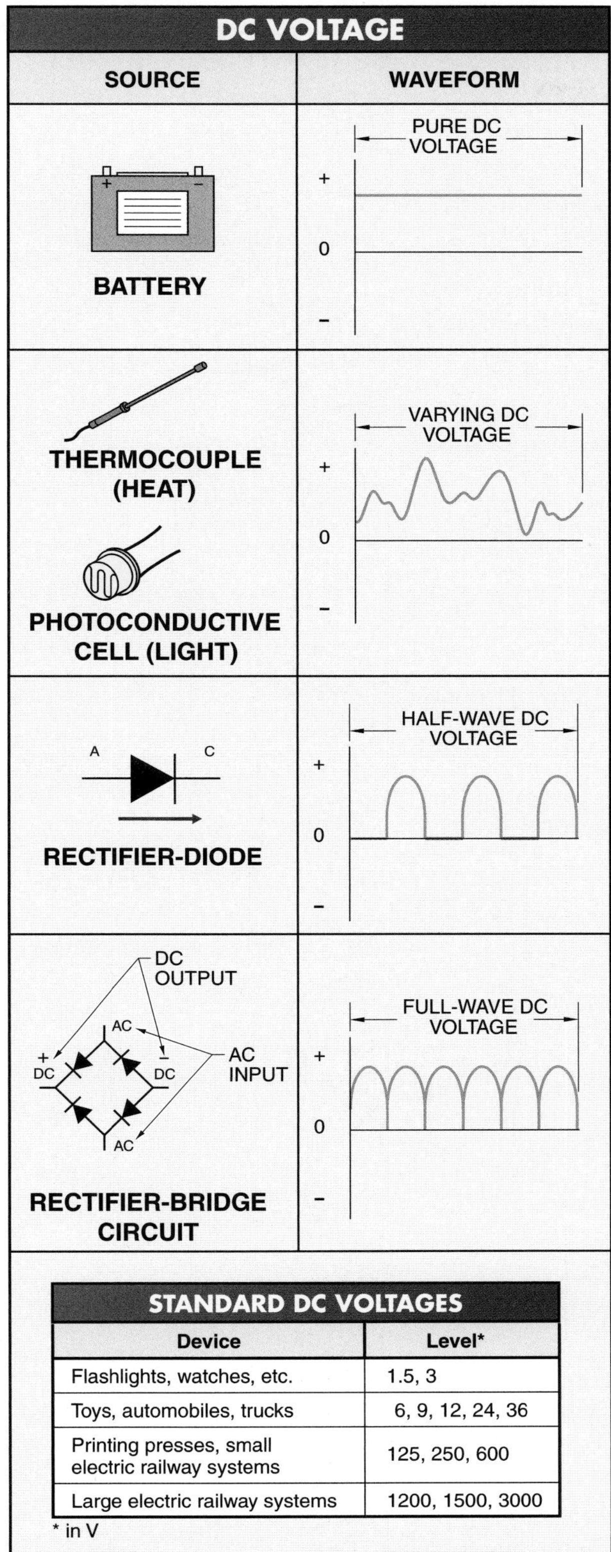

STANDARD DC VOLTAGES	
Device	**Level***
Flashlights, watches, etc.	1.5, 3
Toys, automobiles, trucks	6, 9, 12, 24, 36
Printing presses, small electric railway systems	125, 250, 600
Large electric railway systems	1200, 1500, 3000

* in V

Figure 1-7. DC voltage is produced from batteries, photocells, and rectified AC voltage supplies, and can vary from almost pure DC voltage to half-wave DC voltage.

DC Voltage Measurements. Exercise caution when measuring DC voltages over 60 V.

Warning: Ensure that no body parts contact any part of a live circuit, including the metal contact points at the tip of the test leads.

A standard procedure is followed when taking DC voltage measurements. **See Figure 1-8.** To measure DC voltages with a digital multimeter (DMM), apply the procedure:

1. Set the function switch to DC voltage. If the DMM has more than one voltage position or if the circuit voltage is unknown, select a setting high enough to measure the highest possible circuit voltage.
2. Plug the black test lead into the common jack.
3. Plug the red test lead into the voltage jack.
4. Discharge any capacitors.
5. Connect the DMM test leads to the circuit. Connect the black test lead to circuit ground and the red test lead to the point at which the voltage is under test. Reverse the black and red test leads if a negative sign appears in front of the reading on the DMM.
6. Read the voltage displayed.

AC Voltage

AC voltage is the most common voltage used to produce work. AC voltage is produced by generators, which create AC sine waves as they rotate. An *AC sine wave* is a symmetrical waveform that contains 360 electrical degrees. The wave reaches its peak positive value at 90°, returns to 0 V at 180°, increases to its peak negative value at 270°, and returns to 0 V at 360°.

A *cycle* is one complete positive and negative alternation of a wave form. An *alternation* is half of a cycle. A sine wave has one positive alternation and one negative alternation per cycle. **See Figure 1-9.**

AC voltage is either single-phase (1ϕ) or three-phase (3ϕ). Single-phase AC voltage contains only one alternating voltage waveform. Three-phase AC voltage is a combination of three alternating voltage waveforms, each displaced 120 electrical degrees (one-third of a cycle) apart. Three-phase voltage is produced when three coils are simultaneously rotated in a generator.

Low AC voltages (6 V to 24 V) are used for doorbells and security systems. Medium AC voltages (110 V to 120 V) are used in residential applications for lighting, heating, cooling, cooking, running motors, etc. High AC voltages (208 V to 480 V) are used in commercial/residential applications for cooking, heating, cooling, etc. High AC voltages are also used in industrial applications to convert raw materials into usable products, in addition to providing lighting, heating, and cooling for plant personnel.

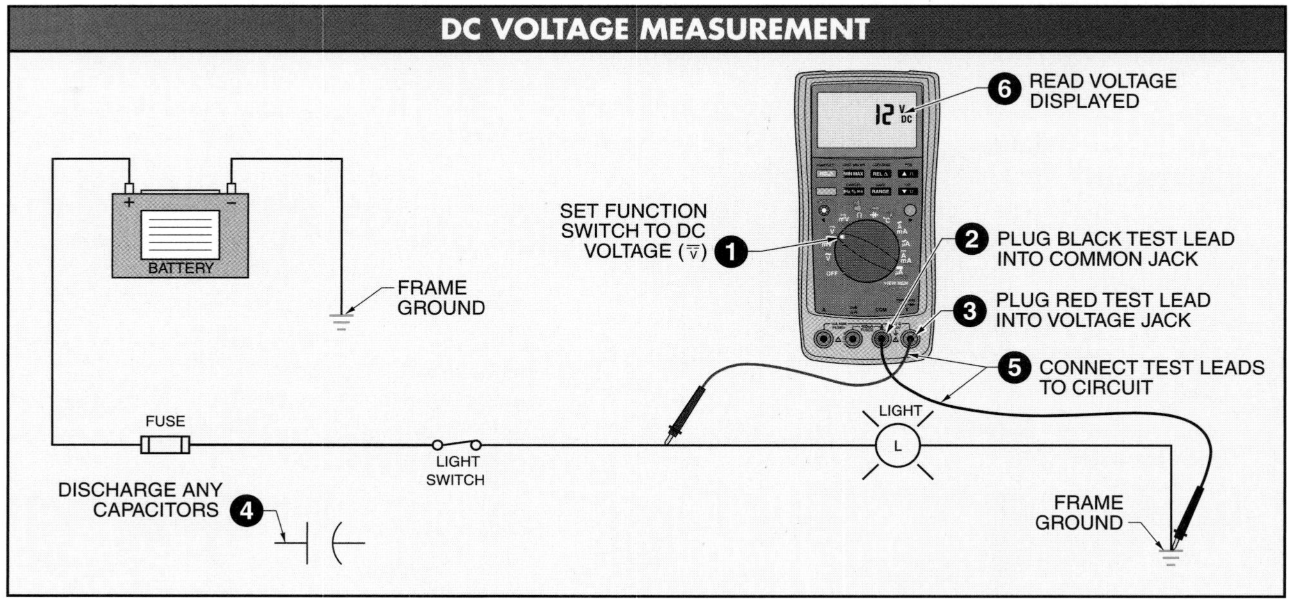

Figure 1-8. Exercise caution when measuring DC voltages over 60 V.

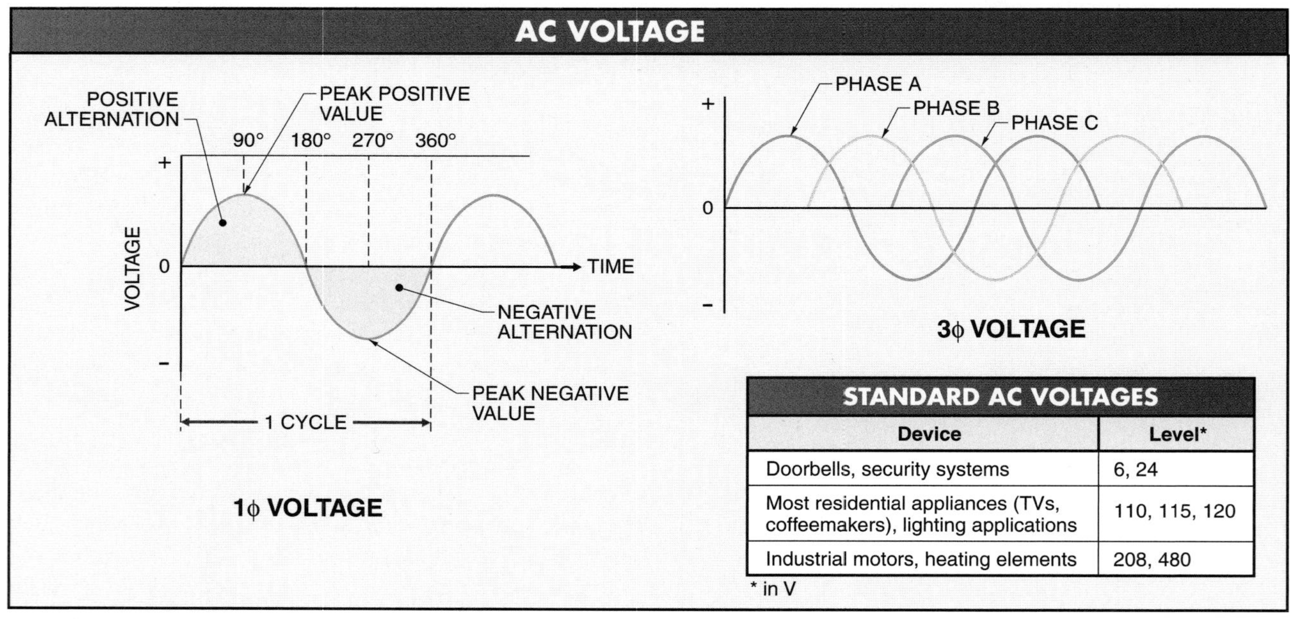

STANDARD AC VOLTAGES

Device	Level*
Doorbells, security systems	6, 24
Most residential appliances (TVs, coffeemakers), lighting applications	110, 115, 120
Industrial motors, heating elements	208, 480

* in V

Figure 1-9. AC voltage is either single-phase (1ϕ) or three-phase (3ϕ).

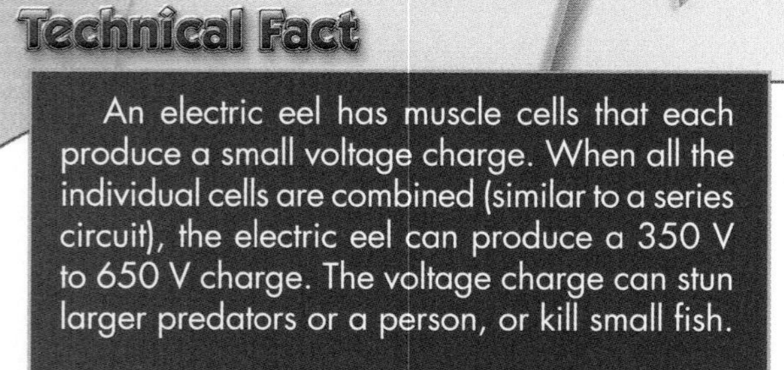

Technical Fact

An electric eel has muscle cells that each produce a small voltage charge. When all the individual cells are combined (similar to a series circuit), the electric eel can produce a 350 V to 650 V charge. The voltage charge can stun larger predators or a person, or kill small fish.

AC Voltage Measurements. Exercise caution when measuring AC voltages over 24 V.

Warning: Ensure that no body parts contact any part of a live circuit, including the metal contact points at the tip of the test leads.

A standard procedure is followed when testing AC voltage measurements. **See Figure 1-10.** To measure AC voltages, apply the procedure:

1. Set the function switch to AC voltage. If the DMM has more than one voltage position or if the circuit

voltage is unknown, select a setting high enough to measure the highest possible circuit voltage.
2. Plug the black test lead into the common jack.
3. Plug the red test lead into the voltage jack.
4. Connect the DMM test leads to the circuit. The position of the test leads is arbitrary. Common industrial practice is to connect the black test lead to the grounded (neutral) side of the AC voltage.
5. Read the voltage displayed.

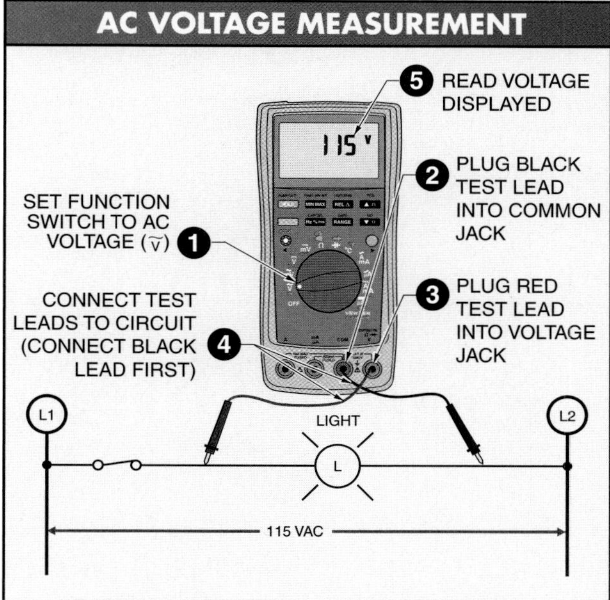

Figure 1-10. Exercise caution when measuring AC voltages over 24 V.

CURRENT

Current flows through a circuit when a source of power is connected to a device that uses electricity. *Current (I)* is the amount of electrons flowing through an electrical circuit. Current is measured in amperes (A). An *ampere* is the number of electrons passing a given point in one second. The more power a load requires, the larger the amount of current flow. For example, a 10 horsepower (HP) motor draws approximately 28 A when wired for 230 V. A 20 HP motor draws approximately 54 A when wired for 230 V.

Current Levels

Different voltage sources produce different amounts of current. For example, standard AAA, AA, A, C, and D size batteries all produce 1.5 V, but each size is capable of delivering a different amount of current. Size AAA batteries are capable of delivering the smallest amount of current, and size D batteries are capable of delivering the highest amount of current. For this reason, a load connected to a size D battery operates longer than the same load connected to a size AAA battery.

Current may be direct current or alternating current. *Direct current (DC)* is current that flows in only one direction. Direct current flows in any circuit connected to a power supply producing a DC voltage. *Alternating current (AC)* is current that reverses its direction of flow at regular intervals. Alternating current flows in any circuit connected to a power supply producing an AC voltage.

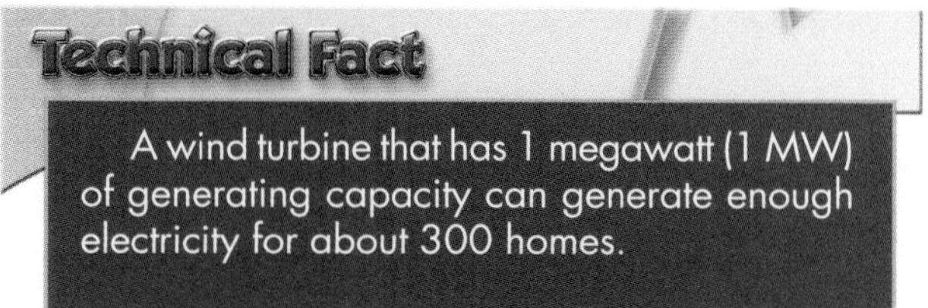

Current Flow

Early scientists believed that electrons flowed from positive (+) to negative (–). Later, when atomic structure was studied, electron flow from negative to positive was introduced. *Conventional current flow* is current flow from positive to negative. *Electron current flow* is current flow from negative to positive. Both current theories are still used. The current flow theory used depends on the industry. For example, the automobile industry usually uses the conventional current flow theory when explaining electricity. The electrical and electronics industry usually uses the electron current flow theory when explaining electricity.

In-Line Current Measurement Procedures

Care is required to protect the DMM, the circuit, and the user when taking in-line AC or DC current measurements. To observe standard safety precautions when taking in-line current measurements, do the following:

- Follow recommended procedures when testing DMM fuses.
- Ensure that the expected load current measurement is less than the current setting (limit) of the DMM. Start with the highest current-measuring range if the load current is unknown. If the current measurement may exceed the limit of the DMM setting, use a clamp-on ammeter or do not take the measurement.
- Ensure that the DMM function switch is set to the proper setting for measuring current (AC or DC). Most DMMs include more than one current level, such as A and mA, or μA.
- Ensure that the test leads are connected to the proper jacks for measuring current. Most DMMs include more than one current jack.

Warning: Always ensure that the function switch position matches the connection of the test leads. The DMM can be damaged if the test leads are connected to measure current and the function switch is set for a different measurement such as voltage or resistance. Some DMMs have Input Alert™, which provides a constant audible warning (beep) if the test leads are connected in the current jacks and a non-current mode is selected.

- Ensure that power to the test circuit is OFF before connecting and disconnecting test leads. If necessary, take a voltage measurement to ensure that the voltage is OFF.
- Do not change the function switch position on the DMM while the circuit under test is energized.
- Turn power to the DMM and the circuit OFF before changing any settings.
- Connect the DMM in series with the load(s) to be measured. Never connect a DMM in parallel with the load(s) to be measured.

Many DMMs include a fuse in the current-measuring circuit to prevent damage caused by excessive current. Before using a DMM, check to see if the DMM is fused on the current range being used. The DMM is marked as fused or not fused at the test lead current terminals. In-line current measurements are not recommended if the DMM is not fused. In-line AC or DC measurements are taken using a standard procedure. **See Figure 1-11.** To take in-line current measurements, apply the procedure:

Warning: Ensure that no body parts contact any part of the live circuit, including the metal contact points at the tip of the test leads.

1. Set the function switch to the proper position for measuring the AC or DC and current level (A and mA, or μA). If the DMM has more than one position, select a setting high enough to measure the highest possible circuit current.
2. Plug the black test lead into the common jack.
3. Plug the red test lead into the current jack. The current jack may be marked A and mA, or μA.
4. Turn the power to the circuit or device under test OFF and discharge all capacitors if possible.
5. Open the circuit at the test point and connect the test leads to each side of the opening. For DC current, the black (negative) test lead is connected to the negative side of the opening, and the red (positive) test lead is connected to the positive side of the opening. Reverse the black and red test leads if a negative sign appears to the left of the measurement displayed.
6. Turn the power ON to the circuit under test.
7. Read the current measurement displayed.
8. Turn the power OFF and remove the DMM from the circuit.

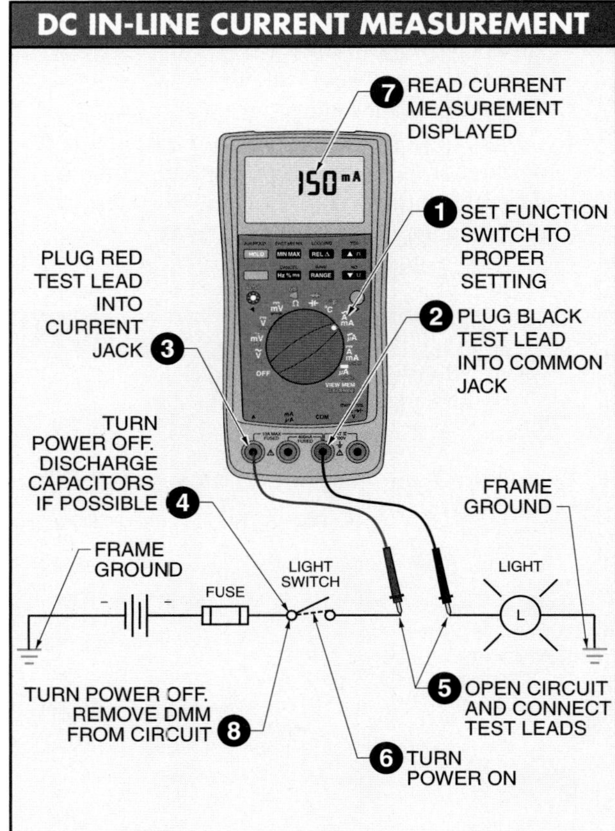

Figure 1-11. A DMM used to take in-line current measurements becomes part of the circuit being tested.

Clamp-on Ammeter Measurement Procedures

Clamp-on ammeters measure current in a circuit by measuring the strength of the magnetic field around a single conductor. Care should be taken to ensure that the meter does not pick up stray magnetic fields. Whenever possible, separate conductors under test from other surrounding conductors by a few inches. If this is not possible, take several readings at different locations along the same conductor.

AC or DC measurements using a clamp-on ammeter or a DMM with a clamp-on current probe accessory follow standard procedures. **See Figure 1-12.** To measure current using a clamp-on ammeter, apply the procedure:

1. Determine if AC or DC current is to be measured.
2. Select the ammeter required to measure the circuit current (AC or DC). If both AC and DC measurements are required, select an ammeter that can measure both AC and DC.
3. Determine if the ammeter range is high enough to measure the maximum current that may exist in the test circuit. If the ammeter range is not high enough,

select an accessory that has a high enough current rating, or select an ammeter with a higher range. If the ammeter includes fused current terminals, check to ensure that the ammeter fuses are good.

4. Set the function switch to the proper current setting (600 A, 200 A, 10 A, 400 mA, etc.). If there is more than one current position or if the circuit current is unknown, select a setting greater than the highest possible circuit current.
5. If required, plug the clamp-on current probe accessory into the DMM. The black test lead of the clamp-on current probe accessory is plugged into the common jack. The red test lead is plugged into the mA jack for current measurement accessories that produce a current output. The red test lead is plugged into the voltage (V) jack for current measurement accessories that produce a voltage output. The current measurement accessories that produce a current output are designed to measure AC only and deliver 1 mA to the DMM for every 1 A of measured current (1 mA/A). Current accessories that produce a voltage output are designed to measure AC or DC and deliver 1 mV to the DMM for every 1 A of measured current (1 mV/A).
6. Open the jaws by pressing against the trigger.
7. Enclose one conductor in the jaws. Ensure that the jaws are completely closed before taking readings.
8. Read the current measurement displayed.

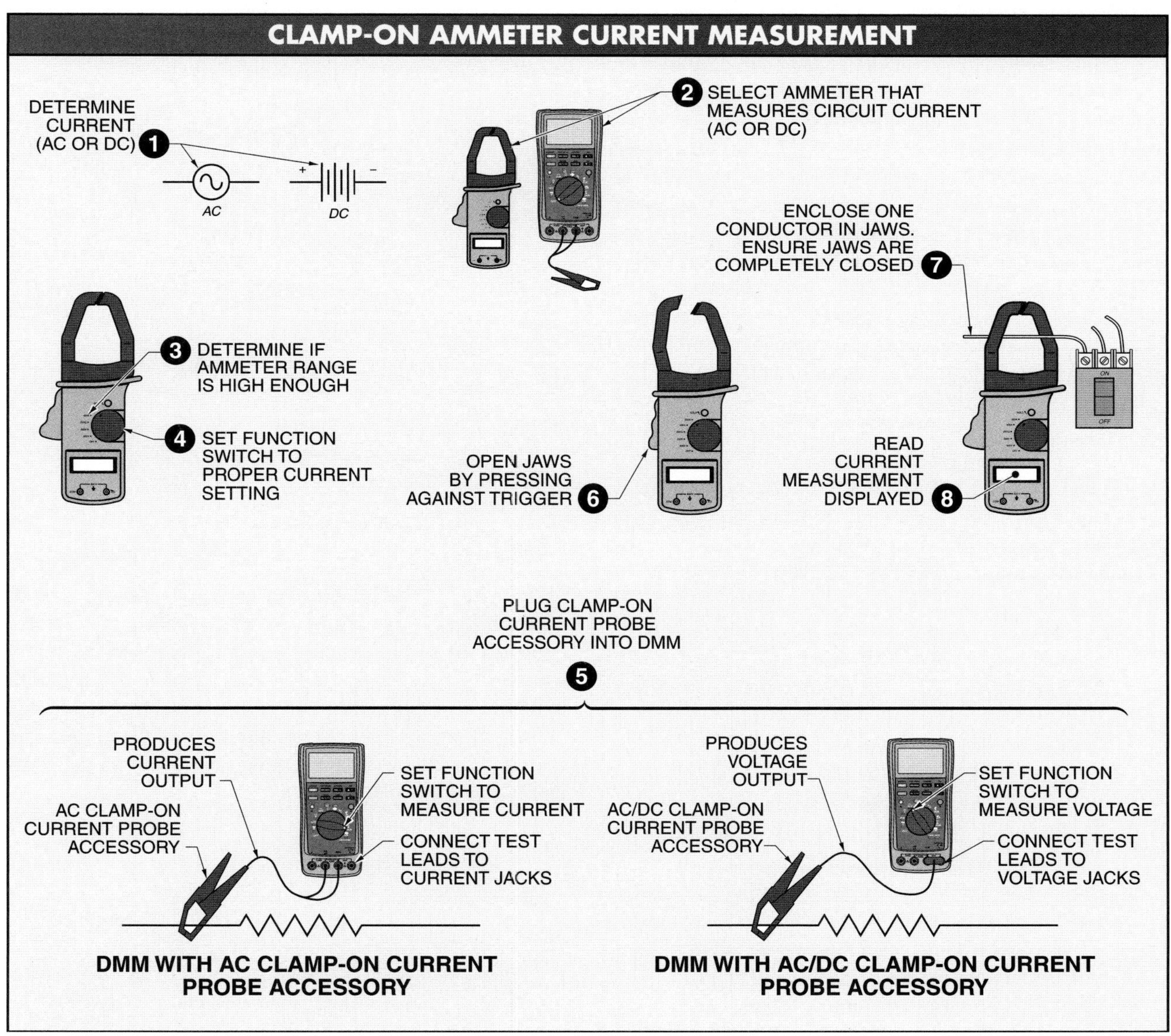

Figure 1-12. Clamp-on ammeters measure current by the strength of the magnetic field around a single conductor.

RESISTANCE

Resistance (R) is the opposition to the flow of electrons. Resistance is measured in ohms. The Greek letter omega (Ω) is used to represent ohms. Higher resistance measurements are expressed using prefixes, as in kilohms (kΩ) and megohms (MΩ).

Resistance limits the flow of current in an electrical circuit. The higher the resistance, the lower the current flow. Likewise, the lower the resistance, the higher the current flow. Components designed to insulate, such as rubber or plastic, should have a very high resistance. Components designed to conduct, such as conductors (wires) or switch contacts, should have a very low resistance. The resistance of insulators decreases when they are damaged by moisture and/or overheating. The resistance of conductors increases when they are damaged by burning and/or corrosion. Factors that affect the resistance of conductors are the size of the wire, length of the wire, conductor material, and temperature.

A conductor with a large cross-sectional area has less resistance than a conductor with a small cross-sectional area. A large conductor may also carry more current. The longer the conductor, the greater the resistance. Short conductors have less resistance than long conductors of the same size. Copper (Cu) is a better conductor (less resistance) than aluminum (Al), and may carry more current for a given size. Temperature also affects resistance. For metals, the higher the temperature, the greater the resistance.

Resistance Measurements

Ensure that no voltage is present in the circuit or component under test before taking resistance measurements. Low voltage applied to a DMM set to measure resistance causes inaccurate readings. High voltage applied to a DMM set to measure resistance causes meter damage. **See Figure 1-13.** Check for voltage using a voltmeter. To measure resistance using a DMM, apply the procedure:

1. Turn power to the circuit or component under test OFF. Remove component if possible.
2. Set the function switch to the resistance position.
3. Plug the black test lead into the common jack.
4. Plug the red test lead into the resistance jack.
5. Check the battery. The battery symbol is displayed when the batteries are low.
6. Connect the DMM leads across the component under test. Ensure that contact between the test leads and the circuit is good. Dirt, solder flux, oil, and other foreign substances greatly affect resistance readings.
7. Read the resistance displayed. Check the circuit schematic for parallel paths. Parallel paths with the resistance under test cause reading errors. Do not touch exposed metal parts of the test leads during the test. Resistance of a person's body can cause reading errors.
8. Turn the meter OFF after measurements are taken to save battery life.

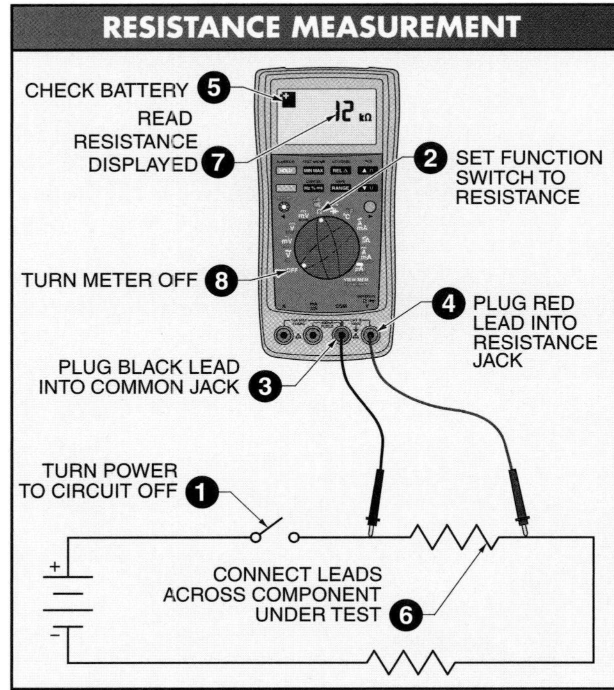

Figure 1-13. Ensure that no voltage is present in the circuit or component under test before taking resistance measurements.

OHM'S LAW

Ohm's law is the relationship between voltage, current, and resistance in a circuit. Ohm's law states that current in a circuit is proportional to the voltage and inversely proportional to the resistance. Any value in this relationship can be found when the other two values are known. The relationship between voltage, current, and resistance may be visualized by presenting Ohm's law in pie chart form. **See Figure 1-14.**

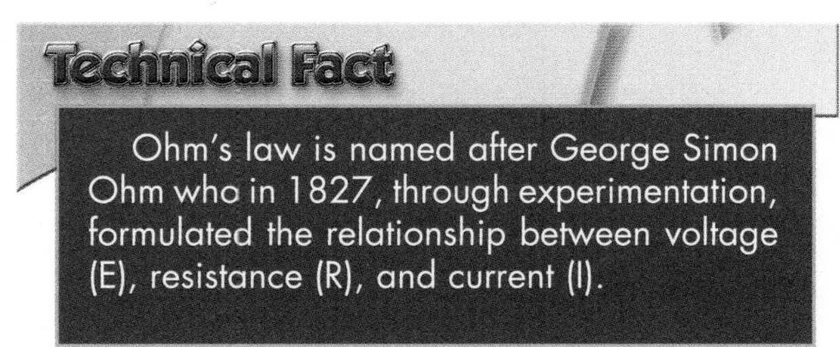

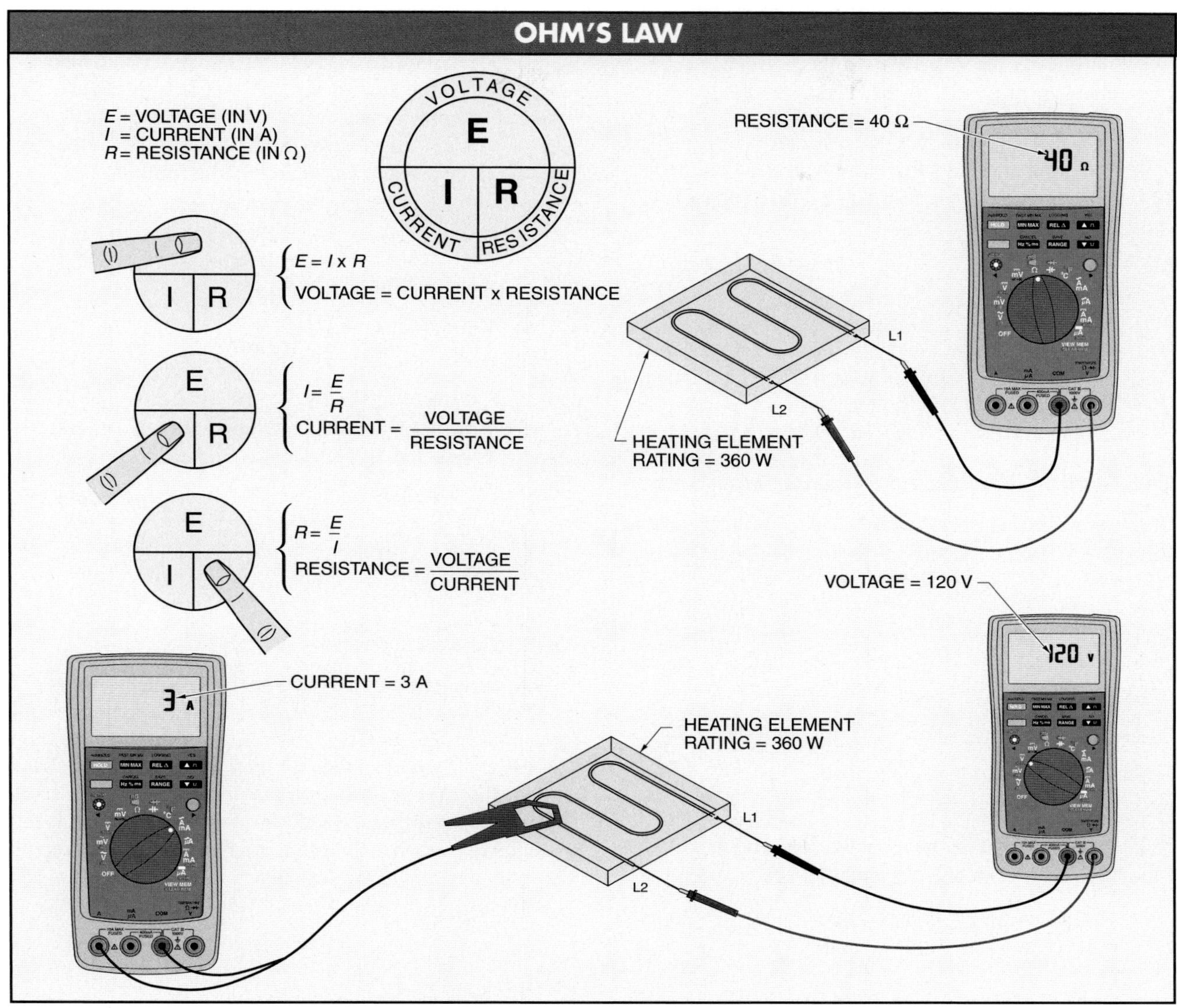

Figure 1-14. Ohm's law is the relationship between voltage (E), current (I), and resistance (R) in a circuit.

Calculating Voltage Using Ohm's Law

Ohm's law states that voltage (E) in a circuit is equal to current (I) times resistance (R). To calculate voltage using Ohm's law, apply the formula:

$$E = I \times R$$

where

E = voltage (in V)

I = current (in A)

R = resistance (in Ω)

For example, what is the voltage in a circuit that includes a 40 Ω heating element that draws 3 A?

$$E = I \times R$$

$$E = 3 \times 40$$

$$E = \mathbf{120\ V}$$

Calculating Current Using Ohm's Law

Ohm's law states that current (I) in a circuit is equal to voltage (E) divided by resistance (R). To calculate current using Ohm's law, apply the formula:

$$I = \frac{E}{R}$$

where

I = current (in A)

E = voltage (in V)

R = resistance (in Ω)

For example, what is the current in a circuit with a 40 Ω heating element connected to a 120 V supply?

$$I = \frac{E}{R}$$

$$I = \frac{120}{40}$$

$$I = \mathbf{3\ A}$$

Calculating Resistance Using Ohm's Law

Ohm's law states that resistance (R) in a circuit is equal to voltage (E) divided by current (I). To calculate resistance using Ohm's law, apply the formula:

$$R = \frac{E}{I}$$

where

R = resistance (in Ω)

E = voltage (in V)

I = current (in A)

For example, what is the resistance of a circuit in which a load that draws 3 A is connected to a 120 V supply?

$$R = \frac{E}{I}$$

$$R = \frac{120}{3}$$

$$R = \mathbf{40\ \Omega}$$

SERIES CIRCUITS

Fuses, switches, loads, and other electrical components can be connected in series. A *series connection* is a connection that has two or more components connected so there is only one path for current flow. Opening the circuit at any point stops the flow of current. Current stops flowing any time a fuse blows, a circuit breaker trips, or a switch or load opens. An example of a series connection is a DC series motor. **See Figure 1-15.** A *DC series motor* is a DC motor that has the series field coils connected in series with the armature. The armature wires are marked A1 and A2. The series coil wires are marked S1 and S2.

Resistance in Series Circuits

The total resistance in a circuit containing series-connected loads equals the sum of the resistances of all loads. The resistance in the circuit increases if loads are added in series and decreases if loads are removed. To calculate total resistance of a series circuit, apply the formula:

$$R_T = R_1 + R_2 + R_3 + \ldots$$

where

R_T = total resistance (in Ω)

R_1 = resistance 1 (in Ω)

R_2 = resistance 2 (in Ω)

R_3 = resistance 3 (in Ω)

For example, what is the total resistance of a circuit that has 2 Ω, 4 Ω, and 6 Ω resistors connected in series?

$$R_T = R_1 + R_2 + R_3$$

$$R_T = 2 + 4 + 6$$

$$R_T = \mathbf{12\ \Omega}$$

Voltage in Series Circuits

The total voltage applied across loads connected in series is divided across the individual loads. Each load drops a set percentage of the applied voltage. The exact voltage drop across each load depends on the resistance of that load. The voltage drops across any two loads are the same if the resistance values are the same. To calculate total voltage of a series circuit when the voltage across each load is known or measured, apply the formula:

$$E_T = E_1 + E_2 + E_3 + \ldots$$

where

E_T = total applied voltage (in V)

E_1 = voltage drop across load 1 (in V)

E_2 = voltage drop across load 2 (in V)

E_3 = voltage drop across load 3 (in V)

For example, what is the total applied voltage of a circuit containing 4 V, 8 V, and 12 V drops across three loads?

$$E_T = E_1 + E_2 + E_3$$

$$E_T = 4 + 8 + 12$$

$$E_T = \mathbf{24\ V}$$

Technical Fact

A voltage rating of a fuse, circuit breaker, or switch can be higher than the circuit's actual voltage, but never lower.

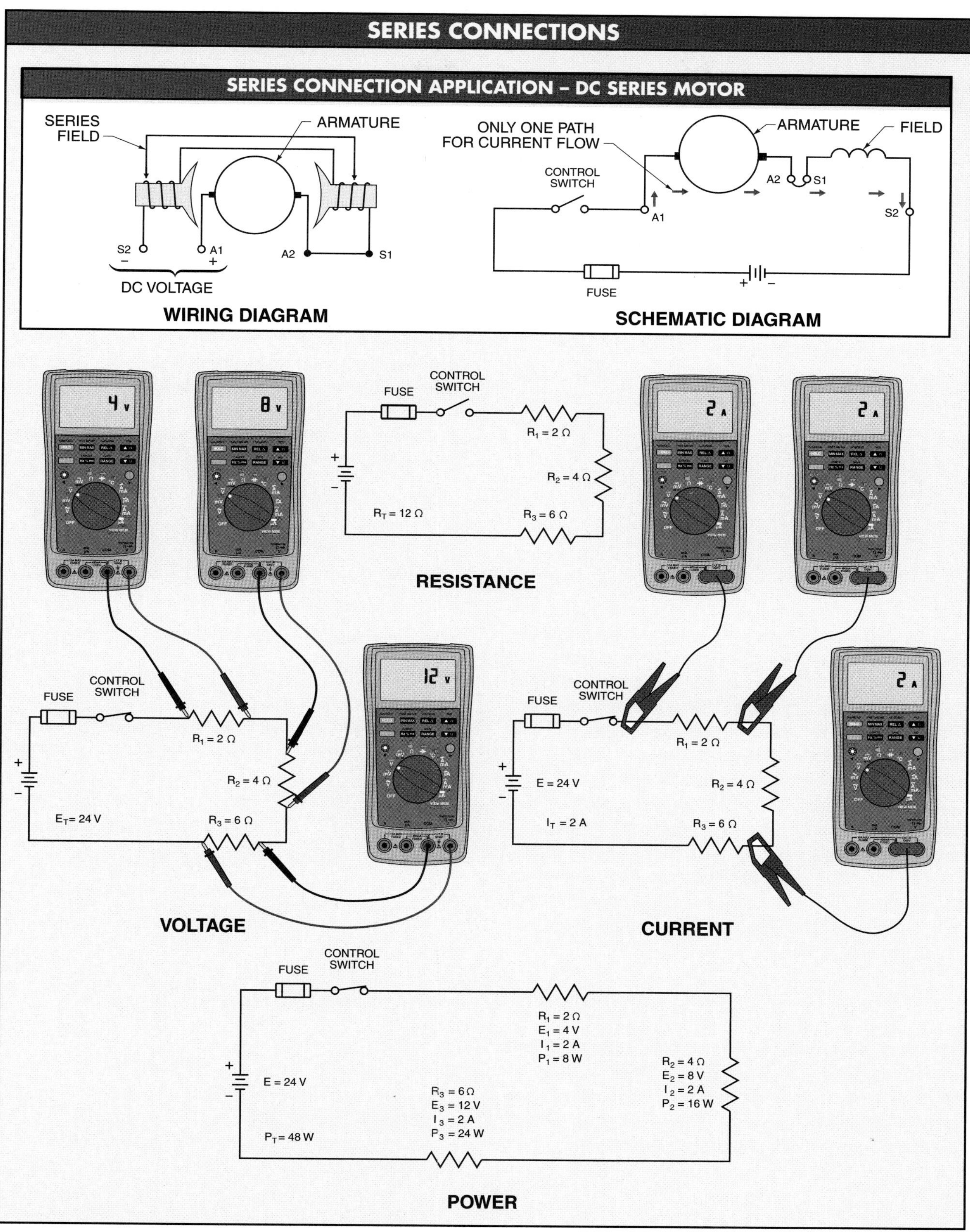

Figure 1-15. A series connection has two or more components connected so there is only one path for current flow.

Current in Series Circuits

The current in a circuit containing series-connected loads is the same throughout the circuit. The current in the circuit decreases if the circuit resistance increases and the current increases if the circuit resistance decreases. To calculate total current of a series circuit, apply the formula:

$I_T = I_1 = I_2 = I_3 = \ldots$

where

I_T = total circuit current (in A)

I_1 = current through load 1 (in A)

I_2 = current through load 2 (in A)

I_3 = current through load 3 (in A)

For example, what is the total current through a series circuit if the current measured at each load is 2 A?

$I_T = I_1 = I_2 = I_3$

$I_T = 2 = 2 = 2$

$I_T = \mathbf{2\ A}$

PARALLEL CIRCUITS

Fuses, switches, loads, and other components can be connected in parallel. A *parallel connection* is a connection that has two or more components connected so there is more than one path for current flow. An example of a parallel connection is a DC shunt motor. **See Figure 1-16.** A *DC shunt motor* is a DC motor that has the field connected in shunt (parallel) with the armature. The armature wires are marked A1 and A2. The parallel (shunt) coil wires are marked F1 and F2.

Care must be taken when working with parallel circuits because current can be flowing in one part of the circuit even though another part of the circuit is OFF. Understanding and recognizing parallel-connected components and circuits enables a technician or troubleshooter to take proper measurements, make circuit modifications, and troubleshoot the circuit.

Resistance in Parallel Circuits

The total resistance in a circuit containing parallel-connected loads is less than the smallest resistance value. The total resistance decreases if loads are added in parallel and increases if loads are removed. To calculate total resistance in a parallel circuit containing two resistors, apply the formula:

$$R_T = \frac{R_1 \times R_2}{R_1 + R_2}$$

where

R_T = total resistance (in Ω)

R_1 = resistance 1 (in Ω)

R_2 = resistance 2 (in Ω)

For example, what is the total resistance in a circuit containing resistors of 16 Ω and 24 Ω connected in parallel?

$$R_T = \frac{R_1 \times R_2}{R_1 + R_2}$$

$$R_T = \frac{16 \times 24}{16 + 24}$$

$$R_T = \frac{384}{40}$$

$R_T = \mathbf{9.6\ \Omega}$

To calculate total resistance in a parallel circuit with three or more resistors, the formula for two resistors can be used by solving the problem for two resistors at a time. In addition, to calculate the total resistance in a parallel circuit with three or more resistors, apply the formula:

$$R_T = \frac{1}{\frac{1}{R_1} + \frac{1}{R_2} + \frac{1}{R_3} + \ldots}$$

where

R_T = total resistance (in Ω)

R_1 = resistance 1 (in Ω)

R_2 = resistance 2 (in Ω)

R_3 = resistance 3 (in Ω)

For example, what is the total resistance in a circuit containing resistors of 16 Ω, 24 Ω, and 48 Ω connected in parallel?

$$R_T = \frac{1}{\frac{1}{R_1} + \frac{1}{R_2} + \frac{1}{R_3} + \ldots}$$

$$R_T = \frac{1}{\frac{1}{16} + \frac{1}{24} + \frac{1}{48}}$$

$$R_T = \frac{1}{.06250 + .04166 + .02083}$$

$R_T = \mathbf{8\ \Omega}$

Technical Fact

Since Ohm's law was formulated in 1827, numerous deviations from it have been discovered.

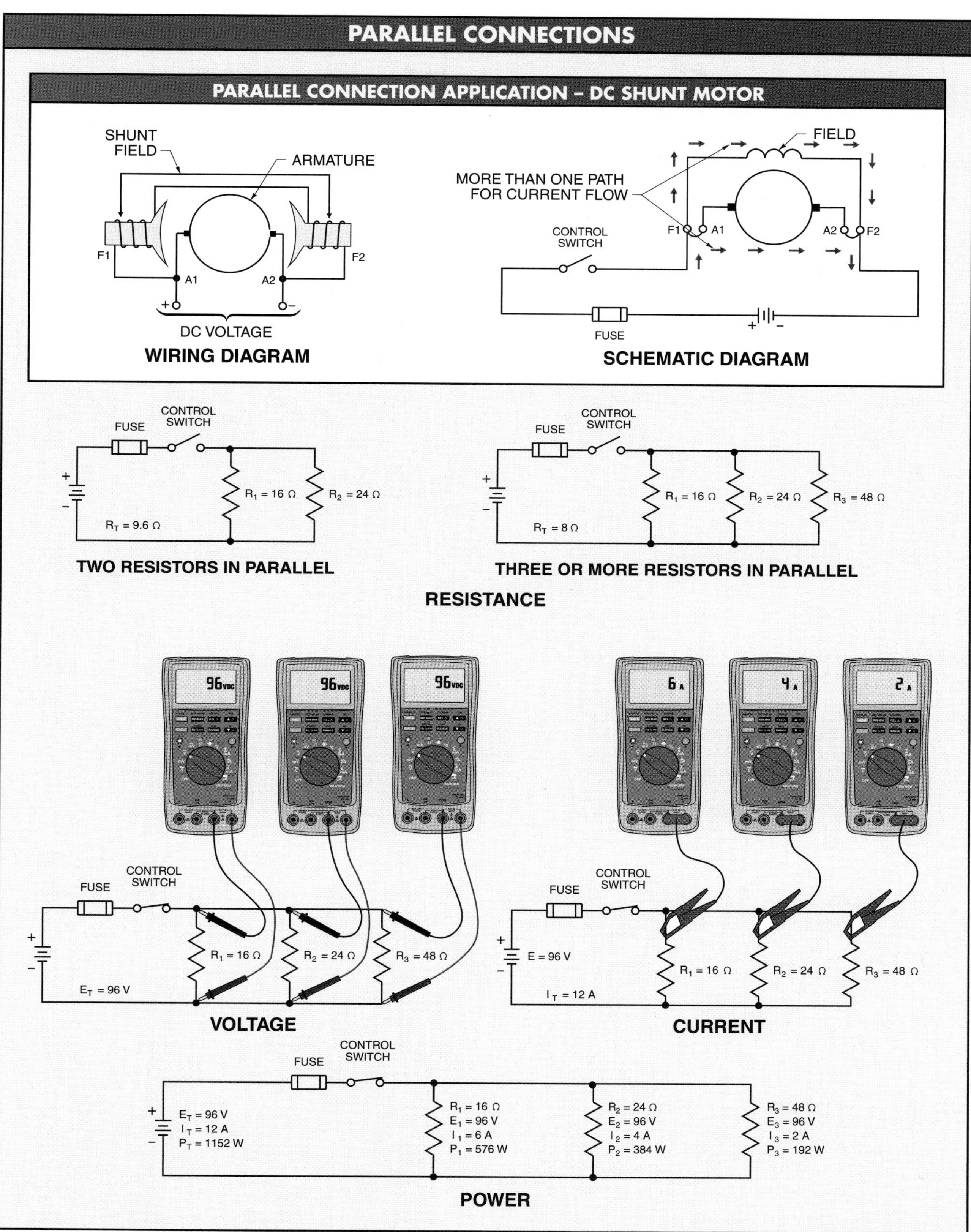

Figure 1-16. A parallel connection has two or more components connected so there is more than one path for current flow.

Voltage in Parallel Circuits

The voltage across each load is the same when loads are connected in parallel. The voltage across each load remains the same if parallel loads are added or removed. To calculate total voltage in a parallel circuit when the voltage across a load is known or measured, apply the formula:

$$E_T = E_1 = E_2 = E_3 = \ldots$$

where

E_T = total applied voltage (in V)

E_1 = voltage across load 1 (in V)

E_2 = voltage across load 2 (in V)

E_3 = voltage across load 3 (in V)

For example, what is the total applied voltage if the voltage across three parallel-connected loads is 96 VDC?

$$E_T = E_1 = E_2 = E_3$$

$$E_T = 96 = 96 = 96$$

$$E_T = \mathbf{96\ VDC}$$

Current in Parallel Circuits

Total current in a circuit containing parallel-connected loads equals the sum of the current through all the loads. Total current increases if loads are added in parallel and decreases if loads are removed. To calculate total current in a parallel circuit, apply the formula:

$$I_T = I_1 + I_2 + I_3 + \ldots$$

where

I_T = total circuit current (in A)

I_1 = current through load 1 (in A)

I_2 = current through load 2 (in A)

I_3 = current through load 3 (in A)

For example, what is the total current in a circuit containing three loads connected in parallel if the current through the three loads is 6 A, 4 A, and 2 A?

$$I_T = I_1 + I_2 + I_3$$

$$I_T = 6 + 4 + 2$$

$$I_T = \mathbf{12\ A}$$

SERIES/PARALLEL CIRCUITS

Fuses, switches, loads, and other components can be connected in a series/parallel connection. A *series/parallel connection* is a combination of series- and parallel-connected components. An example of a series/parallel connection is a DC compound motor. **See Figure 1-17.** A *DC compound motor* is a DC motor with the field connected in both series and shunt with the armature. The armature wires are marked A1 and A2. The parallel (shunt) coil wires are marked F1 and F2. The series coil is marked S1 and S2.

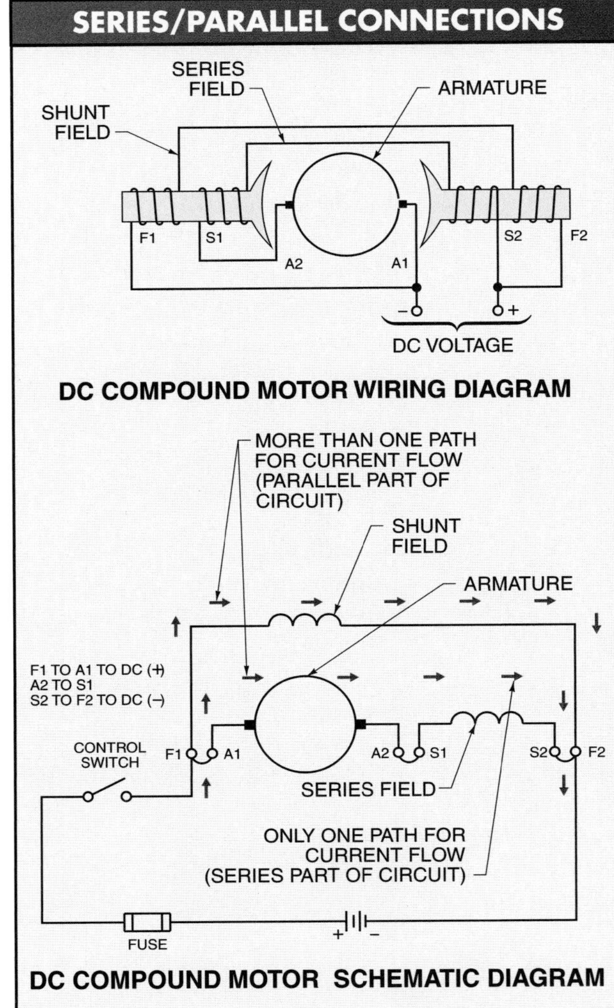

Figure 1-17. A series/parallel connection is a combination of series- and parallel-connected components.

Technical Fact

A basic series/parallel circuit has many practical applications and is used in a circuit that includes a 12 V lamp to indicate when a 120 V load is ON. In this circuit, a resistor used to drop 108 V is connected in series with the 12 V lamp. The series lamp/resistor combination is then connected in parallel with the load.

Resistance in Series/Parallel Circuits

A series/parallel circuit can contain any number of individual resistors (loads) connected in any number of different series/parallel circuit combinations. The series/parallel combination is always equal to one combined total resistance value. The total resistance in a circuit containing series/parallel connected resistors equals the sum of the series loads and the equivalent resistance of the parallel combinations. To calculate total resistance in a series/parallel circuit that contains two resistors in series connected to two resistors in parallel, apply the formula:

$$R_T = \left(\frac{Rp_1 \times Rp_2}{Rp_1 + Rp_2}\right) + Rs_1 + Rs_2$$

where

R_T = total resistance (in Ω)

Rp_1 = parallel resistance 1 (in Ω)

Rp_2 = parallel resistance 2 (in Ω)

Rs_1 = series resistance 1 (in Ω)

Rs_2 = series resistance 2 (in Ω)

For example, what is the total resistance of a 150 Ω and 50 Ω resistor connected in parallel with a 25 Ω and 100 Ω resistor connected in series?

$$R_T = \left(\frac{Rp_1 \times Rp_2}{Rp_1 + Rp_2}\right) + Rs_1 + Rs_2$$

$$R_T = \left(\frac{150 \times 50}{150 + 50}\right) + 25 + 100$$

$$R_T = \left(\frac{7500}{200}\right) + 125$$

$R_T = 37.5 + 125$

$R_T =$ **162.5 Ω**

Current in Series/Parallel Circuits

The total current and current in individual parts of a series/parallel circuit follow the same laws of current as in a basic series and a basic parallel circuit. Current is the same in each series part of the series/parallel circuit. Current is equal to the sum of each parallel combination in each parallel part of the series/parallel circuit.

Voltage in Series/Parallel Circuits

The total voltage applied across resistors (loads) connected in a series/parallel combination is divided across the individual resistors (loads). The higher the resistance of any one resistor or equivalent parallel resistance, the higher the voltage drop. Likewise, the lower the resistance of any one resistor or equivalent parallel resistance, the lower the voltage drop.

Application of Series and Parallel Circuits

Principles of series and parallel circuits can be used to produce several different heat outputs (wattages) in heating element circuits. Heat is produced any time electricity passes through a wire that has resistance. This principle is used to produce heat in devices such as toasters, portable space heaters, hair dryers, electric ovens, coffee makers, irons, and electric hot water heaters. Industrial heating applications also apply this principle to heat solids, gases, liquids, surfaces, and pipes.

Heating elements are made from special resistance wire that is capable of withstanding the temperature produced by the electricity flowing through the element. The amount of heat a wire produces is proportional to the resistance of the wire. The lower the resistance, the greater the current flow in a wire (Ohm's law). The greater the current flow in a wire, the higher the temperature of the wire (power formula). Likewise, the higher the resistance, the smaller the current flow (Ohm's Law). The smaller the current flow, the lower the temperature of the wire (power formula).

By varying the total resistance of a heating element, the heat output (wattage) of the heating element can be varied. The total resistance can be varied by using several individual heating elements that can be connected in series, parallel, or a series/parallel combination. **See Figure 1-18.**

In circuit 1, the four 20 Ω heating elements are connected in series and produce 180 W of power. In circuit 2, the four 20 Ω heating elements are connected in parallel and produce 2880 W of power. In circuit 3, two 20 Ω heating elements are connected in parallel, and then connected in series with the other two 20 Ω heating elements that are also connected in parallel. These heating elements produce 720 W of power. In circuit 4, two 20 Ω heating elements are connected in series, and then in parallel with the other two 20 Ω heating elements that are also connected in series. These heating elements also produce 720 W of power.

Technical Fact

Doubling the area of a conductor reduces the conductor's resistance by half.

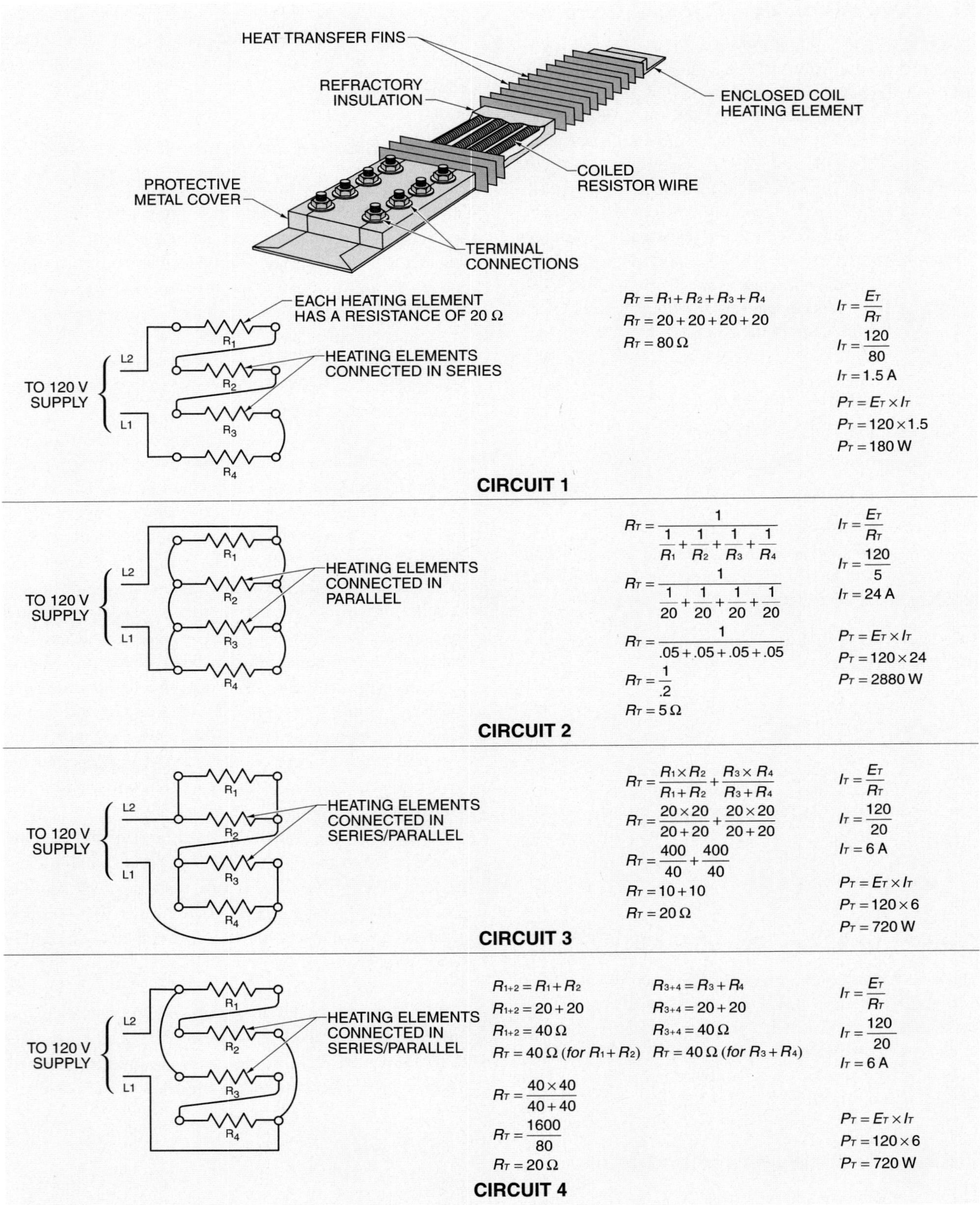

Figure 1-18. Principles of series and parallel circuits can be used to produce several different heat outputs (wattages) in heating element circuits.

POWER FORMULA

Power is the rate of doing work or using energy. The *power formula* is the relationship between power (P), voltage (E), and current (I) in an electrical circuit. Any value in this relationship may be found using the power formula when the other two values are known. The relationship between power, voltage, and current may be visualized by presenting the power formula in pie chart form. **See Figure 1-19.**

Calculating Power Using Power Formula

The power formula states that power (P) in a circuit is equal to voltage (E) times current (I). To calculate power using the power formula, apply the formula:

$P = E \times I$

where

P = power (in W)

E = voltage (in V)

I = current (in A)

For example, what is the power of a load that draws 5 A when connected to a 120 V supply?

$P = E \times I$

$P = 120 \times 5$

$P = \mathbf{600\ W}$

Calculating Voltage Using Power Formula

The power formula states that voltage (E) in a circuit is equal to power (P) divided by current (I). To calculate voltage using the power formula, apply the formula:

$$E = \frac{P}{I}$$

where

E = voltage (in V)

P = power (in W)

I = current (in A)

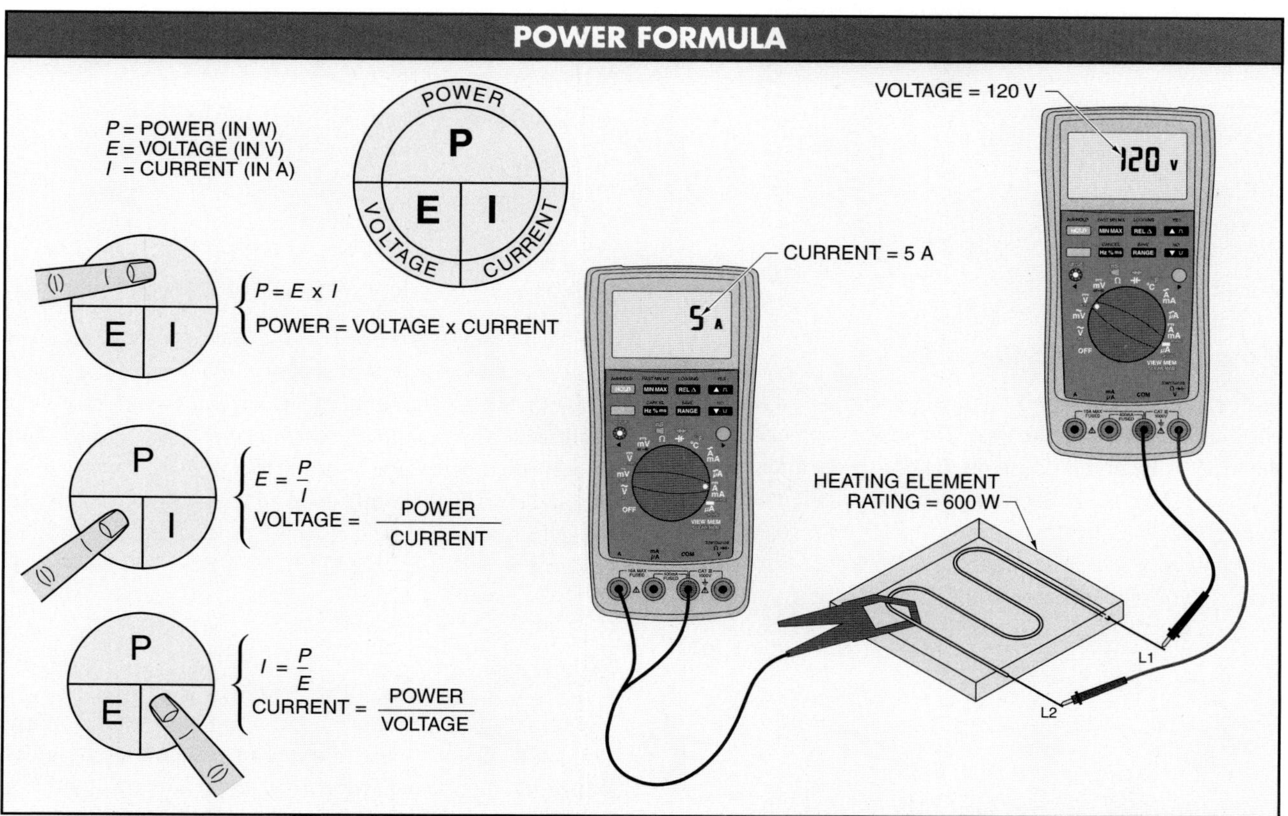

Figure 1-19. The power formula is the relationship between power (P), voltage (E), and current (I) in an electrical circuit.

For example, what is the voltage in a circuit in which a 600 W load draws 5 A?

$$E = \frac{P}{I}$$

$$E = \frac{600}{5}$$

$$E = \mathbf{120\ V}$$

Calculating Current Using Power Formula

The power formula states that current (I) in a circuit is equal to power (P) divided by voltage (E). To calculate current using the power formula, apply the formula:

$$I = \frac{P}{E}$$

where

I = current (in A)

P = power (in W)

E = voltage (in V)

For example, what is the current in a circuit in which a 600 W load is connected to a 120 V supply?

$$I = \frac{P}{E}$$

$$I = \frac{600}{120}$$

$$I = \mathbf{5\ A}$$

Power in Series Circuits

Power is produced when voltage is applied to a load and current flows through the load. The lower the resistance of the load or the higher the applied voltage, the more power produced. The higher the resistance of the load or the lower the applied voltage, the less power produced. To calculate total power in a series circuit when the power across each load is known or measured, apply the formula:

$$P_T = P_1 + P_2 + P_3 + \ldots$$

where

P_T = total applied power (in W)

P_1 = power drop across load 1 (in W)

P_2 = power drop across load 2 (in W)

P_3 = power drop across load 3 (in W)

For example, what is the total power in a series circuit if three loads are connected in series and load 1 equals 8 W, load 2 equals 16 W, and load 3 equals 24 W?

$$P_T = P_1 + P_2 + P_3$$

$$P_T = 8 + 16 + 24$$

$$P_T = \mathbf{48\ W}$$

Power in Parallel Circuits

Power is produced when voltage is applied to a load and current flows through the load. Total power produced in a parallel circuit is equal to the sum of the power produced by each load. To calculate total power in a parallel circuit when the power across each load is known or measured, apply the formula:

$$P_T = P_1 + P_2 + P_3 + \ldots$$

where

P_T = total circuit power (in W)

P_1 = power of load 1 (in W)

P_2 = power of load 2 (in W)

P_3 = power of load 3 (in W)

For example, what is the total circuit power if three loads are connected in parallel and the loads produce 576 W, 384 W, and 192 W?

$$P_T = P_1 + P_2 + P_3$$

$$P_T = 576 + 384 + 192$$

$$P_T = \mathbf{1152\ W}$$

Power in Series/Parallel Circuits

Power is produced when current flows through any load or component that has resistance. The lower the resistance or higher the amount of current, the more power produced. The higher the resistance or lower the amount of current, the less power produced. As with any series or parallel circuit, the total power in a series/parallel circuit is equal to the sum of the power produced by each load or component.

POWER

Electrical energy is converted into another form of energy any time current flows in a circuit. Electrical energy is converted into sound (speakers), rotary motion (motors), light (lamps), linear motion (solenoids), and heat (heating elements).

Power is the rate of doing work or using energy. *True power (P_T)* is the actual power used in an electrical circuit. True power is the power that is converted into work for use by devices, such as sound produced by speakers, rotary motion produced by motors, light produced by lamps, linear motion produced by solenoids, and heat produced by heating elements. True power is measured in watts (W), kilowatts (kW), or megawatts (MW). In DC circuits or AC circuits in which voltage and current are in phase, such as resistive loads, true power is equal to the voltage (E) times the current (I). **See Figure 1-20.** Heating elements are resistive loads and are rated in true power (watts).

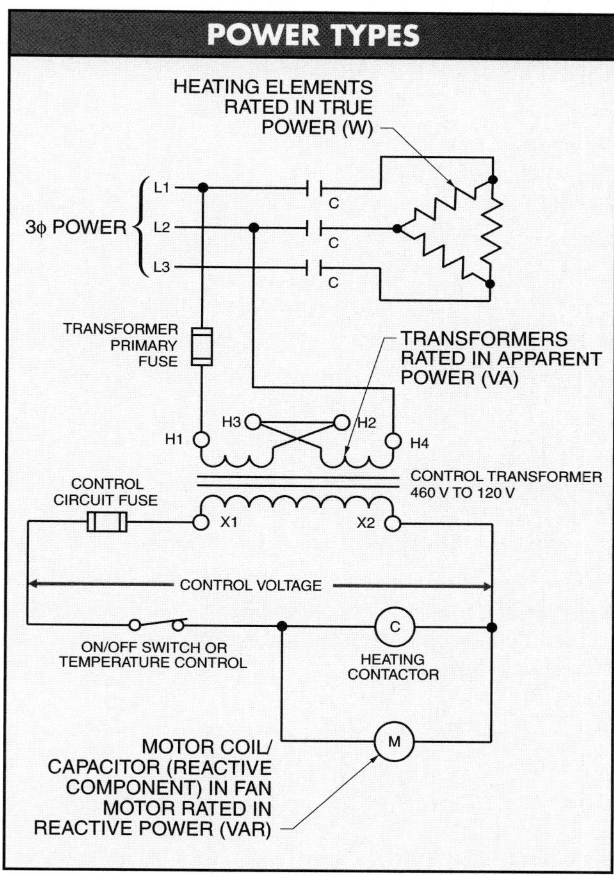

Figure 1-20. Power can be stated as true (W), reactive (VAR), or apparent (VA).

Reactive power (VAR) is power supplied to a reactive load (capacitor or coil). The unit of reactive power is volt-amps reactive (VAR). True power represents a pure resistive component or load and VAR represents a pure reactive (capacitor or coil) component or load. The capacitor on a motor uses reactive power (VAR) to keep the capacitor charged. The capacitor uses no true power because it performs no actual work such as producing light, heat, sound, linear motion, or rotary motion.

In an AC circuit in which voltage and current are in phase, such as a circuit containing only resistance, the power in the circuit is true power. However, almost all AC circuits include capacitive reactance (from capacitors) and/or inductive reactance (from coils). Inductive reactance is the most common, because all motors, transformers, solenoids, and coils have inductive reactance.

Apparent power (P_A) is the product of the voltage and current in a circuit calculated without considering the phase shift that may be present between the voltage and the current in a circuit. True power represents a pure resistive component or load in which voltage and current are in phase. Reactive power represents a pure inductive load or capacitor load in which voltage and current are out of phase. Apparent power represents a load or circuit that includes both true power and reactive power.

Apparent power is expressed in volt amps (VA), kilovolt amps (kVA), or megavolt amps (MVA). Apparent power is a measure of component or system capacity because apparent power considers circuit current regardless of how it is used. For this reason, transformers are sized in volt amps rather than in watts. Transformers have volt amp or kilovolt amp nameplate ratings.

True power is always less than apparent power in any circuit in which there is a phase shift between voltage and current. **See Figure 1-21.** For example, a 1ϕ, ¼ HP AC motor (resistive/reactive load) is required to lift a 60 lb load 30′ in 15 sec. To lift the load, the motor must deliver 186.5 W (true power). The motor nameplate lists motor current at 5 A and voltage at 115 V. The rated current (5 A) multiplied by the rated voltage (115 V) equals 575 VA (W). The difference between the true power and the apparent power exists because the coil in the motor must produce a rotating magnetic field for the motor to perform the work. Reactive power and true power are required from the power source because the motor coil is a reactive load. In small 1ϕ AC motor circuits, apparent power is much higher than true power.

Technical Fact

Power is used to describe the operation of any device or apparatus in which a flow of energy occurs. In many areas of design, the power, rather than the total work to be done, determines the size of the component used. Any device can do a large amount of work performing slowly at low power. However, if a large amount of work must be done rapidly, a high-power device is needed.

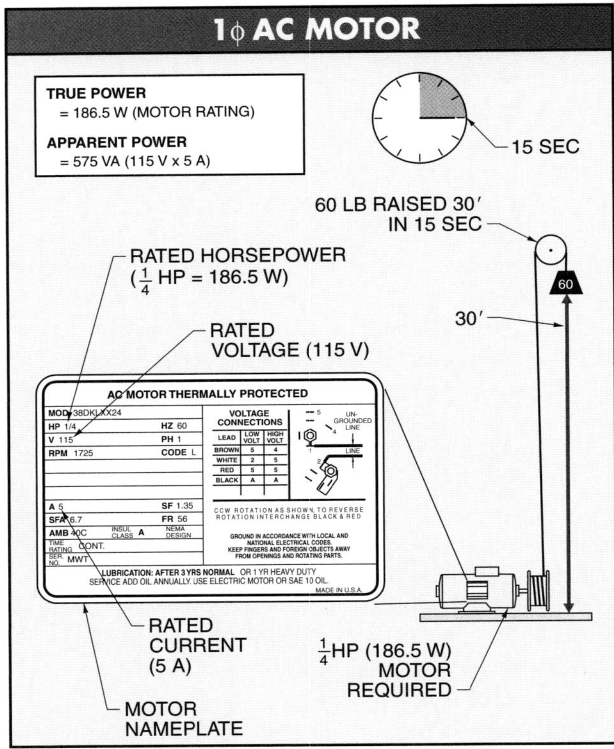

Figure 1-21. True power is always less than apparent power in a circuit with a phase shift between voltage and current.

Phase Shift

Some electrical circuits include only one load type, such as a heating element. However, most electrical circuits include several different load types, such as motors, solenoids, and lamps in addition to heating elements.

Each load affects the electrical system in a different way. For example, some loads cause a phase shift in the circuit. *Phase shift* is the state in which voltage and current in an AC circuit do not reach their maximum amplitude and zero level simultaneously. *In-phase* is the state in which voltage and current in an AC circuit reach their maximum amplitude and zero level simultaneously. **See Figure 1-22.** There is little to no phase shift in AC circuits that contain resistive devices. There is a phase shift in AC circuits that contain devices causing inductance or capacitance.

Resistive Circuits

A *resistive circuit* is a circuit that contains only resistance, such as heating elements and incandescent lamps. All electrical circuits include some resistance because all conductors, switch contacts, connections, and/or loads have resistance. AC voltage and current are in-phase in resistive circuits.

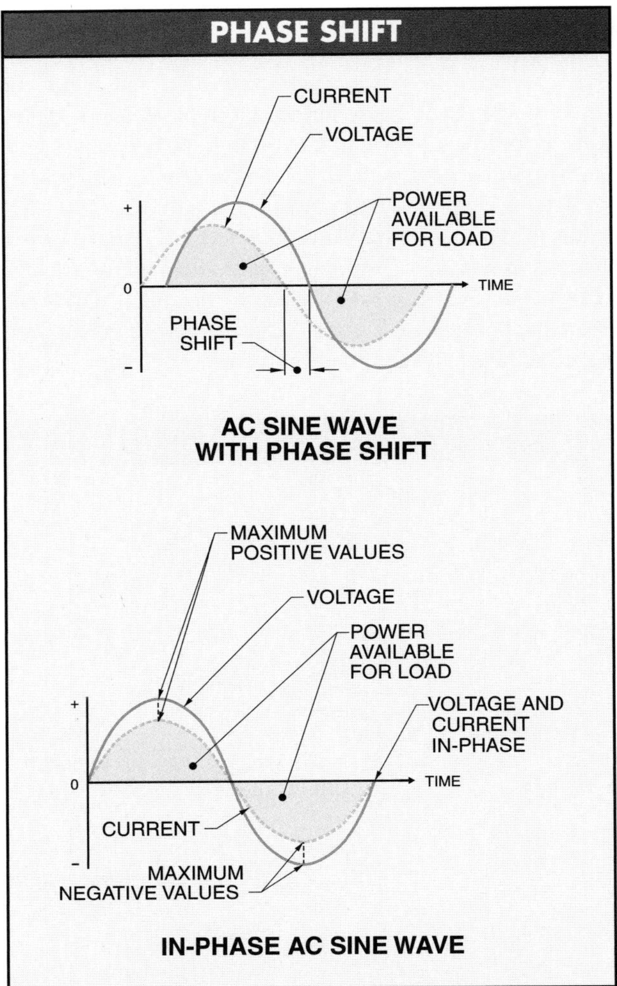

Figure 1-22. Phase shift occurs when voltage and current in an AC circuit do not reach their maximum amplitude and zero level simultaneously.

Inductive Circuits

Inductance (L) is the property of a circuit that causes it to oppose a change in current due to energy stored in a magnetic field. The opposition to a change in current is the result of energy stored in the magnetic field of the coil. Inductance is normally stated in henrys (H), millihenrys (mH), or microhenrys (μH).

Any time current flows through a conductor, a magnetic field is produced around the conductor. A conductor formed into a coil produces a strong magnetic field when current flows through the coil. AC current flow produces an alternating magnetic field around the coil. DC current flow produces a constant magnetic field around the coil. **See Figure 1-23.**

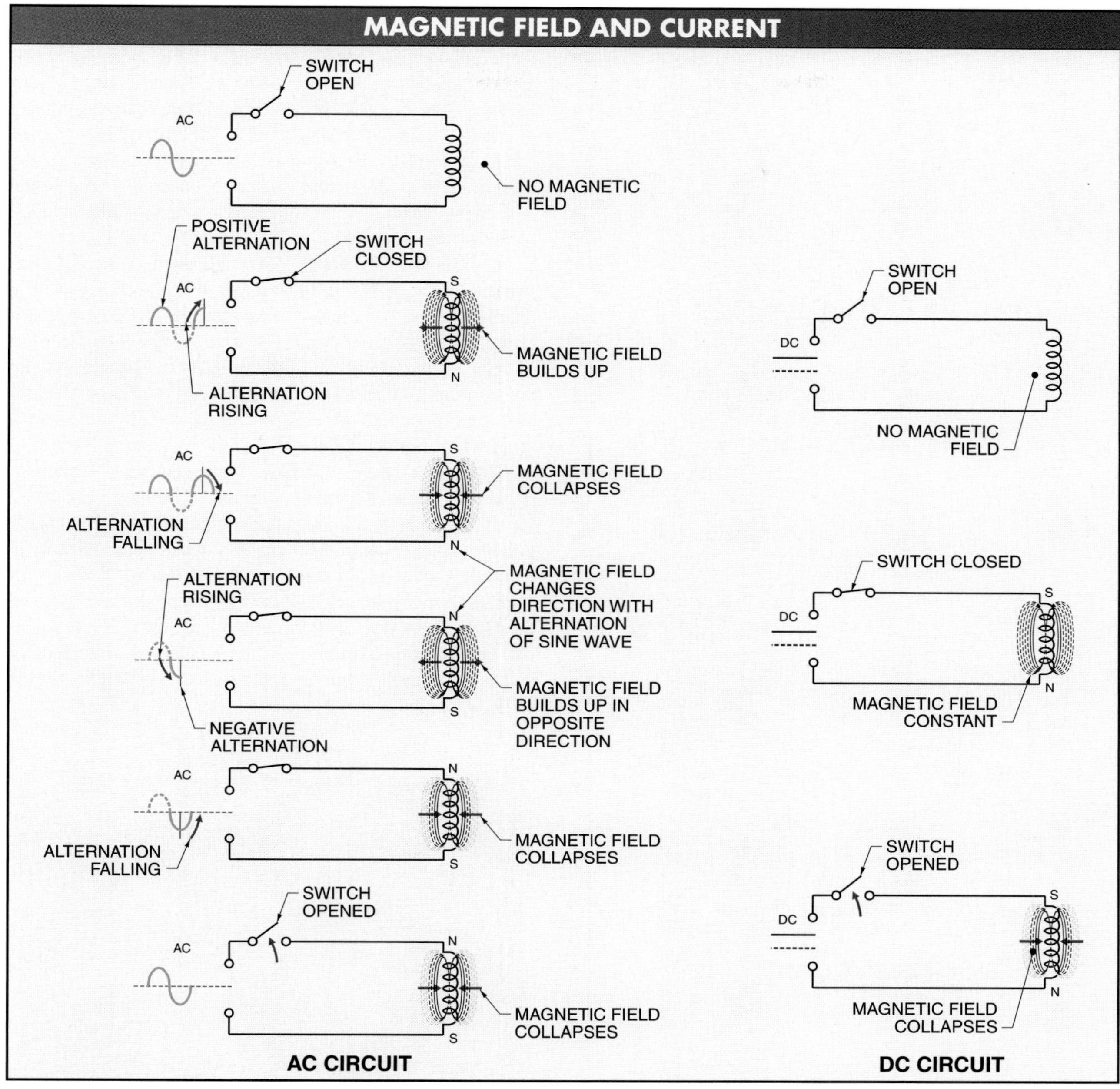

Figure 1-23. A conductor formed into a coil produces a strong magntetic field around the coil when current flows through the coil.

In an AC circuit, the magnetic field is continuously building and collapsing until the circuit is opened. The magnetic field also changes direction with each change in sine wave alternation. In a DC circuit, a magnetic field is created and remains at maximum potential until the circuit (switch) is opened. Once the circuit is opened, the magnetic field collapses.

Coils commonly found in motor windings, transformers, and solenoids create inductance in an electrical circuit. A phase shift occurs between alternating voltage and current in an inductive circuit. An *inductive circuit* is a circuit in which current lags voltage. The greater the inductance in a circuit, the larger the phase shift. **See Figure 1-24.**

Inductive reactance (X_L) is the opposition of an inductor to alternating current. Like resistance, inductive reactance is measured in ohms (Ω). The amount of inductive reactance in a circuit depends on the amount of inductance (in henrys) of the coil (inductor) and the frequency of the current. Inductance is normally a fixed amount. Frequency may be a fixed or variable amount.

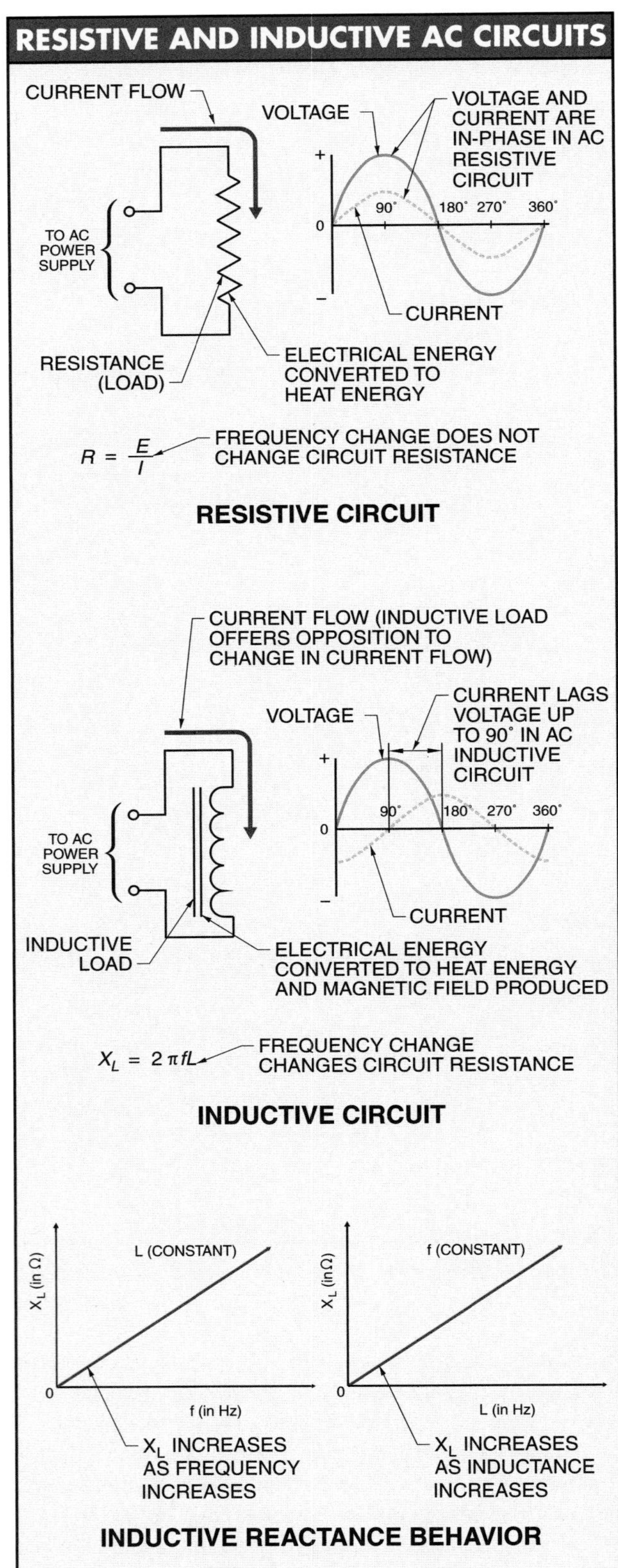

Figure 1-24. Voltage and current are in phase in AC resistive circuits. Current lags voltage in AC inductive circuits.

Capacitive Circuits

Capacitance (C) is the ability of a component or circuit to store energy in the form of an electrical charge. A *capacitor* is an electric device that stores electrical energy by means of an electrostatic field. The unit of capacitance is the farad (F). The farad is too large a unit to express capacitance for most electrical applications. Capacitance and capacitor values are normally stated in microfarads (μF) or picofarads (pF).

A capacitor consists of two conductors (plates) separated by an insulator (dielectric), which allows the development of an electrostatic charge. The electrostatic charge becomes a source of stored energy. The strength of the charge depends on the applied voltage, size of the conductors, and quality of the insulation. The dielectric may be air, paper, oil, ceramic, mica, or any other nonconductive material.

Capacitors create capacitance in an electrical circuit. A phase shift occurs between voltage and current in a capacitive circuit. A *capacitive circuit* is a circuit in which current leads voltage (voltage lags current). The greater the capacitance in a circuit, the larger the phase shift. **See Figure 1-25.**

Capacitive reactance (X_C) is the opposition to current flow by a capacitor. Like inductive reactance (X_L), capacitive reactance is expressed in ohms. Capacitors offer opposition to the flow of current in a circuit when connected to an AC power supply.

Fluke Corporation

A DMM set to measure output voltage will help indicate the condition of a capacitor bank in an electrical motor drive.

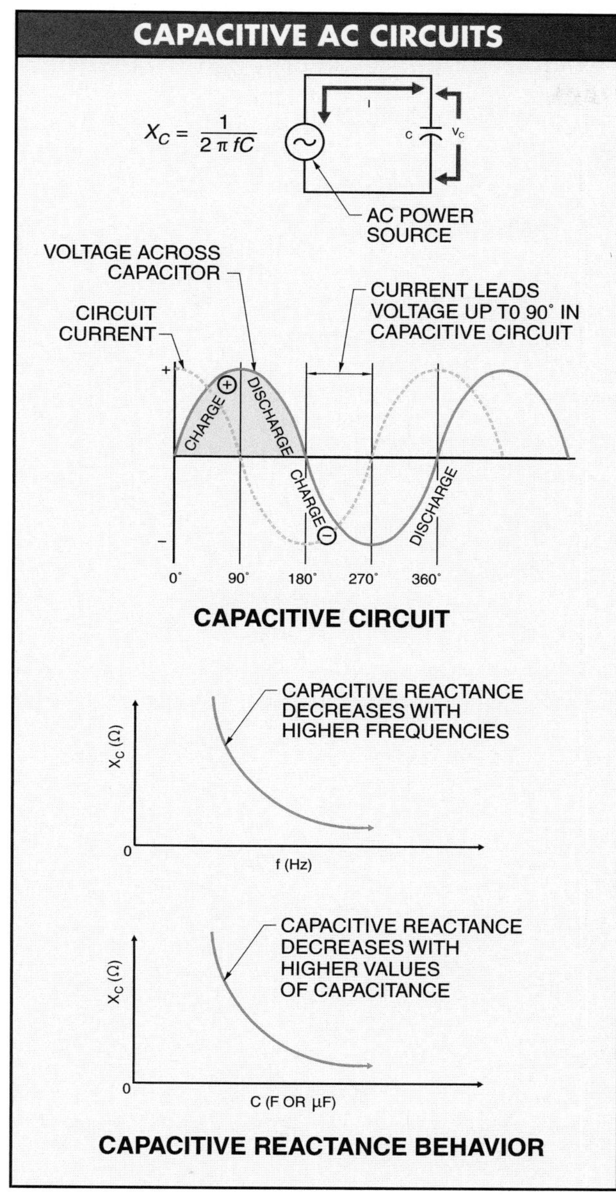

Figure 1-25. Current leads voltage in AC capacitive circuits.

Technical Fact

Capacitance added to a power circuit increases true power when a circuit includes resistive and inductive loads.

Power Factor

Power factor (PF) is the ratio of true power used in an AC circuit to apparent power delivered to the circuit. Power factor is commonly expressed as a percentage. True power equals apparent power only when the power factor is 100%. When the power factor is less than 100%, the circuit is less efficient and has a high operating cost because not all current is performing work. To calculate power factor, apply the formula:

$$PF = \frac{P_T}{P_A} \times 100$$

where

PF = power factor (in %)

P_T = true power (in W)

P_A = apparent power (in VA)

100 = constant (to convert decimal to percent)

For example, what is the power factor when the apparent power of an electrical circuit is 575 VA (.575 kVA) and the true power is 187 W (.187 kW)?

$$PF = \frac{P_T}{P_A} \times 100$$

$$PF = \frac{187}{575} \times 100$$

$$PF = .325 \times 100$$

$$PF = \mathbf{32.5\%}$$

Ohm's Law and Impedance

Ohm's law is limited to circuits in which electrical resistance is the only significant opposition to the flow of current. This limitation includes all DC circuits and any AC circuits that do not contain a significant amount of inductance and/or capacitance. AC circuits that do not contain inductance and/or capacitance include devices such as heating elements and incandescent lamps. AC circuits that include inductance are circuits that include a coil as the load. AC circuits that include capacitance are circuits that include one or more capacitors.

In DC circuits and AC circuits that do not contain a significant amount of inductance and/or capacitance, the opposition to the flow of current is resistance (R). In circuits that contain inductance (X_L) or capacitance (X_C), the opposition to the flow of current is reactance (X). In circuits that contain resistance and reactance, the combined opposition to the flow of current is impedance (Z). Impedance is stated in ohms.

Ohm's law is used in circuits that contain impedance; however, Z is substituted for R in the formula. Z represents the total resistive force (resistance and reactance) opposing current flow. The relationship between voltage (E), current (I), and impedance (Z) may be visualized by presenting the relationship in pie chart form. **See Figure 1-26.**

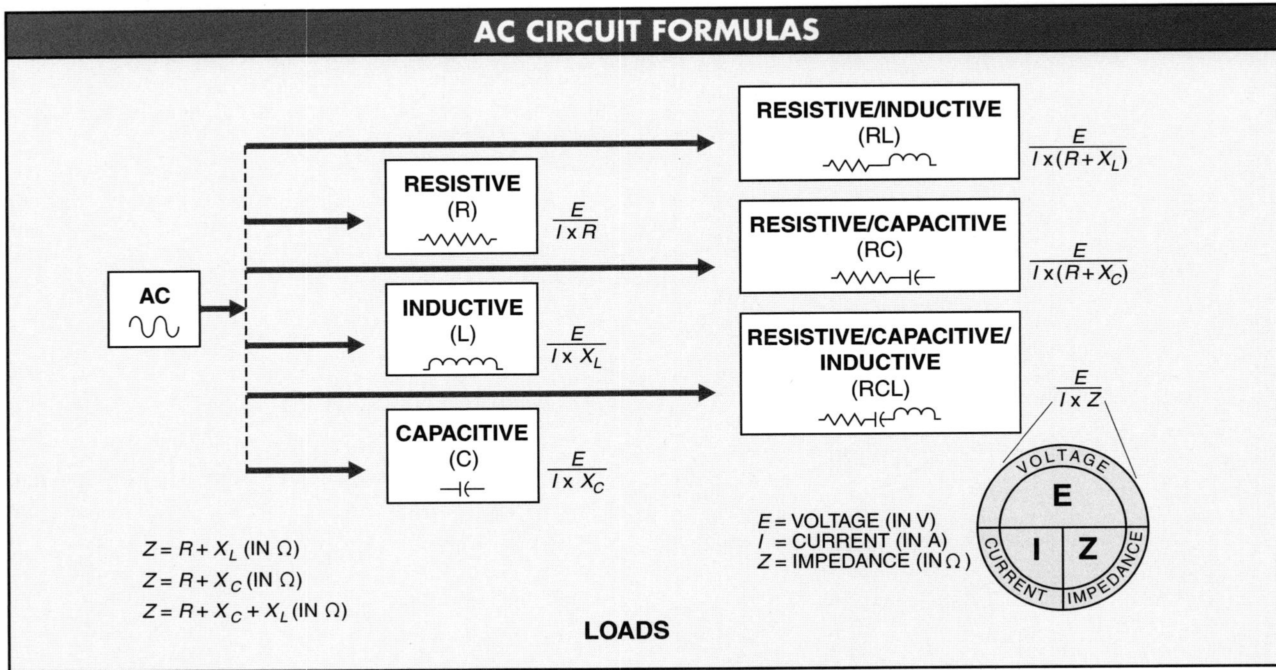

Figure 1-26. Ohm's law can be used on circuits with impedance by substituting Z (impedance) for R (resistance) in the formula.

Review Questions

1. What are the two forms of energy?
2. Which energy source is used to produce most of the generated electrical power?
3. What electrical device is used to change voltage from one level to another?
4. What is the unit of measurement and abbreviation used when measuring voltage?
5. What is the unit of measurement and abbreviation used when measuring current?
6. What is the unit of measurement and abbreviation used when measuring resistance?
7. What is the unit of measurement and abbreviation used when measuring power?
8. What is the unit of measurement and abbreviation used when measuring frequency?
9. Which power is the actual power used in an electrical circuit?
10. Which power is the power supplied to a capacitor or coil?
11. Which power is the product of the voltage and current in a circuit calculated without considering the phase shift that may be present between the voltage and current?
12. What is the property of a circuit that causes it to oppose a change in current due to energy stored in a magnetic field?
13. What is the ratio of true power used in an AC circuit to apparent power delivered to the circuit?
14. What does current times resistance equal in an electrical circuit?
15. What does voltage divided by current equal in an electrical circuit?
16. What does voltage times current equal in an electrical circuit?
17. What does Z stand for?
18. If an additional resistor (resistance) is added into a series-connected electrical circuit, does the total circuit resistance increase or decrease?
19. If an additional resistor (resistance) is added into a parallel-connected electrical circuit, does the total circuit resistance increase or decrease?
20. In which type of circuit is voltage the same across each of the loads?
21. If an additional resistor (resistance) is added into a series-connected electrical circuit, does the total circuit current increase or decrease?
22. If an additional resistor (resistance) is added into a parallel-connected electrical circuit, does the total circuit current increase or decrease?

chapter 2

electrical motor controls *for Integrated Systems*

Electrical Tools and Test Instruments

The proper tool or test instrument must be selected for each job. Tools and test instruments must be used for the tasks for which they were designed. Tools and test instruments must be organized and readily available for use. Consult the operator's manual for correct operation and use before using any new tool or test instrument.

TOOLS

Various hand and power tools are used by electricians for the maintenance, troubleshooting, and installation of electrical equipment. Different tools are designed for the efficient and safe completion of a specific job. Proper use of tools is required for safe and efficient electrical work.

Hand-Operated Tools

Electricians use hand tools for twisting, turning, bending, cutting, stripping, attaching, pulling, and securing operations. Noncutting operations involve tools such as screwdrivers, pliers, wrenches, hammers, fuse pullers, vises, fish tape, cable tie guns, and conduit benders. Cutting operations involve tools such as wire stripper/crimper/cutter tools, pipe cutters, hacksaws, and knives.

Screwdrivers. A *screwdriver* is a hand tool with a tip designed to fit into a screw head for fastening operations. Electricians use screwdrivers in many installation, troubleshooting, and maintenance activities to secure and remove various threaded fasteners. Various types of screwdrivers are available. The two main types of screwdrivers are the flathead and Phillips. Flathead and Phillips screwdrivers are available as standard, offset, and screwholding. **See Figure 2-1.**

Technical Fact

Iron or steel hand tools may produce sparks that can be an ignition source around flammable substances. Spark-resistant tools made of nonferrous materials such as brass or copper should be used where flammable gases, highly volatile liquids, and other explosive substances are stored or used.

Standard screwdrivers are used for the installation and removal of threaded fasteners. Offset screwdrivers provide a means for reaching difficult screws. A screwholding screwdriver is used to hold screws in place when working in tight spots. Once started, the screw is released and tightened with a standard screwdriver. Screwdrivers are available with square shanks to which a wrench can be applied for the removal of stubborn screws. Screwdrivers may also have a thin shank to reach and drive screws in deep, counterbored holes.

When using a screwdriver, ensure that the tip fits the slot of the screw snugly and does not project beyond the screw head. A screwdriver should never be used as a cold chisel or punch. A screwdriver should not be used near energized electrical wires and should never be exposed to excessive heat. A worn tip should be redressed with a file to regain a good, straight edge. A screwdriver that has a worn or broken handle should be discarded.

Pliers. *Pliers* are a hand tool with opposing jaws for gripping and/or cutting. Pliers are used by electricians for various gripping, turning, cutting, positioning, and bending operations. Common pliers include slip-joint, tongue-and-groove, long-nose, diagonal-cutting, side-cutting, end-cutting, and locking pliers. **See Figure 2-2.**

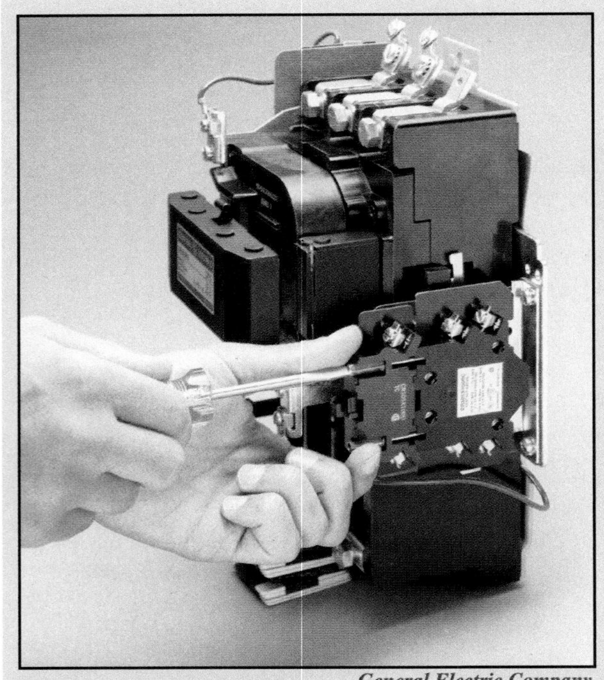

General Electric Company

Screwdrivers are used on electrical devices to add or remove components and make adjustments to the device.

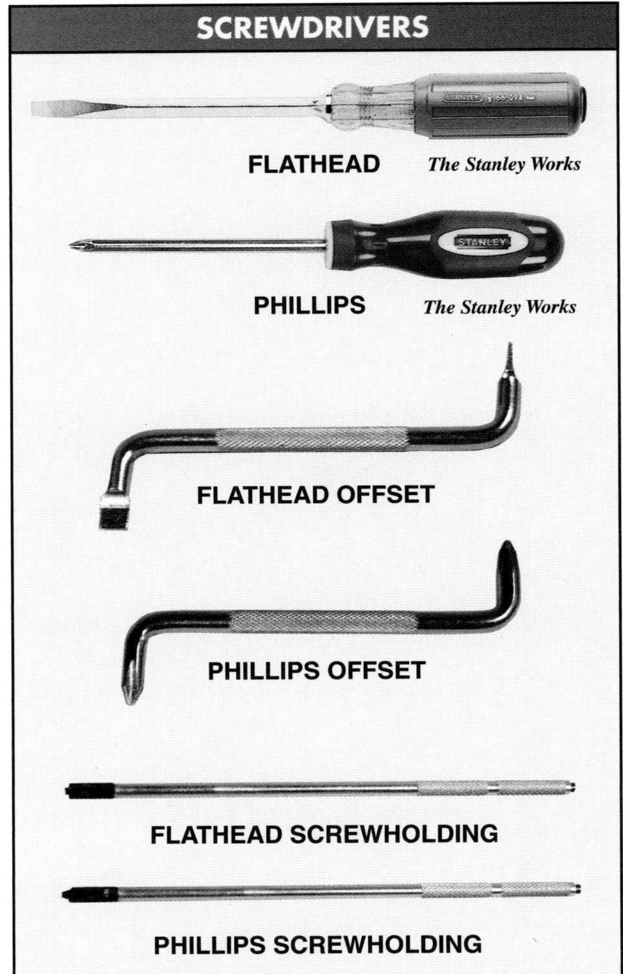

Figure 2-1. Electricians use screwdrivers in many installation, troubleshooting, and maintenance activities to secure and remove threaded fasteners.

Slip-joint pliers are used to tighten box connectors, lock nuts, and small-size conduit couplings. Tongue-and-groove pliers are used for a wide range of applications involving gripping, turning, and bending. The adjustable jaws of tongue-and-groove pliers enable a wide range of sizes. Long-nose pliers are used for bending and cutting wire and positioning small components. Diagonal-cutting pliers are used for cutting cables and wires too difficult to cut with side-cutting pliers.

Side-cutting (lineman's) pliers are used for cutting cable, removing knockouts, twisting wire, and deburring conduit. End-cutting pliers are used for cutting wire, nails, rivets, etc., close to the workpiece. Locking pliers (such as Vise-Grip™ pliers) are used to lock on to a workpiece. Locking pliers can be adjusted to lock at any size with any desired amount of pressure.

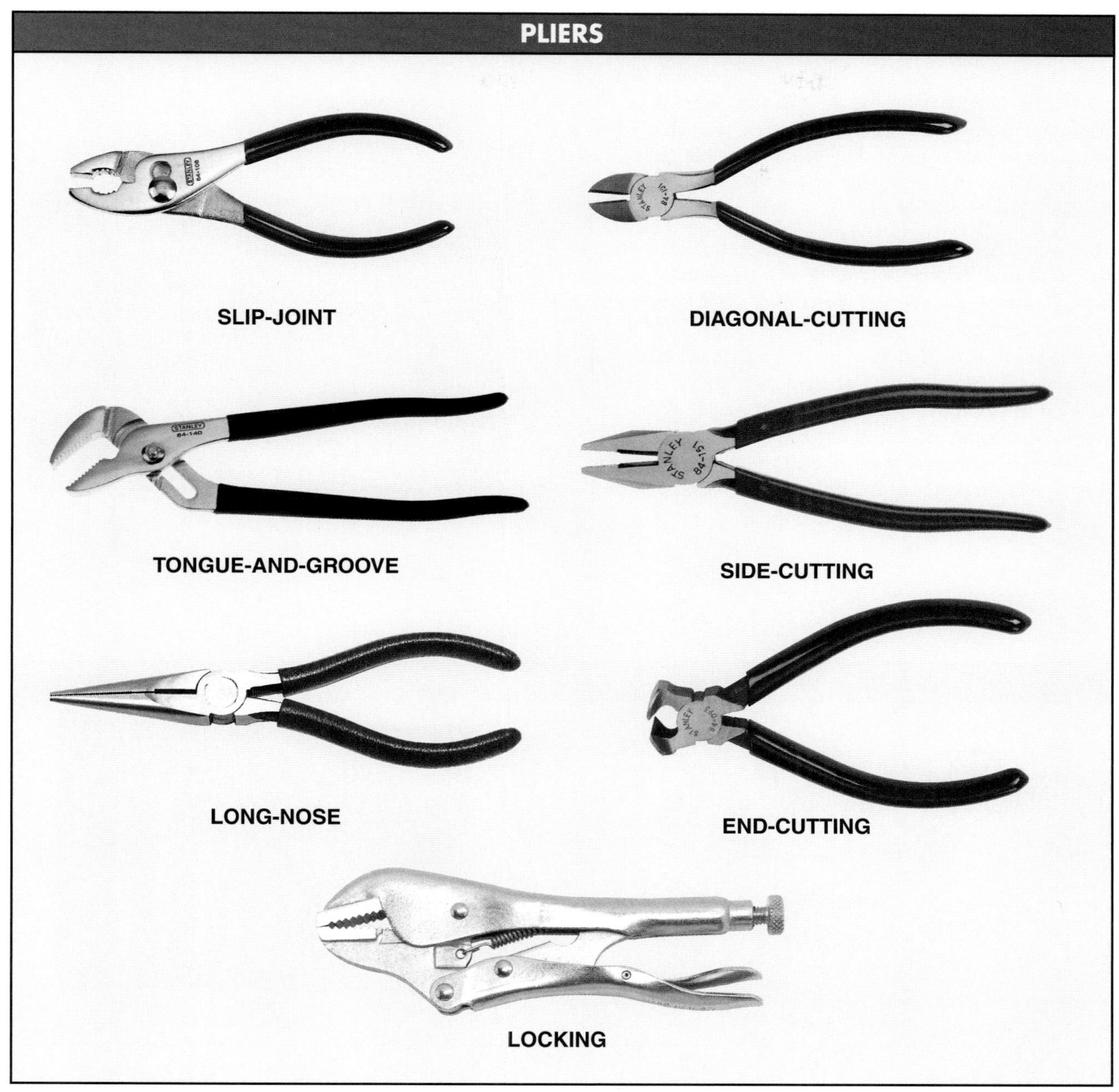

The Stanley Works

Figure 2-2. Pliers are used by electricians for various gripping, turning, cutting, positioning, and bending operations.

Wrenches. A *wrench* is a hand tool with jaws at one or both ends that is designed to turn bolts, nuts, or pipes. Common wrenches include socket, adjustable, hex key, combination, and pipe. **See Figure 2-3.**

Socket wrenches are used to tighten a variety of items, such as hex head lag screws, bolts, and various electrical connectors. Adjustable wrenches are used to tighten items such as hex head lag screws, bolts, and large conduit couplings. Hex key wrenches are used for tightening hex head bolts. A combination wrench is a hand tool with an open-end wrench on one end and a closed-end box wrench on the other. Pipe wrenches may be straight, offset, strap, or chain. Pipe wrenches are used to tighten and loosen pipes and large conduit.

When using wrenches, a pipe extension or other form of "cheater" should never be used to increase the leverage of the wrench. A wrench is selected with an opening which corresponds to the size of the nut to be turned. Too large an opening can spread the jaws of an open-end wrench and can batter the points of a box or socket wrench. Care should be taken to select inch wrenches for inch fasteners and metric wrenches for metric fasteners. If possible, always pull on a wrench handle. The safest wrench is a box or socket wrench because they cannot slip and injure the worker. Always use a straight handle rather than an offset handle if conditions permit.

Hammers. A *hammer* is a striking or splitting tool with a hardened head fastened perpendicular to a handle. Common hammers include the electrician's, ball peen, and sledgehammer. **See Figure 2-4.**

An electrician's hammer is used to mount electrical boxes and drive nails. An electrician's hammer may also be used to determine the height of receptacle boxes because most hammers are 12″ in length from head to end of handle, or can be so marked. Ball peen hammers of the proper size are designed for striking chisels and punches. Ball peen hammers may also be used for riveting, shaping, and straightening unhardened metal. Medium-sized sledgehammers (5 lb to 8 lb) are used for driving stakes and other heavy-duty pounding.

Technical Fact

The *Occupational Safety and Health Act of 1970* encourages states to develop and operate their own job safety and health plans. OSHA approves and monitors these plans. There are currently 26 state plans. States and territories with their own OSHA-approved plans must adopt and enforce standards identical to the federal standards and provide extensive programs of voluntary compliance and technical assistance.

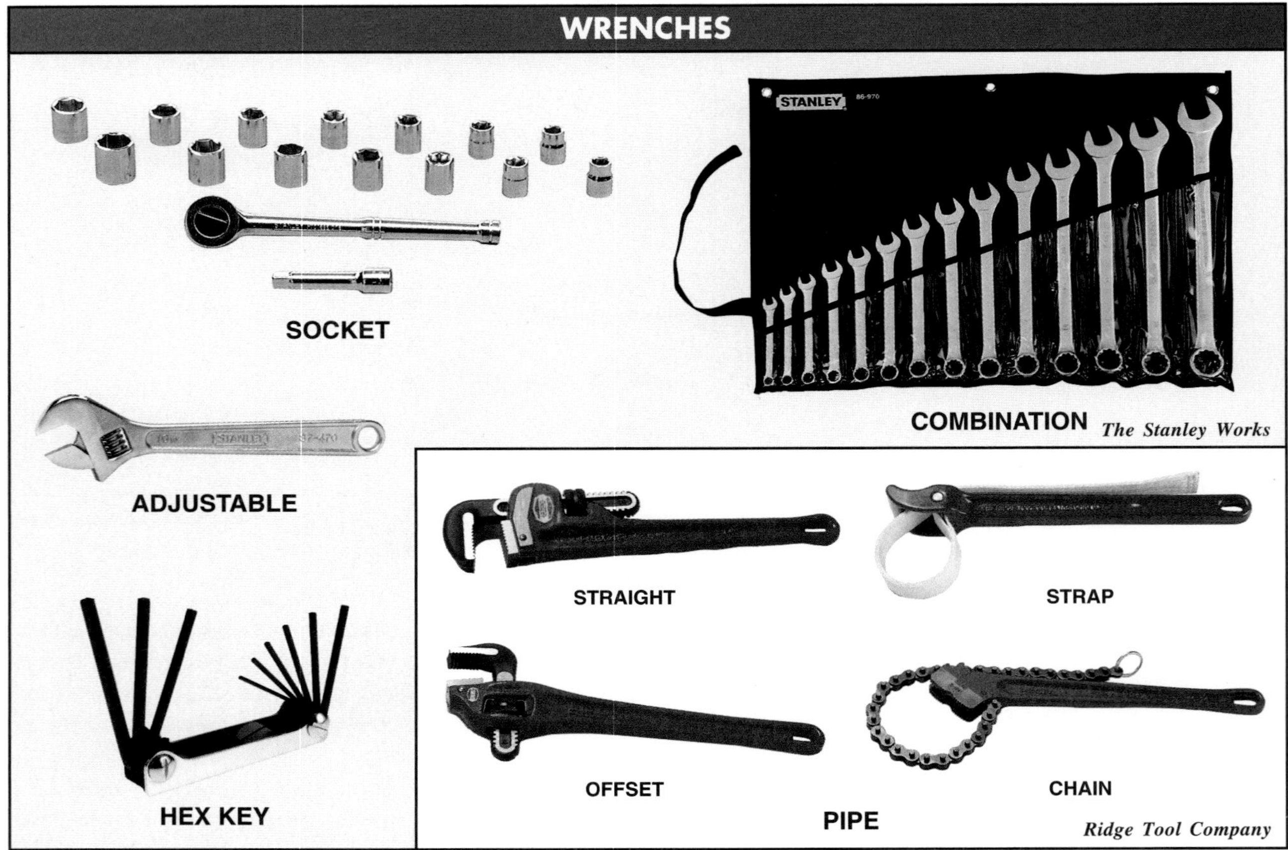

Figure 2-3. Common wrenches include socket, adjustable, hex key, combination, and pipe.

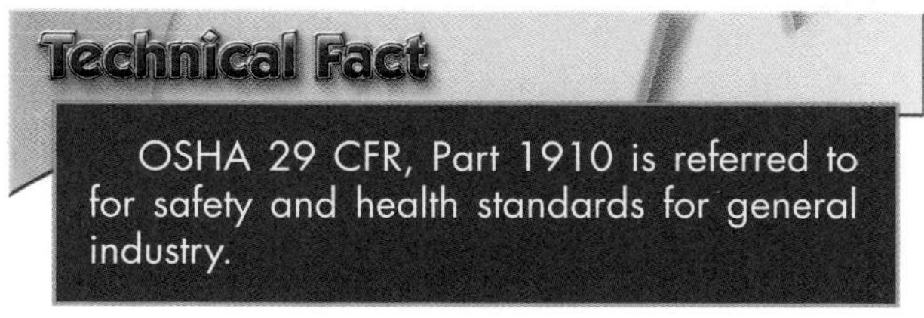

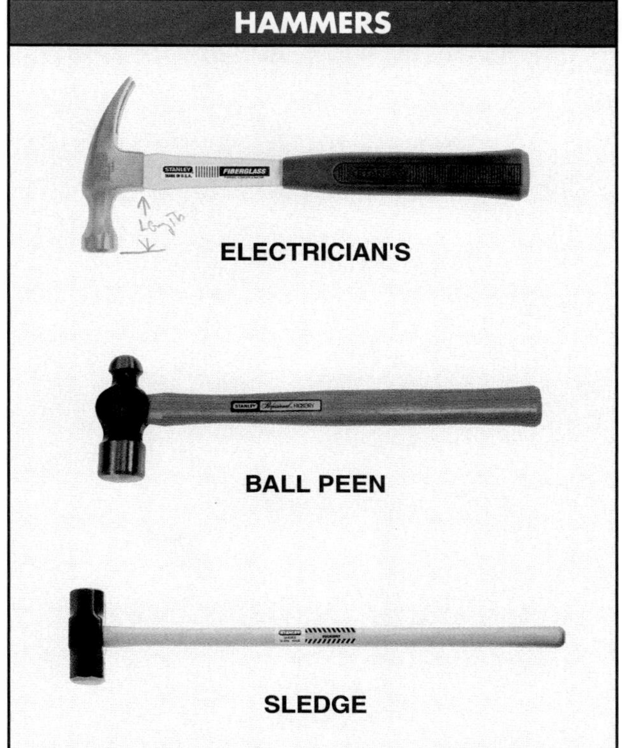

The Stanley Works

Figure 2-4. Common hammers include the electrician's, ball peen, and sledge.

Fuse Pullers. A *fuse puller* is a device that is used for the safe removal of fuses from electrical boxes and cabinets. **See Figure 2-5.** A fuse puller can be used to remove either operational or nonoperational (blown) fuses from an electrical box safely, preventing the electrician from coming in contact with an energized circuit. Fuse pullers have solid-grip jaws, are usually constructed of nylon or laminated fiber, and can be used on a variety of different size cartridge-type fuses.

Wire Stripper/Crimper/Cutter Tools. A *wire stripper/crimper/cutter* is a device used for the removal of insulation from small-diameter wire. **See Figure 2-6.** Most wire strippers strip stranded wire from American Wire Gauge (AWG) size 22 to AWG size 10 and solid wire from AWG size 18 to AWG size 8. A wire stripper/crimper/cutter is also used to crimp terminals from AWG size 22 to AWG size 10. Newer models are also designed with a wire cutter in the nose of the tool and a small-diameter bolt cutter near the handle. Most stripper/crimper/cutter tools shear bolts ranging from 4-40 to 10-32. Wire strippers are also available without the crimper and bolt-shear functions. Wire strippers without these functions are smaller and easier to handle than a stripper/crimper/cutter tool.

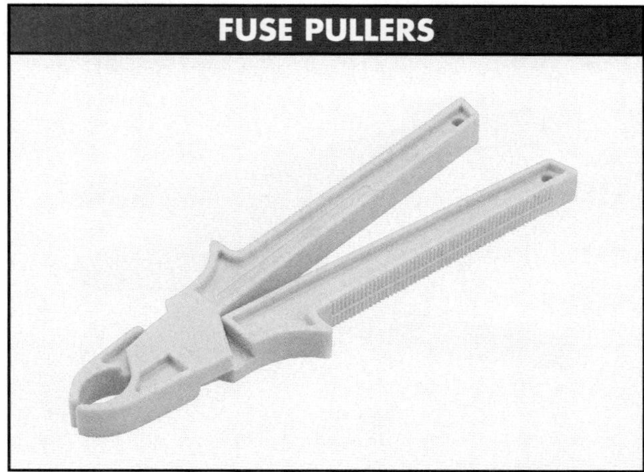

Ideal Industries, Inc.

Figure 2-5. A fuse puller is used to safely remove cartridge fuses from electrical boxes and cabinets.

Ideal Industries, Inc.

Holes saws are used by electricians for making clean, circular cuts in walls, floors, and ceilings.

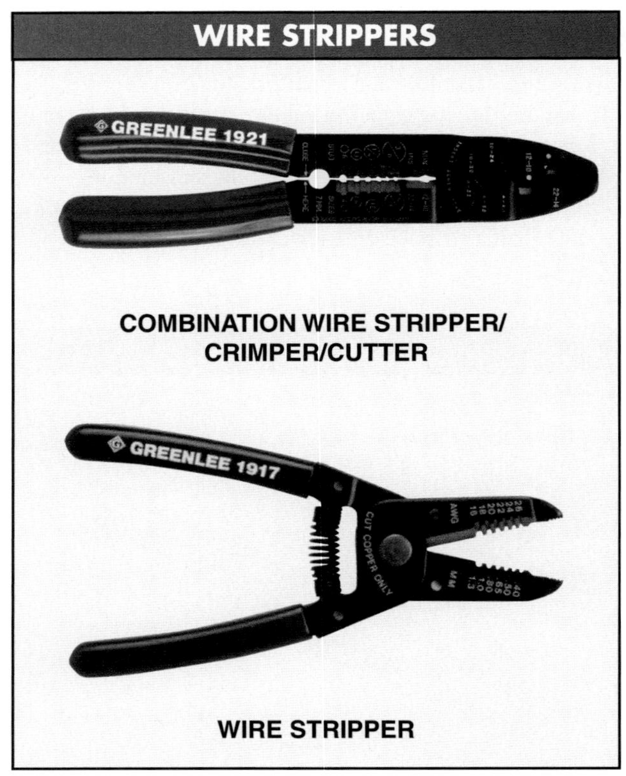

Greenlee Textron Inc.

Figure 2-6. Wire strippers and stripper/crimper/cutter tools are used to remove insulation from electrical wires, crimp terminals, and shear small-diameter bolts.

Pipe Vises. A *vise* is a portable or stationary clamping device used to firmly hold work in place. Pipe vises include yoke and chain vises. **See Figure 2-7.** A pipe vise has a hinge at one end and a hook at the opposite end. This allows the vise to be opened so that the pipe does not need to be threaded through the vise jaws to be worked on. A yoke pipe vise is bolted to a workbench. A clamp kit vise is a vise that contains a clamp for mounting. A clamp kit vise can be temporarily mounted for light-duty work without drilling holes. The yoke pipe vise and chain vise are available in portable workbench models.

Fish Tape. A *fish tape* is a retractable tape, usually of a rectangular cross section, that is pushed through an inaccessible space such as a run of conduit or a partition in order to draw in wires. **See Figure 2-8.** Fish tapes are constructed of either nylon, tempered steel, stainless steel, fiberglass, or multi-strand steel and range in length from 25′ to 200′. Steel fish tapes are coated with nylon to prevent corrosion and reduce friction and jamming. Nylon-coated and fiberglass fish tapes are recommended for applications where the fish tape may come in contact with energized electrical wires.

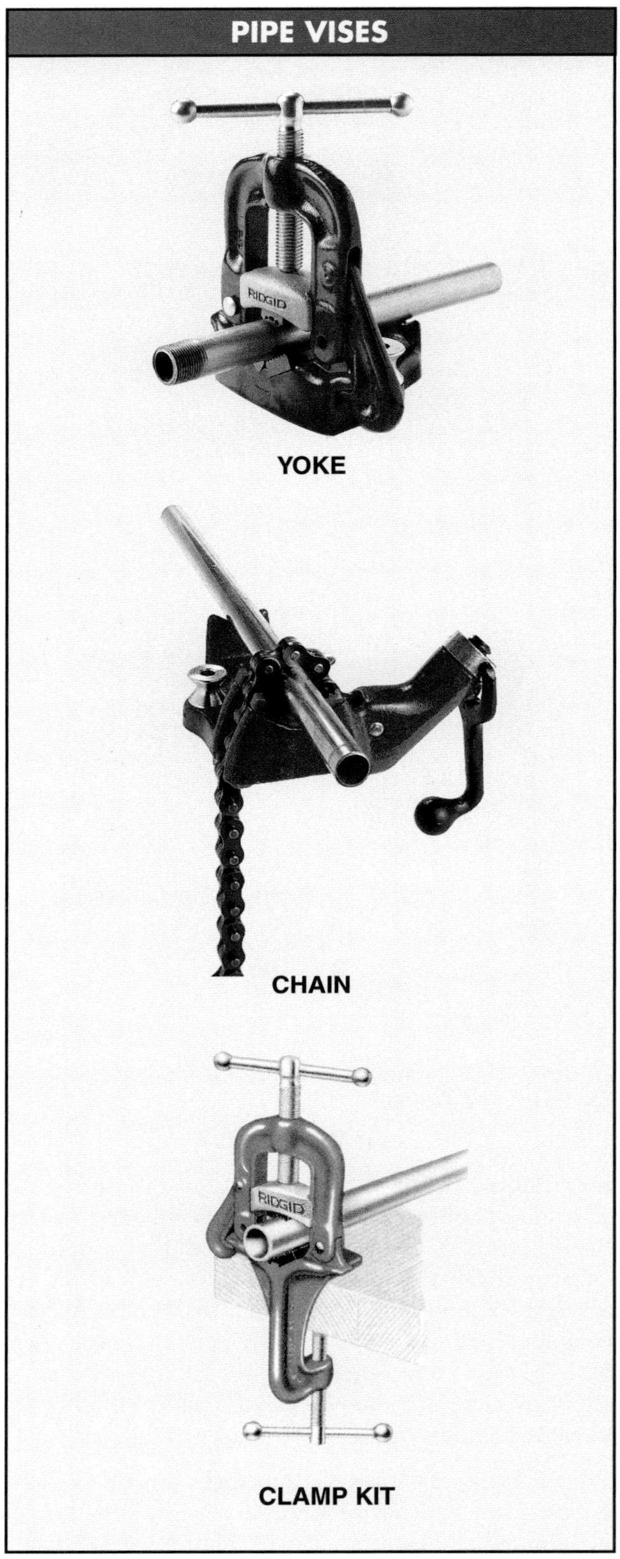

Ridge Tool Company

Figure 2-7. A vise is a portable or stationary clamping device used to firmly hold work in place.

Greenlee Textron Inc.

Figure 2-8. A fish tape is a retractable tape used to pull wires through conduit and other inaccessible spaces.

Ratcheting PVC Pipe Cutters. A *PVC pipe cutter* is a handheld tool designed to cut up to 2″ diameter PVC pipe, polyethylene pipe, and hose quickly and accurately without the use of a vise. **See Figure 2-9.** The workpiece is held in place by a hooked jaw. Alloy steel blades are ratcheted through the pipe to be cut. Alloy steel is used to retain the sharpness of the blades and increase the time period between blade replacement.

Ideal Industries, Inc.

Figure 2-9. A ratcheting PVC pipe cutter is used to cut PVC pipe up to 2″ in diameter without the use of a vise.

Cable Tie Guns. A *cable tie gun* is a handheld device that is used to hold and tighten several plastic or steel cable ties. **See Figure 2-10.** Bundles of wire can be quickly tied together in multiple locations by squeezing the trigger handle of the cable tie gun along the length of the wires to be bundled. Cable tie guns automatically tension and cut off the excess tie material. Cable tie guns are designed to reduce fatigue and increase the productivity of the operator.

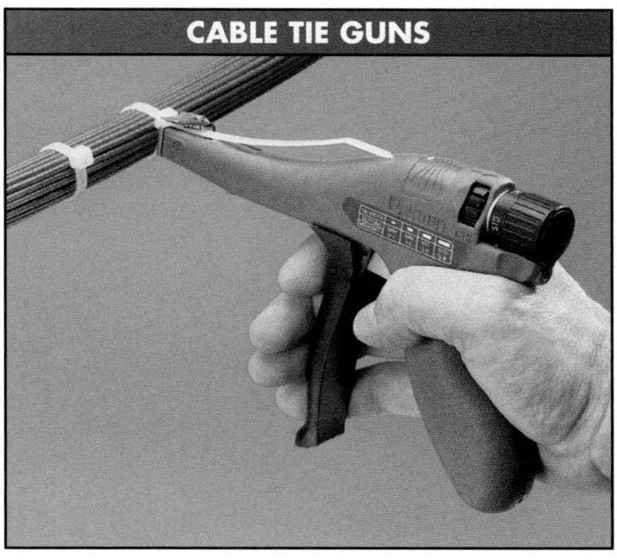

Panduit Corp.

Figure 2-10. A cable tie gun is used to quickly and efficiently tie bundles of small-diameter wires or cables together.

Conduit Benders. A *conduit bender* is a device used to radius (bend) electrical metallic tubing, intermediate metallic conduit, and rigid steel and aluminum conduit in sizes ranging from ½″ to 1½″ diameter. **See Figure 2-11.** Conduit is bent to clear obstructions when machinery or equipment is initially installed or relocated. Conduit can be bent properly using a hand conduit bender (hickey), mechanical conduit bender, or electric conduit bender. All conduit benders have bending shoes with high supporting sidewalls to prevent flattening or kinking of the conduit and are designed for quick, efficient conduit bending.

A hand conduit bender is used by contractors and maintenance personnel who occasionally need to bend conduit. Hand conduit benders are normally made from heat-treated aluminum or iron. Some hand conduit benders are available with a flared handle that can be used as a conduit straightening tool.

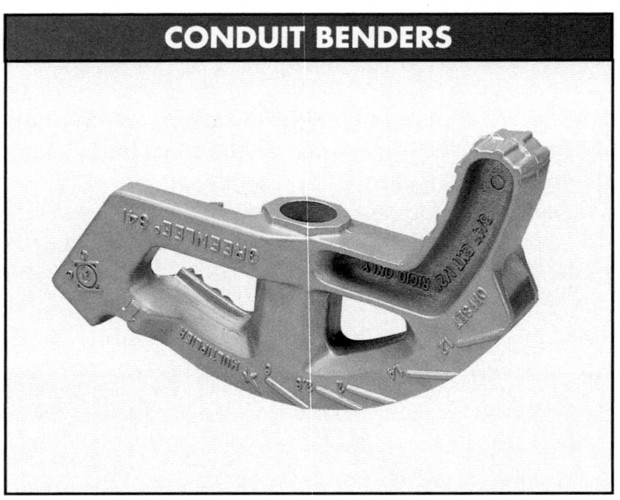

Greenlee Textron Inc.

Figure 2-11. A conduit bender is used to bend and shape metallic tubing from ½″ to 1½″ diameter.

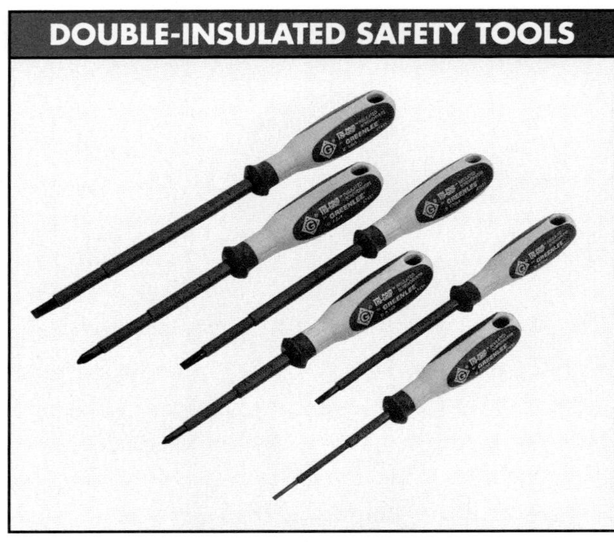

Greenlee Textron Inc.

Figure 2-12. Screwdrivers are double insulated to reduce the chance of electrical shock while working on energized electrical equipment and circuits.

A mechanical conduit bender uses a ratchet device to bend conduit. Some mechanical conduit benders are only used on certain types of conduit, while others have changeable bending shoes that allow one bender to be used on all sizes and types of conduit. A mechanical conduit bender is mounted on a frame and may have wheels so it can be moved easily. As the ratchet is pulled, a series of teeth rotate the bending shoe, bending the conduit. Mechanical conduit benders are generally used by contractors or maintenance personnel who are required to bend conduit frequently but in small volumes.

An electric conduit bender uses an electric motor that powers a chain-and-sprocket drive. High-volume conduit installers generally use electric conduit benders. Electric conduit benders have interchangeable bending shoes that allow the bender to be used on all types of conduit.

Double Insulated (1000 V 70E) Electric Safety Tools. Double insulated electric safety tools have a recommended maximum use of 1000 V and are designed to meet National Fire Protection Association (NFPA) 70E and Occupational Safety and Health Administration (OSHA) requirements for persons working with electrical hazards to be protected from electrical shock. The handles of double insulated electric safety tools are covered with a double-bonded, flame- and impact-resistant insulation material. Certain tools, such as pliers, also have finger guards to prevent accidental metal contact. Wrenches, hex keys, nutdrivers, screwdrivers, pliers, wire strippers, and socket wrenches are available as double insulated electric safety tools. **See Figure 2-12.**

Miscellaneous Hand Tools. Additional hand tools that are used by electricians include reamers, hand threaders, cable strippers, skinning knives, electrician's knives, nutdrivers, and hacksaws. An electrician's knife is similar in design to a pocket knife and is used for removing insulation and servicing conductors. **See Figure 2-13.**

Hand Tool Safety

Tools must be used properly to prevent injury. Tools should not be forced or used beyond their rated capacity. The proper tool does the job it was designed for quickly and safely. The time spent searching for the correct tool or the expense of purchasing it is less costly than a serious accident.

Tools should be kept in good working condition. Periodic inspection of tools helps to keep them in good condition. Always inspect a tool for worn, chipped, or damaged surfaces prior to use. Do not use a tool which is in poor or faulty condition. Tool handles should be free of cracks and splinters and should be fastened securely to the working part. Damaged tools are dangerous and less productive than those in good working condition, and should not be used. Repair or replace a tool immediately when inspection shows a dangerous condition.

Cutting tools should be sharp and clean. Dull tools are dangerous. The extra force exerted while using dull tools often results in losing control of the tool. Dirt, oil, or debris on a tool may cause slippage while being used and cause injury.

Figure 2-13. A wide variety of tools are used by electricians when doing electrical work.

Tools should be kept in a safe place. Any tool can be dangerous when left in the wrong place. Many accidents are caused by tools falling off ladders, shelves, and scaffolds.

Each tool should have a designated place in a toolbox, chest, or cabinet. Do not carry tools in clothing pockets unless the pocket is designed to carry a specific tool. Keep pencils in a pocket designed for them.

Keep all tools away from the edge of a bench or work area. Brushing against the tool may cause it to fall and injure a leg or foot. Carry sharp-edged and pointed tools with the cutting edge or point down and away from the body. Avoid placing work tools in environments such as solvents, prolonged moisture, or excessive heat which could cause permanent damage.

Technical Fact

Power tools are classified according to the type of power source: electric, pneumatic, liquid fuel, hydraulic, or powder actuated. All types must be fitted with appropriate guards and safety switches.

Power-Operated Tools

While much of the work performed by an electrician can be accomplished with hand tools, many tasks require the use of power tools. These tasks include pulling, cutting, drilling, sawing, and grinding. Typical power tools used by electricians include reciprocating saws, battery-powered cable cutters, laser distance estimators, power cable pullers, and hammer drills.

Reciprocating Saws. A *reciprocating saw* is a multipurpose cutting tool in which the blade reciprocates (quickly moves back and forth) to create the cutting action. Reciprocating saw blades can be plunged directly into walls, floors, ceilings, and other resilient material. Reciprocating saws operate at 1700 to 2800 strokes per minute (no load) and are used by electricians to cut holes in drywall, plywood, and hard-surface flooring in order to install electrical boxes and conduit. Reciprocating saws are typically used after the framing is done and additional cuts need to be made. **See Figure 2-14.**

RECIPROCATING SAWS

Milwaukee Electric Tool Corporation

Figure 2-14. Reciprocating saws are used to cut holes in walls, floors, and ceilings to allow placement of electrical boxes and conduit.

Battery-Powered Cable Cutters. A *battery-powered cable cutter* is a tool designed to cut various diameters of electrical and fiber-optic cables. Battery-powered cable cutters are similar to ratcheting PVC pipe cutters. Rather than using a ratcheting action, a battery-powered cable cutter receives power from a rechargeable battery pack and is ideal for large projects that require numerous cuts. Battery-powered cable cutters allow an electrician to reduce fatigue on nerves, muscles, and tendons and increase productivity. **See Figure 2-15.**

POWER CABLE CUTTERS

Greenlee Textron Inc.

Figure 2-15. A battery-powered cable cutter cuts various diameters of cable with minimal effort from the operator.

Laser Distance Estimators. A *laser distance estimator* is a device that uses sonar to take measurements. Laser distance estimators are ideal for pinpointing conduit, pipe, and ductwork locations and distances. The laser is used for aiming purposes only. An ultrasonic circuit performs the measurement. **See Figure 2-16.**

LASER DISTANCE ESTIMATORS

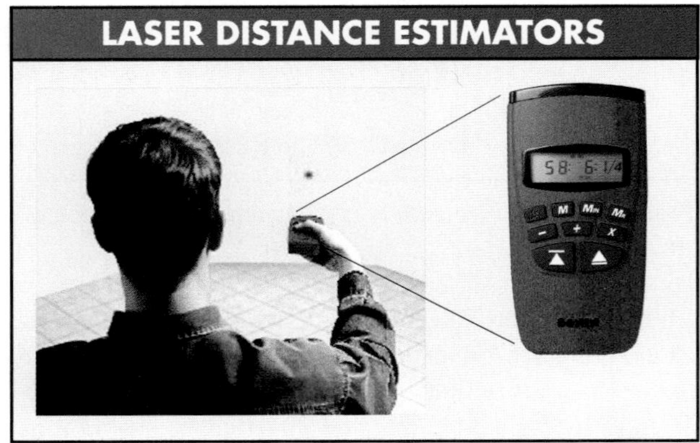

Sonin Incorporated

Figure 2-16. A laser distance estimator is used to take measurements without the use of a tape measure.

Power Cable Pullers. A *power cable puller* is a device used to pull large cables and wires into place. Power cable pullers are similar to fish tapes but use a gas- or electric-powered motor to pull wire. This reduces the strain and fatigue of the operator while increasing productivity. Power cable pullers can pull cables up to distances of 200′ and use up to 2000 lb of force. **See Figure 2-17.**

Greenlee Textron Inc.

Figure 2-17. A power cable puller uses a motor to pull large-diameter cables and wires into place.

Hammer Drills. A *hammer drill* is a drill that rotates and drives simultaneously. Hammer drills are used to perform multiple drilling operations in masonry, concrete, metal, and wood. Hammer drills can drill at speeds up to 3000 rpm and simultaneously hammer into the material at up to 50,000 blows per minute. Hammer drills are used by electricians to drill holes in concrete walls, floors, and ceilings for the attachment of conduit and electrical boxes. **See Figure 2-18.**

Milwaukee Electric Tool Corporation

Figure 2-18. A hammer drill is used to perform multiple drilling operations in masonry, concrete, wood, and metal.

Power Tool Safety

Power tools should only be used after gaining knowledge of their principles of operation, methods of use, and safety precautions. Always obtain authorization from supervisors before using any power tool. Full safety and health standards for the safe operation of power and hand tools are published by OSHA in Title 29, Code of Federal Regulations (CFR), Part 1910, Subpart P.

All power tools should be grounded unless they are approved double insulated. Power tools must have a grounded three-wire cord. A three-prong plug connects into a grounded electrical outlet (receptacle). Approved receptacles may be locking or nonlocking. Consult OSHA, the National Electrical Code® (NEC)®, and local codes for proper grounding requirements. **See Figure 2-19.**

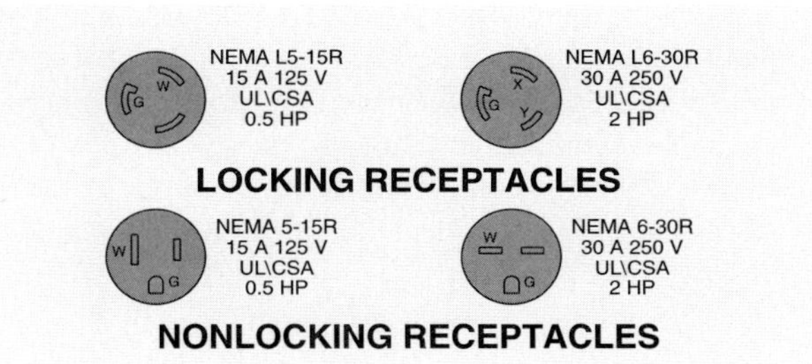

Figure 2-19. Approved receptacles may be locking or nonlocking.

It is dangerous to use an adapter to plug a three-prong plug into a nongrounded receptacle unless a separate ground wire or strap is connected to an approved ground. The ground ensures that any short circuit trips the circuit breaker or blows the fuse. **Warning:** An ungrounded power tool can cause fatal accidents.

Double insulated tools have two prongs and have a notation on the specification plate that they are double insulated. Electrical parts in the motor of a double insulated tool are surrounded by extra insulation to help prevent electrical shock. For this reason, the tool is not required to be grounded. Both the interior and exterior should be kept clean of grease and dirt that may conduct electricity. Safety rules must be followed when using power tools. Electrical power tool safety rules include the following:

- Review and understand all manufacturer safety recommendations.
- Read the owner's manual before using any power tool.
- Ensure that all safety guards are properly in place and in working order.

- Wear safety goggles at all times and a dust mask when required.
- Ensure that the workpiece is free of obstructions and securely clamped.
- Ensure that the tool switch is in the OFF position before connecting a tool to the power source.
- Keep attention focused on the work.
- A change in sound during tool operation normally indicates trouble. Investigate immediately.
- Power tools should be inspected and serviced by a qualified repair person at regular intervals as specified by the manufacturer or by OSHA.
- Inspect electrical cords regularly to ensure that they are in good condition.
- Shut OFF the power when work is completed. Wait until all movement of the tool stops before leaving a stationary tool or laying down a portable tool.
- Clean and lubricate all tools after use.
- Remove all defective power tools from service. Alert others to the situation.
- Take extra precautions when working on damp or wet surfaces. Use additional insulation to prevent any body part from coming into contact with a wet or damp surface.
- Always work with at least one coworker in hazardous or dangerous locations.

Tool Organization

Tools should be marked so that they can be easily identified as belonging to an individual or to a department in a company. To be effective, tools must be available when needed. Tools must not be damaged by abuse. An organized tool system provides both a central location and a means of protecting tools.

Tools can be organized in several ways depending on where and how frequently they are used. A pegboard may be appropriate if the tools are used at a repair bench. An electrician's leather pouch may be used if tools are used only at a construction site. A portable toolbox is normally best when tools are used at a bench and on a jobsite.

Pegboards. Pegboard is available in 4′ × 8′ sheets at most lumberyards. Heavy-duty tempered board is normally best for tools. Once the pegboard is mounted, outlines of the tools can be made to maintain tool inventory. The outlines may be painted on the board or cut out of self-adhesive vinyl paper.

Tool Pouches. Tool pouches are used to safely transport and store many small electrical hand tools and instruments. Tool pouches are normally made of heavy-duty leather or fabric and vary in design and size. Tool pouches are chosen based on specific needs and comfort. Some tool pouches hold only a few tools, while others hold a wide selection of tools. The pouch selected depends on the work required. **See Figure 2-20.**

Toolboxes, Chests, and Cabinets. Toolboxes, chests, and cabinets are used by many electricians to store, organize, and carry tools. A well-designed toolbox can be locked and helps keep tools clean and dry. Toolboxes may have a lift-out tray or may contain lever-operated trays that open automatically when the cover is lifted.

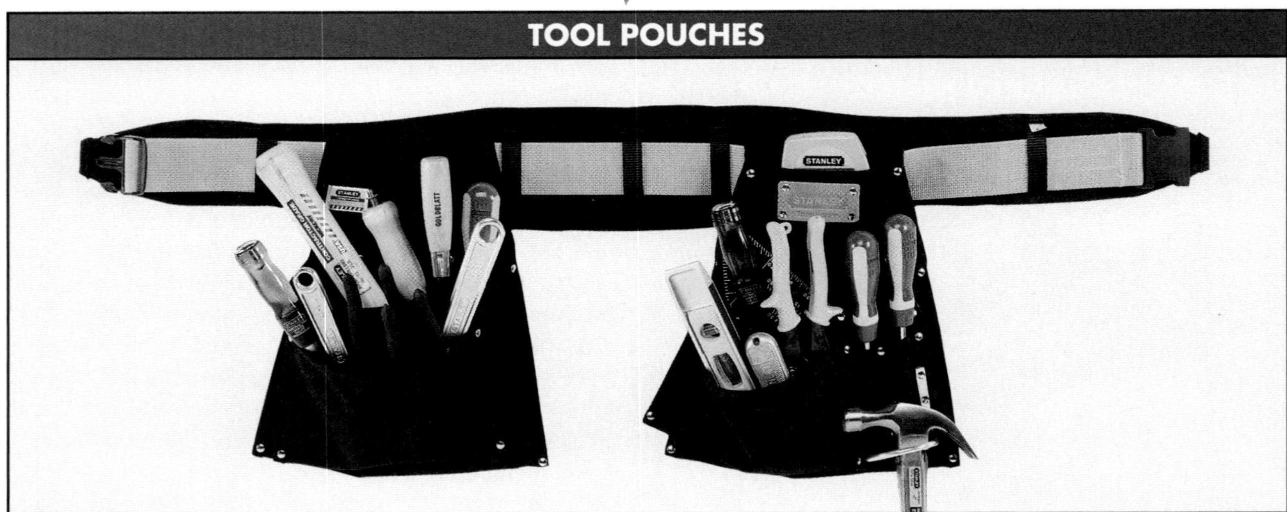

The Stanley Works

Figure 2-20. Tool pouches are used to conveniently and safely transport and store many small electrical hand tools and instruments.

Tool chests are more substantial than toolboxes. A tool chest may have from two to 10 drawers. A tool cabinet is used to organize and store a variety of tools and materials in one location. Tool cabinets are always mounted on casters. Tool chests can be added on top of the cabinet. A list of all tools should be kept in a toolbox, chest, or cabinet to ensure a complete inventory after each job. **See Figure 2-21.**

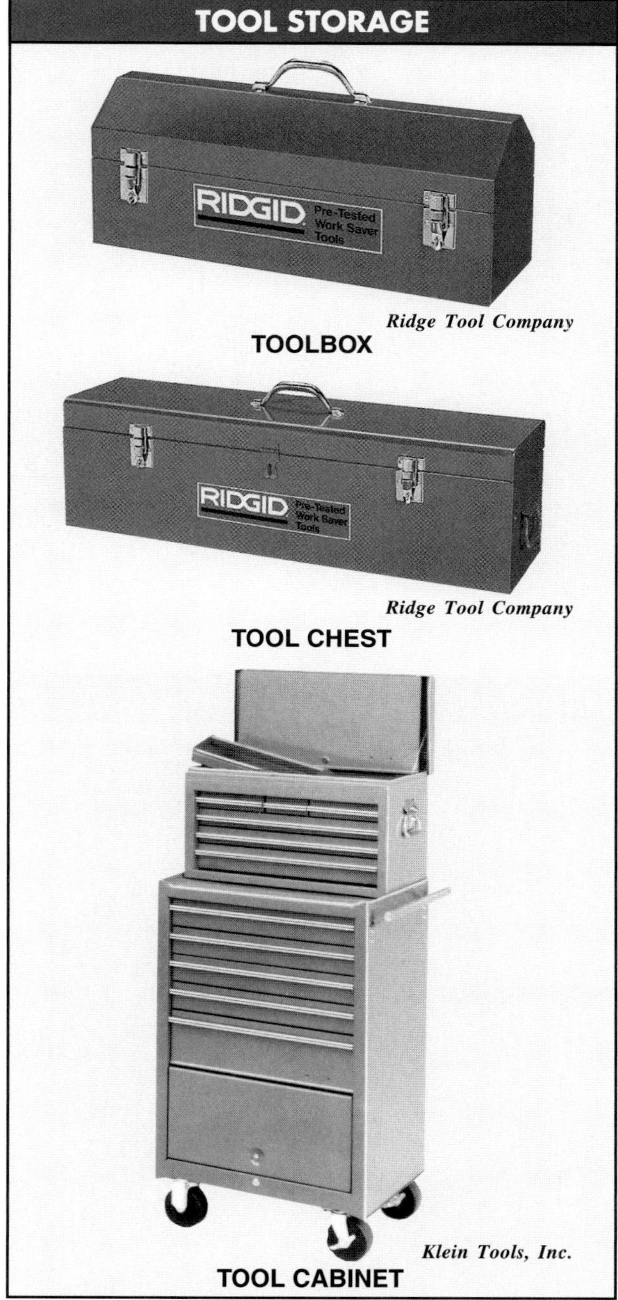

Figure 2-21. Toolboxes, chests, and cabinets are used by many electricians to store, organize, and carry tools.

An organized tool system ensures that clean, dry tools can be found when they are needed. Several manufacturers produce rust and moisture inhibitors that can be placed in a toolbox, chest, or cabinet. Unless protected, steel and some other metals rapidly oxidize (rust). The presence of moisture, even in small quantities, hastens oxidation.

ELECTRICAL TEST INSTRUMENTS

Electrical test instruments are used by electricians to aid in the taking of various electrical measurements. Care must be taken when using electrical test instruments because damage to the instrument or personal injury may result from improper or unsafe usage. The owner's manual must be consulted and all functions of a test instrument fully understood before using it.

Many types of test instruments are used by electricians. Most test instruments are available in either analog or digital versions and are designed to perform specific functions. Electrical test instruments include multimeters, clamp-on ammeters, megohmmeters, oscilloscopes, continuity testers, voltage testers, digital logic probes, phase sequence indicators, ground resistance testers, and receptacle testers.

Multimeters

A *multimeter* is a meter that is capable of measuring two or more electrical quantities. Multimeters can be used to measure electrical functions such as voltage, current, continuity, resistance, capacitance, frequency, and duty cycle. Multimeters may be analog or digital. **See Figure 2-22.**

Analog Multimeters. An *analog multimeter* is a meter that can measure two or more electrical properties and displays the measured properties along calibrated scales using a pointer. Analog multimeters use electromechanical components to display measured values. Most analog multimeters have several calibrated scales which correspond to the different selector switch settings (AC, DC, and R) and placement of the test leads (mA jack and 10 A jack). When reading a measurement on an analog multimeter, the correct scale must be used. The most common measurements made with analog multimeters are voltage, resistance, and current. Analog multimeters may also include scales for measuring decibels (dB) and checking batteries.

Reading Analog Displays. An *analog display* is an electromechanical device that indicates a value by the position of a pointer on a scale. Analog scales may be linear or nonlinear. A *linear scale* is a scale that is divided into equally spaced segments. A *nonlinear scale* is a scale that is divided into unequally spaced segments. **See Figure 2-23.**

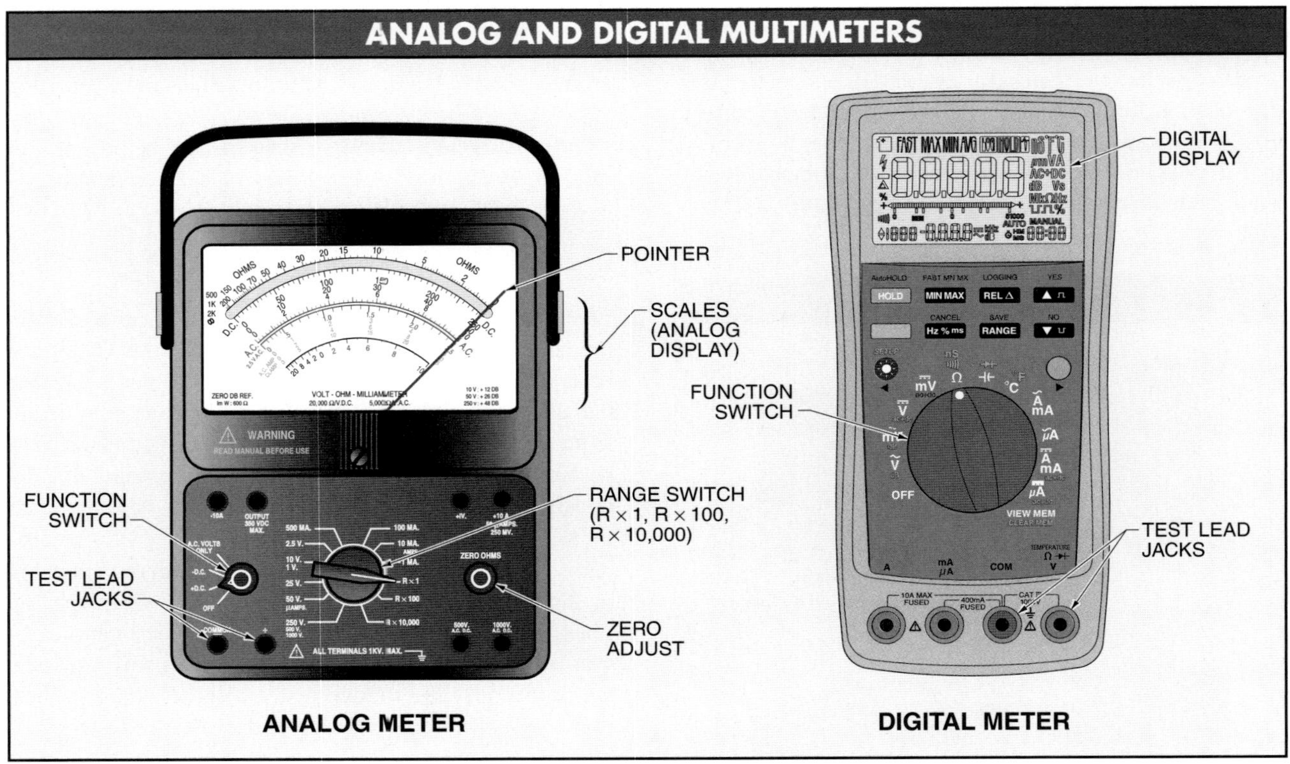

Figure 2-22. Multimeters are capable of measuring two or more electrical quantities and may be analog or digital meters.

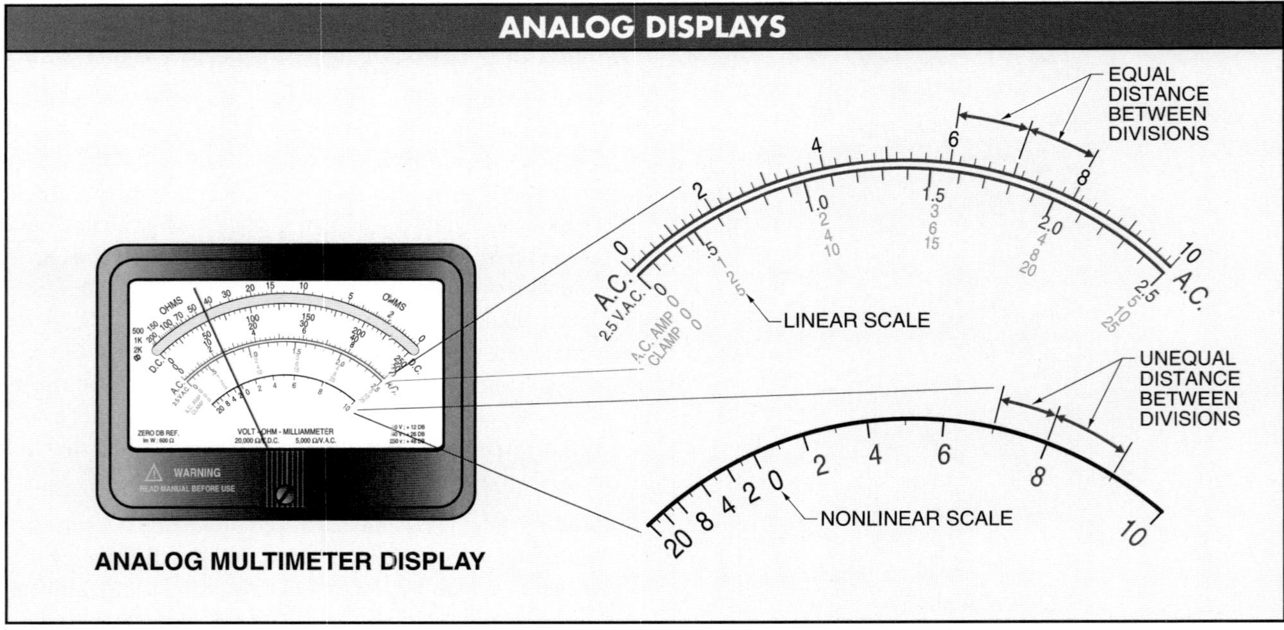

Figure 2-23. An analog display is an electromechanical device that indicates a value by the position of the pointer on a scale.

Analog scales are divided using primary divisions, secondary divisions, and subdivisions. A *primary division* is a division with a listed value. A *secondary division* is a division that divides primary divisions in halves, thirds, fourths, fifths, etc. A *subdivision* is a division that divides secondary divisions in halves, thirds, fourths, fifths, etc.

Secondary divisions and subdivisions do not have listed numerical values. When reading an analog scale, the primary, secondary, and subdivision readings are added. **See Figure 2-24.** To read an analog scale, apply the following procedure:

1. Read the primary division.
2. Read the secondary division if the pointer moves past a secondary division. *Note:* This may not occur with very low readings.
3. Read the subdivision if the pointer is not directly on a primary or secondary division. Round the reading to the nearest subdivision if the pointer is not directly on a subdivision. Round the reading to the next highest subdivision if rounding to the nearest subdivision is unclear. The primary division, secondary division, and subdivision readings are added to obtain the analog reading.

Analog meters are less susceptible to electrical noise than digital meters. Disadvantages of analog meters are that analog scales can be misread, especially if the meter has multiple settings (1 V, 10 V, 25 V, 50 V, 250 V) used with the same scale. In addition, analog meters are less accurate than digital meters (typical voltage measurements are within 1% to 5%, depending on meter specifications).

Digital Multimeters. A *digital multimeter (DMM)* is a meter that can measure two or more electrical properties and displays the measured properties as numerical values. Basic digital multimeters can measure voltage, current, and resistance. Advanced digital multimeters can include functions such as measuring capacitance and/or temperature. The main advantages of a digital multimeter over an analog multimeter are the ability of a digital meter to record measurements, and ease in reading the displayed values.

Reading Digital Displays. A *digital display* is an electronic device that displays readings on a meter as numerical values. Digital displays help eliminate human error when taking readings by displaying exact values measured. Errors occur when reading a digital display if the displayed prefixes, symbols, and/or decimal points are not properly applied.

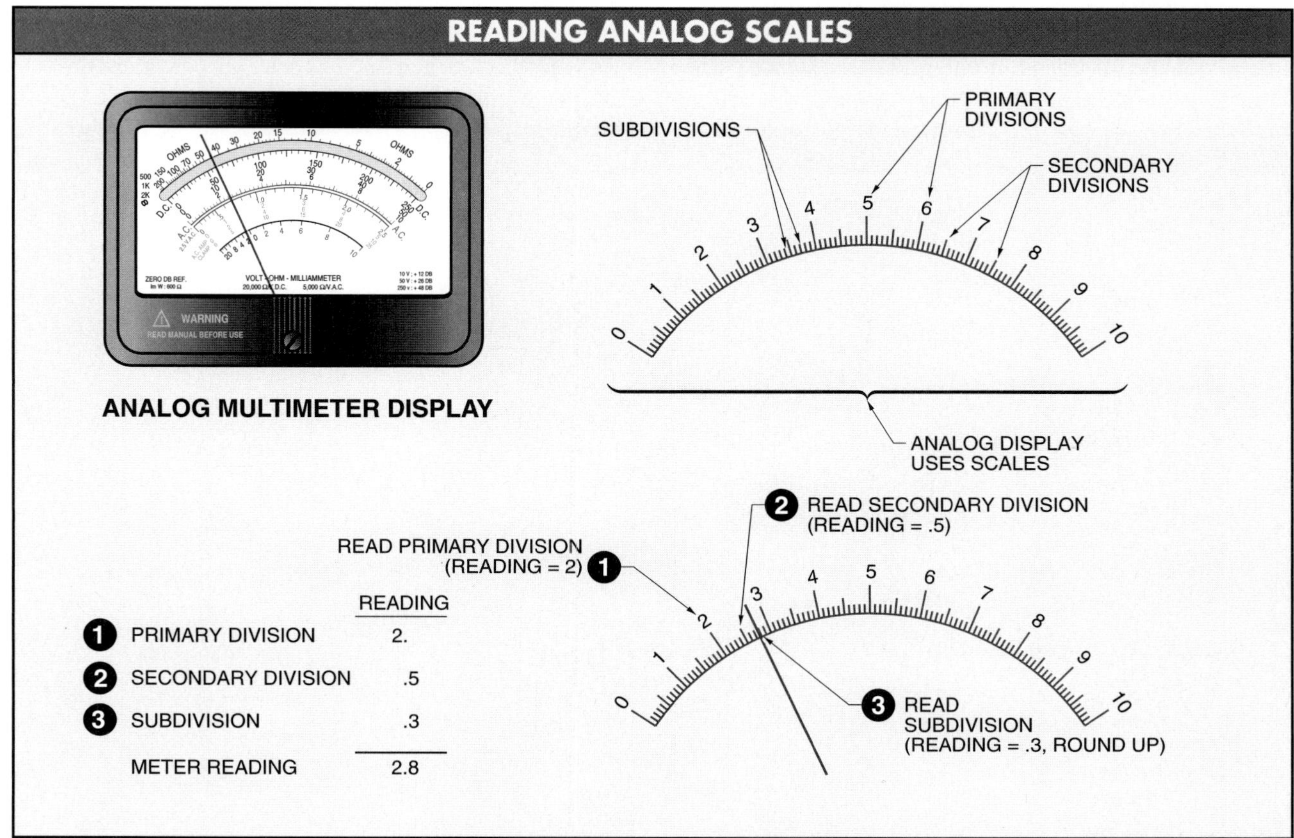

Figure 2-24. When reading an analog scale, the primary, secondary, and subdivision readings are added.

Digital displays display values using either a light-emitting diode (LED) display or a liquid crystal display (LCD). Light-emitting diode displays are easier to read than LCD displays but use more power. Most portable digital meters use liquid crystal displays. The exact value on a digital display is determined from the numbers displayed and the position of the decimal point. A selector switch (range switch) determines the placement of the decimal point.

Typical voltage ranges on a digital display are 3 V, 30 V, and 300 V. The highest possible reading with the range switch on 3 V is 2.999 V. The highest possible reading with the range switch on 30 V is 29.99 V. The highest possible reading with the range switch on 300 V is 299.9 V. Accurate readings are obtained by using the range that gives the best resolution without overloading. **See Figure 2-25.**

Bar Graphs. A *graph* is a diagram that shows a variable in comparison to other variables. Most digital displays include a bar graph to show changes and trends in a circuit. A *bar graph* is a graph composed of segments that function as an analog pointer. The displayed bar graph segments increase as the measured value increases and decrease as the measured value decreases. The polarity of test leads should be reversed if a negative sign is displayed at the beginning of a bar graph. **See Figure 2-26.** A *wraparound bar graph* is a bar graph that displays a fraction of the full range on the graph at one time. The pointer wraps around and starts over when the limit of the bar graph is reached.

A bar graph reading is updated 30 times per second. A digital reading is updated four times per second. The bar graph is used when quickly changing signals cause the digital display to flash or when there is a change in the circuit that is too rapid for the digital display to detect. For example, mechanical relay contacts may bounce open when exposed to vibration. Contact bounce causes intermittent problems in electrical equipment. Frequency and severity of contact bounce increase as a relay ages.

A contact's resistance changes momentarily from zero to infinity and back when a contact bounces open. *Infinity* is an unlimited number or amount. A digital display cannot indicate contact bounce because most digital displays require more than 250 milliseconds (ms) to update their reading. The quick response of bar graphs enables detection of most contact bounce problems. The contact bounce is displayed by the movement of one or more segments the moment the contact opens.

Most digital multimeters are autoranging, meaning that once a measuring function (such as VAC) is selected, the meter automatically selects the best meter range for taking the measurement (400 mV, 4 V, 40 V, 400 V, 1000 V range). In addition, digital multimeters have a greater accuracy than analog meters (typical voltage measurements are within .01% to 1.5%, depending on meter specifications).

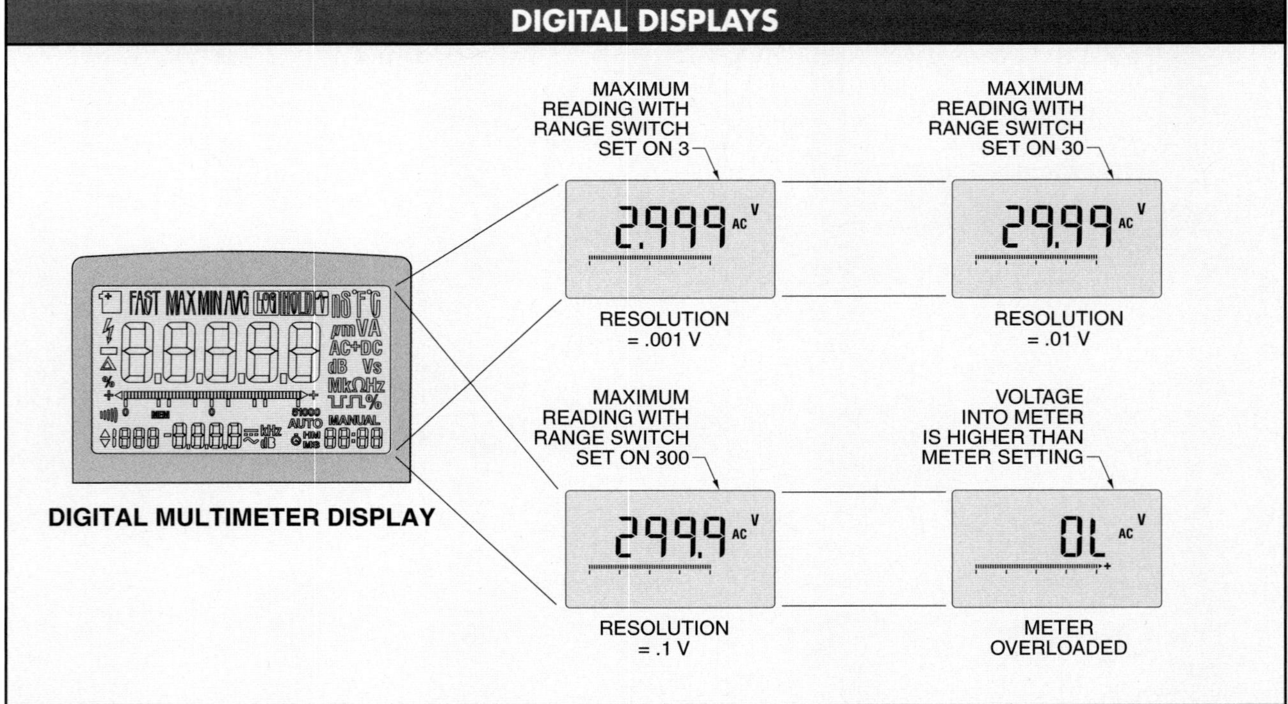

Figure 2-25. Digital displays display values using either a light-emitting diode (LED) display or a liquid crystal display (LCD).

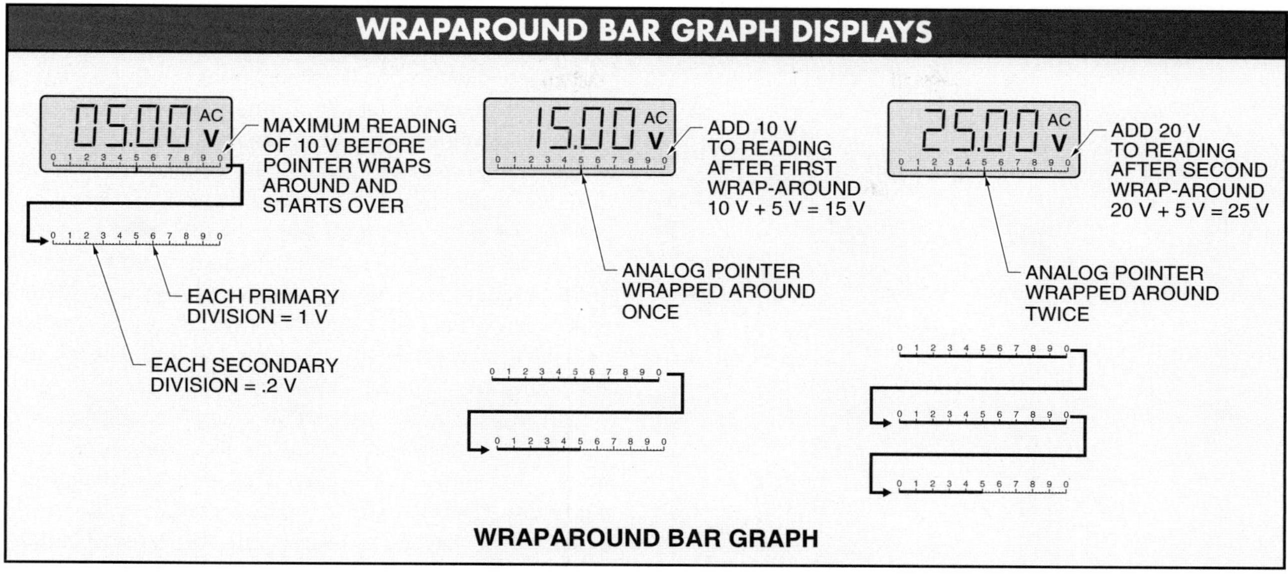

Figure 2-26. A bar graph is composed of segments that function as an analog pointer.

Ghost Voltages. A meter set to measure voltage may display a reading before the meter is connected to a powered circuit. The displayed voltage is a ghost voltage that appears as changing numbers on a digital display or as a vibrating analog display. A *ghost voltage* is a voltage that appears on a meter not connected to a circuit.

Ghost voltages are produced by the magnetic fields generated by current-carrying conductors, fluorescent lighting, and operating electrical equipment. Ghost voltages enter a meter through the test leads because test leads not connected to a circuit act as antennae for stray voltages. **See Figure 2-27.**

Ghost voltages do not damage a meter. Ghost voltages may be misread as circuit voltages when a meter is connected to a circuit that is believed to be powered. A circuit that is not powered can also act as an antenna for stray voltages. To ensure true circuit voltage readings, a meter is connected to a circuit for a long enough time that the meter displays a constant reading.

Clamp-on Ammeters

A *clamp-on ammeter* is a meter that measures the current in a circuit by measuring the strength of the magnetic field around a single conductor. Clamp-on ammeters measure currents from .01 A or less to 1000 A or more. A clamp-on ammeter is normally used to measure current in a circuit with over 1 A of current and in applications in which current can be measured by easily placing the jaws of the ammeter around one of the conductors. Most clamp-on ammeters can also measure voltage and resistance. To measure voltage and resistance, the clamp-on ammeter must include test leads and a voltage and resistance mode. **See Figure 2-28.**

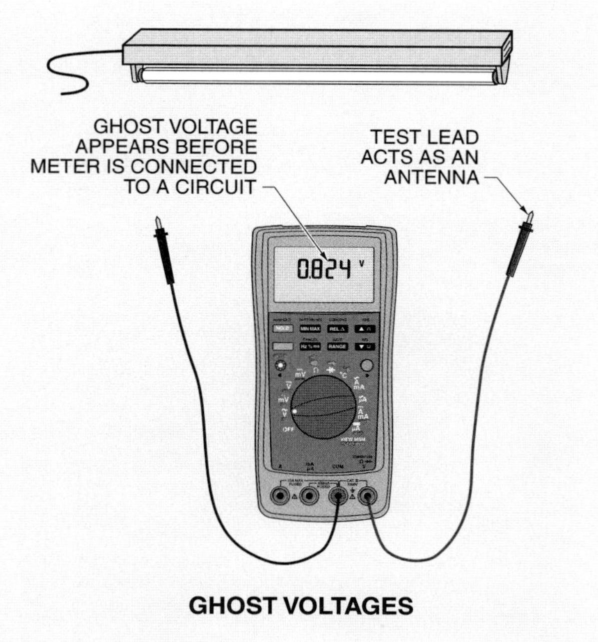

Figure 2-27. Ghost voltages are produced by the magnetic fields generated by current-carrying conductors, fluorescent lighting, and operating electrical equipment.

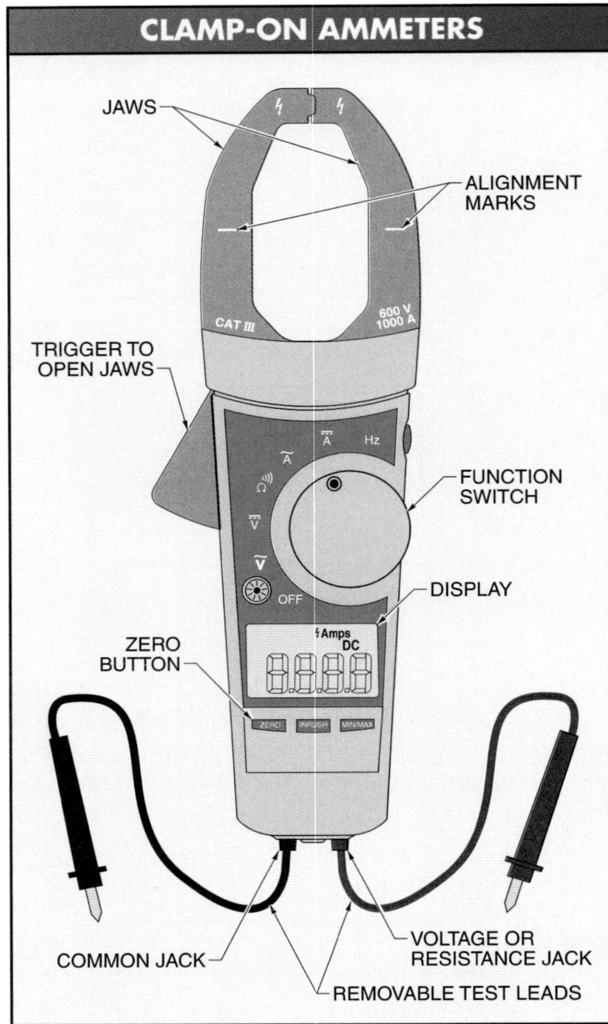

Figure 2-28. A clamp-on ammeter measures the current in a circuit by measuring the strength of the magnetic field around a single conductor.

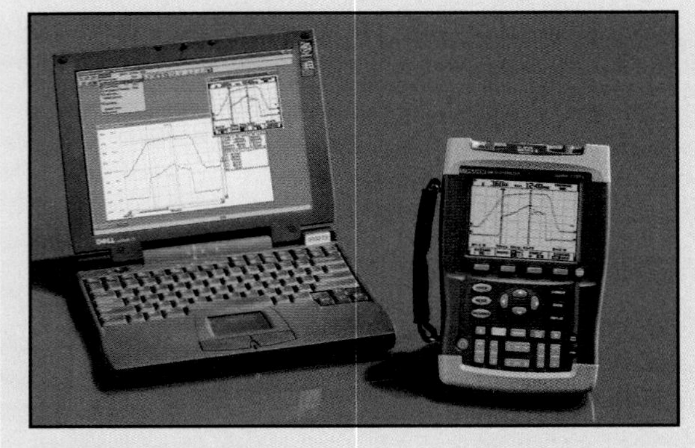

A portable computer is often interfaced with an electrical test instrument to store and analyze data.

Current is a common troubleshooting measurement because only a current measurement can be used by an electrician to determine how much a circuit is loaded or if a circuit is operating correctly. Current measurements vary because current can vary at different points in parallel or series/parallel circuits. Current in series circuits is constant throughout the circuit. The largest amount of current in a parallel circuit is at a point closest to the power source; current decreases as the system distributes current to each of the parallel loads. Any variation that is excessively high must be investigated because the current measurement may indicate that a partial short exists on one of the lines and a small amount of current is flowing to ground.

Electricians must ensure that clamp-on ammeters do not pick up stray magnetic fields by separating conductors being tested as much as possible from other conductors during testing. If stray magnetic fields are possibly affecting a measurement, several measurements at different locations along the same conductor must be taken.

Megohmmeters

A *megohmmeter* is a device that detects insulation deterioration by measuring high resistance values under high test voltage conditions. Megohmmeter test voltages range from 50 V to 5000 V. A megohmmeter detects insulation failure or potential failure of insulation caused by excessive moisture, dirt, heat, cold, corrosive vapors or solids, vibration, and aging. **See Figure 2-29.**

Some insulation, such as that found on conductors used to wire branch circuits, has a thick insulation and is harder to damage or break down. Other insulation, such as the insulation used on motor windings, is very thin (to save weight and space) and breaks down much more easily. Megohmmeters are used to test for insulation breakdown in long wire runs or motor windings.

Insulation allows conductors to stay separated from each other and from earth ground. Insulation must have a high resistance to prevent current from leaking through the insulation. All insulation has a resistance value that is less than infinity, which allows some current leakage to occur. *Leakage current* is current that flows through insulation. Under normal operating conditions, the amount of leakage current is so small (only a few microamperes) that the leaking current has no effect on the operation or safety of a circuit.

Oscilloscopes

Testing electronic circuits and equipment often requires measuring or observing electrical signals. An *oscilloscope* is an instrument that displays an instantaneous voltage. **See Figure 2-30.** An oscilloscope is used to display the shape of a voltage waveform when bench testing electronic circuits. *Bench testing* is testing performed when equipment under test is brought to a designated service area. Oscilloscopes show the shape of a circuit's voltage and allow the voltage level, frequency, and phase to be measured.

Oscilloscopes are used to troubleshoot digital circuits, communication circuits, TVs, VCRs, DVD players, computers, and other types of electronic circuits and equipment. Oscilloscopes are available in basic and specialized types that can display different waveforms simultaneously.

Continuity Testers

A *continuity tester* is a test instrument that tests for a complete path for current to flow. A continuity tester is an economical tester that is used to test switches, fuses, and grounds, and is also used for identifying individual conductors in a multiwire cable. Continuity testers give an audible indication when there is a complete path. Most multimeters have a built-in continuity test mode. **See Figure 2-31.**

The continuity test mode is commonly used to test components such as switches, fuses, electrical connections, and individual conductors. In the continuity test mode, a multimeter emits an audible response (beep) when there is a complete path. Indication of a complete path can be used to determine the condition of a component as open or closed. For example, a good fuse should have continuity, whereas a bad fuse does not have continuity. The main advantage of using the continuity test mode of a multimeter is that an audible response is often more desirable than reading a resistance measurement. An audible response allows an electrician to concentrate on the testing procedures without looking at the display.

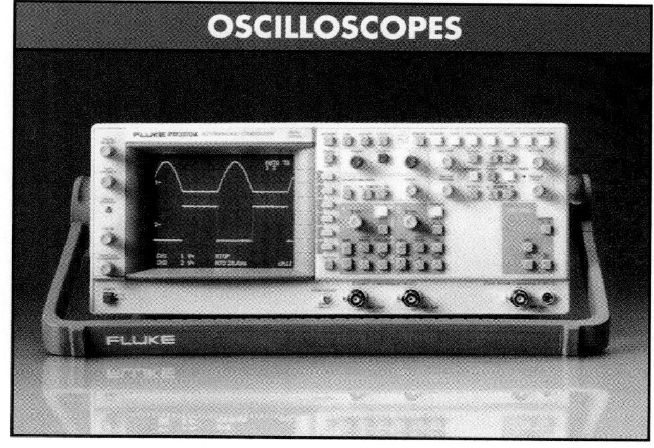

Fluke Corporation

Figure 2-30. An oscilloscope is an instrument that displays an instantaneous voltage.

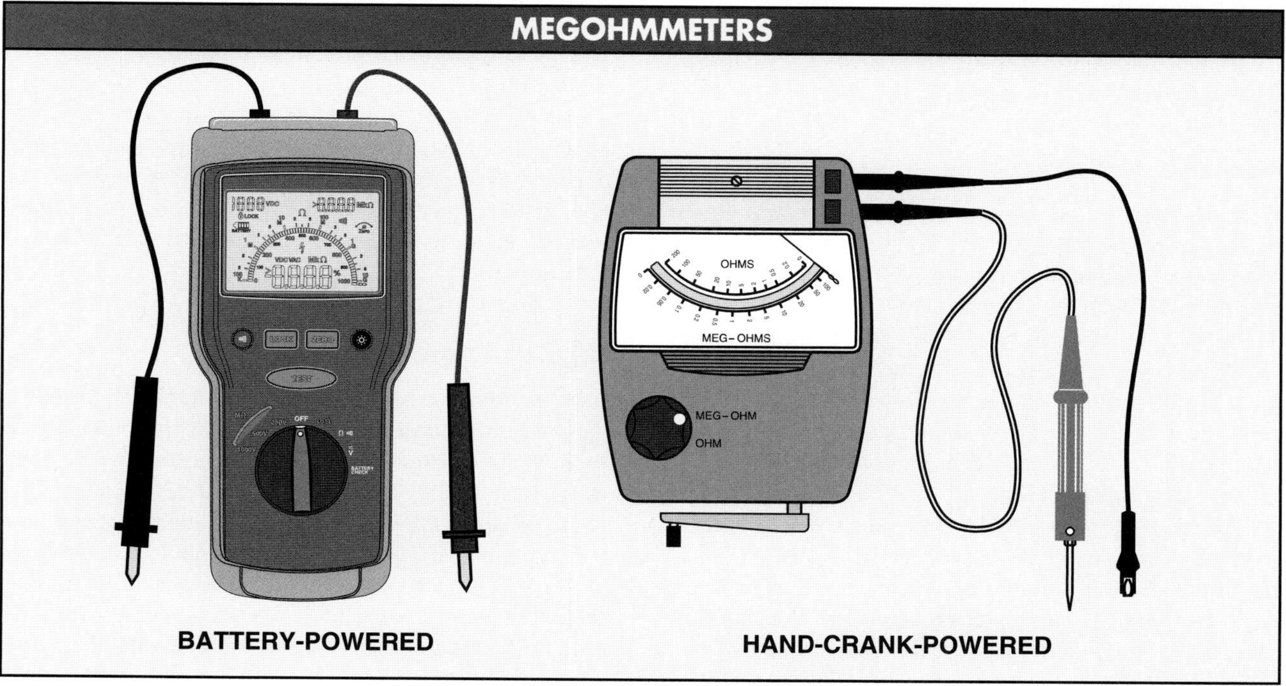

Figure 2-29. A megohmmeter detects insulation deterioration by measuring high resistance values under high test voltage conditions.

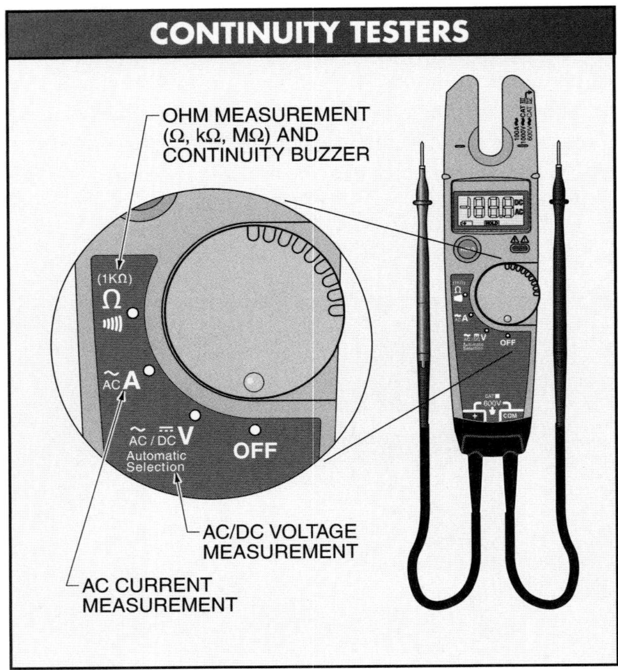

Figure 2-31. A continuity tester tests for a complete path for current to flow.

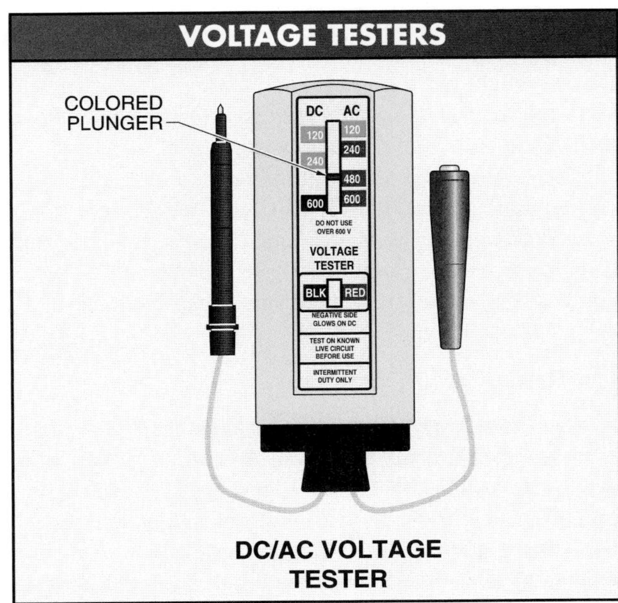

Figure 2-32. A voltage tester indicates approximate voltage level and type (AC or DC) by the movement and vibration of a pointer on a scale.

Warning: A continuity tester must only be used on de-energized circuits or components. Voltage applied to a continuity tester can cause damage to the test instrument and/or harm to the electrician. Always check a circuit for voltage before taking a continuity test.

Voltage Testers

A *voltage tester* is a device that indicates approximate voltage level and type (AC or DC) by the movement and vibration of a pointer on a scale. Voltage testers contain a scale marked 120 VAC, 240 VAC, 480 VAC, 600 VAC, 120 VDC, 240 VDC, and 600 VDC. Some voltage testers include a colored plunger to indicate the polarity of test leads. If the red indicator is up, the red test lead is connected to the positive DC voltage. If the black indicator is up, the black test lead is connected to the positive DC voltage. **See Figure 2-32.**

Voltage testers are used to take voltage measurements anytime voltage of a circuit to be tested is within the rating of the tester and an exact voltage measurement is not required. Exact voltage measurements are not required when testing to determine if a receptacle is energized (hot), if a system is grounded, if fuses or circuit breakers are good or bad, or if a circuit is a 115 VAC, 230 VAC, or 460 VAC circuit.

Before using a voltage tester or any voltage measuring instrument, always check the voltage tester on a known energized circuit that is within the voltage rating of the voltage tester to verify proper operation.

Warning: If a voltage tester does not indicate a voltage, a voltage that can cause an electrical shock may still be present. The voltage tester may also have been damaged during the test (from too high a voltage). Always retest a voltage tester on a known energized circuit after testing a suspect circuit to ensure the voltage tester is still operating correctly.

Digital Logic Probes

A *digital logic probe* is a special DC voltmeter that detects the presence or absence of a signal. Displays on a digital logic probe include logic high, logic low, pulse light, memory, and TTL/CMOS. **See Figure 2-33.** The high light-emitting diode (LED) lights when the logic probe detects a high logic level (1). The low LED lights when the logic probe detects a low logic level (0). The pulse LED flashes relatively slowly when the probe detects logic activity present in a circuit. Logic activity indicates that the circuit is changing between logic levels. The pulse light displays the changes between logic levels because the changes are usually too fast for the high and low LEDs to display.

DIGITAL LOGIC PROBES

Figure 2-33. A digital logic probe is a special DC voltmeter that detects the presence or absence of a signal.

The memory switch sets the logic probe to detect short pulses, usually lasting a few nanoseconds. Any change from the original logic level causes the memory LED to light and remain ON. The memory LED is the pulse LED switch in the memory position. The memory switch is manually moved to the pulse position and back to the memory position to reset the logic probe.

The TTL/CMOS switch selects the logic family of integrated circuits (ICs) to be tested. *Transistor-transistor logic (TTL) ICs* are a broad family of ICs that employ a two-transistor arrangement. The supply voltage for TTL ICs is 5.0 VDC, ±.25 V.

Complementary metal-oxide semiconductor (CMOS) ICs are a group of ICs that employ MOS transistors. CMOS ICs are designed to operate on a supply voltage ranging from 3 VDC to 18 VDC.

Digital circuits fail because the signal is lost somewhere between the circuit input and output stages. Finding the point where the signal is missing and repairing that area usually solves the problem. Repairing normally involves replacing a component, section, or an entire PC board.

Phase Sequence Indicators

A *phase sequence indicator* is a device used to determine phase sequence and open phases. Phase sequence indicators help protect motors, generators, and other equipment from damage due to incorrect motor rotation. Phase sequence indicators are available with either a rotating disk with a colored dot or with two or three LED lamps to indicate phase. **See Figure 2-34.**

Ground Resistance Testers

A *ground resistance tester* is a device used to measure ground connection resistance of electrical installations such as power plants, industrial plants, high-tension towers, and lightning arrestors. Ground resistance testers make routine ground tests as specified by NEC Article 230.95(C). **See Figure 2-35.**

PHASE SEQUENCE INDICATORS

Greenlee Textron Inc.

Figure 2-34. A phase sequence indicator is a device used to determine phase sequence and open phases.

GROUND RESISTANCE TESTERS

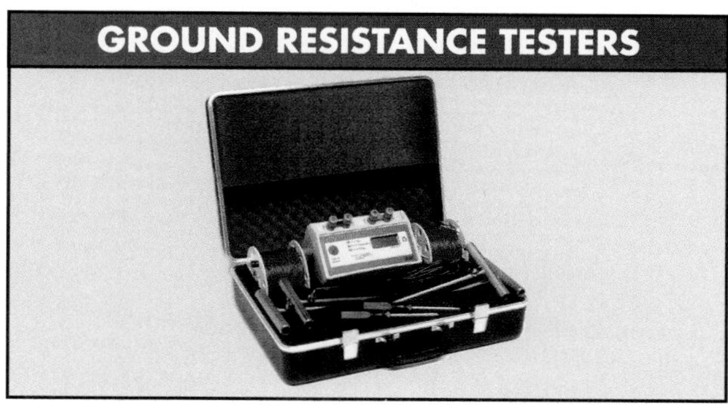

AEMC Instruments

Figure 2-35. A ground resistance tester is a device used to measure ground connection resistance of electrical installations.

Receptacle Testers

A *receptacle tester* is a device that is plugged into a standard receptacle to determine if the receptacle is properly wired and energized. Some receptacle testers include a GFCI (ground fault circuit interrupter) test button that allows the receptacle tester to be used on GFCI receptacles.

When testing a receptacle, the indicator light code indicates whether the receptacle is wired correctly. The situation of having the hot and neutral wires reversed is a safety hazard and must be corrected. Improper grounds are also a safety hazard and must be corrected.

Review Questions

1. What are three of the different methods used to organize electrical tools?
2. What are some of the basic rules to follow for proper and safe tool usage?
3. Why should all power tools be grounded before using them?
4. Why is it dangerous to use a screwdriver around a live electrical wire?
5. Is it safe to use a pipe extension to increase the leverage of a wrench?
6. What tool should be used for bending and cutting small-diameter wire?
7. What tool should be used for cutting cable and twisting wire?
8. What tool should be used for removing insulation from small-diameter wire?
9. What tool should be used to pull wire through conduit?
10. What tool should be used to bend conduit?
11. What tool should be used to remove cartridge-type fuses?
12. What is an analog display?
13. What is a digital display?
14. What type of switch determines the placement of the decimal on a digital multimeter?
15. How do digital displays display values?
16. What is a bar graph?
17. What is a ghost voltage?
18. What is the purpose of a bar graph on a digital multimeter (DMM)?
19. What test instrument should be used for testing a receptacle?
20. What test instrument should be used to ensure a motor is not damaged due to incorrect rotation?
21. What instrument can be used to measure voltage in a circuit?
22. What instrument can be used to measure current in a circuit?

Electrical Safety

A technician must be knowledgeable of the dangers associated with electrical power and the potential hazards on the job. Safe work habits, the proper personal protective equipment, and proper procedures minimize the possibility of personal injury.

ELECTRICAL SAFETY

Technicians must work safely at all times. Electrical safety rules must be followed when working with electrical equipment to help prevent injuries from electrical energy sources. Electrical safety has been advanced by the efforts of the National Fire Protection Association (NFPA), Occupational Safety and Health Administration (OSHA), and state safety laws. *The National Fire Protection Association (NFPA)* is a national organization that provides guidance in assessing the hazards of the products of combustion. The NFPA sponsors the development of the National Electrical Code® (NEC®).

National Electrical Code®

The National Electrical Code® is one of the most widely used and recognized consensus standards in the world. The purpose of the NEC® is to protect people and property from hazards that arise from the use of electricity. Improper procedures when working with electricity can cause permanent injury or death. Many city, county, state, and federal agencies use the NEC® to set requirements for electrical installations. Article 430 of the NEC® covers requirements for motors, motor circuits, and controllers. **See Figure 3-1.** The NEC® is updated every three years.

Electrical safety rules include the following:

- Always comply with the NEC®, state, and local codes.
- Use UL® approved equipment, components, and test equipment.
- Before removing any fuse from a circuit, be sure the switch for the circuit is open or disconnected. When removing fuses, use an approved fuse puller and break contact on the line side of the circuit first. When installing fuses, install the fuse first into the load side of the fuse clip, then into the line side.
- Inspect and test grounding systems for proper operation. Ground any conductive component or element that is not energized.

- Turn OFF, lock out, and tag out any circuit that is not required to be energized when maintenance is being performed.
- Always use personal protective equipment and safety equipment.
- Perform the appropriate task required during an emergency situation.
- Use only a Class C rated fire extinguisher on electrical equipment. A Class C fire extinguisher is identified by the color blue inside a circle.
- Always work with another individual when working in a dangerous area, on dangerous equipment, or with high voltages.
- Do not work when tired or taking medication that causes drowsiness unless specifically authorized by a physician.
- Do not work in poorly lighted areas.
- Ensure there are no atmospheric hazards such as flammable dust or vapor in the area.
- Use one hand when working on a live circuit to reduce the chance of an electrical shock passing through the heart and lungs.
- Never bypass fuses, circuit breakers, or any other safety device.

The Occupational Safety and Health Administration (OSHA) requires employers to provide a safe work environment for employees.

Technical Fact

The National Electrical Code® is adopted for use by local (village, town, city, etc.), county or parish, and state authorities. These authorities are generally represented by building officials who issue permits for jobs and make periodic inspections of work in progress. Additionally, these authorities certify that the completed building meets all applicable codes before occupancy occurs.

Qualified Persons

To prevent an accident, electrical shock, or damage to equipment, all electrical work must be performed by qualified persons. A *qualified person* is a person who is trained and has special knowledge of the construction and operation of electrical equipment or a specific task, and is trained to recognize and avoid electrical hazards that might be present with respect to the equipment or specific task. NFPA 70E Part II *Safety-Related Work Practices*, Chapter 1 *General*, Section 1-5.4.1 *Qualified Persons* provides additional information regarding the definition of a qualified person. A qualified person does the following:

- Determines the voltage of energized electrical parts
- Determines the degree and extent of hazards and uses the proper personal protective equipment and job planning to perform work safely on electrical equipment by following all NFPA, OSHA, equipment manufacturer, state, and company safety procedures and practices
- Performs the appropriate task required during an accident or emergency situation
- Understands electrical principles and follows all manufacturer procedures and approach distances specified by the NFPA
- Understands the operation of test equipment and follows all manufacturer procedures
- Informs other technicians and operators of tasks being performed and maintains all required records

Safety Labels

A *safety label* is a label that indicates areas or tasks that can pose a hazard to personnel and/or equipment. Safety labels appear in several ways on equipment and in equipment manuals. Safety labels use signal words to communicate the severity of a potential problem. The three most common signal words are danger, warning, and caution.

MOTORS—ARTICLE 430

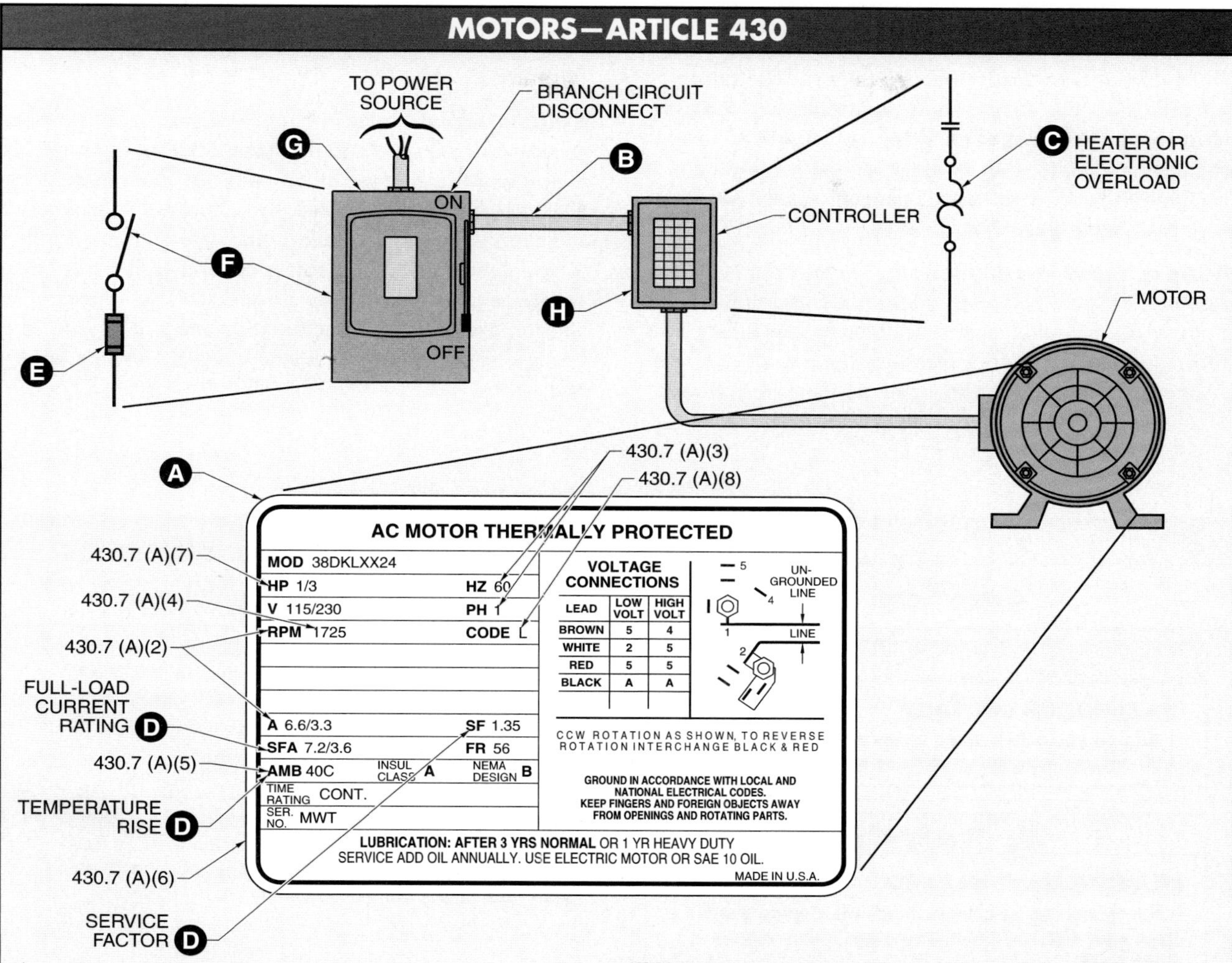

A. **430.7** Motors shall be marked with specific information.

B. **430.22** The ampacity of branch-circuit conductors shall be not less than 125% of the motor current rating.

C. **430.32(A)(1)** Continuous-duty motors over 1 HP shall be protected against overload by a separate overload device rated at not more than 125% of the full-load current rating for motors with a service factor not less than 1.15 or a temperature rise not over 40°C, and at not over 115% for all other motors.

430.32(D)(2)(a) Continuous-duty motors of 1 HP or less which are not permanently installed and are manually started and within sight from the controller are permitted to be protected by the branch-circuit, short-circuit, and ground-fault protective device.

430.32(B)(1), Ex. Any motor in 430.30(b)(1) is permitted on a 20 A, 120 V branch circuit.

430.32(B)(1) Continuous-duty motors of 1 HP or less which are automatically started shall be protected against overload by a separate overload device rated at not more than 125% of the full-load current rating for motors with a service factor not less than 1.15 or a temperature rise not over 40°C, and at not over 115% for all other motors.

D. **430.33** Any motor applications shall be considered as continuous duty unless the driven apparatus is such that the motor cannot operate continuously.

430.32 (C) Where the overcurrent relay selected does not allow the motor to start, the next higher size overload relay is selected provided the trip current does not exceed 140% of the motor's full-load current rating for motors with a service factor of not less than 1.15 or a temperature rise not over 40°C, and 130% for all other motors.

430.35(A) A running overcurrent device may be shunted at starting of a manual-start motor if no hazard is introduced and the branch-circuit device of not over 400% is operative in the circuit during the starting period.

430.35(b) Shunting is not permitted if the motor is automatically started. See Exception.

430.40 Thermal cutouts and overload relays for motor-running protection not capable of opening short circuits shall be protected per 430.52 unless approved for group installation and marked with the maximum size required protection.

E. **430.52** The motor branch-circuit overcurrent device shall be able to carry the starting current (150% – 300% per Table 430.52; absolute maximum 400% with NTDFs and Class CC fuses, and 225% with TDF's).

430.102(A) A disconnecting means shall be in sight from the controller location and shall disconnect the controller. See ex. 1 and 2.

F. **430.102(B)** A disconnecting means shall be in sight from the motor location and the driven machinery location. See Exception.

430.107 One of the disconnecting means shall be readily accessible.

430.108 All disconnecting means shall comply with 430.109 and 430.110.

G. **430.109** The disconnecting means shall be a type specified in 430.109(A) unless otherwise permitted in 430.109(B) through 430.109(G), under the conditions specified.

430.110(A) All disconnecting means shall have an ampere rating of at least 115% of the motor's FLC, taken from FLC tables per 430.6(a).

H. **430.111** A suitable switch or CB may serve as both the disconnecting means and the controller.

Figure 3-1. Article 430 of the NEC® covers requirements for motors, motor circuits, and controllers.

Danger Signal Word. A *danger signal word* is a word used to indicate an imminently hazardous situation which, if not avoided, results in death or serious injury. The information indicated by a danger signal word indicates the most extreme type of potential situation, and must be followed. The danger symbol is an exclamation mark enclosed in a triangle followed by the word "danger" written boldly in a red box. **See Figure 3-2.**

Warning Signal Word. A *warning signal word* is a word used to indicate a potentially hazardous situation which, if not avoided, could result in death or serious injury. The information indicated by a warning signal word indicates a potentially hazardous situation and must be followed. The warning symbol is an exclamation mark enclosed in a triangle followed by the word "warning" written boldly in an orange box.

Caution Signal Word. *A caution signal word* is a word used to indicate a potentially hazardous situation which, if not avoided, may result in minor or moderate injury. The information indicated by a caution signal word indicates a potential situation that may cause a problem to people and/or equipment. A caution signal word also warns of problems due to unsafe work practices. The caution symbol is an exclamation mark enclosed in a triangle followed by the word "caution" written boldly in a yellow box.

SAFETY LABELS

Safety Label	Box Color	Symbol	Significance
DANGER HAZARDOUS VOLTAGE • Ground equipment using screw provided. • Do not use metallic conduits as a ground conductor.	red		**DANGER** – Indicates an imminently hazardous situation which, if not avoided, will result in death or serious injury
WARNING MEASUREMENT HAZARD When taking measurements inside the electric panel, make sure that only the test lead tips touch internal metal parts.	orange		**WARNING** – Indicates a potentially hazardous situation which, if not avoided, could result in death or serious injury
CAUTION MOTOR OVERHEATING Use of a thermal sensor in the motor may be required for protection at all speeds and loading conditions. Consult motor manufacturer for thermal capability of motor when operated over desired speed range.	yellow		**CAUTION** – Indicates a potentially hazardous situation which, if not avoided, may result in minor or moderate injury, or damage to equipment. May also be used to alert against unsafe work practices
WARNING Disconnect electrical supply before working on this equipment.	orange		**ELECTRICAL WARNING** – Indicates a high voltage location and conditions that could result in death or serious injury from an electrical shock
WARNING Do not operate the meter around explosive gas, vapor, or dust.	orange	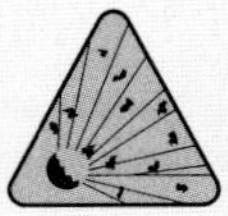	**EXPLOSION WARNING** – Indicates location and conditions where exploding electrical parts may cause death or serious injury

Figure 3-2. Safety labels are used to indicate a situation with different degrees of likelihood of death or injury to personnel.

Other signal words may also appear with danger, warning, and caution signal words used by manufacturers. ANSI Z535.4, *Product Safety Signs and Labels,* provides additional information concerning safety labels. Additional signal words may be used alone or in combination on safety labels.

Electrical Warning Signal Word. *Electrical warning signal word* is a word used to indicate a high-voltage location and conditions that could result in death or serious personal injury from an electrical shock if proper precautions are not taken. An electrical warning safety label is usually placed where there is a potential for coming in contact with live electrical wires, terminals, or parts. The electrical warning symbol is a lightning bolt enclosed in a triangle. The safety label may be shown with no words or may be preceded by the word "warning" written boldly.

Explosion Warning Signal Word. *Explosion warning signal word* is a word used to indicate locations and conditions where exploding parts may cause death or serious personal injury if proper precautions and procedures are not followed. The explosion warning symbol is an explosion enclosed in a triangle. The safety label may be shown with no words or may be preceded by the word "warning" written boldly.

Electrical Shock

An *electrical shock* is a shock that results any time a body becomes part of an electrical circuit. Electrical shock effects vary from a mild sensation, to paralysis, to death. Also, severe burns may occur where current enters and exits the body. The severity of an electrical shock depends on the amount of electric current in milliamps (mA) that flows through the body, the length of time the body is exposed to the current flow, the path the current takes through the body, and the physical size and condition of the body through which the current passes. **See Figure 3-3.**

ELECTRICAL SHOCK EFFECTS

Approximate Current*	Effect On Body†
over 20	Causes severe muscular contractions, paralysis of breathing, heart convulsions
15-20	Painful shock May be frozen or locked to point of electrical contact until circuit is de-energized
8-15	Painful shock Removal from contact point by natural reflexes
8 or less	Sensation of shock but probably not painful

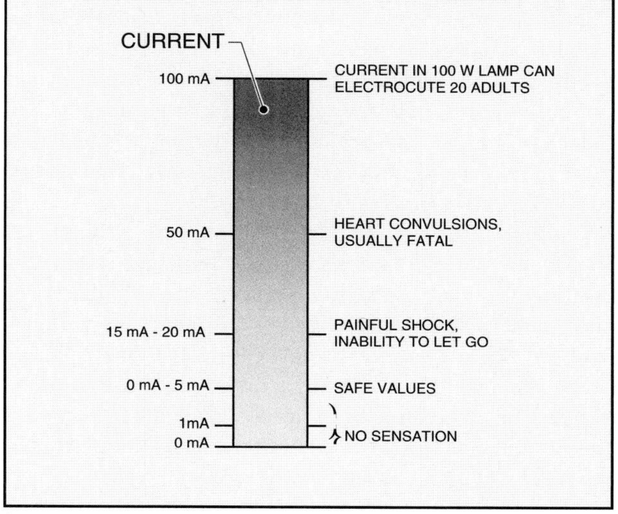

* in mA
† effects vary depending on time, path, amount of exposure, and condition of body

Figure 3-3. Electrical shock is a condition that results any time a body becomes part of an electrical circuit.

Prevention is the best medicine for electrical shock. Anyone working on electrical equipment should have respect for all voltages, have a knowledge of the principles of electricity, and follow safe work procedures. All technicians should be encouraged to take a basic course in cardiopulmonary resuscitation (CPR) so they can aid a coworker in emergency situations.

To reduce the chance of electrical shock, always ensure portable electric tools are in safe operating condition and contain a third wire on the plug for grounding. If electric power tools are grounded and an insulation breakdown occurs, the fault current should flow through the third wire to ground instead of through the body of the operator to ground.

During an electrical shock, the body of a person becomes part of an electrical circuit. The resistance the body of a person offers to the flow of current varies. Sweaty hands have less resistance than dry hands. A wet floor has less resistance than a dry floor. The lower the resistance, the greater the current flow. As the current flow increases, the severity of the electrical shock increases.

If a person is receiving an electrical shock, power should be removed as quickly as possible. If power cannot be removed quickly, the victim must be removed from contact with live parts. Action must be taken quickly and cautiously. Delay may be fatal. An individual must keep themselves from also becoming a casualty while attempting to rescue another person. If the equipment circuit disconnect switch is nearby and can be operated safely, shut OFF the power. Excessive time should not be spent searching for the circuit disconnect. In order to remove the energized part, insulated protective equipment such as a hot stick, rubber gloves, blankets, wood poles, plastic pipes, etc., can be used if such items are accessible.

After the victim is free from the electrical hazard, help is called and first aid (CPR, etc.) begun as needed. The injured individual should not be transported unless there is no other option and the injuries require immediate professional attention.

Grounding. *Grounding* is the connection of all exposed non-current-carrying metal parts to the earth. Grounding provides a direct path to the earth for unwanted fault current without causing harm to persons or equipment. Electrical circuits are grounded to safeguard equipment and personnel against the hazards of electrical shock. Proper grounding of electrical tools, motors, equipment, enclosures, and other control circuitry helps prevent hazardous conditions. On the other hand, improper electrical wiring or misuse of electricity causes destruction of equipment and fire damage to property as well as personal injury.

Grounding is accomplished by connecting the circuit to a metal underground water pipe, the metal frame of a building, a concrete-encased electrode, or a ground ring in accordance with the NEC®. To prevent problems, a grounding path must be as short as possible and of sufficient size as recommended by the manufacturer (minimum 14 AWG copper), never be fused or switched, be a permanent part of the electrical circuit, and be continuous and uninterrupted from the electrical circuit to the ground.

A ground is provided at the main service equipment or at the source of a separately derived system (SDS). A *separately derived system (SDS)* is a system that supplies electrical power derived (taken from) transformers, storage batteries, solar photovoltaic systems, or generators. The majority of separately derived systems are produced by the secondary of the distribution transformer.

The neutral ground connection must be made at the transformer or at the main service panel only. The neutral ground connection is made by connecting the neutral bus to the ground bus with a main bonding jumper. A *main bonding jumper (MBJ)* is a connection at the service equipment that connects the equipment grounding conductor, the grounding electrode conductor, and the grounded conductor (neutral conductor). **See Figure 3-4.**

An *equipment grounding conductor (EGC)* is an electrical conductor that provides a low-impedance ground path between electrical equipment and enclosures within the distribution system. A *grounding electrode conductor (GEC)* is a conductor that connects grounded parts of a power distribution system (equipment grounding conductors, grounded conductors, and all metal parts) to the NEC®-approved earth grounding system. A *grounded conductor* is a conductor that has been intentionally grounded.

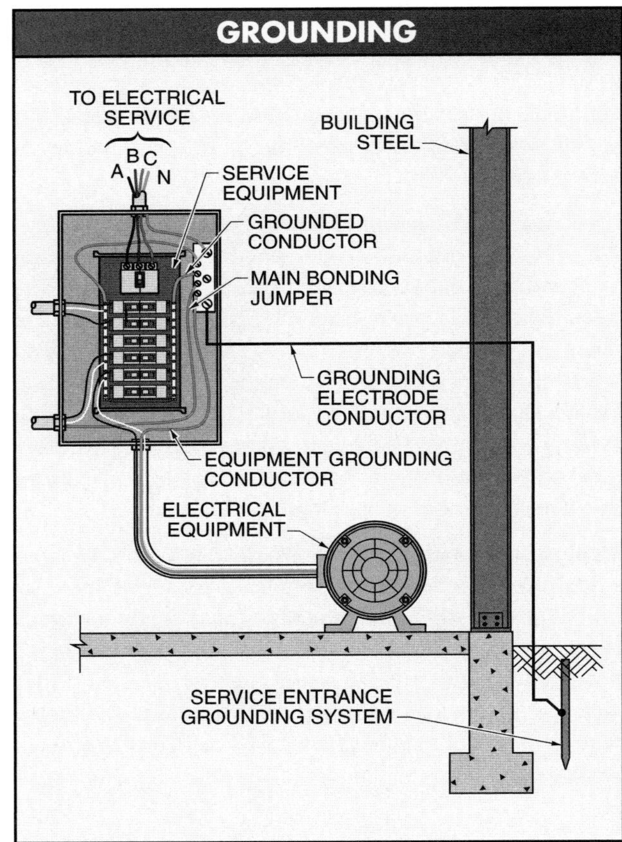

Figure 3-4. Grounding provides a direct path for unwanted (fault) current to the earth without causing harm to individuals or equipment.

Ground Fault Circuit Interrupters. A *ground fault circuit interrupter (GFCI)* is a device that protects against electrical shock by detecting an imbalance of current in the normal conductor pathways and opening the circuit. When current in the two conductors of an electrical circuit varies by more than 5 mA, a GFCI opens the circuit. A GFCI is rated to trip quickly enough (1/40 of a second) to prevent electrocution. **See Figure 3-5.**

A potentially dangerous ground fault is any amount of current above the level that may deliver a dangerous shock. Any current over 8 mA is considered potentially dangerous depending on the path the current takes, the physical condition of the person receiving the shock, and the amount of time the person is exposed to the shock. Therefore, GFCIs are required in such places as dwellings, hotels, motels, construction sites, marinas, receptacles near swimming pools and hot tubs, underwater lighting, fountains, and other areas in which a person may experience a ground fault.

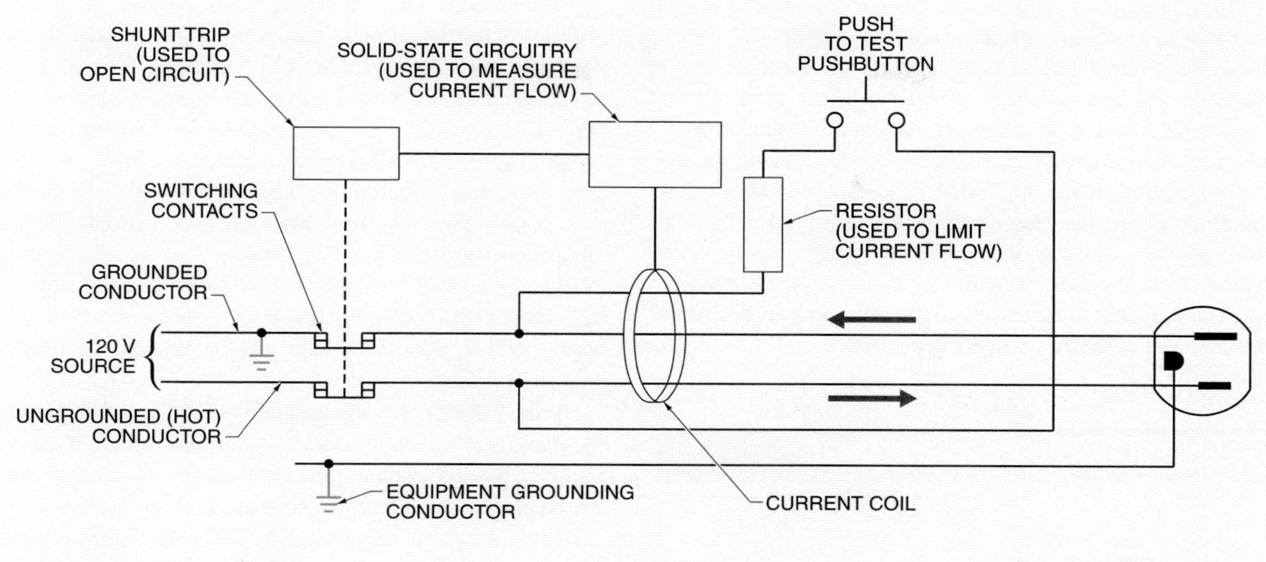

Figure 3-5. A GFCI compares the amount of current in the ungrounded (hot) conductor with the amount of current in the neutral conductor.

A GFCI compares the amount of current in the ungrounded (hot) conductor with the amount of current in the neutral conductor. If the current in the neutral conductor becomes less than the current in the hot conductor, a ground fault condition exists. The amount of current that is missing is returned to the source by some path other than the intended path (fault current).

GFCI protection may be installed at different locations within a circuit. Direct-wired GFCI receptacles provide ground fault protection at the point of installation. GFCI receptacles may also be connected to provide protection at all other receptacles installed downstream on the same circuit. GFCI circuit breakers, when installed in a load center or panelboard, provide GFCI protection and conventional circuit overcurrent protection for all branch-circuit components connected to the circuit breaker.

Plug-in GFCIs provide ground fault protection for devices plugged into them. These plug-in devices are often used by personnel working with power tools in an area that does not include GFCI receptacles.

Portable GFCIs are designed to be easily moved from one location to another. Portable GFCIs commonly contain more than one receptacle outlet protected by an electronic circuit module. Portable GFCIs should be inspected and tested before each use. GFCIs have a built-in test circuit to ensure that the ground fault protection is operational.

A GFCI protects against the most common form of electrical shock hazard, the ground fault. A GFCI does not protect against line-to-line contact hazards, such as a technician holding two hot wires or a hot and a neutral wire in each hand. GFCI protection is required in addition to NFPA grounding requirements.

International Electrotechnical Commission (IEC) 1010 Safety Standard

The *International Electrotechnical Commission (IEC)* is an organization that develops international safety standards for electrical equipment. IEC standards reduce safety hazards that can occur from unpredictable circumstances when using electrical test equipment such as DMMs. For example, voltage surges on a power distribution system can cause a safety hazard. A *voltage surge* is a higher-than-normal voltage that temporarily exists on one or more power lines. Voltage surges vary in voltage amount and time present on power lines. One type of voltage surge is a transient voltage. A *transient voltage* (voltage spike) is a temporary, unwanted voltage in an electrical circuit. Transient voltages typically exist for a very short time, but are often larger in magnitude than voltage surges and very erratic. Transient voltages occur due to lightning strikes, unfiltered electrical equipment, and power being switched ON and OFF. High transient voltages may reach several thousand volts. A transient voltage on a 120 V power line can reach 1000 V (1 kV) or more.

High transient voltages exist close to a lightning strike or when large (high-current) loads are switched OFF. For example, when a large motor (100 HP) is turned OFF, a transient voltage can move down the power distribution system. If a DMM is connected to a point along the system in which the high transient voltage is present, an arc can be created inside the DMM. Once started, the arc can cause a high-current short in the power distribution system even after the original high transient voltage is gone. The high-current short can turn into an arc blast. An *arc blast* is an explosion that occurs when the surrounding air becomes ionized and conductive. **See Figure 3-6.**

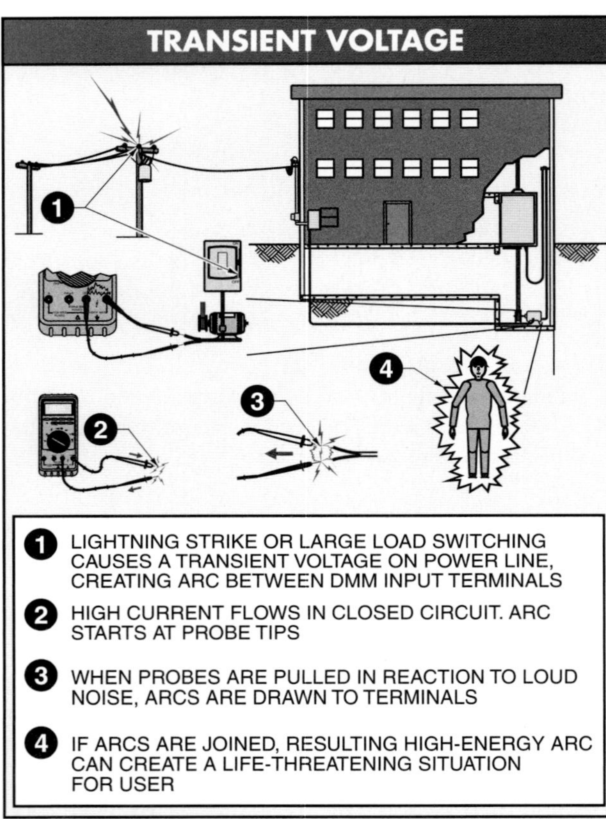

Figure 3-6. When taking measurements in an electrical circuit, transient voltages can cause electrical shock and/or damage to equipment.

The amount of current drawn and potential damage caused depends on the specific location of the power distribution system. All power distribution systems have current limits set by fuses and circuit breakers along the system. The current rating (size) of fuses and circuit breakers decreases farther away from the main distribution panel. The farther away from the main distribution panel, the less likely the high transient voltage is to cause damage.

Overvoltage Installation Categories. The IEC 1010-1 standard defines four overvoltage installation categories in which a DMM may be used (Category I – Category IV). These categories are typically abbreviated as CAT I, CAT II, CAT III, and CAT IV. They determine what magnitude of transient voltage a DMM or other electrical appliance has to withstand when used on the power distribution system. For example, a DMM or other electrical appliance used in a CAT III environment must withstand a 6000 V transient voltage without causing an arc. If the DMM or other appliance is operated on voltages above 600 V, then the DMM must withstand an 8000 V transient voltage.

If the DMM can withstand the voltage, the DMM may be damaged but an arc does not start and no arc blast occurs. To protect against transient voltages, protection must be built into the test equipment used. Over the years, the industry standard followed was IEC 348. This standard has been replaced by IEC 1010. A DMM designed to the IEC 1010 standard offers a higher level of protection. A higher CAT number indicates an electrical environment with higher power available, larger short-circuit current available, and higher energy transients. For example, a DMM designed to the CAT III standard is resistant to higher energy transients than a DMM designed to the CAT II standard. **See Figure 3-7.**

Power distribution systems are divided into categories because a dangerous high-energy transient voltage such as a lightning strike is attenuated (lessened) or dampened as it travels through the impedance (AC resistance) of the system and the system grounds. Within an IEC 1010 standard category, a higher voltage rating denotes a higher transient voltage withstanding rating. For example, a CAT III-1000 V (steady-state) rated DMM has better protection compared to a CAT III-600 V (steady-state) rated DMM. Between categories, a higher voltage rating (steady-state) might not provide higher transient voltage protection. For example, a CAT III-600 V DMM has better transient protection compared to a CAT II-1000 V DMM. A DMM should be chosen based on the IEC overvoltage installation category first and voltage second.

PERSONAL PROTECTIVE EQUIPMENT

Personal protective equipment (PPE) is clothing and/or equipment worn by a technician to reduce the possibility of injury in the work area. The use of personal protective equipment is required whenever work may occur on or near energized exposed electrical circuits. The National Fire Protection Association standard *NFPA 70E, Standard for Electrical Safety in the Workplace,* addresses "electrical safety requirements for employee workplaces that are necessary for the safeguarding of employees in their pursuit of gainful employment." For maximum safety, personal protective equipment must be used as specified in NFPA 70E, OSHA Standard Part 1910 *Subpart I – Personal Protective Equipment (1910.132 through 1910.138),* and other applicable safety mandates.

Per NFPA 70E, "Only qualified persons shall perform testing work on or near live parts operating at 50 V or more." All personal protective equipment and tools are selected for at least the operating voltage of the equipment or circuits to be worked on or near. Equipment, device, tool, or test equipment must be suited for the work to be performed. Personal protective equipment includes protective clothing, head protection, eye protection, ear protection, hand protection, foot protection, back protection, knee protection, and rubber insulated matting. **See Figure 3-8.**

IEC 1010 OVERVOLTAGE INSTALLATION CATEGORIES

Category	In Brief	Examples
CAT I	Electronic	• Protected electronic equipment • Equipment connected to (source) circuits in which measures are taken to limit transient overvoltage to an appropriately low level • Any high-voltage, low-energy source derived from a high-winding-resistance transformer, such as the high-voltage section of a copier
CAT II	1ϕ receptacle-connected loads	• Appliances, portable tools, and other household and similar loads • Outlets and long branch circuits • Outlets at more than 30′ (10 m) from CAT III source • Outlets at more than 60′ (20 m) from CAT IV source
CAT III	3ϕ distribution, including 1ϕ commercial lighting	• Equipment in fixed installations, such as switchgear and polyphase motors • Bus and feeder in industrial plants • Feeders and short branch circuits and distribution panel devices • Lighting systems in larger buildings • Appliance outlets with short connections to service entrance
CAT IV	3ϕ at utility connection, any outdoor conductors	• Refers to the origin of installation, where low-voltage connection is made to utility power • Electric meters, primary overcurrent protection equipment • Outside and service entrance, service drop from pole to building, run between meter and panel • Overhead line to detached building

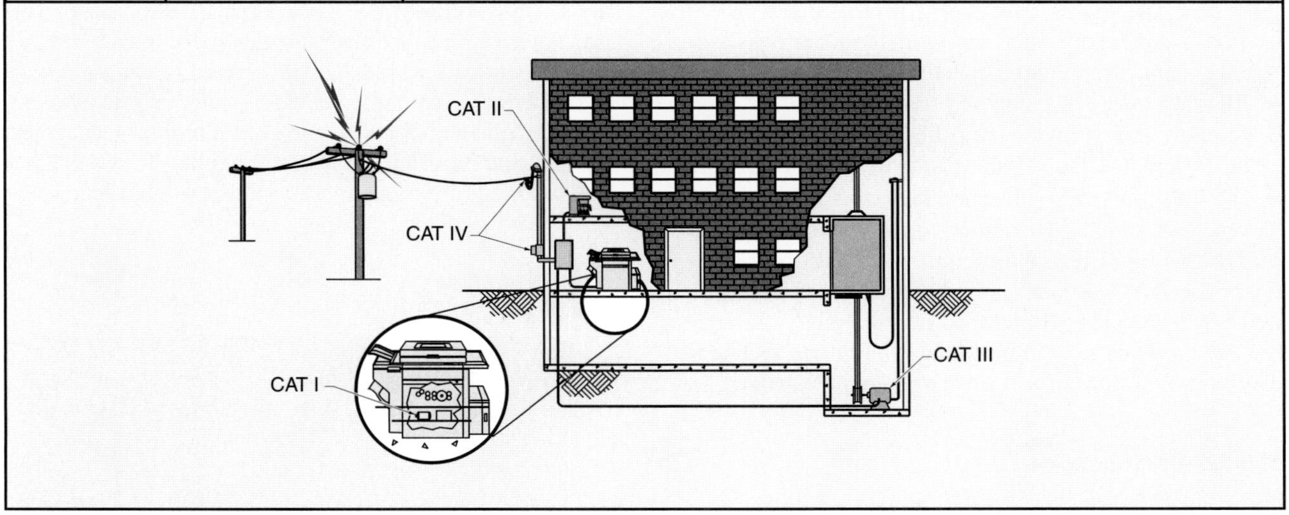

Figure 3-7. The applications in which a DMM may be used are classified by the IEC 1010 standard into four overvoltage installation categories.

Figure 3-8. Personal protective equipment is used to reduce the possibility of an injury.

Protective Clothing

Protective clothing is clothing that provides protection from contact with sharp objects, hot equipment, and harmful materials. Protective clothing made of durable material such as denim should be snug, yet allow ample movement. Clothing should fit snugly to avoid danger of becoming entangled in moving machinery. Pockets should allow convenient access but should not snag on tools or equipment. Soiled protective clothing should be washed to reduce the flammability hazard.

Arc-resistant clothing must be used when working with live high-voltage electrical circuits. Arc-resistant clothing is made of materials such as Nomex®, Basofil®, and/or Kevlar® fibers. The arc-resistant fibers can be coated with PVC to offer weather resistance and to increase arc resistance. Arc-resistant clothing must meet three requirements:

- Clothing must not ignite and continue to burn.
- Clothing must provide an insulating value to dissipate heat throughout the clothing and away from the skin.
- Clothing must provide resistance to the break-open forces generated by the shock wave of an arc.

The National Fire Protection Association specifies boundary distances where arc protection is required. All personnel working within specified boundary distances require arc-resistant clothing and equipment. Boundary distances vary depending on the voltage involved.

Head Protection

Head protection requires using a protective helmet. A *protective helmet* is a hard hat that is used in the workplace to prevent injury from the impact of falling and flying objects, and from electrical shock. Protective helmets resist penetration and absorb impact force. Protective helmet shells are made of durable, lightweight materials. A shock-absorbing lining keeps the shell away from the head to provide ventilation. Protective helmets are identified by class of protection against specific hazardous conditions.

Class G, E, and C helmets are used for construction and industrial applications. Class G (general) protective helmets protect against impact and voltage up to 2200 V, and are commonly used in construction and manufacturing facilities. Class E (electrical) protective helmets protect against impact and voltage up to 20,000 V. Class C (conductive) protective helmets are manufactured with lighter materials yet provide adequate impact protection.

Eye Protection

Eye protection must be worn to prevent eye or face injuries caused by flying particles, contact arcing, and radiant energy. Eye protection must comply with OSHA 29 CFR 1910.133, *Eye and Face Protection*. Eye protection standards are specified in ANSI Z87.1, *Occupational and Educational Eye and Face Protection*. Eye protection includes safety glasses, face shields, and goggles. **See Figure 3-9.**

Safety glasses are an eye protection device with special impact-resistant glass or plastic lenses, reinforced frames, and side shields. Plastic frames are designed to keep the lenses secured in the frame if an impact occurs and minimize the shock hazard when working with electrical equipment. Side shields provide additional protection from flying objects. Tinted-lens safety glasses protect against low-voltage arc hazards.

A *face shield* is an eye and face protection device that covers the entire face with a plastic shield, and is used for protection from flying objects. Tinted face shields protect against low-voltage arc hazards. *Goggles* are an eye protection device with a flexible frame that is secured on the face with an elastic headband. Goggles fit snugly against the face to seal the areas around the eyes, and may be used over prescription glasses. Goggles with clear lenses protect against small flying particles or splashing liquids. Tinted goggles are used to protect against low-voltage arc hazards.

Technical Fact

The NFPA 70E *Standard for Electrical Safety Requirements for Employee Workplaces* requires every employer to perform an electrical arc hazard assessment. No garment should be used in a hazard greater than its stated rating.

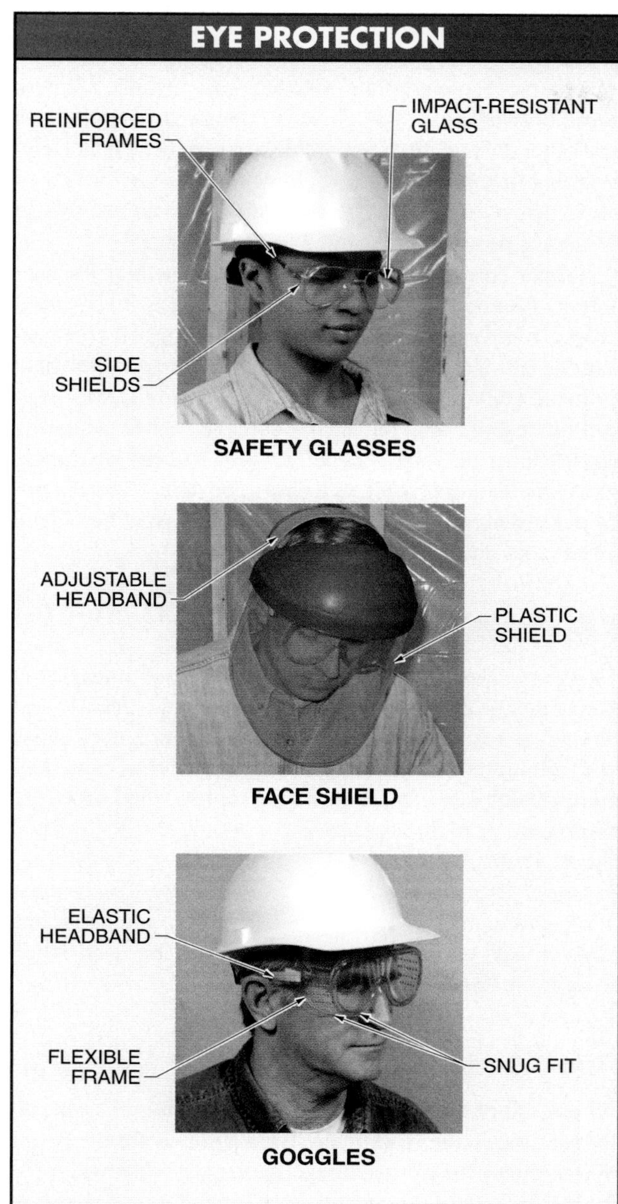

Figure 3-9. Eye protection must be worn to prevent eye or face injuries caused by flying particles, contact arcing, or radiant energy.

Safety glasses, face shields, and goggle lenses must be properly maintained to provide protection and clear visibility. Lens cleaners are available that clean without risk of lens damage. Pitted or scratched lenses reduce vision and may cause lenses to fail on impact.

Ear Protection

Ear protection is any device worn to limit the noise entering the ear and includes earplugs and earmuffs. An *earplug* is an ear protection device made of moldable rubber, foam, or plastic and inserted into the ear canal. An *earmuff* is an ear protection device worn over the ears. A tight seal around an earmuff is required for proper protection.

Power tools and equipment can produce excessive noise levels. Technicians subjected to excessive noise levels may develop hearing loss over a period of time. The severity of hearing loss depends on the intensity and duration of exposure. Noise intensity is expressed in decibels. A *decibel (dB)* is a unit of measure used to express the relative intensity of sound.

Ear protection devices are assigned a noise reduction rating (NRR) number based on the noise level reduced. For example, an NRR of 27 means that the noise level is reduced by 27 dB when tested at the factory. To determine approximate noise reduction in the field, 7 dB is subtracted from the NRR. For example, an NRR of 27 provides a noise reduction of approximately 20 dB in the field.

Hand Protection

Hand protection includes gloves worn to prevent injuries to hands caused by cuts or electrical shock. The appropriate hand protection required is determined by the duration, frequency, and degree of the hazard to the hands. *Rubber insulating gloves* are gloves made of latex rubber and are used to provide maximum insulation from electrical shock. Rubber insulating gloves are stamped with a working voltage range such as 500 V – 26,500 V. *Leather protectors* are gloves worn over rubber insulating gloves to prevent penetration of the rubber insulating gloves and provide added protection against electrical shock. Safety procedures for the use of rubber insulating gloves and leather protectors must be followed at all times. **See Figure 3-10.**

The primary purpose of rubber insulating gloves and leather protectors is to insulate hands and lower arms from possible contact with live conductors. Rubber insulating gloves offer a high resistance to current flow to help prevent an electrical shock. Leather protectors help protect rubber insulating gloves and add additional insulation.

Warning: Rubber insulating gloves are designed for specific applications. Leather protectors are required for protecting rubber insulating gloves and should not be used alone. Rubber insulating gloves offer the highest resistance and greatest insulation. Serious injury or death can result from improper use or using outdated and/or the wrong type of gloves for the application.

The proper care of leather protectors is essential to user safety. Leather protectors should be inspected when inspecting rubber insulating gloves. Metal particles or any substance that could physically damage rubber insulating gloves must be removed from a leather protector before it is used.

RUBBER INSULATING GLOVE CLASSES		
Class	Maximum Use Voltage*	Color of Label
00	500	Beige
0	1000	Red
1	7500	White
2	17,000	Yellow
3	26,500	Green
4	36,000	Orange

* in V

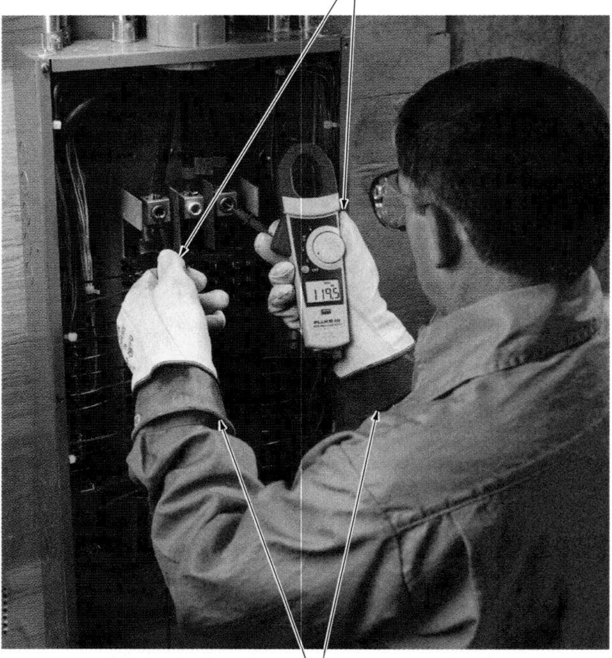

Fluke Corporation

Figure 3-10. Hand protection includes gloves worn to prevent injuries to hands caused by cuts or electrical shock.

The entire surface of rubber insulating gloves must be field tested (visual inspection and air test) before each use. In addition, rubber insulating gloves should also be laboratory tested by an approved laboratory every six months. Visual inspection of rubber insulating gloves is performed by stretching a small area (particularly fingertips) and checking for defects such as punctures or pin holes, embedded or foreign material, deep scratches or cracks, cuts or snags, or deterioration caused by oil, heat, grease, insulating compounds, or any other substance which may harm rubber.

Rubber insulating gloves must also be air tested when there is cause to suspect damage. The entire surface of the glove must be inspected by rolling the cuff tightly toward the palm in such a manner that air is trapped inside the glove, or by using a mechanical inflation device. When using a mechanical inflation device, care must be taken to avoid overinflation. The glove is examined for punctures and other defects. Puncture detection may be enhanced by listening for escaping air by holding the glove to the face or ear to detect escaping air. Gloves failing the air test should be tagged unsafe and returned to a supervisor.

Proper care of leather protectors is essential for user safety. Leather protectors are checked for cuts, tears, holes, abrasions, defective or worn stitching, oil contamination, and any other condition that might prevent them from adequately protecting rubber insulating gloves. Any substance that could physically damage rubber insulating gloves must be removed before use. Rubber insulating gloves or leather protectors found to be defective shall not be discarded or destroyed in the field, but shall be tagged unsafe and returned to a supervisor.

Foot Protection

Foot protection is shoes worn to prevent foot injuries that are typically caused by objects falling less than 4′ and having an average weight of less than 65 lb. Safety shoes with reinforced steel toes protect against injuries caused by compression and impact. Insulated rubber-soled shoes are commonly worn during electrical work to prevent electrical shock. Protective footwear must comply with ANSI Z41, *Personal Protection—Protective Footwear*. Thick-soled work shoes may be worn for protection against sharp objects such as nails. Rubber boots may be used when working in damp locations.

Back Protection

A back injury is one of the most common injuries resulting in lost time in the workplace. Back injuries are the result of improper lifting procedures. Back injuries are prevented through proper planning and work procedures. Assistance should be sought when moving heavy objects. When lifting objects from the ground, ensure the path is clear of obstacles and free of hazards. When lifting objects, the knees are bent and the object is grasped firmly. The object is lifted by straightening the legs and keeping the back as straight as possible. Keep the load close to the body and keep the load steady.

Long objects such as conduit may not be heavy, but the weight might not be balanced. Long objects should be carried by two or more people whenever possible. When carried on the shoulder by one person, conduit should be transported with the front end pointing downward to minimize the possibility of injury to others when walking around corners or through doorways.

Knee Protection

A *knee pad* is a rubber, leather, or plastic pad strapped onto the knees for protection. Knee pads are worn by technicians who spend considerable time working on their knees or who work in close areas and must kneel for proper access to equipment. Knee pads are secured by buckle straps or Velcro® closures. **See Figure 3-11.**

KNEE PROTECTION

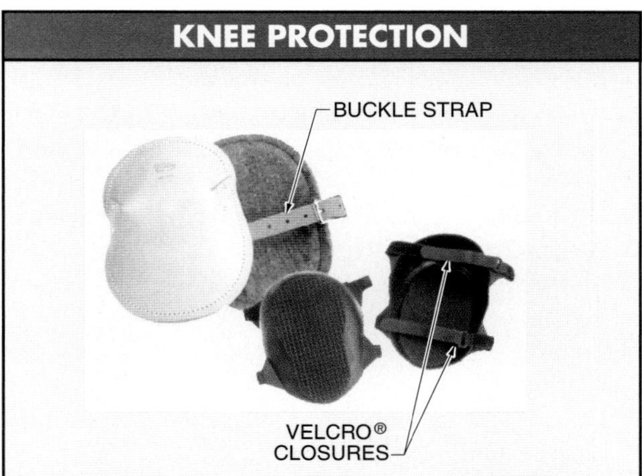

The Stanley Works

Figure 3-11. Knee pads are used to provide protection and comfort to technicians who spend considerable time on their knees.

Rubber Insulating Matting

Rubber insulating matting is a floor covering that provides technicians protection from electrical shock when working on live electrical circuits. Dielectric black fluted rubber matting is specifically designed for use in front of open cabinets or high-voltage equipment. Matting is used to protect technicians when voltages are over 50 V. Two types of matting that differ in chemical and physical characteristics are designated as Type I natural rubber and Type II elastomeric compound matting. **See Figure 3-12.**

LOCKOUT/TAGOUT

Electrical power must be removed when electrical equipment is inspected, serviced, or repaired. To ensure the safety of personnel working with the equipment, power is removed and the equipment must be locked out and tagged out. *Lockout* is the process of removing the source of electrical power and installing a lock which prevents the power from being turned ON. To ensure the safety of personnel working with equipment, all electrical, pneumatic, and hydraulic power is removed and the equipment must be locked out and tagged out. *Tagout* is the process of placing a danger tag on the source of electrical power, which indicates that the equipment may not be operated until the danger tag is removed. Per OSHA standards, equipment is locked out and tagged out before any installation or preventive maintenance is performed. **See Figure 3-13.**

A danger tag has the same importance and purpose as a lock and is used alone only when a lock does not fit the disconnect device. A danger tag shall be attached at the disconnect device with a tag tie or equivalent and shall have space for the technician's name, craft, and other company-required information. A danger tag must withstand the elements and expected atmosphere for the maximum period of time that exposure is expected.

RUBBER INSULATING MATTING RATINGS

Safety Standard	Material Thickness		Material Width (in.)	Test Voltage	Maximum Working Voltage
	Inches	Millimeters			
BS921*	.236	6	36	11,000	450
BS921*	.236	6	48	11,000	450
BS921*	.354	9	36	15,000	650
BS921*	.354	9	48	15,000	650
VDE0680†	.118	3	39	10,000	1000
ASTM D178‡	.236	6	24	25,000	17,000
ASTM D178‡	.236	6	30	25,000	17,000
ASTM D178‡	.236	6	36	25,000	17,000
ASTM D178‡	.236	6	48	25,000	17,000

* BSI–British Standards Instituite
† VDE–Verband Deutscher Elektrotechniker Testing and Certification Institute
‡ ASTM International

Figure 3-12. Rubber insulating matting provides protection from electrical shock when working on live electrical circuits.

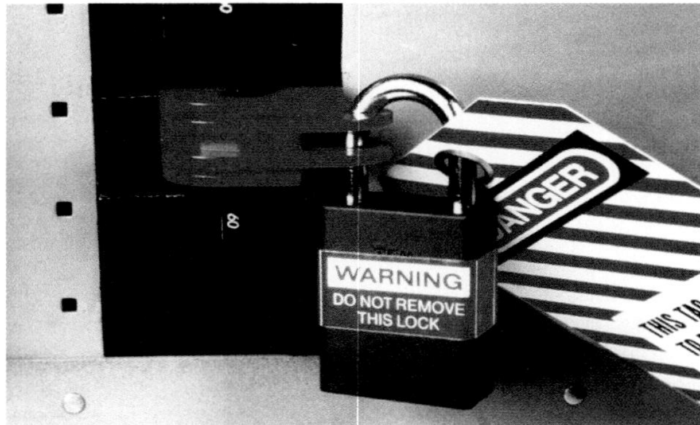

Panduit Corp.

Figure 3-13. Equipment must be locked out and/or tagged out before installation, preventive maintenance, or servicing is performed.

Lockout/tagout is used when:

- power is not required to be ON to a piece of equipment to perform a task
- machine guards or other safety devices are removed or bypassed
- the possibility exists of being injured or caught in moving machinery
- jammed equipment is being cleared
- the danger exists of being injured if equipment power is turned ON

Lockout and tagouts do not by themselves remove power from a machine or its circuitry. OSHA provides a standard procedure for equipment lockout/tagout. Lockout is performed and tagouts are attached only after the equipment is turned OFF and tested. OSHA's procedure is:

1. Prepare for machinery shutdown.
2. Machinery or equipment shutdown.
3. Machinery or equipment isolation.
4. Lockout or tagout application.
5. Release of stored energy.
6. Verification of isolation.

Warning: Personnel should consult OSHA Standard 29 CFR 1910.147 – *The Control of Hazardous Energy (Lockout/Tagout)* for industry standards on lockout/tagout.

A lockout/tagout must not be removed by any person other than the authorized person who installed the lockout/tagout, except in an emergency. In an emergency, only supervisory personnel may remove a lockout/tagout, and only upon notification of the authorized person. A list of company rules and procedures is given to authorized personnel and any person who may be affected by a lockout/tagout. When using a lockout/tagout:

- Use a lockout and tagout when possible.
- Use a tagout when a lockout is impractical. A tagout is used alone only when a lock does not fit the disconnect device.
- Use a multiple lockout when individual employee lockout of equipment is impractical.
- Notify all employees affected before using a lockout/tagout.
- Remove all power sources including primary and secondary.
- Use a DMM set to measure voltage to ensure that the power is OFF.

When more than one technician is required to perform a task on a piece of equipment, each technician shall place a lockout and/or tagout on the energy-isolating device(s). A multiple lockout/tagout device (hasp) must be used because energy-isolating devices typically cannot accept more than one lockout/tagout. A *hasp* is a multiple lockout/tagout device.

LOCKOUT DEVICES

Lockout devices are lightweight enclosures that allow the lockout of standard control devices. Lockout devices are available in various shapes and sizes that allow for the lockout of ball valves, gate valves, and electrical equipment such as plugs, disconnects, etc.

Lockout devices resist chemicals, cracking, abrasion, and temperature changes. They are available in colors to match ANSI pipe colors. Lockout devices are sized to fit standard size industry control devices. **See Figure 3-14.**

Locks used to lock out a device may be color-coded and individually keyed. The locks are rust-resistant and are available with various size shackles.

Danger tags provide additional lockout and warning information. Various danger tags are available. Danger tags may include warnings such as "Do Not Start" or "Do Not Operate," or may provide space to enter worker, date, and lockout reason information. Tag ties must be strong enough to prevent accidental removal and must be self-locking and nonreusable.

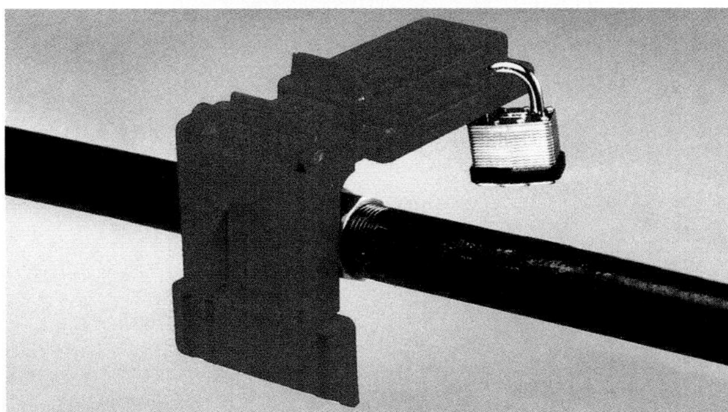

Panduit Corp.

Figure 3-14. Lockout devices are available in various shapes and sizes that allow for the lockout of standard control devices.

Lockout/tagout kits are also available. A lockout/tagout kit contains items required to comply with OSHA lockout/tagout standards. Lockout/tagout kits contain reusable danger tags, multiple lockouts, locks, magnetic signs, and information on lockout/tagout procedures. **See Figure 3-15.** A lockout/tagout should be checked to ensure power is removed when returning to work after leaving a job for any reason or when a job cannot be completed in the same day.

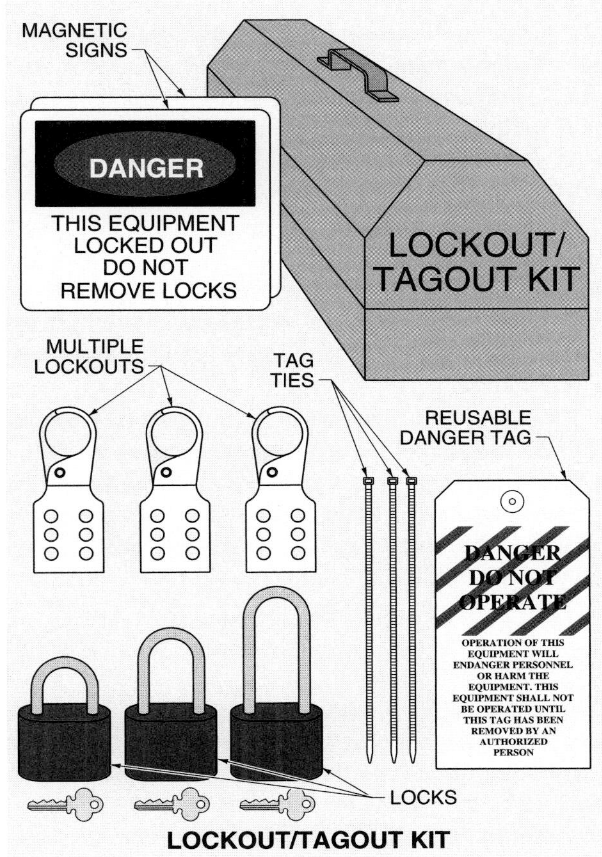

Figure 3-15. Lockout/tagout kits comply with OSHA lockout/tagout standards.

Restoring Equipment to Service. After servicing and/or maintenance work is completed on locked out or tagged out equipment and the equipment is ready to resume normal operation, the following steps must be taken before the lockout/tagout devices are removed:

- Ensure that all tools and nonessential items have been removed from the equipment and that all machine guards, components, etc., have been properly reinstalled.
- Perform a thorough visual check of the area around the equipment to ensure that all individuals are safely positioned or removed from the area and from equipment, circuits, etc., that are about to be reenergized.
- Notify all affected individuals in the area that lockout and/or tagout devices will be removed and the time frame for removal.
- Ensure that only the authorized individual who applied the lockout/tagout removes locks and/or tags from each energy-isolating device.

Note: If the authorized individual who applied the lock and/or tag is unavailable, only the supervisor of the individual may remove the lock and/or tag devices after necessity for removal has been positively established and all of the following conditions have been met:

- A removal of lockout device form has been completed.
- Verification has been made that the individual who applied the lock and/or tag device(s) is not at the facility or location.
- All reasonable efforts have been made to contact the individual who applied the lockout/tagout to inform the individual that their lockout and/or tagout device(s) will be removed.
- The individual's direct supervisor is certain that removal of the lock and/or tag will not endanger anyone.
- Prior to resuming work within the facility or location, the individual that placed the lockout/tagout device shall be notified that his/her lock and/or tag has been removed in the individual's absence.

FIRE SAFETY

Fire safety requires established procedures to reduce or eliminate conditions that could cause a fire. Guidelines in assessing hazards of the products of combustion are provided by the NFPA. Prevention is the best strategy to ward against potential fire hazards. Technicians must take responsibility in preventing conditions that could result in a fire. This includes proper use and storage of lubricants, oily rags, and solvents, and immediate cleanup of combustible spills.

The chance of fire is greatly reduced by good housekeeping. Rags containing oil, gasoline, alcohol, shellac, paint, varnish, lacquer, or other solvents may spontaneously combust and should be kept in a covered metal container. A self-closing steel container specially designed for the disposal of rags containing oil, grease, and flammable liquids is recommended. **See Figure 3-16.** To reduce the possibility of a fire, debris must be kept in a designated area away from the building.

In the event of a fire, a technician must act quickly to minimize injury and damage. An alarm is sounded if a fire occurs, all workers are alerted, and the fire department called. Before starting any work, all individuals should be advised of the location of the nearest telephone and fire alarm reporting station for summoning emergency medical assistance. The telephone shall be reasonably close to the workplace, readily accessible, and functional throughout the work period. When the nearest telephone does not satisfy these requirements, two-way radios or some other positive means of rapid communication must be employed. Cellular telephones can be used only if they are checked to make sure they are operational in the area and approved by the supervisor. Cellular telephones are prohibited by law in an explosionproof environment. A procedure to evacuate the premises and account for all personnel after the fire department is called should be in place and practiced on a regular basis.

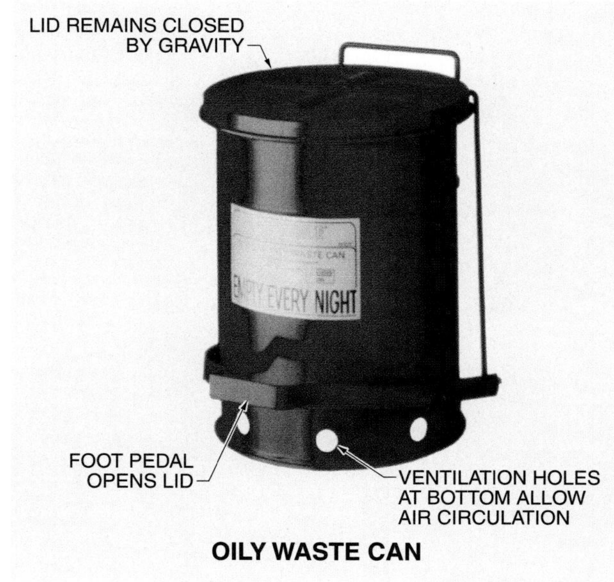

Justrite Manufacturing Company

Figure 3-16. An oily waste can seals out oxygen to prevent spontaneous combustion.

All facilities must have a fire safety plan. A fire safety plan establishes procedures that must be followed if a fire occurs. The fire safety plan lists the locations of the main electrical breaker, fire main, exits, fire alarms, and fire extinguishers for each area of a facility.

Classes of Fire

The five classes of fires are Class A, Class B, Class C, Class D, and Class K. Class A fires include burning wood, paper, textiles, and other ordinary combustible materials containing carbon. Class B fires include burning oil, gas, grease, paint, and other liquids that convert to a gas when heated. Class C fires include burning electrical devices, motors, and transformers. Class D is a specialized class of fires including burning metals such as zirconium, titanium, magnesium, sodium, and potassium. Class K fires include grease in commercial cooking equipment. Fire extinguishers are selected for the class of fire based on the combustibility of the material. **See Figure 3-17.**

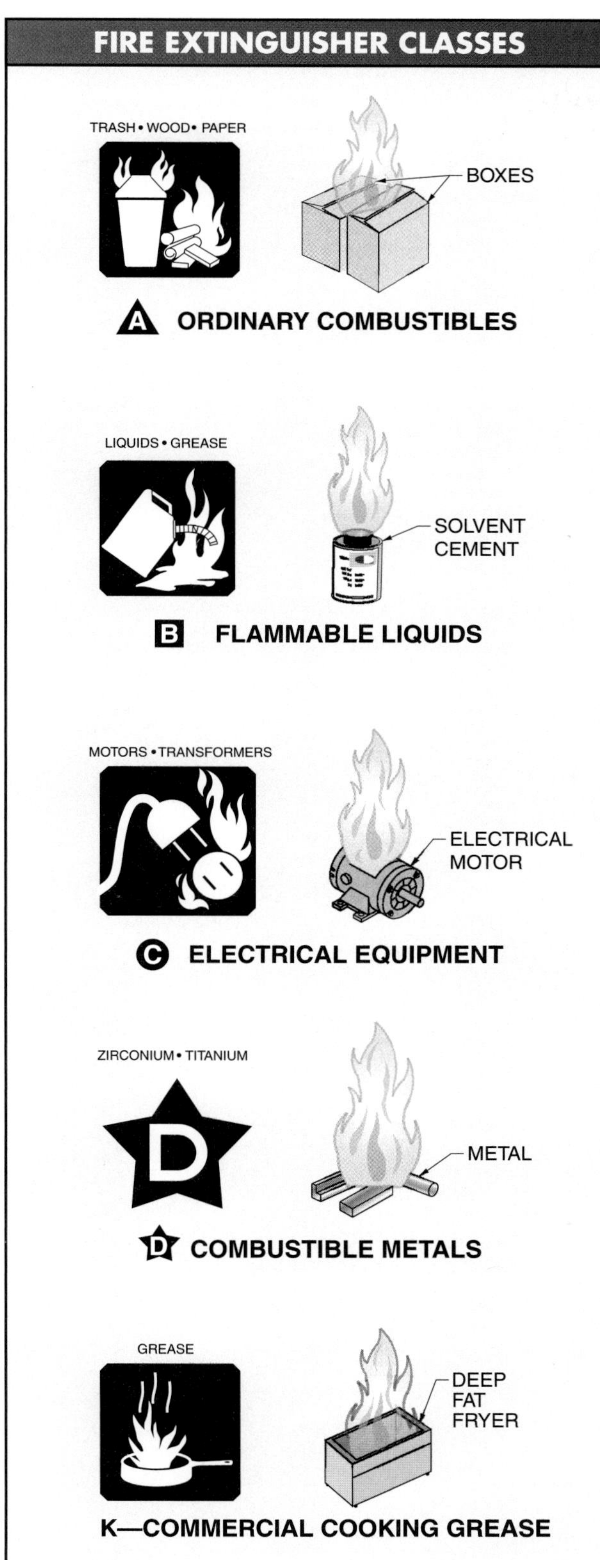

Figure 3-17. Fire extinguisher classes are based on the combustibility of the material.

Fuel, heat, and oxygen are required to start and sustain a fire. A fire goes out when any one of the three is taken away. Fire extinguishing equipment does not take the place of plant fire protection personnel or the local fire department. Proper authorities must be notified whenever there is a fire in the plant. Technicians must know the locations of all fire extinguishing equipment in a facility and be ready to direct firefighters to the location of the fire. In addition, technicians should be able to inform firefighters of any special problems or conditions that exist, such as downed electrical wires, leaks in gas lines, locations of gasoline or propane tanks, and locations of flammable materials.

Fire extinguishing equipment, such as fire extinguishers, water hoses, and sand buckets, must be routinely checked according to plant procedures. The instructions for use should be read before using a fire extinguisher, and the correct fire extinguisher must be used for the class of fire. Fire extinguishers are normally painted red but could also be painted yellow or be made of stainless steel. Fire extinguishers may be located on a red background with a bright red arrow directly above the location so that they can be located easily.

In-Plant Training

All personnel should be acquainted with all fire extinguisher types and sizes available in a plant or a specific work area. Training should include a tour of the facility indicating special fire hazard operations and should be practiced on a routine basis.

In addition, it is helpful to periodically discharge each type of extinguisher. Such practice is essential in learning how to activate each type, knowing the discharge ranges, realizing which types are affected by winds and drafts, familiarizing oneself with discharge duration, learning where to aim the discharge, and learning of any precautions to take as noted on the nameplate.

Fire extinguishers should be inspected at least once a month. It is common to find units that are missing, damaged, or empty. Consider contracting for such a service. Contract for annual maintenance with a qualified service agency. Never attempt to make repairs to fire extinguishers.

Hazardous Locations

The use of electrical equipment in areas where explosion hazards are present can lead to an explosion and fire. This danger exists in the form of escaped flammable gases such as naphtha, benzene, propane, and others. Coal, grain, and other dust suspended in the air can also cause an explosion. Article 500 of the National Electrical Code® (NEC®) and Article 440 of the NFPA cover hazardous locations. **See Figure 3-18.** Any hazardous location requires the maximum in safety and adherence to local, state, and federal guidelines and laws, as well as in-plant safety rules. Hazardous locations are indicated by Class, Division, and Group.

HAZARDOUS LOCATIONS—ARTICLE 500

Hazardous Location – A location where there is an increased risk of fire or explosion due to the presence of flammable gases, vapors, liquids, combustible dusts, or easily-ignitable fibers or flyings.

Location – A position or site.

Flammable – Capable of being easily ignited and of burning quickly.

Gas – A fluid (such as air) that has no independent shape or volume but tends to expand indefinitely.

Vapor – A substance in the gaseous state as distinguished from the solid or liquid state.

Liquid – A fluid (such as water) that has no independent shape but has a definite volume. A liquid does not expand indefinitely and is only slightly compressible.

Combustible – Capable of burning.

Ignitable – Capable of being set on fire.

Fiber – A thread or piece of material.

Flyings – Small particles of material.

Dust – Fine particles of matter.

Classes	Likelihood that a flammable or combustible concentration is present
I	Sufficient quantities of flammable gases and vapors present in air to cause an explosion or ignite hazardous materials
II	Sufficient quantities of combustible dust are present in air to cause an explosion or ignite hazardous materials
III	Easily ignitable fibers or flyings are present in air, but not in a sufficient quantity to cause an explosion or ignite hazardous materials

Divisions	Location containing hazardous substances
1	Hazardous location in which hazardous substance is normally present in air in sufficient quantities to cause an explosion or ignite hazardous materials
2	Hazardous location in which hazardous substance is not normally present in air in sufficient quantities to cause an explosion or ignite hazardous materials

Class I Division I:

Spray booth interiors

Areas adjacent to spraying or painting operations using volatile flammable solvents

Open tanks or vats of volatile flammable liquids

Drying or evaporation rooms for flammable vents

Areas where fats and oils extraction equipment using flammable solvents is operated

Cleaning and dyeing plant rooms that use flammable liquids that do not contain adequate ventilation

Refrigeration or freezer interiors that store flammable materials

All other locations where sufficient ignitable quantities of flammable gases or vapors are likely to occur during routine operations

Class II Division I:

Grain and grain products

Pulverized sugar and cocoa

Dried egg and milk powders

Pulverized spices

Starch and pastes

Potato and wood flour

Oil meal from beans and seeds

Dried hay

Any other organic materials that may produce combustible dusts during their use or handling

Class III Division I:

Portions of rayon, cotton, or other textile mills

Manufacturing and processing plants for combustible fibers, cotton gins, and cotton seed mills

Flax processing plants

Clothing manufacturing plants

Woodworking plants

Other establishments involving similar hazardous processes or conditions

HAZARDOUS LOCATIONS		
Class	**Group**	**Material**
I	A	Acetylene
	B	Hydrogen, butadiene, ethylene oxide, propylene oxide
	C	Carbon monoxide, ether, ethylene, hydrogen sulfide, morpholine, cyclopropane
	D	Gasoline, benzene, butane, propane, alcohol, acetone, ammonia, vinyl chloride
II	E	Metal dusts
	F	Carbon black, coke dust, coal
	G	Grain dust, flour, starch, sugar, plastics
III	No groups	Wood chips, cotton, flax, and nylon

Figure 3-18. Article 500 of the NEC® covers hazardous locations.

When working with energized electrical equipment, it is recommended that a work permit procedure be followed. This practice documents all electrical work performed on the premises and requires the signatures of supervisors and electricians involved in the work being performed. **See Figure 3-19.** Although a work permit procedure is not an NEC® or NFPA requirement, both agencies recommend the procedure. Most commercial insurance companies require such policies be in place as a prerequisite for coverage.

CONFINED SPACES

A *confined space* is a space large enough and so configured that an employee can physically enter and perform assigned work, that has limited or restricted means for entry and exit, and is not designed for continuous employee occupancy. Confined spaces have a limited means of egress and are subject to the accumulation of toxic or flammable contaminants or an oxygen-deficient atmosphere. Confined spaces include, but are not limited to, storage tanks, process vessels, bins, boilers, ventilation or exhaust ducts, sewers, underground utility vaults, tunnels, pipelines, and open top spaces more than 4′ in depth such as pits, tubes, ditches, and vaults.

Confined spaces cause entrapment hazards and life-threatening atmospheres through oxygen deficiency, combustible gases, and/or toxic gases. Oxygen deficiency is caused by the displacement of oxygen by leaking gases or vapors, the combustion or oxidation process, oxygen absorbed by the vessel or product stored, and/or oxygen consumed by bacterial action. Oxygen-deficient air can result in injury or death. **See Figure 3-20.**

Combustible gases in a confined space are commonly caused by leaking gases or gases produced in the space such as methane, carbon monoxide, carbon dioxide, and hydrogen sulfide. Air normally contains 21% oxygen. An increase in the oxygen level increases the explosive potential of combustible gases. Finely ground materials including carbon, grain, fibers, metals, and plastics can also cause explosive atmospheres.

Warning: Confined space procedures vary in each facility. For maximum safety, always refer to specific facility procedures and applicable federal, state, and local regulations.

Confined Space Permits

Confined space permits are required for work in confined spaces based on safety considerations for workers. A *permit-required confined space* is a confined space that has specific health and safety hazards associated with it. OSHA Standard 29 CFR 1910.146 – *Permit-Required Confined Spaces* contains the requirements for practices and procedures to protect workers from the hazards of entry into permit-required confined spaces. These spaces are grouped into the categories of containing or having a potential to contain a hazardous atmosphere, containing a material that has the potential for engulfing an entrant, having an internal configuration such that an entrant could be trapped or asphyxiated by inwardly converging walls or a floor that slopes downward and tapers into a smaller cross-section, or containing any other recognized safety or health hazard.

Permit-required confined spaces require assessment of procedures in compliance with OSHA standards prior to entry. **See Figure 3-21.** A *non-permit confined space* is a confined space that does not contain or, with respect to atmospheric hazards, have the potential to contain any hazards capable of causing death or serious physical harm. These conditions can change with tasks such as welding, painting, or solvent use in the confined space.

Employers must evaluate the workplace to determine if spaces are permit-required confined spaces. If confined spaces exist in the workplace, the employer must inform exposed technicians of the existence, location, and danger posed by the spaces. This is accomplished by posting danger signs or by other equally effective means. In addition, the employer must develop a written permit-required confined space program. A written permit-required confined space program specifies procedures, identification of hazards in each permit-required confined space, restriction of access to authorized personnel, control of hazards, and monitoring of permit-required confined spaces during entry.

The Sinco Group, Inc.

Per OSHA 29 CFR 1910.146 – Permit Required Confined Spaces, at least one attendant must be provided outside a permit-required confined space into which entry is authorized for the duration of entry operations.

ELECTRICAL WORK PERMIT

SEC I: WORK REQUESTED (TO BE COMPLETED BY REQUESTER) Use additional sheets if necessary

1. Description of equipment & location: ______

______

2. Description of work to be performed: ______

______

3. Reason for request: ______

Requested by: ______ Date: ______

SEC II: JOB PROCEDURE (TO BE COMPLETED BY ELECTRICIAN)

1. Detailed description of procedure to be used in performing above work: ______

______

2. Safe work practice description: ______

3. Shock hazard analysis results: ______

4. Electrical shock/flash hazard protection boundary: ______

5. Flash-hazard analysis results: ______

6. PPE requirements: ______

7. Access restriction requirements: ______

8. Pre-work meeting documentation: ______

9. Can above job be performed safely? yes ______ no ______ (If no, return to requester)

Electrician: ______ Date: ______

Electrician: ______ Date: ______

SEC III: MANAGEMENT APPROVALS

______	______
Manufacturing Manager	Maintenance Manager
Safety Manager	Electrician Supervisor
Plant Manager	General Manager
	Date

White Copy: Office
Yellow Copy: Safety Manager
Pink Copy: Maintenance Manager

Figure 3-19. An energized electrical work permit documents all electrical work performed on the premises.

POTENTIAL EFFECTS OF OXYGEN-DEFICIENT ATMOSPHERES*	
Oxygen Content†	**Effects and Symptoms‡**
19.5	Minimum permissible oxygen level
15 – 19.5	Decreased ability to work strenuously. May impair condition and induce early symptoms in persons with coronary, pulmonary, or circulatory problems
12 – 14	Respiration exertion and pulse increases. Impaired coordination, perception, and judgment
10 – 11	Respiration further increases in rate and depth, poor judgment, lips turn blue
8 – 9	Mental failure, fainting, unconsciousness, ashen face, blue lips, nausea, and vomiting
6 – 7	Eight minutes, 100% fatal; 6 minutes, 50% fatal; 4 – 5 minutes, recovery with treatment
4 – 5	Coma in 40 seconds, convulsions, respiration ceases, death

*Values are approximate and vary with state of health and physical activities
† % by volume
‡ at atmospheric pressure

Bacharach, Inc.

Figure 3-20. Oxygen-deficient atmospheres in confined spaces can cause life-threatening conditions.

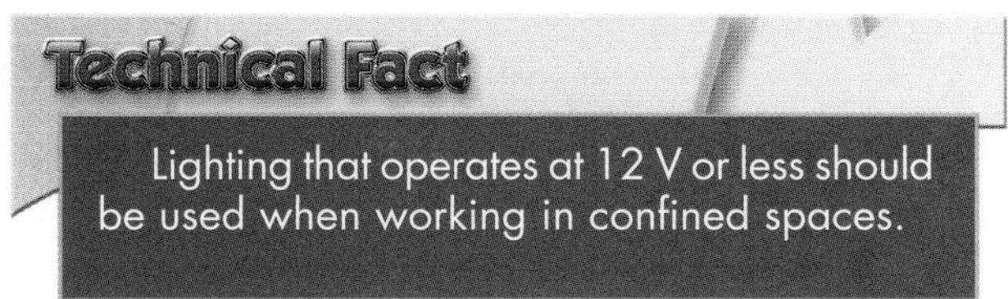

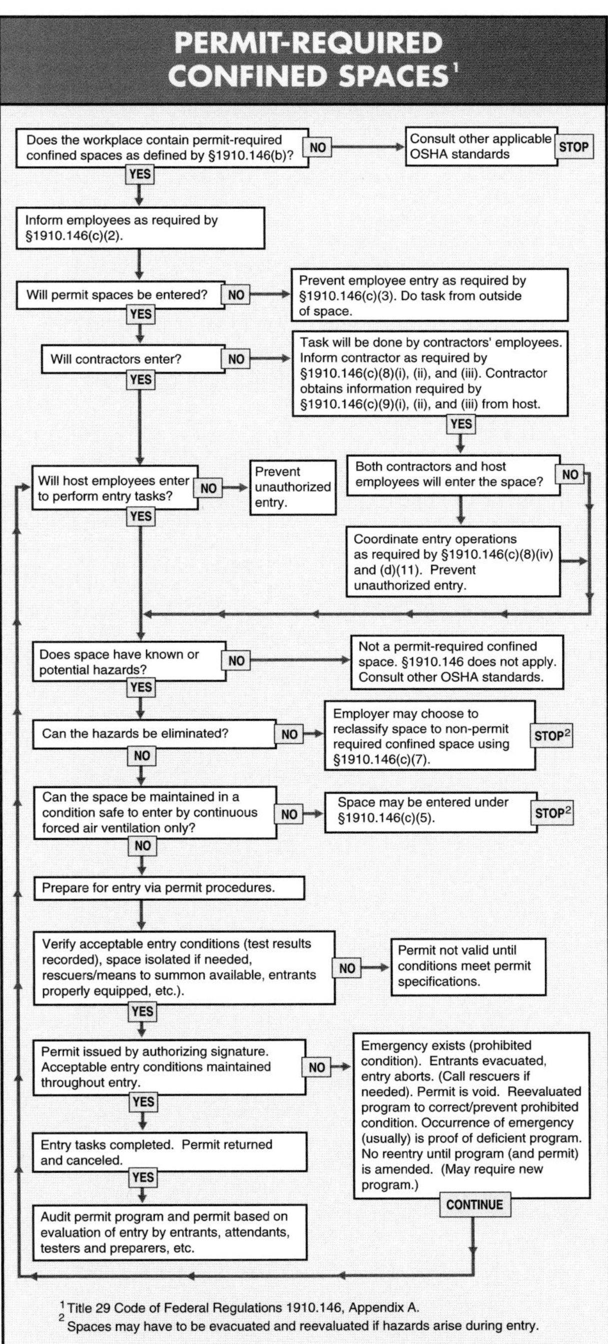

Figure 3-21. For maximum safety, procedures for entering a confined space must follow established OSHA standards.

Entry Permit Procedures

An entry permit must be posted at confined space entrances or otherwise made available to entrants before entering a permit-required confined space. The permit is signed by the entry supervisor and verifies that pre-entry preparations have been completed and that the space is safe to enter. **See Figure 3-22.** A permit-required confined space must be isolated before entry. This prevents hazardous energy or materials from entering the space. Plant procedures for lockout/tagout of permit-required confined spaces must be followed.

The duration of entry permits must not exceed the time required to complete an assignment. The entry supervisor must terminate entry and cancel permits when an assignment has been completed or when new conditions exist. New conditions are noted on the canceled permit and used in revising the permit-required confined space program. All canceled entry permits must be filed for at least one year.

ENTRY PERMIT

✓ CONFINED SPACE ✓ HAZARDOUS AREA

PERMIT VALID FOR 8 HOURS ONLY. ALL COPIES OF PERMIT WILL REMAIN AT JOB SITE UNTIL JOB IS COMPLETED

SITE LOCATION and DESCRIPTION Bunker Fuel Oil Tank #2

PURPOSE OF ENTRY Routine Maintenance/Inspection

SUPERVISOR(S) in charge of crews. Type of Crew Phone #

Michael Green Maintenance Shift II - X5924

* BOLD DENOTES MINIMUM REQUIREMENTS TO BE COMPLETED AND REVIEWED PRIOR TO ENTRY*

REQUIREMENTS COMPLETED	DATE	TIME	REQUIREMENTS COMPLETED	DATE	TIME
Lock Out/De-energize/Try-out	10/2	09:00	**Full Body Harness w/"D" ring**	10/4	08:00
Line(s) Broken-Capped-Blanked	10/2	11:00	**Emergency Escape Retrieval Equip**	10/4	08:00
Purge-Flush and Vent	10/3	09:00	**Lifelines**	10/4	08:00
Ventilation	10/3	10:00	Fire Extinguishers	10/4	08:00
Secure Area (Post and Flag)	10/2	08:00	Lighting (Explosive Proof)	10/4	08:00
Breathing Apparatus	10/4	08:00	Protective Clothing	10/4	08:00
Resuscitator - Inhalator	10/4	08:00	Respirator(s) (Air Purifying)	10/4	08:00
Standby Safety Personnel	10/4	08:00	Burning and Welding Permit	N/A	N/A

Note: Items that do not apply enter N/A in the blank.

** RECORD CONTINUOUS MONITORING RESULTS EVERY 2 HOURS

CONTINUOUS MONITORING** TEST(S) TO BE TAKEN	Permissible Entry Level		10/4							
PERCENT OF OXYGEN	**19.5% to 23.5%**		20.5	20.6	20.7	20.5	20.5			
LOWER FLAMMABLE LIMIT	**Under 10%**		5	5	5	5	6			
CARBON MONOXIDE	**+35 PPM**		0	0	0	0	0			
Aromatic Hydrocarbon	+ 1 PPM	* 5PPM	2	1	2	1	1			
Hydrogen Cyanide	(Skin)	* 4PPM	N/A							
Hydrogen Sulfide	+10 PPM	*15PPM	N/A							
Sulfur Dioxide	+ 2 PPM	* 5PPM	3	2	2	2	2			
Ammonia		* 35PPM	N/A							

* Short-term exposure limit: Employee can work in the area up to 15 minutes.

+ 8 hr. Time Weighted Avg.: Employee can work in area 8 hrs (longer with appropriate respiratory protection).

REMARKS:

GAS TESTER NAME & CHECK #	INSTRUMENT(S) USED	MODEL &/OR TYPE	SERIAL &/OR UNIT #
Marty James	Combination Gas Meter	Industrial Scientific	15A

SAFETY STANDBY PERSON IS REQUIRED FOR ALL CONFINED SPACE WORK

SAFETY STANDBY PERSON(S)	CHECK #	NAME OF SAFETY STANDBY PERSON(S)	CHECK #
Kate Washington	3312		
Tony Linder	3318		

SUPERVISOR AUTHORIZING ENTRY ALL ABOVE CONDITIONS SATISFIED Michael Green

AMBULANCE 2800 Safety 4901 FIRE 2900 Gas Coordinator 4529/5387

Figure 3-22. Confined space entry permit forms document preparations, procedures, and required equipment.

Technical Fact

Permits are required for confined spaces that (1) have a hazardous atmosphere, (2) contain hazardous materials with the potential to engulf an entrant, (3) have an internal configuration such that an entrant could be trapped or asphyxiated by inwardly converging walls or by a floor that slopes downward and tapers to a smaller cross section, or (4) contains any other recognized serious safety or health hazard.

Training is required for all technicians who are required to work in or around permit-required confined spaces. A certificate of training includes the technician's name, the signature or initials of trainer(s), and the dates of training. The certificate must be available for inspection by authorized officials.

OVERHEAD POWER LINE SAFETY

People are killed every day from accidental contact with overhead power lines. *Overhead power lines* are electrical conductors designed to deliver electrical power and that are located in an above-ground aerial position. Overhead power lines are suspended from ceramic insulators which are attached to wood utility poles or metal structures. Overhead power lines are generally owned and operated by an electric utility company. Overhead power line conductors 600 V or higher are usually bare (uninsulated), while low-voltage systems such as service drops to buildings consist of insulated conductors.

Electrical power lines should be located far enough overhead or out of reach as to not pose an electrical hazard. Electrical equipment such as transformers and power panels are also isolated by fences, locked in buildings, or buried underground. Entrances to electrical rooms and other guarded locations containing exposed energized electrical parts must be marked "High Voltage – Do Not Enter," forbidding unqualified persons to enter.

Utility company electrical workers and linemen are skilled workers who have received extensive and specific training to safely work on and near energized overhead power lines and are equipped with the proper personal protection equipment and tools. Workers in other occupations, including residential/commercial/industrial electricians, technicians, engineers, and supervisors, are unqualified (unless trained) to approach overhead power lines closer than an established safe distance. Per NFPA 70E, if the line voltage exceeds 50 kV, the minimum overhead line clearance for all nonqualified individuals is 10′ plus 4″ for every 10 kV over 50 kV.

Scaffolds

A *scaffold* is a temporary or movable platform and structure for workers to stand on when working at a height above the floor. Any person or item on a scaffold must also maintain a safe distance from power lines at all times including during the erection, use, and dismantling of scaffolds. All scaffolds, persons, and items on scaffolds must maintain the minimum distance from power lines.

Review Questions

1. What is grounding?
2. What is the purpose of a GFCI?
3. What are three situations where a lockout/tagout is used?
4. When should rubber boots be used?
5. What is a Class A fire?
6. What is a Class B fire?
7. What is a Class C fire?
8. What is a Class D fire?
9. How often should an unused fire extinguisher be inspected?
10. Who is authorized to sign an entry permit for a permit-required confined space?
11. When should face shields be used?
12. What is the minimum distance a nonutility worker should maintain from a 50,000 V power line?
13. What is the effect of 50 mA of current on the human body?
14. What is the effect of 8 mA of current on the human body?
15. Per NFPA 70 E, what is the minimum overhead line clearance for all nonqualified individuals?

chapter 4

electrical motor controls for Integrated Systems

Electrical Symbols and Diagrams

A basic understanding of electrical symbols, abbreviations, and diagrams is required when working in the electrical industry. The ability to identify commonly used electrical symbols, recognize electrical abbreviations, and read electrical circuit diagrams is required for efficient understanding, wiring, and troubleshooting of electrical circuits.

LANGUAGE OF CONTROL

All trades have a certain language that must be understood in order to transfer information efficiently. This language may include symbols, drawings or diagrams, words, phrases, or abbreviations. Work in the electrical industry requires an understanding of this language, an understanding of the function of electrical components, and an understanding of the relationship between each component in a circuit. With this understanding, an electrician is able to read drawings and diagrams, understand circuit operation, and troubleshoot problems. Drawings and diagrams used to convey electrical information include pictorial drawings, wiring diagrams, schematic diagrams, and line diagrams.

Pictorial Drawings

A *pictorial drawing* is a drawing that shows the length, height, and depth of an object in one view. Pictorial drawings show physical details of an object as seen by the eye. **See Figure 4-1.**

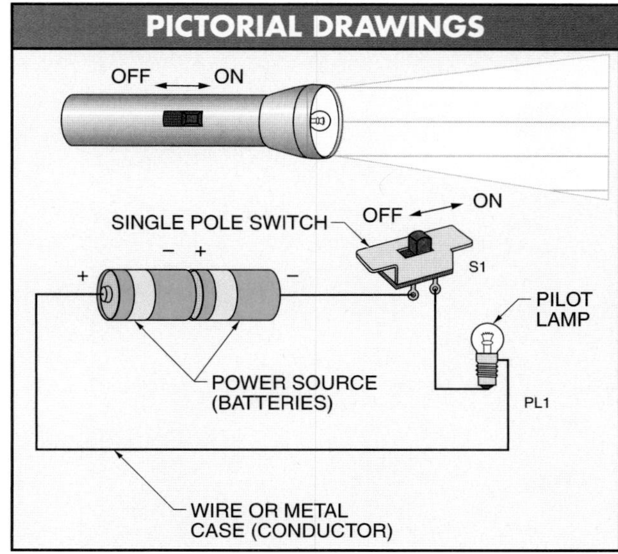

Figure 4-1. A pictorial drawing shows the physical details of components as seen by the eye.

Electrical Symbols and Abbreviations

A *symbol* is a graphic element that represents a quantity or unit. Symbols are used to represent electrical components on electrical and electronic diagrams. An *abbreviation* is a letter or combination of letters that represents a word. **See Figure 4-2. See Appendix.**

ELECTRICAL SYMBOLS . . .

ELECTRICAL POWER SOURCES

Device	Abbr	Symbol
BATTERIES (CHEMICAL)	BAT Pos = (+) Neg = (–)	OPTIONAL; POSITIVE TERMINAL; NEGATIVE TERMINAL SINGLE CELL
	BAT	+ – MULTIPLE CELL
ALTERNATING CURRENT (MAGNETIC)	AC VAC	WAVEFORM REPRESENTS AC
THERMOCOUPLES (HEAT)	TC	OPEN REPRESENTS TERMINAL CONNECTIONS SOLID REPRESENTS THERMOCOUPLE ELEMENT
PHOTOCONDUCTIVE CELLS (LIGHT)	PSC	LIGHT ENERGY + POSITIVE TERMINAL; NEGATIVE TERMINAL –

ELECTRICAL CONDUCTORS

Device	Abbr	Symbol
WIRING WIRE SIZE AND TYPE USUALLY LISTED SIZE = CURRENT CAPACITY TYPE = WHERE IT CAN BE USED	Al = Aluminum Cu = Copper	POWER
		CONTROL
		WIRE NOT CONNECTED
		DOT INDICATES CONNECTION WIRE CONNECTED
	GND	GROUND

ELECTRICAL CONTROL DEVICES

Device	Abbr	Symbol
SWITCH	SPST	BREAK BOTH SIDES SINGLE-POLE SINGLE-THROW, DOUBLE-BREAK
LIMIT SWITCHES (MECHANICAL)	LS	MECHANICAL OPERATOR NORMALLY OPEN
	LS	OPERATOR IN CLOSED POSITION NORMALLY OPEN, HELD CLOSED
	LS	NORMALLY CLOSED
	LS	OPERATOR IN OPEN POSITION NORMALLY CLOSED, HELD OPEN
FOOT SWITCHES	FTS	FOOT OPERATOR NORMALLY OPEN
	FTS	NORMALLY CLOSED
DISCONNECT SWITCHES ON OFF	SPST	TERMINALS FOR CONNECTING WIRE; KNIFE SWITCH SINGLE-POLE, SINGLE-THROW
	DPST	MECHANICALLY TIED TOGETHER, BUT NOT ELECTRICALLY DOUBLE-POLE, SINGLE-THROW
	3PST	THREE-POLE, SINGLE-THROW

Figure 4-2 . . .

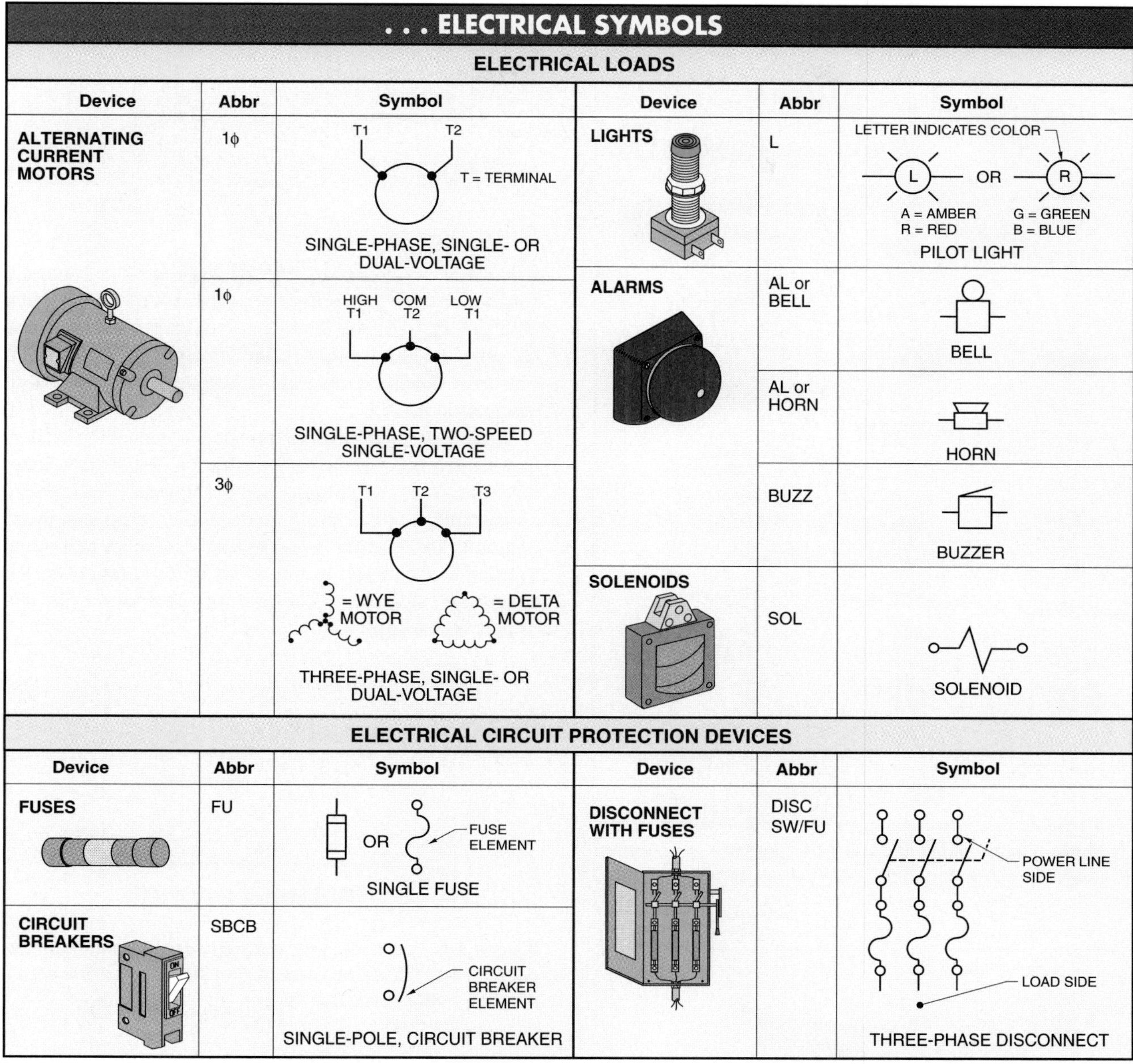

. . . Figure 4-2. Symbols are used to conveniently represent electrical components in diagrams of most electrical and electronic circuits.

Wiring Diagrams

A *wiring diagram* is a diagram that shows the connection of all components in a piece of equipment. Wiring diagrams show, as closely as possible, the actual location of each component in a circuit. Wiring diagrams often include details of the type of wire and the kind of hardware by which wires are fastened to terminals. **See Figure 4-3.**

A wiring diagram is similar to a pictorial drawing except that the components are shown as rectangles or circles. The location or layout of the parts is accurate for the particular equipment. All connecting wires are shown connected from one component to another. Wiring diagrams are used widely by electricians when constructing electronic equipment, and by technicians when maintaining such equipment.

Technical Fact

In industrial wiring, black is used for the hot wire but represents negative (ground) on electronic prints.

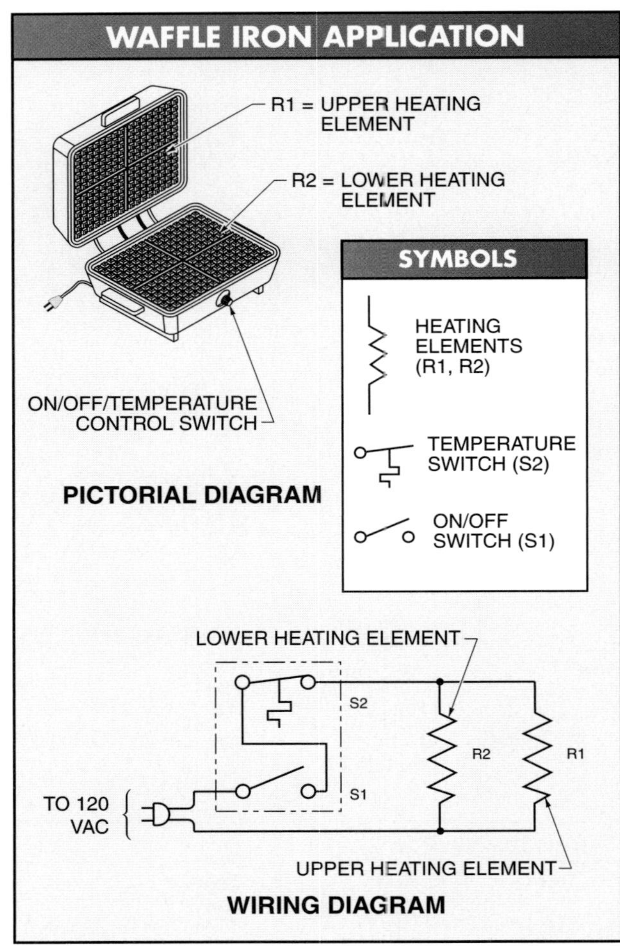

Figure 4-3. In a wiring diagram, the location of components is generally shown as close to the actual circuit configuration as possible.

Bunn-O-Matic

Wiring diagrams are used during the design and assembly of electrical devices and circuits to show the location and relationship of components.

Technical Fact

Standard practices for drawing electrical and electronic diagrams are detailed in *Electrical and Electronic Diagrams, ANSI Y14.15.*

Schematic Diagrams

A *schematic diagram* is a diagram that shows the electrical connections and functions of a specific circuit arrangement with graphic symbols. Schematic diagrams do not show the physical relationship of the components in a circuit. The term schematic diagram is normally associated with electronic circuits.

Schematic diagrams are intended to show the circuitry that is necessary for the basic operation of a device. Schematic diagrams are not intended to show the physical size or appearance of the device. In troubleshooting, schematic diagrams are essential because they enable an individual to trace a circuit and its functions without regard to the actual size, shape, or location of the component, device, or part. **See Figure 4-4.**

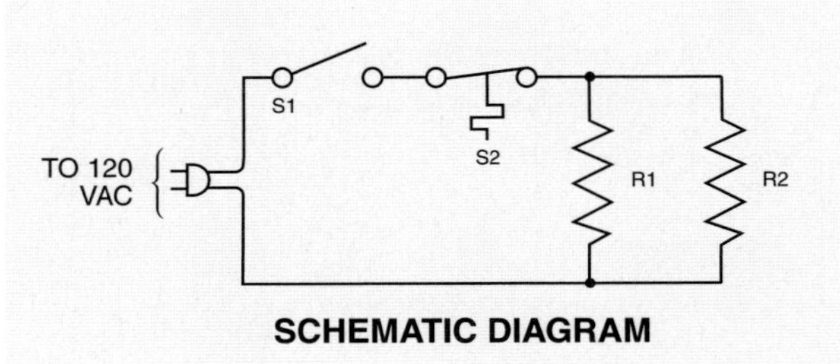

Figure 4-4. In a schematic diagram, components are laid out so the circuit is easily read rather than to show the actual position of the components.

Line Diagrams

A *line (ladder) diagram* is a diagram that shows the logic of an electrical circuit or system using standard symbols. A line diagram is used to show the relationship between circuits and their components but not the actual location of the components. Line diagrams provide a fast, easy understanding of the connections and use of components. **See Figure 4-5.**

The arrangement of a line diagram should promote clarity. Graphic symbols, abbreviations, and device designations are drawn per standards. The circuit should be shown in the most direct path and logical sequence. Lines between symbols can be horizontal or vertical, but should be drawn to minimize line crossing. **See Figure 4-6.**

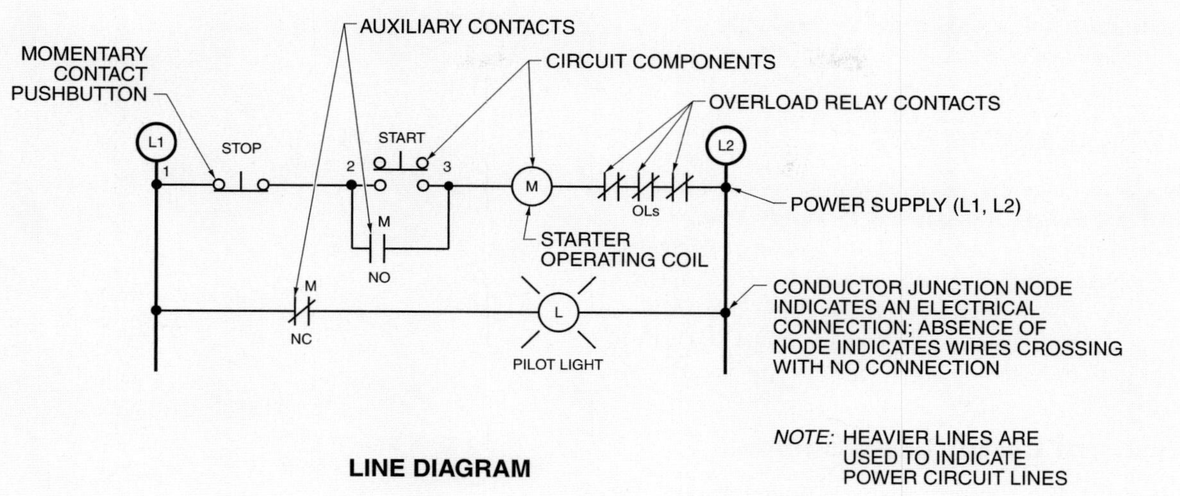

Figure 4-5. A line (ladder) diagram consists of a series of symbols interconnected by lines that are laid out like rungs on a ladder to indicate the flow of current through the various components of a circuit.

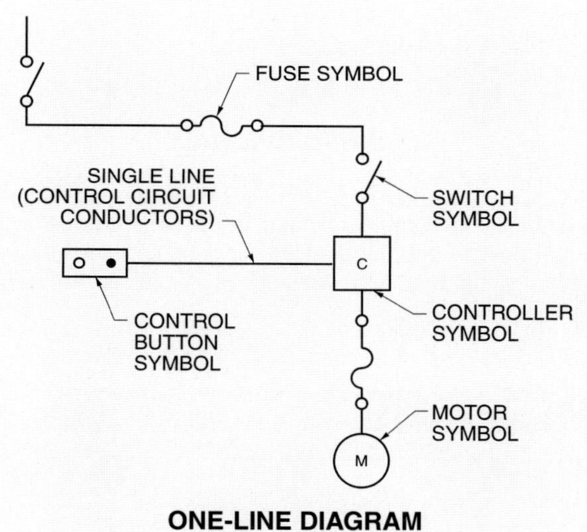

Figure 4-6. A one-line diagram is a diagram that uses single lines and graphic symbols to indicate the path and components of an electrical circuit.

Line diagrams are often incorrectly referred to as one-line diagrams. A *one-line diagram* is a diagram that uses single lines and graphic symbols to indicate the path and components of an electrical circuit. One-line diagrams have only one line between individual components. A line diagram, on the other hand, often shows multiple lines leading to or from a component (parallel connections).

Care must be taken when using electrical symbols to design or communicate electrical circuit operation. Electrical circuit operation may be changed and hazardous situations may be created by using incorrect electrical symbols. One problem that occurs is that limit switch operation is commonly misinterpreted when using electrical circuit diagrams.

For example, a circuit contains four limit switches that are used to control four lamps. **See Figure 4-7.** Lamp 1 is controlled by limit switch 1 (LS1). Limit switch 1 includes a normally open (NO) contact. Thus, lamp 1 is not energized (not turned ON) until an object presses on the limit switch operator and closes the NO limit switch contacts. Lamp 2 is controlled by limit switch 2 (LS2). Limit switch 2 also includes a NO contact. However, the NO contacts are shown in their held closed position. This is often done when a switch would normally be found in the held position, such as a limit switch that detects when a door is closed. Anytime the door is closed, the limit switch NO contacts are held closed and lamp 2 is energized. Thus, for any limit switch symbol, the limit switch contacts are always NO when the moving part of the symbol is drawn below the terminal connections.

Likewise, lamp 3 is controlled by limit switch 3 (LS3). Limit switch 3 includes a normally closed (NC) contact. Thus, lamp 3 is energized (turned ON) before an object presses on the limit switch operator. Lamp 4 is controlled by limit switch 4 (LS4). Limit switch 4 also includes a NC contact. However, the NC contacts are shown in their held open position. This is often done when a switch would normally be found in the held position. Thus, for any limit switch symbol, the limit switch contacts are always NC when the moving part of the symbol is drawn above the terminal connections. Line diagrams are designed to show circuit operation and include switches in their "normal" position and their "held" (actuated) position.

ELECTRICAL CIRCUITS

An *electrical circuit* is an assembly of conductors and electrical devices through which current flows. When an electrical circuit is complete (closed circuit), current makes a complete trip through the circuit. If the circuit is not complete (open circuit), current does not flow. A broken wire, a loose connection, or a switch in the OFF position stops current from flowing in an electrical circuit.

Electrical Circuit Components

All electrical circuits include five basic components. Electrical circuits must include a load that converts electrical energy into some other usable form of energy such as light, heat, or motion; a source of electricity; conductors to connect the individual components; a method of controlling the flow of electricity (switch); and a protection device (fuse or circuit breaker) to ensure that the circuit operates safely and within electrical limits.

Electrical circuit components may be shown using line diagrams, pictorial drawings, and/or wiring diagrams. For example, an automobile interior lighting circuit includes the five components of a typical electrical circuit. **See Figure 4-8.** The source of electricity is the battery, the conductors may be the chassis wires or the car frame, the control device is the plunger-type door switch, the load is the interior light, and the fuse is the protection device.

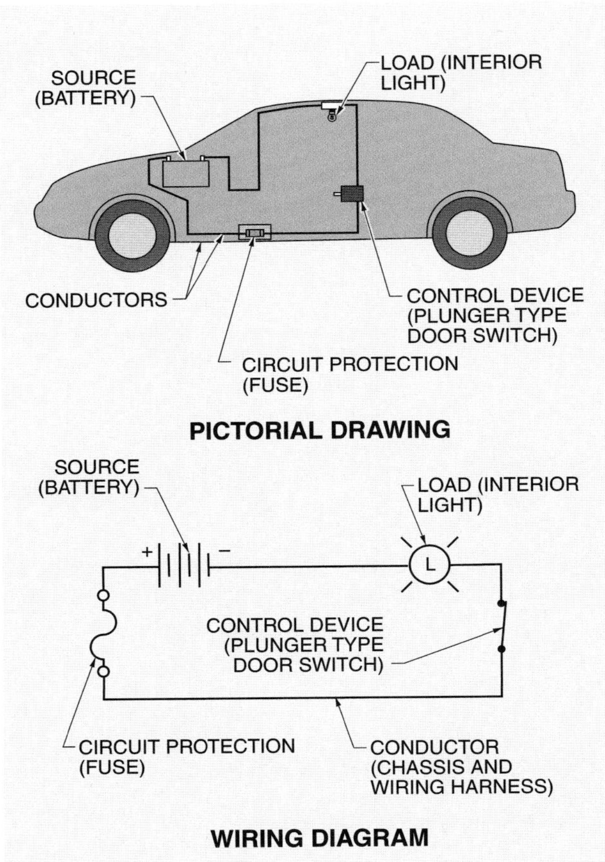

Figure 4-8. All electrical circuits include the source, load, control device, and conductors. Most circuits also include fuses or circuit breakers to provide protection to the circuit.

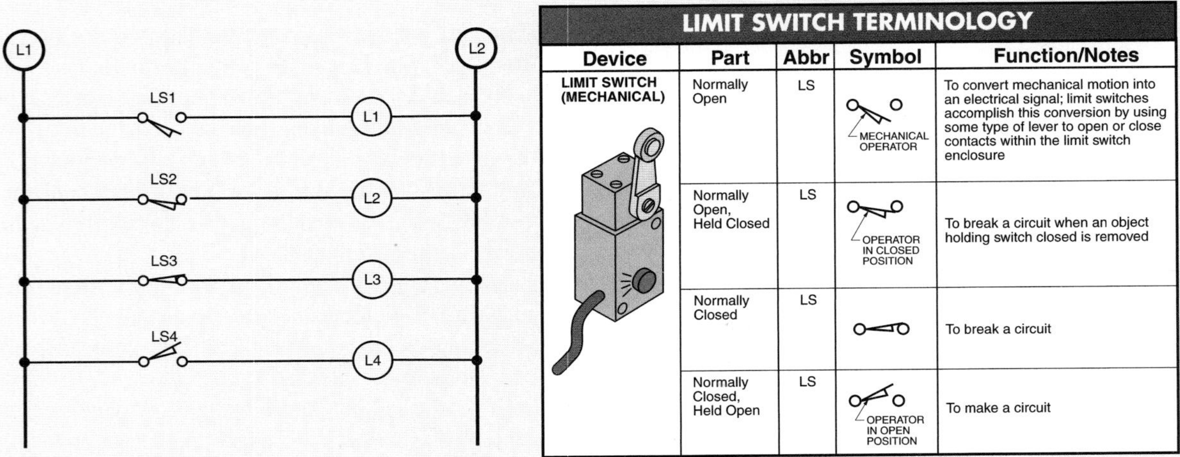

LIMIT SWITCH TERMINOLOGY				
Device	**Part**	**Abbr**	**Symbol**	**Function/Notes**
LIMIT SWITCH (MECHANICAL)	Normally Open	LS	MECHANICAL OPERATOR	To convert mechanical motion into an electrical signal; limit switches accomplish this conversion by using some type of lever to open or close contacts within the limit switch enclosure
	Normally Open, Held Closed	LS	OPERATOR IN CLOSED POSITION	To break a circuit when an object holding switch closed is removed
	Normally Closed	LS		To break a circuit
	Normally Closed, Held Open	LS	OPERATOR IN OPEN POSITION	To make a circuit

Figure 4-7. Care should be taken when using electrical symbols to design or communicate electrical circuit operations because electrical circuit operations may be changed

A *power source* is a device that converts various forms of energy into electricity. The power source in an electrical circuit is normally the point at which to start when reading or troubleshooting a diagram. The components in electrical circuits are connected using conductors. A *conductor* is a material that has very little resistance to current flow and permits electrons to move through it easily. Copper is the most commonly used conductor material.

A *control switch* is a switch that controls the flow of current in a circuit. Switches can be activated manually, mechanically, or automatically. A *load* is any device that converts electrical energy to motion, heat, light, or sound. Common loads include lights, heating elements, speakers, and motors.

An *overcurrent protection device (OCPD)* is a disconnect switch with circuit breakers (CBs) or fuses added to provide overcurrent protection for the switched circuit. A *fuse* is an overcurrent protection device with a fusible link that melts and opens the circuit when an overload condition or short circuit occurs. A *circuit breaker* is an overcurrent protection device with a mechanical mechanism that may manually or automatically open the circuit when an overload condition or short circuit occurs.

Manual Control Circuits

A *manual control circuit* is any circuit that requires a person to initiate an action for the circuit to operate. A line diagram may be used to illustrate a manual control circuit of a pushbutton controlling a pilot light. In a line diagram, the lines labeled L1 and L2 represent the power circuit. **See Figure 4-9.** The voltage of the power circuit is normally indicated on the circuit near these lines. In this circuit, the voltage is 115 VAC, but may be 12 VAC, 18 VAC, 24 VAC, or some other voltage. A common control voltage for many motor control circuits is 24 VAC. The control voltage may also be DC when DC components are used.

The dark black nodes on a circuit indicate an electrical connection. If a node is not present, the wires only cross each other and are not electrically connected.

Line diagrams are read from left (L1) to right (L2). In this circuit, pressing pushbutton 1 (PB1) allows current to pass through the closed contacts of PB1, through pilot light 1 (PL1) and on to L2, forming a complete circuit that activates PL1. Releasing PB1 opens the PB1 contacts, stopping the current flow to the pilot light, and turning the pilot light (PL1) OFF.

A line diagram may be used to illustrate the control and protection of a 1ϕ motor using a manual starter with overload protection. **See Figure 4-10.** The manual starter is represented in the line diagram by the set of normally open (NO) contacts S1 and by the overload contacts OL1. The line diagram is drawn for ease of reading and does not indicate where the devices are physically located. For this reason, the overloads are shown between the motor and L2 in the line diagram but are physically located in the manual starter.

In this circuit, current passes through contacts S1, the motor, the overloads, and on to L2 when the manual starter (S1 NO contacts) is closed. This starts the motor. The motor runs until contacts S1 are opened, a power failure occurs, or the motor experiences an overload. In the case of an overload, the OL1 contacts open and the motor stops. The motor cannot be restarted until the overload is removed and the overload contacts are reset to their normally closed (NC) position. *Note:* For consistency, the overload symbol is often drawn in a line diagram after the motor. In the actual circuit, the overload is located before the motor. It does not matter that the overload is shown after the motor in the line diagram because the overload is a series device and opens the motor control circuit in either position.

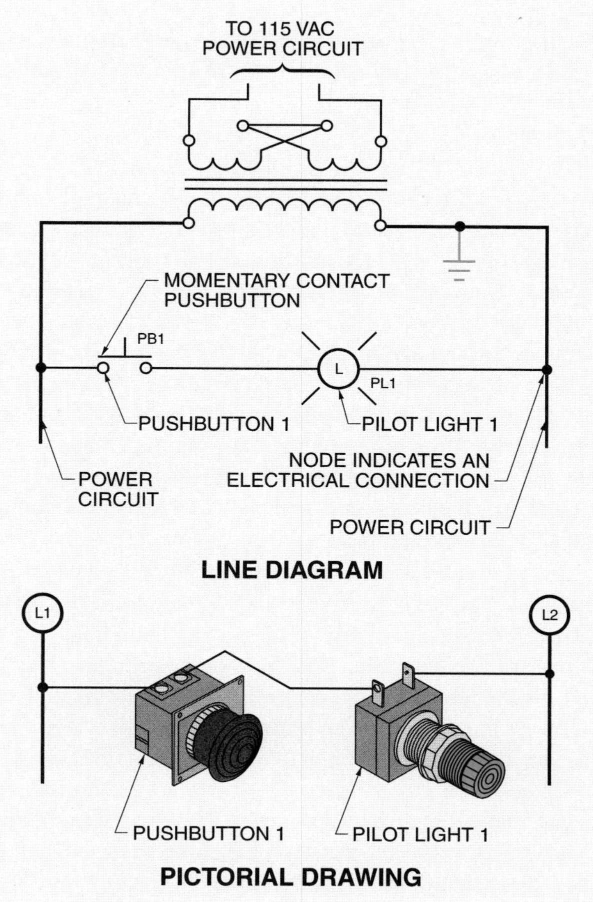

Figure 4-9. A line diagram may be used to illustrate a manual control circuit of a pushbutton controlling a pilot light.

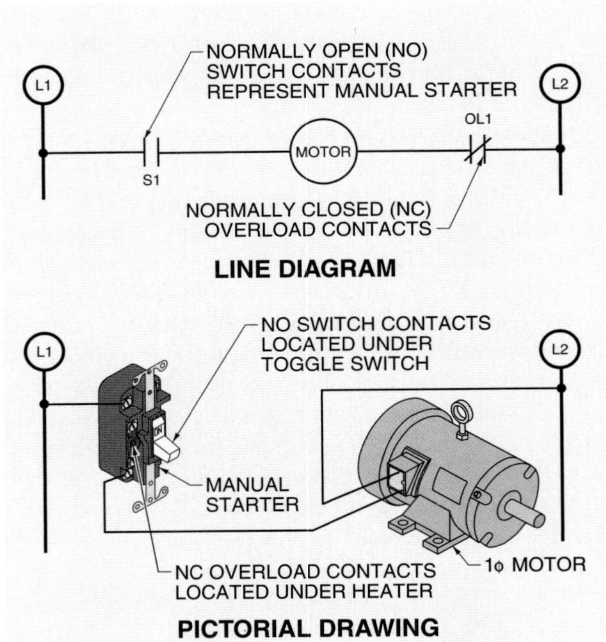

Figure 4-10. A line diagram may be used to illustrate the control and protection of a 1ϕ motor using a manual starter with overload protection.

Automatic Control Circuits

Automatically controlled devices have replaced many functions that were once performed manually. As a part of automation, control circuits are intended to replace manual devices. Any manual control circuit may be converted to automatic operation. For example, an electric motor on a sump pump can be turned ON and OFF automatically by adding an automatic control device such as a float switch. **See Figure 4-11.** This control circuit is used in basements to control a sump pump to prevent flooding. When water reaches a predetermined level, the float switch senses the change in water level and automatically starts the pump, which removes the water.

In this circuit, float switch contacts FS1 determine if current passes through the circuit when switch contacts S1 are closed. Current passes through contacts S1 and float switch contacts FS1, the motor, and the motor overload contacts to L2 when the float switch contacts FS1 are closed. This starts the pump motor. The pump motor pumps water until the water level drops enough to open contacts FS1 and shut OFF the pump motor. A motor overload, a power failure, or the manual opening of contacts S1 would prevent the pump motor from automatically pumping water even after the water reaches the predetermined level.

Control devices such as float switches are normally designed with NO and NC contacts. Variations in the application of the float switch are possible because the NO and NC contacts can close or open when changes in liquid level occur. For example, the NC contacts of the float switch may be used for a pump operation to maintain a certain water level in a livestock water tank. **See Figure 4-12.** The NC contacts close and start the pump motor when the water level drops due to evaporation or drinking. The float opens the NC contacts, shutting the pump motor OFF when the water level rises to a predetermined level.

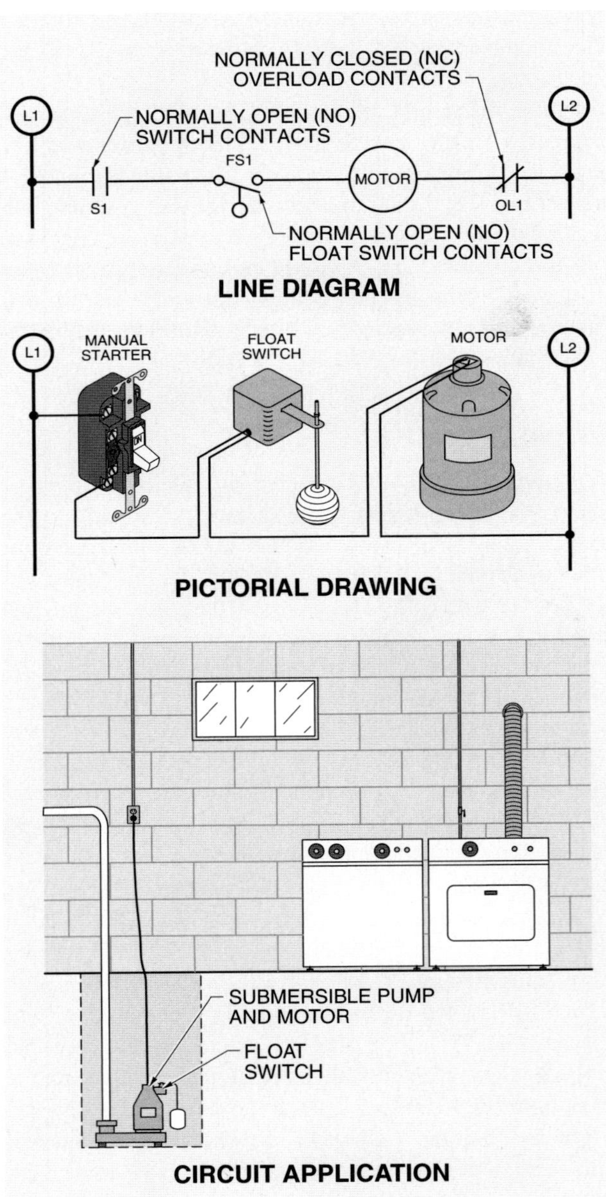

Figure 4-11. An electric motor on a sump pump can be turned ON and OFF by using an automatic control device such as a float switch.

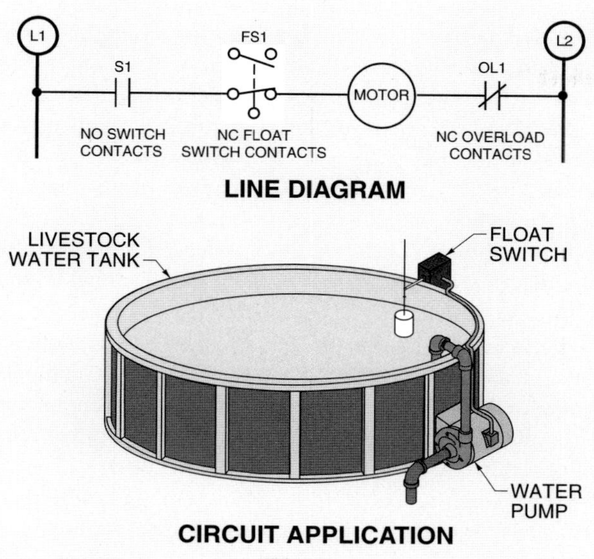

Figure 4-12. The NC contacts of a float switch may be used for a pump operation to maintain a certain level of water in a livestock water tank.

In this circuit, when contacts S1 and float switch contacts FS1 are closed, current passes from L1 through the S1 and FS1 contacts, the motor, and the overloads to L2. This starts the pump motor. The pump motor pumps water until the water level rises high enough to open contacts FS1 and shut OFF the pump. A motor overload, a power failure, or the manual opening of contacts S1 would prevent the pump from automatically filling the tank to a predetermined level.

Magnetic Control Circuits

Although manual controls are compact and sometimes less expensive than magnetic controls, industrial and commercial installations often require that electrical control equipment be located in one area while the load device is located in another. Solenoids, contactors, and magnetic motor starters are used for remote control of devices.

Solenoids. A *solenoid* is an electric output device that converts electrical energy into a linear mechanical force. A solenoid consists of a frame, plunger, and coil. **See Figure 4-13.** A magnetic field is set up in the frame when the coil is energized by an electric current passing through it. This magnetic field causes the plunger to move into the frame. The result is a straight-line force, normally a push or pull action.

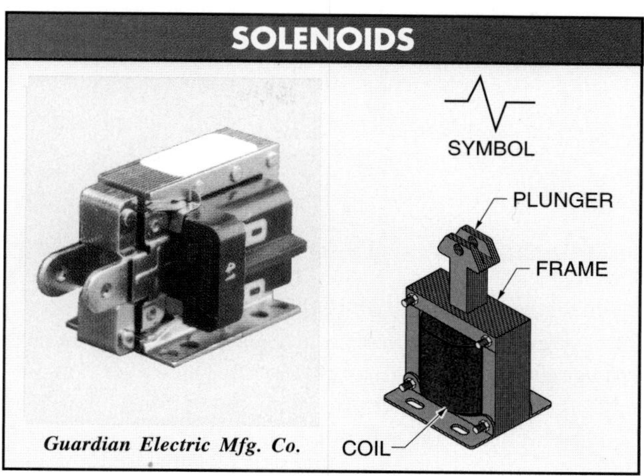

Figure 4-13. A solenoid is an electric output device that converts electrical energy into a linear mechanical force.

A solenoid may be used to control a door lock that opens only when a pushbutton is pressed. **See Figure 4-14.** In this circuit, pressing pushbutton 1 allows an electric current to flow through the solenoid, creating a magnetic field. The magnetic field, depending on the solenoid construction, causes the plunger to push or pull. In this circuit, the door may open as long as the pushbutton is pressed. The door is locked when the pushbutton is released. This circuit provides security access to a building or room.

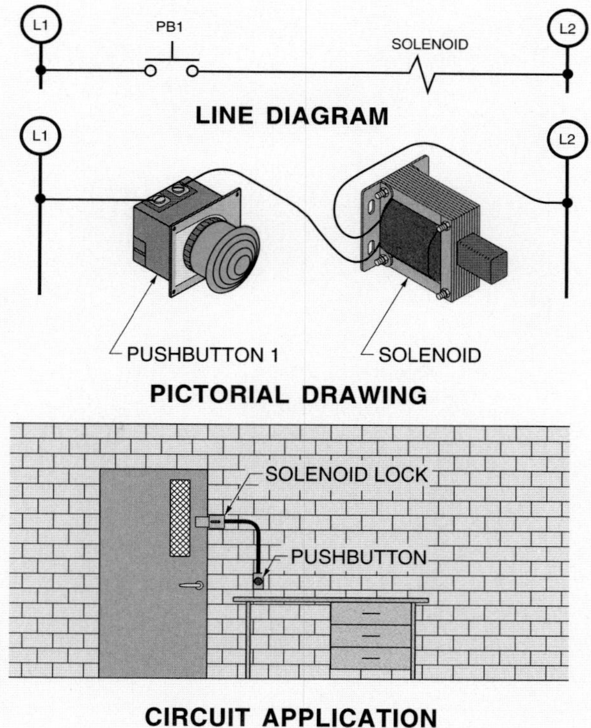

Figure 4-14. A solenoid may be used to control a door lock that is opened only when a pushbutton is pressed.

Technical Fact

Contactors are used to turn ON and OFF high-power (greater than 1500 W), nonmotor loads such as large heating elements and large solenoid-operated valves.

Contactors. A *contactor* is a control device that uses a small control current to energize or de-energize the load connected to it. A contactor does not include overload protection. A contactor is constructed, and operates, similarly to a solenoid. **See Figure 4-15.** Like a solenoid, a contactor has a frame, plunger, and coil. The action of the plunger, however, is directed to close (or open) sets of contacts. The closing of the contacts allows electrical devices to be controlled from remote locations.

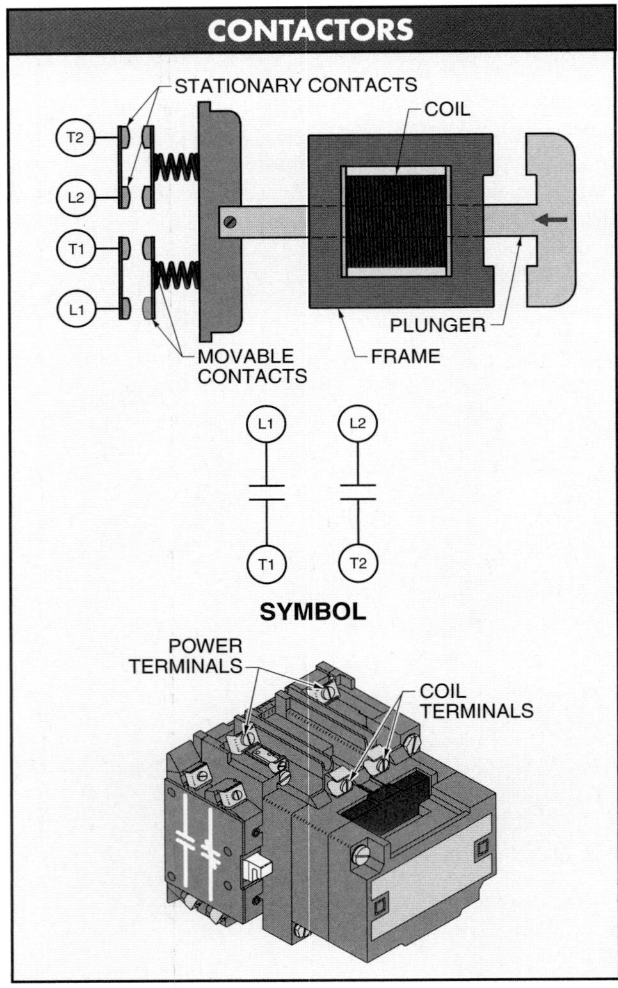

Figure 4-15. A contactor is a control device that uses a small control current to energize or de-energize the load connected to it.

The electrical operation of a contactor can be shown using a line diagram, a pictorial drawing, and/or a wiring diagram. **See Figure 4-16.** In this circuit, pressing pushbutton 1 (PB1) allows current to pass through the switch contacts and the contactor coil (C1) to L2. This energizes the contactor coil (C1). The activation of C1 closes the power contacts of the contactor. The contactor contains power contacts that close each time the control circuit is activated. These contacts are not normally shown in the line diagram but are shown in the pictorial drawing and wiring diagram.

Releasing PB1 stops the flow of current to the contactor coil and de-energizes the coil. The power contacts return to their NO condition when the coil de-energizes. This shuts OFF the lights or other loads connected to the power contacts.

This circuit works well in turning ON and OFF various loads remotely. It does, however, require someone to hold the contacts closed (pressing PB1) if the coil must be continuously energized. Auxiliary contacts may be added to a contactor to form an electrical holding circuit and to eliminate the necessity of someone holding the pushbutton continuously. **See Figure 4-17.** The auxiliary contacts are attached to the side of the contactor and are opened and closed with the power contacts as the coil is energized or de-energized. These contacts are shown on the line diagram because they are part of the control circuit, not the power circuit.

General Electric Company

A contactor may be used to control fractional horsepower motors that contain overload protection.

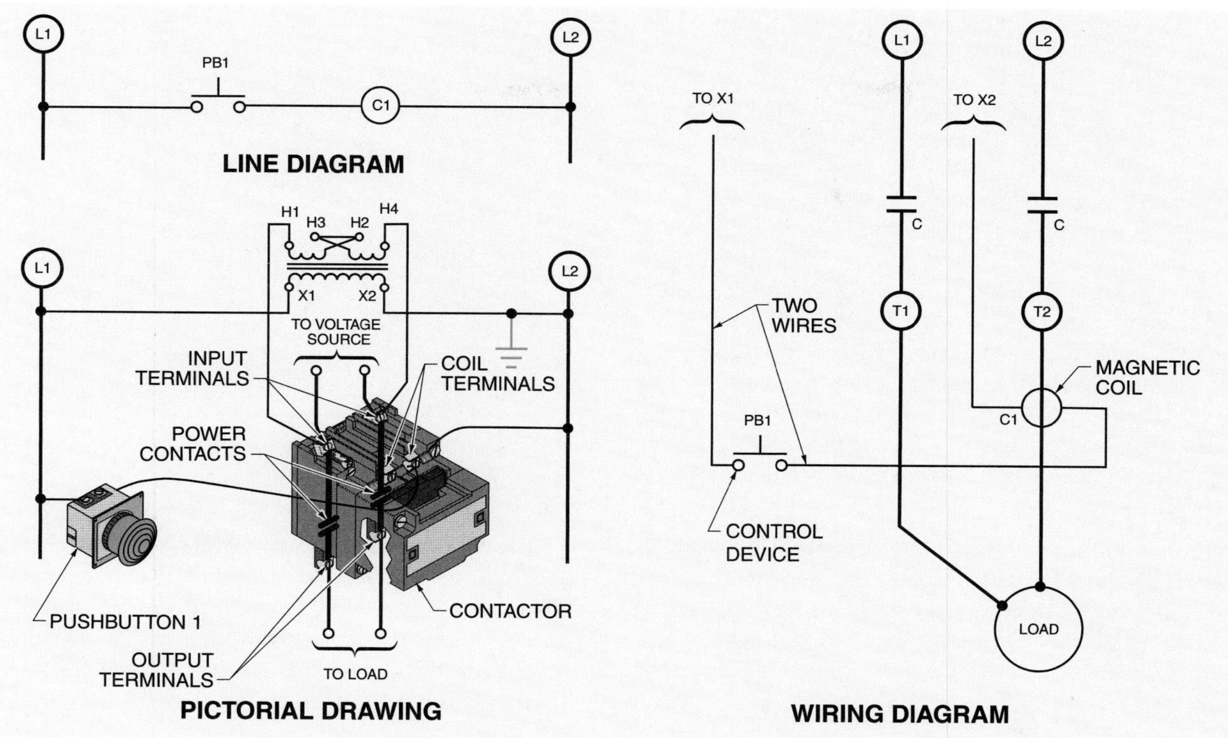

Figure 4-16. The electrical operation of a contactor can be shown using a line diagram, a pictorial drawing, and/or a wiring diagram.

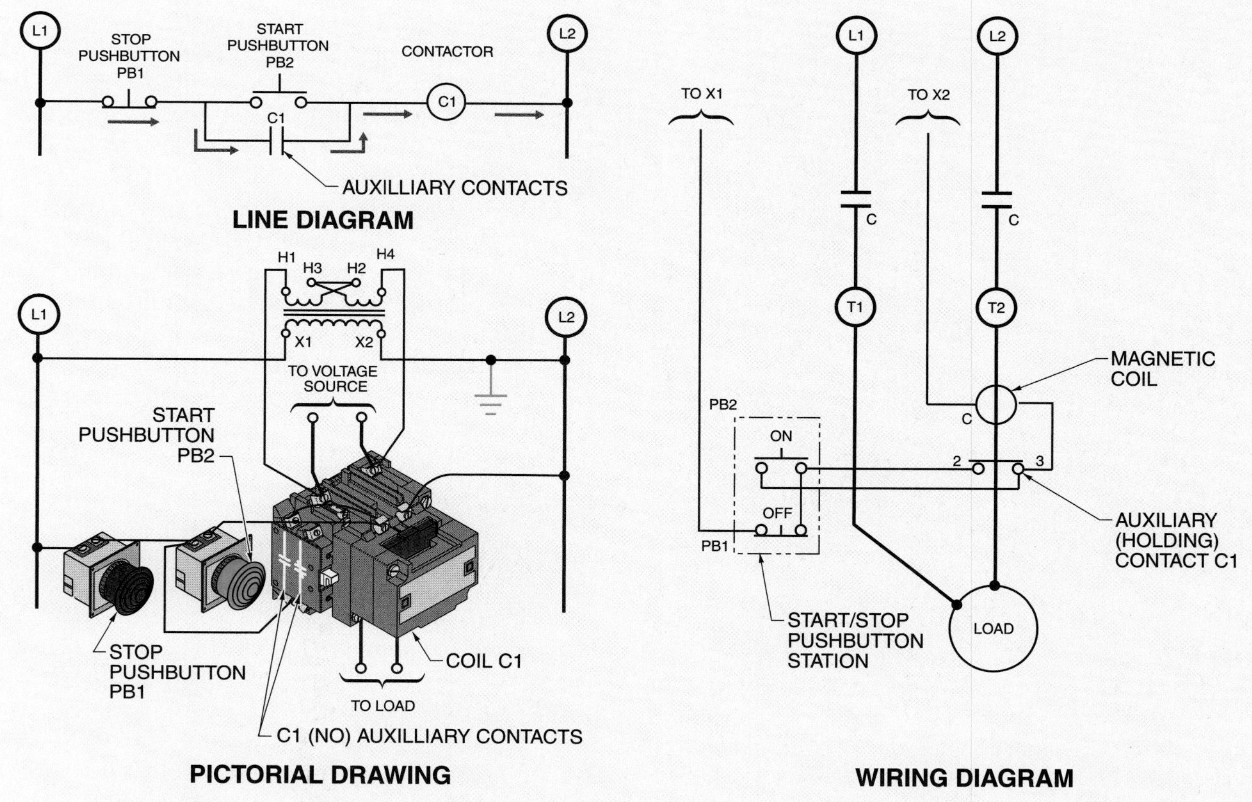

Figure 4-17. Auxiliary contacts may be added to a contactor to form an electrical holding circuit.

A line diagram may be used to show the logic of the electrical holding circuit. In this circuit, pressing the start pushbutton (PB2) allows current to pass through the closed contacts of the stop pushbutton (PB1), through the closed contacts of the start pushbutton, and through coil C1 to L2. This energizes coil C1. With coil C1 energized, auxiliary contacts C1 close and remain closed as long as the coil is energized. This forms a continuous electrical path around the start pushbutton (PB2) so that, even if the start pushbutton is released, the circuit remains energized because the coil remains energized.

The circuit is de-energized by a power failure or by pressing the NC stop pushbutton (PB1). In either case, the current flow to coil C1 stops and the coil is de-energized, causing auxiliary contacts C1 to return to their NO position. The start pushbutton (PB2) must be pressed to re-energize the circuit.

Technical Fact

Magnetic motor starters are sized depending on the horsepower requirements of the circuit in which they are used. National Electrical Manufacturers Association (NEMA) sizes 00 through 3 are available for 1ϕ motors and NEMA sizes 00 through 6 are available for 3ϕ motors. Voltages used with 1ϕ motors include 115 V and 230 V. Voltages used with 3ϕ motors include 200 V, 230 V, 460 V, and 575 V.

Magnetic Motor Starters. A *magnetic motor starter* is an electrically operated switch (contactor) that includes motor overload protection. Magnetic motor starters are used to start and stop motors. Magnetic motor starters are identical to contactors except that they have overloads attached to them. **See Figure 4-18.**

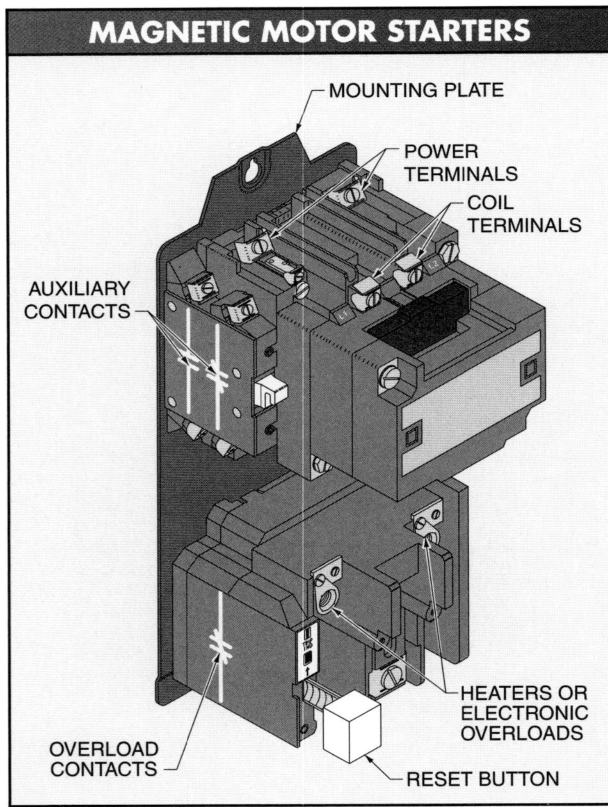

Figure 4-18. A magnetic motor starter is an electrically operated switch (contactor) that includes motor overload protection.

The magnetic motor starter overloads have heaters or electronic overloads (located in the power circuit) that sense excessive current flow to the motor. The heaters open the NC overload contacts (located in the control circuit) when the overload becomes dangerous to the motor.

The electrical operation of a magnetic motor starter may be shown using a line diagram, a pictorial drawing, and/or a wiring diagram. **See Figure 4-19.** The only difference between the drawings and diagrams of a contactor circuit and the drawings and diagrams of a magnetic motor starter circuit is the addition of the heaters or electronic overloads. In this circuit, pressing the start pushbutton allows current to pass through coil M1 and the overload contacts. This energizes coil M1. With coil M1 energized, auxiliary contacts M1 close and the circuit remains energized even if the start pushbutton is released.

This circuit is de-energized if the stop pushbutton is pressed, a power failure occurs, or any one of the overloads senses a problem in the power circuit. Coil M1 de-energizes, causing auxiliary contacts M1 to return to their NO condition if one of these situations occurs. When a motor stops because of an overload, the overload must be removed, the overload device reset, and the start pushbutton pressed to restart the motor.

Troubleshooting Control Circuit Wiring

When troubleshooting electrical control circuits, the actual circuit must be checked against the circuit diagram. Incorrect wiring causes a circuit to malfunction and loads to be energized at the wrong time or not at all. Circuit diagrams can be used to identify the problem in a circuit. All connections must be verified against circuit diagrams, starting from simple to complex. For example, incorrectly numbered wires or swapped wires, such as 6 for 9, are checked against the circuit diagram and specifications.

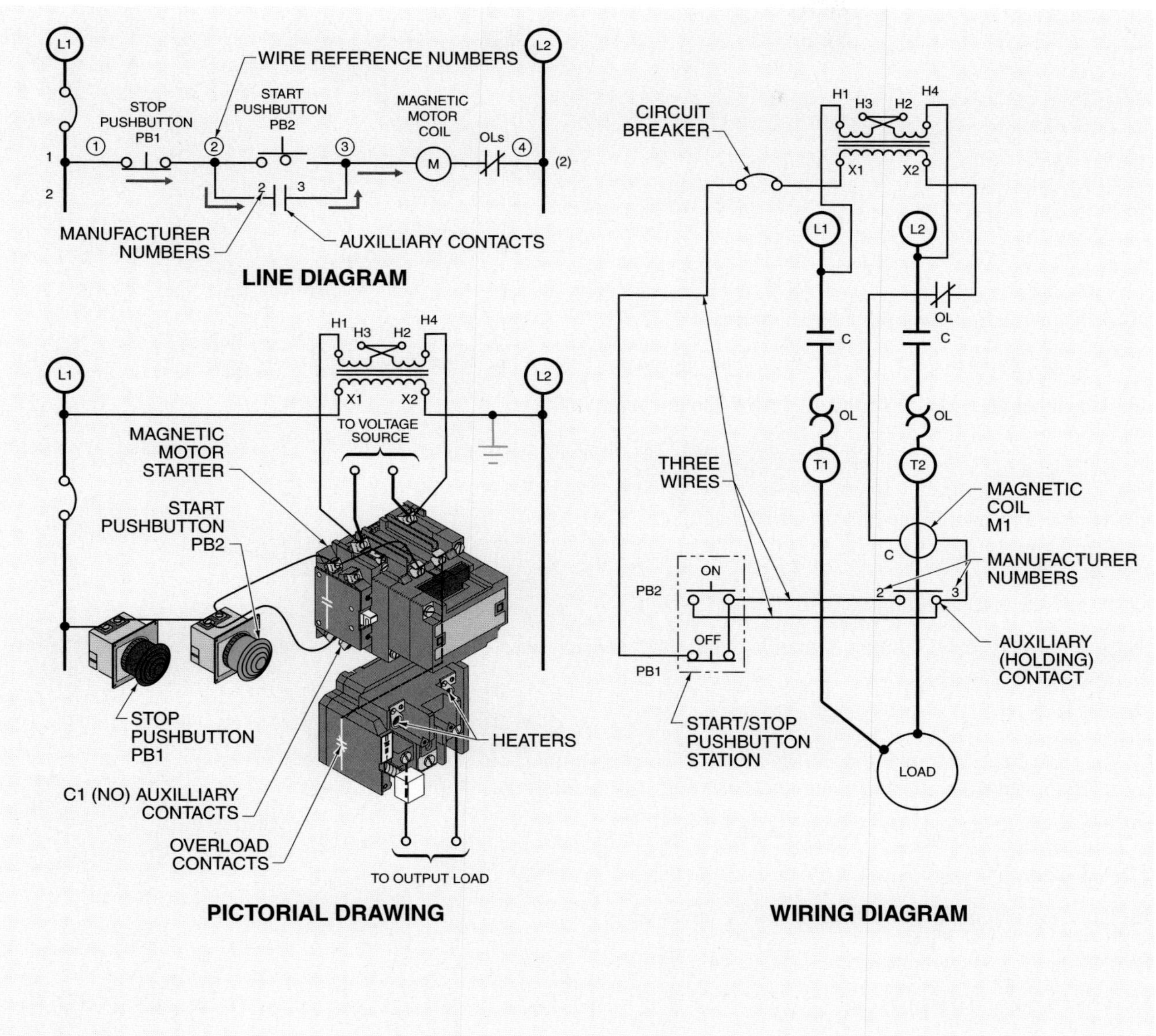

Figure 4-19. The electrical operation of a motor starter can be shown using a line diagram, a pictorial drawing, and/or a wiring diagram.

Review Questions

1. What is the function of the line diagram?
2. How are wires that are electrically connected illustrated on a line diagram?
3. Where are the overload contacts drawn in a line diagram?
4. Are normally open (NO) or normally closed (NC) contacts used when a float switch is used to maintain a predetermined level?
5. How is a solenoid illustrated in a line diagram?
6. How are coils illustrated in a line diagram?
7. What are auxiliary contacts?
8. Are normally open (NO) or normally closed (NC) contacts used when using auxiliary contacts to maintain an electrical holding circuit?
9. What is a magnetic motor starter?
10. In a line diagram showing auxiliary contacts forming a holding circuit around the start pushbutton, how can the power be removed from a magnetic motor starter coil after the start pushbutton is pressed?
11. What is a manual motor control circuit?
12. What are the three major parts of a solenoid?
13. How are magnetic motor starters similar to contactors?
14. What is control language?
15. How are line diagrams read?
16. What type of diagram is used to indicate the path and components in a circuit without showing detail of the actual wire connections?
17. What type of diagram is used to show as closely as possible the actual location of each component in a circuit?
18. If the moving part of a limit switch symbol is drawn below the terminal connections, is the switch NO or NC?
19. If the moving part of a limit switch symbol is drawn above the terminal connections, is the switch NO or NC?
20. What is a symbol?
21. What are the five basic components of all electrical circuits?
22. What is a power source?
23. What is an abbreviation?
24. What is an overcurrent protection device?
25. What is an electrical circuit?

Logic Applied to Line Diagrams

All electrical circuits can be grouped into three basic sections: the signal section, decision section, and action section. Within electrical circuits, certain logic functions can be developed. The most common logic functions are AND, OR, NOT, NOR, and NAND. Combinations of these circuits can be used to develop complex automated decision-making circuits.

BASIC RULES OF LINE DIAGRAMS

The electrical industry has established a universal set of symbols and rules on how line diagrams (circuits) are laid out. By applying these standards, an electrician establishes a working practice with a language in common with all electricians.

One Load per Line

No more than one load should be placed in any circuit line between L1 and L2. A pilot light can be connected into a circuit with a single-pole switch. **See Figure 5-1.** In this circuit, the power lines are drawn vertically on sides of the drawing. The lines are marked L1 and L2. The space between L1 and L2 represents the voltage of the control circuit. This voltage appears across pilot light PL1 when switch S1 is closed. The pilot light glows when current flows through S1 and PL1 because the voltage between L1 and L2 is the proper voltage for the pilot light.

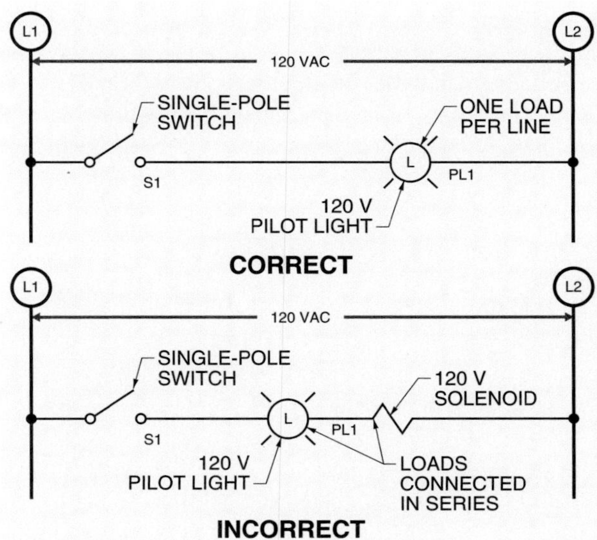

Figure 5-1. No more than one load should be placed in any circuit line between L1 and L2.

Two loads must not be connected in series in one line of a line diagram. If the two loads are connected in series, then the voltage between L1 and L2 must divide across both loads when S1 is closed. The result is that neither device receives the entire 120 V necessary for proper operation.

The load that has the highest resistance drops the highest voltage. The load that has the lowest resistance drops the lowest voltage.

Loads must be connected in parallel when more than one load must be connected in the line diagram. **See Figure 5-2.** In this circuit, there is only one load for each line between L1 and L2, even though there are two loads in the circuit. The voltage from L1 and L2 appears across each load for proper operation of the pilot light and solenoid. This circuit has two lines, one for the pilot light and one for the solenoid.

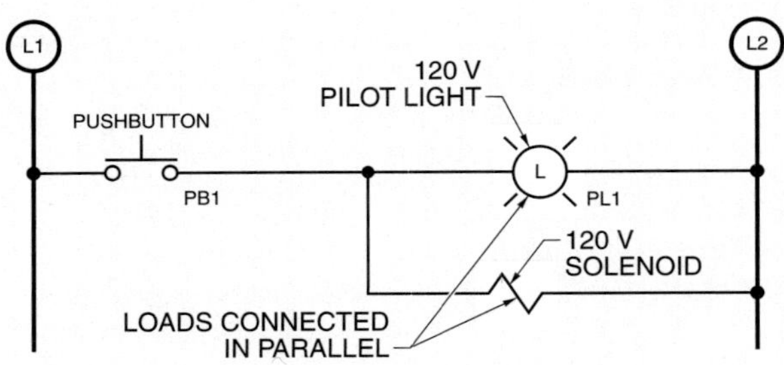

Figure 5-2. Loads must be connected in parallel when more than one load must be connected in the line diagram.

Load Connections

A *load* is any device that converts electrical energy to motion, heat, light, or sound. A load is the electrical device in a line diagram that uses the electrical power from L1 to L2. Control relay coils, solenoids, and pilot lights are loads that are connected directly or indirectly to L2. **See Figure 5-3.**

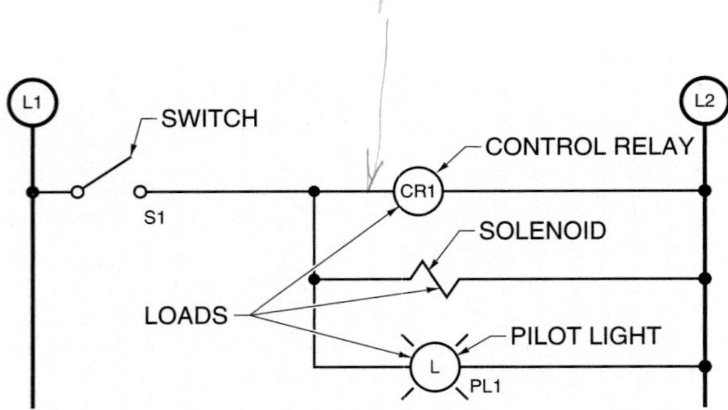

Figure 5-3. Control relays, solenoids, and pilot lights are loads that are connected directly or indirectly to L2.

Magnetic motor starter coils are connected to L2 indirectly through normally closed (NC) overload contacts. **See Figure 5-4.** An overload contact is normally closed and opens only if an overload condition exists in the motor. The number of NC overload contacts between the starter coil and L2 depends on the type of starter and power that are used in the circuit.

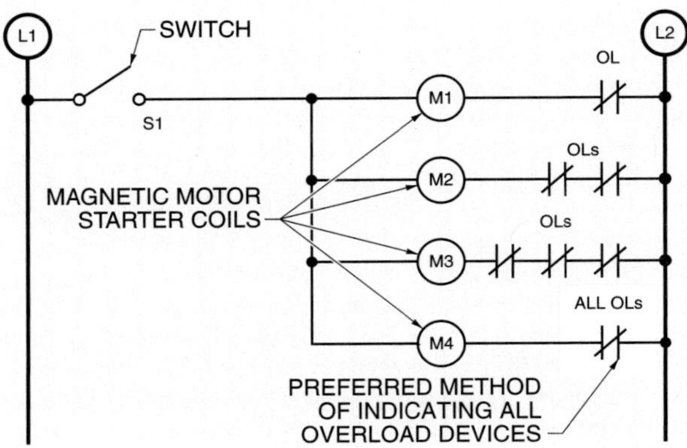

Figure 5-4. Magnetic motor starter coils are connected to L2 indirectly through NC overload contacts.

One to three NC overload contacts may be shown between the starter and L2 in all line diagrams. One to three NC overload contacts are shown because starters may include one, two, or three overload contacts, depending on the manufacturer and motor used. Early starters often included three overload contacts, one for each heater in the starter. Modern starters include only one overload contact. To avoid confusion, it is common practice to draw one set of NC overload contacts and mark these contacts all overloads (OLs). An overload marked this way indicates that the circuit is correct for any motor or starter used. The electrician knows to connect all the NC overload contacts that the starter is designed for in series if there is more than one on the starter.

Control Device Connections

Control devices are connected between L1 and the operating coil (or load). Operating coils of contactors and starters are activated by control devices such as pushbuttons, limit switches, and pressure switches. **See Figure 5-5.**

Each line includes at least one control device. The operating coil is ON all the time if no control device is included in a line. A circuit may contain as many control devices as are required to make the operating coil function as specified. These control devices may be connected in series or parallel when controlling an operating coil. Although a circuit may include any number of loads, the total number of loads determines the required wire size and rating of the incoming power supply (typically a transformer). As loads are added to the circuit, total current increases.

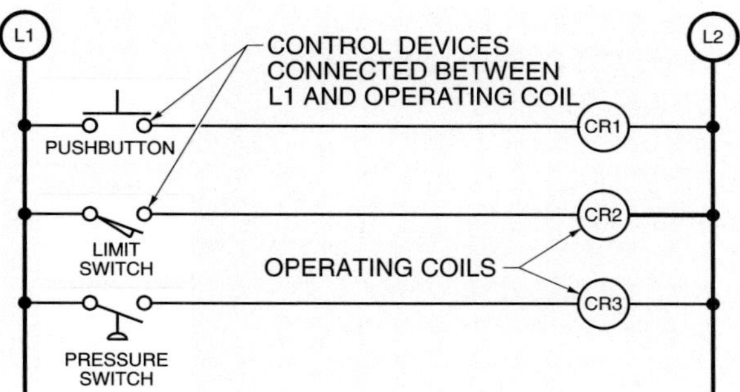

Figure 5-5. Control devices are connected between L1 and the operating coil.

Two control devices (a flow switch and a temperature switch) can be connected in series to control a coil in a magnetic motor starter. The flow switch and temperature switch must close to allow current to pass from L1, through the control device, the magnetic starter coil, and the overloads, to L2. Two control devices (a pressure switch and a foot switch) can be connected in parallel to control a coil in a magnetic motor starter. **See Figure 5-6.** Either the pressure switch or the foot switch can be closed to allow current to pass from L1 through the control device, the magnetic starter coil, and the overloads to L2. Regardless of how the control devices are arranged in a circuit, they must be connected between L1 and the operating coil (or load). The contacts of the control device may be either normally open (NO) or normally closed (NC). The contacts used and the way the control devices are connected into a circuit (series or parallel) determines the function of the circuit.

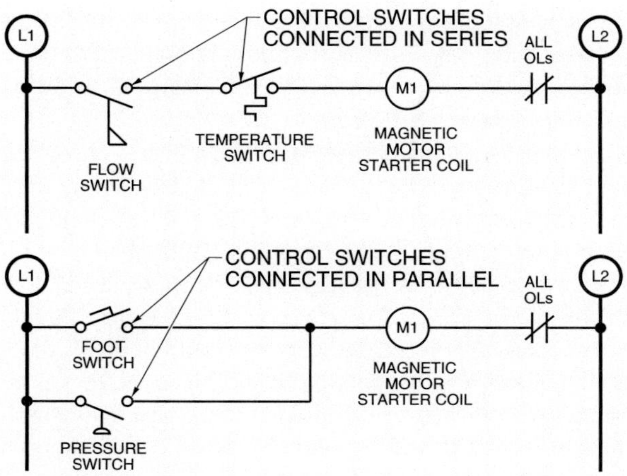

Figure 5-6. Two control devices may be connected in series or parallel to control a coil in a magnetic motor starter.

Line Number Reference

Each line in a line diagram should be numbered starting with the top line and reading down. **See Figure 5-7.**

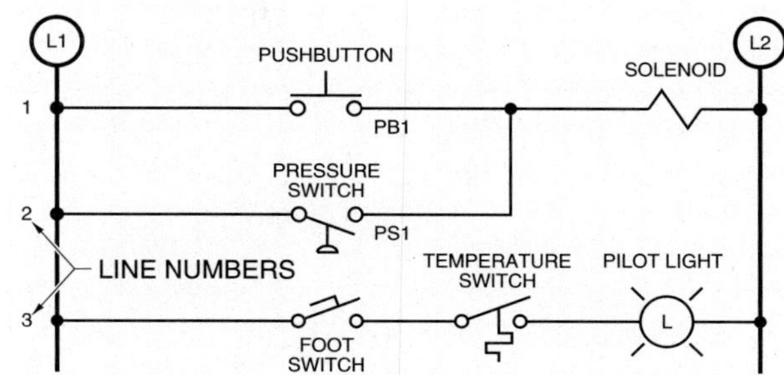

Figure 5-7. Each line in a line diagram should be numbered starting with the top line and reading down.

Line 1 connects PB1 to the solenoid to complete the path from L1 to L2. Line 2 connects PS1 to the solenoid to complete the path from L1 to L2. PB1 and PS1 are marked as two separate lines even though they control the same load, because either the pushbutton or the pressure switch completes the path from L1 to L2. Line 3 connects a foot switch and a temperature switch to complete the path from L1 to L2. The foot switch and temperature switch both appear in the same line because it takes both the foot switch and the temperature switch to complete the path to the pilot light.

Numbering each line simplifies the understanding of the function of a circuit. The importance of this numbering system becomes clear as circuits become more complex and lines are added.

Numerical Cross-Reference Systems

Numerical cross-reference systems are required to trace the action of a circuit in complex line diagrams. Common rules help to simplify the operation of complex circuits.

Numerical Cross-Reference System (NO Contacts). Relays, contactors, and magnetic motor starters normally have more than one set of auxiliary contacts. These contacts may appear at several different locations in the line diagram. Numerical cross-reference systems quickly identify the location and type of contacts controlled by a given device. A numerical cross-reference system consists of numbers in parentheses to the right of the line diagram. NO contacts are represented by line numbers. The line numbers refer to the line on which the NO contacts are located. **See Figure 5-8.**

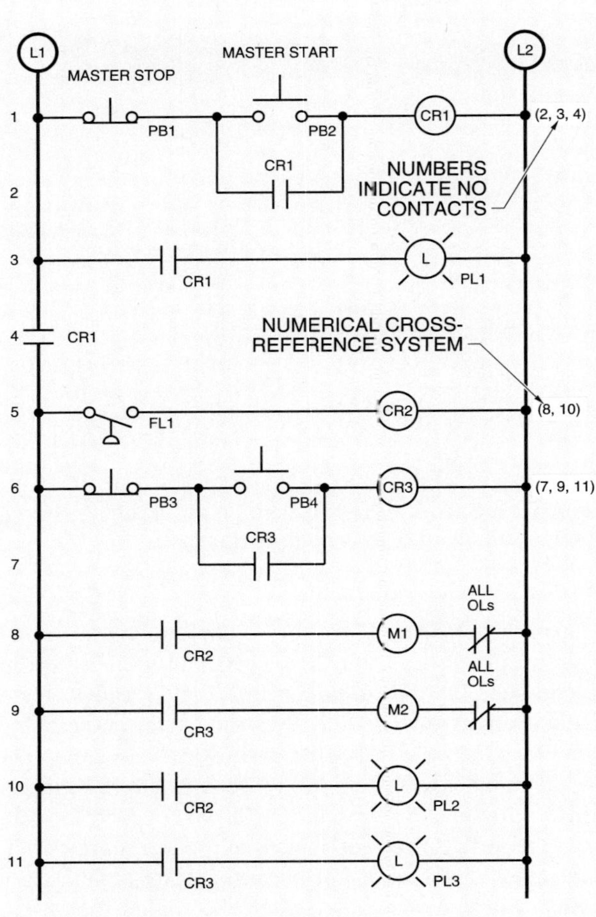

Figure 5-8. The locations of normally open contacts controlled by a device are determined by the numbers on the right side of the line diagram.

In this circuit, pressing master start pushbutton PB2 energizes control relay coil CR1. Control relay coil CR1 controls three sets of NO contacts. This is shown by the numerical codes (2, 3, 4) on the right side of the line diagram. Each number indicates the line in which the NO contacts are located.

In line 2, the NO contacts form the holding circuit (memory) for maintaining the coil CR1 after master start pushbutton PB2 is released. In line 3, the NO contacts energize pilot light PL1, indicating that the circuit has been energized. In line 4, the NO contacts allow the remainder of the circuit to be activated by connecting L1 to the remainder of the circuit. The numerical cross-reference system shows the location of all contacts controlled by coil CR1 as well as the effect each has on the operation of the circuit.

In line 5, control relay CR2 energizes if float switch FL1 closes. Control relay CR2 closes the NO contacts located in lines 8 and 10 as indicated by the numerical codes (8, 10). The magnetic motor starter controlled by coil M1 is energized when the NO contacts of line 8 close. Pilot light PL2 turns ON, indicating the motor has started, when the NO contacts in line 10 close.

In line 6, several NO contacts located in lines 7, 9, and 11 are used to control other parts of the circuit through control relay CR3. In line 7, the NO contacts form the memory circuit for maintaining the circuit to control relay coil CR3 after pushbutton PB4 is released. The NO contacts in line 9 close, energizing the magnetic motor starter controlled by coil M2 when coil CR3 is energized. Simultaneously, the NO contacts in line 11 close, causing pilot light PL3 to light as an indicator that the motor has started.

The numerical cross-reference system allows the simplification of complex line diagrams. Each NO contact must be clearly marked because each set of NO contacts is numbered according to the line in which they appear.

Numerical Cross-Reference System (NC Contacts). In addition to NO contacts, there are also NC contacts in a circuit. To differentiate between NO and NC, NC contacts are indicated as a number that is underlined. The underlined number refers to the line on which the NC contacts are located. **See Figure 5-9.** For example, lines 9 and 11 contain devices that control NC contacts in lines 12 and 13 as indicated by the underlined numbers (12, 13) to the right of the line diagram.

In this circuit, pressing master start pushbutton PB2 energizes control relay coil CR1. Control relay coil CR1 controls three sets of NO contacts. This is shown by the numerical codes (2, 3, 4) on the right side of the line diagram. In line 2, the NO contacts form the holding circuit (memory) for maintaining the coil CR1 after master start pushbutton PB2 is released. In line 3, the NO contacts energize pilot light PL1, indicating that the circuit has been energized. In line 4, the NO contacts allow the remainder of the circuit to be activated by connecting L1 to the remainder of the circuit.

In line 5, when control relay coil CR2 is energized, the NO contacts in lines 8, 9, and 12 close. Closing these contacts energizes coils M1 and M2 and completes the circuit going to the pilot light in line 12. The pilot light in line 12 does not glow because the NC contacts controlled by coil M2 in line 12 are opened at the same time that the NO contacts are closed, leaving the circuit open. With the NO contacts of CR2 closed and the NC contacts of M2 open, the light stays OFF unless something happens to shut down line 9, which contains coil M2. For example, if coil M2 represents a safety cooling fan protecting the motor controlled by M1, the light would indicate the loss of cooling.

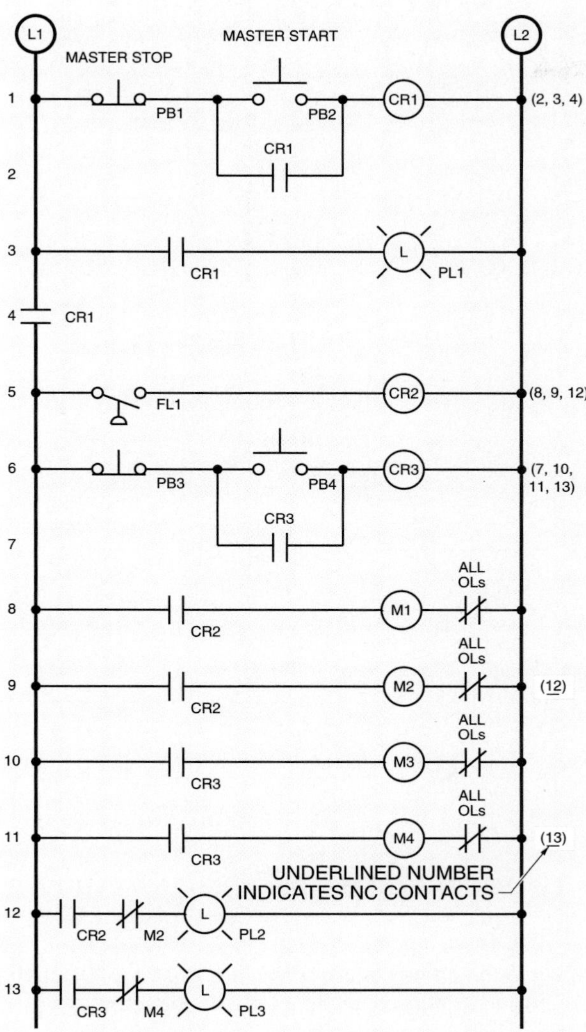

Figure 5-9. Normally closed contacts are indicated by numbers which are underlined to distinguish them from NO contacts.

A similar sequence of events took place when line 6 was energized. In this case, pressing pushbutton PB4 energizes control relay coil CR3, which closes NO contacts in line 7, 10, 11, and 13. A memory circuit is formed in line 7, coils M3 and M4 are energized in lines 10 and 11, and part of the circuit to pilot light PL3 in line 13 is completed. Because coil M4 is energized, the NC contacts it controls open in line 13, forming a similar alarm circuit to the one in line 12.

If coil M4 in line 11 drops out for any reason, the NC contacts in line 13 return to their NC position and the pilot light alarm signal in line 13 is turned ON. This circuit could be used where it is extremely important for the operator to know when something is not functioning.

Wire Reference Numbers

Each wire in a control circuit is assigned a reference point (number) on a line diagram to keep track of the different wires that connect the components in the circuit. Each reference point is assigned a reference number. Reference numbers are normally assigned from the top left to the bottom right. This numbering system can apply to any control circuit such as single-station, multistation, or reversing circuits. **See Figure 5-10.**

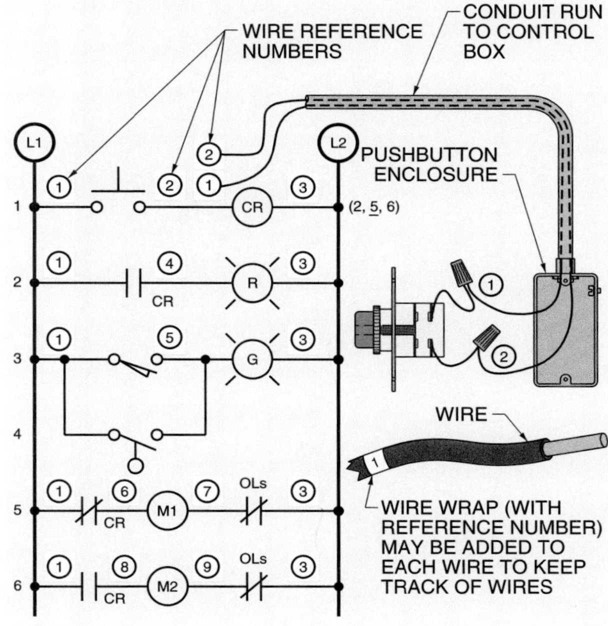

Figure 5-10. Each wire in a control circuit is assigned a reference point on a line diagram to keep track of the different wires that connect the components in the circuit.

Any wire that is always connected to a common point is the same electrically and assigned the same number. The wires that are assigned a number vary from 2 to the number required by the circuit. Any wire that is prewired when the component is purchased is normally not assigned a reference number. The exact numbering system used varies for each manufacturer or design engineer. One common method used is to circle the wire reference numbers. Circling the wire reference numbers helps separate them from other numbering systems.

Manufacturer's Terminal Numbers

Manufacturers of electrical relays, timers, counters, etc., include numbers on the terminal connection points. These terminal numbers are used to identify and separate the different component parts (coil, NC contacts, etc.) included on the individual pieces of equipment. Manufacturer's terminal numbers are often added to a line diagram after the specific equipment to be used in the control circuit is identified. **See Figure 5-11.**

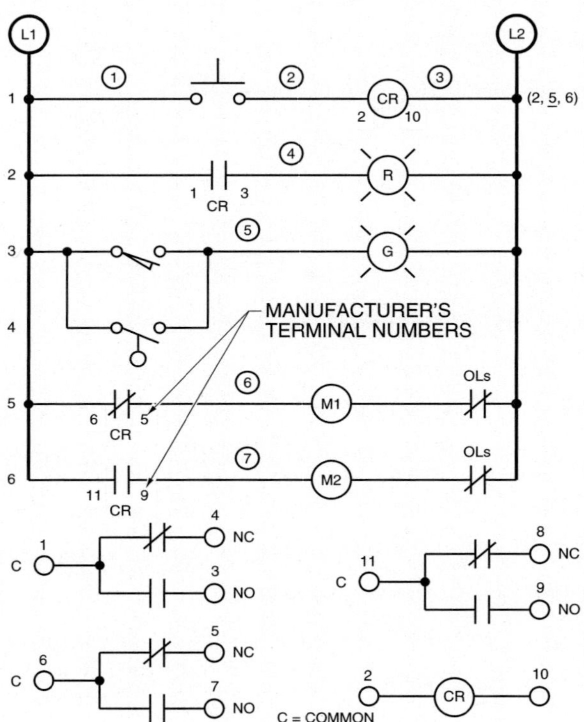

Figure 5-11. Manufacturers include terminal numbers to identify and separate the different component parts included on individual pieces of equipment.

Cross-Referencing Mechanically Connected Contacts

Control devices such as limit switches, flow switches, temperature switches, liquid level switches, and pressure switches normally have more than one set of contacts operating when the device is activated. These devices normally have at least one set of NO contacts and one set of NC contacts that operate simultaneously. For all practical purposes, the multiple contacts of these devices normally do not control other devices in the same lines of a control circuit.

The two methods used to illustrate how contacts found in different control lines belong to the same control switch are the dashed line method and the numerical cross-reference method. **See Figure 5-12.**

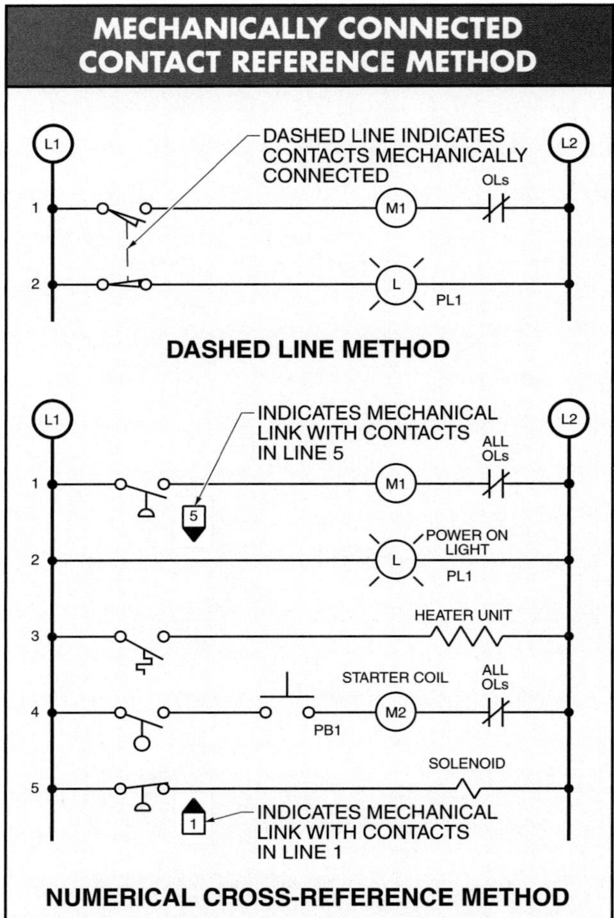

Figure 5-12. Contacts found in different control lines that belong to the same control switch are illustrated using the dashed line or numerical cross-reference method.

In the dashed line method, the dashed line between the NO and NC contacts indicates that both contacts move from the normal position when the arm of the limit switch is moved. In this circuit, pilot light PL1 is ON and motor starter coil M1 is OFF. After the limit switch is actuated, pilot light PL1 turns OFF and the motor starter coil M1 turns ON.

The dashed line method works well when the control contacts are close together and the circuit is relatively simple. If a dashed line must cut across many lines, the circuit becomes hard to follow.

The numerical cross-reference method is used on complex line diagrams where a dashed line cuts across several lines. In this circuit, a pressure switch with an NO contact in line 1 and an NC contact in line 5 is used to control a motor starter and a solenoid. The NO and NC contacts of the pressure switch are simultaneously actuated when a predetermined pressure is reached. A solid arrow pointing down is drawn by the NO contact in line 1 and is marked with a 5 to show the mechanical link with the contact in line 5. A solid arrow pointing up is drawn by the NC contact in line 5 and is marked with a numeral 1 to show the mechanical linkage with the contact in line 1. This cross-reference method eliminates the need for a dashed line cutting across lines 2, 3, and 4. This makes the circuit easier to follow and understand. This system may be used with any type of control switch found in a circuit.

LINE DIAGRAMS—SIGNALS, DECISIONS, AND ACTION

The concept of control is to accomplish specific work in a predetermined manner. A circuit must respond as designed, without any changes. To accomplish this consistency, all control circuits are composed of three basic sections: the signals, the decisions, and the action sections. **See Figure 5-13.** Complete understanding of these sections enables easy understanding of any existing industrial control circuit, as well as those that are created as systems become more mechanized and automated.

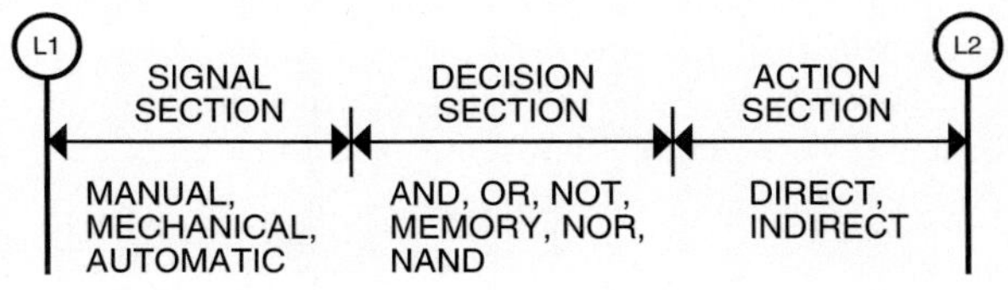

Figure 5-13. All control circuits are composed of signal, decision, and action sections.

Signals

A signal starts or stops the flow of current by closing or opening the control device's contacts. Current is allowed to flow through the control device if the contacts are closed. Current is not allowed to flow through the control device if the contacts are opened. Pushbuttons, limit switches, flow switches, foot switches, temperature switches, and pressure switches may be used as the signal section of a control circuit.

All signals depend on some condition that must take place. This condition can be manual, mechanical, or automatic. A manual condition is any input into the circuit by a person. Foot switches and pushbuttons are control devices that respond to a manual condition. A mechanical condition is any input into the circuit by a mechanically moving part. A limit switch is a control device that responds to a mechanical condition. When a moving object, such as a box, hits a limit switch, the limit switch normally has a lever, roller, ball, or plunger actuator that causes a set of contacts to open or close. An automatic condition is any input that responds automatically to changes in a system. Flow switches, temperature switches, and pressure switches respond to automatic conditions. These devices automatically open and close sets of contacts when a change in the flow of a liquid is created, when a change in temperature is sensed, or when pressure varies. The signal accomplishes no work by itself; it merely starts or stops the flow of current in that part of the circuit.

Decisions

The decision section of a circuit determines what work is to be done and in what order the work is to occur. The decision section of a circuit adds, subtracts, sorts, selects, and redirects the signals from the control devices to the load. For the decision part of the circuit to perform a definite sequence, it must perform in a logical manner. The way the control devices are connected into the circuit gives the circuit logic. The decision section of the circuit accepts informational inputs (signals), makes logical decisions based on the way the control devices are connected into the circuit, and provides the output signal that controls the load.

Action

Once a signal is generated and the decision has been made within a circuit, some action (work) should result. In most cases it is the operating coil in the circuit that is responsible for initiating the action. This action is direct when devices such as motors, lights, and heating elements are turned ON as a direct result of the signal and the decision. This action is indirect when the coils in solenoids, magnetic starters, and relays are energized. The action is indirect because the coil energized by the signal and the decision may energize a magnetic motor starter, which actually starts the motor. Regardless of how this action takes place, the load causes some action (direct or indirect) in the circuit and, for this reason, is the action section of the circuit.

LOGIC FUNCTIONS

Control devices such as pushbuttons, limit switches, and pressure switches are connected into a circuit so that the circuit can function in a predetermined manner. All control circuits are basic logic functions or combinations of logic functions. Logic functions are common to all areas of industry. This includes electricity, electronics, hydraulics, pneumatics, math, and other routine activities. Logic functions include AND, OR, AND/OR, NOT, NOR, and NAND.

AND Logic

AND logic is used in industry when two normally open pushbuttons are connected in series to control a solenoid. **See Figure 5-14.**

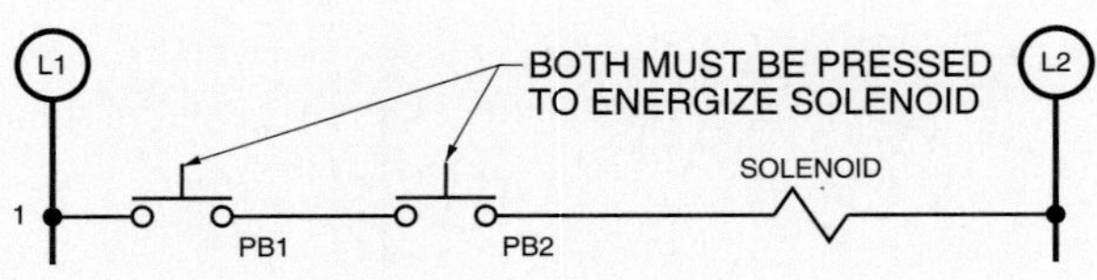

Figure 5-14. In AND logic, the load is ON if both of the control signal's contacts are closed.

PB1 and PB2 must be pressed before the solenoid is energized. The logic function that makes up the decision section of this circuit is AND logic. The reason for using the AND function could be to build in safety for the operator of this circuit.

If the solenoid were operating a punch press or shear, the pushbuttons could be spaced far enough apart so that the operator would have to use both hands to make the machine operate. This ensures that the operator's hands are not near the machine when it is activated. With AND logic, the load is ON only if all the control signal contacts are closed. As with any logic function, the signals may be manually, mechanically, or automatically controlled. Any control device such as limit switches, pressure switches, etc., with NO contacts can be used in developing AND logic. The NO contacts of each control device must be connected in series for AND logic.

A simple example of AND logic takes place whenever an automobile that has an automatic transmission is started. The ignition switch must be turned to the start position and the transmission selector must be in the park position before the starter is energized. Before the action (load ON) in the automobile circuit can take place, the control signals (manual) must be performed in a logical manner (decision).

OR Logic

OR logic is used in industry when a normally open pushbutton and a normally open temperature switch are connected in parallel. **See Figure 5-15.** In this circuit, the load is a heating element that is controlled by two control devices.

The logic of this circuit is OR logic because the pushbutton or the temperature switch energizes the load. The temperature switch is an example of an automatic control device that turns the heating element ON and OFF to maintain the temperature setting for which the temperature switch is set. The manually controlled pushbutton could be used to test or turn ON the heating element when the temperature switch contacts are open.

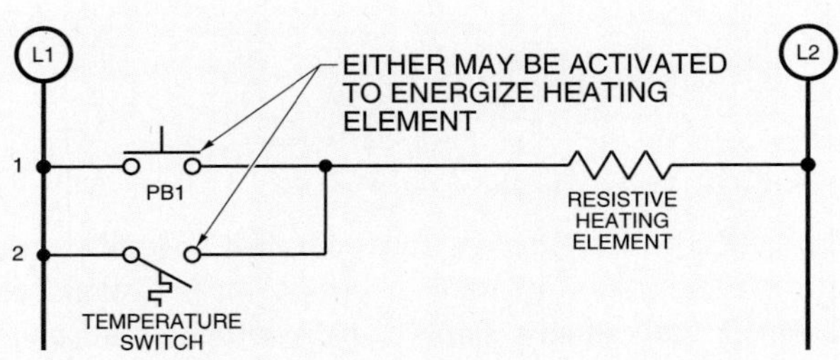

Figure 5-15. In OR logic, the load is ON if any one of the control signal's contacts is closed.

In OR logic, the load is ON if any one of the control signal's contacts is closed. The control devices are connected in parallel. Series and parallel refer to the physical relationship of each control device to other control devices or components in the circuit. This series and parallel relationship is only part of what determines the logic function of any circuit.

An example of OR logic is in a dwelling that has two pushbuttons controlling one bell. The bell (load) may be energized by pressing (signal ON) either the front or the back pushbutton (control device). Here, as in the automobile circuit, the control devices are connected to respond in a logical manner.

AND/OR Logic Combination

The decision section of any circuit may contain one or more logic functions. **See Figure 5-16.** In this circuit, both pressure and flow must be present in addition to the pushbutton or the foot switch being engaged to energize the starter coil (load). This provides the circuit with the advantage of both AND logic and OR logic. The machine is protected because both pressure and flow must exist before it is started, and there is a choice between using a pushbutton or a foot switch for final operation. The action taking place in this circuit is energizing a coil in a magnetic motor starter. The signal inputs for this circuit have to be two automatic and at least one manual.

Each control device responds to its own input signal and has its own decision-making capability. When multiple control devices are used in combination with other control devices making their own decisions, a more complex decision can be made through the combination of all control devices used in the circuit. All industrial control circuits consist of control devices capable of making decisions in accordance with the input signals received.

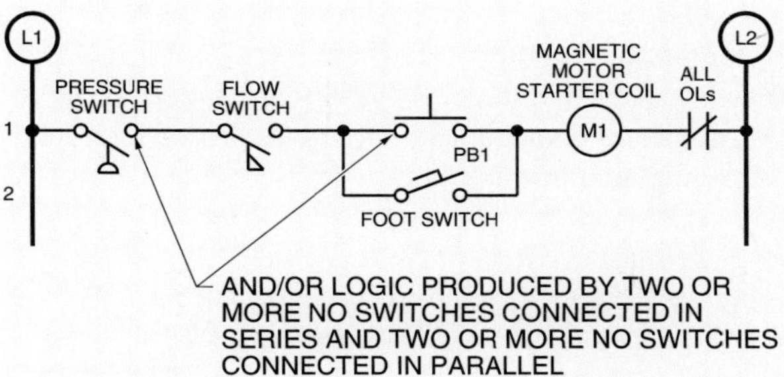

Figure 5-16. The decision section of any circuit may contain one or more logic functions.

NOT Logic

NOT logic has an output if the control signal is OFF. For example, replacing NO contacts on a pushbutton with NC contacts energizes the solenoid and pilot light without pressing the pushbutton. **See Figure 5-17.** Pressing the pushbutton in this circuit de-energizes the loads. There must not be a signal if the loads are to remain energized. With NOT logic, the output remains ON only if the control signal contacts remain closed.

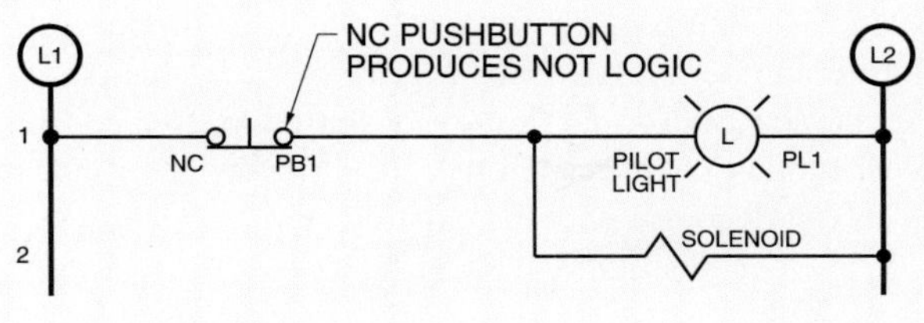

Figure 5-17. In NOT logic, the load is ON only if the control signal contacts are closed.

An example of NOT logic is the courtesy light in a refrigerator. The light is ON if the control signal is OFF. The control signal is the door of the refrigerator. Any time the door is open (signal OFF), the load (courtesy light) is ON. The condition that controls the signal can be manual, mechanical, or automatic. With the refrigerator door, the condition is mechanical.

NOR Logic

NOR logic is an extension of NOT logic in that two or more NC contacts in series are used to control a load. **See Figure 5-18.** In this circuit, additional operator safety is provided by adding several emergency stop pushbuttons (NOT logic) to the control circuit. Pressing any emergency stop pushbutton de-energizes the load (coil M1). By incorporating NOR logic, each machine may be controlled by one operator, but any operator or supervisor can have the capability of turning OFF all the machines on the assembly line to protect individual operators or the entire system. With the knowledge of NOR logic, the electrician can readily add stop pushbuttons by wiring them in series to perform their necessary function.

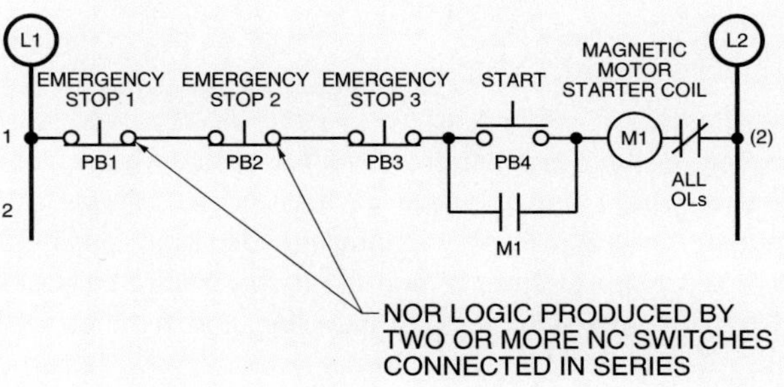

Figure 5-18. NOR logic is an extension of NOT logic in that two or more NC contacts in series are used to control a load.

NAND Logic

NAND logic is an extension of NOT logic in which two or more NC contacts are connected in parallel to control a load. **See Figure 5-19.** In this circuit, two interconnected tanks are filled with a liquid. When pushbutton PB3 is pressed, coil M1 is energized and auxiliary contacts M1 close until both tanks are filled. Both tanks fill to a predetermined level because the float switches in tank 1 and tank 2 do not open until both tanks are full. Every NOT must be open (signal OFF) to stop the filling process based on the input of the float switches. NOR logic is also present in this circuit because the emergency stops (NOT) at tank 1 or tank 2 may stop the process if an operator at either of the tanks sees a problem.

An example of a NAND circuit is the courtesy light in an automobile. In an automobile, the courtesy lights are ON if the control signal (door switches) is OFF (normally closed). This circuit is different from a refrigerator door in that an automobile may have two or more door switches, any of which will turn ON the courtesy lights.

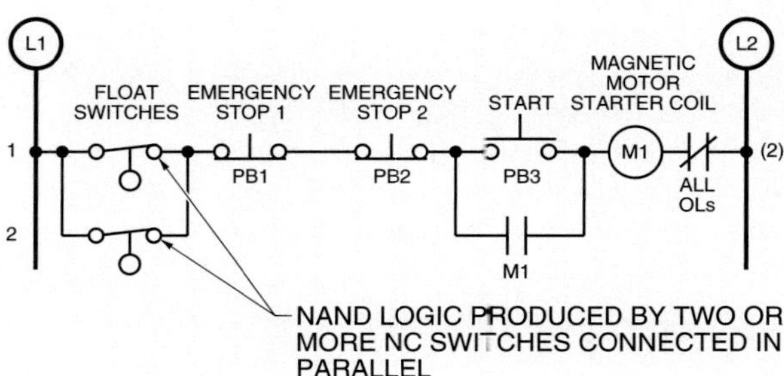

Figure 5-19. NAND logic is an extension of NOT logic in which two or more NC contacts are connected in parallel to control a load.

Memory

Many of today's industrial circuits require their control circuits to not only make logic decisions such as AND, OR, and NOT, but also to be capable of storing, memorizing, or retaining the signal inputs to keep the load energized even after the signals are removed. A switch that controls house lights from only one location is an example of a memory circuit. Memory circuits are also known as holding or sealing circuits. When the memory circuit is ON, it remains ON until it is turned OFF, and remains OFF until it is turned ON. It performs a memory function because the output corresponds to the last input information until new input information is received to change it. In the case of the house light switch, the memory circuit is accomplished by a switch that mechanically stays in one position or another.

In industrial control circuits, it is more common to find pushbuttons with return spring contacts (momentary contacts) than those that mechanically stay held in one position (maintained contacts). Auxiliary contacts are added to provide memory to circuits with pushbuttons. **See Figure 5-20.** Once coil M1 of the magnetic motor starter is energized, it causes coil contacts M1 to close and remain closed (memory) until the coil is de-energized.

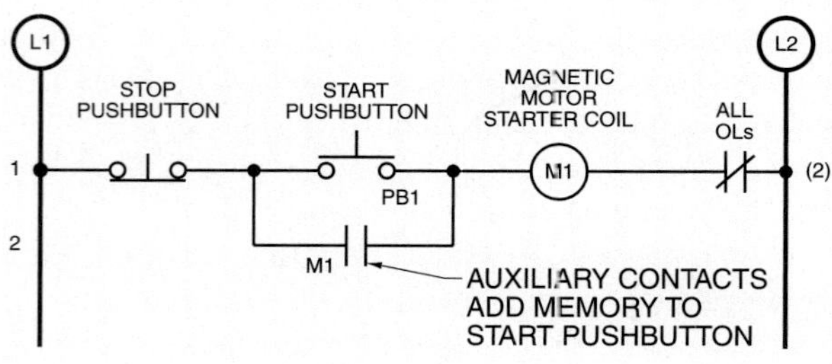

Figure 5-20. Auxiliary contacts are added to give circuits with pushbuttons memory.

Technical Fact

The level of control in industrial circuits varies from simple manual ON/OFF control to automatic control of hundreds of parameters.

NOT logic may be added to memory logic to create a common start/stop control circuit. **See Figure 5-21.** When stop pushbutton PB1 is activated, current to coil M1 stops and contacts M1 open, returning the circuit to its original condition.

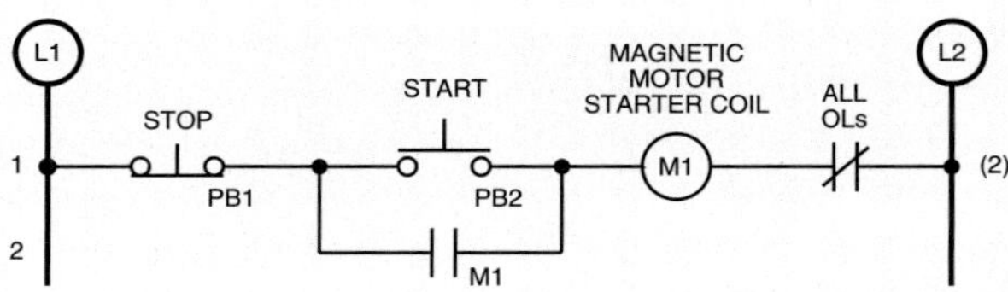

Figure 5-21. A common start/stop control circuit is created by adding the NOT logic of a stop pushbutton to the memory logic of magnetic coil contacts.

COMMON CONTROL CIRCUITS

Various control circuits are commonly used in commercial and industrial electrical circuits. An electrician must understand the entire circuit operation to begin wiring or troubleshooting the circuit.

Start/Stop Stations Controlling Magnetic Starters

A load is often required to be started and stopped from more than one location. **See Figure 5-22.** In this circuit, the magnetic motor starter may be started or stopped from two locations. Additional stop pushbuttons are connected in series (NOR logic) with the existing stop pushbuttons.

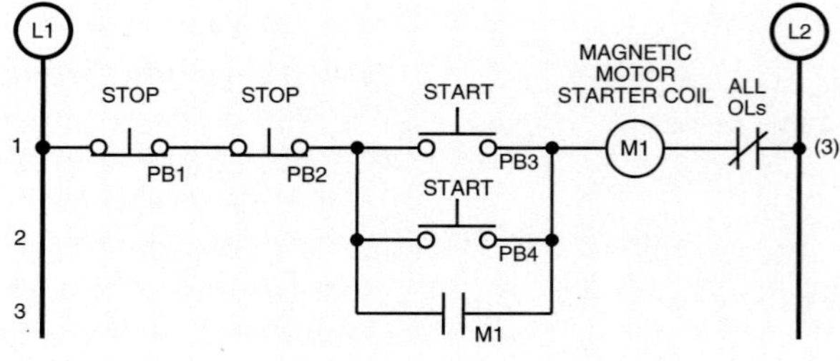

Figure 5-22. Two stop pushbuttons connected in series and two start pushbuttons connected in parallel are used to control a motor from two locations.

Additional start pushbuttons are connected in parallel (OR logic) with the existing start pushbuttons. Pressing any one of the start pushbuttons (PB3 or PB4) causes coil M1 to energize. This causes auxiliary contacts M1 to close, adding memory to the circuit until coil M1 is de-energized. Coil M1 may be de-energized by pressing stop pushbuttons PB1 or PB2, by an overload that would activate the OLs, or by a loss of voltage to the circuit. In the case of an overload, the overload has to be removed and the circuit overload devices reset before the circuit would return to normal starting condition.

Two Magnetic Starters Operated by Two Start/Stop Stations with Common Emergency Stop

In almost all electrical systems, several devices can be found running off a common supply voltage. Two start/stop stations may be used to control two separate magnetic motor starter coils with a common emergency stop protecting the entire system. **See Figure 5-23.** Pressing start pushbutton PB3 causes coil M1 to energize and seal in auxiliary contacts M1. Pressing start pushbutton PB5 causes coil M2 to energize and seal in auxiliary contacts M2. Once the entire circuit is operational, emergency stop pushbutton PB1 can shut down the entire circuit or the individual stop pushbuttons PB2 or PB4 can de-energize the coils in their respective circuits. Each circuit is overload protected and does not affect the other when one magnetic motor starter experiences a problem.

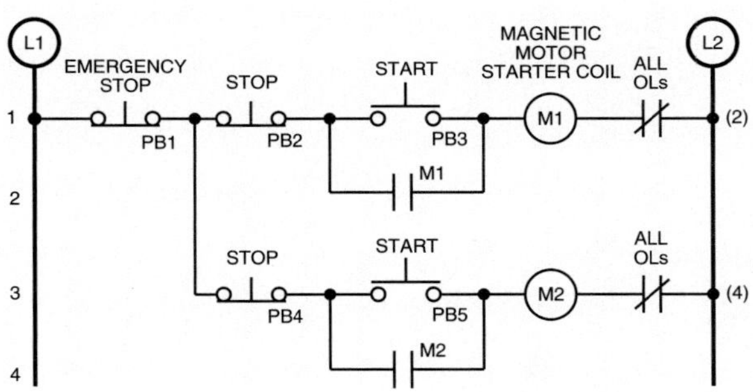

Figure 5-23. Two start/stop stations are used to control two separate magnetic motor starter coils with a common emergency stop protecting the entire system.

Start/Stop Station Controlling Two or More Magnetic Starters

Steel mills, paper mills, bottling plants, and canning plants are industries that require simultaneous operation of two or more motors. In each industry, products or materials are spread out over great lengths but must be started together to prevent product separation or stretching. To accomplish this, two motors can be started almost simultaneously from one location. **See Figure 5-24.**

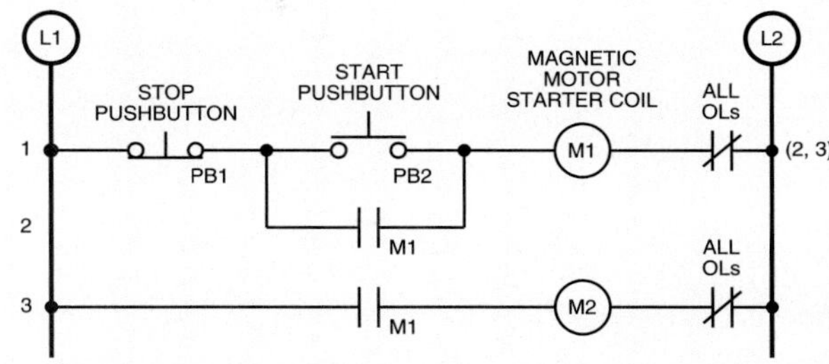

Figure 5-24. Two motors can be started almost simultaneously from one location to prevent product separation or stretching.

In this circuit, pressing start pushbutton PB2 energizes coil M1 and seals in both sets of auxiliary contacts M1. *Note:* It is acceptable to have more than one set of auxiliary contacts controlled by one coil. When both sets of contacts close, the first set of M1 contacts (line 2) provides memory for the start pushbutton and completes the circuit to energize coil M1. The second set of M1 contacts (line 3) completes the circuit to coil M2, energizing coil M2. The motors associated with these magnetic motor starters start almost simultaneously because both coils energize almost simultaneously. Pushing the stop pushbutton breaks the circuit (line 1), de-energizing coil M1. When coil M1 drops out, both sets of auxiliary contacts are deactivated. The motors associated with these magnetic motor starters stop almost simultaneously because both coils de-energize almost simultaneously. An overload in magnetic motor starter M2 affects only the operation of coil M2. The entire circuit is shut down if an overload exists in motor starter M1. The entire circuit stops because de-energizing coil M1 also affects both sets of auxiliary contacts M1. This protection might be used where a machine such as an industrial drill would be damaged if the cooling liquid pump shut OFF while the drill was still operating.

Pressure Switch with Pilot Light Indicating Device Activation

Pilot lights are manufactured in a variety of colors, shapes, and sizes to meet the needs of industry. The illumination of these lights signals an operator that any one of a sequence of events may be taking place. A pilot light may be used with a pressure switch to indicate when a device is activated. **See Figure 5-25.**

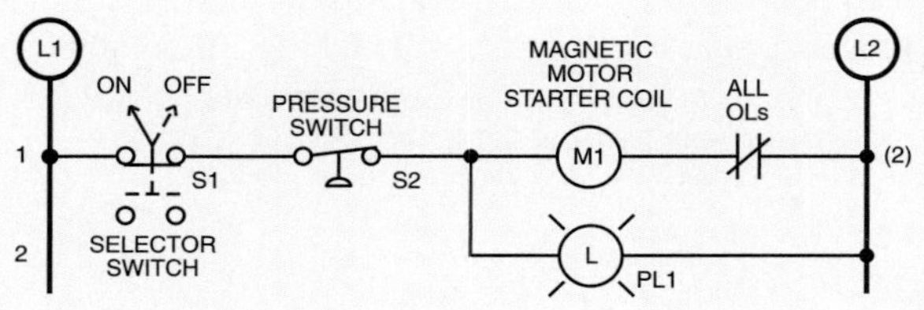

Figure 5-25. A pilot light is used with a pressure switch to indicate when a device is activated.

In this circuit, pressure switch S2 has automatic control over the circuit when switch S1 is closed. When the pressure to switch S2 drops, the switch closes and activates coil M1 which controls the magnetic starter of the compressor motor, starting the compressor. At the same time, contacts M1 close and pilot light PL1 turns ON. The compressor continues to run and the pilot light stays ON as long as the motor runs. When pressure builds sufficiently to open pressure switch S2, coil M1 de-energizes and the magnetic motor starter drops out, stopping the compressor motor. The pilot light goes out because contact M1 controlled by coil M1 opens. The pilot light is ON only when the compressor motor is running. This circuit might be used in a garage to let the owner know when the air compressor is ON or OFF.

A pilot light may be used with a start/stop station to indicate when a device is activated. **See Figure 5-26.** In this circuit, pressing start pushbutton PB2 energizes coil M1, causing auxiliary contacts M1 to close. Closing contacts M1 provides memory for start pushbutton PB2 and maintains an electrical path for the pilot light. As long as coil M1 is energized, the pilot light stays ON. Pressing stop pushbutton PB1 de-energizes coil M1, opening contacts M1 and turning OFF the pilot light. An overload in this circuit also de-energizes coil M1, opening contacts M1 and turning OFF the pilot light. A circuit like this can be used as a positive indicator that some process is taking place. The process may be in a remote place such as in a pump well or in another building.

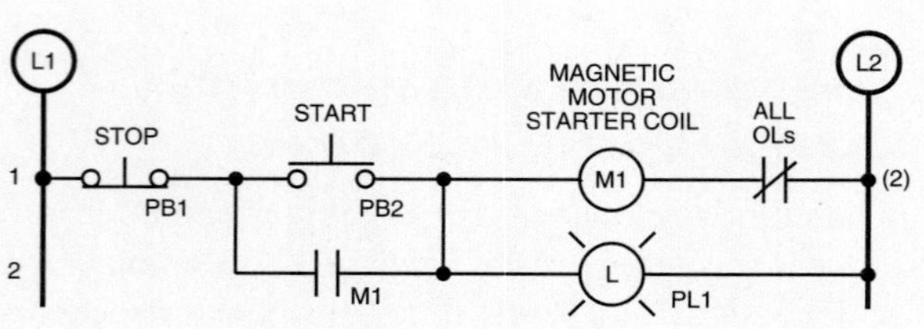

Figure 5-26. A pilot light is used with a start/stop station to indicate when a device is activated.

Start/Stop Station with Pilot Light Indicating NO Device Activation

Pilot lights may be used to show when an operation is stopped as well as when it is started. NOT logic is used in a circuit when a pilot light is used to show that an operation has stopped. NOT logic is established by placing one set of NC contacts in series with a device. **See Figure 5-27.**

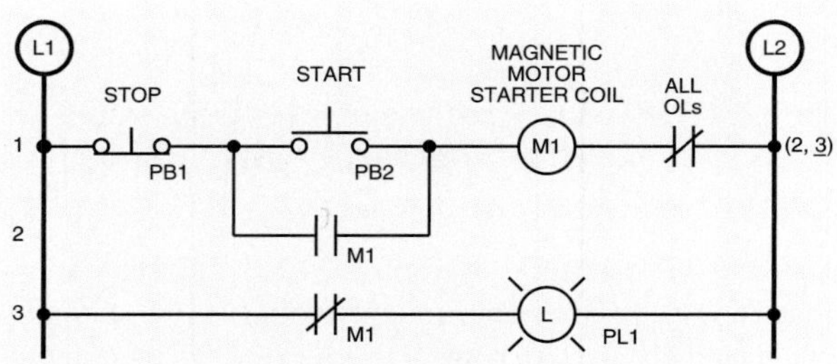

Figure 5-27. NOT logic is used to indicate when a device is not operating.

In this circuit, pressing start pushbutton PB2 energizes coil M1, causing both sets of auxiliary contacts M1 to energize. NO contacts M1 (line 2) close, providing memory for PB2, and NC contacts M1 (line 3) open, disconnecting pilot light PL1 from the line voltage, causing the light to turn OFF. Pressing stop pushbutton PB1 de-energizes coil M1, causing both sets of contacts to return to their normal positions. NO contacts M1 (line 2) return to their NO position, and NC contacts M1 (line 3) return to their NC position, causing the pilot light to be reconnected to the line voltage and causing it to turn ON. The pilot light is ON only when the coil to the magnetic motor starter is OFF. A bell or siren could be substituted for the pilot light to serve as a warning device. A circuit like this is used to monitor critical operating procedures such as a cooling pump for a nuclear reactor. When the cooling pump stops, the pilot light, bell, or siren immediately calls attention to the fact that the process has been stopped.

Pushbutton Sequence Control

Conveyor systems often require one conveyor system to feed boxes or other materials onto another conveyor system. If one conveyor is feeding a second conveyor, a circuit is needed to prevent the pileup of material on the second conveyor if the second conveyor is stopped. A sequence control circuit does not let the first conveyor operate unless the second conveyor has started and is running. **See Figure 5-28.**

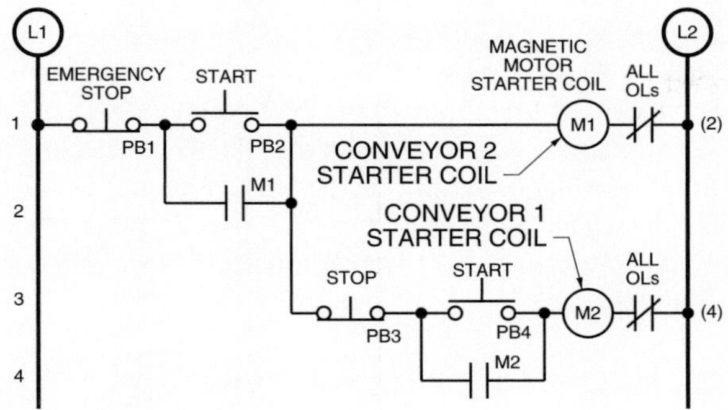

Figure 5-28. A sequence control circuit does not let the first conveyor operate unless the second conveyor has started and is running.

In this circuit, pressing start pushbutton PB2 energizes coil M1 and causes auxiliary contacts M1 to close. With auxiliary contacts M1 closed, PB2 has memory and provides an electrical path to allow coil M2 to be energized when start pushbutton PB4 is pressed.

With start pushbutton PB4 pressed, coil M2 energizes and closes contacts M2, providing memory for start pushbutton PB4 so that both conveyors run. Conveyor 1 (coil M2) cannot start unless conveyor 2 (coil M1) is energized. Both conveyors shut down if an overload occurs in the circuit with coil M1 or if stop pushbutton PB1 is pressed. Only conveyor 1 shuts down if conveyor 1 (coil M2) experiences an overload. A problem in conveyor 1 does not affect conveyor 2. This type of control is also known as cascade control or cascade protection.

Jogging with a Selector Switch

Jogging is the frequent starting and stopping of a motor for short periods of time. Jogging is used to position materials by moving the materials small distances each time the motor starts. A selector switch is used to provide a common industrial jog/run circuit. **See Figure 5-29.** The selector switch (two-position switch) is used to manually open or close a portion of the electrical circuit. In this circuit, the selector switch determines if the circuit is a jog circuit or run circuit. With selector switch S1 in the open (jog) position, pressing start pushbutton PB2 energizes coil M1, causing the magnetic motor starter to operate. Releasing start pushbutton PB2 de-energizes coil M1, causing the magnetic motor starter to stop. With selector switch S1 in the closed (run) position, pressing start pushbutton PB2 energizes coil M1, closing auxiliary contacts M1 and providing memory so that the magnetic starter operates and continues to operate until stop pushbutton PB1 is pressed.

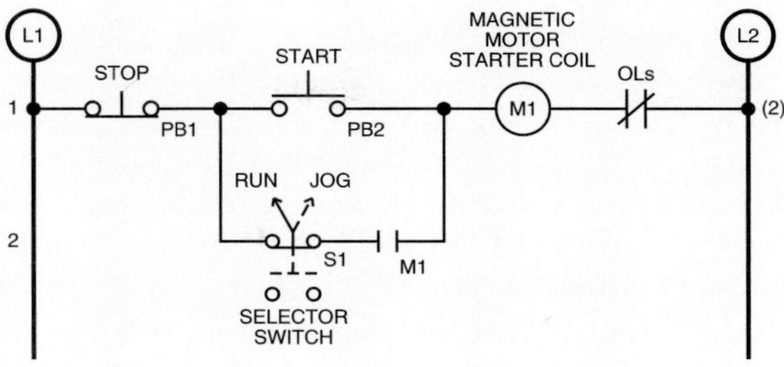

Figure 5-29. A selector switch is used to provide a common industrial jog/run circuit.

When stop pushbutton PB1 is pressed, coil M1 de-energizes and all circuit components return to their original condition. The overloads may also open the circuit and must be reset after the overload is removed to return the circuit to normal operation. This circuit may be found where an operator may run a machine continuously for production, but may stop it at any time for small adjustments or repositioning. Jogging may also be accomplished by other types of circuits. The common feature of all jog circuits is that they prevent the holding circuit from operating.

CONTROL CIRCUIT TROUBLESHOOTING

Troubleshooting is the systematic elimination of the various parts of a system, circuit, or process to locate a malfunctioning part. Troubleshooting electrical control circuits requires an organized, sequenced approach. Troubleshooting requires the use of electrical test equipment, drawings and diagrams, and manufacturer specifications.

Before troubleshooting an electrical circuit, an individual must understand the operation of the circuit, the sequence of events, timing or counting functions, and devices used to energize and de-energize the circuit. A line diagram shows the logic of an electrical circuit using single lines and symbols. Along with a line diagram, a DMM can be used to troubleshoot components in electrical circuits. Common electrical problems include open circuits and short circuits. The most common troubleshooting method is the tie-down troubleshooting method.

Tie-Down Troubleshooting Method

The *tie-down troubleshooting method* is a testing method in which one DMM probe is connected to either the L2 (neutral) or L1 (hot) side of a circuit and the other DMM probe is moved along a section of the circuit to be tested. The tie-down troubleshooting method allows a troubleshooter to work quickly on a familiar circuit that is small enough for the test probes to reach across the test points.

When using the tie-down troubleshooting method, one DMM test lead should be placed (tied down) on the L2 (neutral) conductor, and the other lead should be moved through the circuit starting with L1 (hot conductor). **See Figure 5-30.** If the correct voltage is not measured at L1 and L2, there is a power problem and the main power must be checked (for a possible fuse, circuit breaker, or main switch problem). If the proper voltage is present between L1 and L2, the DMM lead connected to L1 is moved along the circuit until the meter lead is directly at the load. If voltage is measured at the load but the load is not operating, the problem is the load on any circuit in which the load is connected directly to L2.

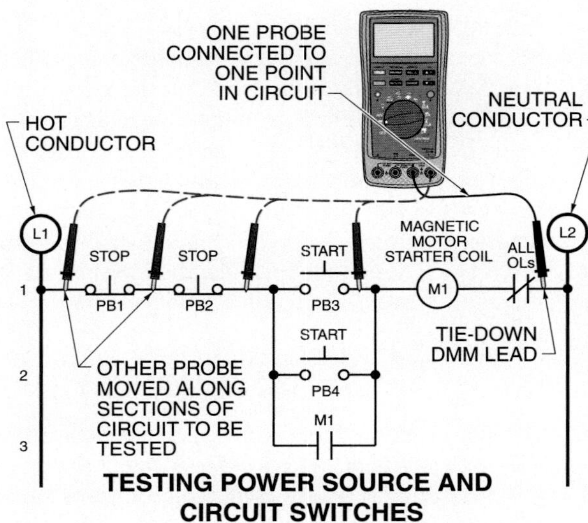

Figure 5-30. When using the tie-down troubleshooting method, one DMM test lead should be placed (tied down) on the L2 (neutral conductor) and the other lead moved through the circuit starting with L1 (hot conductor).

All loads are connected directly to L2 except when a magnetic motor starter overload contact is connected between the starter coil and L2. When a magnetic motor starter overload contact is used in a circuit, the DMM lead connected to L2 can be moved to the other side of the overload (side connected directly to the starter coil) to check if the overloads are open. However, caution must be exercised when doing this because one DMM lead is still connected to L1 (hot conductor). This means the tip of the other DMM lead (the one being moved) can cause an electrical shock if touched and there is a complete path to ground through the troubleshooter's body. **See Figure 5-31.**

Troubleshooting Open Circuits

An *open circuit* is an electrical circuit that has an incomplete path that prevents current flow. An open circuit represents a very high resistance path for current and is usually regarded as having infinite resistance. An open circuit in a series circuit de-energizes the entire circuit. Open circuits may be caused intentionally or unintentionally. An open circuit is caused intentionally when a switch is used to open a circuit. An open circuit may be caused unintentionally when the wiring between parts in a circuit is broken, when a component or device in a circuit malfunctions, or when a fuse blows.

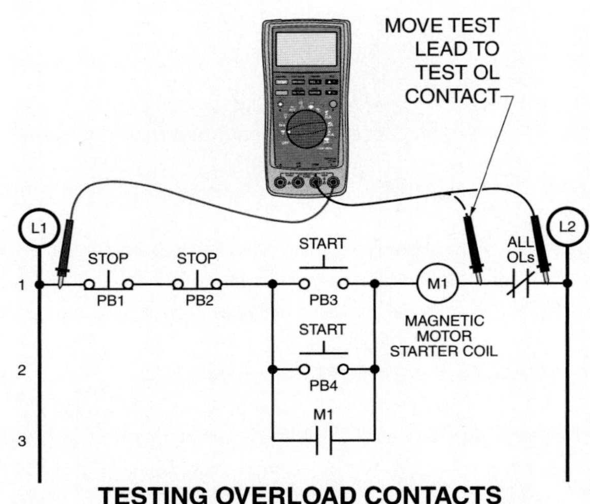

Figure 5-31. When a magnetic motor starter overload contact is used in a circuit, the DMM lead connected to L2 can be moved to the other side of the overload (side connected directly to the starter coil) to check if the overload is open.

A switch in the OFF position is an open circuit. Switches are tested by toggling the switch to check if the contacts open and close. A DMM set to measure voltage can be used to test a mechanical switch. A good switch indicates source voltage when open and 0 V when closed. A faulty switch indicates source voltage both when open and when closed. **See Figure 5-32.**

The proper operation of a switch must be known to determine when it is not operating properly because not all switches operate in the same manner. For example, a good solid-state switch indicates source voltage when open and a slight voltage drop when closed. This is normal due to the construction of the solid-state switch. **See Figure 5-33.**

A switch may also be checked with a jumper wire. A jumper wire is placed in parallel around the switch and the circuit is energized. The jumper wire closes the circuit, energizing the load.

Warning: Jumper wires can cause equipment to start unexpectedly and must be removed from the circuit when no longer needed for testing.

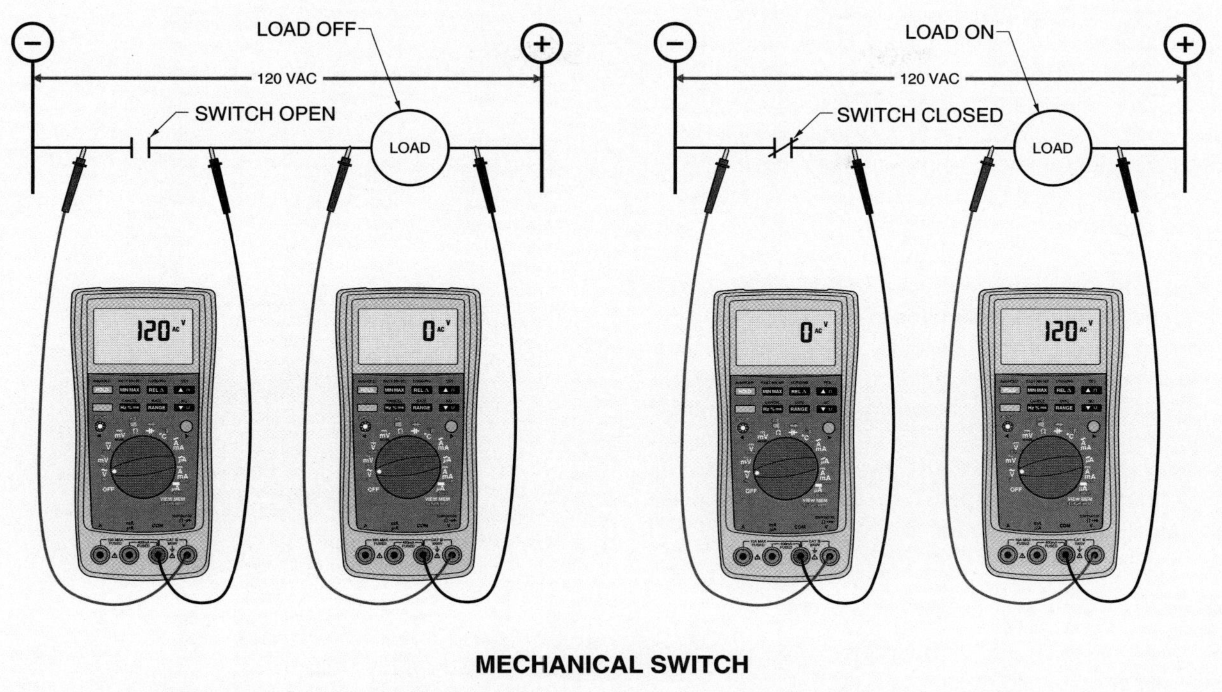

Figure 5-32. A good mechanical switch indicates source voltage when open and 0 V when closed.

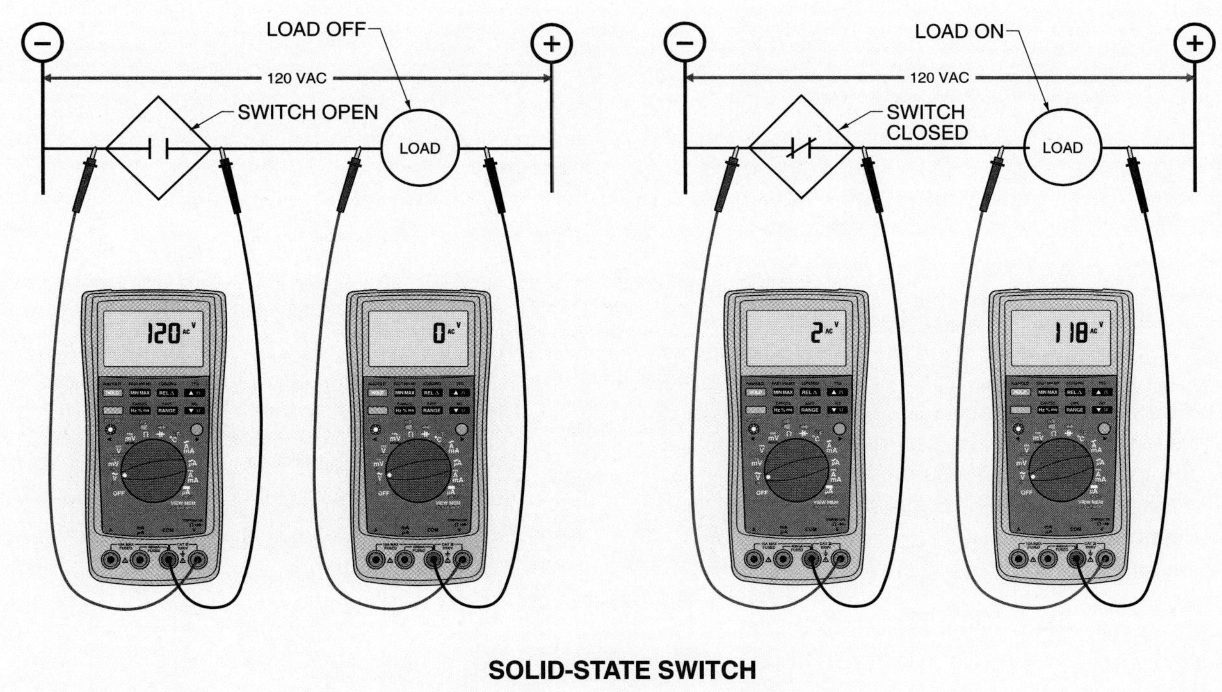

Figure 5-33. A good solid-state switch indicates source voltage when open and a slight voltage drop when closed.

An open circuit may occur unintentionally by a break in the wire of a circuit, a malfunctioning component, or a blown fuse. When a wire breaks, the path for current is interrupted and current flow stops. A broken wire in an individual line of a circuit de-energizes that line only. That branch of the circuit can be tested using a DMM set to measure voltage. For example, a DMM may be placed across a section of wire to determine if there is a break in that part of the wire. This test is often taken across wire connection points. The DMM indicates 0 V if the wire has no break. The DMM indicates source voltage if the wire is broken and there is no other open in the branch.

A faulty electrical component may also cause an unintentional open circuit. For example, a break in the conducting path of an electrical component, such as a burnt-out filament of a light bulb, also breaks the path for current and opens the circuit. In addition, when a fuse blows, the current flow in the circuit increases to a level that opens the conducting path inside the fuse.

Troubleshooting Short Circuits

A *short circuit* is a circuit in which current takes a shortcut around the normal path of current flow. In a short circuit, current leaves the normal current-carrying path and goes around the load and back to the power source or to ground. The low-resistance path can be due to failure of circuit components or failure in the wiring of the circuit. For example, if two pieces of wire accidentally contact each other, the wires produce a dead short across the circuit. A *dead short* is a short circuit that opens the circuit as soon as the circuit is energized or when the section of the circuit containing the short is energized. **See Figure 5-34.**

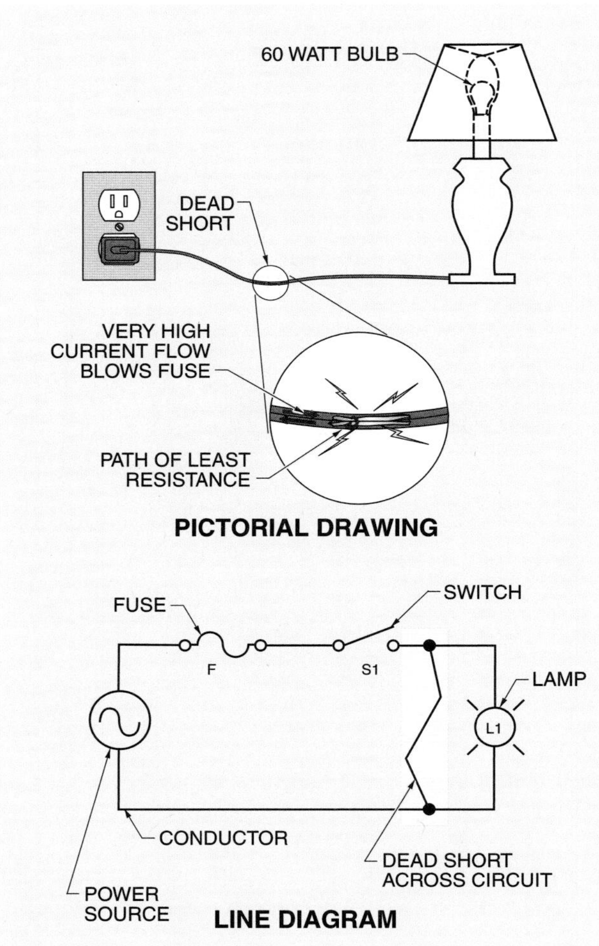

Figure 5-34. When two pieces of wire touch because of damaged insulation, the wires produce a dead short across the circuit.

A dead short reduces the resistance of the short-circuited part of a circuit to nearly 0 Ω. A dead short produces a surge of current in the circuit, resulting in an overload device such as a fuse being blown or circuit breaker being tripped. In a circuit with a dead short, the fuse must be replaced or the circuit breaker reset. The circuit is inspected for the location of the short if the fuse blows or circuit breaker trips again when the circuit is energized. The location of the short is usually indicated by signs of overheating, such as burn marks or discolored insulation. The location of a short can be determined using a continuity tester or a DMM set to measure resistance.

Warning: Ensure that the circuit is de-energized when measuring resistance.

A continuity tester can be used to test for short circuits. **See Figure 5-35.** A continuity tester uses its own power (usually a battery) to power the circuit to determine if a short circuit exists between a wire and its housing. Once the short circuit is located, the shorted wire must be replaced. Ensure that the circuit is disconnected from its power source before testing using a continuity tester.

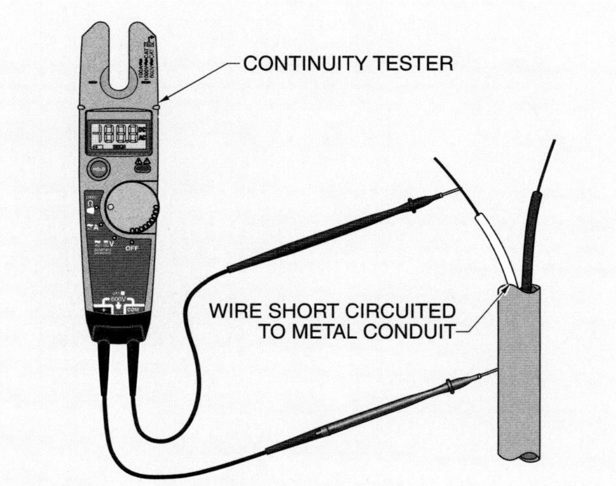

Figure 5-35. A continuity tester can be used to test for short circuits.

A DMM set to measure resistance can also be used to test for short circuits. A circuit is tested for a short circuit with all open contacts closed. In a good circuit, a DMM reads total circuit resistance when all open contacts are closed. In a circuit with a dead short, a DMM reads near 0 Ω. **See Figure 5-36.** To test each branch of a circuit, each branch is isolated by disconnecting a wire from the branch. The branch does not contain a short circuit if this produces no change in the DMM resistance reading. The branch is reconnected after the resistance reading is taken. This process is continued by isolating each branch in succession. A branch contains a short if, when the branch is disconnected, the DMM resistance reading jumps from 0 Ω to a high resistance. This branch is inspected for signs of overheating and crossed, frayed, or loose wires. Further inspection is required to find the exact cause of the short. Large, complex circuits are tested one section at a time to determine which section contains the short. The individual branches of the section are then tested to find the exact location of the short.

Warning: Power must always be removed from the circuit before a resistance check can be made.

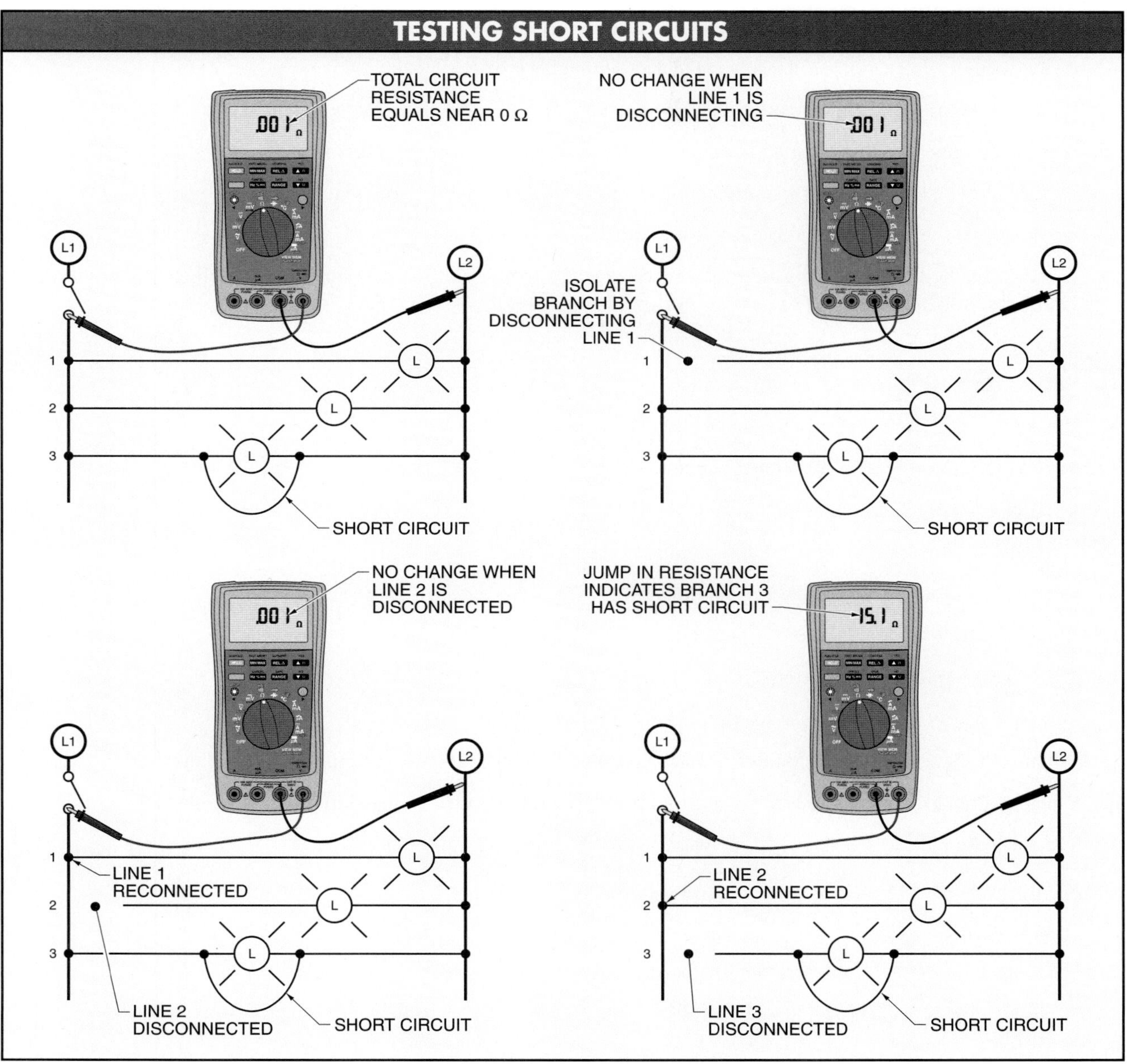

Figure 5-36. In a good circuit, a DMM reads total circuit resistance when all open contacts are closed. In a circuit with a short, a DMM reads near 0 Ω.

Review Questions

1. How many electrical loads can be placed in the control circuit of any one line between L1 and L2?
2. How are the loads connected if more than one load must be connected in a line diagram?
3. Loads such as control relay coils, solenoids, and pilot lights are connected directly or indirectly to which power line?
4. Where are control devices connected in a line diagram?
5. How is each line in a line diagram marked to distinguish that line from all other lines?
6. How are NO and NC contacts identified when using the numerical cross-reference system?
7. In what order are wire reference numbers normally assigned?
8. What are the two methods used to illustrate how contacts found in different control lines belong to the same control switch, such as a limit switch?
9. What are the three basic sections of all control circuits?
10. What is the signal section of a control circuit?
11. What is the decision section of a control circuit?
12. What is the action section of a control circuit?
13. What is AND logic as applied to control circuits?
14. What is OR logic as applied to control circuits?
15. What is NOT logic as applied to control circuits?
16. What is NOR logic as applied to control circuits?
17. What is NAND logic as applied to control circuits?
18. What is memory logic as applied to control circuits?
19. When additional stops are added to a control circuit, should they be connected in series or parallel?
20. When additional starts are added to a control circuit, should they be connected in series or parallel?
21. What is jogging?
22. What is troubleshooting?
23. What is a dead short?
24. What is the tie-down troubleshooting method?

Solenoids, DC Generators, and DC Motors

Solenoids are used to control devices such as valves, relays, and other industrial machinery. A generator is a machine that converts mechanical energy into electrical energy by electromagnetic induction. A direct current (DC) motor is a motor that uses direct current connected to the field and armature to produce shaft rotation.

MAGNETISM

Magnetism was first discovered by the Greeks when they noticed that a certain type of stone attracted bits of iron. This stone was first found in Asia Minor in the province of Magnesia. The stone was named magnetite after this province.

Magnets

A *magnet* is a substance that produces a magnetic field and attracts iron. Magnets are either permanent or temporary. A *permanent magnet* is a magnet that can retain its magnetism after the magnetizing force has been removed. Permanent magnets include natural magnets (magnetite) and manufactured magnets. **See Figure 6-1.**

A *temporary magnet* is a magnet that retains trace amounts of magnetism after the magnetizing force has been removed. **See Figure 6-2.** Temporary magnets have a low retentivity. A magnet with low retentivity has very little residual magnetism (leftover magnetism) remaining once the magnetizing force has been removed.

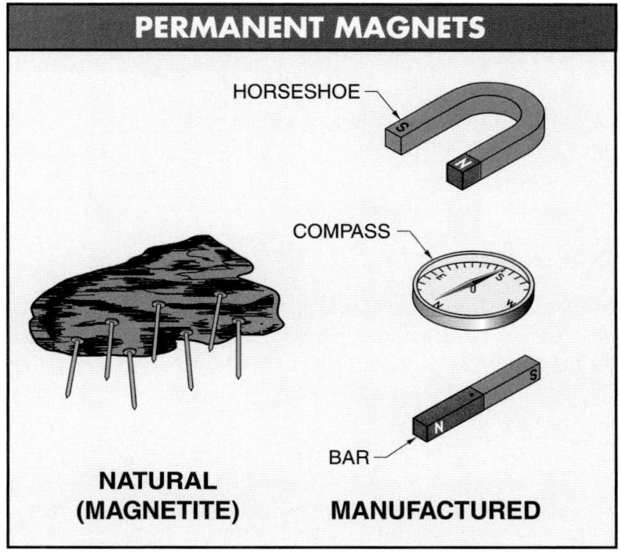

Figure 6-1. Permanent magnets include natural magnets (magnetite) and manufactured magnets.

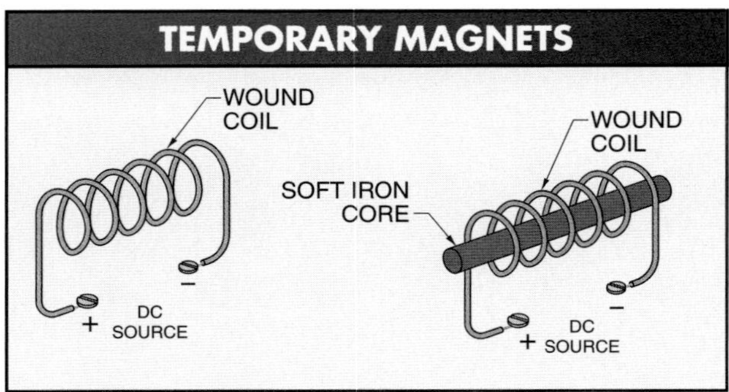

Figure 6-2. Temporary magnets are magnets that retain trace amounts of magnetism after the magnetizing force has been removed.

Molecular Theory of Magnetism

The *molecular theory of magnetism* is the theory that states that all substances are made up of an infinite number of molecular magnets that can be arranged in either an organized or disorganized manner. **See Figure 6-3.** A material is demagnetized if it has disorganized molecular magnets. A material is magnetized if it has organized molecular magnets.

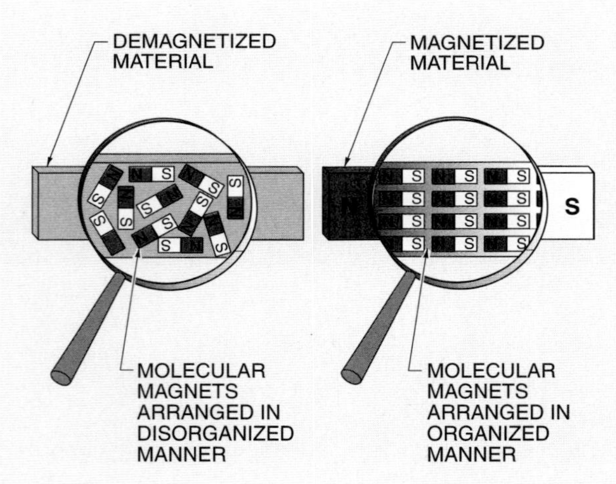

Figure 6-3. The molecular theory of magnetism states that all substances are made up of an infinite number of molecular magnets that can be arranged in either an organized or disorganized manner.

The molecular theory of magnetism explains how certain materials used in control devices react to magnetic fields. For example, it explains why hard steel is used for permanent magnets, while soft iron is used for the temporary magnets found in control devices.

The dense molecular structure of hard steel does not easily disorganize once a magnetizing force has been removed. Hard steel is difficult to magnetize and demagnetize, making it a good permanent magnet. Hard steel is considered to have high retentivity. However, permanent magnets may be demagnetized by a sharp blow or by heat.

The loose molecular structure of soft iron can be magnetized and demagnetized easily. Soft iron is ideal for use as a temporary magnet in control devices because it does not retain residual magnetism very easily. Soft iron is considered to have low retentivity.

ELECTROMAGNETISM

In 1819, the Danish physicist Hans C. Oersted discovered that a magnetic field is created around an electrical conductor when electric current flows through the conductor. *Electromagnetism* is the magnetism produced when electric current passes through a conductor. **See Figure 6-4.**

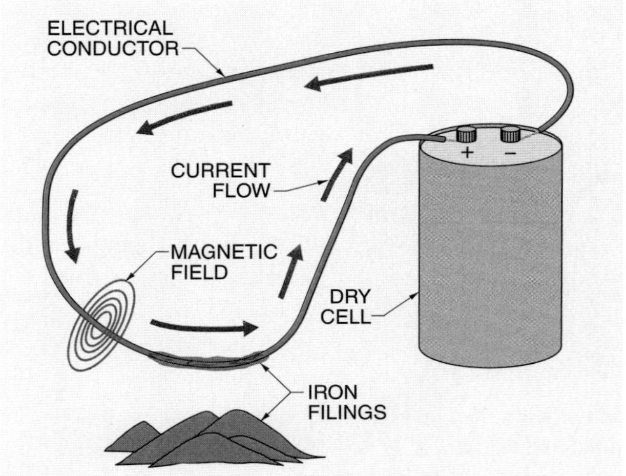

Figure 6-4. In 1819, the Danish physicist Hans C. Oersted discovered that a magnetic field is created around an electrical conductor when electric current flows through the conductor.

The direction in which current flows through a conductor determines the direction of the magnetic field around it. Lines of force (lines of induction) are present all along the full length of the conductor. **See Figure 6-5.**

One line of force is called a maxwell, and the total number of lines is called flux. The total number of lines of force in a space of 1 centimeter (cm) equals the flux density (in gauss) of the field. For example, 16 lines of force in 1 cm equal 16 gauss. **See Figure 6-6.**

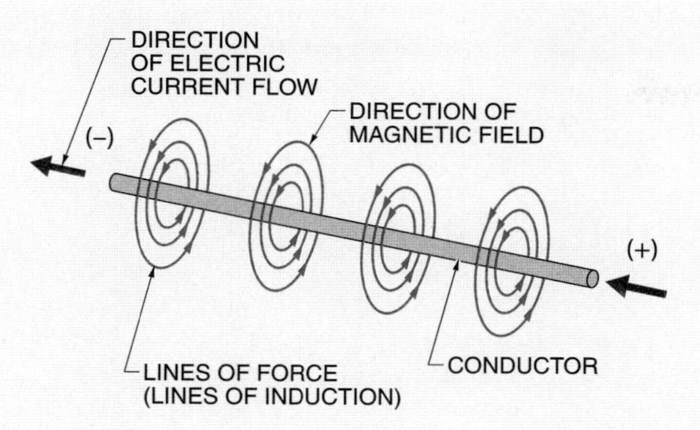

Figure 6-5. The lines of force (lines of induction) are present along the full length of a conductor.

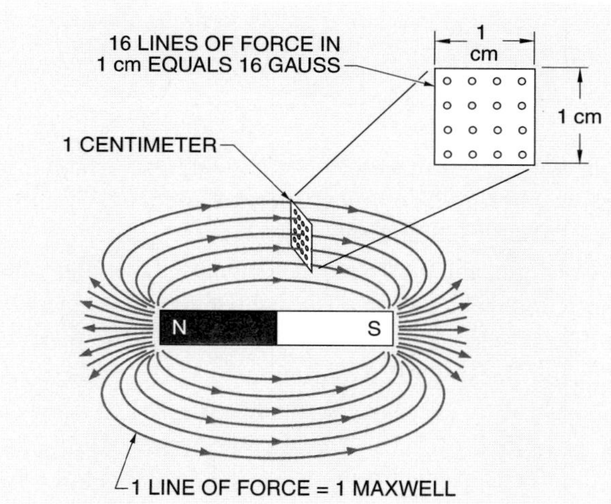

Figure 6-6. The total number of lines of force (maxwells) in a 1 cm section of a magnetic field equals the flux density of the field (gauss).

If a conductor is bent to form a loop, all of the lines of force circling the conductor enter one side of the loop and leave from the other side of the loop. Thus, a north pole is created on one side of the loop and a south pole is created on the other side of the loop. The side of the loop into which the lines of force enter is the south pole, and the side from which the lines of force leave is the north pole. A loop of wire has poles just like a bar magnet.

If a conductor is wound into multiple loops (a coil), the magnetic lines of force combine. Thus, the magnetic force of a coil with multiple turns is stronger than the magnetic force in a coil with a single loop. **See Figure 6-7.**

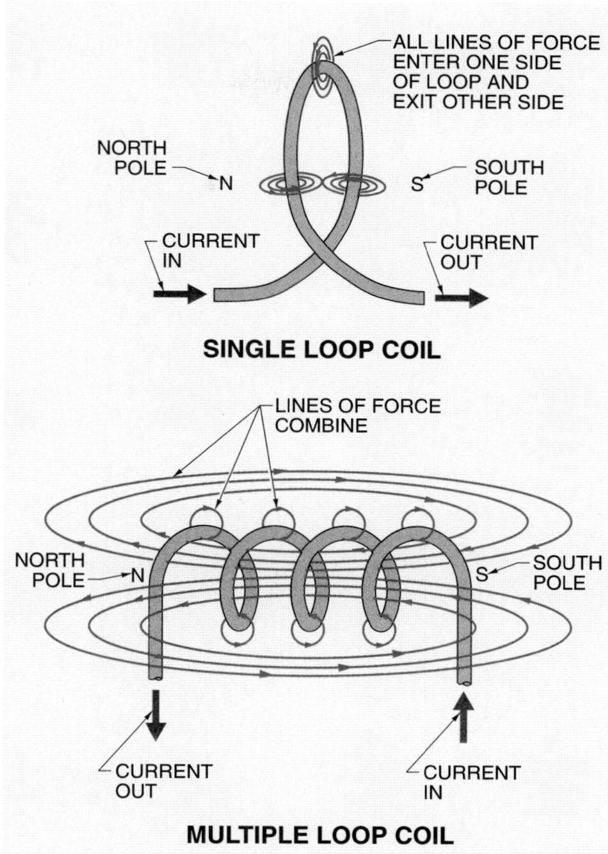

Figure 6-7. If a conductor is formed into a coil, the lines of force combine, forming a stronger field than the lines of force from a single loop.

Oersted tried several other experiments to increase the strength of the magnetic field and found three ways to increase the strength of the magnetic field in a coil: increase the amount of current by increasing the voltage; increase the number of turns in the coil; and insert an iron core through the coil. **See Figure 6-8.** These early experiments have led to the development of a huge control industry, which depends on magnetic coils to convert electrical energy into usable magnetic energy.

Electromagnets

An *electromagnet* is a magnet whose magnetic energy is produced by the flow of electric current. Some electromagnets are so large and powerful that they can lift tons of scrap metal at one time. Other electromagnets used in some electrical and electronic circuits are very small, such as those found in solenoids and relays.

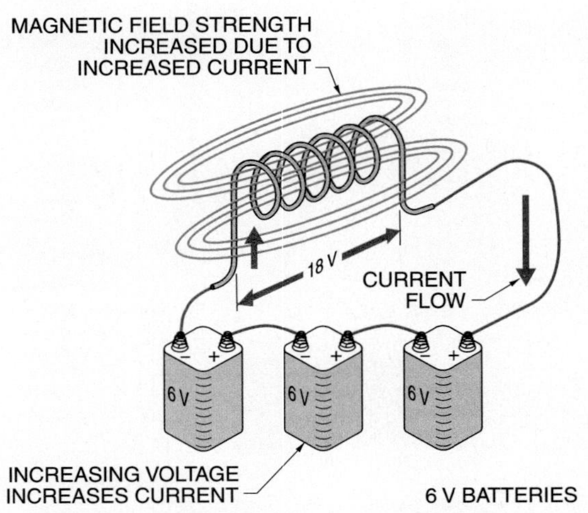

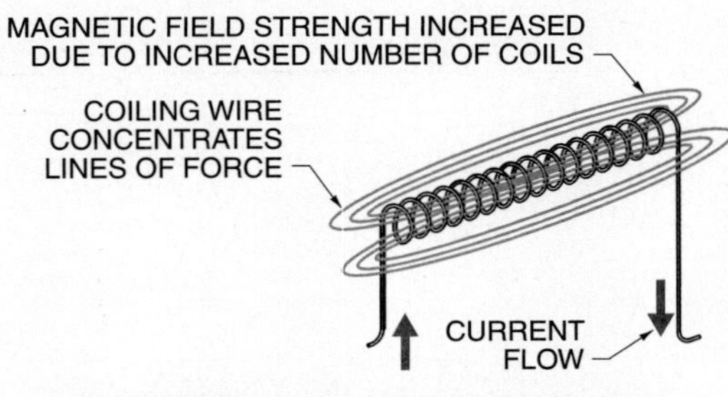

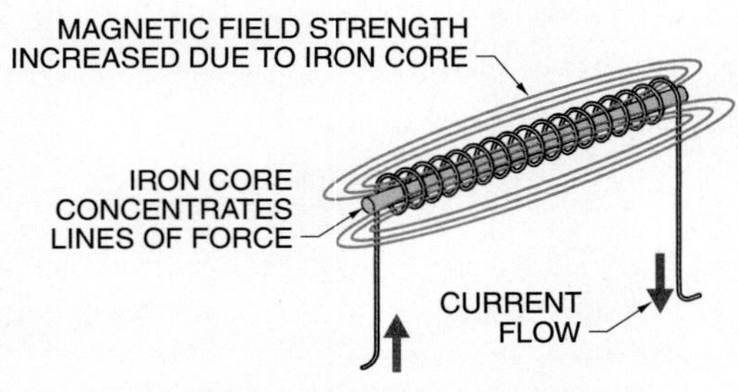

Figure 6-8. The strength of a magnetic field produced by a conductor may be increased by increasing the voltage, increasing the number of coils, or inserting an iron core through the coil.

An electromagnet consists of an iron core inserted into a coil. The iron core concentrates the lines of force produced by the coil. With the core in place and the coil energized, the polarity of the magnet can be determined using the left-hand rule. The left-hand rule states that if the fingers of the left hand are wrapped in the direction of the current flow in a coil, the left thumb points to the magnetic north pole of the coil. **See Figure 6-9.**

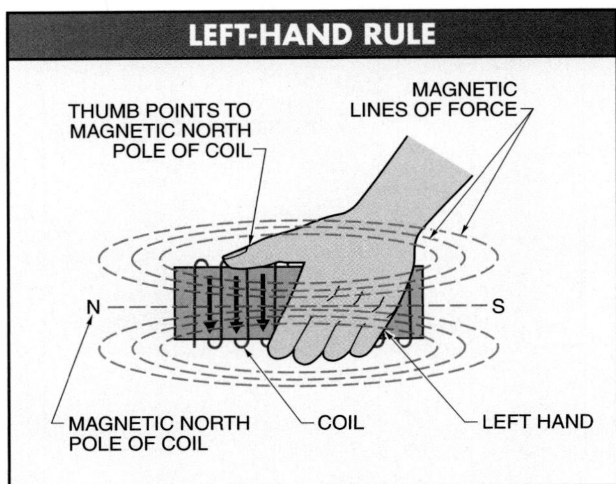

Figure 6-9. The left-hand rule states that if the fingers of the left hand are wrapped in the direction of the current flow in a coil, the left thumb points to the magnetic north pole of the coil.

Permeability. Permeability is the ability of a material to carry magnetic lines of force. The symbol for permeability is the Greek letter mu (μ).

The permeability of a magnetic material is comparable to the conductance of an electrical conductor. Permeability refers to the ability of a specified material to conduct magnetic lines of force as compared to the ability of a standard material (usually vacuum or air) to conduct magnetic lines of force. The reference permeability of a vacuum is 1. Typical reference values are available for a number of materials. **See Figure 6-10.** Pure iron and silicon steel are often used as cores in electromagnets due to their high permeability.

TYPICAL PERMEABILITY VALUES

MATERIAL	PERMEABILITY*
COPPER	0.999991
VACUUM	1
AIR	1.0000004
NICKEL	400 TO 1000
SILICON STEEL	5000 TO 10,000
PURE IRON	6000 TO 8000

* in μ

Figure 6-10. Permeability values refer to the ability of a specified material to conduct magnetic lines of force.

The advantages of electromagnets are that they can be made stronger than permanent magnets, and the magnetic strength can be easily controlled by regulating the electric current. The main characteristics of an electromagnet include the following:

- When electricity flows through a conductor, a magnetic field is created around that conductor.
- The field is stronger close to the wire and weaker further away.
- The strength of the magnetic field and the current are directly related: more current, the stronger the magnetic field; less current, the weaker the magnetic field.
- The direction of the magnetic field is determined by the direction of the current flowing through the conductor.
- The more permeable the core, the greater the concentration of magnetic lines of force.

SOLENOIDS

A *solenoid* is an electric output device that converts electrical energy into a linear mechanical force. The magnetic attraction of a solenoid may be used to transmit force. Solenoids may be combined with an armature, which transmits the force created by the solenoid into useful work. An *armature* is the movable part of a solenoid.

Solenoid Configurations

Solenoids are configured in various ways for different applications and operating characteristics. The five-solenoid configurations are clapper, bell-crank, horizontal-action, vertical-action, and plunger. **See Figure 6-11.**

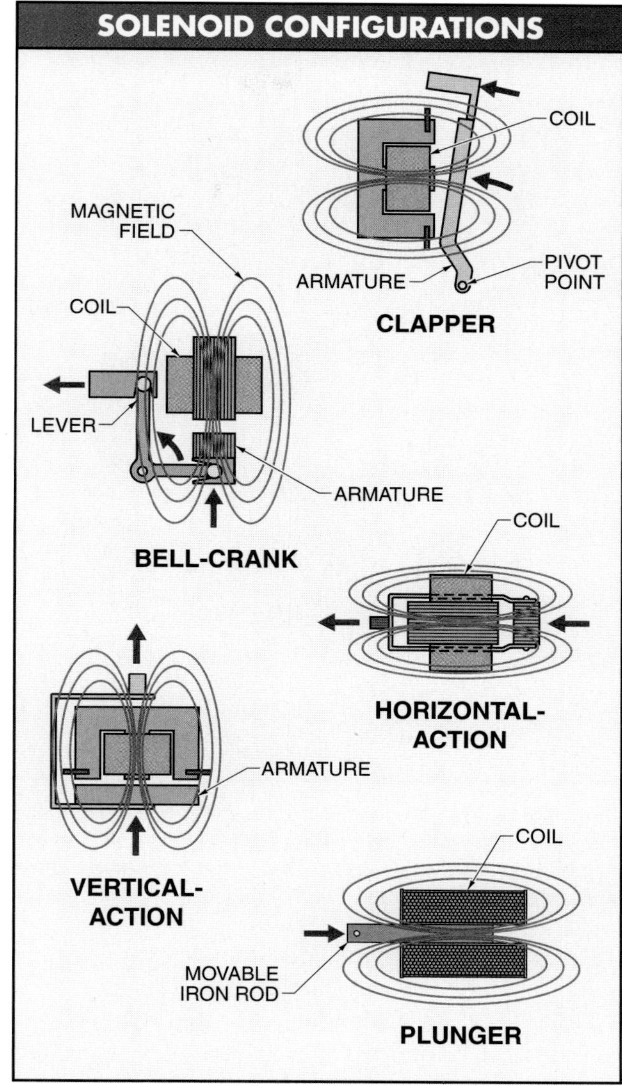

Figure 6-11. The five-solenoid configurations are clapper, bell-crank, horizontal-action, vertical-action, and plunger.

A clapper solenoid has the armature hinged on a pivot point. As voltage is applied to the coil, the magnetic effect produced pulls the armature to a closed position so that it is picked up (sealed in). A bell-crank solenoid uses a lever attached to the armature to transform the vertical action of the armature into a horizontal motion. The use of the lever allows the shock of the armature to be absorbed by the lever and not transmitted to the end of the lever. This is beneficial when a soft but firm motion is required in the controls.

A horizontal-action solenoid is a direct-action device. The movement of the armature moves the resultant force in a straight line. Horizontal-action solenoids are one of the most common solenoid configurations. A vertical-action solenoid also uses a mechanical assembly but transmits the vertical action of the armature in a straight-line motion as the armature is picked up.

A plunger solenoid contains only a moving iron cylinder. A movable iron rod placed within the electrical coil tends to equalize or align itself within the coil when current passes through the coil. The current causes the rod to center itself so that the rod ends line up with the ends of the solenoid, if the rod and solenoid are of equal length.

In a plunger solenoid, a spring is used to move the rod a short distance from its center in the coil. The rod moves against the spring tension to recenter itself in the coil when the current is turned ON. The spring returns the rod to its off-center position when the current is turned OFF. The motion of the rod is used to operate any number of mechanical devices. **See Figure 6-12.**

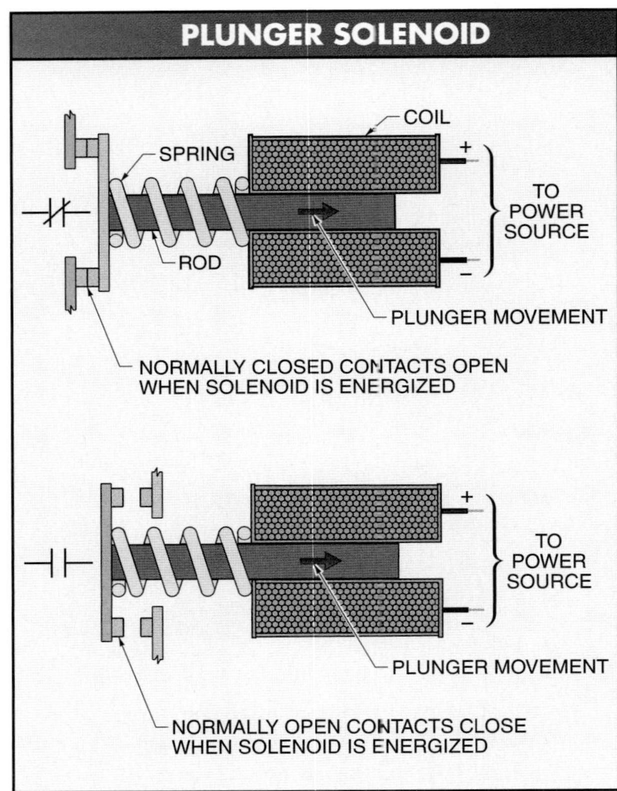

Figure 6-12. In a plunger solenoid, a spring is used to move the rod a short distance from its center in the coil. The rod moves against the spring tension to recenter itself when the current is turned ON.

Solenoid Construction

Solenoids are constructed of many turns of wire wrapped around a magnetic laminate assembly. Passing electric current through the coil causes the armature to be pulled toward the coil. Devices may be attached to the solenoid to accomplish tasks like opening and closing contacts.

Eddy Current. *Eddy current* is unwanted current induced in the metal structure of a device due to the rate of change in the induced magnetic field. Strong eddy currents are generated in solid metal when used with alternating current. In AC solenoids, the magnetic assembly and armature consist of a number of thin pieces of metal laminated together. The thin pieces of metal reduce the eddy current produced in the metal. **See Figure 6-13.** Eddy current is confined to each lamination, thus reducing the intensity of the magnetic effect and subsequent heat buildup. For DC solenoids, a solid core is acceptable because the current is in one direction and continuous.

Armature Air Gap. To avoid chattering, solenoids are designed so that the armature is attracted to its sealed-in position so that it completes the magnetic circuit as completely as possible. To ensure this, both the faces on the magnetic laminate assembly and those on the armature are machined to a very close tolerance.

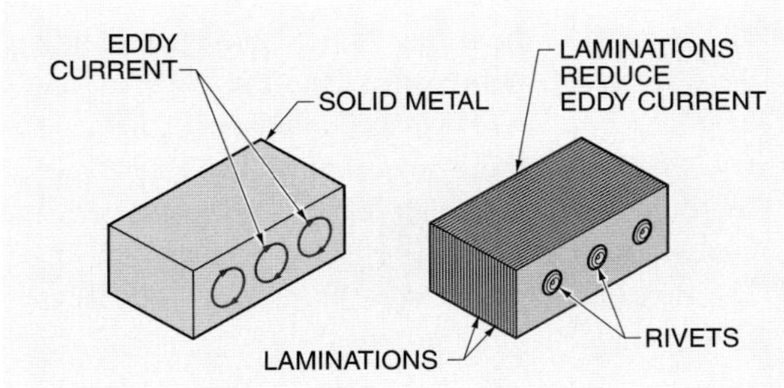

Figure 6-13. In AC solenoids, the magnetic assembly and armature consist of a number of thin pieces of metal laminated together.

As the coil is de-energized, some magnetic lines of force (residual magnetism) are always retained and could be enough to hold the armature in the sealed position. To eliminate this possibility, a small air gap is always left between the armature and the magnetic laminate assembly to break the magnetic field and allow the armature to drop away freely when de-energized. **See Figure 6-14.**

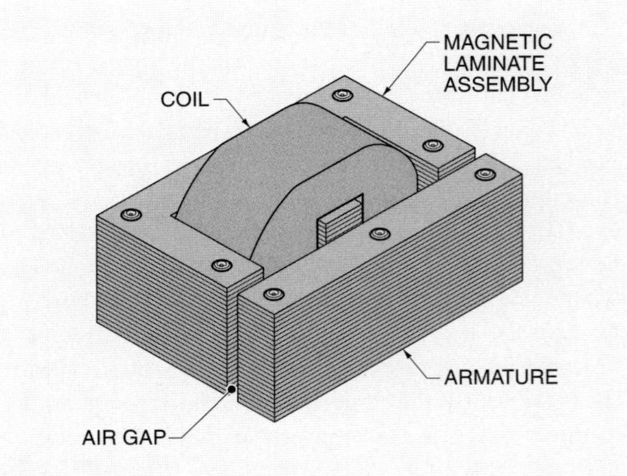

Figure 6-14. A small air gap is left in the magnetic laminate assembly to break the magnetic field and allow the armature to drop away freely after being de-energized.

Shading Coil. A *shading coil* is a single turn of conducting material (normally copper or aluminum) mounted on the face of the magnetic laminate assembly or armature. **See Figure 6-15.** A shading coil sets up an auxiliary magnetic field which helps hold in the armature as the main coil magnetic field drops to zero in an AC circuit.

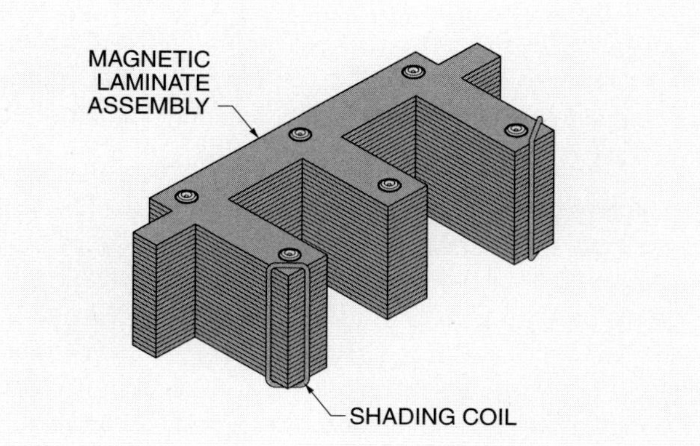

Figure 6-15. A shading coil sets up an auxiliary magnetic field, which is out of phase with the main coil magnetic field.

The magnetic field generated by alternating current periodically drops to zero. This makes the armature drop out or chatter. The attraction of the shading coil adds enough pull to the unit to keep the armature firmly seated. Without the shading coil, excessive noise, wear, and heat builds up on the armature faces, reducing the armature life expectancy.

SOLENOID CHARACTERISTICS

The two primary characteristics of a solenoid are the amount of voltage applied to the coil and the amount of current allowed to pass through the coil. Solenoid voltage characteristics include pick-up voltage, seal-in voltage, and drop-out voltage. Solenoid current characteristics include coil inrush current and sealed current.

Coils

Magnetic coils are normally constructed of many turns of insulated copper wire wound on a spool. The mechanical life of most coils is extended by encapsulating the coil in an epoxy resin or glass-reinforced alkyd material. **See Figure 6-16.** In addition to increasing mechanical strength, these materials greatly increase the moisture resistance of the magnetic coil. Because magnetic coils are encapsulated and cannot be repaired, they must be replaced when they fail.

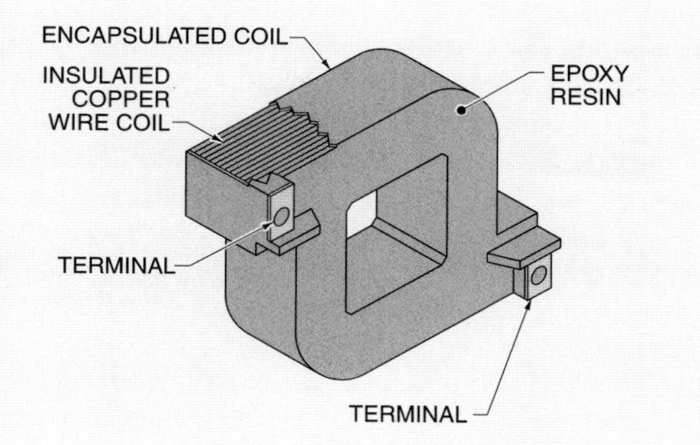

Figure 6-16. The mechanical life of most coils is extended by encapsulating the coil in an epoxy resin or glass-reinforced alkyd material.

Coil Inrush and Sealed Currents. Solenoid coils draw more current when first energized than is required to keep them running. **See Figure 6-17.** In a solenoid coil, the inrush current is approximately six to 10 times the sealed current. After the solenoid has been energized for some time, the coil becomes hot, causing the coil current to fall and stabilize at approximately 80% of its value when cold. The reason for such a high inrush current is that the basic opposition to current flow when a solenoid is energized is only the resistance of the copper coil. Upon energizing, however, the armature begins to move iron into the core of the coil. The large amount of iron in the magnetic circuit increases the magnetic opposition of the coil and decreases the current through the coil. This magnetic opposition is referred to as inductive reactance or total impedance. The heat produced by the coil further reduces current flow because the resistance of copper wire increases when hot, which limits some current flow.

Coil Inrush and Sealed Current Ratings. Magnetic coil data is normally given in volt amperes (VA). For example, a solenoid with a 120 V coil rated at 600 VA inrush and 60 VA sealed has an inrush current of 5 A ($^{600}/_{120}$ = 5 A) and a sealed current of 0.5 A ($^{60}/_{120}$ = 0.5 A). The same solenoid with a 480 V coil draws only 1.25 A ($^{600}/_{480}$ = 1.25 A) inrush current and 0.125 A ($^{60}/_{480}$ = 0.125 A) sealed current. The VA rating helps determine the starting and energized current load drawn from the supply line.

Coil Voltage Characteristics. All solenoids develop a magnetic field in their coil when voltage is applied. This magnetic field produces a force on the armature and tries to move it. The applied voltage determines the amount of force produced on the armature. The voltage applied to a solenoid should be ±10% of the rated solenoid value. A solenoid overheats when the voltage is excessive. The heat destroys the insulation on the coil wire and burns out the solenoid. The solenoid armature may have difficulty moving the load connected to it when the voltage is too low.

Pick-up voltage is the minimum voltage that causes the armature to start to move. *Seal-in voltage* is the minimum control voltage required to cause the armature to seal against the pole faces of the magnet. *Drop-out voltage* is the voltage that exists when voltage is reduced sufficiently to allow the solenoid to open. Seal-in voltage can be higher than pick-up voltage because a higher force may be required to seal in the armature than to just move the armature. Drop-out voltage is lower than pick-up voltage or seal-in voltage because it takes more force to hold the armature in place than to release the armature.

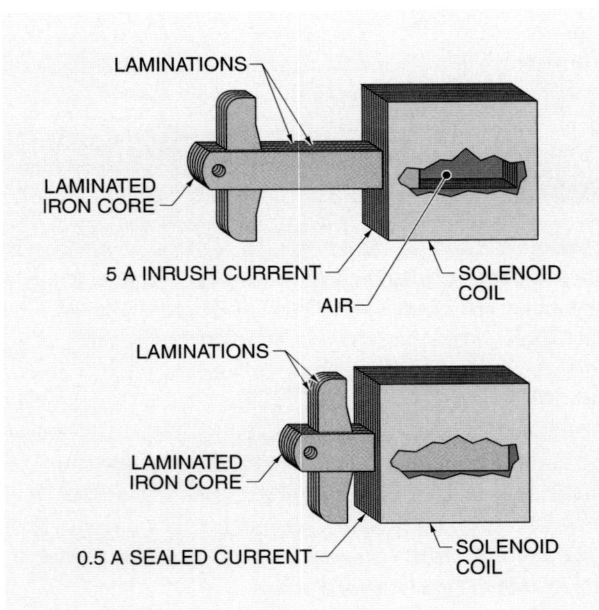

Figure 6-17. Solenoid inrush current is approximately six to 10 times the sealed current.

For most solenoids, the minimum pick-up voltage is about 80% to 85% of the solenoid rated voltage. The seal-in voltage is somewhat higher than the pick-up voltage and should be no less than 90% of the solenoid rated voltage. Drop-out voltage can be as low as 70% of the solenoid rated voltage. The exact pick-up, seal-in, and drop-out voltages depend on the load connected to the solenoid armature and the mounting position of the solenoid. The greater the applied armature load, the higher the required voltage values.

Voltage Variation Effects

Voltage variations are one of the most common causes of solenoid failure. Precautions must be taken to select the proper coil for a solenoid. Excessive or low voltage must not be applied to a solenoid coil.

High Voltage. A coil draws more than its rated current if the voltage applied to the coil is too high. Excessive heat is produced, which causes early failure of the coil insulation. The magnetic pull is also too high and causes the armature to slam in with excessive force. This causes the magnetic faces to wear rapidly, reducing the expected life of the solenoid.

Low Voltage. Low voltage on the coil produces low coil current and reduced magnetic pull. The solenoid may pick up but does not seal in when the applied voltage is greater than the pick-up voltage but less than the seal-in voltage. The greater pick-up current (six to 10 times sealed current) quickly heats up and burns out the coil because it is not designed to carry a high continuous current. The armature also chatters, which creates noise and increases the wear of the magnetic faces.

SELECTING PROPER SOLENOIDS

Solenoids are selected by analyzing the work to be done and selecting the correct solenoid to perform the work. Solenoid application rules are also considered when selecting solenoids.

Solenoid Application Rules

Solenoids are selected for an application based on the loading conditions that give the optimum performance. Rules to determine appropriate solenoid applications include the following:

- Obtain complete data on load requirements. Both the ultimate life of the solenoid and the life of its linkage depend on the loading of the solenoid. The solenoid may not seal correctly, resulting in coil noise, overheating, and eventual burnout if overloaded. An accurate estimate of the required force at specific inch strokes (lb load vs. inch travel) is required to prevent solenoid overload.
- Allow for possible low-voltage conditions of the power supply. Some allowance must be made for low-voltage conditions of the power supply because the pull of the solenoid varies as the square of the voltage (4 lb at 10 V, 16 lb at 20 V, etc.). Solenoids should be applied in accordance with their recommended load. This rating is based on the amount of force the solenoid can develop with 85% of the rated voltage applied to the coil.

- Use the shortest possible stroke. Shorter strokes produce faster operating rates, require less power, produce greater force, and decrease coil heating. Any decrease in heating increases the life expectancy of the coil. The greater force allows a small, low-rated, low-cost solenoid to be used. Also, less destructive mechanical energy is normally available from shorter strokes. This decrease in destructive energy, or impact force, helps to reduce solenoid wear.
- Never use an oversized solenoid. Use of an oversized solenoid is inefficient, resulting in higher initial cost and greater power consumption, and requires a physically larger unit. Any energy not expended in useful work must be absorbed by the solenoid in the form of impact force, because energy produced by a solenoid is constant regardless of the load. This results in reduced mechanical life and subjects the linkage mechanism to unnecessary strain.

Solenoid Selection Methods

After reviewing the basic rules for selecting proper solenoids, specific parameters are analyzed. Solenoids are selected based on the outcome required.

Push or Pull. A solenoid may push or pull, depending on the application. In the case of a door latch, the unit must pull. In a clamping jig, the unit must push.

Length of Stroke. The length of the stroke is calculated after determining whether the solenoid must push or pull. For example, a door latch requires a ½″ maximum stroke length.

Required Force. Manufacturer specification sheets are used to determine the correct solenoid based on the required force. **See Figure 6-18.** For example, an A 100 solenoid is used for an application that requires a horizontal force of 2.7 lb. A solenoid is selected from the Horizontal Force 85% Voltage column with the next highest force if the required force is not given.

Duty Cycle. Solenoid characteristic tables are also used to check the duty cycle requirements of the application against the duty cycle information given for the solenoid. For example, an A 101 solenoid is required for an application requiring 190 operations per minute.

Rofin Sinar

Solenoids are used to control the flow of gas during tube welding.

½″ PULL SOLENOID CHARACTERISTCS								
Solenoid	Seated Force*		Plunger Weight*	Shipping Weight*	VA 100% Voltage Seated	½″ Maximum Stroke		
	85% Voltage	100% Voltage				Horizontal Force 85% Voltage*	VA 100% Voltage	Duty Cycle† 50% Time ON
A 100	7	9	.2	1.3	40	2.7	230	240
A 101	9	12	.3	1.5	50	4.0	322	190
A 102	11	15	.3	1.7	50	4.7	420	180
B 100	11	15	.4	2.3	60	6.2	520	200
B 101	13	18	.5	2.6	70	9.6	790	109

* in lb
† in ops/min

Figure 6-18. Manufacturer specification sheets are used to determine the correct solenoid based on the required force.

Uniform Force Curve. Manufacturers provide specification curves to help determine the overall operating characteristics of a solenoid. The force curve of a solenoid must meet the load throughout its length of travel. See **Figure 6-19.** For example, an A 100 solenoid may be used in an application requiring 2.7 lb of horizontal force over a stroke length of approximately ⅝″.

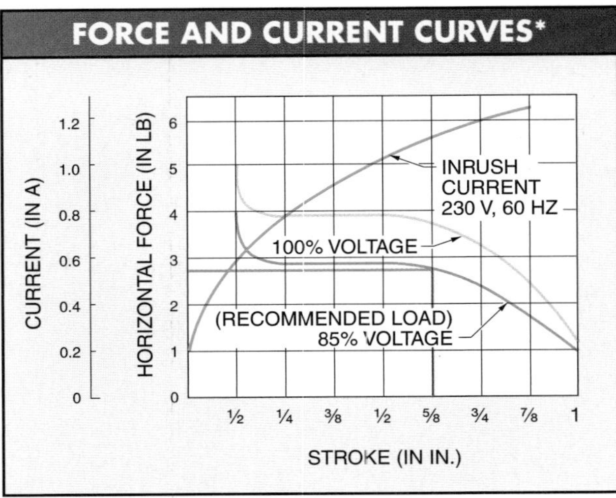

* A 100-60 Hz - ½″ and 1″ stroke solenoid

Figure 6-19. Manufacturers provide specification curves to determine operating characteristics of solenoids.

Heidelberg Harris, Inc.

Large complex systems use solenoids to control product positioning as well as other functions.

Mounting. Manufacturers provide letter or number codes to indicate the solenoid mount. **See Figure 6-20.** For example, an A solenoid is selected for a door latch application because the door latch application requires an end-mounting solenoid.

SOLENOID MOUNTING CODE

Code	Mounting
A	End
B	Right side
C	Throat
D	None (for thru-bolts)
E	Left Side
F	Both Sides

Figure 6-20. Manufacturers provide letter or number codes to indicate the solenoid mount.

Voltage Rating. Manufacturers provide letter or number codes to indicate the voltages that are available for a given solenoid. **See Figure 6-21.** For example, a 2 A solenoid may be used for an application that requires a 115 V coil.

SOLENOID VOLTAGE RATINGS

Number	Voltage (in V)
2 X	115
3 X	230
4 X	460
5 X	575

Figure 6-21. Manufacturers provide letter or number codes to indicate the voltages that are available for a given solenoid.

Other Considerations. Additional background information may be helpful in selecting the proper solenoid for other applications. Most manufacturers provide a specification order sheet that has space for additional information to help select the correct solenoid for an application. **See Figure 6-22.**

SOLENOID APPLICATIONS

Solenoids can be found in a wide range of equipment. In residential equipment, solenoids can be found in doorbells, washing machines, and kitchen appliances. Solenoids are commonly used in commercial and industrial control circuit applications such as hydraulics/pneumatics, refrigeration, combustion, and general-purpose controls.

Hydraulics/Pneumatics

Solenoid-operated valves typically control hydraulic and pneumatic equipment. A solenoid is used to move the valve spool that controls the flow of fluid (air or oil) in a directional control valve. A *directional control valve* is a valve that is used to direct the flow of fluid throughout a fluid power system. Directional control valves are identified by the number of positions and ways and the type of actuators.

Positions. A manual directional control valve is placed in different positions to start, stop, or change the direction of fluid flow. **See Figure 6-23.** A *position* is the number of locations within the valve in which the spool is placed to direct fluid through the valve. A directional control valve normally has two or three positions.

Technical Fact

The plunger, stop, and casing of solenoids used in AC service are laminated to reduce eddy current losses.

LINEAR SOLENOID DESIGN DATA SHEET

(1) Type: Push _______ Pull _______

(2) Stroke: _______ inches

(3) Force _______ lb at start _______ lb at end of stroke (for hold unit)

(4) Duty Cycle: Continuous _______ Intermittent _______

(5) Type of Mounting: Horizontal _______ Vertical _______

(6) Cycles per Second _______

(7) Voltage: DC _______ Min _______ Max _______
AC _______ Min _______ Max _______

(8) Ambient Temperature: _______

(9) Body Size: Length _______
Diameter _______

(10) Electrical Connection:
____

(11) Environmental Conditions:
Dust _______ Water _______ Oil _______
Other _______________

Sketch

Description of Solenoid Application Considered: _______________
____
____

Figure 6-22. Most manufacturers provide a specification order sheet that has space for additional information to help select the correct solenoid for an application.

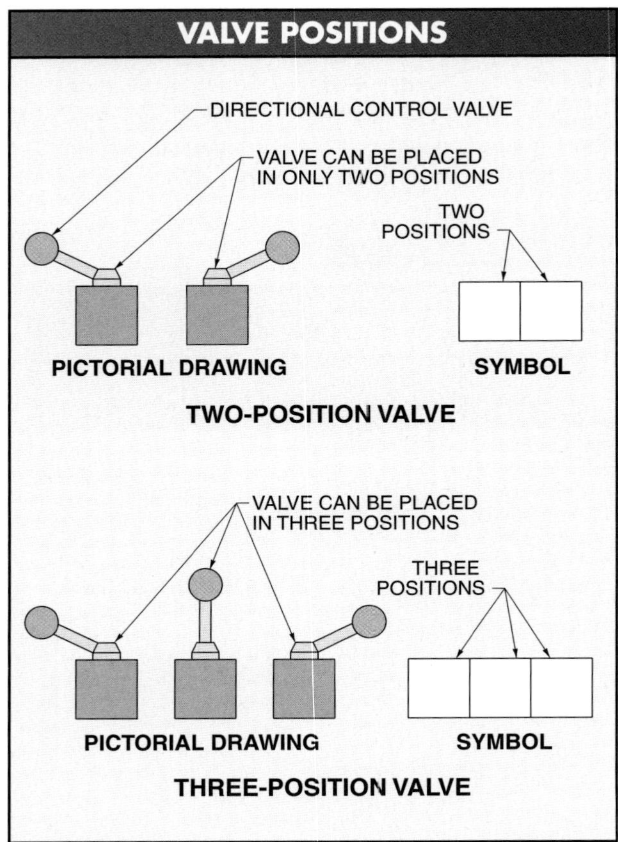

Figure 6-23. Positions are the number of locations within the valve in which the spool is placed to direct fluid through the valve.

Ways. A *way* is a flow path through a valve. Most directional control valves are either two-way or three-way valves. The number of ways required depends on the application. Two-way directional control valves have two main ports that allow or stop the flow of fluid. Two-way valves are used as shutoff, check, and quick-exhaust valves. **See Figure 6-24.**

Valve Actuators. A manual directional control valve uses a handle to change the valve spool position. An electrical control valve uses an actuator to change the position of a valve spool. In an electrical control valve, the solenoid acts as the actuator. **See Figure 6-25.**

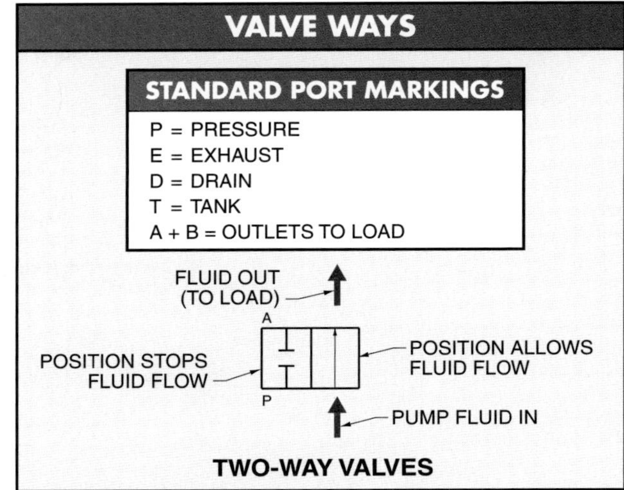

Figure 6-24. A way is a flow path through a valve.

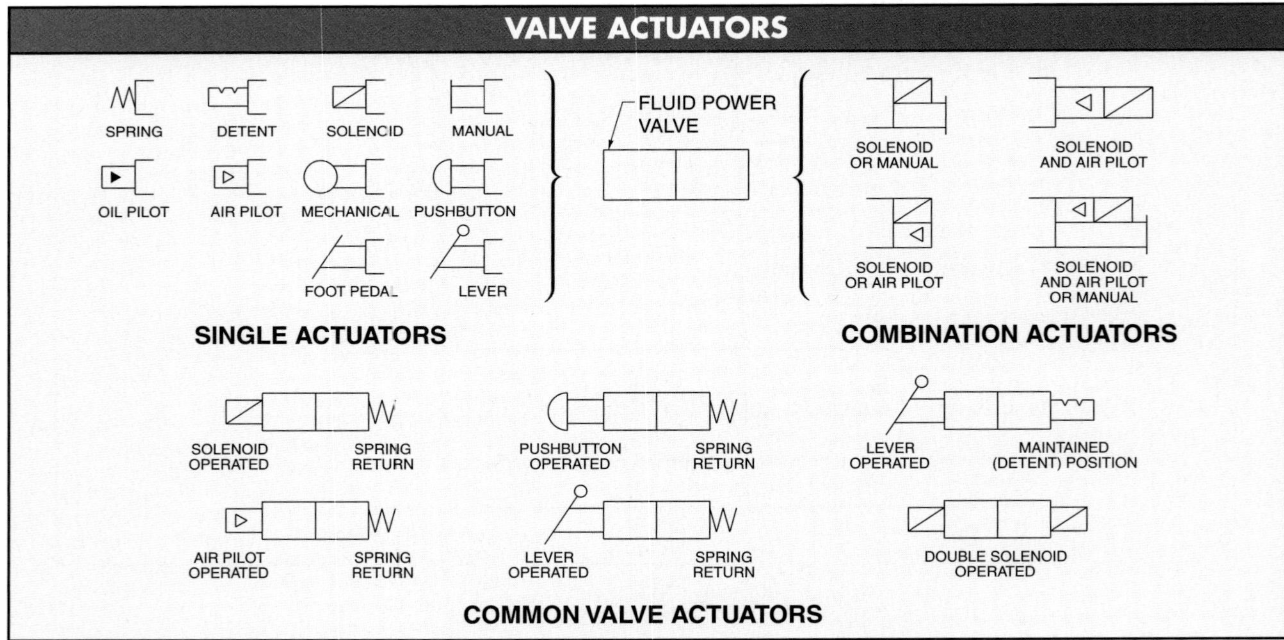

Figure 6-25. In an electrically operated hydraulic control valve, the solenoid acts as the actuator.

Refrigeration

Direct-acting, two-way valves are commonly used in refrigeration equipment. See Figure 6-26. Two-way (shutoff) valves have one inlet and one outlet pipe connection. These units may be constructed as normally open (NO), where the valve is open when de-energized and closed when energized, or they may be constructed as normally closed (NC), where the valve is closed when de-energized and open when energized.

A number of different solenoids may be used in a typical refrigeration system. The liquid line solenoid valves could be operated by two-wire or three-wire thermostats. The hot gas solenoid valve remains closed until the defrost cycle and then feeds the evaporator with hot gas for the defrost operation. **See Figure 6-27.**

Combustion

Solenoids may also be used in an oil-fired single-burner system. **See Figure 6-28.** The solenoids are crucial in the startup and normal operating functions of the system.

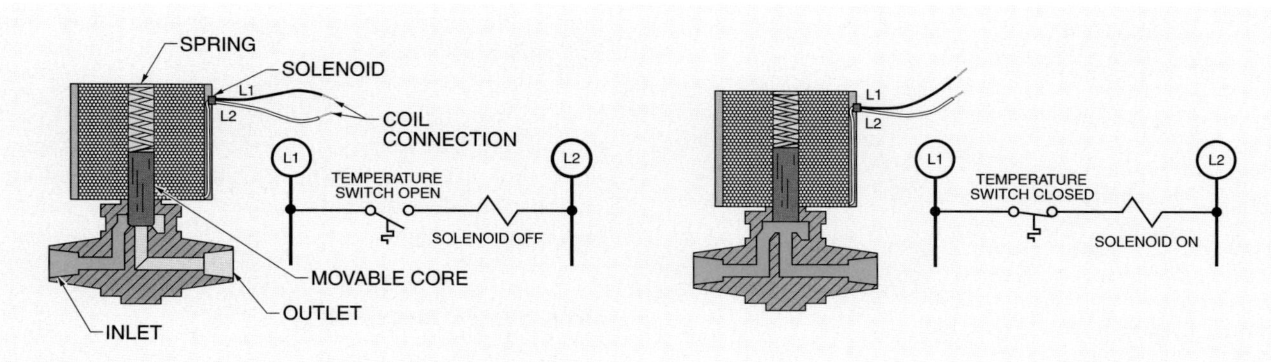

Figure 6-26. Direct-acting two-way valves open and close lines in a refrigeration system.

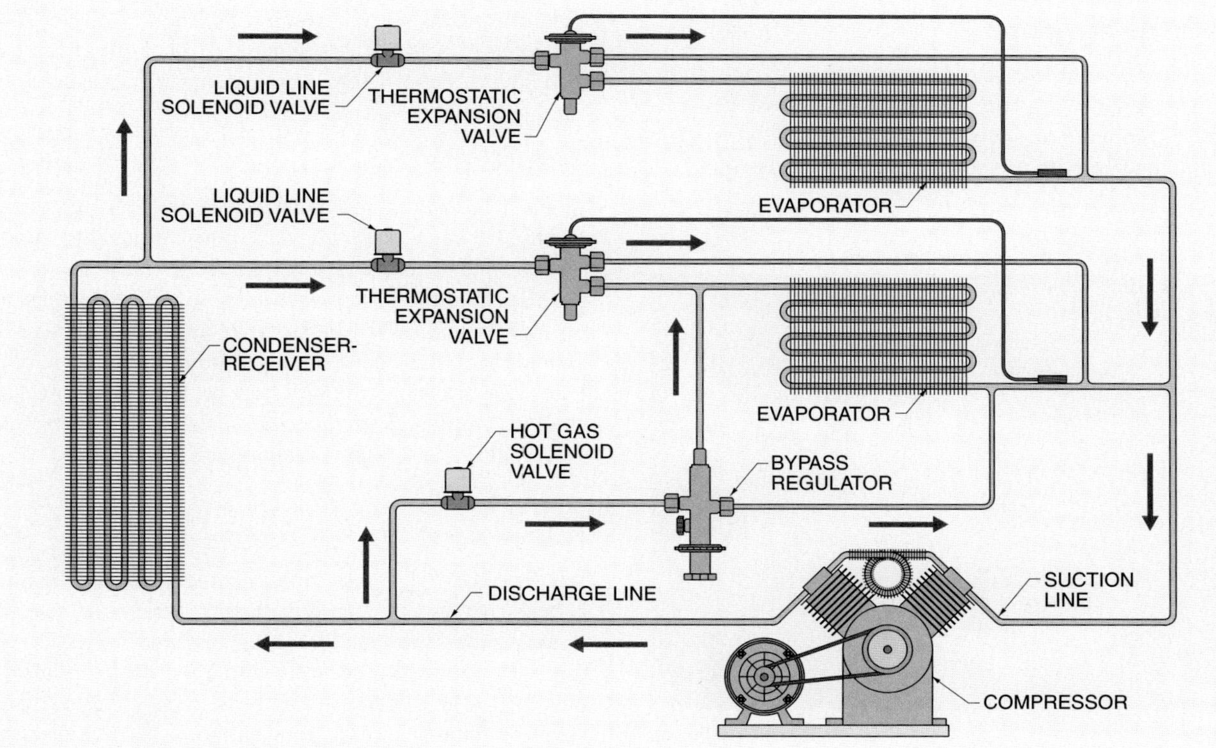

Figure 6-27. Refrigeration systems use solenoid valves to stop, start, and redirect the flow of refrigerant.

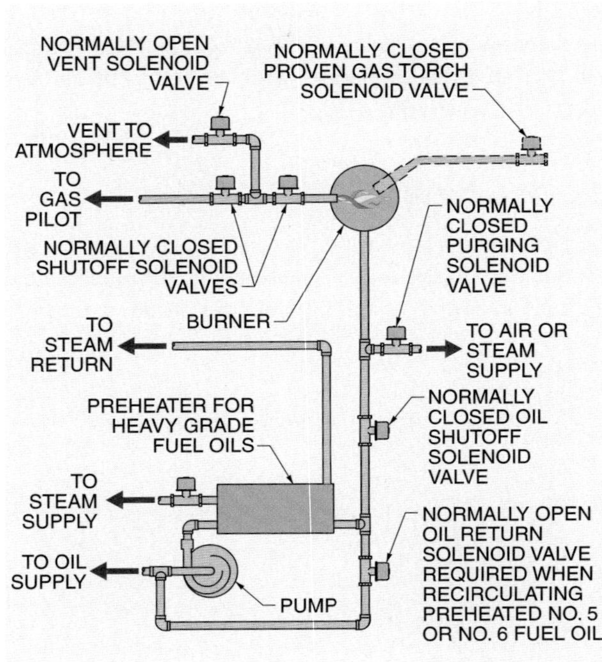

Figure 6-28. Different solenoids are used for the safe operation of an oil-fired single-burner system.

General-Purpose

In addition to commercial and industrial use, solenoids are used for general-purpose applications. Typical general-purpose applications include products such as printing calculators, cameras, and airplanes. **See Figure 6-29.**

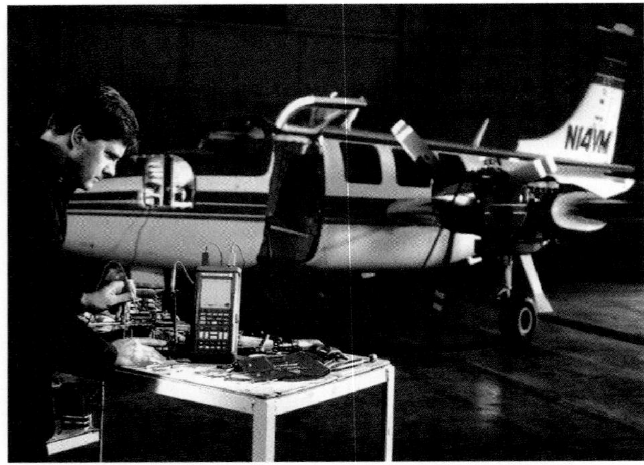

Fluke Corporation

Figure 6-29. Solenoids are used for general-purpose applications, such as in airplanes.

TROUBLESHOOTING SOLENOIDS

Solenoids fail due to coil burnout or mechanical damage. Manufacturer charts are used to help in the determination of the cause of solenoid failure. **See Figure 6-30.**

Incorrect Voltage

The voltage applied to a solenoid should be ±10% of the solenoid rated value. The voltage is measured directly at the valve when the solenoid is energized. A DMM set to measure AC voltage is used for AC solenoids. A DMM set to measure DC voltage is used for DC solenoids. The range setting must be greater than the applied voltage. **See Figure 6-31.**

A solenoid overheats when the voltage is excessive. The heat destroys the insulation on the coil wire and burns out the solenoid. The solenoid has difficulty moving the spool inside the valve when the voltage is too low. The slow operation causes the solenoid to draw its high inrush current longer. Longer high inrush current causes excessive heat.

Incorrect Frequency

Solenoids are available with frequency ratings of 50 Hz or 60 Hz. Solenoids with a frequency rating of 50 Hz are used in Europe, Asia, and much of South America. Solenoids with a frequency rating of 60 Hz are used in the United States, Canada, Mexico, and most of the Caribbean. A solenoid may operate if the frequency is not correct, but may also have a higher failure rate and may produce noise.

Transients

In most industrial applications, the power supplying a solenoid comes from the same power lines that supply electric motors and other solenoids. High transient voltages are placed on the power lines as these inductive loads are turned ON and OFF. Transient voltages may damage the insulation on the solenoid coil, nearby contacts, and other loads. The transient voltages may be suppressed by using snubber circuits. A *snubber circuit* is a circuit that suppresses noise and high voltage on the power lines.

Rapid Cycling

A solenoid draws several times its rated current when first connected to power. This high inrush current produces heat. In normal applications, the heat is low and dissipates over time. Rapid cycling does not allow the heat to dissipate quickly. The heat buildup burns the coil insulation and causes solenoid failure. To eliminate failure, a high-temperature solenoid should be used in applications requiring a solenoid to be cycled more than 10 times per minute.

SOLENOID FAILURE CHARACTERISTICS		
Problem	**Possible Causes**	**Comments**
Failure to operate when energized	Complete loss of power to solenoid	Normally caused by blown fuse or control circuit problem
	Low voltage applied to solenoid	Voltage should be at least 85% of solenoid rated value
	Burned out solenoid coil	Normally evident by pungent odor caused by burnt insulation
	Shorted coil	Normally a fuse is blown and continues to blow when changed
	Obstruction of plunger movement	Normally caused by a broken part, misalignment, or the presence of a foreign object
	Excessive pressure on solenoid plunger	Normally caused by excessive system pressure in solenoid-operated valves
Failure of spring-return solenoids to operate when de-energized	Faulty control circuit	Normally a problem of the control circuit not disengaging the solenoid's hold or memory circuit
	Obstruction of plunger movement	Normally caused by a broken part, misalignment, or the presence of a foreign object
	Excessive pressure on solenoid plunger	Normally caused by excessive system pressure in solenoid-operated valves
Failure of electrically-operated return solenoids to operate when de-energized	Complete loss of power to solenoid	Normally caused by a blown fuse or control circuit problem
	Low voltage applied to solenoid	Voltage should be at least 85% of solenoid rated value
	Burned out solenoid coil	Normally evident by pungent odor caused by burnt insulation
	Obstruction of plunger movement	Normally caused by broken part, misalignment, or presence of a foreign object
	Excessive pressure on solenoid plunger	Normally caused by excessive system pressure in solenoid-operated valves
Noisy operation	Solenoid housing vibrates	Normally caused by loose mounting screws
	Plunger pole pieces do not make flush contact	An air gap may be present, causing the plunger to vibrate. These symptoms are normally caused by foreign matter
Erratic operation	Low voltage applied to solenoid	Voltage should be at least 85% of the solenoid rated voltage
	System pressure may be low or excessive	Solenoid size is inadequate for the application
	Control circuit is not operating properly	Conditions on the solenoid have increased to the point where the solenoid cannot deliver the required force

Figure 6-30. Manufacturer charts are used to help in the determination of the cause of solenoid failure.

Environmental Conditions

A solenoid must operate within its rating and not be mechanically damaged or damaged by the surrounding atmosphere. A solenoid coil is subject to heat during normal operation. This heat comes from the combination of fluid flowing through the valve, the temperature rise from the coil when energized, and the ambient temperature of the solenoid.

Technical Fact

Typical industrial applications of solenoids include the control of door locks, valves dispensing product, movement or stamping of parts, motor/clutch/brake operation, and the operation of fluid power valves.

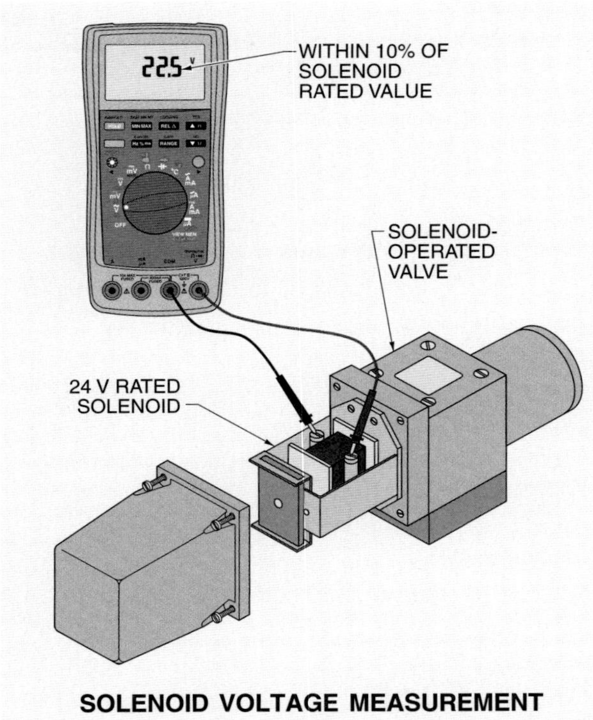

Figure 6-31. The voltage applied to a solenoid should be ±10% of the solenoid rated value.

Solenoid Troubleshooting Procedure

A DMM set to measure voltage and resistance is required when troubleshooting a solenoid. **See Figure 6-32.** To troubleshoot a solenoid, apply the following procedure:

1. Turn electrical power to solenoid or circuit OFF.
2. Measure the voltage at the solenoid to ensure the power is OFF.
3. Remove the solenoid cover and visually inspect the solenoid. Look for a burnt coil, broken parts, or other problems. Replace the coil when burnt. Replace the broken parts when available. Replace the valve, contactor, starter, or solenoid-operated device when the parts are not available. *Note:* Determine the fault before installing a new coil when a solenoid has failed due to a burnt or shorted coil. Coils will continue to burn out if the fault is not corrected. Always observe solenoid operation after a solenoid is replaced.
4. Disconnect the solenoid wires from the electrical circuit when no obvious problem is observed.
5. Check the solenoid continuity. Connect the DMM leads to the solenoid wires with all power turned OFF. The DMM should indicate a resistance reading of ±15% of the coil's normal reading.

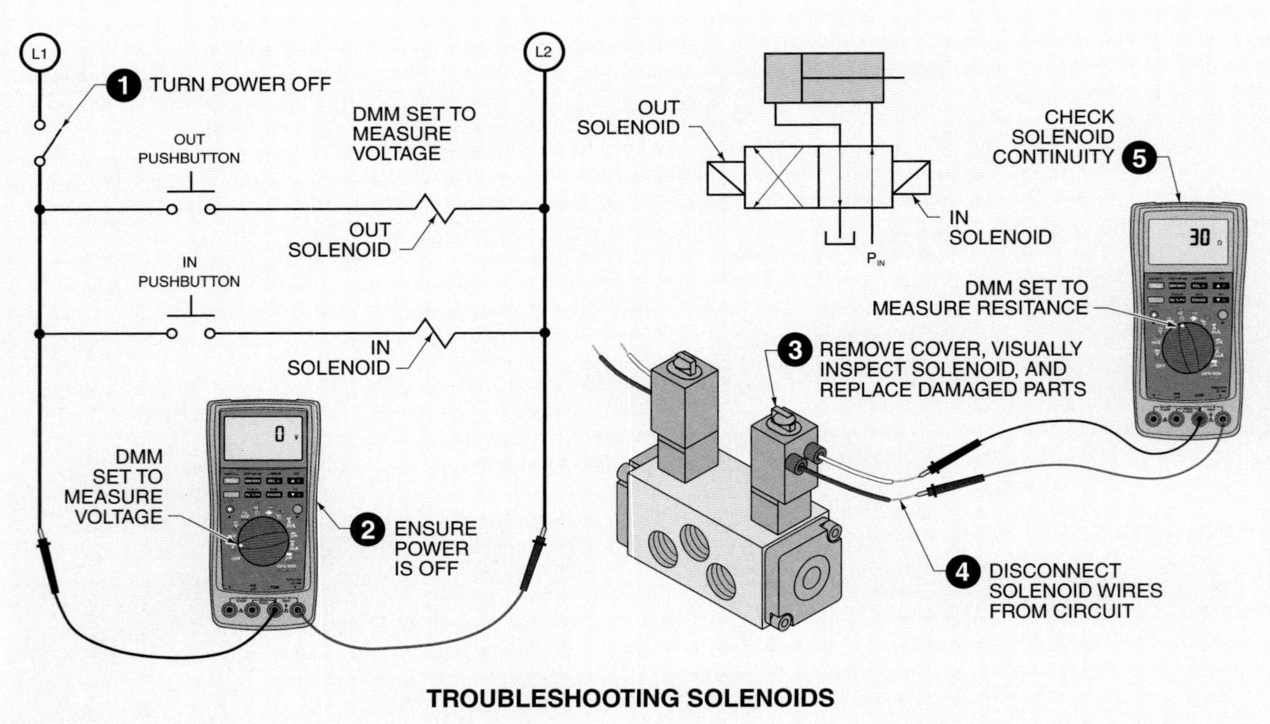

Figure 6-32. A DMM set to measure voltage and resistance is required when troubleshooting a solenoid.

Circuit readings are obtained by testing a good solenoid. A low or zero reading indicates a short or partial short circuit. Replace the solenoid if there is a short circuit. No movement of the needle on an analog meter or infinity resistance on a DMM set to measure resistance indicates the coil is open and defective. Replace the solenoid if the open is not obvious. Manufacturer charts and recommendations are also used to help determine the cause of solenoid failure.

DC GENERATORS

A *generator* is a machine that converts mechanical energy into electrical energy by means of electromagnetic induction. DC generators operate on the principle that when a coil of wire is rotated in a magnetic field, a voltage is induced in the coil. The amount of voltage induced in the coil is determined by the rate at which the coil is rotated in the magnetic field. When a coil is rotated in a magnetic field at a constant rate, the voltage induced in the coil depends on the number of magnetic lines of force in the magnetic field at each given instant of time. DC generators consist of field windings, an armature, a commutator, and brushes. **See Figure 6-33.**

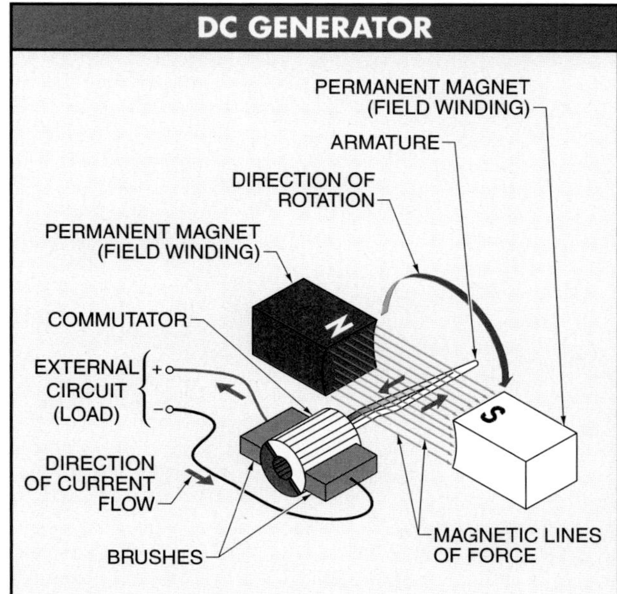

Figure 6-33. DC generators consist of field windings, an armature, commutator, and brushes, and operate on the principle that when a coil of wire is rotated in a magnetic field, a voltage is induced in the coil.

Field Windings

Field windings are magnets used to produce the magnetic field in a generator. The magnetic field used in a generator can be produced by permanent magnets or electromagnets. Permanent magnets are used in very small machines referred to as magnetos. The disadvantages of permanent magnets are that their magnetic lines of force decrease as the age of the magnet increases, and the strength of a permanent magnet cannot be varied for control purposes. Most generators use electromagnets, which must be supplied with current. If the current for the field windings is supplied by an outside source (a battery or another generator), the generator is separately excited. If the generator itself supplies current for the field windings, the generator is referred to as self-excited. DC generators are usually self-excited.

Armature

An *armature* is the movable coil of wire in a generator that rotates through the magnetic field. A DC generator always has a rotating armature and a stationary field (field windings). The rotating armature may consist of many coils. Although increasing the number of coils reduces the ripples (pulsations) in the output voltage, it is impossible to remove the ripples completely.

Commutator

A *commutator* is a ring made of segments that are insulated from one another. Each end of a coil of wire is connected to a segment. A voltage is induced in the coil whenever the coil cuts the magnetic lines of force of a magnetic field. The commutator segments reverse the connections to the brushes every half cycle. This maintains a constant polarity of output voltage produced by the generator. **See Figure 6-34.**

Technical Fact

The theory of operation (production of output voltage) for a DC generator is based upon Faraday's law, which states that when the magnetic field across a coil of wire changes, the resulting induced voltage is proportional to the number of turns in the coil multiplied by the rate of change of the field. The rate of change in a generator may vary from several thousand rotations per minute to several rotations per minute.

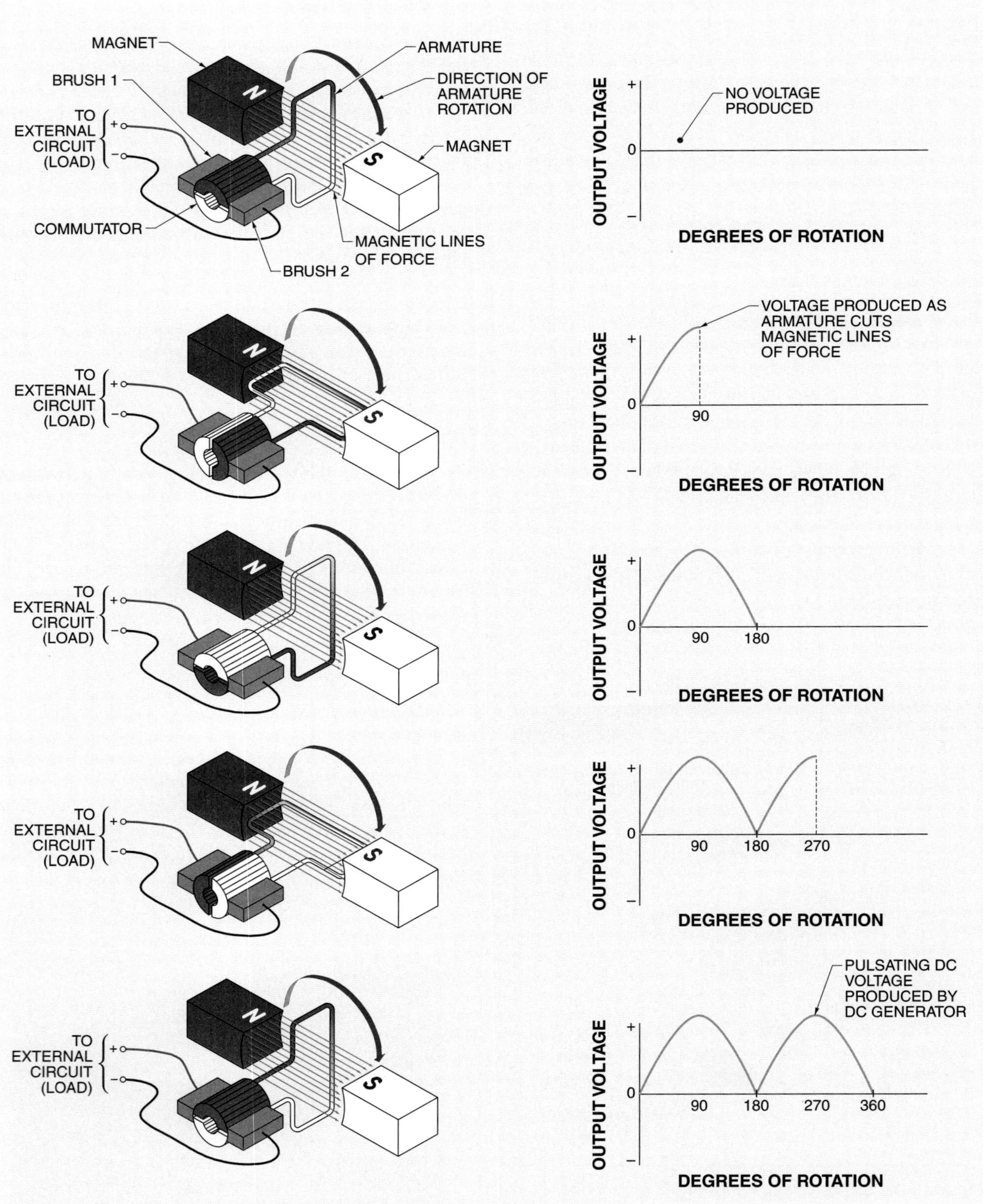

Figure 6-34. A voltage is induced in the coil (armature) of a generator when the coil cuts the lines of force of a magnetic field.

Brushes

A *brush* is the sliding contact that rides against the commutator segments and is used to connect the armature to the external circuit. Brushes are made from soft carbon (natural graphite). Brushes are softer than the commutator bars yet strong enough so that the brushes do not chip or break from vibration. One brush makes contact with each segment of the commutator.

A DC generator is designed so that the brushes ride on the different segments of the commutator each time the current is zero. Therefore, the current in the external circuit (load) always flows in one direction; however, its magnitude varies continuously. The action of reversing the connections to the coil (armature) to obtain a direct current is referred to as commutation. The resulting output voltage of a DC generator is a pulsating DC voltage. The pulsations of the output voltage are known as ripples.

Left-Hand Generator Rule

The *left-hand generator rule* is the relationship between the current in a conductor and the magnetic field existing around the conductor. The left-hand generator rule states that with the thumb, index finger, and middle finger of the left hand set at right angles to each other, the index finger points in the direction of the magnetic field, the thumb points in the direction of the motion of the conductor, and the middle finger points in the direction of the induced current. When using the left-hand generator rule, it is assumed that the magnetic field is stationary and that the conductor is moving through the field. **See Figure 6-35.**

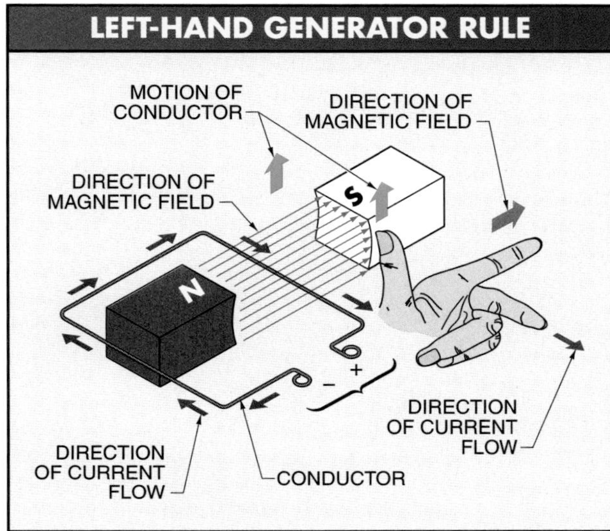

Figure 6-35. The left-hand generator rule expresses the relationship between the conductor, magnetic field, and induced voltage in a generator.

DC Generator Types

The three types of DC generators are series-wound, shunt-wound, and compound-wound generators. The difference between the types is based on the relationship of the field windings to the external circuit.

Series-Wound Generators. A *series-wound generator* is a generator that has its field windings connected in series with the armature and the external circuit (load). **See Figure 6-36.** In a series-wound generator, the field windings consist of a few turns of low-resistance wire because the load current flows through them.

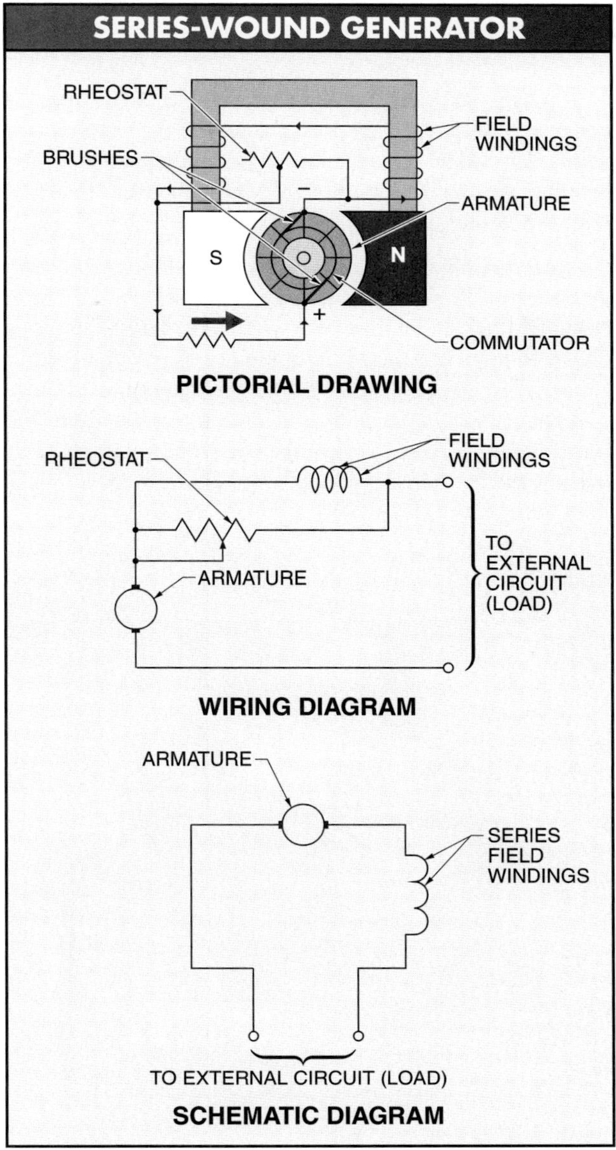

Figure 6-36. A series-wound generator has its field windings connected in series with the armature and load.

The ability of a generator to have a constant voltage output under varying load conditions is referred to as the generator voltage regulation. Series-wound generators have poor voltage regulation. Because of their poor voltage regulation, series-wound DC generators are not used frequently. The output voltage of a series-wound generator may be controlled by a rheostat (variable resistor) connected in parallel with the field windings.

Shunt-Wound Generators. A *shunt-wound generator* is a generator that has its field windings connected in parallel (shunt) with the armature and the external circuit (load). **See Figure 6-37.** Because the field windings are connected in parallel with the load, the current through them is wasted as far as output is concerned. The field windings consist of many turns of high-resistance wire to keep the current flow through them low.

A shunt-wound generator is suitable if the load is constant. However, if the load fluctuates, the voltage also varies. The output voltage of a shunt-wound generator may be controlled by means of a rheostat connected in series with the shunt field.

Compound-Wound Generators. A *compound-wound generator* is a generator that includes series and shunt field windings. In a compound-wound generator, the series field windings and shunt field windings are combined in a manner to take advantage of the characteristics of each. The shunt field is normally the stronger of the two. The series field is used only to compensate for effects that tend to decrease the output voltage. **See Figure 6-38.**

Cummins Power Generation

Industrial generators are used for peak and standby power systems in such locations as airports, hospitals, and factories.

Technical Fact

Per NEC® Article 445.12, constant voltage generators should be protected from overloads by inherent design, circuit breakers, or fuses.

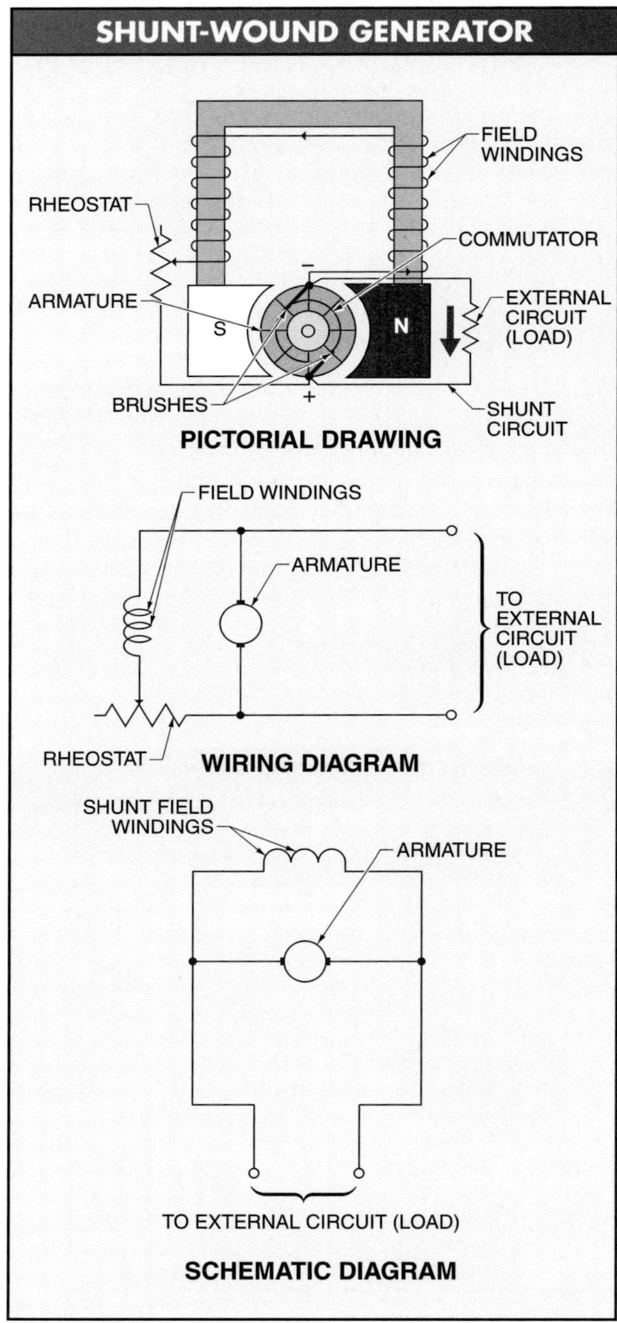

Figure 6-37. A shunt-wound generator has its field windings connected in parallel (shunt) with the armature and load.

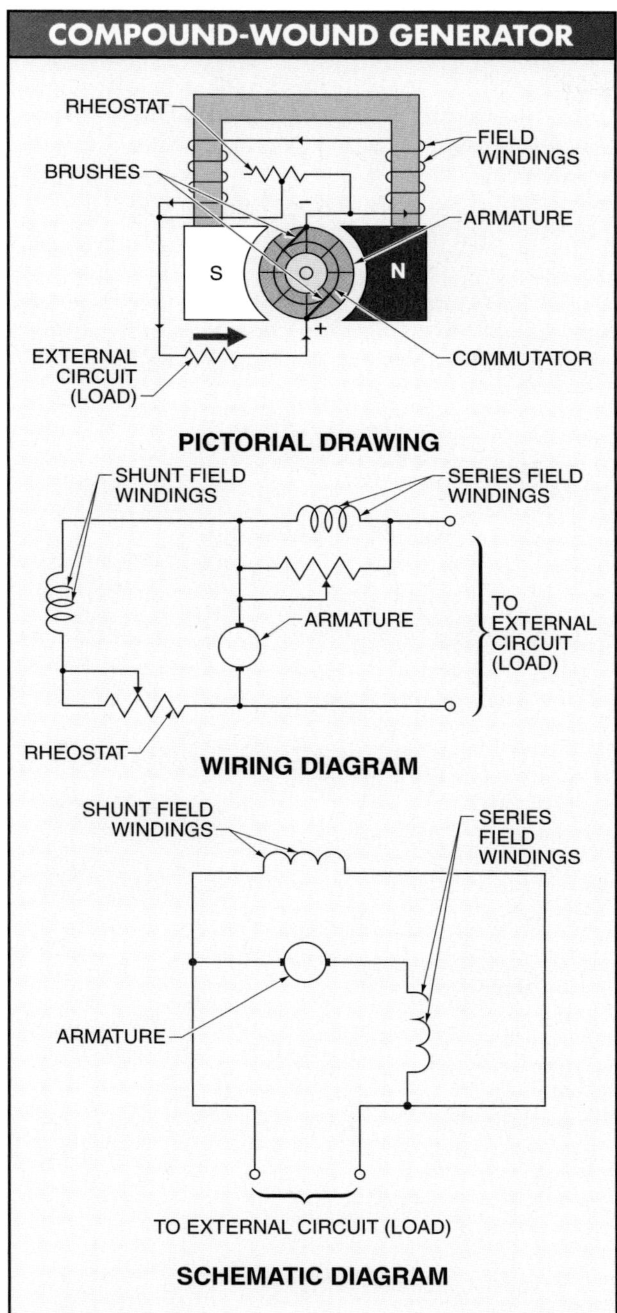

Figure 6-38. A compound-wound generator includes series and shunt field windings.

DC MOTORS

A *motor* is a machine that converts electrical energy into mechanical energy by means of electromagnetic induction. Motors operate on the principle that when a current-carrying conductor is placed in a magnetic field, a force that tends to move the conductor out of the field is exerted on the conductor. The conductor tends to move at right angles to the field. **See Figure 6-39.**

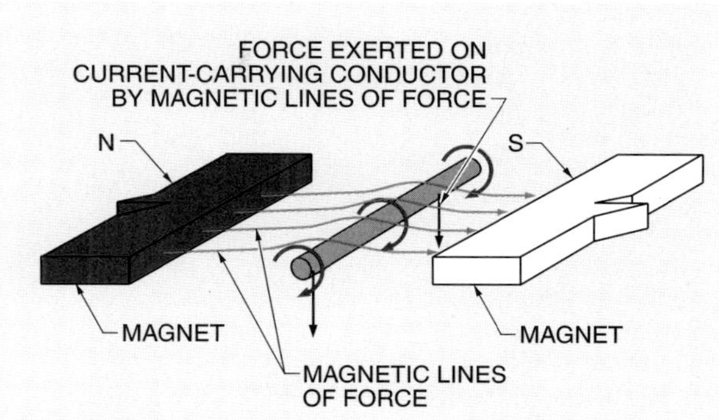

Figure 6-39. When a current-carrying conductor is placed in a magnetic field, a force that tends to move the conductor out of the field is exerted on the conductor.

The direction of the movement of the conductor depends on the direction of the current and the magnetic field. The electron flow motor rule is used to determine the direction of motion of a current-carrying conductor in a magnetic field. **See Figure 6-40.** The electron flow motor rule states that with the thumb, index finger, and middle finger of the right hand set at right angles to each other, the index finger points in the direction of the magnetic field (N to S), the thumb points in the direction of the induced conductor motion, and the middle finger points in the direction of the electron current flow in the conductor.

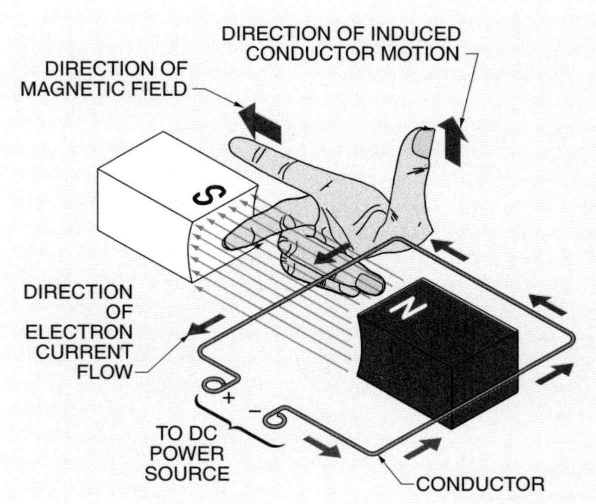

Figure 6-40. The electron flow motor rule is used to determine the direction of motion of a current-carrying conductor in a magnetic field.

Technical Fact

Motors are classified according to power rating as subfractional horsepower, fractional horsepower, or integral horsepower. They are also classified by application as general purpose, definite purpose, and special purpose. In the U.S., definitions of motor terms and standards for dimensions and operating characteristics are set by NEMA and the Small Motors Manufacturers Association (SMMA).

The current in the conductor flows at right angles to the magnetic lines of force of the magnetic field. The force on the conductor is at right angles to both the current in the conductor and the magnetic lines of force. The amount of force on the conductor depends on the intensity of the magnetic field, the current through the conductor, and the length of the conductor. The intensity of the magnetic field and the amount of current in the conductor are normally changed to increase the force on the conductor. However, the amount of force can be increased by increasing any of these three factors.

Torque is developed on a wire loop in a magnetic field. **See Figure 6-41.** Electron current flow must be at a right angle to the magnetic field. This is required for induced motion because no force is exerted on a conductor if the direction of electron current flow through the conductor and direction of the magnetic lines of force are the same (parallel).

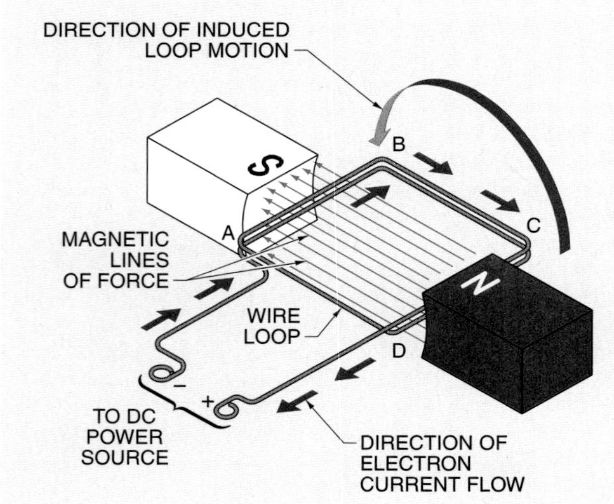

Figure 6-41. Torque is developed on a wire loop in a magnetic field.

Both sections of loop AB and CD have a force exerted on them because the direction of electron current flow in these segments is at right angles to the magnetic lines of force. The exertion of force on AB and CD is opposite in direction because the current flow is opposite in each section of the wire loop.

The result of the two magnetic fields intersecting creates a turning force (torque) on the loop. The magnetic lines of force cause the loop to rotate when they straighten. The left side of the conductor is forced downward and the right side of the conductor is forced upward, causing a counterclockwise rotation. **See Figure 6-42.**

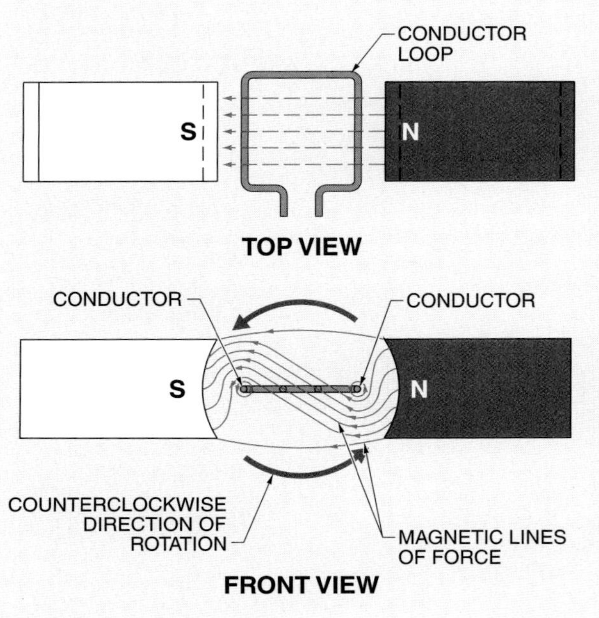

Figure 6-42. The distortion of the magnetic lines of force causes the conductor loop to rotate in a counterclockwise direction.

DC Motor Construction

A *direct current (DC) motor* is a motor that uses direct current connected to the field and armature to produce shaft rotation. A DC motor consists of field windings, an armature, a commutator, and brushes. *Field windings* are the stationary windings or magnets of a DC motor. An *armature* is the rotating part of a DC motor. A magnetic field is produced in the armature by current flowing through the armature coils. The armature magnetic field interacts with the direct current produced by the field windings. The interaction of the magnetic fields causes the armature to rotate.

A commutator is a ring made of segments that are insulated from one another. A commutator is the part of the armature that connects each armature coil to the brushes using copper bars (segments) that are insulated from each other with pieces of mica. The commutator is mounted on the same shaft as the armature and rotates with the shaft.

A brush is the sliding contact that rides against the commutator segments and is used to connect the armature to the external circuit. Brushes are made of carbon or graphite material and are held in place by brush holders. A pigtail connects a brush to the external circuit (power supply). A *pigtail* is an extended, flexible connection or a braided copper conductor. Brushes are free to move up and down in the brush holder. This freedom allows the brush to follow irregularities in the surface of the commutator. A spring placed behind the brush forces the brush to make contact with the commutator. The spring pressure is normally adjustable, as is the entire brush holder assembly. The brushes make contact with successive copper bars of the commutator as the shaft, armature, and commutator rotate. **See Figure 6-43.**

North American Industries, Inc.

Overhead cranes driven by electric motors are used to move heavy metal beams.

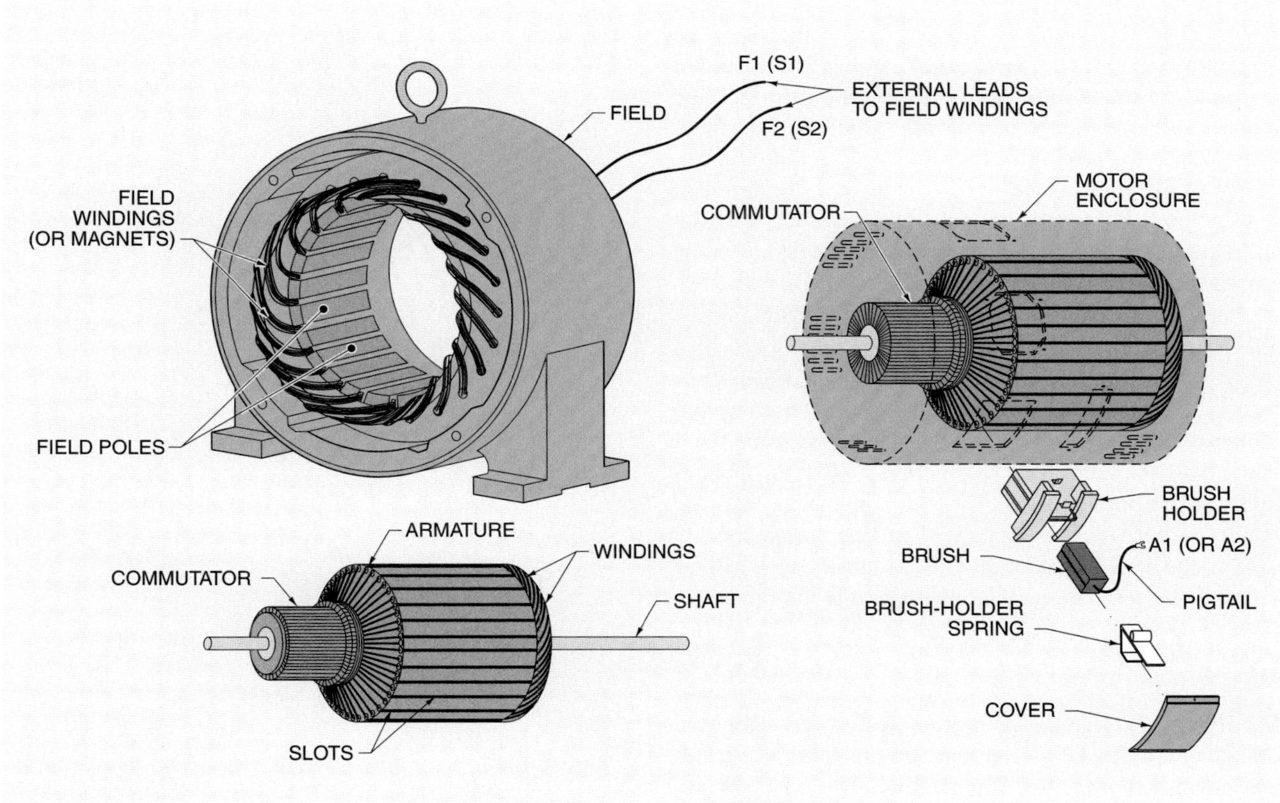

Figure 6-43. DC motors consist of field windings, an armature, commutator, and brushes.

DC power is delivered to the armature coils through the brushes and commutator segments. The armature coils, commutator, and brushes are arranged so that the flow of current is in one direction in the loop on one side of the armature, and the flow of current is in the opposite direction in the loop on the other side of the armature. For example, brush 2 breaks contact with side B of the commutator and makes contact with side A. **See Figure 6-44.** The flow of current through the commutator reverses because the flow of current is at the same polarity on the brushes at all times. This allows the commutator to rotate another 180° in the same direction. After the additional 180° rotation, brush 1 breaks contact with side B of the commutator and makes contact with side A. Likewise, brush 2 breaks contact with side A of the commutator and makes contact with side B. This reverses the direction of current in the commutator again and allows for another 180° of rotation. The armature continues to rotate as long as the commutator winding is supplied with current and there is a magnetic field.

A rotating force is exerted on the armature when it is positioned so that the plane of the armature loop is parallel to the field, and the armature loop sides are at right angles to the magnetic field. No movement takes place if the armature loop is stopped in the vertical (neutral) position. In this position, no further torque is produced because the forces acting on the armature are upward on the top side of the loop and downward on the lower side of the loop. The armature does not stop because of inertia. The armature continues to rotate for a short distance. As it rotates, the magnetic field in the armature is opposite that of the field. This pushes the conductor back in the direction it came, stopping the rotating motion. A method is required to reverse the current in the armature every one-half rotation so that the magnetic fields work together. Brushes and a commutator are added to maintain this positive rotation.

Connecting voltage directly to the field and armature of a DC motor allows the motor to produce higher torque in a smaller frame than AC motors. DC motors provide excellent speed control for acceleration and deceleration with effective and simple torque control. DC motors perform better than AC motors in most traction equipment applications. DC motors do require more maintenance than AC motors because they have brushes that wear. DC motors are used as the drive motor in mobile equipment such as golf carts, quarry and mining equipment, and locomotives.

DC Motor Types

The four basic types of DC motors are DC series motors, DC shunt motors, DC compound motors, and DC permanent-magnet motors. **See Figure 6-45.** These DC motors have similar external appearances, but are different in their internal construction and output performance.

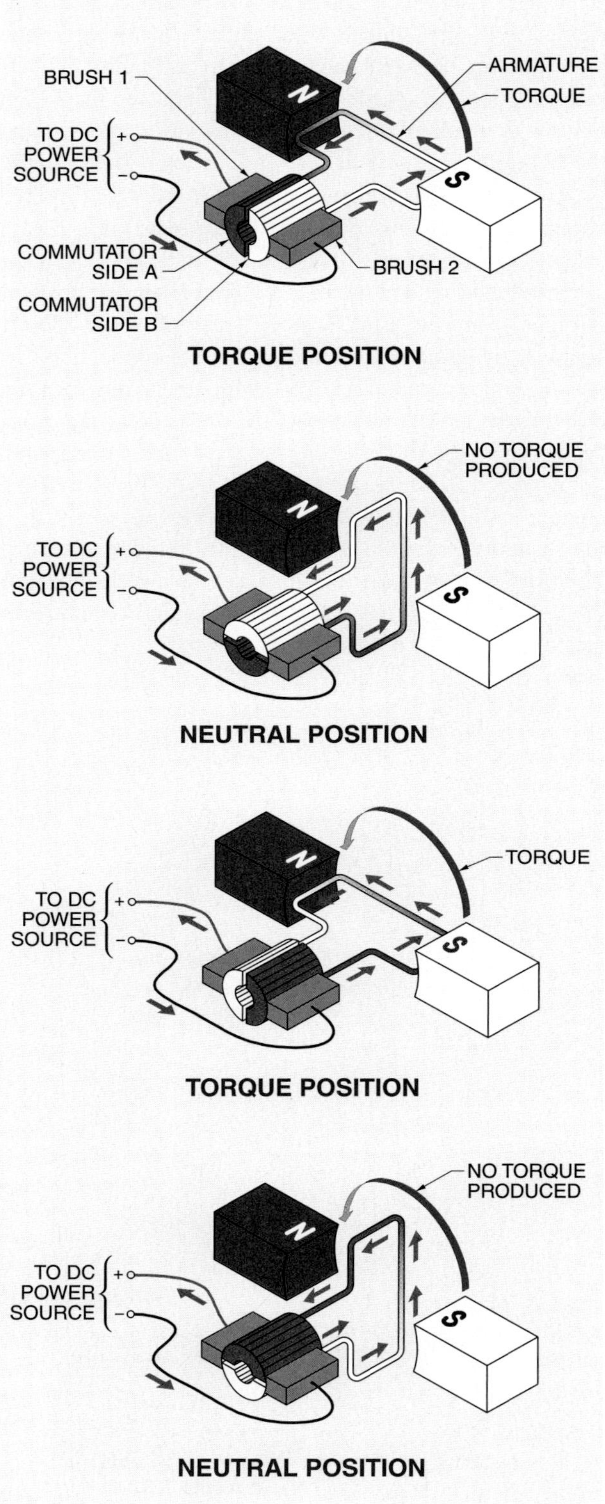

Figure 6-44. A rotating force is exerted on the armature when it is positioned so that the plane of the armature loop is parallel to the field and the armature loop sides are at a right angle to the magnetic field.

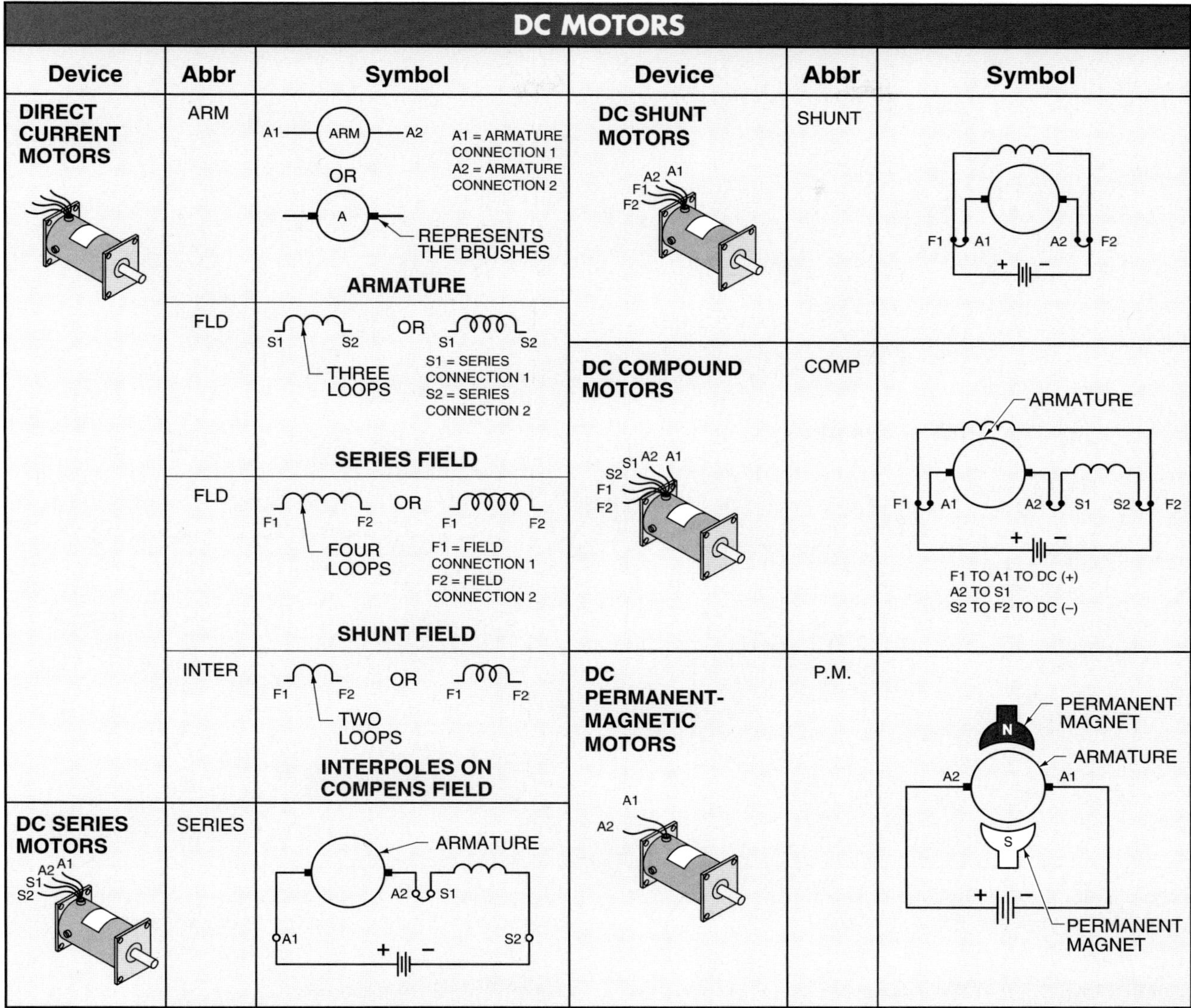

Figure 6-45. The four basic types of DC motors are DC series motors, DC shunt motors, DC compound motors, and DC permanent-magnet motors.

DC Series Motors. A *DC series motor* is a DC motor that has the series field coils connected in series with the armature. The field must carry the load current passing through the armature. The field coil has relatively few turns of heavy-gauge wire. The wires extending from the series coil are marked S1 and S2. The wires extending from the armature are marked A1 and A2. **See Figure 6-46.**

DC series motors are used as traction motors because they produce the highest torque of all DC motors. DC series motors can develop 500% of full-load torque upon starting. Typical applications include traction bridges, hoists, gates, and starters in automobiles.

The speed regulation of a DC series motor is poor. As the mechanical load on the motor is reduced, a simultaneous reduction of current occurs in the field and the armature. If the mechanical load is entirely removed, the speed of the motor increases without limit and may destroy the motor. For this reason, series motors are always permanently connected to the load the motor controls.

Technical Fact

DC motors that have wound poles are called shunt, series, or compound motors. DC motors that use permanent magnets to magnetize their field poles are called PMDC machines.

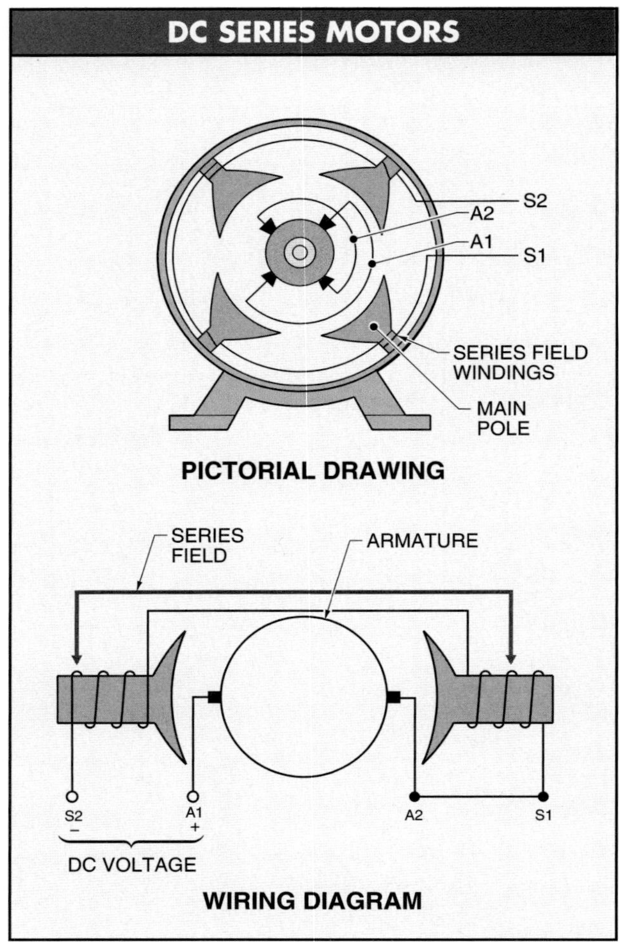

Figure 6-46. A DC series motor is a motor with the field connected in series with the armature.

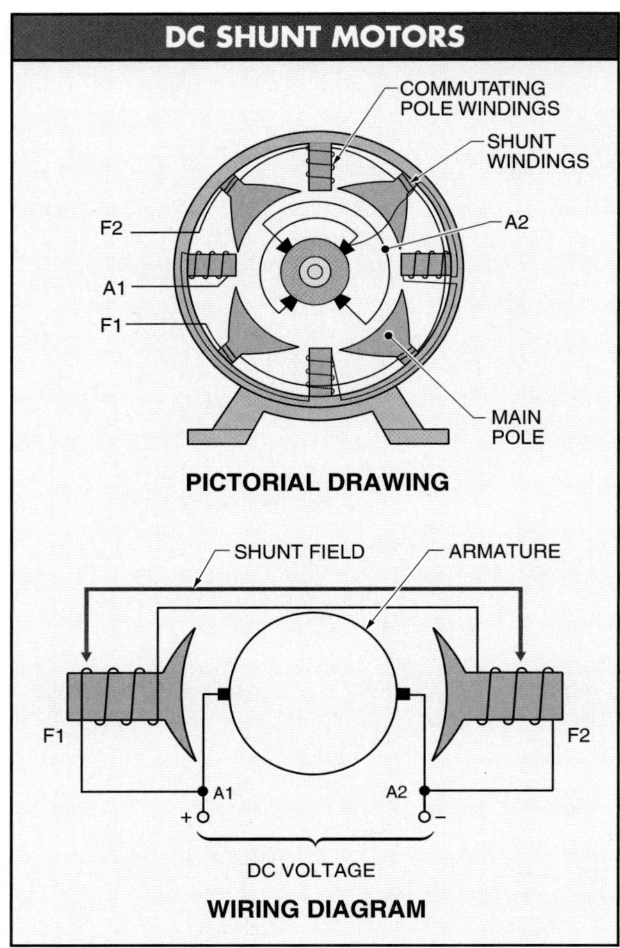

Figure 6-47. A DC shunt motor is a motor with the field connected in shunt (parallel) with the armature.

DC Shunt Motors. A *DC shunt motor* is a DC motor that has the field connected in shunt (parallel) with the armature. The wires extending from the shunt field of a DC shunt motor are marked F1 and F2. The armature windings are marked A1 and A2. **See Figure 6-47.**

The field has numerous turns of wire, and the current in the field is independent of the armature, providing the DC shunt motor with excellent speed control. The shunt field may be connected to the same power supply as the armature or may be connected to another power supply. A *self-excited shunt field* is a shunt field connected to the same power supply as the armature. A *separately excited shunt field* is a shunt field connected to a different power supply than the armature.

DC shunt motors are used where constant or adjustable speed is required and starting conditions are moderate. Typical applications include fans, blowers, centrifugal pumps, conveyors, elevators, woodworking machinery, and metalworking machinery.

DC Compound Motors. A *DC compound motor* is a DC motor with the field connected in both series and shunt with the armature. The field coil is a combination of the series field (S1 and S2) and the shunt field (F1 and F2). **See Figure 6-48.** The series field is connected in series with the armature. The shunt field is connected in parallel with the series field and armature combination. This arrangement gives the motor the advantages of the DC series motor (high torque) and the DC shunt motor (constant speed).

DC compound motors are used when high starting torque and constant speed are required. Typical applications include punch presses, shears, bending machines, and hoists.

DC Permanent-Magnet Motors. A *DC permanent-magnet motor* is a motor that uses magnets, not a coil of wire, for the field windings. DC permanent-magnet motors have molded magnets mounted into a steel shell. The permanent magnets are the field coils. DC power is supplied only to the armature. **See Figure 6-49.**

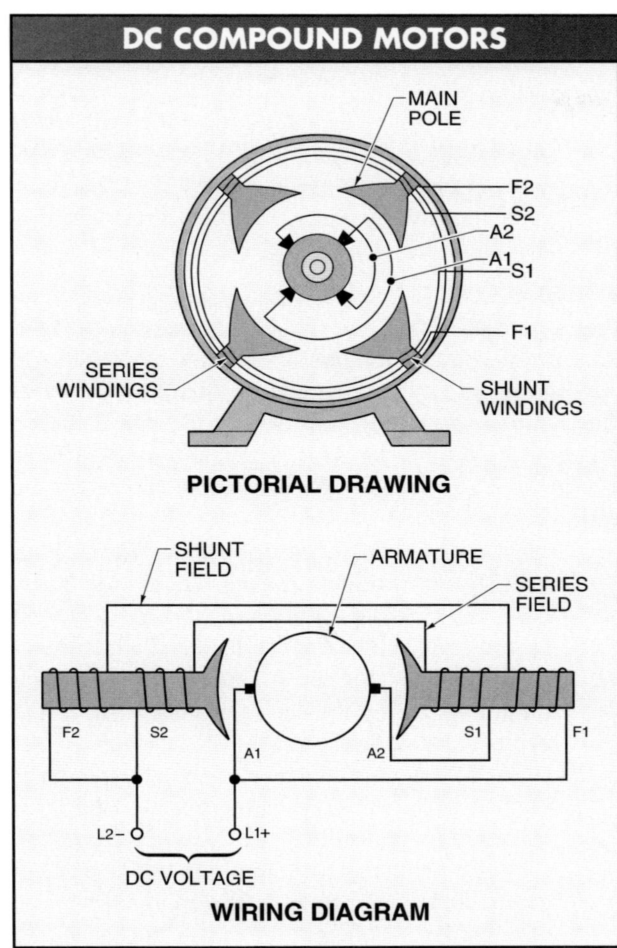

Figure 6-48. A DC compound motor is a motor with the field connected in both series and shunt with the armature.

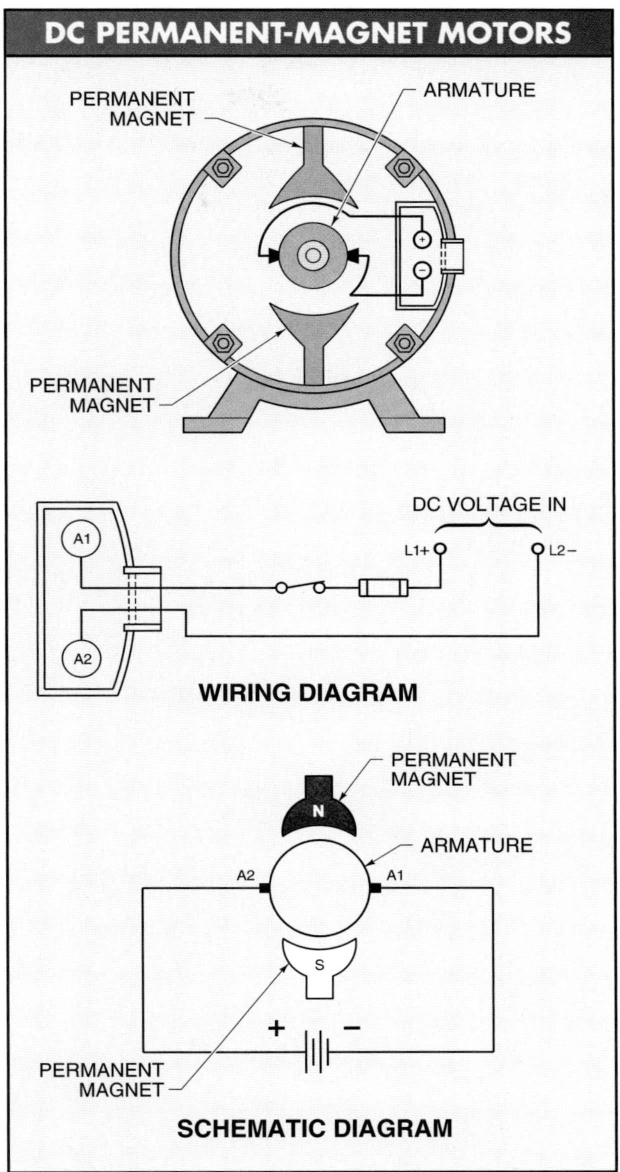

Figure 6-49. A DC permanent-magnet motor uses magnets, not a coil of wire, for the field windings.

DC permanent-magnet motors are used in automobiles to control power seats, power windows, and windshield wipers. DC permanent-magnet motors produce relatively high torque at low speeds and provide some self-braking when removed from power. Not all DC permanent-magnet motors are designed to run continuously because they overheat rapidly. Overheating destroys the permanent magnets.

Troubleshooting DC Motors

DC motors require considerable troubleshooting because of their brushes. The brushes and commutator of a DC motor are subject to wear. The brushes are designed to wear as the motor ages. Most DC motors are designed so that the brushes and the commutator can be inspected without disassembling the motor. Some motors require disassembly for close inspection of the brushes and commutator. Troubleshooting DC motors also includes troubleshooting for grounded, open, or short circuits.

Troubleshooting Brushes. Brushes wear faster than any other component of a DC motor. The brushes ride on the fast-moving commutator. Typically, bearings and lubrication are used to reduce friction when two moving surfaces touch. However, no lubrication is used between the moving brushes and the commutator because the brushes must carry current from the armature. Sparking occurs as the current passes from the commutator to the brushes. Sparking causes heat, burning, and wear of electric parts.

Replacing worn brushes is easier and less expensive than servicing or replacing a worn commutator. When troubleshooting brushes, the brushes are observed as the motor operates. **See Figure 6-50.** The brushes should ride smoothly on the commutator with little or no sparking. There should be no brush noise, such as chattering. Brush sparking, chattering, or a rough commutator indicates service is required. Brushes must be positioned correctly for proper contact with the commutator.

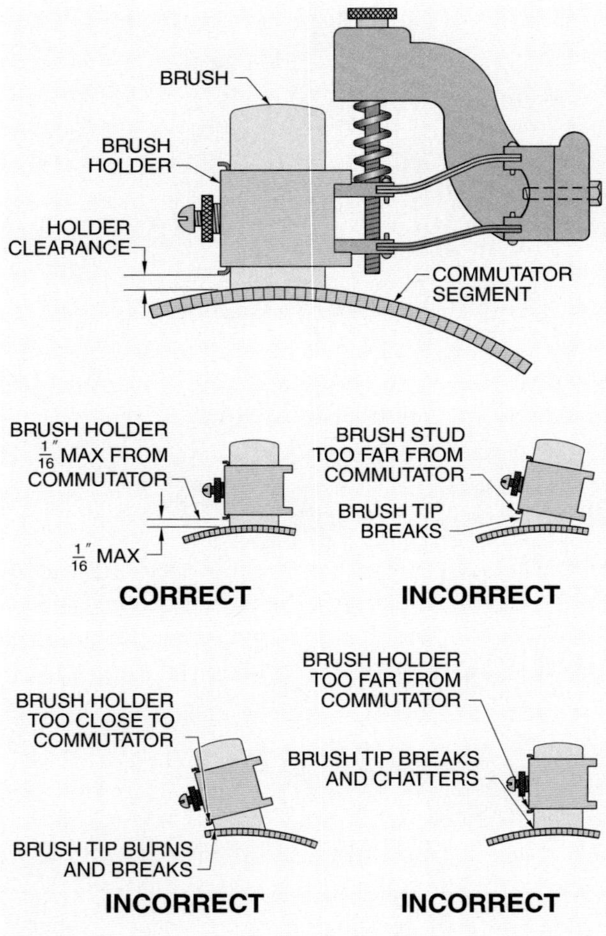

Figure 6-50. Brushes must be positioned correctly to maintain proper contact with the commutator.

To troubleshoot brushes, apply the following procedure:

1. Turn the handle of the safety switch or combination starter OFF. Lock out and tag out the starting mechanism per company policy. **See Figure 6-51.**
2. Measure the voltage at the motor terminals to ensure that the power is OFF.
3. Check the brush movement and tension. Remove the brushes. The brushes should move freely in the brush holder. The spring tension should be approximately the same on each brush.
4. Check the length of the brushes. Brushes should be replaced when they have worn down to about half of their original size. Replace all brushes if any brush is less than half its original length. Never replace only one brush. Always replace brushes with brushes of the same composition. Check manufacturer recommendations for proper brush position and brush pressure.

Troubleshooting Commutators. Brushes wear faster than the commutator. After brushes have been changed once or twice, the commutator usually needs servicing. Any markings on the commutator, such as grooves or ruts, or discolorations other than a polished, brown color where the brushes ride, indicate a problem. **See Figure 6-52.** To troubleshoot commutators, apply the following procedure:

1. Make a visual check of the commutator. The commutator should be smooth and concentric. A uniform, dark, copper oxide-carbon film should be present on the surface of the commutator. This naturally occurring film acts like a lubricant by prolonging the life of the brushes and reducing wear on the commutator surface.
2. Check the mica insulation between the commutator segments. The mica insulation separates and insulates each commutator segment. The mica insulation should be undercut (lowered below the surface) approximately 1/32″ to 1/16″, depending on the size of the motor. The larger the motor, the deeper the undercut. Replace or service the commutator if the mica is raised.

Troubleshooting for Grounded, Open, or Short Circuits. A DC motor is tested for a grounded, open, or short circuit by using a test light. A *grounded circuit* is a circuit in which current leaves its normal path and travels to the frame of the motor. A grounded circuit is caused when insulation breaks down or is damaged, allowing circuit wiring to come in contact with the metal frame of the motor.

An *open circuit* is an electrical circuit that has an incomplete path that prevents current flow. An open circuit is caused when a conductor or connection has physically moved apart from another conductor or connection.

A *short circuit* is a circuit in which current takes a shortcut around the normal path of current flow. A short circuit is caused when the insulation of two conductors fails, allowing different parts of a circuit to come in contact with one another. Short circuits are usually a result of insulation breakdown. Insulation will break down after extended periods of vibration, friction, or abrasion.

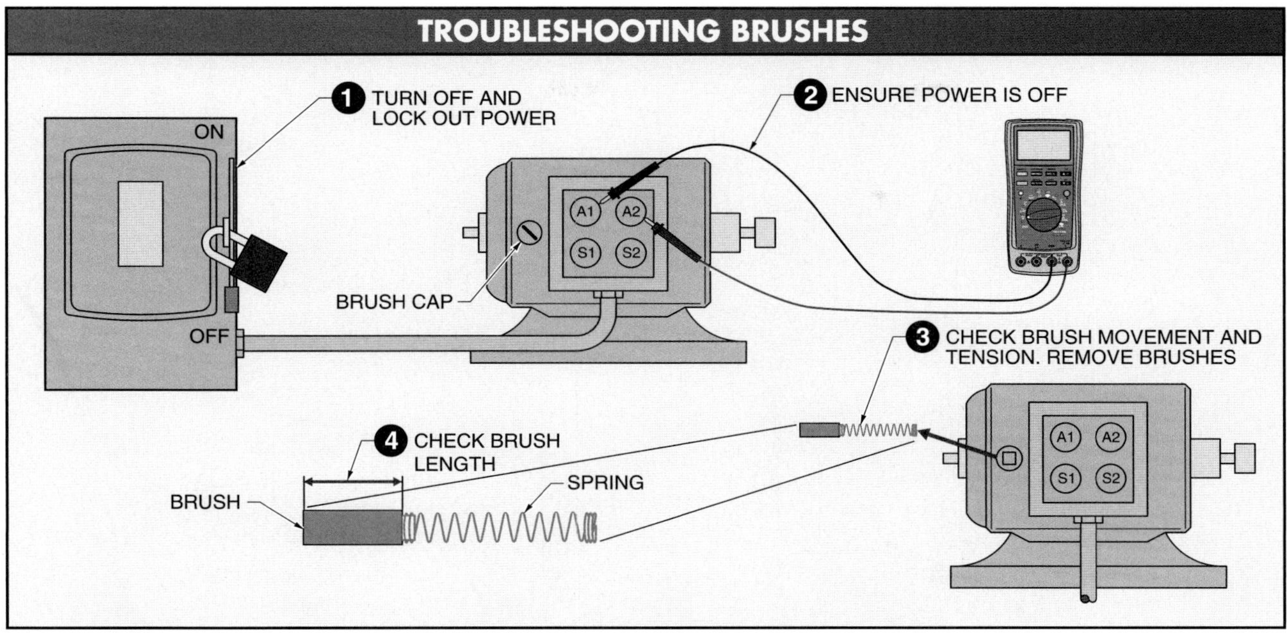

Figure 6-51. Brushes should be replaced when they have worn down to about half of their original size.

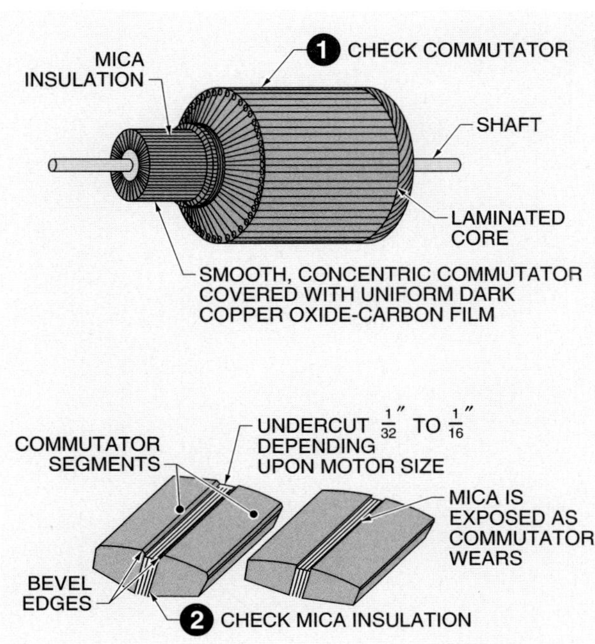

Figure 6-52. Any markings on the commutator, such as grooves or ruts, or discolorations other than a polished, brown color where the brushes ride, indicate a problem.

A continuity tester is preferred for a quick check of a motor. A continuity tester can give results quickly when there is a problem. **See Figure 6-53.** To troubleshoot for a grounded circuit, open circuit, or short circuit, apply the following procedure:

1. Check for a grounded circuit. Connect one lead of the continuity tester to the frame of the motor. Touch the other lead lead from one motor lead to the other. A grounded circuit is present if the continuity tester beeps. Service and repair the motor.
2. Check for an open circuit. Connect the two test leads to the motor field and armature circuits as follows:
   Series motors: A1 to A2 and S1 to S2. Shunt motors: A1 to A2 and F1 to F2. Compound motors: A1 to A2, F1 to F2, and S1 to S2.
   The circuits are complete if the continuity tester beeps. The circuits are open if the continuity tester does not beep. Service and repair the motor.
3. Check for a short between windings. Connect the two leads to the motor field and armature circuits as follows:
   Series motors: A1 to S1, A1 to S2, A2 to S1, and A2 to S2. Shunt motors: A1 to F1, A1 to F2, A2 to F1, and A2 to F2. Compound motors: A1 to F1, A2 to F2; A1 to F2, A2 to S1; A1 to S1, A2 to S2; A1 to S2, F1 to S1; and A2 to F1, F1 to S2.

The circuit is shorted if the continuity tester beeps. The circuit is not shorted if the continuity tester does not beep. Service and repair the motor.

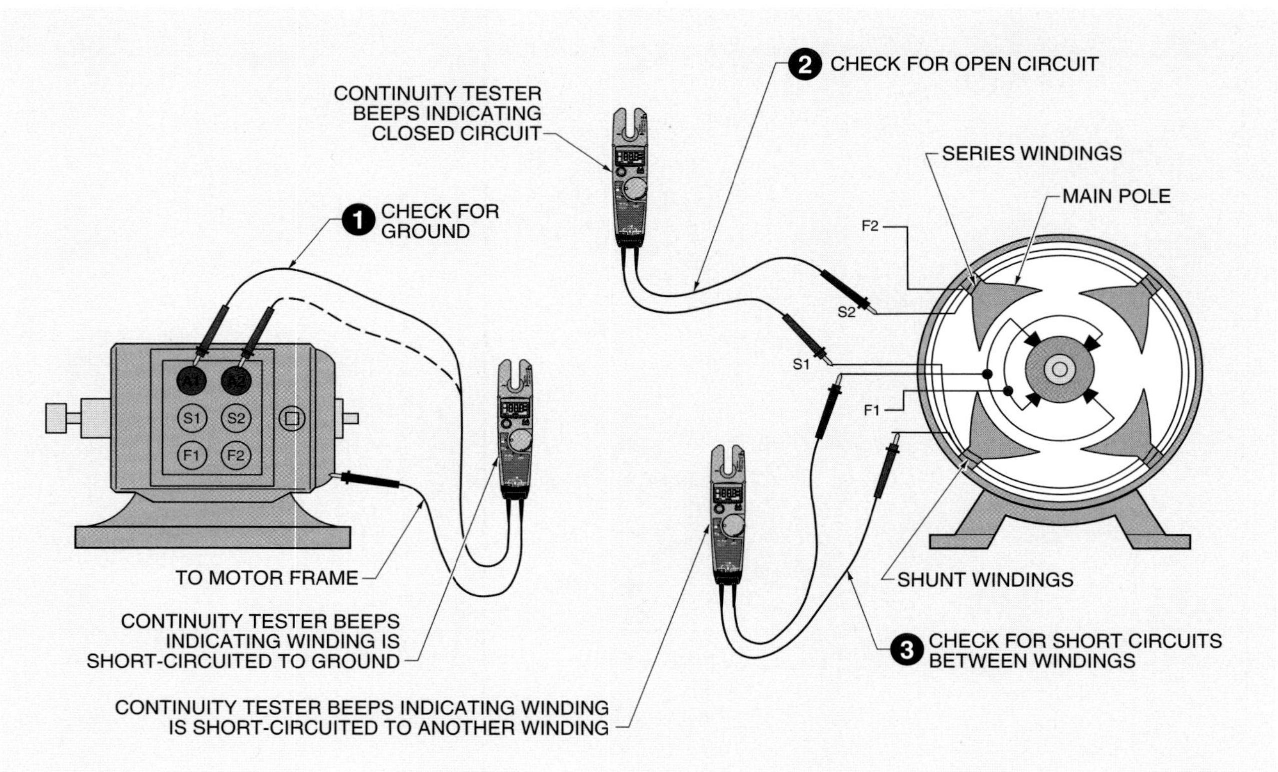

Figure 6-53. A DC motor is tested for a grounded, open, or short circuit by using a continuity tester.

Review Questions

1. What are the two main types of magnets?
2. Why is soft iron more easily magnetized than hard steel?
3. How can the strength of an electromagnet be increased?
4. What are the different types of solenoids used to transmit force?
5. Why is the armature of a solenoid made from a number of thin laminated pieces instead of a solid piece?
6. What is the purpose of having a small air gap on the armature face?
7. What is the function of the shading coil?
8. Why does a solenoid coil have a much higher inrush current than sealed current?
9. What is the effect of a higher-than-rated voltage applied to a solenoid coil?
10. What is the effect of a lower-than-rated voltage applied to a solenoid coil?
11. What are the four basic rules followed when selecting a solenoid?
12. What are the common applications of solenoids?
13. What are the two main causes of solenoid failure?
14. What is the function of a snubber circuit?
15. What are the two primary characteristics of a solenoid?
16. What is permeability?
17. Shading coils are normally made of what material?
18. Magnetic coils are normally made of what type of wire?
19. What are the principal components of a DC generator?
20. What is the function of the commutator of a DC generator?
21. What does the armature of a DC generator consist of?
22. Why do DC motors that use brushes require more repair than motors that do not use brushes?
23. What are three signs that a DC motor requires service?
24. What are signs that the commutator needs servicing?
25. When should brushes be replaced?
26. What is the difference between a grounded circuit and a short circuit?
27. Why is a continuity tester preferred for performing a quick check on a motor?

AC Generators, Transformers, and AC Motors

AC generators convert mechanical energy into AC voltage and current. Transformers are used in electrical distribution systems to increase or decrease the voltage and current safely and efficiently. AC motors use alternating current to produce rotation.

AC GENERATORS

Generators convert mechanical energy into electrical energy by means of electromagnetic induction. AC generators (alternators) convert mechanical energy into AC voltage and current. AC generators consist of field windings, an armature, slip rings, and brushes. **See Figure 7-1.**

Field windings are magnets used to produce the magnetic field in a generator. The magnetic field in a generator can be produced by permanent magnets or electromagnets. Most generators use electromagnets, which must be supplied with current. An *armature* is the movable coil of wire in a generator that rotates through the magnetic field. The armature may consist of many coils. The ends of the coils are connected to slip rings. *Slip rings* are metallic rings connected to the ends of the armature and are used to connect the induced voltage to the brushes. When the armature is rotated in the magnetic field, a voltage is generated in each half of the armature coil. A *brush* is the sliding contact that rides against the slip rings and is used to connect the armature to the external circuit.

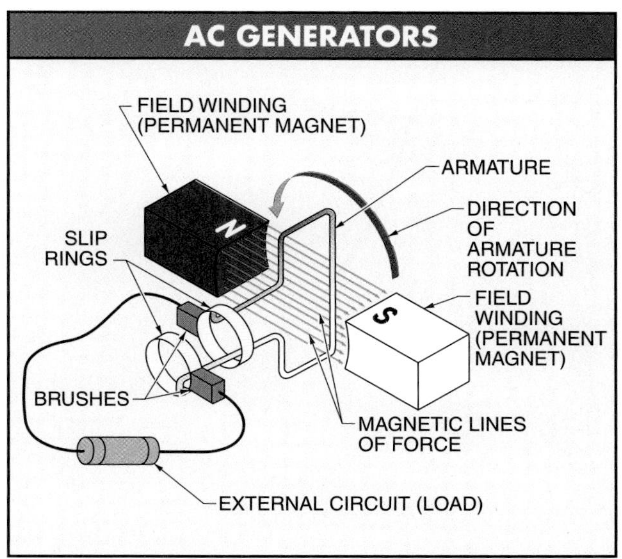

Figure 7-1. AC generators consist of field windings, a coil (armature), slip rings, and brushes.

AC generators are similar in construction and operation to DC generators. The major difference between AC and DC generators is that DC generators contain a commutator which reverses the connections to the brushes every half cycle. This maintains a constant polarity of output voltage produced by the generator. AC generators use slip rings to connect the armature to the external circuit (load). The slip rings do not reverse the polarity of the output voltage produced by the generator. The result is an alternating sine wave output.

As the armature is rotated, each half cuts across the magnetic lines of force at the same speed. Thus, the strength of the voltage induced in one side of the armature is always the same as the strength of the voltage induced in the other side of the armature.

Each half of the armature cuts the magnetic lines of force in a different direction. For example, as the armature rotates in the clockwise direction, the lower half of the coil cuts the magnetic lines of force from the bottom up to the left, while the top half of the coil cuts the magnetic lines of force from the top down to the right. The voltage induced in one side of the coil, therefore, is opposite to the voltage induced in the other side of the coil. The voltage in the lower half of the coil enables current flow in one direction, and the voltage in the upper half enables current flow in the opposite direction.

Cummins Power Generation

Large-scale generators are used in hospitals, airports, and other complexes and campuses such as schools.

Technical Fact

Most electric power is produced by alternating current (AC) generators that are driven by rotating prime movers. Most of the prime movers are steam turbines whose thermal energy comes from steam generators that use either fossil fuel or nuclear energy. Combustion turbines are often used for smaller units and in applications where gas, diesel, or fuel oil is the only available fuel.

However, since the two halves of the coil are connected in a closed loop, the voltages add to each other. The result is that the total voltage of a full rotation of the armature is twice the voltage of each coil half. This total voltage is obtained at the brushes connected to the slip rings, and may be applied to an external circuit.

Single-Phase AC Generators

Each complete rotation of the armature in a single-wire (1ϕ) AC generator produces one complete alternating current cycle. **See Figure 7-2.** In position A, before the armature begins to rotate in a clockwise direction, there is no voltage and no current in the external (load) circuit because the armature is not cutting across any magnetic lines of force (0° of rotation).

As the armature rotates from position A to position B, each half of the armature cuts across the magnetic lines of force, producing current in the external circuit. The current increases from zero to its maximum value in one direction. This changing value of current is represented by the first quarter (90° of rotation) of the sine wave.

As the armature rotates from position B to position C, current continues in the same direction. The current decreases from its maximum value to zero. This changing value of current is represented by the second quarter (91°–180° of rotation) of the sine wave.

As the armature continues to rotate to position D, each half of the coil cuts across the magnetic lines of force in the opposite direction. This changes the direction of the current. During this time, the current increases from zero to its maximum negative value. This changing value of current is shown by the third quarter (181° – 270° of rotation) of the sine wave.

As the armature continues to rotate to position E (position A), the current decreases to zero. This completes one 360° cycle of the sine wave.

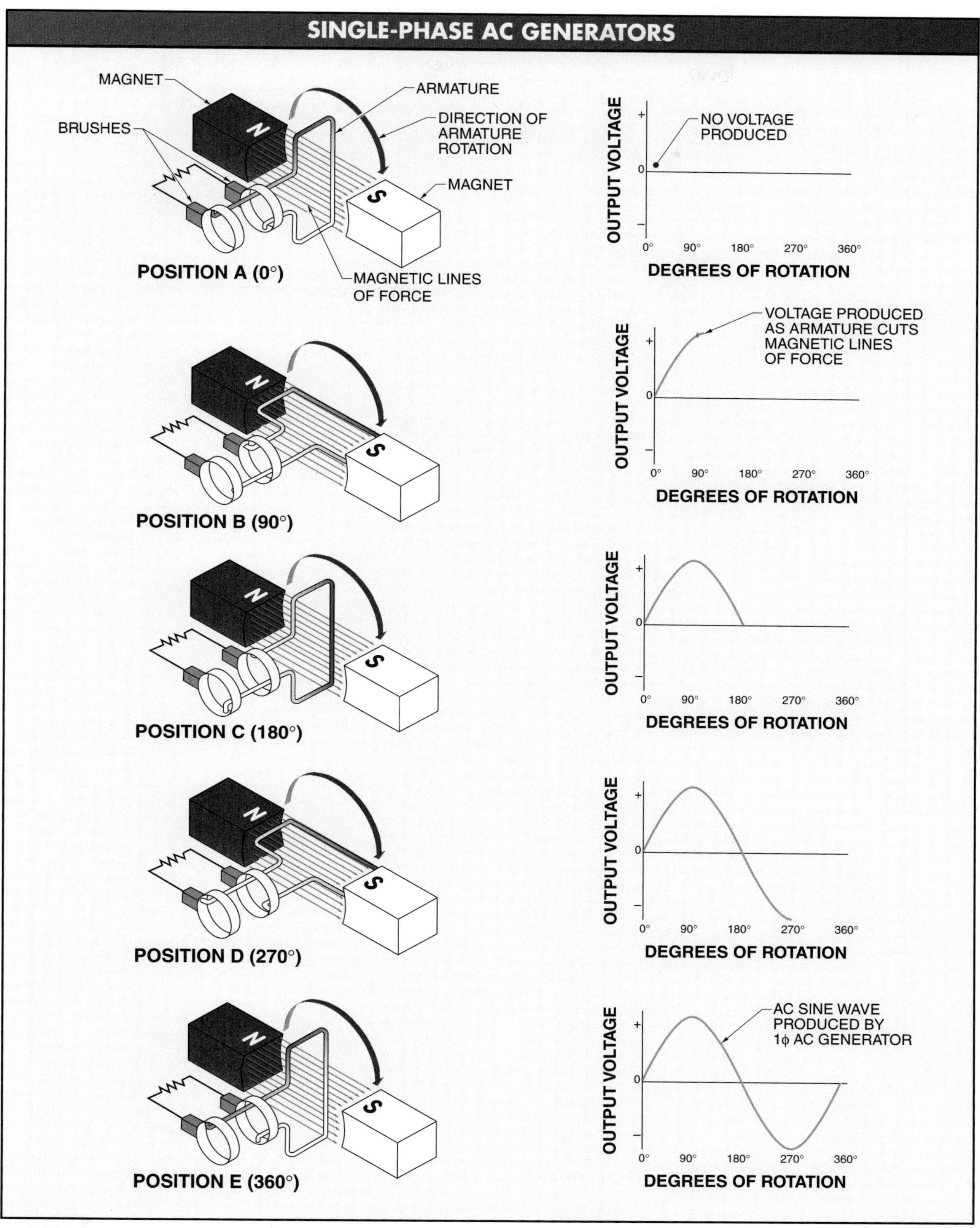

Figure 7-2. In a 1ϕ AC generator, as the armature rotates through 360°of motion, the voltage generated is a continuously changing AC sine wave.

Three-Phase AC Generators

The principles of a 3ϕ generator are the same as a 1ϕ generator except that there are three equally spaced armature windings 120° out of phase with each other. **See Figure 7-3.** The output of a 3ϕ generator results in three output voltages 120° out of phase with each other.

Delta and Wye Connections. A 3ϕ generator has six leads coming from the armature coils. When these leads are brought out from the generator, they are connected so that only three leads appear for connection to the load. Armature coils can be connected in a delta connection or a wye connection. The manner in which the leads are connected determines the electrical characteristics of the generator output. **See Figure 7-4.**

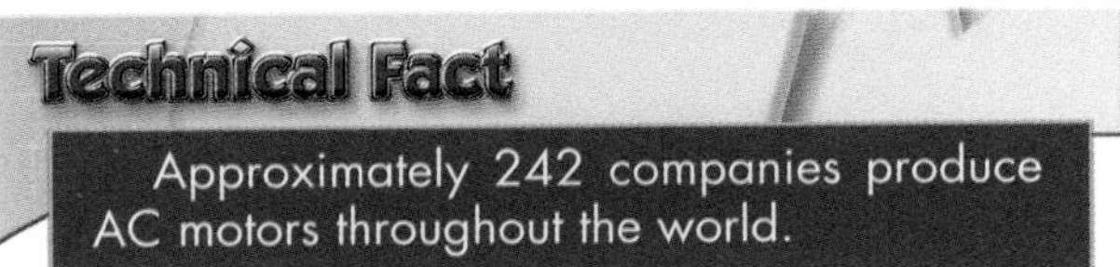

Technical Fact

Approximately 242 companies produce AC motors throughout the world.

THREE-PHASE AC GENERATORS
STAR CONNECTION
THREE LOOPS EQUALLY SPACED
MAGNET
LOAD 1
LOAD 2
DIRECTION OF ARMATURE MOVEMENT
MAGNET
THREE WIRES CONNECTED TO LOAD
BRUSHES
SLIP RINGS
N
S
OUTPUT VOLTAGE
+
0
−
0° 60° 120° 180° 240° 300° 360°
DEGREES OF ROTATION

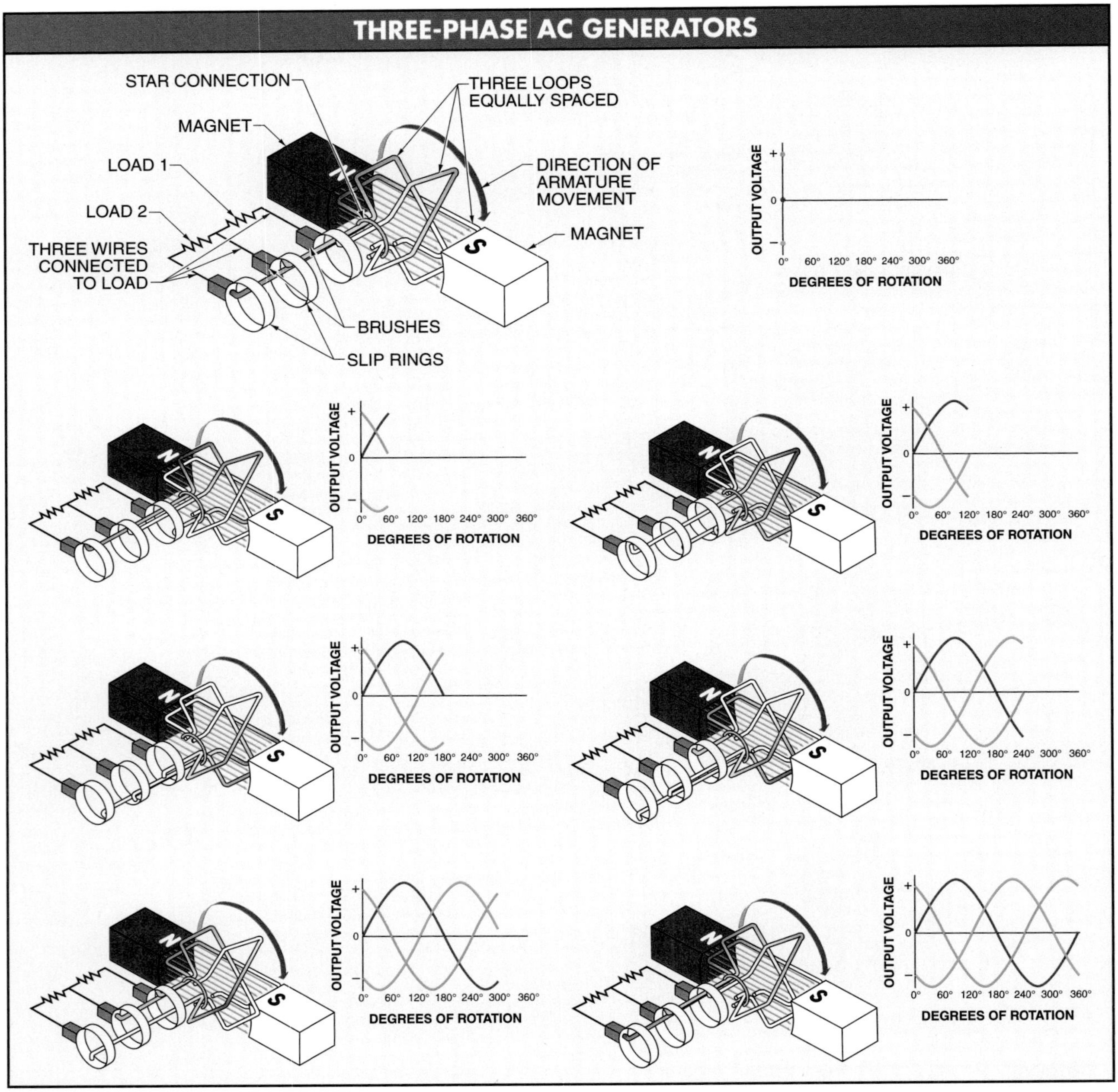

Figure 7-3. A 3ϕ generator has three equally spaced armature windings 120° out of phase with each other.

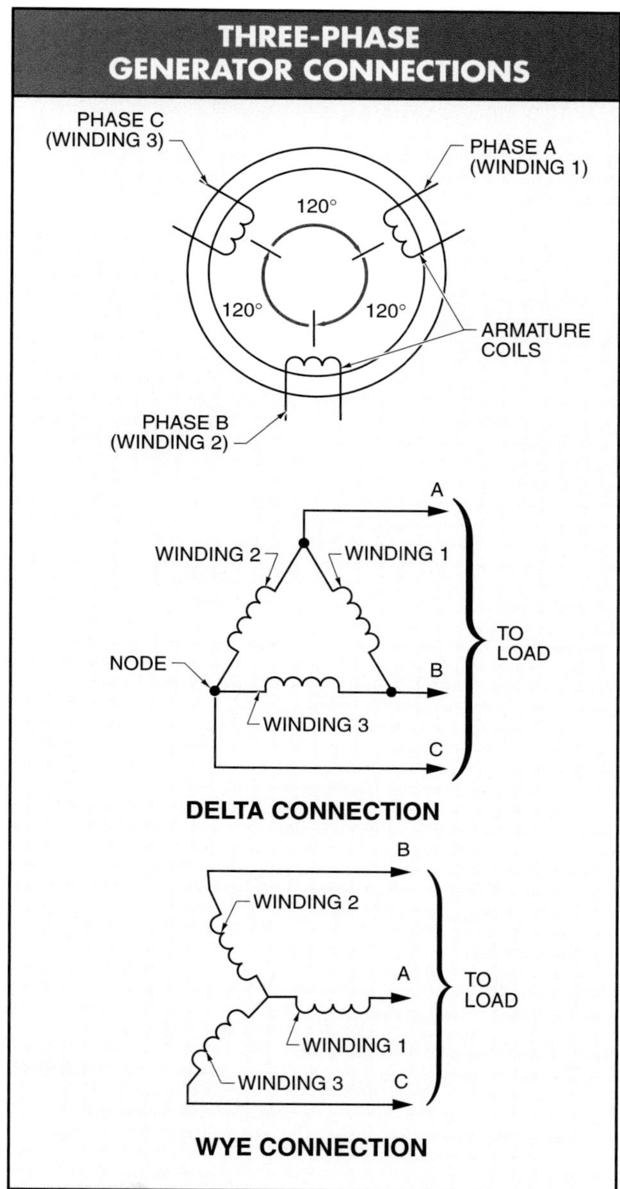

Figure 7-4. When the six leads of a 3ϕ generator are brought out, they are connected in a delta connection or a wye connection.

A *delta connection* is a connection that has each coil end connected end-to-end to form a closed loop. In a delta connection, the three windings are all connected in series and form a closed circuit. The leads are connected to the nodes where two windings are joined. A delta connection appears like the Greek letter delta (Δ).

A *wye connection* is a connection that has one end of each coil connected together and the other end of each coil left open for external connections. In a wye connection, three leads (one from each winding) are connected together while the other three leads are brought out for connecting to the load. A wye connection appears like the letter Y.

Voltage Changes

AC generators are designed to produce a rated output voltage. In addition, all electrical and electronic equipment is rated for operation at a specific voltage. The rated voltage is a voltage range that was normally ±10%. Today, however, with many components derated to save energy and operating cost, the range is normally +5% to –10%. A voltage range is used because an overvoltage is generally more damaging than an undervoltage. Equipment manufacturers, utility companies, and regulating agencies must routinely compensate for changes in system voltage.

Back-up generators are used to compensate for voltage changes. A back-up generator can be powered by a diesel, gasoline, natural gas, or propane engine connected to the generator. If there is any power interruption in the time period between the loss of main utility power and when the generator starts providing power, the generator is usually classified as a standby (emergency) power supply. Voltage changes in a system may be categorized as momentary, temporary, or sustained. **See Figure 7-5.**

Momentary Power Interruption. A *momentary power interruption* is a decrease to 0 V on one or more power lines lasting from .5 cycles up to 3 sec. All power distribution systems have momentary power interruptions during normal operation. Momentary power interruptions can be caused when lightning strikes nearby, by utility grid switching during a problem (short on one line), or during open circuit transition switching. *Open circuit transition switching* is a process in which power is momentarily disconnected when switching a circuit from one voltage supply (or level) to another.

Temporary Power Interruption. A *temporary power interruption* is a decrease to 0 V on one or more power lines lasting for more than 3 sec up to 1 min. Automatic circuit breakers and other circuit protection equipment protect all power distribution systems. Circuit protection equipment is designed to remove faults and restore power. An automatic circuit breaker normally takes from 20 cycles to about 5 sec to close. If the power is restored, the power interruption is only temporary. If power is not restored, a temporary power interruption becomes a sustained power interruption. A temporary power interruption can also be caused by a time gap between power interruptions and when a back-up power supply (generator) takes over, or if someone accidentally opens the circuit by switching the wrong circuit breaker switch.

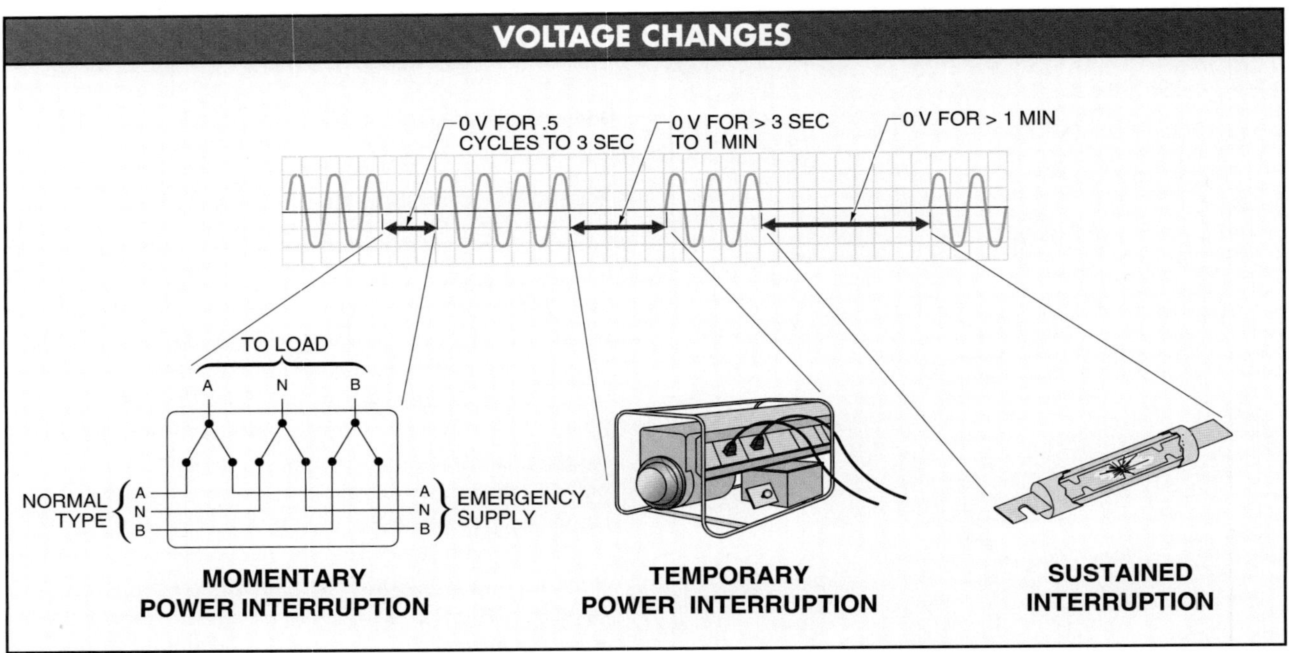

Figure 7-5. Voltage changes in an electrical system may be categorized as momentary, temporary, or sustained.

Sustained Power Interruption. A *sustained power interruption* is a decrease to 0 V on all power lines for a period of more than 1 min. All power distribution systems have a complete loss of power at some time. Sustained power interruptions (outages) are commonly the result of storms, tripped circuit breakers, blown fuses, and/or damaged equipment.

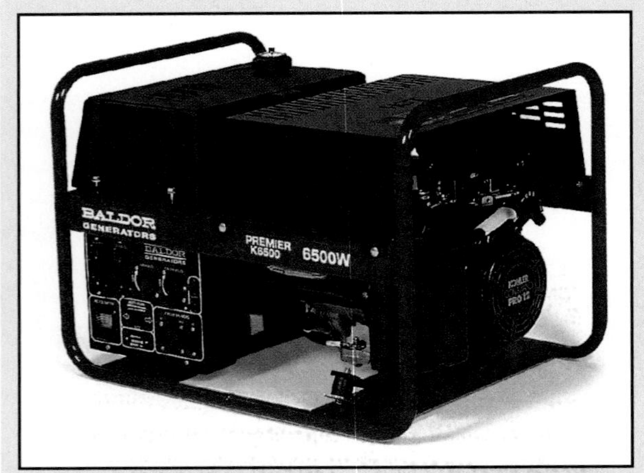

Baldor Electric Co.

Portable generators of various sizes are used to provide power during temporary power interruptions.

The effect of a power interruption on a load depends on the load and the application. If a power interruption could cause equipment, production, and/or security problems that are not acceptable, an uninterruptible power system can be used. An *uninterruptible power system (UPS)* is a power supply that provides constant on-line power when the primary power supply is interrupted. For long-term power interruption protection, a generator/UPS is used. For short-term power interruptions, a static UPS is used.

Transients

A *transient* is a temporary, unwanted voltage in an electrical circuit. Transient voltages are normally erratic, large voltages or spikes that have a short duration and a short rise time. Computers, electronic circuits, and specialized electrical equipment require protection against transient voltages. Protection methods commonly include proper wiring, grounding, shielding of the power lines, and use of surge suppressors. A *surge suppressor* is an electrical device that provides protection from high-level transients by limiting the level of voltage allowed downstream from the surge suppressor. Surge suppressors can be installed at service entrance panels and at individual loads. **See Figure 7-6.**

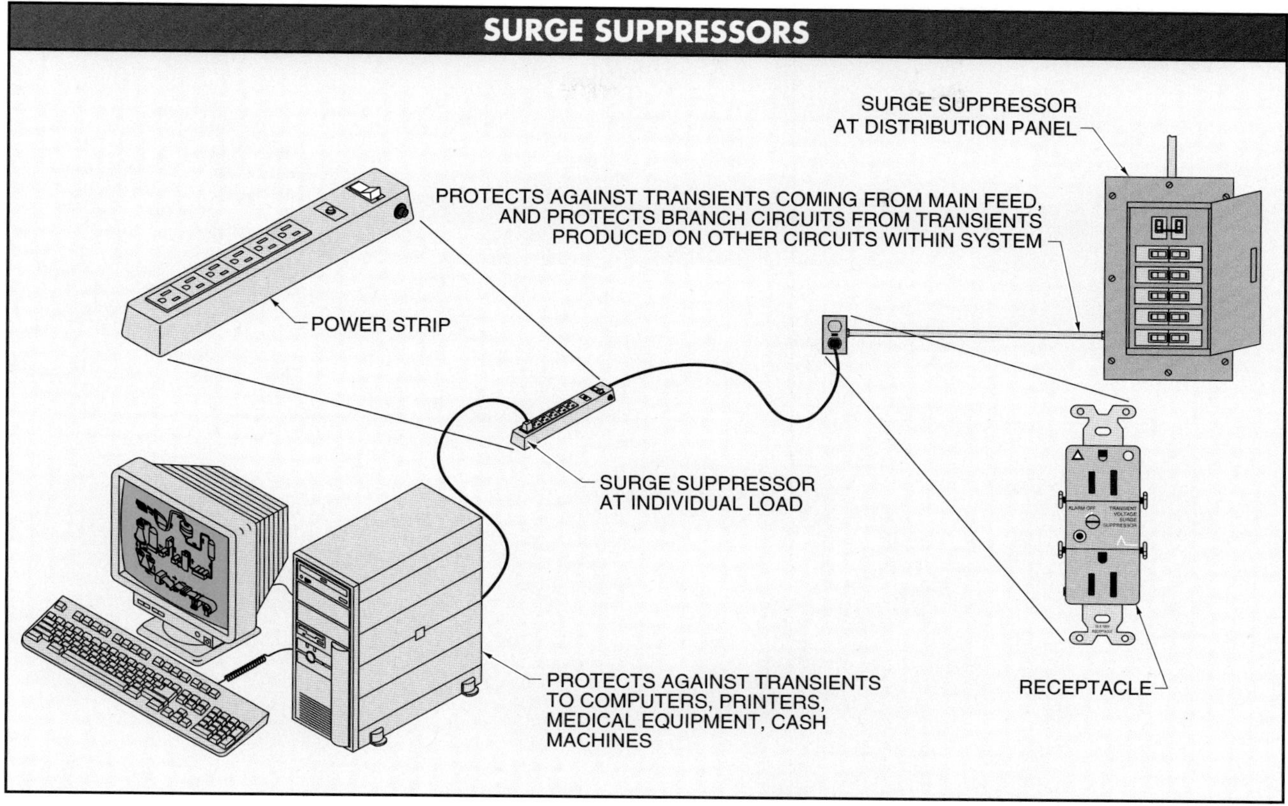

Figure 7-6. A surge suppressor is an electrical device that provides protection from high-level transients by limiting the level of voltage allowed downstream from the surge suppressor.

TRANSFORMERS

A *transformer* is an electric device that uses electromagnetism to change voltage from one level to another or to isolate one voltage from another. Transformers are used in electrical distribution systems to increase or decrease the voltage and current safely and efficiently. For example, transformers are used to increase generated voltage to a high level for transmission across the country and then decrease it to a low level for use by electrical loads. **See Figure 7-7.**

Transformers allow power companies to distribute large amounts of power at a reasonable cost. Large transformers are used for power distribution along city streets and in large manufacturing or commercial buildings. Large transformers are normally maintained by the power company or by workers who have been specifically trained in high-voltage transformer operation and maintenance.

Technicians often work with small control transformers. Control transformers isolate the power circuit from the control circuit, providing additional safety for the circuit operator. Transformers are also used in the power supplies of most electronic equipment to step the power line voltage up or down to provide the required operating voltage for the equipment.

A transformer has a primary winding and a secondary winding wound around an iron core. **See Figure 7-8.** The *primary winding* is the coil of a transformer that draws power from the source. The *secondary winding* is the coil of a transformer that delivers the energy at the transformed or changed voltage to the load.

Technical Fact

In 1819, it was discovered that when an electric current flows in a copper wire (conductor), a magnetic field exists in the space around the conductor. A fixed relationship exists between the direction of the current in a conductor and the magnitude and direction of the resulting magnetic field.

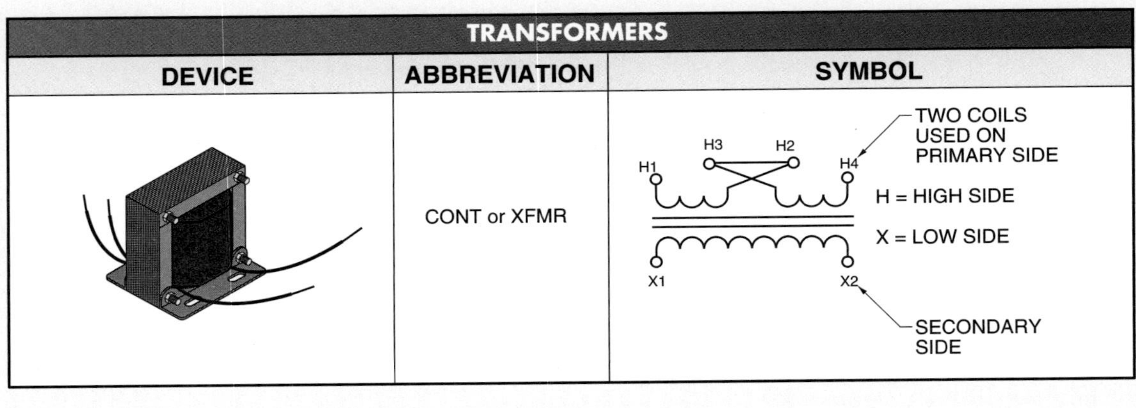

TRANSFORMERS		
DEVICE	ABBREVIATION	SYMBOL
	CONT or XFMR	

ABB Power T&D Company Inc.

LARGE TRANSFORMER

General Electric Company

SMALL TRANSFORMER

Figure 7-7. Transformers are used to increase voltage to a high level for transmission across the country and then decrease it to a low level for use by electrical loads.

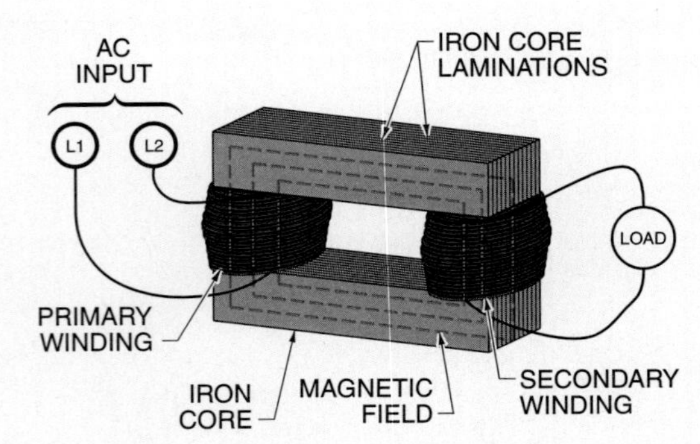

Figure 7-8. A transformer has a primary winding and a secondary winding wound around an iron core.

Transformer Operation

A transformer transfers AC energy from one circuit to another. The energy transfer is made magnetically through the iron core. A magnetic field builds up around a wire when AC is passed through the wire. The magnetic field builds up and collapses each half cycle because the wire is carrying AC. **See Figure 7-9.**

The primary coil of the transformer supplies the magnetic field for the iron core. The secondary coil supplies the load with an induced voltage proportional to the number of turns of a conductor cut by the magnetic field of the core. A transformer is either a step-up or step-down transformer depending on the ratio between the number of turns of the conductor in the primary and secondary sides of the transformer. **See Figure 7-10.**

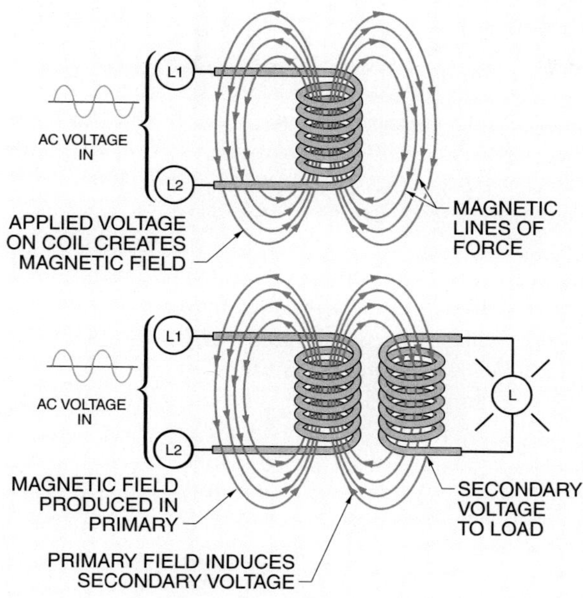

Figure 7-9. In a transformer, magnetic lines of force created by one coil induce a voltage in a second coil.

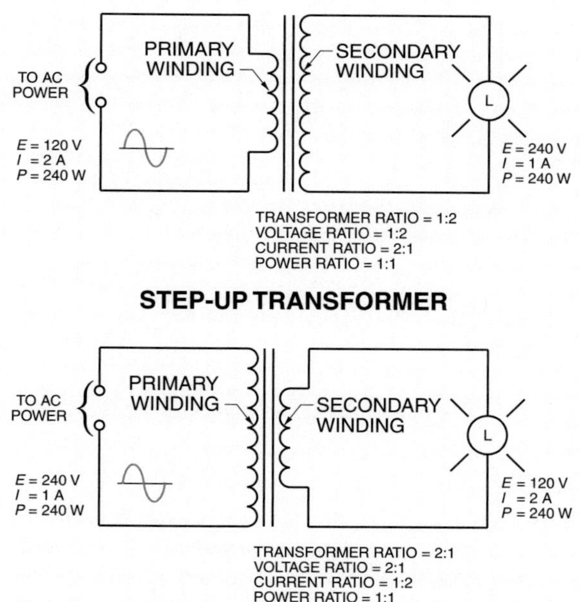

Figure 7-10. Voltage and current change from the primary to secondary winding in step-up and step-down transformers.

If twice as many turns are on the secondary, twice the voltage is induced on the secondary. The ratio of primary to secondary is 1:2, making the transformer a step-up transformer. If only half as many turns are on the secondary, only half the voltage is induced on the secondary. The ratio of primary to secondary is 2:1, making the transformer a step-down transformer.

In a step-up transformer, a ratio of 1:2 doubles the voltage. This may seem like a gain or a multiplication of voltage without any sacrifice. However, the amount of power transferred in a transformer is equal on both the primary and the secondary, excluding small losses within the transformer.

Because power is equal to voltage times current ($P = E \times I$) and power is always equal on both sides of a transformer, the voltage cannot change without changing the current. For example, when voltage is stepped down from 240 V to 120 V in a 2:1 ratio, the current increases from 1 A to 2 A, keeping the power equal on each side of the transformer. By contrast, when the voltage is stepped up from 120 V to 240 V in a 1:2 ratio, the current is reduced from 2 A to 1 A to maintain the power balance. In other words, voltage and current may be changed for particular reasons, but power is constant.

One advantage of increasing voltage and reducing current is that power may be transmitted through smaller gauge wire, thus reducing the cost of power lines. For this reason, the generated voltages are stepped up very high for distribution across large distances, and then stepped back down to meet consumer needs. Although both the voltage and current can be stepped up or down, the terms step up and step down, when used with transformers, always apply to voltage.

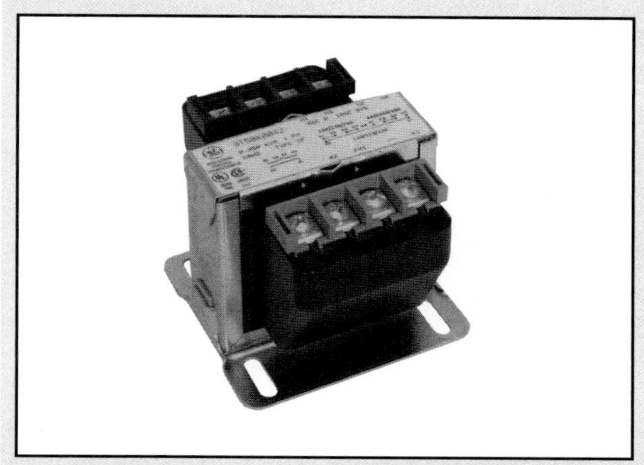

General Electric Company

Type IP core and coil transformers are designed for panelboard, industrial control, machine tool, and general-purpose applications.

Extra care must be taken when working around high-voltage transformers and power lines because of the exposed electrical terminals.

Transformer Losses

Although transformers are very efficient, they are not perfect. Not all of the energy delivered to the primary side by the source is transferred to the secondary load circuit. The majority of the energy lost is lost as heat in the transformer. The three types of losses in an iron core transformer are resistive, eddy current, and hysteresis losses.

Resistive Loss. Resistive loss comes from the resistance of the coil winding. When current passes through a winding, the winding heats up and loses energy that could have been transferred to the secondary.

Eddy Current Loss. Because iron is a fair conductor of electricity, the varying magnetic field which induces a voltage in the secondary of a transformer also induces small voltages in the iron core of the transformer. These small voltages produce eddy currents, which in turn produce heat. This heat also represents a loss because it does no useful work.

Eddy currents are minimized either by making the core out of thin sheets (laminations) which are insulated from each other, or by using powdered-iron cores instead of solid blocks of iron. The insulation between the laminations of a laminated core breaks up the current paths within the core and reduces the eddy currents. This is the same technique used to reduce eddy currents in solenoids.

Hysteresis Loss. Each time the magnetizing force produced by the primary side of a transformer changes, the atoms of the core realign themselves in the direction of the force. The energy required to realign the iron atoms must be supplied by the input power and is not transferred to the secondary load circuit. The realignment of the iron atoms does not follow the magnetizing force instantaneously, but instead lags slightly behind it. This lagging action is called hysteresis. The degree of hysteresis is a measure of the amount of energy required to realign the iron atoms in the core, and this energy loss is called hysteresis loss.

Hysteresis results in heating of the iron core. Because of this and other similarities to mechanical friction, hysteresis is sometimes referred to as magnetic friction. Hysteresis losses are minimized by using high-silicon steel and other alloys in the core.

All of these losses make the typical iron core transformer hot when operating under full load. A transformer may be too hot to touch during normal operation, but there should be no odor of burning insulation or varnish, or signs of discoloration or smoke. Any one of these indicates to the technician that the transformer is overloaded or defective.

Single-Phase Transformer Connections

Electricity is used in residential applications (one-family, two-family, and multifamily dwellings) to provide energy for lighting, heating, cooling, cooking, etc. The electrical service to dwellings is normally 1φ, 120/240 V. The low voltage (120 V) is used for general-purpose receptacles and general lighting. The high voltage (240 V) is used for heating, cooling, cooking, etc.

Residential electrical service may be overhead or lateral. *Overhead service* is electrical service in which service-entrance conductors are run from the utility pole through the air and to the dwelling. *Lateral service* is electrical service in which service-entrance conductors are run underground from the utility service to the dwelling. **See Figure 7-11.**

Three-Phase Transformer Connections

Three 1φ transformers are connected to develop 3φ voltage. The three transformers may be connected in a wye or delta connection. In a wye connection, the end of each coil is connected to the incoming power lines (primary side) or used to supply power to the load or loads (secondary side). Each connecting point in a delta connection is attached to the incoming power lines or used to supply power to the load or loads. The voltage output and type available for the load or loads is determined by whether the transformer is connected in a wye or delta connection.

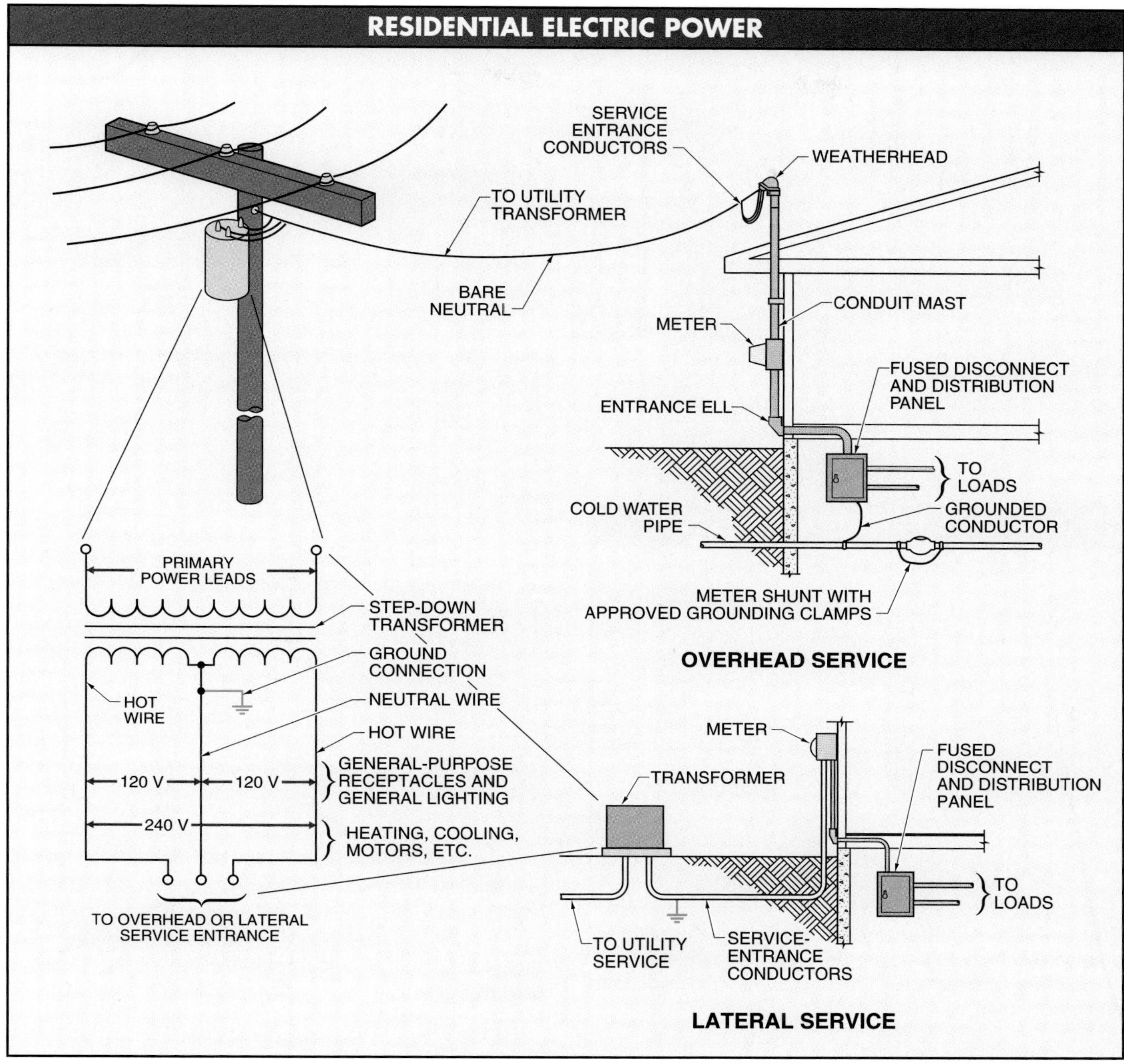

Figure 7-11. Overhead or lateral service may be used to supply power to a residential building.

Transformer Secondary Tap

Many transformers have a secondary coil that has an extra lead (tap) attached to it. A *tap* is a connection brought out of a winding at a point between its endpoints to allow changing the voltage or current ratio. Taps allow different output voltages to be obtained from a transformer. **See Figure 7-12.** For example, the output voltage between leads 1 and 2 is 120 VAC because the turns ratio is 1:1 (100 to 100). The output between the tap and lead 1 is 24 VAC because the turns ratio is approximately 4.17:1 (100 to 24).

A tap that splits a secondary in half is referred to as a center tap. A common application of a transformer with a center tap is a distribution transformer. A distribution transformer is used in residences and businesses to change the high voltage of power company distribution lines to the common 240/120 VAC supply of residences and businesses. **See Figure 7-13.**

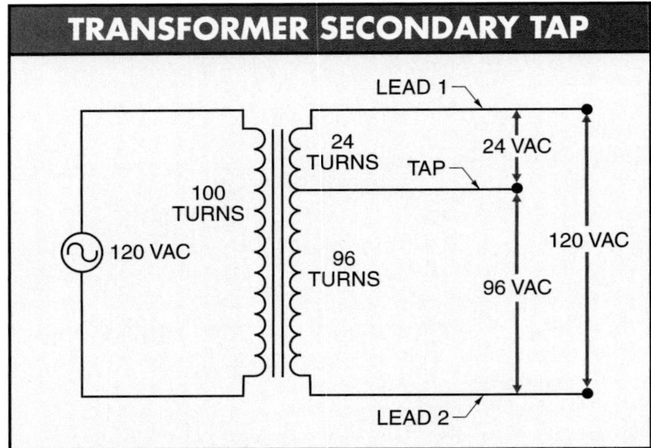

Figure 7-12. Taps allow different output voltages to be obtained from a transformer.

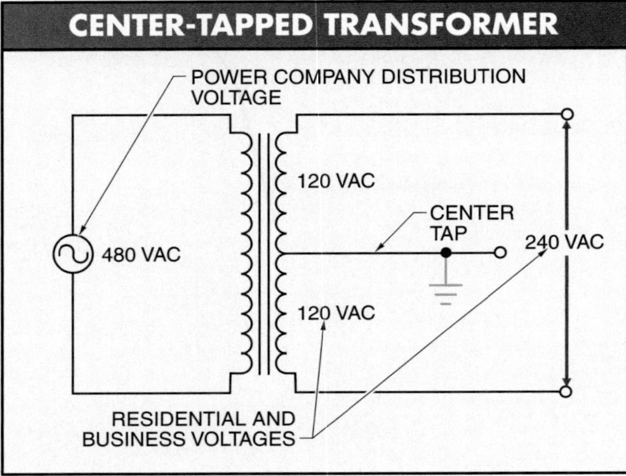

Figure 7-13. A center-tapped transformer is used to change the high voltage of power company distribution lines to the common 240/120 VAC supply of residences and businesses.

The center tap is connected to earth ground and becomes a common conductor. The voltage across the output lines is 240 VAC. However, the voltage measured between either output line and the center tap is 120 VAC.

This circuit is a typical circuit used by the power company to deliver power to a residence. The 240 VAC power is used to supply devices in the residence that require a large amount of operating power, such as a central air conditioner, water heater, clothes dryer, and cooking range. These high-power devices run on 240 VAC to allow smaller conductor wires to deliver power to them. The 120 VAC power is wired to the electrical outlets and lighting system. This provides a much safer level of voltage, which can be used on smaller electrical devices.

Control Transformers

A *control transformer* is a transformer that is used to step down the voltage to the control circuit of a system or machine. The most common control transformers have two primary coils and one secondary coil. **See Figure 7-14.**

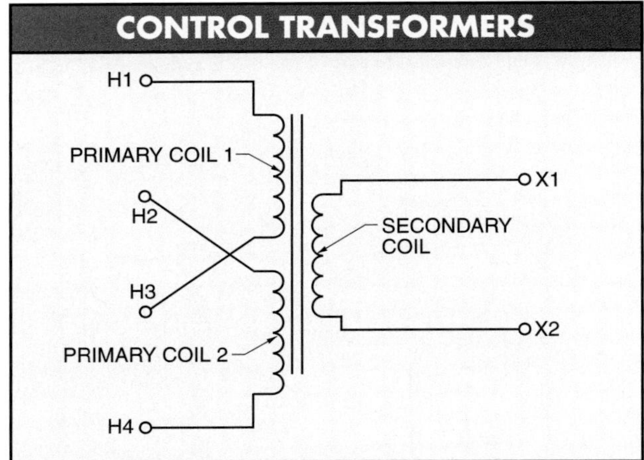

Figure 7-14. The most common control transformers have two primary coils and one secondary coil.

The primary coils of a control transformer are crossed so that metal links can be used to connect the primaries for either 240 VAC or 480 VAC operation. In most applications, a control transformer is used to reduce the main or line voltage of 240 VAC or 480 VAC to a control voltage of 120 VAC.

Technical Fact

Control and signal transformers are self-air-cooled, constant-potential transformers which are used to step down the voltage to supply signal circuits or control circuits of electrically operated switches. Small control transformers operating at 120/24 V are commonplace in small commercial buildings, supplying control voltages for heating and air conditioning units and other large electrical appliances.

240 V Primary. To obtain a control voltage of 120 VAC from a line voltage of 240 VAC, the two primary coils must be connected in parallel. **See Figure 7-15.** If the primary coils are connected in parallel, the effective turns of the two primary coils is 200, the same as if there were only one primary coil. If the secondary has 100 turns, the turns ratio is 2:1. This means an input voltage of 240 VAC produces an output voltage of 120 VAC.

CONTROL TRANSFORMERS-
240 V PRIMARY

Figure 7-15. To obtain a control voltage of 120 VAC from a line voltage of 240 VAC, the two primary coils must be connected in parallel.

480 V Primary. To obtain a control voltage of 120 VAC from a line voltage of 480 VAC, the two primary coils must be connected in series. **See Figure 7-16.** If the primary coils are connected in series, the effective turns of the two primary coils is 400, making the turns ratio 4:1. This means an input voltage of 480 VAC produces an output voltage of 120 VAC.

Transformer Selection

A transformer that is selected for an application must have a higher VA rating than is needed for the application. However, a transformer that has a VA rating that is excessively higher than the required rating should not be selected because the transformer is less efficient in that application. A transformer should be selected that has a rating that is above but close to the required value. The information needed to size a transformer includes input voltage available, output voltage desired, and output current required (both inrush and steady-state). With this information, a catalog specification sheet can be used to select the proper transformer. **See Figure 7-17.**

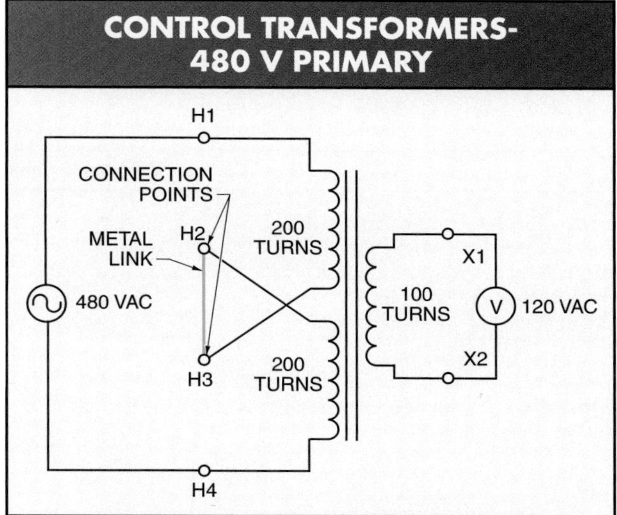

Figure 7-16. To obtain a control voltage of 120 VAC from a line voltage of 480 VAC, the two primary coils must be connected in series.

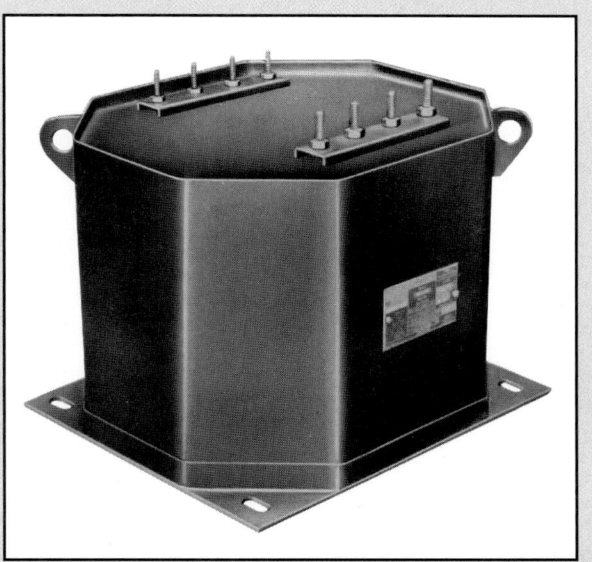

ABB Power T&D Company Inc.

Control transformers are used in small commercial buildings and residential complexes to supply control voltage for HVAC units.

TRANSFORMER ELECTRICAL SPECIFICATIONS AND ORDERING DATA (SUPPLY VOLTAGE 220 VAC)						
VA*	Maximum Inrush VA†	Temperature Rise‡	Dimensions§			Model J201
			H	L	W	
110 V To 120 V Secondary Voltage Rating						
50	180	55	3-5/16	3-3/8	2-1/2	1111
75	218	55	3-9/16	3-3/8	2-7/8	1121
100	273	55	3-3/4	3-3/8	2-7/8	1131
150	660	55	4-5/16	4-1/2	3-13/16	1141
250	1360	55	5	4-1/2	3-13/16	1161
500	1964	115	5-1/2	4-1/2	3-3/4	1191
1000	4014	115	6-3/4	5-1/4	4-3/8	1211
22 V To 24 V Secondary Voltage Rating						
50	180	55	3-5/16	3	2-1/2	1111-824
100	273	55	3-3/4	3-3/8	2-7/8	1131-824
150	660	55	4-5/16	4-1/2	3-13/16	1141-824

* Terminal type
† Capability VA. Refers to maximum inrush VA after calculations are made
‡ 0° C
§ in in.

Figure 7-17. Transformer specification sheets are used to obtain required information when selecting the proper transformer for an application.

The most important guideline used when sizing a transformer is to select a transformer that safely and efficiently provides the maximum current that can be drawn by a load. A common example of sizing a transformer occurs in selecting a transformer to operate a machine. Many machines require a transformer to step down the line voltage (480 VAC or 240 VAC) to the operating voltage of 120 VAC. Machines with motors or other high inrush devices draw their maximum current when the devices are first started. For these machines, this inrush current is the critical value that must be considered when selecting a transformer.

General Electric Company

Autotransformers are used in starting rotating machinery such as synchronous and induction motors.

If a machine does not have devices with high inrush characteristics, inrush current is not as much of a consideration. In this case, the steady-state current is more important. Most machines list both their maximum inrush and steady-state current requirements.

Transformer specification sheets normally list a steady-state volt/amperage (VA). The steady-state volt/amperage is the secondary voltage multiplied by the secondary current ($V_S \times I_S$) load or loads during steady-state current. Another listed value is the maximum inrush VA. This is the secondary voltage multiplied by the secondary current during the inrush period.

Troubleshooting Transformers

After a transformer is installed in a circuit, it may operate without failure for a long time. One reason for this is that transformers have no moving parts. If a transformer does fail, it appears as either a short circuit or an open circuit in one of the coils. The two methods that can be used to determine if a transformer has failed are to measure the input and output voltages and to check the transformer resistance.

Measuring Input and Output Voltages. If a transformer is connected in a circuit, the transformer can be tested by measuring the input and output voltages. The transformer is good if the input and output voltages are reasonably close to the theoretical values. The current levels are tested if the voltage does not stay constant. Although the initial voltage may appear normal, it may not hold up when the transformer is fully loaded.

Checking Transformer Resistance. A DMM set to measure resistance can be used to check for open circuits in coils, short circuits between coils, or coils shorted to the core without power applied to the transformer. **See Figure 7-18.**

- Open circuits in coils. The resistance of each coil is checked with a DMM. The winding is open and the transformer is bad if any of the coils show an infinite resistance reading. Note that very low resistance readings do not indicate a short, just the resistance of the wire.
- Short circuits between primary and secondary coils. A check for short circuits should be made between the primary and secondary coils of the transformer. A DMM should show an infinite resistance reading between the primary and secondary coils.
- Coils shorted to core. A resistance check is made from each transformer coil to the core of the transformer. All coils should show an infinite resistance reading. The transformer should not be used if a resistance is shown between any coil and the core.

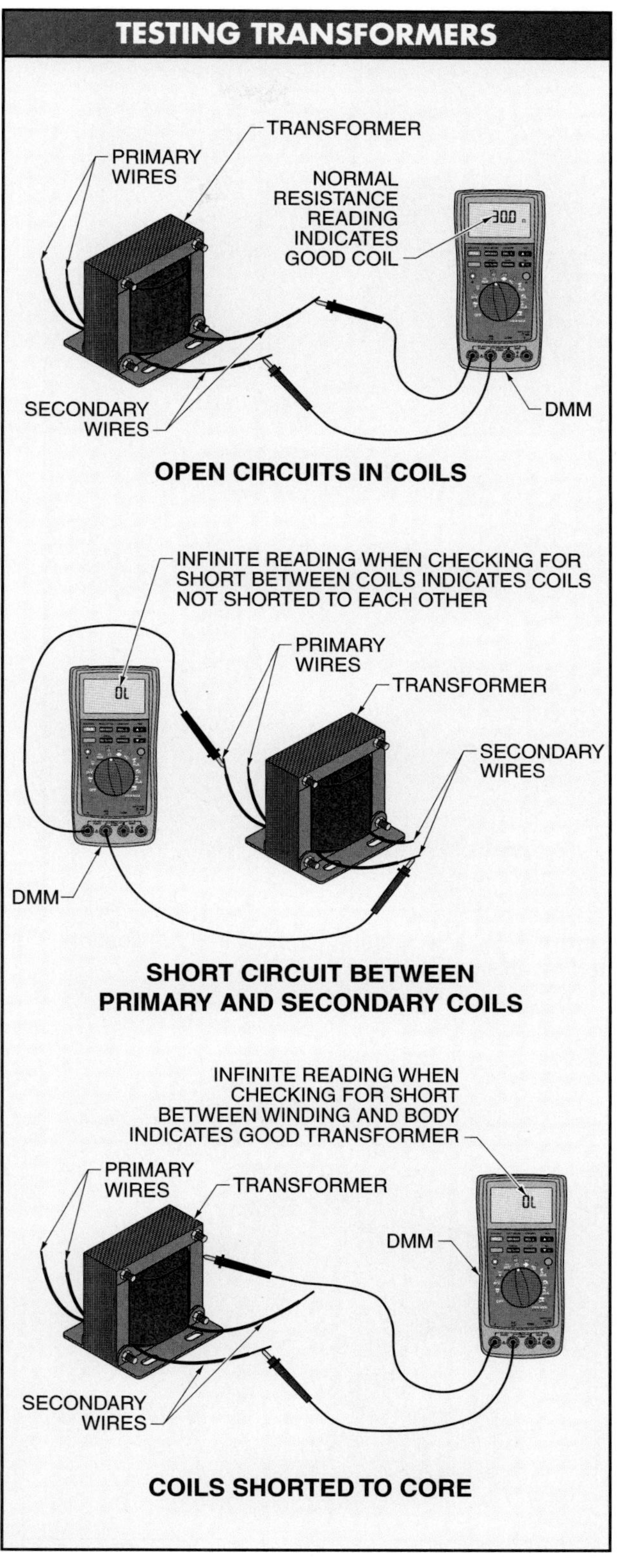

Figure 7-18. Transformers are tested by checking for open circuits in the coils, short circuits between the primary and secondary coils, and coils shorted to the core.

AC MOTORS

An *alternating current (AC) motor* is a motor that uses alternating current to produce rotation. AC motors have several advantages over DC motors. One advantage is that AC motors have only two bearings that can wear. Secondly, there are no brushes to wear because the motor does not have a commutator. For these reasons, maintenance is minimal. Also, no sparks are generated to create a hazard in the presence of flammable materials. The main parts of an AC motor are the rotor and stator. A *rotor* is the rotating part of an AC motor. A *stator* is the stationary part of an AC motor. **See Figure 7-19.** AC motors are 1ϕ or 3ϕ.

Technical Fact

Motors perform work by converting electrical energy to mechanical energy. Depending upon the motor size and design, motors typically convert between 75% and 95% of their electrical power to usable mechanical energy. The mechanical energy is used to produce work. The balance of the electrical power is lost. Lost power adds to the cost of the work produced.

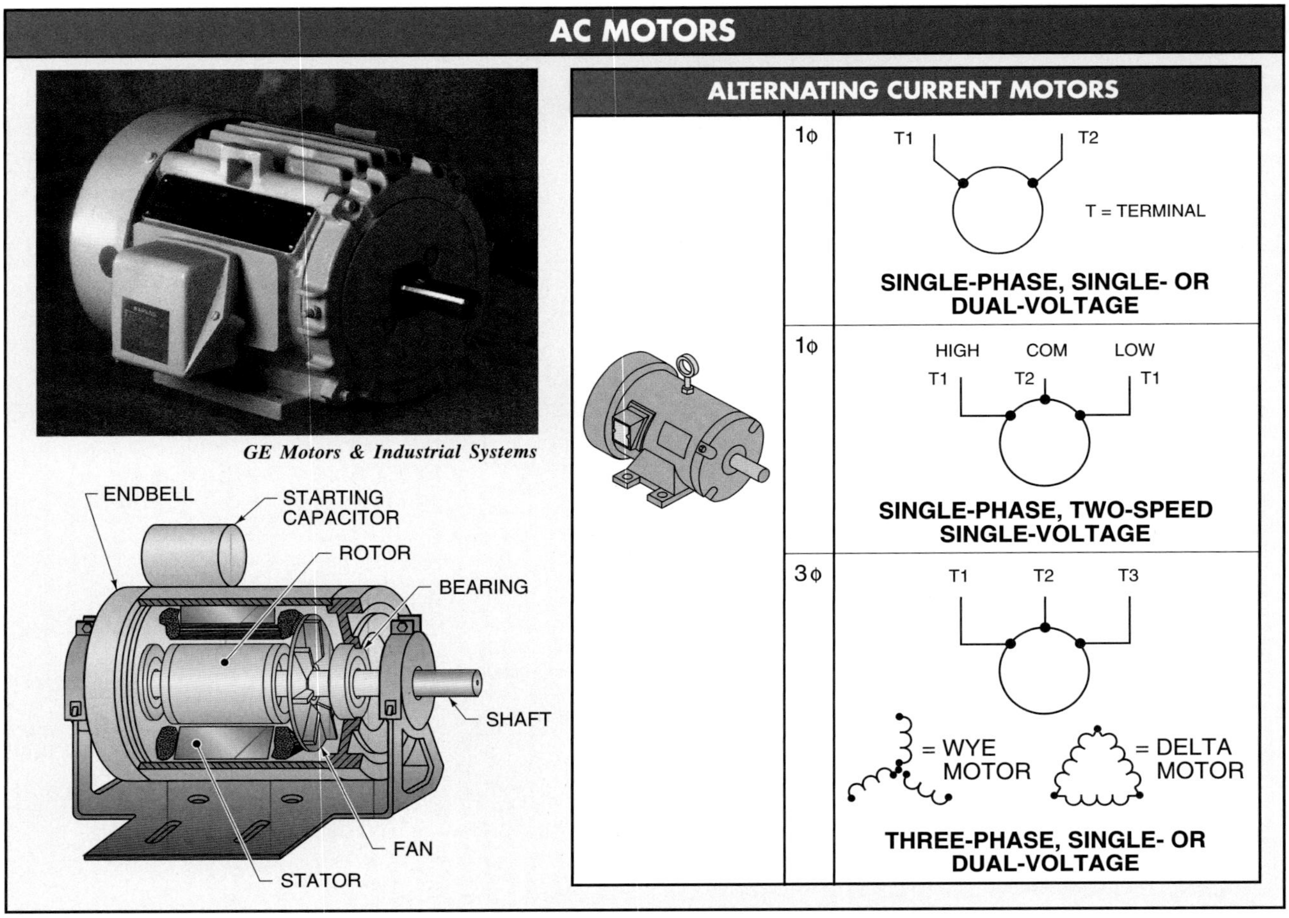

AC MOTORS

GE Motors & Industrial Systems

ENDBELL
STARTING CAPACITOR
ROTOR
BEARING
SHAFT
FAN
STATOR

ALTERNATING CURRENT MOTORS

	Diagram
1φ	T1 T2 T = TERMINAL **SINGLE-PHASE, SINGLE- OR DUAL-VOLTAGE**
1φ	HIGH COM LOW T1 T2 T1 **SINGLE-PHASE, TWO-SPEED SINGLE-VOLTAGE**
3φ	T1 T2 T3 = WYE MOTOR = DELTA MOTOR **THREE-PHASE, SINGLE- OR DUAL-VOLTAGE**

Figure 7-19. The majority of industrial applications normally use AC motors because of their simplicity, ruggedness, and reliability.

Baldor Electric Co.

AC motors with a capacitor start are commonly used in industrial environments.

Single-Phase Motors

Single-phase motors are used in residential applications for AC motor-driven appliances such as furnaces, air conditioners, washing machines, etc. Single-phase motors include shaded-pole, split-phase, and capacitor motors.

Shaded-Pole Motors. A *shaded-pole motor* is a 1φ AC motor that uses a shaded stator pole for starting. Shading the stator pole is the simplest method used to start a 1φ motor. Shaded-pole motors are commonly 1/20 HP or less and have low starting torque. Common applications of shaded-pole motors include small cooling fans found in computers and home entertainment centers. The shaded pole is normally a solid single turn of copper wire placed around a portion of the main pole laminations. **See Figure 7-20.**

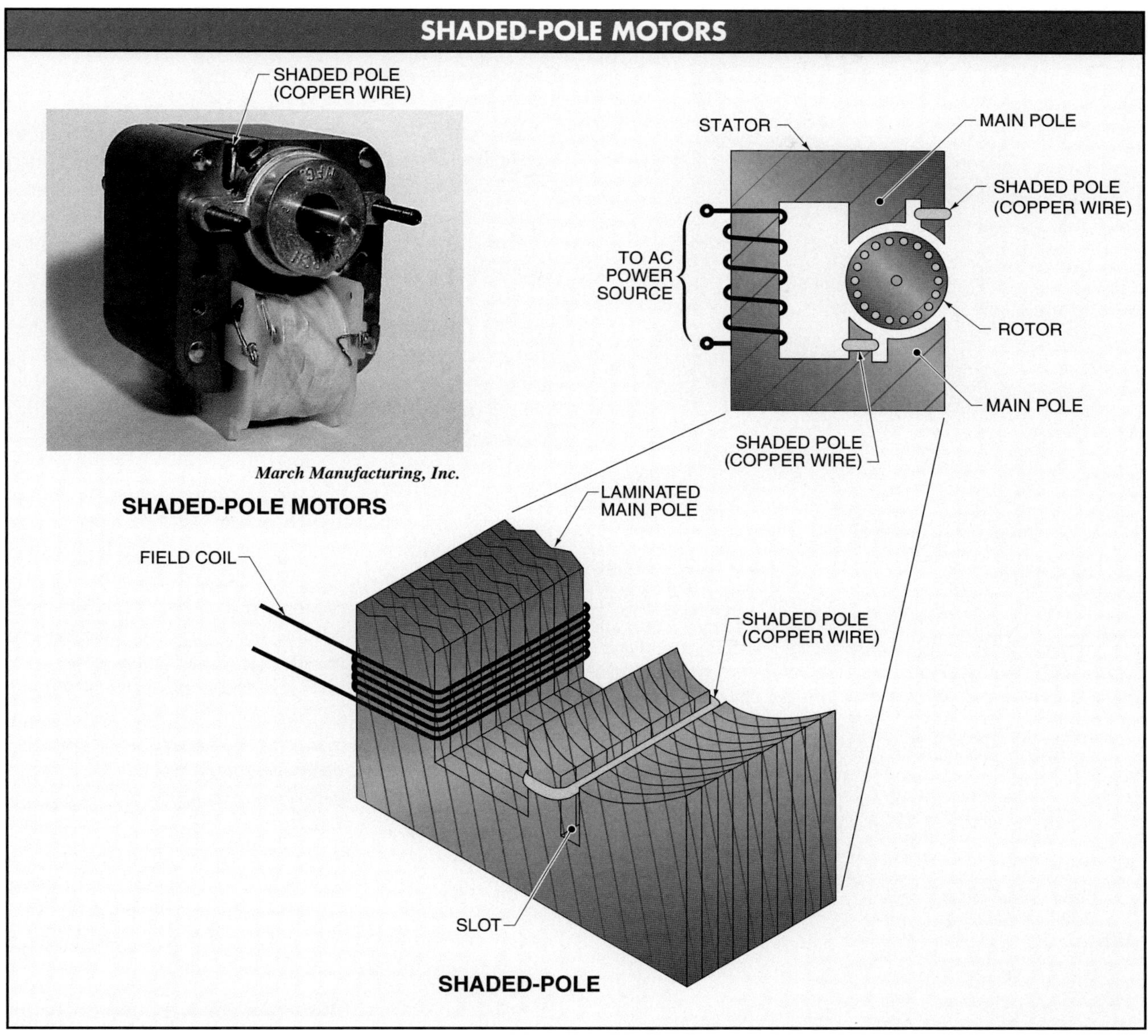

Figure 7-20. A shaded-pole motor uses a shaded stator pole for starting.

The shaded pole delays the magnetic field in the area of the pole that is shaded. Shading causes the magnetic field at the pole area to be positioned approximately 90° from the magnetic field of the main stator pole. The offset magnetic field causes the rotor to move from the main pole toward the shaded pole. This movement determines the starting direction of a shaded-pole motor.

Split-Phase Motors. A *split-phase motor* is a 1ϕ AC motor that includes a running winding (main winding) and a starting winding (auxiliary winding). Split-phase motors are AC motors of fractional horsepower, usually 1⁄20 HP to 1⁄3 HP. Split-phase motors are commonly used to operate washing machines, oil burners, and small pumps and blowers.

A split-phase motor has a rotating part (rotor), a stationary part consisting of the running winding and starting winding (stator), and a centrifugal switch that is located inside the motor to disconnect the starting winding at approximately 60% to 80% of full-load speed. **See Figure 7-21.**

Technical Fact

A continuous-duty motor should be used if the motor must operate at full load for 1 hour or more in a 24-hr period.

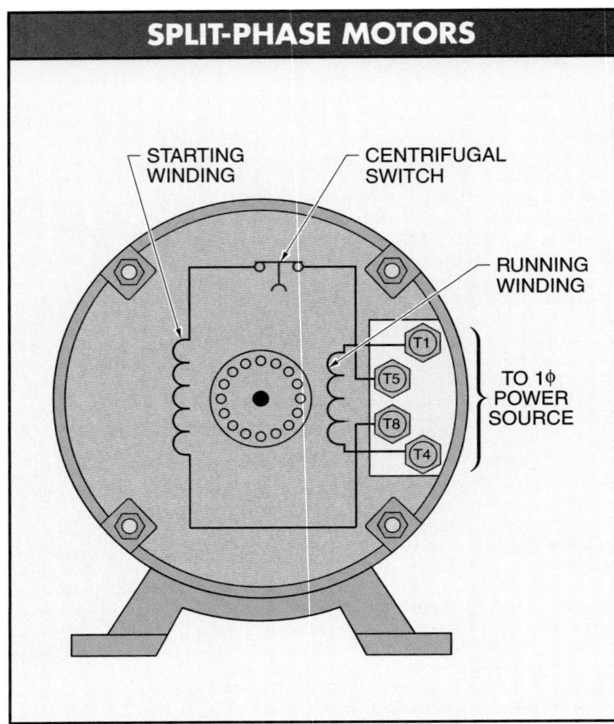

Figure 7-21. A split-phase motor includes a running winding, a starting winding, and a centrifugal switch.

Heidelberg Harris, Inc.

Capacitor motors are used in automated processes where high starting and/or running torque characteristics are required.

When starting, both the running windings and the starting windings are connected in parallel. The running winding is normally made up of heavy insulated copper wire and the starting winding is made of fine insulated copper wire. When the motor reaches approximately 75% of full speed, the centrifugal switch opens, disconnecting the starting winding from the circuit. This allows the motor to operate on the running winding only. When the motor is turned OFF (power removed), the centrifugal switch recloses at approximately 40% of full-load speed.

The running winding is made of larger wire and has a greater number of turns than the starting winding. When the motor is first connected to power, the reactance of the running winding is higher and the resistance is lower than the starting winding. *Reactance* is the opposition to the flow of alternating current in a circuit due to inductance.

The starting winding is made of relatively small wire and has fewer turns than the running winding. When the motor is first connected to power, the reactance of the starting winding is lower and the resistance is higher than the running winding.

When power is first applied, both the running winding and the starting winding are energized. The running winding current lags the starting winding current because of its different reactance. This produces a phase difference between the starting and running windings. A 90° phase difference is required to produce maximum starting torque, but the phase difference is commonly much less. A rotating magnetic field is produced because the two windings are out of phase.

The rotating magnetic field starts the rotor rotating. With the running and starting windings out of phase, the current changes in magnitude and direction, and the magnetic field moves around the stator. This movement forces the rotor to rotate with the rotating magnetic field.

Capacitor Motors. A *capacitor motor* is a 1ϕ AC motor that includes a capacitor in addition to the running and starting windings. Capacitor motor sizes range from ⅛ HP to 10 HP. Capacitor motors are used to operate refrigerators, compressors, washing machines, and air conditioners. The construction of a capacitor motor is similar to that of a split-phase motor, except that in a capacitor motor, a capacitor is connected in series with the starting winding. The addition of a capacitor in the starting winding gives a capacitor motor more torque than a split-phase motor. The three types of capacitor motors are capacitor-start, capacitor-run, and capacitor start-and-run motors.

A capacitor-start motor operates much the same as a split-phase motor in that it uses a centrifugal switch that opens at approximately 60% to 80% of full-load speed. **See Figure 7-22.** In a capacitor-start motor, the starting winding and the capacitor are removed when the centrifugal switch opens. The capacitor used in the starting winding gives a capacitor-start motor high starting torque.

A capacitor-run motor has the starting winding and capacitor connected in series at all times. A lower-value capacitor is used in a capacitor-run motor than in a capacitor-start motor because the capacitor remains in the circuit at full-load speed. This gives a capacitor-run motor medium starting torque and somewhat higher running torque than a capacitor-start motor. **See Figure 7-23.**

A capacitor start-and-run motor uses two capacitors. A capacitor start-and-run motor starts with one value capacitor in series with the starting winding and runs with a different value capacitor in series with the starting winding. Capacitor start-and-run motors are also known as dual-capacitor motors. **See Figure 7-24.**

A capacitor start-and-run motor has the same starting torque as a capacitor-start motor. A capacitor start-and-run motor has more running torque than a capacitor-start motor or capacitor-run motor because the capacitance is better matched for starting and running.

In a typical capacitor start-and-run motor, one capacitor is used for starting the motor and the other capacitor remains in circuit while the motor is running. A large-value capacitor is used for starting and a small-value capacitor is used for running. Capacitor start-and-run motors are used to run refrigerators and air compressors.

Three-Phase Motors

Three-phase motors are the most common motors used in industrial applications. Three-phase motors are used in applications ranging from fractional horsepower to over 500 HP. Three-phase motors are used in most applications because they are simple in construction, require little maintenance, and cost less to operate than 1ϕ or DC motors.

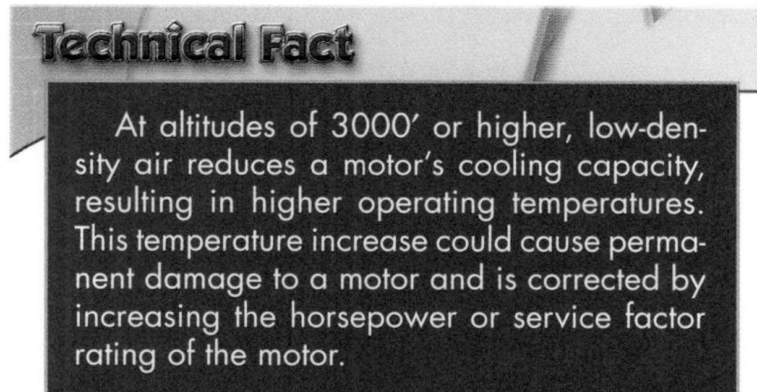

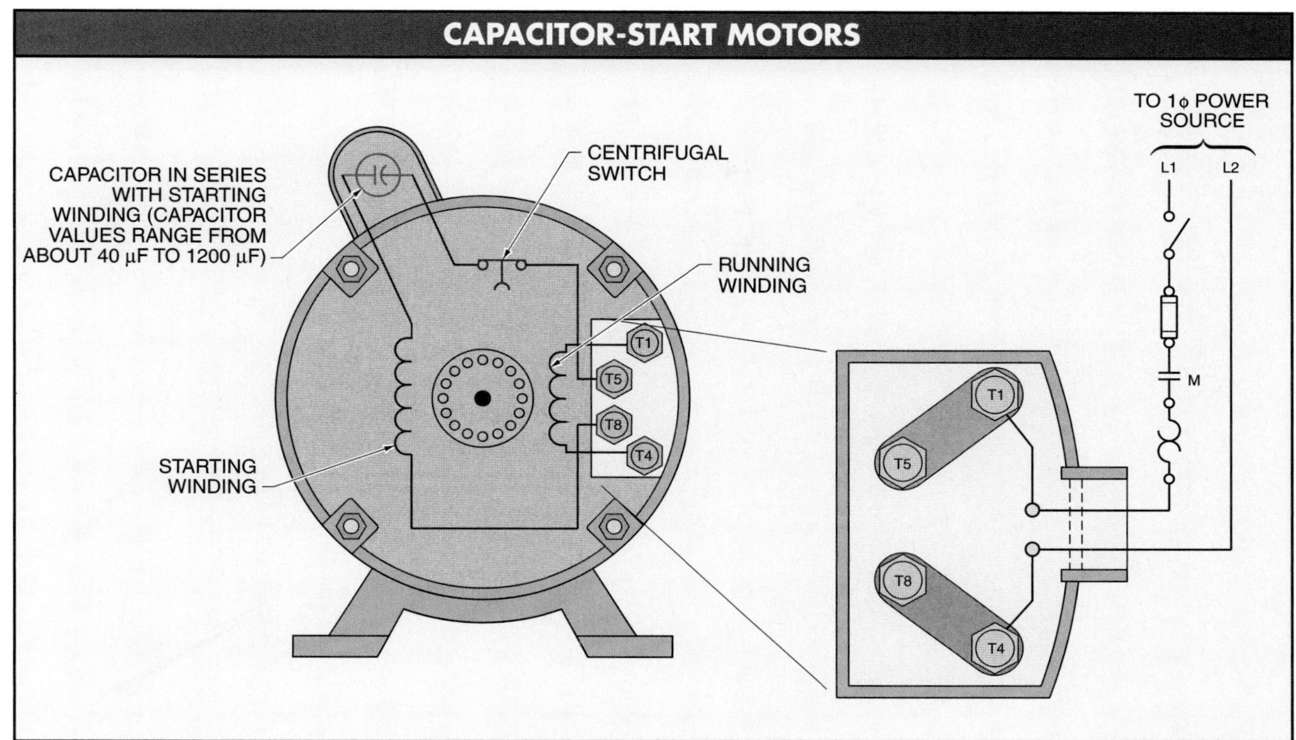

Figure 7-22. A capacitor-start motor has a capacitor in the starting winding, which gives the motor a high starting torque.

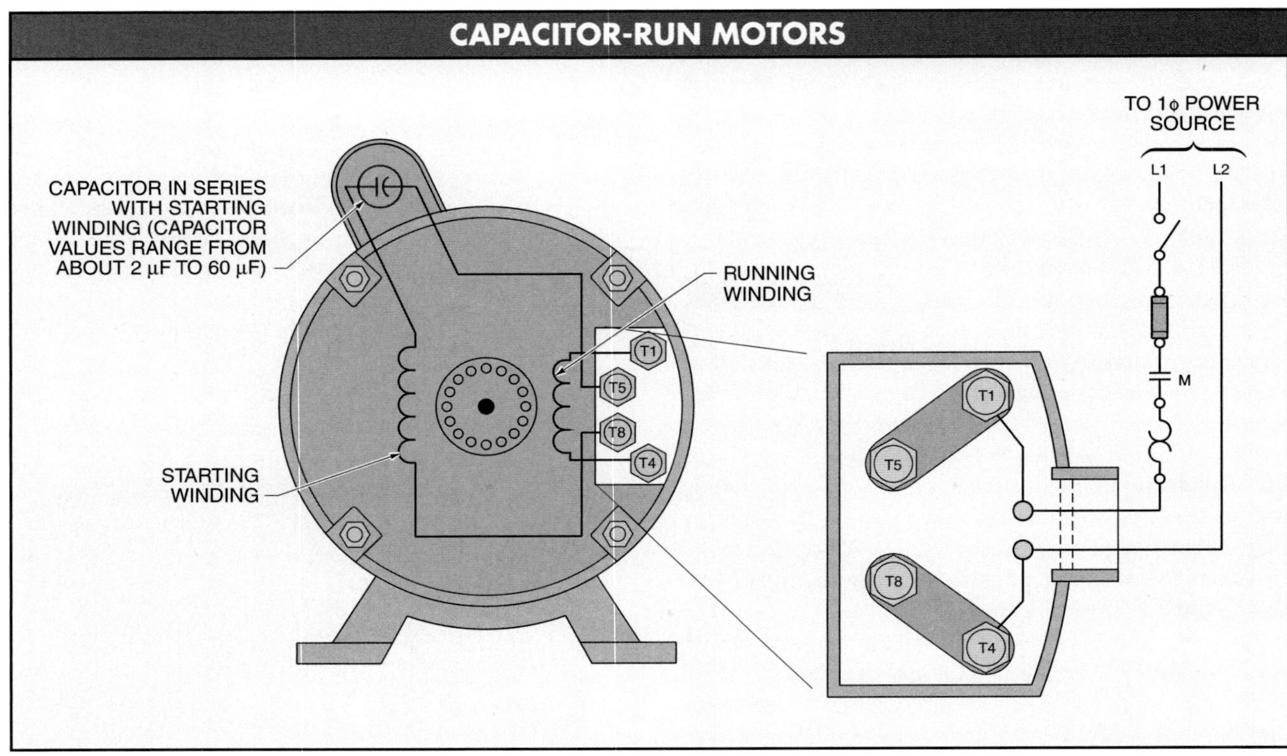

Figure 7-23. A capacitor-run motor has the starting winding and capacitor connected in series at all times.

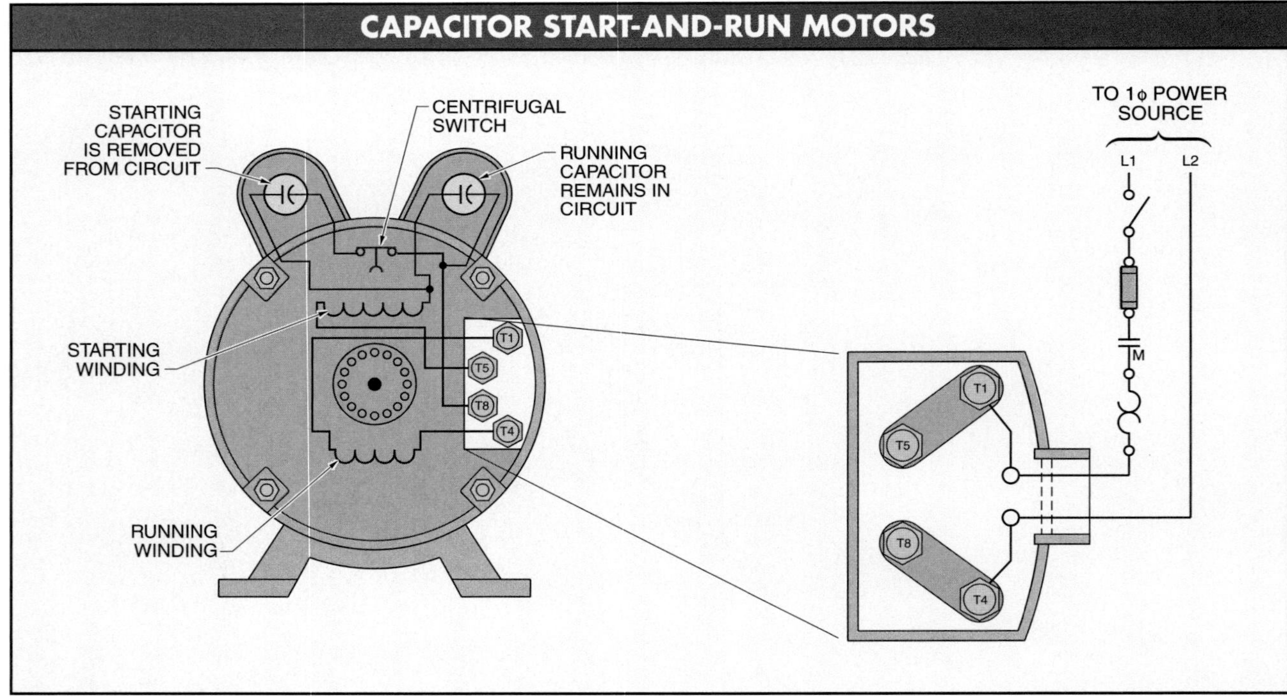

Figure 7-24. In a capacitor start-and-run motor, the starting capacitor is removed when the motor reaches full-load speed, but the running capacitor remains in the circuit.

The most common 3ϕ motor used in most applications is the induction motor. An *induction motor* is a motor that has no physical electrical connection to the rotor. Induction motors have no brushes that wear or require maintenance. Current in the rotor is induced by the rotating magnetic field of the stator.

In a 3ϕ motor, a rotating magnetic field is set up automatically in the stator when the motor is connected to 3ϕ power. The coils in the stator are connected to form three separate windings (phases). Each phase contains one-third of the total number of individual coils in the motor. These composite windings or phases are the A phase, B phase, and C phase. **See Figure 7-25.**

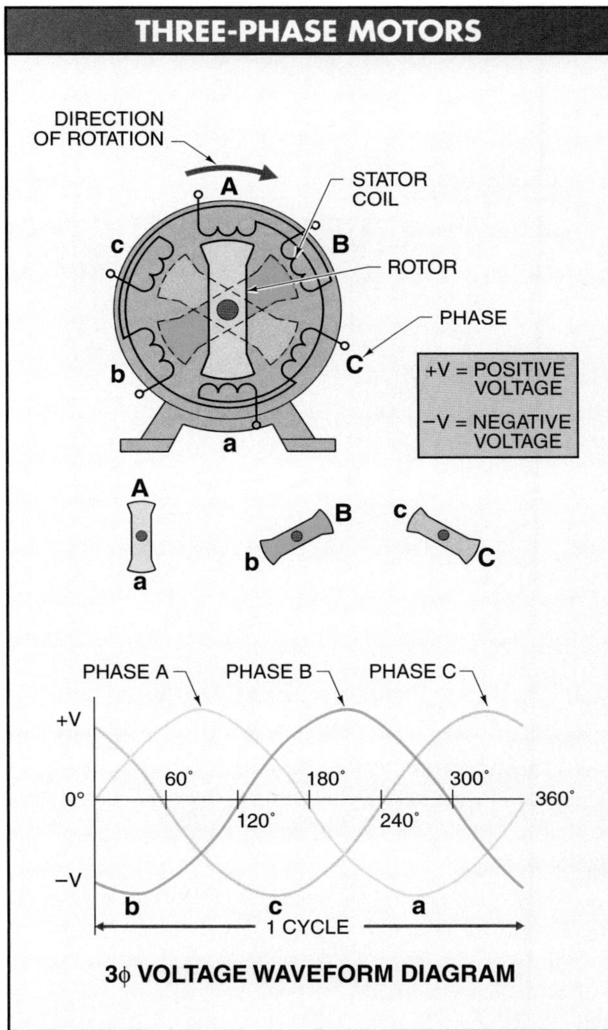

Figure 7-25. The coils in the stator of a 3ϕ motor are connected to form three separate windings (phases).

Each phase is placed in the motor so that it is 120° from the other phases. A rotating magnetic field is produced in the stator because each phase reaches its peak magnetic strength 120° away from the other phases. Three-phase motors are self-starting and do not require an additional starting method because of the rotating magnetic field in the motor.

To develop a rotating magnetic field in a motor, the stator windings must be connected to the proper voltage level. This voltage level is determined by the manufacturer and stamped on the motor nameplate. Three-phase motors are designed as either single-voltage motors or dual-voltage motors.

Single-Voltage, 3ϕ Motors. A *single-voltage motor* is a motor that operates at only one voltage level. Single-voltage motors are less expensive to manufacture than dual-voltage motors, but are limited to locations having the same voltage as the motor. Common single-voltage, 3ϕ motor ratings are 230 V, 460 V, and 575 V. Other single-voltage, 3ϕ motor ratings are 200 V, 208 V, and 220 V. All 3ϕ motors are wired so that the phases are connected in either a wye (Y) or delta (Δ) configuration.

In a single-voltage, wye-connected, 3ϕ motor, one end of each of the three phases is internally connected to the other phases. **See Figure 7-26.** The remaining end of each phase is brought out externally and connected to a power line. The leads that are brought out externally are labeled terminal one (T1), terminal two (T2), and terminal three (T3). When connected, terminals T1, T2, and T3 are matched to the 3ϕ power lines labeled line one (L1), line two (L2), and line three (L3). For the motor to operate properly, the 3ϕ lines supplying power to the wye motor must have the same voltage and frequency as the motor.

GE Motors & Industrial Systems

Industrial 3ϕ motors are commonly used in such manufacturing facilities as paper mills and wood processing plants.

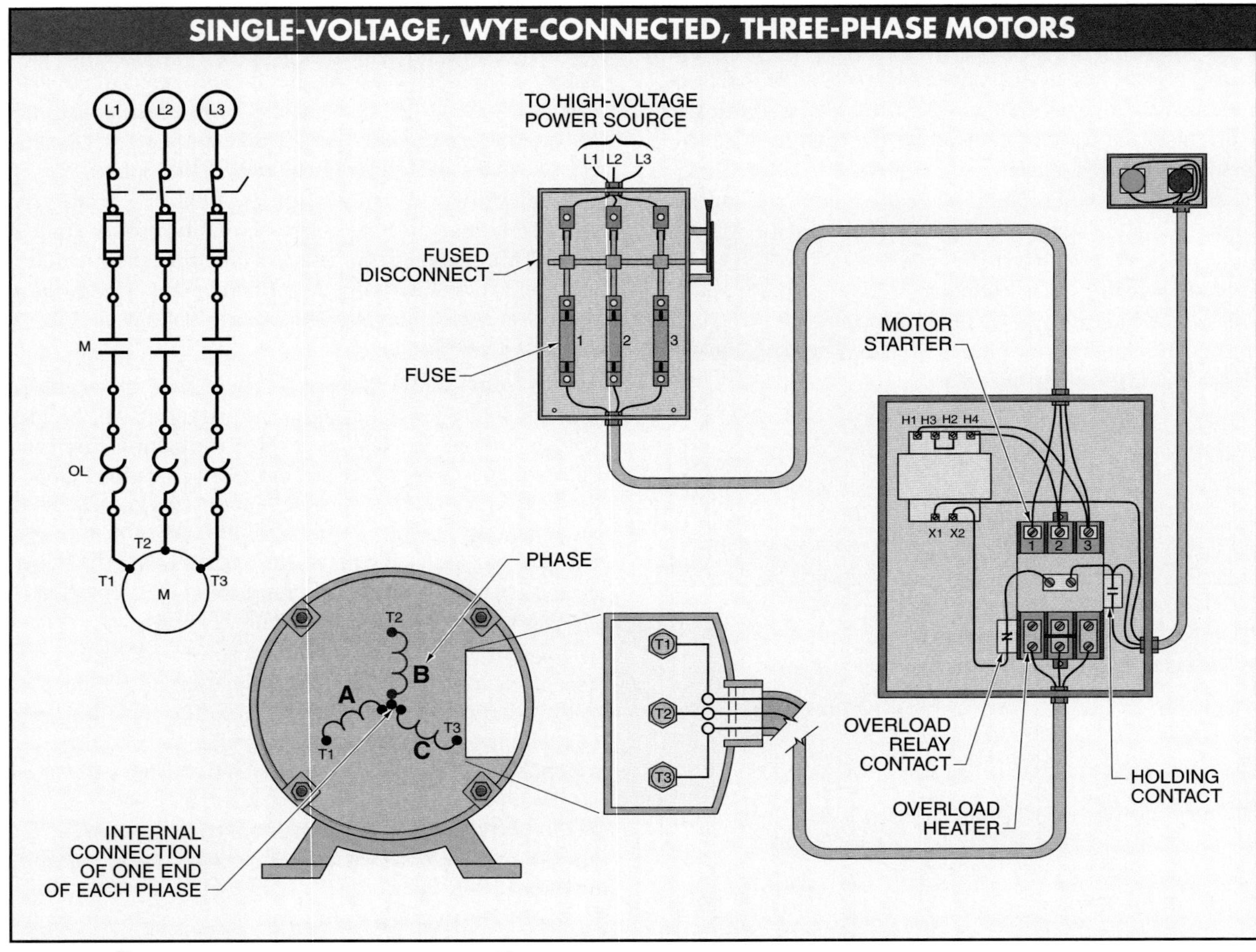

Figure 7-26. In a single-voltage, wye-connected, 3ϕ motor, one end of each phase is internally connected to the other phases.

In a single-voltage, delta-connected, 3ϕ motor, each winding is wired end-to-end to form a completely closed loop circuit. **See Figure 7-27.** At each point where the phases are connected, leads are brought out externally and labeled terminal one (T1), terminal two (T2), and terminal three (T3). These terminals, like those of a wye-connected motor, are attached to power lines one (L1), two (L2), and three (L3). The 3ϕ lines supplying power to the delta motor must have the same voltage and frequency rating as the motor.

Dual-Voltage, 3ϕ Motors. Most 3ϕ motors are made so that they may be connected for either of two voltages. Making motors for two voltages enables the same motor to be used with two different power line voltages. The normal dual-voltage rating of industrial motors is 230/460 V. The motor nameplate should be reviewed for proper voltage ratings.

The higher voltage is preferred when a choice between voltages is available. The motor uses the same amount of power and gives the same horsepower output for either high or low voltage, but as the voltage is doubled (230 V to 460 V), the current is cut in half. Using a reduced current enables the use of a smaller wire size, which reduces the cost of installation. Dual-voltage 3ϕ motors are wired so that the phases are connected in either a wye or delta configuration.

A wiring diagram is used to show the terminal numbering system for a dual-voltage, wye-connected, 3ϕ motor. **See Figure 7-28.** Nine leads are brought out of the motor. These leads are marked T1 through T9 and may be externally connected for either of the two voltages. The terminal connections for high and low voltage are normally provided on the motor nameplate.

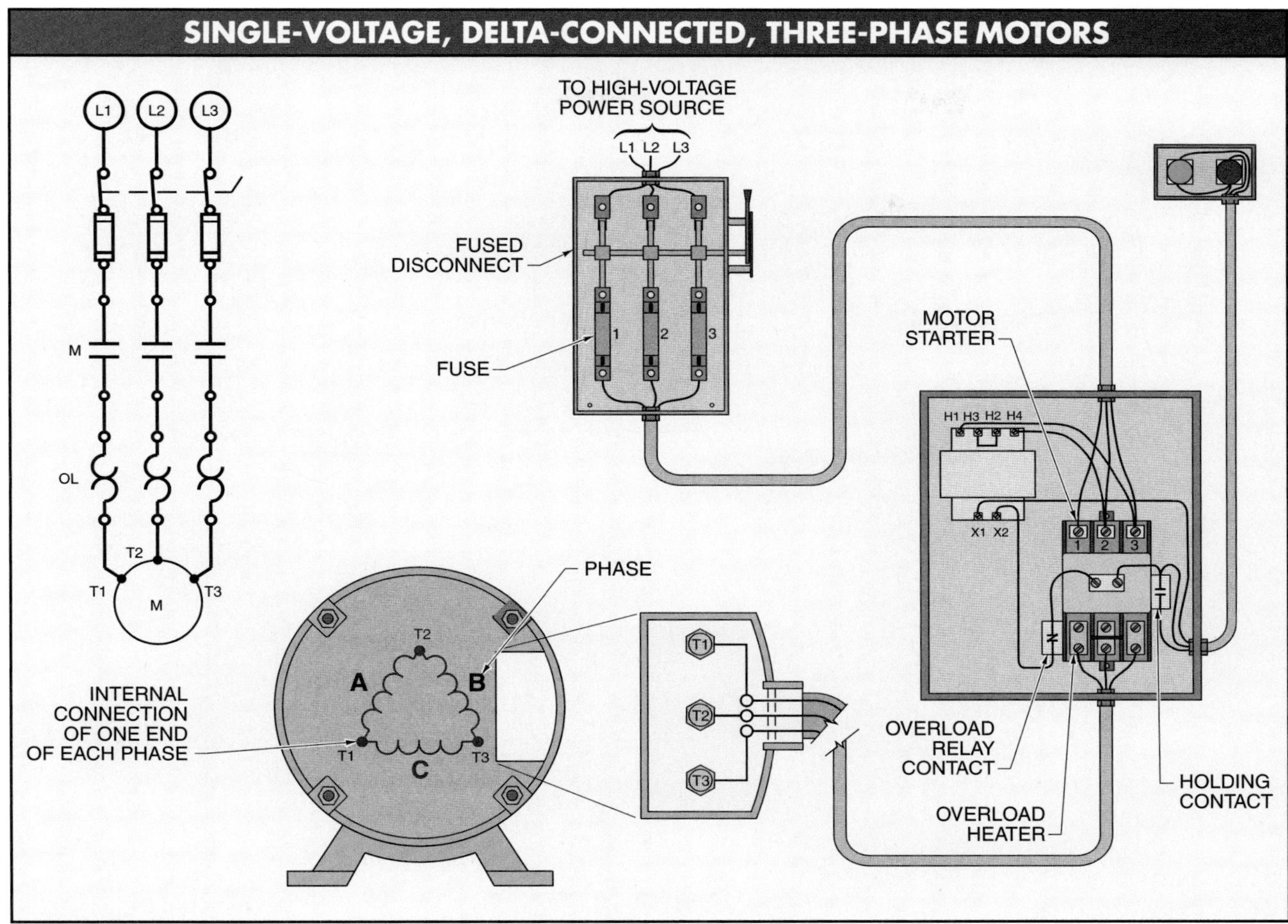

Figure 7-27. In a single-voltage, delta-connected, 3ϕ motor, each phase is wired end-to-end to form a completely closed loop.

The nine leads are connected in either series (high voltage) or parallel (low voltage). To connect a wye-connected motor for high voltage, L1 is connected to T1, L2 to T2, and L3 to T3; T4 is tied to T7, T5 to T8, and T6 to T9. This connects the individual coils in phases A, B, and C in series, each coil receiving 50% of the line-to-neutral point voltage. The neutral point equals the internal connecting point of all three phases.

To connect a wye-connected motor for low voltage, L1 is connected to T1 and T7, L2 to T2 and T8, and L3 to T3 and T9; T4 is tied to T5 and T6. This connects the individual coils in phases A, B, and C in parallel so that each coil receives 100% of the line-to-neutral point voltage.

A wiring diagram is used to show the terminal numbering system for a dual-voltage, delta-connected, 3ϕ motor. **See Figure 7-29.** The leads are marked T1 through T9 and a terminal connection chart is provided for wiring high- and low-voltage operations.

The nine leads are connected in either series or parallel for high or low voltage. In the high-voltage configuration, the coils are wired in series. In the low-voltage configuration, the coils are wired in parallel to distribute the voltage to the individual coil ratings.

Technical Fact

Motor failures are caused by overheating, phase unbalance, voltage unbalance, single phasing, surge voltages, poor ventilation, lack of lubrication, overloads, overcycling, excessive moisture, improper belt tension, misalignment, vibration, loose connections, and incorrect motor selection for the application.

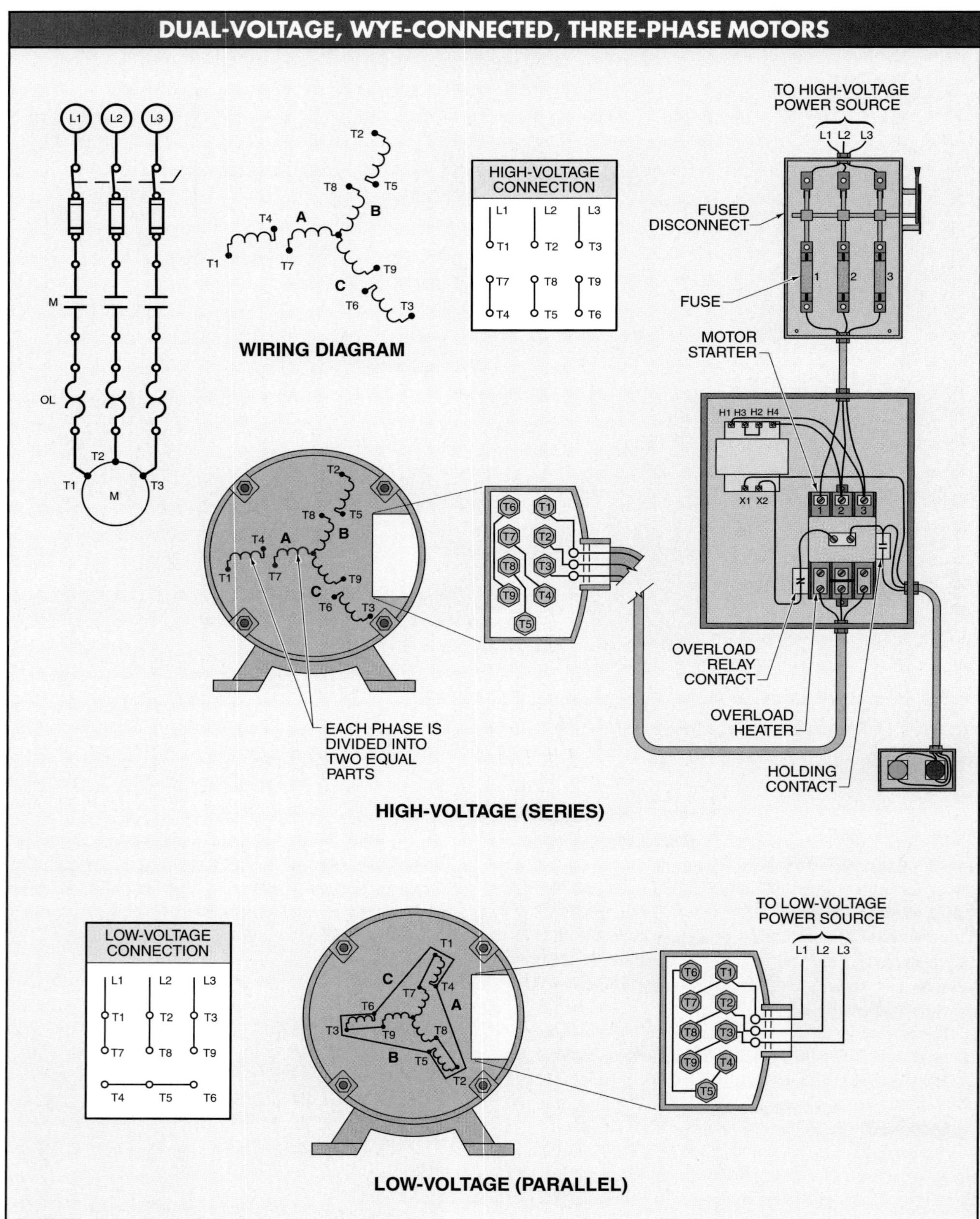

Figure 7-28. In a dual-voltage, wye-connected, 3ϕ motor, each phase coil is divided into two equal parts.

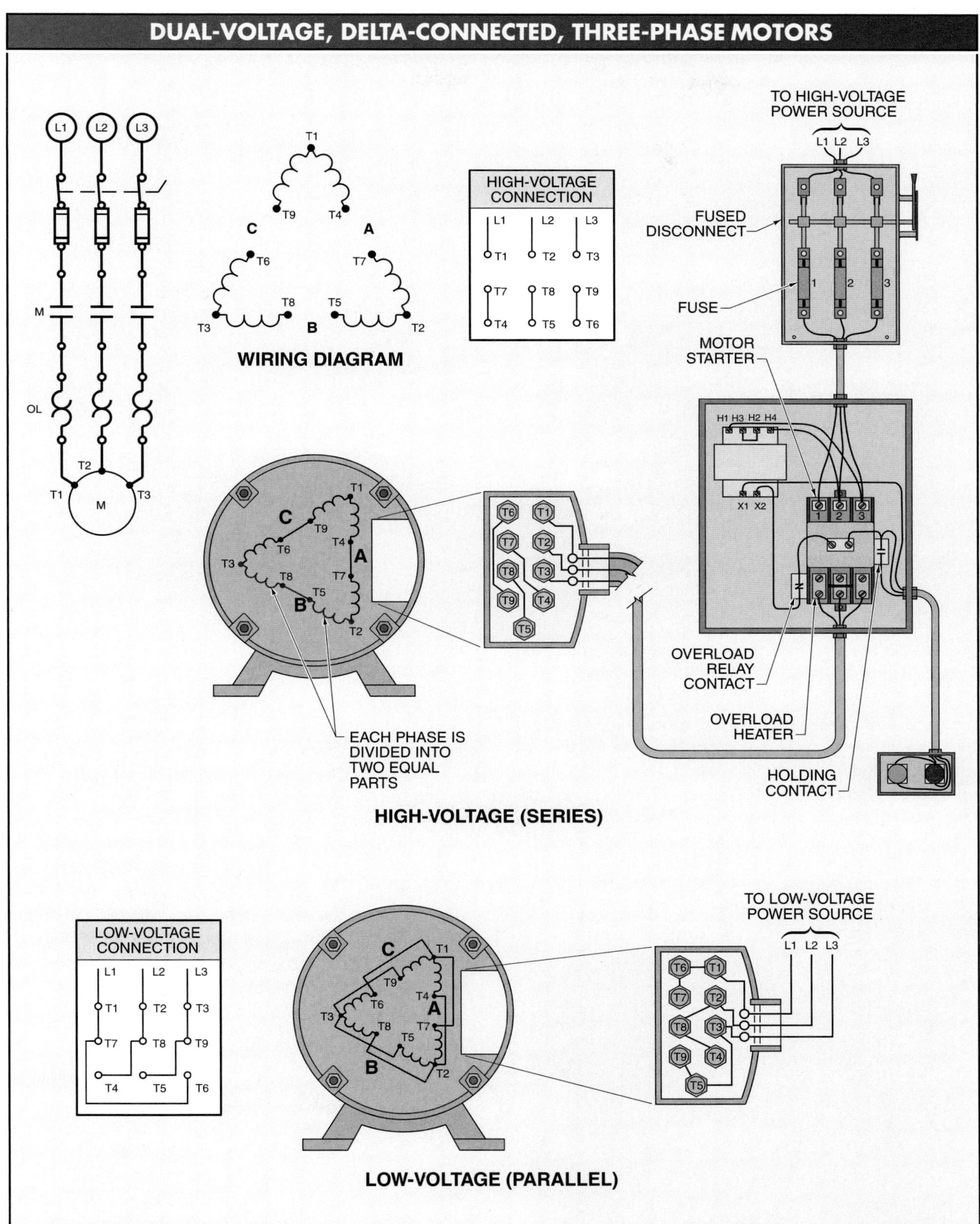

Figure 7-29. In a dual-voltage, delta-connected, 3ϕ motor, each phase coil is divided into two equal parts.

AC Motor Maintenance

Electric motors are very dependable machines. An electric motor gives good service under all operating conditions for which it is designed. For the safest service possible, the information given on the motor nameplate should be checked before putting a motor into operation. **See Figure 7-30.** The nameplate should be checked to ensure that the proper voltage and current are being used.

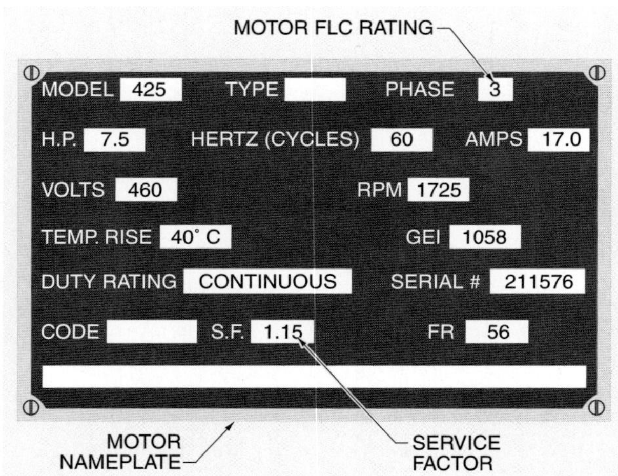

Figure 7-30. Motor nameplate information should be checked to ensure proper voltage and current are being used.

A standard motor should not be operated in very damp locations or where water may enter the motor frame. Motors are designed for use in specific locations. Typical enclosures available are open motor enclosures and totally enclosed motor enclosures. **See Figure 7-31.** When replacing a motor, ensure that the motor enclosure meets the proper specifications.

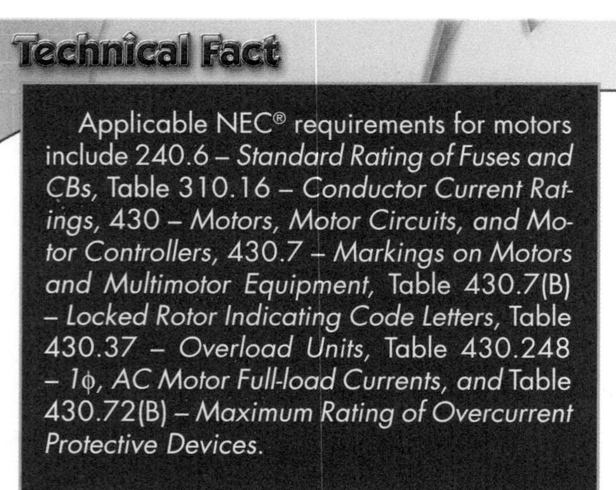

Applicable NEC® requirements for motors include 240.6 – *Standard Rating of Fuses and CBs*, Table 310.16 – *Conductor Current Ratings*, 430 – *Motors, Motor Circuits, and Motor Controllers*, 430.7 – *Markings on Motors and Multimotor Equipment*, Table 430.7(B) – *Locked Rotor Indicating Code Letters*, Table 430.37 – *Overload Units*, Table 430.248 – *1ϕ, AC Motor Full-load Currents*, and Table 430.72(B) – *Maximum Rating of Overcurrent Protective Devices.*

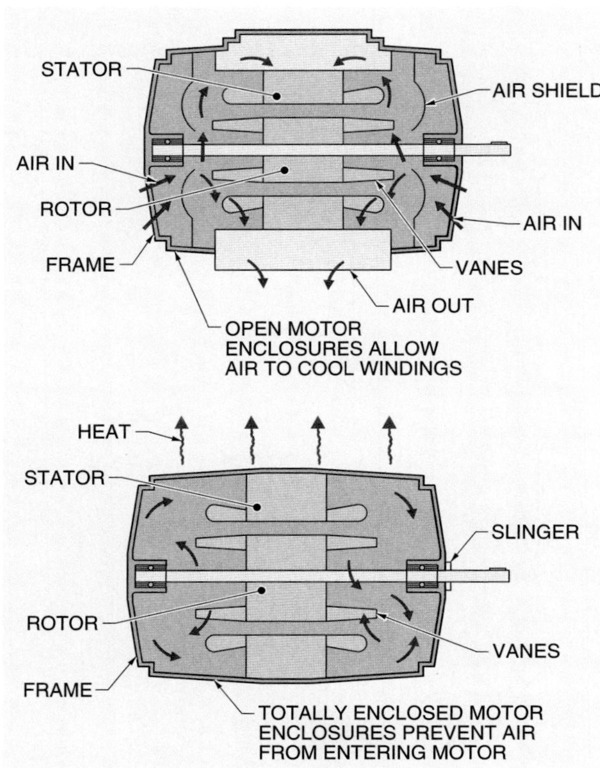

Figure 7-31. An open motor enclosure allows air to flow through the motor to cool the windings. A totally enclosed motor enclosure prevents air from entering the motor.

The frame of a motor should be grounded, especially if the motor is used in a damp location. If a motor shaft does not rotate after the switch has been turned ON, unplug the motor immediately. This prevents the windings from becoming seriously overheated. To prevent an ordinary motor from becoming overheated, keep the air openings on its frame clear at all times. When oiling a motor, apply oil to the bearings only. Excessive oil used on a motor damages the winding insulation and causes the motor to collect an excessive amount of dirt and dust.

Troubleshooting AC Motors

Most problems with 1ϕ AC motors involve the centrifugal switch, thermal switch, or capacitor(s). A motor is usually serviced and repaired if the problem is in the centrifugal switch, thermal switch, or capacitor. A motor is almost always replaced if the motor is less than ⅛ HP. Three-phase motors usually operate for many years without any problems because 3ϕ motors have fewer components that may malfunction than other motors.

Troubleshooting Shaded-Pole Motors. Shaded-pole motors that fail are usually replaced. The reason for the motor failure should be investigated. For example, replacing a motor because it failed due to a jammed load does not solve the problem. **See Figure 7-32.** To troubleshoot a shaded-pole motor, apply the following procedure:

1. Visually inspect the motor after turning power to motor OFF. Replace the motor if it is burned, if the shaft is jammed, or if there is any sign of damage.
2. Check the stator winding. The stator winding is the only electrical circuit that may be tested without taking the motor apart. Measure the resistance of the stator winding. Set a DMM to the lowest resistance scale for taking the reading. The winding is open if the DMM indicates an infinity reading. Replace the motor. The winding is short circuited if the DMM indicates a zero reading. Replace the motor. The winding may still be good if the DMM indicates a low resistance reading. Check the winding with a megohmmeter before replacing the motor.

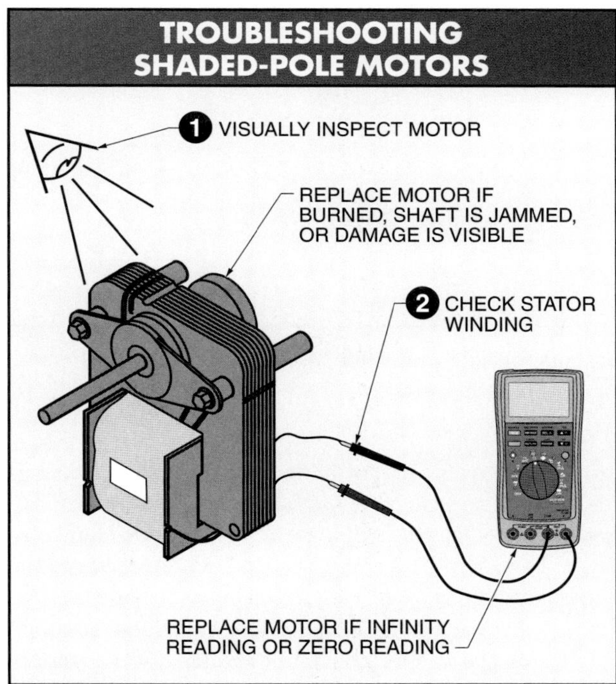

Figure 7-32. Shaded-pole motors that fail are usually replaced.

Troubleshooting Split-Phase Motors. Some split-phase motors include a thermal switch that automatically turns the motor OFF when it overheats. Thermal switches may have a manual reset or an automatic reset. Caution should be taken with any motor that has an automatic reset because the motor may automatically restart at any time. **See Figure 7-33.** To troubleshoot a split-phase motor, apply the following procedure:

1. Visually inspect the motor after turning the power to the motor OFF. Replace the motor if it is burned, if the shaft is jammed, or if there is any sign of damage.
2. Check to determine if the motor is controlled by a thermal switch. Reset the thermal switch and turn the motor ON if the thermal switch is manual.
3. Check for voltage at the motor terminals using a DMM set to measure voltage if the motor does not start. The voltage should be within 10% of the motor listed voltage. Troubleshoot the circuit leading to the motor if the voltage is not correct. If the voltage is correct, turn the power to the motor OFF so the motor may be tested.
4. Turn OFF and lock out and tag out the starting mechanism per company policy.
5. With power OFF, connect a DMM set to measure resistance to the same motor terminals from which the incoming power leads were disconnected. The DMM reads the resistance of the starting and running windings. Their combined resistance is less than the resistance of either winding alone because the windings are connected in parallel. A short circuit is present if the DMM reads zero. An open circuit is present if the DMM reads infinity. Replace the motor. *Note:* Split-phase motors are normally too small for a repair to be cost efficient.
6. Check the centrifugal switch for signs of burning or broken springs. Service or replace the switch if any obvious signs of problems are present. Check the switch using a DMM set to measure resistance if no obvious signs of problems are present. Manually operate the centrifugal switch. The endbell on the switch side may have to be removed. The resistance on the DMM decreases if the motor is good. A problem exists if the resistance does not change.

Troubleshooting Capacitor Motors. Troubleshooting capacitor motors is similar to troubleshooting split-phase motors. The only additional device to be tested is the capacitor. Capacitors have a limited life and are often the problem in capacitor motors. Capacitors may have a short circuit or an open circuit, or may deteriorate to the point where they must be replaced.

Deterioration may also change the value of a capacitor, which causes additional problems. When a capacitor short circuits, the winding in the motor may burn out. When a capacitor deteriorates or opens, the motor has poor starting torque. Poor starting torque may prevent the motor from starting, which usually trips the overloads.

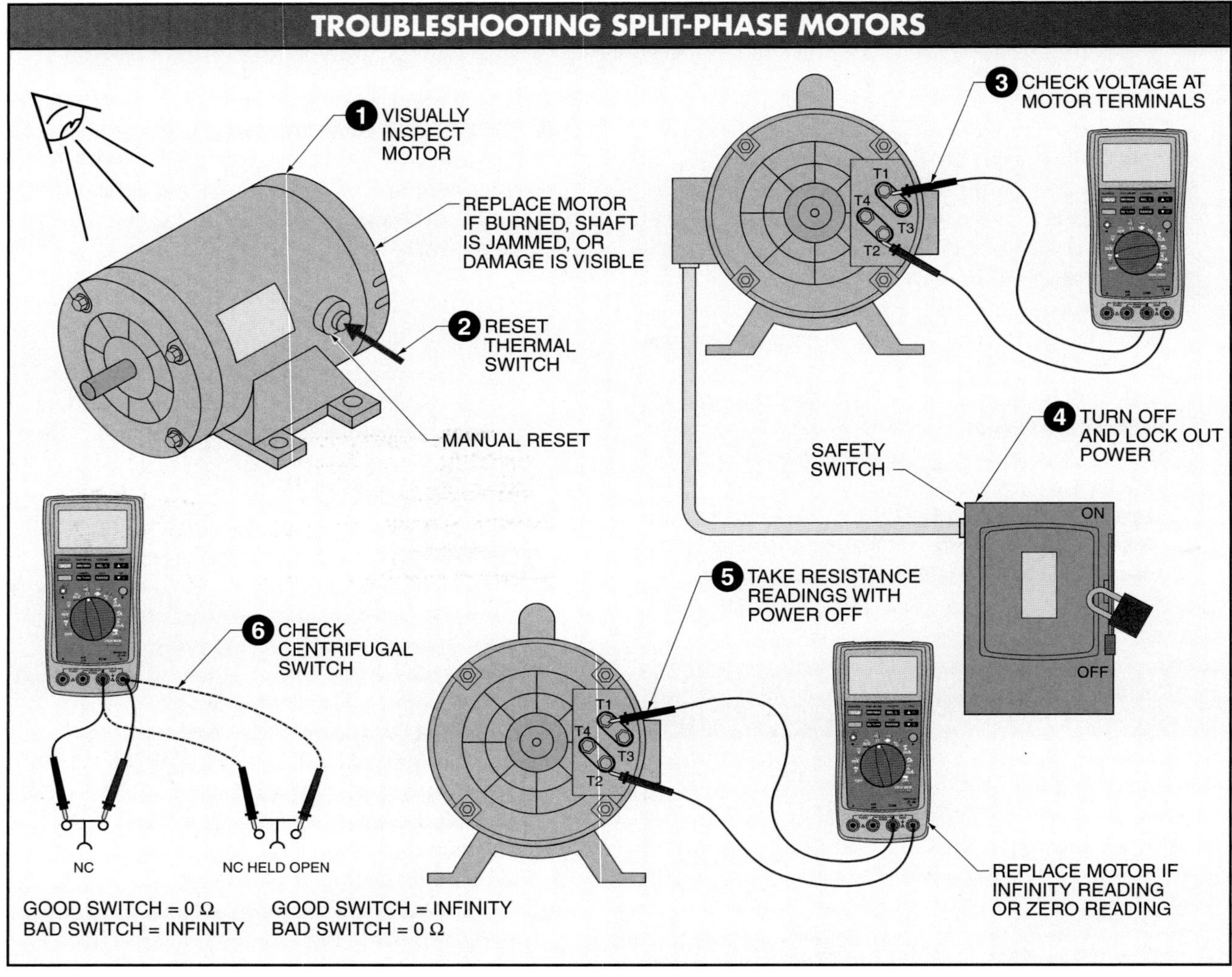

Figure 7-33. Some split-phase motors include a thermal switch that automatically turns the motor OFF when it overheats.

All capacitors are made with two conducting surfaces separated by dielectric material. *Dielectric material* is a medium in which an electric field is maintained with little or no outside energy supply. Dielectric material is used to insulate the conducting surfaces of a capacitor. Capacitors are either oil or electrolytic. Oil capacitors are filled with oil and sealed in a metal container. The oil serves as the dielectric material.

More motors use electrolytic capacitors than oil capacitors. Electrolytic capacitors are formed by winding two sheets of aluminum foil separated by pieces of thin paper impregnated with an electrolyte. An *electrolyte* is a conducting medium in which the current flow occurs by ion migration. The electrolyte is used as the dielectric material. The aluminum foil and electrolyte are encased in a cardboard or aluminum cover. A vent hole is provided to prevent a possible explosion in the event the capacitor is shorted or overheated. AC capacitors are used with capacitor motors. Capacitors that are designed to be connected to AC have no polarity. **See Figure 7-34.** To troubleshoot a capacitor motor, apply the following procedure:

1. Turn the handle of the safety switch or combination starter OFF. Lock out and tag out the starting mechanism per company policy.
2. Use a DMM set to measure voltage at the motor terminals to ensure the power is OFF.
3. Capacitors are located on the outside frame of a motor. Remove the cover of the capacitor. **Caution:** A good capacitor will hold a charge even when power is removed.
4. Visually inspect the capacitor for leakage, cracks, or bulges. If these are present, replace the capacitor.

5. Remove the capacitor from the circuit and discharge it. To safely discharge a capacitor, place a 20,000 Ω, 5 W resistor across the terminals for 5 seconds.
6. After a capacitor is discharged, connect the leads of a DMM set to measure resistance to the capacitor terminals. The DMM indicates the general condition of the capacitor. A capacitor is either good, shorted, or open.

- Good Capacitor. The reading changes from zero resistance to infinity. When the reading reaches the halfway point, remove one of the leads and wait 30 sec. When the lead is reconnected, the reading should change back to the halfway point and continue to infinity. This demonstrates that the capacitor can hold a charge. The capacitor cannot hold a charge and must be replaced when the reading changes back to zero resistance.
- Short Capacitor. The reading changes to zero and does not move. The capacitor is bad and must be replaced.
- Open Capacitor. The reading does not change from infinity. The capacitor is bad and must be replaced.

Troubleshooting 3ϕ Motors. The extent of troubleshooting a 3ϕ motor depends on the motor application. Testing is normally limited to checking the voltage at the motor if a motor is used in an application that is critical to an operation or production. The motor is assumed to be the problem if the voltage is present and correct. Unless it is very large, the motor is normally replaced at this time so production may continue. Further tests may be made to determine the exact problem if time is not a critical factor. **See Figure 7-35.**

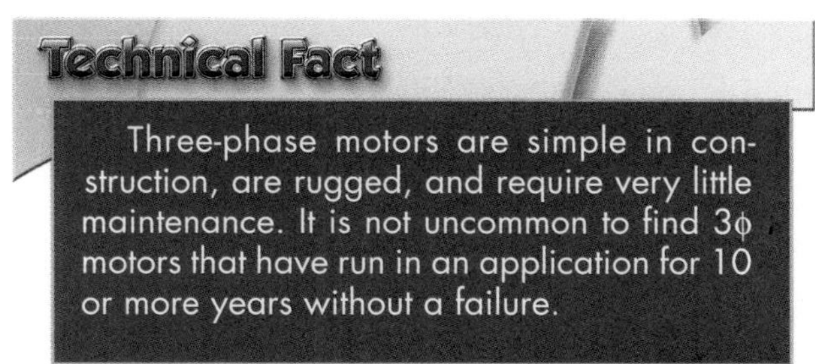

TROUBLESHOOTING CAPACITOR MOTORS

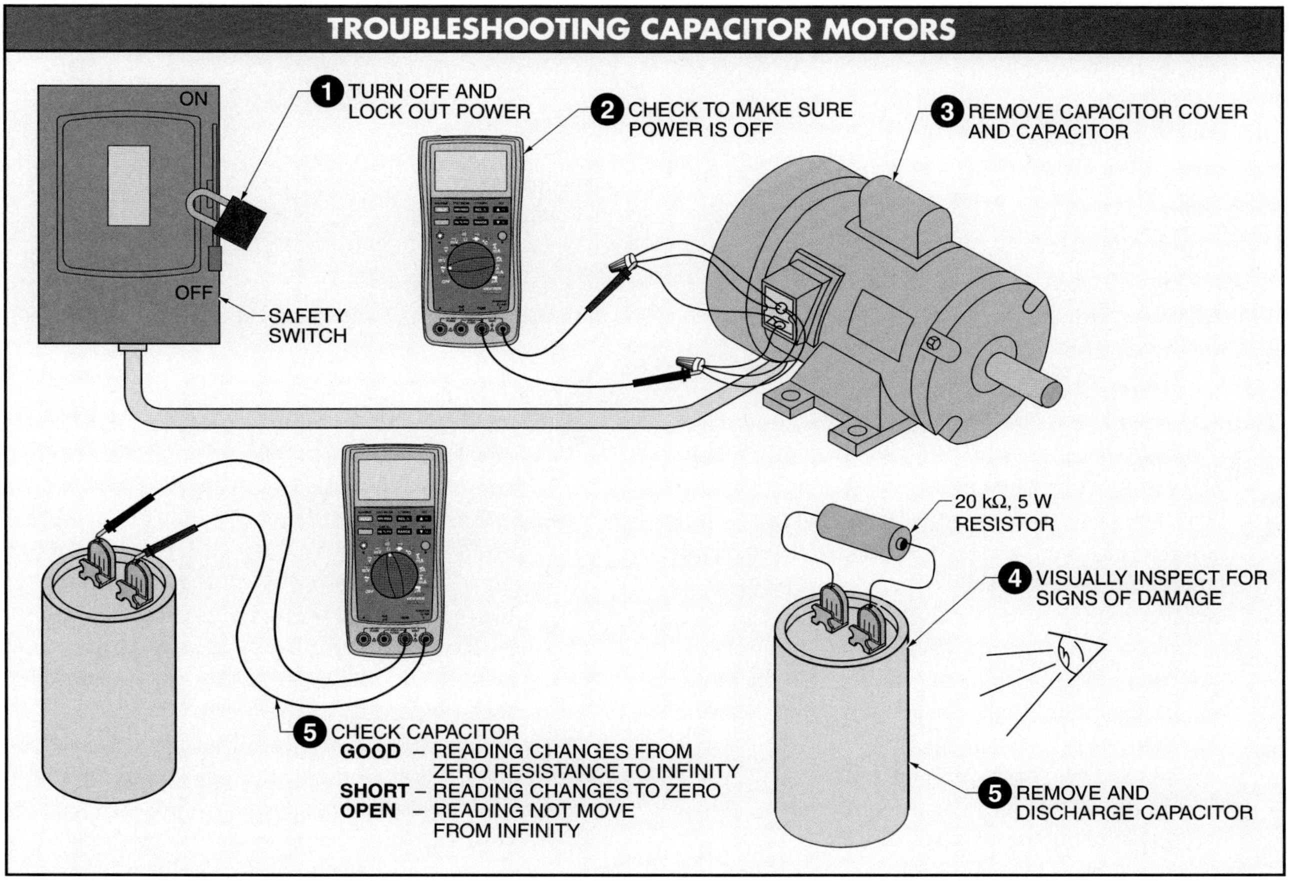

Figure 7-34. Capacitors have a limited life and are often the problem in capacitor motors.

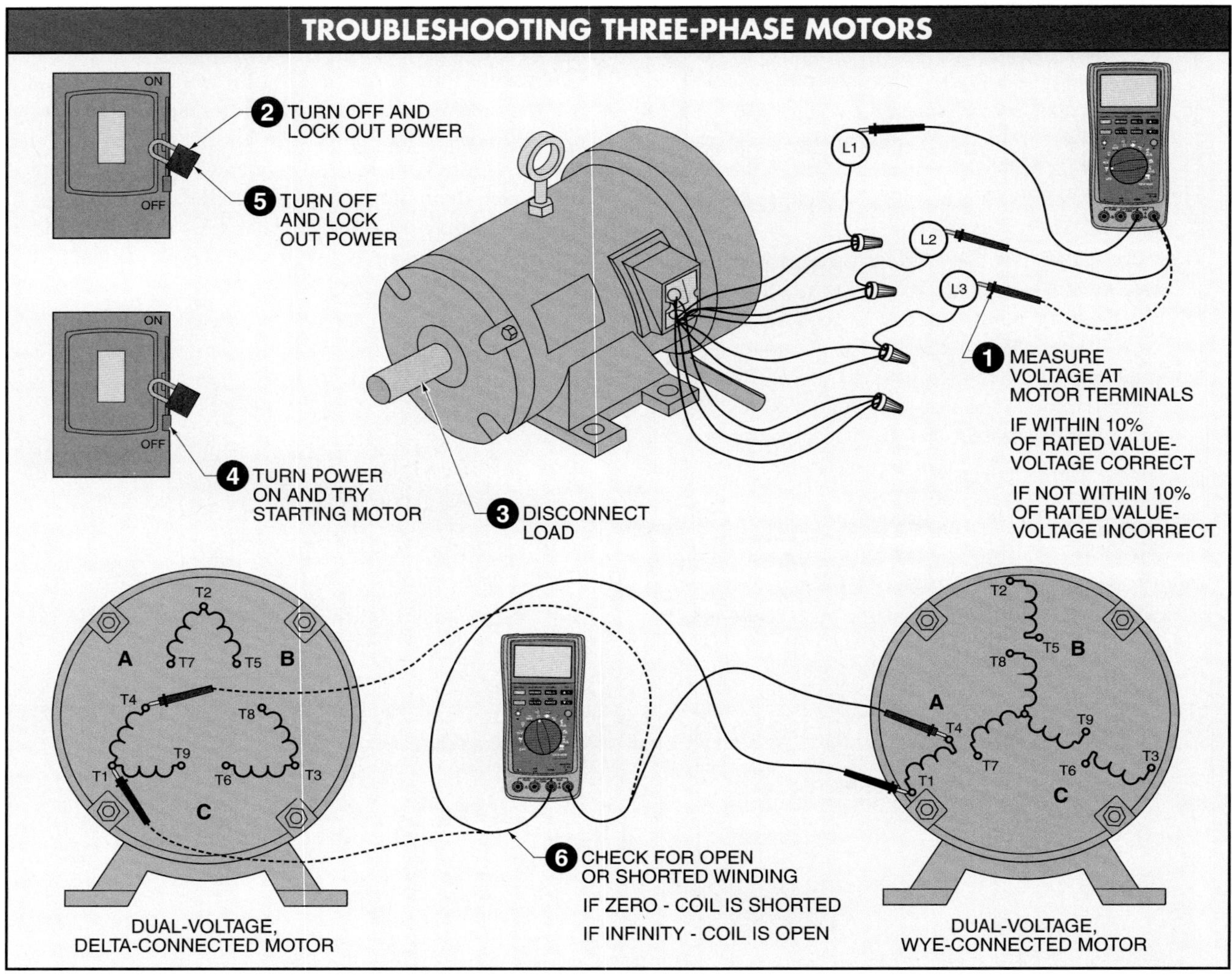

Figure 7-35. Testing of 3ϕ motors is normally limited to checking the voltage at the motor if a motor is used in an application that is critical to an operation or production.

To troubleshoot a 3ϕ motor, apply the following procedure:

1. Measure the voltage at the motor terminals using a DMM set to measure voltage. The motor must be checked if the voltage is present and at the correct level on all three phases. The incoming power supply must be checked if the voltage is not present on all three phases.
2. Turn the handle of the safety switch or combination starter OFF if voltage is present but the motor is not operating. Lock out and tag out the starting mechanism per company policy.
3. Disconnect the motor from the load.
4. Turn power ON to try restarting the motor. Check the load if the motor starts.
5. Turn the motor OFF and lock out the power if the motor does not start.
6. Check the motor windings with a DMM set to measure resistance for any opens or shorts. Take a resistance reading of the T1-T4 coil. The coil must have a resistance reading. The coil is shorted when the reading is zero. The coil is open when the reading is infinity. The resistance is low because the coil winding is made of wire only. However, there is resistance on a good coil winding. The larger the motor, the smaller the resistance reading.

After the resistance of one coil has been found, the basic laws of series and parallel circuits are applied. When measuring the resistance of two coils in series, the total resistance is twice the resistance of one coil. When measuring the resistance of two coils in parallel, the total resistance is one half the resistance of one coil.

Review Questions

1. What is the purpose of the field windings in a generator?
2. What is the major difference between an AC generator and a DC generator?
3. What is a delta connection?
4. What is a wye connection?
5. What is the normal range of the rated voltage of an AC generator?
6. Why are transformers used in power distribution?
7. What is a step-up transformer?
8. What is a step-down transformer?
9. Why aren't transformers 100% efficient?
10. What is a shaded-pole motor?
11. What is a split-phase motor?
12. What is the function of a tap on a transformer?
13. What are the considerations in choosing a transformer?
14. What advantages does an AC motor have over a DC motor?
15. What motor features are used to aid starting?
16. What motor designs are used for special types of locations?
17. What is likely to go wrong with a capacitor motor?
18. What is the most common type of 3ϕ motor?
19. What types of motors are commonly replaced rather than repaired?
20. What kinds of motors are used in household washing machines and refrigerators?

chapter 8

electrical motor controls *for Integrated Systems*

Contactors and Motor Starters

Contactors and motor starters are control devices that use a small control current to energize or de-energize the loads connected to them. A manual contactor manually opens and closes contacts in an electrical circuit. A manual motor starter is a contactor with an added overload protection device.

MANUAL SWITCHING

In the late 1800s, when electric motors were introduced, a method had to be found to start and stop them. This was accomplished through the use of knife switches. **See Figure 8-1.** Knife switches were eventually discontinued as a means of connecting line voltage directly to motor terminals and disconnecting voltage, for three basic reasons. First, the open knife switch had exposed (live) parts that presented an extreme electrical hazard to the operator. In addition, any applications where dirt or moisture were present made the open switch concept vulnerable to problems. Second, the speed of opening and closing contacts was determined solely by the operator. Considerable arcing and pitting of the contacts led to rapid wear and eventual replacement if the operator did not open or close the switch quickly. Finally, most knife switches were made of soft copper, which required replacement after repeated arcing, heat generation, and mechanical fatigue.

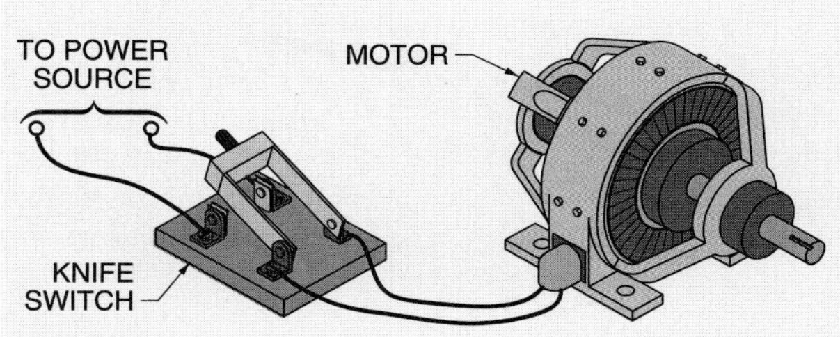

Figure 8-1. Knife switches were the first devices used to start and stop electric motors.

Mechanical Improvements

As industry demanded more electric motors, improvements were made to knife switches to make them more acceptable as control devices. First, the knife switch was enclosed in a steel housing to protect the switch. **See Figure 8-2.** An insulated external handle was added to protect the operator.

Third, an operating spring was attached to the handle to ensure quick opening and closing of the knife blade. The switch handle was designed so that once the handle was moved a certain distance, the tension on the spring forced the contacts to open or close at the same continuous speed each time it was operated.

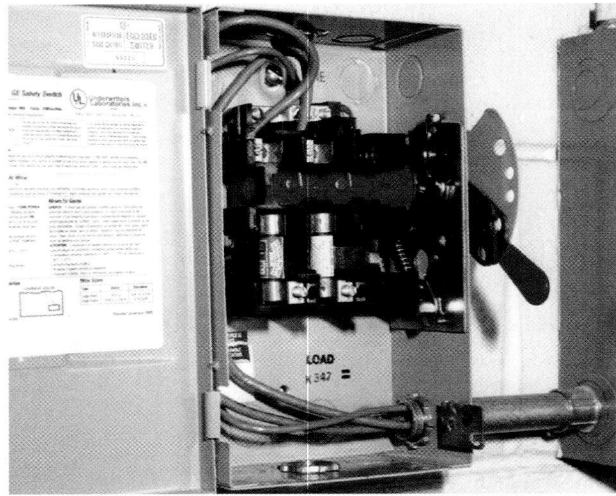

Figure 8-2. A knife switch is enclosed in a steel housing, has an insulated external handle, and includes an operating spring for improved operation and safety.

Even with these improvements, the blade and jaw mechanism of a knife switch had a short mechanical life when the knife switch was used as a direct control device. The knife switch mechanism was discontinued as a means of direct control for motors because of the short life. Knife switches are currently used as electrical disconnects. A *disconnect* is a device used only periodically to remove electrical circuits from their supply source. The mechanical life of the knife switch mechanism is not of major concern because a disconnect is used infrequently. Always follow lockout/tagout procedures when using a disconnect.

MANUAL CONTACTORS

A *manual contactor* is a control device that uses pushbuttons to energize or de-energize the load connected to it. **See Figure 8-3.** A manual contactor manually opens and closes contacts in an electrical circuit. Manual contactors cannot be used to start and stop motors because they have no overload protection built into them. Manual contactors are normally used with lighting circuits, and resistive loads such as heaters or large lamp loads. A fuse or circuit breaker is normally included in the same enclosure with a manual contactor.

Rockwell Automation, Allen-Bradley Company, Inc.

Figure 8-3. A manual contactor uses pushbuttons to energize or de-energize the load connected to it.

Double-Break Contacts

A double-break contact can act as a direct controller. *Double-break contacts* are contacts that break an electrical circuit in two places. **See Figure 8-4.**

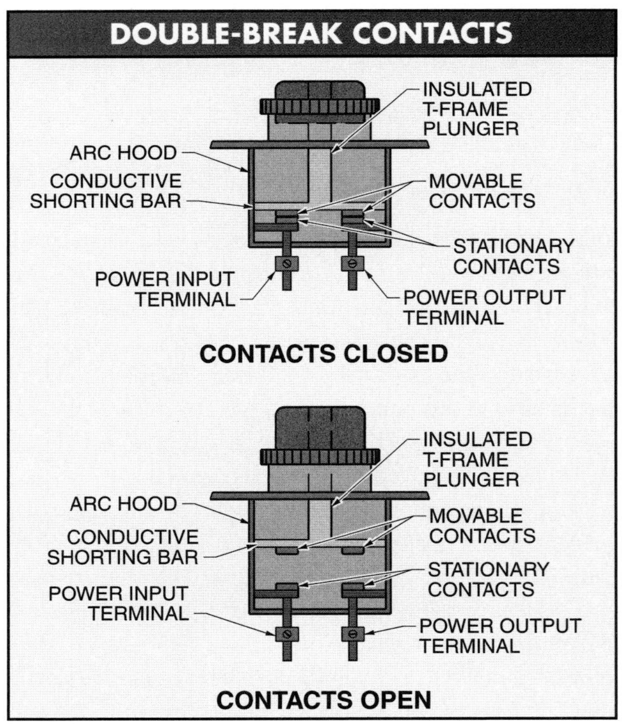

Figure 8-4. Double-break contacts break an electrical circuit in two places.

Double-break contacts allow devices to be designed that have a higher contact rating (current rating) in a smaller space than devices designed with single-break contacts. With double-break contacts, the movable contacts are forced against the two stationary contacts to complete the electrical circuit when a set of normally open (NO) double-break contacts are energized. The movable contacts are pulled away from the stationary contacts and the circuit is open when the manual contactor is de-energized. The procedure is reversed when normally closed (NC) double-break contacts are used.

A 3ϕ manual contactor has three sets of normally open double-break contacts. One set of normally open double-break contacts is used to open and close each phase in the circuit. The movable contacts are located on an insulated T-frame and are provided with springs to soften their impact. The T-frame is activated by a pushbutton mechanism. Similar to a disconnect, the mechanical linkage consistently and quickly makes or breaks the circuits. **See Figure 8-5.**

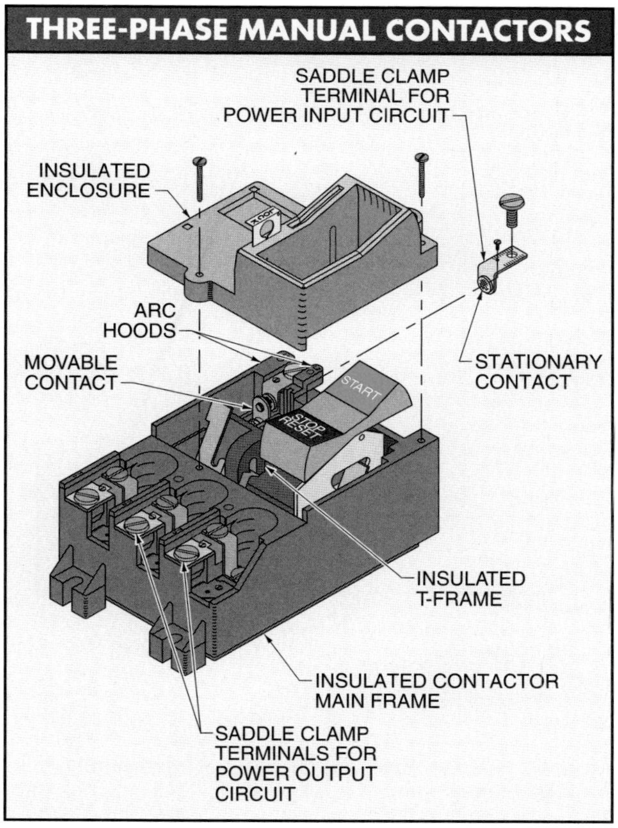

Figure 8-5. A 3ϕ manual contactor has three sets of normally open double-break contacts.

The movable contacts have no physical connection to external electrical wires. The movable contacts move into arc hoods and bridge the gap between a set of fixed contacts to make or break the circuit. All physical electrical connections are made indirectly to the fixed contacts, normally through saddle clamps.

Contact Construction

In the past, a major problem with knife switches was that they were constructed from soft copper. Today, most contacts are made of a low-resistance silver alloy. Silver is alloyed (mixed) with cadmium or cadmium oxide to make an arc-resistant material which has good conductivity (low resistance). In addition, the silver alloy has good mechanical strength, enabling it to endure the continual wear encountered by many openings and closings. Another advantage of silver-alloy contacts is that the oxide that forms on the metal is an excellent conductor of electricity. Even when the contacts appear dull or tarnished, they are still capable of operating normally. **See Figure 8-6.**

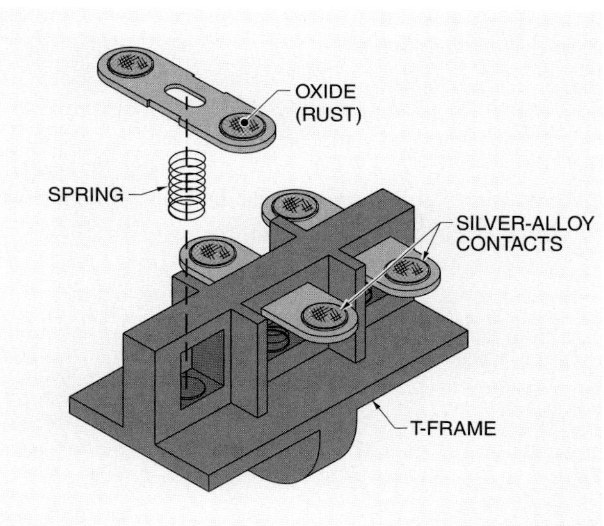

Figure 8-6. The oxide that forms on silver-alloy contacts is an excellent conductor of electricity.

Manual contactors directly control power circuits. Power circuit wiring is shown on a wiring diagram. An understanding of a wiring diagram is required because an electrician may be required to make changes in power circuits as well as in control circuits. **See Figure 8-7.**

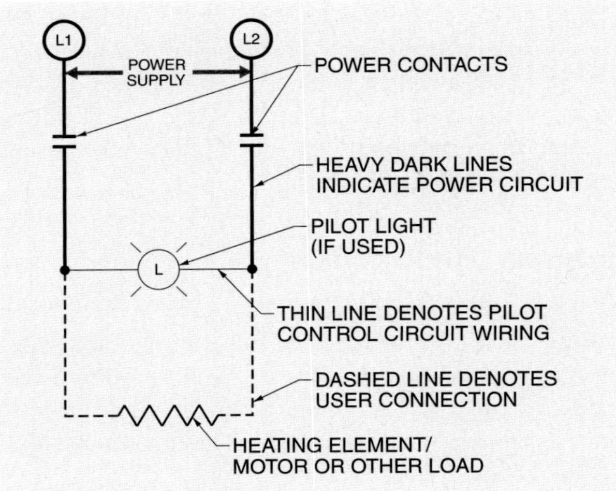

Figure 8-7. A wiring diagram shows the connection of an installation or its component devices or parts.

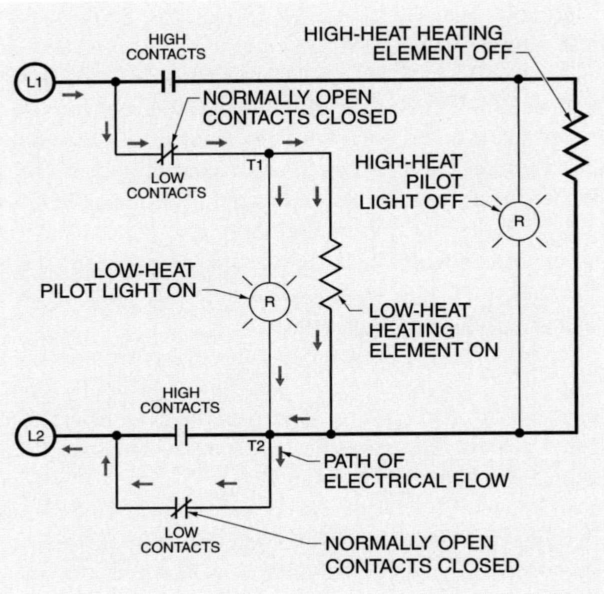

Figure 8-8. In the wiring diagram for a dual-element heater with pilot lights, the low-heat heating element is operated when the low contacts in L1 and L2 are closed so that a connection is made to the low and common terminals of the heater.

The wiring diagram for a double-pole manual contactor and pilot light shows the power contacts and their connection to the load. As in a line diagram, the power circuit is indicated through heavy dark lines and the control circuit is indicated by thin lines. In this circuit, current passes from L1 through the pilot light on to L2, causing the pilot light to glow when the power contacts in L1 and L2 close. At the same time, current passes from L1 through the heating element and on to L2, causing the heating element to be activated. The pilot light and heating element are connected in parallel with each other.

Wiring diagrams may be complex. For example, the wiring diagram for a dual-element heater with pilot lights contains various circuit paths. In this circuit, the low-heat heating element is operated when the low contacts in L1 and L2 are closed so that a connection is made to the low and common terminals of the heater. This allows the low-heat heating element to be energized. **See Figure 8-8.**

To operate the high-heat heating element, the high contacts in L1 and L2 are closed so that a connection is made to the high and common terminals of the heater. This allows the high-heat heating element to be energized. A low-heat pilot light and high-heat pilot light turn ON to indicate each condition because each pilot light is in parallel with the appropriate heating element. **See Figure 8-9.**

One problem that may arise with a dual-element start is that someone may try to energize both sets of elements at the same time. This causes serious damage to the heater. To prevent this problem from occurring, most manual contactors are equipped with a mechanical interlock.

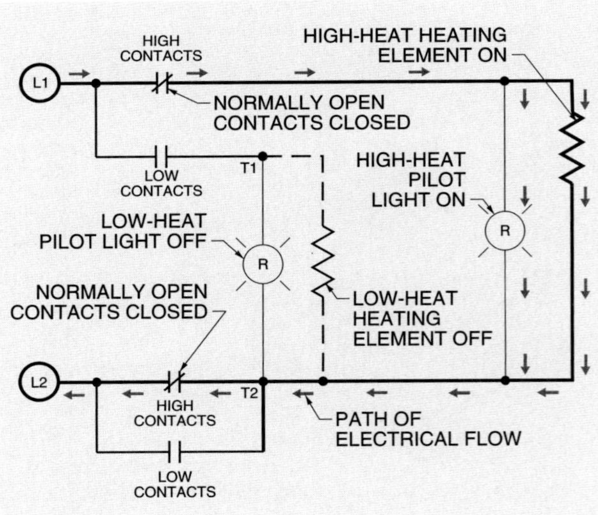

Figure 8-9. In the wiring diagram for a dual-element heater with pilot lights, the high-heat heating element is operated when the high contacts in L1 and L2 are closed so that a connection is made to the high and common terminals of the heater.

A *mechanical interlock* is the arrangement of contacts in such a way that both sets of contacts cannot be closed at the same time. Mechanical interlocking can be established by a mechanism that forces open one set of contacts while the other contacts are being closed. Another method is to provide a blocking bar or holding mechanism that does not allow the first set of contacts to close until the second set of contacts opens. An electrician can determine if a device is mechanically interlocked by consulting the wiring diagram information provided by the manufacturer. This information is normally packaged with the equipment when it is delivered or is attached to the inside of the enclosure.

MANUAL STARTERS

A *manual starter* is a contactor with an added overload protection device. Manual starters are used only in electrical motor circuits. The primary difference between a manual contactor and a manual starter is the addition of an overload protection device. **See Figure 8-10.**

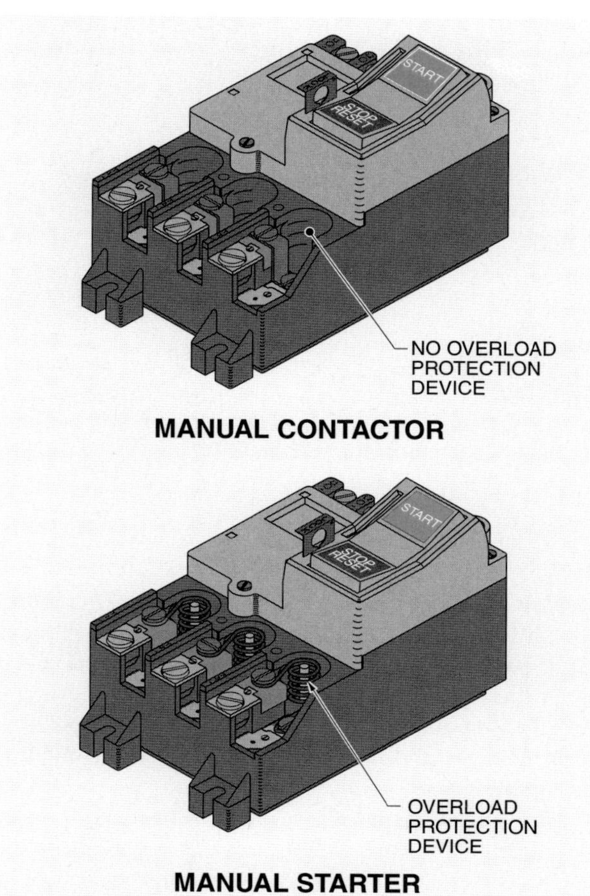

Figure 8-10. A manual starter is a contactor with an added overload protection device.

The overload protection device must be added because the National Electrical Code® (NEC®) requires that a control device shall not only turn a motor ON and OFF, but shall also protect the motor from destroying itself under an overloaded situation, such as a locked rotor. A *locked rotor* is a condition when a motor is loaded so heavily that the motor shaft cannot turn. A motor with a locked rotor draws excessive current and burns up if not disconnected from the line voltage. To protect the motor, the overload device senses the excessive current and opens the circuit.

Overload Protection

A motor goes through three stages during normal operation: resting, starting, and operating under load. **See Figure 8-11.** A motor at rest requires no current because the circuit is open. A motor that is starting draws a tremendous inrush current (normally six to eight times the running current) when the circuit is closed. Fuses or circuit breakers must have a sufficiently high ampere rating to avoid the immediate opening of the circuit caused by the large inrush current required for a motor when starting.

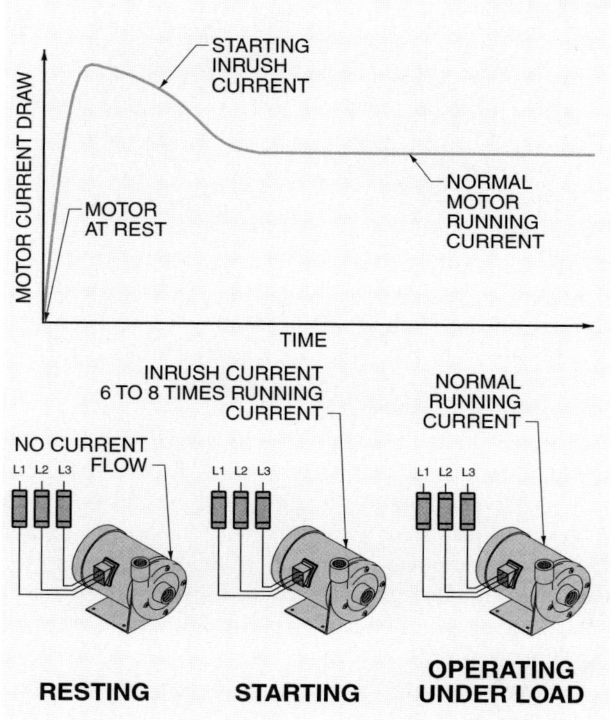

Figure 8-11. The three stages a motor goes through during normal operation include resting, starting, and operating under load.

A motor may encounter an overload while running. While it may not draw enough current to blow the fuses or trip the circuit breakers, it is large enough to produce sufficient heat to burn up the motor. The intense heat concentration generated by excessive current in the windings causes the insulation to fail and burn the motor. It is estimated that every 1°C (1.8°F) rise over normal ambient temperature ratings for insulation can reduce the life expectancy of a motor by almost a year. *Ambient temperature* is the temperature of the air surrounding a motor. The normal rating for many motors is about 40°C (104°F).

Fuses or circuit breakers must protect the circuit against the very high current of a short circuit or a ground fault. An overload relay is required that does not open the circuit while the motor is starting, but opens the circuit if the motor gets overloaded and the fuses do not blow. **See Figure 8-12.**

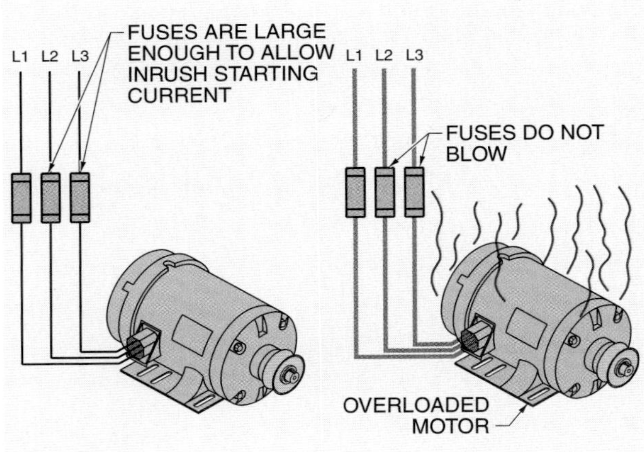

Figure 8-12. An overload relay does not open a circuit while a motor is starting, but opens the circuit if the motor gets overloaded and the fuses do not blow.

To meet motor protection needs, overload relays are designed to have a time delay to allow harmless, temporary overloads without disrupting the circuit. Overload relays must also have a trip capability to open the circuit if mildly dangerous currents that could result in motor damage continue over a period of time. All overload relays have some means of resetting the circuit once the overload is removed.

Melting Alloy Overloads. Heat is the end product that destroys a motor. To be effective, an overload relay must measure the temperature of the motor by monitoring the amount of current being drawn. The overload relay must indirectly monitor the temperature conditions of the motor because the overload relay is normally located at some distance from the motor. One of the most popular methods of providing overload protection is to use a melting alloy overload relay.

A *heater coil* is a sensing device used to monitor the heat generated by excessive current and the heat created through ambient temperature rise. Many different types of heater coils are available. The operating principle of each is the same. A heater coil converts the excess current drawn by a motor into heat, which is used to determine whether the motor is in danger. **See Figure 8-13.**

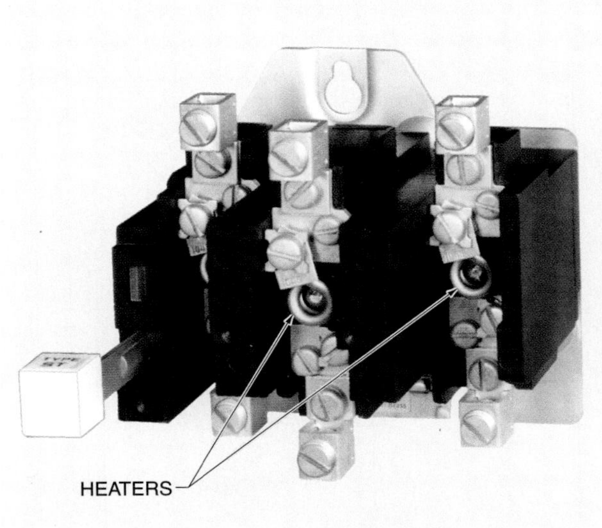

Cutler-Hammer

Figure 8-13. A heater coil is a sensing device used to monitor the heat generated by excessive current and the heat created through ambient temperature rise.

Most manufacturers rely on a eutectic alloy in conjunction with a mechanical mechanism to activate a tripping device when an overload occurs. A *eutectic alloy* is a metal that has a fixed temperature at which it changes directly from a solid to a liquid state. This temperature never changes and is not affected by repeated melting and resetting.

Most manufacturers use a ratchet wheel and eutectic alloy tube combination to activate a trip mechanism when an overload occurs. The eutectic alloy tube consists of an outer tube and an inner shaft connected to a ratchet wheel. The ratchet wheel is held firmly in the tube by the solid eutectic alloy. The inner shaft and ratchet wheel are locked into position by a pawl (locking mechanism) so that the wheel cannot turn when the alloy is cool. **See Figure 8-14.** Excessive current applied to the heater coil melts the eutectic alloy. This allows the ratchet wheel to turn freely.

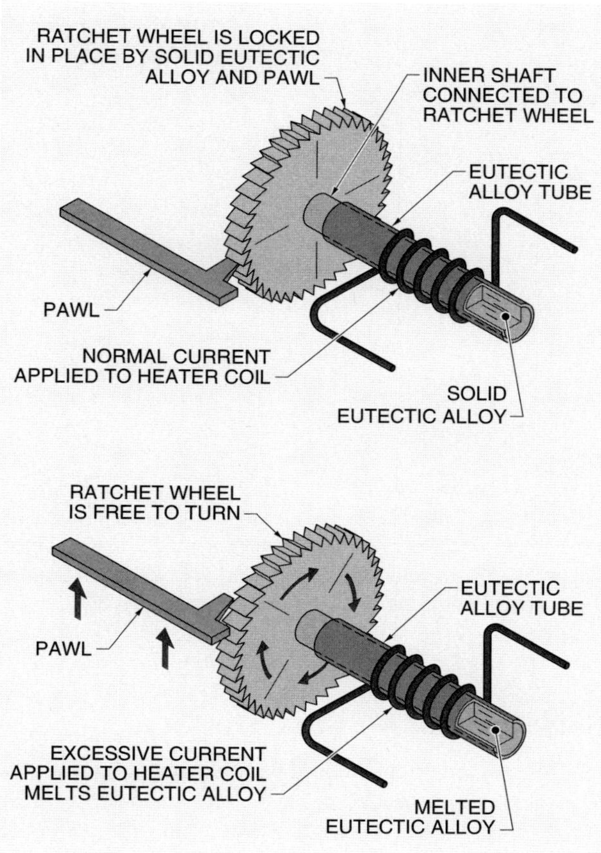

Figure 8-14. Most manufacturers use a ratchet wheel and eutectic alloy combination to activate a trip mechanism when an overload occurs.

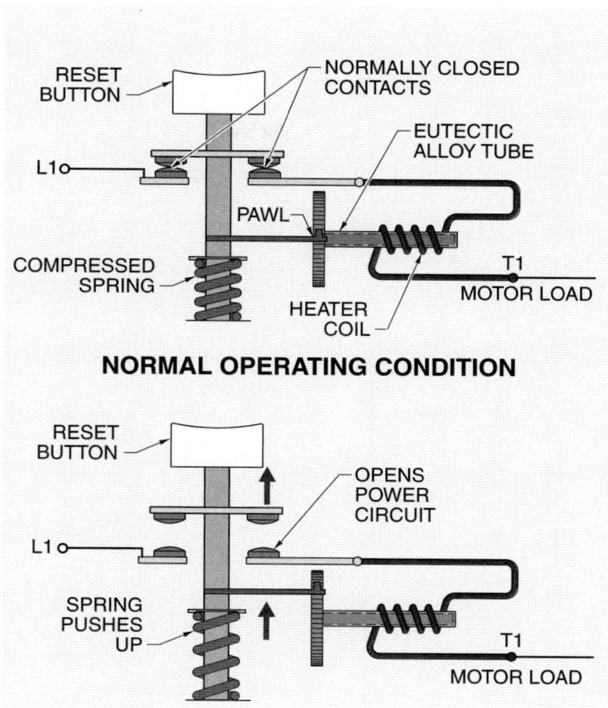

Figure 8-15. In a manual starter overload relay, the compressed spring tries to push the normally closed contacts open under normal operating conditions.

The main device in an overload relay is the eutectic alloy tube. The compressed spring tries to push the normally closed overload contacts open when motor current conditions are normal. The pawl is caught in the ratchet wheel and does not let the spring push up to open the contacts. **See Figure 8-15.**

The heater coil heats the eutectic alloy tube when an overload occurs. The heat melts the alloy, which allows the ratchet wheel to turn. The spring pushes the reset button up, which opens the contacts to the voltage coil of the contactor. The contactor opens the circuit to the motor, which stops the current flow through the heater coil. The heater coil cools, which solidifies the eutectic alloy tube.

Only the normally closed overload contacts open during an overload condition. The normally closed overload contacts can be manually reset to the closed position. The actual heating elements (heaters) installed in the motor starter do not open during an overload. The heaters are only used to produce heat. The higher the current draw of the motor, the more heat produced.

Resetting Overload Devices. The cause of an overload must be found before resetting an overload relay. A relay trips on resetting if the overload is not removed. Once the overload is removed, the device can be reset. The reset button is pushed, forcing the pawl across the ratchet wheel until the contacts are closed and the spring and ratchet wheel are returned to their original condition. The start pushbutton can then be pressed to start the motor. **See Figure 8-16.**

Nothing requires replacement or repair when an overload device trips because the heaters do not open like a fuse would. Once the cause of the overload is removed, the reset button may be pressed. Normally, a few minutes should be allowed for the eutectic alloy to cool.

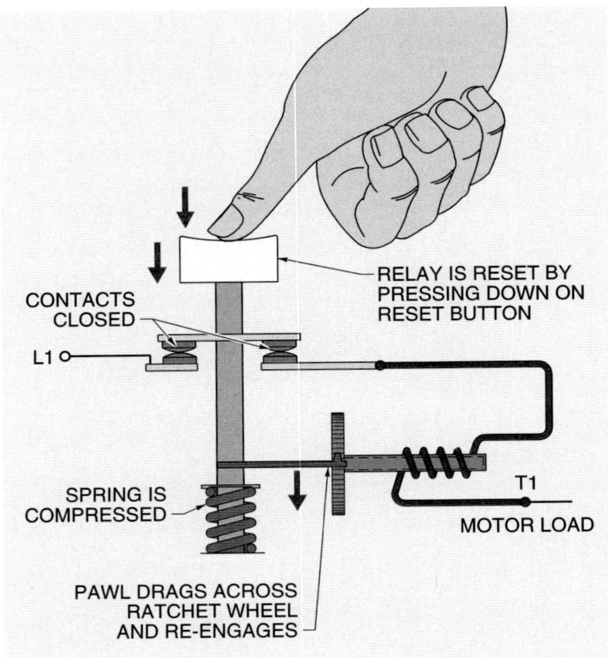

Figure 8-16. The overload relay is reset by pressing the reset button, which forces the pawl across the ratchet wheel until the contacts are closed and the spring and ratchet wheel are returned to their original condition.

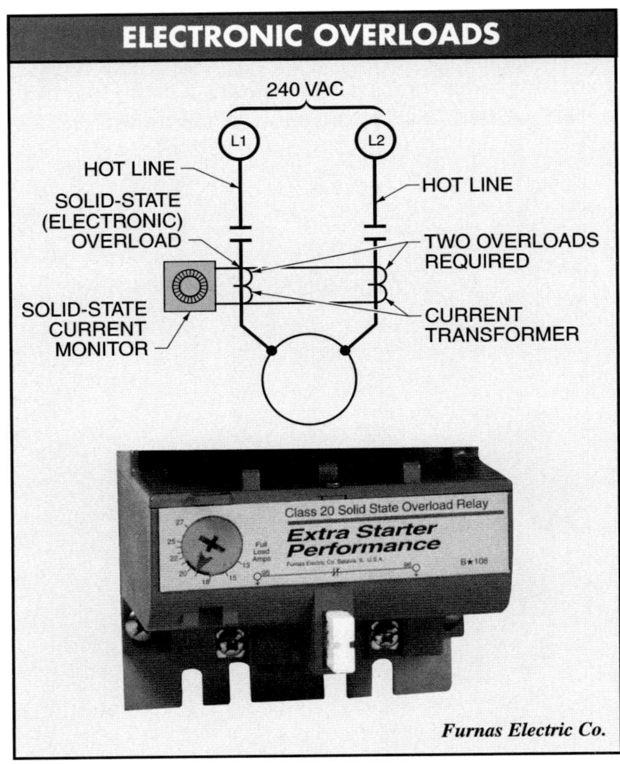

Figure 8-17. An electronic overload has built-in circuitry that senses changes in current and temperature.

The same basic overload relay is used with all sizes of motors. The only difference is that the heater coil size is changed. For small horsepower motors, a small heater coil is used. For large horsepower motors, a large heater coil is used. The NEC® should be checked for selection of appropriate overload heater sizes.

Electronic Overloads

New manual starters normally include an electronic overload instead of heaters. An *electronic overload* is a device that has built-in circuitry to sense changes in current and temperature. An electronic overload monitors the current in the load (motor, heating elements, etc.) directly by measuring the current in the power lines leading to the load. The electronic overload is built directly into the motor starter. **See Figure 8-17.**

An electronic overload measures the strength of the magnetic field around a wire instead of converting the current into heat. The higher the current in the wire leading to a motor, the stronger the magnetic field produced. An electronic circuit is used to activate a disconnecting device that opens the starter power contacts. Electronic overloads have an adjustable range. The setting is based on the nameplate current listed on the motor.

Selecting AC Manual Starters

Electricians are often required to select manual starters for new installations or replace ones that have been severely damaged due to an electrical fire or explosion. In either case, the electrician must specify certain characteristics of the starter to obtain the proper replacement. **See Figure 8-18.** Manual starters are selected based on phasing, number of poles, voltage, starter size, and enclosure type. Starter sizes are given in general motor protection tables. General motor protection tables indicate motor protection device sizes based on motor horsepower, current, fuse classification, and wire size. **See Appendix.**

Technical Fact

Most motor control circuits are powered from step-down transformers to reduce the voltage to the control circuit. A step-down transformer reduces the voltage to the control circuit to a level of 120 V or less.

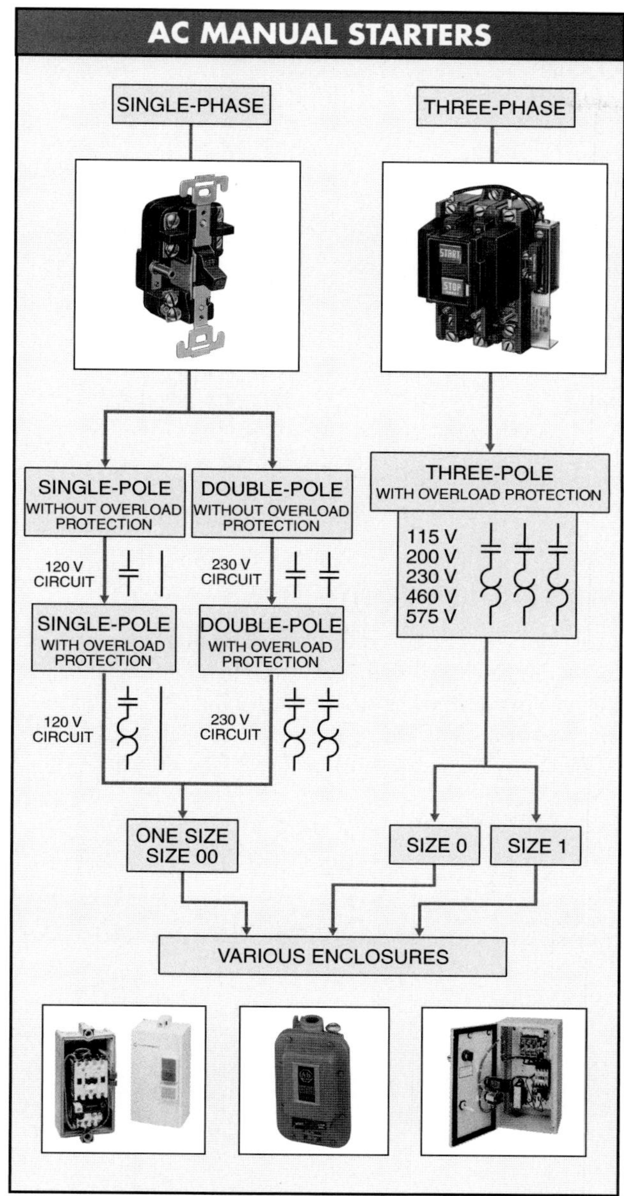

Rockwell Automation, Allen-Bradley Company, Inc.

Figure 8-18. Manual starters are selected considering phasing, number of poles, voltage, starter size, and enclosure type.

Phasing. AC manual starters/contactors can be divided into 1ϕ and 3ϕ contactors. **See Figure 8-19.** A 120 V, 1ϕ power source has one hot wire (ungrounded conductor) and one neutral wire (grounded conductor). A 230 V, 1ϕ power source has two hot wires: L1, L2 (ungrounded conductors), and no neutral. A 3ϕ power source has three hot wires: L1, L2, L3, and no neutral.

Single-phase manual starters are available as single-pole and double-pole devices because the NEC® requires that each ungrounded conductor (hot wire) be open when disconnecting a device. A single-pole device is used on 120 V circuits and a double-pole device is used on 230 V circuits.

Single-phase manual starters have limited horsepower ratings because of their physical size and are normally used as starters for motors of 1 HP or less. Single-phase manual starters are often available in only one size for all motors rated at 1 HP or less. The size established for 1ϕ starters is classified as National Electrical Manufacturers Association (NEMA) size 00. IEC manual starters/contactors are horsepower rated. Single-phase manual contactors and starters are normally used for 1ϕ, 1 HP and under motors where low-voltage protection is not needed. They are also used for 1ϕ motors that do not require a high frequency of operation.

Three-phase manual starters are physically larger than 1ϕ manual starters and may be used for motors of 10 HP or less. Three-phase manual contactors are normally pushbutton-operated instead of toggle-operated like 1ϕ starters.

Motor circuits require a manual starter that has overloads. Contactors, however, can be used in certain applications, such as in lighting circuits, without overload devices. In those cases, the fuse or CB in the main disconnect provides the overload protection.

Three-phase devices are designed with three-pole switching because 3ϕ devices have three hot wires that must be disconnected. Three-phase devices, like 1ϕ, use contacts and have quick-make and quick-break mechanisms. Three-phase contactors and starters are normally designed to be used on circuits from 115 V up to and including 575 V.

Three-phase starters are normally used for 3ϕ, 7.5 HP and under motors operating at 208/230 V, or 10 HP and under operating at 380/575 V. Three-phase starters are also used for 3ϕ motors where low-voltage protection is not needed, motors that do not require a high frequency of operation, and motors that do not need remote operation by pushbuttons or limit switches.

Enclosures. Enclosures provide mechanical and electrical protection for the operator and the starter. **See Appendix.** Although the enclosures are designed to provide protection in a variety of situations (water, dust, oil, and hazardous locations), the internal electrical wiring and physical construction of the starter remain the same.

Consult the NEC® and local codes to determine the proper selection of an enclosure for a particular application. For example, NEMA Type 1 enclosures are intended for indoor use primarily to provide a degree of protection against human contact with the enclosed equipment in locations where unusual service conditions do not exist.

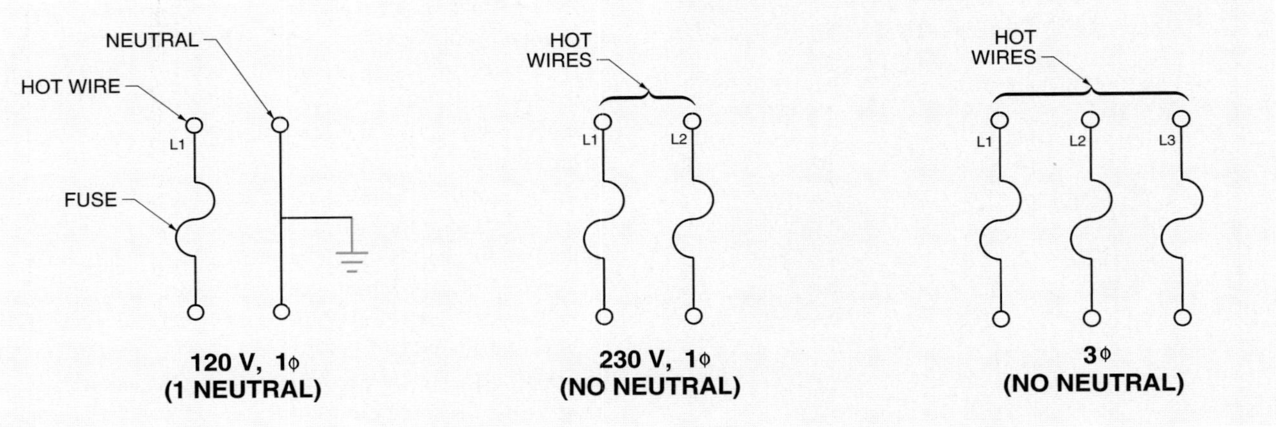

Figure 8-19. AC manual contactors can be divided into 1ϕ and 3ϕ contactors.

Manual Starter Applications

Manual motor starters are used in applications such as conveyor systems and drill presses. **See Figure 8-20.** In most applications, the manual starter provides the means of turning ON and OFF the device while providing motor overload protection.

MAGNETIC CONTACTORS

Contactors may be operated manually or magnetically. Contactors are devices for repeatedly establishing and interrupting an electrical power circuit. Contactors are used to make and break the electrical power circuit to loads such as lights, heaters, transformers, and capacitors. **See Figure 8-21.**

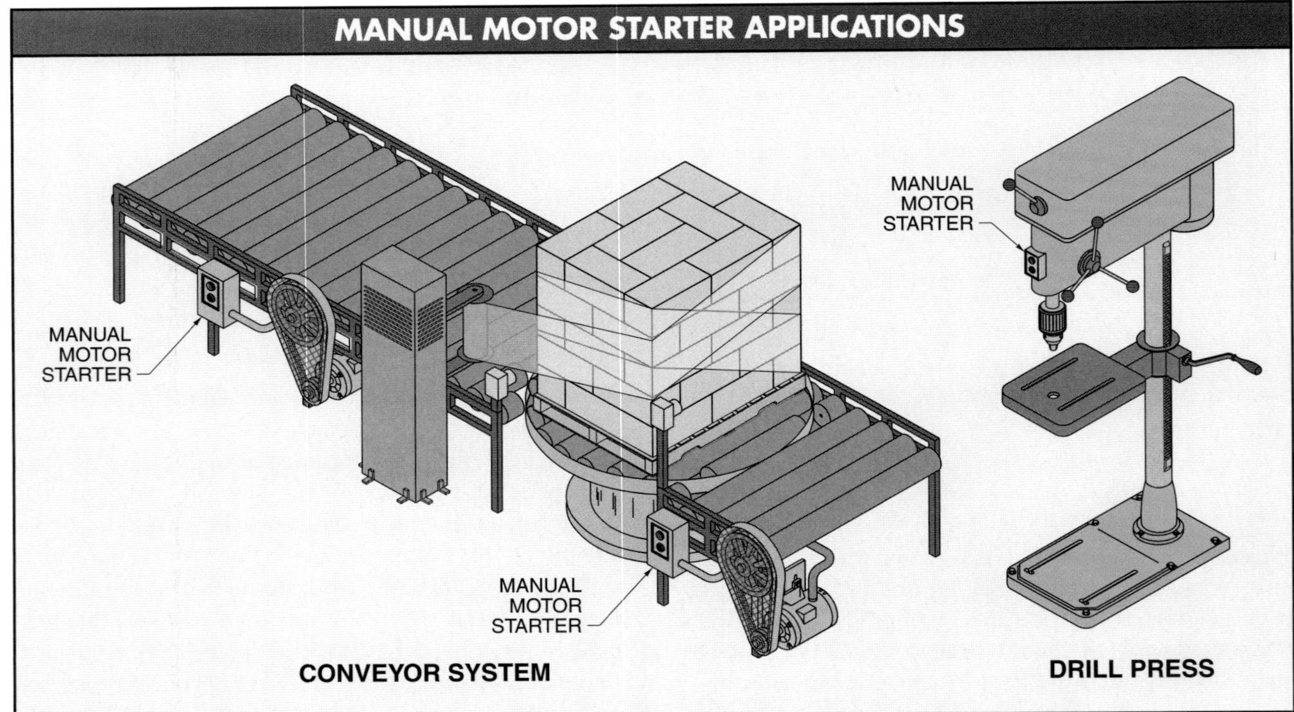

Figure 8-20. Manual motor starters are used in applications such as conveyor systems and drill presses.

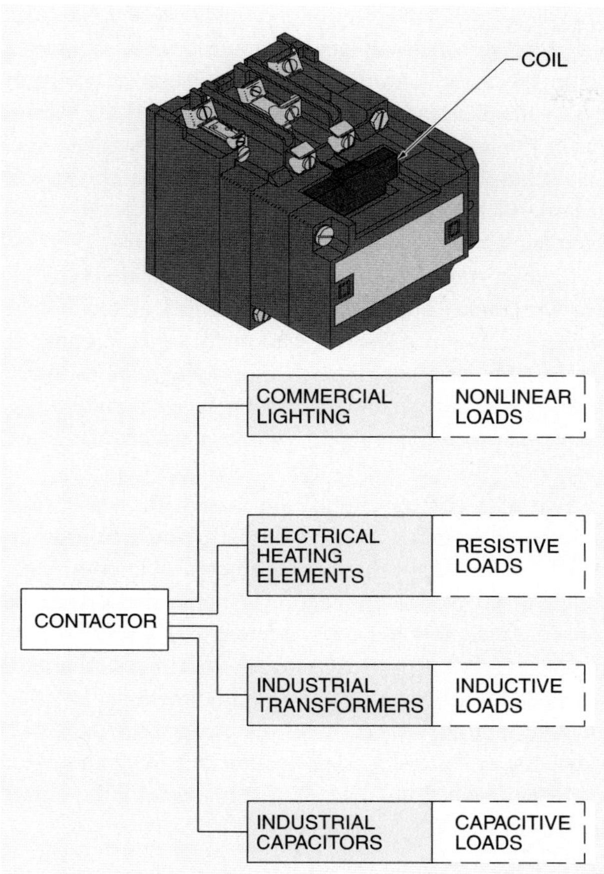

Figure 8-21. Contactors are used to make and break the electrical power circuit to lights, heaters, transformers, and capacitors.

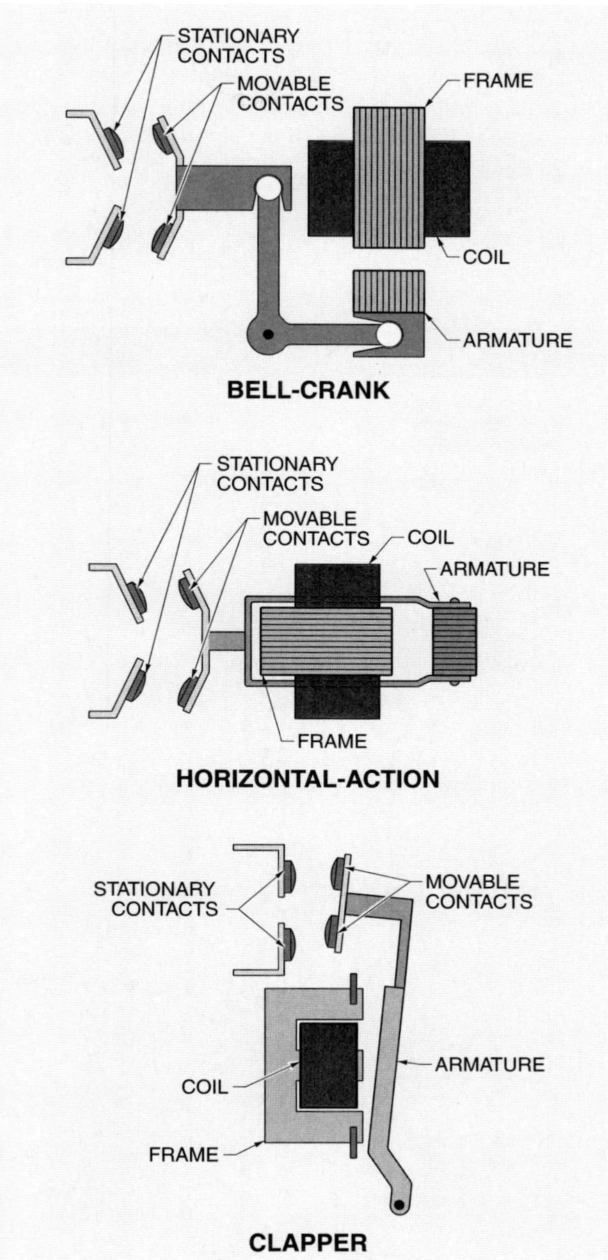

Figure 8-22. Solenoid action is the principal operating mechanism for magnetic contactors.

Magnetic Contactor Construction

Solenoid action is the principal operating mechanism for magnetic contactors. The linear action of a solenoid is used to open and close sets of contacts instead of pushing and pulling levers and valves. **See Figure 8-22.** The use of solenoid action rather than manual input is an advantage of a magnetic contactor over a manual contactor. Remote control and automation, which are impossible with manual contactors, can be designed into a system using magnetic contactors.

Magnetic Contactor Wiring

Control circuits are often referred to by the number of conductors used in the control circuit, such as two-wire and three-wire control. Two-wire control involves two conductors to complete the circuit. Three-wire control involves three conductors to complete the circuit.

Two-Wire Control. Two-wire control has two wires leading from the control device to the contactor or starter. **See Figure 8-23.** The control device could be a thermostat, float switch, or other contact device. When the contacts of the control device close, they complete the coil circuit of the contactor, causing it to energize. This connects the load to the line through the power contacts. The contactor coil is de-energized when the contacts of the control device open. This de-energizes coil C, which opens the contacts that control the load. The contactor functions automatically in response to the condition of the control device without the attention of an operator.

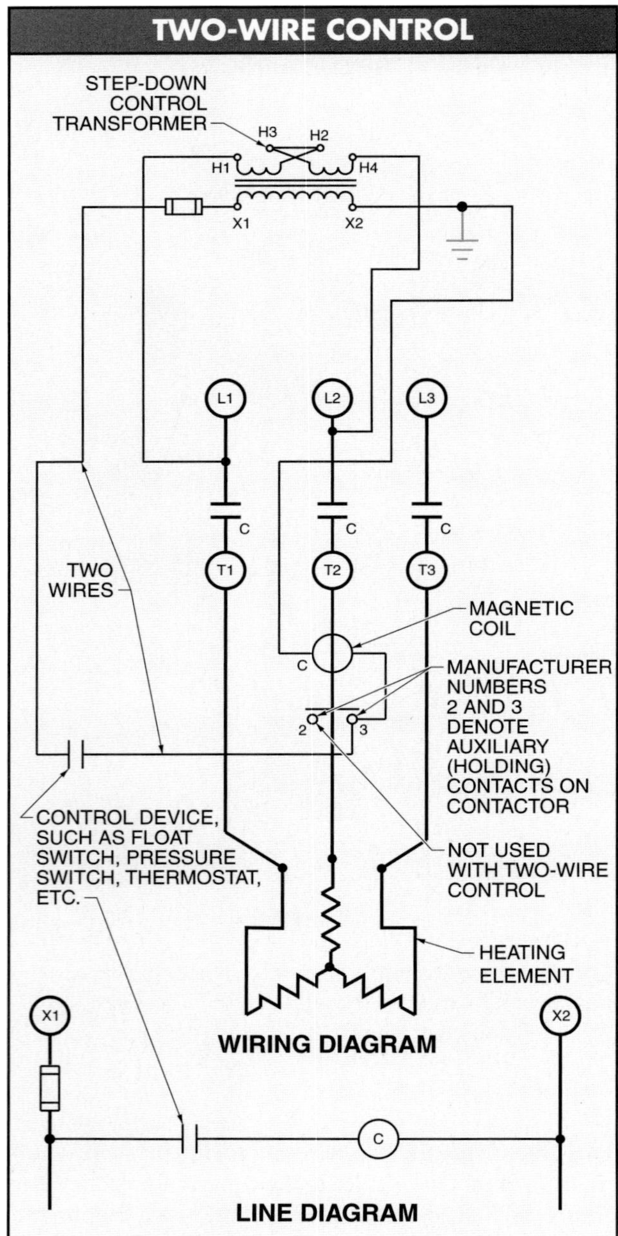

Figure 8-23. In two-wire control, two wires lead from the control device to the contactor or starter.

A two-wire control circuit provides low-voltage release, but not low-voltage protection. In the event of a power loss in the control circuit, the contactor de-energizes (low-voltage release), but also re-energizes if the control device remains closed when the circuit has power restored. Low-voltage protection cannot be provided in this circuit because there is no way for the operator to be protected from the circuit once it has been re-energized.

Caution must be exercised in the use and service of two-wire control circuits because of the lack of low-voltage protection. Two-wire control is normally used for remote or inaccessible installations, such as pumping stations, water or sewage treatment, air conditioning or refrigeration systems, and process line pumps where an immediate return to service after a power failure is required.

Two-wire control circuits are used with motor loads and nonmotor loads. Motor overload protection must be added to a contactor that is used to control a motor load. When motor overload protection is included as part of the contactor assembly, the unit is referred to as a motor starter. Contactors are not used to control motors unless the motor is a small horsepower motor (normally fractional HP) that includes internal protection, or the contactor is used with a separate motor overload protection unit. With nonmotor loads, the contactor is used to directly control the power applied to the load.

Three-Wire Control. Three-wire control has three wires leading from the control device to the starter or contactor. **See Figure 8-24.** The circuit uses a momentary contact OFF pushbutton (NC) wired in series with a momentary contact ON pushbutton (NO) wired in parallel to a set of contacts which form a holding circuit interlock (memory).

When the normally open ON pushbutton is pressed, current flows through the normally closed OFF pushbutton, through the momentarily closed ON pushbutton, through magnetic coil C, and on to L2. This causes the magnetic coil to energize. When energized, the auxiliary holding circuit interlock contacts (memory) close, sealing the path through to the coil circuit even if the start pushbutton is released.

Pressing the OFF pushbutton (NC) opens the circuit to the magnetic coil, causing the contactor to de-energize. A power failure also de-energizes the contactor. The interlock contacts (memory) reopen when the contactor de-energizes. This opens both current paths to the coil, through the ON pushbutton and the interlock.

Three-wire control provides low-voltage release and low-voltage protection. The coil drops out at low or no voltage and cannot be reset unless the voltage returns and the operator presses the start pushbutton.

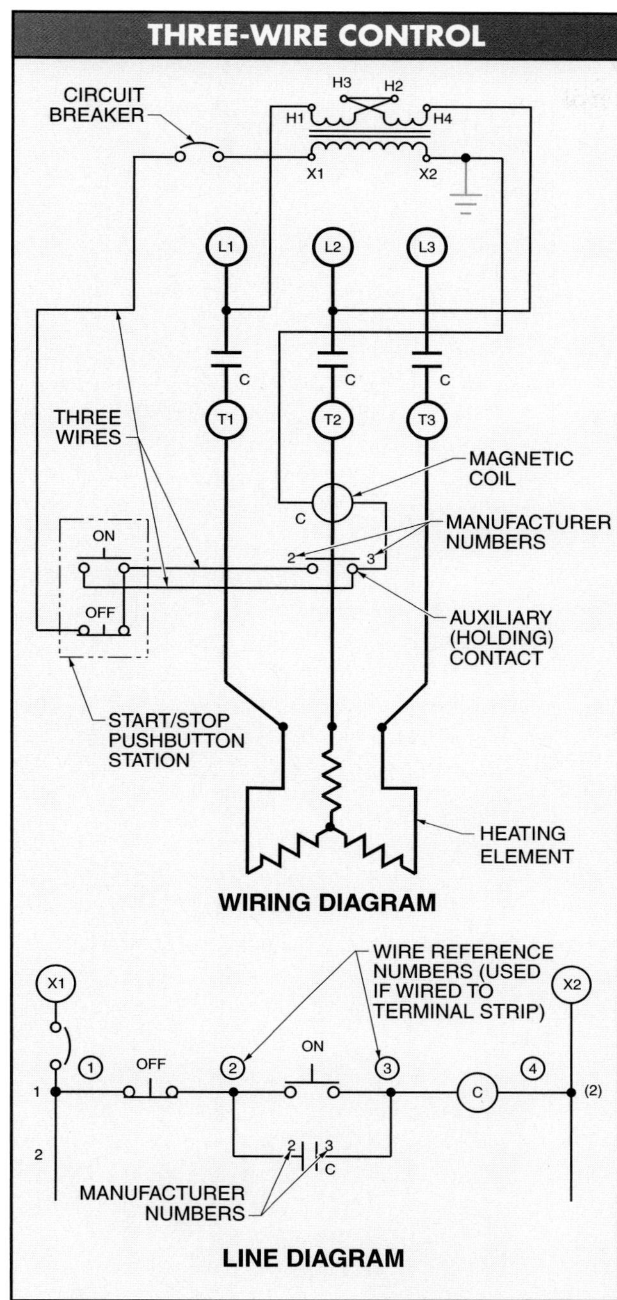

Figure 8-24. In three-wire control, three wires lead from the control device to the starter or contactor.

Control Circuit Voltage

Pushbuttons, limit switches, pressure switches, temperature switches, etc., are used to control the flow of power to the contactor (or motor starter) magnetic coil in the control circuit. When the control circuit is connected to the same voltage level as the load (lamps, heating elements, motors), the control circuit must be rated for the same voltage.

In most circuits in which the load is rated higher than 115 V (normally 208 V, 230 V, 240 V, 460 V, and 480 V), the control circuit is operated at a lower voltage level than the load. A step-down control transformer is used to step down the voltage to the level required in the control circuit. Normally, the secondary of the transformer is rated for 12 V, 24 V, or 120 V. The voltage of a control circuit can be any voltage (AC or DC), but is commonly less than 120 V. **See Figure 8-25.**

AC and DC Contactors

AC contactor assemblies may have several sets of contacts. DC contactor assemblies typically have only one set of contacts. **See Figure 8-26.** In 3ϕ AC contactors, all three power lines must be broken. This creates the need for several sets of contacts. For multiple contact control, a T-bar assembly allows several sets of contacts to be activated simultaneously. In a DC contactor, it is necessary to break only one power line.

AC contactor assemblies are made of laminated steel, while DC assemblies are solid. Laminations are unnecessary in a DC coil because the current travels in one direction at a continuous rate and does not create eddy current problems. The other major differences between AC and DC contactors are the electrical and mechanical requirements necessary for suppressing the arcs created in opening and closing contacts under load.

Products Unlimited

Contactors have higher current ratings than relays because they are used to control high-power loads.

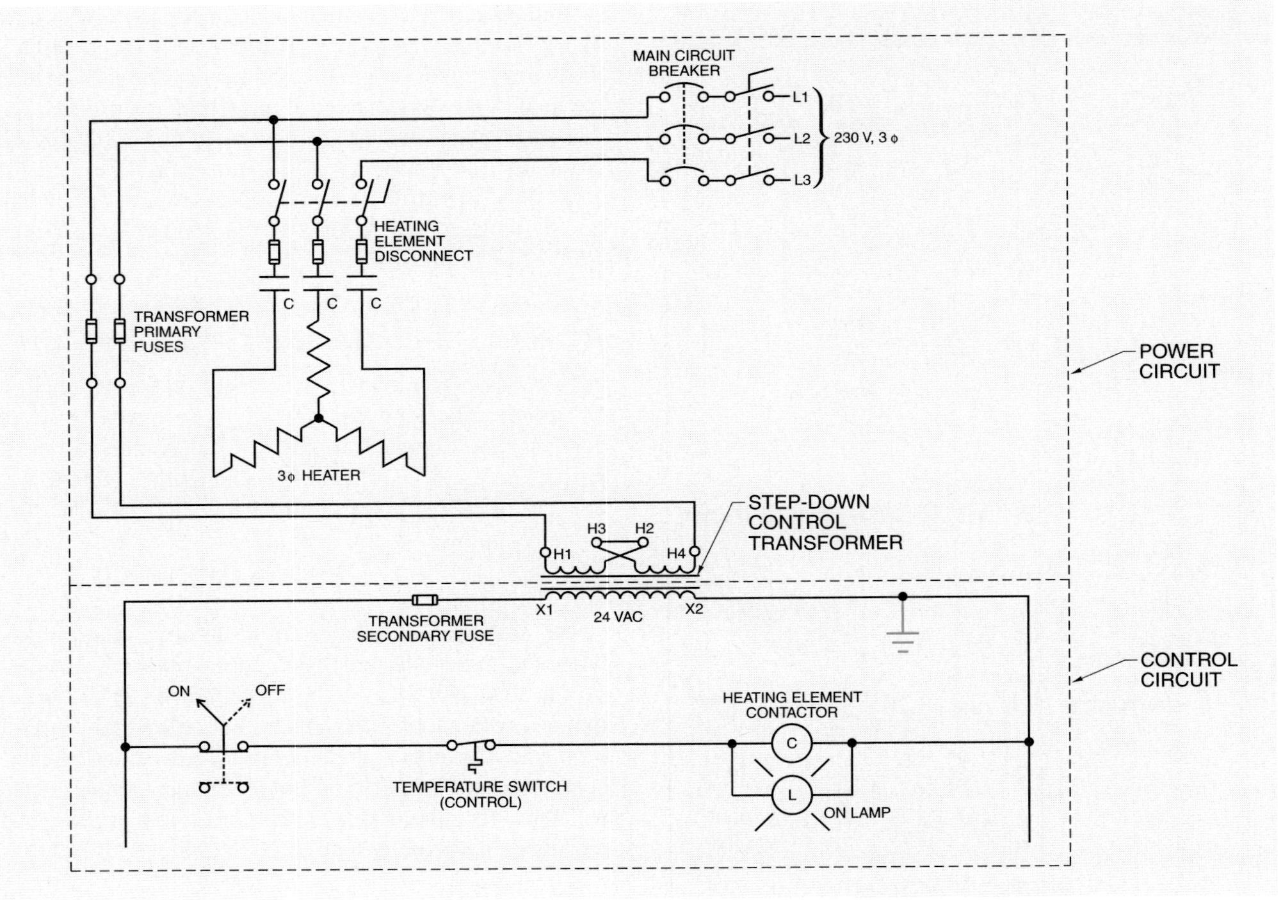

Figure 8-25. A step-down control transformer is used to step down the voltage to the level required in the control circuit.

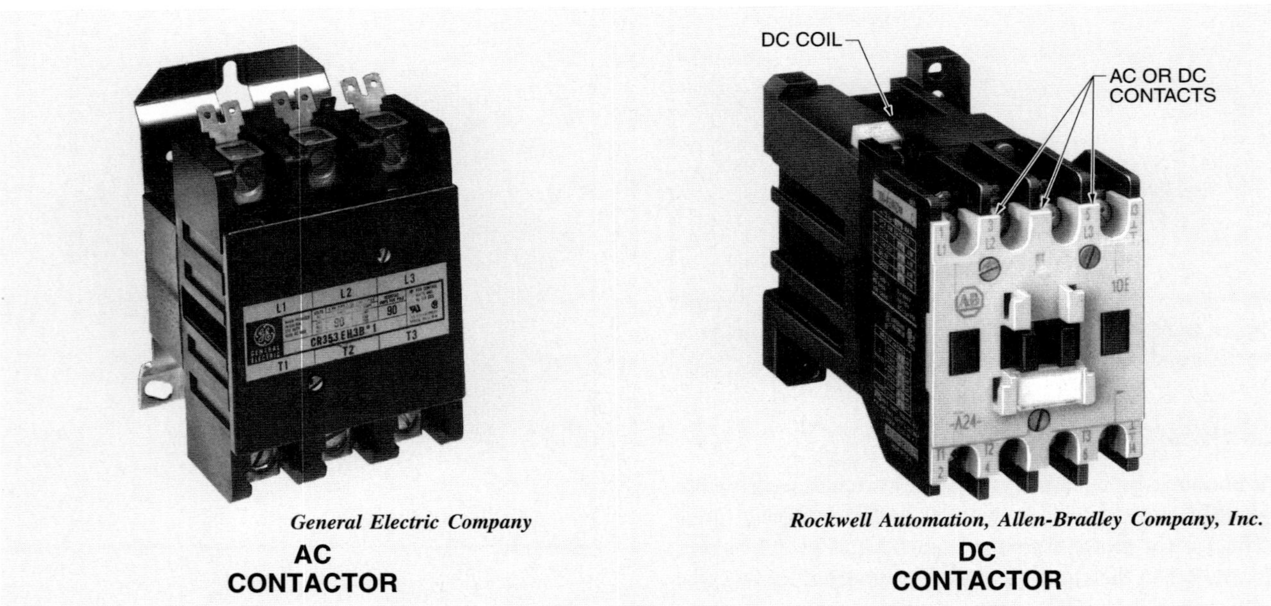

Figure 8-26. Contactors have either an AC coil or a DC coil, but may have either AC or DC contacts.

Arc Suppression

Arc suppression is required on contactors and motor starters. An *arc suppressor* is a device that dissipates the energy present across opening contacts. Without arc suppression, contactors and motors may require maintenance prematurely and could result in excessive downtime.

Opening Contact Arc. A short period of time (a few thousandths of a second) exists when a set of contacts is opened under load during which the contacts are neither fully in touch with each other nor completely separated. **See Figure 8-27.**

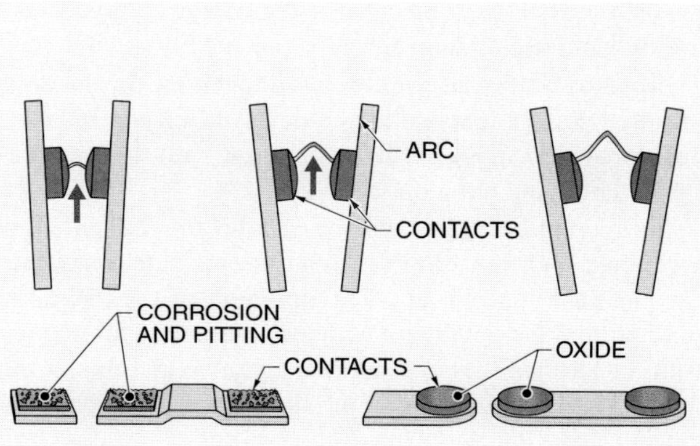

Figure 8-27. An electrical arc is created between contacts as they are opened. Prolonged arcing may result in damage to contact surfaces.

As the contacts continue to separate, the contact surface area decreases, increasing the electrical resistance. With full-load current passing through the increasing resistance, a substantial temperature rise is created on the surface of the contacts. This temperature rise is often high enough to cause the contact surfaces to become molten and emit ions of vaporized metal into the gap between the contacts. This hot ionized vapor permits the current to continue to flow in the form of an arc, even though the contacts are completely separated. The arcs produce additional heat, which, if continued, can damage the contact surfaces. The sooner the arc is extinguished, the longer the life expectancy of the contacts.

DC Arc Suppression. DC arcs are considered the most difficult to extinguish because the continuous DC supply causes current to flow constantly and with great stability across a much wider gap than does an AC supply of equal voltage. To reduce arcing in DC circuits, the switching mechanism must be such that the contacts separate rapidly and with enough of an air gap to extinguish the arc as soon as possible on opening. DC contactors are larger than AC contactors to allow for the additional air gap. In addition, the operating characteristics of DC contactors are faster than AC contactors.

When closing DC contacts, it is necessary to move the contacts together as quickly as possible to avoid some of the same problems encountered in opening them. One disadvantage in rapid closing of DC contactors is that the contacts must be buffered to eliminate contact bounce due to excessive closing force. Contact bounce may be minimized through the use of certain types of solenoid action and springs attached under the contacts to absorb some of the shock.

AC Arc Suppression. An AC arc is self-extinguishing when a set of contacts is opened. In contrast to a DC supply of constant voltage, an AC supply has a voltage that reverses its polarity every $\frac{1}{120}$ of a second when operated on a 60 hertz (Hz) line frequency. The alternation allows the arc to have a maximum duration of no more than a half-cycle. During any half-cycle, the maximum arcing current is reached only once in that half-cycle. **See Figure 8-28.**

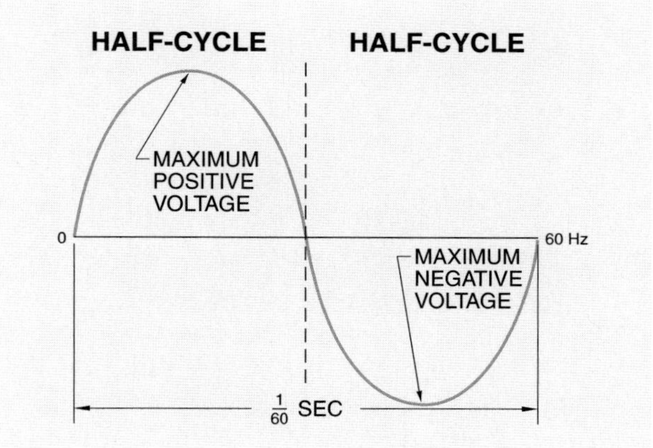

Figure 8-28. The maximum arcing current is reached only once during any half-cycle of AC voltage.

The contacts can be separated more slowly and the gap length may be shortened because an AC arc is self-extinguishing. This short gap keeps the voltage across the gap and the arc energy low. With low gap energy, ionizing gases cool more rapidly, extinguishing the arc and making it difficult to restart. AC contactors need less room to operate and run cooler, which increases contact life.

Arc at Closing. Arcing may also occur on AC and DC contactors when they are closing. The most common arcing occurs when the contacts come close enough that an arc is able to bridge the open space between the contacts.

Arcing also occurs if a whisker or rough edge of the contact touches first and melts, causing an ionized path that allows current to flow. In either case, the arc lasts until the contact surfaces are fully closed. Contactor design is quite similar for both AC and DC devices. The contactor should be designed so that the contacts close as rapidly as possible, without bouncing, to minimize the arc at each closing.

Arc Chutes. An *arc chute* is a device that confines, divides, and extinguishes arcs drawn between contacts opened under load. **See Figure 8-29.** Arc chutes are used to contain large arcs and the gases created by them. Arc chutes employ the de-ion principle which confines, divides, and extinguishes the arc for each set of contacts.

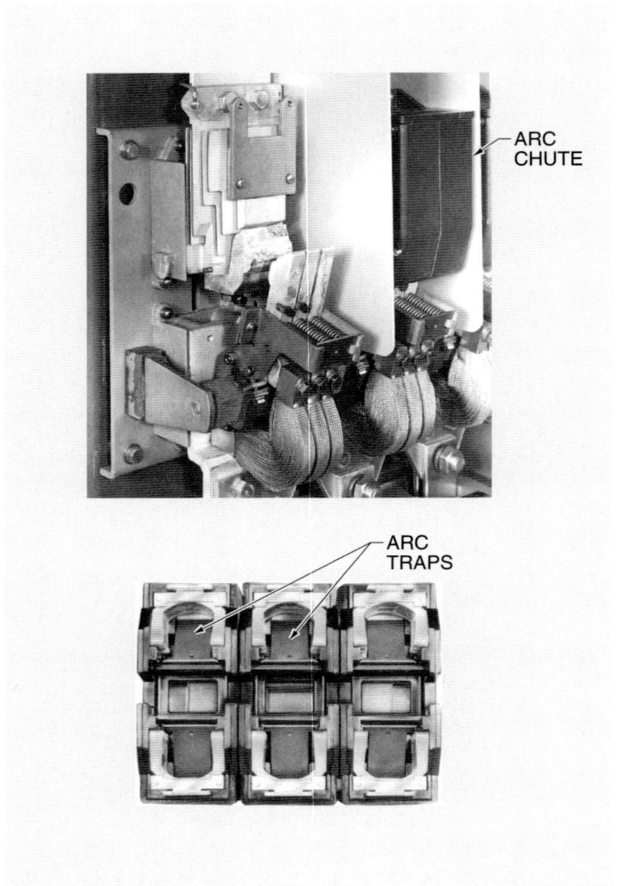

Figure 8-29. Arc chutes and arc traps are used to confine, divide, and extinguish arcs drawn between contacts opened under load.

Arcs may also be extinguished by using special arc traps and arc-quenching compounds. This circuit breaker technique attracts, splits, and quickly cools arcs as well as vents ionized gases. Vertical barriers between each set of contacts, as well as arc covers, confine arcs to separate chambers and quickly quench them.

DC Magnetic Blowout Coils. When a DC circuit carrying large amounts of current is interrupted, the collapsing magnetic field of the circuit current may induce a voltage that helps sustain the arc. Action must be taken to quickly limit the damaging effect of the heavy current arcs because a sustained electrical arc may melt the contacts, weld them together, or severely damage them.

One way to stop the arc quickly is to move the contacts some distance from each other as quickly as possible. The problem is that the contactor has to be large enough to accommodate such a large air gap.

Magnetic blowout coils are used to reduce the distance required and yet quench arcs quickly. Magnetic blowout coils provide a magnetic field that blows out the arc similarly to blowing out a match.

A magnetic field is created around the current flow whenever a current flows through a conductive medium (in this case ionized air). The direction of the magnetic field around the conductor is determined by wrapping the right or left hand around the conductor. When the thumb on the right hand points in the direction of conventional current flow, the wrapping fingers point in the direction of the resulting magnetic field. When the thumb on the left hand points in the direction of electron current flow, the wrapping fingers point in the direction of the resulting magnetic field. **See Figure 8-30.**

The electron flow motor rule states that when a current-carrying conductor (represented by the middle finger) is placed in a parallel magnetic field (represented by the index finger), the resulting force or movement is in the direction of the thumb. This action occurs because the magnetic field around the current flow opposes the parallel magnetic field above the current flow. This makes the magnetic field above the current flow weaker, while aiding the magnetic field below the current flow, making the magnetic field stronger. The net result is an upward push that quickly elongates the arc current so that it breaks (blows out). An electromagnetic blowout coil is often referred to as a puffer because of its blowout ability. **See Figure 8-31.**

Contact Construction

Contact design and materials depend on the size, current rating, and application of the contactor. Double-break contacts are normally made of a silver-cadmium alloy. Single-break contacts in large contactors are frequently made of copper because of the low cost.

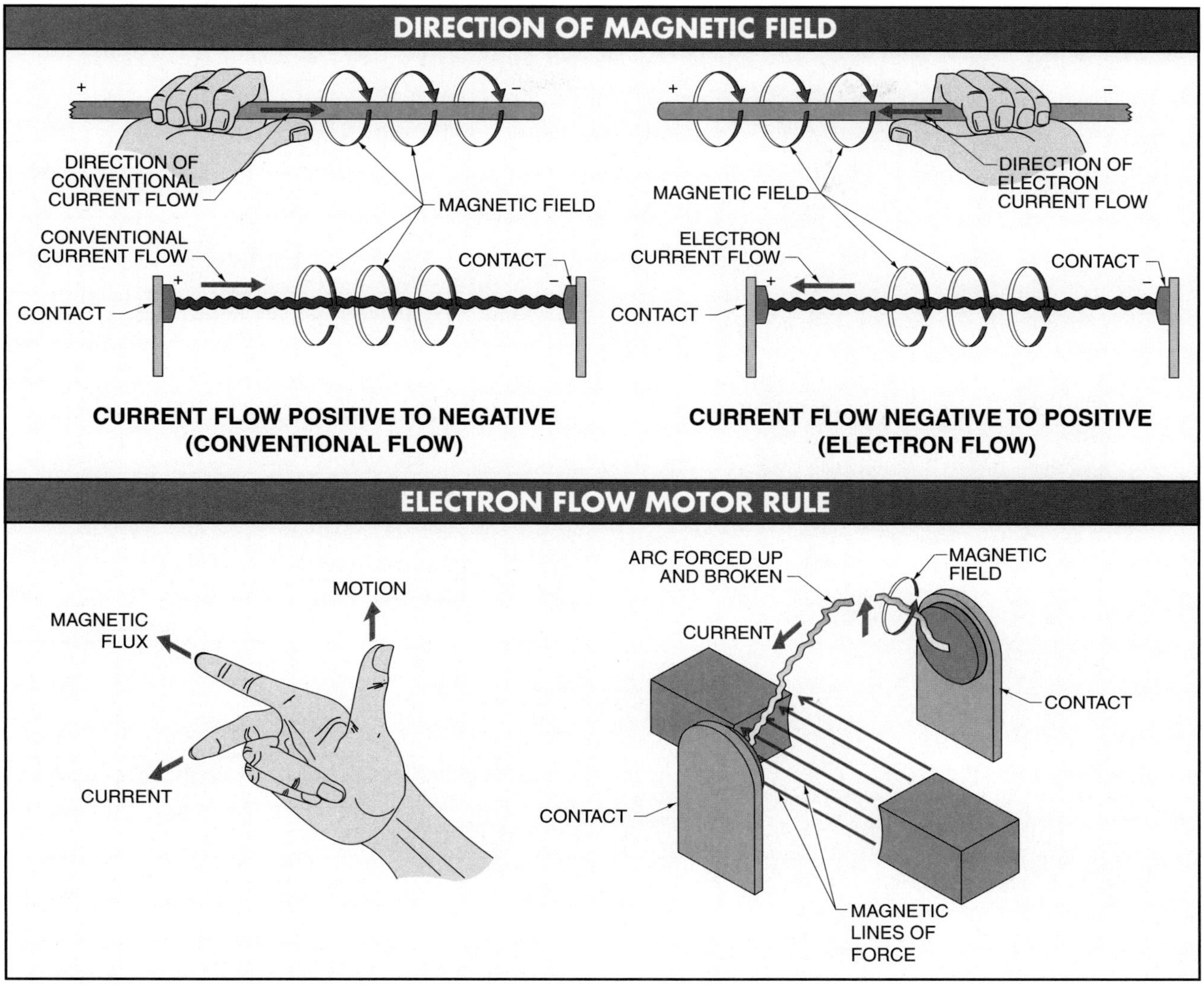

Figure 8-30. The direction of the magnetic field around the conductor is determined by wrapping the right or left hand around the conductor. The electron flow motor rule indicates the motion of an arc cutting through magnetic lines of force.

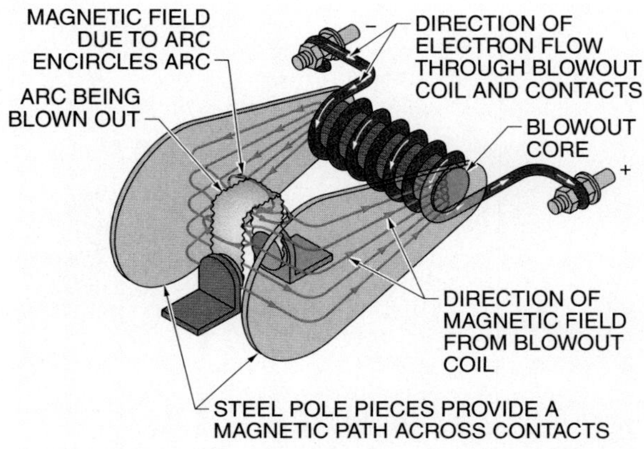

Figure 8-31. Electromagnetic blowout coils rapidly extinguish DC arcs.

Single-break copper contacts are designed with a wiping action to remove the copper oxide film that forms on the copper tips of the contacts. The wiping action is necessary because copper oxide formed on the contacts when not in use is an insulator and must be eliminated for good circuit conductivity.

Technical Fact

Contacts are actuated directly or by a linkage that is driven manually, mechanically, electromagnetically, hydraulically, or pneumatically.

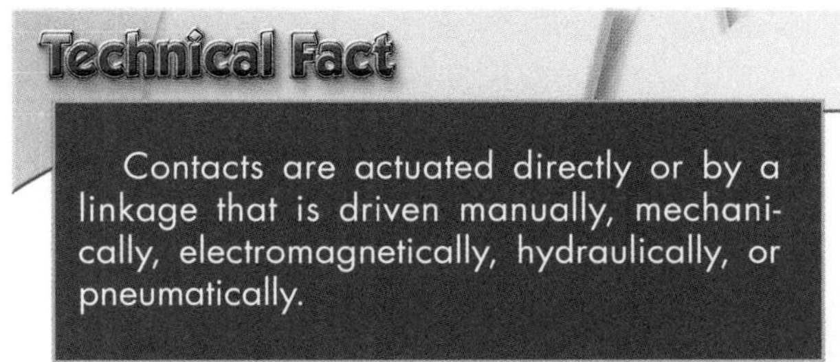

Damaged or burned contacts should be replaced immediately to prevent irreversible damage to motors and other electrical equipment.

In most cases, the slight rubbing action and burning that occur during normal operation keep the contact surfaces clean for proper operation. Copper contacts that seldom open or close, or those being replaced, should be cleaned to reduce contact resistance. High contact resistance often causes serious heating of the contacts.

General-Purpose AC/DC Contactor Sizes and Ratings

Magnetic contactors, like manual contactors, are rated according to the size and type of load by the National Electrical Manufacturers Association (NEMA). Tables are used to indicate the number/size designations and establish the current load carried by each contact in a contactor. **See Figure 8-32.** The rating is for each contact individually, not for the entire contactor. For example, a size 0, three-pole contactor rated at 18 A is capable of, and rated for, switching three separate 18 A loads simultaneously.

Contactor dimensions vary greatly. The range is from inches to several feet in length. Contactors are selected based on type, size, and voltage available. **See Figure 8-33.**

Contactors are also available in a variety of enclosures. The enclosures offer protection ranging from the most basic protection to high levels of protection required in hazardous locations where any spark caused by the closing or opening of the contact could cause an explosion.

60 Hz AC CONTACTOR STANDARD NEMA RATINGS

Size	Rating*	Power Rating†				
		3φ			1φ	
		200 V	230 V	230/460 V	115 V	230 V
00	9	1½	1½	2	⅓	1
0	18	3	3	5	1	2
1	27	7½	7½	10	2	3
2	45	10	15	25	3	7½
3	90	25	30	50	—	—
4	135	40	50	100	—	—
5	270	75	100	200	—	—
6	540	150	200	400	—	—
7	810	—	300	600	—	—
8	1215	—	450	900	—	—
9	2250	—	800	1600	—	—

* in A
† in HP

DC CONTACTOR STANDARD NEMA RATINGS

Size	Rating*	Power Rating†		
		115 V	230 V	550 V
1	25	3	5	—
2	50	5	10	20
3	100	10	25	50
4	150	20	40	.75
5	300	40	75	150
6	600	75	150	300
7	900	110	225	450
8	1350	175	350	700
9	2500	300	600	1200

* in A
† in HP

Figure 8-32. Tables indicate the number/size designations and establish the current load carried by each contact in a contactor.

Technical Fact

Contact forces of up to several hundred pounds are sometimes applied when contacts are in the closed position. For contacts that must interrupt high currents, refractory materials with antiwelding characteristics such as molybdenum and tungsten are used. These materials are essential for contacts that must close in on short-circuit currents and open at a later time under normal operating force.

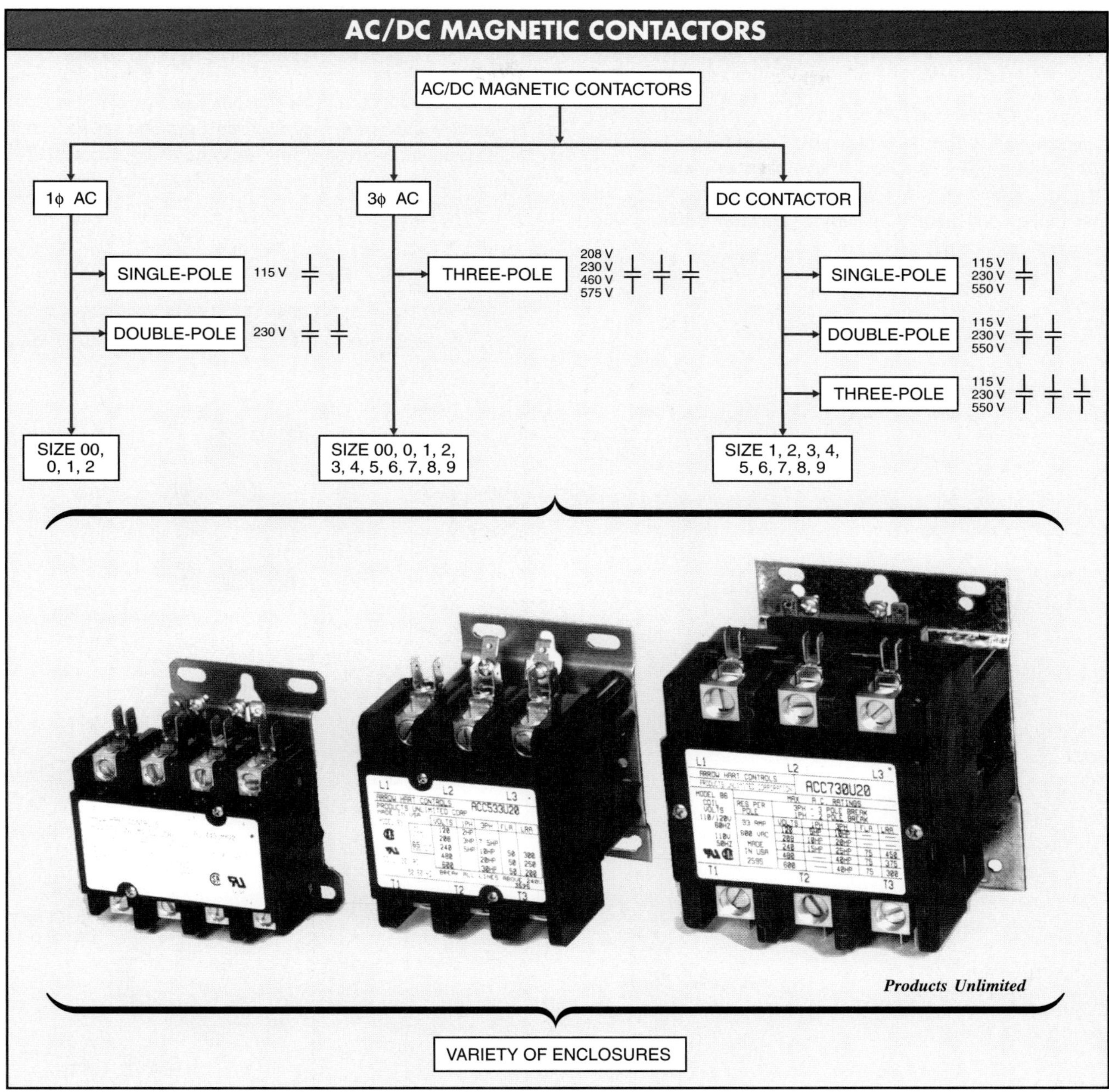

Figure 8-33. Contactor dimensions vary from inches to several feet in length.

MAGNETIC MOTOR STARTERS

A *magnetic motor starter* is an electrically operated switch (contactor) that includes motor overload protection. Magnetic motor starters include overload relays that detect excessive current passing through a motor and are used to switch all types and sizes of motors. Magnetic motor starters are available in sizes that can switch loads of a few amperes to several hundred amperes. **See Figure 8-34.**

Overload Protection

The main difference between the sensing device for a manual motor starter and a magnetic motor starter is that on a manual motor starter a manual overload opens the power contacts on the starter. The overload device on a magnetic motor starter opens a set of contacts to the magnetic coil, de-energizing the coil and disconnecting the power. Overload devices include melting alloy, magnetic, and bimetallic overload relays. The overload unit (heater) does not open, as a fuse or CB does, but only produces the heat required to open the overload contacts.

Melting Alloy Overload Relays. The melting alloy overload relays used in magnetic motor starters are similar to the melting alloy overload relays used in manual motor starters. They consist of a heater coil, eutectic alloy, and mechanical mechanism to activate a tripping device when an overload occurs.

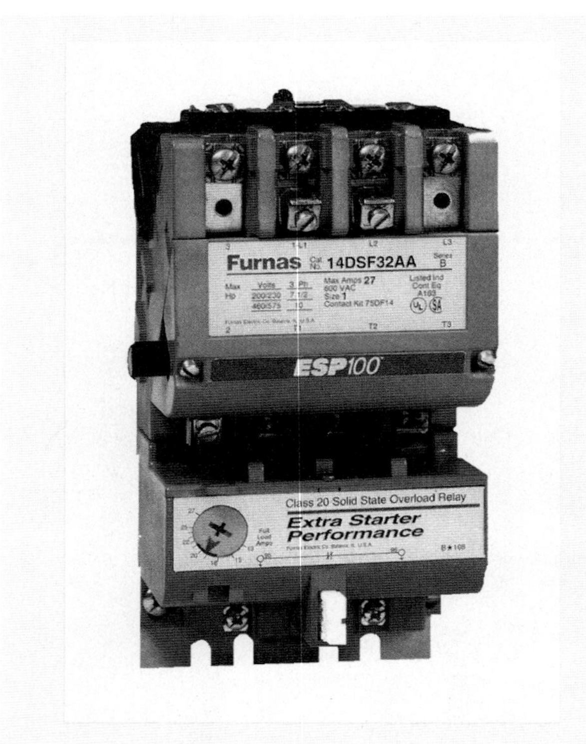

Furnas Electric Co.

Figure 8-34. A magnetic motor starter is a contactor with overload protection added.

Magnetic Overload Relays. Magnetic overload relays provide another means of monitoring the amount of current drawn by a motor. A magnetic overload relay operates through the use of a current coil. At a specified overcurrent value, the current coil acts as a solenoid, causing a set of normally closed contacts to open. This causes the circuit to open and protect the motor by disconnecting it from power. **See Figure 8-35.**

Magnetic overload relays are used in special applications, such as steel mill processing lines or other heavy-duty industrial applications where holding a specified level of motor current is required. A magnetic overload relay is also ideal for special applications, such as slow acceleration motors, high-inrush-current motors, or any use where normal time/current curves of thermal overload relays do not provide satisfactory operation. This flexibility is made possible because the magnetic unit may be set for either instantaneous or inverse time-tripping characteristics. The device may also offer independent adjustable trip time and trip current.

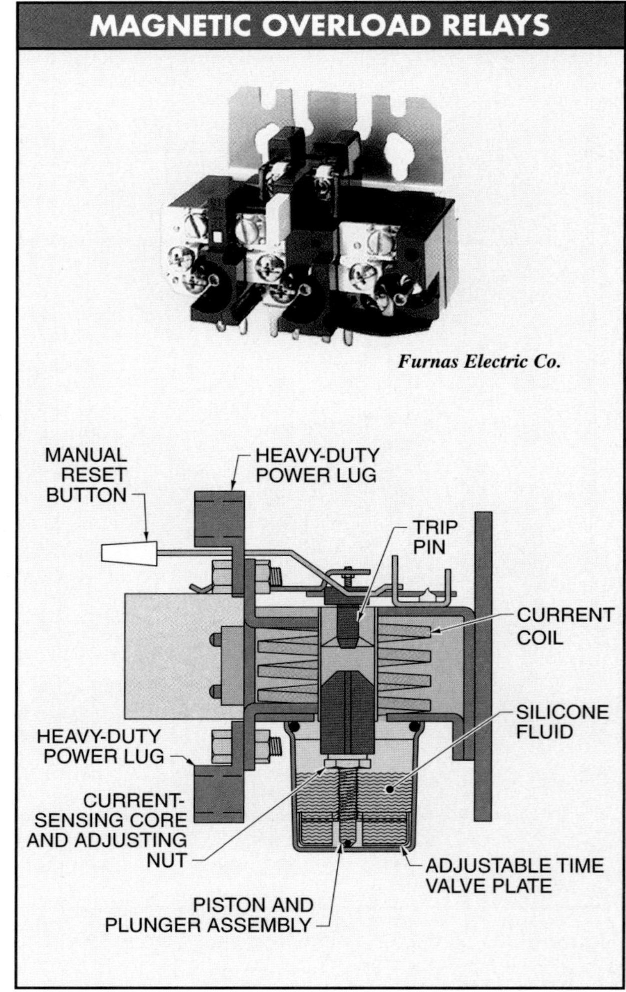

Figure 8-35. Magnetic overload relays use a current coil which, at a specific overcurrent value, acts like a solenoid and causes a set of normally closed contacts to open.

Magnetic overload relays are extremely quick to reset because they do not require a cooling-off period before being reset. Magnetic overload relays are much more expensive than thermal overload relays.

Bimetallic Overload Relays. In certain applications such as walk-in meat coolers, remote pumping stations, and some chemical process equipment, overload relays that reset automatically to keep the unit operating up to the last possible moment may be required. A *bimetallic overload relay* is an overload relay that resets automatically. Bimetallic overload relays operate on the principle of the bimetallic strip. A bimetallic strip is made of two pieces of dissimilar metal that are permanently joined by lamination. Heating the bimetallic strip causes it to warp because the dissimilar metals expand and contract at different rates. The warping effect of the bimetallic strip is used as a means of separating contacts. **See Figure 8-36.**

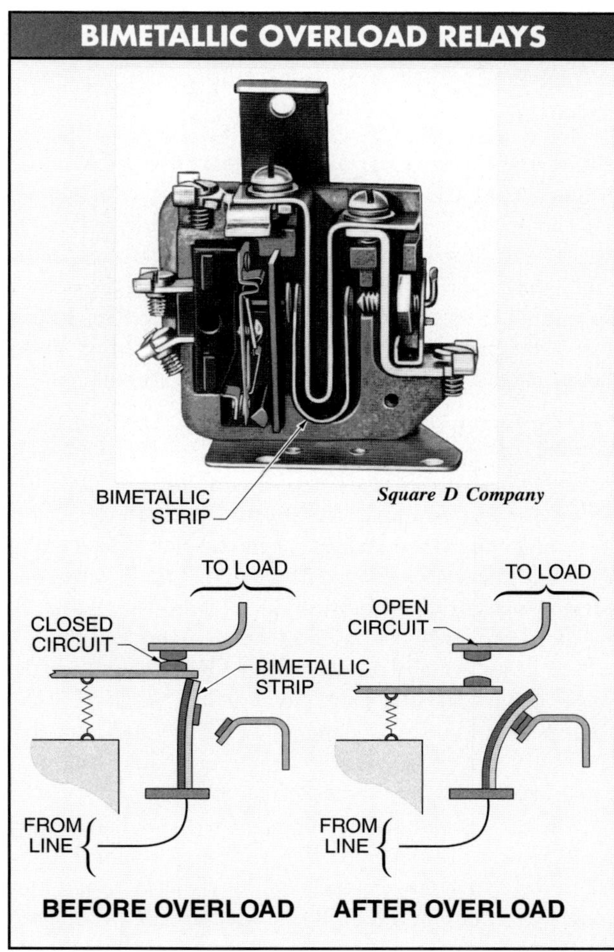

Figure 8-36. The warping effect of a bimetallic strip is used as a means for separating contacts.

Once the tripping action has taken place, the bimetallic strip cools and reshapes itself. In certain devices, such as circuit breakers, a trip lever needs to be reset to make the circuit operate again. In other devices, such as bimetallic overload relays, the device automatically resets the circuit when the bimetallic strip cools and reshapes itself.

The motor restarts even when the overload has not been cleared, and trips and resets itself again at given intervals. Care must be exercised in the selection of a bimetallic overload relay because repeated cycling eventually burns out the motor. The bimetallic strip may be shaped in the form of a U. The U-shape provides a uniform temperature response.

Trip Indicators. Many overload devices have a trip indicator built into the unit to indicate to the operator that an overload has taken place within the device. **See Figure 8-37.** A red metal indicator appears in a window located above the reset button when the overload relay has tripped. The red indicator informs the operator or electrician why the unit is not operating and that it potentially is capable of restarting with an automatic reset.

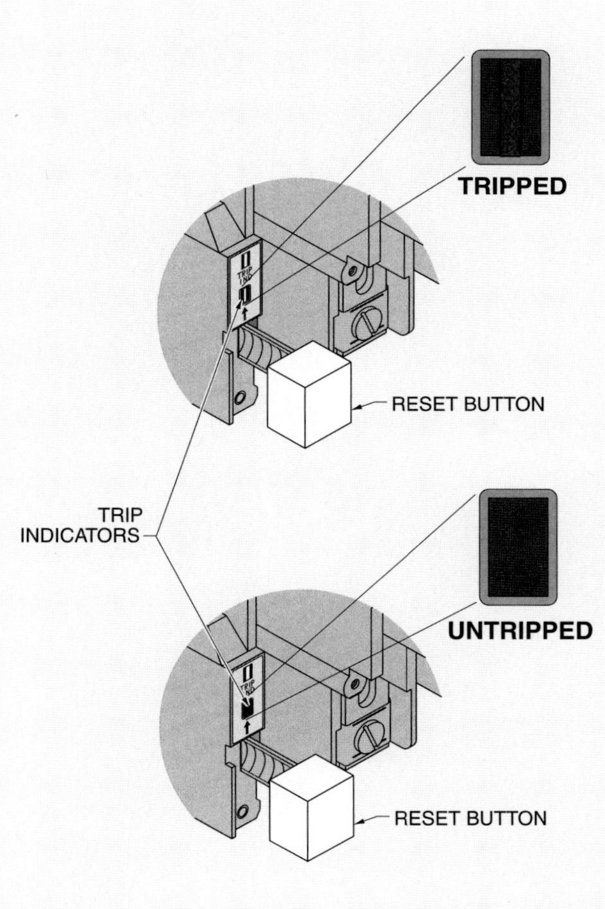

Figure 8-37. Trip indicators indicate that an overload has taken place within the device.

Overload Current Transformers. Large horsepower motors have currents that exceed the values of standard overload relays. To make the overload relays larger would greatly increase their physical size, which would create a space problem in relation to the magnetic motor starter. To avoid such a conflict, current transformers are used to reduce the current in a fixed ratio. **See Figure 8-38.** A current transformer is used to change the amount of current flowing to a motor but reduces the current to a lower value for the overload relay. For example, if 50 A were flowing to a motor, only 5 A would flow to the overload relay through the use of the current transformer. Standard current transformers are normally rated in primary and secondary rated current such as 50/5 or 100/5.

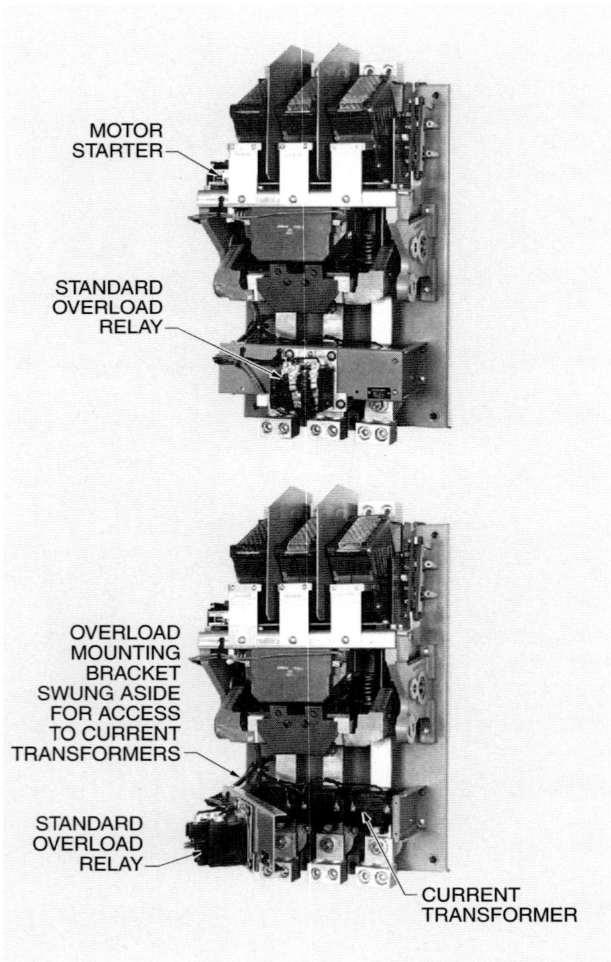

Cutler-Hammer

Figure 8-38. Standard overload relays may be used on very large starters by using current transformers with specific reduction ratios.

Because the ratio is always the same, an increase in the current to a motor also increases the current to the overload relay. If the correct current transformer and overload relay combination is selected, the same overload protection can be provided to a motor as if the overload relay were actually in the load circuit. The overload relay contacts open and the coil to the magnetic motor starter is de-energized when excessive current is sensed. This shuts the motor OFF. Several different current transformer ratios are available to make this type of overload protection easy to provide.

Overload Heater Sizes

Each motor must be sized according to its own unique operating characteristics and applications. Thermal overload heaters are selected based on the full-load current rating (FLC), service factor (SF), and ambient temperature (surrounding air temperature) of the motor when it is operating.

Full-Load Current Rating. Selection of thermal overload heaters is based on the full-load current shown on the motor nameplate or in the motor manufacturer specification sheet. The current value reflects the current to be expected when the motor is running at specified voltages, specified speeds, and normal torque operating characteristics. Heater manufacturers develop current charts indicating which heater should be used with each full-load current.

Service Factor. In most motor applications, there are times when the motor must produce more than its rated horsepower for a short period of time without damage. A *service factor (SF)* is a number designation that represents the percentage of extra demand that can be placed on a motor for short intervals without damaging the motor. Common service factors range from 1.00 to 1.25, indicating that the motor can produce 0% to 25% extra demand over that for which it is normally rated. A 1.00 SF indicates that the motor cannot produce more power than it is rated for and to do so would result in damage. A 1.25 SF indicates that the motor can produce up to 25% more power than it is rated for, but only for short periods of time.

The excessive current that can be safely handled by a given motor for short periods of time is approximated by multiplying the service factor by the full-load current. For example, if a motor is rated at an FLC of 10 A with an SF of 1.15, the excess short-term current equals 11.5 A ($10 \times 1.15 = 11.5$ A). The motor could handle an additional 1.5 A for a short period of time.

Ambient Temperature. A thermal overload relay operates on the principle of heat. When an overload takes place, sufficient heat is generated by the excessive current to melt a metal alloy, produce movement in a current coil, or warp a bimetallic strip and allow the device to trip. The temperature surrounding a thermal overload relay must be considered because the relay is sensitive to heat from any source. The ambient temperature is a factor when considering moving a thermal overload relay from a refrigerated meat packing plant to a location near a blast furnace.

Overload relay devices are normally rated to trip at a specific current when surrounded by an ambient temperature of 40°C (104°F). This standard ambient temperature is acceptable for most control applications. Compensation must be provided for higher or lower ambient temperatures.

Overload Heater Selection

Overload heater coils for continuous-duty motors are selected from manufacturer tables based on the motor nameplate full-load current for maximum motor protection and compliance with Section 430.32 of the NEC®. The class, type, and size information of a magnetic motor starter are found on the nameplate on the face of the starter. **See Figure 8-39.** The phase, service factor, and full-load current of the motor are determined from the motor nameplate. Common applications use 40°C as the ambient temperature. Questionable ambient temperatures should be measured at the job site or determined by some other method.

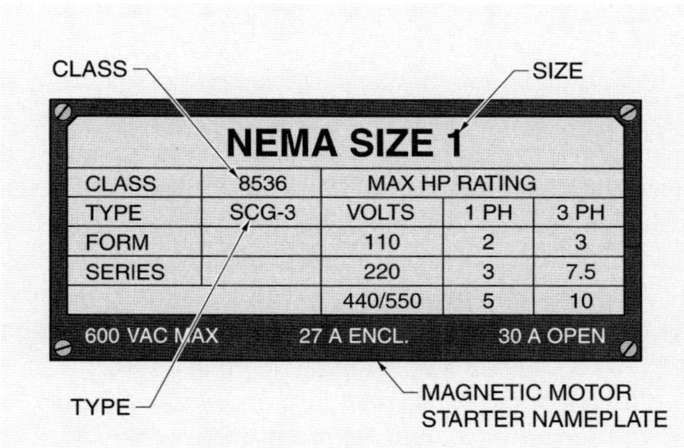

Figure 8-39. The nameplate of a magnetic motor starter includes the class, type, and size of the starter.

Always refer to the manufacturer instructions on thermal overload relay selection to see if any restrictions are placed on the class of starter required. **See Figure 8-40.** For example, unless a class 8198 starter is used, motors with service factors of 1.15 to 1.25 may use 100% of the motor full-load current for thermal overload selection.

Motor and controller in *same ambient temperature*:

a. All starter classes, except Class 8198

1. For 1.15 to 1.25 service factor motors use 100% of motor full-load current for thermal unit selection.
2. For 1.0 service factor motors use 90% of motor full-load current for thermal unit selection.

b. Class 8198 only:

CLASS RESTRICTIONS

1. For 1.0 service factor motors use 100% of motor full-load current for thermal unit selection.
2. For 1.15 to 1.25 service factor motors use 110% of motor full-load current for thermal unit selection.

Figure 8-40. Manufacturer instructions on thermal overload relay selection detail restrictions that are placed on classes of starters.

Manufacturer Heater Selection Charts. Manufacturers provide charts for use in selecting proper thermal overload heaters. The correct chart must be used for the appropriate size starter. **See Figure 8-41.** This information is also found within the enclosure of many motor starters. Each motor starter manufacturer has a chart that applies to their specific brand.

For example, a thermal unit number B2.40 is the correct overload heater for controlling a 3ϕ motor with a full-load current of 1.50 A. Column three is used because all three phases of the 3ϕ motor must have thermal overload protection. The heater must provide protection of approximately 1.5 A (1.44 – 1.62) based on the motor full-load current. Manufacturers have different numbers that relate to their specific heaters, but the selection procedure is similar.

Checking Selections. Section 430.32 of the NEC® indicates that a motor must be protected up to 125% of its full-load current rating. Because the minimum full-load current of a B2.40 overload device is 1.44 A, the device trips at 125% of this value or 1.8 A (1.44 × 1.25 = 1.8 A). Dividing the minimum trip current (1.8 A) by the full-load current of the motor (1.5 A) and multiplying by 100% determines if this range is acceptable (1.8 / 1.5 × 100% = 120%). The heater selection is correct because the trip current is less than the NEC® limit of 125%.

THERMAL UNIT CURRENT RATINGS			
Motor Full-Load Current (Amps)			Thermal Unit Number
1 Unit (HEATER)	2 Units (HEATERS)	3 Units (HEATERS)	
0.29 – 0.31	0.29 – 0.31	0.28 – 0.30	B0.44
0.32 – 0.34	0.32 – 0.34	0.31 – 0.34	B0.51
0.35 – 0.38	0.35 – 0.38	0.35 – 0.37	B0.57
0.39 – 0.45	0.39 – 0.45	0.38 – 0.44	B0.63
0.46 – 0.54	0.46 – 0.54	0.45 – 0.53	B0.71
0.55 – 0.61	0.55 – 0.61	0.54 – 0.59	B0.81
0.62 – 0.66	0.62 – 0.66	0.60 – 0.64	B0.92
0.67 – 0.73	0.67 – 0.73	0.65 – 0.72	B1.03
0.74 – 0.81	0.74 – 0.81	0.73 – 0.80	B1.16
0.82 – 0.94	0.82 – 0.94	0.81 – 0.90	B1.30
0.95 – 1.05	0.95 – 1.05	0.91 – 1.03	B1.45
1.06 – 1.22	1.06 – 1.22	1.04 – 1.14	B1.67
1.23 – 1.34	1.23 – 1.34	1.15 – 1.27	B1.88
1.35 – 1.51	1.35 – 1.51	1.28 – 1.43	B2.10
1.52 – 1.71	1.52 – 1.71	1.44 – 1.62	B2.40
1.72 – 1.93	1.72 – 1.93	1.63 – 1.77	B2.65
1.94 – 2.14	1.94 – 2.14	1.78 – 1.97	B3.00
2.15 – 2.40	2.15 – 2.40	1.98 – 2.32	B3.30
2.41 – 2.72	2.41 – 2.72	2.33 – 2.51	B3.70
2.73 – 3.15	2.73 – 3.15	2.52 – 2.99	B4.15
3.16 – 3.55	3.16 – 3.55	3.00 – 3.42	B4.85
3.56 – 4.00	3.56 – 4.00	3.43 – 3.75	B5.50
4.01 – 4.40	4.01 – 4.40	3.76 – 3.98	B6.25
4.41 – 4.88	4.41 – 4.88	3.99 – 4.48	B6.90
4.89 – 5.19	4.89 – 5.19	4.49 – 4.93	B7.70
5.20 – 5.73	5.20 – 5.73	4.94 – 5.21	B8.20
5.74 – 6.39	5.74 – 6.39	5.22 – 5.84	B9.10
6.40 – 7.13	6.40 – 7.13	5.85 – 6.67	B10.2
7.14 – 7.90	7.14 – 7.90	6.68 – 7.54	B11.5
7.91 – 8.55	7.91 – 8.55	7.55 – 8.14	B12.8
8.56 – 9.53	8.56 – 9.53	8.15 – 8.72	B14.0
9.54 – 10.6	9.54 – 10.6	8.73 – 9.66	B15.5
10.7 – 11.8	10.7 – 11.8	9.67 – 10.5	B17.5
11.9 – 13.2	11.9 – 12.0	10.6 – 11.3	B19.5
13.3 – 14.9	—	11.4 – 12.0	B22.0
15.0 – 16.6	—	—	B25.0
16.7 – 18.0	—	—	B28.0
Following Selections for Size 1 Only			
—	11.9 – 13.2	—	B19.5
—	13.3 – 14.9	11.4 – 12.7	B22.0
—	15.0 – 16.6	12.8 – 14.1	B25.0
16.7 – 18.9	16.7 – 18.9	14.2 – 15.9	B28.0
19.0 – 21.2	19.0 – 21.2	16.0 – 17.5	B32.0
21.3 – 23.0	21.3 – 23.0	17.6 – 19.7	B36.0
23.1 – 25.5	23.1 – 25.5	19.8 – 21.9	B40.0
25.6 – 26.0	25.6 – 26.0	22.0 – 24.4	B45.0
—	—	24.5 – 26.0	B50.0

Figure 8-41. Manufacturers provide charts for use in selecting proper overload heaters.

Ambient Temperature Compensation. As ambient temperature increases, less current is needed to trip overload devices. As ambient temperature decreases, more current is needed to trip overload devices. Most heater manufacturers provide special overload heater selection tables that provide multipliers to compensate for temperature changes above or below the standard temperature of 40°C. The multipliers ensure that the increase or decrease in temperature does not affect the proper protection provided by the overload relay. **See Figure 8-42.**

THERMAL UNIT SELECTION				
Controller Class	Continuous-Duty Motor Service Factor	Melting Alloy and Noncompensated Bimetallic Relays		
		Ambient Temperature of Motor		
		*	†	‡
		Full-Load Current Multiplier		
All Classes except 8198	1.15 – 1.25	1.0	0.9	1.05
	1.0	0.9	0.8	.95
Class 8198	1.15 – 1.25	1.1	1.0	1.15
	1.0	1.0	0.9	1.05

* same as controller ambient
† constant 10°C (18°F) higher than controller ambient
‡ constant 10°C (18°F) lower than controller ambient

Figure 8-42. Special overload heater selection tables provide multipliers to compensate for ambient temperatures above or below the standard temperature of 40°C.

For example, a multiplier of 0.9 is required for an ambient temperature increase of 10°C to 50°C. Multiplying the motor full-load current (1.5 A) by the correction factor (0.9) determines the compensated overload heater current rating of 1.35 A (1.5 A × 0.9 = 1.35 A). Using a heater selection chart, the acceptable current range is 1.28 A – 1.43 A. A B2.10 heater is required based on the increase in ambient temperature. This is one size smaller than the heater required (B2.40) at a 40°C ambient temperature.

The temperature surrounding an overload heater is 30°C if the ambient temperature is decreased 10°C. The correction multiplier is 1.05 for a 10°C decrease in ambient temperature. The corrected current is 1.575 A using a full-load current of 1.5 A (1.5 A × 1.05 = 1.575 A). In a heater selection chart, a range of 1.44 – 1.62 is acceptable. In this case, the same size heater could be used. Always consult manufacturer specifications and tables for proper heater sizing.

In rare instances, such as older installations or severely damaged equipment, it may be impossible to determine a motor full-load current from its nameplate. Manufacturers provide charts listing approximate full-load currents based on average motor full-load currents. **See Figure 8-43.**

AMPERE RATINGS OF 3ϕ, 60 Hz, AC INDUCTION MOTORS															
HP	**rpm Speed**	**200 V**	**230 V**	**380 V**	**460 V**	**575 V**	**2200 V**	**HP**	**rpm Speed**	**200 V**	**230 V**	**380 V**	**460 V**	**575 V**	**2200 V**
1/4	1800	1.09	.95	.55	.48	.38	—	25	3600	69.9	60.8	36.8	30.4	24.3	—
	1200	1.61	1.40	.81	.70	.56			1800	74.5	64.8	39.2	32.4	25.9	
	900	1.84	1.60	.93	.80	.64			1200	75.4	65.6	39.6	32.8	26.2	
									900	77.4	67.3	40.7	33.7	27.0	
1/3	1800	1.37	1.19	.69	.60	.48	—	30	3600	84.4	73.7	44.4	36.8	29.4	—
	1200	1.83	1.59	.92	.80	.64			1800	86.9	75.6	45.7	37.8	30.2	
	900	2.07	1.80	1.04	.90	.72			1200	90.6	78.8	47.6	39.4	31.5	
									900	94.1	81.8	49.5	40.9	32.7	
1/2	18800	1.98	1.72	.99	.86	.69	—	40	3600	111.0	96.4	58.2	48.2	38.5	—
	1200	2.47	2.15	1.24	1.08	.86			1800	116.0	101.0	61.0	50.4	40.3	
	900	2.74	2.38	1.38	1.19	.95			1200	117.0	102.0	61.2	50.6	40.4	
									900	121.0	105.0	63.2	52.2	41.7	
3/4	1800	2.83	2.46	1.42	1.23	.98	—	50	3600	138.0	120.0	72.9	60.1	48.2	—
	1200	3.36	2.92	1.69	1.46	1.17			1800	143.0	124.0	75.2	62.2	49.7	
	900	3.75	3.26	1.88	1.63	1.30			1200	145.0	126.0	76.2	63.0	50.4	
									900	150.0	130.0	78.5	65.0	52.0	
1	3600	3.22	2.80	1.70	1.40	1.12	—	60	3600	164.0	143.0	86.8	71.7	57.3	—
	1800	4.09	3.56	2.06	1.78	1.42			1800	171.0	149.0	90.0	74.5	59.4	
	1200	4.32	3.76	2.28	1.88	1.50			1200	173.0	150.0	91.0	75.0	60.0	
	900	4.95	4.30	2.60	2.15	1.72			900	177.0	154.0	93.1	77.0	61.5	
1½	3600	5.01	4.36	2.64	2.18	1.74	—	75	3600	206.0	179.0	108.0	89.6	71.7	—
	1800	5.59	4.86	2.94	2.43	1.94			1800	210.0	183.0	111.0	91.6	73.2	
	1200	6.07	5.28	3.20	2.64	2.11			1200	212.0	184.0	112.0	92.0	73.5	
	900	6.44	5.60	3.39	2.80	2.24			900	222.0	193.0	117.0	96.5	77.5	
2	3600	6.44	5.60	3.39	2.80	2.24	—	100	3600	266.0	231.0	140.0	115.0	92.2	—
	1800	7.36	6.40	3.87	3.20	2.56			1800	271.0	236.0	144.0	118.0	94.8	
	1200	7.87	6.84	4.14	3.42	2.74			1200	275.0	239.0	145.0	120.0	95.6	
	900	9.09	7.90	4.77	3.95	3.16			900	290.0	252.0	153.0	126.0	101.0	
3	3600	9.59	8.34	5.02	4.17	3.34	—	125	3600	—	292.0	176.0	146.0	116.0	—
	1800	10.8	9.40	5.70	4.70	3.76			1800		293.0	177.0	147.0	117.0	23.6
	1200	11.7	10.2	6.20	5.12	4.10			1200		298.0	180.0	149.0	119.0	24.2
	900	13.1	11.4	6.90	5.70	4.55			900		305.0	186.0	153.0	122.0	24.8
5	3600	15.5	13.5	8.20	6.76	5.41	—	150	3600	—	343.0	208.0	171.0	137.0	—
	1800	16.6	14.4	8.74	7.21	5.78			1800		348.0	210.0	174.0	139.0	29.2
	1200	18.2	15.8	9.59	7.91	6.32			1200		350.0	210.0	174.0	139.0	29.9
	900	18.3	15.9	9.60	7.92	6.33			900		365.0	211.0	183.0	146.0	30.9
7½	3600	22.4	19.5	11.8	9.79	7.81	—	200	3600	—	458.0	277.0	229.0	164.0	—
	1800	24.7	21.5	13.0	10.7	8.55			1800		452.0	274.0	226.0	181.0	34.8
	1200	25.1	21.8	13.2	10.9	8.70			1200		460.0	266.0	230.0	184.0	35.5
	900	26.5	23.0	13.9	11.5	9.19			900		482.0	2.79.0	241.0	193.0	37.0
10	3600	29.2	25.4	15.4	12.7	10.1	—	250	3600	—	559.0	338.0	279.0	223.0	—
	1800	30.8	26.8	16.3	13.4	10.7			1800		568.0	343.0	284.0	227.0	57.5
	1200	32.2	28.0	16.9	14.0	11.2			1200		573.0	345.0	287.0	229.0	58.5
	900	35.1	30.5	18.5	15.2	12.2			900		600.0	347.0	300.0	240.0	60.5
15	3600	41.9	36.4	22.0	18.2	14.5	—	300	1800	—	678.0	392.0	339.0	271.0	69.0
	1800	45.1	39.2	23.7	19.6	15.7			1200		684.0	395.0	342.0	274.0	70.0
	1200	47.6	41.4	25.0	20.7	16.5									
	900	51.2	44.5	26.9	22.2	17.8									
20	3600	58.0	50.4	30.5	25.2	20.1	—	400	1800	—	896.0	518.0	448.0	358.0	91.8
	1800	58.9	51.2	31.0	25.6	20.5		500	1800	—	1110.0	642.0	555.0	444.0	116.0
	1200	60.7	52.8	31.9	26.4	21.1									
	900	63.1	54.9	33.2	27.4	21.9									

Figure 8-43. Most manufacturers provide charts for approximating full-load current when motor nameplate information is not available.

These charts should be used only as a last resort. This technique is not suggested as a standard procedure because the average rating could be higher or lower for a specific motor and, therefore, selection on this basis always involves risk. For fully reliable motor protection, select heat coils based on the motor full-load current rating shown on the motor nameplate. The full-load current of a motor stated on charts should be used in the selection of a heater using the same procedure as if it were the motor nameplate information. These charts provide approximately the same information that may be found on the motor nameplate, but should be used only if motor nameplate information is not available.

Inherent Motor Protectors

An *inherent motor protector* is an overload device located directly on or in a motor to provide overload protection. Certain inherent motor protectors base their sensing element on the amount of heat generated or the amount of current consumed by a motor. Inherent motor protectors directly or indirectly (using contactors) trip a circuit that disconnects the motor from the power circuit based on what the motor protector senses. Bimetallic thermodiscs and thermistor overload devices are inherent motor protectors.

Bimetallic Thermodiscs. A bimetallic thermodisc operates on the same principle as a bimetallic strip. The differences between these devices are the shape of the device and its location. A thermodisc has the shape of a miniature dinner plate and is located within the frame of a motor. **See Figure 8-44.** A bimetallic thermodisc warps and opens the circuit when a motor is overloaded. Bimetallic thermodiscs are normally used on small horsepower motors to disconnect the motor directly from the power circuit. Bimetallic thermodiscs may be tied into the control circuit of a magnetic contactor coil where they can be used as indirect control devices.

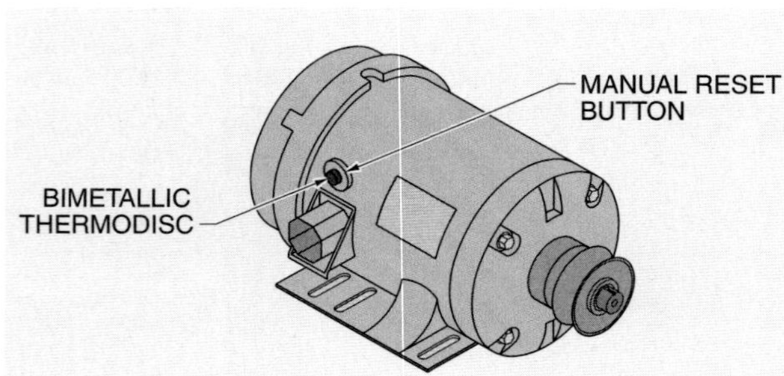

Figure 8-44. Bimetallic thermodiscs are normally used on small horsepower motors to directly disconnect the motor from the power circuit.

Always ensure power to the motor is turned OFF before resetting a manual-reset thermodisc. This prevents a potential hazard when the motor restarts.

Thermistor Overload Devices. A thermistor-based overload is a sophisticated form of inherent motor protection. A thermistor overload device combines a thermistor, solid-state relay, and contactor into a custom-built overload protector. **See Figure 8-45.**

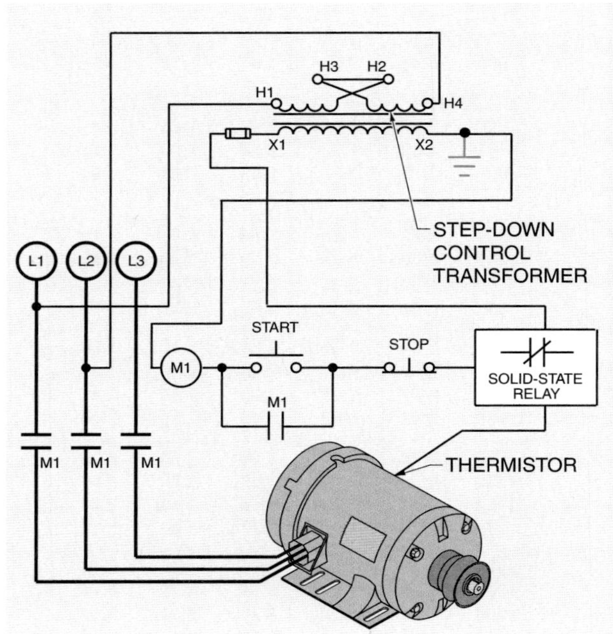

Figure 8-45. A thermistor overload device combines a thermistor, solid-state relay, and contactor into a custom-built overload protector.

A thermistor is similar to a resistor in that its resistance changes with the amount of heat applied to it. As the temperature increases, the resistance of the thermistor decreases and the amount of current passing through the thermistor increases. The changing signal must be amplified before it can do any work, such as triggering a relay, because the thermistor is a low-power device (normally in the thousandths of an ampere range). When a thermistor overload device is amplified, a relay may open a set of contacts in the control circuit of a magnetic motor starter, de-energizing the power circuit of the motor.

The major drawback to thermistor overload devices is that they require a close coordination between the user and the manufacturer to customize the design. Custom-designed overload protectors are more costly than standard, off-the-shelf overload protectors. With the exception of special and high-priced motors requiring extensive protection, custom-designed overload protectors are uneconomical and are not recommended.

Electronic Overload Protection

An electronic overload is a device that has built-in circuitry to sense changes in current and temperature. Because electronic devices may include amplifiers, small changes can be responded to before mechanical devices can be activated. An electronic overload monitors the current in the motor directly by measuring the current in the power lines of a motor.

CONTACTOR AND MAGNETIC MOTOR STARTER MODIFICATIONS

Certain devices may be added to basic contactors or motor starters to expand their capability. These devices include additional electrical contacts, power poles, pneumatic timers, transient suppression modules, and control circuit fuse holders. **See Figure 8-46.**

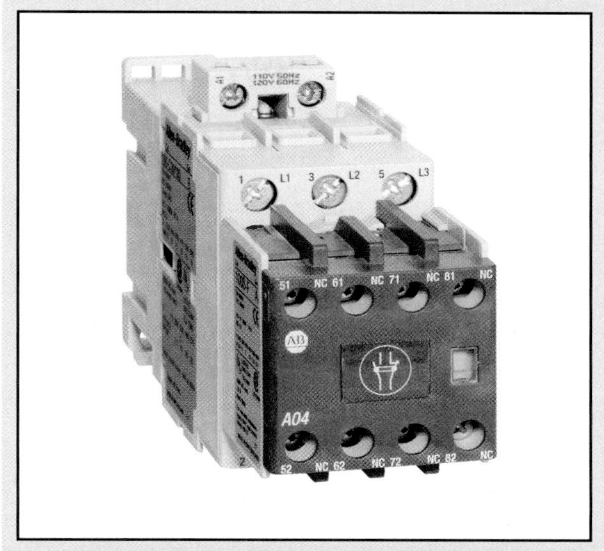

Rockwell Automation, Allen-Bradley Company, Inc.

Safety control contactors provide positively guided contacts at various amperage levels which are required in feedback circuits for modern safety applications.

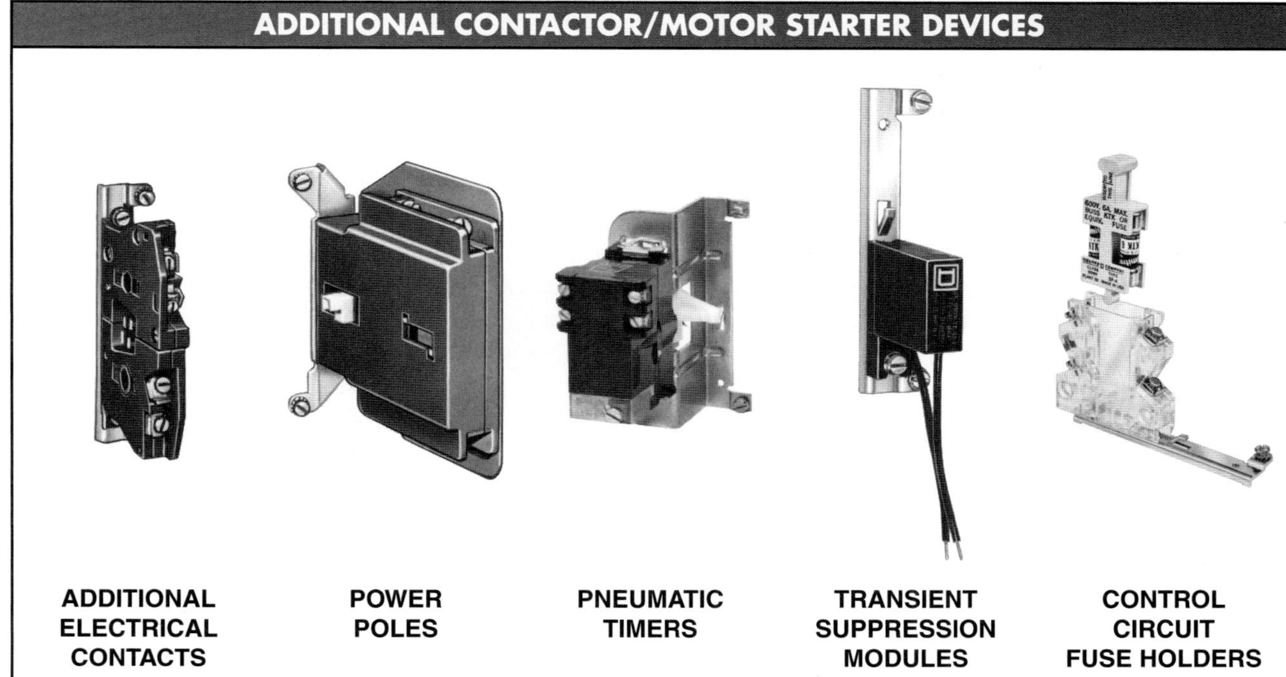

Square D Company

Figure 8-46. The devices that may be added to basic contactors or magnetic motor starters to expand their capability include additional electrical contacts, power poles, pneumatic timers, transient suppression modules, and control circuit fuse holders.

Additional Electrical Contacts

Most contactors and motor starters have the ability to control several additional electrical contacts if the additional contacts are added to existing auxiliary contacts. The additional contacts may be used as extra auxiliary contacts. Both normally open and/or normally closed contacts may be wired to control additional loads. Normally closed contacts are used to turn additional loads ON anytime the contactor or starter is OFF, as well as to provide electrical interlocking. Normally open contacts are used to turn additional loads ON anytime a contactor or starter is ON.

Power Poles

In certain cases, additional power poles (contacts capable of carrying a load) may be added to a contactor. The power poles are available with normally open or normally closed contacts. Normally, only one power pole unit with one or two contacts is added per contactor or motor starter.

In certain cases with large-sized contactors or motor starters, it may be necessary to replace the coil to handle the additional load created when energizing the additional poles. Most power poles are factory or field installed.

Pneumatic Timers

A mechanically operated pneumatic timer can be mounted on some sizes of contactors and motor starters for applications requiring the simultaneous operation of a timer and a contactor. The use of mechanically operated timers results in considerable savings in panel space over a separately mounted timer. Available in time delay after de-energization (OFF-delay) or time delay after energization (ON-delay), the timer attachment has an adjustable timing period over a specified range.

Most manufacturers provide units that are field convertible from ON-delay to OFF-delay (or vice versa) without additional parts. The pneumatic timers are ordered either fixed or variable. Most timers mount on the side of the contactor and are secured firmly. One single-pole, double-throw contact is provided.

Transient Suppression Modules

Transient suppression modules are designed to be added where the transient voltage generated when opening the coil circuit interferes with the power operation of nearby components and solid-state control circuits. Transient suppression modules normally consist of resistance/capacitance (RC) circuits and are designed to suppress the voltage transients to approximately 200% of peak coil supply voltage.

In certain cases, a voltage transient is generated when switching the integral control transformer that powers the coil control circuit. A transient suppression module, when used with devices wired for common control, is connected across the 120 V transformer secondary. The transient suppression module is not connected across the control coil.

Control Circuit Fuse Holders

Control circuit fuse holders can be attached to contactors or starters when either one or two control circuit fuses may be required. The fuse holder helps satisfy the NEC® requirements in Section 430.72.

INTERNATIONAL STANDARDS

The International Electrotechnical Commission (IEC), headquartered in Geneva, Switzerland, is primarily associated with equipment used in Europe. The National Electrical Manufacturers Association (NEMA), headquartered in Washington, D.C., is primarily associated with equipment used in North America. The IEC and NEMA rate contactors and motor starters. This causes confusion because ratings are different for the same horsepower. IEC devices are smaller in size for the equivalent-rated contactor. **See Figure 8-47.**

IEC devices are built with materials required for average applications. NEMA devices are built for a high level of performance in a variety of applications. IEC devices are less expensive, but more application sensitive. NEMA devices are more costly, but less application sensitive. IEC devices are commonly used in original equipment manufacturer (OEM) machines where machine specifications are known and do not change. NEMA devices are commonly used where machine requirements and specifications may vary.

TROUBLESHOOTING CONTACTORS AND MOTOR STARTERS

Contactors or motor starters are the first devices checked when troubleshooting a circuit that does not work or has a problem. Contactors or motor starters are checked first because they are the point where the incoming power, load, and control circuit are connected. Basic voltage readings are taken at a contactor or motor starter to determine where the problem lies. The same basic procedure used to troubleshoot a motor starter works for contactors because a motor starter is a contactor with added overload protection.

IEC AND NEMA DEVICE COMPARISON		
Considerations	**IEC**	**NEMA**
Size	Smaller per horsepower rating than NEMA	Larger per horsepower rating than IEC
Cost	Lower cost per horsepower	Higher cost per horsepower
Performance	Electrical life of 1,000,000 operations is acceptable	Electrical life 2.5 to 4 times higher than equivalent IEC device
Applications	Application sensitive with greater knowledge and care necessary	Application easier with fewer parameters to consider
Overloads	Fixed heaters that are adjustable to match different motors at same horsepower. Heaters are not field changeable	Field-changeable heaters allow adjustment to motors of different horsepowers
Additional Information	Reset/stop dual-function operation mechanism typical	Reset-only mechanism typical
	Hand/auto reset typical	Hand reset-only typical
	Typically designed for use with fast-acting, current-limiting European fuses	Designed for use with domestic time delay fuses and circuit breakers
	DIN-rail mountable	

Figure 8-47. The difference between IEC and NEMA devices is based on size, cost, performance, application, and overloads.

The tightness of all terminals and busbar connections is checked when troubleshooting control devices. Loose connections in the power circuit of contactors and motor starters cause overheating. Overheating leads to equipment malfunction or failure. Loose connections in the control circuit cause control malfunctions. Loose connections of grounding terminals lead to electrical shock and cause electromagnetic-generated interference.

The power circuit and the control circuit are checked if the control circuit does not correctly operate a motor. The two circuits are dependent on each other, but are considered two separate circuits because they are normally at different voltage levels and always at different current levels. **See Figure 8-48.** To troubleshoot a motor starter, apply the following procedure:

1. Inspect the motor starter and overload assembly. Service or replace motor starters that show heat damage, arcing, or wear. Replace motor starters that show burning. Check the motor and driven load for signs of an overload or other problem.
2. Reset the overload relay if there is no visual indication of damage. Replace the overload relay if there is visual indication of damage.
3. Observe the motor starter for several minutes if the motor starts after resetting the overload relay. The overload relay continues to open if an overload problem continues to exist.
4. Check the voltage into the starter if resetting the overload relay does not start the motor. Check circuit voltage ahead of the starter if the voltage reading is 0 V. The voltage is acceptable if the voltage reading is within 10% of the motor voltage rating. The voltage is unacceptable if the voltage reading is not within 10% of the motor voltage rating.
5. Energize the starter and check the starter contacts if the voltage into the starter is present and at the correct level. The starter contacts are good if the voltage reading is acceptable. Open the starter, turn the power OFF, and replace the contacts if there is no voltage reading.
6. Check the overload relay if voltage is coming out of the starter contacts. Turn the power OFF and replace the overload relay if the voltage reading is 0 V. The problem is downstream from the starter if the voltage reading is acceptable and the motor is not operating.

Technical Fact

IEC-rated contactors are applied when electrical parameters are defined. NEMA-rated contactors are applied when electrical parameters vary.

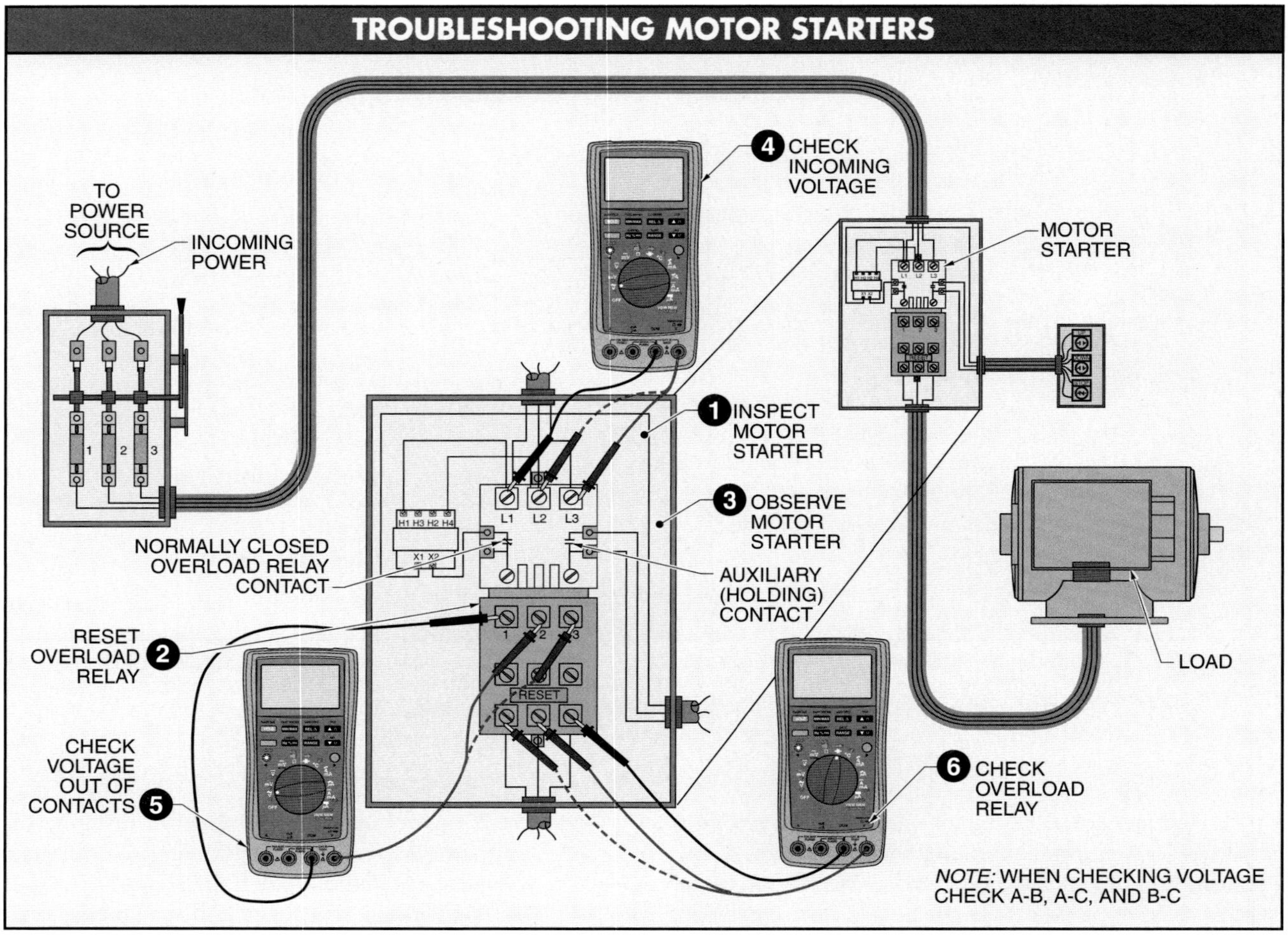

TROUBLESHOOTING MOTOR STARTERS

TO POWER SOURCE
INCOMING POWER
❹ CHECK INCOMING VOLTAGE
MOTOR STARTER
❶ INSPECT MOTOR STARTER
❸ OBSERVE MOTOR STARTER
NORMALLY CLOSED OVERLOAD RELAY CONTACT
AUXILIARY (HOLDING) CONTACT
RESET OVERLOAD RELAY ❷
LOAD
CHECK VOLTAGE OUT OF CONTACTS ❺
❻ CHECK OVERLOAD RELAY
NOTE: WHEN CHECKING VOLTAGE CHECK A-B, A-C, AND B-C

Figure 8-48. A contactor or motor starter is the first device checked when troubleshooting a circuit that does not work or has a problem.

General Electric Company

Safety precautions must be followed when troubleshooting motor control circuits because the circuit must be energized when troubleshooting.

Troubleshooting Guides

A troubleshooting guide is used when troubleshooting contactors and motor starters. The guide states a problem, its possible cause(s), and corrective action(s) that may be taken. **See Figure 8-49.**

MOTOR DRIVES

Contactors and motor starters are being replaced in many applications by motor drives. A *motor drive* is an electronic unit designed to control the speed of a motor using solid-state components. Motor drives may be AC motor drives (AC drives) or DC motor drives (DC drives). AC drives are the most common. AC drives may also be referred to as adjustable speed drives, variable frequency drives, or inverters. **See Figure 8-50.**

CONTACTOR AND MOTOR STARTER TROUBLESHOOTING GUIDE		
Problem	**Possible Cause**	**Corrective Action**
Humming noise H-M-M-M-M	Magnetic pole faces misaligned	Realign. Replace magnet assembly if realignment is not possible
	Too low voltage at coil	Measure voltage at coil. Check voltage rating of coil. Correct any voltage that is 10% less than coil rating
	Pole face obstructed by foreign object, dirt, or rust	Remove any foreign object and clean as necessary. Never file pole faces
Loud buzz noise	Shading coil broken	Replace coil assembly
Controller fails to drop out POWER IN POWER OUT	Voltage to coil not being removed	Measure voltage at coil. Trace voltage from coil to supply. Search for shorted switch or contact if voltage is present
	Worn or rusted parts causing binding	Replace worn parts. Clean rusted parts
	Contact poles sticking	Check for burning or sticky substance on contacts. Replace burned contacts. Clean dirty contacts
	Mechanical interlock binding	Check to ensure interlocking mechanism is free to move when power is OFF. Replace faulty interlock
Controller fails to pull in POWER IN POWER OUT	No coil voltage	Measure voltage at coil terminals. Trace voltage loss from coil to supply voltage if voltage is not present
	Too low voltage	Measure voltage at coil terminals. Correct voltage level if voltage is less than 10% of rated coil voltage. Check for a voltage drop as large loads are energized
	Coil open	Measure voltage at coil. Remove coil if voltage is present and correct but coil does not pull in. Measure coil resistance for open circuit. Replace if open
	Coil shorted	Shorted coil may show signs of burning. The fuse or breakers should trip if coil is shorted. Disconnect one side of coil and reset if tripped. Remove coil and check resistance for short if protection device does not trip. A shorted coil has zero or very low resistance. Replace shorted coil. Replace any coil that is burned
	Mechanical obstruction	Remove any obstructions
Contacts badly burned or welded	Too high inrush current	Measure inrush current. Check load for problem of higher-than-rated load current. Change to larger controller if load current is correct but excessive for controller
	Too fast load cycling	Change to larger controller if load cycles ON and OFF repeatedly
	Too large overcurrent protection device	Size overcurrent protection to load and controller
	Short circuit	Check fuses or circuit breakers. Clear any short circuit
	Insufficient contact pressure	Check to ensure contacts are making good connection
Nuisance tripping	Incorrect overload size	Check size of overload against rated load current. Adjust size as permissible per NEC®
	Lack of temperature compensation	Check setting of overload if controller and load are at different ambient temperatures
	Loose connections	Check for loose terminal connection

Figure 8-49. A troubleshooting guide used when troubleshooting contactors and motor starters states a problem, its possible cause or causes, and corrective actions that may be taken.

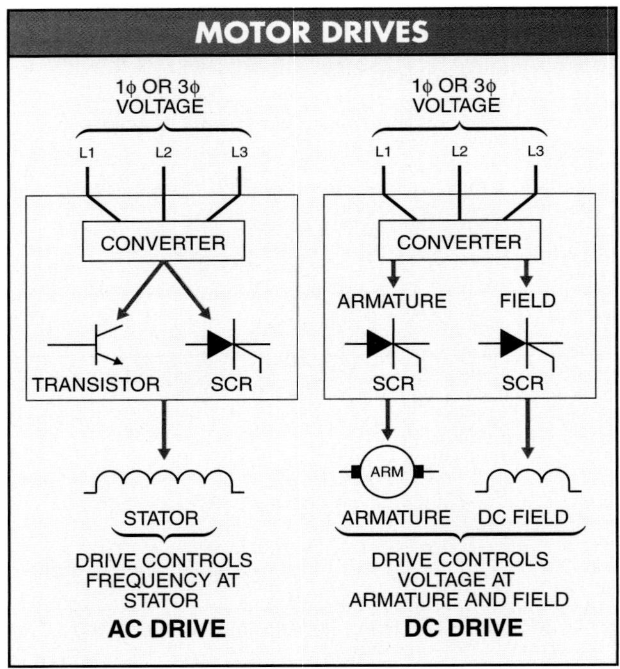

Figure 8-50. Motor drives are designed to control the speed of a motor using solid-state components and may be AC drives or DC drives.

A motor drive performs the same function as a motor starter, but can also vary motor speed, reverse a motor, provide additional protection features, display operating information, and be interfaced with other electrical equipment. Motor drives can be used to control almost any size motor from fractional horsepower to hundreds of horsepower.

AC motor drives control motor speed by converting incoming AC to DC and then converting the DC back to a variable frequency AC. Varying the frequency (Hz) to the motor can change the speed of an AC motor. For example, a standard 60 Hz AC motor runs at full speed when connected to 60 Hz, half speed when connected to 30 Hz, and one-quarter speed when connected to 15 Hz.

DC motor drives control motor speed by controlling and monitoring the DC output voltage to the motor by varying the amount of voltage and current on the motor field and armature. AC power can also be connected to DC drives that convert incoming AC to DC.

The section of a drive that converts AC to DC is referred to as the converter and the section that converts the DC back to AC is referred to as the inverter. A *converter* is an electronic device that changes AC voltage into DC voltage. An *inverter* is an electronic device that changes DC voltage into AC voltage. **See Figure 8-51.** Silicon controlled rectifiers (SCRs) are used to convert the incoming AC voltage into DC. In addition to converting AC into DC, SCRs can also control the level of the DC voltage.

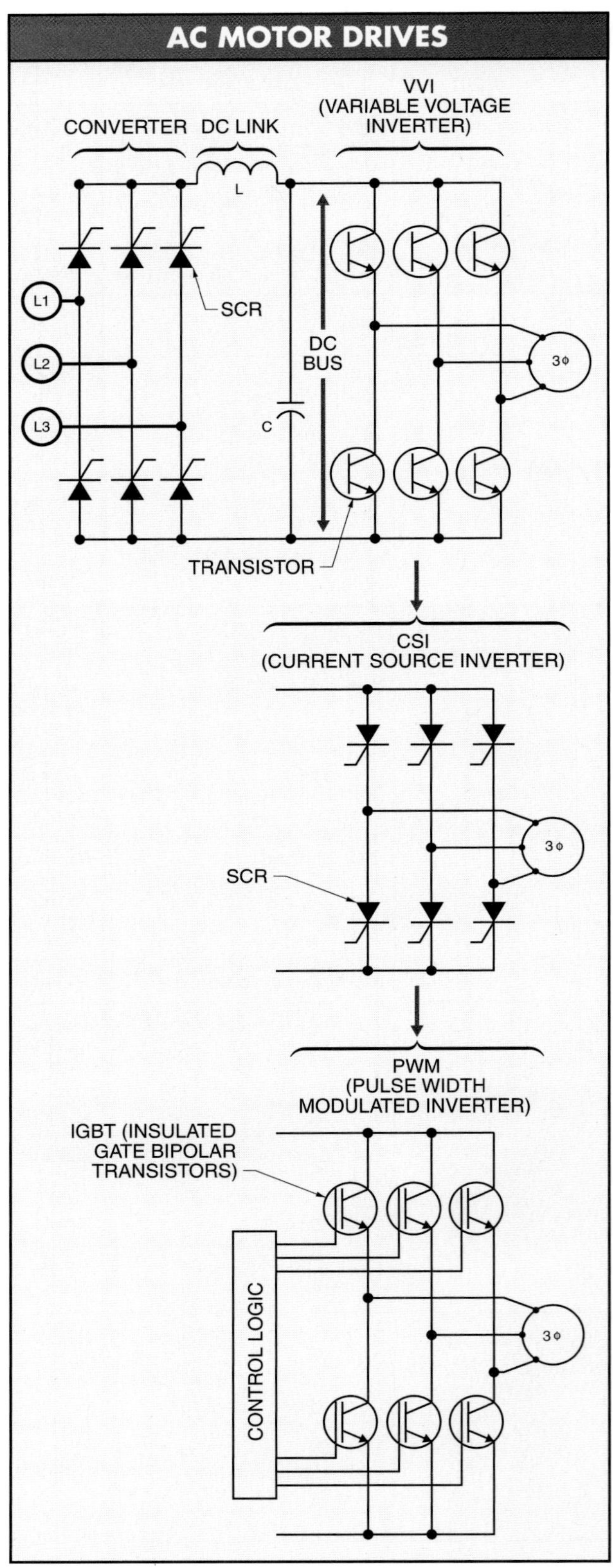

Figure 8-51. AC drives include a converter, DC link, and an inverter.

On all AC drives, the inverter is used to switch the DC voltage ON and OFF to produce simulated AC. The simulated AC is actually pulsating DC. The pulsating DC can be used to drive an AC motor and control the speed by varying the voltage and frequency applied to the motor. AC drives control the voltage and frequency to a motor by switching DC ON and OFF. Electronic switching is used because of the fast switching required. A microprocessor circuit located within the drive controls the switching.

Inverters are referred to according to the method used to change the frequency of the incoming voltage. Inverters include variable voltage inverters (VVI), current source inverters (CSI), and pulse width modulated (PWM) inverters.

Motor Drive Programming

All motor drives must be programmed to ensure proper motor operation. A properly programmed motor drive gives maximum system performance. An improperly programmed motor drive can cause damage to the motor, operator, and other system components.

Programming a motor drive is done at the main programming module, which is referred to as the human interface module (HIM). A *human interface module (HIM)* is a manually operated input control unit that includes programming keys, system operating keys, and normally a status display. The status display may be a liquid crystal display (LCD) or light emitting diode (LED) display. Liquid crystal displays or LED displays are used to present programming information, drive status conditions, and diagnostic data during system operation. **See Figure 8-52.**

The human interface module normally has several levels of information that are used to display system information and programming. For example, a common motor drive human interface module may have an operational level, mode level, group level, and parameter level. These levels are accessed with the programming keys and displayed on the LCD. **See Figure 8-53.**

The operational level is normally the first level seen when viewing an LCD. The operational level indicates current drive operational status. As the system is operating, the operational level displays the condition of the motor (stopped, accelerating, etc.) and any system faults that have occurred.

The mode level provides additional information not found at the operational level. The mode level is the starting point for making system changes. The mode level includes operating modes such as display, process, program, search, and control status. The display mode is used to view system parameters without making any modifications to parameter settings. The process mode is used to display application-specific status information instead of the standard status information. The program mode is used to view system

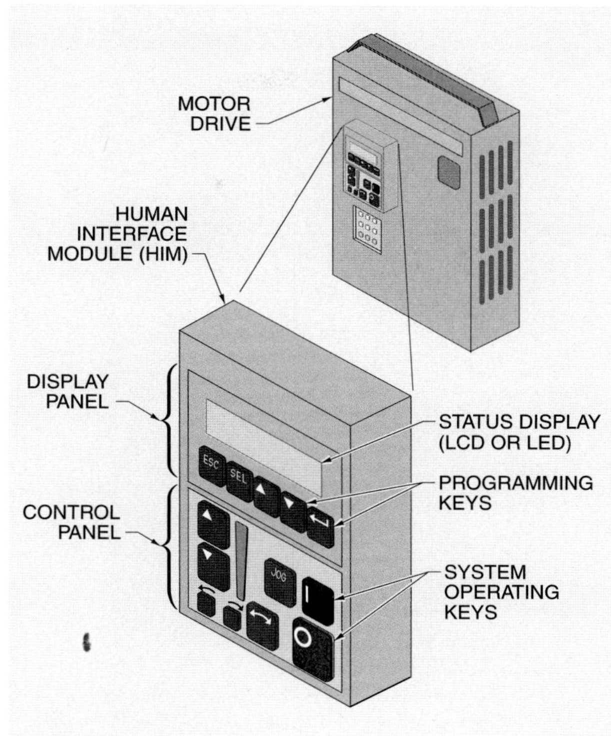

Figure 8-52. A human interface module includes programming keys, system operating keys, and normally a status display (LCD).

parameters and enable changes in parameter settings. The search mode searches for parameters that are not at their factory-set (default) value. The control status mode permits the given parameter to be disabled or enabled.

The group level is used to select the groups of parameters to be displayed or programmed when using the display mode or program mode. In the display mode, the parameters can only be viewed. In the program mode, the parameters can be viewed or changed.

The parameter level is the level used to view or program individual system parameter values. For example, parameter values such as output current, output power, and output voltage can be viewed. Parameters such as motor acceleration time may be viewed or reprogrammed for a longer (or shorter) length of time.

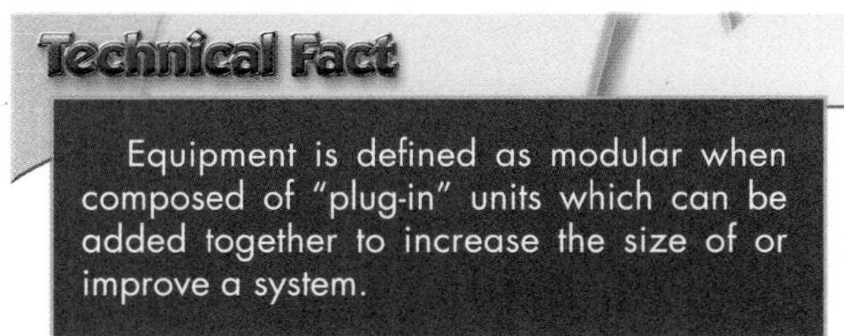

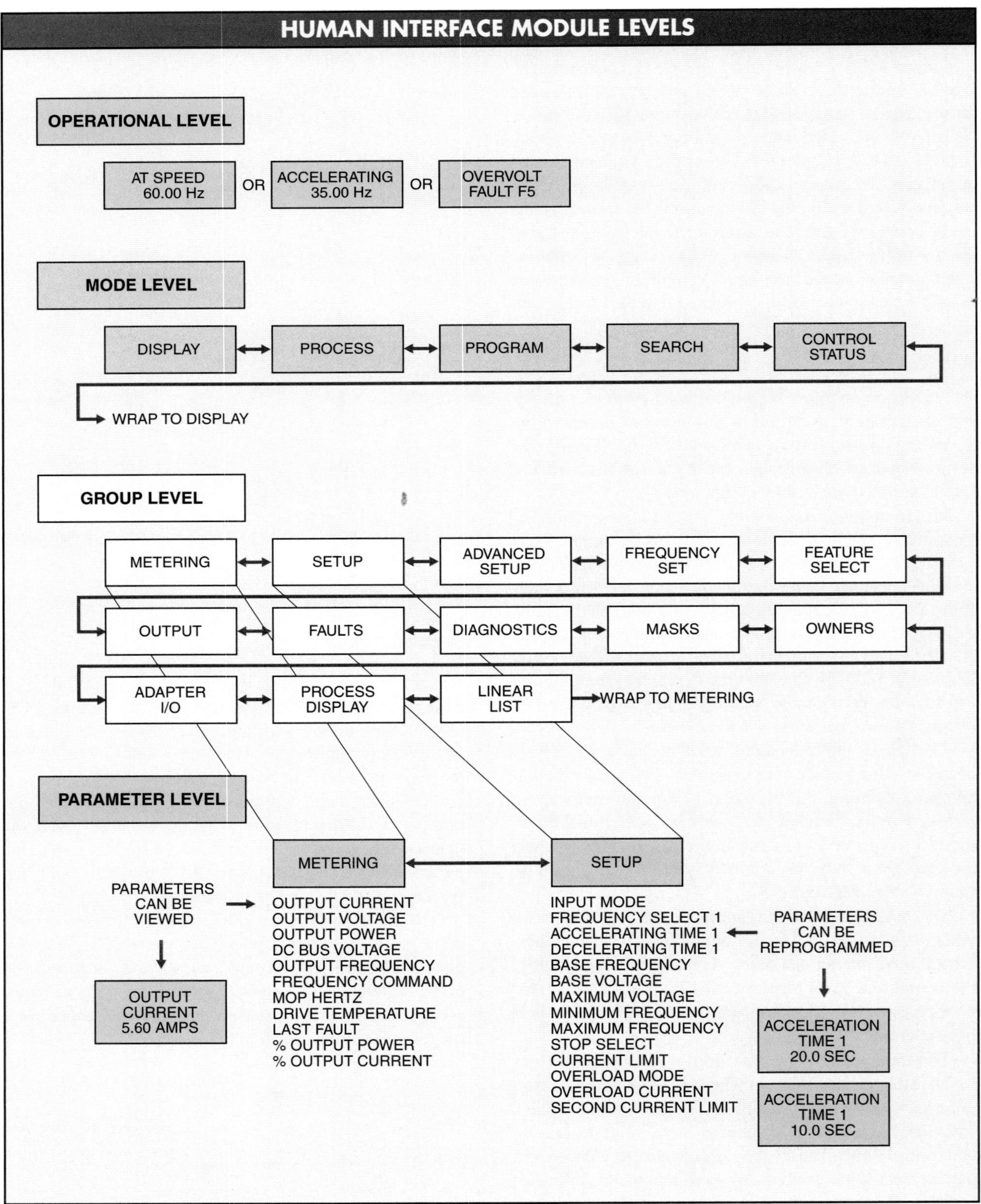

Figure 8-53. A motor drive human interface module normally has several levels, such as operational, mode, group, and parameter levels, that are used for displaying system information and programming.

Programming Overload Protection

When a magnetic motor starter is used to turn ON a motor, the motor starter must start the motor and provide overload protection. Overload protection is used to protect a motor when running. Motor starters provide overload protection through heaters or electronic overloads that are included with the motor starter. The heaters are sized and electronic overloads are set based on the current rating of the motor. Motor drives must also provide motor overload protection. Overload protection is programmed into the drive using the human interface module.

Article 430 of the NEC® covers the design and installation of electrical systems that contain motors, motor circuits, and motor controllers. Motor drive installation and operation must conform to Article 430 because a motor drive is a motor controller. Part III of Article 430 covers motor overload (running) protection requirements. Motor drives can be used as the motor overload protection device if properly programmed. **See Figure 8-54.**

Overload protection devices are designed to protect the windings of a running motor from excessive heating. Overload protection is different from overcurrent protection. Overcurrent protection is designed to protect a motor from short circuits and when starting. Fuses and circuit breakers are used for overcurrent protection. Overcurrent and ground-fault protection is covered in Part IV of Article 430.

The amount of current a motor draws from the power lines while running depends primarily on the load. Full-load current (FLC) is drawn when a motor is connected to the maximum load the motor is designed to drive. The full-load current of a motor is the current listed on the motor nameplate. A motor draws less than its full-load current if it is connected to a light load. A motor draws its full-load current or more if it is connected to a heavy load.

A motor is protected up to the maximum current allowed when its full-load current is used to size overload (running) protection. A properly sized (or programmed) overload protection device removes the motor from the circuit if an overload is present for a long enough period of time to damage the motor.

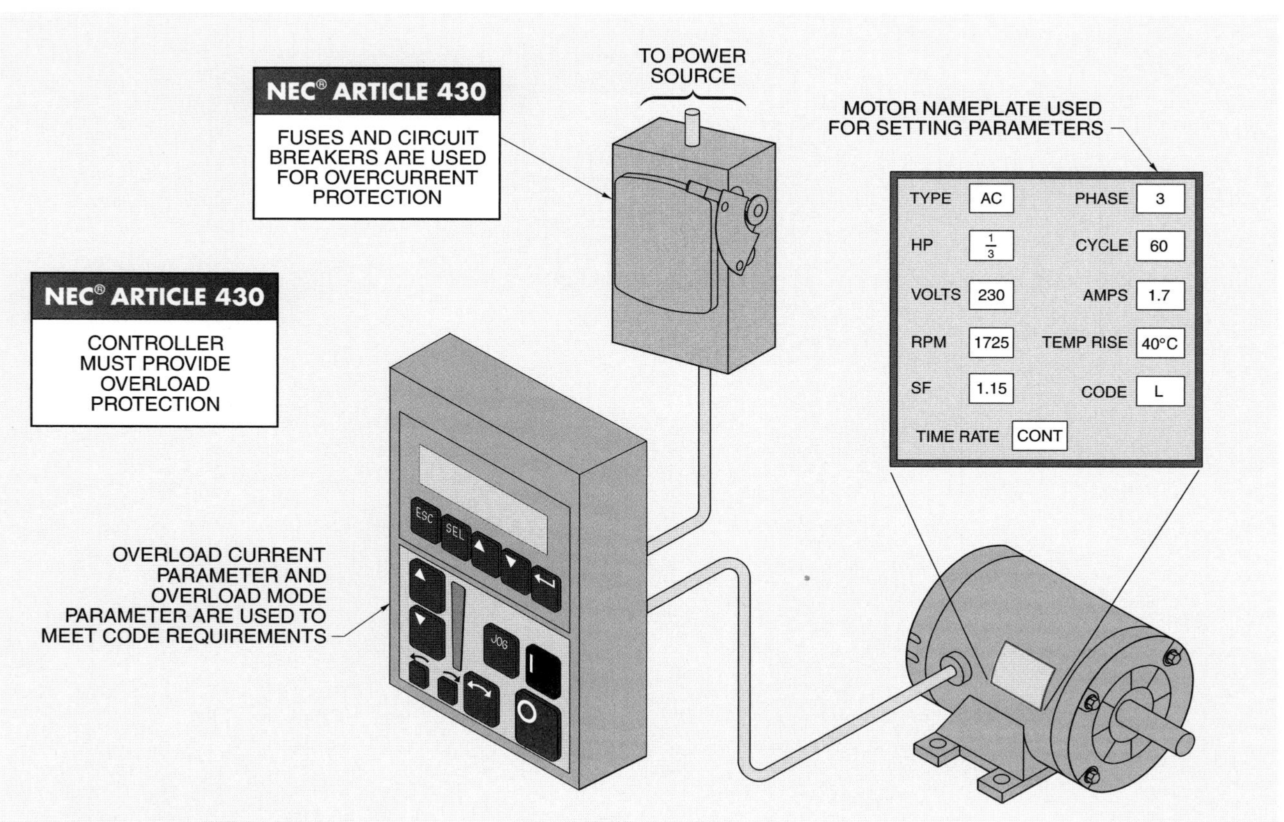

Figure 8-54. Motor drives must be programmed for overload protection to meet NEC® Article 430.32 requirements.

Overload protection devices are set to open a circuit at a maximum of 115% or 125% of the motor full-load current. **See Figure 8-55.** Per NEC® Article 430.32, the percentage above full-load current depends on the motor temperature rise and service factor. Motors with a marked service factor not less than 1.15 require a maximum overload protection of 125% times the motor full-load current. Motors with a marked temperature rise not over 40°C also require a maximum overload protection of 125% times the motor full-load current. All other motors require a maximum overload protection of 115% times the motor full-load current.

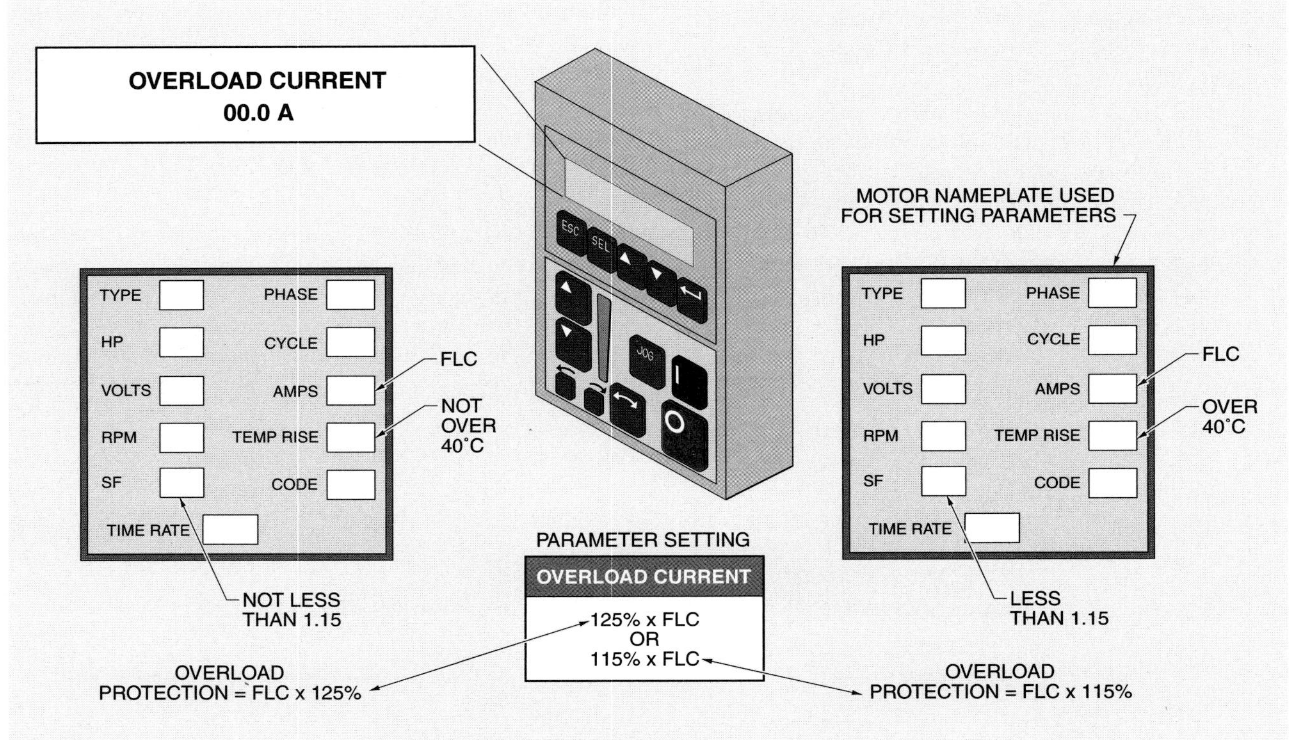

Figure 8-55. Overload protection devices are set to open a circuit at a maximum of 115% or 125% of the motor full-load current.

Transformers Connected for a Delta-to-Delta Installation. Figure 9-20 illustrates how three single-phase transformers are connected for a Delta-to-Delta step-down transformer bank. Remember from our previous discussion that the line voltage is equal to the coil voltage and that the line current is equal to 1.73 times the coil current in a balanced Delta system.

As shown in Figure 9-20, the Delta connected secondary, with one coil centered tapped at the mid-point, provides for three different types of service: (1) a three-phase 240 volt service for the three-phase motor load; (2) single-phase 120 volt service for the lighting loads; and (3) single-phase 240 volts for the 240 single-phase motor. Not illustrated is the high-phase of line C to N. This is because the high-phase is not to be used.

As with the Wye transformer bank, a closed Delta bank would deliver a total KVA output equal to the sum of the individual transformer ratings. However, we have mentioned that one advantage of a Delta system is that if one transformer is damaged or removed from service, the other two can be connected in an open-delta connection (also called the V-connection). This type of connection enables power to be maintained at a reduced level. This level is reduced to 57.7 percent of a full Delta-connected transformer bank not to 66.6%, as might be thought in a reduction of 3 to 2 transformers.

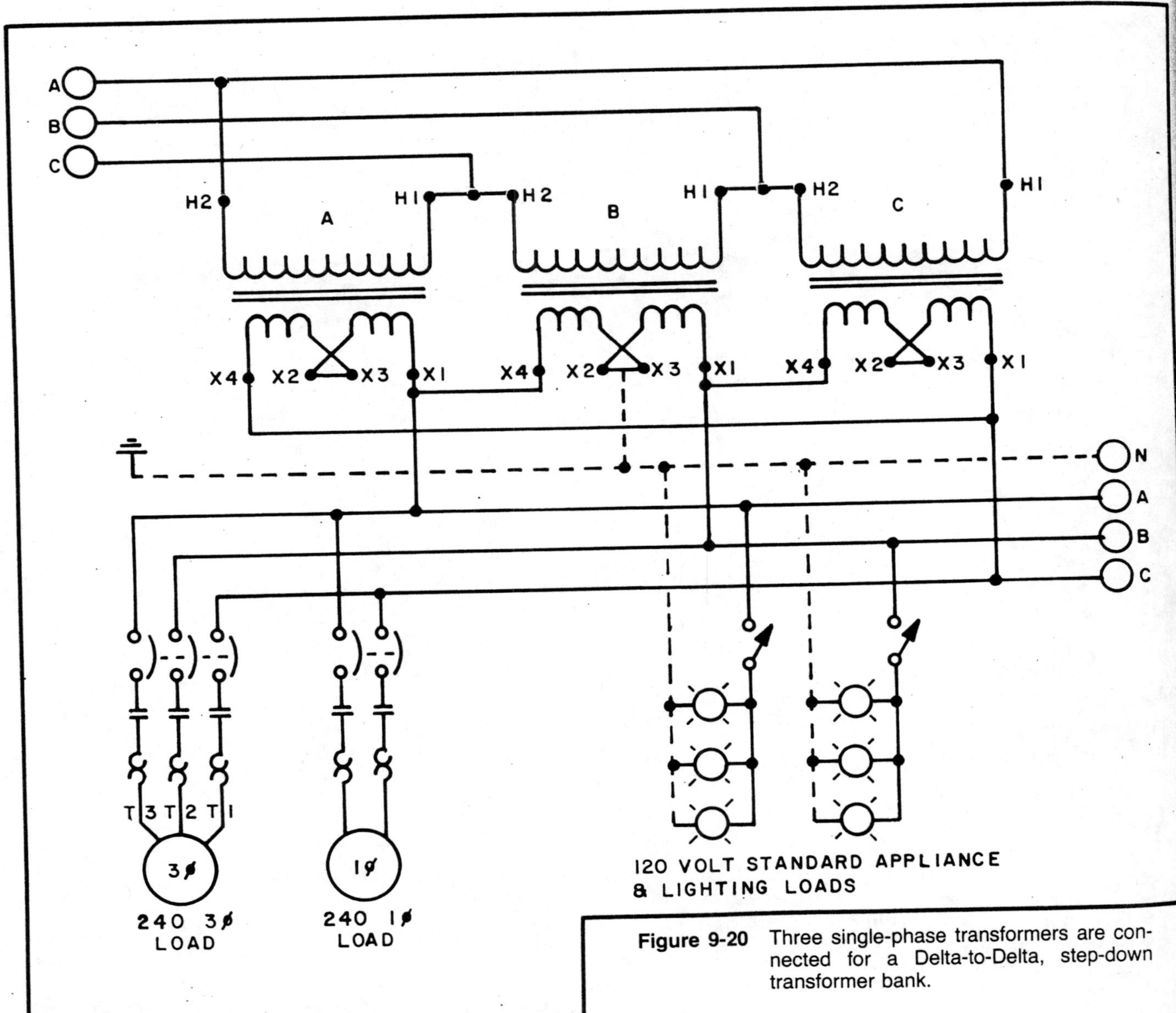

Figure 9-20 Three single-phase transformers are connected for a Delta-to-Delta, step-down transformer bank.

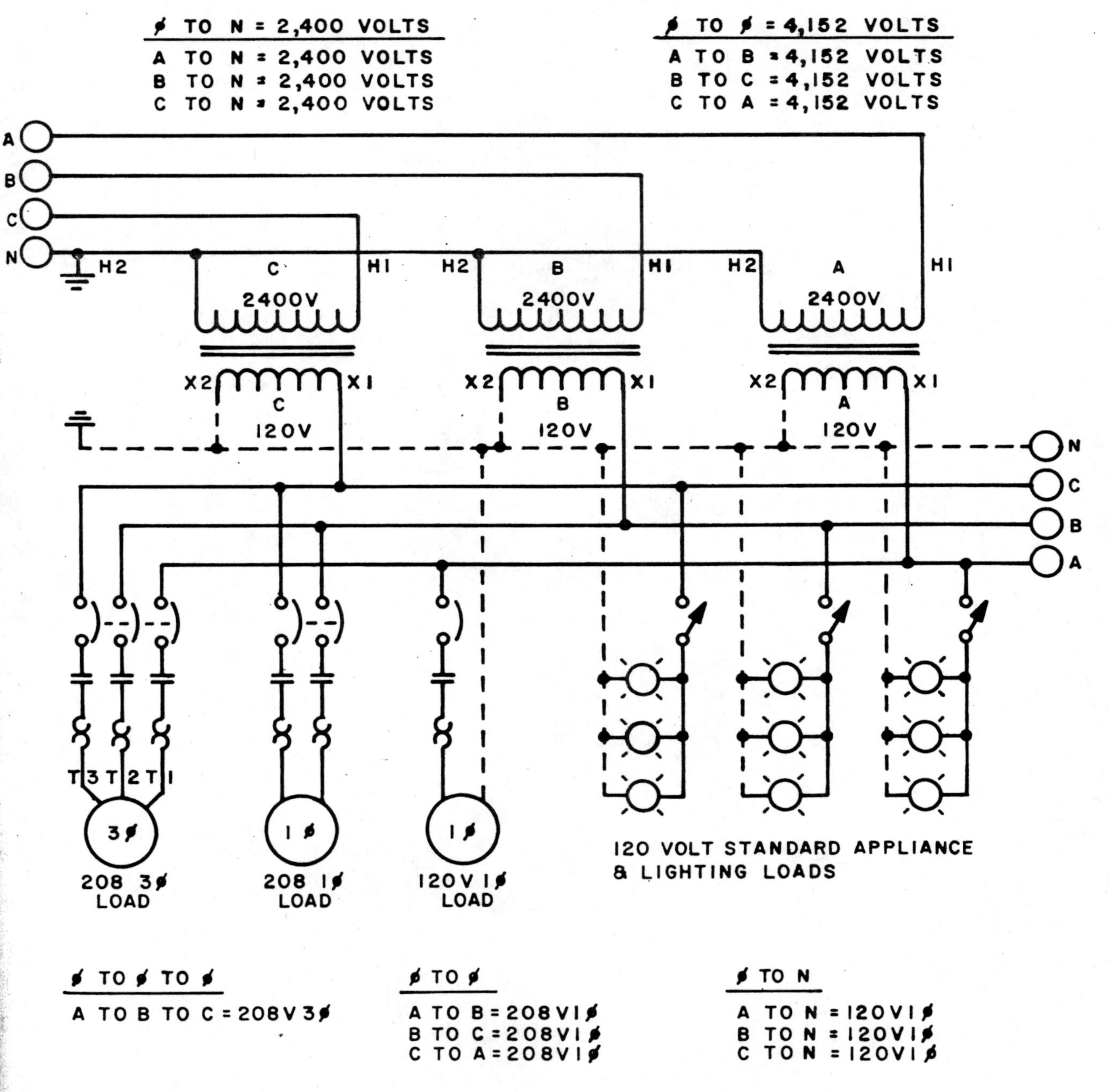

Figure 9-19 Three single-phase transformers are connected for a Wye-to-Wye step-down transformer bank.

Review Questions

1. What are the disadvantages of using a knife switch for starting and stopping electric motors?
2. What is a normally open double-break contact?
3. What is the function of arc hoods?
4. Why is a silver alloy used on switching contacts?
5. What is the function of a mechanical interlock?
6. How does a manual starter differ from a manual contactor?
7. Why must a motor be protected by an overload relay and a fuse or breaker?
8. What is ambient temperature?
9. What are the basic requirements of an overload relay?
10. What is a contactor?
11. What are the differences between AC and DC contactors?
12. Why is arc suppression needed?
13. Why is it harder to extinguish an arc on contacts passing DC than on contacts passing AC?
14. How are AC and DC contactors rated?
15. What is a magnetic motor starter?
16. What general factors must be considered when selecting overloads?
17. What is the service factor rating of a motor?
18. What is ambient temperature compensation?
19. What is an inherent motor protector?
20. How does an AC motor drive change motor speed?
21. How does a DC motor drive change motor speed?

chapter 9

electrical motor controls for Integrated Systems

Control Devices

Control devices range from simple pushbutton switches to complex solid-state sensors. Manufacturer's specification sheets detail required amperage, voltage, and sizing information for control devices. Installation of control devices requires proper position and location for safety and function in the intended environment.

INDUSTRIAL PUSHBUTTONS

Pushbuttons are the most common control switches used on industrial equipment. Almost all industrial machines and processes have a manually controlled position, even if the machine or process is designed to operate automatically. An industrial pushbutton consists of a legend plate, an operator, and one or more contact blocks (electrical contacts). **See Figure 9-1.**

Legend Plates

A *legend plate* is the part of a switch that includes the written description of the switch's operation. A legend plate indicates the pushbutton's function in the circuit. Legend plates are available indicating common circuit operations such as start, stop, jog, up, down, ON, OFF, reset, and run, or are available blank. The lettering on legend plates is normally uppercase for clarity and visibility. Legend plates are also available in different colors. The color red is normally used for such circuit functions as stop, OFF, and emergency stop. The color black with white lettering is used for most other circuit functions. However, different colored legend plates can be used along with colored operators to highlight different circuit functions. When color is used, red is normally used to indicate a stop or OFF function, green is used to indicate an ON or open function, and amber is used to indicate a manual override or reset function.

Technical Fact

Switches may have more than two poles or two throws. A single- or double-pole switch can have up to five or more throw positions.

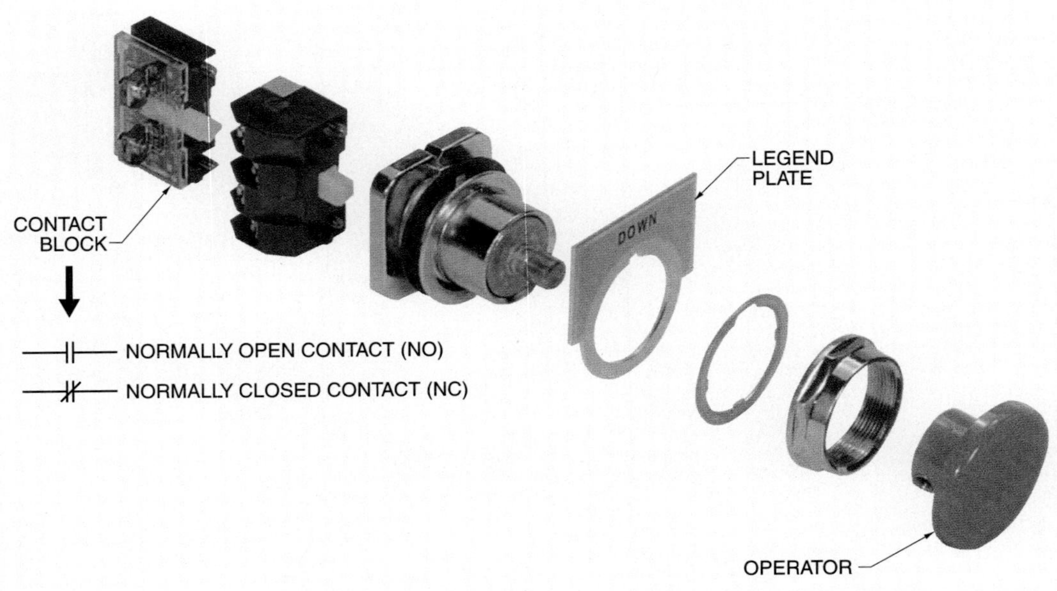

Figure 9-1. An industrial pushbutton consists of a legend plate, an operator, and one or more contact blocks (electrical contacts).

Operators

An *operator* is the device that is pressed, pulled, or rotated by the individual operating the circuit. An operator activates the pushbutton's contacts. Operators are available in many different colors, shapes, and sizes. Standard pushbutton operators include the flush, half-shrouded, extended, and jumbo mushroom button. The operator used depends on the application. **See Figure 9-2.**

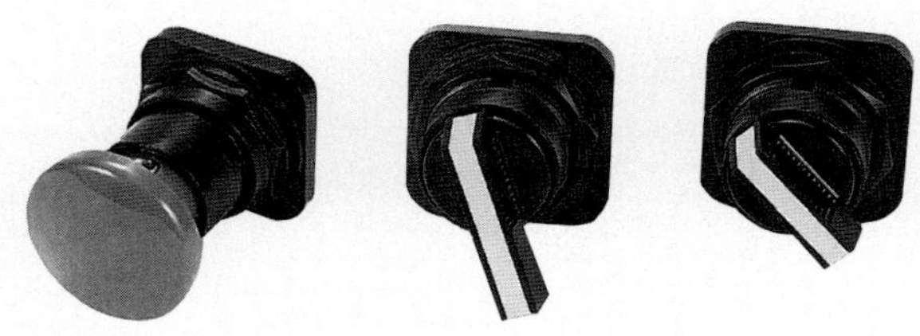

Figure 9-2. An operator is the device that is pressed, pulled, or rotated by the individual operating the circuit.

Rockwell Automation, Allen-Bradley Company, Inc.

The operator of a selector switch is rotated to control the operation of an electrical circuit.

Flush Button Operators. A *flush button operator* is a pushbutton with a guard ring surrounding the button that prevents accidental operation. The flush button operator is the most common operator used in applications in which accidental turn-ON may create a dangerous situation.

Half-Shrouded Button Operators. A *half-shrouded button operator* is a pushbutton with a guard ring that extends over the top half of the button. The guard ring helps prevent accidental operation, but allows for easier operation with the thumb. The half-shrouded button operator is used where avoiding accidental operation is preferred, but where the operator may be wearing gloves. Wearing gloves makes depressing a flush button operator difficult.

Extended Button Operators. An *extended button operator* is a pushbutton that has the button extended beyond the guard. An extended button operator is easily accessible and the color of the operator may be seen from all angles. The extended button operator is the most common operator used in applications in which an accidental start is not dangerous, such as when turning on lights.

Jumbo Mushroom Button Operators. A *jumbo mushroom button operator* is a pushbutton that has a large curved operator extending beyond the guard. A jumbo mushroom button operator is easily seen because of its large size. It can be operated from any angle and is used in applications that require fast operation such as emergency stops, motor stops, and valve shutoffs.

Contact Blocks

A *contact block* is the part of the pushbutton that is activated when the operator is pressed. A contact block includes the switching contacts of the pushbutton. Contact blocks include normally open (NO), normally closed (NC), or both NO and NC contacts. The most common contact block includes one NO and one NC contact. NO contacts make the circuit when the pushbutton operator is pressed and are used mainly for start or ON functions. NC contacts break the circuit when the pushbutton operator is pressed and are used mainly for stop or OFF functions. More than one contact block may be added to an operator. **See Figure 9-3.**

Pushbuttons are housed in pushbutton stations. A *pushbutton station* is an enclosure that protects the pushbutton, contact block, and wiring from dust, dirt, water, and corrosive fluids. Enclosures are available in various sizes and with a number of punched holes for mounting the operators. Every basic NEMA and IEC enclosure size is available because pushbutton stations need to be mounted where they can be conveniently operated.

A pushbutton must be placed in the proper enclosure for continuous and safe operation. Pushbuttons are often required to operate in environments where dust, dirt, oil, vibration, corrosive material, extreme variations of temperature and humidity, as well as other damaging factors are present. Always match the correct components and enclosure to the environment in which they will operate. **See Figure 9-4. See Appendix.**

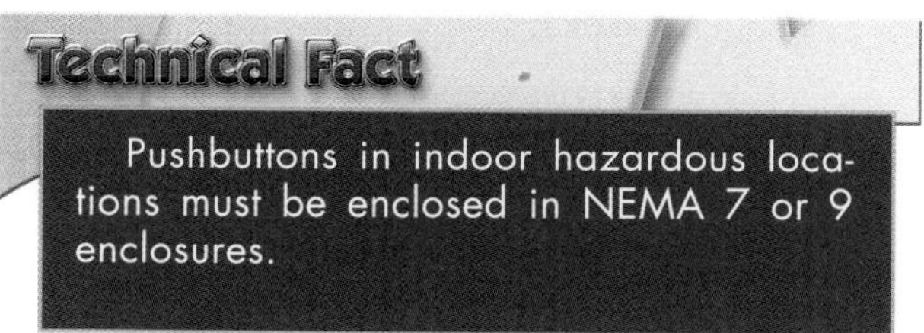

Pushbuttons in indoor hazardous locations must be enclosed in NEMA 7 or 9 enclosures.

Square D Company

Pushbuttons and pushbutton stations are designed to allow the use of interchangeable elements to meet various control application needs.

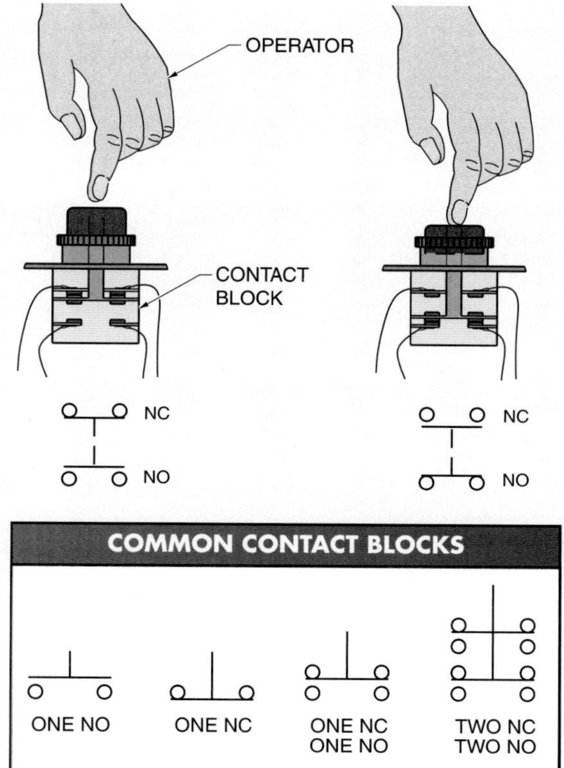

Figure 9-3. Contact blocks include normally open (NO), normally closed (NC), or both NO and NC contacts.

NEMA ENCLOSURE CLASSIFICATION					
Type	**Use**	**Service Conditions**	**Tests**	**Comments**	**Type**
1	Indoor	No unusual	Rod entry, rust resistance		1
3	Outdoor	Windblown dust, rain, sleet, and ice on enclosure	Rain, external icing, dust, and rust resistance	Does not provide protection against internal condensation or internal icing	
3R	Outdoor	Falling rain and ice on enclosure	Rod entry, rain, external icing, and rust resistance	Does not provide protection against dust, internal condensation, or internal icing	
4	Indoor/outdoor	Windblown dust and rain, splashing water, hose-directed water, and ice on enclosure	Hosedown, external icing, and rust resistance	Does not provide protection against internal condensation or internal icing	4
4X	Indoor/outdoor	Corrosion, windblown dust and rain, splashing water, hose-directed water, and ice on enclosure	Hosedown, external icing, and corrosion resistance	Does not provide protection against internal condensation or internal icing	4X
6	Indoor/outdoor	Occasional temporary submersion at a limited depth			
6P	Indoor/outdoor	Prolonged submersion at a limited depth			
7	Indoor locations classified as Class I, Groups A, B, C, or D, as defined in the NEC ®	Withstand and contain an internal explosion of specified gases, contain an explosion of specified gases, contain an explosion sufficiently so an explosive gas-air mixture in the atmosphere is not ignited	Explosion, hydrostatic, and temperature	Enclosed heat-generating devices shall not cause external surfaces to reach temperatures capable of igniting explosive gas-air mixtures in the atmosphere	7
9	Indoor locations classified as Class II, Groups E or G, as defined in the NEC ®	Dust	Dust penetration, temperature, and gasket aging	Enclosed heat-generating devices shall not cause external surfaces to reach temperatures capable of igniting explosive gas-air mixtures in the atmosphere	9
12	Indoor	Dust, falling dirt, and dripping noncorrosive liquids	Drip, dust, and rust resistance	Does not provide protection against internal condensation	12
13	Indoor	Dust, spraying water, oil, and noncorrosive coolant	Oil explosion and rust resistance	Does not provide protection against internal condensation	

IEC ENCLOSURE CLASSIFICATION

IEC Publication 529 describes standard degrees of protection that enclosures of a product must provide when properly installed. The degree of protection is indicated by two letters, IP, and two numerals. International Standard IEC 529 contains descriptions and associated test requirements to define the degree of protection that each numeral specifies. The following table indicates the general degrees of protection. For complete test requirements refer to IEC 529.

FIRST NUMERAL *†	SECOND NUMERAL*†
Protection of persons against access to hazardous parts and protection against penetration of solid foreign objects.	Protection against liquids‡ under test conditions specified in IEC 529.
0 Not protected	**0** Not protected
1 Protection against objects greater than 50 mm in diameter (hands)	**1** Protection against vertically falling drops of water (condensation)
2 Protection against objects greater than 12.5 mm in diameter (fingers)	**2** Protection against falling water with enclosure tilted 15°
3 Protection against objects greater than 2.5 mm in diameter (tools, wires)	**3** Protection against spraying of falling water with enclosure tilted 60°
4 Protection against objects greater than 1.0 mm in diameter (tools, small wires)	**4** Protection against splashing water
5 Protection against dust (dust may enter during test but must not interfere with equipment operation or impair safety)	**5** Protection against low-pressure water jets
6 Dusttight (no dust observable inside enclosure at end of test)	**6** Protection against powerful water jets
	7 Protection against temporary submersion
	8 Protection against continuous submersion

Example: IP41 describes an enclosure that is designed to protect against the entry of tools or objects greater than 1 mm in diameter, and to protect against vertically dripping water under specified test conditions.

* All first and second numerals up to and including numeral 6 imply compliance with the requirements of all preceding numerals in their respective series. Second numerals 7 and 8 do not imply suitability for exposure to water jets unless dual coded; e.g., IP_5/IP_7

† The IEC permits use of certain supplementary letters with the characteristic numerals. If such letters are used, refer to IEC 529 for an explanation.

‡ The IEC test requirements for degrees of protection against liquid ingress refer only to water

Figure 9-4. A pushbutton station is an enclosure that protects the pushbutton, contact block, and wiring from dust, dirt, water, and corrosive fluids.

SELECTOR SWITCHES

A *selector switch* is a switch with an operator that is rotated (instead of pushed) to activate the electrical contacts. Selector switches select one of several different circuit conditions. They are normally used to select either two or three different circuit conditions. However, selector switches are available that have more than three positions.

Two-Position Selector Switches

A two-position selector switch allows the operator to select one of two circuit conditions. For example, a two-position selector switch may be used to place a heating circuit in the manual (HAND) or automatic (AUTO) condition. Only the manual control switch can turn the heating contactor ON or OFF when the selector switch is placed in the HAND position. The heating contactor controls the high-power heating elements. Only the temperature switch can turn the heating contactor ON or OFF when the selector switch is placed in the AUTO position. Circuit conditions controlled by two-position selector switches include ON/OFF, left/right, manual/automatic, up/down, slow/fast, run/stop, forward/reverse, jog/run, and open/close conditions. **See Figure 9-5.**

Three-Position Selector Switches

A three-position selector switch allows the operator to select one of three circuit conditions. For example, a three-position selector switch may be used to place a heating circuit in the manual, automatic, or OFF position. The OFF position is added for safety. In the OFF position, the heating contactor (or other machine being controlled) cannot be energized by the manual or automatic switch. Circuit conditions controlled by three-position selector switches include manual/OFF/automatic, heat/OFF/cool, forward/OFF/reverse, jog/OFF/run, slow/stop/fast, and up/stop/down conditions. **See Figure 9-6.**

Truth Tables

Contact position on a selector switch may be illustrated using truth tables (target tables) or solid lines, dashed lines, and a series of small circles. **See Figure 9-7.** In truth tables, each contact on the line diagram is marked A, B, etc., and each position of the selector switch is marked 1, 2, etc. The truth table is made and positioned near the switch to illustrate each position and each contact.

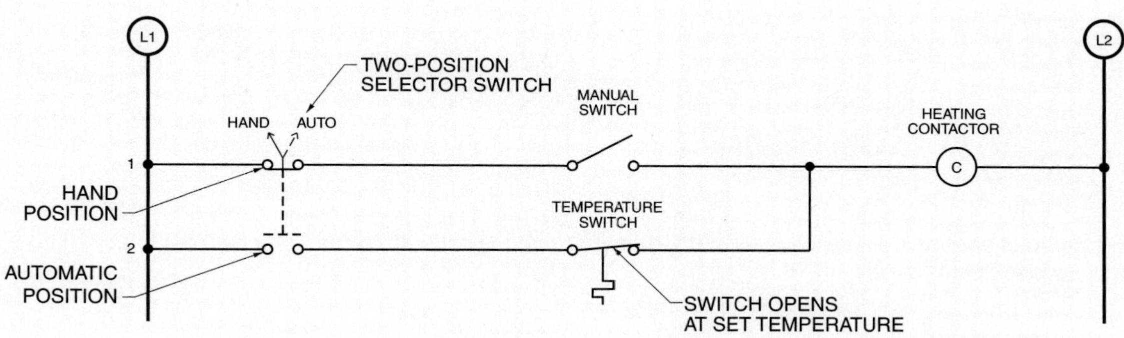

Figure 9-5. A two-position selector switch allows the operator to select one of two circuit conditions.

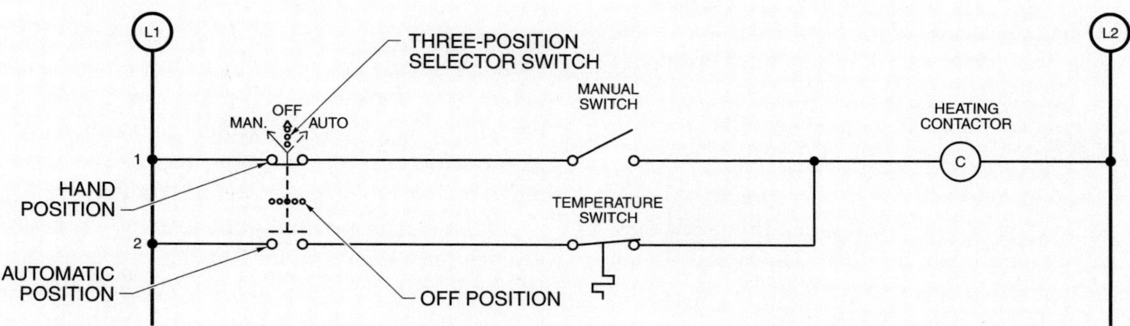

Figure 9-6. A three-position selector switch allows the operator to select one of three circuit conditions.

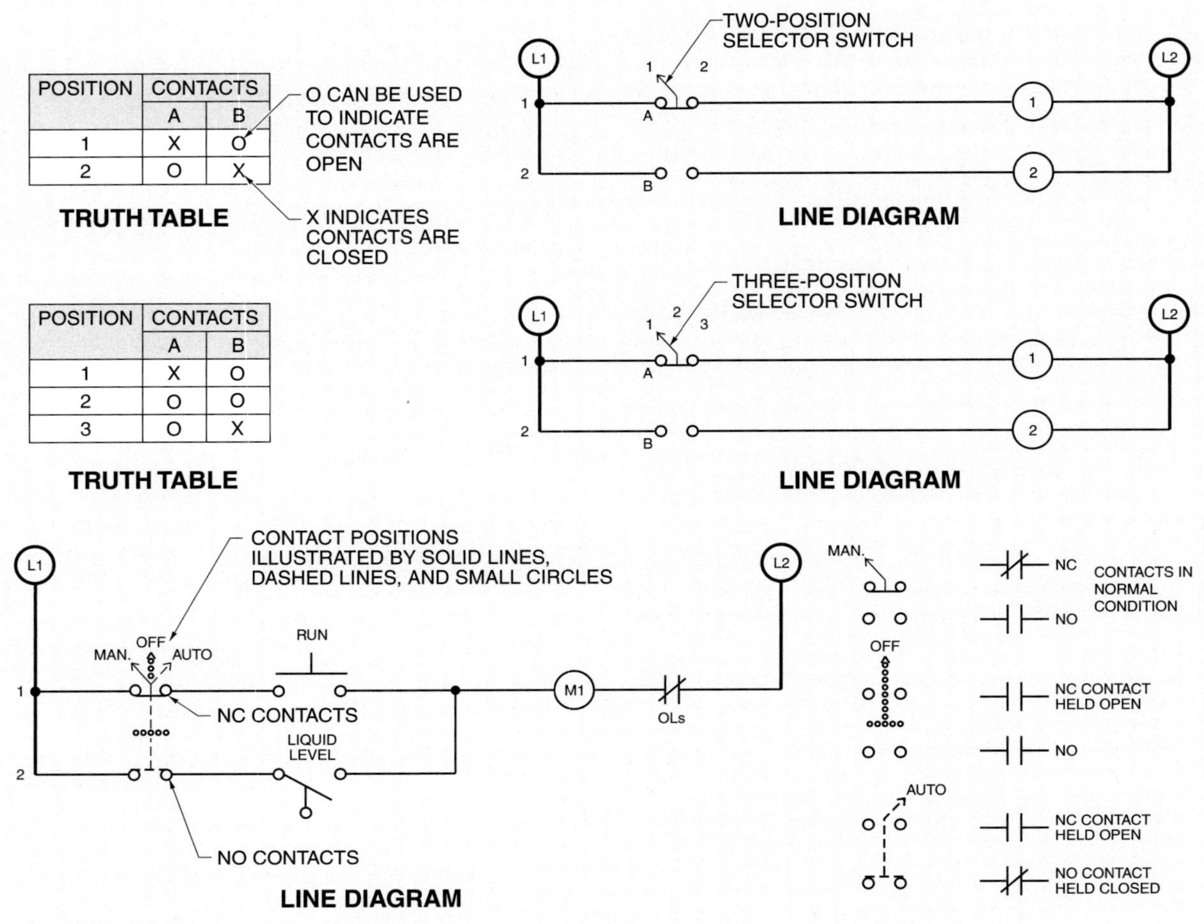

Figure 9-7. Contact position on a selector switch may be illustrated using truth tables (target tables) or solid lines, dashed lines, and a series of small circles.

An X is placed in the table if a contact is closed in any position. The table is easily read as to what contacts are closed in what positions. Truth tables illustrate the selector switch contacts more clearly than the method of using solid and dashed lines and small circles when a selector switch has more than two contacts or more than three positions.

JOYSTICKS

A *joystick* is an operator that selects one to eight different circuit conditions by shifting the joystick from the center position into one of the other positions. The most common joysticks can move from the center position into one of four different positions (up, down, left, or right).

The advantage of a joystick is that a person may control many operations without removing their hand from the joystick and they do not have to take their eyes off the operation performed by the circuit.

The most common circuit condition controlled by a joystick is in controlling a hoist (or crane) in the raise, lower, left, right, or OFF position. **See Figure 9-8.** In the hoist application, two reversing motors move the hoist and pulleys. One forward and reversing motor starter controls the hoist drive motor, and another forward and reversing motor starter controls the pulley motor. The joystick can turn only one motor starter ON at a time.

Two methods are used to indicate which position the joystick must be placed in to operate the contacts. In the first method, a dot is placed in the symbol of the joystick to indicate the position the joystick must be in to switch the contacts. The NO contacts close and the NC contacts open when the contacts are switched.

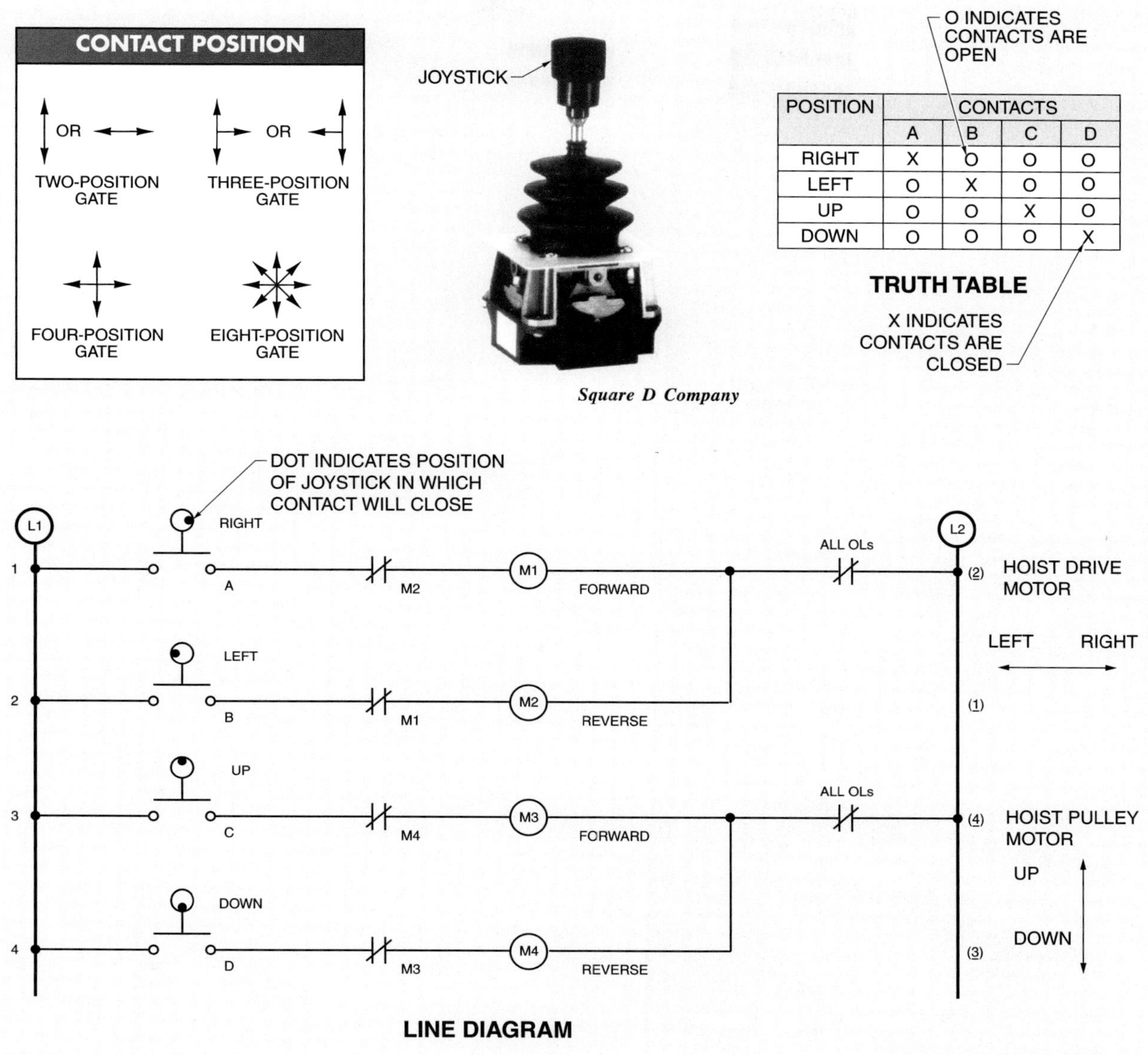

Figure 9-8. A joystick is used to control many different circuit operations from one location.

In the second method, a truth table is used to indicate which contacts are switched in each position. In the truth table, an X indicates when the contact is closed. Truth tables are normally given in manufacturer's catalogs showing joystick operation, and a dot in the symbol is normally used on the line diagram.

LIMIT SWITCHES

A *limit switch* is a mechanical input that requires physical contact of the object with the switch actuator. The physical contact is obtained from a moving object that comes in contact with the limit switch. The mechanical motion physically opens or closes a set of contacts within the limit switch enclosure. The contacts start or stop the flow of current in the electrical circuit. The contacts start, stop, operate in forward, operate in reverse, recycle, slow, or speed an operation. **See Figure 9-9.**

For example, a limit switch is used to automatically turn ON a light in a refrigerator or prevent a microwave oven from operating with the door open. In a washing machine, a limit switch is used to automatically turn OFF the washer if the load is not balanced. In an automobile, limit switches are used to automatically turn ON lights when a door is opened, and prevent overtravel of automatically operated windows. In industry, limit switches are used to limit the travel of machine parts, sequence operations, detect moving objects, monitor an object's position, and provide safety by, for example, detecting guards in place.

Limit switch contacts are normally snap-acting switches, which quickly change position to minimize arcing at the contacts. Limit switch contacts may be NO, NC, or any combination of NO and NC contacts. Most limit switches include one NO contact and one NC contact. Contacts are rated for the maximum current and voltage they can safely control.

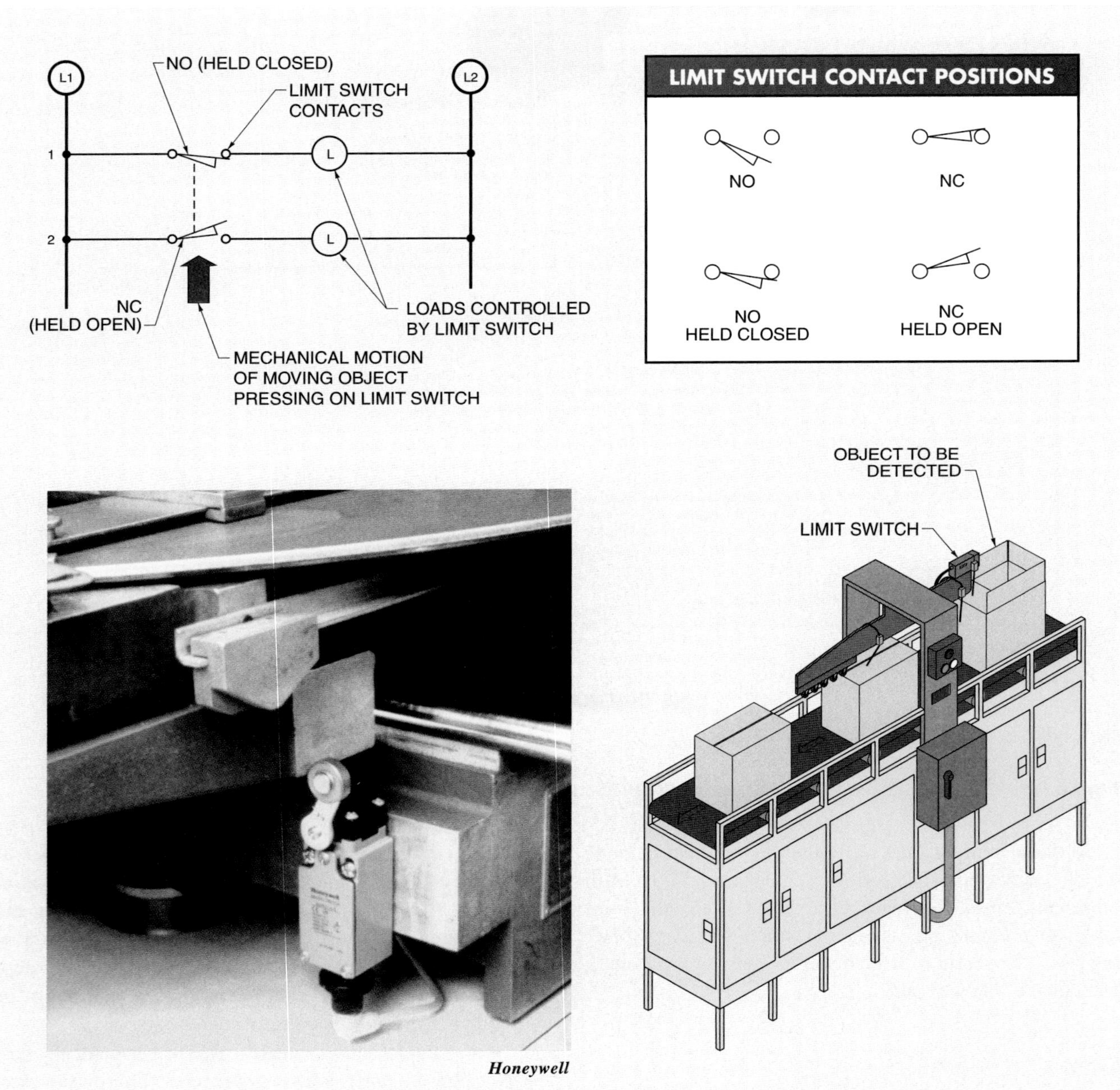

Figure 9-9. Limit switches are used to convert a mechanical motion into an electrical signal.

Limit switch contacts must be connected to the proper polarity. There is no arcing between the contacts when the contacts energize and de-energize the load as long as the contacts are at the same polarity. Arcing or welding of the contacts may occur from a possible short circuit if the contacts are connected to opposite polarity. **See Figure 9-10.** Contacts must be selected according to proper voltage and current size according to the load and manufacturer specifications.

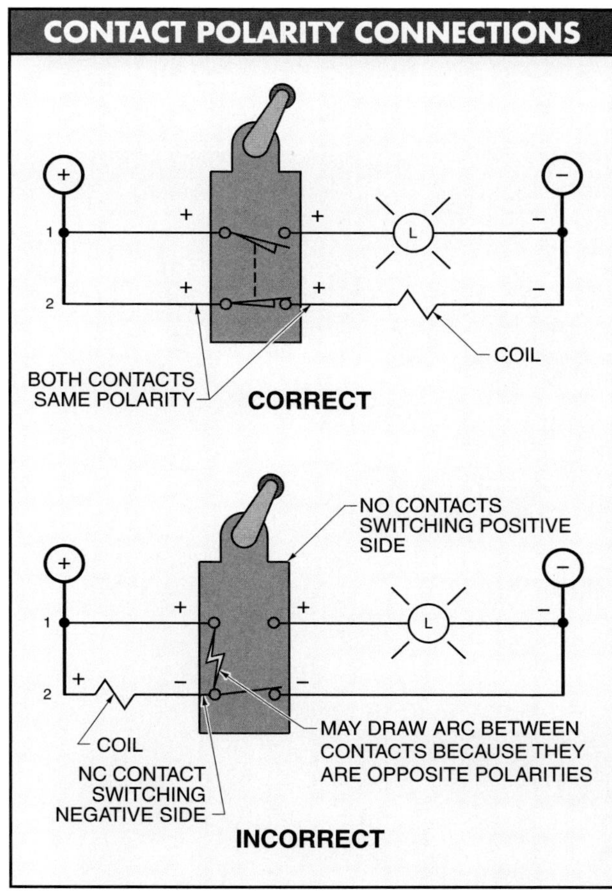

Figure 9-10. Arcing or welding of the contacts may occur from a possible short circuit if the contacts are connected to opposite polarity.

Technical Fact

Limit switches are the most cost-effective switch for detecting objects that can be touched. Limit switches are available in a variety of sizes, enclosures, actuation methods, and electrical ratings.

A relay, contactor, or motor starter must be used to interface the limit switch with the load if the load current exceeds the contact rating. **See Figure 9-11.**

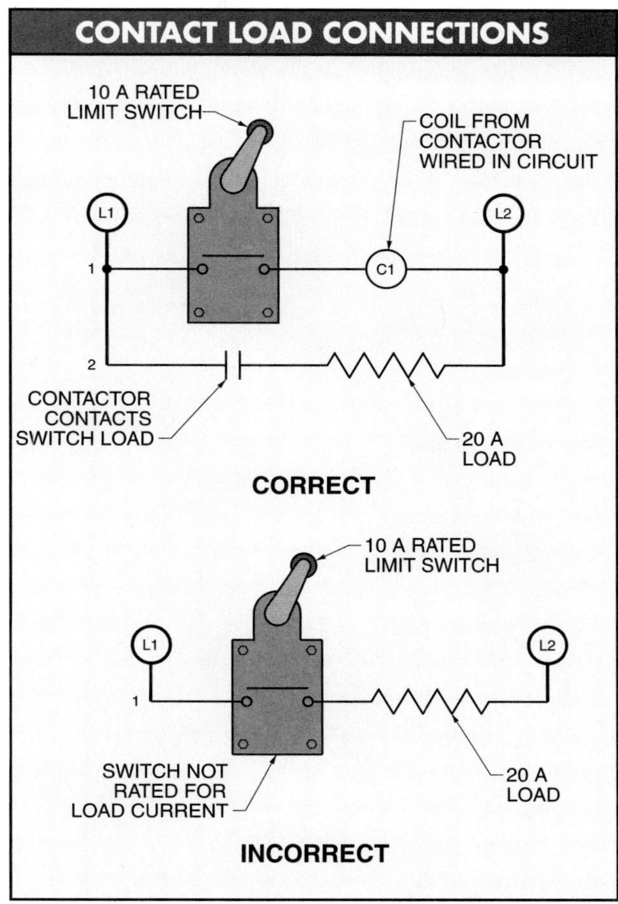

Figure 9-11. A relay, contactor, or motor starter must be used to interface the limit switch with the load if the load current exceeds the contact rating.

Limit Switch Actuators

An *actuator* is the part of a limit switch that transfers the mechanical force of the moving part to the electrical contacts. The basic actuators used on limit switches include lever, fork lever, push roller, and wobble stick. Most manufacturers offer several variations in addition to the basic actuators. **See Figure 9-12.**

Small limit switches are available with one fixed actuator. The fixed actuator may be any one of several different types, but is neither removable nor interchangeable. Large limit switches are available with a knurled shaft that allows different actuators to be attached. The actuator used depends on the application.

Levers. A *lever actuator* is an actuator operated by means of a lever that is attached to the shaft of the limit switch. The lever actuator includes a roller on the end that helps to prevent wear. The length of the lever may be fixed or adjustable. The adjustable lever is used in applications in which the length of the arm, or actuator travel, may require adjustment. A lever actuator may be operated from either direction but is normally used in applications in which the actuating object is moving in only one direction. A typical application is on an assembly line conveyor system.

Fork Levers. A *fork lever actuator* is an actuator operated by either one of two roller arms. Fork lever actuators are used where the actuating object travels in two directions. A typical application is a grinder that automatically alternates back and forth.

Push Rollers. A *push-roller actuator* is an actuator operated by direct forward movement into the limit switch. A direct thrust with very limited travel is accomplished. Push-roller actuators are commonly used to prevent overtravel of a machine part or object. The switch contacts stop the forward movement of the object when the machine part comes in contact with the limit switch. A typical application is on a milling-machine or an automatic turret lathe, where the travel of the work surface would need to be monitored.

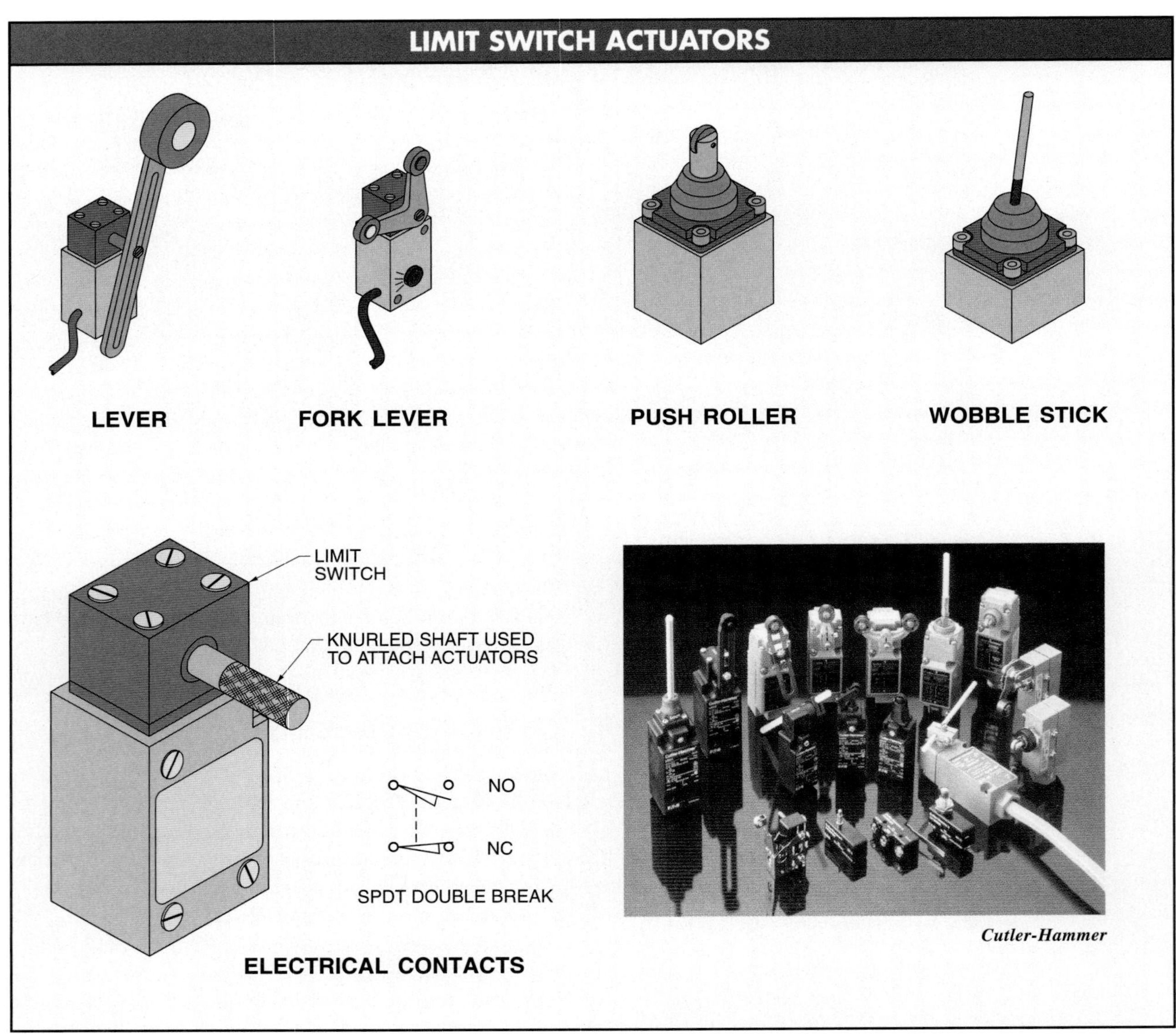

Figure 9-12. An actuator is the part of a limit switch that transfers the mechanical force of the moving part to the electrical contacts.

Wobble Sticks. A *wobble-stick actuator* is an actuator operated by means of any movement into the switch, except a direct pull. The wobble-stick actuator normally has a long arm that may be cut to the required length. Wobble-stick actuators are used in applications that require detection of a moving object from any direction such as in the robotics section of an automated manufacturing facility.

Limit Switch Installation

Limit switches are actuated by a moving part. Limit switches must be placed in the correct position in relationship to the moving part. Limit switches should not be operated beyond the manufacturer's recommended travel specifications. **See Figure 9-13.**

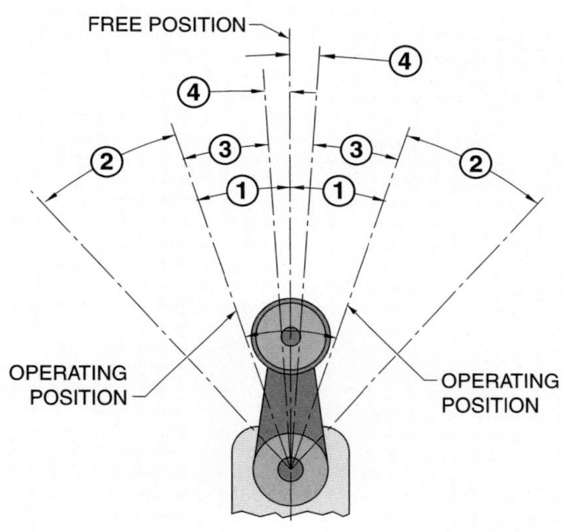

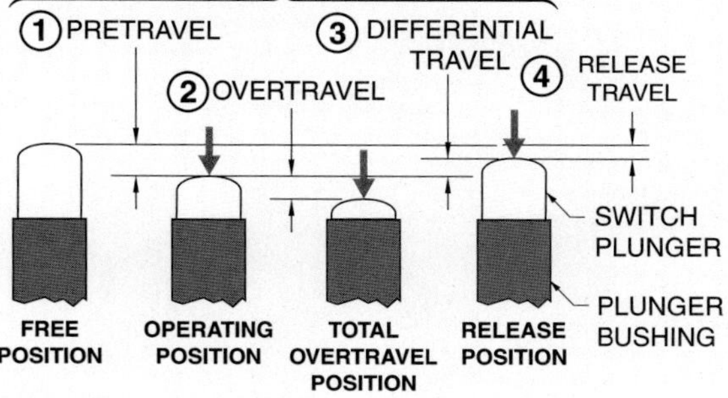

Figure 9-13. Limit switches should not be operated beyond the manufacturer's recommended travel specifications.

Limit switch contacts do not operate if the actuating object does not force the limit switch actuator to move far enough. Limit switch contacts operate, but may return to their normal position if the actuating object forces the actuator of the limit switch to move too far. Overtravel may also damage the limit switch or force it out of position.

A rotary cam-operated limit switch must be installed according to manufacturer recommendations. A push-roller actuator should not be allowed to snap back freely. The cam should be tapered to allow a slow release of the lever. This helps to eliminate roller bounce and switch wear, and allows for better repeat accuracy. **See Figure 9-14.**

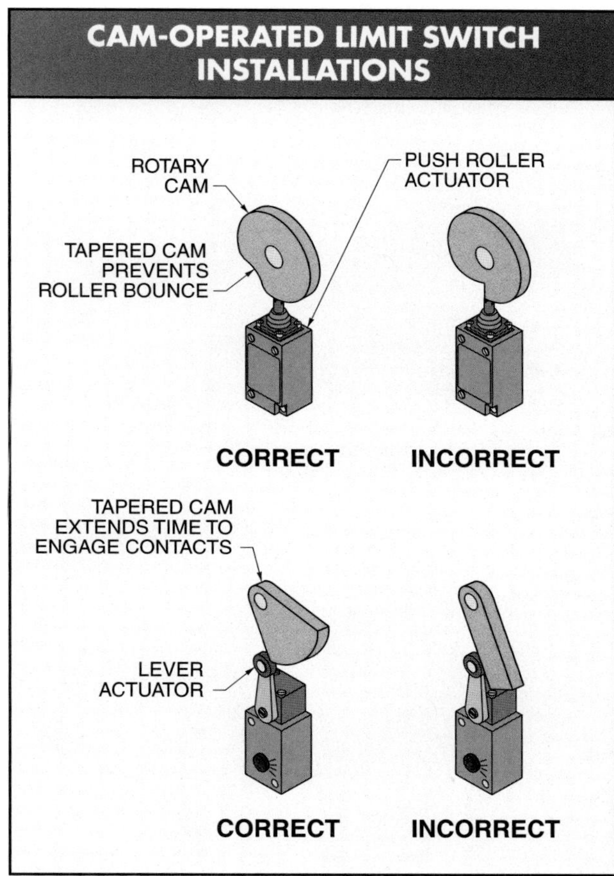

Figure 9-14. A cam-operated limit switch must be installed to prevent severe impact and allow a slow release of the lever.

Limit switches installed where relatively fast motions are involved must be installed so that the limit switch's lever does not receive a severe impact. The cam should be tapered to extend by the time it takes to engage the electrical contacts. This prevents wear on the switch and allows the contacts a longer closing time, ensuring that the circuit is complete.

Limit switches using push-roller actuators must not be operated beyond their travel in emergency conditions. A lever actuator is used instead of a push-roller actuator in applications where an override may occur. **See Figure 9-15.** A lever actuator has an extended range, which prevents damage to the switch and mounting in case of an overtravel condition. A limit switch should never be used as a stop. A stop plate should always be added to protect the limit switch and its mountings from any damage due to overtravel.

Limit switches are designed to be used as automatic controllers that are mechanically activated. Care should be taken to avoid any human error. Limit switches should be mounted so that an operator cannot accidentally activate the limit switch. **See Figure 9-16.**

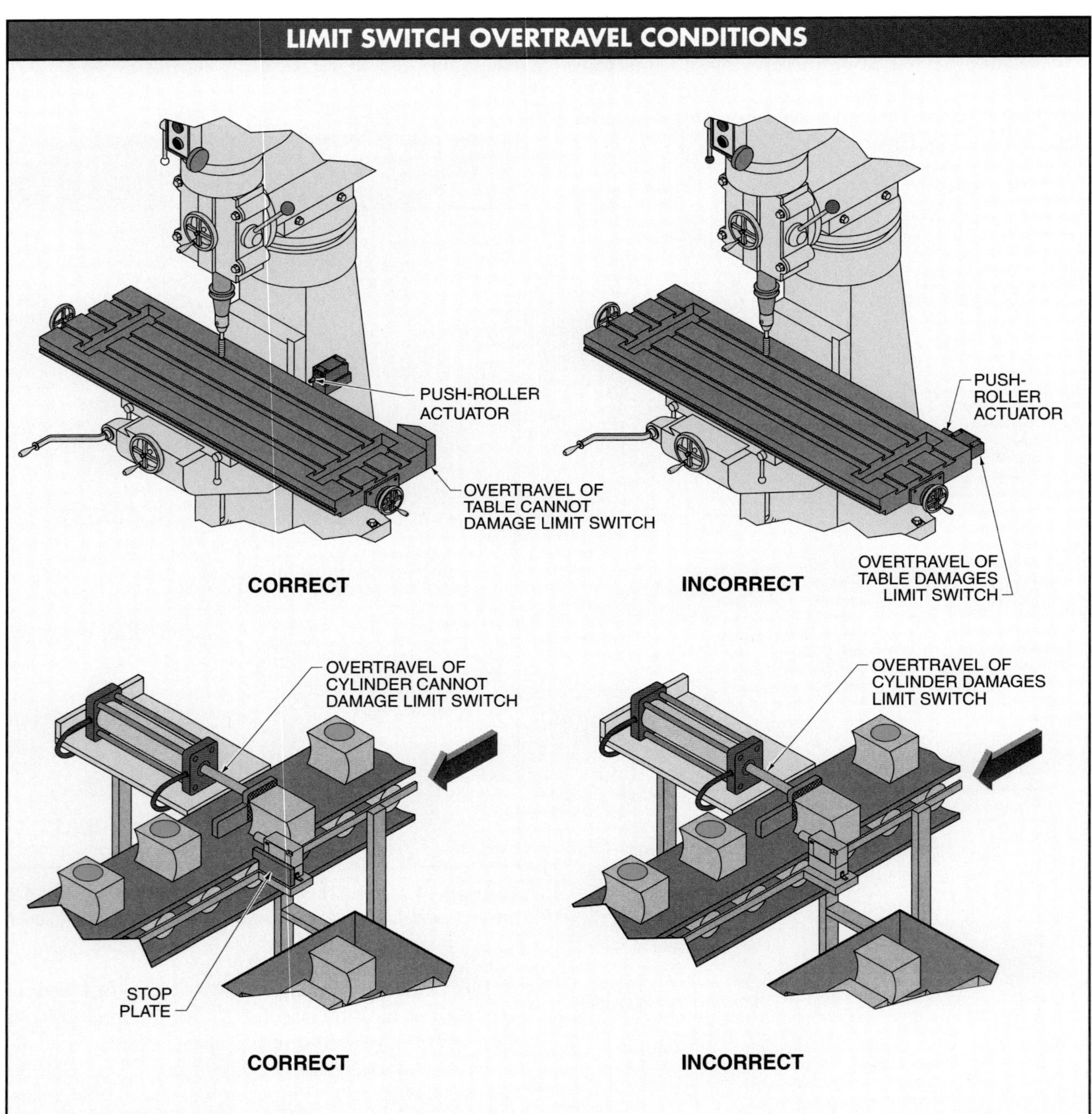

Figure 9-15. Limit switches using push-roller actuators must not be operated beyond their travel limit.

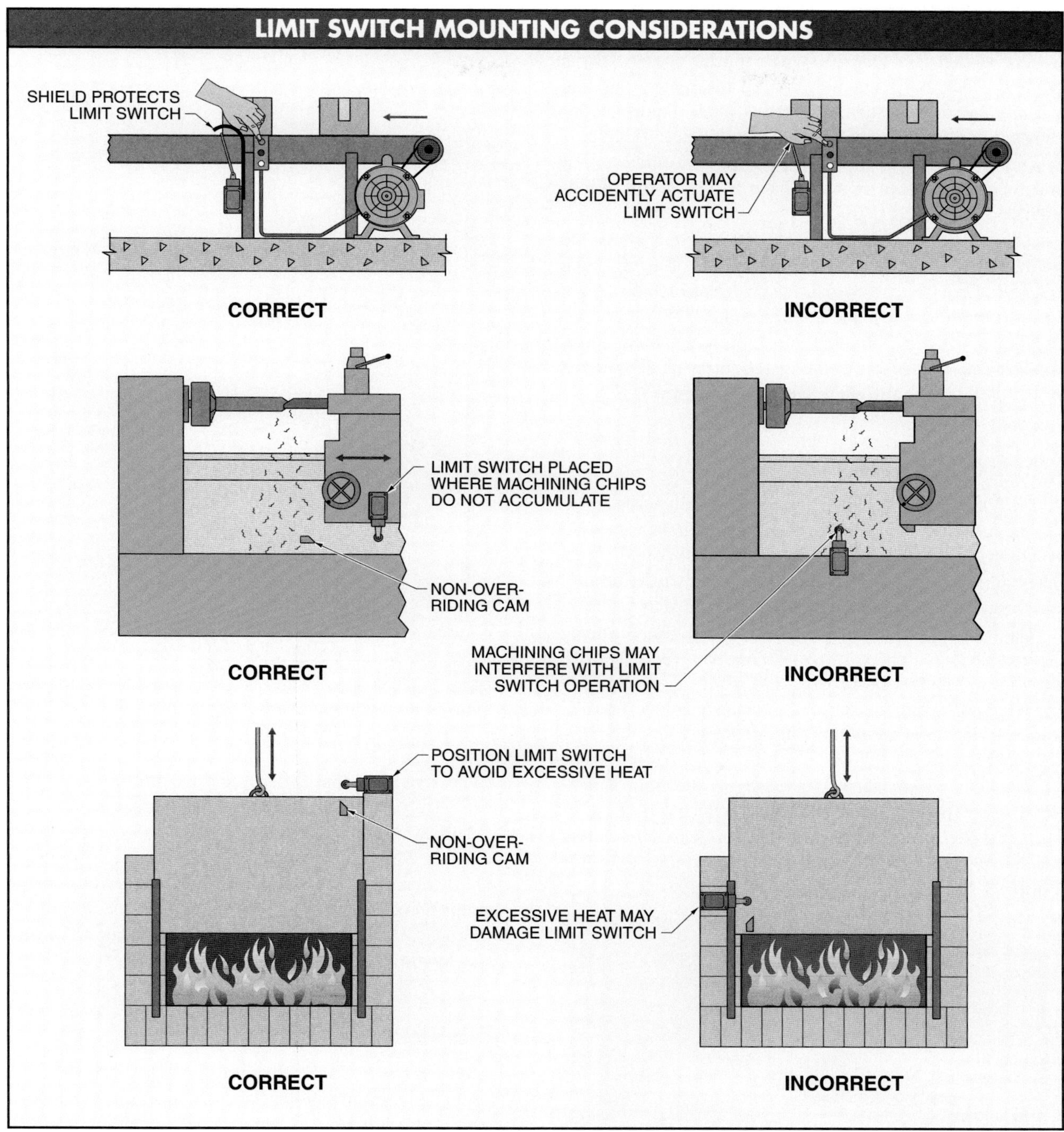

Figure 9-16. Limit switches should be mounted to avoid accidental activation, accumulated materials, and excessive heat.

The atmosphere and surroundings of the limit switch must be considered when mounting a limit switch. A limit switch must be mounted in a location where machining chips or other materials do not accumulate. These could interfere with the operation of the limit switch and cause circuit failure. Submerging the limit switch or splashing it with oils, coolants, or other liquids must be avoided. Heat levels above the specified limits of the switch must also be avoided. Always position a limit switch to avoid any excessive heat.

Foot Switches

A *foot switch* is a control switch that is operated by a person's foot. A foot switch is used in applications that require a person's hands to be free or that require an additional control point. Foot switch applications include sewing machines, drill presses, lathes, and other similar machines. Most foot switches have two positions, a toe-operated position and an OFF position.

The OFF position is normally spring-loaded so that the switch automatically returns to the OFF position when released. Foot switches with three positions include a pivot on a fulcrum to allow toe or heel control. Like the two-position foot switch, the three-position foot switch is normally spring-loaded so that the switch automatically returns to the OFF position when released. **See Figure 9-17.**

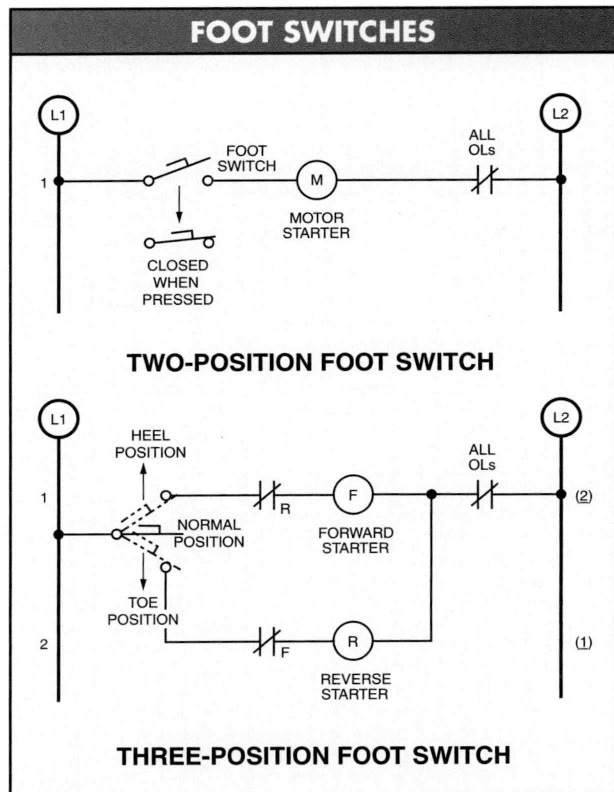

Figure 9-17. A foot switch is used to allow hands-free control or an additional control point.

DAYLIGHT SWITCHES

A *daylight switch* is a switch that automatically turns lamps ON at dusk and OFF at dawn. Daylight switches are used to control outdoor lamps such as streetlights and signs. They are also used to provide safety and security around buildings and other areas that require lighting at night.

Daylight switches use a sensor that changes resistance with a change in light intensity. The higher the light source, the lower the sensor's resistance. Current flows through the relay coil when the sensor's resistance is low. The NC contacts open and the lamp is turned OFF when the relay coil energizes.

The sensor's resistance increases as the light source decreases. The increased resistance reduces the flow of current to the point that the relay de-energizes. This causes the NC contacts to close and the lamp to turn ON. **See Figure 9-18.**

Ruud Lighting, Inc.

Low-voltage track lighting systems are available with a photocontrol/time clock option, which turns the lights ON at dusk and keeps them ON for a set time period.

The sensor of the switch must be positioned so that artificially produced light from the lamp that is being controlled, as well as from other lamps, does not fall on the sensor. Most daylight switches have a sensitivity adjustment that changes the amount of light required to switch a lamp ON or OFF. Most daylight switches include an approximate 30 sec time delay to prevent nuisance switching caused by automobile headlights.

PRESSURE SWITCHES

Pressure is force exerted over a surface divided by its area. The exerted force always produces a deflection or change in the volume or dimension of the area to which it is applied. Pressure is expressed in pounds per square inch (psi). Low pressures are expressed in inches of water column (in. WC). One psi equals 27.68 in. WC.

A *pressure switch* is a switch that detects a set amount of force and activates electrical contacts when the set amount of force is reached. The contacts may be activated by positive, negative (vacuum), or differential pressures. Differential pressure switches are connected to two different system pressures.

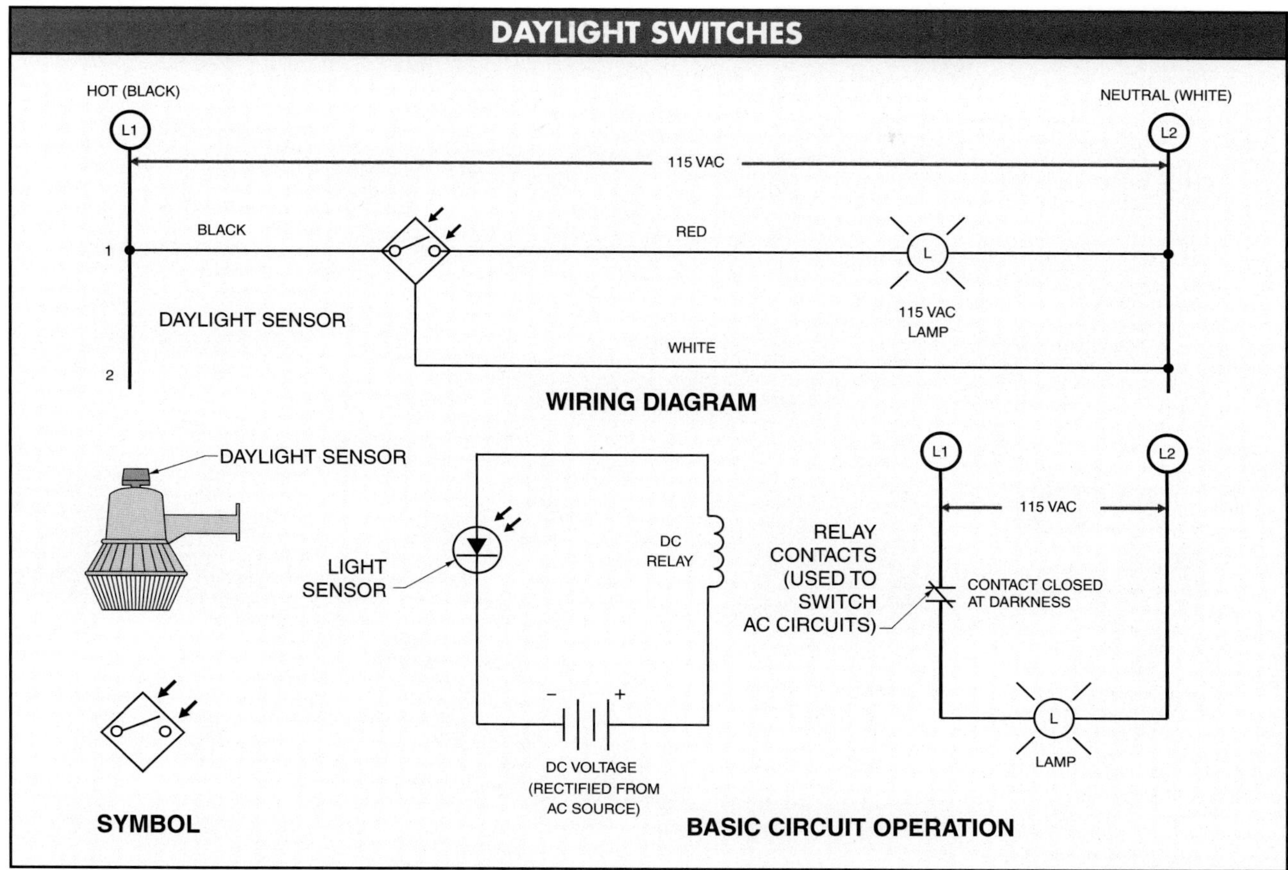

Figure 9-18. A daylight switch is a switch that automatically turns lamps ON at dusk and OFF at dawn.

NC or NO contacts are used, depending on the application. NC contacts are used to maintain system pressure. The closed contacts energize a pump motor until system pressure is reached. **See Figure 9-19.**

When system pressure is reached, the contacts open and the pump motor is turned OFF. NO contacts are used to signal an overpressure condition. An alarm is sounded when the open contacts close. The alarm remains ON until the pressure is reduced.

Pressure switches use different sensing devices to detect the amount of pressure. The pressure switch used depends on the application and system pressure. Most pressure switches use a diaphragm, bellows, or piston sensing device.

Diaphragm and bellows sensing devices are used for low-pressure applications. Piston sensing devices are used for high-pressure applications. **See Figure 9-20.**

Diaphragms

A *diaphragm* is a deflecting mechanism that moves when a force (pressure) is applied. One side of the diaphragm is connected to the pressure to be detected (source pressure) and the other side is vented to the atmosphere.

The diaphragm moves against a spring switch mechanism that operates electrical contacts when the source pressure increases. The spring tension is adjustable to allow for different pressure settings. A diaphragm pressure switch is used with pressures of less than 200 psi, but some are designed to detect several thousand pounds of pressure.

Bellows

A *bellows* is a cylindrical device with several deep folds that expand or contract when pressure is applied. One end of the bellows is closed and the other end is connected to the source pressure. The expanding bellows moves against a spring switch mechanism that operates electrical contacts when the source pressure increases.

The spring tension is adjustable to allow for different pressure settings. A bellows pressure switch is used with pressures up to 500 psi, but some are designed for higher pressures.

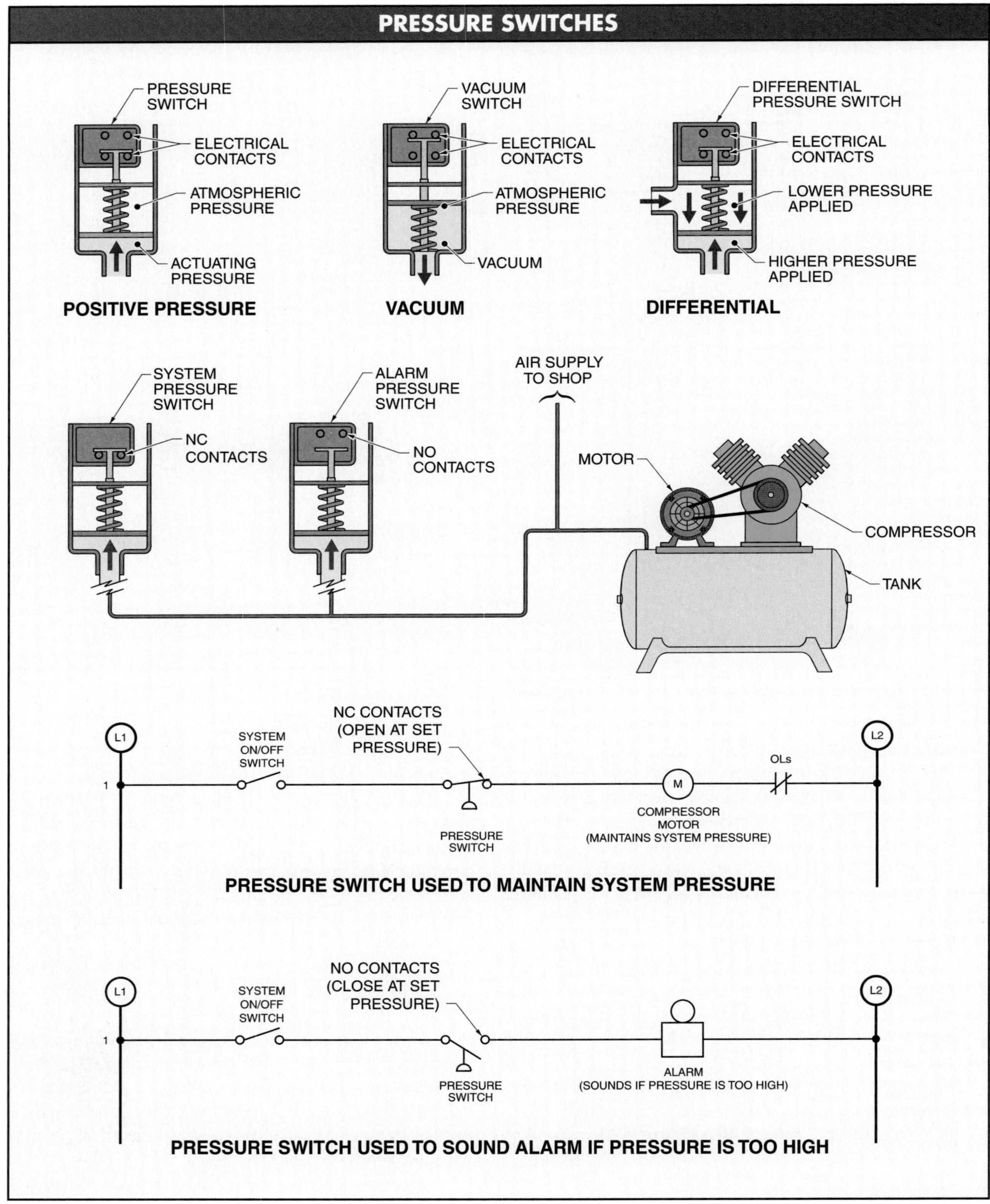

Figure 9-19. A pressure switch is a control switch that detects a set amount of force and activates electrical contacts when the set amount of force is reached.

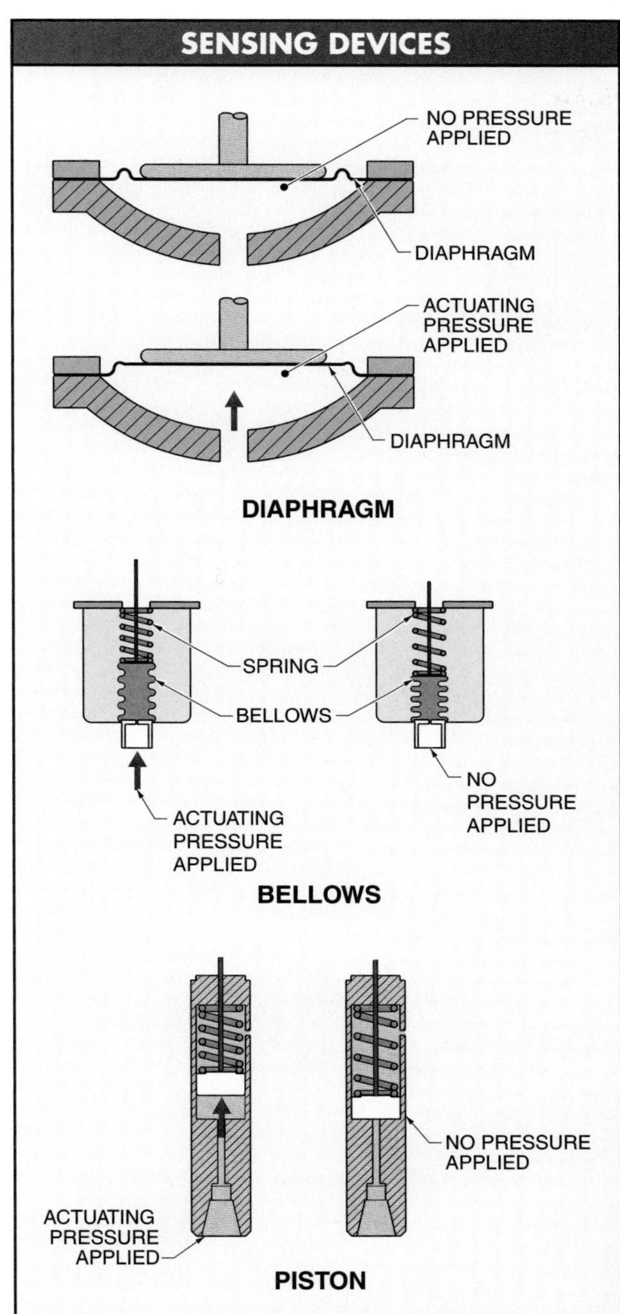

Figure 9-20. Pressure switches use different sensing devices to detect the amount of pressure.

Pistons

A *piston* is a cylinder that is moved back and forth in a tight-fitting chamber by the pressure applied in the chamber. A piston sensing device (pressure switch) uses a stainless steel piston moving against a spring tension to operate electrical contacts. The piston moves a switch mechanism that operates electrical contacts when the source pressure increases.

The spring tension is adjustable to allow for different pressure settings. Piston sensing devices (pressure switches) are designed for high-pressure applications of 10,000 psi or more.

Deadband

When a change in pressure occurs causing the diaphragm to move far enough to actuate the switch contacts, some of the pressure must be removed before the switch resets for another cycle. *Deadband (differential)* is the amount of pressure that must be removed before the switch contacts reset for another cycle after the setpoint has been reached and the switch has been actuated. **See Figure 9-21.**

Deadband is inherent in all pressure, temperature, level, and flow switches, and most automatically actuated switches. Deadband is not a fixed amount, but is different at each setpoint. Deadband is minimum when the setpoint is at the low end of the switch range. Deadband is maximum when the setpoint is at the high end of the switch range.

Deadband may be beneficial or detrimental. Without a deadband range, or too small of one, electrical contacts chatter ON and OFF as a pressure switch approaches the setpoint. However, a large deadband is detrimental in applications that require the pressure to be maintained within a very close range. Different switches have different deadband ratings. Always check the amount of listed deadband when using pressure switches in different applications.

Ruud Lighting, Inc.

Daylight switches increase the security and convenience of landscape lighting.

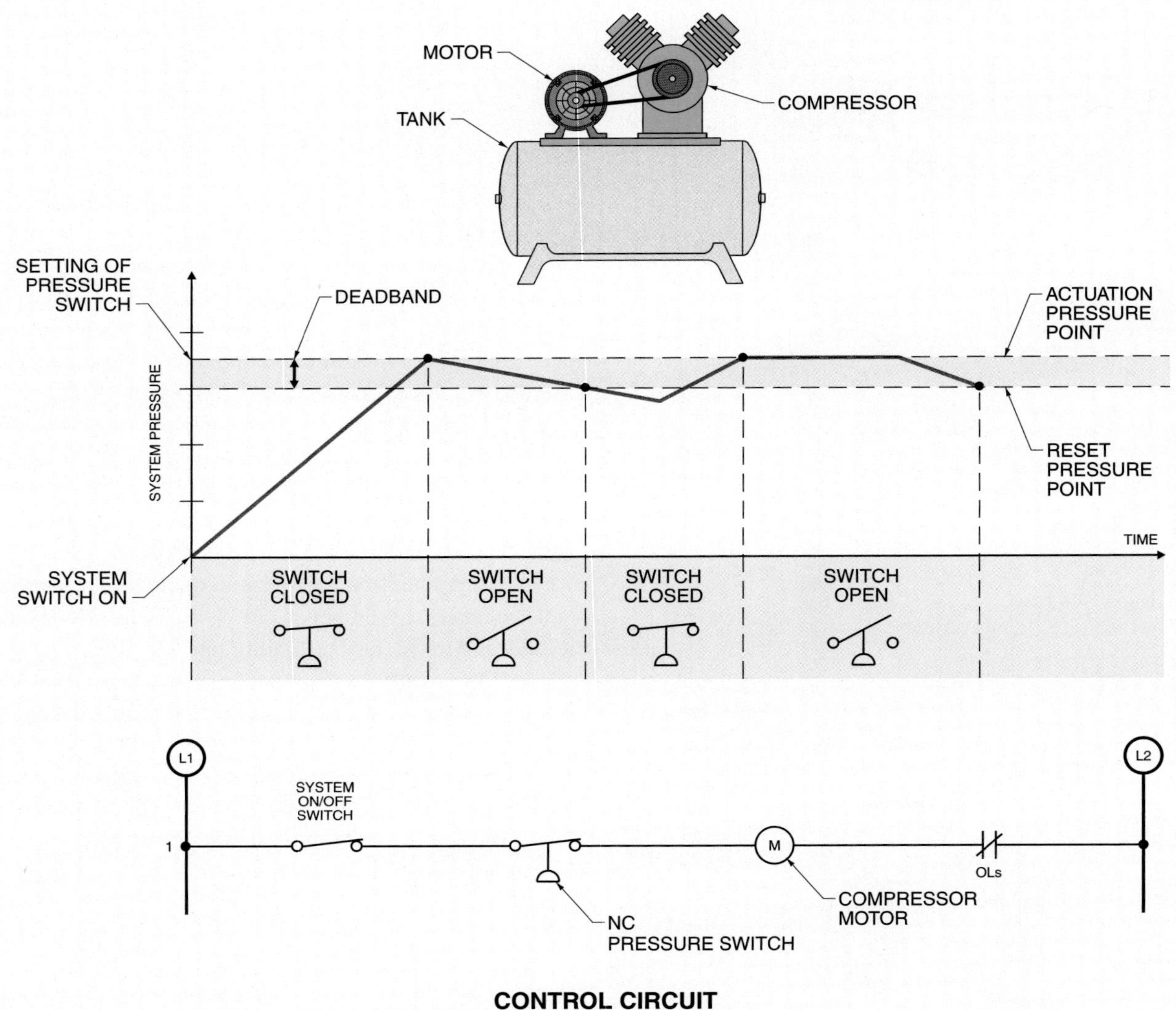

Figure 9-21. Deadband is the amount of pressure that must be removed before the switch contacts reset for another cycle after the setpoint has been reached and the switch has been actuated.

Pressure Switch Applications

Most pressure switches are used to maintain a predetermined pressure in a tank or reservoir. Pressure switches may also be used to sequence the return of pneumatic or hydraulic cylinders. **See Figure 9-22.** The two-position, four-way, directional control valve solenoid is energized when the operator presses the start pushbutton. This changes the directional control valve from the spring position to the solenoid-actuated position. The cylinder advances because the flow of pressure is changed in the cylinder. The control relay energizes and its NO contacts close because it is in parallel with the solenoid. This adds memory to the circuit. The operator releases the pushbutton and the cylinder continues to advance. The cylinder advances until the preset pressure is reached on the pressure switch, which signals the return of the cylinder by de-energizing the solenoid and relay. The de-energizing of the solenoid returns the directional control valve to the spring position. The return of the directional control valve to the spring position reverses the flow in the cylinder that returns it. The return of the cylinder occurs until the pushbutton is pressed. The setting of the pressure switch depends on the application of the cylinder. This setting may be low for packing fragile materials or high for forming metals.

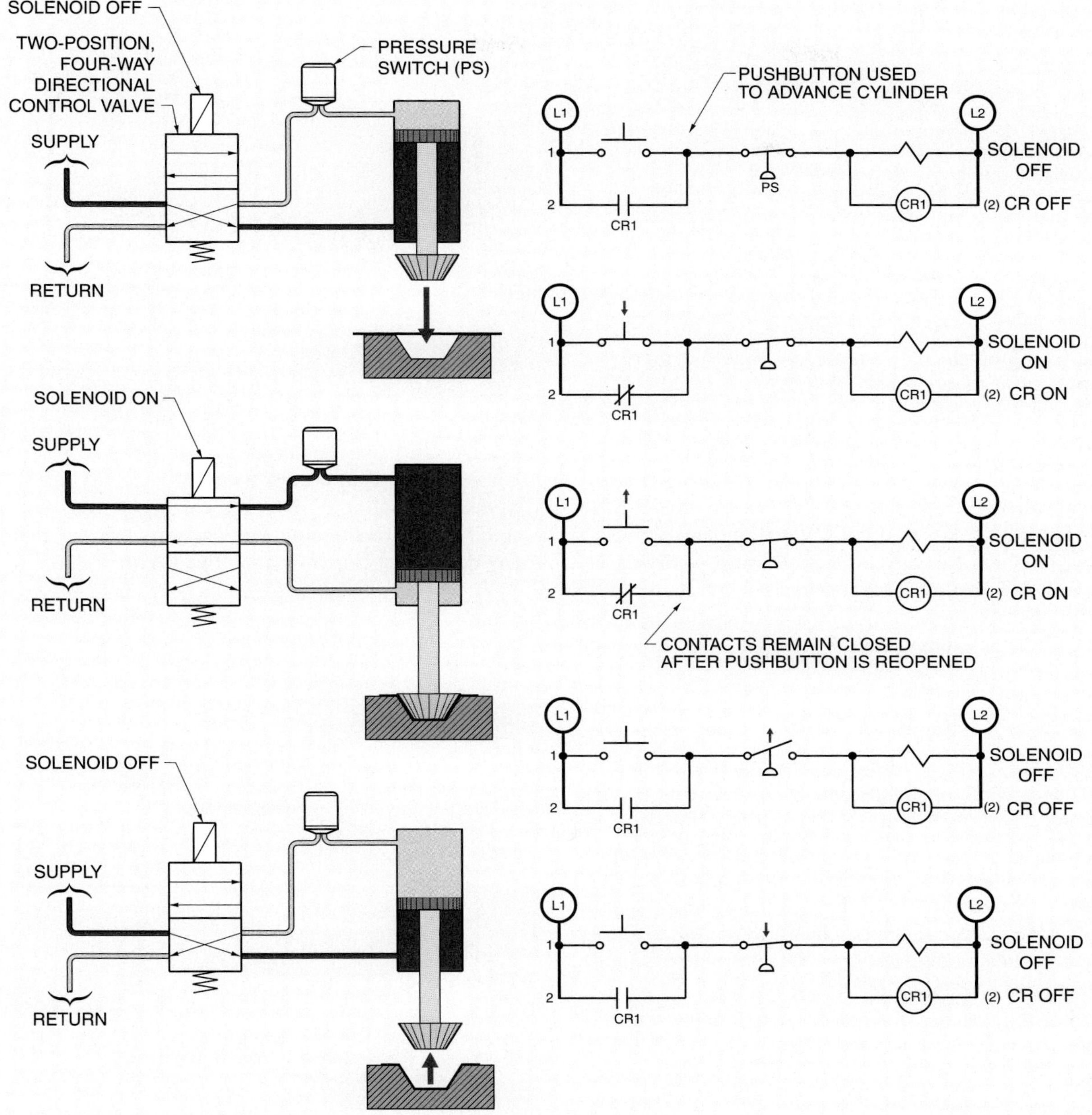

Figure 9-22. Pressure switches are used to maintain a predetermined pressure in a tank or reservoir, or sequence the return of pneumatic or hydraulic cylinders.

The advantage of using a pressure switch over a pushbutton for returning the cylinder is that the load always receives the same amount of pressure before the cylinder returns. An emergency stop could be added to the control circuit for manual return of the cylinder.

A low-range pressure switch may be used with a metal tubing arrangement in a fluidic sensor. **See Figure 9-23.** In this application, a constant low-pressure stream of air is directed at a sheet of material through the metal tubing. As long as the material in process is present, the air stream is deflected. The stream of air is sensed in the receiver tube if the material breaks. The pressure switch would signal corrective action through the control relay.

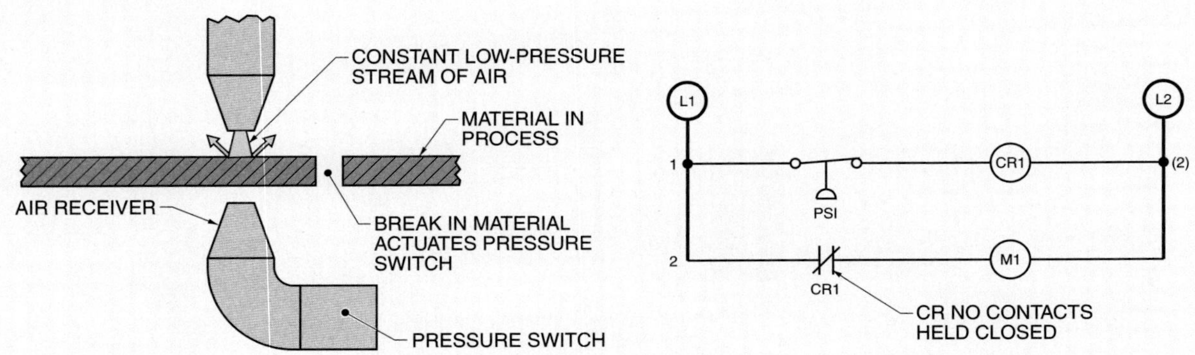

Figure 9-23. A low-range pressure switch may be used with a metal tubing arrangement in a fluidic sensor to determine the presence of a material in process.

The fluidic sensor has certain advantages over using a photoelectric sensor. For example, the air stream flowing over the material in normal operation may perform a second function such as cooling, cleaning, or drying the material. A fluidic sensor is also inexpensive compared to a photoelectric sensor because the only cost is the metal tube and the pressure switch.

TEMPERATURE SWITCHES

Temperature switches are control devices that react to heat intensity. Temperature switches are used in heating systems, cooling systems, fire alarm systems, process control systems, and equipment/circuit protection systems. In most applications, temperature switches react to rising or falling temperatures. Cooling systems, alarm systems, and protection systems use temperature switches that react to rising temperatures. Heating systems use temperature switches that react to falling temperatures.

A heating system maintains a set temperature when the ambient temperature drops. In a heating system, as ambient temperature drops, the switch contacts close and turn ON the heat-producing device. The heat-producing device may be an electric coil, gas furnace, heat pump, or any device that produces heat. See **Figure 9-24.**

The temperature switch energizes a heating contactor. The heating contactor energizes the heat-producing coils. By having the temperature switch control a contactor, the high current required by the heating coils does not pass through the contacts of the temperature switch. This reduces the size of the required temperature switch and increases the life of the contacts. A temperature switch may also sound an alarm if the temperature rises too high.

A cooling system maintains a set temperature when the ambient temperature rises. In a cooling system, as the ambient temperature rises, the switch contacts close and turn ON a cooling device. The cooling device may be a standard air conditioning unit, cooling tower, radiator, or any device that is used to cool. **See Figure 9-25.**

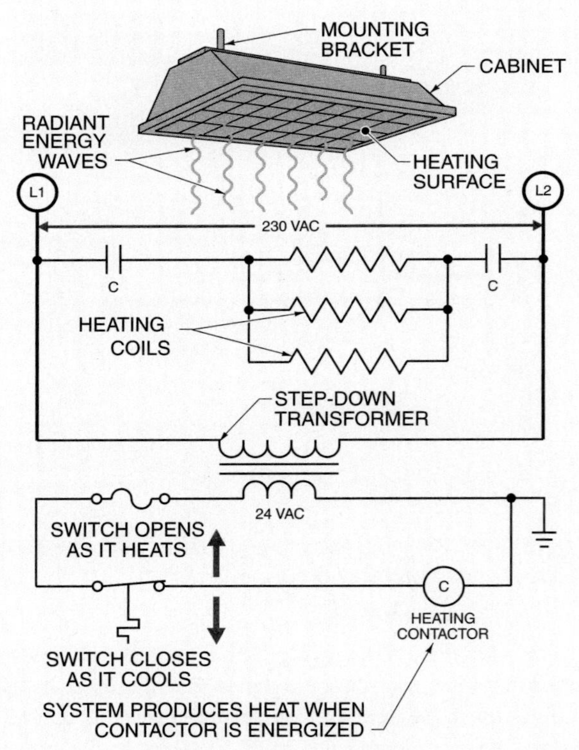

Figure 9-24. In a heating system, heat is produced when the temperature switch contacts cool and close.

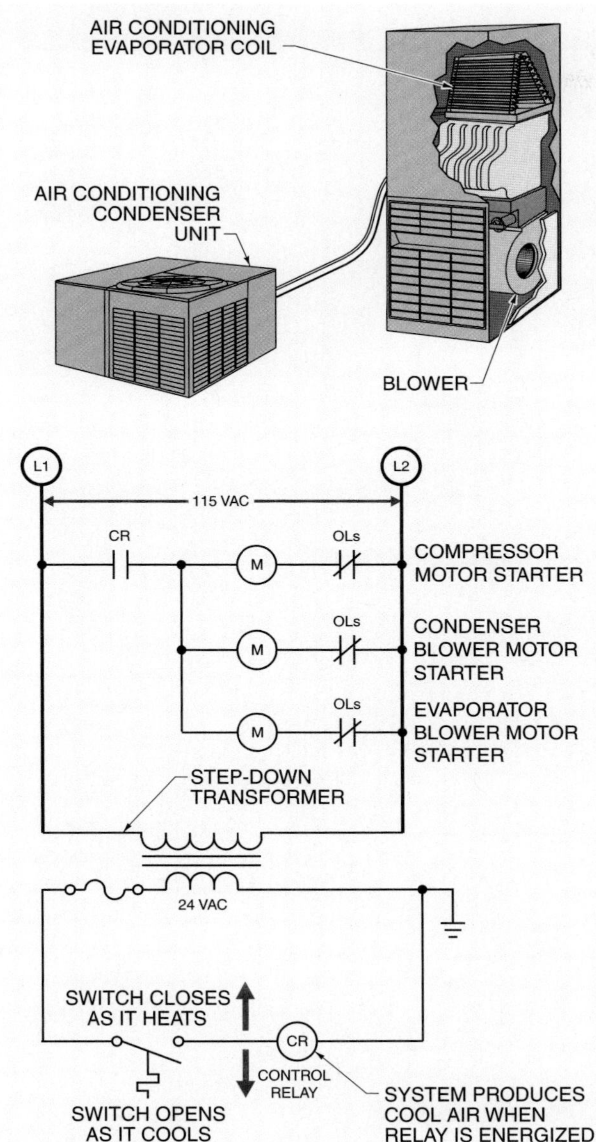

Figure 9-25. In a cooling system, cool air is produced when the temperature switch contacts heat and close.

The temperature switch energizes a control relay. The control relay energizes the motor starters. The motor starters control the motors that produce cooling and provide overload protection for the motors. The control relay may also open a valve that allows water (or coolant) to circulate over the heated area.

Temperature switches may be activated by several different sensors. The different sensors used to activate electrical contacts in response to temperature changes include bimetallic, capillary tube, thermistor, and thermocouple. **See Figure 9-26.**

Bimetallic

A *bimetallic sensor* is a sensor that bends or curls when the temperature changes. It is made of two different metals bonded together. The two metals expand at different rates when heated. This causes the metal strip to bend into an arc, which is used to open and close electrical contacts. To improve performance, most bimetallic sensors use a strip wound into a coil. The coil moves a bulb filled with liquid mercury. When the coil expands, the mercury moves across a set of electrical contacts, completing an electrical circuit.

Capillary Tubes

A *capillary tube sensor* is a sensor that changes internal pressure with a change in temperature. A capillary tube sensor uses a tube filled with a temperature-sensitive liquid. The pressure in the tube changes in proportion to the temperature surrounding the tube. As the temperature rises, the pressure in the tube increases. As the temperature falls, the pressure in the tube decreases. The pressure change inside the tube is transmitted to a bellows inside the temperature control through a capillary tube. The movement of the bellows activates electrical contacts, completing an electrical circuit.

Thermistors

A *thermistor* is a temperature-sensitive resistor whose resistance changes with a change in temperature. Thermistors have either a positive temperature coefficient (PTC) or a negative temperature coefficient (NTC). PTC thermistors increase in resistance when the temperature increases. NTC thermistors increase in resistance when the temperature decreases. The thermistor is placed at the point where the temperature is to be measured, and is connected to an electronic circuit which is set to respond to the changes in resistance. The electronic circuit activates a relay at the set temperature, completing an electrical circuit.

Technical Fact

Temperature switches are used to turn ON a cooling system when the temperature in a building rises to a predetermined setpoint and turn ON a heating system when the temperature in a building falls to a predetermined setpoint. Temperature switches are also wired into furnace control circuits to protect the system from overheating.

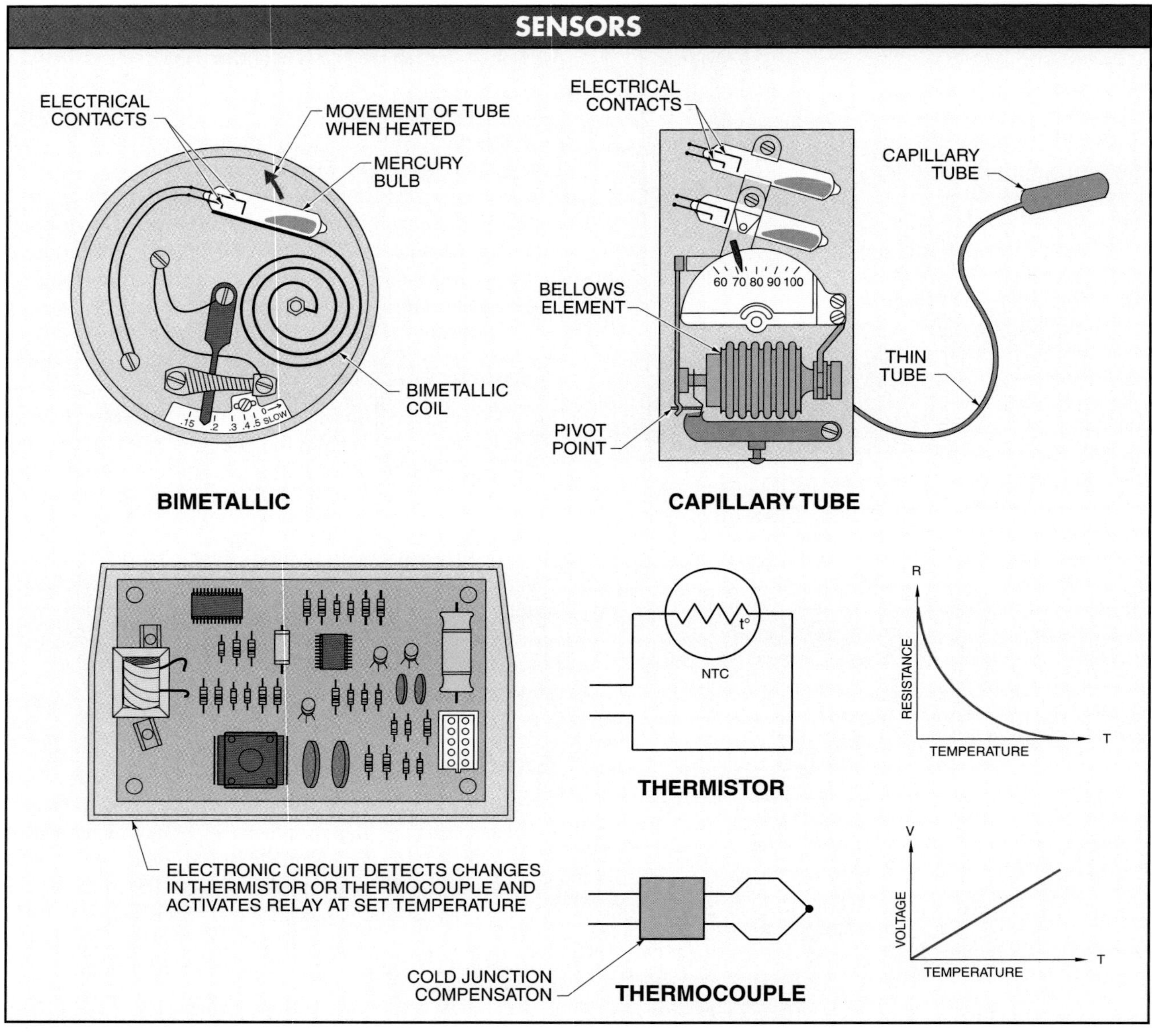

Figure 9-26. The different sensors used to activate electrical contacts in response to temperature changes include bimetallic, capillary tube, thermistor, and thermocouple.

Liquid, Air, or Surface Temperature Control Applications. Thermistors are available that monitor the temperature of liquid or air, or a surface temperature. **See Figure 9-27.** Any one or all may be used depending on the application.

A thermistor is used to monitor and control liquids in many applications. This is one of the most common process controls used in industry. The temperature of many liquids must be at a set point before a mixing or fill process can start. In other applications, a process may have to be stopped if the liquid is too cool or too hot. Thermistors are available in a wide range of temperatures with extreme ranges from –400°F to 3200°F (–240°C to 1742°C).

Thermistors that sense air temperatures are available in a wide range of temperatures, with the most common in the –30°F to 150°F range. These sensors are used mostly in heating and air conditioning control to maintain a desired temperature of a room or a storage unit.

Thermistors that sense surface temperatures are normally designed to attach to a metallic surface. They can be used to detect an ice buildup in an air conditioning system or heat buildup in many other processes. The advantage of surface temperature sensors is that they are easy to install on pipes without having to open the system.

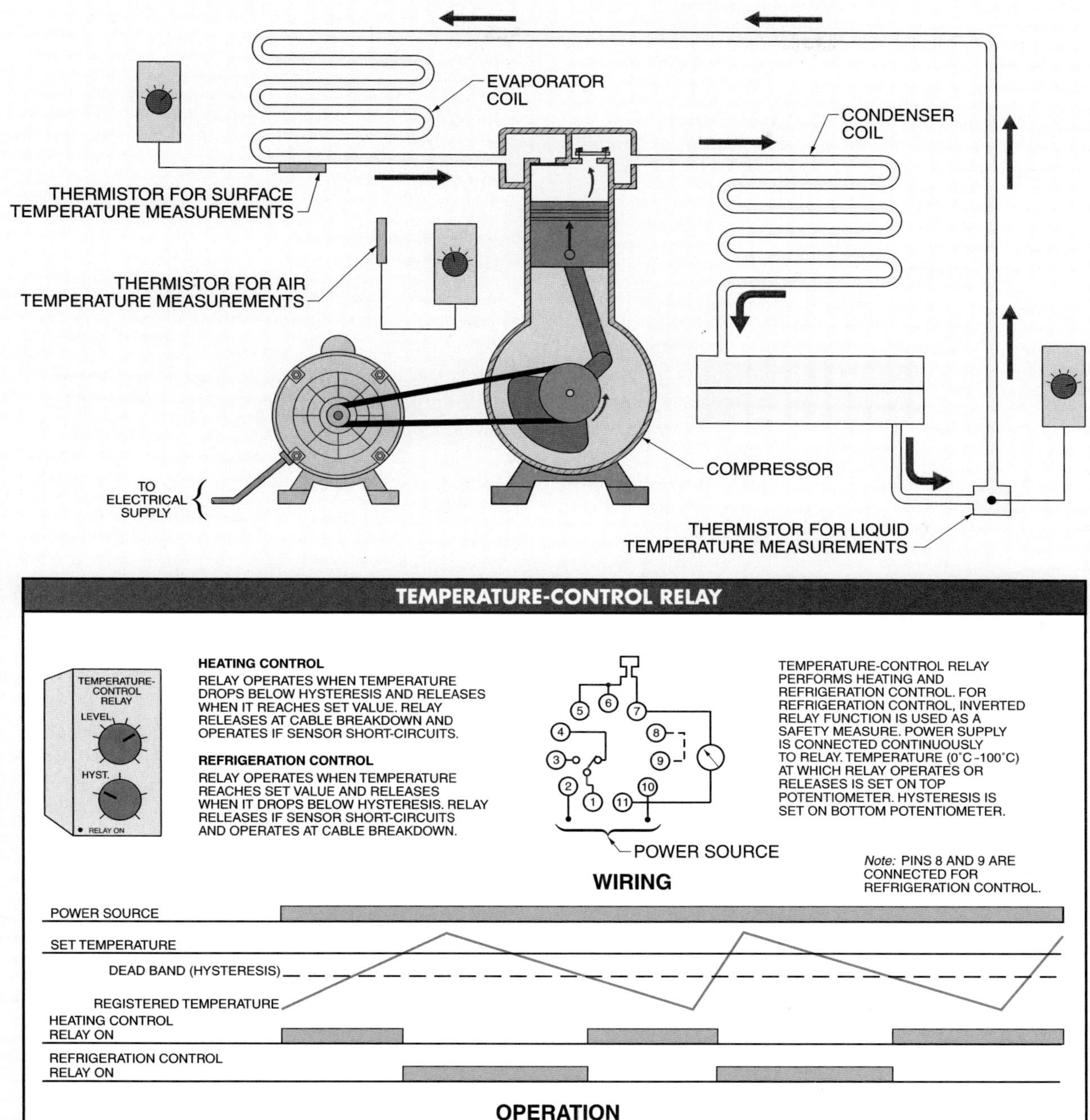

Figure 9-27. Thermistors are available that monitor the temperature of liquid, air, or a surface temperature.

Thermocouples

A *thermocouple* is a temperature sensor that consists of two dissimilar metals joined at the end where heat is to be measured (hot junction), and that produces a voltage output at the other end (cold junction) proportional to the measured temperature. The *hot junction* (measuring junction) is the joined end of a thermocouple that is exposed to the process where the temperature measurement is desired. The *cold junction* (reference junction) is the end of a thermocouple that is kept at a constant temperature in order to provide a reference point. Thermocouples work on the principle that two different metals connected together produce a voltage between the hot junction and cold junction when the hot junction is heated. If heat is applied to the hot junction, the voltage increases across the cold junction. If the hot junction is cooled, the voltage decreases across the cold junction. A thermocouple is connected to an electronic circuit, which is set to respond to the changes in voltage. The electronic circuit activates a relay at the set temperature.

FLOW SWITCHES

Flow is the travel of fluid in response to a force caused by pressure or gravity. The fluid may be air, water, oil, or some other gas or liquid. Most industrial processes depend on fluids flowing from one location to another. Problems may occur if the flow is stopped or slowed. Flow may be stopped by a frozen pipe, a clogged pipe, or an improperly closed valve (manual or automatic). A *flow switch* is a control switch that detects the movement of a fluid. **See Figure 9-28.**

Applications that use flow switches to detect the presence or absence of flow include:

- Boilers
- Cooling lines
- Air compressors
- Fluid pumps
- Food processing systems
- Machine tools
- Sprinkler systems
- Water treatment systems
- Heating processes
- Refrigeration systems
- Chemical processing and refining

Flow switches use different methods to detect if the fluid is flowing. The methods used to detect if the fluid is flowing include the paddle and transmitter/receiver methods. In the paddle method, a paddle extends into the pipe or duct. The paddle moves and actuates electrical contacts when the fluid flow is sufficient to overcome the spring tension on the paddle. The spring tension is adjustable on many flow switches, allowing for different flow rate adjustments.

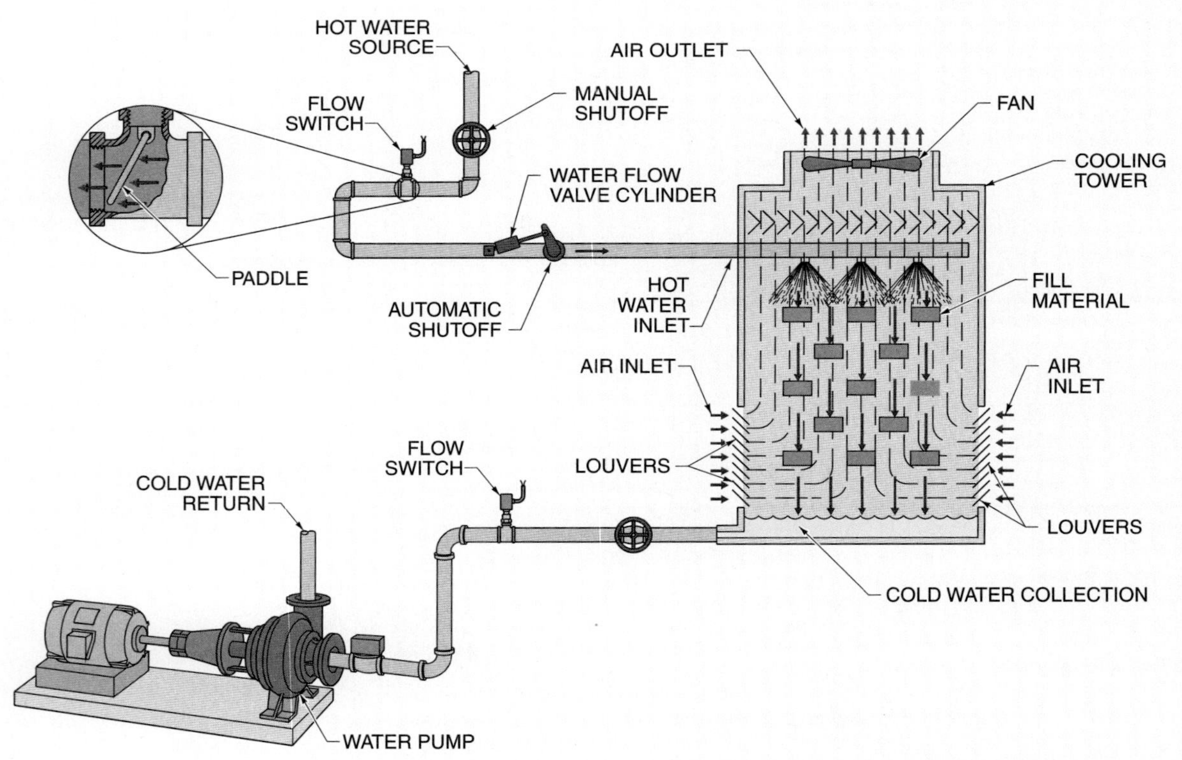

Figure 9-28. A flow switch is a control switch that detects the movement of a fluid.

In the transmitter/receiver method, a transmitter sends a signal through the pipe. The receiver picks up the transmitted signal. The strength of the signal changes when the product is flowing. One common transmitter/receiver method uses a sound transmitter to produce sound pulses through the fluid. Moving solids or bubbles in the fluid reflect a distorted sound back to the receiver. The transmitter/receiver unit is adjustable for detecting different flow rates.

Both NO and NC electrical contacts are used with flow switches. **See Figure 9-29.** In some applications, a flow of fluid indicates a problem. For example, in an automatic sprinkler system used as fire protection, the flow of water indicates a problem. In this application, an NO contact on the flow switch could be used to sound an alarm. When a fire (or high heat) opens the sprinkler head, the water starts flowing through the pipe. The flow of water closes the NO contacts and the alarm sounds. The alarm sounds as long as the water is flowing.

An NC contact is used to signal when a fluid is not flowing. The NC contact may be used to sound an alarm if fluid stops flowing. When fluid is flowing, the NC contacts are held open by the fluid flow.

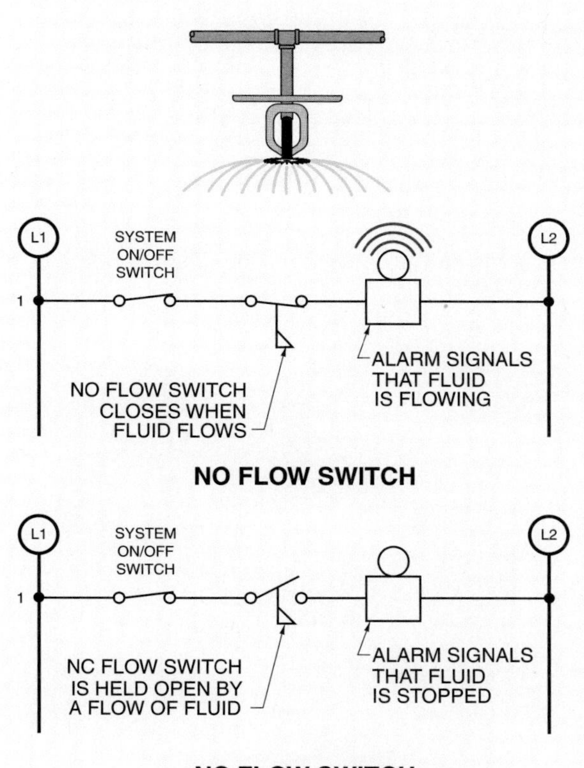

Figure 9-29. Both NO and NC electrical contacts are used with flow switches.

The Foxboro Company

An electronic differential pressure transmitter is commonly used for flow measurment.

A flow switch may also be used to detect airflow across the heating elements of an electric heater. **See Figure 9-30.** The heating elements burn out if sufficient airflow is not present. The flow switch is used as an economical way to turn the heater OFF anytime there is not enough airflow. This circuit can also be applied to an air conditioning or refrigeration system. In this circuit, the flow switch is used to detect insufficient airflow over the refrigeration coils. The restricted airflow is normally caused by the icing of the coils, which blocks the airflow. In this case, the flow switch would automatically start the defrost cycle of the refrigeration unit.

Flow switches may also be used to detect the proper airflow in a ventilation system. **See Figure 9-31.** The ventilation system may be directing dangerous gases away from the operator. Poisonous gases could overcome the operator or damage could occur to the process involved if there is insufficient airflow.

Flow switches are often used to protect the large motion picture projector used in theaters, where poor airflow would cause heat buildup and reduce the life expectancy of the expensive bulbs used in projectors. Airflow may be restricted from a large draft caused by high winds outside the building or by clogged air filters in the intake system.

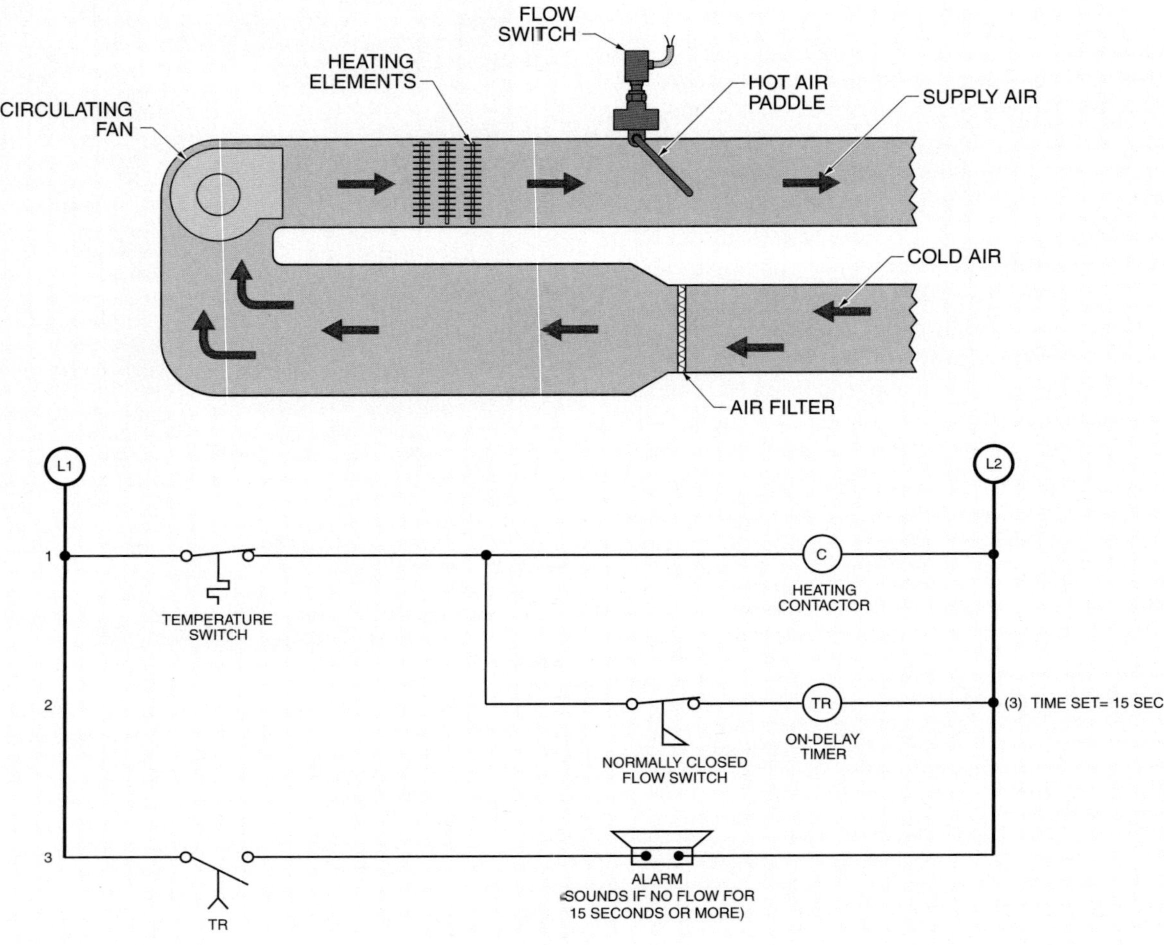

Figure 9-30. A flow switch may be used to determine if sufficient air is flowing across the heating elements of an electric heater.

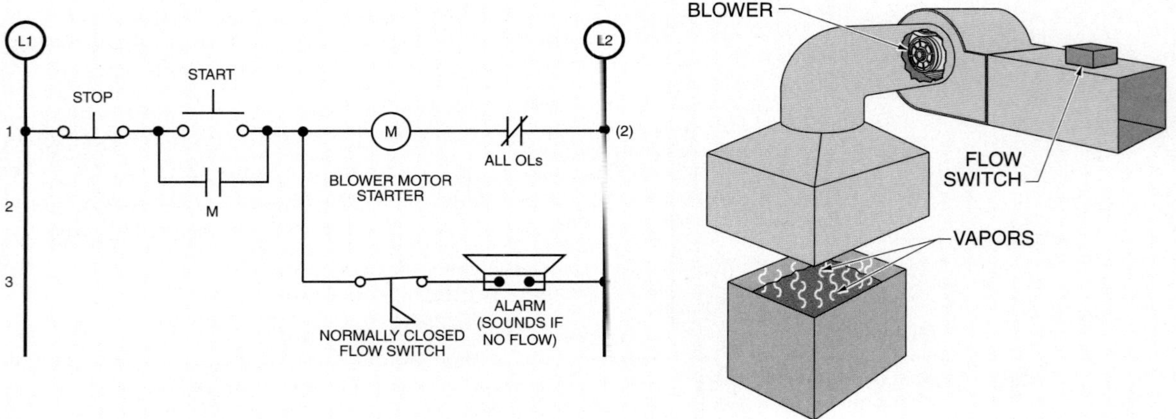

Figure 9-31. A flow switch may be used to maintain a critical ventilation process.

A flow switch may be used to advance a clogged filter based on restricted airflow. **See Figure 9-32.** The flow switch is used to start a gear-reduced motor that slowly advances the roll of filter material until sufficient airflow is present.

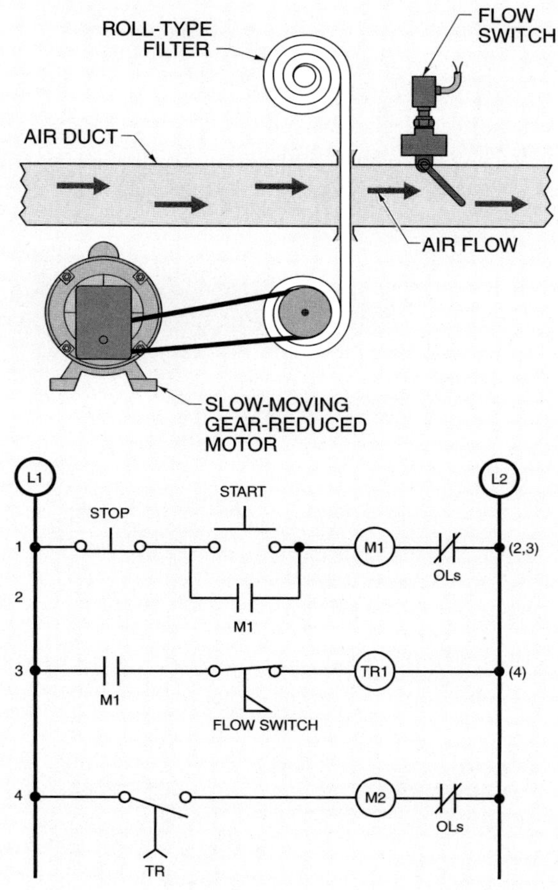

Figure 9-32. A flow switch may be used to advance a clogged filter when restricted airflow is sensed.

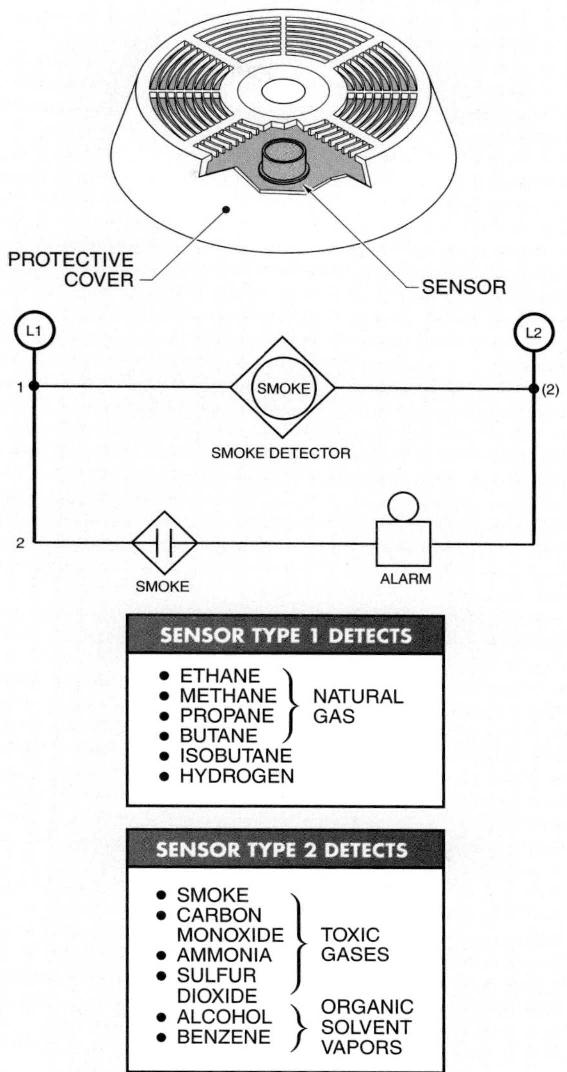

Figure 9-33. A smoke switch (smoke detector) is a switch that detects a set amount of smoke caused by smoldering or burning material and activates a set of electrical contacts. A gas switch (gas detector) is a switch that detects a set amount of a specified gas and activates a set of electrical contacts.

SMOKE/GAS SWITCHES

Smoke/gas switches detect vapor. A *vapor* is a gas that can be liquefied by compression without lowering the temperature. A *smoke switch (smoke detector)* is a switch that detects a set amount of smoke caused by smoldering or burning material and activates a set of electrical contacts. **See Figure 9-33.** Smoke switches are used as early-warning devices in fire protection systems. They are available as self-contained units such as the common units used in most houses, or as industrial units used as part of a large fire protection system.

A *gas switch (gas detector)* is a switch that detects a set amount of a specified gas and activates a set of electrical contacts. Many gas detectors have interchangeable sensor units that can detect different groups of gases. For example, one common gas sensor is designed to detect gases such as propane and butane, but not carbon monoxide and smoke. This gas detector is used in applications such as detecting leaks in areas that have (or fill) propane tanks. Another sensor is designed to detect toxic gases such as carbon monoxide and ammonia. This gas detector is used in applications such as detecting high carbon monoxide levels in an area with operating combustion engines.

LEVEL SWITCHES

In most industrial plants there are tanks, vessels, reservoirs, and other containers in which process water, wastewater, raw materials, or product must be stored or mixed. The level of the product must be controlled. A *level switch* is a switch that detects the height of a liquid or solid (gases cannot be detected by level switches) inside a tank.

Systems that use level switches include processing systems for products such as milk, water, oil, beer, wine, solvents, plastic granules, coal, grains, sugar, chemicals, and many other products. Different level switches are used to detect each product. Factors that determine the correct level switch to use for an application include:

- Motion – Turbulence causes some level switches to chatter ON and OFF or actuate falsely.
- Corrosiveness – Level switches are made of different materials such as stainless steel, copper, plastic, etc. Always use a level switch made of a material that is compatible with the product to be detected.
- Density – All solid materials have a certain density. Capacitive level switches are designed to detect different amounts of density.
- Physical state – The physical state of a liquid depends on its type, temperature, and condition. Any liquid or solid may be detected if the correct level switch is used. Different level switches are designed to operate at different temperatures and to detect different types and thicknesses of liquids.
- Movement – A moving product may require a special level switch. A product that is stationary for a long period may cause certain mechanical level switches to stick.
- Conductivity – Some level switches with metal probes placed in a liquid depend on the liquid to be a conductor for proper operation.
- Abrasiveness – Noncontact level switches should be used with abrasive products.
- Sensing distance – Some level switches are designed to detect short distances and others are designed to detect long distances.

All level switches are designed to detect a certain range of materials. Some level switches can only detect liquids; others can detect both liquids and solids. Some level switches must come in direct contact with the product to be detected; others do not have to make contact. The level switch used depends on the application, cost, life expectancy, and product to be detected. The different level switches include mechanical, magnetic, conductive probe, capacitive, optical, and ultrasonic.

Mechanical

Mechanical level switches were the first level switches used and are still one of the most common. *Mechanical level switches* are level switches that use a float that moves up and down with the level of the liquid and activates electrical contacts at a set height. **See Figure 9-34.** Mechanical level switches may be used with many different liquids because the float is the only part of the switch that is in contact with the liquid. Mechanical level switches work well with water (even dirty water) and any other liquid that dries without leaving a crust. Mechanical limit switches are not used with paint because the paint builds up on the float as the paint dries. The dried paint weighs the float down and affects the operation of the switch.

One of the most common applications of a mechanical level switch is in sump pumps found in most houses with basements. In a sump application, the level switch turns ON a pump when the level reaches a set height.

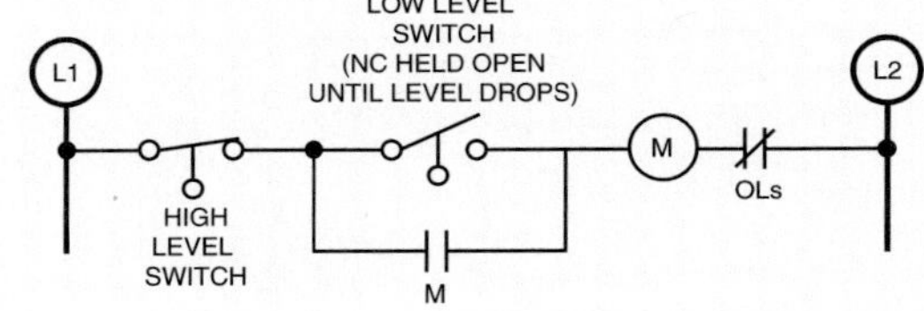

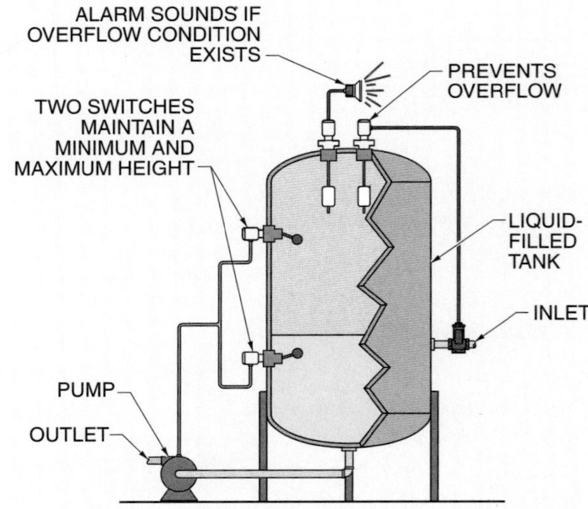

Figure 9-34. A mechanical level switch uses a float, which moves up and down at the level of the liquid and activates electrical contacts at a set height.

Magnetic

A *magnetic level switch* is a switch that contains a float, a moving magnet, and a magnetically operated reed switch to detect the level of a liquid. The float moves with the level of the liquid. A permanent magnet inside the float moves up and down with the liquid. The magnet passes alongside a magnetically operated reed switch as it moves with the level of the liquid. The reed switch contacts change position when the magnetic field is present. The reed switch contacts return to their normal position when the magnetic field moves away. An advantage of using a magnetic level switch is that several individual switches may be placed on one housing. **See Figure 9-35.**

Conductive Probe

A *conductive probe level switch* is a level switch that uses liquid to complete the electrical path between two conductive probes. The voltage that is applied to the probes is 24 V or less. The liquid must be a fair to good electrical conductor. All water-based solutions conduct electricity to some degree. The conductance of water increases as salts and acids are added to the water (resistance decreases).

A fluid with an electrical resistance of less than 25 kΩ allows a sufficient amount of current to pass through it to actuate the relay inside the conductive probe level control. The electrical resistance is high and no current passes through the probes when the conductive liquid is no longer between the two probes. **See Figure 9-36.**

The number of probes used and their length depends on the application. Two probes of the same length may be used to detect a liquid at a given height. The relay inside the level control is activated when the liquid reaches the probes. The relay is no longer activated when the liquid is no longer in contact with the probes.

Two probes of different lengths and a ground may be used to detect a liquid at different heights. The ground is connected to the conductive tank. The relay is activated when the liquid reaches the highest probe. The relay is not deactivated until the liquid no longer is in contact with the lowest probe. The probes may be any distance apart. Three probes of different lengths are used if the tank that holds the liquid is made of a nonconductive material. The lowest probe is connected to the ground wire of the level control.

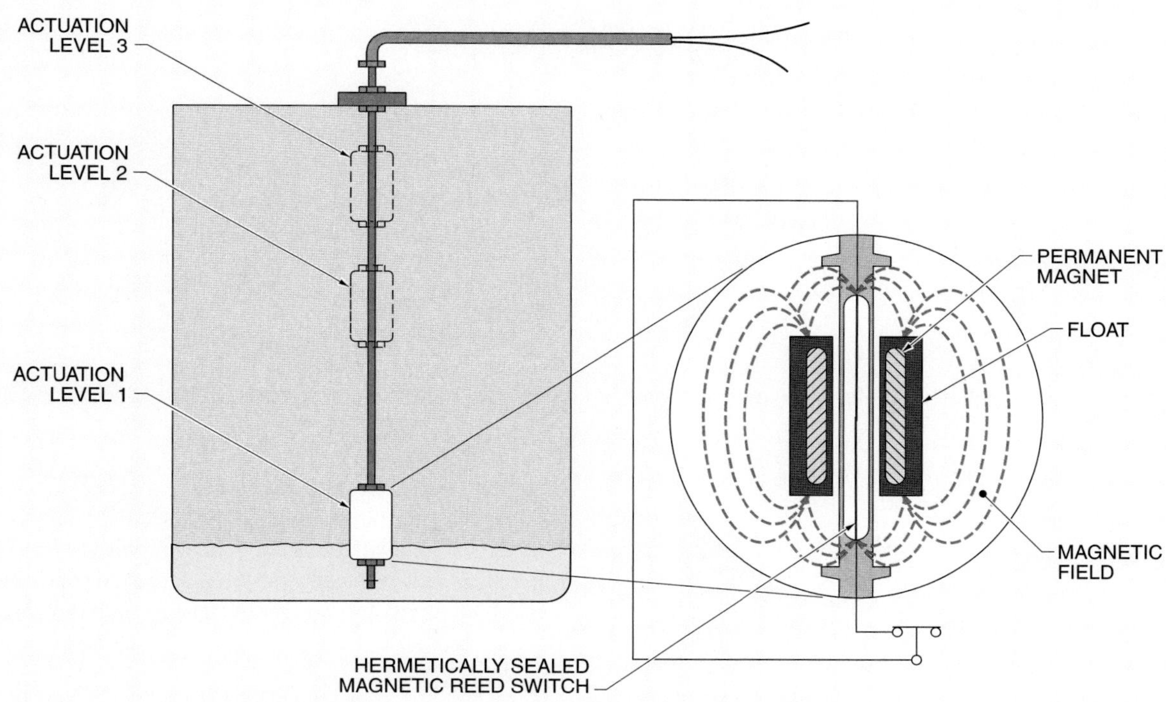

Figure 9-35. A magnetic level switch uses a magnetically operated reed switch and a moving magnet to detect the level of a liquid.

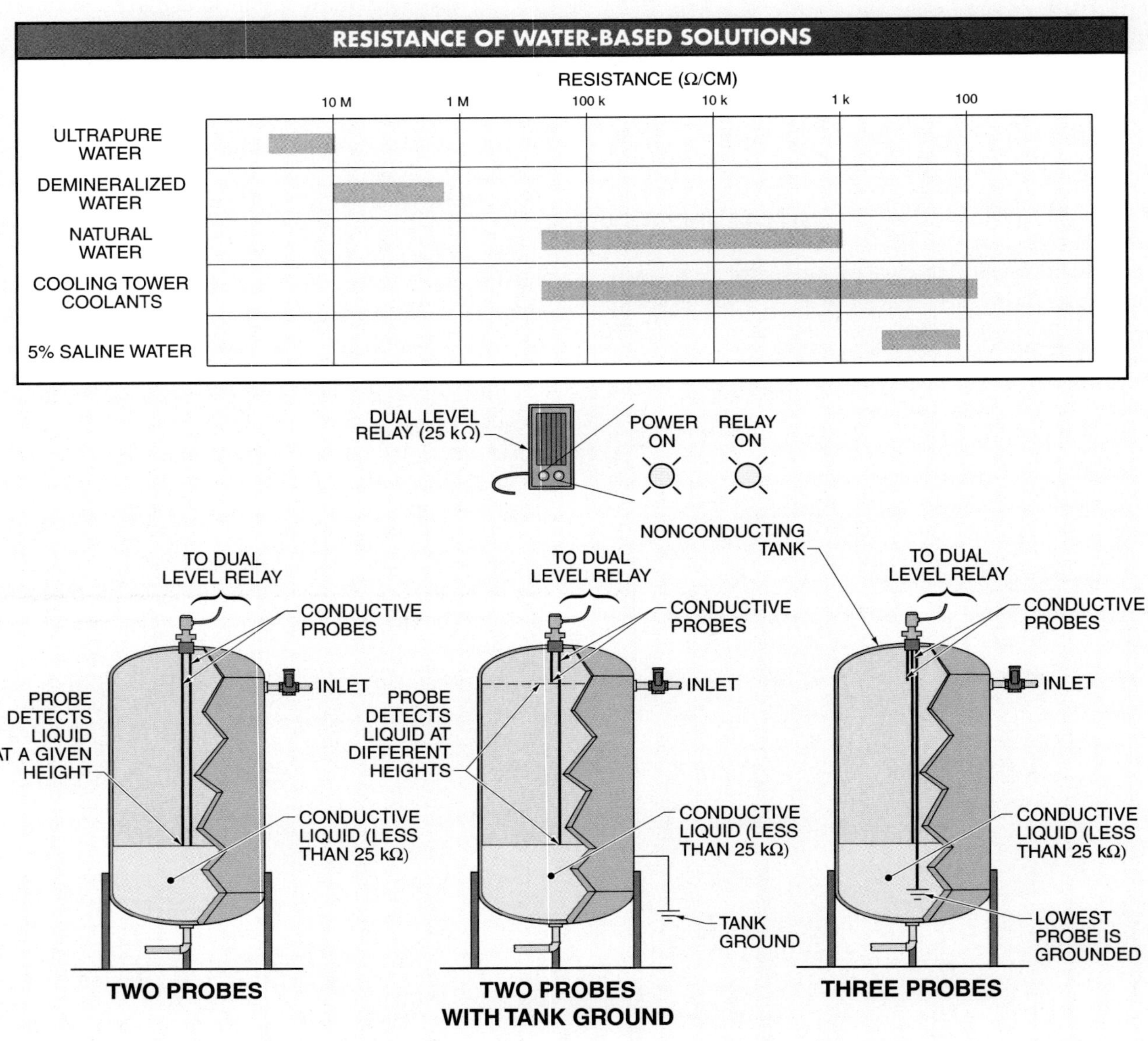

Figure 9-36. A conductive probe level switch uses liquid to complete the electrical path between two conductive probes.

Capacitive

A *capacitive level switch* is a level switch that detects the dielectric variation when the product is in contact (proximity) with the probe and when the product is not in contact with the probe. *Dielectric variation* is the range at which a material can sustain an electric field with a minimum dissipation of power. Capacitive level switches are used to detect solids or granules such as sand, sugar, grain, and chemicals in addition to some liquids. The capacitance of the sensor is changed when the product comes in proximity with the sensor. Materials with a dielectric constant of 1.2 or greater can be sensed. Some capacitive sensors are adjustable to allow only certain products to be sensed. Capacitive level switches are available that work well with hard-to-detect products such as plastic granules, shredded paper, copying machine toner, and fine powders. **See Figure 9-37.**

Optical

Optical level switches are level switches that use a photoelectric beam to sense the liquid. These switches are normally enclosed in a corrosion-resistant housing that makes them ideal for certain liquids. They are used to detect liquids such as oil, gas, beer, wine, milk, alcohol, and many acids. Because of the housing material (usually stainless steel or a high-temperature thermoplastic), the sensor can be safely washed down with very hot (up to 212°F) water or solvents.

Figure 9-37. A capacitive level switch detects the dielectric variation when the product is in contact with the probe and when the product is not in contact with the probe.

The tip of the sensor forms a 90° angle that acts as a prism, reflecting the transmitted light beam to the receiver through the air. The beam is reflected into the liquid and a relay inside the level switch is activated if the sensor tip is immersed in a liquid having a refractive index different from that of air. **See Figure 9-38.**

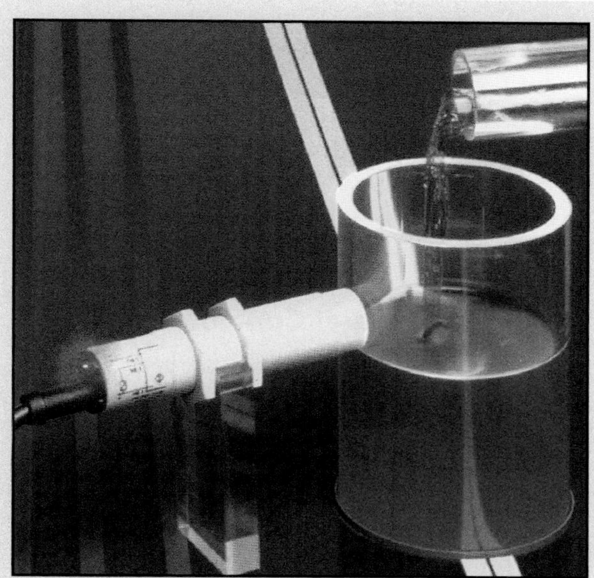

Carlo Gavazzi Inc. Electromatic Business Unit

Capacitive level switches are used in applications that require liquid detection through a container wall.

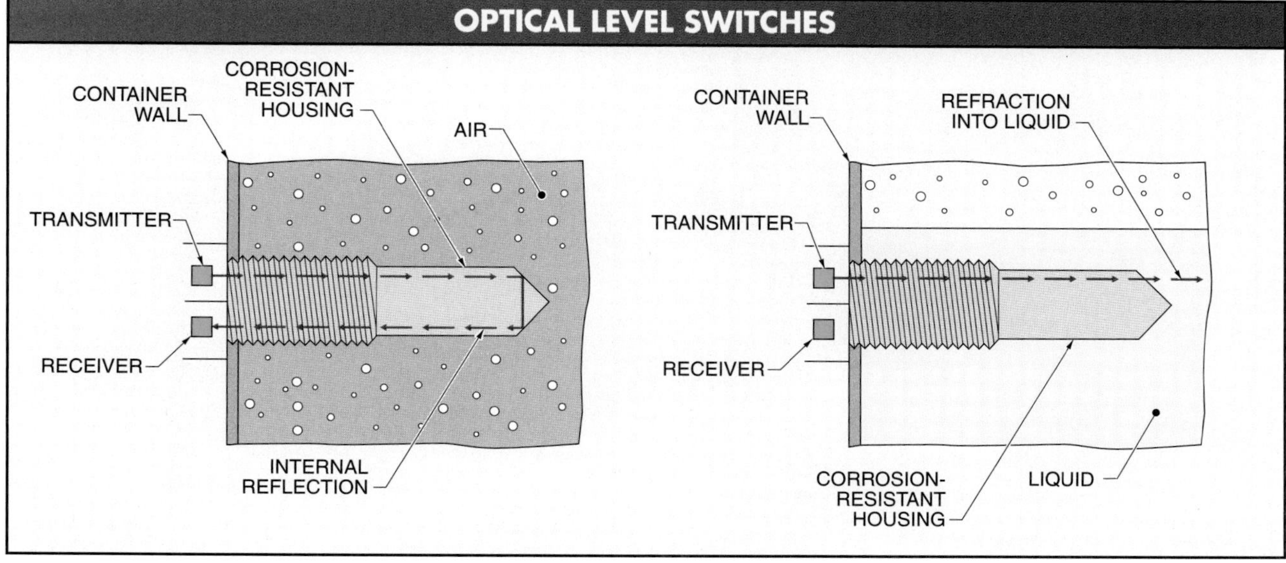

Figure 9-38. Optical level switches use a photoelectric beam to sense the liquid.

Charging and Discharging

Level switches detect and respond to the level of a material in a tank. The response is normally to charge or discharge the tank. Charging a tank is also known as pump control and discharging a tank is also known as sump control.

In a charging application, the level in a tank is maintained. As liquid is removed from the tank, the level switch signals the circuit to add liquid. Liquid may be added by opening a valve or starting a pump motor. The liquid is added until the level switch detects the correct height. Flow is stopped when the level switch detects the correct height. Liquid is added when the level switch is no longer in contact with the liquid. **See Figure 9-39.**

In a discharging application, the liquid in a tank is removed once it reaches a predetermined level. Liquid may be removed by a pump or through gravity when a valve is opened. The liquid is removed until the level switch detects the tank is empty.

One- or Two-Level Control

In level control applications, the distance between the high and low level must be considered. This distance may be small or large. In applications using a one-level switch, the distance is small. In applications using a two-level switch, the distance may be any length. **See Figure 9-40.** Although it may appear that a small distance maintained in a system is the best, this may not be true because the smaller the distance to be maintained, the greater the number of times the pump motor must cycle ON and OFF.

Level controls are required when a certain level must be maintained in a tank. For example, an explosion could occur if the water level in a boiler falls too low, or if the pressure rises too high.

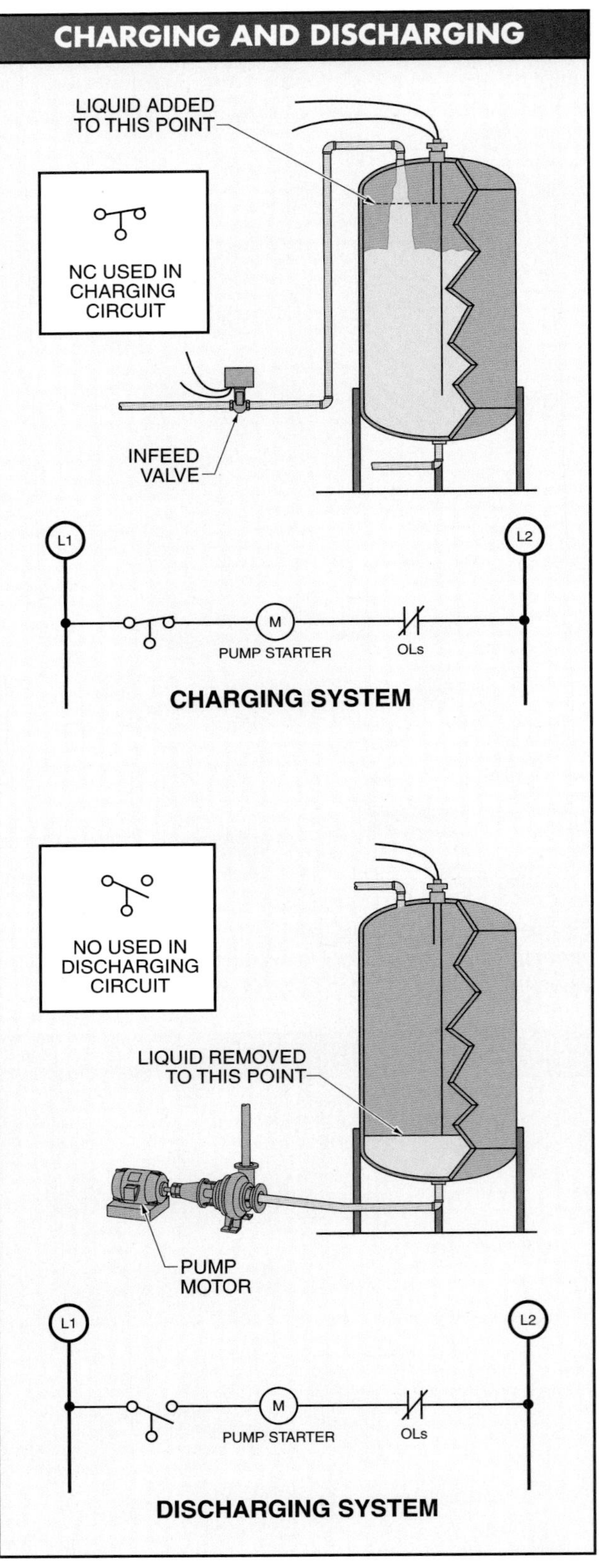

Figure 9-39. Level switches detect and respond to the level of a material in a tank.

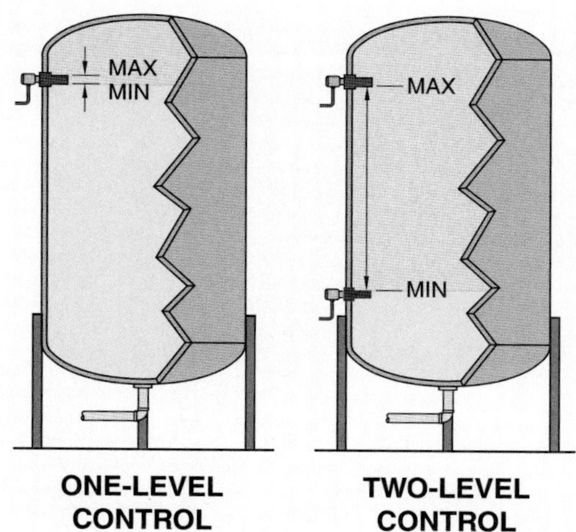

Figure 9-40. The distance controlled in one-level control is small. Any distance may be controlled in two-level control.

Since motors draw much more current when starting than when running, excess heat is produced in a motor that must turn ON and OFF frequently. The faster the level in the tank drops, the faster the motor must cycle.

While it may appear that two-level registration is the best because the pump does not have to cycle as often, what is best for the motor may not be best for the total system. For example, if common house paint is to be maintained in a fill tank, problems develop if the length of time between the high level and low level is excessive. As the paint dries on the inside of the tank, it accumulates layer by layer. This causes skin to form, which may clog or impede the pump or fill action if the skin falls into the product.

Therefore, product type must be taken into consideration when determining the distance and time between the high and low level. In general, one-level control is best when the liquid is emptied very slowly from the tank. Two-level control is best when the liquid is emptied at a fast rate.

Temperature and Level Control Combinations

Temperature and level controls may be combined in some applications. **See Figure 9-41.** A contactor is used in the power circuit to control the heating element, and a temperature control relay is used in the control circuit. A magnetic motor starter is used to control the pump motor. A contactor can be used if a solenoid is used to open and close a valve that fills the tank. Circuit interlocking may be required to prevent the heating element from turning ON unless the level in the tank is at the maximum point.

Solar Heating Control

Solar energy is plentiful and can provide substantial amounts of energy that can be used to replace other forms of more expensive or less available fuels. Although solar energy has been used in the past for many applications, such as supplying power in space, its main use is in space heating and hot water heating. Temperature control is required in each application.

In controlling a solar heating system, the temperature controllers must be able to measure the temperature inside the solar collector and at the water storage area, and circulate the heat accordingly. This circulation of heat must be accomplished based on the temperature setting and differential setting of the controller. To meet these requirements, the temperature controller must be able to take two separate signal inputs (collector and heated area) and make a decision that provides the correct heat circulation.

Solar Heating System Control Circuit. A solar heating control system needs two signal inputs (temperature sensor 1 and 2), a decision circuit, and an action circuit to start or stop the pump motor. Sensor 1 monitors the temperature at the solar collector. Sensor 2 monitors the temperature at the heating unit. The decision circuit compares temperatures at each sensor and starts or stops the pump based on demand. The decision circuit can be a logic module that has only the input and output connected to the module. The action circuit is a starter coil that energizes the pump motor based on demand.

Heidelberg Harris, Inc.

Air pollution control is required in any facility that contains highly sensitive manufacturing or production systems.

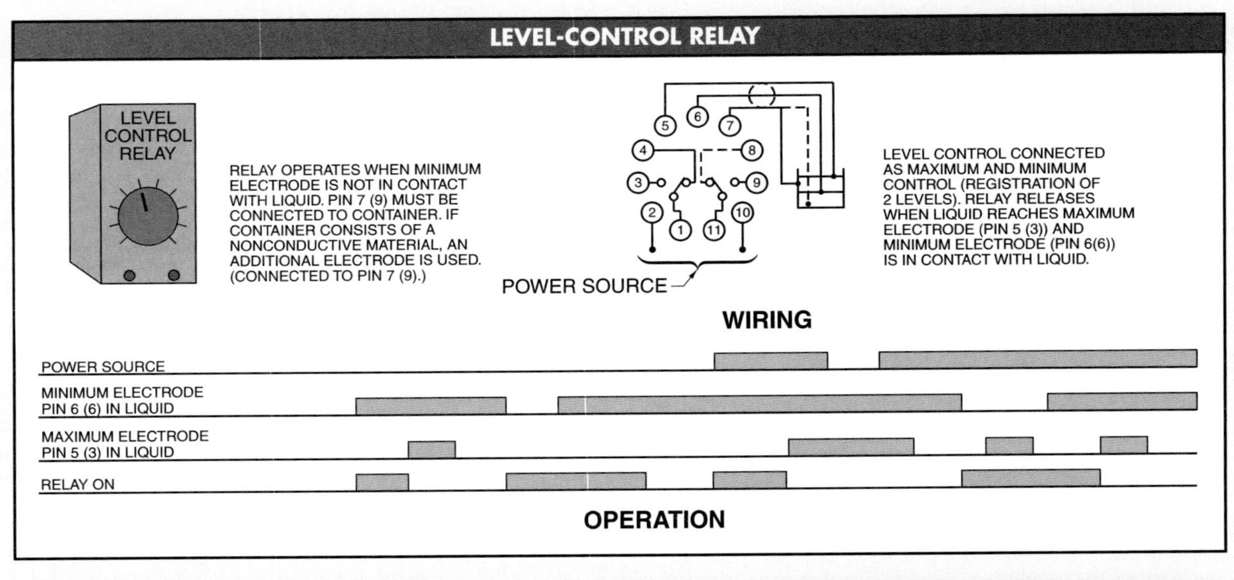

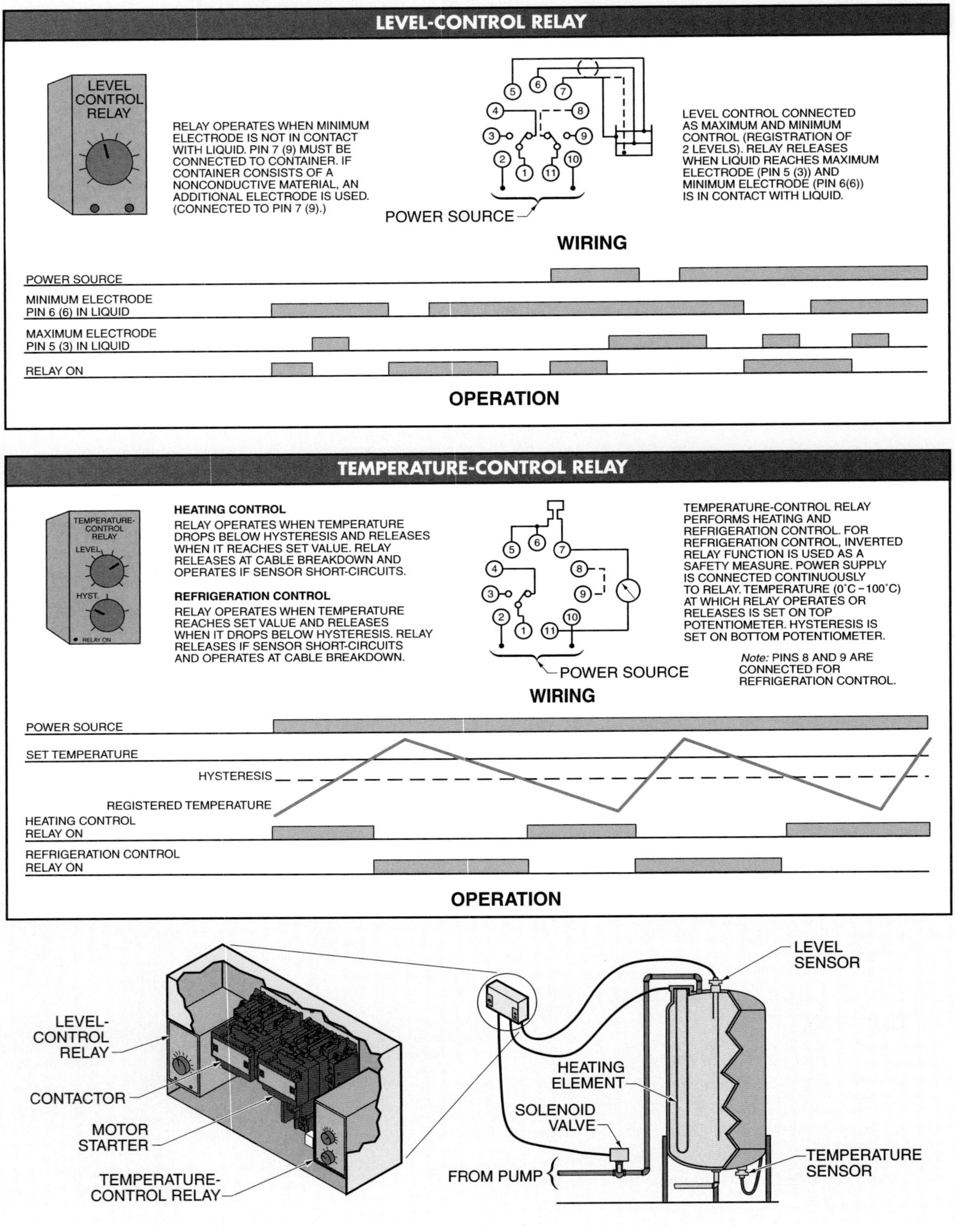

TEMPERATURE-CONTROL RELAY

TEMPERATURE-CONTROL RELAY
LEVEL
HYST.
RELAY ON

HEATING CONTROL
RELAY OPERATES WHEN TEMPERATURE DROPS BELOW HYSTERESIS AND RELEASES WHEN IT REACHES SET VALUE. RELAY RELEASES AT CABLE BREAKDOWN AND OPERATES IF SENSOR SHORT-CIRCUITS.

REFRIGERATION CONTROL
RELAY OPERATES WHEN TEMPERATURE REACHES SET VALUE AND RELEASES WHEN IT DROPS BELOW HYSTERESIS. RELAY RELEASES IF SENSOR SHORT-CIRCUITS AND OPERATES AT CABLE BREAKDOWN.

POWER SOURCE
WIRING

TEMPERATURE-CONTROL RELAY PERFORMS HEATING AND REFRIGERATION CONTROL. FOR REFRIGERATION CONTROL, INVERTED RELAY FUNCTION IS USED AS A SAFETY MEASURE. POWER SUPPLY IS CONNECTED CONTINUOUSLY TO RELAY. TEMPERATURE (0°C – 100°C) AT WHICH RELAY OPERATES OR RELEASES IS SET ON TOP POTENTIOMETER. HYSTERESIS IS SET ON BOTTOM POTENTIOMETER.

Note: PINS 8 AND 9 ARE CONNECTED FOR REFRIGERATION CONTROL.

POWER SOURCE
SET TEMPERATURE
HYSTERESIS
REGISTERED TEMPERATURE
HEATING CONTROL RELAY ON
REFRIGERATION CONTROL RELAY ON
OPERATION

LEVEL-CONTROL RELAY
CONTACTOR
MOTOR STARTER
TEMPERATURE-CONTROL RELAY
LEVEL SENSOR
HEATING ELEMENT
SOLENOID VALVE
FROM PUMP
TEMPERATURE SENSOR

Figure 9-41. A temperature and level control may be combined to control the temperature and level of a fluid.

Solar Heating System Temperature Control Circuit. In a solar heating system, the circulation pump is activated by the temperature differential between T1 and T2. **See Figure 9-42.** In this application, a water storage tank is used to hold hot water. A solar heat collector is used to absorb the sun's heat, and a pump motor is used to circulate the water.

This system does not include an auxiliary heating system, such as gas or electric, to supplement the solar system. A wiring diagram is used to properly install the control circuit of the solar heating system. **See Figure 9-43.**

The temperature relay is operated in conjunction with two semiconductor temperature sensors. These temperature sensors are available for use in liquids or gases, or may be constructed to measure the temperature of a metallic surface. These sensors are available in either a semiconductor (thermistor) or bulb-type construction. The relay reacts to certain temperature differences as set on the adjustable knob.

The relay operates when the temperature in the solar panel (measuring point T1) exceeds the temperature in the water tank used to accumulate the heat (measuring point T2) by a predetermined value, which is set on the relay.

The smallest differential temperature (T1 – T2) at which the relay operates can be set to any desired value between 3°C and 10°C. The relay does not release until the temperature difference is reduced to 2°C (T1 – T2 = 2°C) as set by the factory. For example, if the relay is set for a 6°C temperature difference between T1 and T2, the pump turns ON at a 6°C temperature change between the two measuring points. The pump circulates the hot water from the solar collector to the storage tank. This circulation continues until there is a 2°C temperature difference between the two measuring points, at which point the relay turns OFF the pump.

A temperature meter may be added to the circuit to indicate the temperatures. A temperature meter with a neutral position in the middle of the scale must be used in cases where negative temperature differences may be displayed.

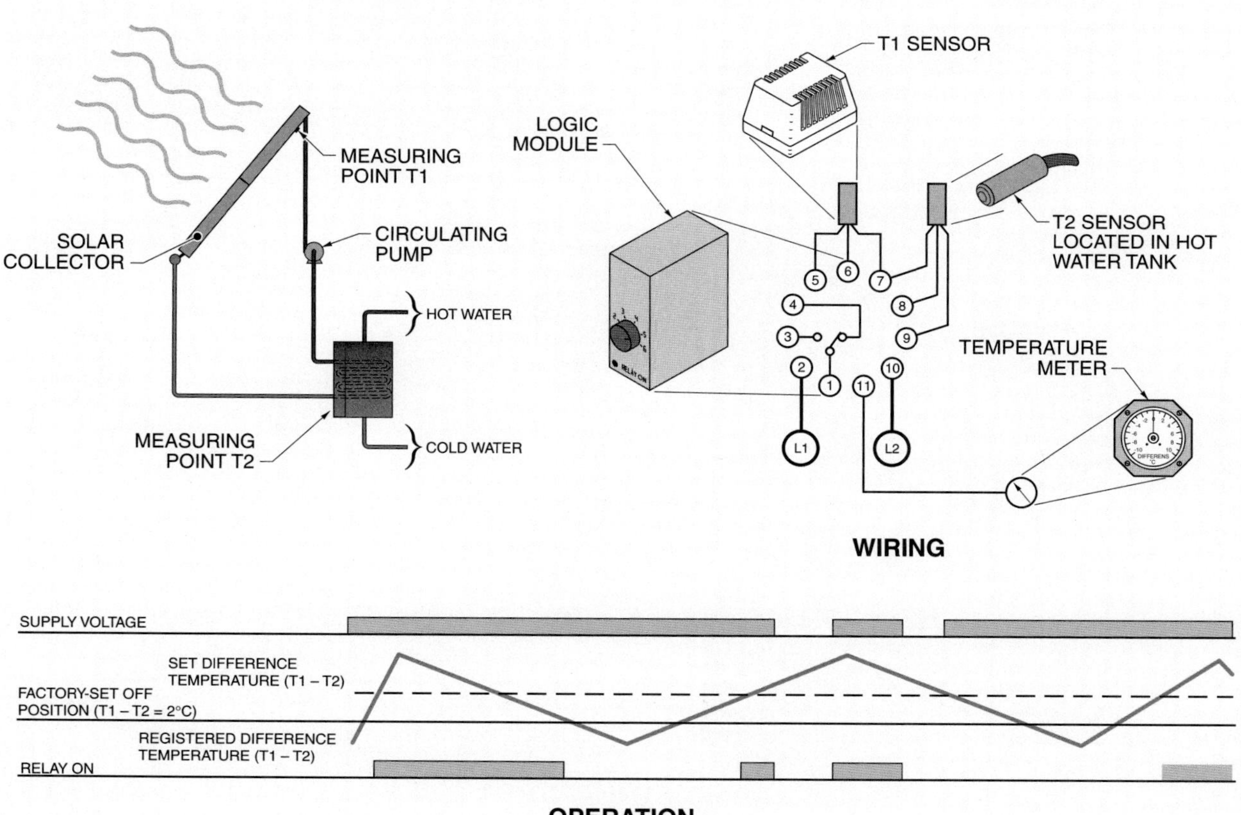

Figure 9-42. In a solar heating system, the circulation pump is activated by the temperature differential between T1 and T2.

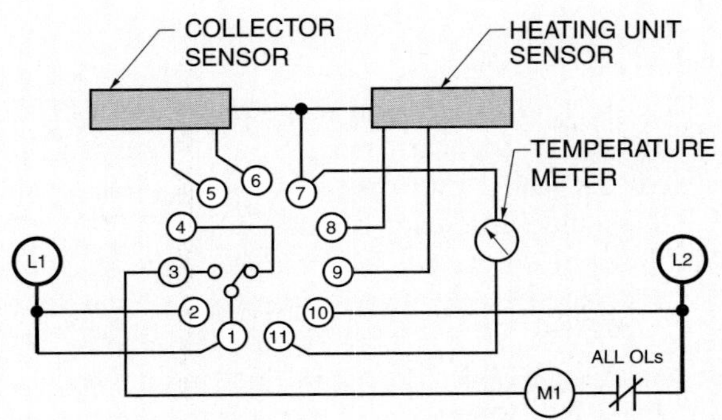

Figure 9-43. A wiring diagram is used to properly install the control circuit of the solar heating system.

Air Pollution Control

The major emphasis on air pollution control today helped develop automatic air pollution control devices. The air pollution control devices detect and react to gases and smoke in the air. Such a system is similar to residential smoke detectors. Industrial applications often require the detection of other pollutants in addition to smoke. These applications may require the detection of gases or carbon in the air. Such detection may be used for measuring pollutants to meet recommended safety levels.

The air pollution control device must be able to measure the pollutants in the air and compare them to a set level. It must also take appropriate action if the level is exceeded. This action may be the sounding of an alarm or some corrective measure such as turning ON an exhaust fan. In meeting these requirements, the system must include a pollution detector (signal input), a logic module or control circuit that can react to the detector (decision), and an output (action). **See Figure 9-44.** The circuit uses an air pollution sensor and matching relay to detect flammable gases, smoke, and carbon in the air.

The relay is used with an air pollution detector that detects very small concentrations of all reducing or flammable gases such as hydrogen, carbon dioxide, methane, propane, butane, acetylene, and sulfur dioxide. Even very small concentrations, such as .02% (200 parts per million [ppm]), are detected. The detector also reacts to a smoke-filled carbonaceous atmosphere.

Technical Fact

Air pollution is caused by particles released into the air from burning fuels and by the release of gases such as sulfur dioxide, carbon monoxide, and chemical vapors into the atmosphere. These gases chemically react in the atmosphere and form smog and acid rain.

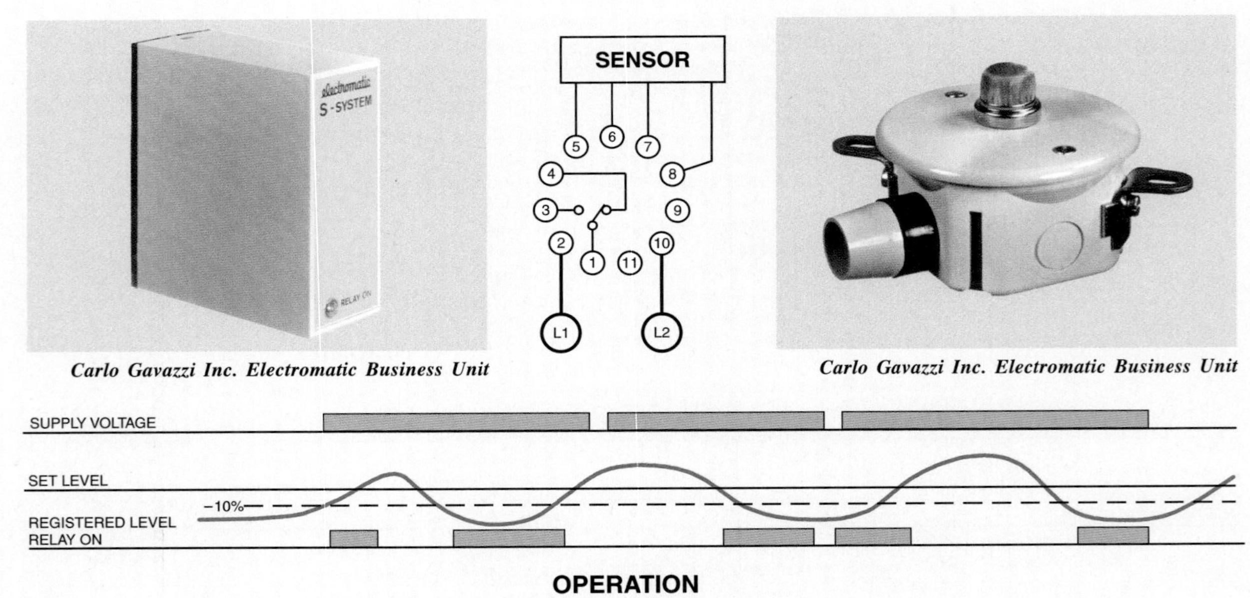

Figure 9-44. An air pollution detector uses a logic module and an air pollution sensor to measure the pollutants in the air and compare them to a set level.

The relay immediately operates when the supply voltage is applied. The relay releases when the registered pollution level exceeds the set level. The relay does not operate again until the registered level is about 10% lower than the set level. The activation of the relay can be used to signal an alarm or record the time and duration of high levels of pollution.

WIND METERING

Windmills are used to harness the wind to generate power. A typical windmill consists of one or two generators, depending on the size. A windmill with two generators commonly has a small generator with a synchronous speed of 1000 rpm and a large generator with a synchronous speed of 1500 rpm. The small generator is first connected to the system. The large generator is connected after a short time delay when the maximum power of the small generator is reached.

On a windmill, a control device (anemometer) is required to measure and react to wind velocity. On some windmill applications, a control device (wind vane) is also required to determine relative wind direction. **See Figure 9-45.**

An anemometer is used to stop the windmill at too low and too high a wind velocity. Stopping the windmill at low wind velocities prevents the constant connection and disconnection of the small generator. Stopping the windmill at high wind velocities helps to protect the windmill against damage and wear from the high wind speeds. To stop the windmill, the anemometer and wind velocity relay control a mechanical brake inside the windmill. The specifications for the anemometer and wind velocity relay are given in manufacturer data sheets. **See Figure 9-46.**

Bergey Windpower Co., Inc.

Wind turbine systems are designed for use based on the average wind speed of the location.

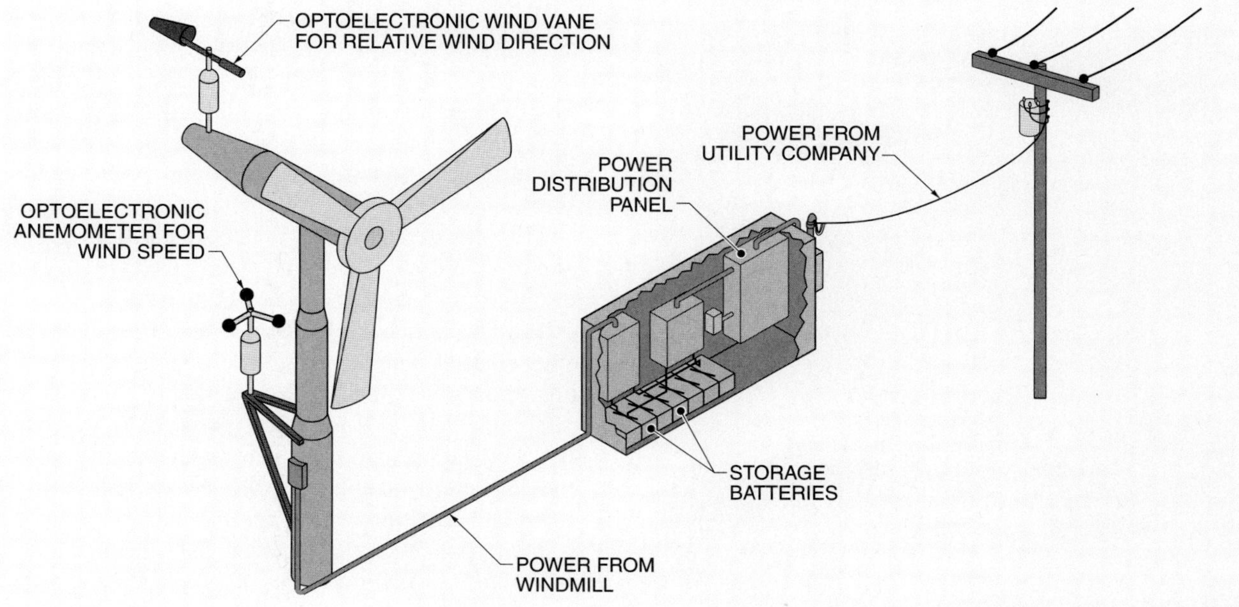

Figure 9-45. Windmills are used to harness the wind to generate power.

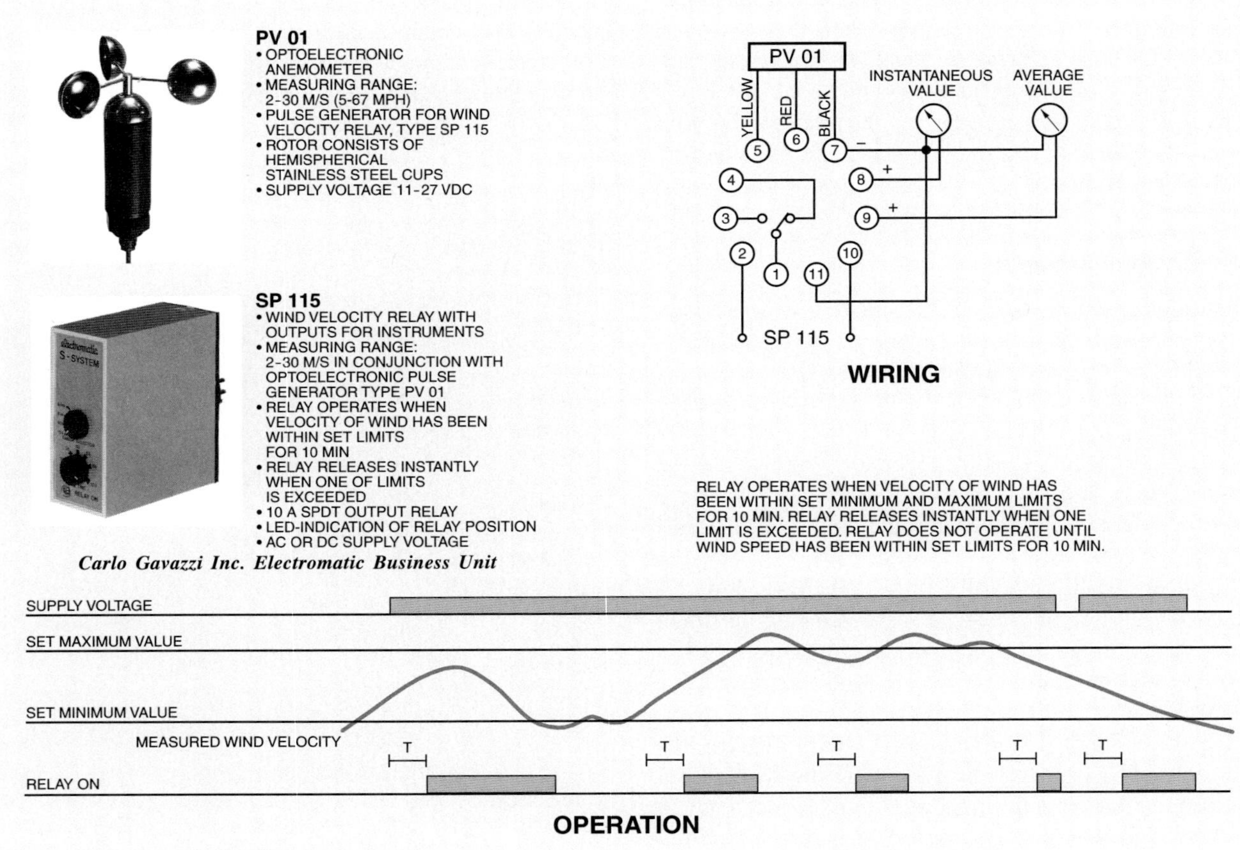

Figure 9-46. Anemometer and wind velocity relay specifications are given in manufacturer data sheets.

Bergey Windpower Co., Inc.

Wind turbines are used for supplying power for pumping, battery charging, and residential power demands.

A wind vane is used to control the yawing function of the windmill. *Yawing* is a side-to-side movement. This is required to turn the windmill into the direction of the wind for windmills that are not freely allowed to turn. To do this, the wind direction relay incorporates a relay with a neutral center position. In this position, the top of the windmill is kept still while the two working positions of the contacts turn to either the right or left of the windmill. The specifications for wind vanes and wind direction relays are given in manufacturer's data sheets. **See Figure 9-47.**

In addition to wind speed and wind direction controls, other controls may be required depending on the application of the generated power. In a simple windmill application, the generated power can be used directly without synchronization with other power. An example of this is an application that connects the generated power to a set of heating elements.

In an application that connects the generated power to utility lines, synchronization of frequency and voltages must be built in. A rectifier circuit is required when connecting the generated power to a set of batteries.

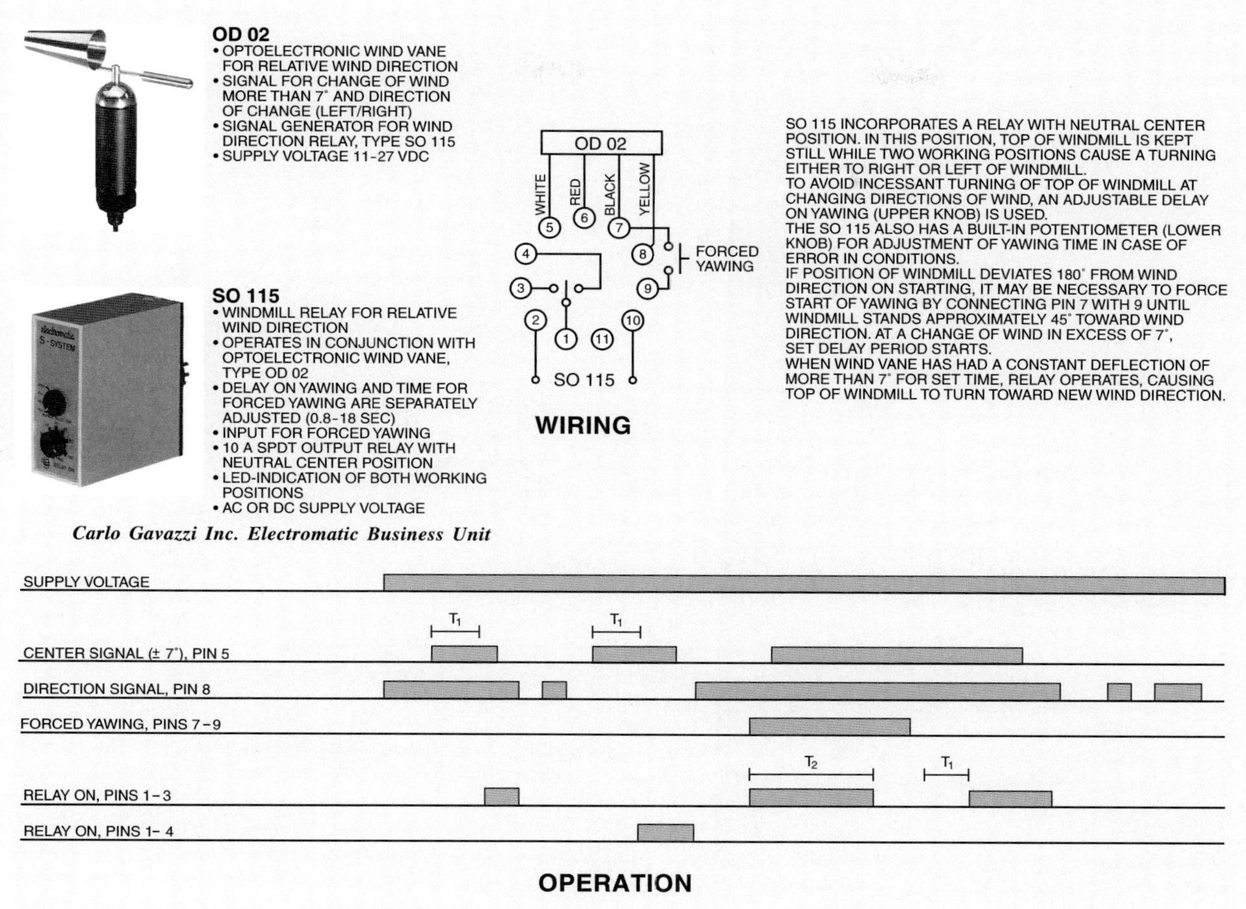

Figure 9-47. A wind vane and a wind direction relay are used to control the yawing function of a windmill.

AUTOMATED SYSTEMS

An automated system includes manual, mechanical, and automatic control devices. These control devices are interconnected to provide the required inputs to make the system function as designed. **See Figure 9-48.** This system could be used for paint, food products, beverages, or other products put in a container.

In this system, manual inputs such as pushbuttons and selector switches are required at the individual and main control stations. Automatic inputs such as pressure, temperature, flow, and level controls are required to control each step of the process from start to finish. Mechanical inputs are used to detect position as required.

All of these basic control devices provide a method for controlling the product or operation. Each control device must be selected and installed as if the entire operation depends on that one control input. In an automated system, the failure of any one control device could shut down the entire process.

Preventing Problems When Installing Control Devices

All electrical circuits must be controlled. For this reason, control devices are used in every type of control application. A control device must be properly protected and installed to ensure that the control device operates properly for a long time. Proper protection means that the switching contacts are operated within their electrical rating and are not subjected to destructive levels of current or voltage. Proper installation means that the control device is installed in such a manner as to ensure it operates as designed.

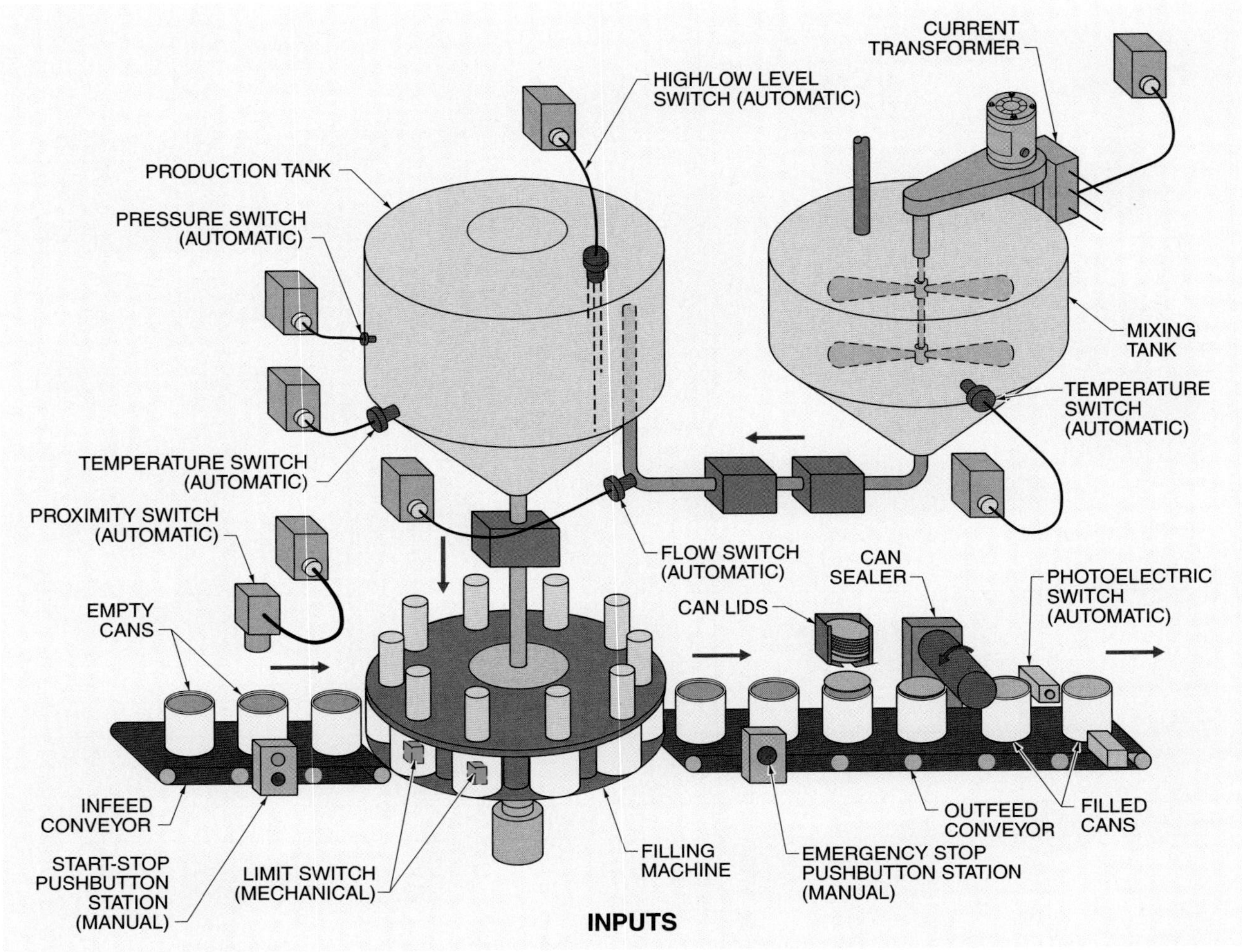

Figure 9-48. An automated system includes manual, mechanical, and automatic control devices that are interconnected to provide the required inputs to make the system function as designed.

Protecting Switch Contacts

Control devices are used to switch ON or OFF or redirect the flow of current in an electrical circuit. The control devices switch contacts that are rated for the amount of current they can safely switch. The switch rating is normally specified for switching a resistive load, such as small heating elements. Resistive loads are the least destructive loads to switch. However, most loads that are switched are inductive loads, such as solenoids and motor starter coils. Inductive loads are the most destructive loads to switch because of the collapsing magnetic field present due to counter EMF when the contacts are opened.

A large induced voltage appears across the switch contacts when inductive loads are turned OFF. The induced voltage is the opposite polarity of the applied voltage. The induced voltage causes arcing at the switch contacts. Arcing may cause the contacts to burn, stick, or weld together. Contact protection should be added when frequently switching inductive loads to prevent or reduce arcing. **See Figure 9-49.**

A diode is added in parallel with the load to protect contacts that switch DC. The diode conducts only when the switch is open, providing a path for the induced voltage in the load to dissipate.

A resistor and capacitor (RC network or snubber) are connected across the switch contacts to protect contacts that switch AC. The capacitor acts as a high-impedance (resistor) load at 60 Hz, but becomes a short circuit at the high frequencies produced by the induced voltage of the load. This allows the induced voltage to dissipate across the resistor when the load is switched OFF.

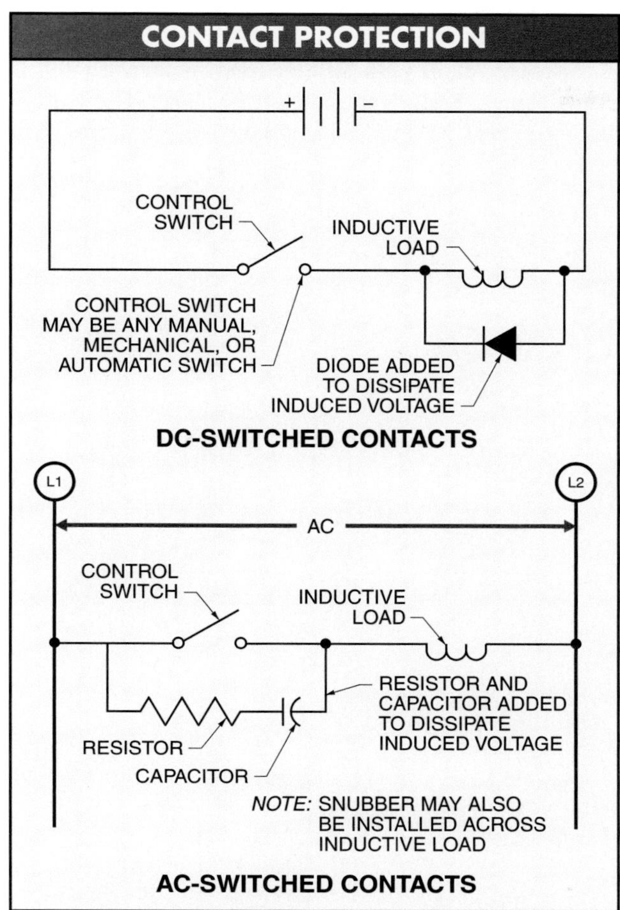

Figure 9-49. Contact protection may be added when switching large DC and AC inductive loads to prevent or reduce arcing at the switch contacts.

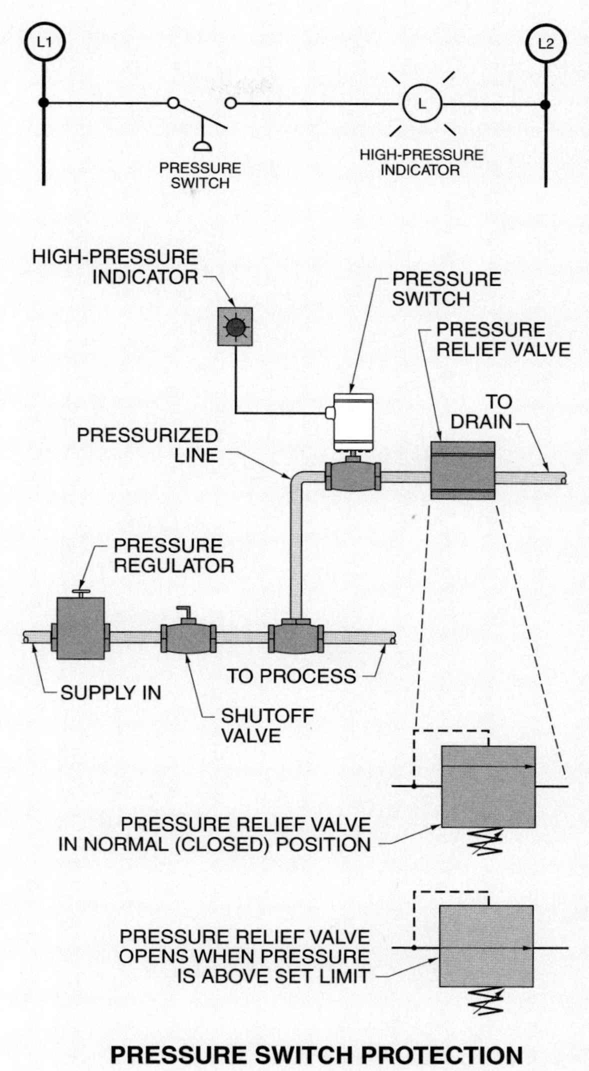

Figure 9-50. A pressure relief valve may be added to a circuit to protect a pressure switch from excessive pressure.

Protecting Pressure Switches

A pressure switch is a switch that detects a set amount of force and activates electrical contacts when the set amount of force is reached. Pressure switches are designed to activate their contacts at a preset pressure. A pressure switch is rated according to its operating pressure range. A pressure switch may be damaged if its maximum pressure limit is exceeded.

Protection for a pressure switch should be added in any system in which a higher pressure than the maximum limit is possible. A pressure relief valve is installed to protect the pressure switch. A pressure relief valve should be set just below the pressure switch's maximum limit. The valve opens when the system pressure increases to the setting of the relief valve.

Caution: The output of the relief valve must be connected to a proper drain (or return line) if the product under pressure is a gas or a fluid. **See Figure 9-50.**

Installing Flow Switches

A flow switch is a switch that detects the movement of a fluid. Most flow switches use a paddle to detect the movement of the product. The paddle is designed to detect the product movement with the least possible pressure drop across the switch. A flow switch must be installed correctly to ensure it does not interfere with the movement of the product. Most flow switches are designed to operate in the horizontal position. There is a great deal of turbulence in flowing product at a distance of at least three pipe inside diameters (ID) on each side of the flow switch. **See Figure 9-51.** For example, the minimum horizontal distance of straight pipe required on each side of a flow switch is 4½″ (1½″ × 3 = 4½″) when used in an application that moves a product through a 1½″ diameter pipe.

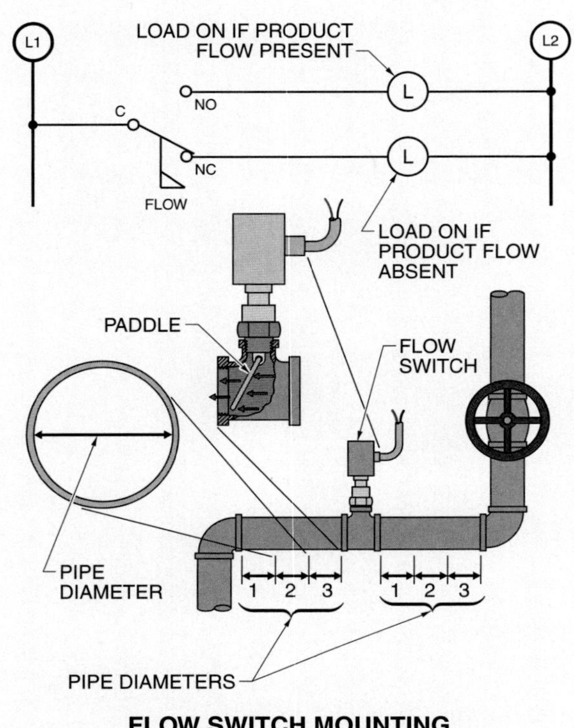

Figure 9-51. Allow a distance of at least three pipe inside diameters (ID) on each side of the flow switch when mounting a flow switch.

TROUBLESHOOTING CONTROL DEVICES

Control devices use mechanical or solid-state switches to control the flow of current. A *mechanical switch* is any switch that uses silver contacts to start and stop the flow of current in a circuit. Silver contacts can be used to switch either AC or DC loads. A solid-state switch has no moving parts (contacts). A *solid-state switch* is a switch that uses a triac, SCR, current sink (NPN) transistor, or current source (PNP) transistor to perform the switching function. The triac output is used for switching AC loads. The SCR output is used for switching high-power DC loads. The current sink and current source outputs are used for switching low-power DC loads.

Testing Electromechanical Switches

A suspected fault with an electromechanical switch is tested using a DMM set to measure voltage. The voltage setting on the DMM is used to test the voltage flowing into and out of the switch. **See Figure 9-52.**

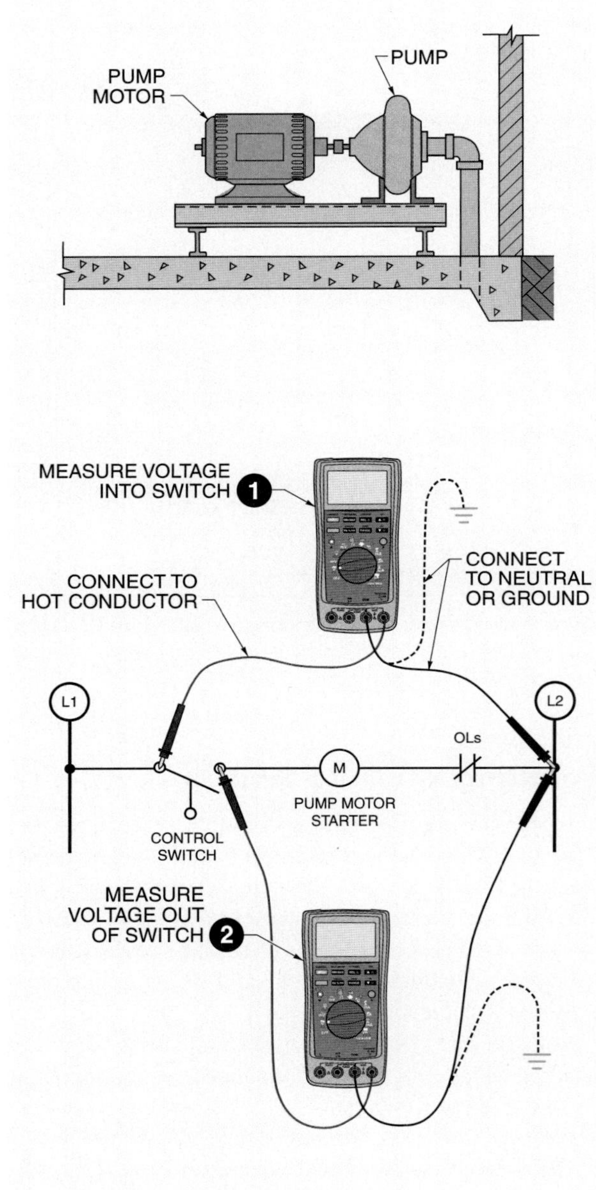

Figure 9-52. A DMM set to measure voltage is used to test the operation of a switch.

To test a mechanical switch, apply the procedure:

1. Measure the voltage into the switch. Connect the DMM between the neutral and hot conductor feeding the switch. Set the DMM to the voltage setting. When working on a grounded system, the DMM lead may be connected to ground instead of neutral if the neutral conductor is not available in the same box in which the switch is located. The problem is located upstream from the switch when there is no voltage present or the voltage is not at the correct level. The problem may be a blown fuse or open circuit. Voltage must be reestablished to the switch before the switch may be tested.
2. Measure the voltage out of the switch. There should be a voltage reading when the switch contacts are closed. There should not be a voltage reading when the switch contacts are open. The switch has an open and must be replaced if there is no voltage reading in either switch position. The switch has a short and must be replaced if there is a voltage reading in both switch positions.

Warning: Always ensure power is OFF before changing a control switch. Use a DMM set to measure voltage to ensure the power is OFF.

A clamp meter can be used to troubleshoot solid-state devices by testing for voltage, current, or resistance.

Testing Solid-State Switches

A suspected fault in a two-wire solid-state switch may be tested using a DMM set to measure voltage. The voltage setting on the DMM is used to test the voltage into and out of the switch. **See Figure 9-53.**

To test a two-wire solid-state switch, apply the following procedure:

1. Measure the voltage into the switch. The problem is upstream from the switch when no voltage is present or the voltage is not at the correct level. The problem may be a blown fuse or open circuit. Voltage must be reestablished before the switch may be tested.
2. Measure the voltage out of the switch. The voltage should equal the supply voltage minus the voltage drop (3 V to 8 V) when the switch is conducting (load ON). Replace the switch if the voltage output is not correct.

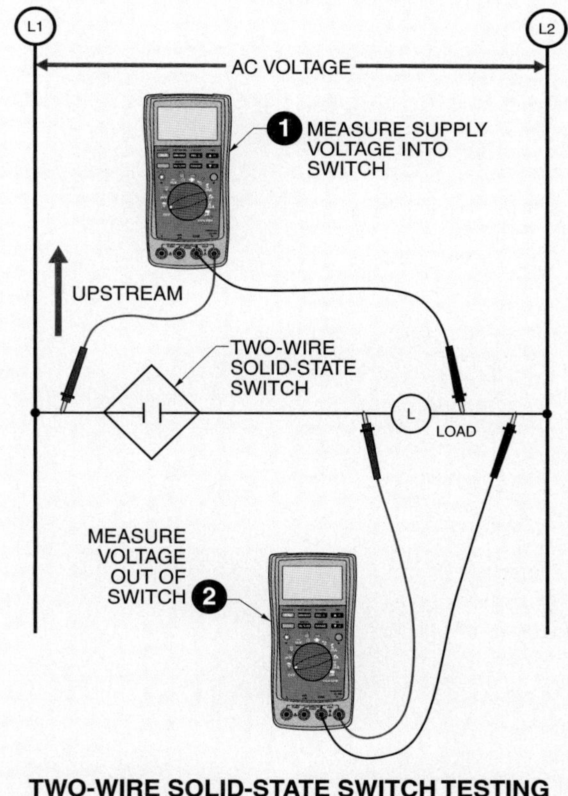

Figure 9-53. A DMM set to measure voltage is used to measure the voltage into and out of a two-wire solid-state switch.

Technical Fact

When a DMM is used to measure voltage across an open circuit, the DMM completes the circuit. The voltage is dropped across the DMM because the resistance of the DMM is greater than that of the circuit.

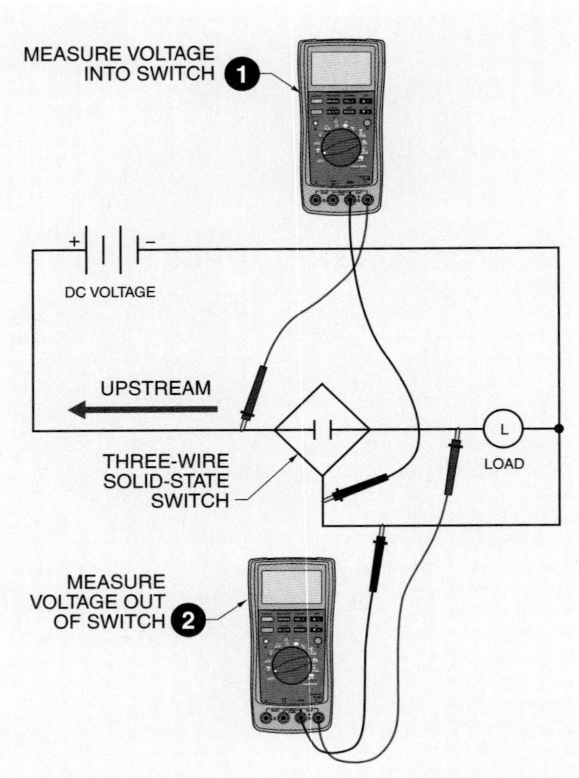

Figure 9-54. A DMM set to measure voltage is used to measure the voltage into and out of a three-wire solid-state switch.

Advanced Assembly Automation, Inc.

Smart input devices are used with programmable controllers to provide multiple control functions to a circuit as in a conveyor system that must have the ability to start and stop automatically based on the materials present.

Warning: Always ensure power is OFF before changing a control switch. Use a DMM set to measure voltage to ensure the power is OFF.

A suspected fault with a three-wire solid-state switch may be tested using a DMM set to measure voltage. The DMM is used to test the voltage into and out of the switch. **See Figure 9-54.**

To test a three-wire solid-state switch, apply the following procedure:

1. Measure the voltage into the switch. The problem is upstream from the switch when no voltage is present or the voltage is not at the correct level. The problem may be a blown fuse or open circuit. Voltage must be reestablished before the switch may be tested.
2. Measure the voltage out of the switch. The voltage should equal the supply voltage when the switch is conducting (load ON). Replace the switch if the voltage output is not correct.

Warning: Always ensure power is OFF before changing a control switch. Use a DMM set to measure voltage to ensure the power is OFF.

SMART (INTELLIGENT) INPUT DEVICES

In hardwired circuits, inputs such as limit switches and temperature switches are wired into the system using several different wires for each device. At least one set of wires must go to each input device. **See Figure 9-55.** Wires that share a common terminal number can be connected together. For example, one wire can be used for the master stop and master start in the car wash system because both are wired to terminal number 5. However, even if some of the wires can be shared, hard wiring requires a large number of control wires to interconnect a system.

Smart (intelligent) devices were developed to reduce wiring and increase control and monitoring of a system. A *smart (intelligent) device* is a device that includes an electronic circuit (chip) that provides communication and diagnostic capabilities to the device. A smart device uses a communication protocol (convention) that allows individual devices (limit switches, photoelectric switches, motor starters, etc.) to communicate with each other over a single cable. **See Figure 9-56.**

Smart devices replace a large number of wires with a single cable. Each smart device monitors itself and any other device on the system. This self-diagnostic reduces downtime and aids in troubleshooting. The software used to control the system can be based on standard line diagram format, flowcharts, Microsoft® Windows–based instructions, or other control languages. The software can be operated by a central control device such as a programmable controller (PLC), personal computer (PC), or human machine interface (HMI).

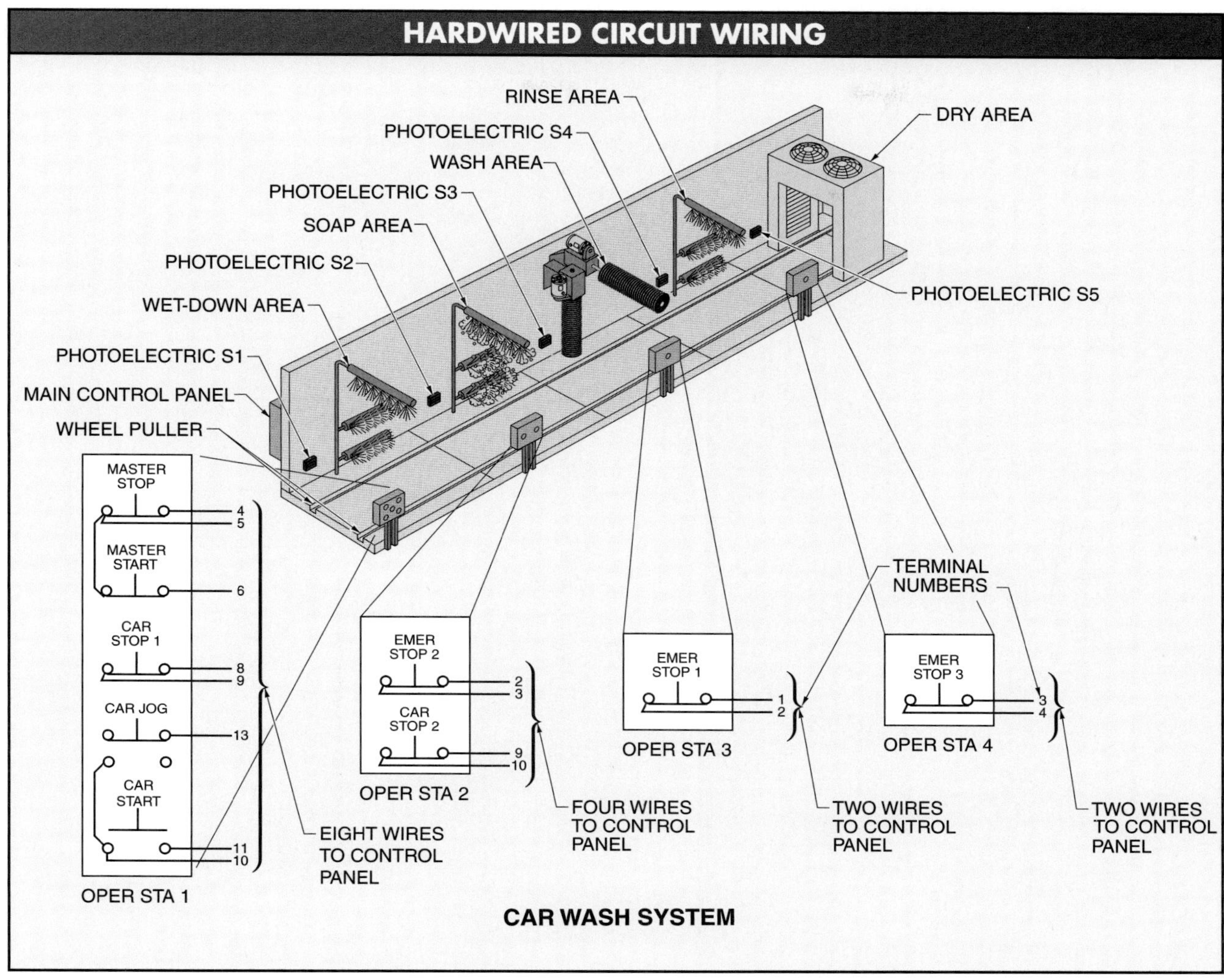

Figure 9-55. Hard wiring requires a large number of control wires to interconnect a system.

A *programmable controller (PLC)* is a solid-state control device that is programmed and reprogrammed to automatically control an industrial process or machine. PLCs are reliable and familiar to electricians, and are often already a part of most electrical systems. However, their software is proprietary and differs between manufacturers. Personal computers are reliable in a clean environment, familiar to most people, flexible, generally available, and interchangeable. For these reasons, smart devices are used with PLCs and personal computers as the main system control device.

In a smart device system, each device is assigned an address such as 1, 2, 25, etc. The power supply allows data and control power to be transmitted on the same line. For this reason, a standard power supply cannot be used. The power supply voltage is normally 30 VDC or less. Interface modules can be added that allow standard (non-smart) inputs and outputs to also be connected into the same system. Any smart device (limit switch, etc.) can be directly connected to the cable.

Technical Fact

PLCs are designed for trouble-free operation in a plant environment. To further minimize the possibility of system malfunction, the connections to the I/O modules should be periodically checked to ensure that all plugs, terminals strips, sockets, and modules have good electrical connections.

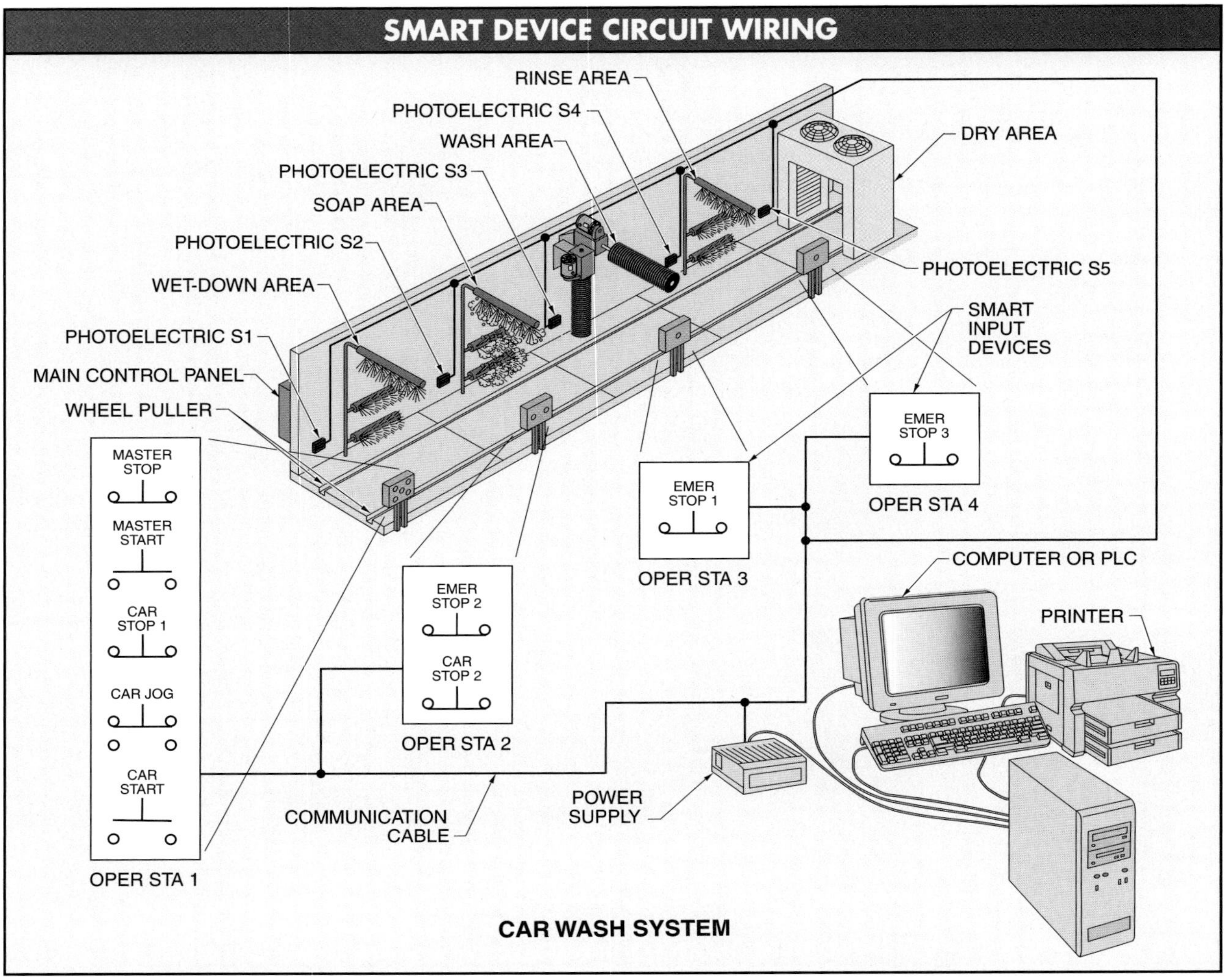

Figure 9-56. A smart device uses a communication protocol that allows individual devices to communicate with each other over a single cable.

Data and control power are carried over the same two-wire cable. This cable is normally an unshielded cable that includes approximately five wires. The cable is normally AWG No. 14 or No. 16 wire. The exact size of the cable depends on the required system current and cable length. The cable system is designed so new branches can be added where required. The smart devices may be connected using screw connectors or quick connectors. **See Figure 9-57.**

Another advantage of smart (intelligent) devices is the ability to print out all activity on the system. For example, the exact time any input is energized (turned ON) or de-energized (turned OFF) can be printed and documented. This information can be used for quality control, scheduling preventive maintenance, scheduling production, and compliance with regulatory agencies.

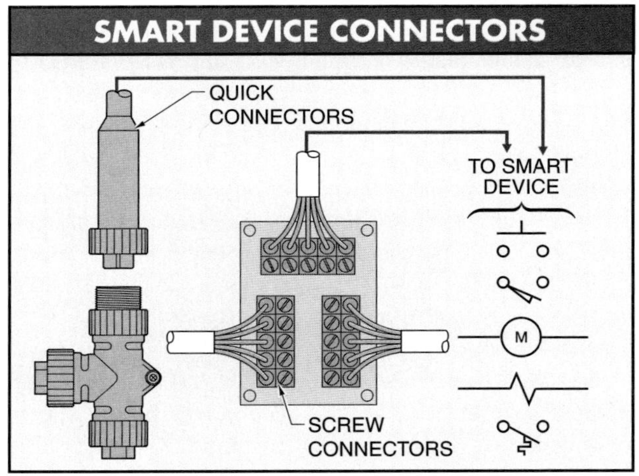

Figure 9-57. Smart devices may be connected using screw connectors or quick connectors.

Review Questions

1. What is the most common contact configuration used on pushbuttons?
2. What are the different operators that are available on pushbuttons?
3. What operator is best suited for an emergency stop pushbutton?
4. What NEMA enclosures are available for pushbutton stations?
5. What is the purpose of selector switches?
6. What is the purpose of a truth table?
7. What is a joystick?
8. How many positions can a joystick have?
9. What are the two methods used to illustrate the different positions of a joystick?
10. What types of contacts are normally included with limit switches?
11. What is an actuator?
12. What are the basic actuators available for use with limit switches?
13. What is a daylight switch?
14. What is a pressure switch?
15. What types of pressure switch contacts are used to signal an overpressure condition?
16. What is the deadband setting of a pressure switch?
17. What may happen if the pressure differential setting of a pressure switch is too small?
18. What are the uses of most pressure switches?
19. What is a temperature switch?
20. What are the different methods used to switch electrical contacts in response to a temperature change?
21. What is a flow switch?
22. Can gas detection be made to detect different groups of gas?
23. Can level switches only sense liquids?
24. What other two names are charging systems and discharging systems known by?
25. How many single temperature inputs does a solar heating system need?

chapter 10

electrical motor controls *for Integrated Systems*

Reversing Motor Circuits

Reversing circuits provide the means to safely and conveniently reverse the direction of a motor. Reversing the electrical connections to a motor may be accomplished with manual reversing starters, drum switches, magnetic reversing starters, PLCs, or motor drives.

REVERSING MOTORS USING MANUAL STARTERS

A manual starter is a contactor with an added overload protective device. Two manual starters are connected together to create a manual reversing starter. Manual reversing starters are used in pairs to change the direction of rotation of 3ϕ, 1ϕ, and DC motors. **See Figure 10-1.**

Manual reversing starters are used to operate low horsepower motors, such as those found on fans, small machines, pumps, and blowers, in forward and reverse directions. Individual manual starters are marked start/stop instead of forward/stop or reverse/stop. This is common when two manual starters are placed in the same enclosure to make up a manual reversing starter. The electrician must correctly label the unit once it is properly wired.

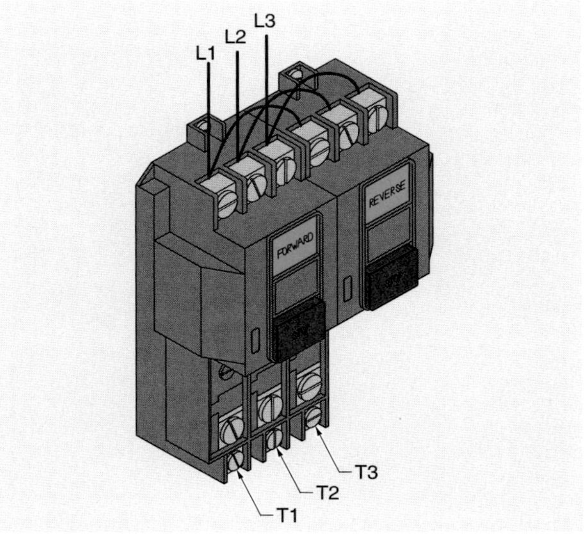

Figure 10-1. Manual starters are used in pairs to reverse 3ϕ, 1ϕ, and DC motors.

Since a motor cannot run in forward and reverse simultaneously, some means must be included to prevent both starters from energizing at the same time. A manual reversing starter uses a mechanical interlock to separate the starter contacts. A *mechanical interlock* is the arrangement of contacts in such a way that both sets of contacts cannot be closed at the same time. Mechanical interlock devices are inserted between the two starters to ensure that both switching mechanisms cannot be energized at the same time. Crossing dashed lines are used between the manual starters in the wiring diagram to indicate a mechanical interlock. **See Figure 10-2.** An electrician must ensure that the interlock is provided if the unit is not preassembled.

Figure 10-2. Mechanical interlock devices are inserted between the two starters to ensure that both switching mechanisms cannot be energized at the same time.

Leeson Electric Corporation

Electric motors are available in a wide range of sizes and types for any industrial application.

Reversing 3ϕ Motors Using Manual Starters

A wiring diagram illustrates the electrical connections necessary to properly reverse a 3ϕ motor using a manual reversing starter. **See Figure 10-3.** Only one set of overloads is required.

Reversing the direction of rotation of 3ϕ motors is accomplished by interchanging any two of the three main power lines to the motor. Although any two lines may be interchanged, the industry standard is to interchange L1 and L3. This standard is true for all 3ϕ motors including three-, six-, and nine-lead wye- and delta-connected motors.

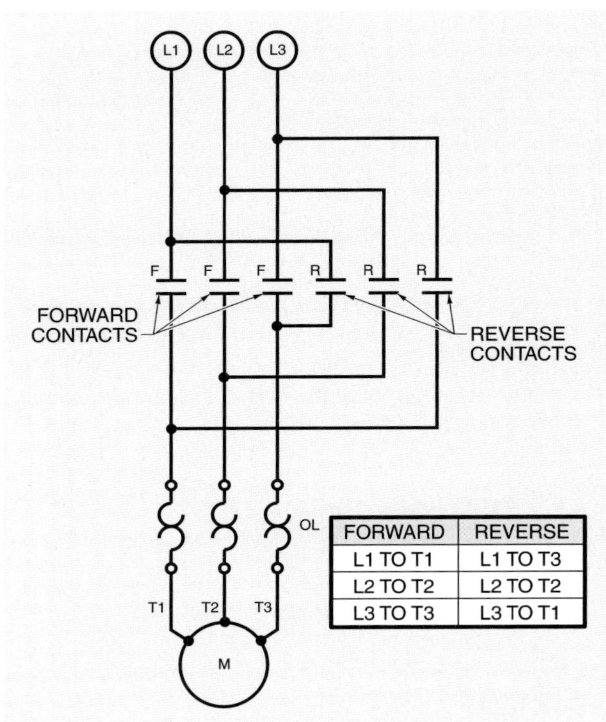

Figure 10-3. A wiring diagram illustrates the electrical connections necessary to properly reverse a 3ϕ motor using a manual reversing starter.

Regardless of the type of 3ϕ motor, when using a manual reversing starter to reverse a 3ϕ motor, L1 is connected to T1, L2 is connected to T2, and L3 is connected to T3 for forward rotation (when the forward contacts close). **See Figure 10-4.**

Line 1 is connected to T3, L2 is connected to T2, and L3 is connected to T1 for reverse rotation (when the reverse contacts close and the forward contacts open). The motor changes direction each time forward or reverse is depressed because it is necessary to interchange only two leads on a 3ϕ motor to reverse rotation. If a 3ϕ motor has more than three leads coming out, these leads are connected according to the motor wiring diagram.

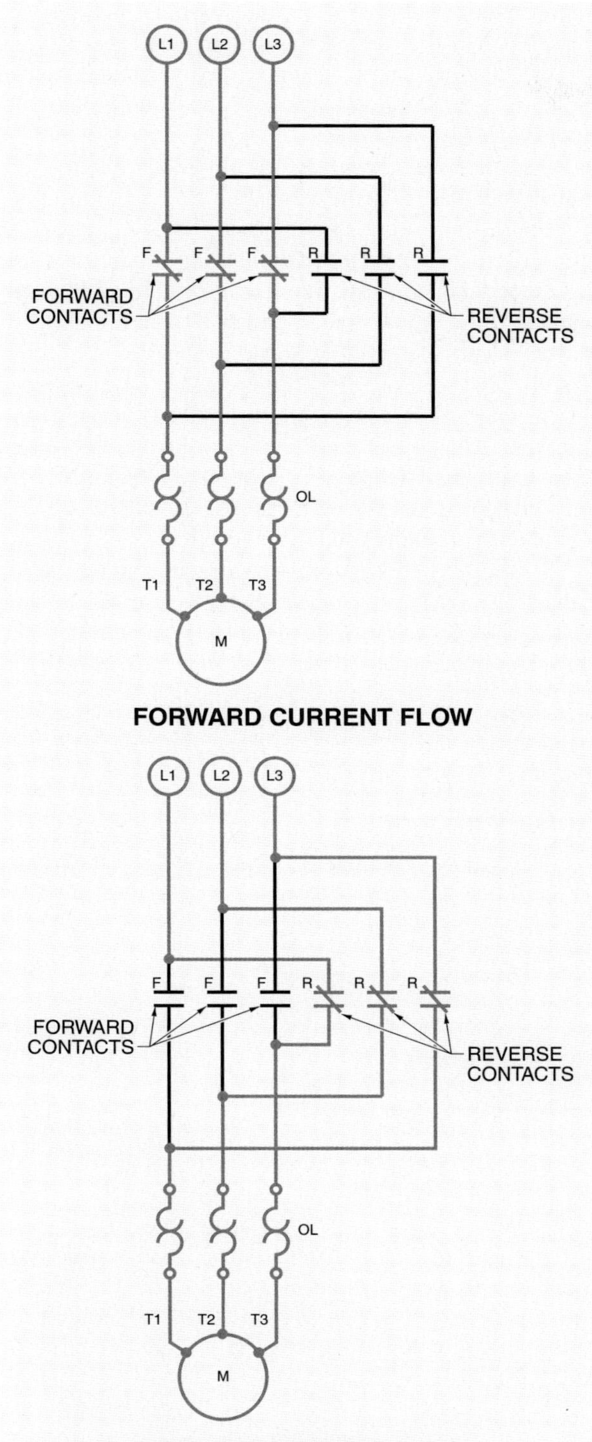

Figure 10-4. When using a manual reversing starter to reverse a 3ϕ motor, L1 is connected to T1, L2 is connected to T2, and L3 is connected to T3 for forward rotation. Line 1 is connected to T3, L2 is connected to T2, and L3 is connected to T1 for reverse rotation.

Interchanging L1 and L3 is standard for safety reasons. When first connecting a motor, the direction of rotation is not usually known until the motor is started. A motor may be temporarily connected to determine the direction of rotation before making permanent connections. Motor lead temporary connections are not taped. By always interchanging L1 and L3, L2 can be permanently connected to T2, creating an insulated barrier between L1 and L3.

Reversing 1ϕ Motors Using Manual Starters

Reversing the rotation of 1ϕ motors is accomplished by interchanging the leads of the starting or running windings. The manufacturer wiring diagram is used to determine the exact wires to interchange to properly reverse a 1ϕ motor using a manual reversing starter. **See Figure 10-5.** *Note:* Always check the manufacturer wiring diagrams for proper reversal of 1ϕ motors. An electrician can measure the resistance of the starting winding and running winding to determine which leads are connected to which windings if manufacturer information is not available. The running winding is made of a heavier gauge wire than the starting winding, so the running winding has a much lower resistance than the starting winding.

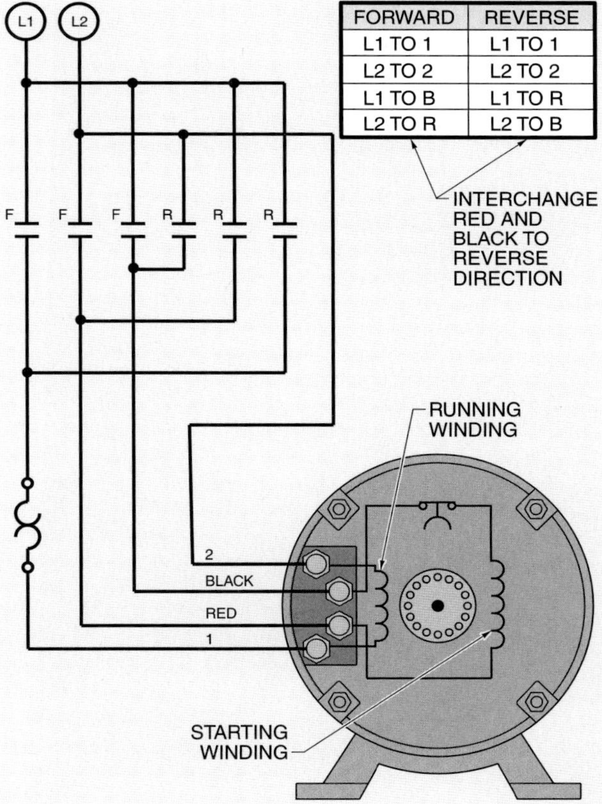

FORWARD	REVERSE
L1 TO 1	L1 TO 1
L2 TO 2	L2 TO 2
L1 TO B	L1 TO R
L2 TO R	L2 TO B

Figure 10-5. A wiring diagram illustrates the electrical connections necessary to properly reverse a 1ϕ motor using a manual reversing starter.

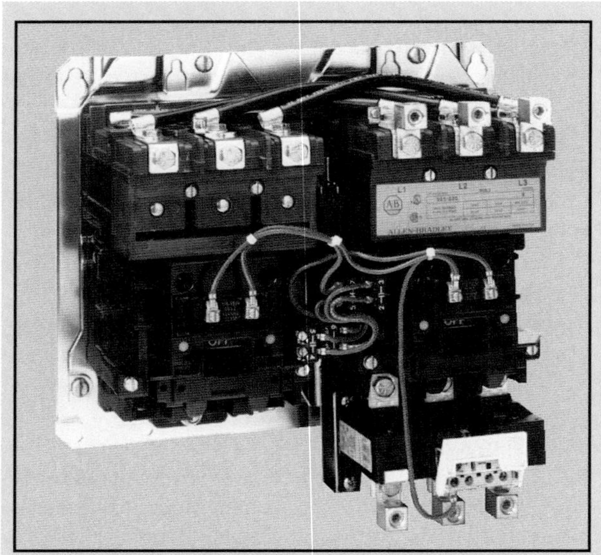

Rockwell Automation, Allen-Bradley Company, Inc.

Full-voltage reversing starters may allow auto/manual reset; communication capability; and phase loss, ground fault, and jam protection.

When the forward contacts are closed, L1 is connected to the black lead of the starting winding and side 1 of the running winding, and L2 is connected to the red lead of the starting winding and side 2 of the running winding. **See Figure 10-6.**

When the reverse contacts are closed and the forward contacts open, L1 is connected to the red lead of the starting winding and side 1 of the running winding, and L2 is connected to the black lead of the starting winding and side 2 of the running winding. The starting windings are interchanged while the running windings remain the same.

The motor changes direction each time forward or reverse is depressed because it is necessary to interchange only the starting windings on a 1ϕ motor to reverse rotation. *Note:* Always check the manufacturer wiring diagram when reversing 1ϕ motors to determine which leads are connected to the starting winding. The red and black wires are the ones normally used for reversal.

The direction of rotation of a capacitor motor can be changed by reversing the connections to the starting or running windings. Whenever possible, the manufacturer wiring diagram should be checked for the exact wires to interchange.

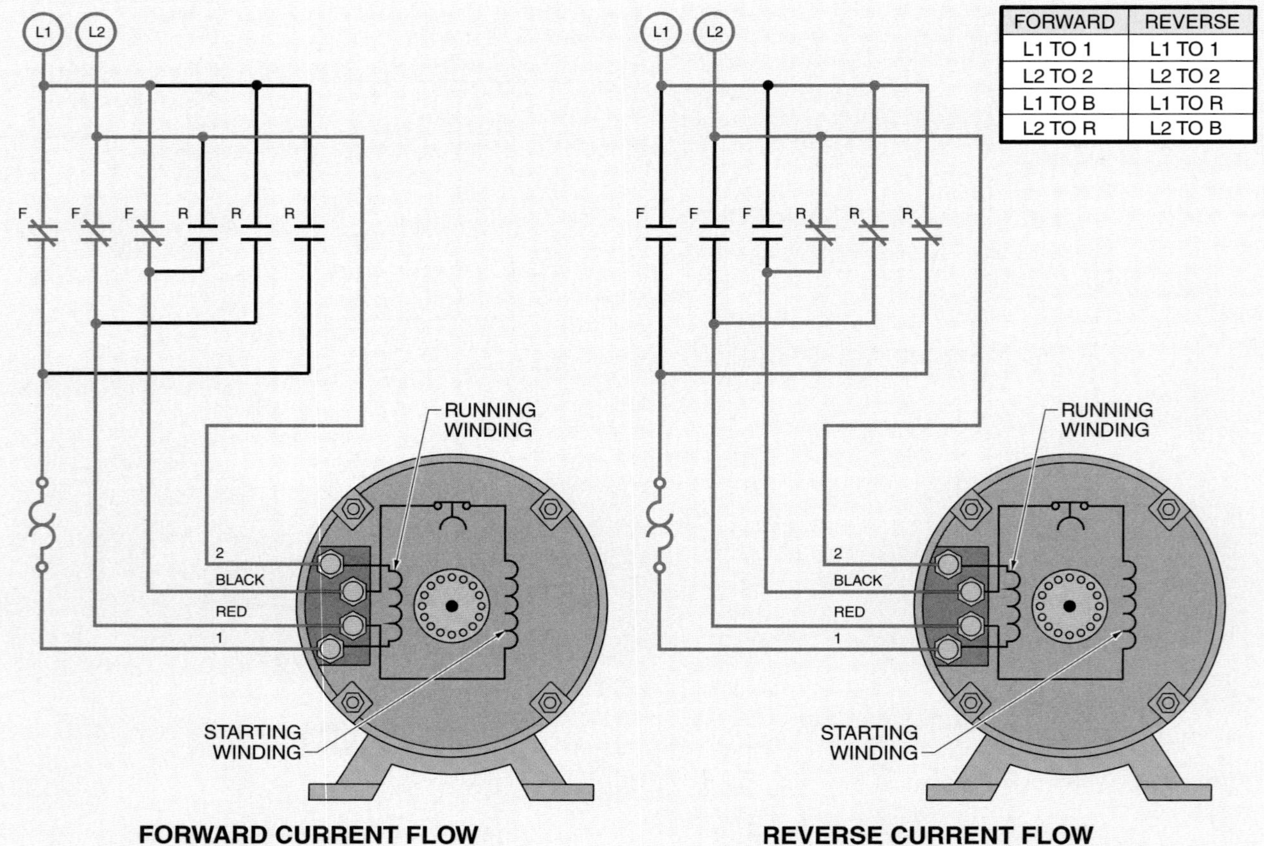

FORWARD	REVERSE
L1 TO 1	L1 TO 1
L2 TO 2	L2 TO 2
L1 TO B	L1 TO R
L2 TO R	L2 TO B

Figure 10-6. To reverse the direction of a 1ϕ motor, the direction of current through the starting winding is reversed.

Reversing DC Motors Using Manual Starters

A manual starter can be used to reverse the direction of current flow through the armature of all DC motors. The motor is wired to the starter so that the polarity of the applied DC voltage on the field remains the same in either direction, but the polarity on the armature is opposite for each direction. The direction of rotation of DC series, shunt, and compound motors may be reversed by reversing the direction of the current through the field without changing the direction of the current through the armature, or by reversing the direction of the current through the armature, but not both. The industrial standard is to reverse the current through the armature.

A wiring diagram is used to properly wire a DC series motor for reversing. A DC series motor is wired to the starter so that A2 is positive and A1 is negative when the forward contacts are closed, and A2 is negative and A1 is positive when the reverse contacts are closed. **See Figure 10-7.** Regardless of whether the forward contacts or reverse contacts are closed, S2 is always positive and S1 is always negative. The motor reverses direction for each position of the starter because only the polarity of the armature reverses direction.

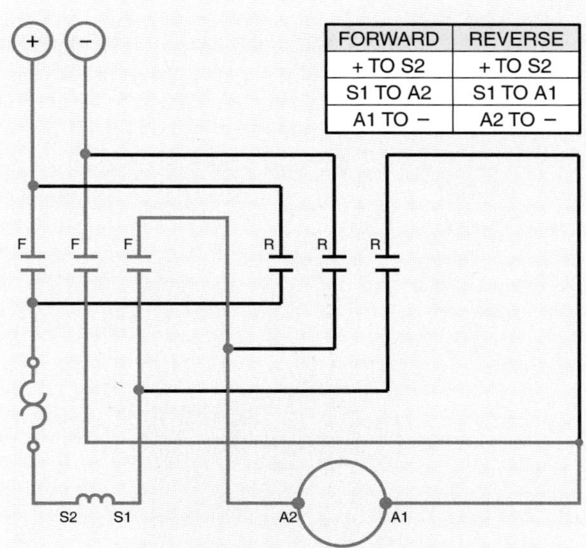

Figure 10-7. A DC series motor is wired to a starter so that A2 is positive and A1 is negative when the forward contacts are closed, and A2 is negative and A1 is positive when the reverse contacts are closed.

A wiring diagram is used to properly wire a DC shunt motor for reversing. A DC shunt motor is wired to the starter so that A2 is positive and A1 is negative when the forward contacts are closed, and A2 is negative and A1 is positive when the reverse contacts are closed. **See Figure 10-8.** Regardless of whether the forward contacts or reverse contacts are closed, F2 is always positive and F1 is always negative. The motor reverses direction for each position of the starter because only the polarity of the armature reverses direction.

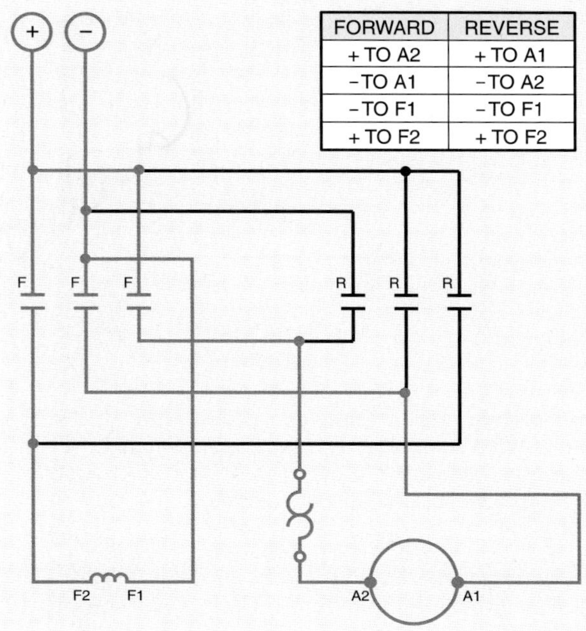

Figure 10-8. A wiring diagram is used to properly wire a DC shunt motor for reversing.

A wiring diagram is used for properly wiring a DC compound motor for reversing. **See Figure 10-9.** A DC compound motor is wired to the starter so that A2 is positive and A1 is negative when the forward contacts are closed, and A2 is negative and A1 is positive when the reverse contacts are closed. Regardless of whether the forward contacts or reverse contacts are closed, S2 and F2 are always positive and S1 and A1 are always negative. The motor reverses direction for each position of the starter because only the polarity of the armature reverses direction.

In a DC compound motor, the series and shunt field relationship to the armature must be left unchanged. The shunt field must be connected in parallel with the armature and the series field must be connected in series with the armature. Reversal is accomplished by reversing the armature connections only. If the motor has commutating pole windings, these windings are considered a part of the armature circuit and the current through them must be reversed when the current through the armature is reversed.

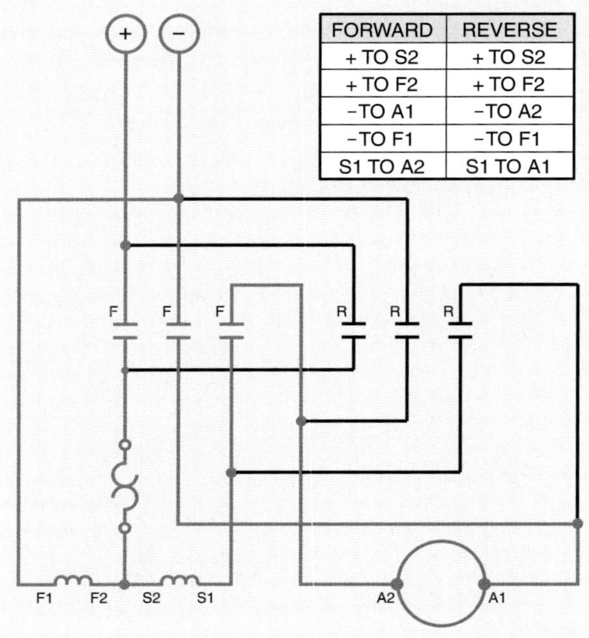

Figure 10-9. A DC compound motor is wired to a starter so that A2 is positive and A1 is negative when the forward contacts are closed, and A2 is negative and A1 is positive when the reverse contacts are closed.

A wiring diagram is used for properly wiring a DC permanent-magnet motor for reversing. **See Figure 10-10.** A DC permanent-magnet motor is wired to the starter so that A2 is positive and A1 is negative when the forward contacts are closed and A2 is negative and A1 is positive when the reverse contacts are closed. A permanent-magnet field never reverses its direction of polarity regardless of the polarity to which the armature is connected. The direction of rotation of a DC permanent-magnet motor is reversed by reversing the direction of the current through the armature only, since there are no field connections available.

Technical Fact

The wiring diagram on a motor nameplate must always be checked to see if the motor is designed for dual-voltage use. If a motor is a dual-voltage motor, care must be taken when reversing the motor to ensure that the wires are attached to the correct voltages. If the wiring is connected incorrectly, performance will differ from nameplate values and from applicable standards.

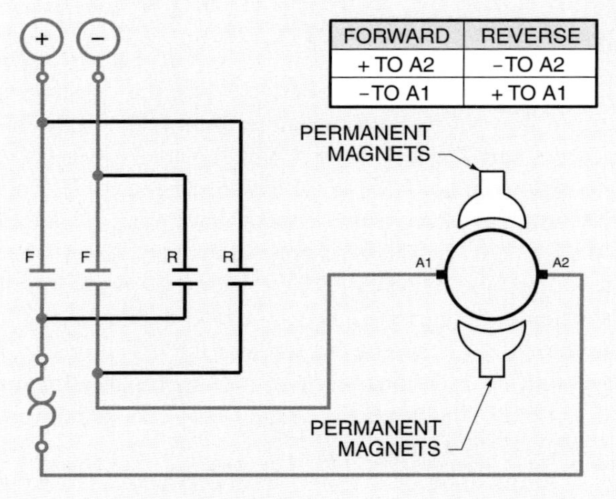

Figure 10-10. A wiring diagram is used to properly wire a DC permanent-magnet motor for reversing.

REVERSING MOTORS USING DRUM SWITCHES

A *drum switch* is a manual switch made up of moving contacts mounted on an insulated rotating shaft. **See Figure 10-11.** The moving contacts make and break contact with stationary contacts within the switch as the shaft is rotated.

Drum switches are totally enclosed and an insulated handle provides the means for moving the contacts from point to point. Drum switches are available in several sizes and can have different numbers of poles and positions. Drum switches are usually used where an operator's eyes must remain on a particular operation such as a crane raising and lowering a load.

Furnas Electric Co.

Figure 10-11. A drum switch is a manual switch with moving contacts mounted on an insulated rotating shaft.

A drum switch may be purchased with maintained contacts or spring-return contacts. In either case, when the motor is not running in forward or reverse, the handle is in the center (OFF) position. To reverse a running motor, the handle must first be moved to the center position until the motor stops and then moved to the reverse position.

Drum switches are not motor starters because they do not contain protective overloads. Separate overload protection is normally provided by placing a nonreversing starter in line before the drum switch. This provides the required overload protection and acts as a second disconnecting means. A drum switch is used only as a means of controlling the direction of a motor by switching the leads of the motor.

Reversing 3ϕ Motors Using Drum Switches

A 3ϕ motor may be connected to the contacts of a drum switch to change the direction of rotation from forward to reverse. **See Figure 10-12.** Charts are used to show the internal operation of a drum switch and the resulting motor connections for forward and reverse. Line 1 and L3 are interchanged as the drum switch is moved from the forward to the reverse position. The motor changes direction each time the drum switch is moved to forward or reverse because only the two leads on a 3ϕ motor must be interchanged to reverse rotation.

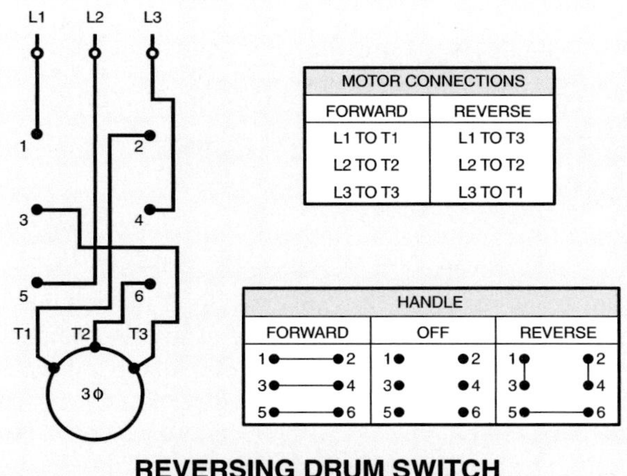

MOTOR CONNECTIONS	
FORWARD	REVERSE
L1 TO T1	L1 TO T3
L2 TO T2	L2 TO T2
L3 TO T3	L3 TO T1

HANDLE		
FORWARD	OFF	REVERSE
1●—●2	1● ●2	1● ●2
3●—●4	3● ●4	3● ●4
5●—●6	5● ●6	5●—●6

REVERSING DRUM SWITCH INTERNAL SWITCHING

Figure 10-12. A 3ϕ motor may be connected to the contacts of a drum switch to change the direction of rotation from forward to reverse.

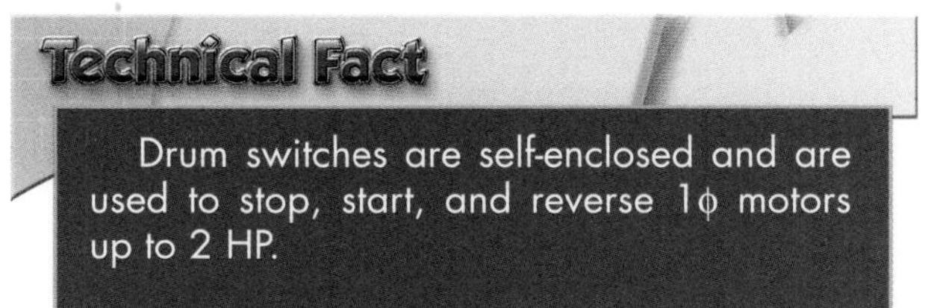

Drum switches are self-enclosed and are used to stop, start, and reverse 1ϕ motors up to 2 HP.

Reversing 1ϕ Motors Using Drum Switches

A 1ϕ motor may be connected to the contacts of a drum switch to change the direction of rotation from forward to reverse. **See Figure 10-13.** Charts are used to show the internal operation of a drum switch. Always consult the manufacturer wiring diagram to ensure proper wiring.

The motor changes direction each time the drum switch is moved to forward or reverse. This occurs because only the starting windings must be interchanged on a 1ϕ motor to reverse rotation.

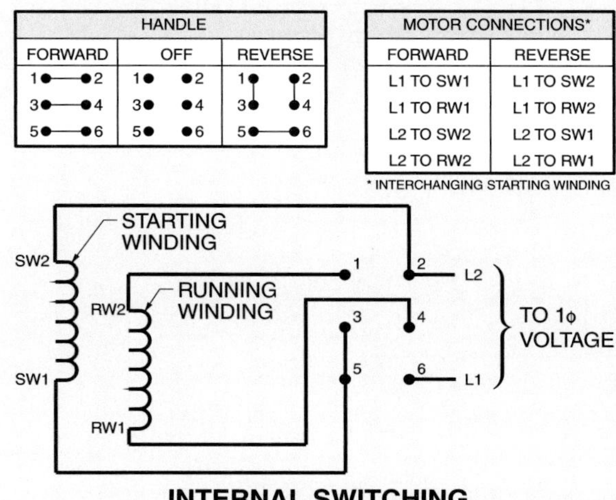

HANDLE		
FORWARD	OFF	REVERSE
1●—●2	1● ●2	1● ●2
3●—●4	3● ●4	3● ●4
5●—●6	5● ●6	5●—●6

MOTOR CONNECTIONS*	
FORWARD	REVERSE
L1 TO SW1	L1 TO SW2
L1 TO RW1	L1 TO RW2
L2 TO SW2	L2 TO SW1
L2 TO RW2	L2 TO RW1

* INTERCHANGING STARTING WINDING

INTERNAL SWITCHING

Figure 10-13. A 1ϕ motor may be connected to the contacts of a drum switch to change the direction of rotation from forward to reverse.

Reversing DC Motors Using Drum Switches

The direction of rotation of any DC series, shunt, compound, or permanent-magnet motor may be reversed by reversing the direction of the current through the fields without changing the direction of the current through the armature, or by reversing the direction of the current through the armature without changing the direction of the current through the fields. The industrial standard is to reverse the direction of current through the armature. A drum switch may be connected to change the direction of rotation of any DC series, shunt, compound, or permanent-magnet motor. **See Figure 10-14.** In each circuit, the current through the armature is changed. Some DC motors have commutating windings (interpoles) that are used to prevent sparking at the brushes in the motor. For this reason, the armature circuit (armature and commutating windings) must be reversed on all DC motors with commutating windings to reverse the direction of motor rotation.

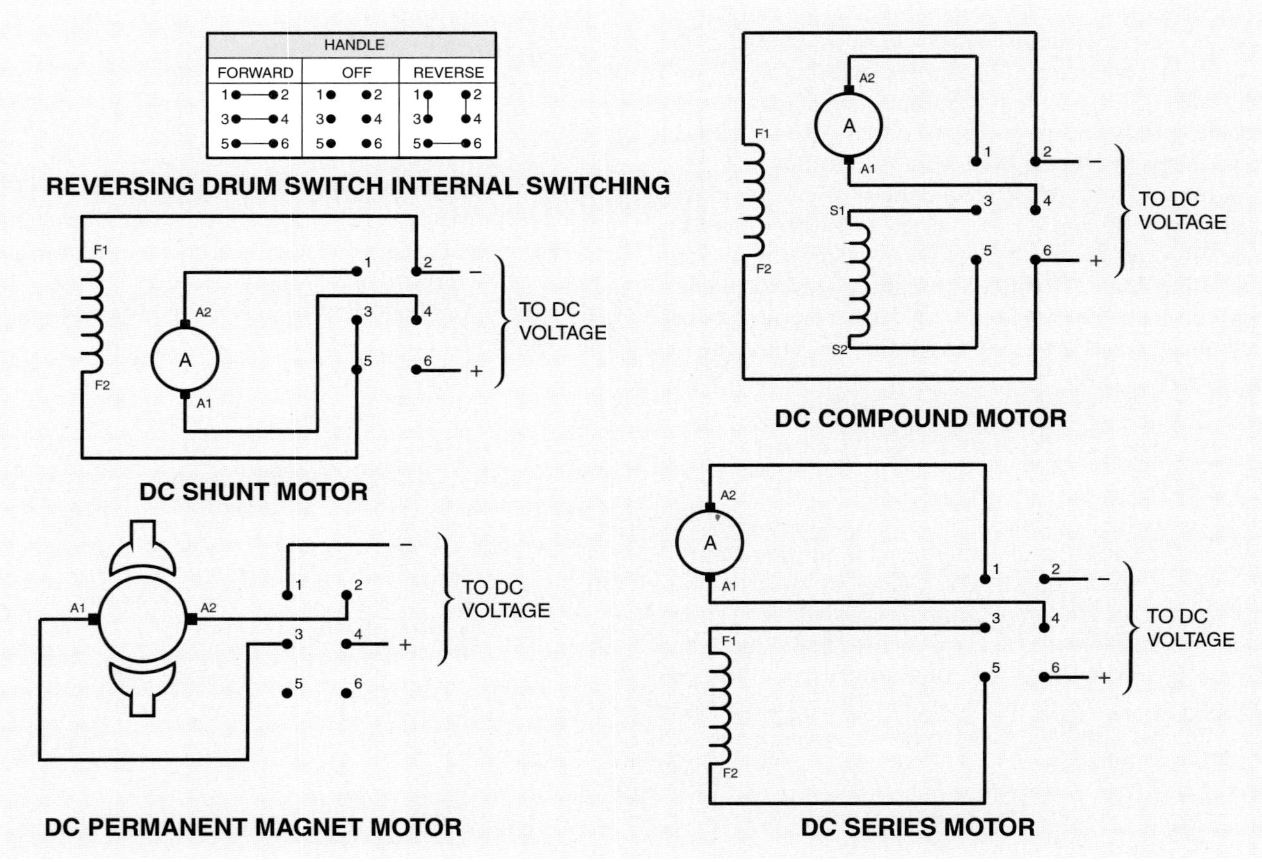

Figure 10-14. A drum switch may be connected to change the direction of rotation of any DC series, shunt, compound, or permanent-magnet motor.

Atlas Technologies Inc.

Multiple forward and reverse circuits are required on large industrial machine operations.

REVERSING MOTORS USING MAGNETIC STARTERS

A magnetic reversing starter performs the same function as a manual reversing starter. The only difference between manual and magnetic reversing starters is the addition of forward and reverse coils and the use of auxiliary contacts. **See Figure 10-15.** The forward and reverse coils replace the pushbuttons of a manual starter and the auxiliary contacts provide additional electrical protection and circuit flexibility. The reversing circuit is the same for both manual and magnetic starters.

Mechanical Interlocking

A magnetic reversing starter may be controlled by forward and reverse pushbuttons. **See Figure 10-16.** *Note:* A line diagram does not show the power contacts. The power contacts are found in the wiring diagram. The broken lines running from the forward coil to the reverse coil indicate that the coils are mechanically interlocked like those of a manual reversing starter. This mechanical interlock is normally factory-installed by the manufacturer.

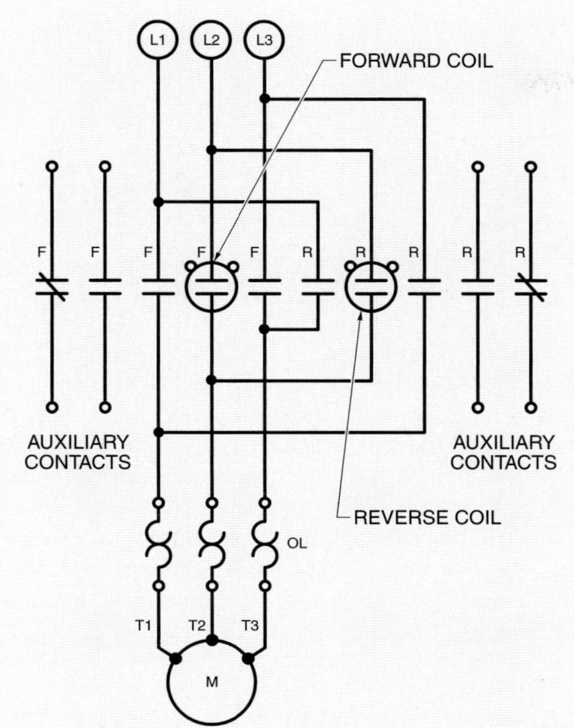

Figure 10-15. A magnetic reversing starter has forward and reverse coils, which replace the pushbuttons of a manual starter, and auxiliary contacts that provide additional electrical protection and circuit flexibility.

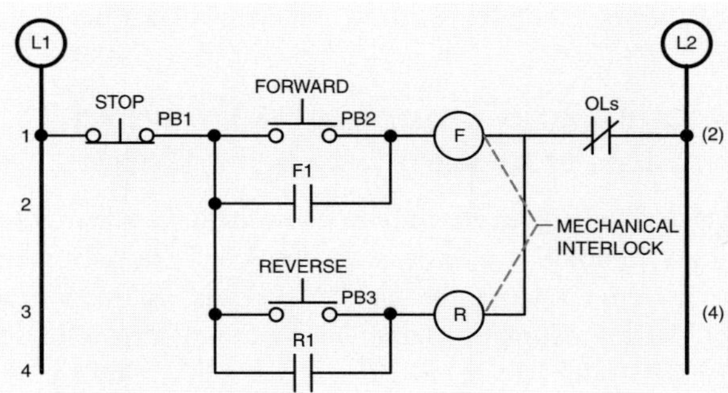

Figure 10-16. A magnetic reversing starter may be controlled by forward and reverse pushbuttons.

In this circuit, depressing forward pushbutton PB2 completes the forward coil circuit from L1 to L2, energizing coil F. Coil F energizes auxiliary contacts F1, providing memory. Mechanical interlocking keeps the reversing circuit from closing. Depressing stop pushbutton PB1 opens the forward coil circuit, causing coil F to de-energize and contacts F1 to return to their NO position. Depressing reverse pushbutton PB3 completes the reverse coil circuit from L1 to L2, energizing coil R. Coil R energizes auxiliary contacts R1, providing memory. Mechanical interlocking keeps the forward circuit from closing. Depressing stop pushbutton PB1 opens the reverse coil circuit, causing coil R to de-energize and contacts R1 to return to their NO position. Overload protection is provided in forward and reverse by the same set of overloads.

Auxiliary Contact Interlocking

Although most magnetic reversing starters provide mechanical interlock protection, some circuits are provided with a secondary backup or safety backup system that uses auxiliary contacts to provide electrical interlocking. **See Figure 10-17.**

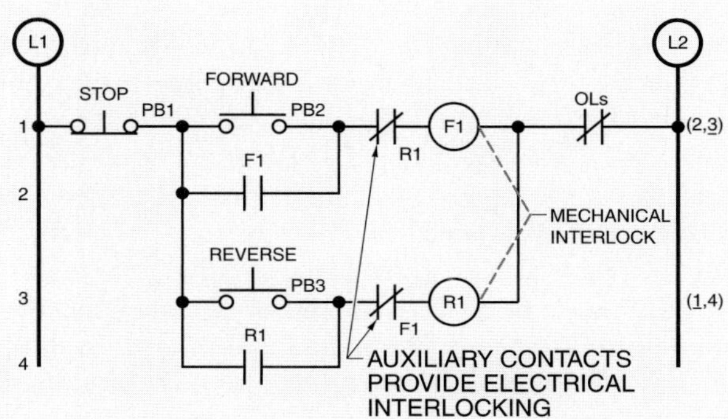

Figure 10-17. Although most magnetic reversing starters provide mechanical interlock protection, some circuits are provided with a secondary backup system that uses auxiliary contacts to provide electrical interlocking.

In this circuit, one NO set and one NC set of contacts are activated when the forward coil circuit is energized. The NO contacts close, providing memory, and the NC contacts open, providing electrical isolation in the reverse coil circuit. When the forward coil circuit is energized, the reverse coil circuit is automatically opened or isolated from the control voltage. Even if the reverse pushbutton is closed, no electrical path is available in the reverse circuit. For the reverse circuit to operate, the stop pushbutton must be pressed so that the forward circuit de-energizes and returns the NC contacts to their normal position. Depressing the reverse pushbutton provides the same electrical interlock for the reverse circuit when the forward contacts are in their normal position.

Pushbutton Interlocking

Pushbutton interlocking may be used with either or both mechanical and auxiliary interlocking. Pushbutton interlocking uses both NO and NC contacts mechanically connected on each pushbutton. **See Figure 10-18.**

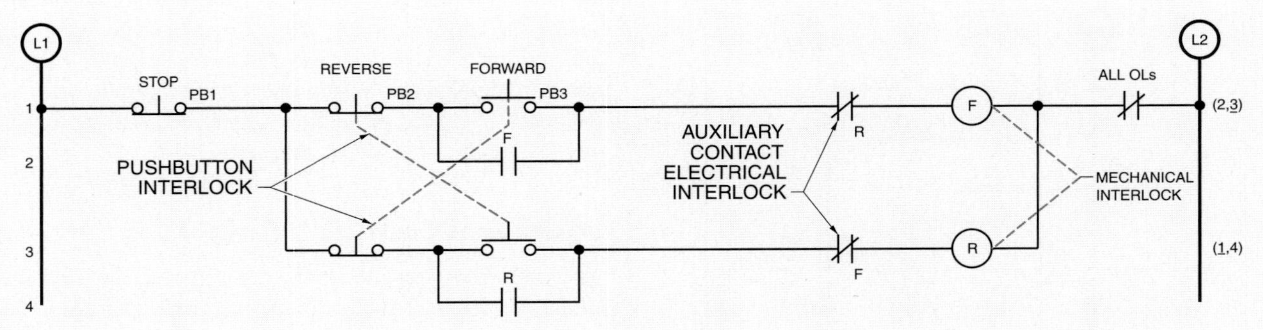

Figure 10-18. Pushbutton interlocking uses both NO and NC contacts mechanically connected on each pushbutton.

In this circuit, the NC contacts wired into the R coil circuit open, providing electrical isolation, when NO contacts on the forward pushbutton close to energize the F coil circuit. Conversely, the NC contacts wired into the F coil circuit open, providing electrical isolation, when the NO contacts on the reverse pushbutton close to energize the R coil circuit. Mechanical and auxiliary contact electrical interlocking is also provided in the circuit.

Caution: In many cases, motors or the equipment they are powering cannot withstand a rapid reversal of direction. Care must be exercised to determine the equipment that can be safely reversed under load. Also consider the braking that must be provided to slow the machine to a safe speed before reversal.

Reversing Power and Control Circuits

A power circuit and a control circuit are required when using motor starters or motor drives. The power circuit includes the incoming circuit main power, the motor starter (or drive), and the motor. The control circuit includes the required circuit inputs (pushbuttons, limit switches, etc.), motor starter coils, motor starter auxiliary contacts, overload contacts, timers, counters, and any other device designed to operate in the control circuit.

The control circuit is normally operated at a lower voltage than the power circuit. The low control circuit voltage is obtained by using a step-down transformer. When used with motor control circuits, this transformer is often referred to as the control transformer.

Technical Fact

Motor windings may be at line voltage even when the motor is not rotating. Use caution when troubleshooting or servicing a motor.

Although the power circuit and control circuit operate together to control the motor, they are electrically isolated from each other through the transformer. This electrical isolation allows individual control circuits to control different motor types (1ϕ motors, DC motors, and 3ϕ motors). **See Figure 10-19.**

MAGNETIC REVERSING STARTER APPLICATIONS

Many applications can be built around a basic magnetic reversing starter because magnetic reversing starters are controlled electrically. These circuits include the functions of starting and stopping motors in forward and reverse and controlling the motors with various control devices.

Starting and Stopping in Forward and Reverse with Indicator Lights

Operators are often required to know the direction of rotation of a motor at a given moment. A start/stop/forward/reverse circuit with indicator lights enables an operator to know the direction of rotation of a motor at any time. **See Figure 10-20.** An example is a motor controlling a crane which raises and lowers a load. The line diagram is capable of indicating, through lights, the direction the motor is operating. If an electrician adds nameplates, these lights could indicate up and down directions of the hoist.

In this circuit, pressing the momentary contact forward pushbutton causes the NO and NC contacts to move simultaneously. The NO contacts close, energizing coil F while the pushbutton is depressed. Coil F causes the memory contacts F to close and the NC electrical interlock to open, isolating the reversing circuit. The forward pilot light turns ON when holding contacts F are closed. For the period of time the pushbutton is depressed, the NC contacts of the forward pushbutton open and isolate the reversing coil R.

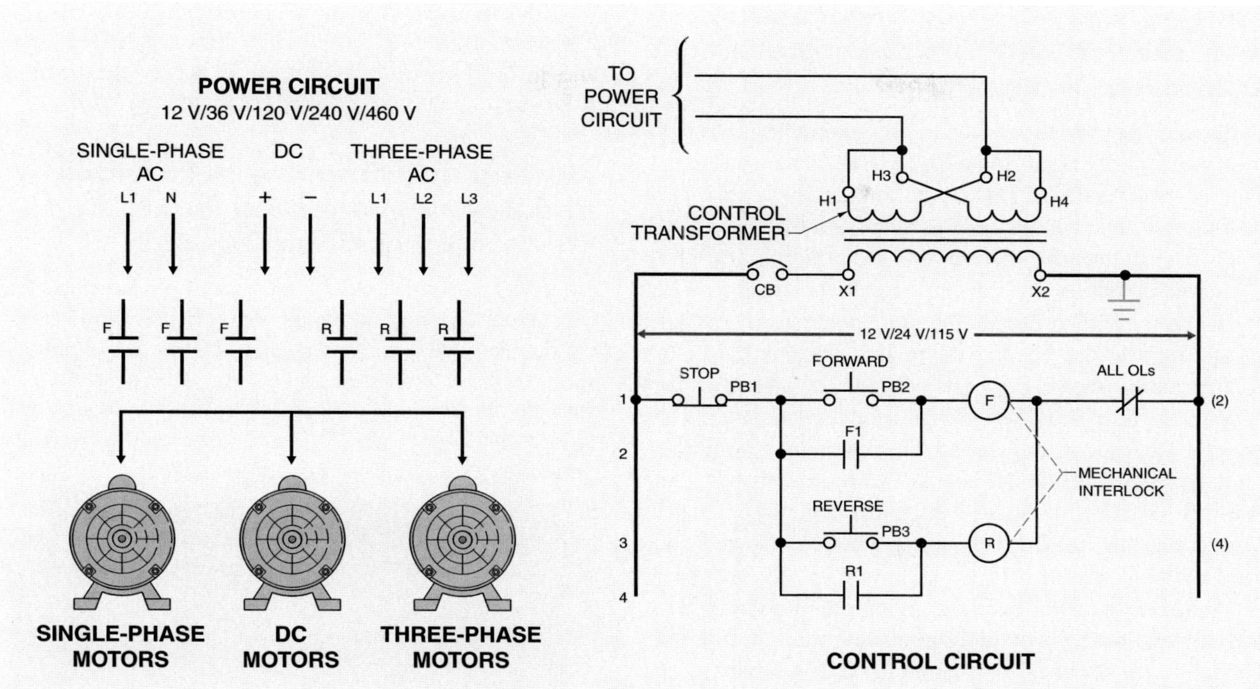

Figure 10-19. A control circuit can be used to control different motor types, such as 1ϕ motors, DC motors, and 3ϕ motors.

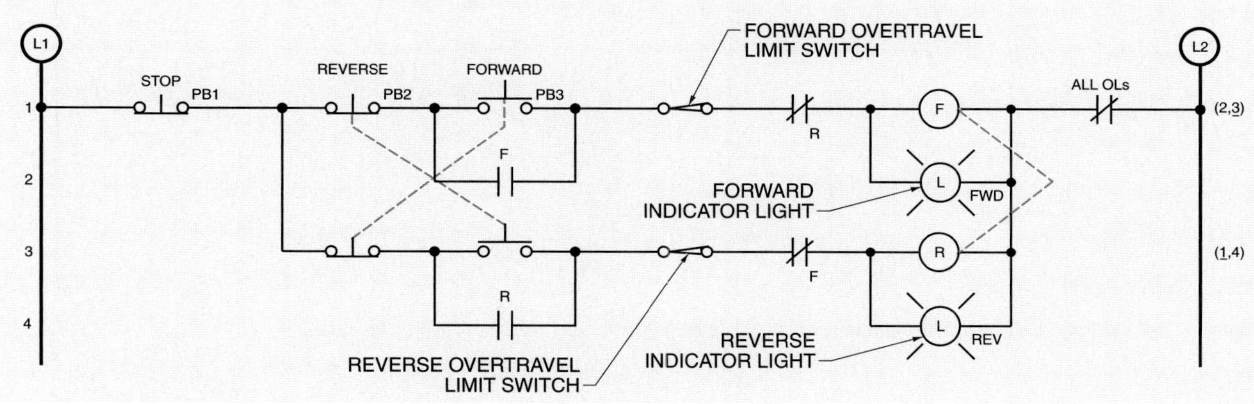

Figure 10-20. A start/stop/forward/reverse circuit with indicator lights enables an operator to know the direction of rotation of a motor at a given moment.

Pressing the momentary contact reverse pushbutton causes the NO and NC contacts to move simultaneously. The opening of the NC contacts de-energizes coil F. With coil F de-energized, the memory contacts F open and the electrical interlock F closes. The closing of the NO contacts energizes coil R. Coil R causes the holding contacts R to close and the NC electrical interlock to open, isolating the forward circuit. The reverse pilot light turns ON when the memory contacts R close. Pressing the stop pushbutton with the motor running in either direction stops the motor and causes the circuit to return to its normal state.

Overload protection for the circuit is provided by heater coils. Operation of the overload contacts breaks the circuit, opening the overload contacts. The motor cannot be restarted until the overloads are reset and the forward or reverse pushbutton is pressed.

This circuit provides protection against low voltage or a power failure. A loss of voltage de-energizes the circuit and hold-in contacts F or R open. This design prevents the motor from starting automatically after the power returns.

Starting and Stopping in Forward and Reverse with Limit Switches Controlling Reversing

Limit switches may be used to provide automatic control of reversing circuits. **See Figure 10-21.** This circuit uses limit switches and a control relay to automatically reverse the direction of a machine at predetermined points. This circuit could control the table of an automatic grinding machine where the operation must be periodically reversed.

In this circuit, pressing the start pushbutton causes control relay CR to become energized. The auxiliary CR contacts close when control relay CR is energized. One set of contacts form the holding circuit, and the other contacts connect the limit switch circuit. The motor runs when the limit switch circuit is activated. The motor runs in the forward direction if the forward limit switch is closed. The motor runs in the opposite direction if the reverse limit switch is closed.

Overload protection for the circuit is provided by heater coils. Operation of the overload contacts breaks the circuit. The motor cannot be restarted until the overloads are reset and the start pushbutton is pressed.

This circuit provides protection against low voltage or a power failure. A loss of voltage de-energizes the circuit and the hold-in contacts CR open. This prevents the motor from starting automatically after the power returns.

Starting and Stopping in Forward and Reverse with Limit Switch as Safety Stop in Either Direction

For safety reasons, it may be necessary to ensure that a load controlled by a reversing motor does not go beyond certain operating points in the system. For example, a hydraulic lift should not rise too high. Limit switches are incorporated to shut the operation down if a load travels far enough to be unsafe. **See Figure 10-22.** The circuit provides overtravel protection through the use of limit switches.

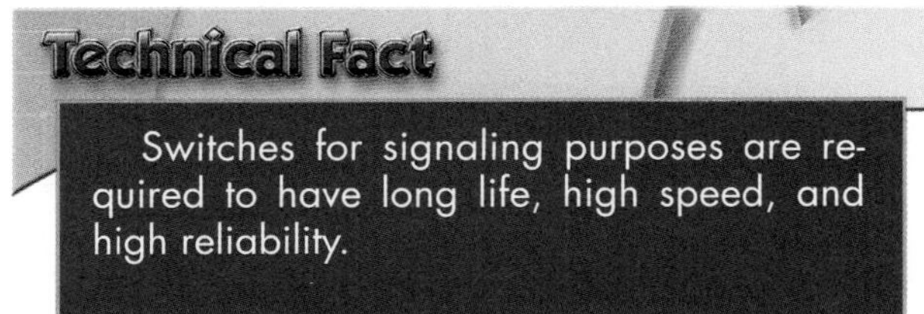

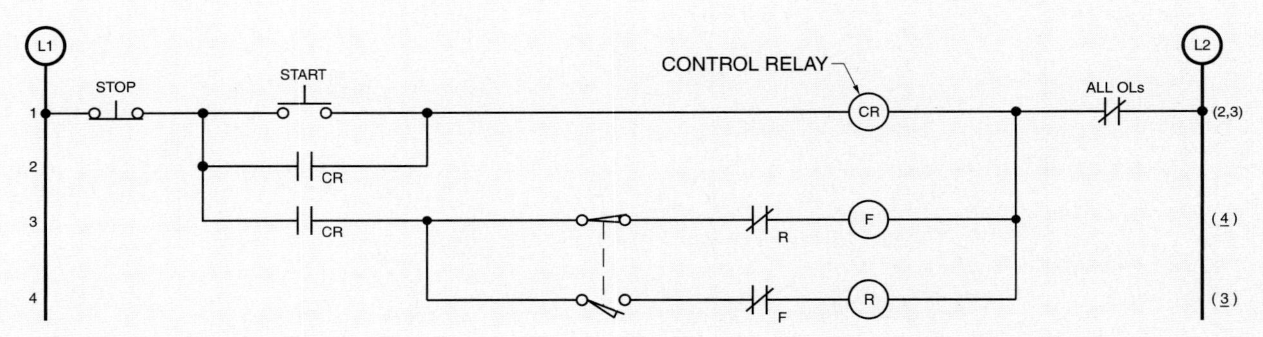

Figure 10-21. Limit switches may be used to provide automatic control of reversing circuits.

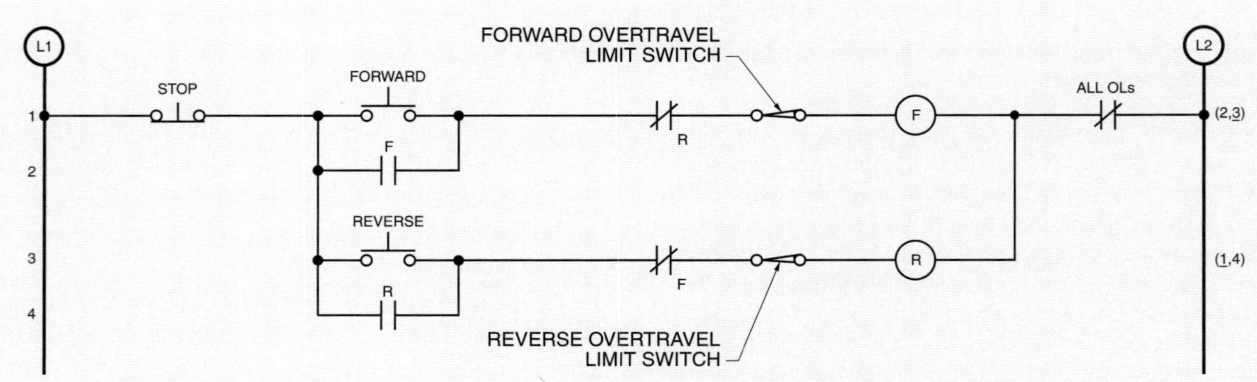

Figure 10-22. Limit switches may be used in a circuit to provide overtravel protection.

In this circuit, pressing the forward pushbutton activates coil F. Coil F pulls in the holding contacts F and opens the electrical interlock F, isolating the reversing circuit. The motor runs in the forward direction until either the stop pushbutton is pressed or the limit switch is activated. The circuit is broken and the holding contacts and electrical interlock return to their normal state if either control is activated.

Pressing the reverse pushbutton activates coil R. Coil R pulls in the holding contacts R and opens the electrical interlock R, isolating the forward circuit. The motor runs in the reverse direction until either the stop pushbutton is pressed or the limit switch is activated. The circuit is broken and the holding contacts and electrical interlock return to their normal state if either control is activated. The circuit may still be reversed to clear a jam or undesirable situation if either limit switch is opened. This allows the operator to operate the motor in the direction opposite the tripped overtravel limit switch to reset the tripped limit switch.

Overload protection for the circuit is provided by heater coils. Operation of the overload contacts breaks the circuit. The motor cannot be restarted until the overloads are reset and the forward pushbutton is pressed.

This circuit provides protection against low voltage or a power failure in that a loss of voltage de-energizes the circuit and holding contacts F or R open. This prevents the motor from starting automatically after the power returns.

Selector Switch Used to Determine Direction of Motor Travel

A selector switch and a basic start/stop station can be used to reverse a motor. **See Figure 10-23.** The motor can be run in either direction, but the desired direction must be set by the selector switch before starting.

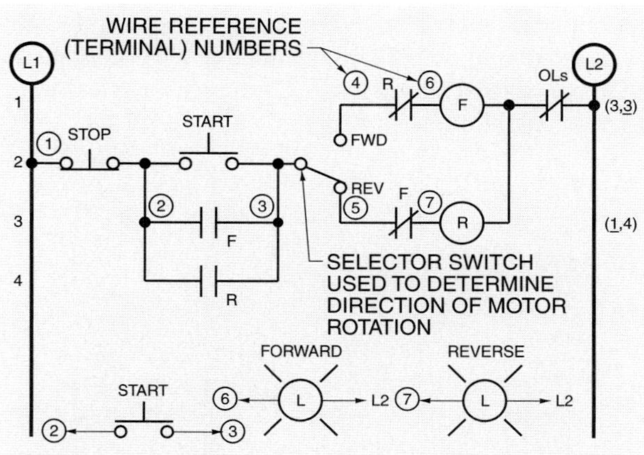

Figure 10-23. A selector switch and a basic start/stop station can be used to reverse a motor.

Troubleshooting of motor control circuits is normally performed at the motor control center.

In this circuit, pressing the start pushbutton with the selector switch in the forward position energizes coil F. Coil F closes the holding contacts F and opens the electrical interlock F, isolating the reversing circuit. Pressing the stop pushbutton de-energizes coil F, which releases the holding contacts and the electrical interlock. Pressing the start pushbutton with the selector switch in the reverse position energizes coil R. Coil R closes the holding contacts R and opens the electrical interlock R, isolating the forward circuit.

Overload protection for the circuit is provided by heater coils. Operation of the overload contacts breaks the circuit. The motor cannot be restarted until the overloads are reset and the start pushbutton is pressed.

This circuit provides protection against low voltage or a power failure in that a loss of voltage de-energizes the circuit and hold-in contacts F or R open. This prevents the motor from starting automatically after the power returns.

This circuit also illustrates the proper connections for adding forward and reverse indicator lights. The forward indicator light is connected to wire 6 and L2. The reverse indicator light is connected to wire 7 and L2. Additional start pushbuttons are connected to wires 2 and 3. It is standard industrial practice to mark the NO memory contacts 2 and 3. It is also standard industrial practice to mark the wire coming from the forward coil and leading to the NC reverse contact (used for interlocking) as wire 6. Likewise, the wire coming from the reverse coil and leading to the NC forward contact is marked 7. These numbers are usually printed on the magnetic starters to help in wiring the circuit.

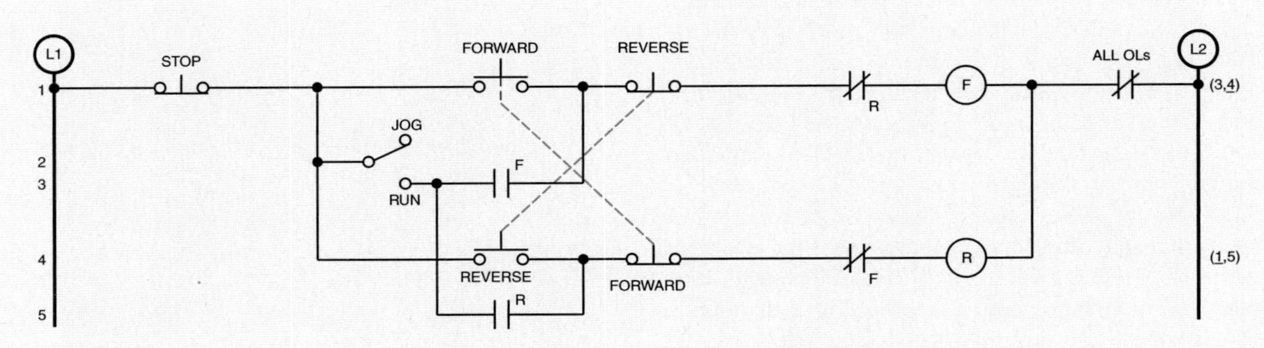

Figure 10-24. A jogging circuit allows the operator to start a motor for short times without memory.

Starting, Stopping, and Jogging in Forward and Reverse with Jogging Controlled through a Selector Switch

In certain industrial operations, it may be necessary to reposition equipment a little at a time for small adjustments. A jogging circuit allows the operator to start a motor for short times without memory. **See Figure 10-24.** *Jogging* is the frequent starting and stopping of a motor for short periods of time.

In this circuit, small adjustments may be made in forward and reverse motor rotation or in continuous operation, depending on the position of the selector switch. Pressing the forward pushbutton with the selector switch in the run position activates coil F. Coil F pulls in the NO holding contacts F and opens the NC electrical interlock F, isolating the reversing circuit. The motor starts and continues to run. Pressing the reverse pushbutton with the selector switch in the run position activates coil R. Coil R pulls in the NO holding contacts R and opens the NC electrical interlock R, isolating the forward circuit. The motor starts in the reverse direction and continues to run.

Pressing the stop pushbutton in either direction breaks the circuit and returns the circuit contacts to their normal positions. Pressing the forward pushbutton with the selector switch in the jog position activates coil F and the motor only for the period of time that the forward pushbutton is depressed. In addition, the NC electrical interlock F opens and isolates the reversing circuit. Pressing the reverse pushbutton with the selector switch in the jog position activates coil R and the motor only for the period of time that the reverse pushbutton is depressed. In addition, the NC electrical interlock R opens and isolates the forward circuit.

Overload protection is provided by heater coils. Operation of the overload contacts breaks the circuit. The motor cannot be restarted until the overloads are reset and the start pushbutton is pressed.

This circuit provides protection against low voltage or a power failure in that a loss of voltage de-energizes the circuit and hold-in contacts F or R open. This prevents the motor from starting automatically after the power returns.

PLC REVERSING CIRCUITS

A PLC can be used to control a forward and reversing motor application. The advantage of using a PLC is that the PLC simplifies the circuit by eliminating much of the wiring and required components. For example, a PLC can eliminate the need for normally closed (NC) auxiliary contacts on the starter that are used for interlocking. **See Figure 10-25.**

In a hard-wired reversing circuit, auxiliary contact interlocking is accomplished by wiring an NC forward contact in series with the reverse starter coil and wiring an NC reversing contact in series with the forward starter coil. This requires additional wires to be connected into the system. In a PLC reversing circuit, no auxiliary contacts are required on the starters. Interlocking is accomplished by programming an NC output into the control circuit. The NC contacts exist only in the PLC program so no additional wiring is required.

Technical Fact

When troubleshooting a PLC, the first step is to identify the problem and the source of the problem. The source of the problem can usually be identified as the I/O hardware, the wiring, the processor module, the machine inputs or outputs, or the logic program.

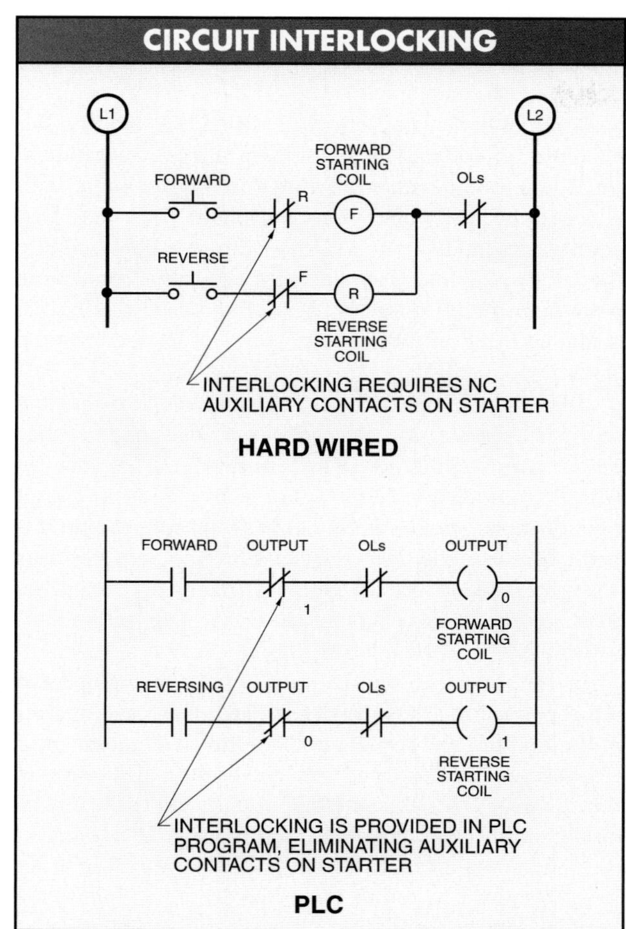

Figure 10-25. A PLC can eliminate the need for normally closed (NC) auxiliary contacts on the starter that are used for interlocking.

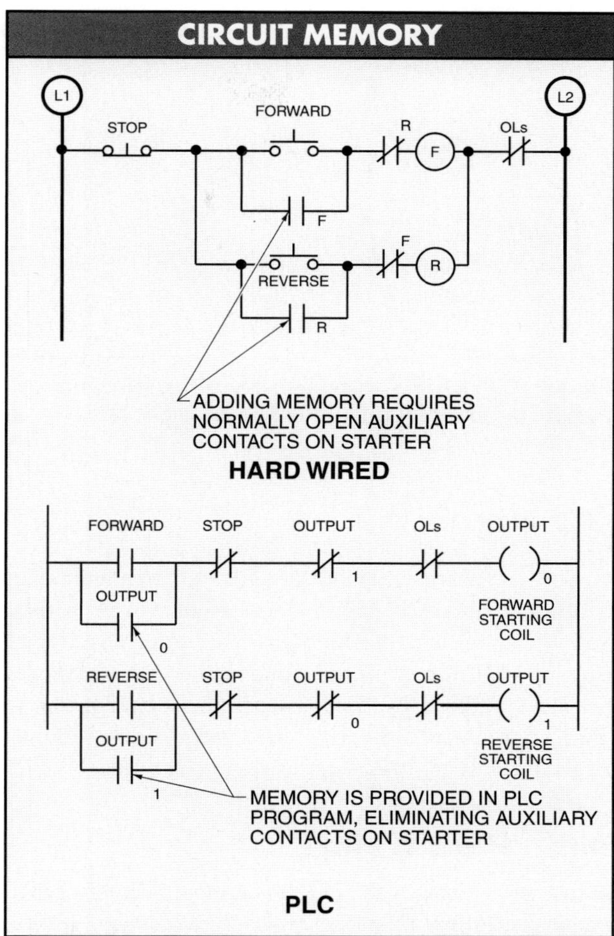

Figure 10-26. A PLC can eliminate the need for NO auxiliary contacts on the starter that are used for developing circuit memory.

Likewise, to produce circuit memory in a forward and reversing circuit, normally open (NO) auxiliary contacts are required. The auxiliary contacts must be included on the starter and wired into the control circuit. A PLC can eliminate the need for NO auxiliary contacts on the starter that are used for developing circuit memory. **See Figure 10-26.**

In a hard-wired reversing circuit, memory is accomplished by wiring an NO forward contact in parallel with the forward pushbutton and wiring an NO reversing contact in parallel with the reverse pushbutton. This requires additional wires to be connected into the system. In a PLC reversing circuit, no auxiliary contacts are required on the starters. Memory is accomplished by programming an NO output into the control circuit. The NO contacts exist only in the PLC program so no additional wiring is required.

The wiring of the input pushbuttons and the output starting coils is simplified because the PLC program eliminates the need for auxiliary contacts on the starter. The stop, forward, and reverse pushbuttons are wired to the input section of the PLC. The forward and reverse starting coils are wired to the output section of the PLC. **See Figure 10-27.**

MOTOR CONTROL WIRING METHODS

A motor must have a method of control in order to operate safely and efficiently. Motor control circuits vary from simple to complex. Reversing motor control circuits, like nonreversing motor control circuits, can be wired using manual controls (manual starters, drum switches), magnetic controls (magnetic starters), motor drives, or PLCs to control the operation of a motor.

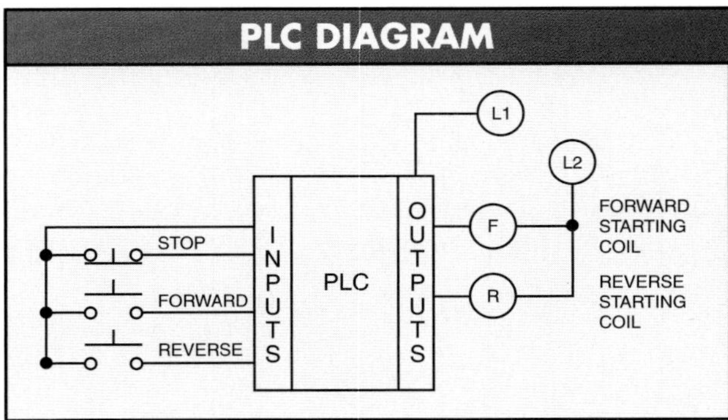

Figure 10-27. A PLC simplifies the wiring of inputs and outputs by eliminating the need for auxiliary contacts on the starter.

Several different methods of wiring a motor and motor control circuit are available. These methods can be used individually or in combination to control the operation of a motor. Each motor control wiring method has advantages and disadvantages. The four basic methods of motor control wiring are direct hard wiring, hard wiring using terminal strips, PLC wiring, and electric motor drive wiring.

Direct Hard Wiring

Direct hard wiring is the oldest and most straightforward motor control wiring method used. In direct hard wiring, the power circuit and the control circuit are wired point-to-point. **See Figure 10-28.** *Point-to-point wiring* is wiring in which each component in a circuit is connected (wired) directly to the next component as specified on the wiring and line diagrams. For example, the transformer X1 terminal is connected directly to the fuse, the fuse is connected directly to the stop pushbutton, the stop pushbutton is connected directly to the reverse pushbutton, the reverse pushbutton is connected directly to the forward pushbutton, and so on until the final connection from the overload (OL) contact is made back to the transformer X2 terminal.

A direct hard wired circuit may operate properly for a period of time. The disadvantage of a direct hard wired circuit is that circuit troubleshooting and circuit modification are time consuming.

For example, when a problem occurs in a direct hard wired circuit, the circuit operation must be understood, measurements taken, and the problem identified. Circuit operation can be understood from a wiring diagram. Without a wiring diagram, the circuit wiring is determined by tracing each wire throughout the circuit. The circuit problem can eventually be found; however, tracing each wire in a circuit to find the wire with a problem is time consuming. Time is saved as experience is gained from working on a circuit several times and understanding its operation and components.

A direct hard wired circuit is difficult to modify. For example, if a forward indicator lamp and a reverse indicator lamp are to be added to a motor control circuit, their exact connection points must be found. Once the exact connection points are found, the lamps can be wired into the control enclosure. Even when the exact connection points are found, problems may arise when making the actual connection (not enough room under the terminal screw, etc.).

Some circuit modifications, such as adding forward and reverse indicator lamps, may not be a problem because they only require adding new wires. In this modification, old wires do not need to be moved or removed. Some circuit modifications, such as adding limit switches, are more difficult. For example, if forward and reverse limit switches are to be added to a circuit, some wiring must be removed from the circuit. In addition, the new wiring for the limit switches must be added. **See Figure 10-29.**

In this circuit, before the limit switches are added, the wires connecting the normally closed interlock contacts of the forward and reverse coils to the pushbuttons have to be removed (or opened) and the limit switch wired in the opening. To do this, the individual making the circuit modification must have a wiring diagram of the circuit (or understand the circuit from past experience) in order to know which wires to open and where to locate the limit switches.

Baldor Electric Co.

Conveyor systems used to move material from one location to another are typically direct hard wired systems because they do not require the memory and decision-making capabilities of a PLC system.

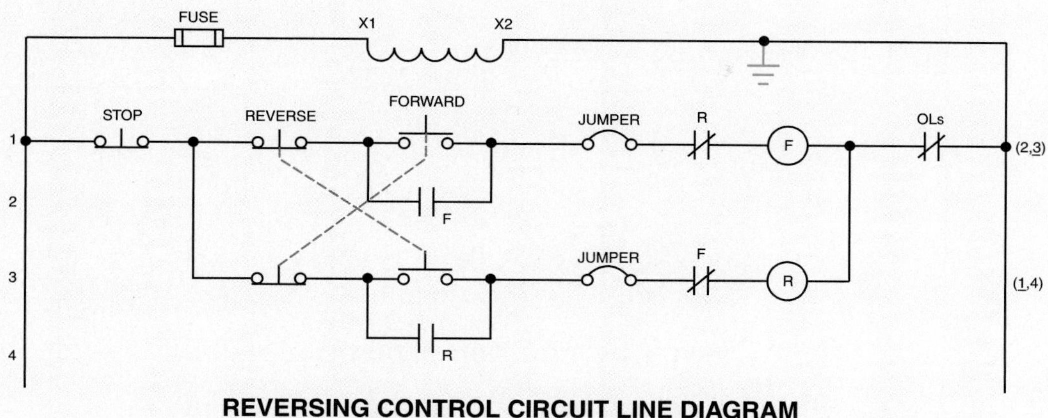

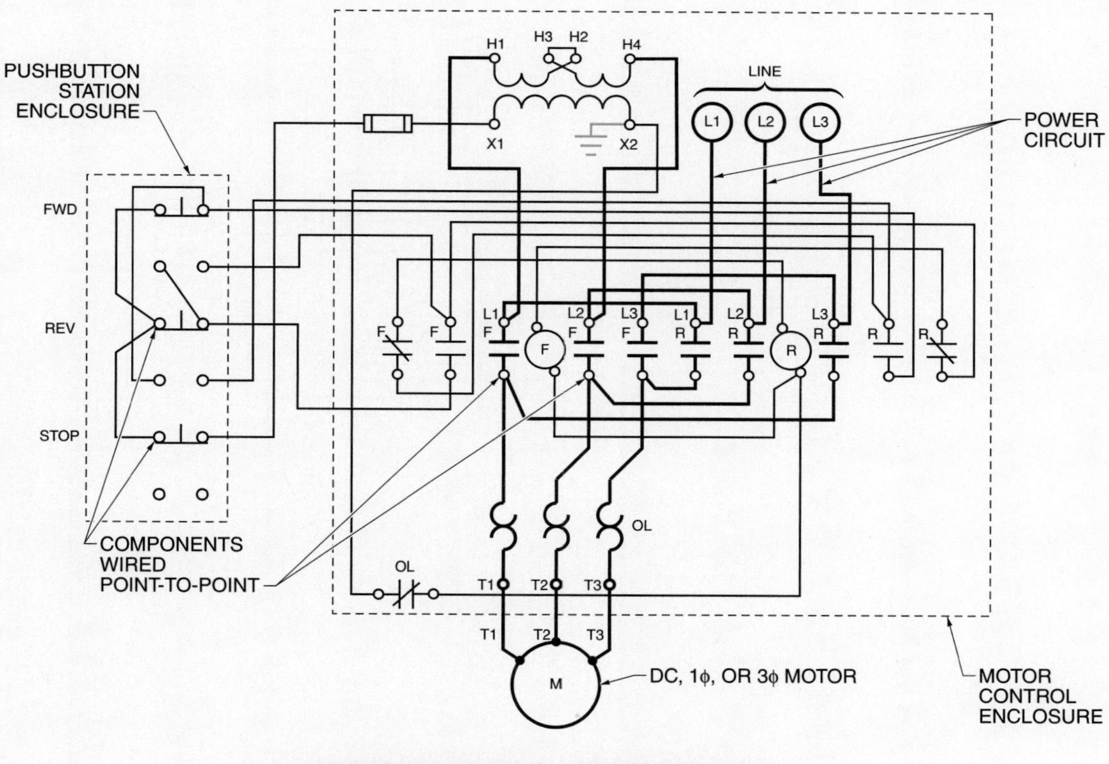

Figure 10-28. In direct hard wiring, the power circuit and the control circuit are wired point-to-point.

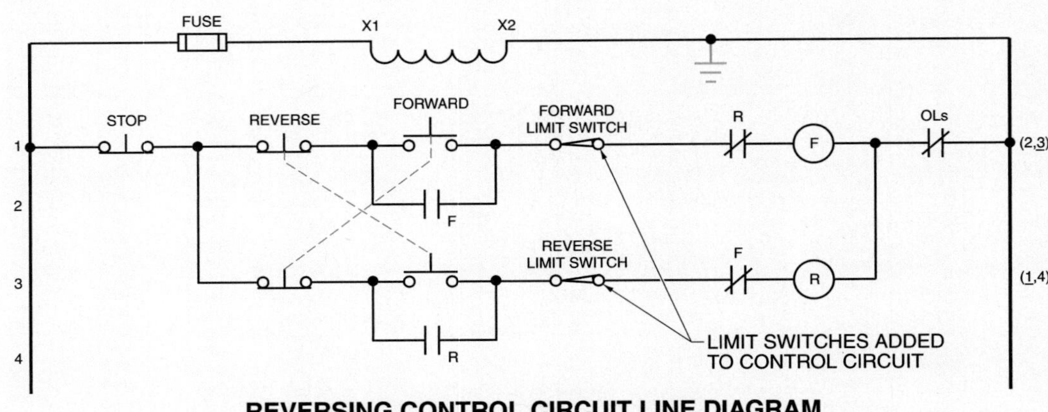

REVERSING CONTROL CIRCUIT LINE DIAGRAM

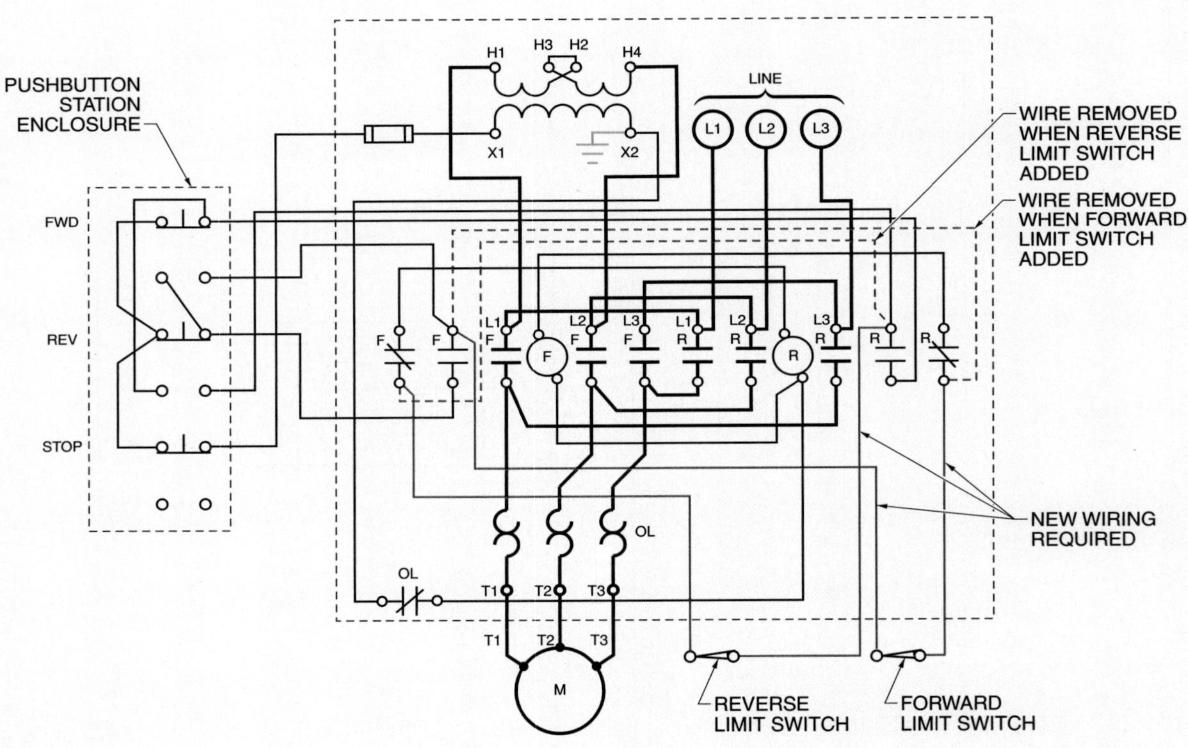

REVERSING CIRCUIT WIRING DIAGRAM

Figure 10-29. In direct hard wired circuits, circuit modifications may require the removal and/or addition of circuit wiring.

Hard Wiring Using Terminal Strips

Hard wiring to a terminal strip allows for easy circuit modification and simplifies circuit troubleshooting. When wiring using a terminal strip, each wire in the control circuit is assigned a reference point on the line diagram to identify the different wires that connect the components in the circuit. Each reference point is assigned a wire reference number. **See Figure 10-30.**

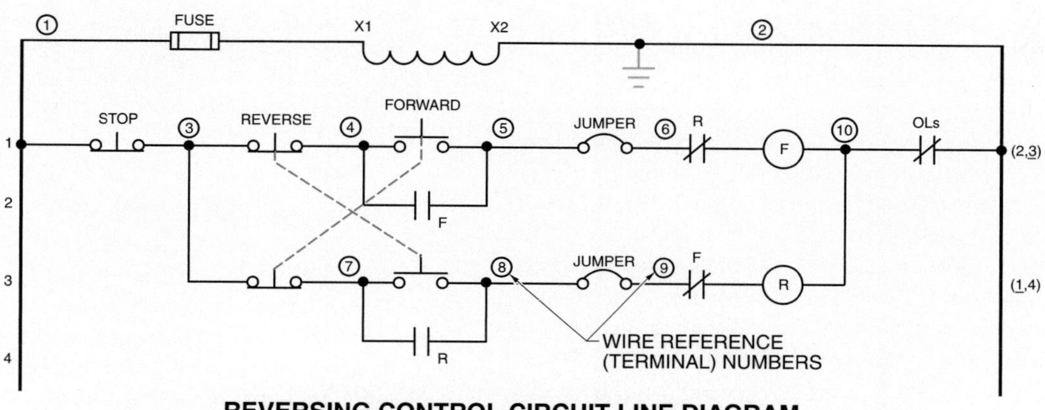

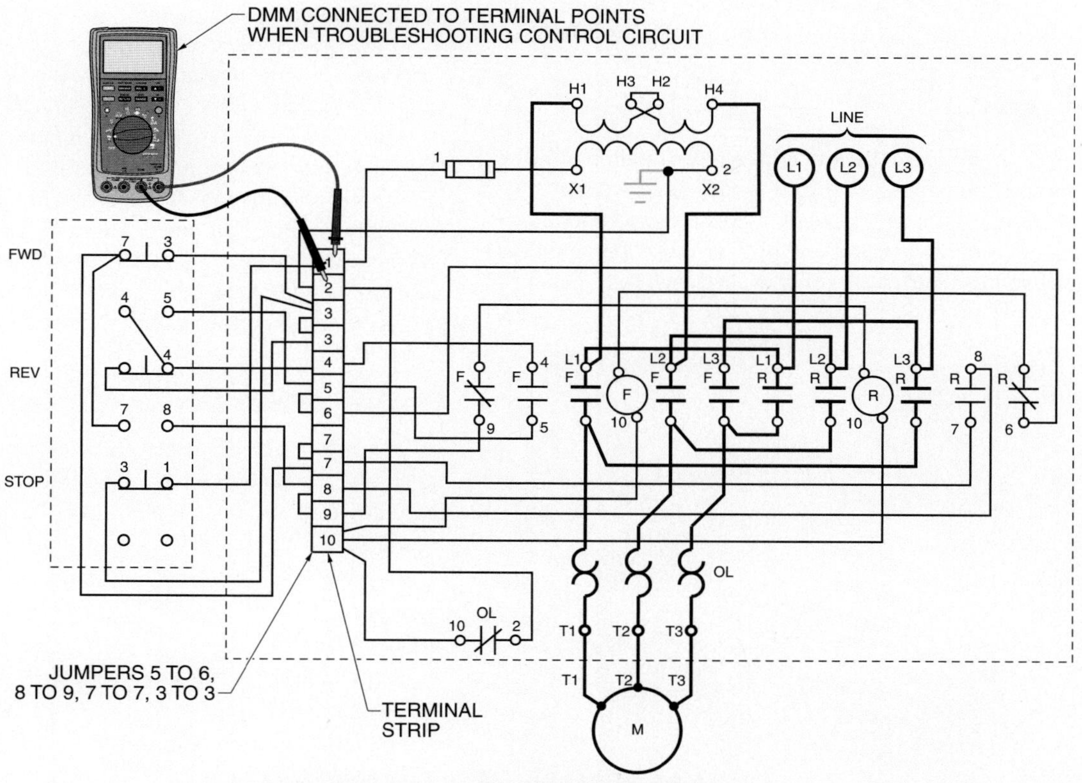

Figure 10-30. When hard wiring a circuit using a terminal strip, each wire in the control circuit is assigned a reference point on the diagram to identify the different wires that connect the components in the circuit.

Wire reference numbers were commonly assigned from the top left to the bottom right. However, most new diagrams have the power line on the left (usually L1 or X1) assigned the number 1, and the power line on the right (usually L2 or X2) number 2. This way the control circuit voltage can always be found at terminal 1 and at terminal 2. This aids when troubleshooting a circuit. If several connections of a given number are required, jumpers can be added to the terminal strip to provide multiple connection points to one given terminal number.

When troubleshooting a circuit with a terminal strip, the troubleshooter can go directly to the terminal strip and take measurements to help isolate the problem. The DMM is first placed on terminals 1 and 2. If the voltage is not correct at that point, the problem is located on the primary side of the transformer. If the voltage is correct at terminals 1 and 2, one DMM lead is left on terminal 2 and the other lead is moved to different terminals until the problem is located.

In addition to the terminal strip and wire reference numbers being an aid when troubleshooting, they also make circuit modification easier. This is because most, if not all, of the wires required to make the change are disconnected and reconnected at the terminal strip. **See Figure 10-31.**

PLC Wiring

Using a PLC for motor control allows for greater flexibility and circuit monitoring of a motor and control circuit. A PLC can monitor and control all motor control functions, but cannot directly monitor and display motor parameters such as voltage, current, frequency, and power.

Omron Electronics, Inc.

Programmable controllers can be designed as rack systems to provide multiple control options for the circuit.

Technical Fact

The difference between hard wired logic and programmable logic is that hard wired logic is fixed. Changes can be made only by altering the way devices are connected. Programmable logic is based on basic logic functions, which are programmable and can be easily changed.

When using a PLC to control a circuit, the PLC is connected into the control circuit. The power circuit does not change. What does change is that the control circuit inputs (pushbuttons, limit switches, and overload contacts) are wired to the PLC input module and the control circuit outputs (motor starter coils and indicator lamps) are wired to the PLC output module. **See Figure 10-32.**

The circuit operation (logic) is programmed using the PLC software and the circuit is downloaded to the PLC. The PLC software can monitor and display the condition (ON or OFF) of the circuit inputs and outputs. If changes in the control circuit are required, they can be reprogrammed and downloaded without changing the circuit wiring.

When using a PLC to program inputs, the actual input type (normally open or normally closed) and the way the input is programmed must be considered. This is because when using a PLC, an input can be wired normally open and programmed either normally open or normally closed. Likewise, an input can be wired normally closed and programmed either normally closed or normally open.

Electric Motor Drive Wiring

An electric motor drive can be used to control various functions of a motor. These functions normally include the following:

- Starting
- Stopping
- Jogging
- Speed control
- Motor direction control
- Acceleration time
- Deceleration time
- Overload protection
- Braking force
- Programmable output contacts
- Voltage, current, power, and frequency metering and display
- Preselection of multiple remote controls

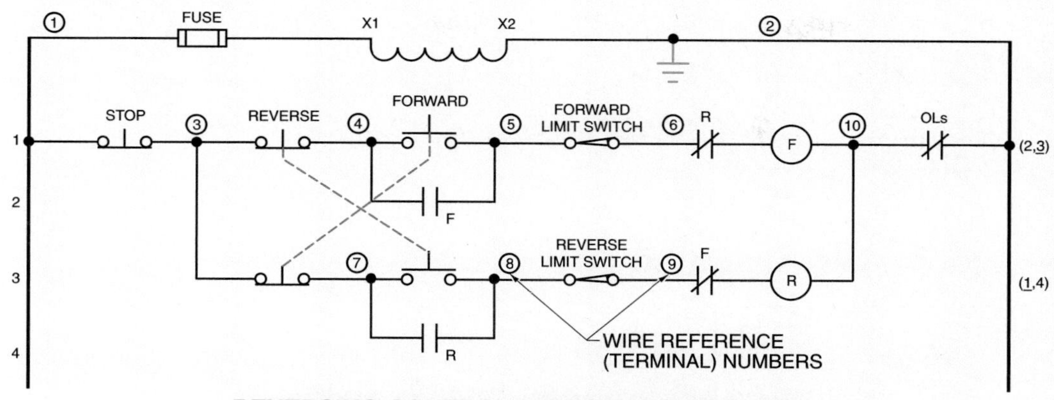

REVERSING CONTROL CIRCUIT LINE DIAGRAM

NOTE: REMOVE THIS JUMPER WHEN WIRING FORWARD LIMIT SWITCH

NOTE: REMOVE THIS JUMPER WHEN WIRING REVERSE LIMIT SWITCH

JUMPERS 5 TO 6, 8 TO 9, 7 TO 7, 3 TO 3

OPTIONAL WIRING

REMOVE JUMPERS 5 TO 6 AND 8 TO 9

FORWARD LIMIT SWITCH

REVERSE LIMIT SWITCH

REVERSING CIRCUIT WIRING DIAGRAM

Figure 10-31. Terminal strips and wire reference numbers enable easy circuit modification because most wires required to make a change are disconnected and reconnected at the terminal strip.

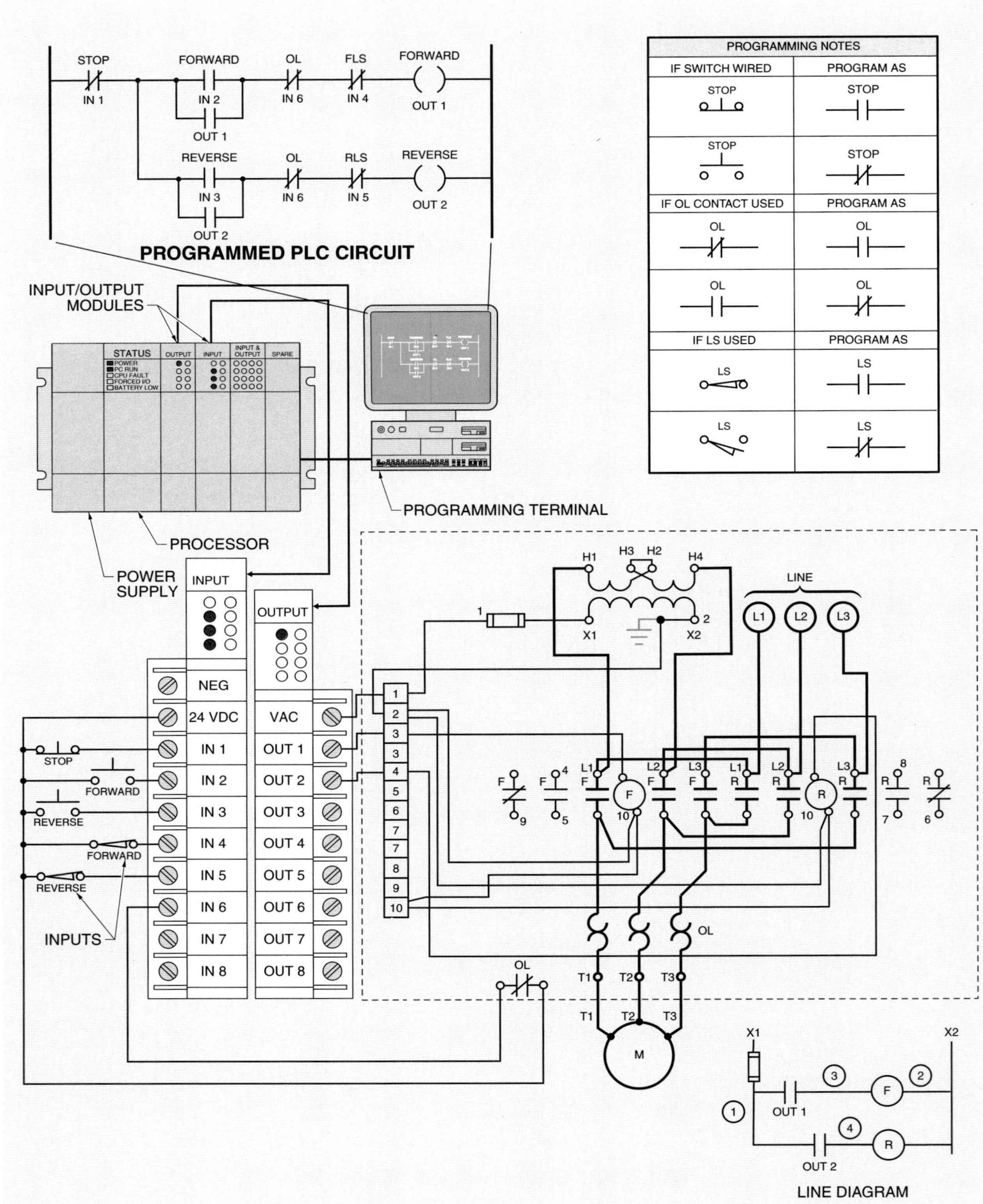

Figure 10-32. When using a PLC to control a circuit, the control circuit inputs are wired to the PLC input module and the control circuit outputs are wired to the PLC output module.

Motor drives eliminate the need for forward and reversing starters because the motor drive can be used to select motor direction. The direction of the motor can be selected using the keypad on the motor drive or external pushbuttons connected to the drive input terminals. **See Figure 10-33.** The motor drive internal circuit and parameter settings can be used to prevent changing motor direction before the motor has come to a full stop.

When wiring the power circuit using a motor drive, the incoming power is connected to L1, L2, and L3 (R, S, and T). The motor is connected to T1, T2, and T3 (U, V, and W). The motor ground, drive ground, and power supply ground are all connected to form a common ground. The power circuit is simplified when using a motor drive because there is no need for a reversing part of the power circuit. All reversing functions are performed internally within the drive. **See Figure 10-34.**

Pandjiris, Inc.

Motor drives control the direction of a motor, such as on a positioner for a welding machine, which is used to position parts to be welded.

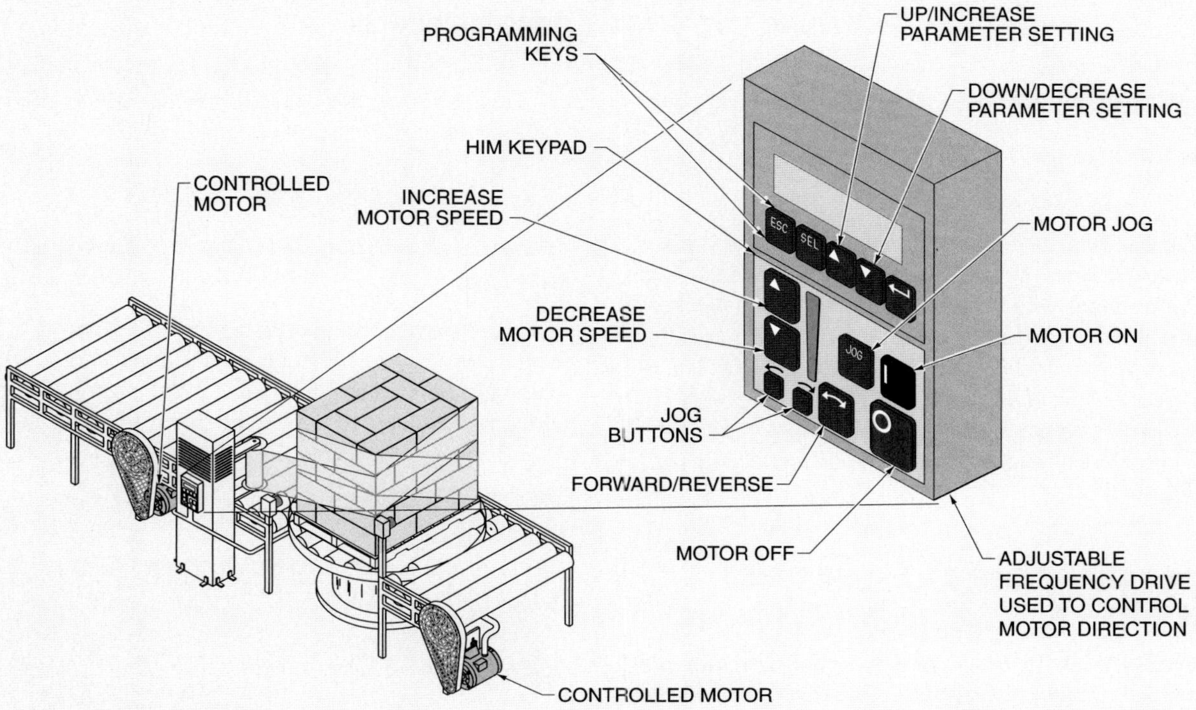

Figure 10-33. A motor drive may be used to select the direction of a motor.

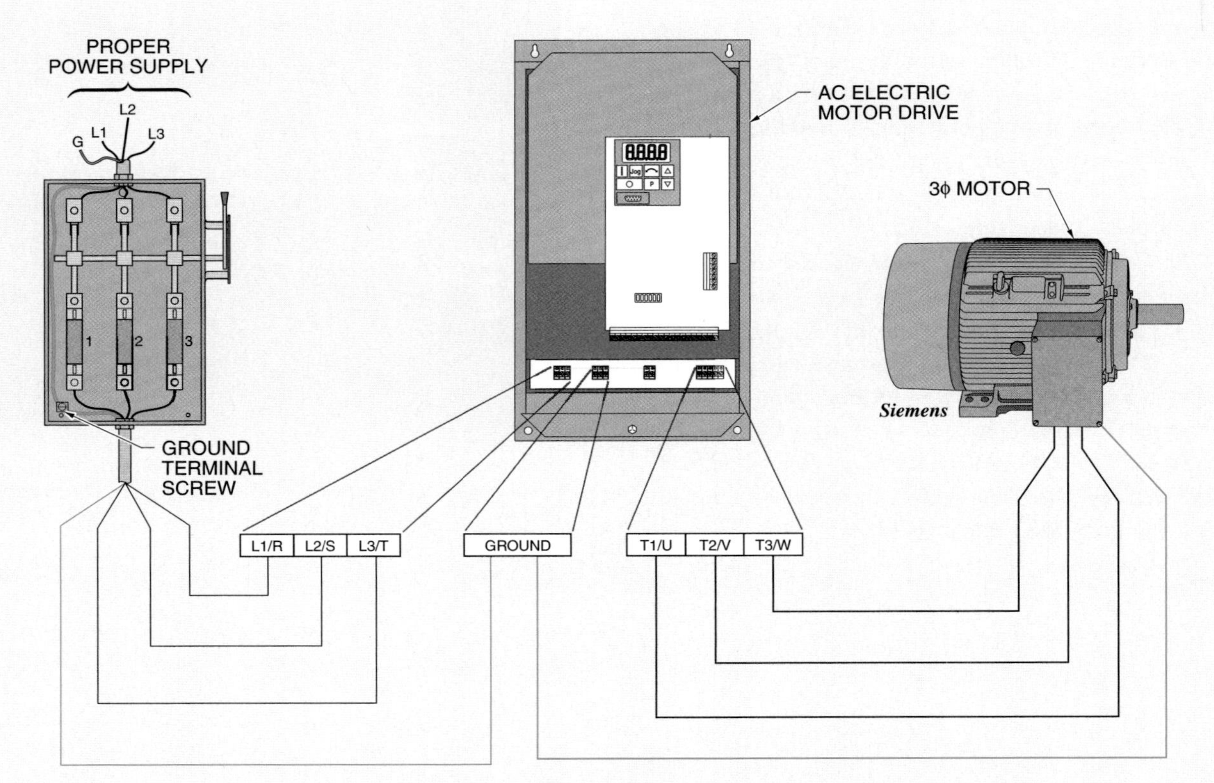

Figure 10-34. A motor drive simplifies a reversing power circuit because all reversing functions are performed internally within the drive.

When wiring the control circuit using a motor drive, the control devices (pushbutton, etc.) are wired to the drive control terminal strip. The power at the terminal strip is usually already stepped down to less than 30 V, making the control circuit safe and simple to wire. A PLC can be used to control the motor through the motor drive, either by using the PLC output contact or by using direct PLC/drive communication through a designated port (serial port, etc.). **See Figure 10-35.**

TROUBLESHOOTING REVERSING CIRCUITS

When a reversing motor circuit does not operate properly, the problem may be electrical or mechanical. The control circuit and power circuit are tested using a DMM to check for proper electrical operation. Troubleshooting starts inside the control cabinet when testing reversing control circuits or power circuits.

Troubleshooting Reversing Control Circuits

When troubleshooting reversing control circuits, a line diagram is used to illustrate circuit logic, and a wiring diagram is used to locate the actual test points at which a DMM is connected to the circuit. **See Figure 10-36.**

Technical Fact

Reversing the starting winding is the industry standard for reversing the direction of rotation for a 1φ motor. The direction of rotation of 3φ motors is reversed by interchanging any two of the 3φ power lines to the motor. The industrial standard is to interchange T1 and T3. This standard applies to all 3φ motors.

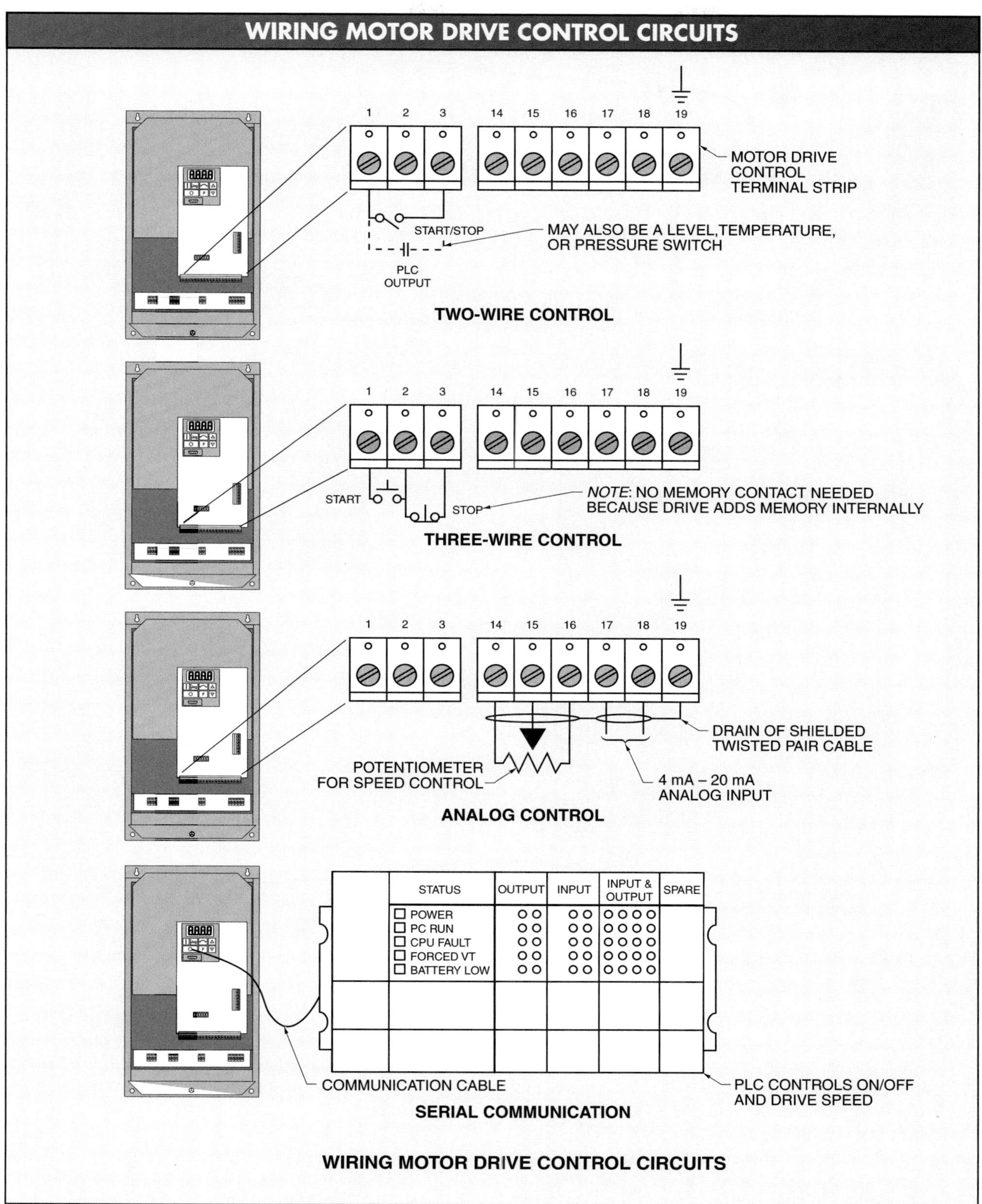

Figure 10-35. A PLC can be used to control a motor through a motor drive by either using the PLC output contact or by using direct PLC/drive communication through a designated port.

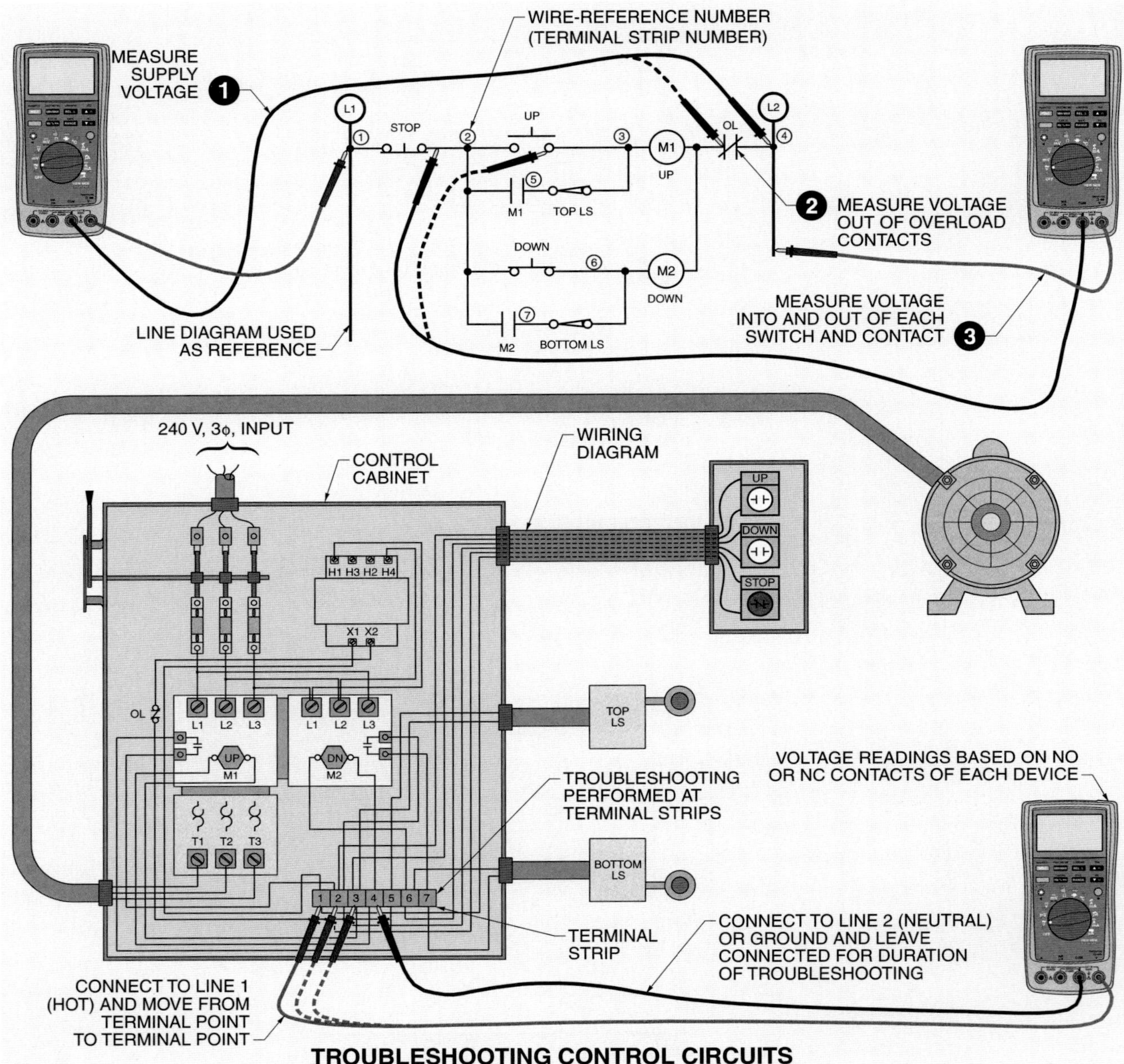

Figure 10-36. When troubleshooting reversing control circuits, a line diagram is used to illustrate circuit logic, and a wiring diagram is used to locate the actual test points at which a DMM is connected.

To troubleshoot a reversing control circuit, apply the following procedure:

1. Measure the supply voltage of the control circuit by connecting a DMM set to measure voltage between line 1 (hot conductor) and line 2 (neutral conductor). The voltage must be within 10% of the control circuit rating. Test the power circuit if the voltage is not correct. The control circuit voltage rating is determined by the voltage rating of the loads used in the control circuit (motor starter coils, etc.).
2. Measure the voltage out of the overload contacts to ensure the contacts are closed. The contacts are tripped or are faulty if no voltage is present. Reset the overloads if tripped. Overloads are installed to protect the motor during operation. The control circuit does not operate when the overloads are tripped.
3. Measure the voltage into and out of the control switch or contacts. Normally closed switches (stop pushbuttons, etc.) should have a voltage output before they are activated. Normally open switches (start pushbuttons, memory contacts, etc.) should have a voltage output only after they are activated.

Troubleshooting Reversing Power Circuits

A *power circuit* is the part of an electrical circuit that connects the loads to the main power lines. Troubleshooting reversing power circuits normally involves determining the point in the system where power is lost. **See Figure 10-37.** To troubleshoot a reversing power circuit, apply the following procedure:

1. Measure the incoming voltage between each pair of power leads. Incoming voltage must be within 10% of the voltage rating of the motor. Measure the voltage at the main power panel feeding the control cabinet if no voltage is present or if the voltage is not at the correct level.
2. Measure the voltage out of each fuse or circuit breaker. The fuse or breaker is open if no voltage reading is obtained. Replace any blown fuse or tripped circuit breaker.

Warning: Use caution when manually operating starter contacts because loads may start or stop without warning.

3. Measure the voltage out of the motor starter. The voltage should be present when either the forward power contacts or reverse power contacts are closed. The contacts can be closed manually at most motor starters if the power contacts cannot be closed by using the control circuit pushbuttons. Disconnect the incoming power and check the motor starter contacts for burning or wear if the voltage is not at the correct level.
4. Measure the voltage at the motor terminals. The voltage must be within 10% of the motor rating and equal on each power line. There is a problem with the motor or mechanical connection if the voltage is correct and the motor does not operate.

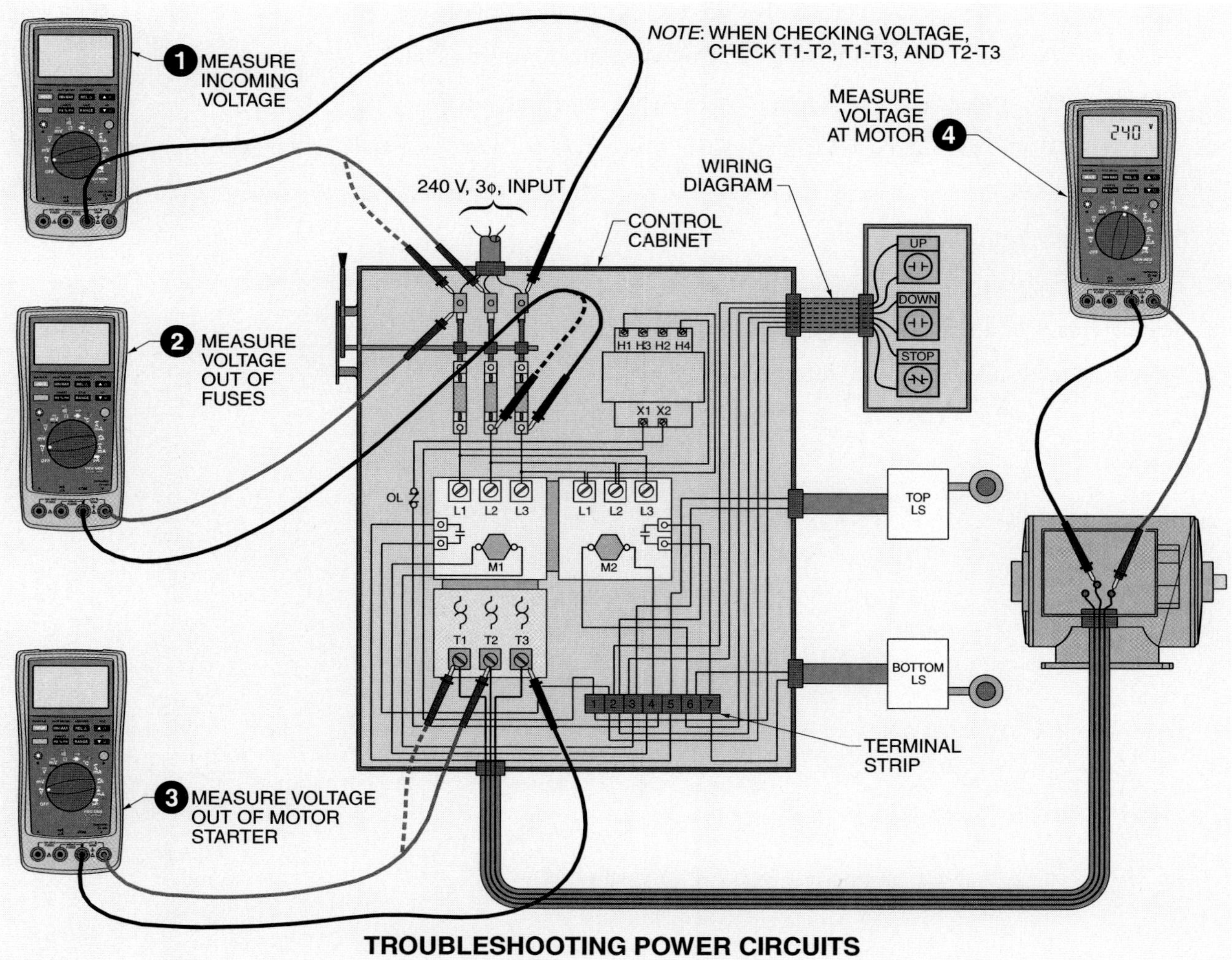

Figure 10-37. Troubleshooting reversing power circuits normally involves determining the point in the system where power is lost.

Review Questions

1. How is a 3ϕ motor reversed?
2. How is a 1ϕ motor reversed?
3. How is a capacitor-start motor reversed?
4. How are DC permanent-magnet motors reversed?
5. Why is mechanical interlocking used on forward and reversing starter combinations?
6. What type of interlocking uses a normally closed contact on the starter to lock out the other starter?
7. What kind of contacts must a pushbutton have to use pushbutton interlocking?
8. Why are indicator lights often used with forward and reversing circuits?
9. Are NO or NC contacts used when a limit switch is used to stop a motor that is running in one direction?
10. Why is a drum switch not considered a motor starter even though it can start and stop a motor in either direction?
11. What is a power circuit?
12. How can interlocking be accomplished when using a PLC to control the forward and reverse circuit of a motor?
13. What are some of the advantages and disadvantages of direct hard wiring?
14. Why is a direct hard wired circuit difficult to modify?

POWER DISTRIBUTION SYSTEMS

The power distribution system from a generating source to a plant and within a plant must be in good working order and properly maintained. *Power distribution* is the process of delivering electrical power to where it is needed. Power control, protection, transformation, and regulation must occur before any power can be delivered to the end user.

Power is transmitted and distributed from the power generating plant to the customer service-entrance equipment. A utility company power transmission and distribution system delivers power to industrial, commercial, and residential customers. **See Figure 11-1.** A power transmission and distribution system commonly includes the following:

- Step-up transformers. The generated voltage is stepped up to transmission voltage level. The transmission voltage level is normally between 12.47 kV and 245 kV.
- Power plant transmission lines. The 12.47 kV to 245 kV power plant transmission lines deliver power to the transmission substations.
- Transmission substations. The voltage is transformed to a lower primary (feeder) voltage. The primary voltage level is normally between 4.16 kV and 34.5 kV.
- Primary transmission lines. The 4.16 kV to 34.5 kV primary transmission lines deliver power to the distribution substations and heavy industry.
- Distribution substations. The voltage is transformed down to utilization voltages. Utilization voltage levels range from 480 V to 4.16 kV.
- Distribution lines. Power is carried from the distribution substation along the street or rear lot lines to the final step-down transformers.
- Final step-down transformers. Voltage is transformed to the required voltage, such as 480 V or 120/240 V. The final step-down transformers may be installed on poles, on grade-level pads, or in underground vaults. The secondary of the final step-down transformer is connected to service-entrance cables that deliver power to service-entrance equipment.

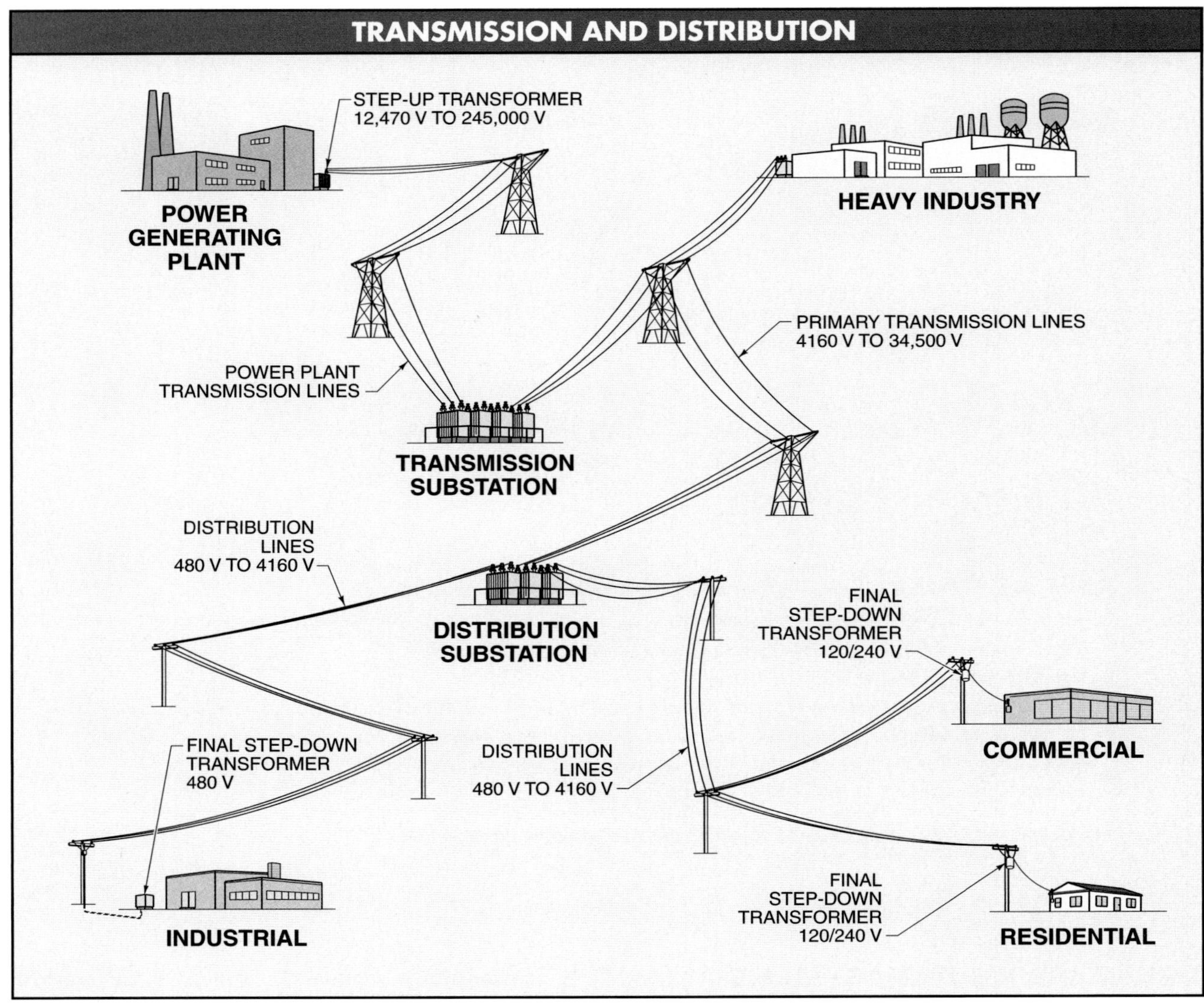

Figure 11-1. A utility company power transmission and distribution system delivers power to industrial, commercial, and residential customers.

Technical Fact

The success of the transformer is due largely to George Westinghouse and William Stanley. Westinghouse worked to secure two English patents in 1885 and turned them over to Stanley to develop. In 1886, a transformer that Stanley designed was able to light several stores in Great Barrington, MA. The transformer he designed was capable of stepping down 500 V on a line from a generator located a half-mile away.

The number and size of transformers used to step down (reduce) the voltage before the customer power distribution system depends on customer power requirements. For example, in a typical heavy industrial facility, the electricity may be delivered directly from a transmission substation to an outside transformer vault. Service-entrance conductors are routed from the outside transformer vault through an outdoor busway to a metered switchboard. A *busway* is a metal-enclosed distribution system of busbars available in prefabricated sections. A *switchboard* is a piece of equipment into which a large block of electric power is delivered from a substation and broken down into smaller blocks for distribution throughout a building. Power is then fed through circuit breakers in the panelboard and routed through busways to power distribution panels and busways with plug-in sections to the points of use. A *panelboard* is a wall-mounted distribution cabinet containing a group of overcurrent and short-circuit protection devices for lighting, appliance, or power distribution branch circuits. **See Figure 11-2.**

Depending on customer needs, the power distribution system delivers power at standard voltage levels and fixed current ratings to set points such as receptacles. Common voltage levels include 110 V, 115 V, 120 V, 208 V, 220 V, 240 V, 277 V, 440 V, 460 V, and 480 V.

A power distribution system and its components can be represented on a one-line diagram. A one-line diagram uses single lines and symbols to show flow path, voltage values, disconnects, overcurrent protection devices, transformers, and panelboards in a circuit. A one-line diagram is helpful when troubleshooting a power system and can show the entire distribution system or specific parts of a system. **See Figure 11-3.**

For example, a one-line diagram may show a 13.8 kV feed into a building and the transformers used for the distribution of specific voltages. High voltages are used for distribution of large amounts of power using small conductor sizes. The high voltage is then stepped down to low voltage levels and delivered to distribution panels. The distribution panels route power to individual loads such as industrial equipment, motors, lamps, and computers.

Generator (Alternator) Phase Connections

In a 3ϕ generator (alternator), three individual phases are present. Six wires extend from the alternator because each phase coil has a beginning and an end. The three coils with six wires may be connected internally or externally in wye (star) or delta connections.

Cutler-Hammer

Power distribution equipment is used to measure, step down, and distribute all incoming power.

POWER DISTRIBUTION TO POINT OF USE

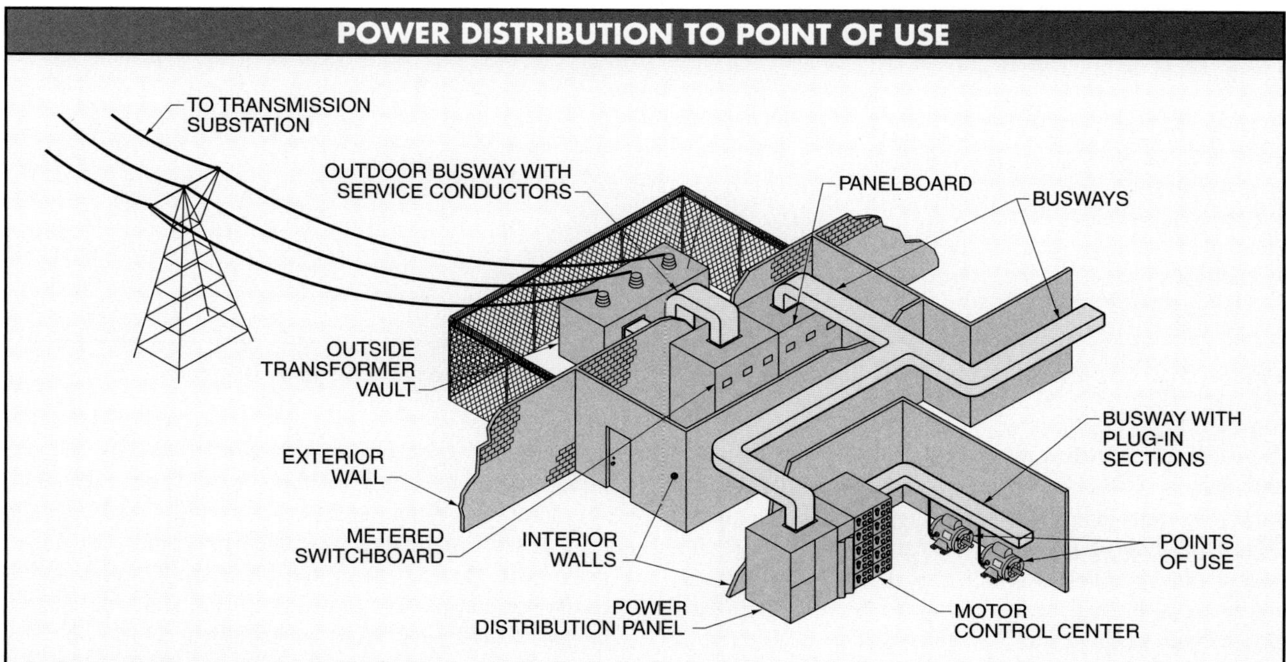

Figure 11-2. Power is fed through circuit breakers in the panelboard and routed through busways to power distribution panels and busways with plug-in sections to the points of use.

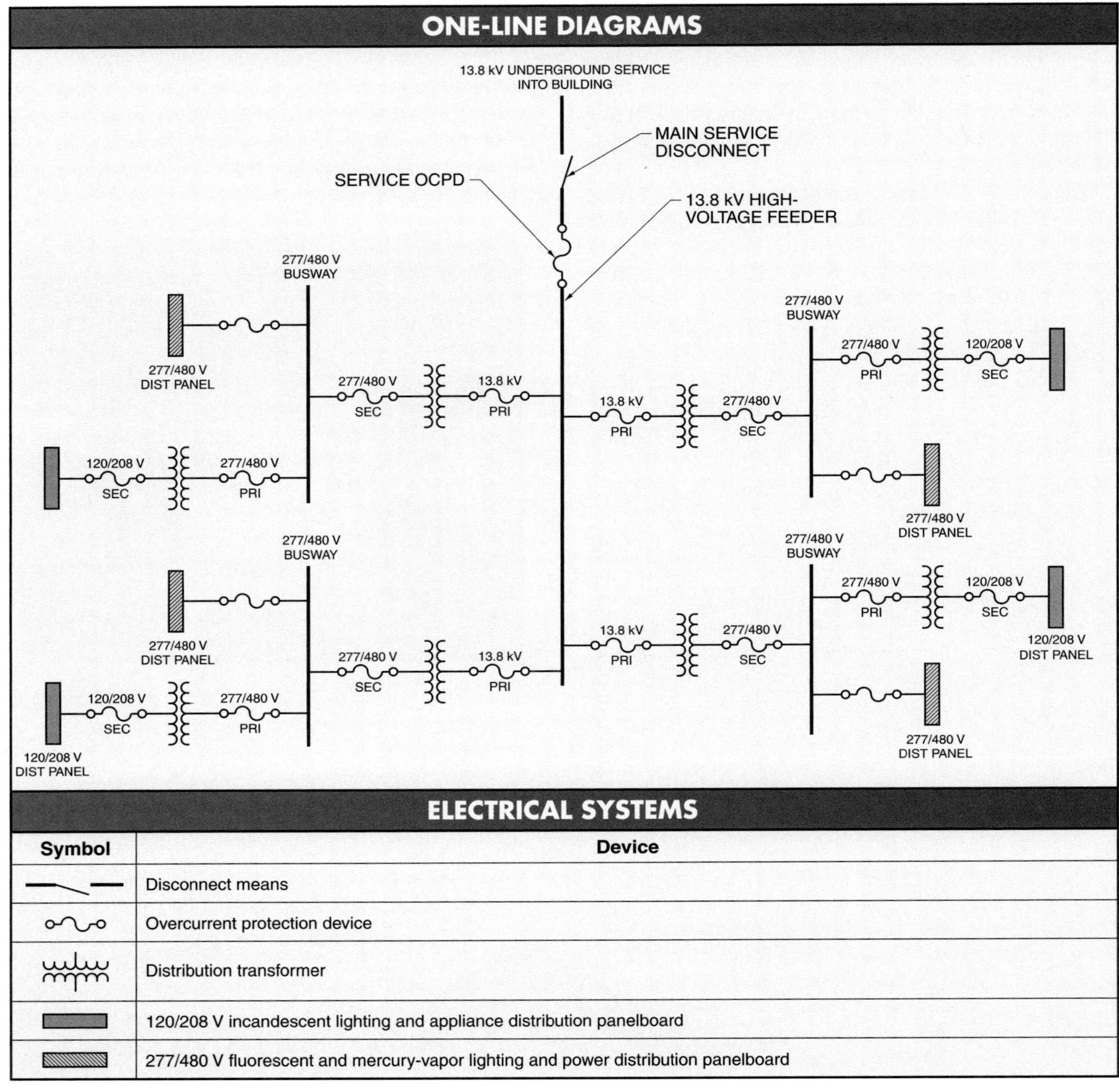

Symbol	Device
	Disconnect means
	Overcurrent protection device
	Distribution transformer
	120/208 V incandescent lighting and appliance distribution panelboard
	277/480 V fluorescent and mercury-vapor lighting and power distribution panelboard

Figure 11-3. One-line diagrams use single lines and symbols to show system components and operation.

Wye (Y) Connections. A wye connection is a connection that has one end of each coil connected together and the other end of each coil left open for external connections. Three lights may be connected to each separate phase of a wye-connected alternator. **See Figure 11-4.** Each light illuminates from the generated 1ϕ power delivered from each phase. The A2, B2, and C2 wires return to the alternator together. This circuit can be simplified by using only one wire and connecting it to the A2, B2, and C2 phase ends. This common wire is the neutral wire.

The three ends can be safely connected at the neutral point because no voltage difference exists between them. As phase A is maximum, phases B and C are opposite to A. If the equal opposing values of B and C are added vectorially, the opposing force of B and C combined is exactly equal to A. **See Figure 11-5.** For example, if three people are pulling with the same amount of force on ropes tied together at a single point, the resulting forces cancel each other and the resultant force is zero in the center (neutral point).

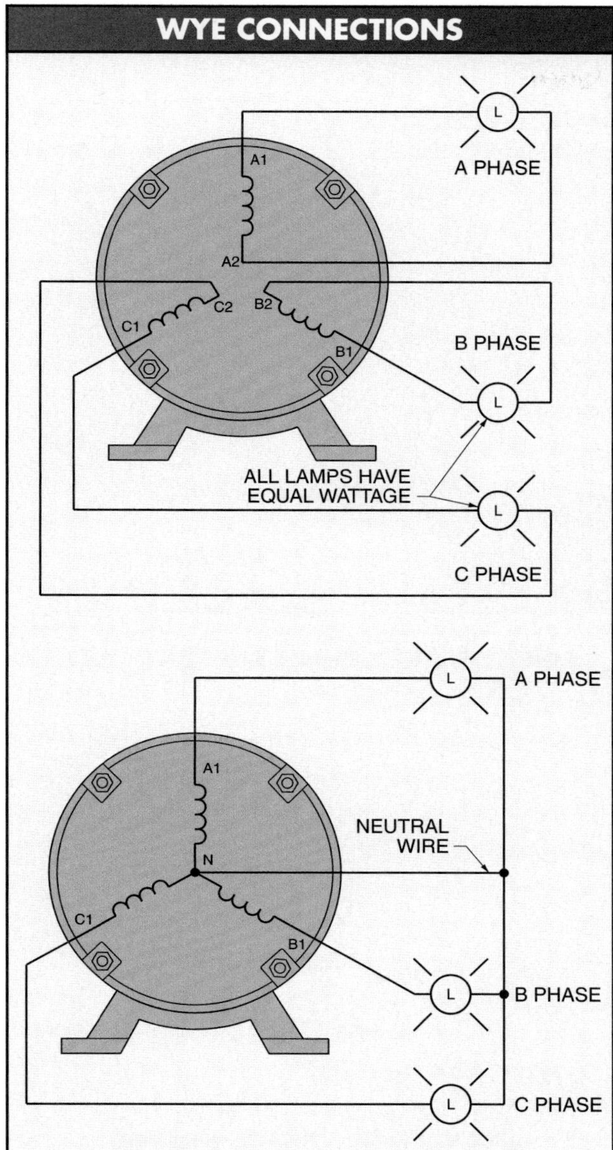

Figure 11-4. A common neutral wire can safely connect the internal leads of a wye-connected alternator to form a common return for lighting loads.

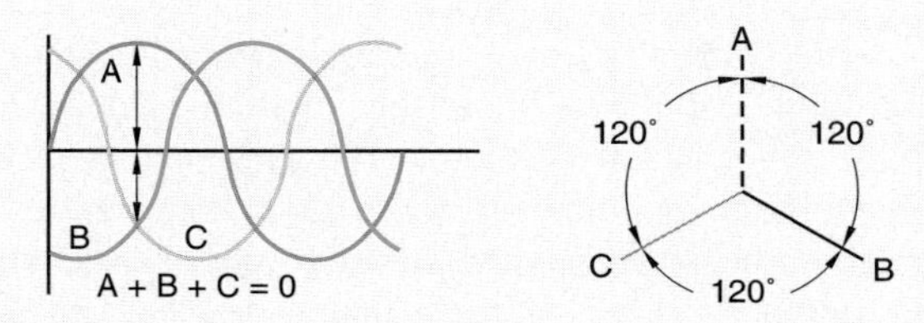

Figure 11-5. The 3ϕ voltages of a wye-connected alternator effectively cancel each other at the neutral point, allowing the three leads of the alternator to be connected.

The net effect is a large voltage (pressure) difference between the A1, B1, and C1 coil ends, but no pressure difference between the A2, B2, and C2 coil ends. In a 3 wye-connected lighting circuit, the 3ϕ circuit is balanced because the loads are all equal in power consumption. In a balanced circuit, there is no current flow in the neutral wire because the sum of all the currents is zero.

All large power distribution systems are designed as 3ϕ systems with the loads balanced across the phases as closely as possible. The only current that flows in the neutral wire is the unbalanced current. This is normally kept to a minimum because most systems can be kept fairly balanced. The neutral wire is normally connected to a ground such as the earth. **See Figure 11-6.** The voltages available from a wye-connected system are phase-to-neutral, phase-to-phase, or phase-to-phase-to-phase.

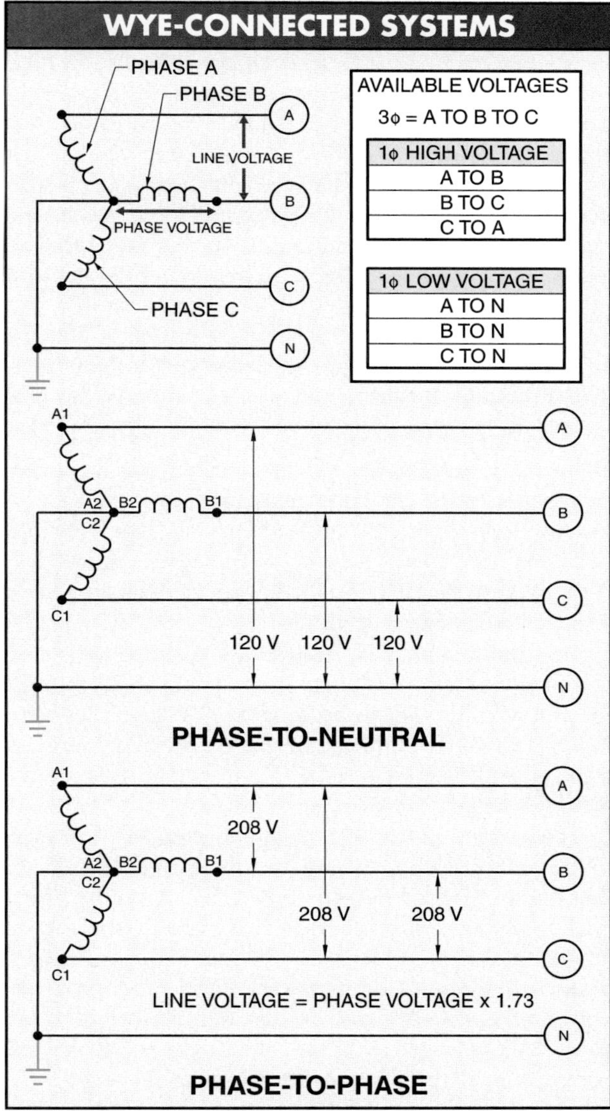

Figure 11-6. In wye-connected systems, the neutral wire is connected to ground and has various available voltages.

In a 3ϕ wye-connected system, the phase-to-neutral voltage is equal to the voltage generated in each coil. For example, if an alternator produces 120 V from A1 to A2, the equivalent 120 V is present from B1 to B2 and C1 to C2. Thus, in a 3ϕ wye-connected system, the output voltage of each coil appears between each phase and the neutral.

In a 3ϕ wye-connected system, the voltage values must be added vectorially because the coils are set 120 electrical degrees apart. In such an arrangement, the phase-to-phase voltage is obtained by multiplying the phase-to-neutral voltage by 1.73.

Similarly, on large wye-connected systems, a phase-to-neutral voltage of 2400 V creates a 4152 V line-to-line voltage, and a phase-to-neutral voltage of 7200 V creates a 12,456 V line-to-line voltage. One of the benefits wye-connected systems bring to the utility company is that even though its alternators are rated at 2400 V or 7200 V per coil, they can transmit at a higher phase-to-phase voltage with a reduction in losses and can provide better voltage regulation. This is because the higher the transmitted voltage, the less the voltage losses. This is especially important in long rural power lines.

In a wye-connected system, the neutral connection point is grounded and a fourth wire is carried along the system and grounded at every distribution transformer location. This solidly grounded system is regarded as the safest of all distribution systems.

In a 3ϕ wye-connected system, the current in the line is the same as the current in the coil (phase) windings. This is because the current in a series circuit is the same throughout all parts of the circuit.

Delta (Δ) Connections. A delta connection is a connection that has each coil end connected end-to-end to form a closed loop. Alternator coil windings of a 3ϕ system can also be connected as a delta connection. **See Figure 11-7.** As in a wye-connected system, the coil windings are spaced 120 electrical degrees apart.

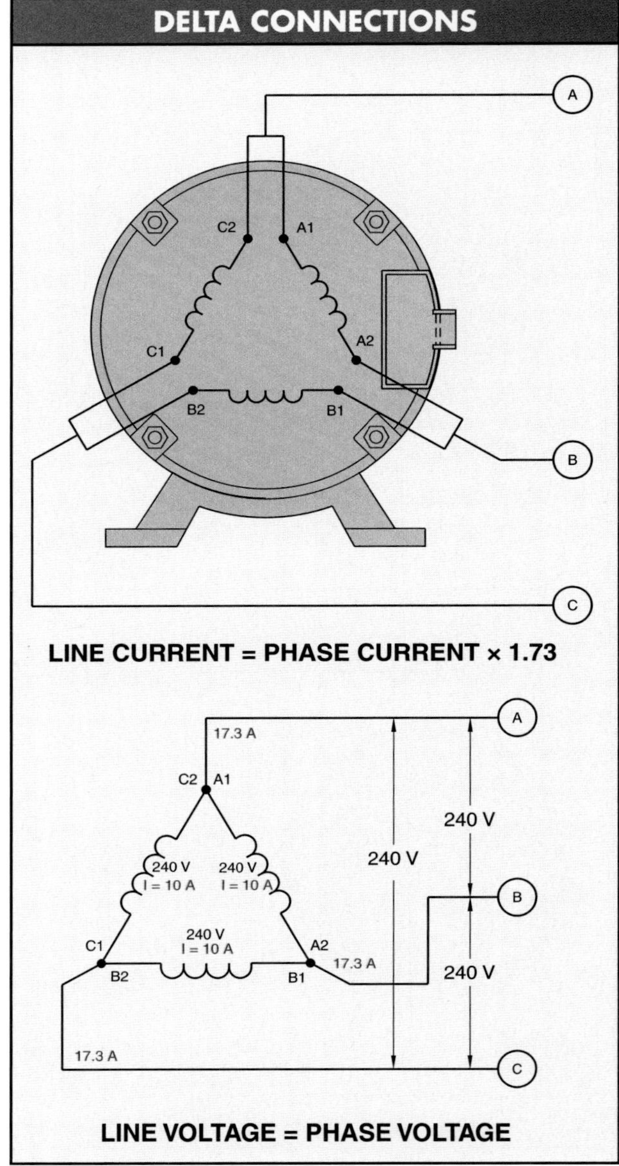

Figure 11-7. A delta connection has each coil end connected end-to-end to form a closed loop.

In a delta-connected system, the voltage measured across any two lines is equal to the voltage generated in the coil winding. This is because the voltage is measured directly across the coil winding. For example, if the generated coil voltage is equal to 240 V, the voltage between any two lines equals 240 V.

Following any line in a delta-connected system back to the connection point shows that the current supplied to that line is supplied by two coils. Phase A can be traced back to connection point A1, C2. However, as in a wye-connected system, the coils are 120 electrical degrees apart. Therefore, the line current is the vector sum of the two coil currents. In a balanced system, the phase currents are equal. In a balanced 3ϕ delta-connected system, the line current is equal to 1.73 times the current in one of the coils. For example, if each coil current is equal to 10 A, the line current is equal to 17.3 A (1.73 × 10 = 17.3 A).

In a delta-connected system, only three wires appear in the system. None of the three wires are normally connected to ground. However, when a delta-connected system is not grounded, it is possible for one phase to accidentally become grounded without anyone being aware of this. This problem is not apparent until another phase also grounds. For this reason, some plants deliberately ground one corner of the delta-connected system so that inadvertent faults on the other two phases cause a fuse or breaker to trip. Some plants also may make a ground in a delta-connected system by grounding the midpoint of one of the phases.

A delta-connected system permits different voltage possibilities. **See Figure 11-8.** Three-phase power (240 V) is available between A, B, and C. Single-phase power (120 V) is available from A to N and from C to N. Single-phase power (240 V) is available from A to B, B to C, and C to A. Also available is approximately 195 V, 1ϕ power from B to N. This voltage is the high/dangerous voltage, and should be avoided because it could damage equipment. This phase (phase B) is also known as the high leg (stinger). In a 120/240 V, 3ϕ delta system, phase B is never used for 120 VAC loads.

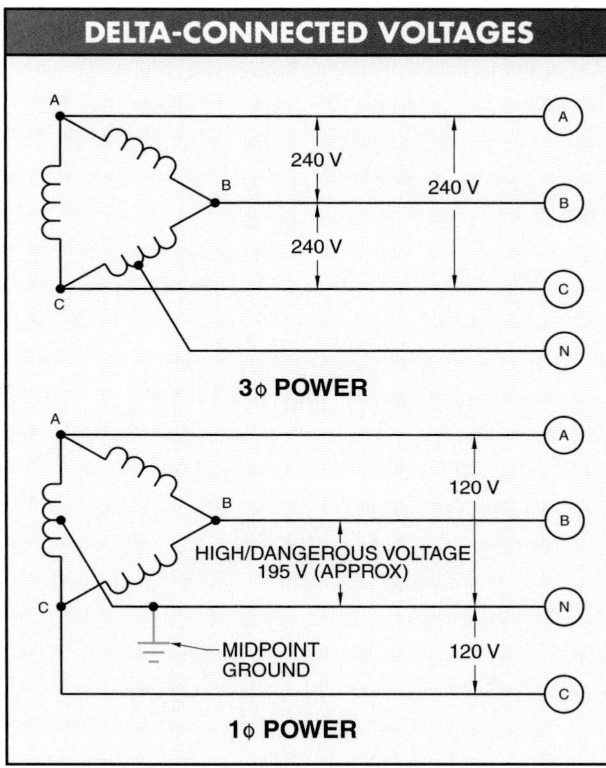

Figure 11-8. A delta-connected system permits different voltage possibilities.

An open-delta connection makes 3ϕ power available anywhere along the distribution line with only two transformers rather than the usual three. Although this system delivers only 57.7% of the nominal full-load capability of a full bank of three transformers, it has the advantage of lower initial cost if the extra power is not presently needed. Another advantage is that, in an emergency, this system allows reduced 1ϕ and 3ϕ power at a location where one transformer has burned out.

Transformers Connected for Wye and Delta Distribution Systems. Large amounts of power are generated using a 3ϕ system. The generated voltage is stepped up and down many times before it reaches the loads in industrial, commercial, and residential buildings. The transformation can be accomplished by using wye- or delta-connected transformers or a combination of both wye and delta transformers with differing voltage ratio transformers.

Three 1ϕ transformers may be connected for a wye-to-wye step-down transformer bank. **See Figure 11-9.** The line voltage is equal to 1.73 times the coil voltage and the line current and coil current are equal in a balanced wye system.

The wye-connected secondary provides three different types of service: a 208 V, 3ϕ service for 3ϕ motor loads; a 208 V, 1ϕ service for 208 V, 1ϕ motor loads; and a 120 V, 1ϕ service for lighting loads. A wye-connected system is commonly used in schools, commercial stores, and offices.

On the primary side of the transformer bank, the grounded neutral wire is connected to the common points of all three high-voltage primary coil windings. These coil windings are marked H1 and H2 on each transformer to indicate the high-voltage side of the transformer. The low-voltage side is marked X1 and X2. The voltage from the neutral to any phase of the three power lines is 2400 V. The voltage across the three power lines is 4152 V (1.73 × 2400 V = 4152 V).

As on the primary side, the grounded neutral is connected to the common points of all three low-voltage secondary coil windings. This allows for a 120 V output on each of the secondary coils. The voltage across the three secondary power lines is 208 V (1.73 × 120 V = 208 V).

To help maintain a balanced transformer bank, the loads should be connected to evenly distribute them among the three transformers. This naturally occurs when connecting a 3ϕ motor because a 3ϕ motor draws the same amount of current from each line. Care should be taken to balance the 1ϕ loads. Three 1ϕ transformers of the same power rating are used in most wye-to-wye systems. The capacity of transformers is rated in kilovolt-amperes (kVA). The total kVA capacity of a transformer bank is found by adding the individual kVA ratings of each transformer in the bank. For example, if each transformer is rated at 50 kVA, the total capacity of the bank is 150 kVA (50 kVA + 50 kVA + 50 kVA = 150 kVA).

Technical Fact

The sheet steel used in modern transformers contains a sufficient amount of silicon to prevent the steel from aging. Aging is the characteristic of the magnetic circuit which increases the iron loss as the transformer continues to operate over a period of time.

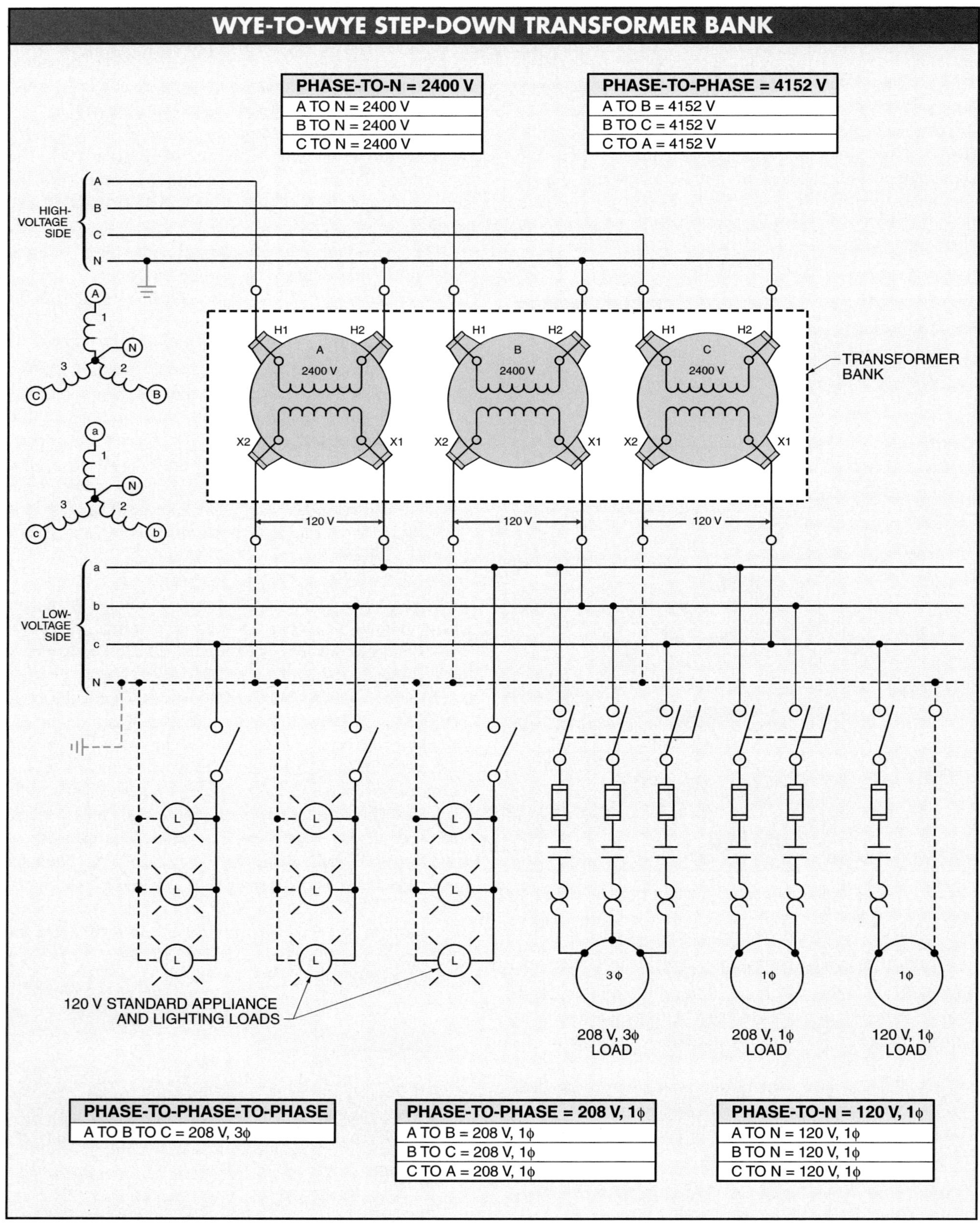

Figure 11-9. Three 1ϕ transformers may be connected for a wye-to-wye, step-down transformer bank.

Three 1ϕ transformers may be connected for a delta-to-delta step-down transformer bank. **See Figure 11-10.** The line voltage is equal to the coil voltage and the line current is equal to 1.73 times the coil current in a balanced delta system.

The delta-connected secondary, with one coil center-tapped, provides three different types of service: a 240 V, 3ϕ service for 3ϕ motor loads; a 120 V, 1ϕ service for lighting loads; and a 240 V, 1ϕ service for 240 V, 1ϕ motor loads. The high phase of line B to N is not used.

As with the wye-connected transformer bank, a closed delta bank delivers a total kVA output equal to the sum of the individual transformer ratings. One advantage of a delta system is that if one transformer is damaged or removed from service, the other two can be connected in an open-delta connection. This type of connection enables power to be maintained at a reduced level. The level is reduced to 57.7% of a full delta-connected transformer bank.

Transformer Installation

Extreme care must be taken when working around transformers because of the high voltage present. Ensure that proper protective equipment is used and all plant safety procedures are followed. All transformer installations should follow National Electrical Code® (NEC®) and National Fire Protection Association (NFPA) requirements. For example, NEC® Section 450.21 covers the installation of indoor dry-type transformers. **See Figure 11-11.** National Fire Protection Association (NFPA) Section 420.10 covers the installation of all transformers. Following these requirements protects persons working around the transformer, helps prevent fires, and reduces the chance of an electrical shock. Always follow NEC® and NFPA requirements when installing transformers.

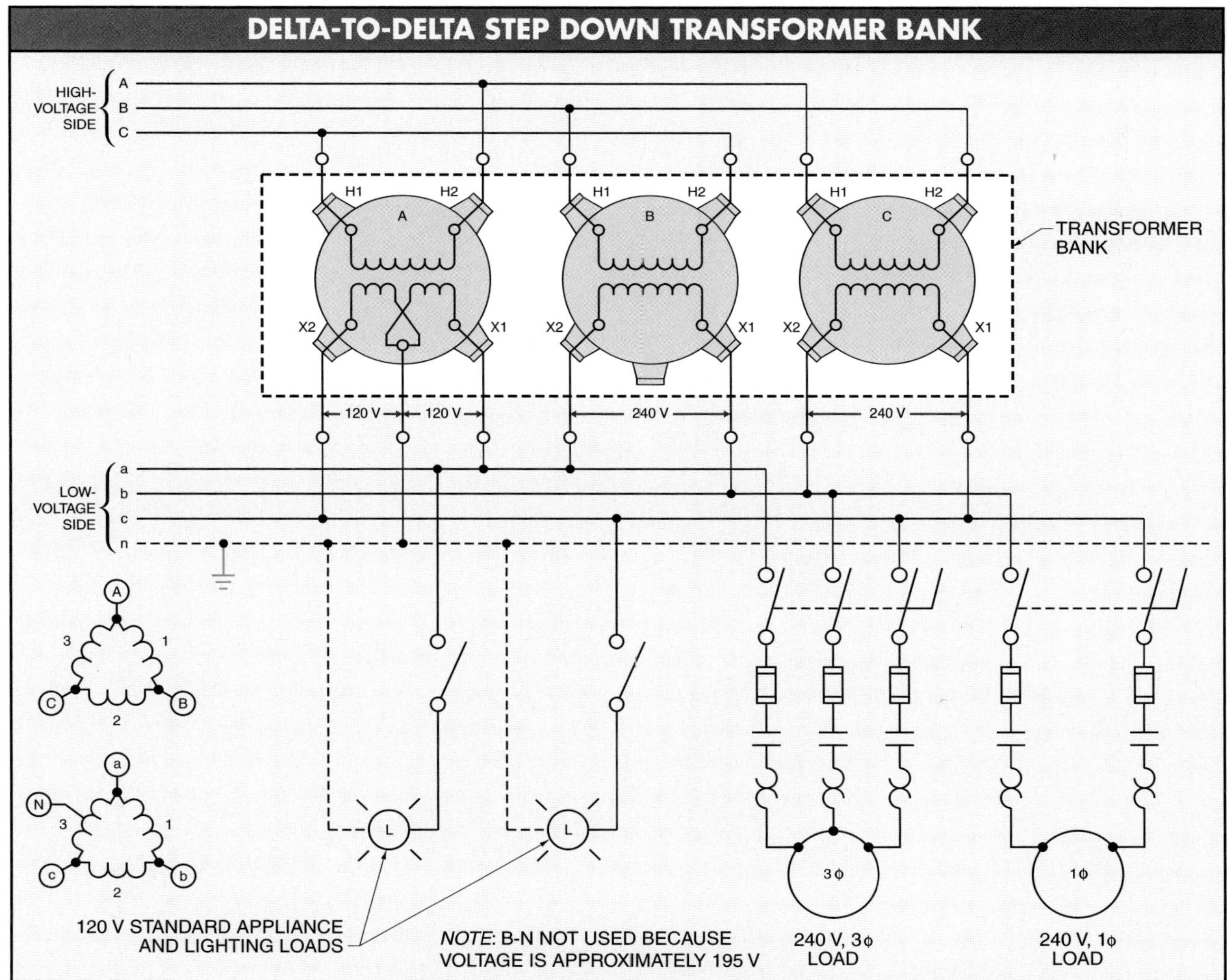

Figure 11-10. Three 1ϕ transformers may be connected for a delta-to-delta, step-down transformer bank.

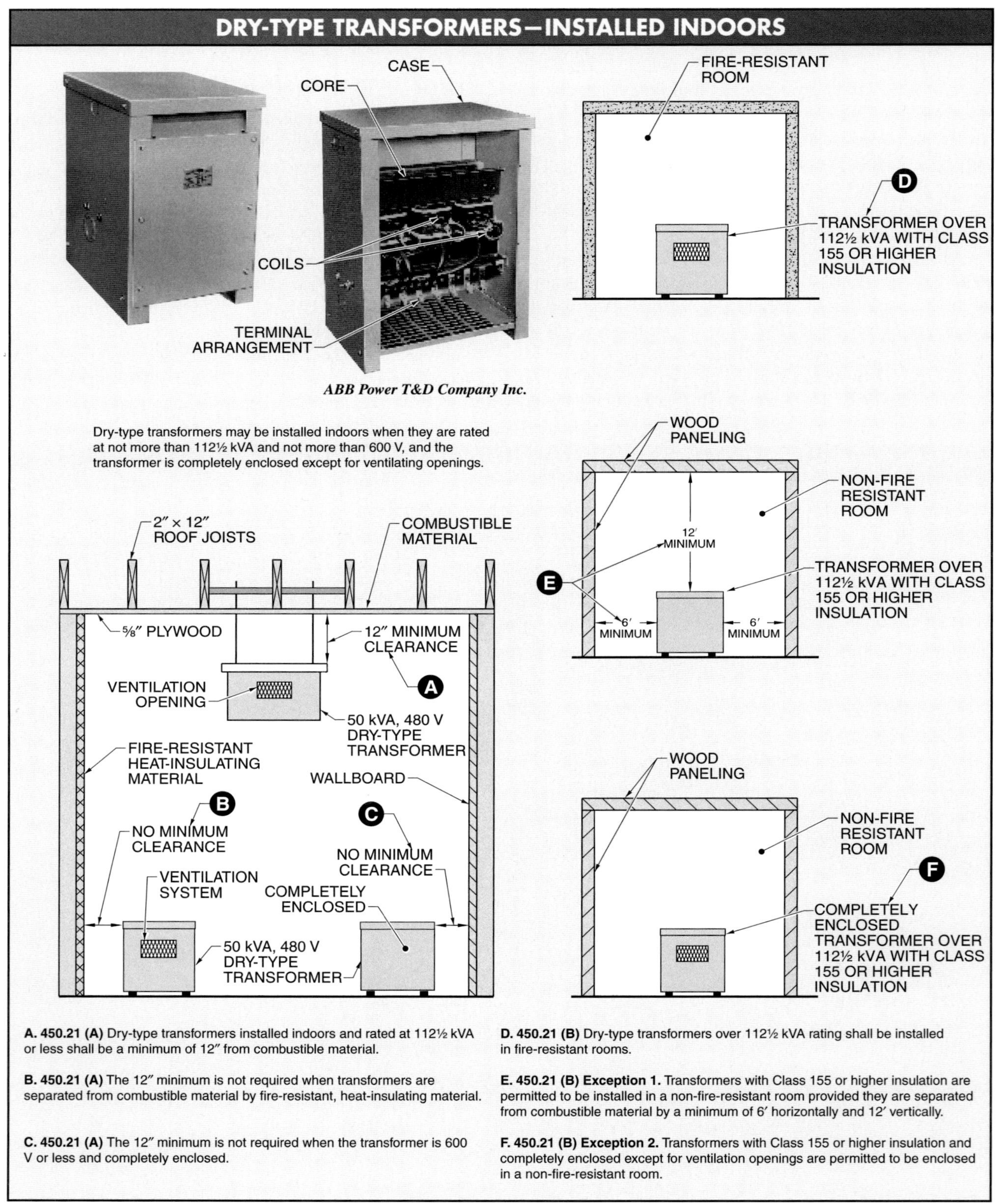

Figure 11-11. NEC® Section 450.21 contains requirements for the installation of indoor dry-type transformers which must be followed to protect persons working around the transformer.

Substations

Substations serve as a source of voltage transformation and control along the distribution system. Their function includes:

- Receiving voltage generated and increasing it to a level appropriate for transmission
- Receiving the transmitted voltage and reducing it to a level appropriate for customer use
- Providing a safe point in the distribution system for disconnecting the power in the event of problems
- Providing a place to adjust and regulate the outgoing voltage
- Providing a convenient place to take measurements and check the operation of the distribution system
- Providing a switching point where different connections may be made between various transmission lines

Substations have three main sections: primary switchgear, transformer, and secondary switchgear sections. **See Figure 11-12.** Depending on the function of the substation (step-up or step-down voltage), the primary or secondary switchgear section may be the high-voltage or low-voltage section. In step-up substations, the primary switchgear section is the low-voltage section and the secondary switchgear section is the high-voltage section. In step-down substations, the primary switchgear section is the high-voltage section and the secondary switchgear section is the low-voltage section. The substation sections normally include breakers, junction boxes, and interrupter switches.

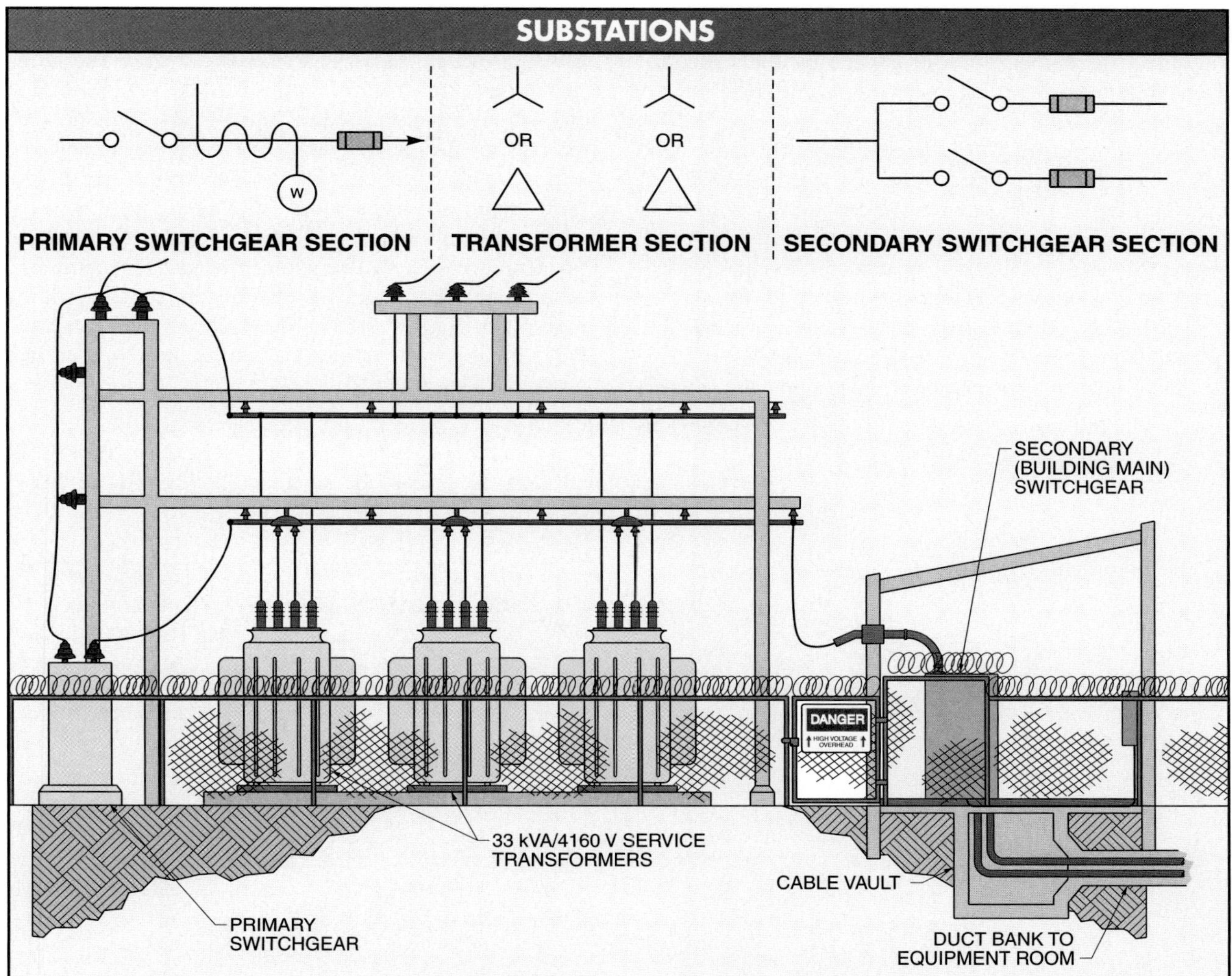

Figure 11-12. The three main sections of a substation are the primary switchgear, transformer, and secondary switchgear sections.

Substations may be entirely enclosed in a building or totally in the open, as in the case of outdoor substations located along a distribution system. The location for a substation is generally selected so that the station is as near as possible to the area to be served.

Substations can be built to order or purchased from factory-built, metal-enclosed units. The purchased units are unit substations. A unit substation offers standardization and flexibility for future changes when quick replacements are needed.

A transformer's function in a substation is the same as that of any transformer. Transformers are broadly classified as wet or dry.

In wet types, oil or some other liquid serves as a heat-transfer medium and insulation. Dry types use air or inert gas in place of the liquid. Fans may be used for forced-air cooling on transformers to provide additional power for peak demand periods where the surrounding air is hot and the transformers are not able to handle their full-rated load without exceeding their recommended temperature. The normal loads are handled by natural circulation.

Most transformers include a voltage regulator. The voltage regulator on a transformer has taps that allow for variable output. These taps are needed because transmission lines rarely deliver the transformer input voltage for which they are rated because of transmission line losses, line loading, and other factors. **See Figure 11-13.**

The taps are normally provided at 2.5% increments for adjusting above and below the rated voltage. The adjustment may be manual or automatic. In an automatic voltage regulator, a control circuit automatically changes the tap setting on the transformer's windings. This allows the outgoing voltage to be kept nearly constant even though the incoming primary voltage or load demands may vary.

Switchboards

Electrical power is delivered to industrial, commercial, and residential buildings through a distribution and transmission system. Once the power is delivered to a building, it is up to the building electrician to further distribute the power to where it is required within the building.

A switchboard is the link between the power delivered to a building (property) and the start of the local power distribution system in a building. The switchboard is the last point on the power distribution system for the power company and the beginning of the distribution system for the property owner's electrician.

A switchboard is a piece of equipment which a large block of electric power is delivered to from a substation and broken down into smaller blocks for distribution throughout a building. **See Figure 11-14.** Switchboards are rated by the manufacturer for a maximum voltage and current output. For example, a switchboard may have a 600 V rating and a bus rating up to 5000 A.

In addition to dividing the incoming power, a switchboard may contain all the equipment needed for controlling, monitoring, protecting, and recording the functions of the substation. Switchboards are designed for use in three categories: service-entrance, distribution, and service-entrance/distribution.

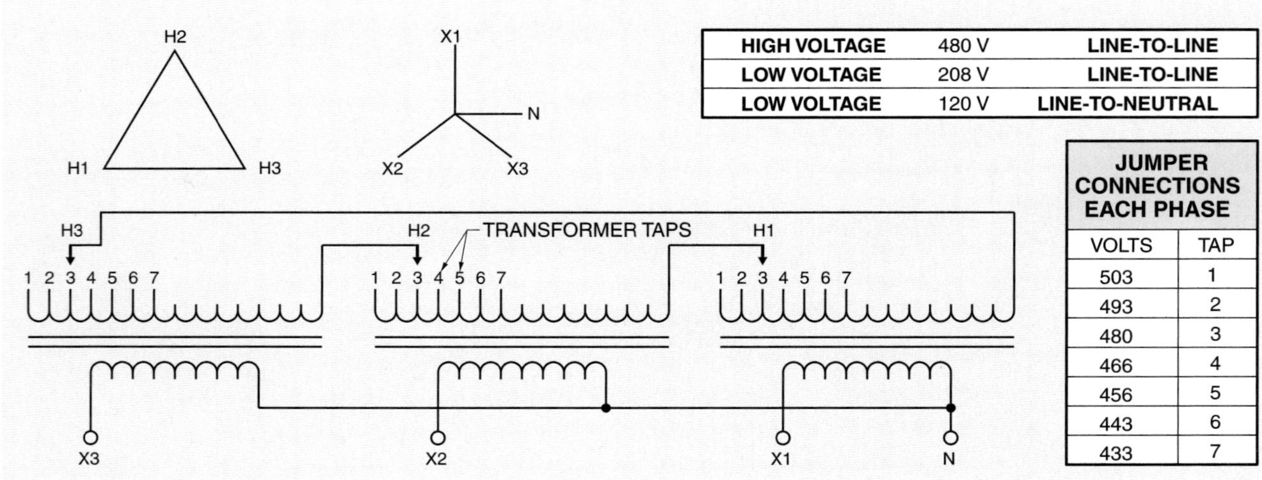

HIGH VOLTAGE	480 V	LINE-TO-LINE
LOW VOLTAGE	208 V	LINE-TO-LINE
LOW VOLTAGE	120 V	LINE-TO-NEUTRAL

JUMPER CONNECTIONS EACH PHASE	
VOLTS	TAP
503	1
493	2
480	3
466	4
456	5
443	6
433	7

Figure 11-13. Taps are built into a transformer to compensate for voltage differences.

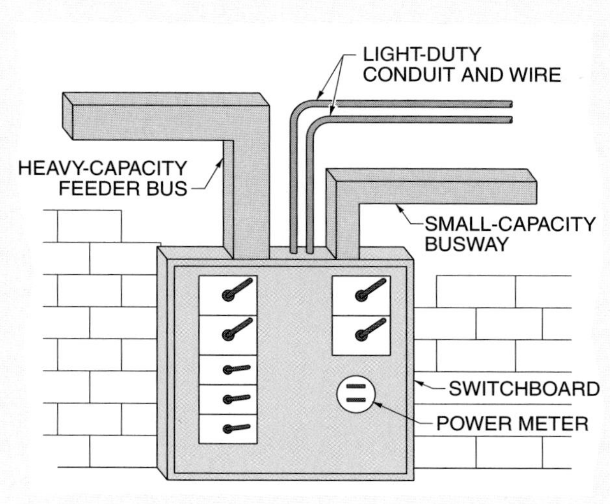

Figure 11-14. A switchboard is used to divide incoming power into smaller branch circuits.

A service-entrance switchboard has space and mounting provisions for metering equipment (as required by the local power company), for overcurrent protection, and for a means of disconnect for the service conductors. **See Figure 11-15.** Provision for grounding the service neutral conductor when a ground is needed is also provided. A distribution switchboard contains the protective devices and feeder circuits required to distribute the power throughout a building. A distribution switchboard may contain either circuit breakers or fused switches.

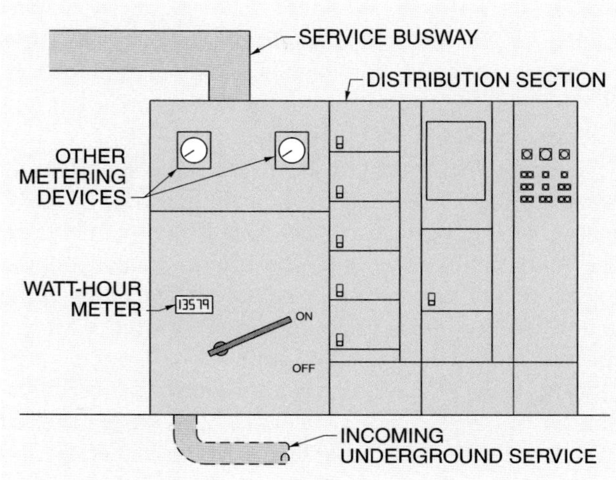

Figure 11-15. A distribution switchboard contains the protective devices and feeder circuits required to distribute the power throughout the building.

A distribution switchboard has the space and mounting provisions required by the local power company. Building power distribution systems are used to deliver the required type (DC, 1ϕ, or 3ϕ) and level (120 V, 230 V, 460 V, etc.) of power to the loads connected to the system. Metering equipment for the power used by the building's tenants is also installed at this location. To meter the incoming power, the switchboard must have a watt-hour meter to measure power usage. Metering is always located on the incoming line side of the disconnect. The compartment cover is sealed to prevent tapping power ahead of the power company metering equipment.

Other meters and indicator lights, such as ammeters and voltmeters, may also be built into the meter compartment. In most cases, these are not requirements but options, depending on the application and the plant requirements. A voltmeter is used to indicate to the maintenance personnel the various incoming and outgoing voltages. An ammeter is used to indicate the various current levels throughout the system. A wattmeter is used to indicate the power used throughout the system. Each of these instruments can be of the indicating type, recording type, or both. A recording instrument is used to keep track of the various values over a period of time.

In addition to measuring the voltage, current, and power of a system, a distribution switchboard also controls the power. Control is achieved through the use of switches and overcurrent and overvoltage relays that are used to disconnect the power. These devices protect the distribution system in the event of a fault.

Switchboards that have more than six switches or circuit breakers must include a main switch to protect or disconnect all circuits. Switchboards with more than one but not more than six switches or circuit breakers do not require a main switch. In a switchboard with more than six switches or breakers, the service-entrance section of the switchboard may have any number of feeder circuits added to the rated capacity of the main. A switchboard with a main section can easily contain more than one distribution system. The system requirement depends on the number of feeder circuits entering the building. Please refer to NEC® Article 240 or NFPA Article 410.9 for more information.

In addition to distributing the power throughout a building, the distribution section of a switchboard may contain provisions for motor starters and other control devices. **See Figure 11-16.** The addition of starters and controls to the switchboard allows for motors to be connected to the switchboard. This combination can be used when the motors to be controlled are located near the switchboard. This combination allows for high-current loads such as motors to be connected to the source of power without further power distribution.

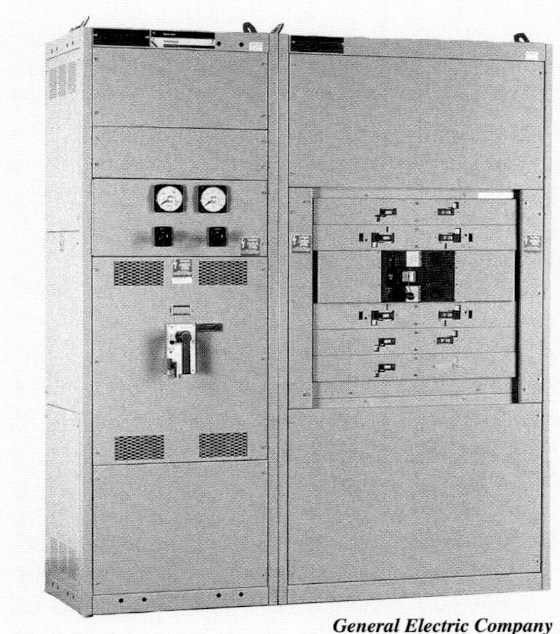

General Electric Company

Figure 11-16. A switchboard may contain provisions for motor starters and other devices.

Panelboards and Branch Circuits

A panelboard is a wall-mounted distribution cabinet containing a group of overcurrent and short-circuit protection devices for lighting, appliance, or power distribution branch circuits. The wall-mounted feature distinguishes the panelboard from a switchboard, which is normally freestanding. **See Figure 11-17.**

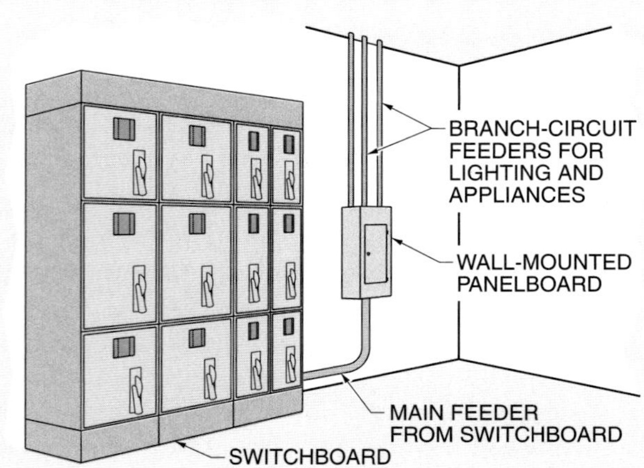

Figure 11-17. A panelboard is a wall-mounted distribution cabinet containing overcurrent and short-circuit protection devices.

General Electric Company

Type IP core and coil transformers are designed for panelboard, industrial control, machine tool, and general-purpose applications.

A panelboard is normally supplied from a switchboard and further divides the power distribution system into smaller parts. Panelboards are the part of the distribution system that provides the last centrally located protection for the final power run to the load and its control circuitry. Panelboards are classified according to their use in the distribution system.

A panelboard provides the required circuit control and overcurrent protection for all circuits and power-consuming loads connected to the distribution system. **See Figure 11-18.** The panelboards are located throughout a plant or building, providing the necessary protection for the branch circuits feeding the loads.

A branch circuit is the portion of a distribution system between the final overcurrent protection device and the outlet or load connected to it. The basic requirements for panelboards and overcurrent protection devices are given in Article 240 of the National Electrical Code® and Article 420.4 of the National Fire Protection Association and must be met for individual applications. In addition, local power company, city, and county regulations should also be met. Note: Per NEC® 408.15, not more than 42 overcurrent devices are allowed in a panelboard.

Overcurrent protection devices used for protecting branch circuits include fuses or circuit breakers. Overcurrent protection devices must provide for proper overload and short-circuit protection. The size (in amperes) of the overcurrent protection device is based on the rating of the panelboard and load. The overcurrent protection device must protect the load and be within the rating of the panelboard. If the overcurrent protection device exceeds the ampacity of the busbars in the panelboard, the panelboard is undersized for the load(s) that are to be connected.

A panelboard may be compared to a load center found in most residential dwellings. The load center in residential dwellings contains the fuses or breakers that control the individual branch circuits throughout the dwelling.

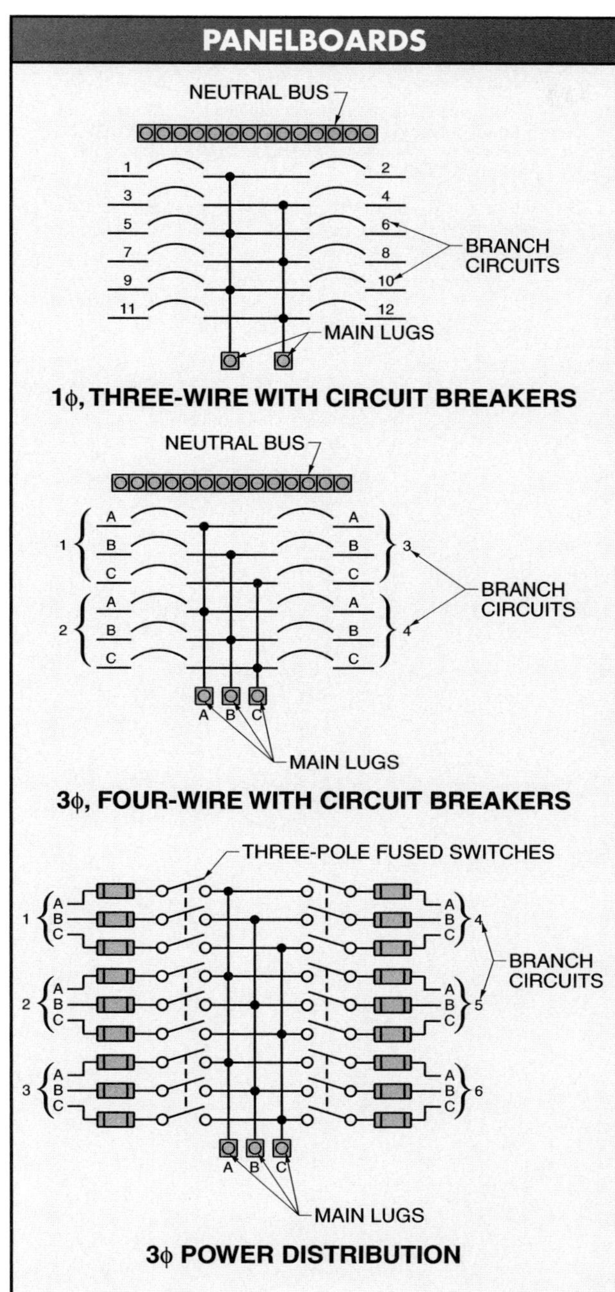

Figure 11-18. A panelboard provides the required circuit control and overcurrent protection for all circuits and power-consuming loads connected to the distribution system.

Although panelboards and load centers perform the same function, panelboards for commercial and industrial applications have distinct features. Panelboard features not shared by load centers include:

- A box fabricated of Underwriters Laboratories Inc.® (UL) approved corrosion-resistant galvanized steel
- A minimum of a 4″ wiring gutter on all sides
- Combination catch and lock in addition to hinges
- Busbars listed to 1200 A (load center main busbars are generally 200 A maximum)
- Enclosure depth to accommodate 2½″ or greater conduit
- Main and branch terminal lugs

Panelboard Installation

Panelboards are the main place where most in-house electrical circuits start. The panelboard must have a sufficient current and voltage rating and must be properly installed and grounded. An electrical shock or fire is possible if the panelboard is not properly installed and grounded. In addition, the system can be overloaded and a fire can result if the panelboard is not properly sized. All panelboard installations should follow NEC® and NFPA requirements. For example, NEC® Article 408 and NFPA Article 420.4 cover the installation of switchboards and panelboards. **See Figure 11-19.** Following these requirements protects persons working around panelboards, helps prevent fires, and reduces the chance of an electrical shock.

Motor Control Centers

In a power distribution system, many different kinds of loads are connected to the system. The loads vary considerably from application to application, as does their degree of control. For example, a light may be connected to a system requiring only a switch for control (along with proper protection). However, other loads, such as motors, may require complicated and lengthy control and protection circuits. The more complicated a control circuit becomes, the more difficult it is to wire into the system.

UE Systems, Inc.

A motor control center provides a central location for the wiring, control, and troubleshooting of motor control circuits.

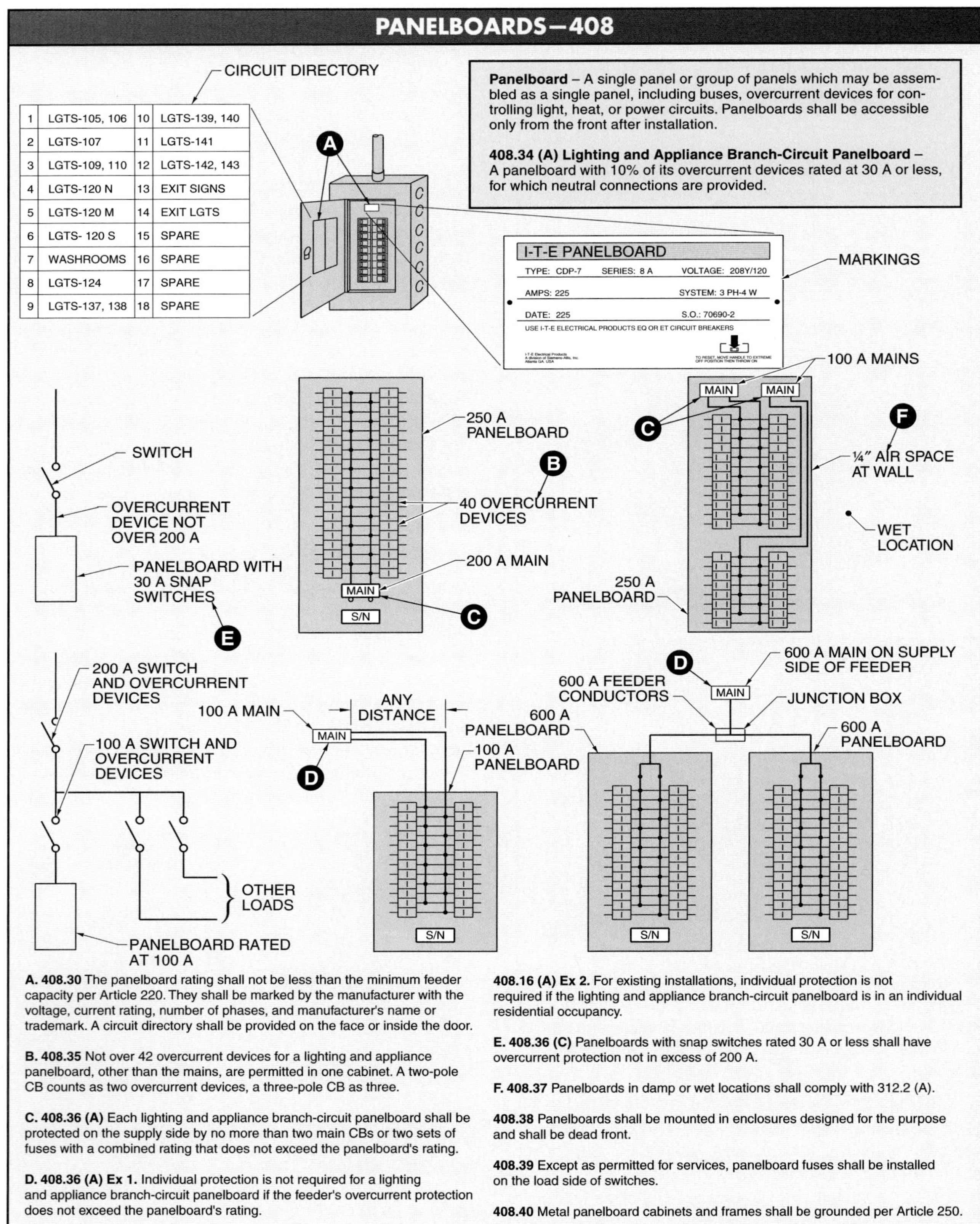

Figure 11-19. NEC® Article 408 covers the installation of panelboards.

The most common loads requiring simple and complex control are electric motors. Simplifying and consolidating motor control circuits is required because an electric motor is the backbone of almost all production and industrial applications. To do this, a control center takes the incoming power, control circuitry, required overload and overcurrent protection, and any transformation of power, and combines them into one convenient motor control center.

A motor control center combines individual control units into standard modular structures joined on formed sills. Power for a motor control center is normally supplied from a panelboard or switchboard. A motor control center is different from a switchboard containing motor panels in that the motor control center is a modular structure designed specifically for plug-in control units and motor control. **See Figure 11-20.**

Figure 11-20. A motor control center combines the incoming power, control circuitry, overload and overcurrent protection, and any transformation of power into one convenient location.

A motor control center receives the incoming power and delivers it to the control circuit and motor loads. The motor control center provides space for the control and load wiring in addition to providing required control components. The control inputs into the motor control center are the control devices such as pushbuttons, liquid level and limit switches, and other devices that provide a signal. The output of the motor control center is the wire connecting the motors. All other control devices are located in the motor control center. These control devices include relays, control transformers, motor starters, overload and overcurrent protection devices, timers, counters, and any other required control devices.

One advantage of a motor control center is that it provides one convenient place for installing and troubleshooting control circuits. This is especially useful in applications that require individual control circuits to be related to other control circuits. An example includes assembly lines in which one machine feeds the next.

A second advantage is that individual units can be easily removed, replaced, added to, and interlocked at one central location. Manufacturers of motor control centers produce factory-preassembled units to meet all the standard motor functions, such as start/stop, reversing, reduced-voltage starting, and speed control. This leaves only the connecting of the control devices (pushbuttons, limit switches, level switches, pressure switches, etc.) and the motors to the motor control center.

Common preassembled motor control center panels are available from the factory, along with their schematic diagrams. **See Figure 11-21.** The only required wiring by the electrician is the connection to control inputs, terminal blocks, and the motor.

The motor is connected to T1, T2, and T3. The control inputs are connected to the terminal blocks marked 1, 2, and 3. If a two-wire control, like a liquid level switch, is connected to the circuit, it is connected to terminals 1 and 3 only. Also provided on each unit are predrilled holes to allow for easy additions to the circuit. These holes match the manufacturer's standard devices, and most manufacturers provide templates for easy layout and circuit design. Motor control units should not be removed or installed unless the disconnect for that unit is in the OFF position. *Note:* Always verify that power is off, by using a DMM set to measure voltage, before removing or installing any electrical equipment.

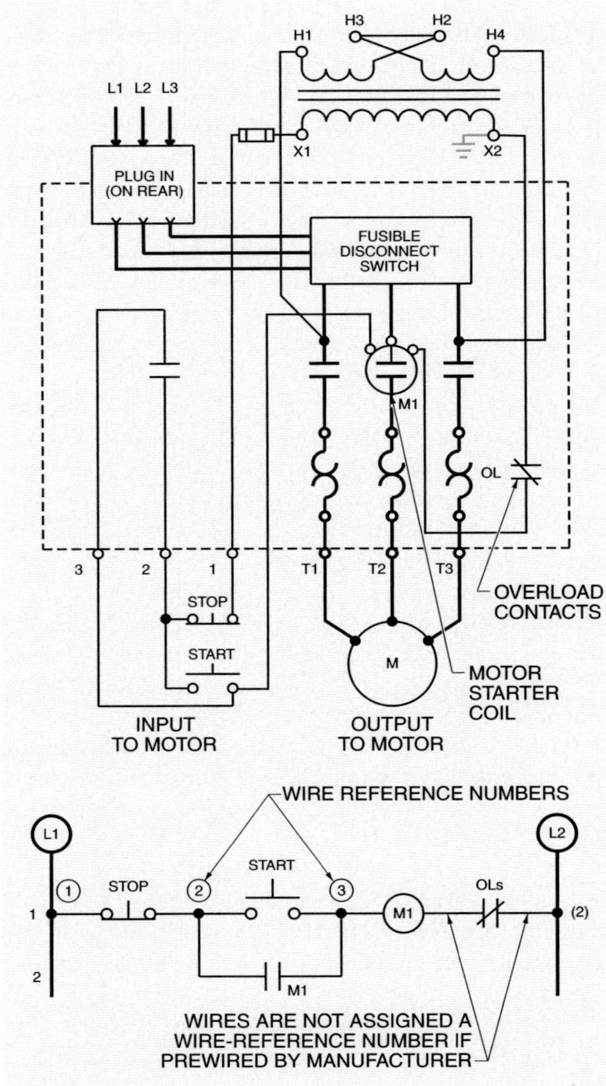

Figure 11-21. Common preassembled motor control center panels are available from the factory, along with corresponding schematic diagrams.

Feeders and Busways

The electrical distribution system in a plant must transport the electrical power from the source of supply to the loads. In today's plants, this may consist of distribution over large areas with many different electrical requirements. **See Figure 11-22.** In many cases where shifting of production machinery is common, the distribution system must be changed from time to time. A busway is a metal-enclosed distribution system of busbars available in prefabricated sections. Prefabricated fittings, tees, elbows, and crosses simplify the connecting and reconnecting of the distribution system. By bolting sections together, the electrical power is available at many locations and throughout the system.

A busway does not have exposed conductors. This is because the power in a plant distribution system is at a high level. To offer protection from the high voltage, the conductors of a busway are supported with insulating blocks and covered with an enclosure to prevent accidental contact. A typical busway distribution system provides for fast connection and disconnection of machinery. Busways enable manufacturing plants to be retooled or re-engineered without major changes in the distribution system.

The most common length of busways is 10′. Shorter lengths are used as needed. Prefabricated elbows, tees, and crosses make it possible for the electrical power to run up, down, and around corners, and to be tapped off from the distribution system. This allows the distribution system to have maximum flexibility with simple and easy connections when working on installations.

The two basic types of busways are feeder and plug-in busways. **See Figure 11-23.** Feeder busways deliver the power from the source to a load-consuming device. Plug-in busways serve the same function, but also allow load-consuming devices to be conveniently added along the bus structure. A plug-in power module is used on a plug-in busway system.

The three general types of plug-in power panels used with busways are fusible switches, circuit breakers, and specialty plugs (duplex receptacles with circuit breakers, twistlock receptacles, etc.). The conduit and wire is run to a machine or load from the fusible switches and circuit breaker plug-in panels. Generally, power cords may be used only for portable equipment.

The loads connected to the power distribution system are often portable or unknown at the time of installation. For this reason, the power distribution system must often terminate in such a manner as to provide for a quick connection of a load at some future time. To do this, an electrician installs receptacles throughout the building or plant to serve the loads as required. With these receptacles, different loads can be connected easily.

Because the distribution system wiring and protection devices determine the size of the load that can be connected to it, a method is required for distinguishing the rating in voltage and current of each termination. This is especially true in industrial applications that require a variety of different currents, voltages, and phases.

The National Electrical Manufacturers Association (NEMA) has established a set of standard plug and receptacle configurations that clearly indicate the type of termination. **See Appendix.** The standard configurations enable the identification of the voltage and current rating of any receptacle or plug simply by looking at the configuration. Plug-in bus disconnect switches must not be removed or installed unless the disconnect switch is in the OFF position.

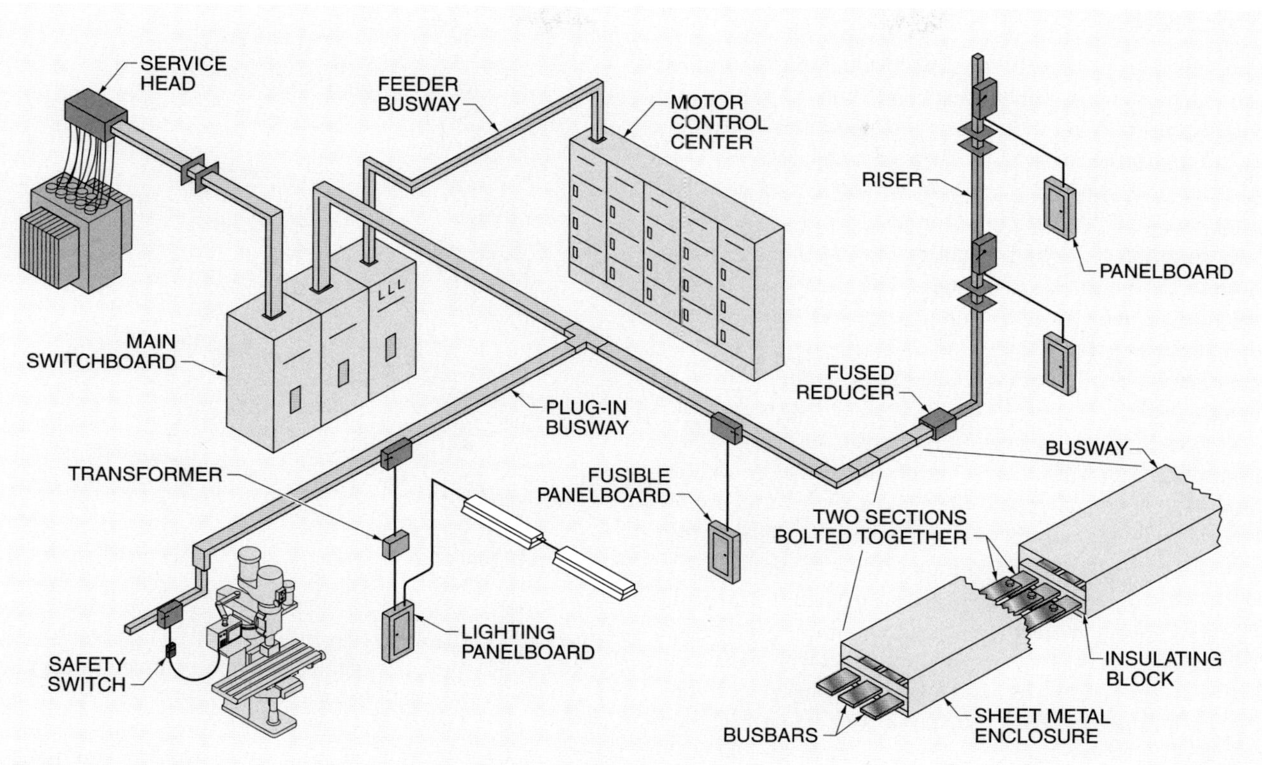

Figure 11-22. The electrical distribution system in a plant must transport the electrical power from the source of supply to the loads.

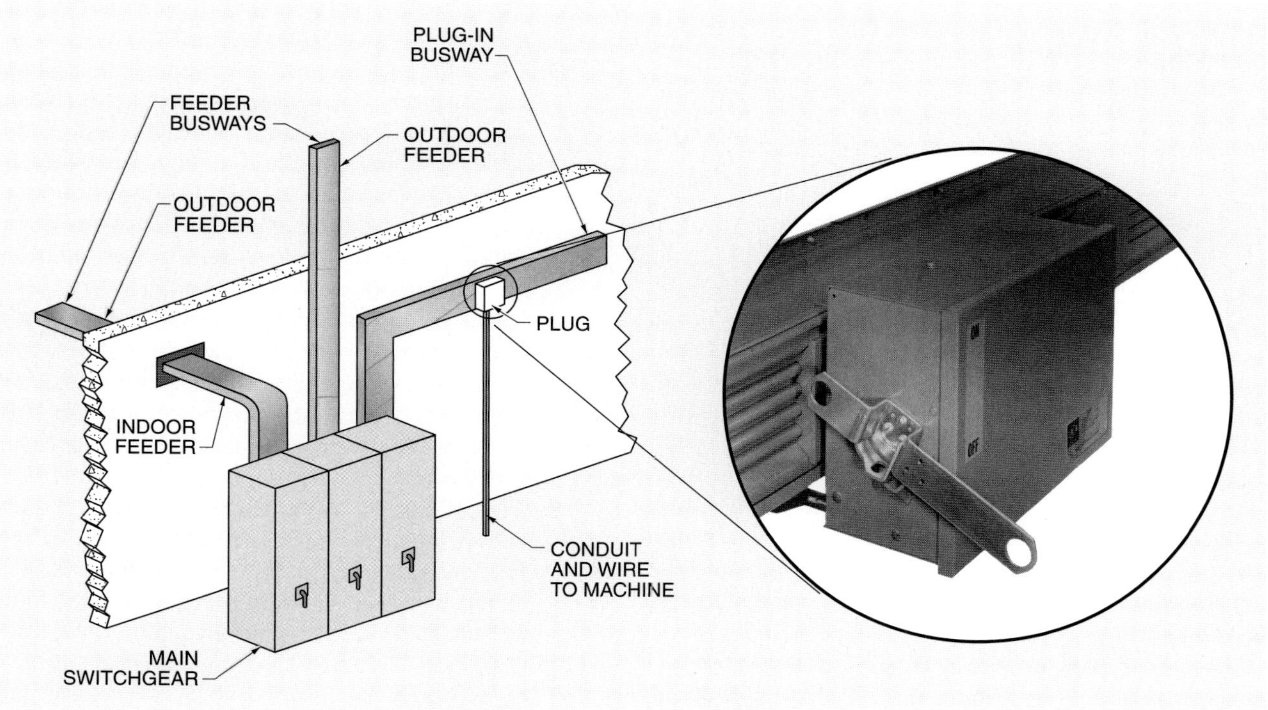

Figure 11-23. The two basic types of busways are feeder and plug-in busways.

Grounding

Equipment grounding is required throughout the entire distribution system. This means connecting to ground all non-current-carrying metal parts including conduit, raceways, transformer cases, and switchgear enclosures. The objective of grounding is to limit the voltage between all metal parts and the earth to a safe level.

Grounding is accomplished by connecting the non-current-carrying metal to a ground bus with an approved grounding conductor and fitting. A ground bus is a network that ties solidly to grounding electrodes. A *ground electrode* is a conductor embedded in the earth to provide a good ground.

A ground bus should surround the transmission station or building. This bus must be connected to the grounding electrodes in several spots. The size of the ground bus is determined by the amount of current that flows through the grounding system and the length of time it flows.

In addition to grounding all non-current-carrying metal, lightning arresters may be needed. A *lightning arrester* is a device that protects transformers and other electrical equipment from voltage surges caused by lightning. A lightning arrester provides a path over which the surge can pass to ground before it has a chance to damage electrical equipment.

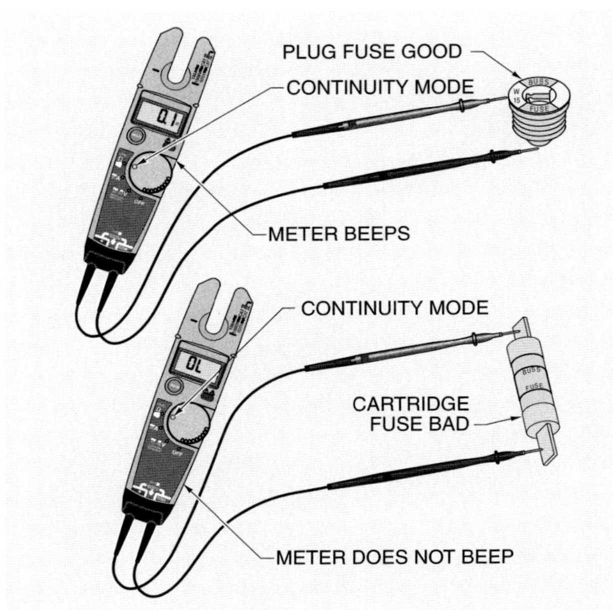

Figure 11-24. Fuses can be checked using a continuity tester set to measure continuity placed across a fuse that has been removed from a circuit.

Troubleshooting Fuses

A *fuse* is an overcurrent protection device (OCPD) with a fusible link that melts and opens the circuit on an overcurrent condition. Fuses are connected in series with a circuit to protect the circuit from overcurrents or short circuits. Fuses may be one-time or renewable. One-time fuses are fuses that cannot be reused after they have opened. One-time fuses are the most common. Renewable fuses are OCPDs designed so that the fusible link can be replaced. Fuses can be checked using a continuity tester or a DMM set to measure voltage.

Fuses can be checked using a continuity tester placed across a fuse that has been removed from a circuit. **See Figure 11-24.** When testing a good fuse, the continuity tester beeps because the circuit inside the fuse is closed, has a low resistance, and lets current flow through the fuse and continuity test circuit. When testing a bad fuse, the continuity tester does not beep because the circuit inside the fuse is open, has an infinite resistance (OL), and does not allow current to flow through the fuse and continuity test circuit.

Fuses can also be checked using a DMM set to measure voltage. **See Figure 11-25.** To troubleshoot fuses using a DMM set to measure voltage, apply the following procedure:

1. Turn the handle of the safety switch or combination starter to the OFF position.
2. Open the door of the safety switch or combination starter. The operating handle must be capable of opening the switch. If the operating handle is not working properly, replace the switch.
3. Check the enclosure and interior parts for deformation, displacement of parts, and burning. Such damage may indicate a short circuit, fire, or lightning strike. Deformation requires replacement of the part or complete device. Any indication of arcing damage or overheating, such as discoloration or melting of insulation, requires replacement of the damaged part(s).
4. Check the incoming voltage between each pair of power leads. Incoming voltage should be within 10% of the voltage rating of the motor. A secondary problem exists if voltage is not within 10%. This secondary problem may be the reason the fuses have blown.
5. Test the enclosure for grounding if voltage is present and at the correct level. To test for grounding, connect one side of a DMM set to measure voltage to an unpainted metal part of the enclosure and touch the other side to each of the incoming power leads. A voltage difference is indicated if the enclosure is properly grounded. The line-to-ground voltage probably does not equal the line-to-line voltage reading taken in Step 4.

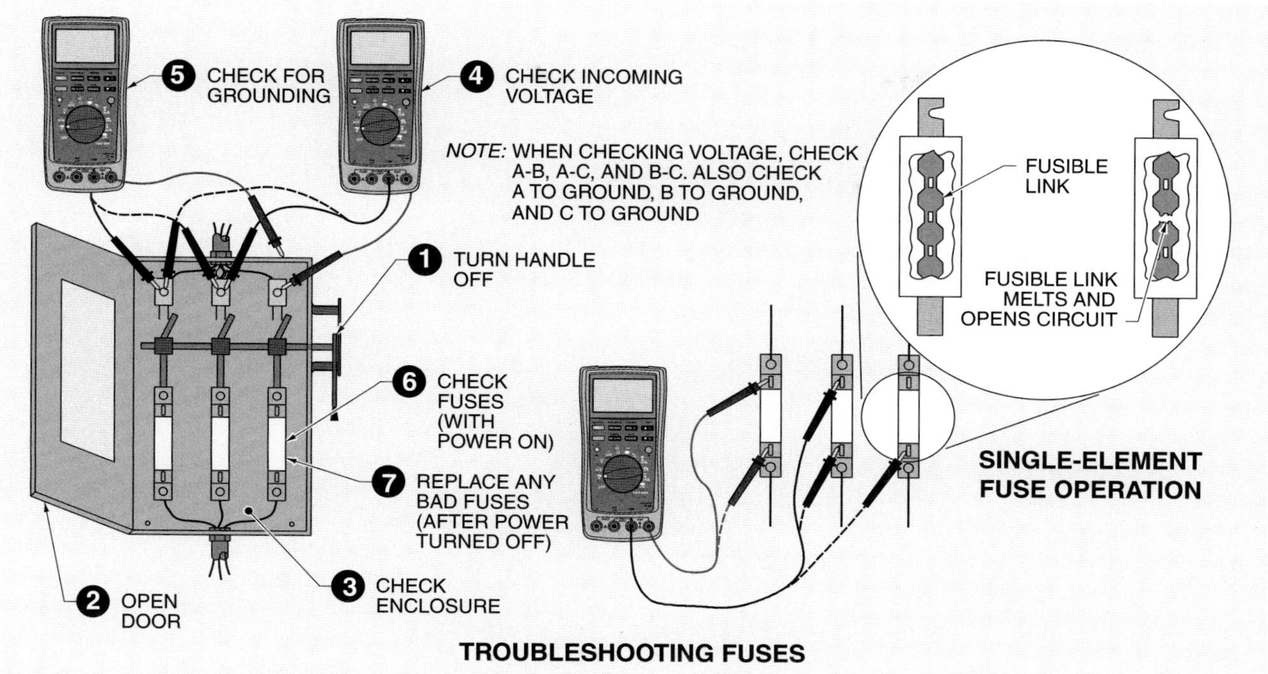

Figure 11-25. Fuses can be checked using a DMM set to measure voltage.

6. Check fuses. Turn the handle of the safety switch or combination starter to the ON position to test the fuses. **Warning:** When turning a circuit breaker, disconnect, etc. ON, always stand to the side and look away. One side of a DMM set to measure voltage is connected to one side of an incoming power line at the top of one fuse. The other side of the DMM is connected to the bottom of each of the remaining fuses. A voltage reading indicates the fuse is good. If no voltage reading is obtained, the fuse is open and no voltage passes through. The fuse must be replaced (not at this time). Repeat this procedure for each fuse. When testing the last fuse, the DMM is moved to a second incoming power line.
7. Replace bad fuses. Turn the handle of the safety switch or combination starter to the OFF position to replace the fuses. Use a fuse puller to remove bad fuses. Replace all bad fuses with the correct type and size replacement. Close the door on the safety switch or combination starter and turn the circuit to the ON position.

Troubleshooting Circuit Breakers

A *circuit breaker (CB)* is an overcurrent protection device with a mechanical mechanism that may manually or automatically open the circuit when an overload condition or short circuit occurs. CBs are connected in series with the circuit. They protect a circuit from overcurrents or short circuits. CBs are thermally or magnetically operated and are reset after an overload. A DMM set to measure voltage is used to test CBs. Circuit breakers perform the same function as fuses and are tested the same way. **See Figure 11-26.** To troubleshoot CBs, apply the following procedure:

1. Turn the handle of the safety switch or combination starter to the OFF position.
2. Open the door of the safety switch or combination starter. The operating handle must be capable of opening the switch. Replace the operating handle if it does not open the switch.
3. Check the enclosure and interior parts for deformation, displacement of parts, and burning.
4. Check the incoming voltage between each pair of power leads. Incoming voltage should be within 10% of the voltage rating of the motor.
5. Test the enclosure for grounding if voltage is present and at the correct level.
6. Examine the CB. It is in one of three positions: ON, TRIPPED, or OFF.
7. Reset the CB. If no evidence of damage is present, reset the CB by moving the handle to the OFF position and then back to the ON position. CBs must be cooled before they are reset. CBs are designed so they cannot be held in the ON position if an overload or short is present. Check the voltage of the reset CB if resetting the CB does not restore power. Replace all faulty CBs. Never try to service a faulty CB.

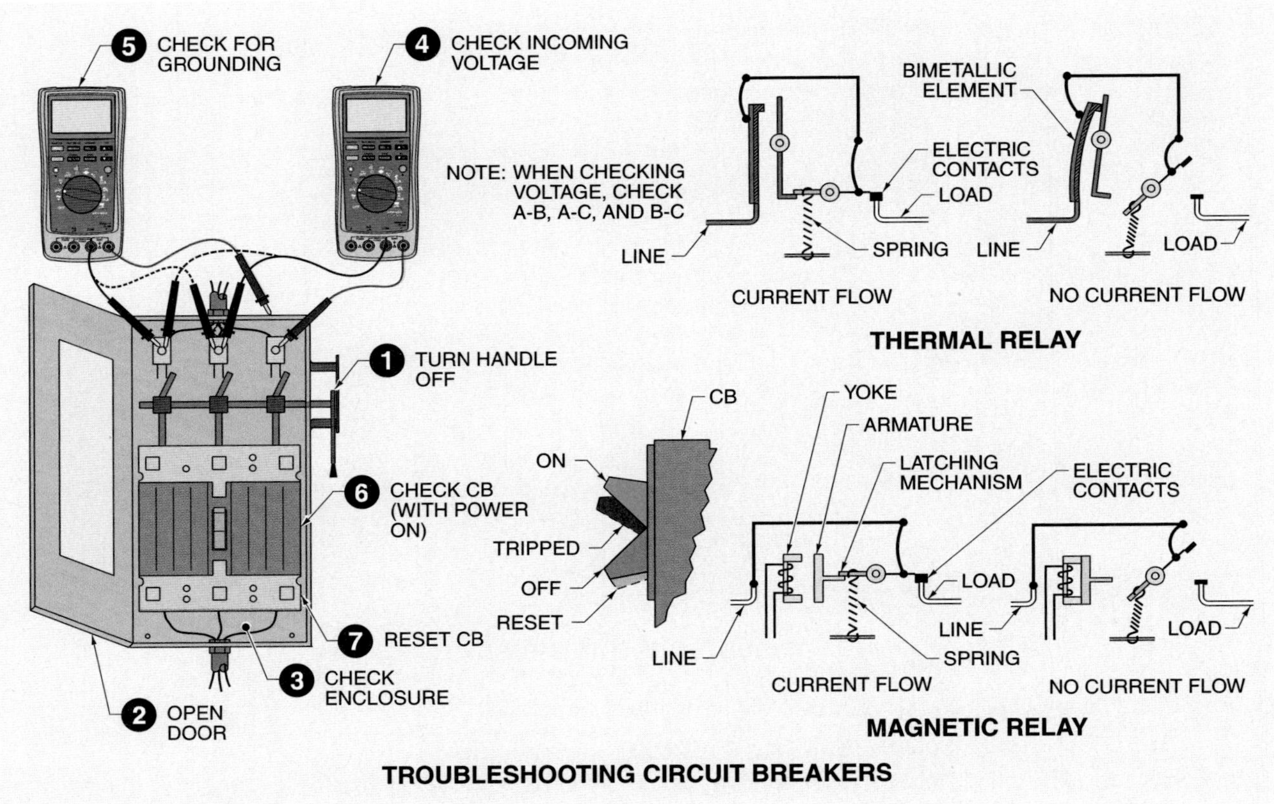

Figure 11-26. Circuit breakers (CBs) perform the same function as fuses and are tested in the same manner.

Testing Control Transformers

A control transformer is used in a circuit to step down the supply voltage to provide a safe voltage level for the control circuit. A control transformer should be checked if there is a problem in a control circuit that may be related to the power supply. **See Figure 11-27.**

All transformers are capable of delivering a limited current output at a given voltage. The power limit of a transformer equals the current times the voltage. This power limit is listed on the nameplate of the transformer as its kilovoltampere (kVA) rating. This rating indicates the apparent power the transformer can deliver. The transformer overheats and the control circuit does not function properly if this limit is exceeded. The transformer comes closer to reaching its limit when loads are added. To test a control transformer, apply the following procedure:

1. Check the input and output voltages of the transformer with the power supply energized. The input and output voltages should be within 5% of the transformer nameplate rating (10% max). The transformer is good if the voltage is within the rating or proportionally low.
2. Measure the current drawn by the transformer with a clamp-on ammeter. The apparent power drawn by the control circuit is determined by multiplying the current reading by the voltage reading. A larger transformer is required if the voltamperes (VA) drawn are more than the rating of the transformer.
3. Check the transformer ground. A ground test should be performed on new transformer installations or if a ground problem is suspected. Connect one lead of a DMM set to measure voltage to the metal frame of the transformer. Do not connect it to a painted or varnished surface. Connect the second lead to each lead of the transformer on the secondary. Under normal circumstances, if X2 is grounded, the DMM displays a voltage when connected to X1. Under normal circumstances, the DMM will also read a voltage when connected to H1 (if hot side of primary) and no voltage on H2 (if neutral side of primary).

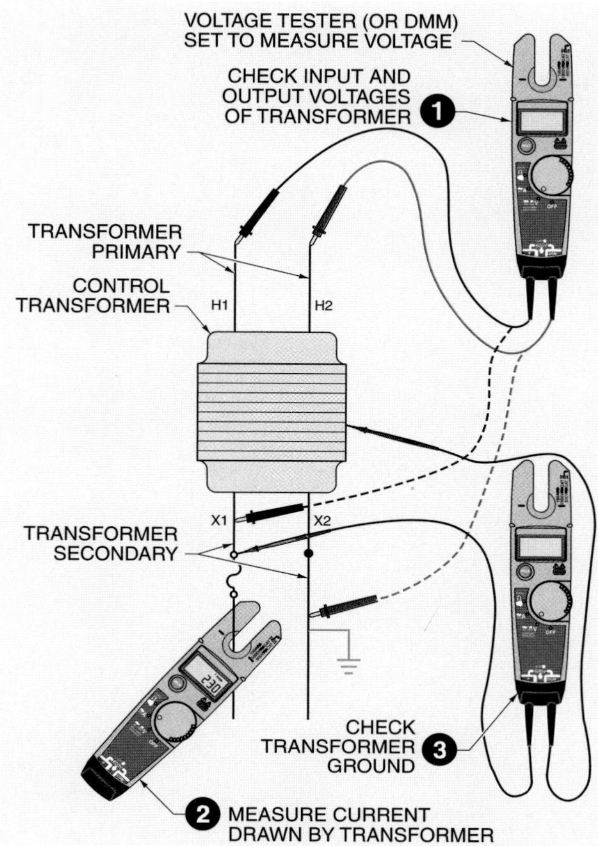

Figure 11-27. A control transformer should be checked if there is a problem in a control circuit that may be related to the power supply.

Locating Circuits in Switchboards, Panelboards, or Load Centers

A technician must often locate one circuit in a switchboard, panelboard, or load center to turn OFF the power before troubleshooting or working on a circuit. Switchboards, panelboards, and load centers are often crowded with wires that are not marked or that are mismarked. A technician cannot start turning OFF each circuit until the correct circuit is found because this disconnects all loads connected to that circuit. Timers, counters, clocks, starters, and other control devices must be reset, or critical equipment such as alarms and safety circuits may be stopped. A flashing lamp and a clamp-on ammeter may be used to isolate a particular circuit. **See Figure 11-28.**

The flashing lamp is plugged into any receptacle on the circuit that is to be disconnected. As the lamp is flashing ON and OFF, a clamp-on ammeter is used to check each circuit. Each circuit displays a constant current reading except the one with the flashing lamp. The circuit with the flashing lamp displays a varying value on the ammeter equal to the flashing time of the lamp. This circuit may then be turned OFF for troubleshooting.

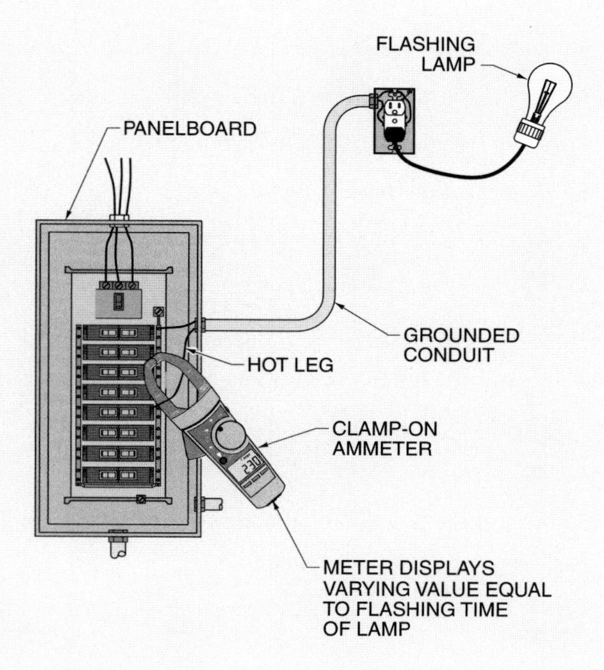

Figure 11-28. A flashing lamp and a clamp-on ammeter may be used to isolate a particular circuit.

Review Questions

1. What type of diagram is used to show flow path, voltage values, disconnects, overcurrent protection devices, transformers, and panelboards in a circuit?
2. What device is installed to protect electrical equipment from lightning surges?
3. What are the voltages available from a wye-connected system with a common neutral?
4. What is the phase-to-phase voltage if the phase-to-neutral voltage is 208 V in a wye-connected system?
5. What are the voltages available from a 240 V delta-connected system without a neutral wire?
6. How is the high side of a transformer identified?
7. How is the low side of a transformer identified?
8. How is a transformer rated for power output?
9. Why is it important to balance a transformer bank?
10. What are the three main parts of a substation?
11. What is the last point on a power distribution system, as far as the power company is concerned?
12. What is the difference between a service-entrance switchboard and a distribution switchboard?
13. What is a panelboard?
14. What is a branch circuit?
15. What is the function of a motor control center in a power distribution system?

chapter 12

electrical motor controls *for Integrated Systems*

Solid-State Devices and System Integration

Semiconductor devices are devices that have electrical conductivity between that of a conductor and that of an insulator. Semiconductor devices include diodes, thermistors, photocells, solar cells, Hall effect sensors, transistors, SCRs, triacs, UJTs, diacs, ICs, operational amplifiers, etc.

ELECTRONIC CONTROL SYSTEMS AND DEVICES

Control circuits can be simplified as consisting of signals, decisions, and actions. **See Figure 12-1.** Although this description holds true for simple circuits, it does not fully explain the integration of complex circuits, especially those using solid-state devices. The development of solid-state devices has greatly increased the capabilities of control circuits.

In the 1970s, solid-state devices began entering the manufacturing environment. This created computer-aided drafting and design (CADD), computer numerical control (CNC) for machine tools such as engine lathes and milling machines, and programmable controllers (PLCs) used to control complete factory processes. The addition of solid-state devices to the manufacturing environment has created an integrated systems technology that enables precise control of these processes.

A single integrated technology system can be comprised of input transducers, amplifiers, processing logic, switching/communication/interfacing, and outputs. **See Figure 12-2.** Inputs can come from a variety of sources such as temperature, light, pressure, flow, and speed. Because some of the input signals are quite small, they require amplification before being processed. The processing logic may be a simple AND and OR logic with memory, or a complex set of instructions such as those found in a PLC.

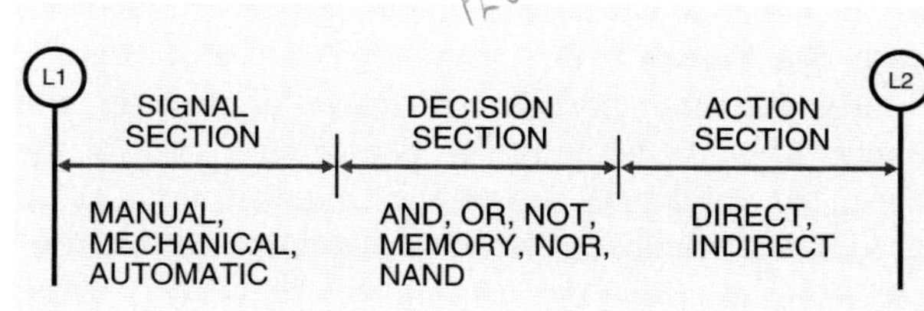

Figure 12-1. All control circuits are composed of signal, decision, and action sections.

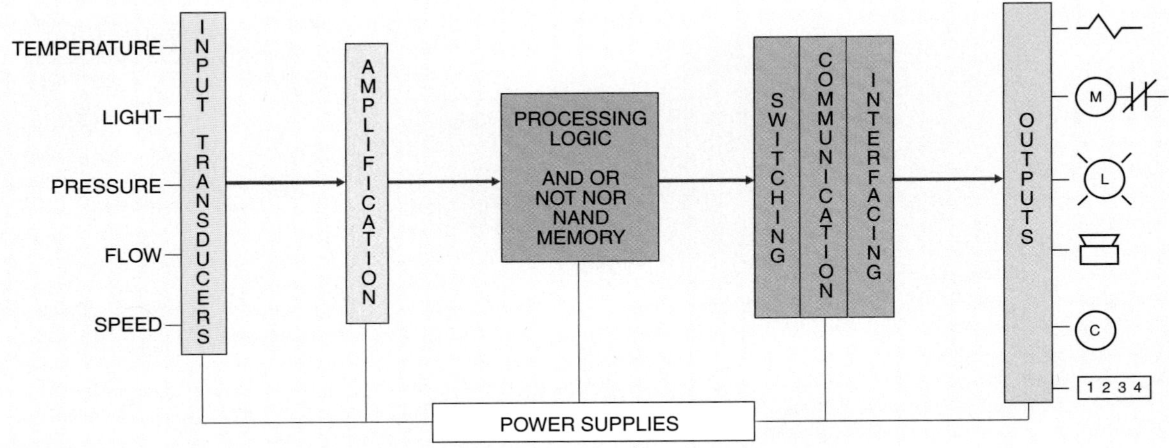

Figure 12-2. An integrated technology system is comprised of input transducers, amplifiers, processing logic, switching/communication/interfacing, and outputs.

> **Technical Fact**
>
> Printed circuit boards are also referred to as PC boards. PC boards are extremely sensitive to static electricity and should only be handled by certified electronics technicians who are attached to the proper grounding equipment. The work area must be free of static electricity and direct contact should be avoided with the components of the PC board.

Once the decision has been made, the signal is communicated to an interface or switching device that controls the outputs or loads. The outputs or loads could be solenoids, motors, lights, alarms, control relays, or display devices. A variety of well-regulated power supplies are required to energize these components or systems.

Solid-state components are also no longer isolated from each other in industrial circuits. Solid-state components may be integrated or connected on a printed circuit (PC) board. **See Figure 12-3.** A *printed circuit (PC) board* is an insulating material such as fiberglass or phenolic with conducting paths laminated to one or both sides of the board. The solid-state components mounted on a single PC board may include input devices, amplifiers, and decision-processing devices. The components on the PC board are referred to as semiconductor devices and are rapidly replacing mechanical devices in control applications where precision and reliability are required.

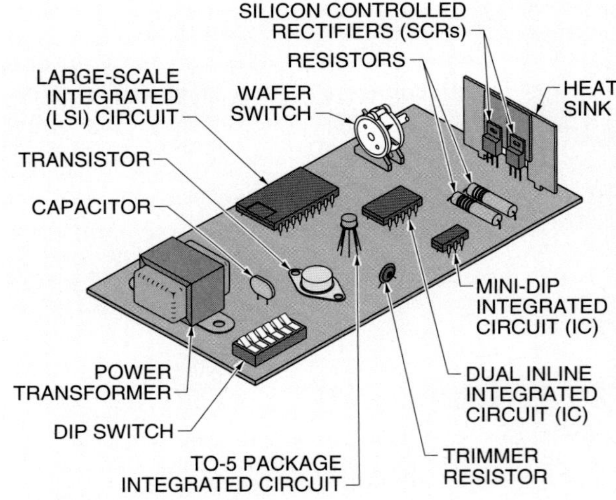

Figure 12-3. Solid-state control devices are normally mounted on a printed circuit (PC) board.

Semiconductor Devices

Semiconductor devices are devices that have electrical conductivity between that of a conductor (high conductivity) and that of an insulator (low conductivity). Semiconductor devices are often mounted on a PC board. PC boards provide electrical paths of sufficient size to ensure a reliable electronic circuit. **See Figure 12-4.** *Pads* are small round conductors to which component leads are soldered. *Traces (foils)* are conducting paths used to connect components on a PC board. Traces are used to interconnect two or more pads. A *bus* is a large trace extending around the edge to provide conduction from several sources.

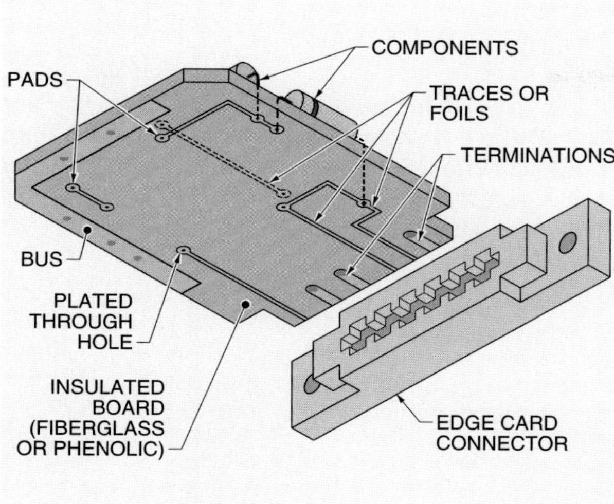

Figure 12-4. A PC board is constructed of an insulating material such as fiberglass or phenolic with conducting paths laminated on one or both sides of the board.

An *edge card* is a PC board with multiple terminations (terminal contacts) on one end. Most edge cards have terminations made from copper, which is the same material as the traces. In some instances the terminations are gold plated, allowing for the lowest possible contact resistance. An *edge card connector* is a connector that allows the edge card to be connected to the system's circuitry with the least amount of hardware. An edge card must not be removed while the card is energized. This could result in permanent damage to the card.

Semiconductor control devices are normally mounted on one side of a PC board. In some cases where space is at a premium, components may be mounted on both sides of the PC board. Component leads extend through the board and are connected to the pads, traces, and bus with solder. PC boards may have markings next to each component to help identify the component in relation to the schematic.

Semiconductor Theory

All matter consists of an organized collection of atoms. An *atom* is the smallest particle that an element can be reduced to and still keep the properties of that element. The three fundamental particles contained in atoms are protons, neutrons, and electrons. Protons and neutrons make up the nucleus, and electrons whirl about the nucleus in orbits or shells.

The *nucleus* is the heavy, dense center of an atom and has a positive electrical charge. A *proton* is a particle contained in the nucleus of an atom that has a positive electrical charge. A *neutron* is a particle contained in the nucleus of an atom that has no electrical charge. The nucleus is surrounded by one or more electrons. *Electrons* are negatively charged particles whirling around the nucleus at great speeds in shells. Each shell can hold a specific number of electrons. The innermost shell can hold two electrons. The second shell can hold eight electrons. The third shell can hold 18 electrons, etc. The shells are filled starting with the inner shell and working outward, so that when the inner shells are filled with as many electrons as they can hold, the next shell is started. Electrons and protons have equal amounts of opposite charges. There are as many electrons as there are protons in an atom, which leaves the atom electrically neutral.

Technical Fact

Semiconductors with impurities that have a shortage of electrons are called P-type semiconductors. Semiconductors with a surplus of electrons are called N-type semiconductors. Together, N-type and P-type semiconductors are the basic building blocks of most solid-state electronic devices.

Valence Electrons. Most elements do not have a completed outer shell with the maximum allowable number of electrons. *Valence electrons* are electrons in the outermost shell of an atom. Valence electrons determine the conductive or insulative value of a material. Conductors normally have only one or two valence electrons in their outer shell. **See Figure 12-5.** Insulators normally have several electrons in their outer shell, which is either almost or completely filled with electrons.

Semiconductor materials fall between the low resistance offered by a conductor and the high resistance offered by an insulator. Semiconductors are made from materials that have four valence electrons.

Doping

The basic material used in most semiconductor devices is either germanium or silicon. In their natural state, germanium and silicon are pure crystals. These pure crystals do not have enough free electrons to support a significant current flow. To prepare these crystals for use as a semiconductor device, their structure must be altered to permit significant current flow.

Doping is the addition of impurities to the crystal structure of a semiconductor. In doping, some of the atoms in the crystal are replaced with atoms of other elements. The addition of new atoms in the crystal structure creates N-type material and P-type material.

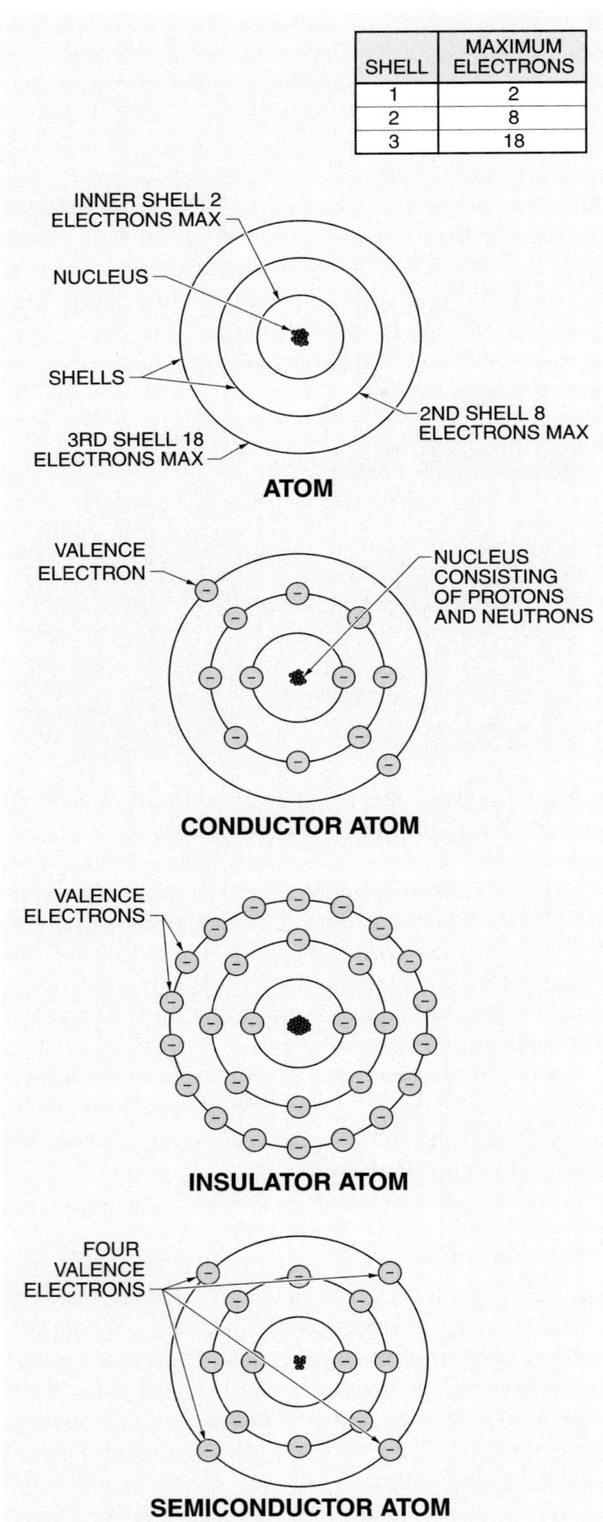

SHELL	MAXIMUM ELECTRONS
1	2
2	8
3	18

Figure 12-5. Valence electrons determine the amount of conductivity or insulating characteristics of a given material.

N-Type Material. *N-type material* is material created by doping a region of a crystal with atoms of an element that has more electrons in its outer shell than the crystal. Adding these atoms to the crystal results in more free electrons. Free electrons (carriers) support current flow. Current flows from negative to positive through the crystal when voltage is applied to N-type material. The material is N-type material because electrons have a negative charge. **See Figure 12-6.**

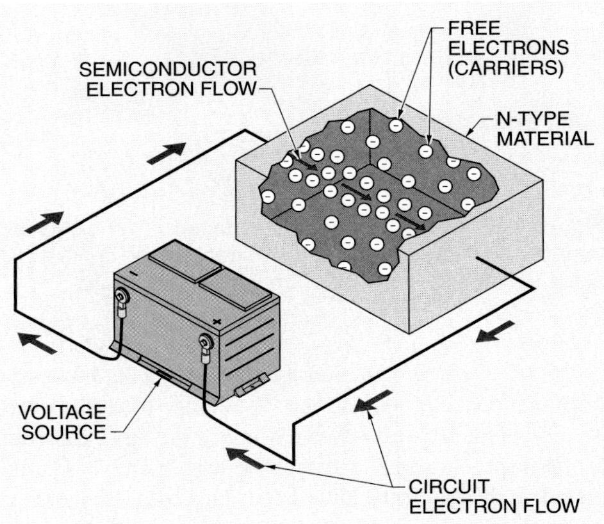

Figure 12-6. Current flows from negative potential to positive potential and is assisted by free electrons when voltage is applied to N-type material.

Elements commonly used for creating N-type material are arsenic, bismuth, and antimony. The quantity of doping material used ranges from a few parts per billion to a few parts per million. By controlling these small quantities of impurities in a crystal, the manufacturer controls the operating characteristics of the semiconductor.

P-Type Material. *P-type material* is material with empty spaces (holes) in its crystal structure. To create P-type material, a crystal is doped with atoms of an element that has fewer electrons in its outer shell than the crystal. *Holes* are the missing electrons in the crystal structure. The holes are represented as positive charges.

In P-type material, the holes act as carriers. The holes are filled with free electrons when voltage is applied, and the free electrons move from negative potential to positive potential through the crystal. **See Figure 12-7.** Movement of the electrons from one hole to the next makes the holes appear to move in the opposite direction. Hole flow is equal to and opposite of electron flow. Typical elements used for doping a crystal to create P-type material are gallium, boron, and indium.

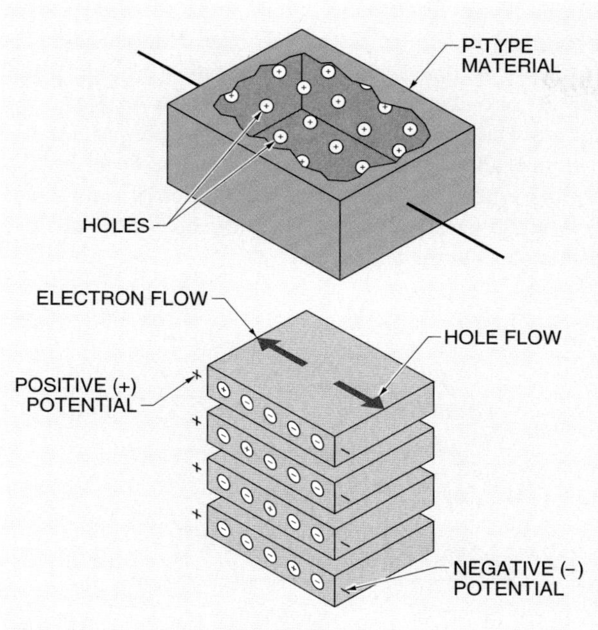

Figure 12-7. When voltage is applied to P-type material, the holes are filled with free electrons that move from the negative potential to the positive potential through the crystal.

DIODES, RECTIFICATION, AND POWER SUPPLIES

When semiconductors are placed in a circuit, they must receive the proper voltage level. Minor variations in voltage can reduce the output of solid-state devices. Major variations in voltage can quickly destroy solid-state devices. DC power supplies include a transformer, diode rectifier, filter, and voltage regulator. **See Figure 12-8.**

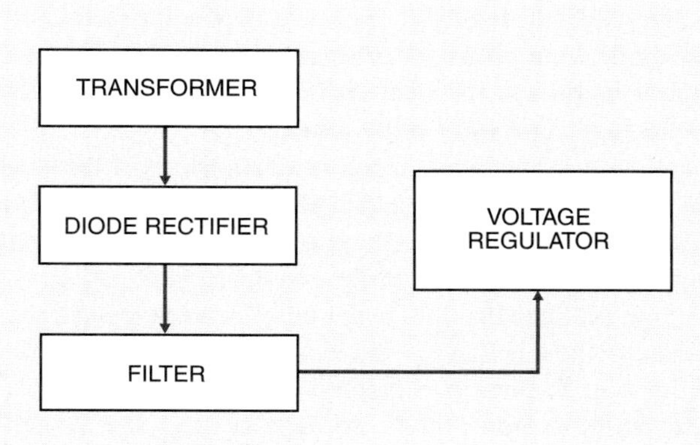

Figure 12-8. A DC power supply includes a transformer, diode rectifier, filter, and voltage regulator.

Technical Fact

Diodes are devices that conduct electricity in one direction only and are sometimes referred to as PN (positive-negative) devices because they are made of a single semiconductive crystal with a P-type (positive terminal or anode) area and an N-type area (negative terminal or cathode).

Diodes

A *diode* is an electronic component that allows current to pass through it in only one direction. This is made possible by the doping process, which creates N-type material and P-type material on the same component. The P-type and N-type materials exchange carriers at the junction of the two materials, creating a thin depletion region. **See Figure 12-9.**

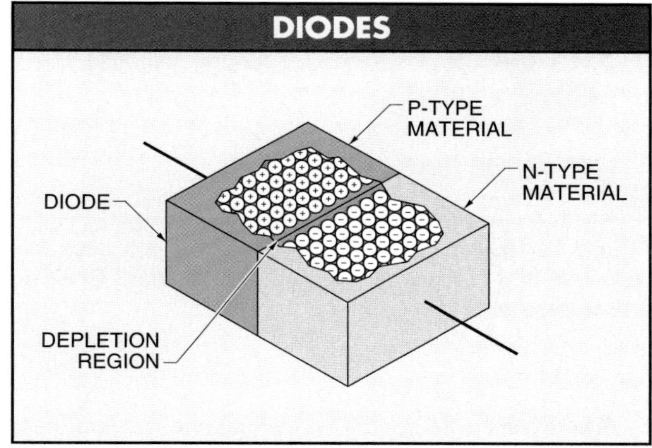

Figure 12-9. In a diode, P-type and N-type materials exchange carriers at the junction of the two materials, creating a thin depletion region.

The thin depletion region responds rapidly to voltage changes. The operating characteristics of a specific diode can be determined through the use of its characteristic curve. **See Figure 12-10.**

When voltage is applied to a diode, the action occurring in the depletion region either blocks current flow or passes current. *Forward-bias voltage* is the application of the proper polarity to a diode. Forward bias results in forward current. *Reverse-bias voltage* is the application of the opposite polarity to a diode. Reverse bias results in a reverse current, which should be very small (normally 1 mA).

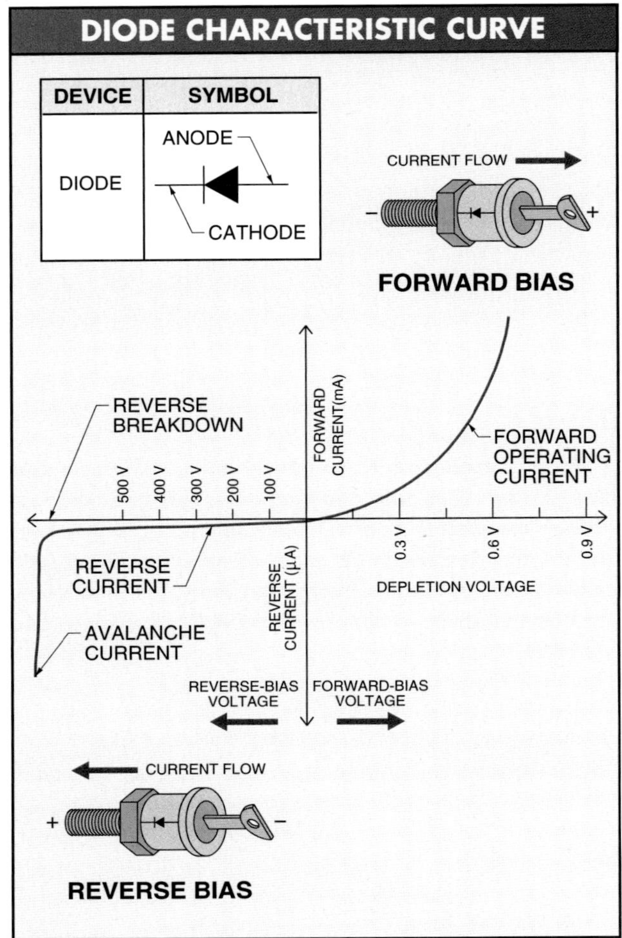

Figure 12-10. A diode characteristic curve indicates the response of a diode when subjected to different forward- and reverse-bias voltages.

Peak inverse voltage (PIV) is the maximum reverse bias voltage that a diode can withstand. The PIV ratings for most diodes used in industry range from a few volts to several thousand volts. A diode breaks down and passes current freely if the reverse bias applied to the diode exceeds its PIV rating. *Avalanche current* is current passed when a diode breaks down. Avalanche current can destroy diodes. Diodes with the correct voltage rating must be used to avoid avalanche current.

Rectification of Alternating Current

Alternating current (AC) power is more efficiently and economically generated and transmitted than direct current (DC) power. Alternating current must be changed to DC because machinery and other loads often need DC to operate. *Rectification* is the changing of AC to DC.

Single-Phase Rectifiers. A half-wave rectifier is used to convert AC to pulsating DC. **See Figure 12-11.** A *half-wave rectifier* is a circuit containing a diode which permits only the positive half-cycles of the AC sine wave to pass. Half-wave rectification is accomplished because current is allowed to flow only when the anode terminal is positive with respect to the cathode. Current is not allowed to flow through the rectifier when the cathode is positive with respect to the anode.

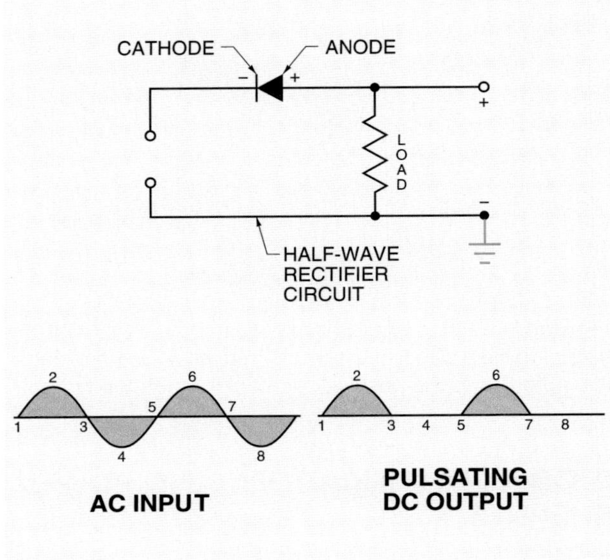

Figure 12-11. A half-wave rectifier converts AC to pulsating DC.

The output voltage of a half-wave rectifier is considered pulsating DC with half of the AC sine wave cut off. A half-wave rectifier passes either the positive or negative half-cycle of the input AC sine wave, depending on the way the diode is connected into the circuit. Half-wave rectification is inefficient for most applications because one-half of the input sine wave is not used.

A full-wave rectifier circuit uses both halves of the input AC sine wave. Full-wave rectification may be obtained from a single-phase AC source by using two diodes with a center-tapped transformer, or by using a bridge rectifier circuit.

In a circuit using two diodes and a center-tapped transformer, when voltage is induced in the secondary from point A to B, point A is positive with respect to point N. Current flows from A to N, through the load, and through diode 1 (D1). **See Figure 12-12.** Diode 1 conducts current and diode 2 (D2) blocks current because A is positive with respect to N.

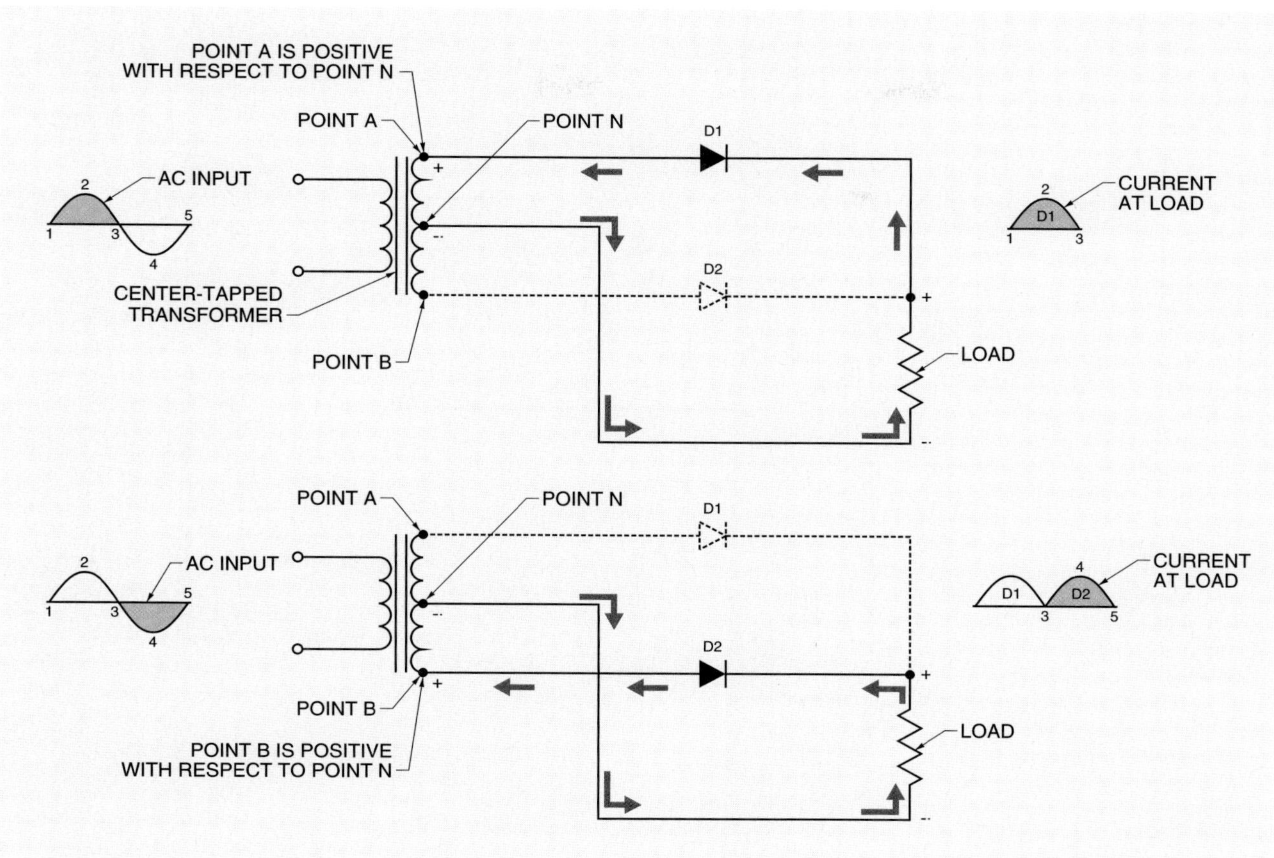

Figure 12-12. Full-wave rectification may be obtained from a 1ϕ AC source by using two diodes with a center-tapped transformer.

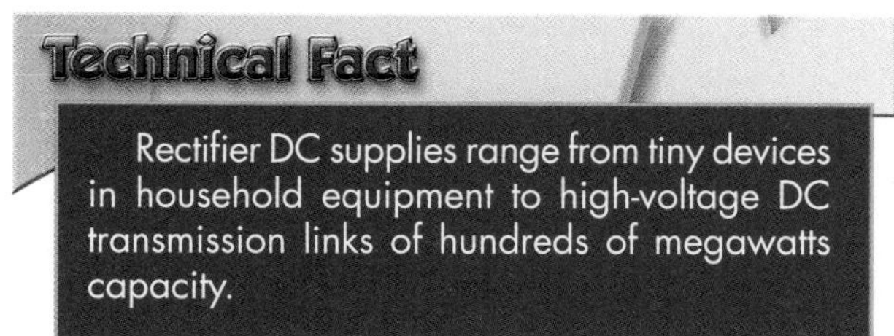

When the voltage across the secondary reverses during the negative half-cycle of the AC sine wave, point B is positive with respect to point N. Current then flows from B to N, through the load, and through D2. Diode 2 conducts and diode 1 blocks current because B is positive with respect to N. This is repeated every cycle of the AC sine wave, producing a full-wave DC output.

A bridge rectifier circuit produces the same full-wave DC output. **See Figure 12-13.** A bridge rectifier circuit requires four diodes and eliminates the need for a transformer. A bridge rectifier circuit is more efficient than a center-tapped circuit because each diode blocks only half as much reverse voltage for the same output voltage.

In this circuit, when the AC supply voltage is positive at point A and negative at point B, current flows from point B, through D2, the load, D1, and to point A. When the AC supply voltage is positive at point B and negative at point A, current flows from point A, through D4, the load, D3, and to point B.

The output of a full-wave rectifier is pulsating DC and must be filtered or smoothed out before it can be used in most electronic equipment. This filtering is done by a filter circuit connected to the output of the rectifier circuit. This filter circuit normally consists of one or more capacitors, inductors, or resistors connected in different combinations. The choice of a filter circuit is determined by the load (how much ripple it can take), cost, and available space.

Filtered DC eliminates pulsations and provides DC at a constant level. **See Figure 12-14.** This is accomplished because the pulsating voltage no longer drops to zero at the end of each pulsation. This results in the average voltage delivered by the rectifier circuit being higher. The purpose of a filter is to smooth and increase the DC voltage output of the circuit.

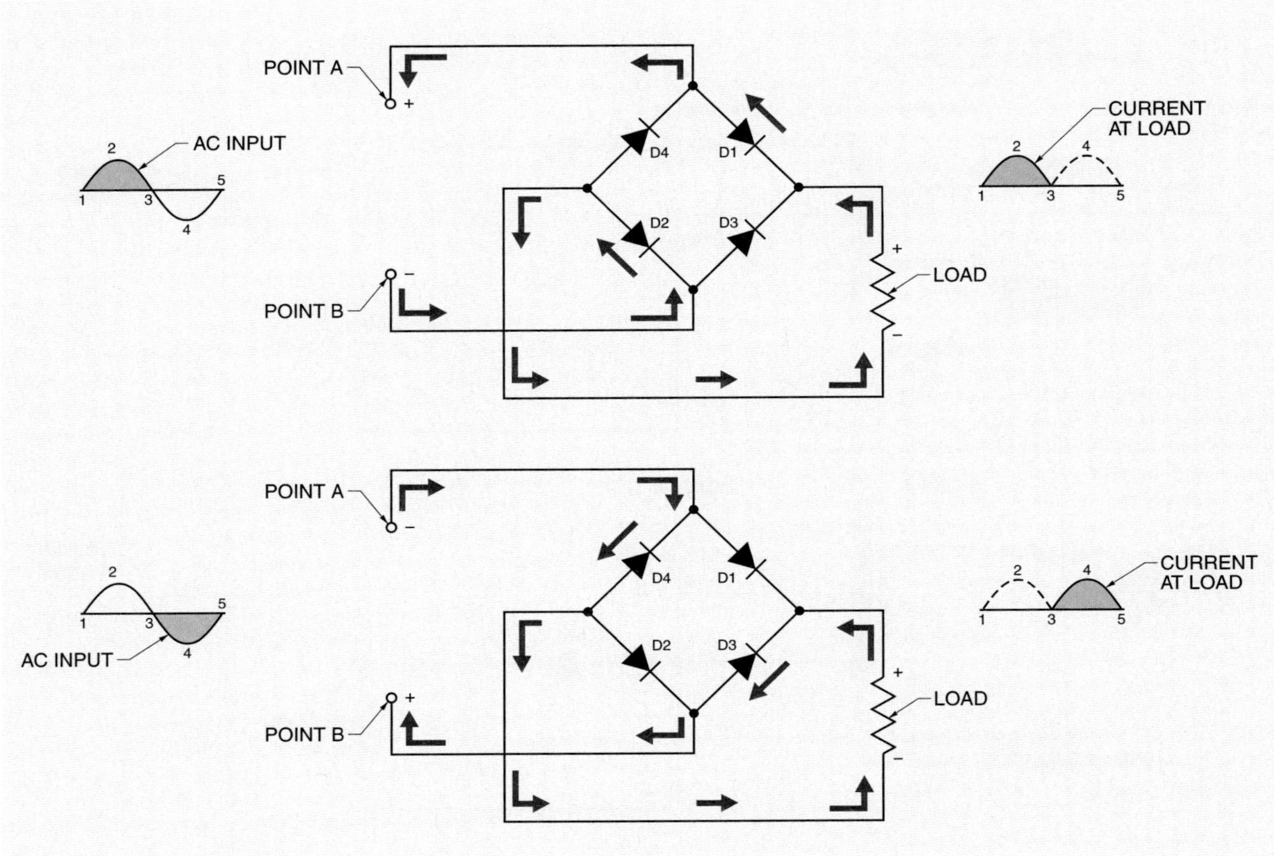

Figure 12-13. A bridge rectifier circuit is more efficient than a center-tapped circuit because each diode blocks only half as much reverse voltage for the same output voltage.

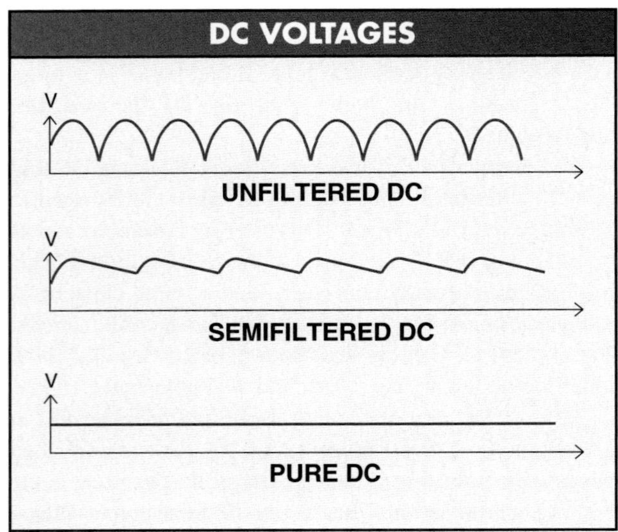

Figure 12-14. Filtered DC eliminates pulsations and provides DC at a constant level.

Three-Phase Rectifiers. A DC output can also be supplied from a three-phase power source. The advantage of using three-phase power is that it is possible to obtain a smooth DC output without the use of a filter circuit. This is possible in a three-phase circuit because, when any one phase becomes negative, at least one of the other phases becomes positive. The result is a relatively smooth output without any filtering.

A three-phase rectifier circuit uses six diodes connected to a wye circuit with a neutral tap. **See Figure 12-15.** Each diode conducts in succession while the remaining two are blocking. The output voltage never goes below a certain voltage level. This circuit delivers the same smooth DC output as a filtered single-phase bridge rectifier circuit.

Zener Diodes

A *zener diode* is a silicon PN junction that differs from a rectifier diode in that it operates in the reverse breakdown region. A *PN junction* is the area on a semiconductor material between the P-type and N-type material. A zener diode acts as a voltage regulator either by itself or in conjunction with other semiconductor devices. A zener diode symbol differs from a standard diode symbol in that the normally vertical cathode line is bent slightly at each end. Standard diodes normally conduct in forward bias and can be destroyed if the reverse bias is exceeded. A zener diode is often referred to as an avalanche diode because it normally operates in reverse breakdown.

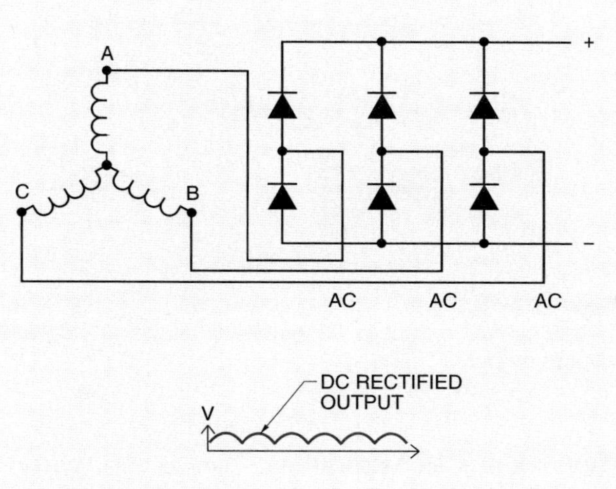

Figure 12-15. A three-phase rectifier circuit uses six diodes connected to a wye circuit to produce DC.

Zener Diode Operation. The forward breakover voltage and current characteristics are similar to a standard diode when a source voltage is applied to a zener diode in the forward direction. **See Figure 12-16.** When a source voltage is applied to a zener diode in the reverse direction, the current remains low until the reverse voltage reaches reverse breakdown (zener breakdown). The zener diode conducts heavily (avalanche breakdown) at zener breakdown. Reverse current flow through a zener diode must be limited by a resistor or other device to prevent diode destruction. The maximum current that may flow through a zener diode is determined by diode size. Like the forward voltage drop of a standard diode, the reverse voltage drop (zener voltage) of a zener diode remains essentially constant despite large current fluctuations.

A zener diode is capable of being a constant voltage source because of the resistance changes that take place within the PN junction. The resistance of the PN junction remains high and should produce leakage current in the microampere range when a source of voltage is applied to the zener diode in the reverse direction. However, as the reverse voltage is increased, the PN junction reaches a critical voltage and the zener diode avalanches. As the avalanche voltage is reached, the normally high resistance of the PN junction drops to a low value and the current increases rapidly. The current is normally limited by a circuit resistor or load resistance.

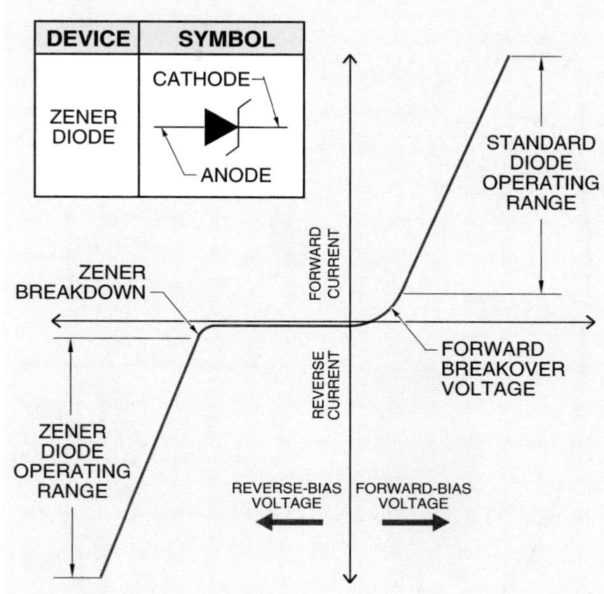

Figure 12-16. A zener diode is often referred to as an avalanche diode because it normally operates in reverse breakdown.

SOLID-STATE POWER SOURCES

A *solid-state power source* is a semiconductor device that controls electrons, electric fields, and magnetic fields in a solid material and typically has no moving parts. Photovoltaic cells (also referred to as solar cells) are commonly used as a solid-state power source.

Photovoltaic Cells

A *photovoltaic cell (solar cell)* is a device that converts solar energy to electrical energy. A solar cell is sensitive to light and produces a voltage without an external source. Several different solar cells are available. A solar cell is equivalent to a single-cell voltage source similar to that found in standard 6 V or 9 V batteries.

The use of solar cells as a remote power source is becoming more popular. Many manufacturers are designing solar cells into their products in individual and multi-cell applications. For example, most handheld calculators are powered by solar cells and do not require batteries.

Photovoltaic Cell Operation. A solar cell generates energy by using a PN junction to convert light energy into electrical energy. **See Figure 12-17.** Solar cells produce a potential difference between a pair of terminals only when exposed to light.

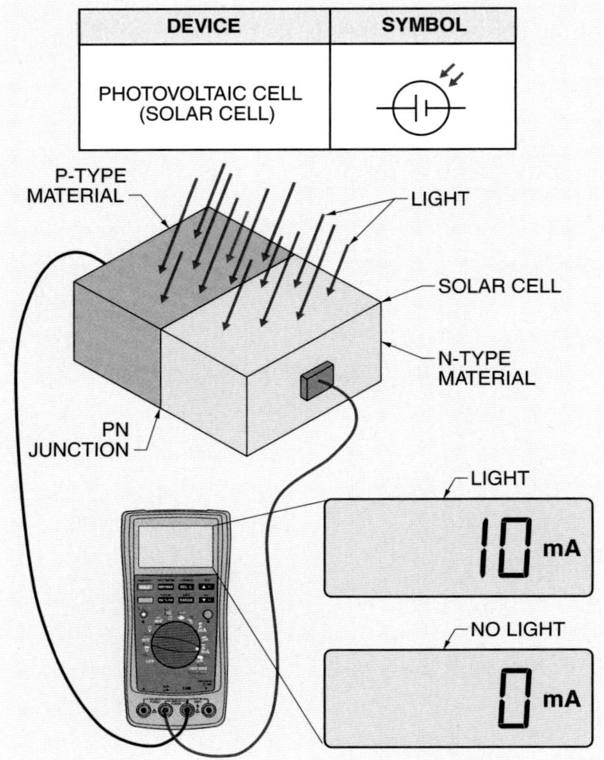

Figure 12-17. A solar cell generates energy by using a PN junction to convert light energy into electrical energy.

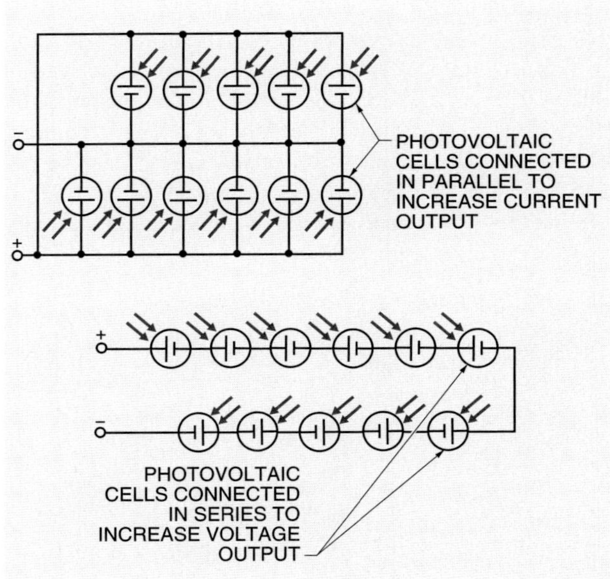

Figure 12-18. Solar cells are connected in parallel to increase current output or connected in series to increase voltage output.

At the junction of N-type material and P-type material, some recombination of the electrons and holes occurs, but the junction itself acts as a barrier between the two charges. The electrical field at the junction maintains the negative charges on the N-type material side and the positive charges in the holes on the P-type material side.

Current flows, with light acting as a generator, if a load is connected across the PN junction. When current flows through the load, the electron-hole pairs formed by light energy recombine and return to the normal condition prior to the application of light. Consequently, there is no loss or addition of electrons to the silicon during the process of converting light energy to electrical energy. A solar cell should have no limit to its life span, provided it is not damaged.

Photovoltaic Cell Output. Photovoltaic cells are rated by the amount of energy they convert. Most manufacturers rate the output in terms of volts (V) and milliamps (mA). Photovoltaic cells may operate with up to 0.5 V and as high as 40 mA (.040 A) per cell. To increase the current output from a set of photovoltaic cells, they should be connected in parallel. To increase the voltage output, they should be connected in series. **See Figure 12-18.**

Photovoltaic Cell Applications. The electrical systems of satellites are powered by solar energy. The solar cells convert light energy from the sun into electrical energy. Communication satellites, which operate almost indefinitely in space, have hundreds of solar cells, constantly converting solar energy into electrical energy. Satellites use the electrical energy to receive and transmit television and radio signals to and from the surface of the Earth. Weather satellites also use solar cells to produce electrical energy. **See Figure 12-19.** Satellite communication enables systems to be integrated anywhere in the world.

INPUT DEVICES

Input devices are often referred to as transducers. A *transducer* is a device used to convert physical parameters, such as temperature, pressure, and weight, into electrical signals. Transducers provide a link in integrating industrial control systems because they convert mechanical, magnetic, thermal, electrical, optical, and chemical variations into electrical voltage and current signals. These voltage and current signals are used directly or indirectly to drive other control systems. Because of the variety of solid-state devices available, many types of electromechanical transducers are being replaced with solid-state transducers. Solid-state transducers include thermistors, photoconductive cells (photocells), photoconductive diodes (photodiodes), Hall effect sensors, and pressure sensors.

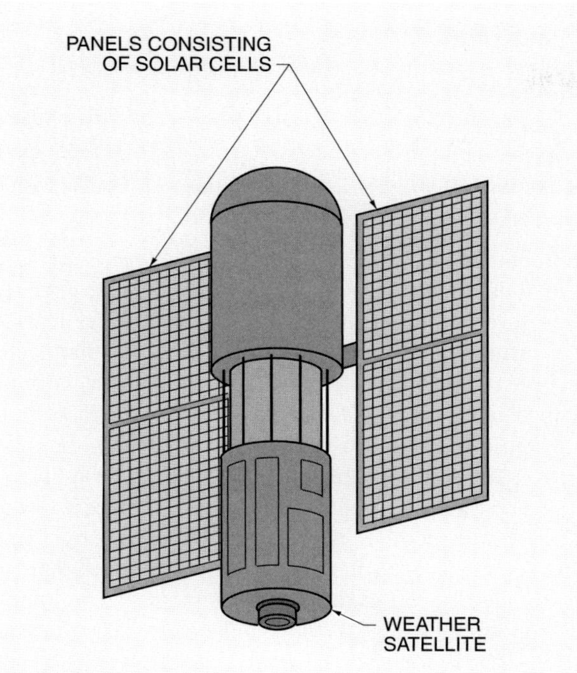

Figure 12-19. Weather satellites use solar cells to produce electrical energy.

Thermistors

A *thermistor* is a temperature-sensitive resistor whose resistance changes with a change in temperature. **See Figure 12-20.** The operation of a thermistor is based on the electron-hole theory. As the temperature of the semiconductor increases, the generation of electron-hole pairs increases due to thermal agitation. Increased electron-hole pairs cause a drop in resistance.

Thermistors are popular because of their small size, allowing them to be mounted in places that are inaccessible to other temperature-sensing devices. Thermistors may be directly heated or indirectly heated.

Controlling a fan motor is a typical application of a thermistor. As the thermistor is heated, its resistance decreases and more current flows through the circuit. When enough current flows through the circuit, a solid-state relay turns ON. The solid-state relay is used to switch ON a fan motor at high temperatures. Such a circuit can be used to automatically reduce heat in attics or to circulate warm air.

Directly heated thermistors are used in voltage regulators, vacuum gauges, and electronic time-delay circuits. Indirectly heated thermistors are used for precision temperature measurement and temperature compensation. Each type of thermistor is represented by a separate schematic symbol.

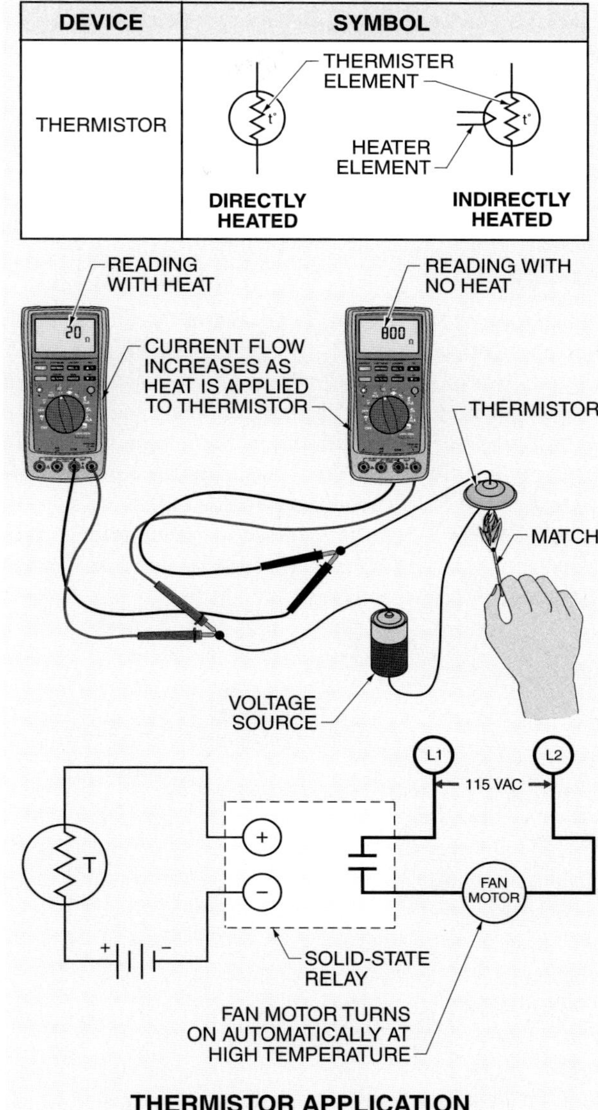

Figure 12-20. A thermistor is a temperature-sensitive resistor whose resistance changes with a change in temperature.

PTC and NTC Thermistors. Two classes of thermistors are positive temperature coefficient (PTC) and negative temperature coefficient (NTC). Although most thermistors are negative temperature coefficient thermistors, some applications require positive temperature coefficient thermistors. With a PTC thermistor, an increase in temperature causes the resistance of the thermistor to increase. With an NTC thermistor, an increase in temperature causes the resistance of the thermistor to decrease. The resistance of both thermistors returns to its original state (resistance value) when the heat is removed.

Cold and Hot Resistance. Cold resistance and hot resistance refer to the operating resistance of a thermistor at extreme temperatures. Cold resistance is measured at 25°C (room temperature). However, some manufacturers specify lower temperatures. The specification sheet should always be checked for the correct temperature specification. Hot resistance is the resistance of a heated thermistor. In a directly heated thermistor, heat is generated from the ambient temperature, the current, and the heating element of the thermistor.

Thermistor Applications. A fire alarm circuit is a common application of an NTC thermistor. **See Figure 12-21.** The purpose of this circuit is to detect a fire and activate an alarm. In normal operating environments, the resistance of the thermistor is high because ambient temperatures are relatively low. The high resistance keeps the current to the control circuit low. The alarm remains OFF. However, in the presence of a fire, the increased ambient temperature lowers the resistance of the thermistor. The lower resistance allows current flow, activating the alarm.

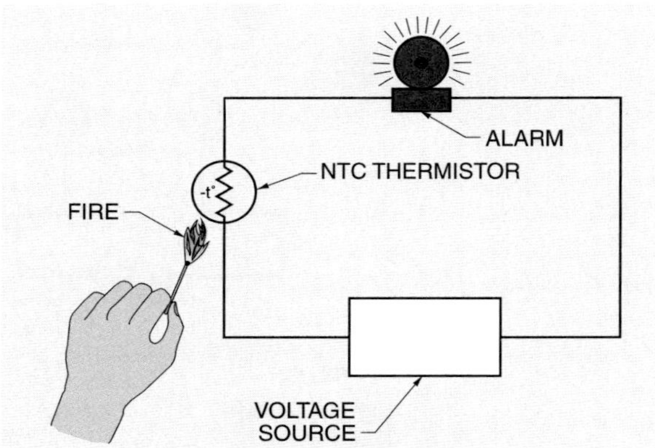

Figure 12-21. In the presence of fire, the increase in temperature lowers the resistance of an NTC thermistor, which increases current and activates an alarm.

Photoconductive Cells

A *photoconductive cell (photocell)* is a device which conducts current when energized by light. Current increases with the intensity of light because resistance decreases. A photocell is, in effect, a variable resistor. A photocell is formed with a thin layer of semiconductor material such as cadmium sulfide (CdS) or cadmium selenide (CdSe) deposited on a suitable insulator. Leads are attached to the semiconductor material and the entire assembly is hermetically sealed in glass. The transparency of the glass allows light to reach the semiconductor material. For maximum current-carrying capacity, the photocell is manufactured with a short conduction path having a large cross-sectional area.

Technical Fact

Dark current is the residual amount of current that flows through a photocell when it is OFF.

Photoconductive Cell Applications. Photocells are used when time response is not critical. Photocells are not used when several thousand responses per second are needed to transmit accurate data. Applications where photocells can be used efficiently include slow-responding electromechanical equipment such as pilot lights or street lights.

Photocells are used to control the pilot light on a gas furnace. **See Figure 12-22.** In this application, the light level determines if the pilot light (flame) on the furnace is ON or OFF. When a pilot light is present, the light from the flame reduces the resistance of the photocell. Current is allowed to pass through the cell and activate a control relay. The control relay allows the main gas valve to be energized when the thermostat calls for heat. The same procedure would be used for similar applications such as gas-powered water heaters, clothes dryers, and ovens (commercial or residential), as well as similar electrical applications.

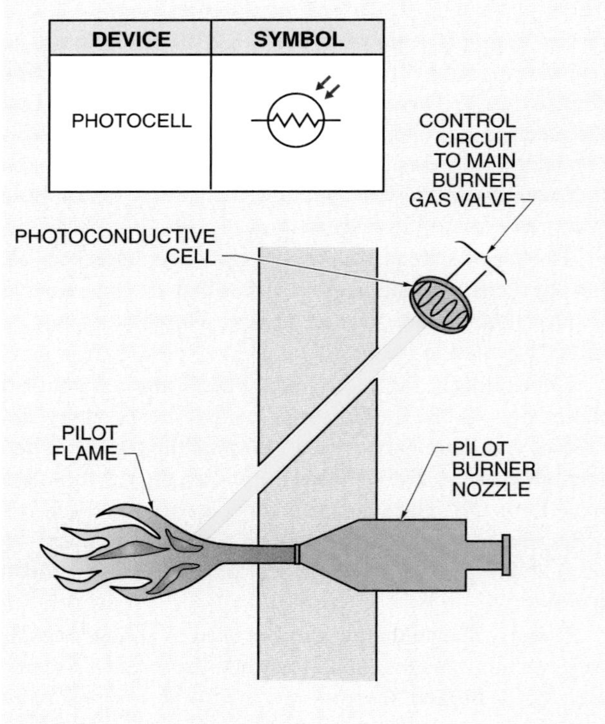

Figure 12-22. A photocell is used to determine if the pilot light on a gas furnace is ON or OFF.

A photocell may also be used in a street light circuit. **See Figure 12-23.** In this circuit, an increase of light at the photocell results in a decrease in resistance and current flow through the solid-state relay. The increased current in the relay causes the normally closed (NC) contacts to open, and the light turns OFF. With darkness, the resistance increases, causing the normally closed contacts to return to their original position, turning the light ON.

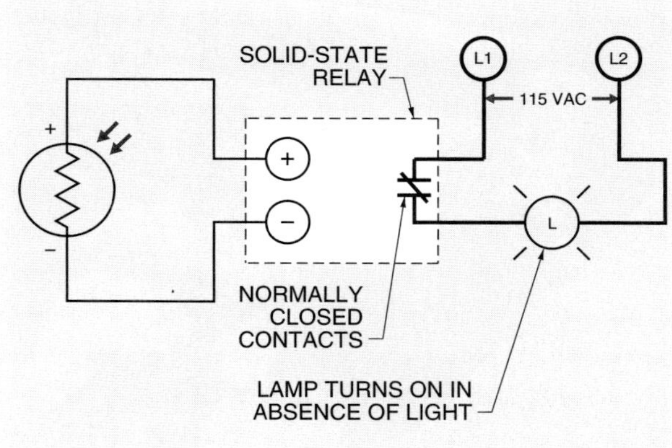

Figure 12-23. A photocell is used to determine when a street light should turn ON or OFF.

Photoconductive Diodes

A *photoconductive diode (photodiode)* is a diode which is switched ON and OFF by light. A photodiode is similar internally to a regular diode. The primary difference is the addition of a lens in the housing for focusing light on the PN junction. **See Figure 12-24.**

Photoconductive Diode Operation. In a photodiode, the conductive properties change when light strikes the surface of the PN junction. Without light, the resistance of the photodiode is high. The resistance is reduced proportionately when a photodiode is exposed to light.

Photoconductive Diode Applications. Photodiodes respond much faster than photocells, and are usually more rugged. Photodiodes are found in movie equipment, conveyor systems, and other equipment requiring a rapid response time. Photodiodes can be used for positioning an object and turning functions ON and OFF, as is done with a filling machine. **See Figure 12-25.**

A constant light source is placed across the conveyor from the photodiode so that cartons can move between the light source and the photodiode. The photodiode is energized as long as there are no cartons in the way to prevent light from passing. The photodiode de-energizes when a carton passes between the light source and the photodiode. The PLC records the response and stops the conveyor, placing the carton in the correct position and filling the carton. This arrangement eliminates the need for slow mechanical equipment.

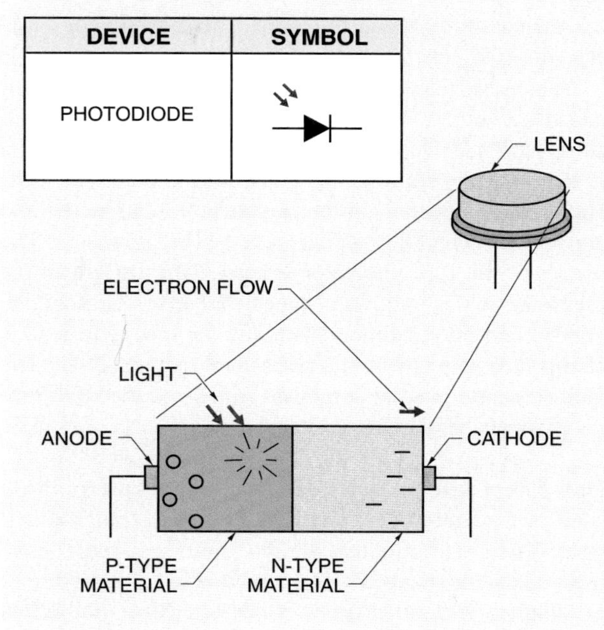

Figure 12-24. A photodiode is a diode which is switched ON and OFF by a light.

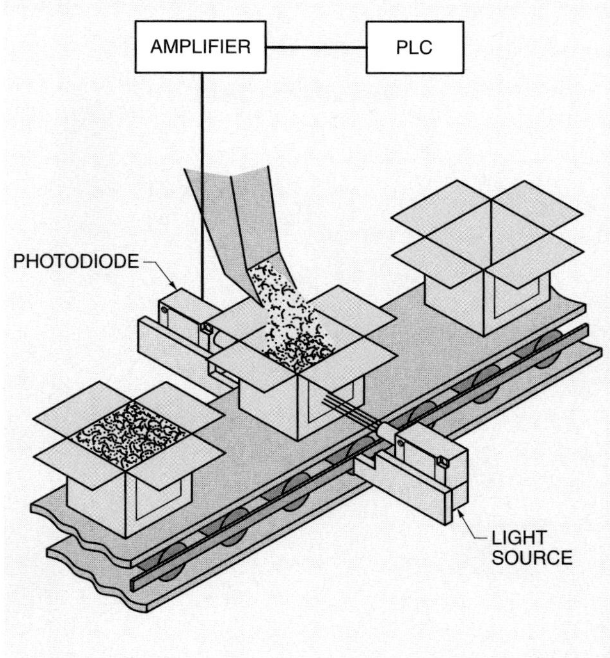

Figure 12-25. Photodiodes are used to position objects and turn machine functions ON and OFF.

Hall Effect Sensors

A *Hall effect sensor* is a sensor that detects the proximity of a magnetic field. The output of a Hall generator depends on the presence of a magnetic field and the current flow in the Hall generator. **See Figure 12-26.** A constant control current passes through a thin strip of semiconductor material (the Hall generator).

When a permanent magnet is brought near, a small voltage (Hall voltage) appears at the contacts that are placed across the narrow dimension of the strip. As the magnet is removed, the Hall voltage is reduced to zero. Thus, the Hall voltage depends on the presence of a magnetic field and on the current flowing through the Hall generator. The output of the Hall generator is zero if the current or the magnetic field is removed. In most Hall effect sensors, the control current is held constant and the magnetic field is changed by movement of a permanent magnet. *Note:* The Hall generator must be combined with associated electronics to form a Hall effect sensor.

Hall Effect Sensor Applications. Hall effect sensors are used in a variety of commercial and industrial applications. Their size, minimal weight, and ruggedness make them ideal for many sensing jobs that are impossible to accomplish with other types of devices. Most Hall effect sensor applications use either a digital or linear output transducer. The choice depends on the output requirements of each application. For example, Hall effect sensors can be embedded in the human heart to serve as a timing element. Other Hall effect sensor applications include speed detection, automobile transmission control, shaft rotation and bottle counting, camera shutter positioning, flow-rate metering, and use as a low-liquid warning sensor or magnetic card reader.

Fluke Corporation

A digital multimeter is used to check the rectifiers in automobiles that change the generated AC from the alternator into DC.

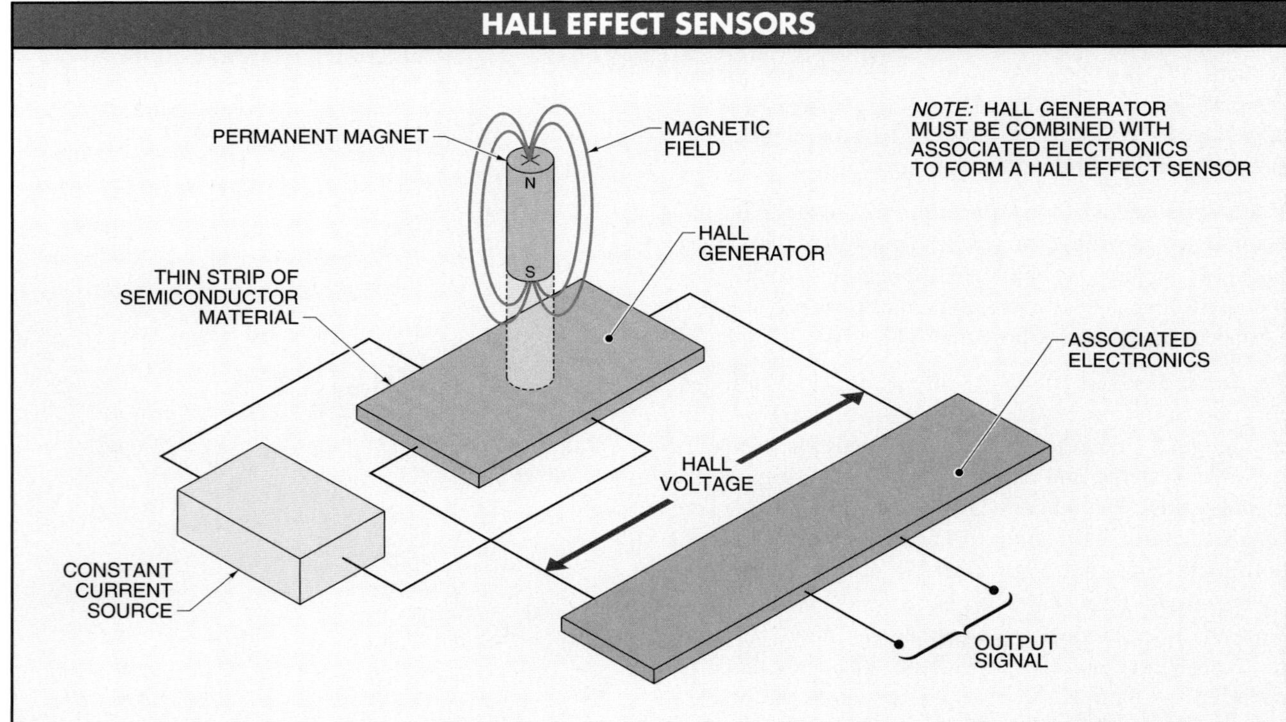

Figure 12-26. A Hall effect sensor produces a voltage depending on the strength of the magnetic field applied to the sensor.

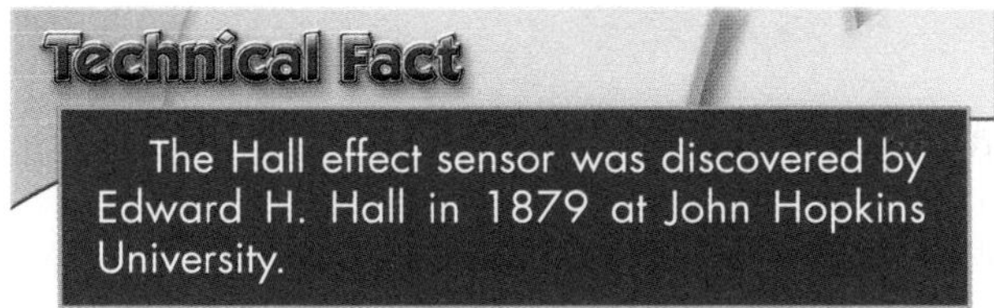

Sensing the speed of a shaft is one of the most common applications of a Hall effect sensor. The magnetic field required to operate the sensor may be furnished by individual magnets mounted on the shaft or hub, or by a ring magnet. Each change in polarity results in an output signal. **See Figure 12-27.**

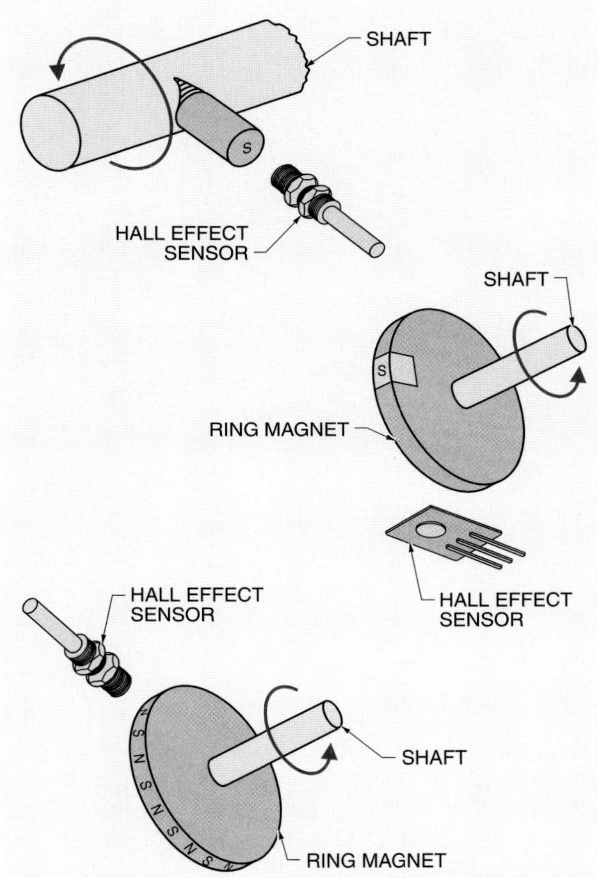

Figure 12-27. Each change in polarity results in an output from a Hall effect sensor used in a shaft speed sensor application.

Another application of a Hall effect sensor is as a low-liquid warning sensor. A low-liquid warning sensor measures and responds to the level of a liquid in a tank. One method used to determine the level in a tank uses a notched tube with a cork floater inserted into the tank. The magnet is mounted in the float assembly, which is forced to move in one plane (vertically). As the liquid level goes down, the magnet passes the digital output sensor Hall effect sensor. The liquid level is indicated when the sensor is actuated. **See Figure 12-28.**

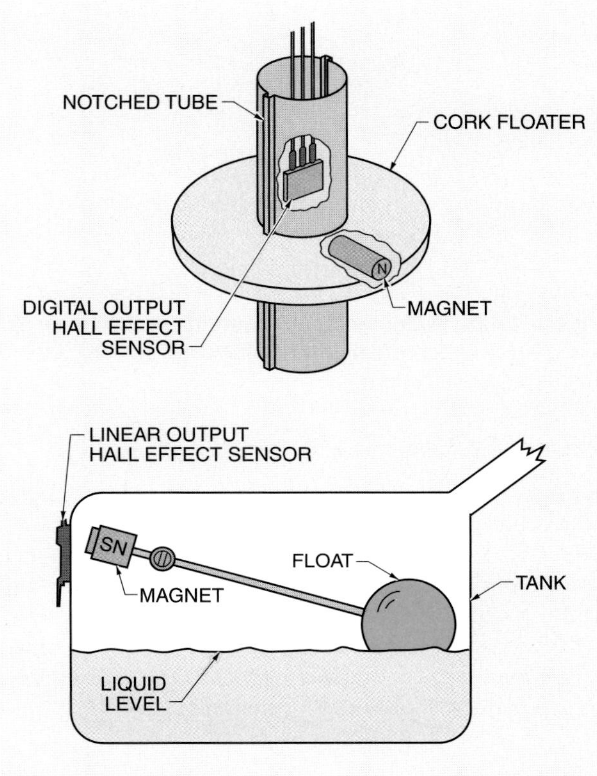

Figure 12-28. A Hall effect sensor can be used to measure the level of liquid in a tank.

A linear output Hall effect sensor may also be used to indicate the liquid level in a tank. As the liquid level falls, the magnet moves closer to the sensor, causing an increase in output voltage. This method allows measurement of liquid levels without any electrical connections in the interior of the tank. Common applications include fuel tanks, transmission fluid reservoirs, and stationary tanks used in food and chemical processing plants.

A door-interlock security system can be designed using a Hall effect sensor, a magnetic card, and associated electronic circuitry. **See Figure 12-29.** In this circuit, the magnetic card slides by the sensor and produces an analog output signal. This analog signal is converted to a digital signal to provide a crisp signal to energize the door latch relay. When the solenoid of the relay pulls in, the door is unlocked. For systems that require additional security measures, a series of magnets may be molded into the card.

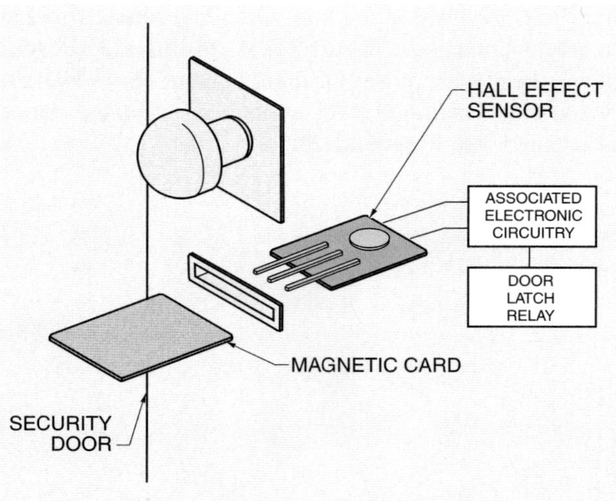

Figure 12-29. A door-interlock system can be designed using a Hall effect sensor, a magnetic card, and associated electronic circuitry.

Pressure Sensors

A *pressure sensor* is a transducer that changes resistance with a corresponding change in pressure. **See Figure 12-30.** A pressure sensor is designed to activate or deactivate when its resistance reaches a predetermined value. A pressure sensor is used for high- or low-pressure control, depending on the switching circuit design. Pressure sensors are suited for a wide variety of pressure measurements on compressors, pumps, and other similar equipment.

DEVICE	SYMBOL
PRESSURE SENSOR	

PRESSURE-SENSING ELEMENT

Figure 12-30. A pressure sensor is a transducer that changes resistance with a corresponding change in pressure.

A pressure sensor can detect low pressure or high pressure, or it can trigger a relief valve. A pressure sensor is also used to measure compression in various types of engines because it is extremely rugged.

AMPLIFICATION

Amplification is the process of taking a small signal and making it larger. In control systems, amplifiers are used to increase small signal currents and voltages so they can do useful work. Amplification is accomplished by using a small input signal to control the energy output from a larger source, such as a power supply. Amplification is accomplished by transistors. Transistors can also be used as DC switching devices.

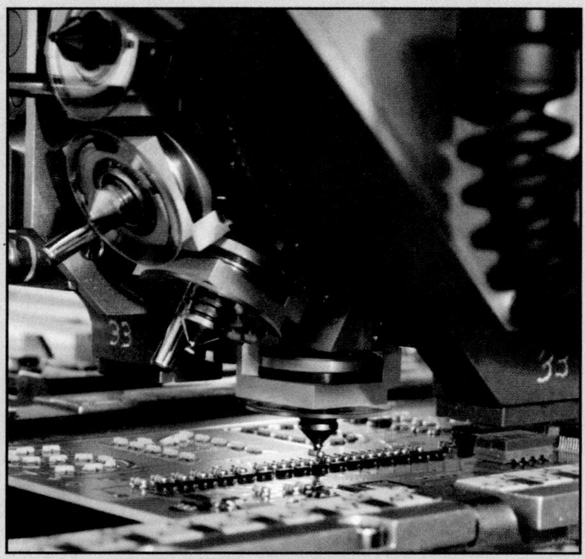

Siemens Corporation

High-speed surface-mount systems are used in modern PC board manufacturing facilities.

Amplifier Gain

The primary objective of an amplifier is to produce gain. *Gain* is a ratio of the amplitude of an output signal to the amplitude of an input signal. In determining gain, the amplifier can be thought of as a black box. A signal applied to the input of the black box gives the output of the box. Mathematically, gain can be found by dividing output by input:

$$\text{Gain} = \frac{\text{Output}}{\text{Input}}$$

Often, a single amplifier does not provide enough gain to increase the output signal to the amplitude needed. In such a case, two or more amplifiers can be used to obtain the gain required. **See Figure 12-31.** For example, amplifier A has a gain of 10, and amplifier B has a gain of 10. The total gain of the two amplifiers is 100 (10 × 10 = 100). If the gains of the amplifiers were 8 and 9, respectively, the total gain would be 72 (8 × 9 = 72). Amplifiers connected in this manner are called cascaded amplifiers. For many amplifiers, gain is in the hundreds or even thousands.

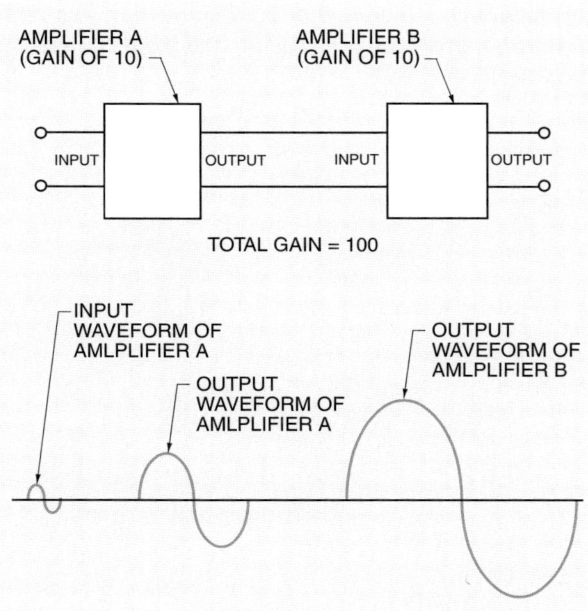

Figure 12-31. Two or more amplifiers can be used to obtain the required gain if a single amplifier does not provide enough gain to increase the output signal to the amplitude needed.

Note: Gain is a ratio of output to input and has no unit of measure, such as volts or amps, attached to it. Therefore, the term gain is used to describe current gain, voltage gain, and power gain. In each case, the output is merely being compared to the input.

Transistors as AC Amplifiers

Transistors may be used as AC amplification devices. A *transistor* is a three-terminal device that controls current through the device depending on the amount of voltage applied to the base. Transistors are bipolar devices. A *bipolar device* is a device in which both holes and electrons are used as internal carriers for maintaining current flow. Transistors may be PNP or NPN transistors. A PNP transistor is formed by sandwiching a thin layer of N-type material between two layers of P-type material. An NPN transistor is formed by sandwiching a thin layer of P-type material between two layers of N-type material. **See Figure 12-32.** Transistor terminals are the emitter (E), base (B), and collector (C). The symbols for PNP and NPN transistors show the emitter, base, and collector in the same places. The difference in the terminals is the direction in which the emitter arrow points. In both cases, the arrow points from the P-type material toward the N-type material.

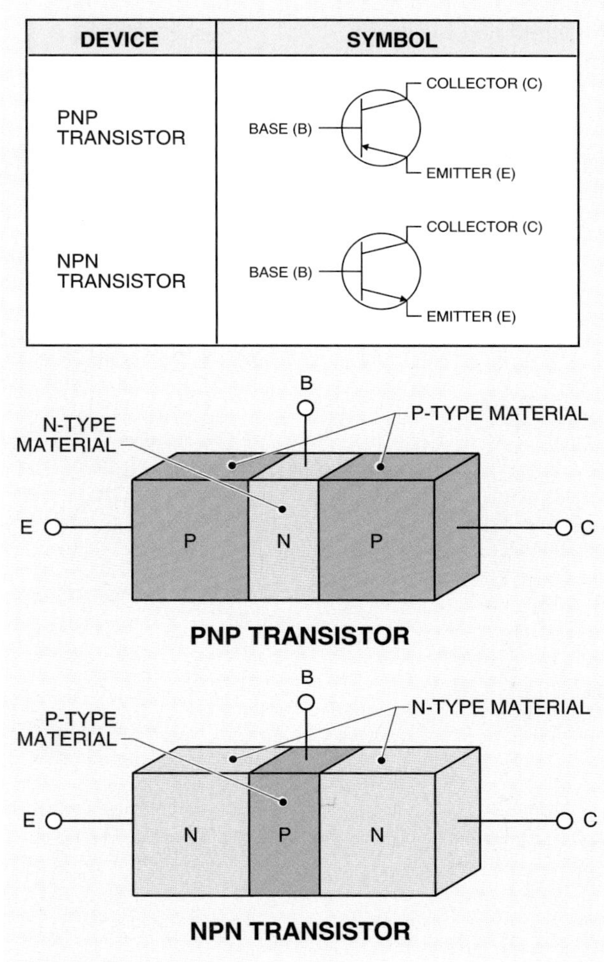

Figure 12-32. A PNP transistor is formed by sandwiching a thin layer of N-type material between two layers of P-type material. An NPN transistor is formed by sandwiching a thin layer of P-type material between two layers of N-type material.

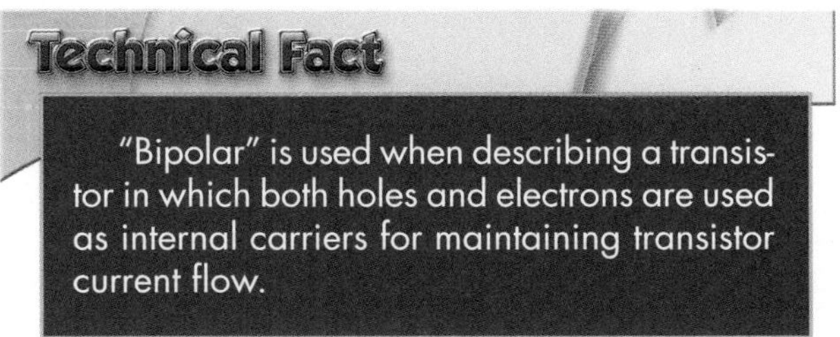

The three basic transistor amplifiers are the common-emitter, common-base, and common-collector. **See Figure 12-33.** Each amplifier is named after the transistor connection that is common to both the input and the load. For example, the input of a common-emitter circuit is across the base and emitter, while the load is across the collector and emitter. Thus, the emitter is common to the input and load.

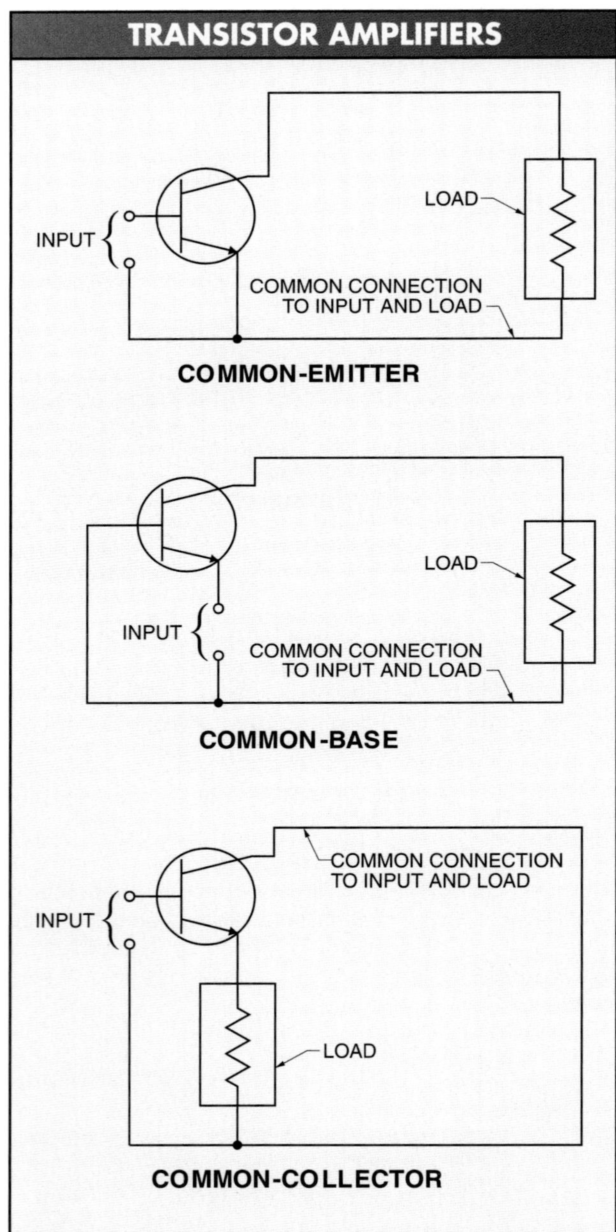

Figure 12-33. The three basic transistor amplifiers are the common-emitter, common-base, and common-collector.

Transistor Terminal Arrangements

Transistors are manufactured with two or three leads extending from their case. A transistor outline (TO) number is used as a reference when a specific-shaped transistor must be used. **See Figure 12-34.** Transistor outline numbers are determined by individual manufacturers. *Note:* The bottom view of transistor TO-3 shows only two leads (terminals). Frequently, transistors use the metal case as the collector-pin lead.

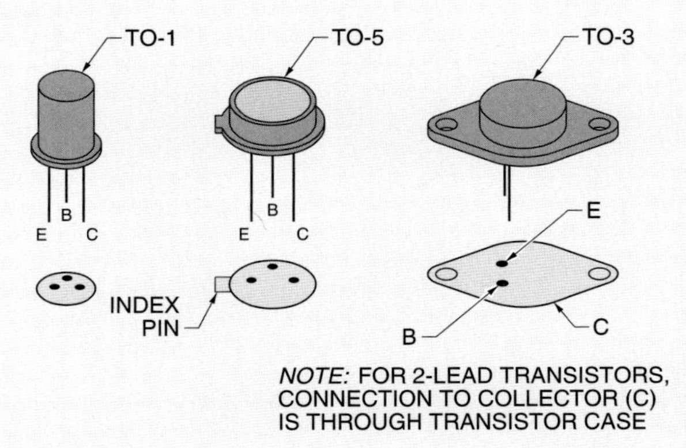

Figure 12-34. Transistors are manufactured with two or three leads extending from their cases.

Spacing can also be used to identify transistor leads. Normally, the emitter and base leads are close together and the collector lead is farther away. The base lead is normally in the middle. A transistor with an index pin must be viewed from the bottom. An *index pin* is a metal extension from the transistor case. The leads are identified in a clockwise direction from the index pin. For example, the loads on TO-5 are identified as E, B, and C. The emitter is closest to the index pin. Refer to a transistor manual or to manufacturer specification sheets for detailed information on transistor construction and identification.

Biasing Transistor Junctions

In any transistor circuit, the base/emitter junction must always be forward biased and the base/collector junction must always be reverse biased. **See Figure 12-35.** The external voltage (bias voltage) is connected so that the positive terminal connects to the P-type material (base) and the negative terminal connects to the N-type material (emitter). This arrangement forward biases the base/emitter junction. Current flows from the emitter to the base. The action that takes place is the same as the action that occurs for a forward biased semiconductor diode.

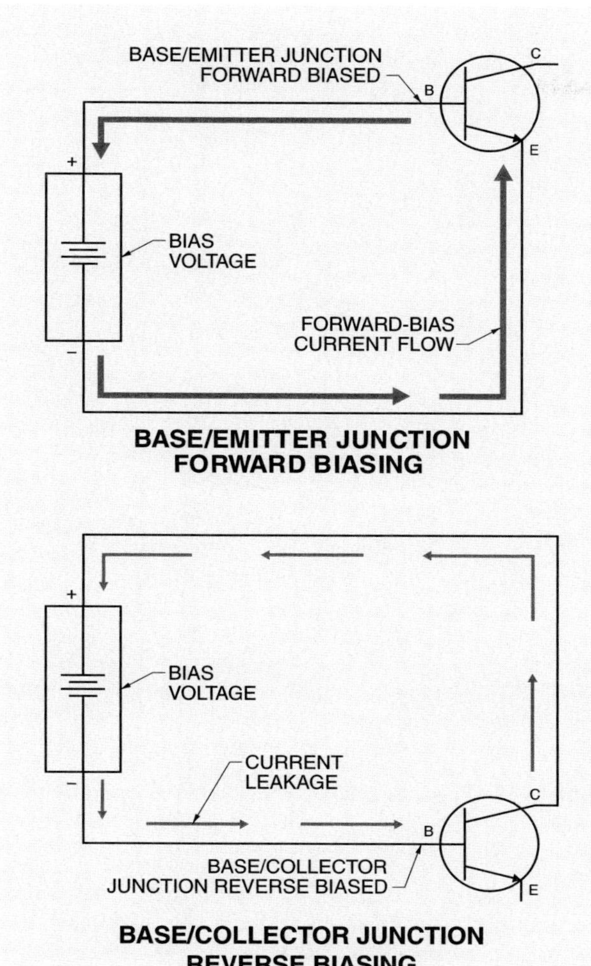

Figure 12-35. In a transistor circuit, the base/emitter junction must always be forward biased and the base/collector junction must always be reverse biased.

Boeing Commercial Airplane Group

Many aircraft systems contain electrical, electronic, and fluid power circuits controlled by solid-state electronic components.

In any transistor circuit, the base/collector junction must always be reverse biased. The external voltage is connected so that the negative terminal connects to the P-type material (base) and the positive terminal connects to the N-type material (collector). This arrangement reverse biases the base/collector junction. Only a very small current (leakage current) flows in the external circuit. The action that takes place is the same as the action that occurs for a semiconductor diode with reverse bias applied.

Transistor Current Flow

Individual PN junctions can be used in combination with two bias arrangements. **See Figure 12-36.** The base/emitter junction is forward biased while the base/collector junction is reverse biased. This circuit arrangement results in an entirely different current path than the path that occurs when the individual circuits are biased separately.

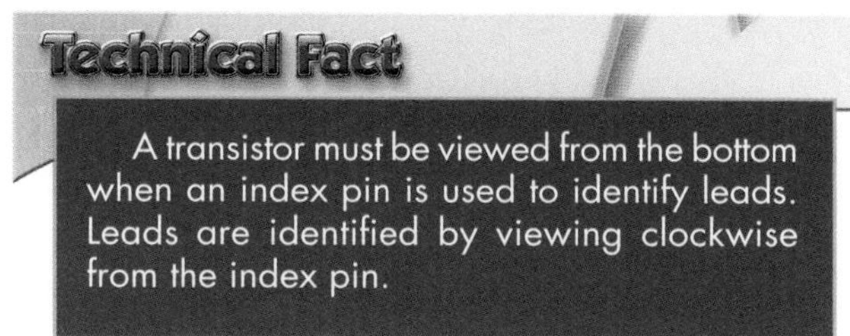

A transistor must be viewed from the bottom when an index pin is used to identify leads. Leads are identified by viewing clockwise from the index pin.

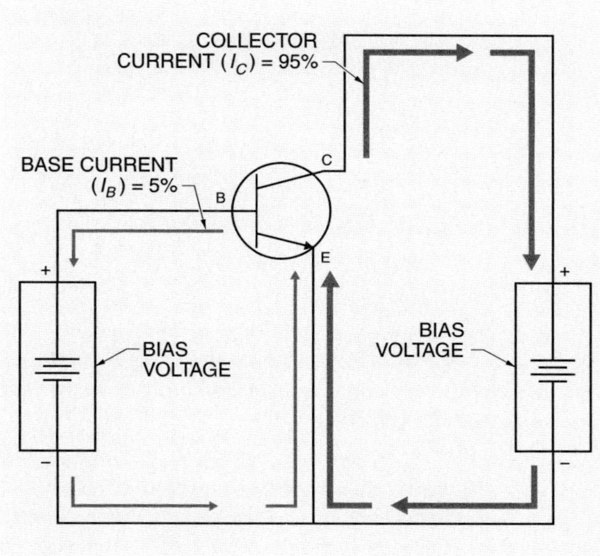

Figure 12-36. In a transistor, an entirely different current path is created when both junctions are biased simultaneously than when each junction is biased separately.

The forward bias of the base/emitter circuit causes the emitter to inject electrons into the depletion region between the emitter and the base. Because the base is less than .001″ thick for most transistors, the more positive potential of the collector pulls the electrons through the thin base. As a result, the greater percentage (95%) of the available free electrons from the emitter pass directly through the base (I_C) into the N-type material, which is the collector of the transistor.

Control of Base Current

The base current (I_B) is a critical factor in determining the amount of current flow in a transistor because the forward biased junction has a very low resistance and could be destroyed by heavy current flow. Therefore, the base current must be limited and controlled.

Fluke Corporation

The proper test equipment must be used when bench testing solid-state components, such as transistors, to prevent damage to the component.

Transistors as DC Switches

Transistors were mainly developed to replace mechanical switches. Transistors have no moving parts and can switch ON and OFF quickly. Mechanical switches have two conditions: open and closed or ON and OFF. Mechanical switches have a very high resistance when open and a very low resistance when closed.

A transistor can be made to operate like a switch. For example, a transistor can be used to turn a pilot light ON or OFF. **See Figure 12-37.** In this circuit, the resistance between the collector (C) and the emitter (E) is determined by the current flow between the base (B) and emitter (E). When no current flows between B and E, the collector/emitter resistance is high, like that of an open switch. The pilot light does not glow because there is no current flow.

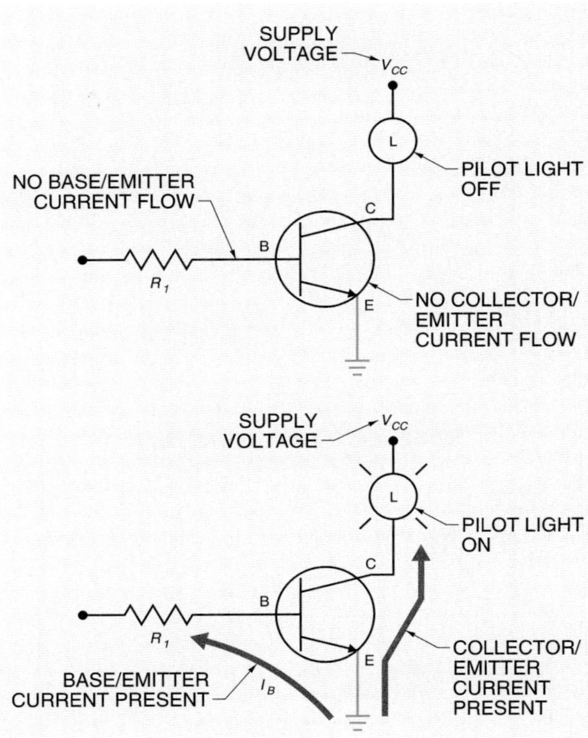

Figure 12-37. A transistor can be made to operate like a switch.

If a small current flows between B and E, the collector/emitter resistance is reduced to a very low value, like that of a closed switch. The pilot light is switched ON. A transistor switched ON is normally operating in the saturation region. The *saturation region* is the maximum current that can flow in a transistor circuit. At saturation, the collector resistance is considered zero and the current is limited only by the resistance of the load.

When the circuit reaches saturation, the resistance of the pilot light is the only current-limiting device in the circuit. When the transistor is switched OFF, it is operating in the cutoff region. The *cutoff region* is the point at which the transistor is turned OFF and no current flows. At cutoff, all the voltage is across the open switch (transistor) and the collector-emitter voltage is equal to the supply voltage V_{CC}.

Technical Fact

Base current is limited in a transistor by using a series-limiting resistor that keeps the base-emitter junction voltage and current low.

Power Dissipation

Because a solid-state device can dissipate only a certain amount of heat (this amount is called the power dissipation factor) before it begins to change operating characteristics, methods must be introduced to remove the heat. If an overload continues for even a short period of time, a solid-state device is destroyed by thermal runaway. Thermal runaway is a condition unique to semiconductors where the increase in current produces more heat, less resistance, and more current.

Transistors and other solid-state devices use heat sinks for thermal protection. The cases of certain power transistors are designed specifically for ease in cooling. Some power transistors use radial fins as part of the transistor design for conducting away heat. Other power transistors are designed for use with heat sinks. **See Figure 12-38.** When transistors are required to operate in high ambient temperatures, forced cooling by a fan or an air conditioner is used. In some precision solid-state equipment, the unit shuts down if the ambient temperature rises past a certain level.

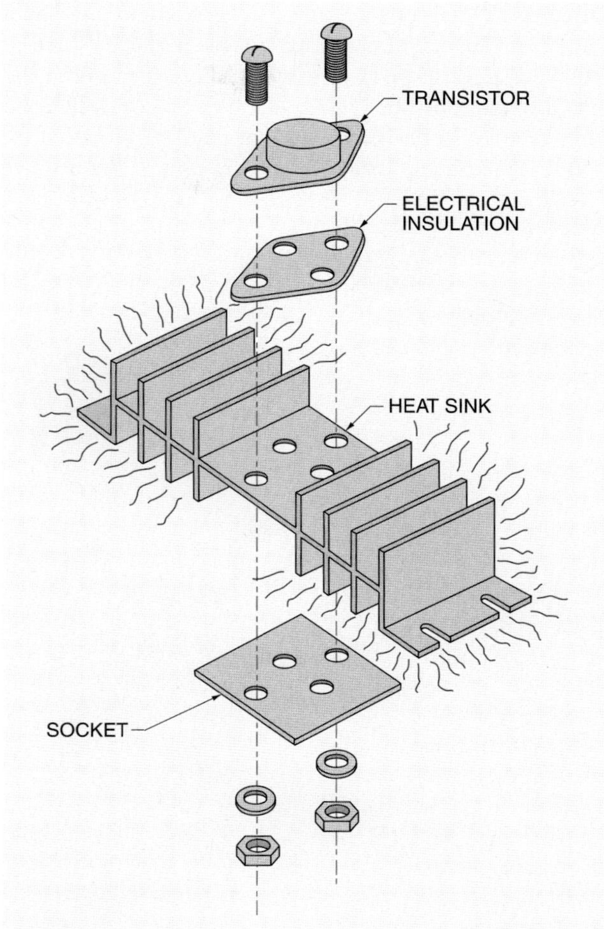

Figure 12-38. Transistors use heat sinks to dissipate excess heat and provide thermal protection.

Integrated Circuits

An *integrated circuit (IC)* is a circuit composed of thousands of semiconductor devices, providing a complete circuit function in one small semiconductor package. An integrated circuit consists of a piece (chip) of silicon or other semiconductor material on which is etched or imprinted a network of electronic components such as transistors, diodes, resistors, and their interconnections. Integrated circuits are popular because they provide a complete circuit function in one package. Integrated circuits are often referred to as chips, although chips are actually a part of the integrated circuit. **See Figure 12-39.** Although many processes have been developed to create these devices, the end result is always a totally enclosed system with specific inputs and outputs.

Because of the nature of integrated circuits, a technician must approach them in an entirely different manner from individual solid-state components. Integrated circuits consist of systems within a system. The entire system must be understood. Data books and manufacturer specification sheets can normally provide this information. The inputs and outputs of the system must be studied by using meters and an oscilloscope when data books and manufacturer specification sheets are not available. Troubleshooting integrated circuits requires knowledge of the system functions and input and output characteristics. Defective integrated circuits must be replaced because they cannot be repaired.

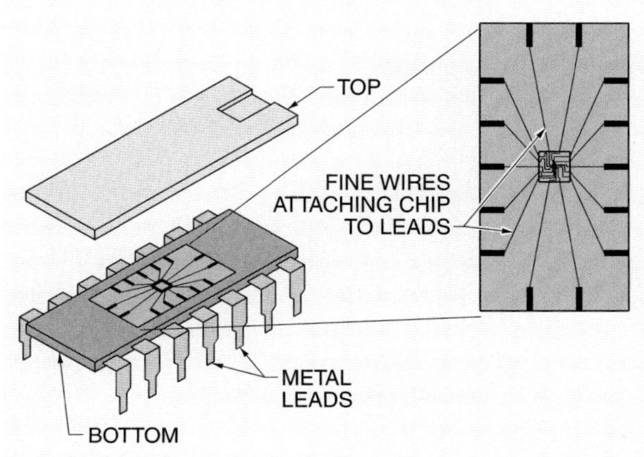

Figure 12-39. Integrated circuits are thousands of semiconductors, providing a complete circuit function in one small semiconductor package.

Integrated Circuit Packages. Integrated circuit shapes and sizes range from standard transistor shapes, such as TO-5 packages, to the latest in large-scale integration (LSI). Metal-oxide substrate (MOS) is a type of LSI. Integrated circuits are also designed with flat-pack construction for applications where space is at a premium. **See Figure 12-40.**

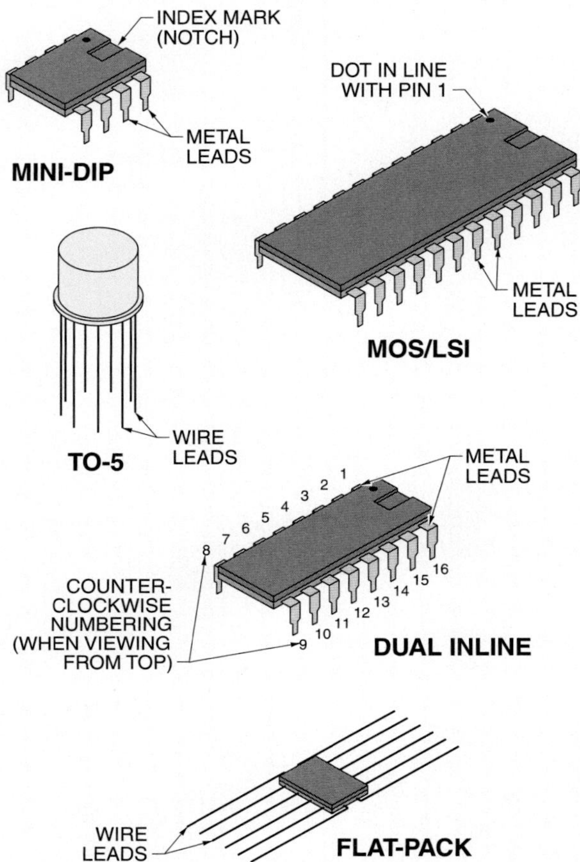

Figure 12-40. Integrated circuit shapes and sizes range from standard transistor shapes, such as the TO-5 package, to the latest in large-scale integration (LSI).

The dual inline package (DIP) with 14, 16, or 24 pins is the most widely used configuration. The mini-DIP is a smaller dual inline package with 8 pins. A modified TO-5 is available with 8, 10, or 12 pins. The housings for integrated circuits may be metal, plastic, or ceramic. Ceramic has inherently high heat resistance and is used in applications where high temperatures may be a factor.

Pin Numbering System. All manufacturers use a standardized pin numbering system for their devices. Manufacturer data sheets should be consulted when unsure about pin numbering patterns.

Dual inline packages and flat packs have index marks and notches at the top for reference. Before removing an integrated circuit, note where the index mark is in relation to the board or socket to aid in installation of the new unit. The numbering of the pins is always the same. The notch is at the top of the chip. To the left of the notch is a dot that is in line with pin 1. The pins are numbered counterclockwise around the chip when viewed from the top.

Technical Fact

Printed circuit boards often have many dual inline package (DIP) switches that are used to configure the board semipermanently.

Operational Amplifiers

An operational amplifier (op-amp) is one of the most widely used integrated circuits. An *op-amp* is a very high gain, directly coupled amplifier that uses external feedback to control response characteristics. An example of this feedback control is gain. The gain of an op-amp can be controlled externally by connecting feedback resistors between the output and input. A number of different amplifier applications can be achieved by selecting different feedback components and combinations. With the right component combinations, gains of 500,000 to 1,000,000 are common.

Fluke Corporation

A graphical multimeter is used to troubleshoot solid-state electrical control devices.

The schematic symbol for an op-amp may be shown in two ways. In each case, the two inputs of the op-amp are the inverting (–) and the noninverting (+). **See Figure 12-41.** The two inputs are normally drawn with the inverting input at the top. The exception to the inverting input being at the top is when it complicates the schematic. In either case, the two inputs should be clearly identified by polarity symbols on the schematic symbol.

Figure 12-41. The schematic symbol for an op-amp which has two inputs should be clearly identified by polarity symbols.

Internal Op-Amp Operation. An op-amp consists of a high-impedance differential amplifier, a high-gain stage, and a low-output-impedance power output stage. The high-impedance differential amplifier provides the wide bandwidth and the high impedance. The high-gain stage boosts the signal. The power output stage isolates the gain stage from the load and provides for the power output.

The operation of the differential amplifier is unique. Current to the emitter-coupled transistors Q1 and Q2 is supplied by the source Q3. The characteristics of Q1 and Q2, along with their biasing resistors R1, R2, and R3, are closely matched to make them as equal as possible. **See Figure 12-42.**

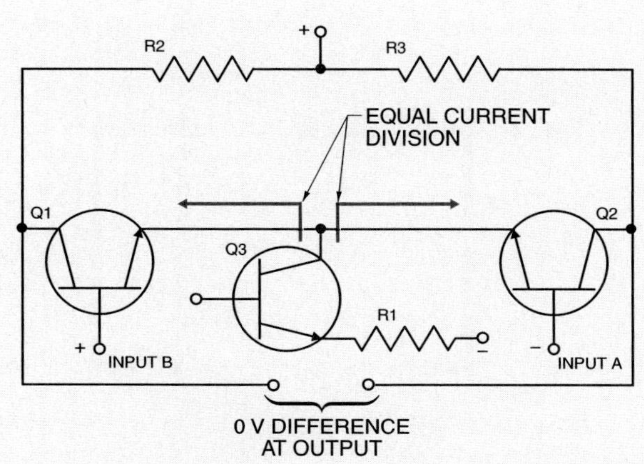

Figure 12-42. An op-amp is a very high gain, directly coupled amplifier that uses external feedback to control response characteristics.

As long as the two input voltages, A and B, are either zero or equal in amplitude and polarity, the amplifier is balanced because the collector currents are equal. Zero voltage difference exists between the two collectors when balanced.

The sum of the emitter currents is always equal to the current supplied by Q3. Thus, if the input to one transistor causes it to draw more current, the current in the other decreases and the voltage difference between the two collectors changes in a differential manner. The differential swing, or output signal, is greater than the simple variation that can be obtained from only one transistor. Each transistor amplifies in the opposite direction so that the total output signal is twice that of one transistor. This swing is amplified through the high-gain stage and matched to the load through the power output stage. By changing op-amps to different configurations, they can be made into oscillators, pulse generators, and level detectors.

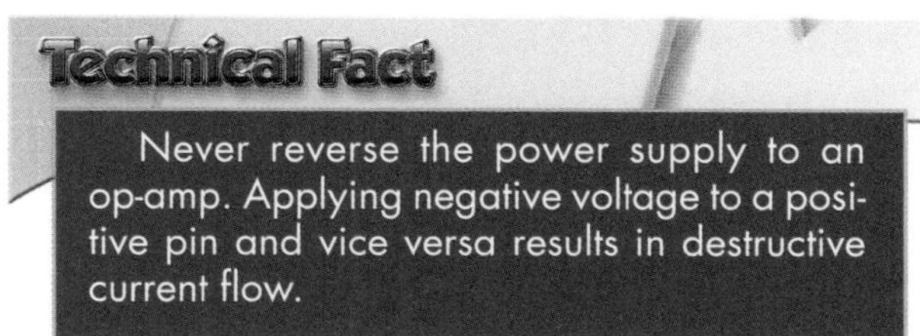

PROCESSING

Processing is an activity or systematic sequence of operations that produces a specified result. Typically, processing is a computer function that consists of or involves procedure code, data storage, and an interface for communicating with other processes. Several types of different processes are used with solid-state devices and system integration.

Digital Integrated Circuits

Electronic signals may be analog or digital. Analog signals (voltage and current) vary smoothly or continuously. Digital signals are a series of pulses that change levels between the OFF or ON state.

The analog and digital processes can be seen by comparing a light dimmer and light switch. A light dimmer varies the intensity of light from fully OFF to fully ON. This is an example of an analog process. A standard light switch has only two positions: fully OFF or fully ON. This is an example of a digital process. Electronic circuits that process these quickly changing pulses are digital or logic circuits. The four most common gates used in digital electronics are the AND, OR, NAND, and NOR gates.

AND Gates. An *AND gate* is a device with an output that is high only when both of its inputs are high. The quad AND gate is one type of integrated circuit chip. **See Figure 12-43.** The manufacturer places four AND gates in one package. By using the numbering system on the chip, any one or all four of the AND gates may be used. In this case, voltage is applied to the circuit at pins 14 and 7.

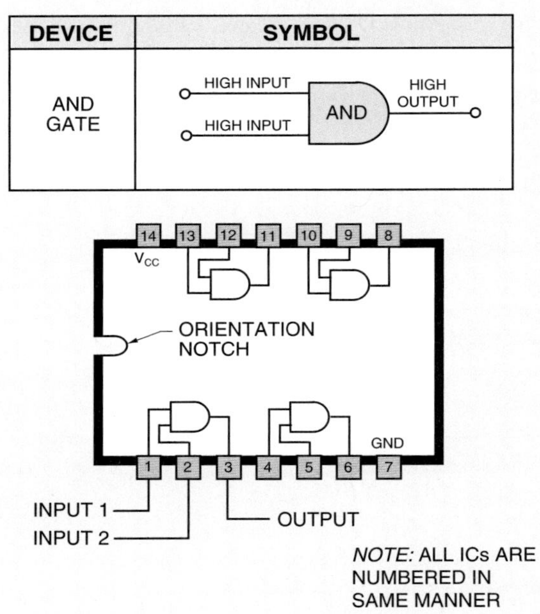

Figure 12-43. An AND gate is a device with an output that is high only when both of its inputs are high.

To connect an external circuit to an AND gate, pins 1, 2, and 3 of the quad AND gate chip could be used. Pins 1 and 2 are the input and pin 3 is the output. An application of an AND gate is in an elevator control circuit. **See Figure 12-44.** The elevator cannot move unless the inner and outer doors are closed. Once both doors are closed, the output of the AND gate could be fed to an op-amp, which fires a triac that starts the elevator motor.

OR Gates. An *OR gate* is a device with an output that is high when either or both inputs are high. **See Figure 12-45.** An application of an OR gate is in a burglar alarm circuit. A signal is sent to the burglar alarm circuit if the front door or the back door is opened. The electrical equivalent of an OR gate is two pushbuttons connected in parallel.

NAND Gates. A NAND (NOT-AND) gate is an inverted AND function. A *NAND gate* is a device that provides a low output when both inputs are high. The NAND gate is represented by the AND symbol followed by a small circle indicating an inversion of the output. **See Figure 12-46.**

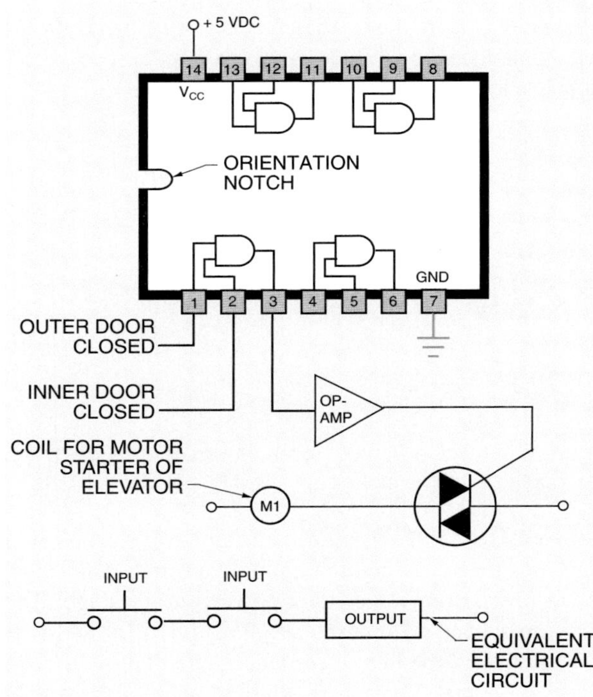

Figure 12-44. An AND gate may be used in an elevator control circuit.

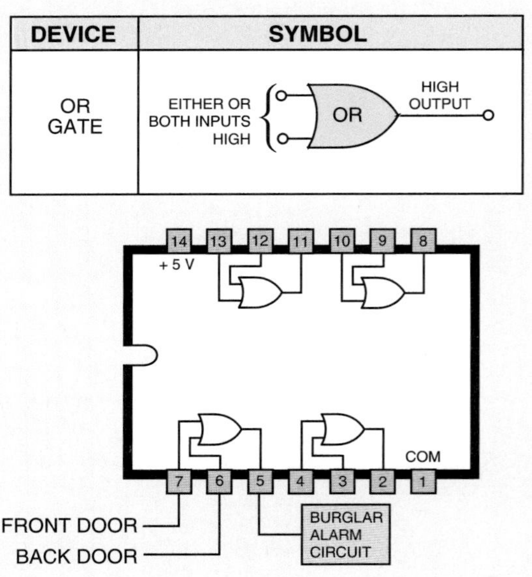

Figure 12-45. An OR gate is a device with an output that is high when either or both inputs are high.

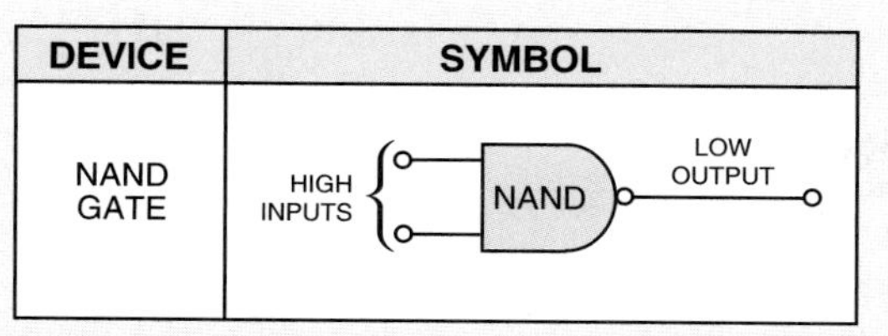

Figure 12-46. A NAND gate is a device that provides a low output when both inputs are high.

A NAND gate is a universal building block of digital logic. NAND gates are normally used in conjunction with other elements to implement more complex logic functions. NAND gates are also available in quad integrated circuit packaging.

NOR Gates. A NOR (NOT-OR) gate is the same as an inverted OR function. A *NOR gate* is a device that provides a low output when either or both inputs are high. A NOR gate is represented by the OR gate symbol followed by a small circle indicating an inversion of the output. **See Figure 12-47.** The NOR gate is a universal building block of digital logic. NOR gates are normally used in conjunction with other elements to implement more complex logic functions. NOR gates are also available in quad integrated circuit packaging.

DEVICE	SYMBOL
NOR GATE	SMALL CIRCLE INDICATES AN INVERSION OF OUTPUT EITHER OR BOTH INPUTS HIGH — NOR — LOW OUTPUT

Figure 12-47. A NOR gate is a device that provides a low output when either or both inputs are high.

Technical Fact

A flip-flop is the electronic equivalent of a toggle switch. It has two outputs, one high and one low. When one switch is in the high output setting, the other will be set in the low output setting, and vice versa.

555 Timers

A *555 timer* is an integrated circuit designed to output timing pulses for control of certain types of circuits. A 555 timer consists of a voltage divider network (R1, R2, and R3), two comparators (Comp 1 and Comp 2), two control transistors (Q1 and Q2), a power output amplifier, and a flip-flop. **See Figure 12-48.** A *flip-flop* is an electronic circuit having two stable states or conditions normally designated set and reset. Flip-flops have two outputs, high and low. When one is high, the other is low, and vice versa.

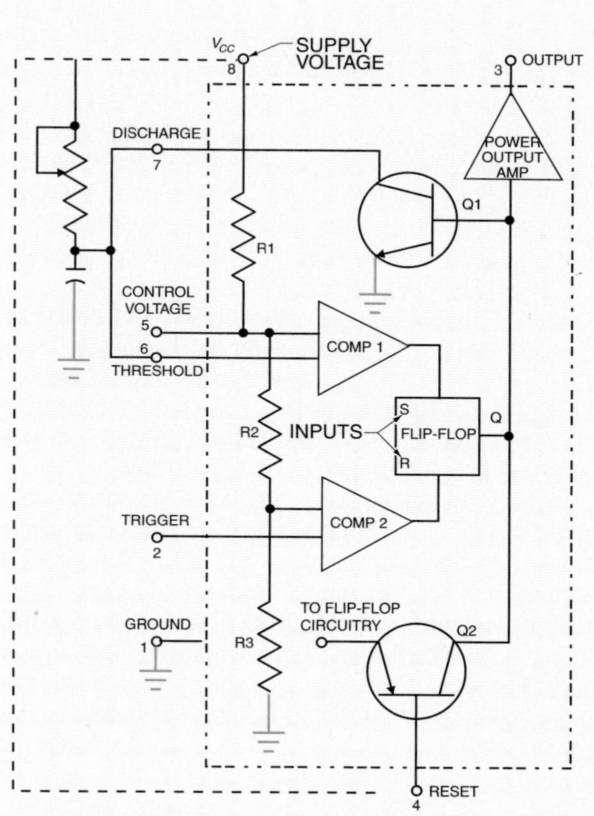

Figure 12-48. A 555 timer is an integrated circuit designed to output timing pulses for control of certain types of circuits.

The comparators compare the input voltages to internal reference voltages that are created by the voltage divider, which consists of resistors R1, R2, and R3. Because the resistors are of equal value, the reference voltage provided by two resistors is two-thirds of the supply voltage (V_{cc}). The other resistor provides one-third of V_{cc}. The value of V may change (9 V, 12 V, 15 V, etc.) from chip to chip. However, the 2/3:1/3 ratio always remains the same.

The comparator goes into saturation and produces a signal that triggers the flip-flop when the input voltage to either one of the comparators is higher than the reference voltage. In this integrated circuit, the flip-flop has two inputs, S and R.

Note: The two comparators feed signals into the flip-flop. Comparator 1 is the threshold comparator and comparator 2 is the trigger comparator. Comparator 1 is connected to the S input of the flip-flop and comparator 2 is connected to the R input of the flip-flop.

The output of the flip-flop is high whenever the voltage at S is positive and the voltage at R is zero. The output of the flip-flop is low whenever the voltage at S is zero and the voltage at R is positive. The output from the flip-flop at point Q is applied to transistors Q1 and Q2 and to the output amplifier simultaneously. Q1 turns ON such that pin 7 (the discharge pin) is grounded through the emitter-collector circuit if the signal is high. Q1 is then in a position to turn ON pin 7 to ground through the emitter-collector circuit. *Note:* Pin 7 is the discharge pin because it is connected to the timing capacitor. When Q1 conducts, pin 7 is grounded and the capacitor can be discharged.

The flip-flop signal is also applied to Q2. A signal to pin 4 can be used to reset the flip-flop. Pin 4 can be activated when a low-level voltage signal is applied. Once applied, this signal overrides the output signal from the flip-flop. The reset pin (pin 4) forces the output of the flip-flop to be low, regardless of the state of the other inputs.

The flip-flop signal is also applied to the power output amplifier. The power output amplifier boosts the signal and the 555 timer delivers up to 200 mA of current when operated at 15 V. The output can be used to drive other transistor circuits and even a small audio speaker. The output of the power output amplifier is always an inverted signal compared to the input. The output is low if the input to the power output amplifier is high. The output is high if the input is low.

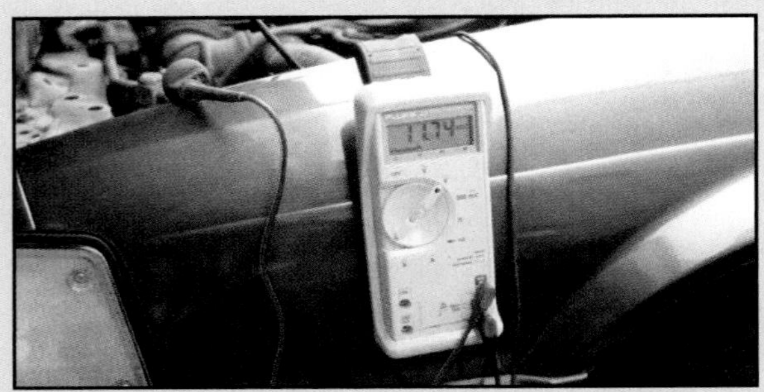

Fluke Corporation

Digital multimeters are used to check the solid-state components in modern automobiles such as diodes in the alternator and SCRs and transistors in electronic ignition systems.

SWITCHING DEVICES

Switching devices are components used for making, breaking, or changing the connections in an electrical circuit. A switch looks at incoming current to determine the destination. Based on that destination, a transmission path is set up through the switching matrix between the incoming and outgoing communications ports and links. Switching devices that are commonly found in integrated solid-state systems include silicon controlled rectifiers (SCRs), triacs, unijunction transistors (UJTs), and diacs.

Silicon Controlled Rectifiers (SCRs)

A *silicon controlled rectifier (SCR)* is a solid-state rectifier with the ability to rapidly switch heavy currents. SCRs use three electrodes for normal operation. **See Figure 12-49.** The three electrodes are the anode, cathode, and gate. The anode and cathode of an SCR are similar to the anode and cathode of a semiconductor diode.

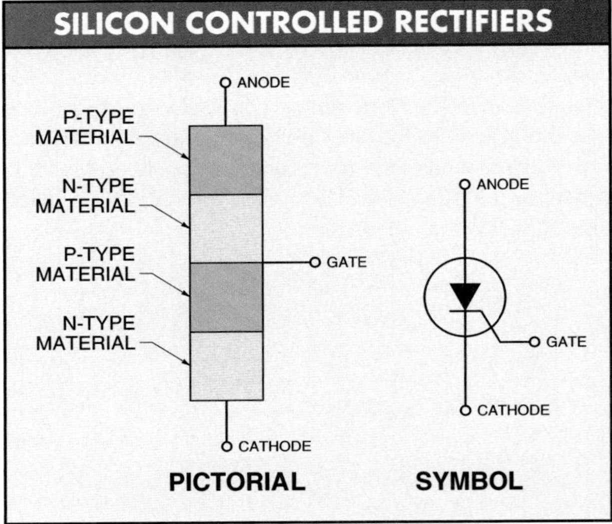

Figure 12-49. An SCR is a four-layer (PNPN) semiconductor device that does not pass significant current, even when forward biased, unless the anode voltage equals or exceeds the forward breakover voltage.

The gate serves as the control point for an SCR. The SCR differs from a semiconductor diode in that it does not pass significant current, even when forward biased, unless the anode voltage equals or exceeds the forward breakover voltage. *Forward breakover voltage* is the voltage required to switch an SCR into a conductive state. The SCR switches ON and becomes highly conductive when the forward breakover voltage is reached. The SCR is unique because the gate current is used to reduce the level of breakover voltage necessary for the SCR to conduct.

Low-current SCRs can operate with an anode current of less than 1 mA. High-current SCRs can handle load currents in the hundreds of amperes. The size of an SCR increases with an increase in its current rating.

SCR Characteristic Curves. The voltage-current characteristic curve of an SCR shows that the SCR operates much like a regular diode in reverse bias. **See Figure 12-50.** With reverse bias, there is a small current until avalanche is reached. After avalanche is reached, the current increases dramatically. This current can cause damage if thermal runaway begins.

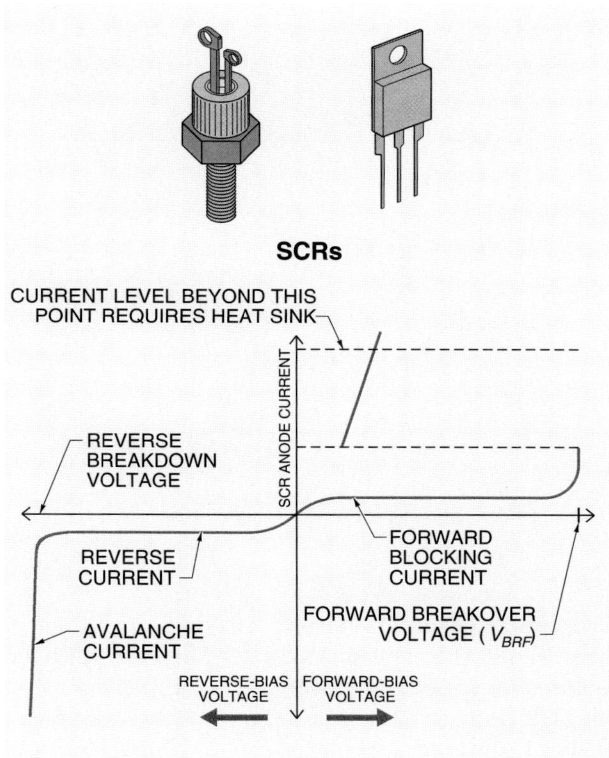

Figure 12-50. The voltage-current characteristic curve of an SCR shows that the SCR operates much like a regular diode in reverse bias.

When an SCR is forward biased, there is also a small forward leakage current (forward blocking current). This current stays relatively constant until the forward breakover voltage is reached. At that point, the current increases rapidly and is often referred to as the forward avalanche region. In the forward avalanche region, the resistance of the SCR is very low. The SCR acts much like a closed switch and the current is limited only by the external load resistance. A short in the load circuit of an SCR can destroy an SCR if overload protection is not adequate.

SCR Operating States. An SCR operates much like a mechanical switch. An SCR is either ON or OFF. An SCR is ON (fires) when the applied voltage is above the forward breakover voltage (V_{BRF}). The SCR remains ON as long as the current stays above the holding current. *Holding current* is the minimum current necessary for an SCR to continue conducting. An SCR returns to its OFF state when voltage across the SCR drops to a value too low to maintain the holding current.

Gate Control of Forward Breakover Voltage. The value of the forward breakover voltage can be reduced when the gate is forward biased and current begins to flow in the gate/cathode junction. Increasing values of forward bias can be used to reduce the amount of forward breakover voltage (V_{BRF}) necessary to get an SCR to conduct.

Once an SCR has been turned ON by the gate current, the gate current loses control of the SCR forward current. Even if the gate current is completely removed, the SCR remains ON until the anode voltage has been removed. The SCR also remains ON until the anode voltage has been significantly reduced to a level where the current is not large enough to maintain the proper level of holding current.

Siemens Corporation

Surface-mount systems include solid-state components that are used for precision control of the heat produced when assembling PC boards.

SCR Applications. SCRs may be used in circuits to provide heat control. For example, an SCR can bring a chemical mixture stored in a vat to a specific temperature and maintain that temperature. **See Figure 12-51.** With the proper circuitry, the temperature of the mixture can be precisely controlled. Using a bridge circuit, the temperature can be maintained within 1°F over a temperature range of 20°F to 150°F.

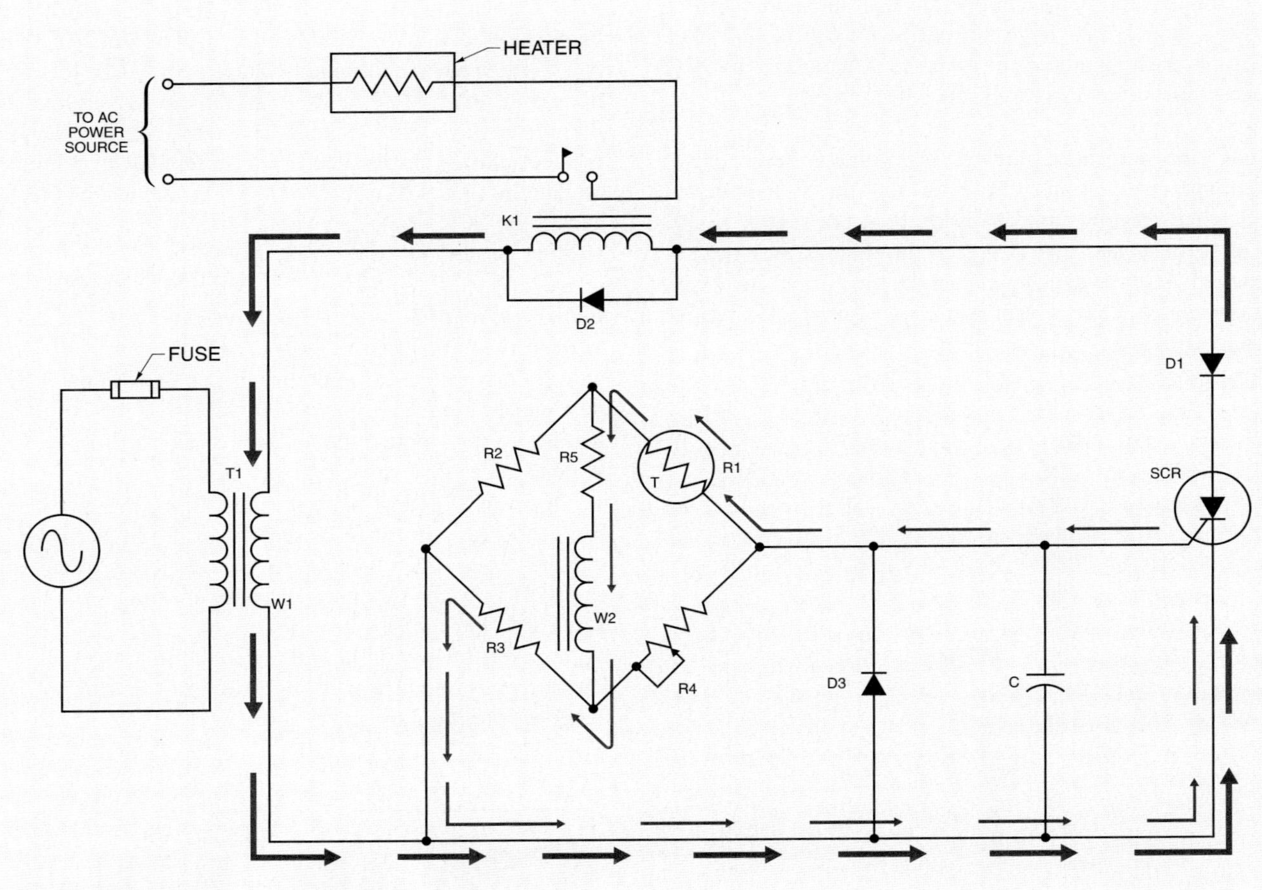

Figure 12-51. SCRs may be used in circuits to provide heat control.

In this circuit, transformer T1 has two secondary windings, W1 and W2. W1 furnishes voltage through the SCR to relay coil K1. W2 furnishes AC voltage to the gate circuit of the SCR. Primary control over this circuit is accomplished through the use of the bridge circuit. The bridge circuit is formed by thermistor R1, fixed resistors R2 and R3, and potentiometer R4. Resistor R5 is a current-limiting resistor used to protect the bridge circuit. The fuse is used to protect the primary of the transformer.

The bridge is balanced when the resistance of R1 equals the resistance setting on R4. None of the AC voltage introduced into the bridge by winding W2 is applied to the gate of the SCR. The relay coil K1 remains de-energized and its normally closed contacts apply power to the heating elements.

The resistance of thermistor R1 decreases if the temperature increases above a preset level. The bridge becomes unbalanced such that a current flows to the gate of the SCR while the anode of the SCR is still positive. This turns ON the SCR and energizes the relay coil K1, thereby switching power from the load through the relay contact. R1 unbalances the bridge in the opposite direction if the temperature falls below the preset temperature setting. A negative signal is applied to the gate of the SCR when the anode of the SCR is positive. The negative signal stops the SCR from conducting and allows current to continue to flow to the heating elements.

Triacs

A *triac* is a three-terminal semiconductor thyristor that is triggered into conduction in either direction by a small current to its gate. Triacs are triggered into conduction in a manner similar to the action of an SCR. Triacs were developed to provide a means for producing improved controls for AC power. Triacs are available in a variety of packaging arrangements. Triacs can handle a wide range of currents and voltages. Triacs normally have relatively low current capabilities compared to SCRs. Triacs are normally limited to less than 50 A and cannot replace SCRs in high-current applications.

Triac Construction. The terminals of a triac are the gate, main terminal 1 (MT1), and main terminal 2 (MT2). There is no designation of anode and cathode. Current may flow in either direction through MT1 and MT2. MT2 is the case- or metal-mounting tab to which the heat sink can be attached. A triac can be considered two NPN switches sandwiched together on a single N-type material wafer.

Triac Operation. A triac blocks current in either direction between MT1 and MT2. A triac can be triggered into conduction in either direction by a momentary pulse in either direction supplied to the gate. A triac operates much like a pair of SCRs connected in a reverse parallel arrangement. The triac conducts if the appropriate signal is applied to the gate.

A triac characteristic curve shows the characteristics of a triac when triggered into conduction. **See Figure 12-52.** The triac remains OFF until the gate is triggered. The trigger circuit pulses the gate and turns ON the triac, allowing current to flow. The trigger circuit can be designed to produce a pulse that varies at any point in the positive or negative half-cycle. Therefore, the average current supplied to the load may vary.

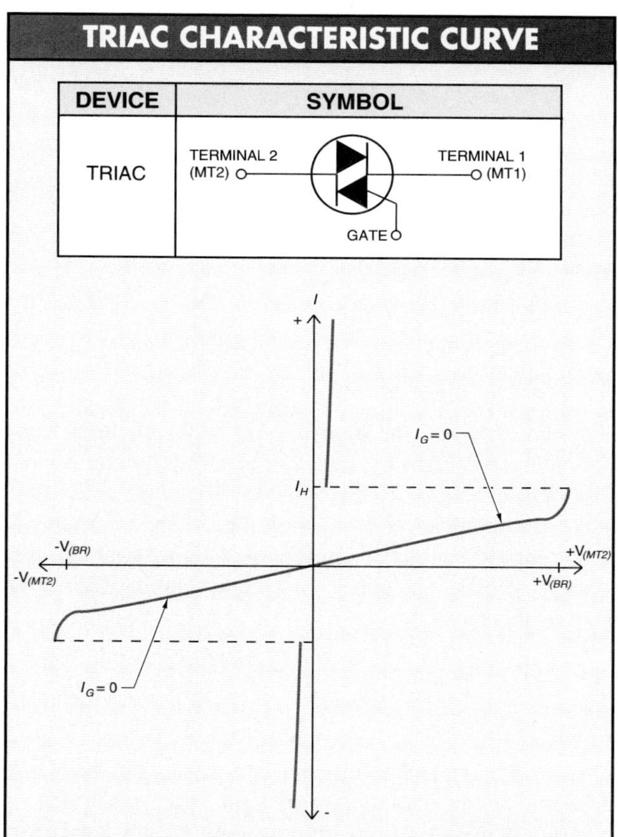

Figure 12-52. A triac characteristic curve shows the characteristics of a triac when triggered into conduction.

> **Technical Fact**
>
> Gate, latching, and holding currents of triacs are easily predictable, once a single parameter is known. The interrelationships (ratios) between SCRs and triacs can be used by engineers and designers for device selection and circuit design and application.

One advantage of a triac is that virtually no power is wasted by being converted to heat. Heat is generated when current is impeded, not when current is switched OFF. A triac is either fully ON or fully OFF. Triacs never partially limit current. Another important feature of a triac is the absence of a reverse breakdown condition of high voltage and high current such as found in diodes and SCRs. A triac turns ON if the voltage across it rises too high. A triac can conduct a reasonably high current when turned ON.

Unijunction Transistors (UJTs)

A *unijunction transistor (UJT)* is a transistor consisting of N-type material with a region of P-type material doped within the N-type material. The N-type material functions as the base and has two leads, base 1 (B1) and base 2 (B2). The lead extending from the P-type material is the emitter (E). **See Figure 12-53.**

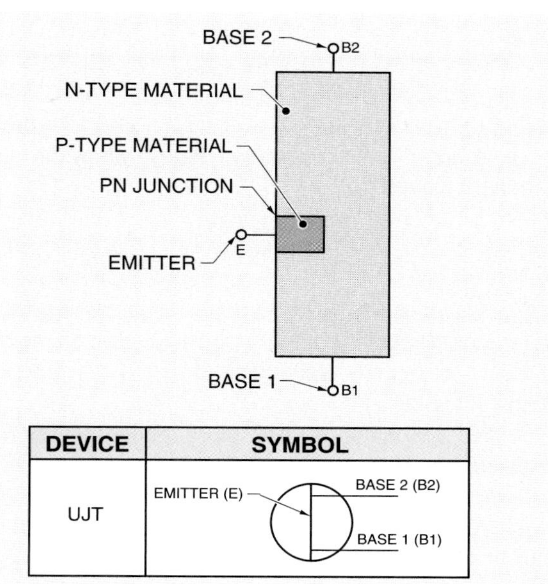

Figure 12-53. A UJT consists of N-type material with a region of P-type material doped within the N-type material.

A UJT is used primarily as a triggering device because it serves as a step-up device between low-level signals and SCRs and triacs. Outputs from photocells, thermistors, and other transducers can be used to trigger UJTs, which fire SCRs and triacs. UJTs are also used in oscillators, timers, and voltage/current-sensing applications.

UJT Biasing. In normal operation, B1 is negative and a positive voltage is applied to B2. The internal resistance between B1 and B2 divides at the emitter (E), with approximately 60% of the resistance between E and B1. The remaining 40% of resistance is between E and B2. The net result is an internal voltage split. This split provides a positive voltage at the N-type material of the emitter junction, creating an emitter junction that is reverse biased. As long as the emitter voltage remains less than the internal voltage, the emitter junction remains reverse biased, even at a very high resistance.

The junction of a UJT is forward biased when the emitter voltage is greater than the internal value. This rapidly drops the resistance between E and B1 to a very low value. A UJT characteristic curve shows the dramatic change in voltage due to this resistance change. **See Figure 12-54.**

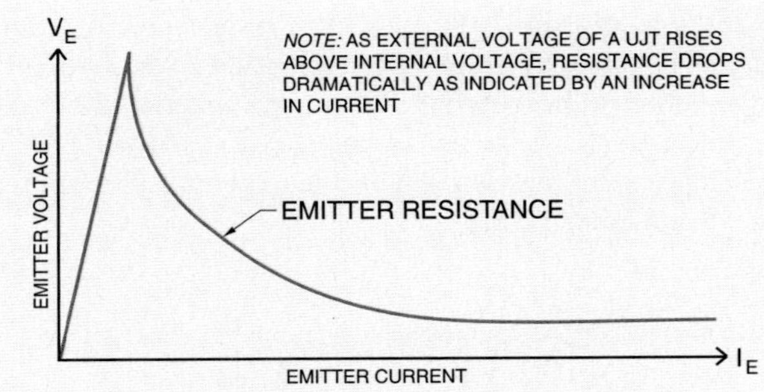

Figure 12-54. A UJT characteristic curve shows the change in voltage due to the resistance change when the device is forward biased.

Diacs

A *diac* is a three-layer bidirectional device used primarily as a triggering device. Unlike a transistor, the two junctions of a diac are heavily and equally doped. Each junction is almost identical to the other.

A diac acts much like two zener diodes that are series connected in opposite directions. Diacs accomplish this through the use of their negative resistance characteristic. *Negative resistance characteristic* is the characteristic that current decreases with an increase in applied voltage. A diac has negative resistance because it does not conduct current until the voltage across it reaches breakover voltage. **See Figure 12-55.**

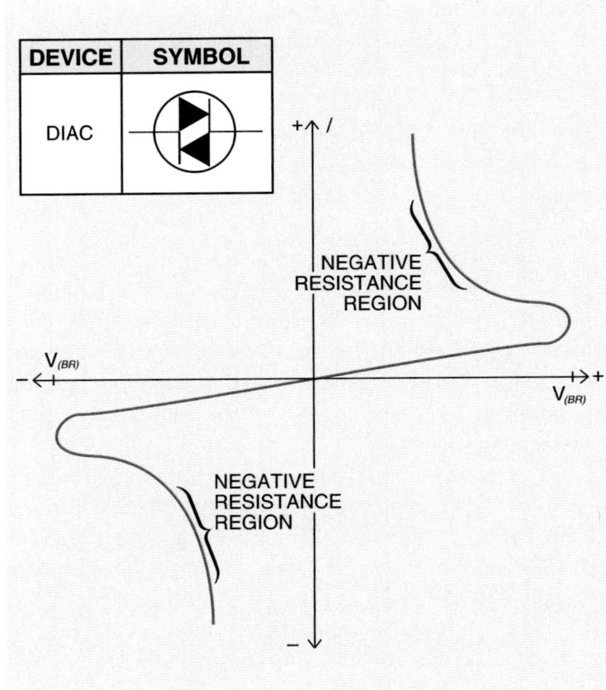

Figure 12-55. A diac rapidly switches from a high-resistance state to a low-resistance state when a positive or negative voltage reaches the breakover voltage.

A diac rapidly switches from a high-resistance state to a low-resistance state when a positive or negative voltage reaches the breakover voltage. Because a diac is a bidirectional device, it is ideal for controlling triacs, which are also bidirectional devices.

COMMUNICATION

Communication is the transmission of information from one point to another by means of electromagnetic waves. Fiber-optic technology has been developed that uses light wave communication to transmit data. This communication is carried out by transmitting modulating light through strands of glass (or plastic in certain applications) over various distances.

Fiber Optics

Fiber optics is a technology that uses a thin, flexible glass or plastic optical fiber to transmit light. Fiber optics is most commonly used as a transmission link. As a transmission link, it connects two electronic circuits consisting of a transmitter and a receiver. **See Figure 12-56.**

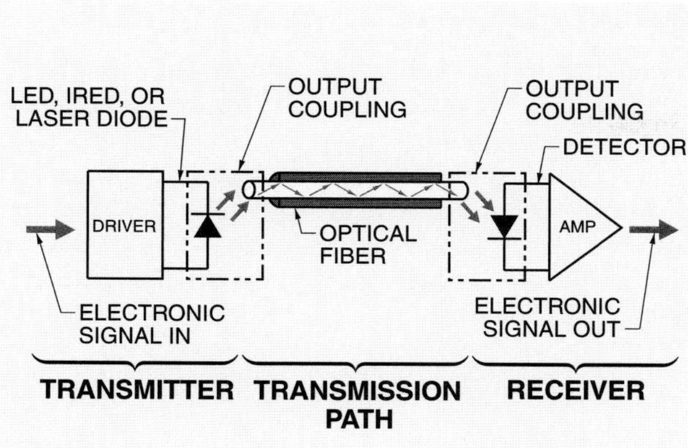

Figure 12-56. Fiber optics uses a thin flexible glass or plastic optical fiber to transmit light.

The central part of the transmitter is its source. The source consists of a light emitting diode (LED), infrared emitting diode (IRED), or laser diode which changes electrical signals into light signals. The receiver normally contains a photodiode that converts light back into electrical signals. The receiver output circuit also amplifies the signal and produces the desired results, such as voice transmission or video signals. Advantages of fiber-optic cables include large bandwidth, low cost, low power consumption, low loss (attenuation), electromagnetic interference (EMI) immunity, small size, light weight, and security (the inability to be tapped by unauthorized users).

Optical Fibers. Optical fibers consist of a core, cladding, and protective jacket. **See Figure 12-57.** The *core* is the actual path for light. The core is normally made of glass but may occasionally be constructed of plastic. *Cladding* is the first layer of protection for the glass or plastic core of the optical fiber cable. A glass or plastic cladding layer is bonded to the core. The cladding is enclosed in a jacket for additional protection.

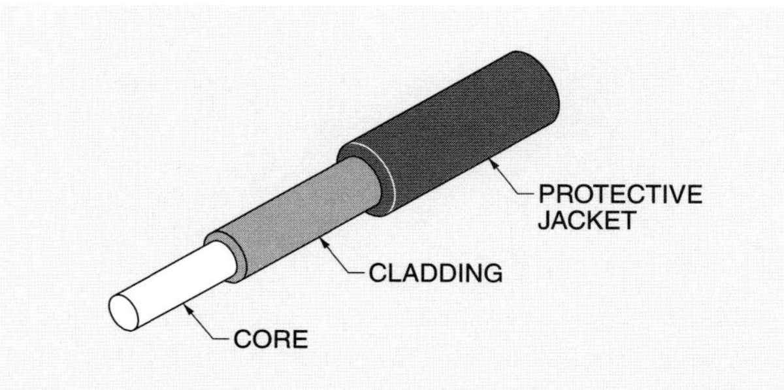

Figure 12-57. Optical fibers consist of a core, cladding, and protective jacket.

Light Source. The light source feeding the cable must be properly matched to the light-activated device for a fiber-optic cable to operate effectively. The source must also be of sufficient intensity to drive the light-activated device.

Laser Diodes. A *laser diode* is a diode similar to an LED but with an optical cavity, which is required for lasing (emitting coherent light) production. The optical cavity is formed by coating opposite sides of a chip to create two highly reflective surfaces. **See Figure 12-58.**

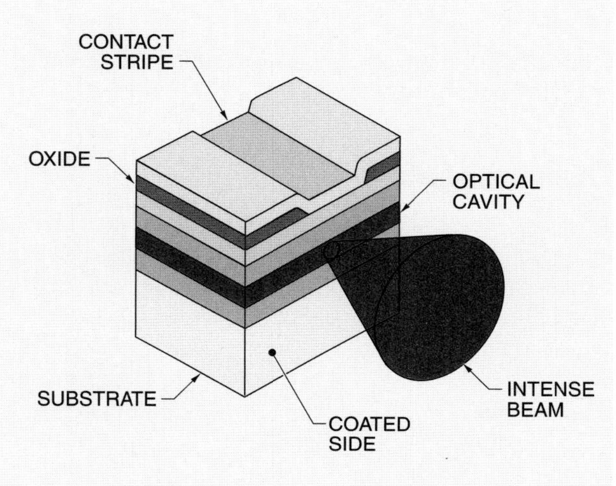

Figure 12-58. A laser diode is similar to an LED but has an optical cavity, which is required for lasing production (emitting coherent light).

Fiber Connectors. The ideal interconnection of one fiber to another is an interconnection that has two fibers that are optically and physically identical. These two fibers are held together by a connector or splice that squarely aligns them on their center axes. The joining of the fibers is so nearly perfect that the interface between them has no influence on light propagation. A perfect connection is limited by variations in fibers and the high tolerances required in the connector or splice. These two factors affect cost and ease of use.

Fiber-Connecting Hardware. Splices and fiber interconnections are often more of a negative factor than poor quality materials because of alignment problems that can arise. The elimination of alignment problems can be accomplished through proper installation of fiber splices, connectors, and couplers. **See Figure 12-59.**

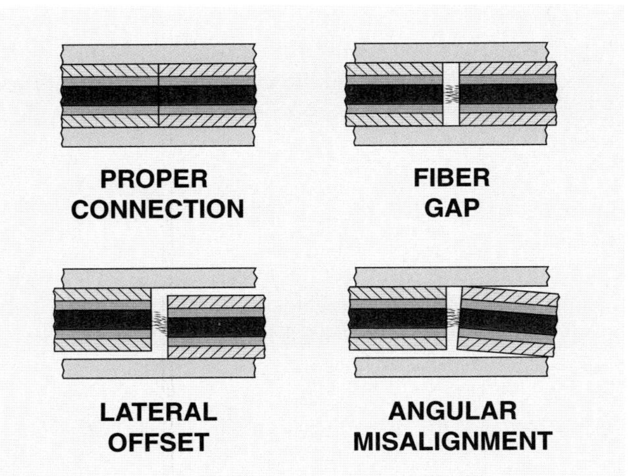

Figure 12-59. Improper connection of fiber-optic cable can result in poor or no transmission.

Optocouplers

An *optocoupler* is a device that consists of an IRED as the input stage and a silicon NPN phototransistor as the output stage. An optocoupler is normally constructed as a dual inline plastic package. **See Figure 12-60.**

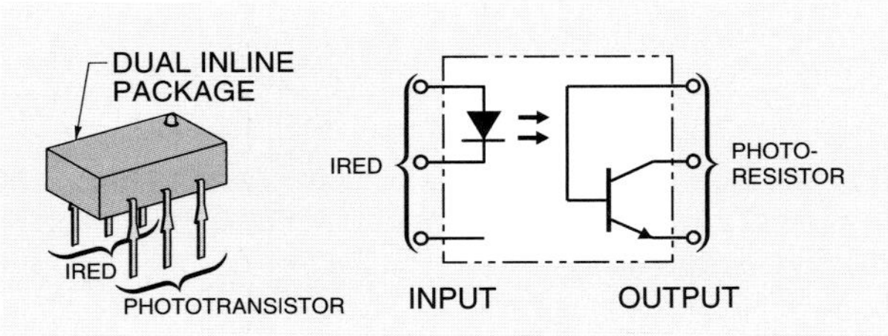

Figure 12-60. An optocoupler consists of an IRED as the input stage and a silicon NPN phototransistor as the output stage.

An optocoupler uses a glass dielectric sandwich to separate input from output. The coupling medium between the IRED and sensor is the infrared transmitting glass. This provides one-way transfer of electrical signals from the IRED to the photodetector (phototransistor) without an electrical connection between the circuitry containing the devices.

Photons emitted from the IRED (emitter) have wavelengths of about 900 nm (nanometers). The detector (transistor) responds effectively to photons with this same wavelength. Input and output devices are always spectrally matched for maximum transfer characteristics. The signal cannot go back in the opposite direction because the emitters and detectors cannot reverse their operating functions.

INTERFACING SOLID-STATE DEVICES

Solid-state devices are usually connected (interfaced) through the use of electrical devices but can also be interfaced through light-driven systems. Once light rays have passed through the optical fiber, they must be detected and converted back into electrical signals. The detection and conversion is accomplished with light-activated devices such as PIN photodiodes, phototransistors, light-activated SCRs, and phototriacs.

PIN Photodiodes

A *PIN photodiode* is a diode with a large intrinsic region sandwiched between P-type and N-type regions. PIN stands for P-type material, insulator, and N-type material. The operation of a PIN photodiode is based on the principle that light radiation, when exposed to a PN junction, momentarily disturbs the structure of the PN junction. The disturbance is due to a hole created when a high-energy photon strikes the PN junction and causes an electron to be ejected from the junction. Thus, light creates electron-hole pairs which act as current carriers. PIN photodiodes are used in gas detectors, spectrometers, and gas analyzers. **See Figure 12-61.**

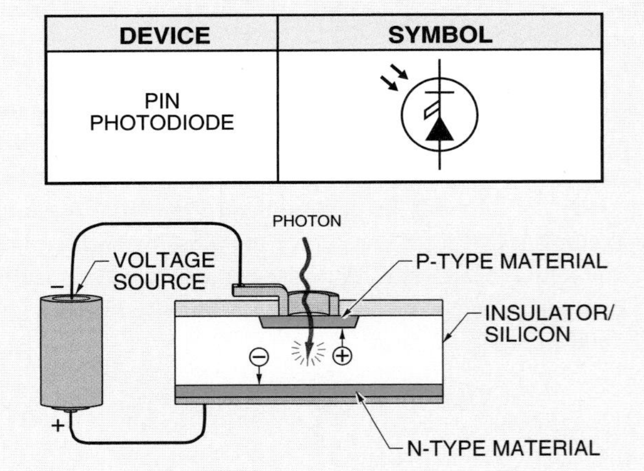

Figure 12-61. A PIN photodiode is a diode with a large intrinsic region sandwiched between P-type and N-type regions.

Phototransistors

A *phototransistor* is a device that combines the effect of a photodiode and the switching capability of a transistor. **See Figure 12-62.** A phototransistor, when connected in a circuit, is placed in series with the bias voltage so that it is forward biased.

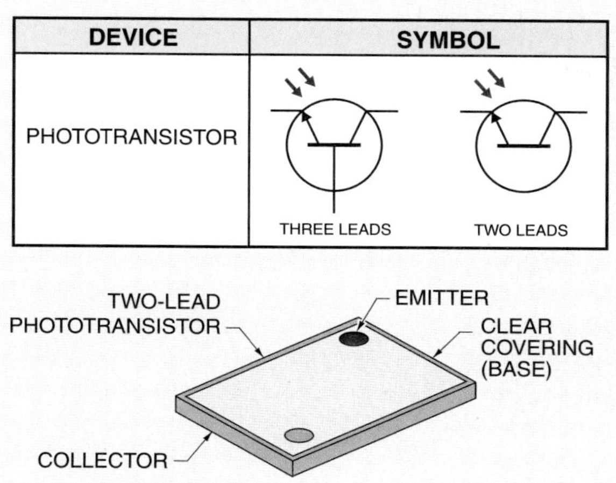

Figure 12-62. A phototransistor is a device that combines the effect of a photodiode and the switching capability of a transistor.

In a two-lead phototransistor, the base lead is replaced by a clear covering. This covering allows light to fall on the base region. Light falling on the base region causes current to flow between the emitter and collector. The collector-base junction is enlarged and works as a reverse biased photodiode controlling the phototransistor. The phototransistor conducts more or less current, depending on the light intensity. If the light intensity increases, resistance decreases and more emitter-to-base current is created. Although the base current is relatively small, the amplifying capability of the small base current is used to control the large emitter-to-collector current. The collector current depends on the light intensity and the DC current gain of the phototransistor. In darkness, the phototransistor is switched OFF with the remaining leakage current (collector dark current).

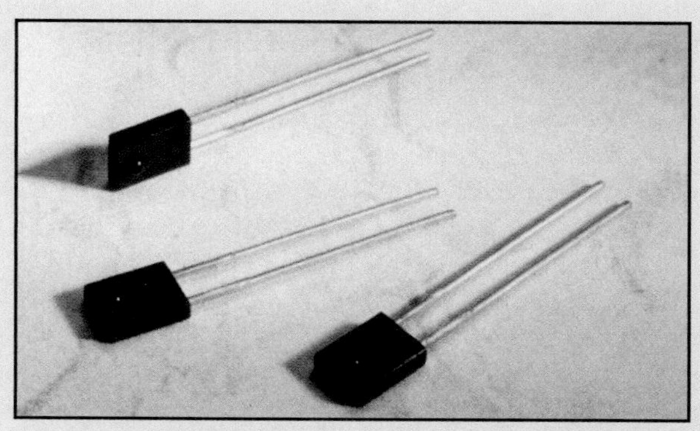

Honeywell

Photodiodes provide a low-power-consumption detection solution when interfaced with battery-powered systems amplifier.

Technical Fact

Additional applications for PIN photodiodes include downhole instruments, where they are used for drilling, formation, and product quality measurement of deep wells in oil and geothermal fields; in turbine engines used in aircraft and stationary power generation; and in internal combustion engines, where they work well in high-temperature environments and help control emissions.

Light-Activated SCRs

A *light-activated SCR (LASCR)* is an SCR that is activated by light. The symbol of a LASCR is identical to the symbol of a regular SCR. The only difference is that arrows are added in the LASCR symbol to indicate a light-sensitive device. **See Figure 12-63.**

DEVICE	SYMBOL
LIGHT-ACTIVATED SCRs	ANODE GATE CATHODE

Figure 12-63. A light-activated SCR is an SCR that is activated by light.

Like a photodiode, current is of a very low level in a LASCR. Even the largest LASCRs are limited to a maximum of a few amps. When larger current requirements are necessary, the LASCR can be used as a trigger circuit for an SCR.

The primary advantage of a LASCR over an SCR is its ability to provide isolation. Because the LASCR is triggered by light, it provides complete isolation between the input signal and the output load current.

Phototriacs

A *phototriac* is a triac that is activated by light. The gate of a phototriac is light sensitive. It triggers the triac at a specified light intensity. **See Figure 12-64.** In darkness, the triac is not triggered. The remaining leakage current is referred to as peak blocking current. A phototriac is bilateral and is designed to switch AC signals.

DEVICE	SYMBOL
PHOTOTRIAC	

Figure 12-64. A phototriac is bilateral and is designed to switch AC signals.

OUTPUT/DISPLAY DEVICES

The visual display of informational outputs is usually in the form of a screen or through an array of illuminated digits. Typical display devices are light emitting diodes and liquid crystal displays. Light emitting diodes use less power than normal incandescent light bulbs but more power than liquid crystal displays.

Light Emitting Diodes

A *light emitting diode (LED)* is a semiconductor diode that produces light when current flows through it. As electrons move across the depletion region, they give up extra kinetic energy. The extra energy is converted to light. An electron must acquire additional energy to get through the depletion region. This additional energy comes from the positive field of the anode. The electron does not get through the depletion region and no light is emitted if the positive field is not strong. **See Figure 12-65.**

Ruud Lighting, Inc.

Outdoor and landscape lighting systems are available with photocells to control the operation of the lighting systems.

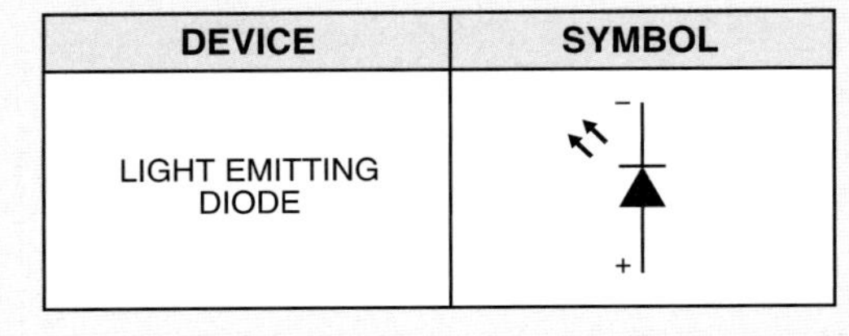

DEVICE	SYMBOL
LIGHT EMITTING DIODE	

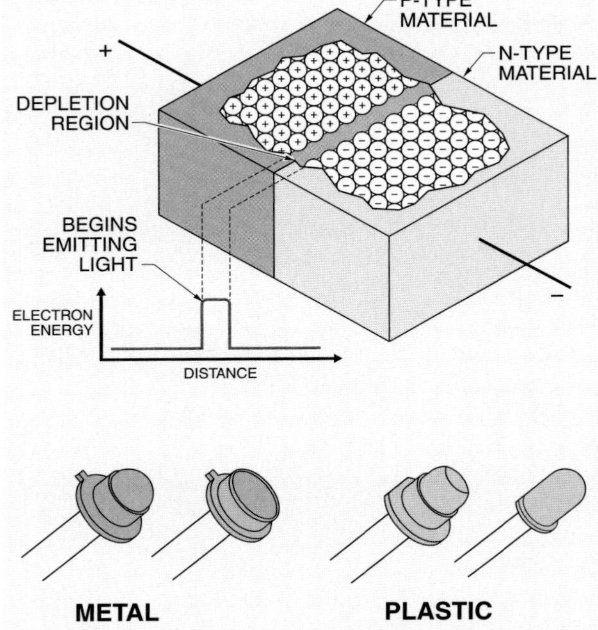

Figure 12-65. A light emitting diode is a diode that produces light when current flows through it.

For a standard silicon diode, a minimum of 0.6 V must be present before the diode conducts. For a germanium diode, 0.3 V must be present before the diode conducts. Most LED manufacturers make a larger depletion region that requires 1.5 V for the electrons to get across.

LED Construction. LED manufacturers normally use a combination of gallium and arsenic with silicon or germanium to construct LEDs. By adding other impurities to the base semiconductor and adjusting them, different wavelengths of light can be produced. LEDs are capable of producing infrared light. *Infrared light* is light that is not visible to the human eye. LEDs may emit a visible red or green light. Colored plastic lenses are available if different colors are desired. As with standard semiconductor diodes, there is a method for determining which end of an LED is the anode and which end is the cathode. The cathode lead is identified by the flat side of the device or it may have a notch cut into the ridge.

A colored plastic lens focuses the light produced at the junction of the LED. Without the lens, the small amount of light produced at the junction is diffused and becomes virtually unusable as a light source. The size and shape of the LED package determines how it is positioned for proper viewing.

The schematic symbol for an LED is exactly like that of a photodiode, but the arrows point away from the diode. The LED is forward biased and a current-limiting resistor is normally present to protect the LED from excessive current.

LED Applications. LEDs are used as status indicators. A *status indicator* is a light that shows the condition of the components in a system. Status indicators may be used to indicate whether a device or system is ON or OFF, or whether it is malfunctioning. In the case of a programmable controller module, the LEDs can be used for diagnostics. **See Figure 12-66.** In this application, the LEDs indicate whether the machine is running properly (RUN) or whether there is a problem in the processor (PROCESSOR), or in the memory (MEMORY). Selecting the diagnostic output is accomplished by using a keylock selector.

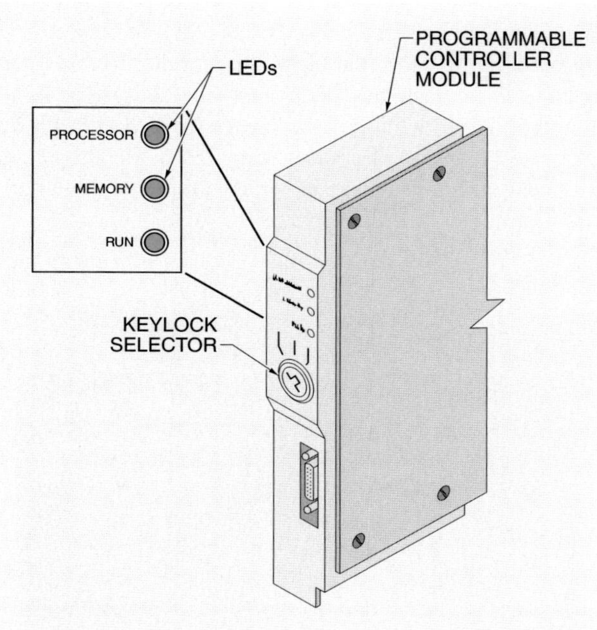

Figure 12-66. When used as status indicators, LEDs can be used to indicate the operating status of a machine or system.

Liquid Crystal Displays

A *liquid crystal display (LCD)* is a display device consisting of a liquid crystal hermetically sealed between two glass plates. An LCD is packaged much like an LED and can produce the same alphanumeric information. Two significant differences between the devices are that an LCD controls light while an LED generates light, and an LCD requires much less power to operate than an LED. Because of these advantages, LCDs are used extensively in new test equipment, clock displays, and calculators.

LCD Construction. An LCD consists of a front and rear piece of glass. These two pieces of glass are separated by a nematic liquid (liquid crystal material). **See Figure 12-67.** Both pieces of glass are coated with a thin, transparent, microscopic layer of metal. The coating is applied to each piece of glass so that it faces the nematic liquid. The layer of metal applied to the front surface of the rear piece of glass covers the entire active area of the display. The layer of metal applied to the rear surface of the front piece of glass is broken into segments. The metal segments are brought through the separator seal to the edges of the display to provide electrical connection points for the driving circuitry.

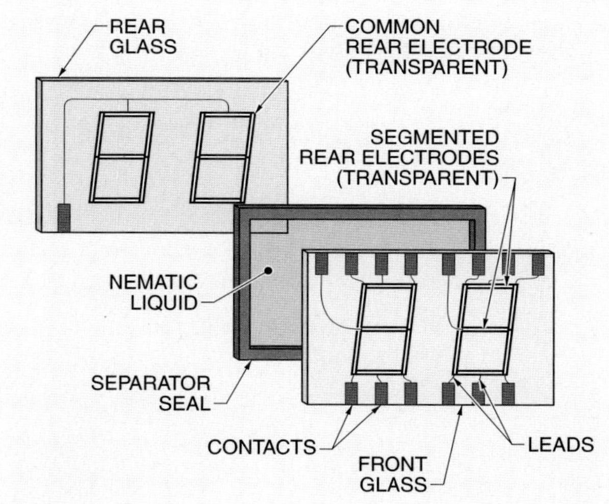

Figure 12-67. A liquid crystal display is composed of two pieces of glass separated by a nematic liquid.

The key to LCD operation is that the nematic liquid fills the space between the front glass and the rear glass. The molecules of the nematic liquid are normally in a parallel alignment. However, when a voltage is applied to the nematic liquid, the molecules turn 90° to alter the light passing through it. **See Figure 12-68.**

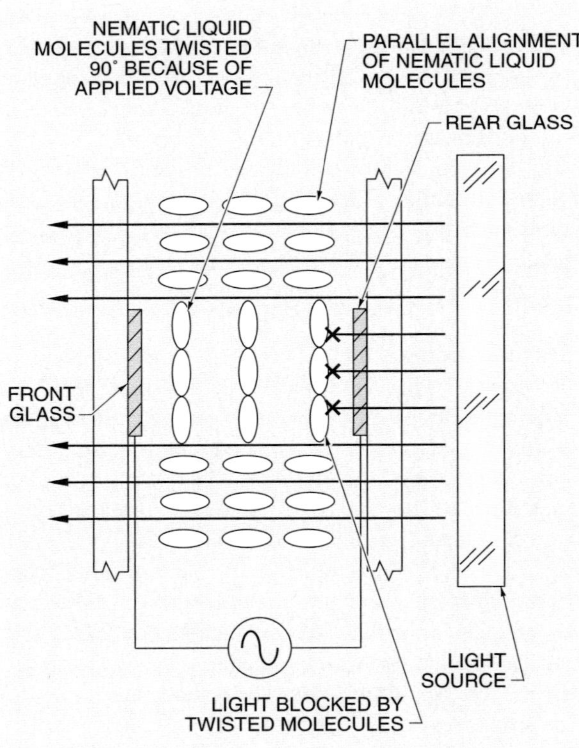

Figure 12-68. The molecules of the nematic liquid are normally aligned in parallel until a voltage is applied to it, twisting the molecules 90° to alter the light passing through.

TROUBLESHOOTING SOLID-STATE DEVICES

High voltages, improper connections, and overheating can damage solid-state devices. An electrician or technician may be responsible for determining the condition of solid-state devices.

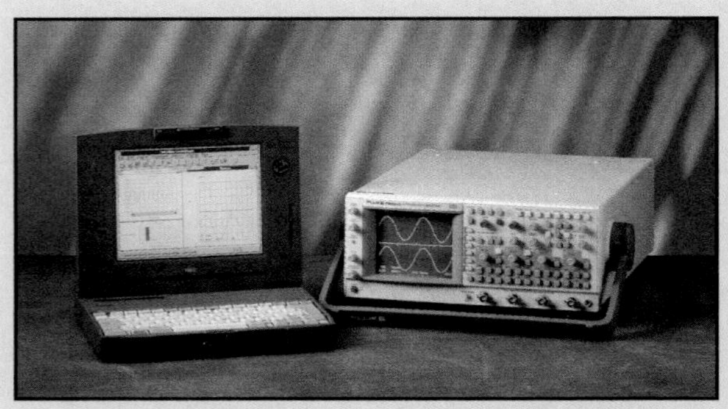

Fluke Corporation

Test equipment that displays voltage and current patterns is used to test solid-state electronic circuits and components.

Testing Diodes

The best way to test a diode is to measure the voltage drop across the diode when it is forward biased. Testing a diode using a DMM set to measure resistance may not indicate whether a diode is good or bad. Testing a diode that is connected in a circuit with a DMM may give false readings because other components may be connected in parallel with the diode under test.

A good diode has a voltage drop across it when it is forward biased and conducting current. The voltage drop is between .5 V and .8 V for the most commonly used silicon diodes. Some diodes are made of germanium and have a voltage drop between .2 V and .3 V.

A DMM in the diode test mode is used to test the voltage drop across a diode. In this position, the DMM produces a small voltage between the test leads. The DMM displays the voltage drop when the leads are connected across a diode. **See Figure 12-69.** To test a diode using the diode test mode on a DMM, apply the following procedure:

1. Ensure that all power in the circuit is OFF. Test for voltage using a DMM set to measure voltage to ensure power is OFF.
2. Set the DMM on the diode test mode.
3. Connect the DMM leads to the diode. Record the reading.
4. Reverse the DMM leads. Record the reading.

The DMM displays a voltage drop between .5 V and .8 V (for a silicon diode) or .2 V and .3 V (for a germanium diode) when a good diode is forward biased. The DMM displays an OL when a good diode is reverse biased. The OL reading indicates that the diode is acting like an open switch. An open (bad) diode does not allow current to flow through it in either direction. The DMM displays an OL reading in both directions when the diode is open. A shorted diode gives the same voltage drop reading in both directions. This reading is normally about .4 V.

Technical Fact

A zener diode acts as a normal rectifier until the voltage applied to it reaches a certain point known as zener (threshold) voltage. At this point, the zener diode either turns ON or OFF, depending on the application. Applications include computer equipment (turn ON), voice-activated devices (turn ON), and surge protectors (turn OFF). Zener diodes are also called "reference diodes."

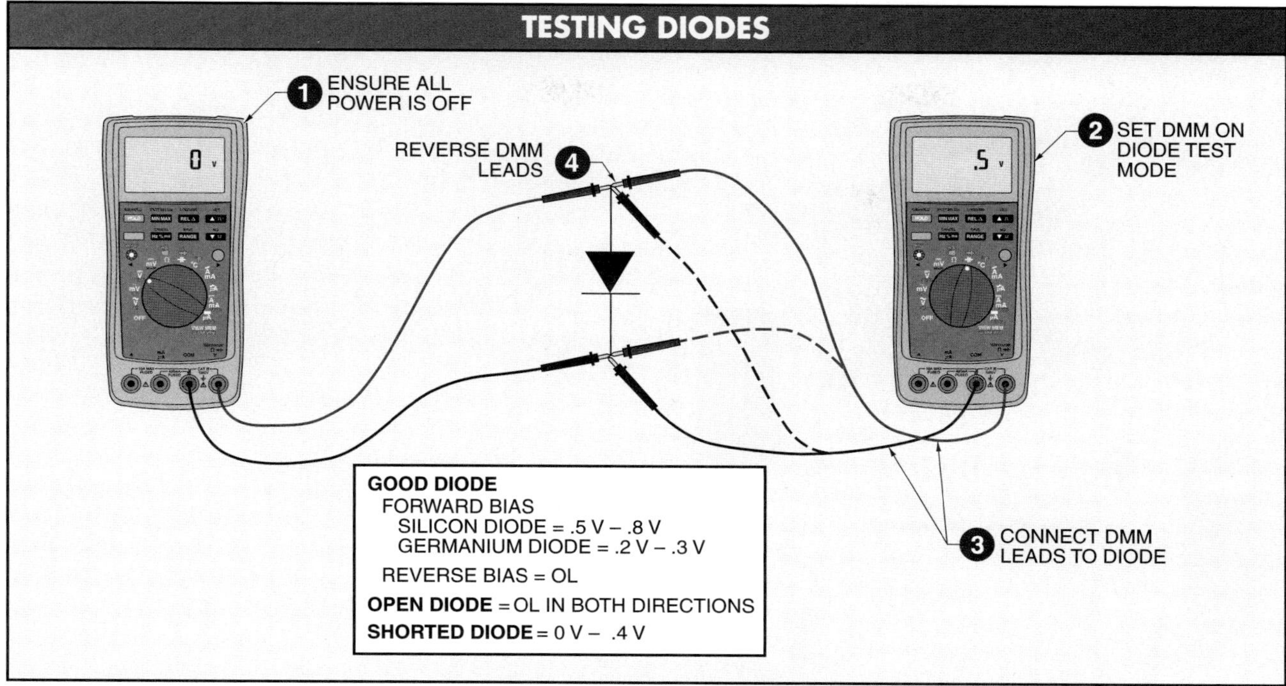

Figure 12-69. A good diode has a voltage drop across it when it is forward biased and conducting current.

Testing Zener Diodes

A zener diode either provides voltage regulation or it fails. A zener diode must be replaced to return the circuit to proper operation if it fails. Occasionally, a zener diode may appear to fail only in certain situations. To check for intermittent failures, a zener diode must be tested while in operation. An oscilloscope is used for testing the characteristics of a zener diode in an operating situation. An oscilloscope displays the dynamic operating characteristics of the zener diode. **See Figure 12-70.**

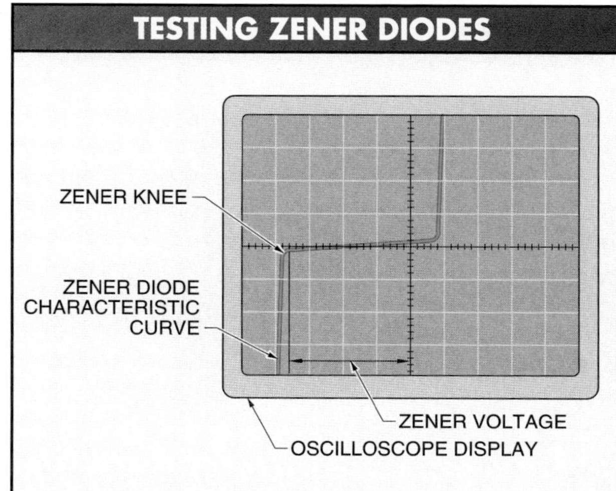

Figure 12-70. An oscilloscope test display indicates if a zener diode is good.

Testing Thermistors

A thermistor must be properly connected to an electronic circuit. Loose or corroded connections create a high resistance in series with the thermistor resistance. The control circuit may sense the additional resistance as a false temperature reading. The hot and cold resistance of a thermistor can be checked with a DMM. **See Figure 12-71.** To test the hot and cold resistance of a thermistor, apply the following procedure:

1. Remove the thermistor from the circuit.
2. Connect the DMM leads to the thermistor leads and place the thermistor and a thermometer in ice water. Record the temperature and resistance readings.
3. Place the thermistor and thermometer in hot water (not boiling). Record the temperature and resistance readings.

Technical Fact

A thermistor is composed of a mixture of metal oxides that exhibit large negative coefficients of resistance changes as the temperature increases. Thermistors can be made in the shape of beads, disks, washers, and rods. The sizes and shapes vary, depending on the type of application.

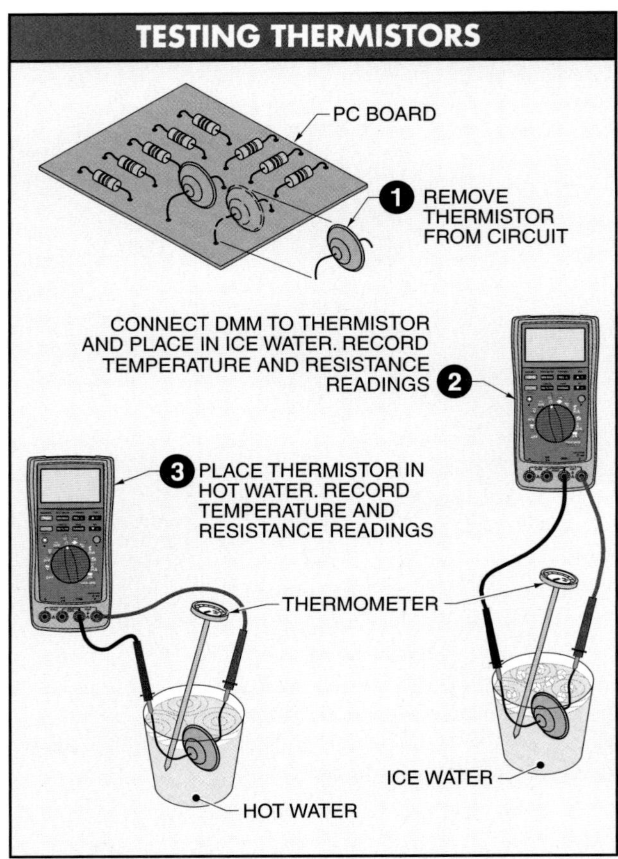

Figure 12-71. The hot and cold resistance of a thermistor can be checked with a DMM.

Compare the hot and cold readings with the manufacturer specification sheet or with a similar thermistor that is known to be good.

Honeywell

Fiber-optic receivers have differential data and signal quality detect outputs, adjustable signal quality detect levels, and 500 Ω output drive capability.

Technical Fact

Pressure sensors have a design suited for difficult physical environments and are ideal for applications that measure compression in various types of engines. In some applications, the output signal of the transducer is applied to an oscilloscope so that the pattern of the oscilloscope provides a visual indication of changes in pressure.

Testing Pressure Sensors

Pressure sensors are tested by checking the resistance of the device at low and high pressure and then comparing the value to manufacturer specification sheets. **See Figure 12-72.** To test a pressure sensor, apply the following procedure:

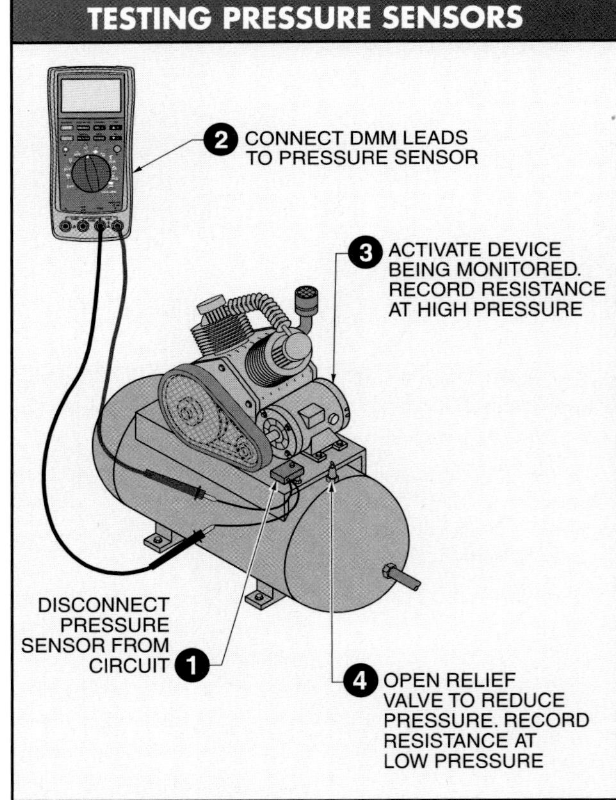

Figure 12-72. Pressure sensors are tested by checking the resistance of the device at low and high pressure and then comparing the value to the manufacturer specification sheets.

Technical Fact

A photoconductor is a transducer material available in sheet, belt, or drum form that changes in electrical conductivity when acted upon by light and relies on the action of light to change the potential of a charged surface. A facsimile machine operates on this principle.

1. Disconnect the pressure sensor from the circuit.
2. Connect the DMM leads to the pressure sensor.
3. Activate the device being monitored (compressor, air tank, etc.) until pressure builds up. Record the resistance of the pressure sensor at the high-pressure setting.
4. Open the relief or exhaust valve and reduce the pressure on the sensor. Record the resistance of the pressure sensor at the low-pressure setting.

Compare the high and low resistance readings with manufacturer specification sheets. Use a replacement pressure sensor that is known to be good when manufacturer specification sheets are not available.

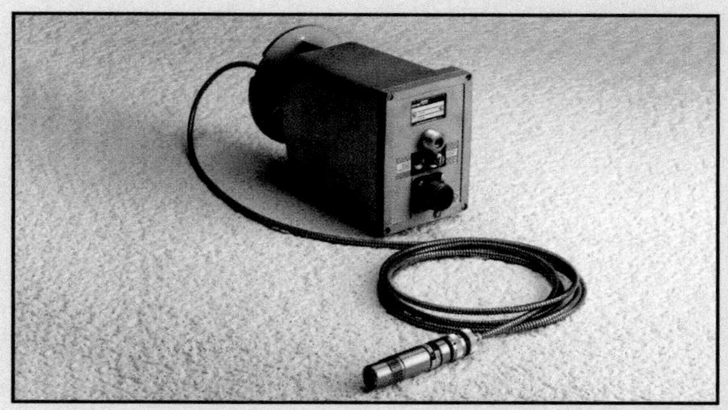

Ircon, Inc.

Fiber-optic infrared thermometers are designed to operate at temperatures between 1300°F and 6500°F.

Testing Photocells

Humidity and contamination are the primary causes of photocell failure. **See Figure 12-73.** The use of quality components that are hermetically sealed is essential for long life and proper operation. Some plastic units are less rugged and more susceptible to temperature changes than glass units. To test the resistance of a photocell, apply the following procedure:

1. Disconnect the photocell from the circuit.
2. Connect the DMM leads to the photocell.
3. Cover the photocell and record dark resistance.
4. Shine a light on the photocell and record light resistance.

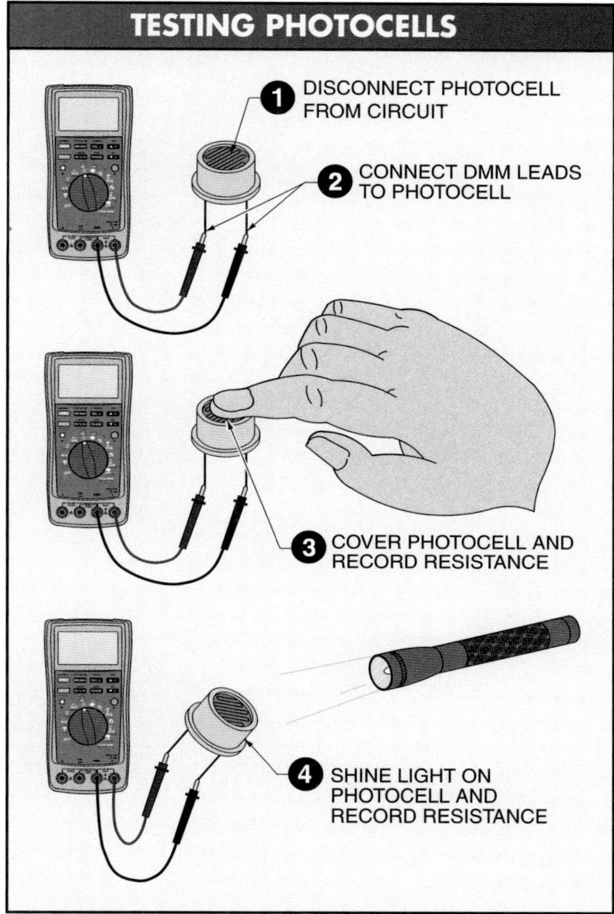

Figure 12-73. Humidity and contamination are the primary causes of photocell failure.

Compare the resistance readings with manufacturer specification sheets. Use a similar photocell that is known to be good when specification sheets are not available. All connections should be tight and corrosion free.

Testing Transistors

A transistor becomes defective from excessive current or temperature. A transistor normally fails due to an open or shorted junction. The two junctions of a transistor may be tested with a DMM set to measure resistance. **See Figure 12-74.** To test an NPN transistor for an open or shorted junction, apply the following procedure:

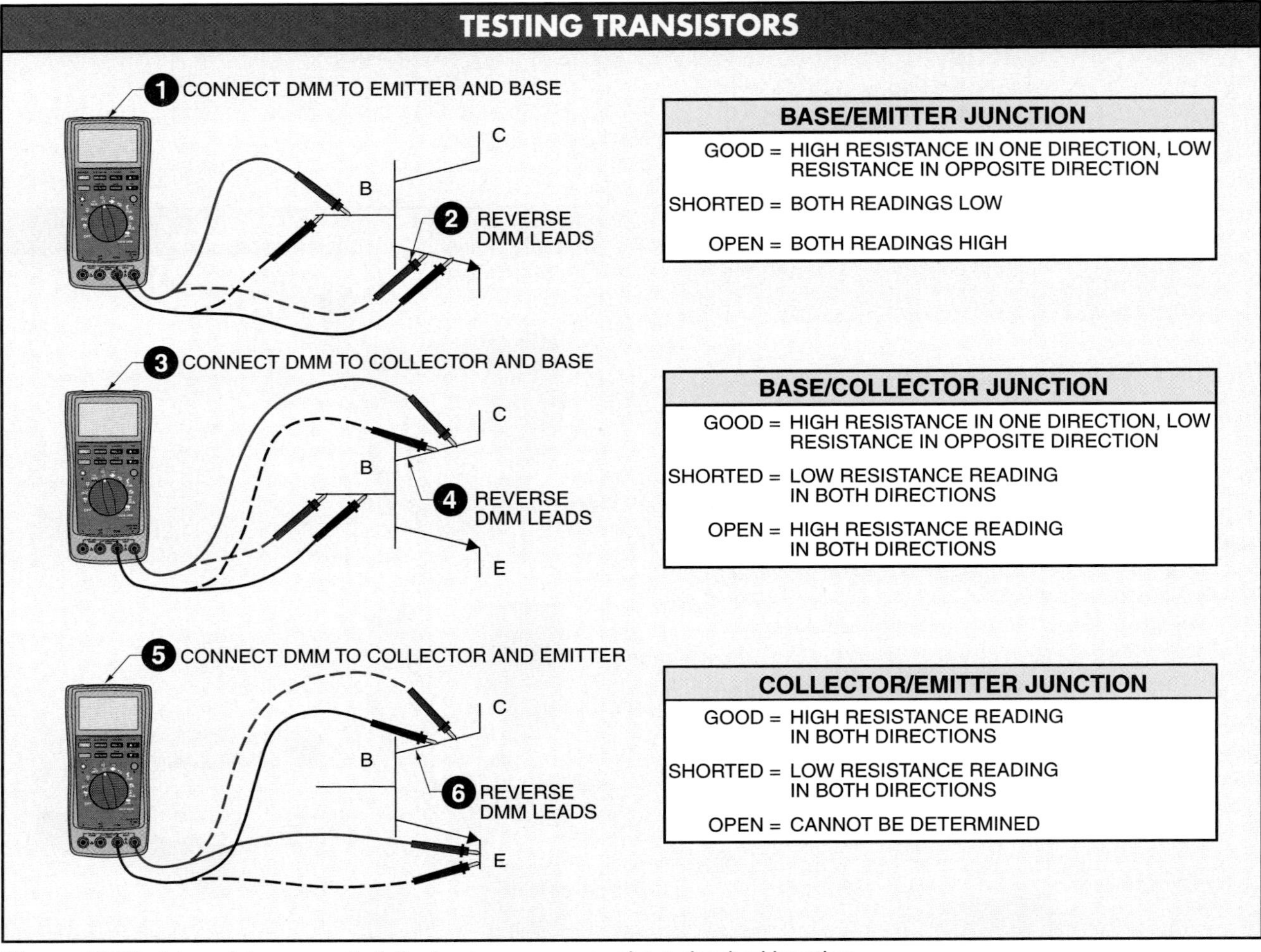

Figure 12-74. A transistor normally fails due to an open or short-circuited junction.

1. Connect a DMM to the emitter and base of the transistor. Measure the resistance.
2. Reverse the DMM leads and measure the resistance. The emitter/base junction is good when the resistance is high in one direction and low in the opposite direction.

   Note: The ratio of high to low resistance should be greater than 100:1. Typical resistance values are 1 kΩ with the positive lead of the DMM on the base, and 100 kΩ with the positive lead of the DMM on the emitter. The junction is shorted when both readings are low. The junction is open when both readings are high.
3. Connect the DMM to the collector and base of the transistor. Measure the resistance.
4. Reverse the DMM leads and measure the resistance. The collector/base junction is good when the resistance is high in one direction and low in the opposite direction.

   Note: The ratio of high to low resistance should be greater than 100:1. Typical resistance values are 1 kΩ with the positive lead of the DMM on the base, and 100 kΩ with the positive lead of the DMM on the collector.
5. Connect the DMM to the collector and emitter of the transistor. Measure the resistance.
6. Reverse the DMM leads and measure the resistance. The collector/emitter junction is good when the resistance reading is high in both directions.

The same test used for an NPN transistor is used for testing a PNP transistor. The difference is that the DMM test leads must be reversed to obtain the same results.

Testing SCRs

An oscilloscope is needed to properly test an SCR under operating conditions. A rough test using a test circuit can be made using a DMM. **See Figure 12-75.** To test an SCR using a DMM, apply the following procedure:

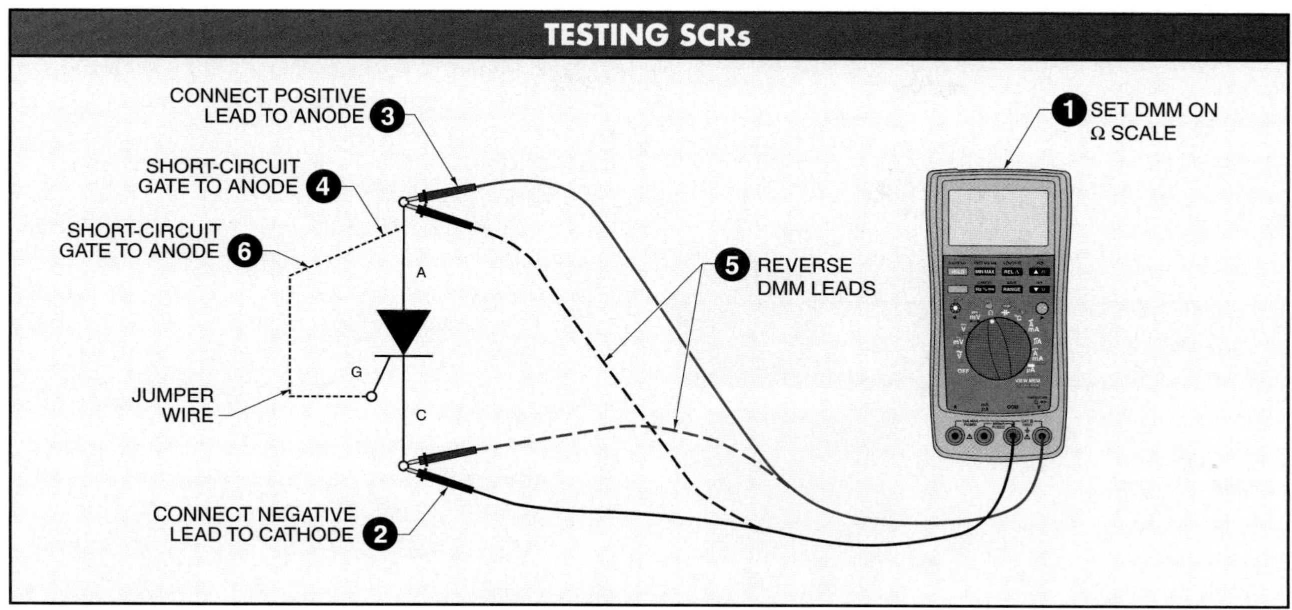

Figure 12-75. A rough test using a test circuit can be made on an SCR using a DMM.

1. Set the DMM on the Ω scale.
2. Connect the negative lead of the DMM to the cathode.
3. Connect the positive lead of the DMM to the anode. The DMM should read infinity.
4. Short-circuit the gate to the anode using a jumper wire. The DMM should read almost 0 Ω. Remove the jumper wire. The low resistance reading should remain.
5. Reverse the DMM leads so that the positive lead is on the cathode and the negative lead is on the anode. The DMM should read almost infinity.
6. Short-circuit the gate to the anode using a jumper wire. The resistance on the DMM should remain high.

Honeywell

Surface-mount emitters and detectors are photodiode or phototransistor detectors used for optical encoders for motion control, computer peripherals, smoke detectors, and medical equipment.

Testing Diacs

A DMM may be used to test a diac for a short circuit. **See Figure 12-76.**

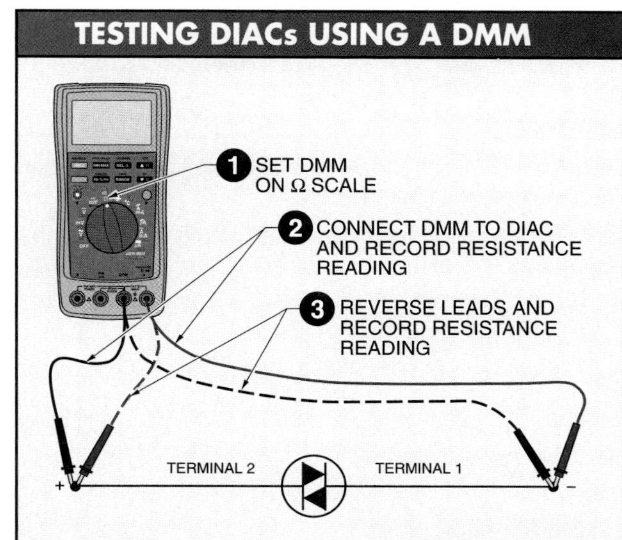

Figure 12-76. A DMM may be used to test a diac for a short circuit.

To test a diac for a short circuit, apply the following procedure:

1. Set the DMM on the Ω scale.
2. Connect the DMM leads to the leads of the diac and record the resistance reading.
3. Reverse the DMM leads and record the resistance reading.

Both resistance readings should show high resistance because the diac is essentially two zener diodes connected in series. Testing a diac in this manner only shows that the component is shunted.

A diac should be tested using an oscilloscope if it is suspected of being open. **See Figure 12-77.**

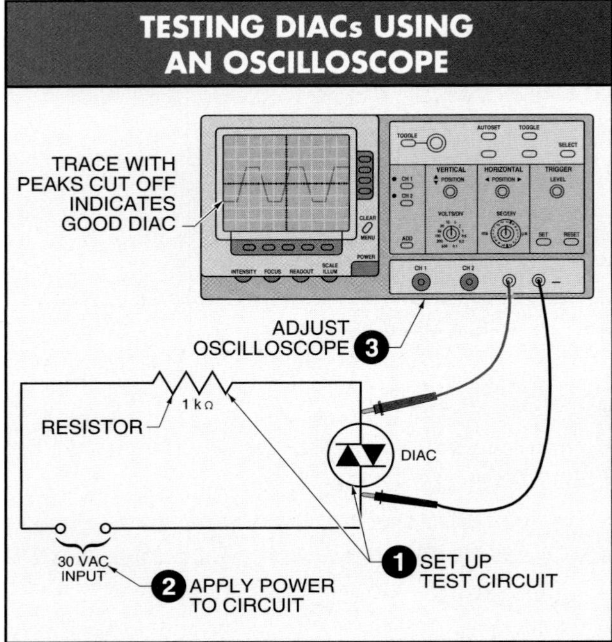

Figure 12-77. A diac should be tested using an oscilloscope if it is suspected of being open.

To test a diac using an oscilloscope, apply the following procedure:

1. Set up the test circuit.
2. Apply power to the circuit.
3. Adjust the oscilloscope.

A trace of an AC sine wave with the peaks cut off indicates that the diac is good.

Technical Fact

Exceeding voltage breakover of SCRs and triacs is not recommended as a turn-on method. Leakage current increases until it exceeds the gate current required to turn ON SCRs and triacs at a small, localized point. Localized heating in a small area will melt the silicon or damage the device if the increasing current is not severely limited.

Testing Triacs

Triacs should be tested under operating conditions using an oscilloscope. A DMM may be used to make a rough test with the triac out of the circuit. **See Figure 12-78.** To test a triac using a DMM, apply the following procedure:

1. Set the DMM on the Ω scale.
2. Connect the negative lead to main terminal 1.
3. Connect the positive lead to main terminal 2. The DMM should read infinity.
4. Short-circuit the gate to main terminal 2 using a jumper wire. The DMM should read almost 0 Ω. The zero reading should remain when the lead is removed.
5. Reverse the DMM leads so that the positive lead is on main terminal 1 and the negative lead is on main terminal 2. The DMM should read infinity.
6. Short-circuit the gate of the triac to main terminal 2 using a jumper wire. The DMM should read almost 0 Ω. The zero reading should remain after the lead is removed.

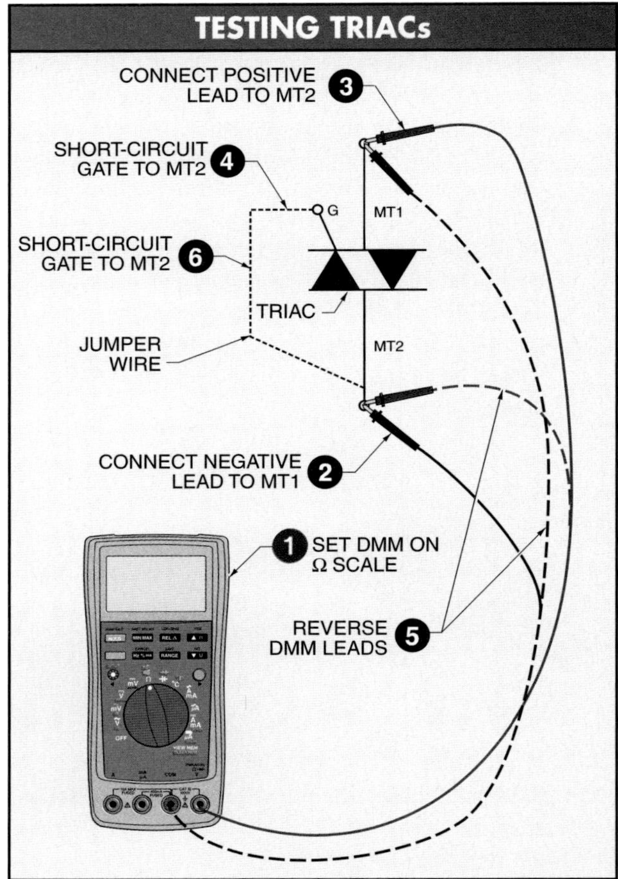

Figure 12-78. A DMM may be used to make a rough test of a triac that is out of the circuit.

Review Questions

1. What determines whether an element is an insulator, conductor, or semiconductor?
2. How is the crystal structure in most semiconductors altered?
3. What is a PC board?
4. What are forward bias and reverse bias in semiconductors?
5. What is a half-wave rectifier?
6. How does a zener diode operate in a circuit?
7. What is a thermistor?
8. How does a photoconductive cell differ from a photovoltaic cell?
9. What is a Hall effect sensor?
10. What is a solid-state pressure sensor?
11. What is an LED?
12. What is a transistor?
13. How is the shape of a transistor identified?
14. What is rectification?
15. What is gain?
16. What are the three types of transistor amplifiers?
17. What is forward breakover voltage?
18. What is a triac?
19. How does a triac function?
20. How does a diac function?
21. What is an integrated circuit?
22. How are pins on an integrated circuit identified?
23. What is an op-amp?
24. What are the two basic types of electronic signals?
25. What are the four most common gates in digital electronics?

chapter

13

electrical motor controls *for Integrated Systems*

Timers and Counters

Timers include dashpot, synchronous clock, solid-state, and programmable timers. Timers are used in any application that requires a time delay at some point of circuit operation. Several different timing functions are available to meet the many different requirements of time-based circuits and applications. Totalizers and counters are used to count events and provide an output.

TIMERS

The four major categories of timers are dashpot, synchronous clock, solid-state, and programmable. Dashpot, synchronous clock, and solid-state timers are stand-alone timers. Stand-alone timers are connected between the input device (limit switch, etc.) and the output device (solenoid, etc.) controlled by the timer. Programmable timers are timing functions that are included in electrical control devices such as programmable controllers (PLCs). **See Figure 13-1.** A *programmable controller (PLC)* is a solid-state control device that is programmed and reprogrammed to automatically control an industrial process or machine.

Dashpot timers are the oldest industrial timers. Dashpot timers can be found on old equipment but are rarely used in new installations. Synchronous clock timers have been installed in millions of control applications and are still specified in some new applications. Solid-state timers are the most common stand-alone timer used in control applications today. However, in most electrical systems that include a PLC, the internal programmable timers of the PLC can be used to replace all stand-alone timers. Although each device accomplishes its task in a different way, all timers have the common ability to introduce some degree of time delay into a control circuit.

Dashpot Timers

A *dashpot timer* is a timer that provides time delay by controlling how rapidly air or liquid is allowed to pass into or out of a container through an orifice (opening) that is either fixed in diameter or variable. **See Figure 13-2.** For example, if the piston of a hand-operated tire pump is forced down, the piston moves down rapidly if the valve opening is unrestricted. However, if the valve opening is restricted, the travel time of the piston increases. The smaller the opening, the longer the travel time.

Figure 13-1. The four major categories of timers are dashpot, synchronous clock, solid-state, and programmable.

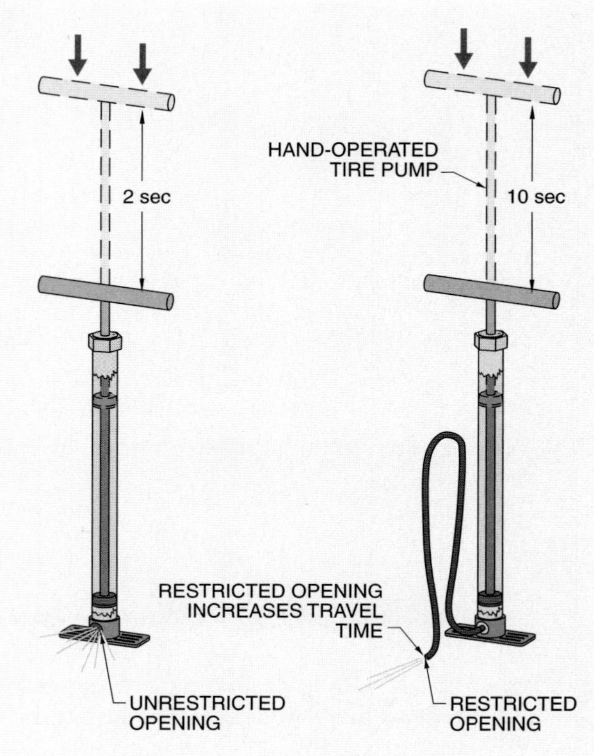

Figure 13-2. A dashpot timer provides time delay by controlling how rapidly air or liquid is allowed to pass into or out of a container.

Synchronous Clock Timers

A *synchronous clock timer* is a timer that opens and closes a circuit depending on the position of the hands of a clock. **See Figure 13-3.** Synchronous clock timers may have one or more contacts through which the circuit may be opened or closed.

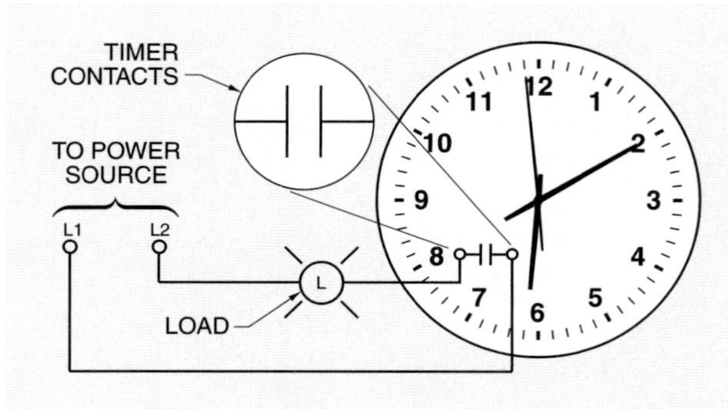

Figure 13-3. A synchronous clock timer opens and closes a circuit depending on the position of the hands of a clock.

The time delay is provided by the speed at which the clock hands move around the perimeter of the face of the clock. In this case, the contacts are closed once every 12 hours. A synchronous clock motor operates the timer. Synchronous clock motors are AC-operated and maintain their speed based on the frequency of the AC power line which feeds them. Synchronous clock timers are accurate timers because power companies regulate the line frequency within strict tolerances.

Solid-State Timers

A *solid-state timer* is a timer whose time delay is provided by solid-state electronic devices enclosed within the timing device. A solid-state timing circuit provides a very accurate timing function at the most economical cost. Solid-state timers can control timing functions ranging from a fraction of a second to hundreds of hours. Most solid-state timers are designed as plug-in modules for quick replacement.

Solid-state timers can replace dashpot and synchronous timers in most applications. Solid-state timers are less susceptible to outside environmental conditions because they, like relay coils, are often encapsulated in epoxy resin for protection. However, because they are encapsulated, they are normally discarded when they fail because they cost less than other timers and are impossible to repair. **See Figure 13-4.**

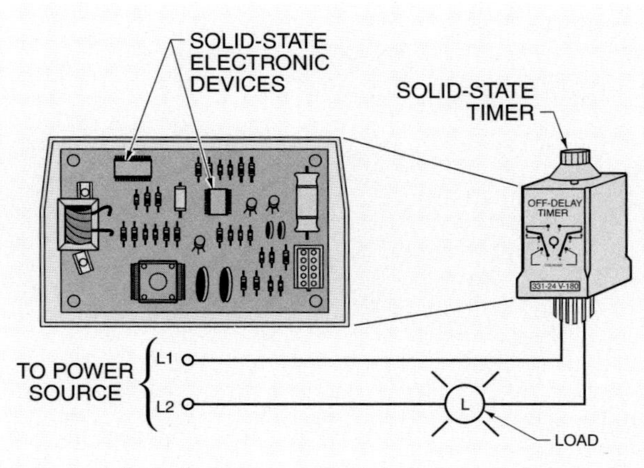

Figure 13-4. A solid-state timer has a time delay provided by solid-state electronic devices enclosed within the timing device.

Programmable Timers

A *programmable timer* is a timer (timing function) included in electrical control devices such as PLCs. When a programmable timer is used to control a circuit, the timers inside the PLC are used to produce the required timing operations.

Programmable timers include retentive and nonretentive timers. A *retentive timer* is a timer that maintains its current accumulated time value when its control input signal is interrupted or power to the timer is removed. A *nonretentive timer* is a timer that does not maintain its current accumulated time value when its control input signal is interrupted or power to the timer is removed. **See Figure 13-5.**

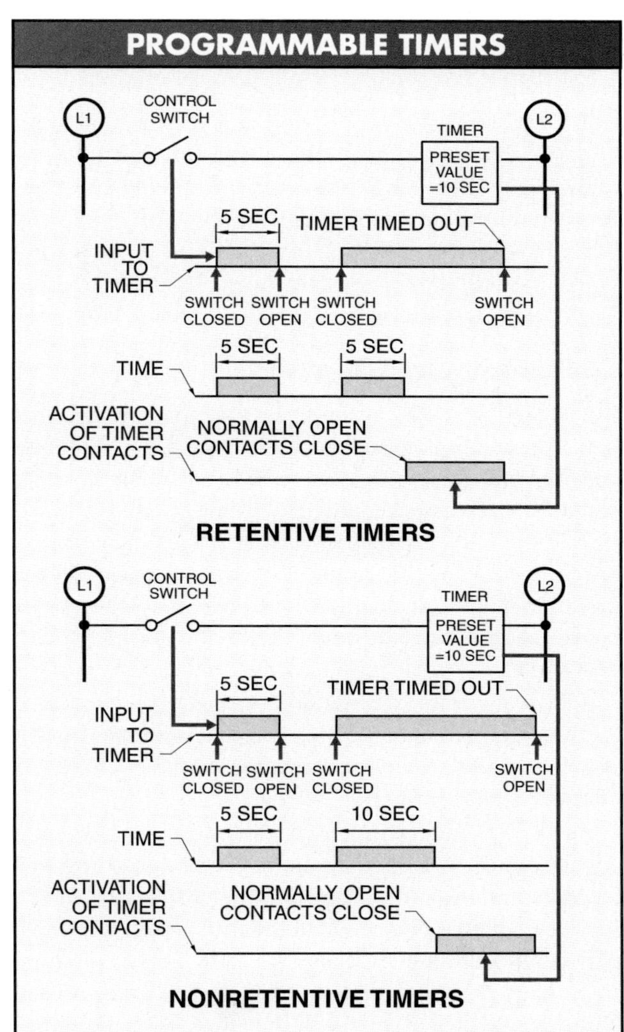

Figure 13-5. Programmable timers include retentive and nonretentive timers.

When a retentive timer is used, the timer retains all accumulated time values if it is interrupted during a procedure. Retentive timers are used in applications where the timer value is to continue where it left off after an interruption. When a nonretentive timer is used, the timer does not retain any accumulated time values if it is interrupted during a procedure. Nonretentive timers reset to the values they had before they began timing. Retentive and nonretentive timers retain all preset values during a power failure.

TIMING FUNCTIONS

Several different timing functions are available to meet the many different requirements of time-based circuits and applications. ON-delay and OFF-delay timing functions were the only two timing functions available when dashpot and synchronous timers were the only timers used.

When solid-state timers became available, they offered ON-delay, OFF-delay, one-shot, and recycle functions. Today's solid-state timers offer dozens of special timing functions in addition to the four basic timing functions because solid-state timing circuits can be easily modified. Several of the special timing functions are normally combined into one multiple-function timer.

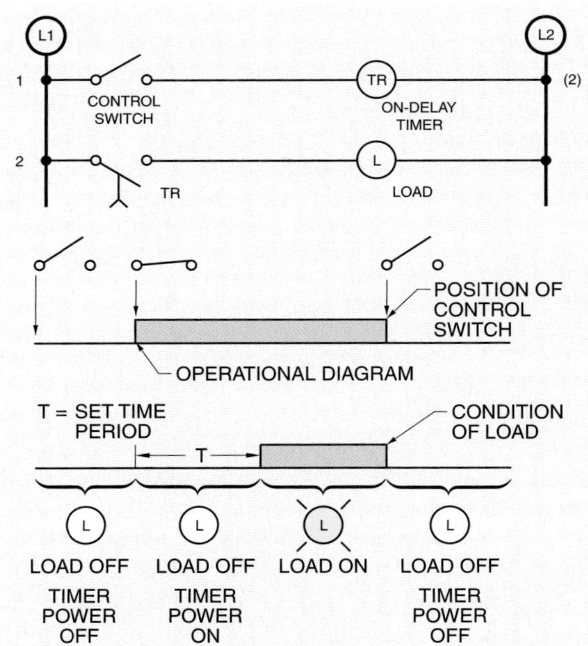

Figure 13-6. ON-delay timer contacts do not change position until the set time period passes after the timer receives power.

ON-Delay Timers

An *ON-delay (delay on operate) timer* is a device that has a preset time period that must pass after the timer has been energized before any action occurs on the timer contacts. Once activated, the timer may be used to turn a load ON or OFF, depending on the way the timer contacts are connected into the circuit. The load energizes after the preprogrammed time delay when a normally open timer contact is used. The load de-energizes after the preset time delay when a normally closed timer contact is used.

ON-delay timer contacts do not change position until the set time period passes after the timer receives power. **See Figure 13-6.** After the preset time has passed, the timer contacts change position.

In the ON-delay timer circuit, the normally open contacts close and energize the load. The load remains energized as long as the control switch remains closed. The load de-energizes the second the control switch is opened. An operational diagram is used to show timer operation. In the operational diagram, the top line shows the position of the control switch and the bottom line shows the condition of the load.

ON-Delay (Timed-Closed). An ON-delay (timed-closed) function may be illustrated using two balloons. **See Figure 13-7.** The solenoid plunger forces air out of balloon A, through orifice B, and into balloon C when control switch S1 is closed. Contacts TR1 close, energizing the circuit to the load after balloon C is filled. This energizes the load. The ON-delay function takes 5 seconds if it takes 5 seconds for balloon C to fill.

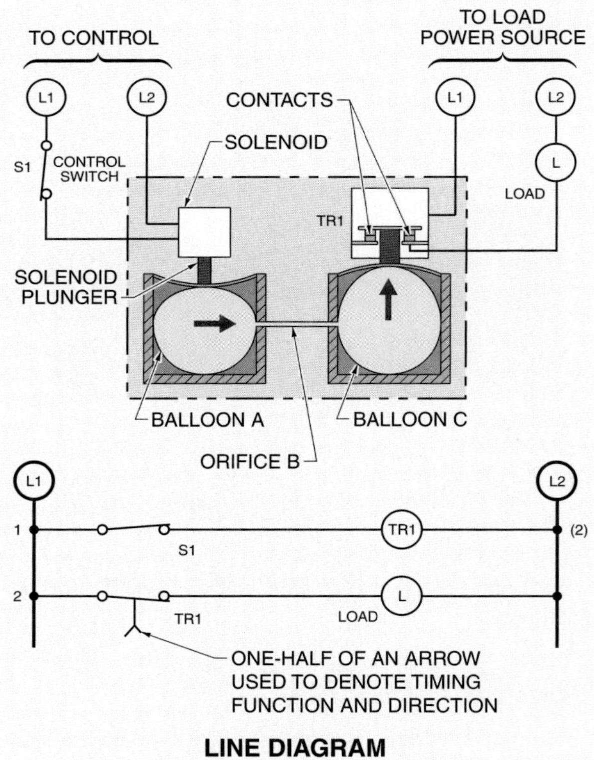

Figure 13-7. With an ON-delay (timed-closed) function, the contacts close after the timing cycle is complete.

One-half of an arrow is used to indicate the direction of time delay of the normally open timing contacts in ON-delay timers. The half arrow points in the direction of ON delay. The operational diagram should be used if an arrow is not used with an ON-delay timer.

ON-Delay (Timed-Open). An ON-delay timer could be designed to open or close a circuit after a predetermined time delay. With an ON-delay (timed-open) function, the balloon forces the contacts open after the timing cycle is complete. **See Figure 13-8.** With control switch S1 closed, the solenoid plunger forces air from balloon A through orifice B, and into balloon C. After 5 seconds, contacts TR1 open the circuit to the load and the load is de-energized. One-half of an arrow is shown in the line diagram. The arrow indicates that the normally closed contacts open after the ON-delay function has taken place. This pneumatically operated timing function is the way dashpot timers operate. A synchronous clock timer or solid-state timer could be substituted for the pneumatic timer. A pneumatic timer is the easiest to understand in terms of mechanical and timing operation.

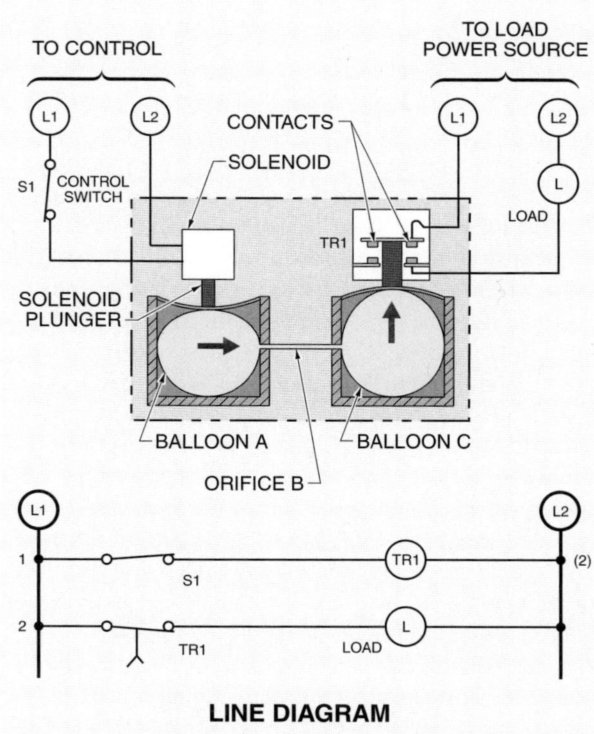

Figure 13-8. With an ON-delay (timed-open) function, the contacts open after the timing cycle is complete.

OFF-Delay Timers

An *OFF-delay (delay on release) timer* is a device that does not start its timing function until the power is removed from the timer. **See Figure 13-9.** In this circuit, a control switch is used to apply power to the timer. The timer contacts change immediately and the load energizes when power is first applied to the timer.

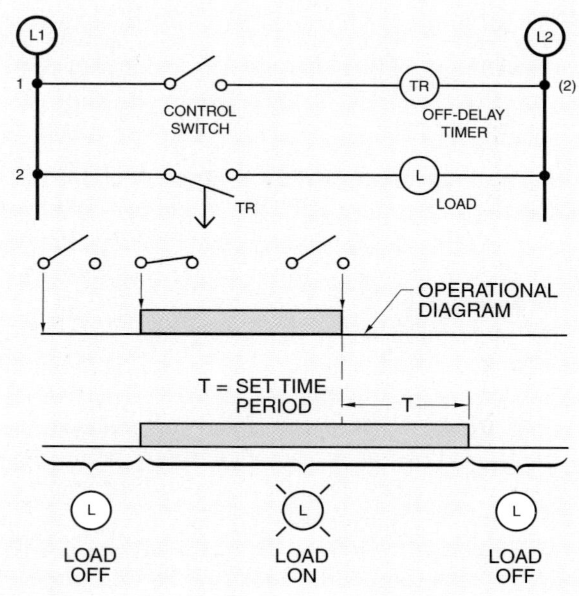

Figure 13-9. An OFF-delay (delay on release) timer is a device that does not start its timing function until the power is removed from the timer.

The timer contacts remain in the changed position and the time period starts when power is removed from the timer. The timer contacts return to their normal position and the load is de-energized on expiration of the set time period.

OFF-Delay (Timed-Open). An OFF-delay (timed-open) contact circuit may be used to provide cooling in a projector once the bulb has been turned OFF but has not had time to cool down. **See Figure 13-10.**

Technical Fact

General-purpose timers are used for quick replacement and space-saving applications, while industrial timers are designed for heavy-duty applications where the timer must have long life and safety circuits.

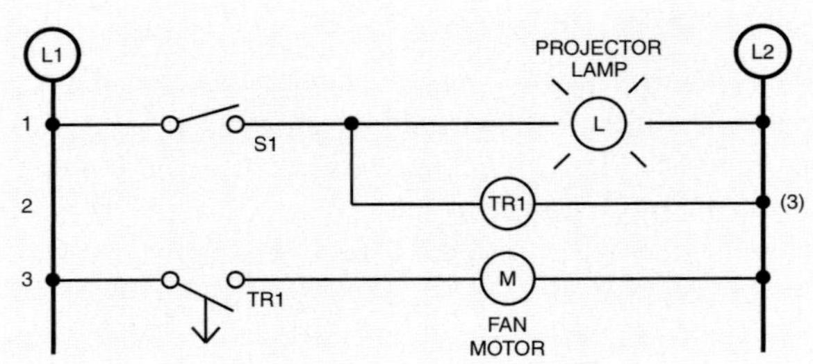

Figure 13-10. An OFF-delay (timed-open) contact circuit may be used to provide cooling in a projector once the bulb has been turned OFF but has not had time to cool down.

In this circuit, closing switch S1 turns ON the projector bulb and activates timer coil TR1. With timer TR1 energized, normally open contacts TR1 immediately close, energizing the fan motor, which controls the cooling of the projector.

The projector bulb and the cooling fan remain ON as long as switch S1 stays closed. When switch S1 is opened, the projector bulb turns OFF and power is removed from the timer. Contacts TR1 remain closed for a predetermined OFF delay and then open, causing the cooling fan to turn OFF. This OFF-delay timed-open circuit is generally set to adequately cool the projector equipment before it shuts OFF. This circuit could also be used for large cooling fan motors by replacing the fan motor in the control circuit with a motor starter. The motor starter could be used to control any size motor.

Rockwell Automation, Allen-Bradley Company, Inc.

All programmable controllers include internal programmable timers that can be used for any required timing function.

OFF-Delay (Timed-Closed). An OFF-delay (timed-closed) contact circuit may be used to provide a pumping system with backspin protection and surge protection on stopping. **See Figure 13-11.**

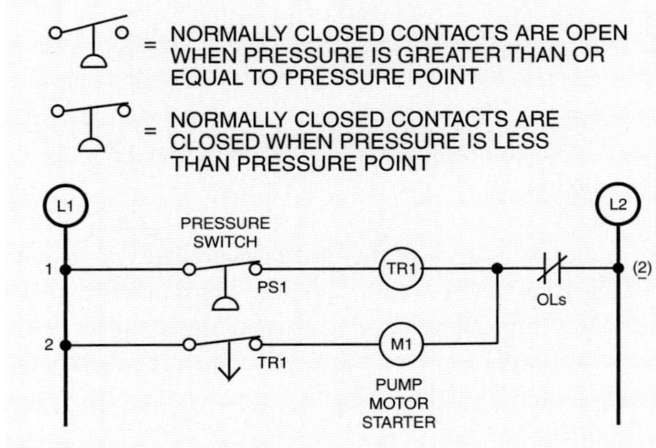

Figure 13-11. An OFF-delay (timed-closed) contact circuit may be used to provide a pumping system with backspin protection and surge protection on stopping.

Surge protection is often necessary when a pump is turned OFF and a high column of water is stopped by a check valve. The force of the sudden stop may cause surges which operate the pressure switch contacts, subjecting the starter to chattering. Backspin is the backward turning of a centrifugal pump when the head of water runs back through the pump just after it has been turned OFF. Starting the pump during backspin might damage the pump motor.

To eliminate the damage resulting from surges and backspin, actuating pressure switch PS1 causes timer relay coil TR1 to energize. With TR1 energized, the normally closed contacts of TR1 immediately open and cause coil M1 to energize, shutting OFF the pump motor. Even if pressure switch PS1 recloses, the coil and motor remain OFF for a predetermined time (to allow surges and backspin to clear) and then restart once the OFF-delay function has taken place. This OFF-delay function prevents the system from operating until TR1 has timed out and its normally closed contacts return to their normally closed position.

A comparison chart may be used to compare the operation of ON-delay and OFF-delay timing functions and contacts. **See Figure 13-12.** To help compare timing functions, instantaneous relay contacts are also included. Some manufacturers also use abbreviations in their catalogs to describe the type of contacts used.

TIMER COMPARISON			
Abbreviation	**Meaning**	**Function**	**Symbols**
NOTC	Normally open, timed-closed	ON-delay (timed-closed) contact – Timer contact normally open. Timed-closed on timer energization. Opens immediately on timer de-energization.	
NCTO	Normally closed, timed-open	ON-delay (timed-open) contact – Timer contact normally closed. Timed-open on timer energization. Closes immediately on timer de-energization.	
NOTO	Normally open, timed-open	OFF-delay (timed-open) contact – Timer contact normally open. Closes immediately on timer energization. Timer contact times open on timer de-energization.	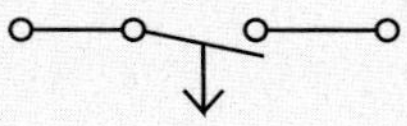
NCTC	Normally closed, timed-closed	OFF-delay (timed-closed) contact – Timer contact normally closed. Opens immediately on timer energization. Timer contact times closed on timer de-energization.	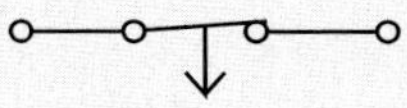
NO	Normally open	Instantaneous contact – Normally open. Contact closes immediately on relay energization. Opens immediately on relay de-energization.	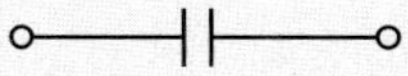
NC	Normally closed	Instantaneous contact – Normally closed. Contact opens immediately on relay energization. Closes immediately on relay de-energization.	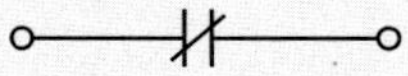

Figure 13-12. A comparison chart may be used to compare the operation of ON-delay and OFF-delay timing functions and contacts.

One-Shot Timers

A *one-shot (interval) timer* is a device in which the contacts change position immediately and remain changed for the set period of time after the timer has received power. **See Figure 13-13.** After the set period of time has passed, the contacts return to their normal position.

One-shot timers are used in applications in which a load is ON for only a set period of time. One-shot timer applications include coin-operated games, dryers, car washes, and other machines. One-shot timing functions have not been available as long as ON-delay and OFF-delay timing functions because the one-shot timing function became available only when solid-state timers became available. For this reason, and the fact that so many other timing functions are now available, no standard symbol was established for any other timer contacts except ON-delay and OFF-delay. Today, the basic normally open and normally closed contacts, along with the timer type and/or operational diagram, are used with all timers that are not ON-delay or OFF-delay.

Recycle Timers

A *recycle timer* is a device in which the contacts cycle open and closed repeatedly once the timer has received power. The cycling of the contacts continues until power is removed from the timer. **See Figure 13-14.** In a recycle timer circuit, the closing of the control switch starts the cycling function. The load continues to turn ON and OFF at regular time intervals as long as the control switch is closed. The cycling function stops when the control switch is opened.

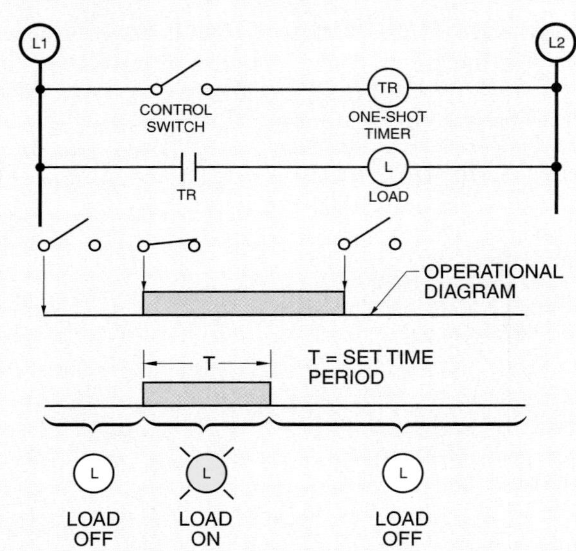

Figure 13-13. A one-shot (interval) timer has contacts which change position and remain changed for the set period of time after the timer has received power.

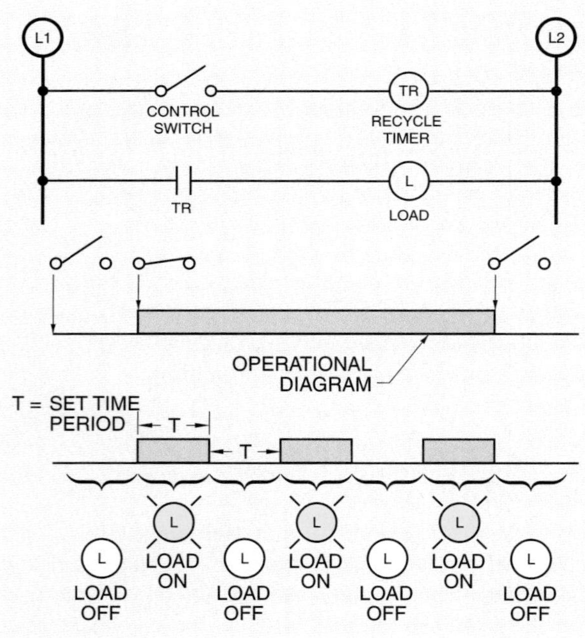

Figure 13-14. A recycle timer is a device in which the contacts cycle open and closed repeatedly once the timer has received power.

Recycle timers may be symmetrical or asymmetrical. A *symmetrical recycle timer* is a timer that operates with equal ON and OFF time periods. An *asymmetrical recycle timer* is a timer which has independent adjustments for the ON and OFF time periods. Asymmetrical timers always have two different time adjustments.

Multiple-Function Timers

ON-delay, OFF-delay, one-shot, and recycle timers are considered monofunction timers. That is, they perform only one timing function, such as ON-delay or OFF-delay. Multiple-function timers are solid-state timers that can perform many different timing functions. Multiple-function timers are normally programmed for different timing functions by the placement of DIP (dual in-line package) switches located on the timer. **See Figure 13-15.**

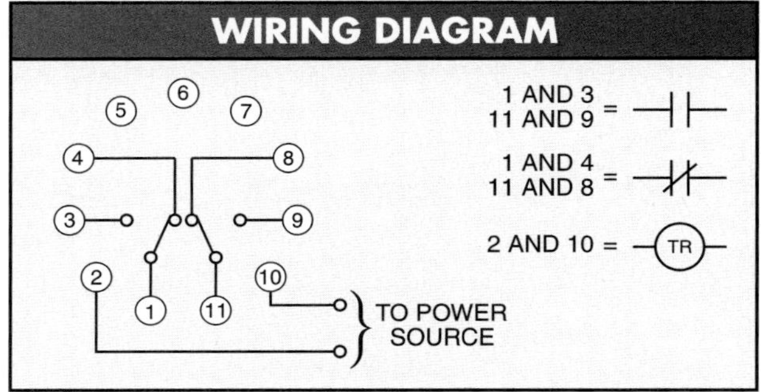

TIME/FUNCTION SETTINGS

SELECTION OF TIME RANGES DIP SWITCH SELECTOR (1 & 2)	SELECTION OF FUNCTION DIP SWITCH SELECTOR (3 & 4)
0.8 sec-15.0 sec	ON-DELAY
3 sec-60 sec	ONE-SHOT
24 sec-480 sec	RECYCLER, OFF-TIME FIRST
3 min-60 min	RECYCLER, ON-TIME FIRST

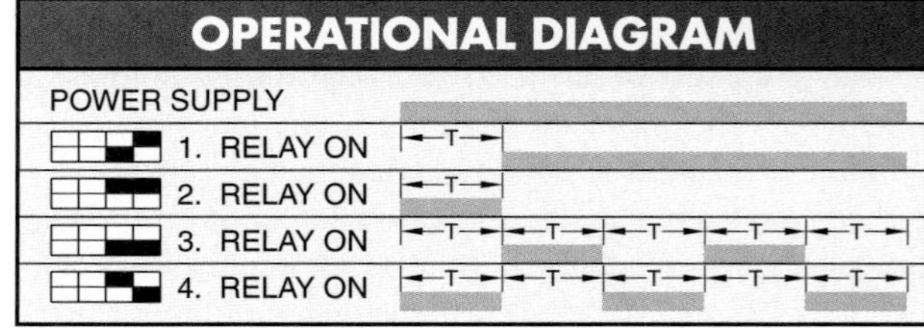

Figure 13-15. A multiple function timer may use the placement of DIP switches to determine the type of timing function and timer setting.

In this timer, four DIP switches are used to set the timer range and function. The first two DIP switches set the time range from .8 sec to 60 min. The last two DIP switches set the timer function. The timer can be set for an ON-delay, one-shot, or recycle time function. The recycle time function can be set to start with the OFF time occurring first or the ON time occurring first.

In addition to some (or all) of the basic timing functions, many multiple-function timers can also be programmed for special timing functions. **See Figure 13-16.** In this multiple-function timer, many timing functions can be programmed with a time range from .15 sec to 220 hr. This timer includes standard timing functions such as an ON-delay (program setting 1) and special timing functions such as a combination of both ON-delay and OFF-delay (program setting 5).

OPERATIONAL DIAGRAM

1 2 3 4 POWER SUPPLY
1. RELAY ON
2. RELAY ON
3. RELAY ON
4. RELAY ON
5. RELAY ON
6. RELAY ON
7. RELAY ON
8. RELAY ON
9. RELAY ON

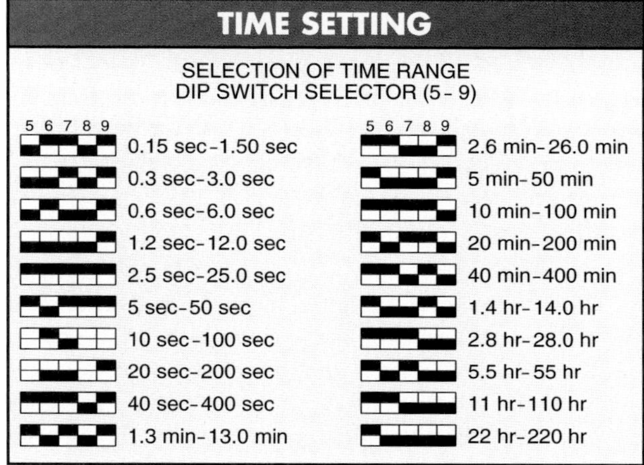

TIME SETTING

SELECTION OF TIME RANGE
DIP SWITCH SELECTOR (5 - 9)

5 6 7 8 9	5 6 7 8 9
0.15 sec-1.50 sec	2.6 min-26.0 min
0.3 sec-3.0 sec	5 min-50 min
0.6 sec-6.0 sec	10 min-100 min
1.2 sec-12.0 sec	20 min-200 min
2.5 sec-25.0 sec	40 min-400 min
5 sec-50 sec	1.4 hr-14.0 hr
10 sec-100 sec	2.8 hr-28.0 hr
20 sec-200 sec	5.5 hr-55 hr
40 sec-400 sec	11 hr-110 hr
1.3 min-13.0 min	22 hr-220 hr

Figure 13-16. Multiple function timers can be programmed for many timing functions of various times.

WIRING DIAGRAMS

In the past, when only dashpot, synchronous clock, and the earliest solid-state timers were in use, timing functions were controlled by applying and removing power from the timer coil (or circuit). This meant that the control switch had to be rated for the same type and level of voltage as the timer coil. This is still true when using dashpot, synchronous clock, and basic solid-state timers. Today, solid-state timers are available that use different methods of controlling the timer. The advantage of the new solid-state timers is that they can be used directly with electronic circuits and other solid-state devices (such as photoelectric, proximity, and temperature sensors), and do not require the control switch to be at the same voltage level as the timer coil.

Supply Voltage Controlled Timers

A *supply voltage timer* is a timer that requires the control switch to be connected so that it controls power to the timer coil. **See Figure 13-17.** In this circuit, the control switch is connected in series with the timer coil. The advantage of this control method is that it is exactly the same as most electrical control circuits used over the years. The disadvantage is that if a standard 115 VAC timer is used, the control switch has to switch the 115 VAC. This means that the control switch has to be installed using standard AWG 14 copper wire and properly enclosed for safety.

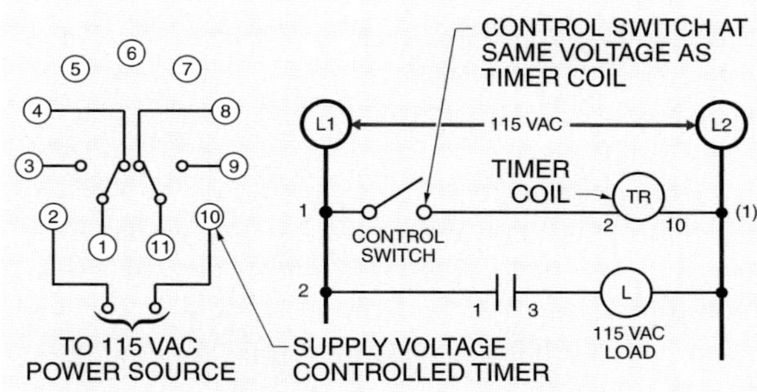

Figure 13-17. A supply voltage timer requires the control switch to be connected so that it controls power to the timer coil.

Contact-Controlled Timers

A *contact-controlled timer* is a timer that does not require the control switch to be connected in line with the timer coil. **See Figure 13-18.** In a contact-controlled timer, the timer supplies the voltage to the circuit in which the control switch is placed. The voltage of the control circuit is normally less than 24 VDC.

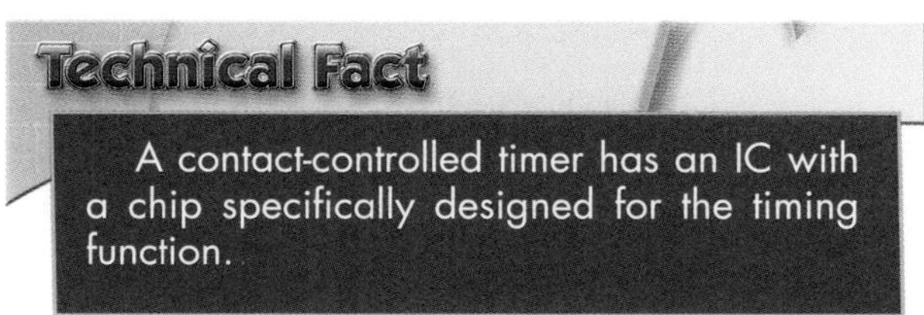

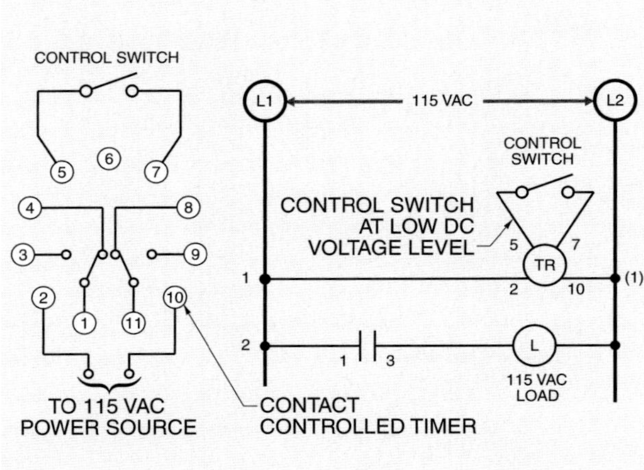

Figure 13-18. A contact-controlled timer does not require the control switch to be connected in line with the timer coil.

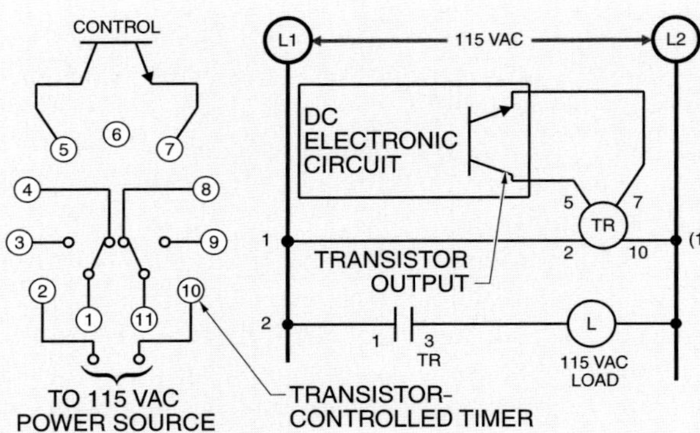

Figure 13-19. A transistor-controlled timer is controlled by an external transistor from a separately powered electronic circuit.

The advantage of this control method is that the control switch can be wired using low-voltage wire (AWG 16, 18, or 20). The control switch contacts can be small and normally require less than a 100 mA rating because the timer control circuit requires little current to pass through the control switch. The disadvantage of this control method is that many electricians are unfamiliar with connecting the control switch outside a standard 115 VAC control circuit. The timer low-voltage circuit (pins 5 and 7) is often connected to the timer 115 VAC circuit (pins 2 and 10). This results in the destruction of the timer.

Transistor-Controlled Timers

Transistor-controlled timers are like contact-controlled timers. A *transistor-controlled timer* is a timer that is controlled by an external transistor from a separately powered electronic circuit. **See Figure 13-19.** Modern industrial control circuits are often connected to DC electronic circuits that include a transistor as their output. Such devices include solid-state temperature controls, counters, computers, and other control devices. A timer that uses a transistor as the control input can be connected to almost any electronic circuit that includes a transistor output. The advantages and disadvantages of a transistor-controlled timer are the same as a contact-controlled timer. Any transistor-controlled timer can use a contact as the control device. However, not all contact-controlled timers can use a transistor as the control device. Check the manufacturer specification sheet when using any unfamiliar timer.

Sensor-Controlled Timers

A sensor-controlled timer is like contact-controlled and transistor-controlled timers that include an additional output from the timer. A *sensor-controlled timer* is a timer controlled by an external sensor in which the timer supplies the power required to operate the sensor. **See Figure 13-20.** The advantage of a sensor-controlled timer is that, by supplying power out of the timer itself, no external power supply is required to operate the control sensor. Sensor-controlled timers are generally used with photoelectric and proximity controls, but may be used with any control that meets the specifications of the timer.

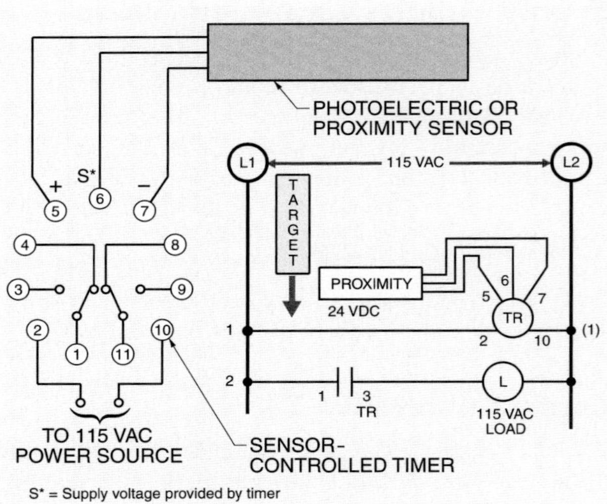

Figure 13-20. A sensor-controlled timer is controlled by an external sensor in which the timer supplies the power required to operate the sensor.

MULTIPLE CONTACT TIMERS

In the past, synchronous clock timer manufacturers included instantaneous and time delay contacts on their timers to meet the many different application requirements of timers. The timers may be used for numerous applications when they have both types of contacts. These timers are still often used and have been updated to include solid-state timing circuits and can be converted from ON-delay to OFF-delay timing functions.

Multiple Contact Timer Wiring Diagrams

Wiring diagrams are required on multiple contact timers because several different connection points must be located and wired to the timer for it to perform properly. **See Figure 13-21.** The wiring diagram for a timer is normally located on the back of the timer. By quickly surveying the diagram, the timer clutch coil, the synchronous motor (or solid-state circuit), and the timer pilot light can be located.

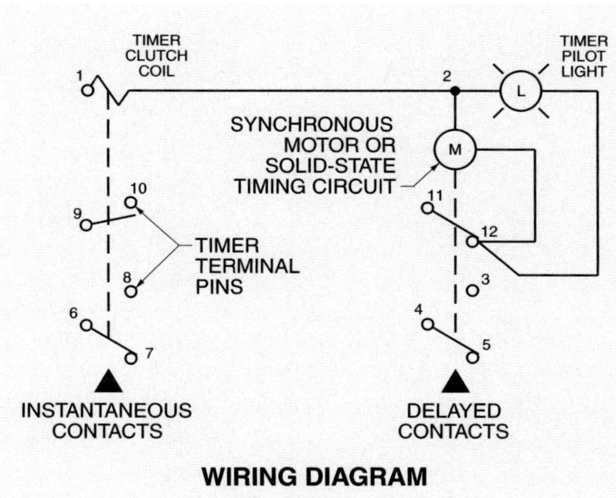

Figure 13-21. Wiring diagrams simplify locating connection points and wiring multiple contact timers.

In this timer, the timer clutch coil engages and disengages the motor. This is similar to that found in an automobile. The timer clutch coil engages and disengages contacts 9 and 10, and 6, 7, and 8. When the clutch is engaged, contacts 9 and 10, and 6 and 8 close instantaneously. Contacts 6 and 7 open simultaneously. In other words, the timer clutch controls two normally open contacts and one normally closed contact instantaneously. The timing motor controls contacts 11 and 12, and 3, 4, and 5 through a time delay. When the motor times out, contacts 4 and 5 and 11 and 12 open and contacts 3 and 4 close.

Contacts 11 and 12 open slightly later than contacts 4 and 5 after the motor times out. The timer pilot light, wired in parallel with the motor, indicates when the motor is timing. Contacts 4 and 5 and 11 and 12 close and contacts 3 and 4 open when the timer is reset.

Wiring Motor-Driven Timers. A motor-driven timer can be wired into L1 and L2 to provide power to the circuit. **See Figure 13-22.** In this circuit, current flows from L1, through the timer clutch coil, and on to L2. Current also flows from terminal 1 through the closed contacts 11 and 12, feeding the parallel circuit provided by the timer motor and timer pilot light, and then on to L2.

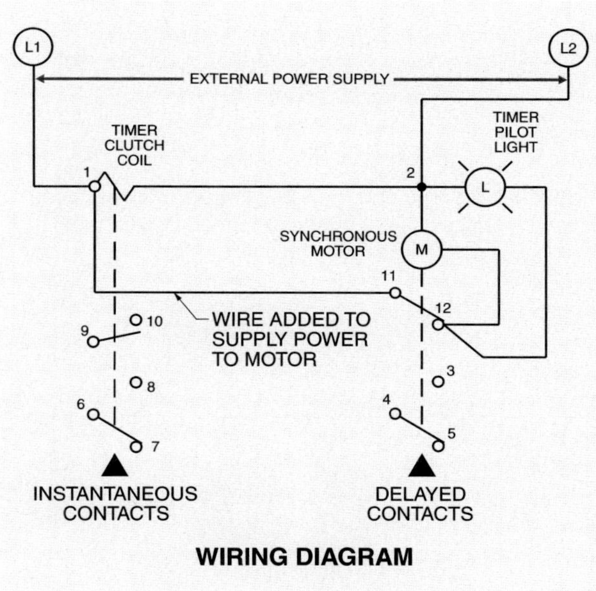

Figure 13-22. A motor-driven timer can be wired into L1 and L2 of a multiple contact timer to provide power.

Operation of Motor-Driven Timers. A timer activated by a limit switch has effects on various loads wired into the circuit. This timer can be controlled by a manual, mechanical, or automatic input. The timer may be used to achieve control using a sustained mechanical input. **See Figure 13-23.** This input (limit switch) must remain closed to energize the timer clutch coil and power the timer motor. The timer is connected for ON-delay, requiring input power to close the clutch and start the timer. The contacts, both instantaneous and time delay, are connected to four loads marked A, B, C, and D. These loads may be any load, such as a solenoid, magnetic motor starter, or light. A code is added above each load to illustrate the sequence during reset, timing, and when timed out.

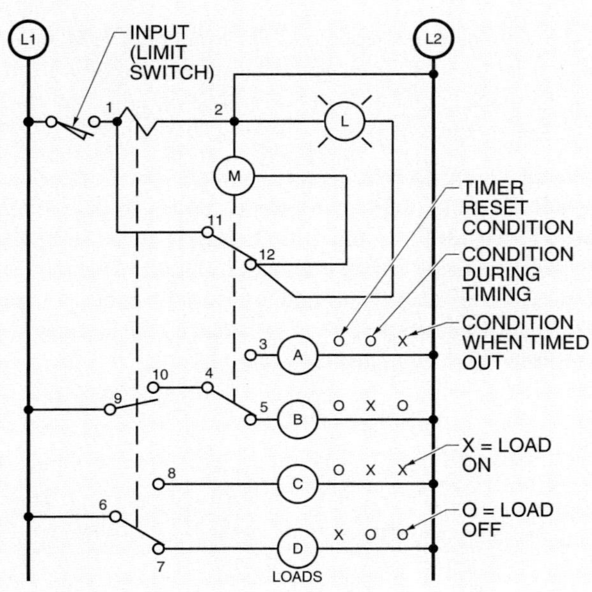

Figure 13-23. Multiple contact timers may be used to achieve control using a sustained mechanical input.

Rockwell Automation, Allen-Bradley Company, Inc.

Repeat cycle (recycle) timers are ideal for applications that require a fixed ON and OFF time period.

The code is used to indicate the condition (ON or OFF) of each load during the three stages of the timer. The three stages include the reset condition (no power applied to timer), the timing condition (time at which the timer is timing, but not timed out), and the timed-out condition. An O indicates when the load is de-energized and an X indicates when the load is energized.

Loads C and D are relay-type responses, using only the instantaneous contacts. Loads A and B use the combined action of the instantaneous and delay contacts to achieve the desired sequence. In this circuit, the timing motor is wired through delay contacts 11 and 12 to ensure motor cutoff after the timer times out. This is required because the limit switch remains closed after timing out. The limit switch otherwise would have to be opened to reset the timer. This also means that a loss of plant power resets the timer because the clutch opens when power is lost.

TIMER APPLICATIONS

Timers are used in any application that requires a time delay at some point of circuit operation. The timer used (ON-delay, OFF-delay, etc.) depends on the application. The timer selected (dashpot, solid-state, synchronous, or programmable) depends on cost, expected usage, type of equipment, operating environment, and personal preference. In general, monofunction solid-state timers are the best choice for most applications in which the circuit is not likely to change in function or time range. For applications that require changing timing functions and/or large time range fluctuations, multiple-function programmable timers are the best choice. Even if most of the machines and equipment in a plant use monofunction timers, multiple-function timers are often stocked in the maintenance shop because they can be used to replace many different types of timers.

ON-Delay Timer Applications

ON-delay timers are the most common timer in use. ON-delay timers may be used to monitor a medical patient's breathing. **See Figure 13-24.** In this application, the timer is used to sound an alarm if a patient does not take a breath within 10 sec. The circuit includes a low-pressure switch built into a patient monitoring system. Pressure switches are available that can activate electrical contacts at pressures less than 1 psi. Pressure switches that react to pressures less than 1 psi are rated in inches of water column (in. WC). For example, the pressure switch used in this application may be rated at 4 in. WC. Approximately 27 in. WC is equal to 1 psi. The circuit is turned ON by the ON/OFF switch once the patient is connected to the monitor.

If the patient does not take a breath, the timer starts timing and continues timing until the patient takes a breath (which resets the timer) or the timer times out. If the timer times out, the timer contacts close, which sounds a warning.

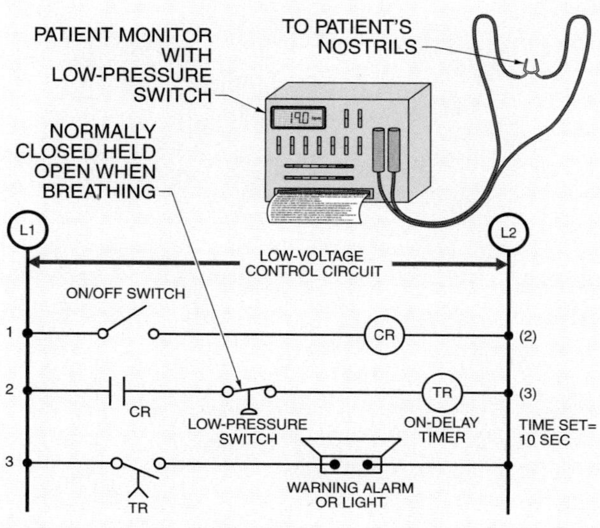

Figure 13-24. ON-delay timers may be used to monitor a patient's breathing.

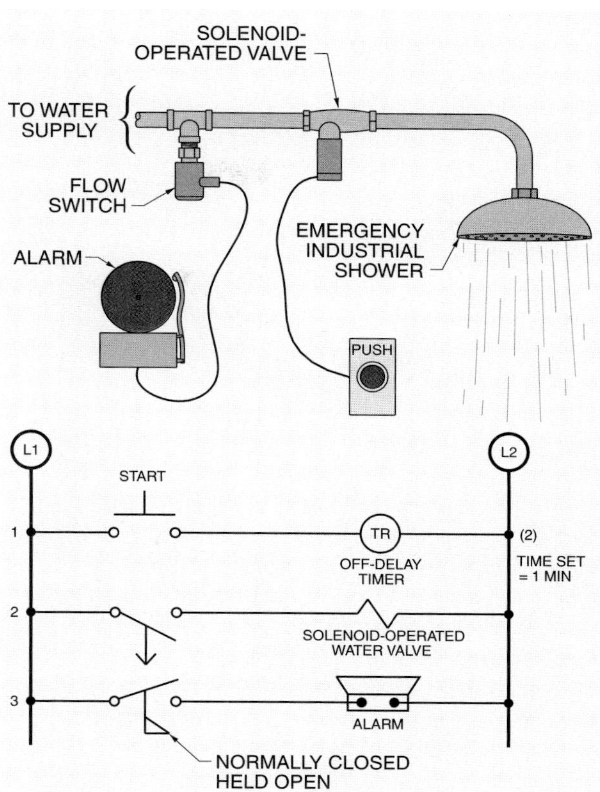

Figure 13-25. OFF-delay timers are used in applications that require a load to remain energized even after the input control has been removed.

OFF-Delay Timer Applications

OFF-delay timers are used in applications that require a load to remain energized even after the input control has been removed. **See Figure 13-25.** In this circuit, the OFF-delay timer is used to keep the water flowing for 1 min after the emergency shower pushbutton is pressed and released.

After the shower pushbutton is pressed, the timer contacts close and the solenoid-operated valve starts the flow of water. The water flows even if the pushbutton is released. A flow switch is used to indicate when water is flowing. The flow switch sounds an alarm that can be used to bring help. The flow switch also sounds the alarm if there is a water break at any point downstream from the switch.

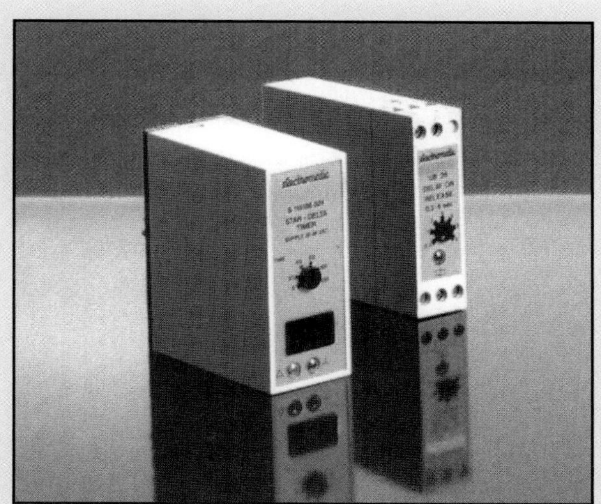

Carlo Gavazzi Inc. Electromatic Business Unit

Timers may include LEDs to indicate when the timer is timing and when the timer is timed out.

One-Shot Timer Applications

One-shot timers are used in applications that require a fixed-timed output for a set period of time. **See Figure 13-26.** In this application, a one-shot timer is used to control the amount of time that plastic wrap is wound around a pallet of cartons.

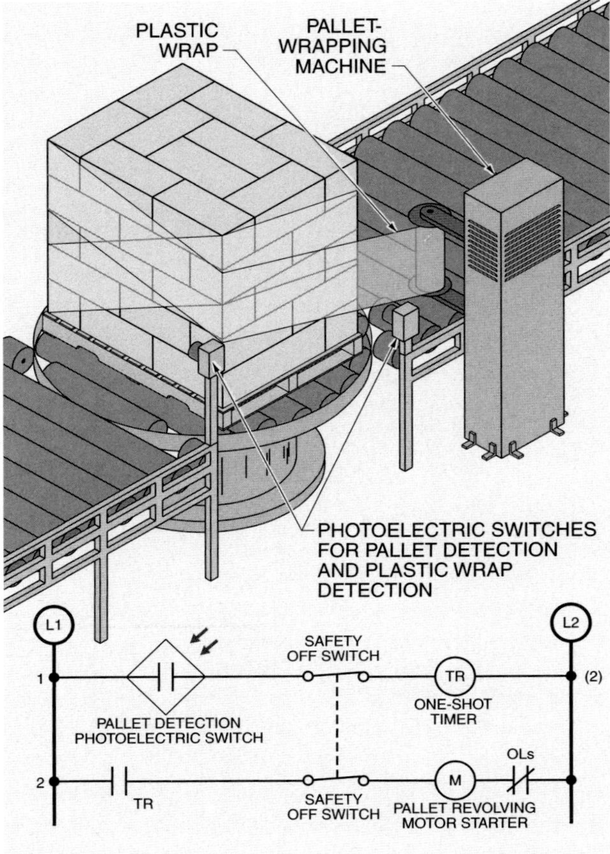

Figure 13-26. One-shot timers are used in applications that require a fixed-timed output for a set period of time.

A photoelectric switch detects a pallet entering the plastic wrap machine. The photoelectric switch energizes the one-shot timer. The one-shot timer contacts close, starting the wrapping process. The wrapping process continues for the setting of the timer. A second photoelectric switch could be used to detect that the plastic wrap is actually being applied. This is helpful in indicating a tear in the plastic or an empty roll.

Recycle Timer Applications

Recycle timers are used in applications that require a fixed ON and OFF time period. **See Figure 13-27.** In this application, a recycle timer is used to keep a product automatically mixed.

Power is applied to the timer when the three-position selector switch is placed in the automatic position. The timer starts recycling for as long as the selector switch is in the automatic position. The recycle timer turns the mixing motor ON and OFF at the set time. In this application, an asymmetrical timer works best. The ON time (mixer motor ON) is set less than the OFF time (mixer motor OFF). For example, the timer may be set to mix the product for 5 min every 2 hr.

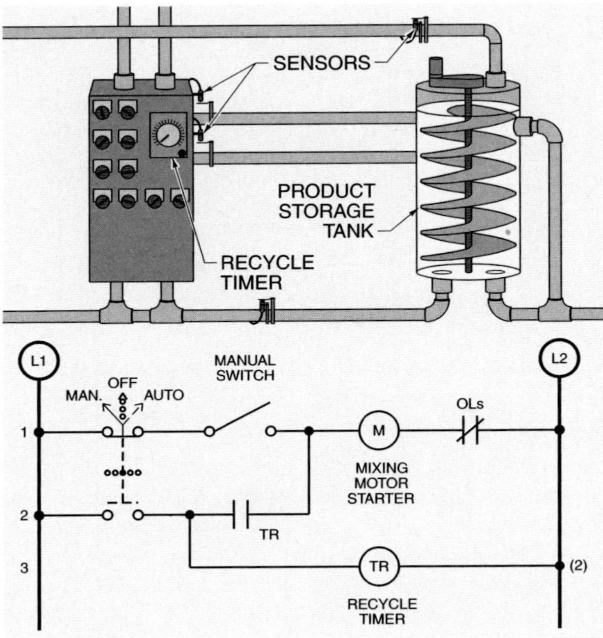

Figure 13-27. Recycle timers are used in applications that require a fixed ON and OFF time period.

Multiple Contact Timer Applications

Multiple contact timers are used in applications where a sustained input is used to control a circuit. **See Figure 13-28.** In this circuit, the cartons coming down the conveyor belt are to be filled with detergent. Each carton must be filled with the same amount of detergent and the process must be automatic. To accomplish this, the timer circuit is used to control the time it takes to fill one carton. A limit switch sustained input is used to detect the carton. A motor drives the conveyor belt and a solenoid opens or closes the hopper full of detergent. The limit switch could be replaced with a photoelectric or proximity switch.

Technical Fact

Adjustable recycle timers can be used for applications such as spot welding, flashing signs, electroplating, and heat-treating applications.

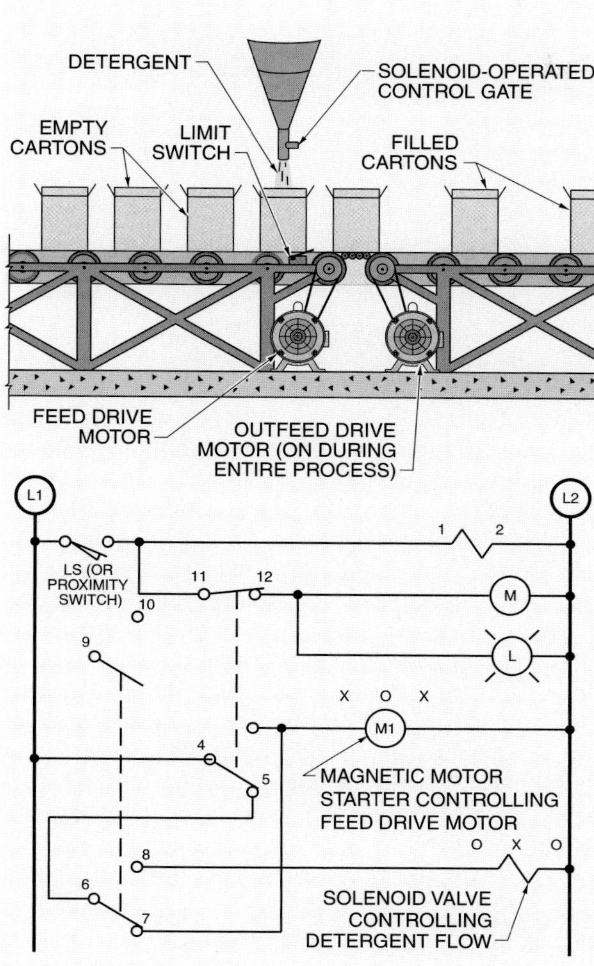

Figure 13-28. Multiple contact timers are used in applications where a sustained input controls a circuit.

As the cartons are coming down the conveyor, the feed drive motor is ON and the solenoid valve is OFF (no detergent fill). As a carton contacts the limit switch, the feed drive motor shuts OFF. This stops the conveyor and energizes the solenoid. The solenoid opens the control gate on the hopper of detergent. After the timer times out, the feed drive motor turns ON and the solenoid valve closes. This removes the filled carton, which opens the limit switch, resetting the timer. The code for each load is added above the respective load. The code illustrates the desired sequence of operation of the loads.

Memory could be added to a circuit if the circuit requires a momentary input, such as a pushbutton, to initiate timing. **See Figure 13-29.** In this circuit, a pushbutton is used as the input signal. As with any memory circuit, a normally open instantaneous contact must be connected in parallel with the pushbutton if memory is to be added into the circuit. To accomplish this, a normally open instantaneous contact 9 and 10 is connected in parallel with the pushbutton so that power is maintained when the control switch is released. A separate reset (stop) switch is used to reset the timer. After the timer has timed out, the timer resets only if the reset pushbutton is pressed. The sequence of each load is illustrated by the code above each load.

Technical Fact

Indoor timer enclosures are classified NEMA 1 and outdoor timer enclosures are classified NEMA 3.

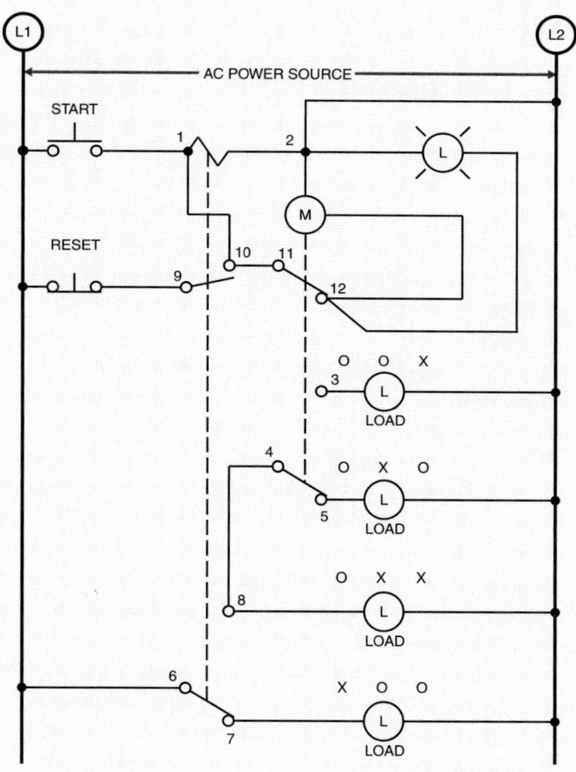

Figure 13-29. Memory could be added to a circuit if the circuit requires a momentary input, such as a pushbutton, to initiate timing.

These diagrams are not in pure line diagram form. Line diagrams are standards used by all manufacturers. Wiring diagrams are the actual diagrams matching the logic of the line diagram to the manufacturer product designed to perform that logic. A wiring diagram is used because this is the actual diagram found on the timer.

TROUBLESHOOTING TIMING CIRCUITS

Troubleshooting timing circuits is a matter of checking power to the timer and checking for proper timer contact operation. The timer is replaced if there is a problem with any part of the timer. The most common problem is contact failure. The current at the contacts is measured if the contacts failed prematurely. The current must not exceed the rating of the contacts. **See Figure 13-30.**

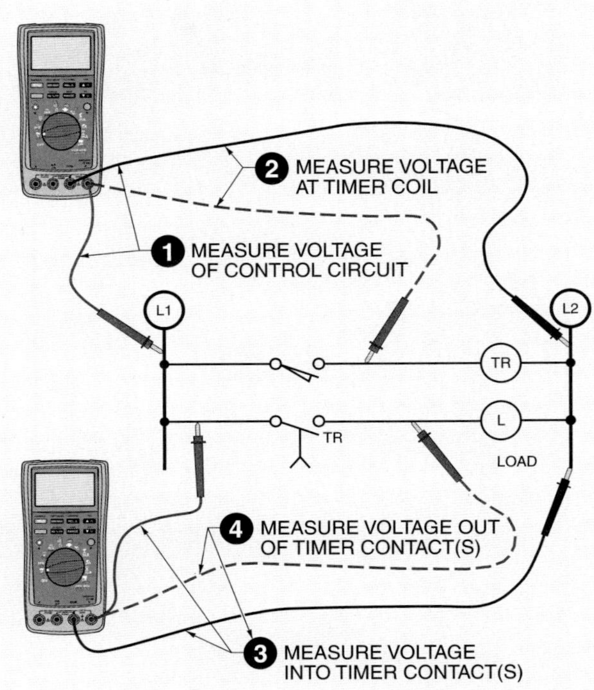

Figure 13-30. Troubleshooting timing circuits is a matter of checking power to the timer and checking for proper timer contact operation.

To troubleshoot timers, apply the following procedure:

1. Measure the voltage of the control circuit. The voltage must be within the specification range of the timer. Correct the voltage problem if the voltage is not within the specification range of the timer. A common problem is an overloaded control transformer that is delivering a low-voltage output.
2. Measure the voltage at the timer coil. The voltage should be the same as the voltage of the control circuit. Check the control switch if the voltage is not the same. Check the voltage at the timer contacts if the voltage is correct.
3. Measure the voltage into the timer contacts. The voltage must be within the range of the load the timer is controlling. Correct the voltage problem if the voltage is not within this range.
4. Measure the voltage out of the timer contacts if the voltage into the timer contacts is correct. The voltage should be the same as the voltage of the control circuit. Check the timer contact connection points and wiring for a bad connection or corrosion if the voltage is not the same.

COUNTERS

In most applications it is necessary to account for the number of events within the system. This may be counting the number of products made, number of products required to fill a carton, number of rejected parts, number of gallons flowing through a pipe, etc. A totalizer can be used if only the total number of events is required to be known. A *totalizer* is a counting device that keeps track of the total number of inputs and displays the counted value.

Counters are also used to count events and provide an output. A *counter* is a counting device that accounts for the total number of inputs entering the counter and can provide an output (mechanical or solid-state contacts) at predetermined counts in addition to displaying the counted value. The two basic types of counters are up counters and up/down counters.

Up Counters

An *up counter* is a device used to count inputs and provide an output (contacts) after the preset count value is reached. An up counter either has one count input or has one count input and one reset input. Removing power to the counter resets a counter with one count input. A counter with two inputs uses one input to add a count into the counter each time it is activated, and a second input to reset the counter to zero counts when it is activated. **See Figure 13-31.**

The count input adds one count each time it is closed and opened. When the preset count value is reached, the counter contacts are activated (normally open contacts close, and normally closed contacts open). If a counter has a total count display in addition to a preset count setting, any additional counts sent into the counter are displayed as part of the total.

Technical Fact

Conveyor-mounted counters are usually designed for either bottle/can or case/carton counting.

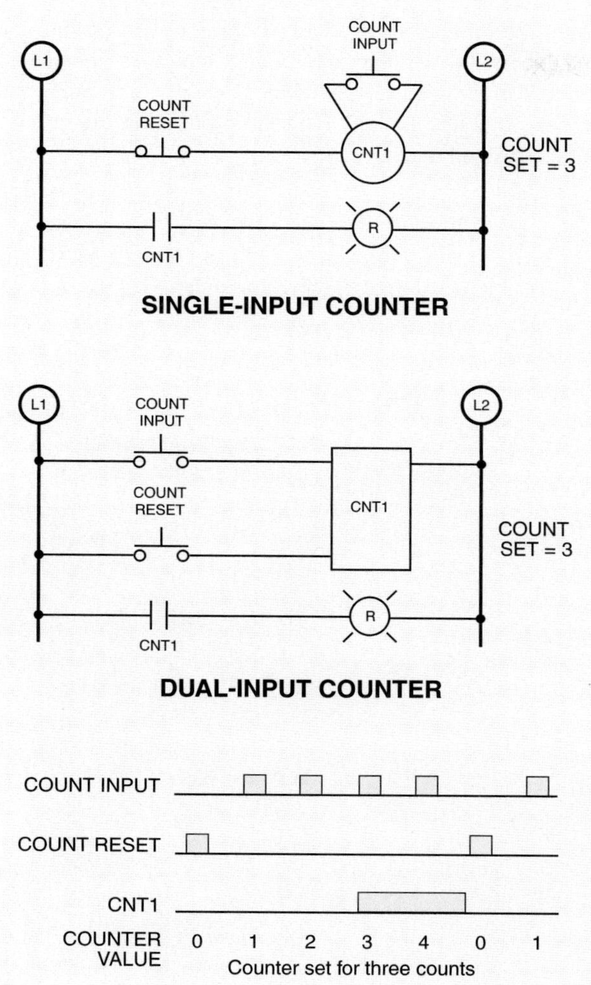

Figure 13-31. Up counters are used in applications that require an output after a fixed number of counts.

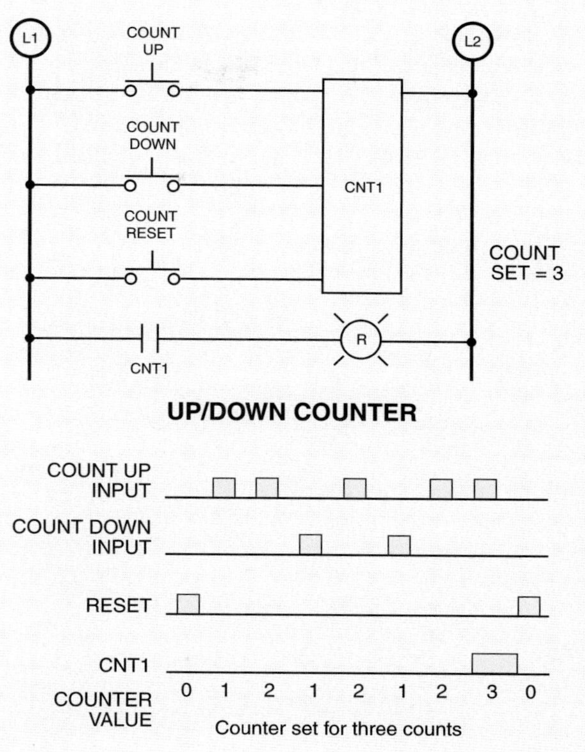

Figure 13-32. Up/down counters are used to keep track of the number of counts when counts are both added to and subtracted from an application.

Up/Down Counters

An *up/down counter* is a device used to count inputs from two different inputs, one that adds a count and the other that subtracts a count. Up/down counters provide an output (contacts) after the preset count value is reached. A third input is used as the reset input, which resets the counter to zero counts when it is activated. **See Figure 13-32.**

Up/down counters are used in applications such as a parking garage. In this application, each time a car enters the garage, one count is added to the count. Each time a car leaves the garage, one count is subtracted from the count. When the garage is full (preset count value reached), the counter contact is used to turn ON a "lot full" sign. If one car leaves, the sign turns OFF.

Totalizers and counters are commonly used in most product production applications. **See Figure 13-33.** For example, in a shrink-wrap application, totalizers and counters can be used to count the number of times a shrink-wrap roller motor has turned ON and OFF.

The motor is turned ON and OFF once for each product packing (detected by a proximity or photoelectric switch). There is only a set number of times the shrink-wrap roller motor can be turned ON and OFF before a new roll of shrink wrap must be placed in the machine. The counter can indicate when a new roll is required by using the counter output contacts to turn ON an alarm or lamp.

The products coming off the production line can be sent to storage area 1 or 2. A counter can be used to count the number of products coming out of the heat machine and send a fixed number (six, 12, 36, etc.) first to storage area 1 and then a new group to storage area 2, allowing control of the distribution of the produced item.

An inspection station could be added to the output of the heat machine that inspects the product and rejects any that do not meet predetermined requirements (weight, size, proper labels, etc.). Items that pass inspection can be sent to storage area 1 and items that fail inspection can be sent to storage area 2. A counter may be used to track the number of products going to storage area 1 and a second counter may be used to track the number of products going to storage area 2. An up/down counter can also be used to count the number of products coming out of production and moving to storage area 1 and subtract the rejected product moving to storage area 2. Finally, a counter can be used to automatically stop the production line after a predetermined count.

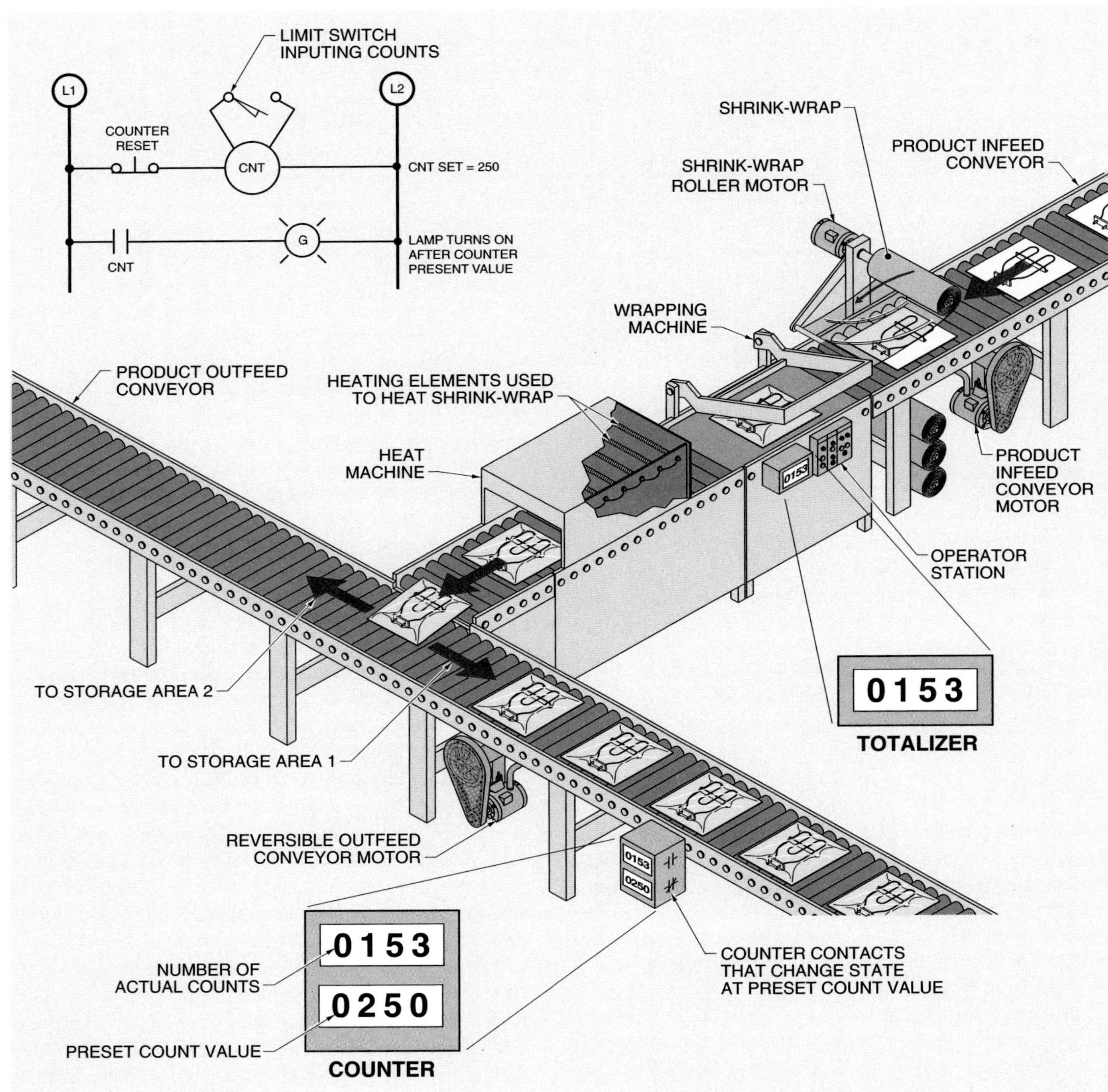

Figure 13-33. Totalizers and counters are commonly used in applications such as counting shrinkable plastic wrap on large quantities of finished goods.

Review Questions

1. What are the four major categories of timers?
2. How does a dashpot timer develop a time delay?
3. How does a synchronous timer develop a time delay?
4. How does a solid-state timer develop a time delay?
5. What is an ON-delay timer?
6. What is another name for ON-delay?
7. What is an OFF-delay timer?
8. What is another name for OFF-delay?
9. What is a one-shot timer?
10. What is another name for one-shot?
11. What is a recycle timer?
12. How is the control switch connected when a supply voltage controlled timer is used?
13. What is the advantage of using a contact-controlled timer?
14. Can a transistor-controlled timer use a contact as the control device?
15. What type of timer provides power to allow photoelectric and proximity control of the timer?
16. When using an X or O code to indicate the condition of a load, how is a de-energized load coded?
17. When using an X or O code to indicate the condition of a load, how is an energized load coded?
18. What type of recycle timer operates with equal ON and OFF time periods?
19. What does an operational diagram for a multiple-function timer show?
20. What is the most common problem in timing circuits?
21. What is the difference between an up counter and an up/down counter?
22. What is a totalizer?

chapter **14**

electrical motor controls *for Integrated Systems*

Relays and Solid-State Starters

Relays are used extensively in machine tool control, industrial assembly lines, and commercial equipment. Relay applications include switching starting coils and turning ON small devices such as pilot lights and audible alarms. The two major types of relays are the electromechanical relay and the solid-state relay.

RELAYS

A *relay* is a device that controls one electrical circuit by opening and closing contacts in another circuit. Depending on design, relays normally do not control power-consuming devices, except for small loads which draw less than 15 A. Relays are used extensively in machine tool control, industrial assembly lines, and commercial equipment. Relays are used to switch starting coils in contactors and motor starters, heating elements, pilot lights, audible alarms, and some small motors (less than ⅛ HP).

A small voltage applied to a relay results in a larger voltage being switched. **See Figure 14-1.** For example, applying 24 V to the relay coils may operate a set of contacts that control a 230/460 V circuit. In this case, the relay acts as an amplifier of the voltage or current in the control circuit because relay coils require a low current or voltage to switch, but can energize larger currents or voltages.

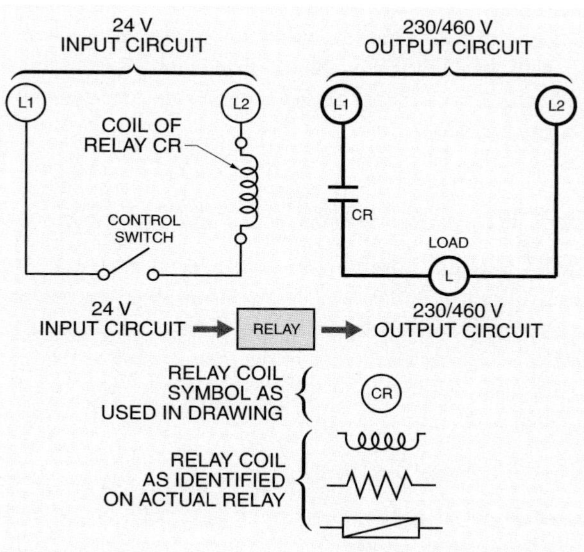

Figure 14-1. Relays may be compared to amplifiers in that small voltage input results in large voltage output.

Another example of a relay providing an amplifying effect is when a single input to the relay results in several other circuits being energized. **See Figure 14-2.** An input may be considered amplified because certain mechanical relays provide eight or more sets of contacts controlled from any one input.

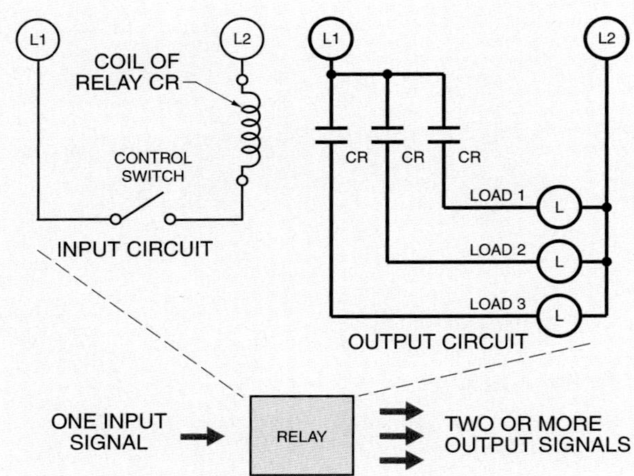

Figure 14-2. Relays may be compared to amplifiers in that a single input may result in multiple outputs.

The two major types of relays are the electromechanical relay and the solid-state relay. An *electromechanical relay (EMR)* is a switching device that has sets of contacts that are closed by a magnetic effect. A *solid-state relay (SSR)* is a switching device that has no contacts and switches entirely by electronic means. A *hybrid relay* is a combination of electromechanical and solid-state technology used to overcome unique problems that cannot be solved by one device or the other alone. Hybrid relays are generally considered EMRs.

ELECTROMECHANICAL RELAYS (EMRs)

EMRs that are common to commercial and industrial applications may be reed, general-purpose, or machine control relays. The major difference between the types of EMRs is their intended use in the circuit, cost, and the life expectancy of the device.

Reed Relays

A *reed relay* is a fast-operating, single-pole, single-throw switch with normally open (NO) contacts hermetically sealed in a glass envelope. **See Figure 14-3.** During the sealing operation, dry nitrogen is forced into the tube, creating a clean inner atmosphere for the contacts. Because the contacts are sealed, they are unaffected by dust, humidity, and fumes. The life expectancy of reed relay contacts is quite long.

A reed relay includes a very low current-rated contact (less than .25 mA) that is activated by the presence of a magnetic field. Reed relays may be activated in a variety of ways, which allows them to be used in circuit applications where other relay types are inappropriate.

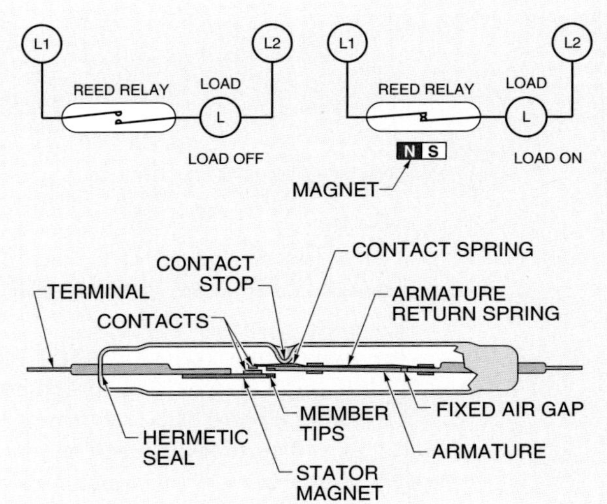

Figure 14-3. A reed relay is a fast-operating, single-pole, single-throw switch that is activated by a magnetic field.

Reed relays are designed to be actuated by an external movable permanent magnet or DC electromagnet. When a magnetic field is brought close to the two reeds, the ferromagnetic (easily magnetized) ends assume opposite magnetic polarity. If the magnetic field is strong enough, the attracting force of the opposing poles overcomes the stiffness of the reed, drawing the contacts together. Removing the magnetizing force allows the contacts to spring open. AC electromagnets are not suitable for reed relays because the reed relay switches so fast that it would energize and de-energize on alternate half-cycles of a standard 60 Hz line.

Reed Contacts. To obtain a low and consistent contact resistance, the overlapping ends of the contacts may be plated with gold, rhodium, silver alloy, or other low-resistance metal. Contact resistance is often under 0.1 Ω when closed. Reed contacts have an open contact resistance of several million ohms.

Most reed contacts are capable of direct switching of industrial solenoids, contactors, and motor starters. Reed relay contact ratings indicate the maximum current, voltage, and volt-amps that may be switched by the relay. Under no circumstances should these values be exceeded.

Reed Relay Actuation

A permanent magnet is the most common actuator for a reed relay. Permanent-magnet actuation can be arranged in several ways depending on the switching requirement. The most commonly used arrangements are proximity motion, rotary motion, shielding, and biasing.

Proximity Motion. The proximity motion arrangement uses the presence of a magnetic field that is brought within a specific proximity (close distance) to the reed relay to close the contacts. The distance for activating any given relay depends on the sensitivity of the relay and the strength of the magnet. A more sensitive relay or stronger magnet needs less distance for actuation. Methods of proximity motion operation are the pivoted motion, perpendicular motion, parallel motion, and front-to-back motion. **See Figure 14-4.** In each method, either the magnet or relay is moved. In some applications, both the magnet and relay are in motion. The contacts operate quickly, with snap action and little wear. The application and switching requirements determine the best method.

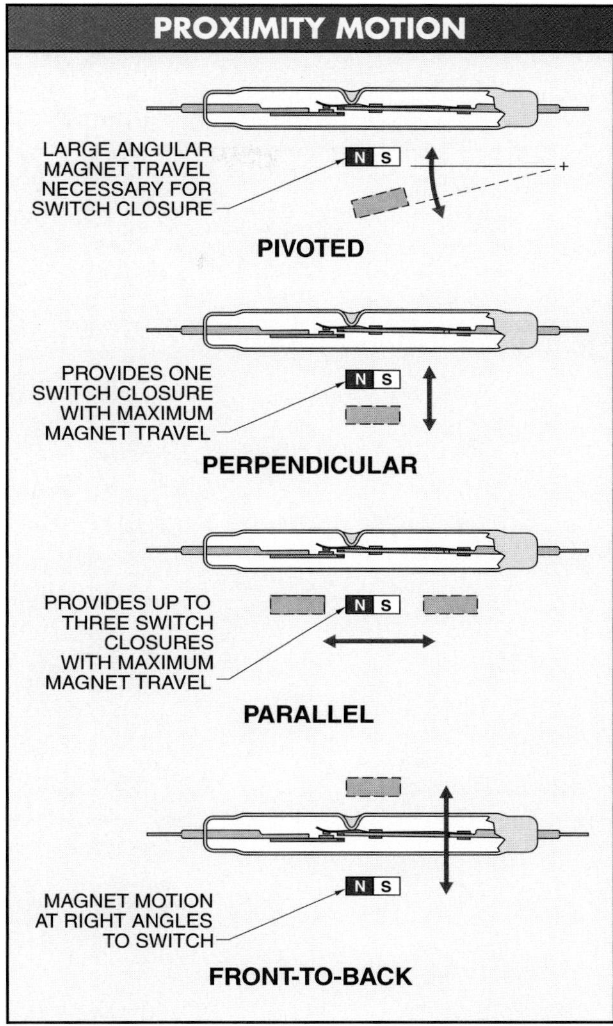

Figure 14-4. The proximity motion arrangement uses the presence of a magnetic field brought within a specific proximity to the reed relay to close the contacts.

Rotary Motion. The rotary motion arrangement involves revolving the magnet or relay, which results in relay contact operation every 180° or two operations every 360°. **See Figure 14-5.** The contacts are closed when the magnet and relay are parallel. The contacts are opened when the magnet and relay are perpendicular. Although the magnetic poles reverse every 180°, they induce a magnetic field with opposite polarity in the relay and close the contacts.

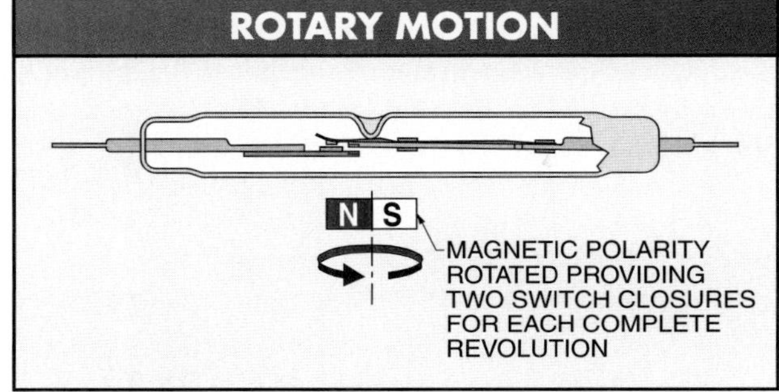

Figure 14-5. A reed relay may be activated by rotary motion.

Shielding. The shielding arrangement involves permanently fixing the magnet and relay so that the relay's contacts are held closed. **See Figure 14-6.** The contacts are open as ferromagnetic material (material with strong magnetic characteristics) is passed between the magnet and relay. The ferromagnetic material acts like a short circuit or shunt for the magnetic field and eliminates the magnetic field holding the contacts. As the shield is removed, the contacts are closed. It makes no difference at what angle the shield is passed between the magnet and relay. This method may be used to signal that a protective shield, such as a cover on a high voltage box, has been removed.

Biasing. In the biasing arrangement, a bias magnet holds the switch closed until an actuating magnet cancels the magnetic field of the bias magnet and opens the switch. **See Figure 14-7.** The actuating magnet approaches the bias magnet with opposite polarity, which cancels the magnetic field of the bias magnet and opens the relay's contacts.

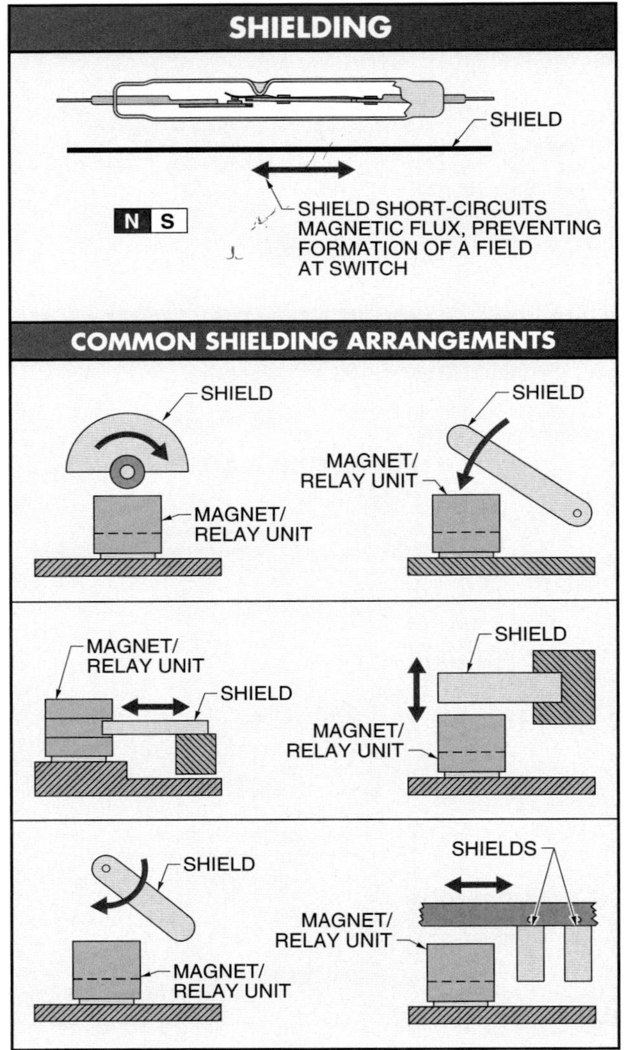

Figure 14-6. The shielding arrangement involves permanently fixing the magnet and relay so that the relay's contacts are held closed. The contacts are opened when a ferromagnetic material passes between the magnet and relay.

The relay's contacts open only if the magnetic fields of the two magnets cancel each other. The correct strength of the magnetic field and distance are required. This application may be used for detecting magnetic polarity or other similar applications.

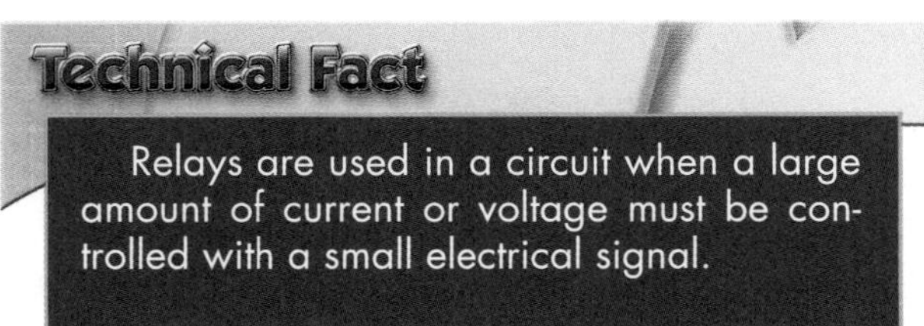

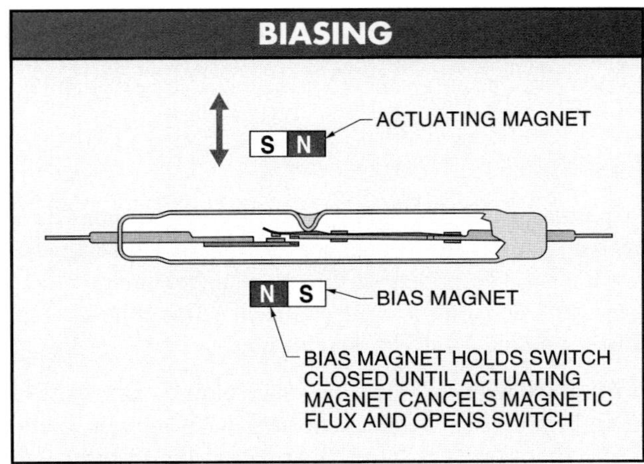

Figure 14-7. In the biasing arrangement, a bias magnet holds the switch closed until an actuating magnet cancels the magnetic field of the bias magnet and opens the switch.

General-Purpose Relays

General-purpose relays are EMRs that include several sets (normally two, three, or four) of nonreplaceable NO and NC contacts (normally rated at 5 A to 15 A) that are activated by a coil. A general-purpose relay is a good relay for applications that can use a throwaway plug-in relay to simplify troubleshooting and reduce costs. Special attention must be given to the contact current rating when using general-purpose relays because the contact rating for switching DC is less than the contact rating for switching AC. For example, a 15 A AC-rated contact normally is only rated for 8 A to 10 A DC.

Several different styles of general-purpose relays are available. These relays are designed for commercial and industrial applications where economy and fast replacement are high priorities. Most general-purpose relays have a plug-in feature that makes for quick replacement and simple troubleshooting. **See Figure 14-8.**

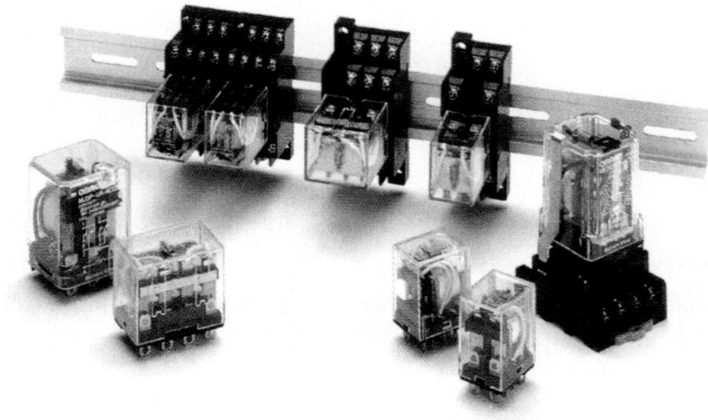

Omron Electronics, Inc.

Figure 14-8. General-purpose relays are EMRs that include several sets of nonreplaceable NO and NC contacts that are activated by a coil.

A *general-purpose relay* is a mechanical switch operated by a magnetic coil. **See Figure 14-9.** General-purpose relays are available in AC and DC designs. These relays are available with coils that can open or close contacts ranging from millivolts to several hundred volts. Relays with 6 V, 12 V, 24 V, 48 V, 115 V, and 230 V coils are the most common. General-purpose relays are available that require as little as 4 mA at 5 VDC, or 22 mA at 12 VDC, making them IC-compatible with transistor-transistor logic (TTL) and complementary metal oxide semiconductor (CMOS) logic gates. These relays are available in a wide range of switching configurations.

Contacts. A *contact* is the conducting part of a switch that operates with another conducting part of the switch to make or break a circuit. Relay contacts switch electrical circuits. The most common contacts are the single-pole, double-throw (SPDT), double-pole, double-throw (DPDT), and the three-pole, double-throw (3PDT) contacts. Relay contacts are described by their number of poles, throws, and breaks. **See Figure 14-10.**

A *break* is the number of separate places on a contact that open or close an electrical circuit. For example, a single-break contact breaks an electrical circuit in one place. A double-break (DB) contact breaks the electrical circuit in two places. All contacts are single break or double break. Single-break (SB) contacts are normally used when switching low-power devices such as indicating lights. Double-break contacts are used when switching high-power devices such as solenoids.

A *pole* is the number of completely isolated circuits that a relay can switch. A single-pole contact can carry current through only one circuit at a time. A double-pole contact can carry current through two circuits simultaneously. In a double-pole contact, the two circuits are mechanically connected to open or close simultaneously and are electrically insulated from each other.

The mechanical connection is represented by a dashed line connecting the poles. Relays are available with one to 12 poles. A *throw* is the number of closed contact positions per pole. A single-throw contact can control only one circuit. A double-throw contact can control two circuits.

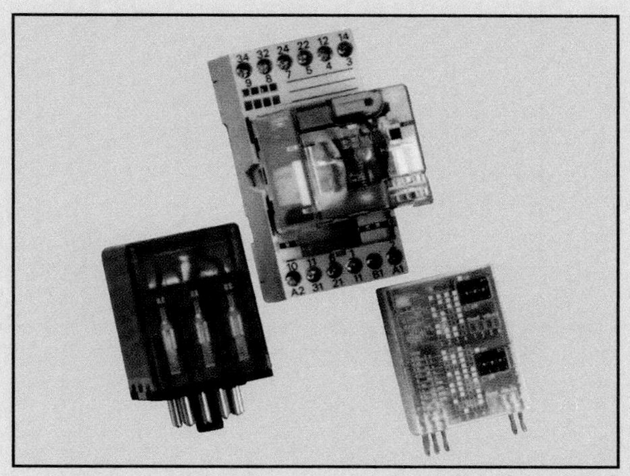

Sprecher + Schuh

Control relays are plug-in units that are commonly available in a standard 8-pin or 11-pin profiles, depending on the number of contacts required for the application.

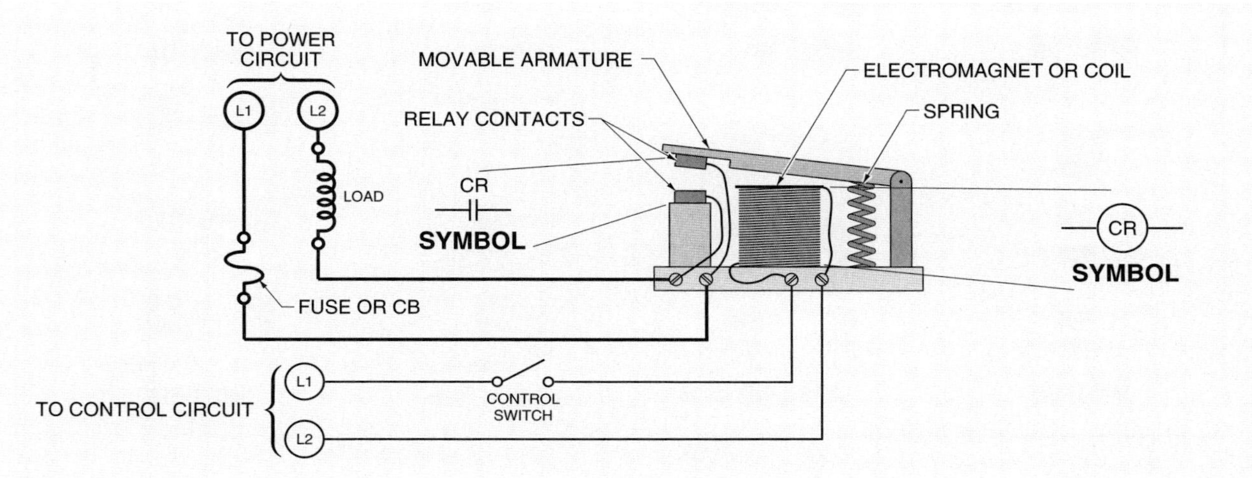

Figure 14-9. A general-purpose relay is a mechanical switch operated by a magnetic coil.

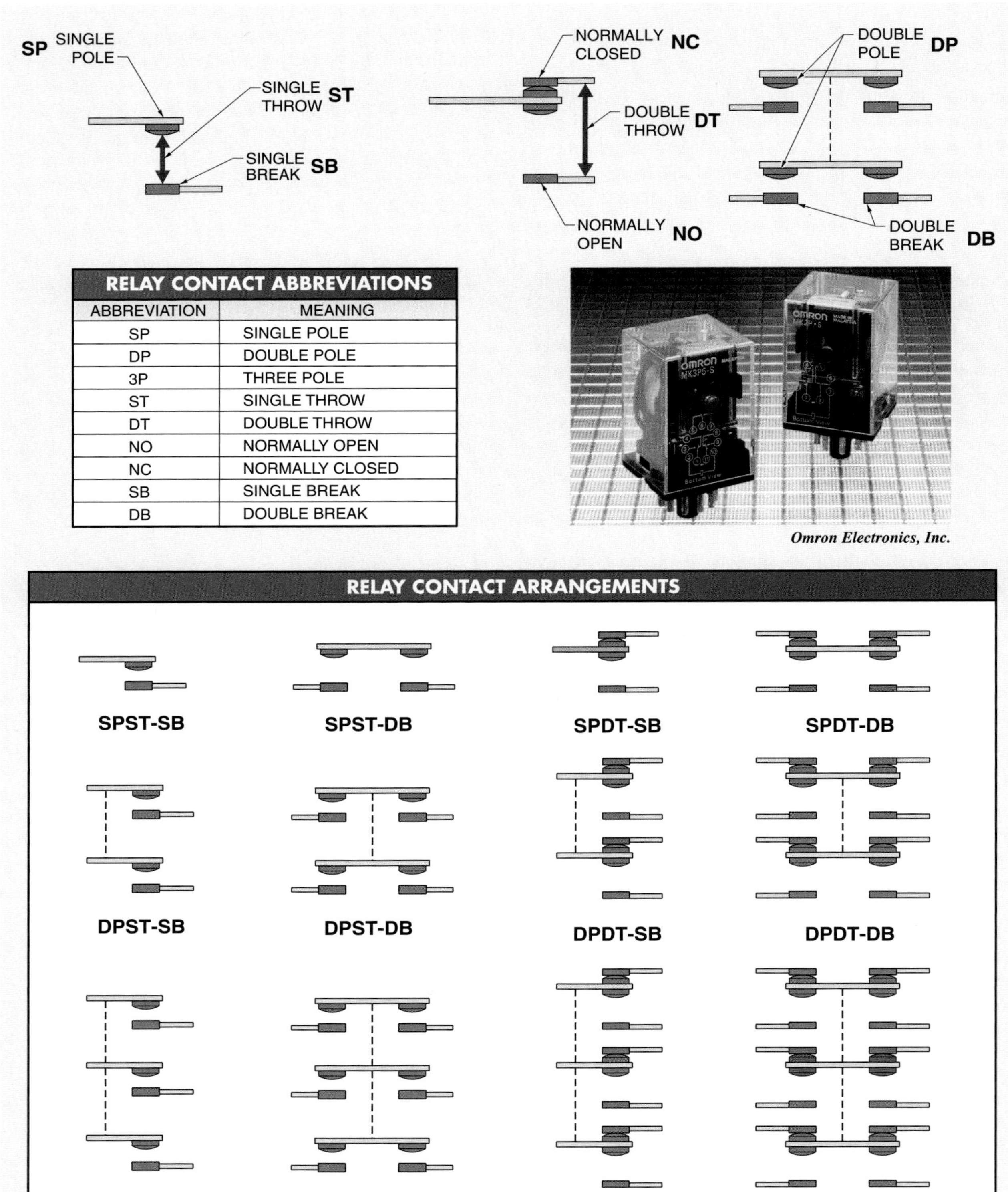

RELAY CONTACT ABBREVIATIONS

ABBREVIATION	MEANING
SP	SINGLE POLE
DP	DOUBLE POLE
3P	THREE POLE
ST	SINGLE THROW
DT	DOUBLE THROW
NO	NORMALLY OPEN
NC	NORMALLY CLOSED
SB	SINGLE BREAK
DB	DOUBLE BREAK

Figure 14-10. Relay contacts are described by their number of poles, throws, and breaks.

Relay manufacturers use a common code to simplify the identification of relays. **See Figure 14-11.** This code uses a form letter to indicate the type of relay. For example, Form A has one contact that is NO, and closes (makes) when the coil is energized. Form B has one contact that is NC, and breaks (opens) when the coil is energized. Form C has one pole that first breaks one contact and then makes a second contact when the coil is energized. Form C is the most common form.

RELAY FORM IDENTIFICATION			
DESIGN	SEQUENCE	SYMBOL	FORM
SPST-NO	MAKE (1)		A
SPST-NC	BREAK (1)		B
SPDT	BREAK (1) MAKE (2)		C
SPDT	MAKE (1) BEFORE BREAK (2)		D
SPDT (B-M-B)	BREAK (1) MAKE (2) BEFORE BREAK (3)		E
SPDT-NO	CENTER OFF		K
SPST-NO (DM)	DOUBLE MAKE (1)		X
SPST-NC (DB)	DOUBLE BREAK (1)		Y
SPDT-NC-NO (DB-DM)	DOUBLE BREAK (1) DOUBLE MAKE (2)		Z

Figure 14-11. Relay manufacturers use a common code (form letter) to simplify the identification of relays.

In some electrical applications, the exact order in which each contact operates (makes or breaks) must be known so the circuit can be designed to reduce arcing. Arcing occurs at any electrical contact that has current flowing through it when the contact is opened.

Technical Fact

In automobile security systems, relays are useful for alarm sensors and triggers, flashing lights, and starter interrupters. They are also used for door locks and courtesy lights that operate on entering or exiting a vehicle.

Machine Control Relays

A *machine control relay* is an EMR that includes several sets (usually two to eight) of NO and NC replaceable contacts (typically rated at 10 A to 20 A) that are activated by a coil. **See Figure 14-12.** Machine control relays are the backbone of electromechanical control circuitry and are expected to have long life and minimum problems. Machine control relays are used extensively in machine tools for direct switching of solenoids, contactors, and starters. Machine control relays provide easy access for contact maintenance and may provide additional features like time delay, latching, and convertible contacts for maximum circuit flexibility. Convertible contacts are mechanical contacts that can be placed in either an NO or NC position. Machine control relays are also known as heavy-duty or industrial control relays.

Sprecher + Schuh

Figure 14-12. Machine control relays are used extensively in machine tools for direct switching of solenoids, contactors, and starters.

In a machine control relay, each contact is a separate removable unit that may be installed to obtain any combination of NO and NC switching. These contacts are also convertible from NO to NC and from NC to NO. **See Figure 14-13.** The unit may be used as either an NO or NC contact by changing the terminal screws and rotating the unit 180°. Relays of one to 12 contact poles are readily assembled from stock parts. Machine control relays may have additional decks (groups of contacts) stacked onto the base unit.

The control coils for machine control relays are easily changed from one control voltage to another and are available in AC or DC standard ratings. Machine control relays have a large number of accessories that may be added to the relay unit. These include indicating lights, transient suppression, latching controls, and time controls.

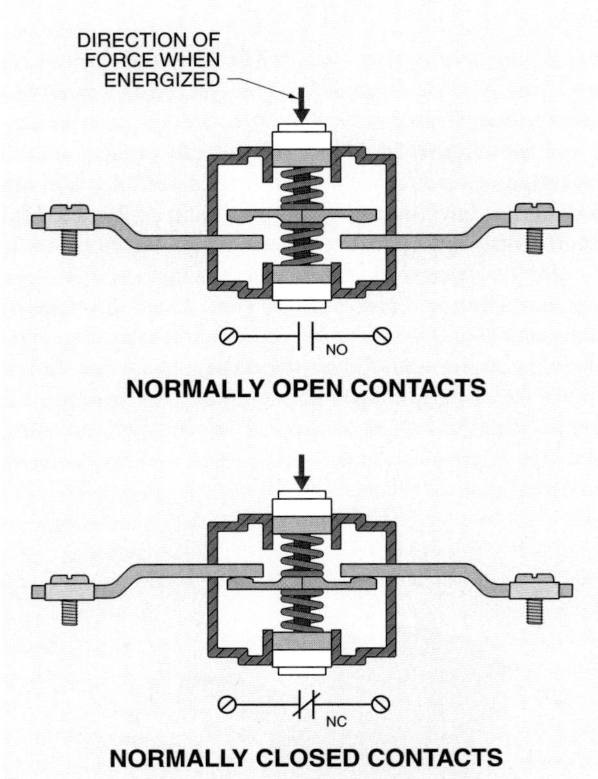

Figure 14-13. Machine control relay contacts are separate, removable units that are installed to obtain any combination of NO and NC switching.

Electromechanical Relay Life

EMR life expectancy is rated in contact life and mechanical life. *Contact life* is the number of times a relay's contacts switch the load controlled by the relay before malfunctioning. Typical contact life ratings are 100,000 to 500,000 operations. *Mechanical life* is the number of times a relay's mechanical parts operate before malfunctioning. Typical mechanical life ratings are 1,000,000 to 10,000,000 operations.

Relay contact life expectancy is lower than mechanical life expectancy because the life of a contact depends on the application. The contact rating of a relay is based on the contact's full rated power. Contact life is increased when contacts switch loads less than their full rated power. Contact life is reduced when contacts switch loads that develop destructive arcs. *Arcing* is the discharge of an electric current across a gap, such as when an electric switch is opened. Arcing causes contact burning and temperature rise. **See Figure 14-14.**

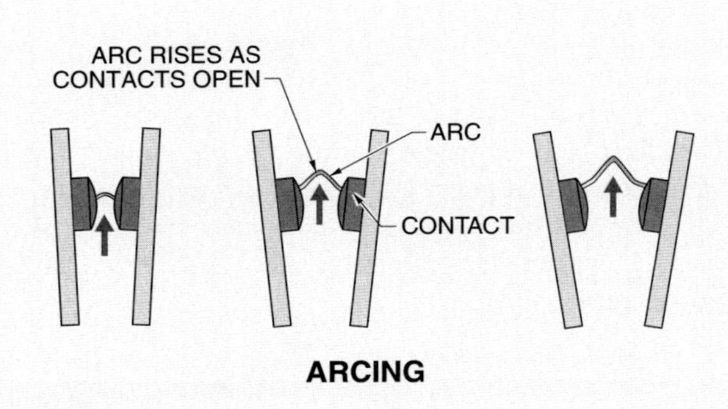

Figure 14-14. Arcing is the discharge of an electric current across a gap, such as when an electric switch is opened.

Arcing is minimized by using an arc suppressor and by using the correct contact material for the application. An arc suppressor is a device that dissipates the energy present across opening contacts. Arc suppression is used in applications that switch arc-producing loads such as solenoids, coils, motors, and other inductive loads.

Arc suppression is also accomplished by using a contact protection circuit. A *contact protection circuit* is a circuit that protects contacts by providing a nondestructive path for generated voltage as a switch is opened. A contact protection circuit may contain a diode, a resistance/capacitance (RC) circuit (snubber), or a varistor.

A diode is used as contact protection in DC circuits. The diode does not conduct electricity when the load is energized. The diode conducts electricity and shorts the generated voltage when the switch is opened. Because the diode shorts the generated voltage, the voltage is dissipated across the diode and not the relay contacts. **See Figure 14-15.**

An RC circuit and varistor are used as contact protection in AC circuits. The capacitor in an RC circuit is a high impedance to the 60 Hz line power and a short circuit to generated high frequencies. The short circuit dissipates generated voltage. A *varistor* is a resistor whose resistance is inversely proportional to the voltage applied to it. The varistor becomes a low-impedance circuit when its rated voltage is exceeded. The low-impedance circuit dissipates generated voltage when a switch is opened.

Contact Material

Relay contacts are available in fine silver, silver-cadmium oxide, gold-flashed silver, and tungsten. Fine silver has the highest electrical conductivity of all metals. However, fine silver sticks, welds, and is subject to sulfidation when used for many applications. Sulfidation is the formation of film on the contact surface. Sulfidation increases the resistance of the contacts. Silver is alloyed with other metals to reduce sulfidation.

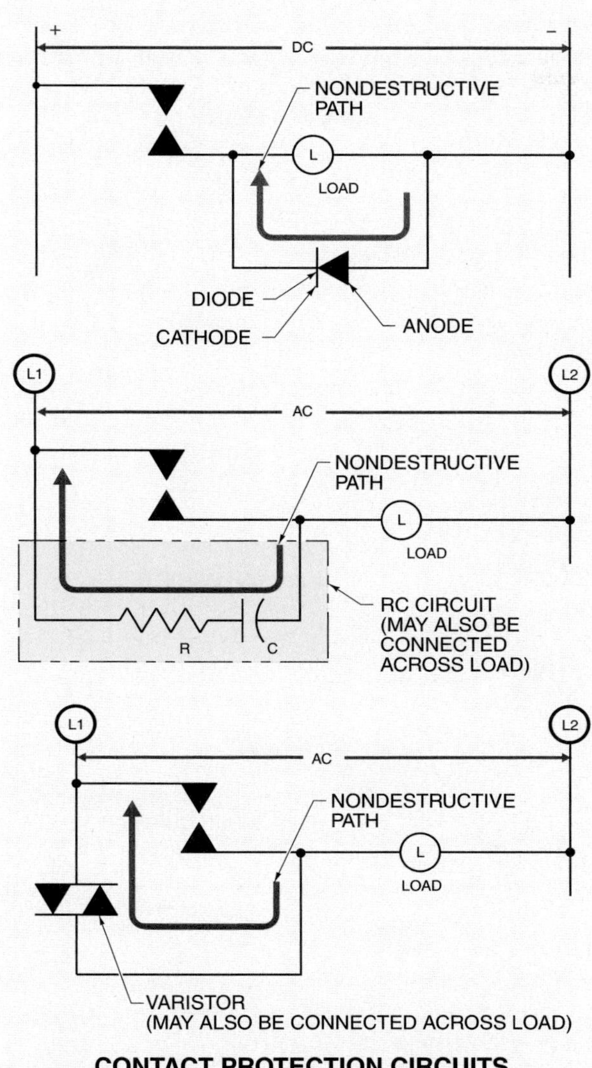

Figure 14-15. A contact protection circuit is a circuit that protects contacts by providing a nondestructive path for generated voltage as a switch is opened.

Silver is alloyed with cadmium to produce a silver-cadmium alloy. Silver-cadmium alloy contacts have good electrical characteristics and low resistance, which helps the contact resist arcing but not sulfidation. Silver or silver-cadmium alloy contacts are used in circuits that switch several amperes at more than 12 V, which burns off the sulfidation.

Sulfidation can damage silver contacts when used in intermittent applications. Gold-flashed silver contacts are used in intermittent applications to minimize sulfidation and provide a good electrical connection. Gold-flashed silver contacts are not used in high-current applications because the gold burns off quickly. Gold-flashed silver contacts are good for switching loads of 1 A or less.

Tungsten contacts are used in high-voltage applications because tungsten has a high melting temperature and is less affected by arcing. Tungsten contacts are used when high repetitive switching is required.

Contact Failure

In most applications, a relay fails due to contact failure. In some low-current applications, the relay contacts may look clean but may have a thin film of sulfidation, oxidation, or contaminants on the contact surface. This film increases the resistance to the flow of current through the contact. Normal contact wiping or arcing usually removes the film. In low-power circuits this action may not take place. In most applications, contacts are oversized for maximum life. Low-power circuit contacts should not be oversized to the extent that they switch just a small fraction of their rated value.

Contacts are often subject to high-current surges. High-current surges reduce contact life by accelerating sulfidation and contact burning. For example, a 100 W incandescent lamp has a current rating of about 1 A. The life of the contacts is reduced if a relay with 5 A contacts is used to switch the lamp because the lamp's filament has a low resistance when cold. When first turned ON, the lamp draws 12 A or more. The 5 A relay switches the lamp, but will not switch it for the rated life of the relay. Contacts are oversized in applications that have high-current surges.

SOLID-STATE RELAYS (SSRs)

The industrial control market has moved to solid-state electronics. Due to declining cost, high reliability, and immense capability, solid-state devices are replacing many devices that operate on mechanical and electromechanical principles. The selection of either a solid-state or electromechanical relay is based on the electrical, mechanical, and cost characteristics of each device and the required application.

SSR Switching Methods

The SSR used in an application depends on the load to be controlled. The different SSRs are designed to properly control certain loads. The four basic SSRs are the zero switching (ZS), instant-ON (IO), peak switching (PS), and analog switching (AS).

Zero Switching. A *zero switching relay* is an SSR that turns ON the load when the control voltage is applied and the voltage at the load crosses zero (or within a few volts of zero). The relay turns OFF the load when the control voltage is removed and the current in the load crosses zero. **See Figure 14-16.**

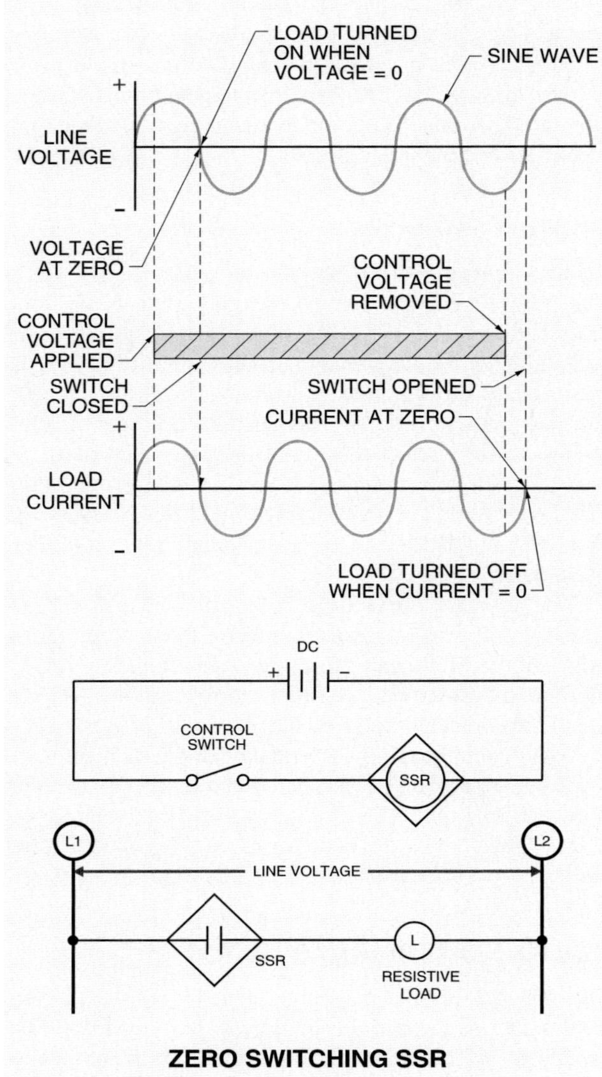

Figure 14-16. A zero switching relay turns ON the load when the control voltage is applied and the voltage at the load crosses zero.

The zero switching relay is the most widely used relay. Zero switching relays are designed to control resistive loads. Zero switching relays control the temperature of heating elements, soldering irons, extruders for forming plastic, incubators, and ovens. Zero switching relays control the switching of incandescent lamps, tungsten lamps, flashing lamps, and programmable controller interfacing.

Instant-ON. An *instant-ON switching relay* is an SSR that turns ON the load immediately when the control voltage is present. This allows the load to be turned ON at any point on the AC sine wave.

The relay turns OFF when the control voltage is removed and the current in the load crosses zero. Instant-ON switching is exactly like electromechanical switching because both switching methods turn ON the load at any point on the AC sine wave. **See Figure 14-17.**

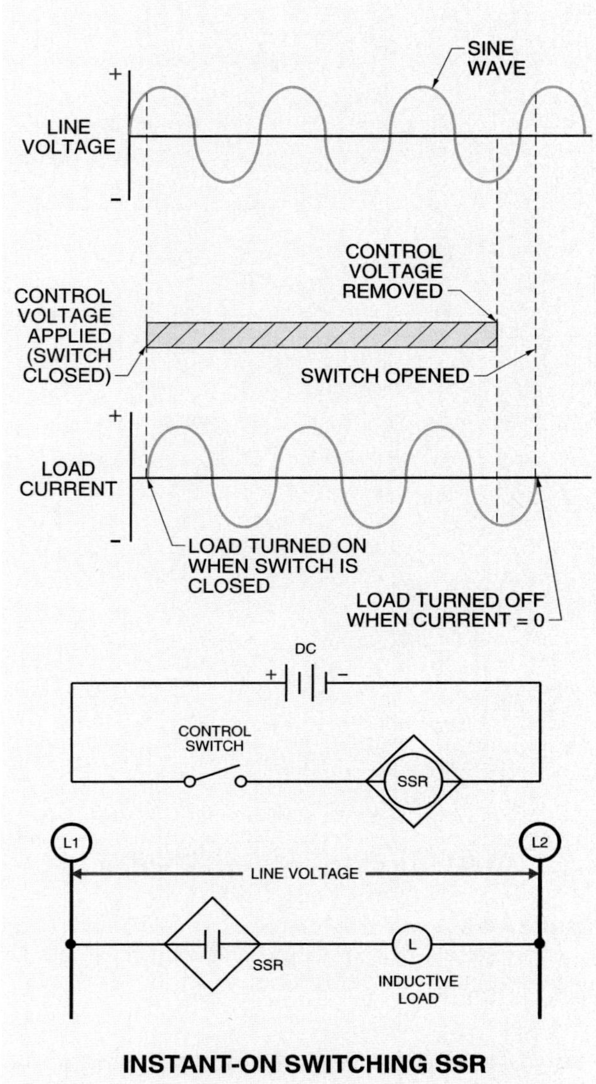

Figure 14-17. An instant-ON switching relay turns ON the load immediately when the control voltage is present.

Instant-ON relays are designed to control inductive loads. In inductive loads, voltage and current are not in phase, and the loads turn ON at a point other than the zero voltage point that is preferred. Instant-ON relays control the switching of contactors, magnetic valves and starters, valve positioning, magnetic brakes, small motors (used for position control), 1ϕ motors, small 3ϕ motors, lighting systems (fluorescent and HID), programmable controller interfaces, and phase control (by pulsing the input).

Peak Switching. A *peak switching relay* is an SSR that turns ON the load when the control voltage is present and the voltage at the load is at its peak. The relay turns OFF when the control voltage is removed and the current in the load crosses zero. Peak switching is preferred when the voltage and the current are about 90° out of phase because switching at peak voltage is switching at close to zero current. **See Figure 14-18.**

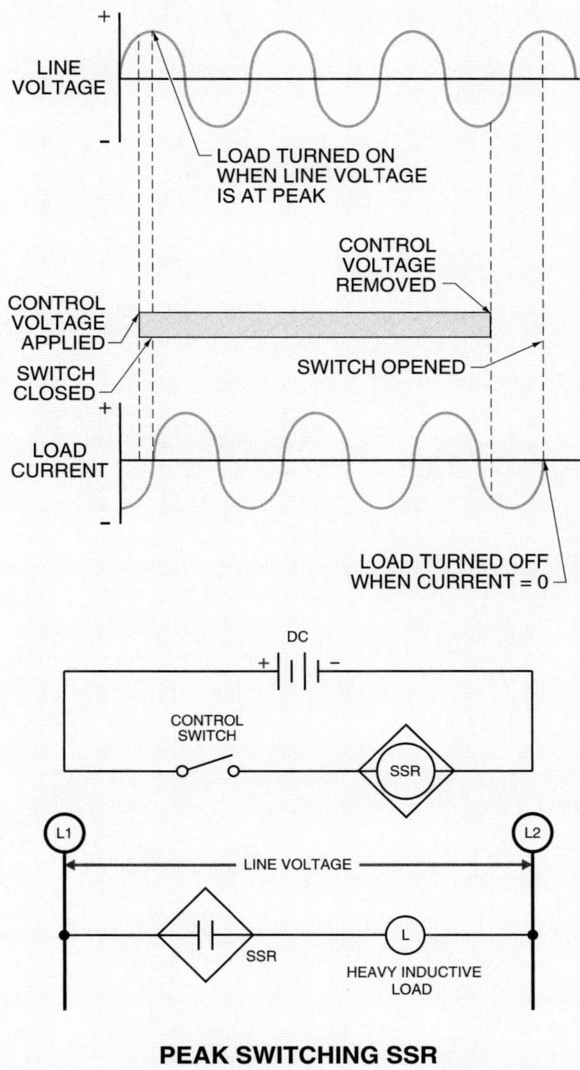

Figure 14-18. A peak switching relay turns ON the load when the control voltage is present and the voltage at the load is at peak.

Peak switching relays control transformers and other heavy inductive loads and limit the current in the first half-period of the AC sine wave. Peak switching relays control the switching of transformers, large motors, DC loads, high inductive lamps, magnetic valves, and small DC motors.

Analog Switching. An *analog switching relay* is an SSR that has an infinite number of possible output voltages within the relay's rated range. An analog switching relay has a built-in synchronizing circuit that controls the amount of output voltage as a function of the input voltage. This allows for a ramp-up function of the load. In a ramp-up function, the voltage at the load starts at a low level and is increased over a period of time. The relay turns OFF when the control voltage is removed and the current in the load crosses zero. **See Figure 14-19.**

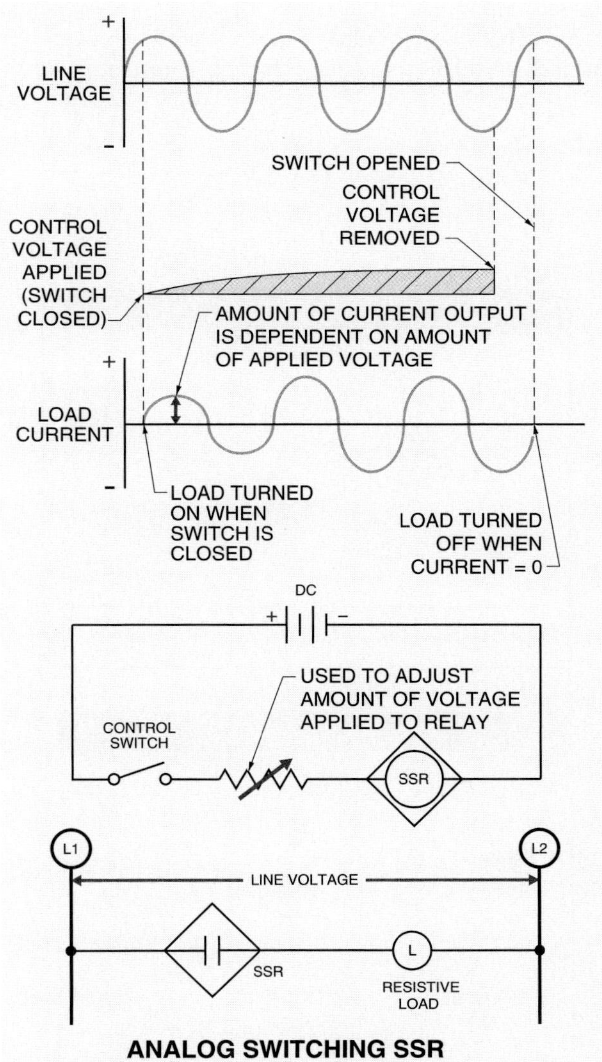

Figure 14-19. An analog switching relay has an infinite number of possible output voltages within the relay's rated range.

A typical analog switching relay has an input control voltage of 0 VDC to 5 VDC; these low and high limits correspond respectively to no switching and full switching on the output load. For any voltage between 0 VDC and 5 VDC, the output is a percentage of the available output voltage. However, the output is normally nonlinear when compared to the input, and the manufacturer's data must be checked.

Analog switching relays are designed for closed-loop applications. A closed-loop application is a temperature control with feedback from a temperature sensor to the controller.

In a closed-loop system, the amount of output is directly proportional to the amount of input signal. For example, if there is a small temperature difference between the actual temperature and the set temperature, the load (heating element) is given low power. However, if there is a large temperature difference between the actual temperature and the set temperature, the load (heating element) is given high power. This relay may also be used for starting high-power incandescent lamps to reduce the inrush current.

SSR Circuits

An SSR circuit consists of an input circuit, a control circuit, and an output (load-switching) circuit. These circuits may be used in any combination, providing many different solid-state switching applications. **See Figure 14-20.**

Input Circuits. The *input circuit* of an SSR is the part of the relay to which the control component is connected. The input circuit performs the same function as the coil of an EMR. The input circuit is activated by applying a voltage to the input of the relay that is higher than the specified pickup voltage of the relay. The input circuit is deactivated when a voltage less than the specified minimum dropout voltage of the relay is applied. Some SSRs have a fixed input voltage rating, such as 12 VDC. Most SSRs have an input voltage range, such as 3 VDC to 32 VDC. The voltage range allows a single SSR to be used with most electronic circuits.

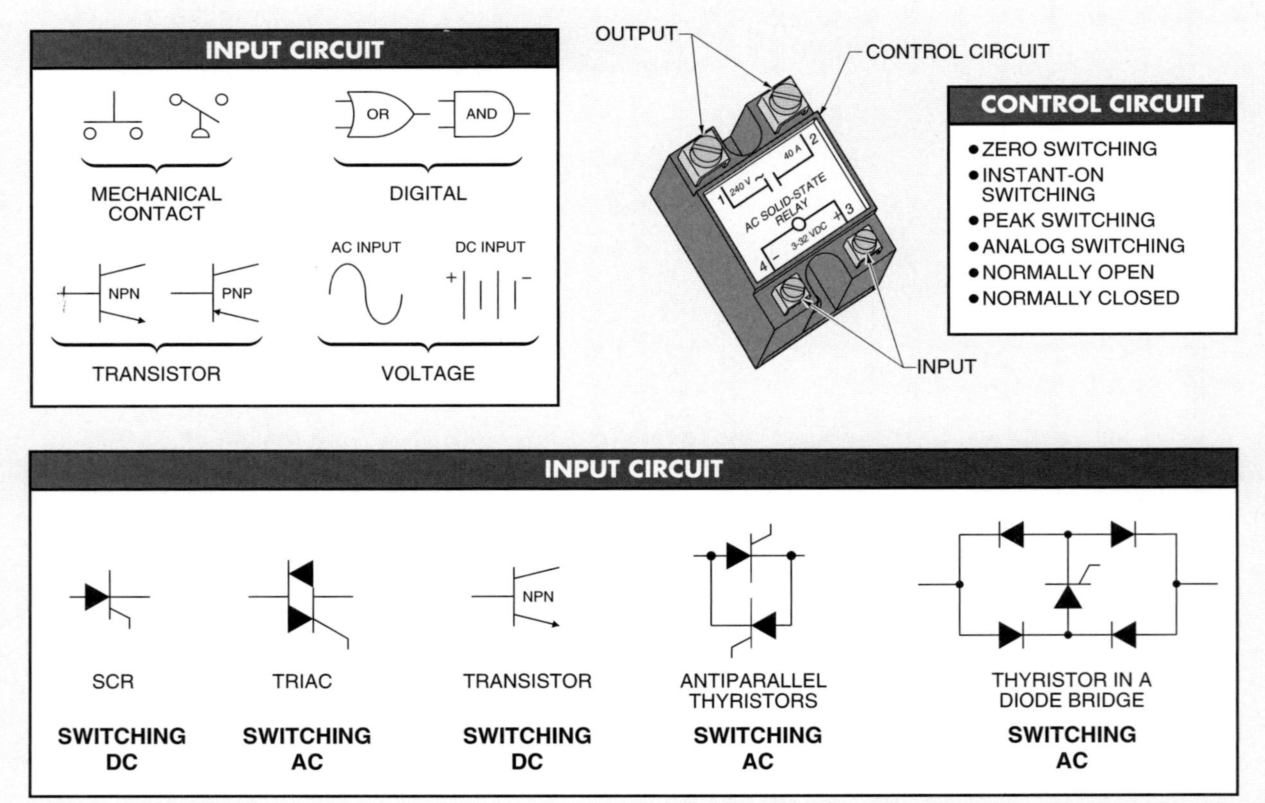

Figure 14-20. An SSR circuit consists of an input circuit, a control circuit, and an output (load-switching) circuit.

The input voltage of an SSR may be controlled (switched) through mechanical contacts, transistors, digital gates, etc. Most SSRs may be switched directly by low-power devices, which include integrated circuits, without adding external buffers or current-limiting devices. Variable-input devices, such as thermistors, may also be used to switch the input voltage of an SSR.

Control Circuits. The *control circuit* of an SSR is the part of the relay that determines when the output component is energized or de-energized. The control circuit functions as the coupling between the input and output circuits. This coupling is accomplished by an electronic circuit inside the SSR. In an EMR, the coupling is accomplished by the magnetic field produced by the coil.

When the control circuit receives the input voltage, the circuit is switched or not switched depending on whether the relay is a zero switching, instant-ON, peak switching, or analog switching relay.

Each relay is designed to turn ON the load-switching circuit at a predetermined voltage point. For example, a zero switching relay allows the load to be turned ON only after the voltage across the load is at or near zero. The zero switching function provides a number of benefits such as the elimination of high inrush currents on the load.

Output (Load-Switching) Circuits. The *output (load-switching) circuit* of an SSR is the load switched by the SSR. The output circuit performs the same function as the mechanical contacts of an electromechanical relay. However, unlike the multiple output contacts of EMRs, SSRs normally have only one output contact.

Most SSRs use a thyristor as the output switching component. Thyristors change from the OFF state (contacts open) to the ON state (contacts closed) very quickly when their gate switches ON. This fast switching action allows for high-speed switching of loads. The output switching device used depends on the type of load to be controlled. Different outputs are required when switching DC circuits than are required when switching AC circuits. Common outputs used in SSRs include:

- SCRs–used to switch high-current DC loads
- Triacs–used to switch low-current AC loads
- Transistors–used to switch low-current DC loads
- Antiparallel thyristors–used to switch high-current AC loads. They are able to dissipate more heat than a triac.
- Thyristors in diode bridges–used to switch low-current AC loads

SSR Circuit Capabilities

An SSR can be used to control most of the same circuits that an EMR is used to control. Because an SSR differs from an EMR in function, the control circuit for an SSR differs from that of an EMR. This difference is in how the relay is connected into the circuit. An SSR performs the same circuit functions as an EMR but with a slightly different control circuit.

Two-Wire Control. An SSR may be used to control a load using a momentary control such as a pushbutton. **See Figure 14-21.** In this circuit, the pushbutton signals the SSR, which turns ON the load. To keep the load turned ON, the pushbutton must be held down. The load is turned OFF when the pushbutton is released. This circuit is identical in operation to the standard two-wire control circuit used with EMRs, magnetic motor starters, and contactors. For this reason, the pushbutton could be changed to any manual, mechanical, or automatic control device for simple ON/OFF operation. The same circuit may be used for liquid level control if the pushbutton is replaced with a float switch.

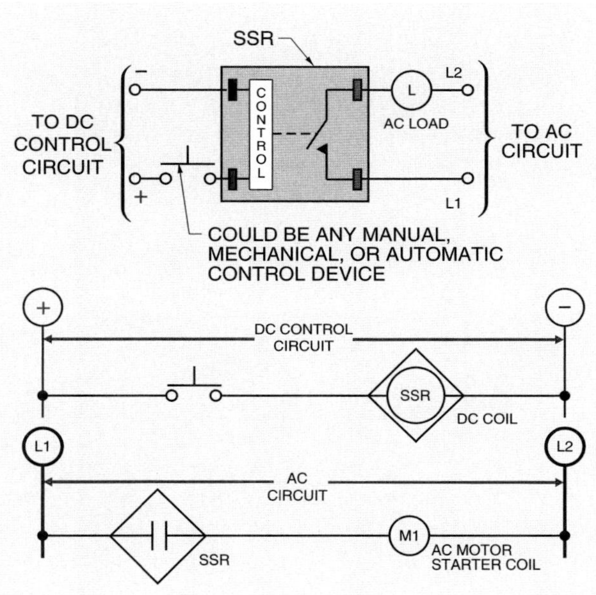

Figure 14-21. An SSR may be used to control a load using a momentary control such as a pushbutton.

Three-Wire Memory Control. An SSR may be used with a silicon-controlled rectifier (SCR) for latching the load ON. **See Figure 14-22.** This circuit is identical in operation to the standard three-wire memory control circuit. An SCR is used to add memory after the start pushbutton is pressed. An SCR acts as a current-operated OFF-to-ON switch. The SCR does not allow the DC control current to pass through until a current is applied to its gate. There must be a flow of a definite minimum current to turn the SCR ON. This is accomplished by pressing the start pushbutton. Once the gate of the SCR has voltage applied, the SCR is latched in the ON condition and allows the DC control voltage to pass through even after the start pushbutton is released. Resistor R1 is used as a current-limiting resistor for the gate, and is determined by gate current and supply voltage.

The circuit must be opened to stop the anode-to-cathode flow of DC current to the SCR. This is accomplished by pressing the stop pushbutton. Additional start pushbuttons are added in parallel with the start pushbutton. Additional stops may be added to the circuit by placing them in series with the stop pushbutton. The additional start/stops may be any manual, mechanical, or automatic control.

Equivalent NC Contacts. An SSR may be used to simulate an equivalent NC contact condition. **See Figure 14-23.**

Technical Fact

An SCR is called a rectifier because it conducts current in only one direction, with a gate controlling forward resistance. Switching of an SCR is normally controlled through gate voltage.

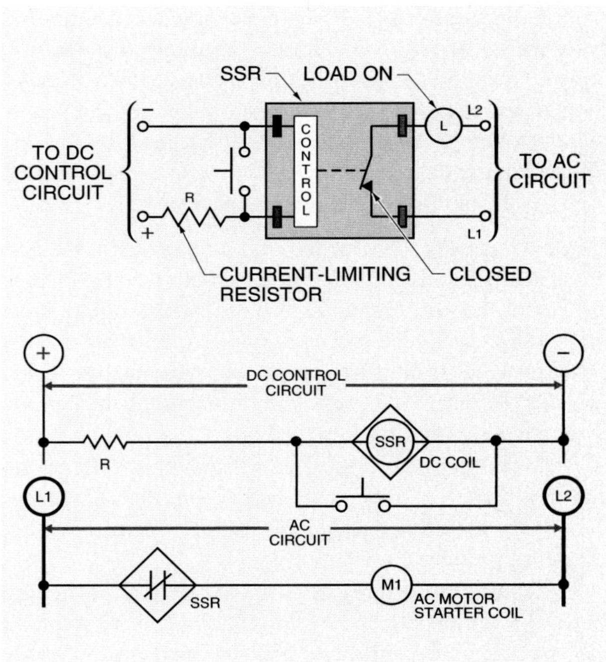

Figure 14-23. An SSR with a load (current limiting) resistor may be used to simulate an equivalent NC contact condition.

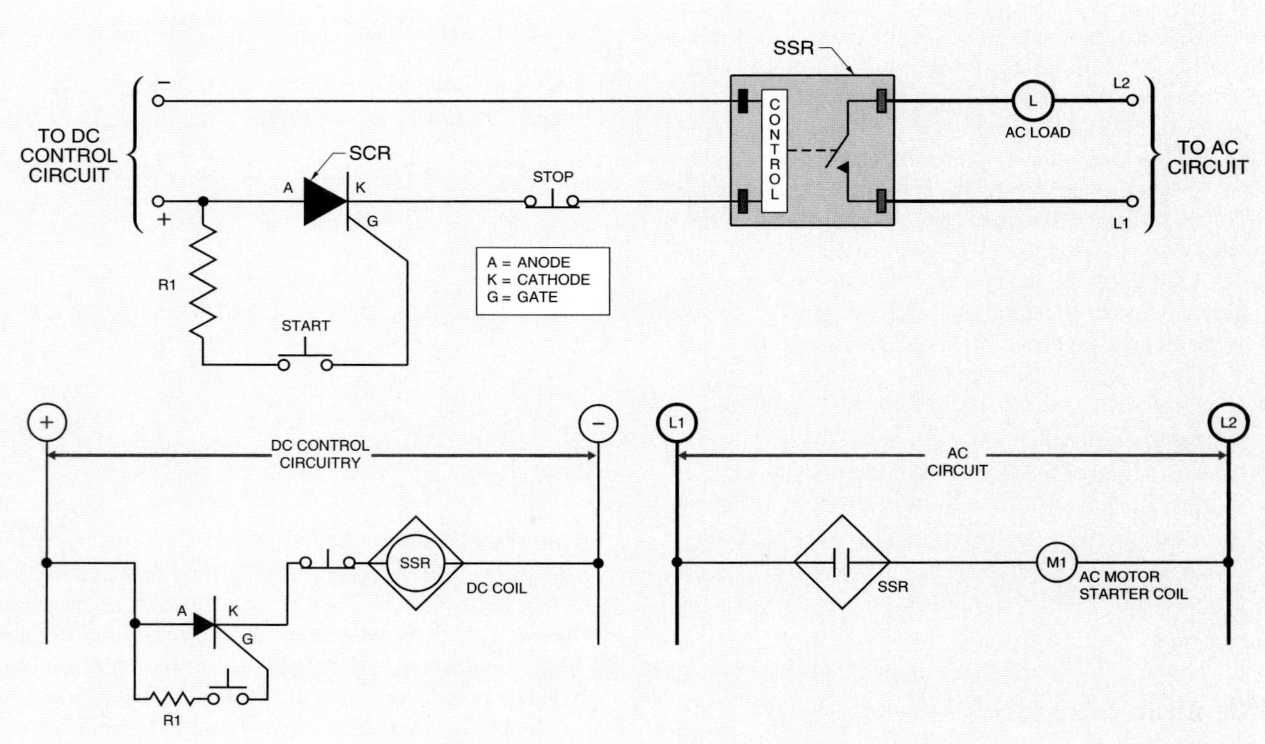

Figure 14-22. An SSR may be used with an SCR to latch a load ON.

An NC contact must be electrically made because most SSRs have the equivalent of an NO contact. This is accomplished by allowing the DC control voltage to be connected to the SSR through a current-limiting resistor (R). The load is held in the ON condition because the control voltage is present on the SSR. The pushbutton is pressed to turn OFF the load. This allows the DC control voltage to take the path of least resistance and electrically remove the control voltage from the relay. This also turns OFF the load until the pushbutton is released.

Transistor Control. SSRs are also capable of being controlled by electronic control signals from integrated circuits and transistors. **See Figure 14-24.** In this circuit, the SSR is controlled through an NPN transistor which receives its signal from IC logic gates, etc. The two resistors (R1 and R2) are used as current-limiting resistors.

Series and Parallel Control of SSRs. SSRs can be connected in series or parallel to obtain multicontacts that are controlled by one input device. Multicontact SSRs may also be used. Three SSR control inputs may be connected in parallel so that, when the switch is closed, all three are actuated. **See Figure 14-25.** This controls the 3ϕ circuit.

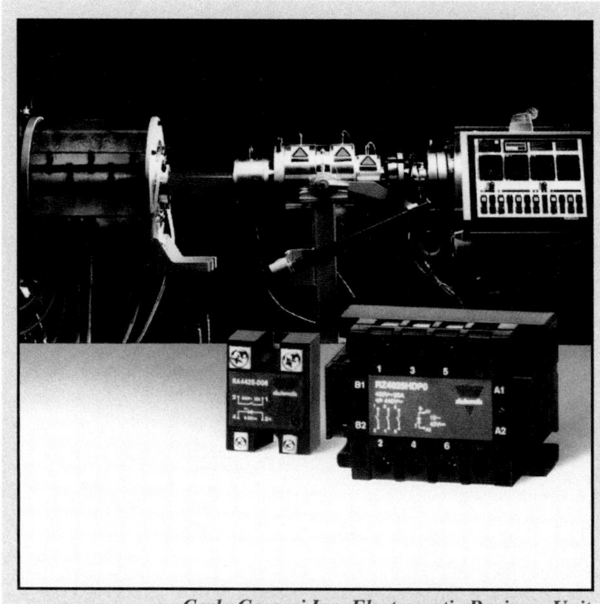

Carlo Gavazzi Inc. Electromatic Business Unit

Three-phase relays may be used for direct and delta switching of motor loads and direct switching of motor loads with shunting by an electromechanical contactor.

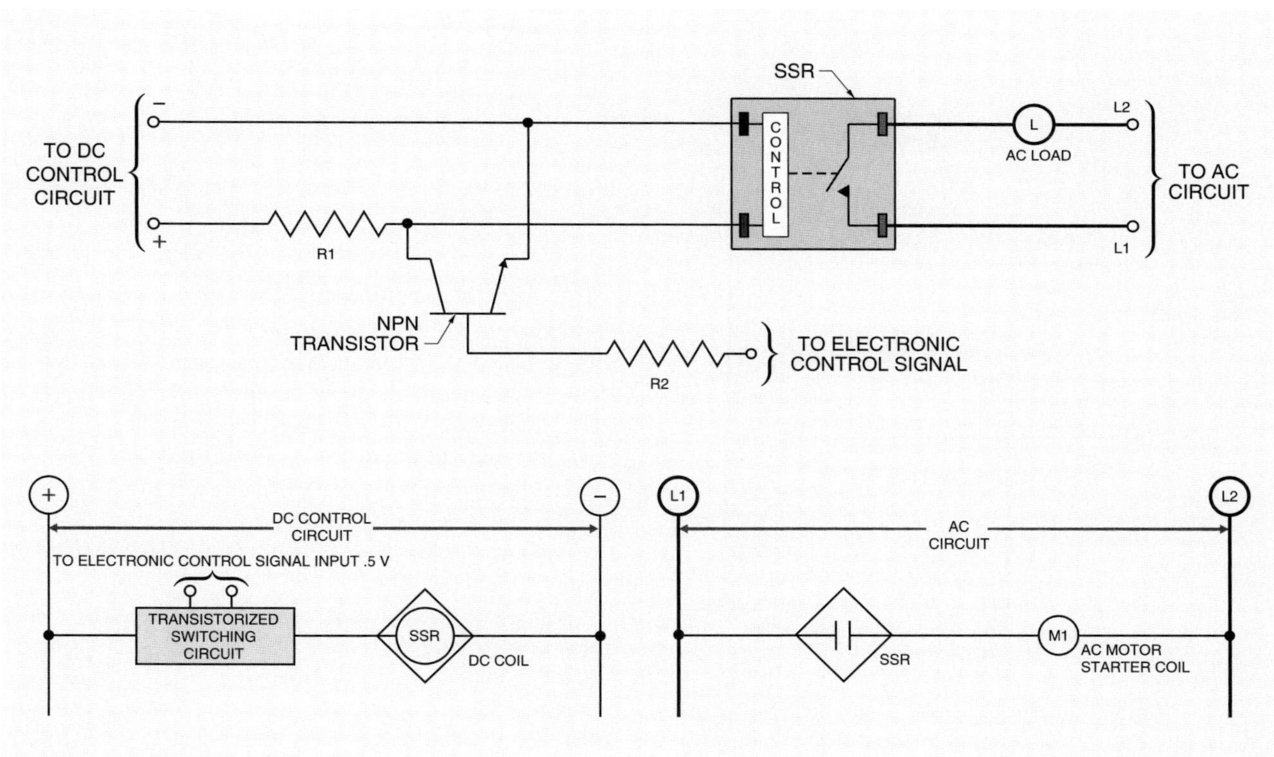

Figure 14-24. SSRs may be controlled by electronic control signals from integrated circuits and transistors.

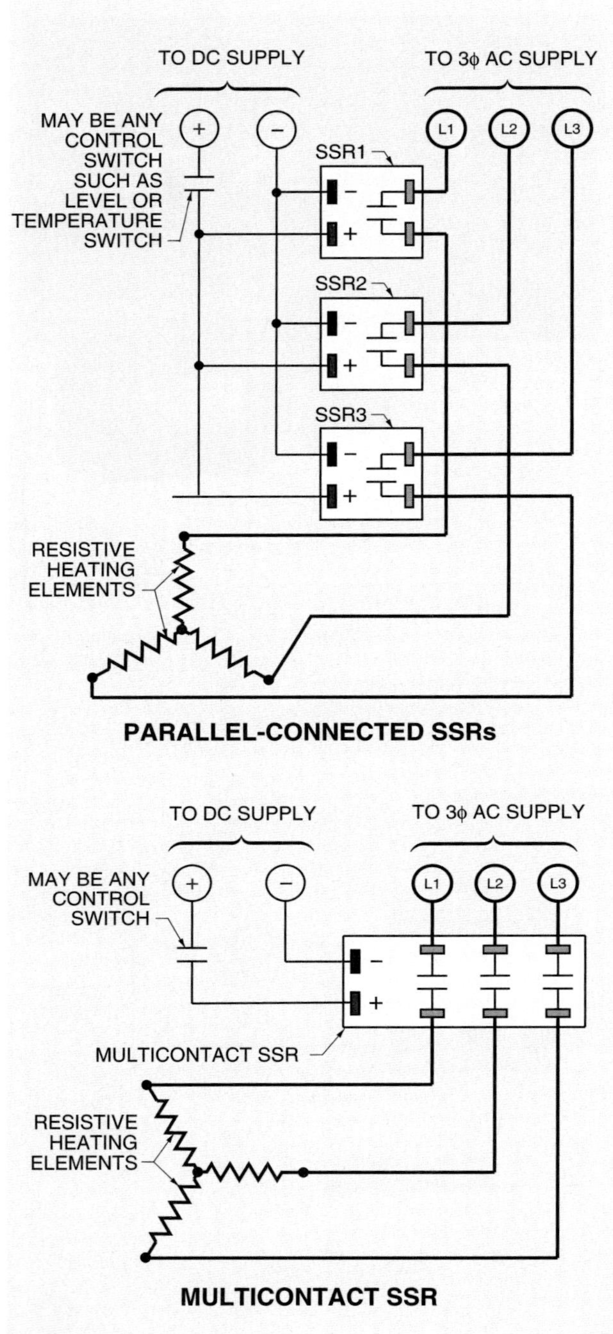

Figure 14-25. Three SSRs may be connected in parallel to control a 3ϕ circuit, or a multicontact SSR may be used.

In this application, the DC control voltage across each SSR is equal to the DC supply voltage because they are connected in parallel. When a multicontact SSR is used, there is only one input that controls all output switches.

SSRs can be connected in series to control a 3ϕ circuit. **See Figure 14-26.** The DC supply voltage is divided across the three SSRs when the switch is closed. For this reason, the DC supply voltage must be at least three times greater than the minimum operating voltage of each relay.

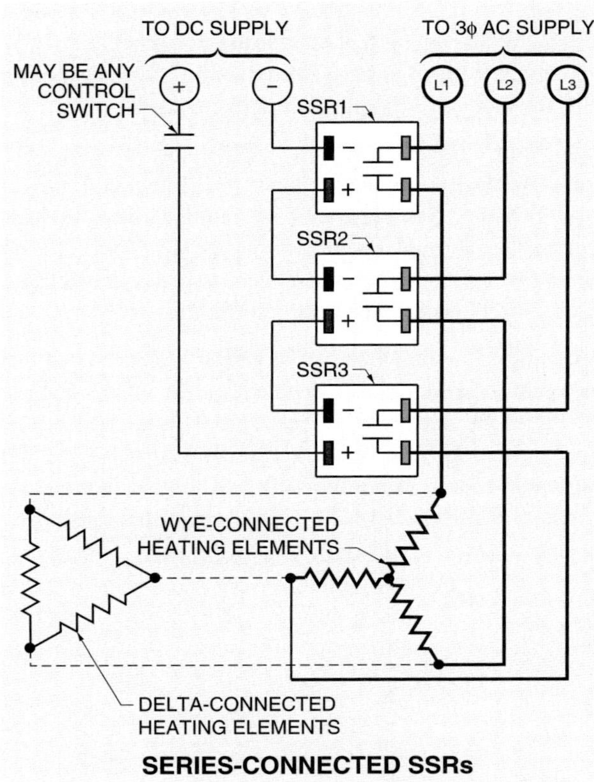

Figure 14-26. Three SSRs may be connected in series to control a 3ϕ circuit.

SSR Temperature Problems

Temperature rise is the largest problem in applications using an SSR. As temperature increases, the failure rate of SSRs increases. As temperature increases, the number of operations of an SSR decreases. The higher the heat in an SSR, the more problems occur. **See Figure 14-27.**

The failure rate of most SSRs doubles for every 10°C temperature rise above an ambient temperature of 40°C. An ambient temperature of 40°C is considered standard by most manufacturers.

Solid-state relay manufacturers specify the maximum relay temperature permitted. The relay must be properly cooled to ensure that the temperature does not exceed the specified maximum safe value. Proper cooling is accomplished by installing the SSR to the correct heat sink. A heat sink is chosen based on the maximum amount of load current controlled.

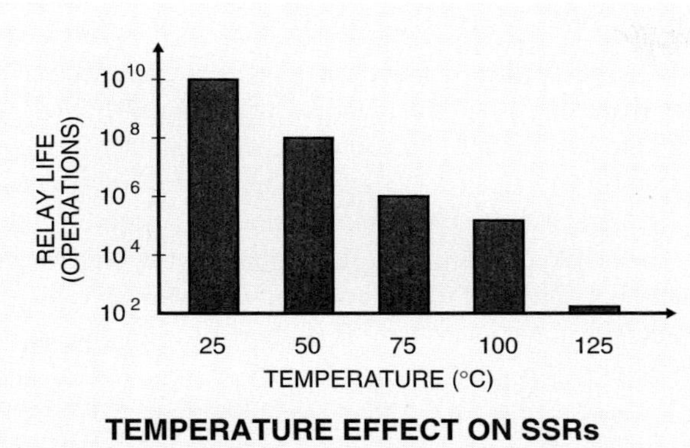

Figure 14-27. As temperature increases, the number of operations of an SSR decreases.

Heat Sinks

The performance of an SSR is affected by ambient temperature. The ambient temperature of a relay is a combination of the temperature of the relay location and the type of enclosure used. The temperature inside an enclosure may be much higher than the ambient temperature of an enclosure that allows good air flow.

The temperature inside an enclosure increases if the enclosure is located next to a heat source or in the sun. The electronic circuit and SSR also produce heat. Forced cooling is required in some applications.

Selecting Heat Sinks

A low resistance to heat flow is required to remove the heat produced by an SSR. The opposition to heat flow is thermal resistance. *Thermal resistance (R_{TH})* is the ability of a device to impede the flow of heat. Thermal resistance is a function of the surface area of a heat sink and the conduction coefficient of the heat sink material. Thermal resistance is expressed in degrees Celsius per watt (°C/W).

Heat sink manufacturers list the thermal resistance of heat sinks. The lower the thermal resistance number, the more easily the heat sink dissipates heat. The larger the thermal resistance number, the less effectively the heat sink dissipates heat. The thermal resistance value of a heat sink is used with an SSR load current/ambient temperature chart to determine the size of the heat sink required. **See Figure 14-28.**

A relay can control a large amount of current when a heat sink with a low thermal resistance number is used. A relay can control the least amount of current when no heat sink (free air mounting) is used. To maximize heat conduction through a relay and into a heat sink:

- Use heat sinks made of a material that has a high thermal conductivity. Silver has the highest thermal conductivity rating. Copper has the highest practical thermal conductivity rating. Aluminum has a good thermal conductivity rating, and is the most cost-effective and widely used heat sink.
- Keep the thermal path as short as possible.
- Use the largest cross-sectional surface area in the smallest space.
- Always use thermal grease or pads between the relay housing and the heat sink to eliminate air gaps and aid in thermal conductivity.

HEAT SINK SELECTIONS

TYPE	H × W × L (mm)	R_{TH}(°C/W)
01	15 × 79 × 100	2.5
02	15 × 100 × 100	2.0
03	25 × 97 × 100	1.5
04	37 × 120 × 100	0.9
05	40 × 60 × 150	0.5
06	40 × 200 × 150	0.4

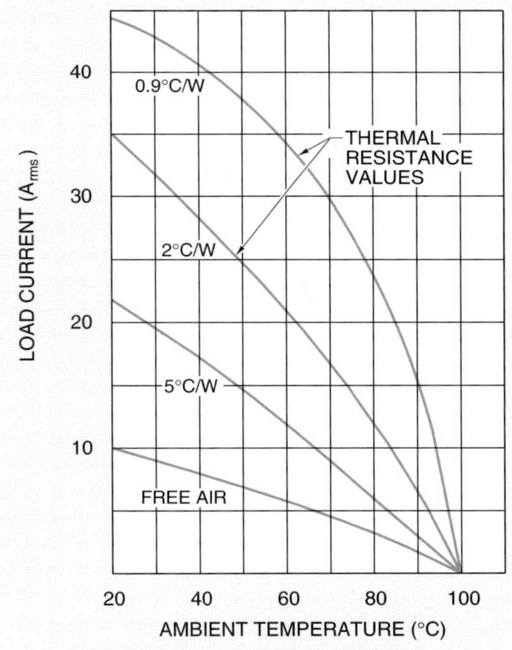

Figure 14-28. Thermal resistance (R_{TH}) is the ability of a device to impede the flow of heat.

Mounting Heat Sinks

A heat sink must be correctly mounted to ensure proper heat transfer. To properly mount a heat sink:

- Choose a smooth mounting surface. The surfaces between a heat sink and a solid-state device should be as flat and smooth as possible. Ensure that the mounting bolts and screws are securely tightened.
- Locate heat-producing devices so that the temperature is spread over a large area. This helps prevent higher temperature areas.
- Use heat sinks with fins to achieve as large a surface area as possible.
- Ensure that the heat from one heat sink does not add to the heat from another heat sink.
- Always use thermal grease between the heat sink and the solid-state device to ensure maximum heat transfer.

Relay Current Problems

The overcurrent passing through an SSR must be kept below the maximum load current rating of the relay. An overload protection fuse is used to prevent overcurrents from damaging an SSR.

An overload protection fuse opens the circuit when the current is increased to a higher value than the nominal load current. The fuse should be an ultrafast fuse used for the protection of semiconductors. **See Figure 14-29.**

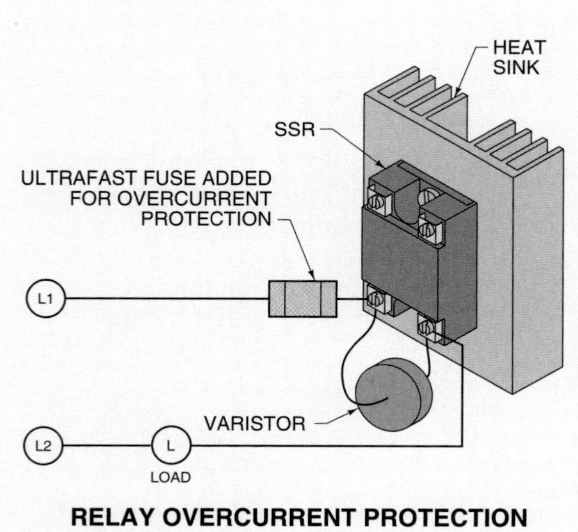

Figure 14-29. An overload protection fuse opens the circuit when the current is increased to a higher value than the nominal load current.

Relay Voltage Problems

Most AC power lines contain voltage spikes superimposed on the voltage sine wave. Voltage spikes are produced by switching motors, solenoids, transformers, motor starters, contactors, and other inductive loads. Large spikes are also produced by lightning striking the power distribution system.

The output element of a relay can exceed its breakdown voltage and turn ON for part of a half period if overvoltage protection is not provided. This short turn-on can cause problems in the circuit.

Varistors are added to the relay output terminals to prevent an overvoltage problem. A varistor should be rated 10% higher than the line voltage of the output circuit. The varistor bypasses the transient current. **See Figure 14-30.**

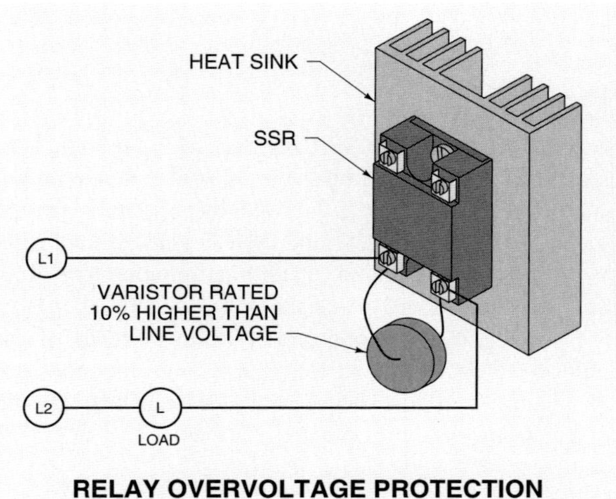

Figure 14-30. Varistors are added to relay output terminals to prevent an overvoltage problem.

Voltage Drop

In all series circuits, the total circuit voltage is dropped across the circuit components. The higher the resistance of any component, the higher the voltage drop. The lower the resistance of any component, the lower the voltage drop. Thus, an open switch that has a meter connected across it shows a very high voltage drop because the meter and open switch have a very high resistance when compared to the load. Conversely, a closed switch that has a meter connected across it shows a very low voltage drop because the meter is closed and closed switches have a very low resistance when compared to the load.

A voltage drop in the switching component is unavoidable in an SSR. The voltage drop produces heat. The larger the current passing through the relay, the greater the amount of heat produced. The generated heat affects relay operation and can destroy the relay if not removed. **See Figure 14-31.**

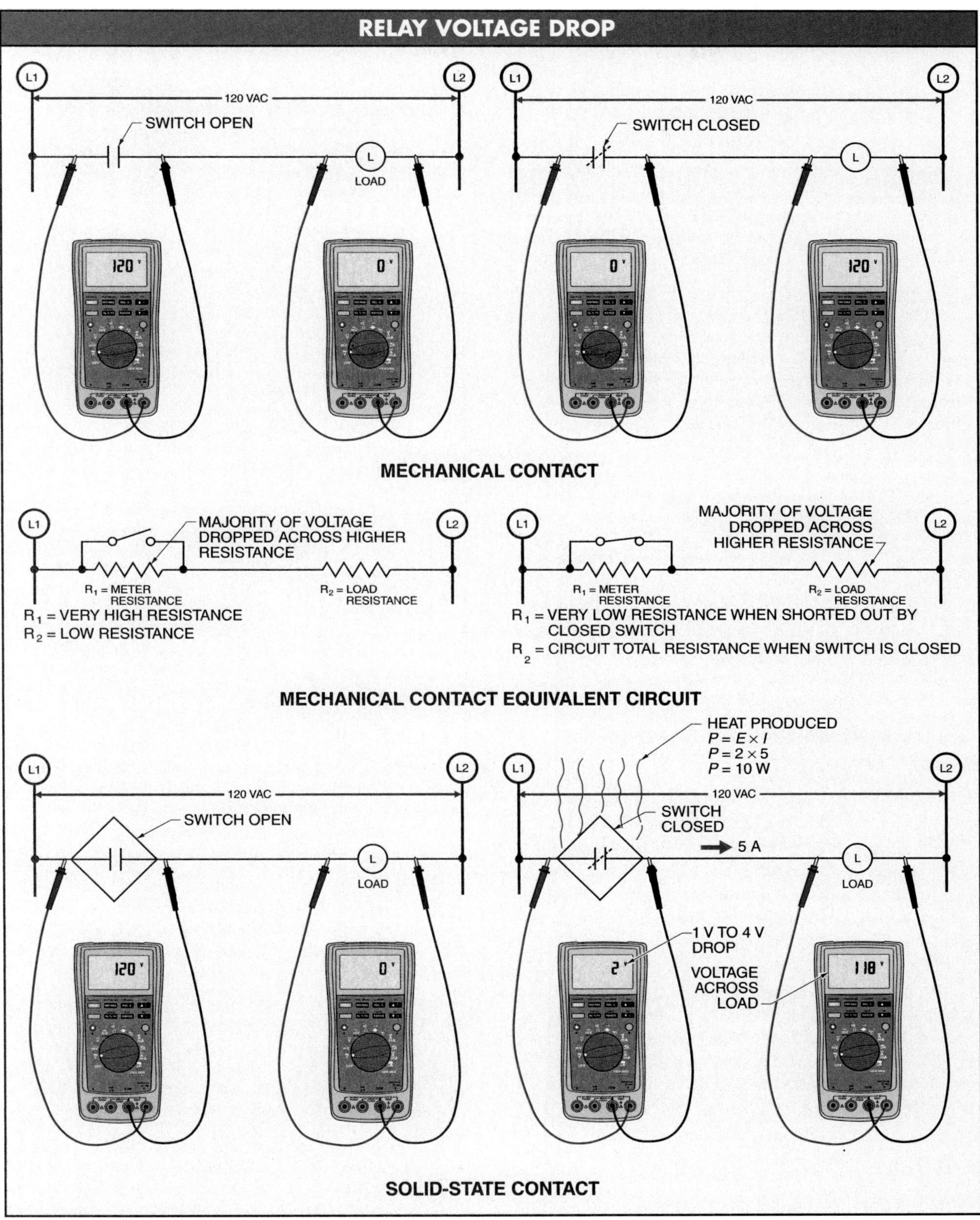

Figure 14-31. The voltage drop in the switching component of an SSR produces heat, which can destroy the relay if not removed.

The voltage drop in an SSR is usually 1 V to 1.6 V, depending on the load current. For small loads (less than 1 A), the heat produced is safely dissipated through the relay's case. High-current loads require a heat sink to dissipate the extra heat. **See Figure 14-32.**

SSR VOLTAGE DROP		
LOAD CURRENT (IN A)	VOLTAGE DROP (IN V)	POWER AT SWITCH (IN V)
1	2	2
2	2	4
5	2	10
10	2	20
20	2	40
50	2	100

Figure 14-32. For small loads (less than 1 A), the heat produced in an SSR is safely dissipated through the relay's case.

For example, if the load current in a circuit is 1 A and the SSR switching device has a 2 V drop, the power generated is 2 W. The 2 W of power generates heat that can be dissipated through the relay's case.

If the load current in a circuit is 20 A and the SSR switching device has a 2 V drop, the power generated in the device is 40 W. The 40 W of power generates heat that requires a heat sink to safely dissipate the heat.

Electromechanical and Solid-State Relay Comparison

EMRs and SSRs are designed to provide a common switching function. An EMR provides switching through the use of electromagnetic devices and sets of contacts. An SSR depends on electronic devices such as SCRs and triacs to switch without contacts. In addition, the physical features and operating characteristics of EMRs and SSRs are different. **See Figure 14-33.**

An equivalent terminology chart is used as an aid in the comparison of EMRs and SSRs. Because the basic operating principles and physical structures of the devices are so different, it is difficult to find a direct comparison of the two. Differences arise almost immediately both in the terminology used to describe the devices and in their overall ability to perform certain functions. **See Figure 14-34.**

Advantages and Limitations

EMRs and SSRs are used in many applications. The relay used depends on the application's electrical requirements, cost requirements, and life expectancy.

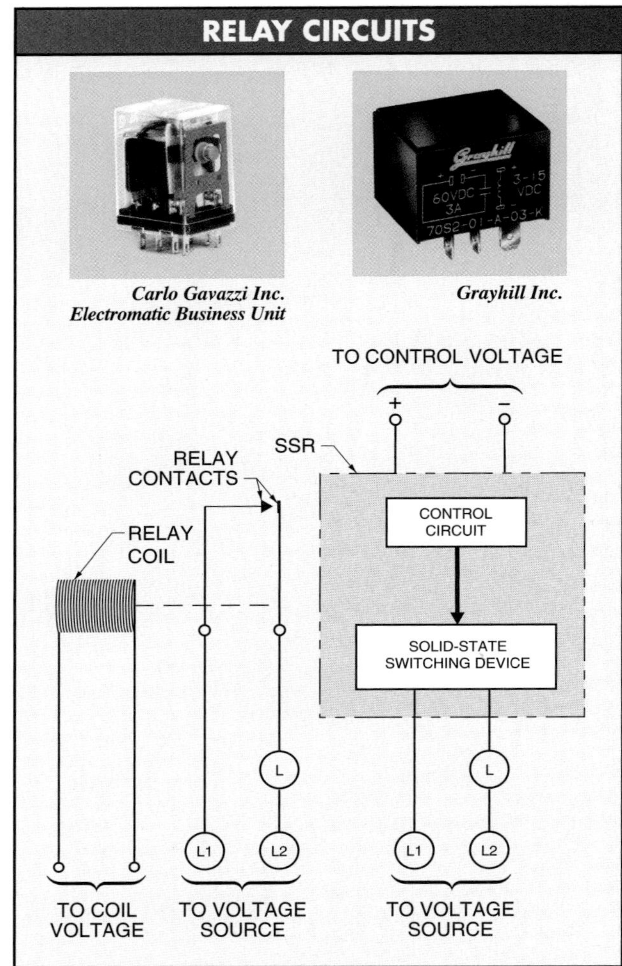

Figure 14-33. An EMR provides switching using electromagnetic devices. An SSR depends on SCRs and triacs to switch without contacts.

Although SSRs are replacing EMRs in many applications, EMRs are still very common. EMRs offer many advantages that make them very cost-effective. However, they do have disadvantages that limit their use in some applications. SSRs provide many advantages such as small size, fast switching, long life, and the ability to handle complex switching requirements. SSRs have some limitations that restrict their use in some applications.

Technical Fact

The first electromagnetic relay used in the U.S. was invented by Samuel Morse and used to regenerate signals on the telegraph line between Baltimore and Washington, D.C., in 1844.

EMR/SSR EQUIVALENT TERMINOLOGY CHART			
EMRs		SSRs	
Term	**Definition**	**Term**	**Definition**
Coil Voltage	Minimum voltage necessary to energize or operate relay. Also referred to as pickup voltage	Control Voltage	Minimum voltage required to gate or activate control circuit of SSR. Generally, a maximum value is also specified
Coil Current	Amount of current necessary to energize or operate relay	Control Current	Minimum current required to turn ON solid-state control circuit. Generally, a maximum value is also specified
Holding Current	Minimum current required to keep a relay energized	Control Current	
Dropout Voltage	Maximum voltage at which the relay is no longer energized	Control Voltage	
Pull-In Time	Amount of time required to operate (open or close) relay contacts after coil voltage is applied	Turn-ON Time	Elapsed time between application of control voltage and application of voltage to load circuit
Dropout Time	Amount of time required for the relay contacts to return to their normal de-energized position after coil voltage is removed	Turn-OFF Time	Elapsed time between removal of control voltage and removal of voltage from load circuit
Contact Voltage Rating	Maximum voltage rating that contacts of relay are capable of safely switching	Load Voltage	Maximum output voltage-handling capability of an SSR
Contact Current Rating	Maximum current rating that contacts of relay are capable of safely switching	Load Current	Maximum output current-handling capability of an SSR
Surge Current	Maximum peak current which contacts on a relay can withstand for short periods of time without damage	Surge Current	Maximum peak current which an SSR can withstand for short periods of time without damage
Contact Voltage Drop	Voltage drop across relay contacts when relay is operating (usually low)	Switch-ON Voltage Drop	Voltage drop across SSR when operating
Insulation Resistance	Amount of resistance measured across relay contacts in open position	Switch-OFF Resistance	Amount of resistance measured across an SSR when turned OFF
No equivalent or comparison		OFF-State Leakage Current	Amount of leakage current through SSR when turned OFF but still connected to load voltage
No equivalent or comparison		Zero Current Turn-OFF	Turn-OFF at zero crossing of load current that flows through an SSR. A thyristor turns OFF only when current falls below minimum holding current. If input control is removed when current is a higher value, turn-OFF is delayed until next zero current crossing
No equivalent or comparison		Zero Voltage Turn-ON	Initial turn-ON occurs at a point near zero crossing of AC line voltage. If input control is applied when line voltage is at a higher value, initial turn-ON is delayed until next zero crossing

Figure 14-34. An equivalent terminology chart is used as an aid in the comparison of EMRs and SSRs.

EMR Advantages

- Normally have multipole, multithrow contact arrangements
- Contacts can switch AC or DC
- Low initial cost
- Very low contact voltage drop, thus no heat sink is required
- Very resistant to voltage transients
- No OFF-state leakage current through open contacts

EMR Limitations

- Contacts wear, thus have a limited life
- Short contact life when used for rapid switching applications or high current loads
- Generate electromagnetic noise and interference on the power lines
- Poor performance when switching high inrush currents

SSR Advantages

- Very long life when properly applied
- No contacts to wear
- No contact arcing to generate electromagnetic interference
- Resistant to shock and vibration because they have no moving parts
- Logic compatible with programmable controllers, digital circuits, and computers
- Very fast switching capability
- Different switching modes (zero switching, instant-ON, etc.)

SSR Limitations

- Normally only one contact available per relay
- Heat sink required due to voltage drop across switch
- Can switch only AC or DC
- OFF-state leakage current when switch is open
- Normally limited to switching only a narrow frequency range such as 40 Hz to 70 Hz

Input Signals. Applying a voltage to the input coil of an electromagnetic device creates an electromagnet that is capable of pulling in an armature with a set of contacts attached to control a load circuit. It takes more voltage and current to pull in the coil than to hold it in due to the initial air gap between the magnetic coil and the armature. The specifications used to describe the energizing and de-energizing process of an electromagnetic device are coil voltage, coil current, holding current, and drop-out voltage.

An SSR has no coil or contacts and requires only minimum values of voltage and current to turn it ON and turn it OFF. The two specifications needed to describe the input signal for an SSR are control voltage and control current.

The electronic nature of an SSR and its input circuit allows easy compatibility with digitally controlled logic circuits. Many SSRs are available with minimum control voltages of 3 V and control currents as low as 1 mA, making them ideal for a variety of current state-of-the-art logic circuits.

Response Time. One of the significant advantages of an SSR over an EMR is its response time (ability to turn ON and turn OFF). An EMR may be able to respond hundreds of times per minute. An SSR is capable of switching thousands of times per minute with no chattering or bounce.

DC switching times for an SSR are in the microsecond range, while AC switching time, with the use of zero-voltage turn-on, is less than 9 ms. The reason for this advantage is that the SSR may be turned ON and turned OFF electronically much more rapidly than a relay may be electromagnetically pulled in and dropped out.

The higher speed of SSRs has become increasingly more important as industry demands higher productivity from processing equipment. The more rapidly the equipment can process or cycle its output, the greater the productivity of the machine.

Voltage and Current Ratings. Electromechanical relays and SSRs have certain limitations, which determine how much voltage and current each device can safely handle. The values vary from device to device and from manufacturer to manufacturer. Data sheets are used to determine if a given device can safely switch a given load. The advantages of SSRs are that they have a capacity for arcless switching, have no moving parts to wear out, and are totally enclosed, and thus are able to be operated in potentially explosive environments without special enclosures.

The advantage of EMRs is the possibility for replacement of contacts when the device receives an excessive surge current. In an EMR, the contacts may be replaced. In an SSR, the complete device must be replaced.

Voltage Drop. When a set of contacts on an EMR closes, the contact resistance is normally low unless the contacts are pitted or corroded. The SSR, however, being constructed of semiconductor materials, opens and closes a circuit by increasing or decreasing its ability to conduct. Even at full conduction, the device presents some residual resistance, which can create a voltage drop of up to approximately 1.5 V in the load circuit. This voltage drop is usually considered insignificant because it is small in relation to the load voltage, and in most cases presents no problems. This unique feature may have to be taken into consideration when load voltages are small. A method of removing the heat produced at the switching device must be used when load currents are high.

Insulation and Leakage. The air gap between a set of open contacts provides an almost infinite resistance through which no current flows. SSRs, because of their unique construction, provide a very high but measurable resistance when turned OFF. SSRs have a switched OFF resistance not found on EMRs.

It is possible for small amounts of current (OFF-state leakage) to pass through an SSR because some conductance is still possible through an SSR even though it is turned OFF. OFF-state leakage current is not found on EMRs.

Off-state leakage current is the amount of current that leaks through an SSR when the switch is turned OFF, normally about 2 mA to 10 mA. The rating of OFF-state leakage current in an SSR is usually determined at 200 VDC across the output and should not usually exceed more than 200 mA at this voltage.

This leakage current normally presents no problem unless the load device is affected by low values of leakage current. For example, small neon indicator lights and some programmable controllers cannot be switched OFF, and remain ON because of the leakage current.

SOLID-STATE MOTOR STARTERS

A motor can be turned ON and OFF using mechanical contacts or solid-state components. Mechanical contacts normally consist of silver-plated contacts that start and stop the flow of current in a circuit. The contacts either allow full current flow (closed contacts), or allow no current flow (open contacts). Solid-state components consist of SCRs or triacs that allow current flow when they are conducting and stop current flow when they are not conducting.

Solid-state motor starters eliminate mechanical contacts by using solid-state components to turn a motor ON and OFF. Solid-state motor starters are connected into a circuit after the disconnect/overcurrent protection device and before the motor. Solid-state motor starters include motor overload protection and are controlled by the same switches (pushbuttons, pressure switches, etc.) as electromechanical starters. However, in addition to their ability to apply power to and remove power from a motor, solid-state motor starters can also control the amount of voltage and current applied to a motor, and torque produced by a motor during starting and stopping. **See Figure 14-35.**

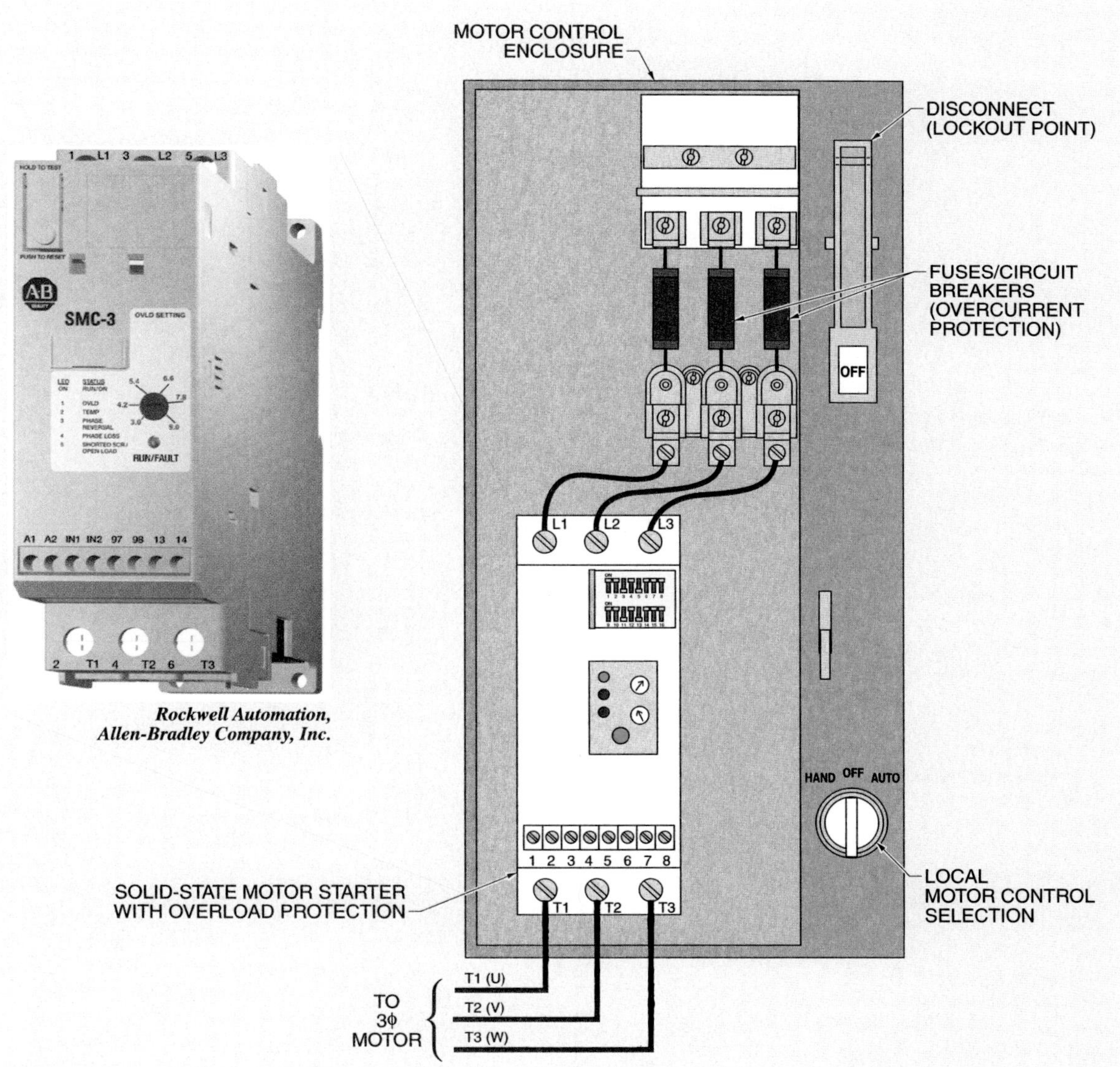

Figure 14-35. Solid-state motor starters eliminate electromechanical components by using solid-state components to turn a motor ON and OFF.

Solid-State Motor Starter Sections

Electromechanical and solid-state motor starters have terminals for connecting the incoming supply power (L1/R, L2/S, L3/T) and terminals for connecting a motor (T1/U, T2/V, T3/W). Solid-state motor starters also include a terminal strip for connecting external inputs (pushbuttons, proximity switches, etc.), a dual inline package (DIP) switchboard for programming starter functions (starting mode/time, stopping mode, etc.), and potentiometers for adjusting motor full-load current (in amps) and trip class. Solid-state motor starters may also include LEDs to provide visual indication of circuit conditions. **See Figure 14-36.**

Wiring Solid-State Motor Starter Power and Control Circuits

The power circuit of a solid-state motor starter is wired by bringing power from the fuses/circuit breaker into the starter. The incoming power must be at the same voltage level for which the motor is rated or wired. The higher the voltage of the power lines, the lower the current for any given motor horsepower rating. **See Figure 14-37.**

Technical Fact

The first transistor was developed in 1947.

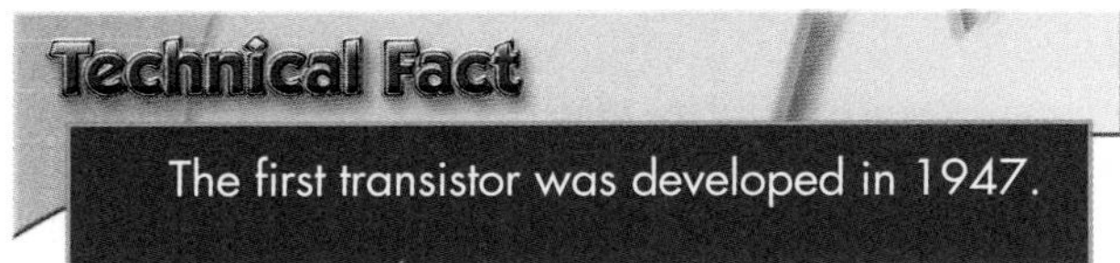

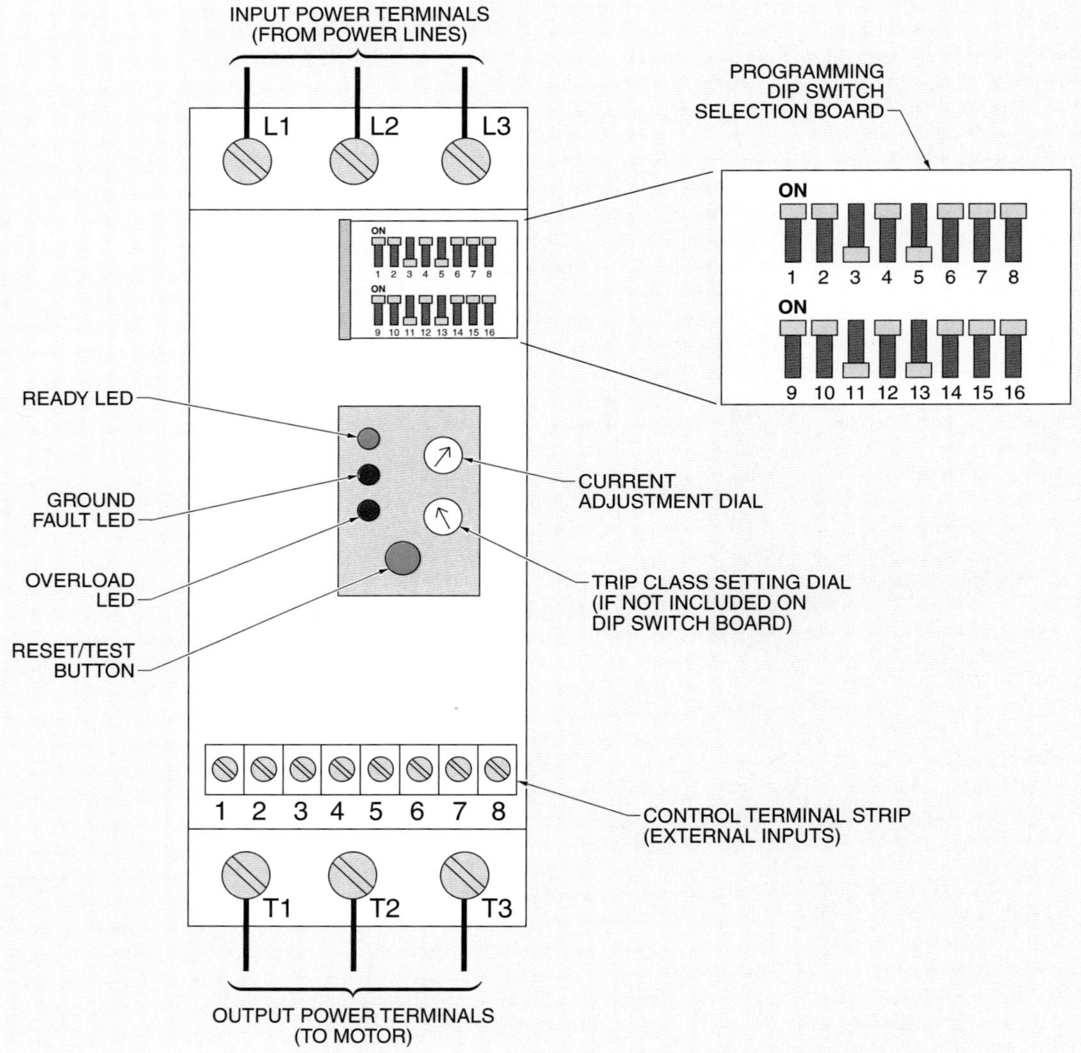

Figure 14-36. Solid-state motor starters have a control terminal strip, input and output power terminals, dials for current and trip class adjustment, and programming DIP switches.

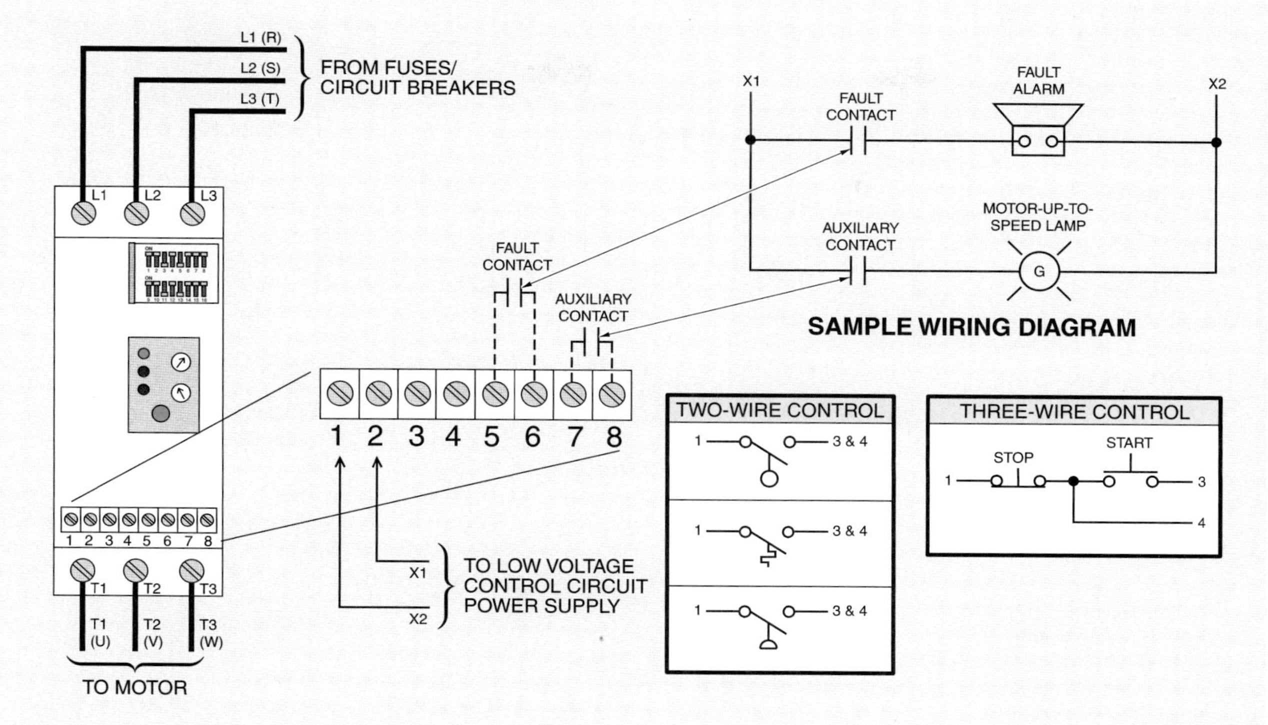

Figure 14-37. The power circuit of a solid-state motor starter is wired by bringing power from the fuses/circuit breaker into the starter. The control circuit is wired to the control terminal strip located on the starter.

The control circuit is wired to the control terminal strip located on the starter. The control circuit voltage is less than the power circuit voltage (typically 12 VDC, 12 VAC, 24 VDC, or 24 VAC). The control terminal strip includes a connection for external control voltage (when required), connections for external control switches (pushbuttons, temperature switches, etc.), and connections for output contacts (alarms, indicating lamps, etc.) that can be used for controlling external loads.

Setting Solid-State Motor Starter Overload Protection

Motors must be protected from overcurrents and overloads. Fuses and circuit breakers (normally located in the motor disconnect) are used to protect a motor from overcurrents (short circuits and high operating currents). Overloads located in the motor starter protect the motor from overload current caused when the load on the motor is greater than the motor design torque rating. Overloads can be thermal overloads (heaters) or solid-state overloads.

Solid-state overloads use a current transformer (CT) to monitor each power line. Solid-state overloads are set by selecting a current limit based on full-load current ratings listed on the motor nameplate, and trip class setting (10, 15, 20, 30). The current limit is set by adjusting the current adjustment dial located on the starter. The *trip class setting* is the length of time it takes for an overload relay to trip and remove power from the motor. The lower the trip class setting, the faster the trip time of the solid-state overload. The higher the trip class setting, the slower the trip time of the solid-state overload.

The trip class setting is based on the motor application (type of load placed on the motor). The trip class setting may be adjusted using a trip class setting dial located near the current adjustment dial, or by using DIP switches. **See Figure 14-38.** *Cold trip* is the trip point from the time the motor starts until the first time the overloads trip (motor operating below nameplate rated current). *Hot trip* is the trip point after the overloads have tripped and have been reset (motor operating near or over nameplate rated current).

Technical Fact

If an overload contact trips due to excessive heater temperature, it is an indication that the motor is nearing its critical temperature and the motor should be allowed to cool down.

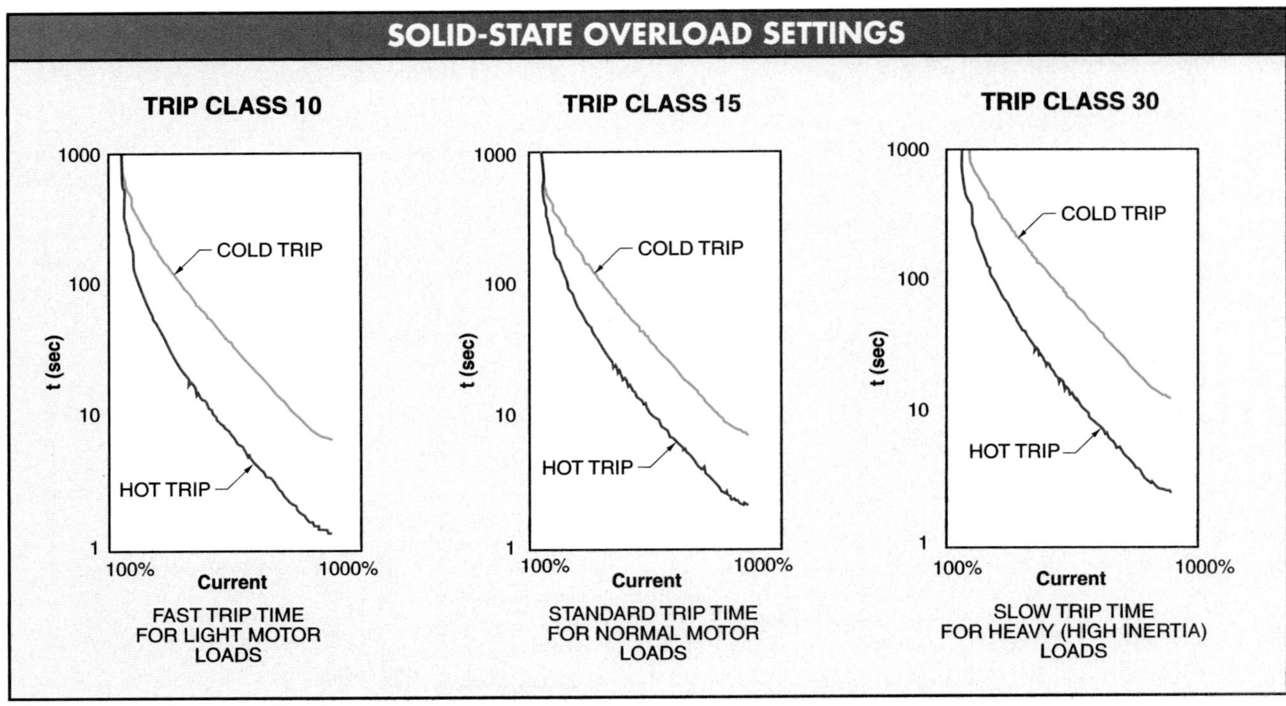

Figure 14-38. The trip class setting of solid-state overloads is based on the motor application (type of load placed on the motor).

Programming Solid-State Motor Starter Operating Functions

A solid-state motor starter must be programmed for proper operation before any power is applied to the starter. A solid-state motor starter is programmed by setting each DIP switch to a predetermined position based on motor and application requirements. The number of DIP switch parameters can range from a few parameters (4 to 6) to numerous parameters. The higher the number of parameters available, the greater the number of applications for which the starter can be used. **See Figure 14-39.**

Technical Fact

A dual in-line package (DIP) switch is a series of tiny switches built into a circuit board to control motor function. DIP switches allow a circuit board to be constructed for many applications. DIP switches have toggle switches that have two possible positions, OFF or ON (0 or 1).

DIP switch parameters include motor starting mode (start time, soft start, start boost, etc.), and the operation of auxiliary contacts (when they are open or closed). Each DIP switch setting must be understood and checked before any power is applied to the starter because some settings can be critical to protecting workers, the motor, and the system. For example, the overload reset function can be placed in a manual or automatic mode. In the manual mode, the reset button on the starter must be pressed before the motor can be restarted manually (by external pushbuttons, etc.). However, in the automatic reset mode, the starter automatically restarts the motor after a short time period if the external control switch (pressure switch, etc.) is still closed. This can cause a safety hazard if the person working on or around the system does not know the motor may automatically restart. For this reason, always refer to the manufacturer literature regarding the setting and meaning of each DIP switch position.

Motor Starting Modes

Electromechanical and solid-state motor starters can be used to start a motor. When an electromechanical motor starter is used, the motor is connected to the full supply voltage. When a motor is connected to full supply voltage, the motor has the highest possible current draw, the highest possible torque applied to the load, and the shortest acceleration time. This operating condition may be acceptable for some loads. However, many loads cannot be started with high starting torque because they control light loads (small parts, etc.) or delicate loads (paper rolls, etc.). High starting current can also damage the power distribution system and trip breakers or blow fuses. Solid-state motor starters can be programmed for different starting modes to help reduce problems caused by full-voltage starting. Motor starting modes include soft start, soft start with start boost, and current limit start. **See Figure 14-40.**

Soft Start. Soft start is the most common solid-state starting method. When a starter is set for soft start, the motor is gradually accelerated over a programmable time period, normally 0 sec to 30 sec. Common start time periods include 2 sec, 4 sec, 6 sec, 8 sec, or 16 sec. The starting torque is adjustable to a percent of the motor's locked rotor torque. Common starting torque settings include 15%, 25%, 50%, or 60%. A soft start helps cushion the stress applied to loads connected to the motor.

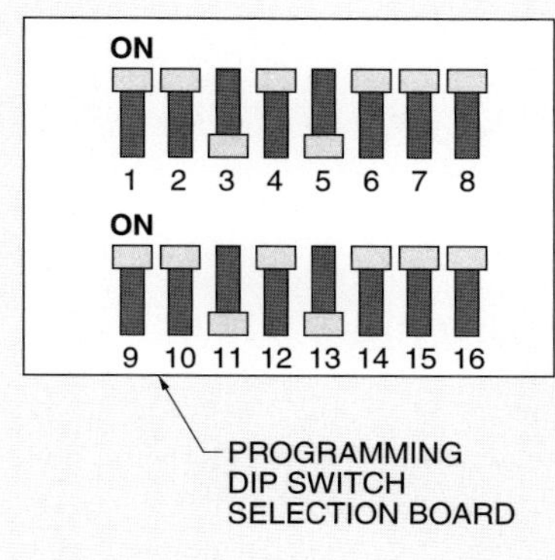

POSITION NUMBER	DESCRIPTION
1	Start time
2	Start time
3	Start mode (current limit or soft start)
4	Current limit start setting (when selected) or soft start initial torque setting (when selected)
5	Current limit start setting (when selected) or soft start initial torque setting (when selected)
6	Soft stop
7	Soft stop
8	Not used
9	Start boost (kick start)
10	Start boost (kick start)
11	Overload class selection
12	Overload class selection
13	Overload reset
14	Auxiliary relay #1 (normal or up-to-speed)
15	Optional auxiliary relay #2 (normal or up-to-speed)
16	Phase rotation check

START TIME		
DIP Switch Number		Time (seconds)
1	2	
OFF	OFF	2
ON	OFF	5
OFF	ON	10
ON	ON	15

START MODE	
DIP Switch Number 3	Setting
OFF	Current limit
ON	Soft start

CURRENT LIMIT START SETTING		
DIP Switch Number		Current Limit % FLA
4	5	
OFF	OFF	150
ON	OFF	250
OFF	ON	350
ON	ON	450

SOFT START INITIAL TORQUE SETTING		
DIP Switch Number		Initial Torque % LRT
4	5	
OFF	OFF	15
ON	OFF	25
OFF	ON	35
ON	ON	65

SOFT STOP		
DIP Switch Number		Setting
6	7	
OFF	OFF	Coast-to-reset
ON	OFF	100% of start time
OFF	ON	200% of start time
ON	ON	300% of start time

START BOOST (KICK START)		
DIP Switch Number		Time (seconds)
9	10	
OFF	OFF	OFF
ON	OFF	.5
OFF	ON	1.0
ON	ON	1.5

OVERLOAD CLASS SELECTION		
DIP Switch Number		Trip Class
11	12	
OFF	OFF	OFF
ON	OFF	10
OFF	ON	15
ON	ON	20

OVERLOAD RESET	
DIP Switch Number 13	Reset
OFF	Manual
ON	Automatic

AUXILIARY RELAY #1	
DIP Switch Number 14	Setting
OFF	Normal
ON	Up-to-speed

OPTIONAL AUXILIARY RELAY #2	
DIP Switch Number 15	Setting
OFF	Normal
ON	Up-to-speed

PHASE ROTATION CHECK	
DIP Switch Number 16	Setting
OFF	Enabled
ON	Disabled

Figure 14-39. A solid-state motor starter is programmed by setting each DIP switch to a predetermined position based on motor and application requirements.

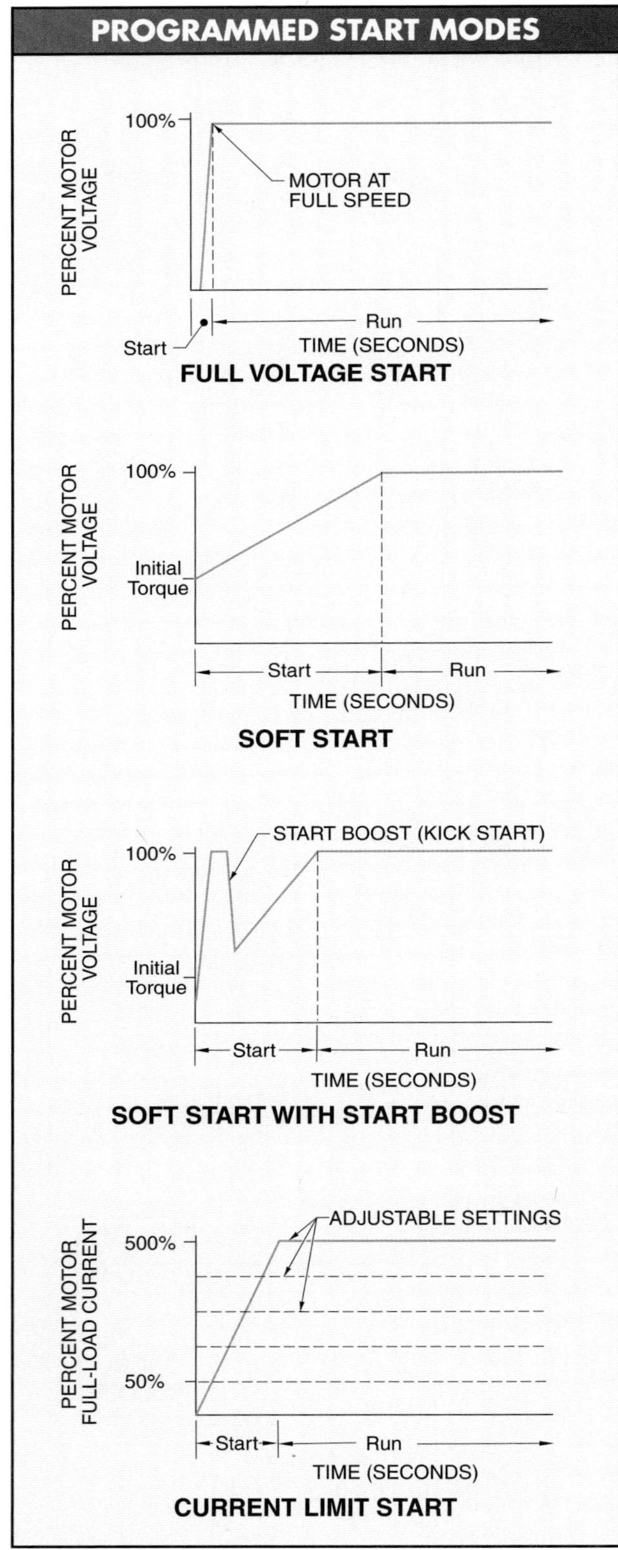

Figure 14-40. Solid-state motor starters can be programmed for different starting modes to help reduce problems caused by full voltage starting.

Soft Start with Start Boost. When a solid-state motor starter is set for soft start with start boost, the motor is given a current pulse during starting to provide additional starting torque for loads that are hard to start. The boost time is usually adjustable from 0 sec to 2 sec. The start boost is normally applied when there is a problem starting a motor using only the soft start starting mode.

Current Limit Start. When a solid-state motor starter is set for current limit starting, the current is limited to a programmed value, normally from 50% to 600% of the motor's full-load current. Common starting current limits include 150%, 250%, 350%, or 450% of full-load current. If the motor is not up to speed after the programmed starting time, the starter applies full voltage to the motor. Current limit starting helps reduce stress on the power distribution system in addition to helping reduce starting torque.

Motor Stopping Modes

Electromechanical and solid-state motor starters can be used to stop a motor when power is removed. When an electromechanical starter is used, the motor coasts to a stop at a rate determined by the load connected to the motor. Solid-state motor starters can be programmed for different stopping modes. This allows greater application flexibility and protection of the motor/load. Motor stopping modes include soft stop, pump control, and brake stop. **See Figure 14-41.**

Soft Stop. Soft stop is the most common solid-state motor starter stopping method. Soft stops allow for an extended controlled stop. In a soft stop, the deceleration time is controlled by the starter, not the load. The soft stop mode is designed for friction loads that tend to stop suddenly when voltage is removed from the motor.

Pump Control. The pump control mode is used to reduce surges that occur when centrifugal pumps are started and stopped. The pump control mode produces smooth acceleration and deceleration of motors and pumps. Common motor and pump starting times range from a few seconds to 30 sec. Common motor and pump stopping times range from a few seconds to 120 sec, depending on the size of the motor and pump.

Brake Stop. Some applications require a fast motor stop. The brake stop mode provides motor braking for a faster stop than a coast stop or soft stop. The amount of braking (and thus braking time) is programmed based on the application requirements. When using the brake stop mode, the longest time is set first and adjusted downward as needed.

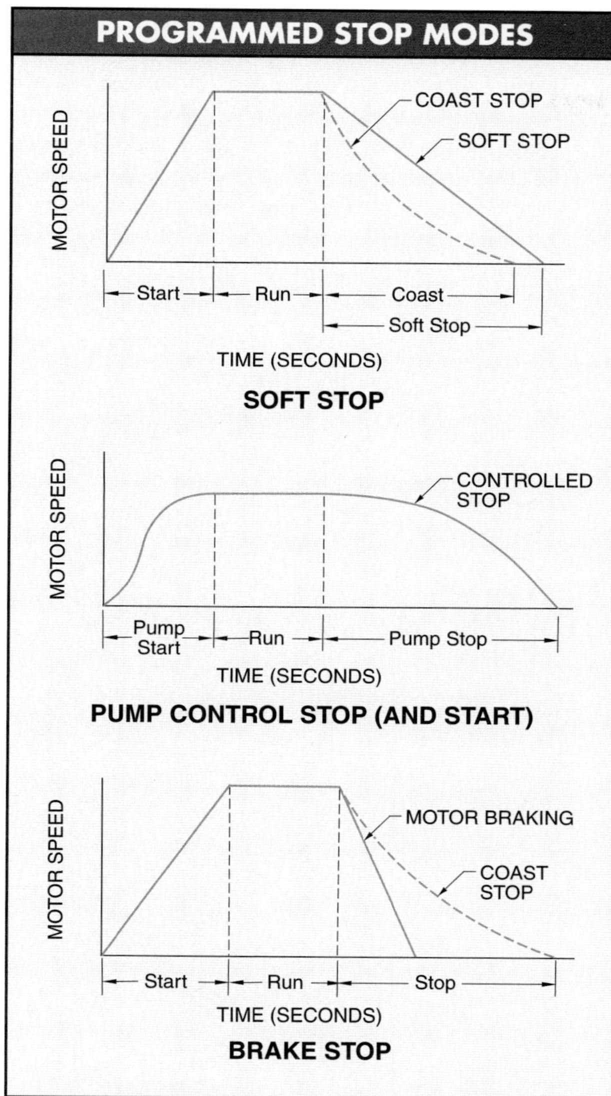

Figure 14-41. Solid-state motor starters can be programmed for different stopping modes to allow greater application flexibility and protection of the motor/load.

TROUBLESHOOTING RELAYS

When troubleshooting EMRs, the input and output of the relay are checked to determine if the circuit on the input side of the relay is the problem, if the circuit on the output side of the relay is the problem, or if the relay itself is the problem. The relay coil and contacts are checked to determine if the relay is the problem. The correct voltage must be applied to the relay's coil before it energizes. The relay contacts are checked by energizing and de-energizing the coil. The contacts should have little or no voltage drop across them when closed. The contacts should have nearly full voltage across them when open.

SSRs require periodic inspection. Dirt, burning, or cracking should not be present on an SSR. Printed circuit (PC) boards should be properly seated. Ensure that the board locking tabs are in place if used. Consider adding locking tabs if a PC board without locking tabs loosens. Check to ensure that any cooling provisions are working and free of obstructions.

Troubleshooting EMRs

Check for contact sticking or binding if the relay is not functioning properly. Tighten any loose parts. Replace any broken, bent, or badly worn parts. Check all contacts for signs of excessive wear and dirt buildup. Contacts are not harmed by discoloration or slight pitting. Vacuum or wipe contacts with a soft cloth to remove dirt. Never use a contact cleaner on relay contacts. Contacts require replacement when the silver surface has become badly worn. Replace all contacts when severe contact wear is evident on any contact. Replacing all contacts prevents uneven and unequal contact closing. Never file a contact.

Relay coils should be free of cracks and burn marks. Replace the coil if there is any evidence of overheating, cracking, melting, or burning. Check the coil terminals for the correct voltage level. Overvoltage or undervoltage conditions of more than 10% should be corrected. Use only replacement parts recommended by the manufacturer when replacing parts of a relay. Using nonapproved parts can void the manufacturer's warranty and may transfer product liability from the manufacturer. Relays are tested by manual operation and by using a digital multimeter (DMM).

Manual Relay Operation

Most relays can be manually operated. Manually operating a relay determines whether the circuit that the relay is controlling (output side) is working correctly. A relay is manually operated by pressing down at a designated area on the relay. This closes the relay contacts. Electromechanical relays may include a push-to-test button. **See Figure 14-42.**

Warning: Use caution when manually operating a relay because loads may start or stop without warning.

When manually operating relay contacts, the circuit controlling the coil is bypassed. Troubleshoot from the relay through the control circuit when the load controlled by the relay operates manually. Troubleshoot the circuit that the relay is controlling if the load controlled by the relay does not operate when the relay is manually operated.

Digital Multimeter Test

A DMM is also used to test an electromechanical relay. A DMM is connected across the input and output side of a relay. Troubleshoot from the input of the relay through the control circuit when no voltage is present at the input side of the relay. The relay is the problem if the relay is not delivering the correct voltage.

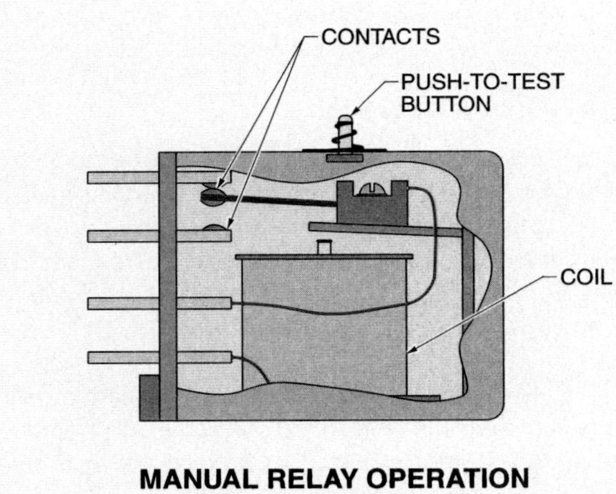

Figure 14-42. Manually operating a relay determines whether the circuit that the relay is controlling (output side) is working correctly.

Troubleshoot from the output of the relay through the power circuit when the relay is delivering the correct voltage. The supply voltage measured across an open contact indicates that the DMM is completing the circuit across the contact. The contacts are not closing and the relay is defective if the voltage measured across the contact remains at full voltage when the coil is energized and de-energized. The contacts are welded closed and the relay is defective if the voltage measured across the contacts remains zero (or very low) when the coil is energized and de-energized. **See Figure 14-43.** To troubleshoot an electromechanical relay, apply the procedure:

1. Measure the voltage in the circuit containing the control relay coil. The voltage should be within 10% of the voltage rating of the coil. The relay coil cannot energize if the voltage is not present. The coil may not energize properly if the voltage is not at the correct level. Troubleshoot the power supply when the voltage level is incorrect.
2. Measure the voltage across the control relay coil. The voltage across the coil should be within 10% of the coil's rating. Troubleshoot the switch controlling power to the coil when the voltage level is incorrect.
3. Measure the voltage in the circuit containing the control relay contacts. The voltage should be within 10% of the rating of the load. Troubleshoot the power supply if the voltage level is incorrect.
4. Measure the voltage across the control relay contacts. The voltage across the contacts should be less than 1 V when the contacts are closed and nearly equal to the supply voltage when open. The contacts have too much resistance and are in need of service if the voltage is more than 1 V when the contacts are closed. Troubleshoot the load when the voltage is correct at the contacts and the circuit does not work.

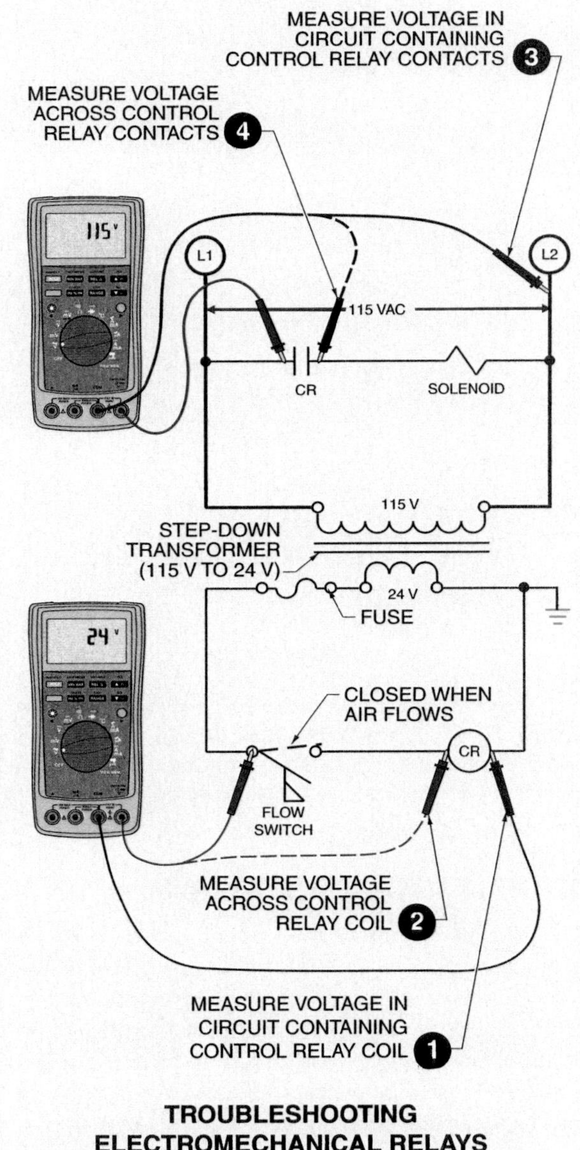

Figure 14-43. A DMM is connected across the input and output side to test an EMR.

Technical Fact

Use caution when troubleshooting solid-state relays because they may fail in the "short circuit" condition on the outputs. This creates an unsafe situation because, if the circuit is shorted, unexpected current is flowing in the circuit, which could cause injury.

Troubleshooting SSRs

Troubleshooting an SSR is accomplished by either the exact replacement method or the circuit analysis method. The *exact replacement method* is a method of SSR replacement in which a bad relay is replaced with a relay of the same type and size. The exact replacement method involves making a quick check of the relay's input and output voltages. The relay is assumed to be the problem and is replaced when there is only an input voltage being switched.

The *circuit analysis method* is a method of SSR replacement in which a logical sequence is used to determine the reason for the failure. Steps are taken to prevent the problem from recurring once the reason for a failure is known. The circuit analysis method of troubleshooting is based on three improper relay operations, which are:

- The relay fails to turn OFF the load
- The relay fails to turn ON the load
- The relay operates erratically

Relay Fails to Turn OFF Load

A relay may not turn OFF the load to which it is connected when a relay fails. This condition occurs either when the load is drawing more current than the relay can withstand, the relay's heat sink is too small, or transient voltages are causing a breakover of the relay's output. A *transient voltage* is a temporary, unwanted voltage in an electrical circuit. Overcurrent permanently shorts the relay's switching device if the load draws more current than the rating of the relay. High temperature causes thermal runaway of the relay's switching device if the heat sink does not remove the heat.

Replace the relay with one of a higher voltage rating and/or add a transient suppression device to the circuit if the power lines are likely to have transients (usually from inductive loads connected on the same line). **See Figure 14-44.** To troubleshoot an SSR that fails to turn OFF a load, apply the procedure:

1. Disconnect the input leads from the SSR. See Step 3 if the relay load turns OFF. The relay is the problem if the load remains ON and the relay is normally open.
2. Measure the voltage of the circuit that the relay is controlling. The line voltage should not be higher than the rated voltage of the relay. Replace the relay with a relay that has a higher voltage rating if the line voltage is higher than the relay's rating. Check to ensure that the relay is rated for the type of line voltage (AC or DC) being used.
3. Measure the current drawn by the load. The current draw must not exceed the relay's rating. For most applications, the current draw should not be more than 75% of the relay's maximum rating.
4. Reconnect the input leads and measure the input voltage to the relay at the time when the control circuit should turn the relay OFF. The control circuit is the problem and needs to be checked if the control voltage is present. The relay is the problem if the control voltage is removed and the load remains ON. Before changing the relay, ensure that the control voltage is not higher than the relay's rated limit when the control circuit delivers the supply voltage. Ensure that the control voltage is not higher than the relay's rated drop-out voltage when the control circuit removes the supply voltage. This condition may occur in some control circuits using solid-state switching.

Relay Fails to Turn ON Load

A relay may fail to turn ON the load to which it is connected when the relay fails. This condition occurs when the relay's switching device receives a very high voltage spike or the relay's input is connected to a higher-than-rated voltage. A high voltage spike blows open the relay's switching device, preventing the load from turning ON. Excessive voltage on the relay's input side destroys the relay's electronic circuit.

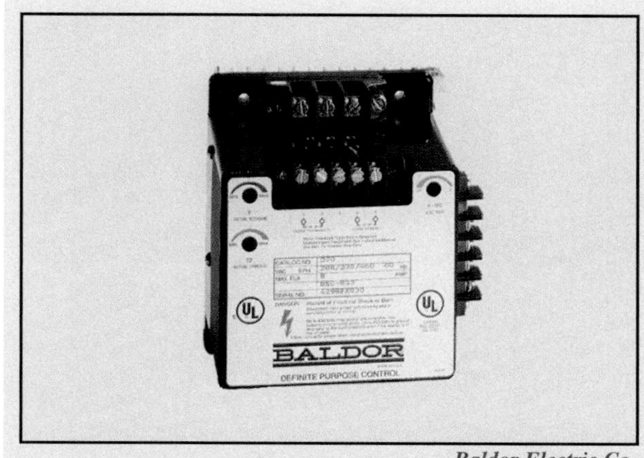

Baldor Electric Co.

A solid-state relay with timer allows the relay to be set so that the start or stop function occurs only after a set time. Timers can be set to energize or de-energize a system.

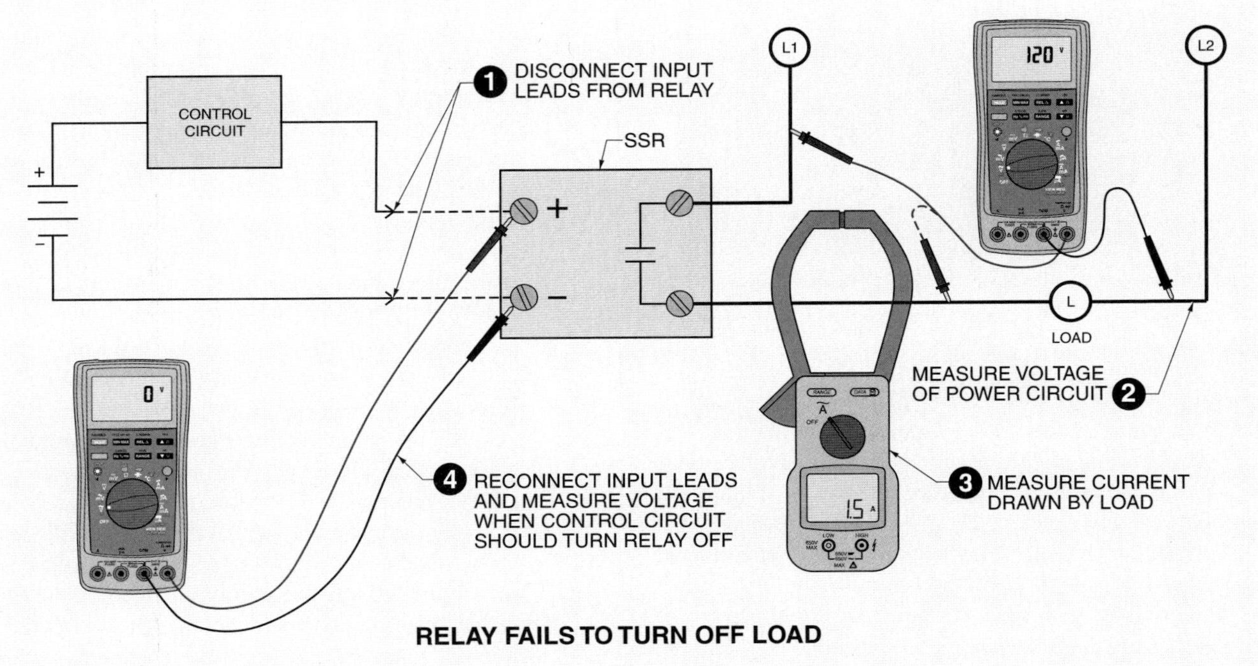

Figure 14-44. A relay may not turn OFF the load to which it is connected when the relay fails.

Baldor Electric Co.

Solid-state motor starters are used on applications such as hoisting equipment for raising and lowering, for horizontal movement, and when soft starts and stops are required.

Replace the relay with one that has a higher voltage and current rating and/or add a transient suppression device to the circuit if the power lines are likely to have high voltage spikes. **See Figure 14-45.** To troubleshoot an SSR that fails to turn ON a load, apply the procedure:

1. Measure the input voltage when the relay should be ON. Troubleshoot the circuit ahead of the relay's input if the voltage is less than the relay's rated pickup voltage. The circuit ahead of the relay is the problem if the voltage is greater than the relay's rated pickup voltage. The higher voltage may have destroyed the relay. The relay may be a secondary problem caused by the primary problem of excessive applied voltage. Correct the high-voltage problem before replacing the relay. The relay or output circuit is the problem if the input voltage is within the pickup limits of the relay.
2. Measure the voltage at the output of the relay. The relay is probably the problem if the relay is not switching the voltage. See Step 3. The problem is in the output circuit if the relay is switching the voltage. Check for an open circuit in the load.
3. Insert a DMM set to measure current in series with the input leads of the relay. Measure the current when the relay should be ON. The relay input is open if no current is flowing. Replace the relay. The relay is bad if the current flow is within the relay's rating. Replace the relay. The control circuit is the problem if current is flowing but is less than that required to operate the relay.

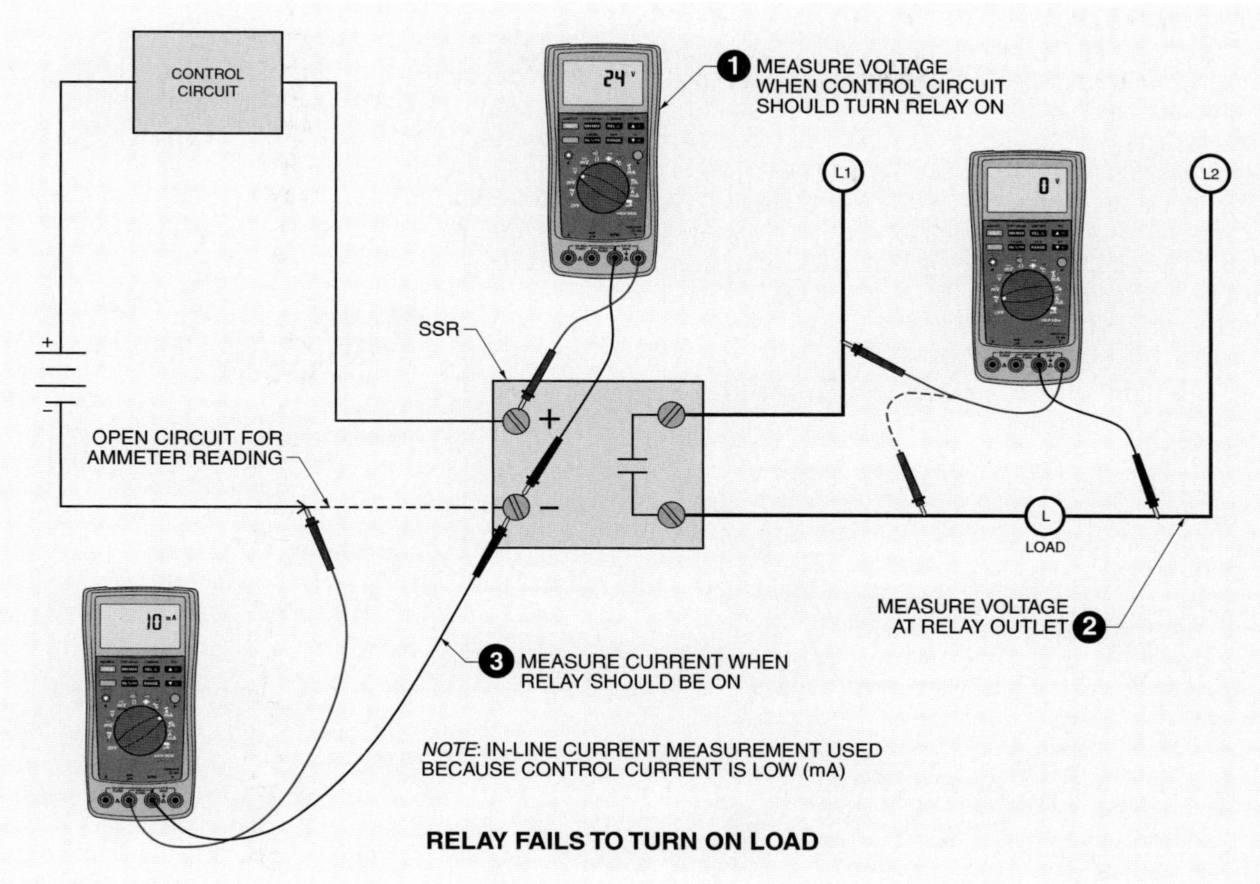

Figure 14-45. A relay may fail to turn ON the load to which it is connected when the relay fails.

Erratic Relay Operation

Erratic relay operation is the proper operation of a relay at times, and the improper operation of the relay at other times. Erratic relay operation is caused by mechanical problems (loose connections), electrical problems (incorrect voltage), or environmental problems (high temperature). **See Figure 14-46.** To troubleshoot erratic relay operation, apply the procedure:

1. Check all wiring and connections for proper wiring and tightness. Loose connections cause many erratic problems. No sign of burning should be present at any terminal. Burning at a terminal usually indicates a loose connection.
2. Ensure that the input control wires are not next to the output line or load wires. The noise carried on the output side may cause unwanted input signals.
3. The relay may be half-waving if the load is a chattering AC motor or solenoid. *Half-waving* is a phenomenon that occurs when a relay fails to turn OFF because the current and voltage in the circuit reach zero at different times. Half-waving is caused by the phase shift inherent in inductive loads. The phase shift makes it difficult for some solid-state relays to turn OFF. Connecting an RC or another snubber circuit across the output load should allow the relay to turn OFF. An *RC circuit* is a circuit in which resistance (R) and capacitance (C) are used to help filter the power in a circuit.

Technical Fact

Solid-state relays have many features which electromechanical relays do not such as long life, shock/vibration resistance, no generation of RFI or EMI, no contact bounce, arcless switching, no acoustic noise, zero voltage switching, IC compatibility, and immunity to prolonged exposure to salt spray, dirt, debris, humidity, and water.

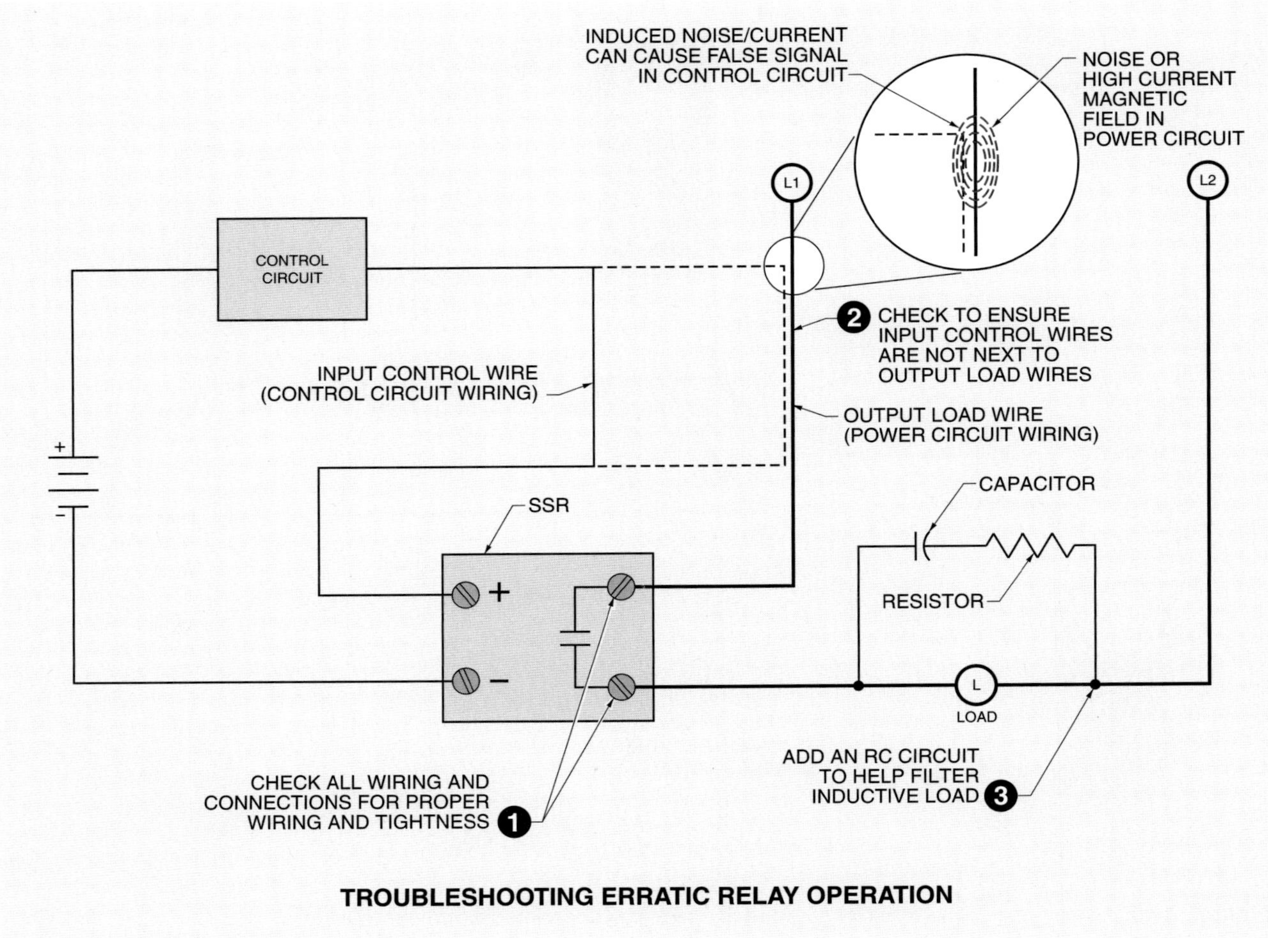

Figure 14-46. Erratic relay operation is the proper operation of a relay at times, and the improper operation of the relay at other times.

Review Questions

1. What are the two major types of relays?
2. What type of relay is activated by the presence of a magnetic field?
3. Do reed relays normally have more than one set of contacts?
4. Do reed relays normally have NC contacts?
5. Do general-purpose relays normally have more than one set of contacts?
6. Do general-purpose relays normally have NC contacts?
7. What does SPDT stand for?
8. What does 3PDT stand for?
9. How many breaks can contacts have?
10. What is the name of a contact that can carry current through two circuits simultaneously?
11. How many circuits can single-throw contacts control?
12. What are convertible contacts?
13. Does an electromechanical relay normally have a higher electrical life rating or a higher mechanical life rating than a solid-state relay?
14. What is sulfidation?
15. What type of SSR is designed to control resistive loads and is the most widely used relay?
16. What type of SSR turns ON a load immediately when the control voltage is present?
17. What is a ramp-up function?
18. What output is used in SSRs to switch high-current DC loads?
19. What output is used in SSRs to switch low-current AC loads?
20. What output is used in SSRs to switch low-current DC loads?
21. What happens to the failure rate of SSRs as the temperature rises?
22. What is thermal resistance?
23. What is OFF-state leakage current?
24. What are transient voltages?
25. What is half-waving?

Sensing Devices and Controls

Photoelectric sensors detect the presence of an object by means of a beam of light. Ultrasonic sensors operate by emitting and receiving high-frequency sound waves. Proximity sensors detect the presence of an object by means of an electronic sensing field. Photoelectric, ultrasonic, and proximity sensors are used in a variety of applications for the detection of almost any object.

PHOTOELECTRIC SENSORS

A *photoelectric sensor (photoelectric switch)* is a solid-state sensor that can detect the presence of an object without touching the object. A photoelectric sensor detects the presence of an object by means of a beam of light. Photoelectric sensors can detect most materials, and have a longer sensing distance than ultrasonic and proximity sensors. Depending on the model, photoelectric sensors can detect objects from several millimeters to over 100′ away.

The maximum sensing distance of any sensor is determined by the size, shape, color, and character of the surface of the object to be detected. Many sensors include an adjustable sensing distance, making it possible to exclude detection of the background of the object. Photoelectric sensors are used in applications in which the object to be detected is excessively light, heavy, hot, or untouchable. **See Figure 15-1.**

Scanning Techniques

Photoelectric sensors are comprised of two separate major components: a light source (phototransmitter) and a photosensor (photoreceiver). The light source emits a beam of light and the photosensor detects the beam of light. The light source and photosensor may be housed in the same enclosure or in separate enclosures. *Scanning* is the process of using the light source and photosensor together to measure a change in light intensity when a target is present in, or absent from, the transmitted light beam.

When the photosensor detects the target, it sends a signal to the control circuit. The control circuit processes the signal and activates a solid-state output switch (thyristor or transistor). The output switch energizes or de-energizes a solenoid, relay, magnetic motor starter, or other load.

The phototransmitter and photoreceiver may be set up for several different scanning techniques. The best technique depends on the particular application.

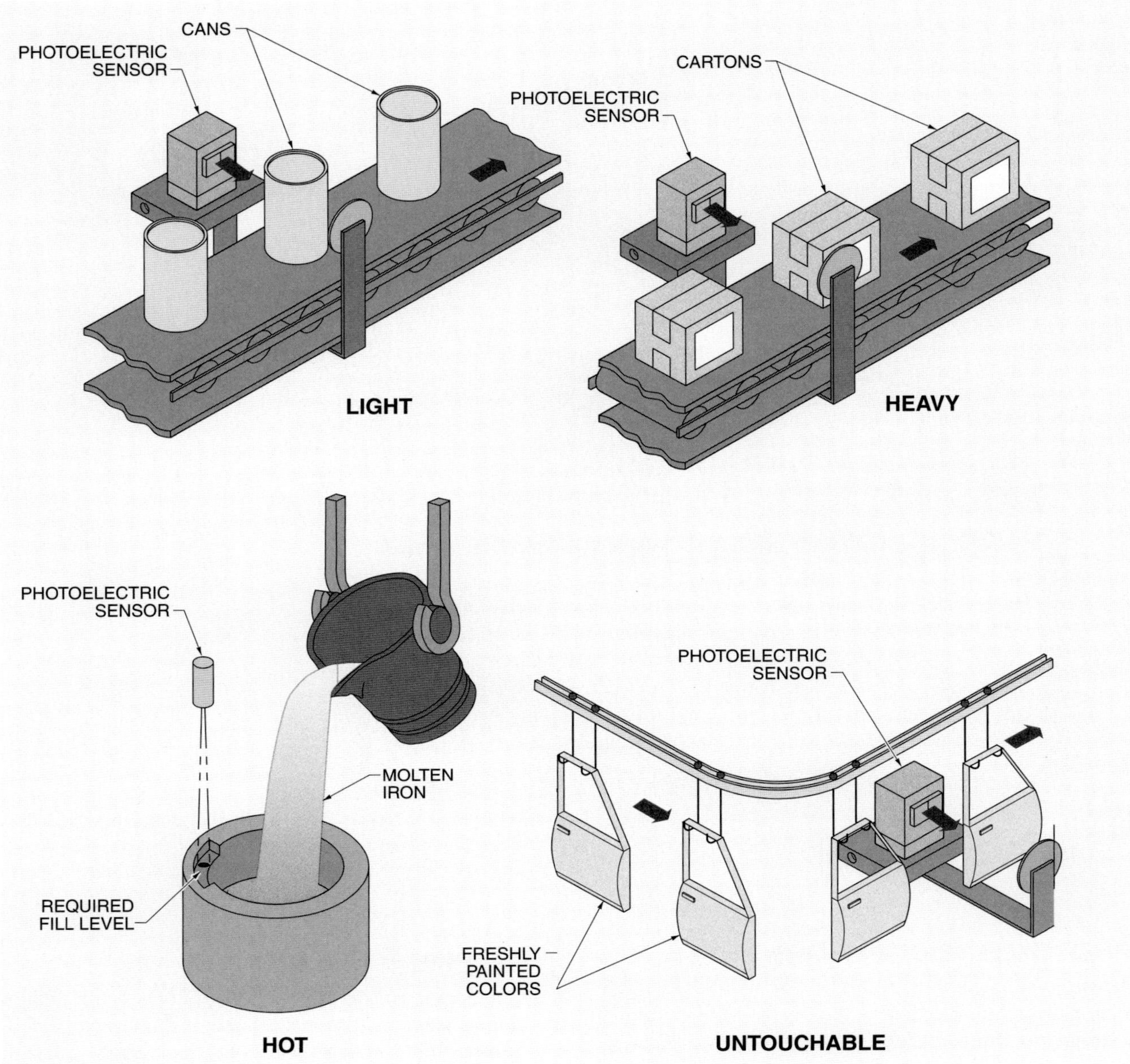

Figure 15-1. Photoelectric sensors are used to detect objects without touching the object.

Factors that determine the best scanning technique include the following:

- Scanning distance. An application may require the target to be a few millimeters or several feet away.
- Size of target. The target may be as small as a needle or as large as a truck.
- Reflectance level of target. All targets reflect the transmitted light. Light targets reflect more light than dark targets.
- Target positioning. Targets may enter the detection area in the same position or in different positions.
- Differences in color and reflective properties between the background and the target. The transmitted light beam is reflected by the background as well as the target.
- Changes in the ambient light intensity. The photosensor may be affected by the amount of natural (or artificial) light at the detection area.
- Condition of the surrounding air. The transmitted light beam is affected by the quality (amount of impurities) of the air which the transmitted light beam must travel through. Impurities reduce the range of the transmitted light beam.

The most common scanning techniques used with photoelectric switches are the direct, retroreflective, polarized, specular, diffuse, and convergent beam scanning methods.

Direct Scan. *Direct scan (transmitted beam, thru-beam, opposed scan)* is a method of scanning in which the transmitter and receiver are placed opposite each other so that the light beam from the transmitter shines directly at the receiver. The target must pass directly between the transmitter and receiver. The target size should be at least 50% of the diameter of the receiver lens to block enough light for detection. **See Figure 15-2.** For very small targets, a special converging lens or aperture may be used.

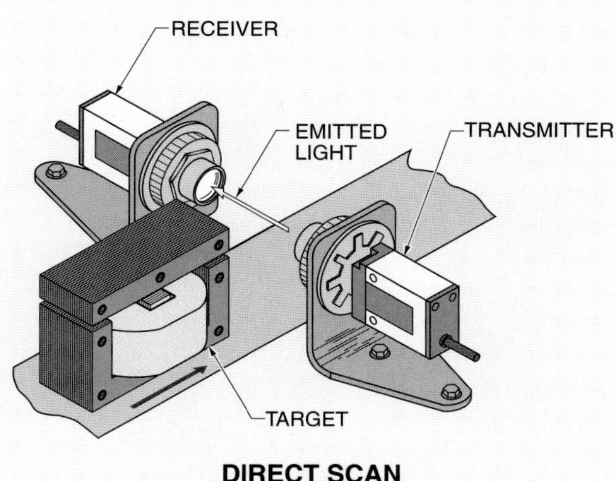

Figure 15-2. Direct scan is a method of scanning in which the target is detected as it passes between the transmitter and receiver.

The direct scan method should generally be the first choice for scanning targets that block most of the light beam. Because the light beam travels in only one direction, direct scan provides long-range sensing and works well in areas of heavy dust, dirt, mist, etc. Direct scan may be used at distances of over 100′.

Retroreflective Scan. *Retroreflective scan (retro scan)* is a method of scanning in which the transmitter and receiver are housed in the same enclosure and the transmitted light beam is reflected back to the receiver from a reflector. The light beam is directed at a reflector, which returns the light beam to the receiver when no target is present. When a target blocks the light beam, the output switch is activated. **See Figure 15-3.**

Alignment is not critical with retroreflective scan. Misalignment of a reflector of up to 15° does not normally affect operation. This makes it a good choice for high vibration applications. Retroreflective scan is used in applications in which sensing is possible from only one side at distances up to about 40′. Retroreflective scan does not work well in applications that require the detection of translucent or transparent materials.

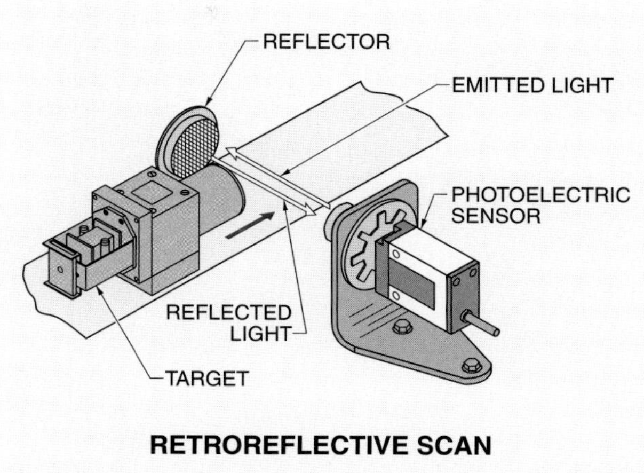

Figure 15-3. Retroreflective scan is a method of scanning in which the target is detected as it passes between the photoelectric sensor and reflector.

Polarized Scan. *Polarized scan* is a method of scanning in which the receiver responds only to the depolarized reflected light from corner cube reflectors or polarized sensitive reflective tape. The light source (emitter) and photoreceiver in a polarized scanner are located on the same side of the object to be detected. A special lens filters the emitter's beam of light so that it is projected in one plane only. **See Figure 15-4.** The receiver ignores the light reflected from most varieties of shrink-wrap materials, shiny luggage, aluminum cans, or common reflective objects. Thus, the receiver picks up the reflection from the reflector, but cannot pick up the reflection from most shiny targets.

Specular Scan. *Specular scan* is a method of scanning in which the transmitter and receiver are placed at equal angles from a highly reflective surface. With reflective surfaces, the angle at which light strikes the reflecting surface equals the angle at which it reflects from the surface. This is similar to billiards in which a ball leaves the cushion at an angle equal to the angle at which it struck the cushion. **See Figure 15-5.**

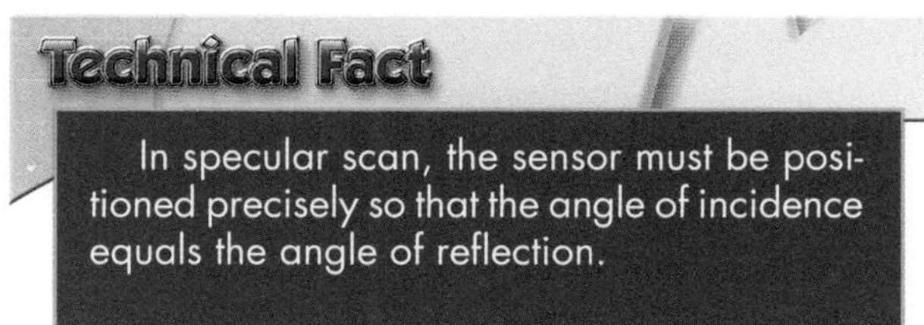

In specular scan, the sensor must be positioned precisely so that the angle of incidence equals the angle of reflection.

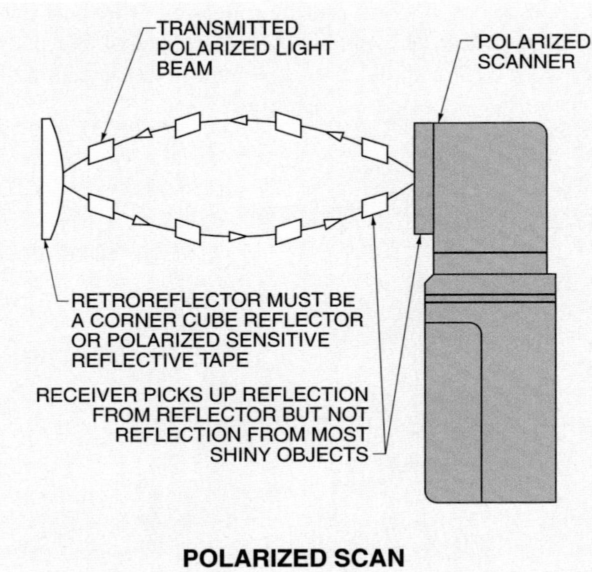

Figure 15-4. Polarized scan uses a special lens that filters the emitter's beam of light so it is projected in one plane only.

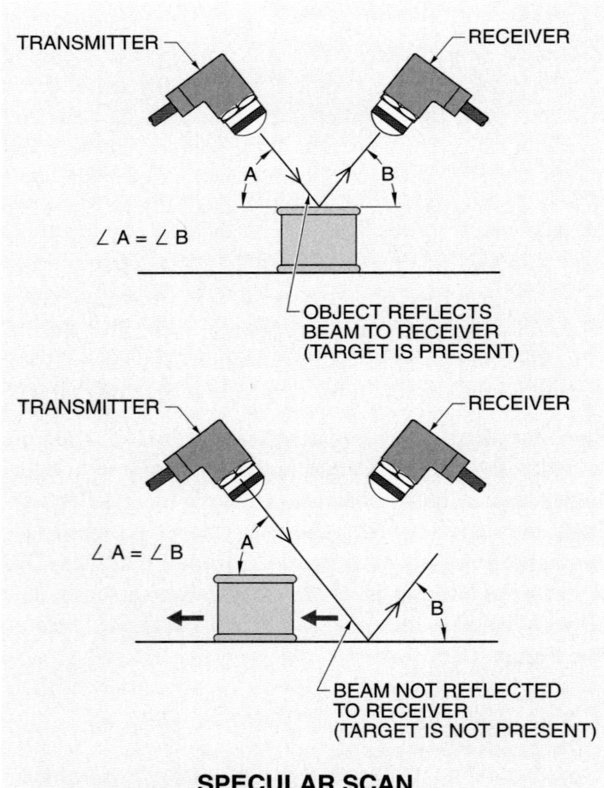

Figure 15-5. Specular scan is a method of scanning in which the transmitter and receiver are placed at equal angles from a highly reflective surface.

Specular scan distinguishes between shiny and nonshiny (matte) surfaces. For example, specular scan may detect a break when printing a newspaper. The newspaper may be moved over a stainless steel plate in which a photoelectric sensor is positioned. When a break occurs, the photoelectric sensor detects the break and stops the press.

Diffuse Scan. *Diffuse scan (proximity scan)* is a method of scanning in which the transmitter and receiver are housed in the same enclosure and a small percentage of the transmitted light beam is reflected back to the receiver from the target. In diffuse scan, the transmitter and the receiver are placed in the same enclosure so the receiver picks up some of the diffused (scattered) light. The target detected may be large or small. **See Figure 15-6.**

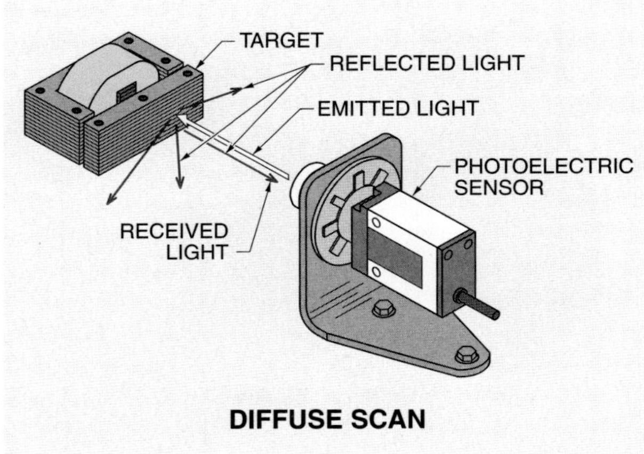

Figure 15-6. Diffuse scan is a method of scanning in which the target is detected when some of the emitted, reflected light is received.

Diffuse scan is used in color mark detection to detect the amount of light that is reflected from a printed surface. Color marks (registration or index marks) are used for registering a specific location on a product. For example, registration marks are used in packaging applications to determine the cutoff point and to identify the point for adding printed material.

The color of a registration mark is selected to provide enough contrast so that the diffuse scanner can detect the difference between the registration mark and the background material. Black marks against a white background provide the best contrast. However, to provide a better selection of sensors, manufacturers offer transmitters with infrared, visible red, green, or white light sources. By using different colors of transmitted light, many different color registration marks may be used with different color backgrounds.

Convergent Beam Scan. *Convergent beam scan* is a method of scanning that simultaneously focuses and converges a light beam to a fixed focal point in front of the photoreceiver. **See Figure 15-7.**

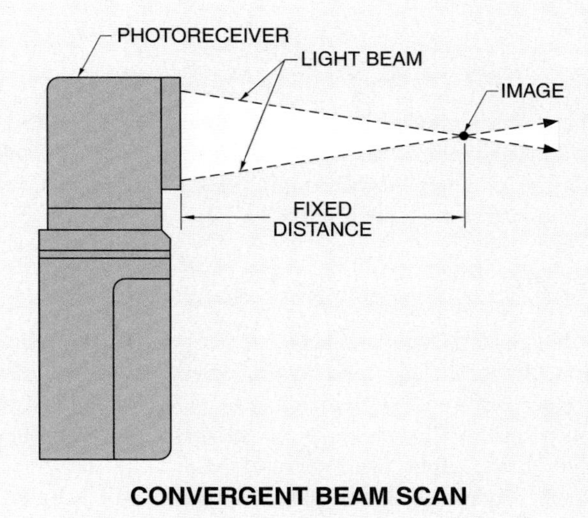

Figure 15-7. Convergent beam scan is a method of scanning which simultaneously focuses and converges a light beam to a fixed focal point in front of the photoreceiver.

Convergent beam scanning is used to detect products that are inches away from another reflective surface. It is a good choice for edge-guiding or positioning clear or translucent materials. The well-defined beam makes convergent beam scanning a good choice for position sensing of opaque materials.

The convergent beam scanner's optical system can only sense light reflected back from an object in its focal point. The scanner is blind a short distance before and beyond the focal point. Operation is possible when highly reflective backgrounds are present. Convergent beam scanning is used for detecting the presence or absence of small objects while ignoring nearby background surfaces.

Parts on a conveyor can be sensed from above while ignoring the conveyor belt. Parts may also be sensed from the side without detecting guides or rails directly in back of the object. Convergent beam scanning can detect the presence of fine wire, resistor leads, needles, bottle caps, pencils, stack height of material, and fill level of clear liquids. It is also capable of sensing bar code marks against a contrasting background.

Fiber Optics

Fiber optics is a technology that uses a thin flexible glass or plastic optical fiber (POF) to transmit light. Optical fiber cables are used with photoelectric sensors to conduct the transmitted light into and out of the sensing area. The optical fiber cables are used as light pipes.

Glass optical fiber cables can withstand much higher temperatures than POF cables. Glass cables can typically withstand temperatures up to 500°F. POF cables are usually limited to about 158°F. POF cable is more suitable than glass optical fiber cable for applications that require severe bending over short distances because POF is less susceptible to breakage than glass optical fiber cables. Although POF is less expensive than glass, glass performs better than POF, offering better transmission quality at higher speeds over longer distances.

The control beam is transmitted through an optical fiber cable and returned to the receiver through a separate cable either combined in the same cable assembly (known as a bifurcated fiber bundle) or within a separate cable assembly. Retroreflective scan and diffuse scan use a bifurcated cable and direct scan uses two separate cables (emitter and receiver). Scan distances commonly vary from 0.4″ to 54″, depending on the scanning technique. **See Figure 15-8.** An optical lens accessory that attaches to certain cable ends significantly increases scan distances.

Fiber-optic controllers are available in different sizes and configurations. **See Figure 15-9.** Combining the optical fiber cables with photoelectric controls enables use in limited mounting spaces and for small parts detection, and detection in applications having high temperature, high vibration, or high electrical noise levels.

Banner Engineering Corp.

Sensors are available in sensing modes including retroreflective, diffuse, convergent beam, and glass fiber optic.

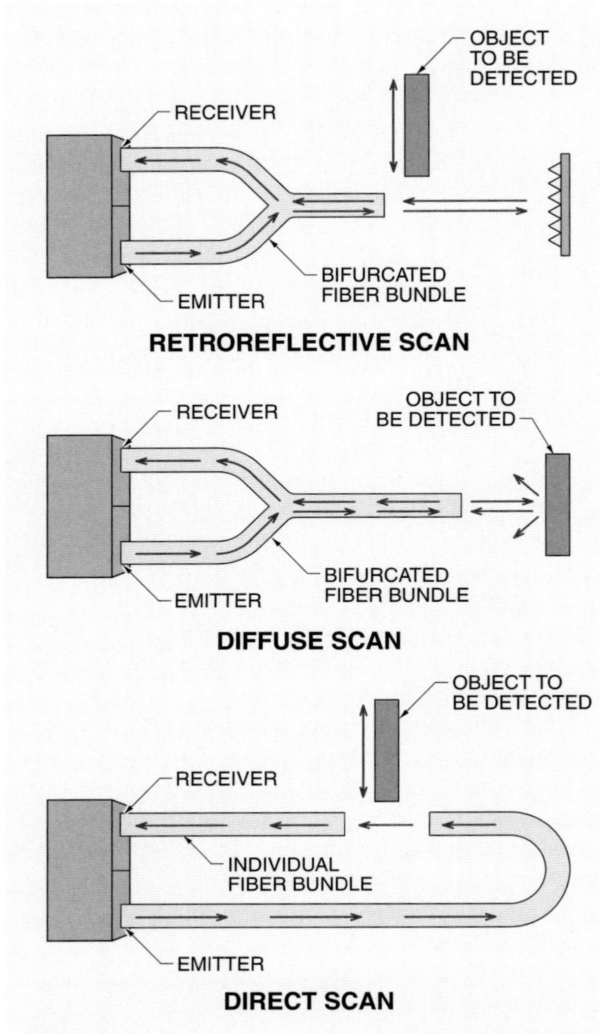

Figure 15-8. Fiber optics use transparent fibers of glass or plastic to conduct and guide light energy.

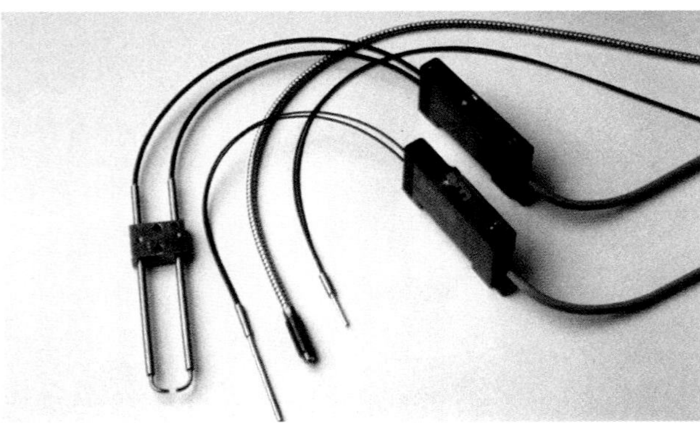

Honeywell

Figure 15-9. Fiber-optic controllers are available in different sizes and configurations.

The light emitting and receiving components in fiber optics are located remotely at the control's housing, and only optical fiber cables are exposed to the severe environmental or hazardous areas. The optical fiber cables are not adversely affected by electrical noise because they carry signals in the form of light.

Selection of Scanning Methods

The scanning method used for an application depends on the environment of the scanning area. In many applications, several methods of scanning work, but normally one method is the best. **See Figure 15-10.**

Modulated and Unmodulated Light

Although some older photoelectric transmitters used white light (unmodulated), modern photoelectric light sources produce infrared light which can be either modulated or unmodulated. Most photoelectric sensors today use modulated infrared light. **See Figure 15-11.**

In a modulated light source, the light source is turned ON and OFF at a very high frequency, normally several kilohertz (kHz). The control responds to this modulated frequency rather than just the intensity of the light. Because the receiver circuitry is tuned to the phototransmitter modulating frequency, the control does not respond to ambient light. This feature also helps to reject other forms of light (noise). A modulated light source should always be considered first when using a photoelectric control.

In an unmodulated light source, the light beam is constantly ON and is not turned ON and OFF. Unmodulated light is considered when the scanning range is very short and when dirt, dust, and bright ambient light conditions are not a problem. Unmodulated light sources are also used for high-speed counting because the beam is continually transmitting and responds quickly. Most manufacturers offer both types of photoelectric controls.

Technical Fact

A fiber-optic system is the combination of a fiber-optic photoelectric sensor with a glass optical fiber cable. Components for fiber-optic systems are assembled in many combinations to fit a particular application. Fiber-optic components are also custom made for many sensing applications. For maximum reliability, many fibers are formed to a specific shape to match a particular object to be sensed.

SCANNING METHODS			
Methods	**Features**		
	Configuration	**Advantages**	**Disadvantages**
Direct	Transmitter on one side sends signal to receiver on other side; object to be detected passes between the transmitter and receiver	Reliable performance in contaminated areas; long range scanning; most well-defined, effective beam of all scanning techniques	Wiring and alignment required for both transmitter and receiver; high installation cost
Retroreflective	Transmitter and receiver are housed in one package and are placed on same side of object to be detected; signal from transmitter is reflected to receiver by retroreflector	Ease of installation in that wiring on only one side is required; alignment need not be exact; more tolerant to vibration	Sensitive to contamination since light source must travel to retroreflector and back; hard to detect transparent or translucent materials; not good for small part detection
Polarized	Transmitter and receiver housed in one package and placed on side of object to be detected; special lens is used to filter light beam to project it in one plane only	Only depolarized light from transmitter is detected, ignoring other unwanted light sources	Detection distance and plane of detection limited
Specular	Transmitter sends signal to receiver by reflecting signal of object to be detected; transmitter and receiver are not housed in same package and receiver must be positioned precisely to receive reflected light	Good for detecting shiny versus dull surfaces; depth of field can be changed by changing transmitter/receiver angle	Wiring required for both transmitter and receiver; proper alignment is important
Diffuse	Transmitter and receiver are housed in one package; object being detected reflects signal back to detector; no retroreflector is used	Ease of installation in that wiring on only one side is required since detected object returns signal; exact alignment is not critical; best scanning technique for transparent or translucent materials	Limited range since object is used to reflect transmitted light; performance changes from one type of object to be detected to another
Convergent Beam	Transmitter and receiver are housed in one package and are placed on side of object to be detected; light beam is focused to a fixed point in front of controller	Detection point is fixed so that objects before or beyond focal point are not detected	Detection point is very small, not allowing for that much variation in distance that may be caused by such factors as vibration

Figure 15-10. The scanning method used for an application depends on the environment of the scanning area.

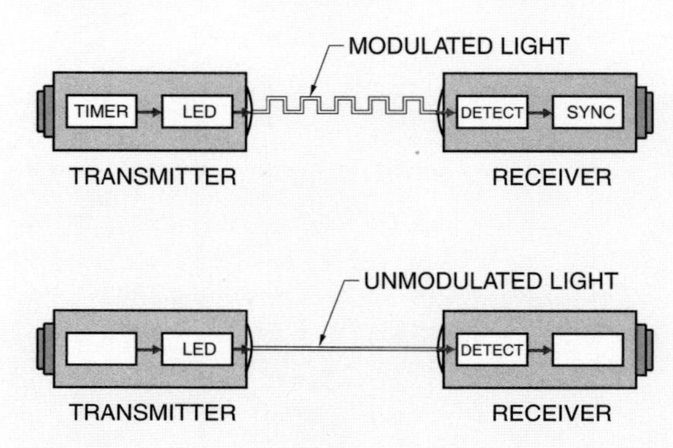

Figure 15-11. Photoelectric controls use modulated or unmodulated light.

Response Time

Response time is the number of pulses (objects) per second a controller can detect. Response time of a photoelectric control must be considered when the object to be detected moves past the beam at a very high speed or when the object to be detected is not much bigger than the effective beam of the controller. This information is listed in the specification sheet of the photoelectric control. For example, a photoelectric control may have an activating frequency of 10 pulses per second. This means the photoelectric control, on average, can detect an object passing by it every 1⁄10 (.1) sec. **See Figure 15-12.**

The beam must be totally blocked before the receiver shuts OFF. The receiver turns ON when the object uncovers an edge of the beam. This has the effect of shortening the size of the object to be detected as seen by the photoelectric control.

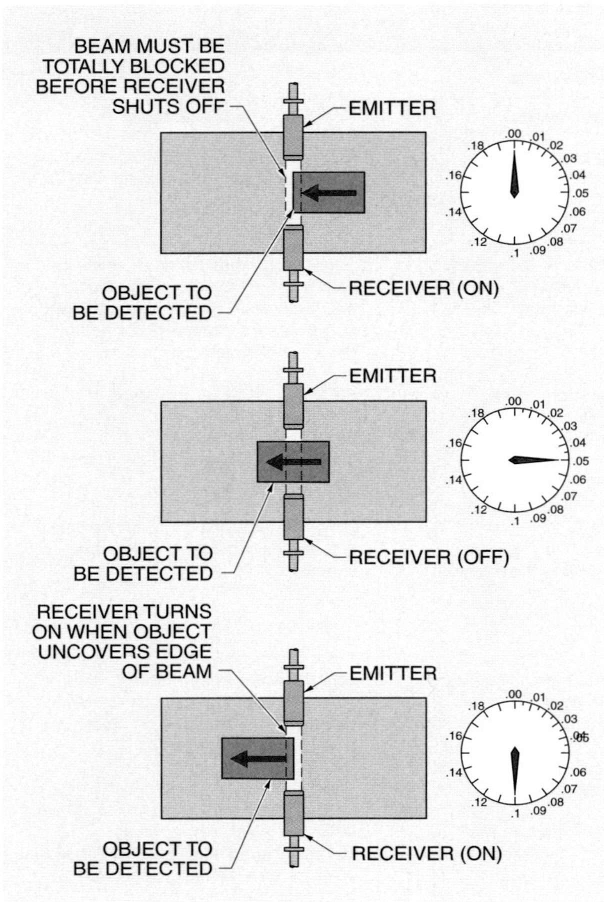

Figure 15-12. Response time of a photoelectric control is the number of pulses per second the controller can detect.

The length of time that an object breaks the beam is found by applying the formula:

$$t = \frac{w - D}{s}$$

where

t = time object takes to break beam (in sec)

w = width of object moving through beam (in in.)

D = effective beam diameter (in in.)

s = speed of object (in in./sec)

Example: Calculating Object Beam Break Time

What is the length of time it takes an object that is 2¼″ wide to pass a ¼″ diameter beam when the object is moving 2″ per second?

$$t = \frac{w - D}{s}$$

$$t = \frac{2.25 - .25}{2}$$

$$t = \frac{2}{2}$$

t = **1 sec**

A photoelectric control rated for 10 pulses per second may be used because the object takes 1 sec to pass the photoelectric control. If the speed of the object is increased to 10″ per second, the length of time it takes the object to move past the photoelectric control is .2 sec ([2.25 – .25] ÷ 10 = .2 sec). A photoelectric control rated for 10 pulses per second may be used because a photoelectric control rated for 10 pulses per second can detect an object passing by it every ¹⁄₁₀ (.1) sec and the object takes .2 sec to pass the photoelectric control.

Sensitivity Adjustment

Many photoelectric and proximity sensors have a sensitivity adjustment that determines the operating point or the intensity of light which triggers the output. **See Figure 15-13.** This adjustment allows the sensitivity to be set between a minimum and maximum range. The adjustment is made after the unit has been installed and the minimum and maximum settings for the application are experimentally determined. The sensitivity adjustment is normally set halfway between the minimum and maximum points. A low sensitivity setting may be desirable, especially when there is bright ambient light or electrical noise interference, or when detecting translucent objects. Reducing the sensitivity may prevent false triggering of the control in these conditions. Some photoelectric sensors include a two-color LED. When the LED is red, the photoelectric sensor is operating at its maximum range. For optimum performance, the photoelectric sensor should be operated when the LED is green.

Technical Fact

The first commercially available, fully integrated wireless sensor was introduced in the fall of 1998. The self-contained device is part of a program for monitoring the condition of rotating machinery and measures temperature, vibration, and other required parameters. The distributed architecture of these systems allows new sensors to be added to the network as needed.

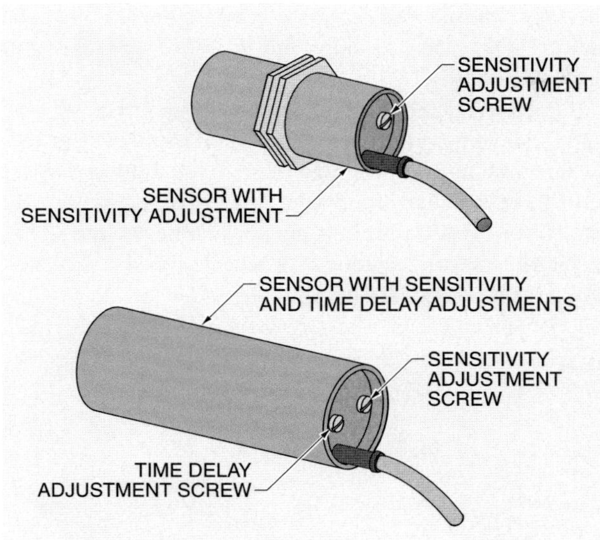

Figure 15-13. Many photoelectric and proximity sensors have adjustments for setting sensitivity of operation and may have a setting to adjust time delay.

Dark-Operated/Light-Operated

Basic photoelectric controls are designed to be dark-operated. A *dark-operated photoelectric control* is a photoelectric control that energizes the output switch when a target is present (breaks the beam). Some photoelectric controls include an optional feature that allows the control to be set in a light-operated mode. A *light-operated photoelectric control* is a photoelectric control that energizes the output switch when the target is missing (removed from the beam). **See Figure 15-14.**

Dark-operated or light-operated mode is usually set using a selector switch located on the photoelectric control. The selector switch is typically marked with an LO/DO (light-operated/dark-operated) position or an INV (invert) position. In a security system, the dark-operated mode is used to activate the switch contacts that sound an alarm when a person walks into the beam. The light-operated mode is used to activate the switch contacts that sound an alarm when a person removes an object (toolbox, painting, etc.) from the light beam.

PHOTOELECTRIC CONTROL APPLICATIONS

Photoelectric controls are used in a variety of applications for the detection of almost any object. Along a production line, photoelectric controls are used for counting, positioning, sorting, and safety. In security systems, they are used to detect the presence or removal of an object. Photoelectric controls are also used to detect the presence of vehicles at tollgates, parking areas, and truck docks. In most applications, photoelectric controls are used as inputs into timers, relays, counters, programmable controllers, and motor control circuits.

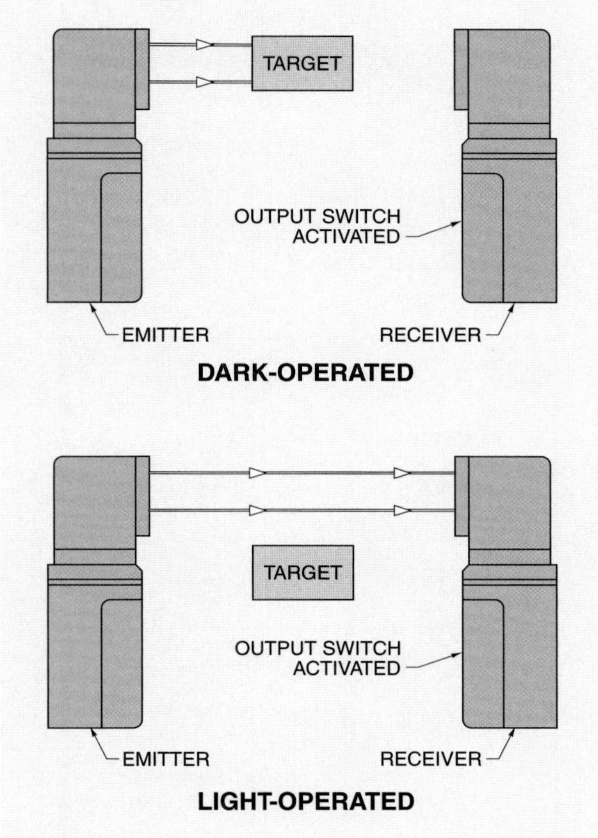

Figure 15-14. A photoelectric control can have a dark-operated or light-operated mode.

Height and Distance Monitoring

Photoelectric controls may be used to monitor a truck loading bay for clearance and distance. **See Figure 15-15.** In this application, a truck in a loading bay is monitored for necessary clearance and distance. Any truck 14′-0″ or larger must be unloaded at another bay. (The dimensions vary depending on particular needs.) In the control circuit, photoelectric control 1 (photo 1) turns ON an alarm in line 2 if the truck is too high. Photoelectric control 2 (photo 2) starts a recycle timer (TR) that flashes a yellow light ON and OFF at a distance of 2′ from the dock. Photoelectric control 3 (photo 3) starts a recycle timer that turns ON a red light at a distance of 6″ from the dock.

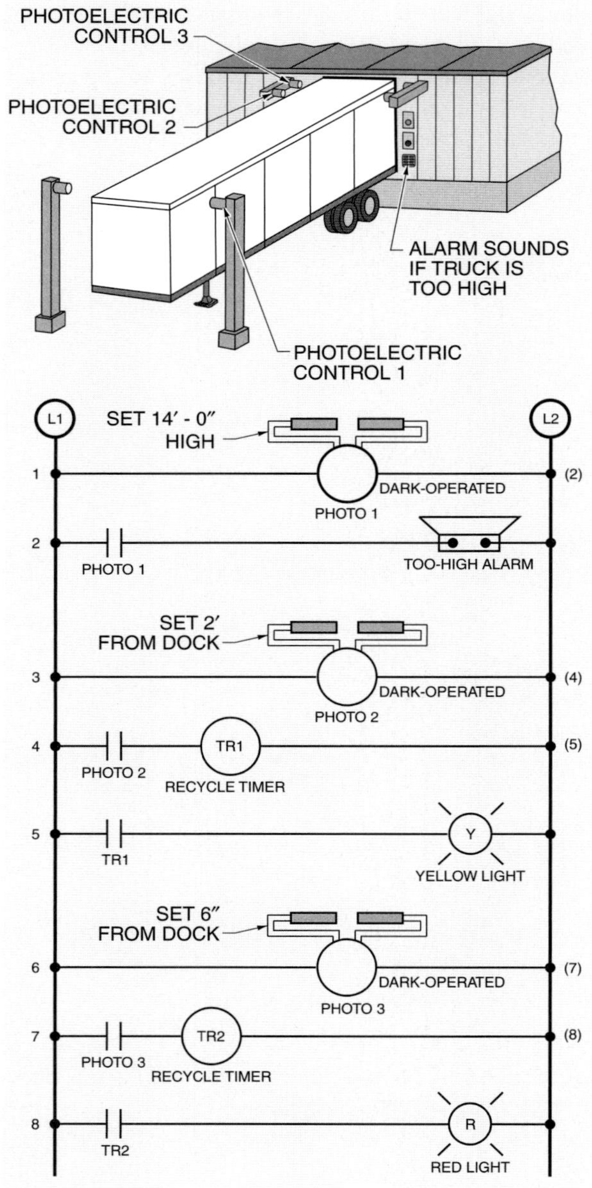

Figure 15-15. Photoelectric controls may be used to monitor a truck loading bay for clearance and distance.

Direct scan using modulated controls is the best scan method for this application. The photoelectric control is connected for dark operation of the controller, allowing for operation only when the truck blocks the beam.

Product Monitoring

Photoelectric controls may be used to detect a backup of a product on a conveyor line. **See Figure 15-16.** In this application, three photoelectric controls are used to turn ON a warning light and turn OFF the conveyor motor if required.

Photoelectric control 1 (photo 1) turns ON a warning light, indicating product is at the end of the conveyor line. At this time an operator may remove the product or wait until more products are on the line. This allows for best use of the worker's time. If the product backs up to photoelectric control 2 (photo 2), a recycle timer is activated which flashes the warning light.

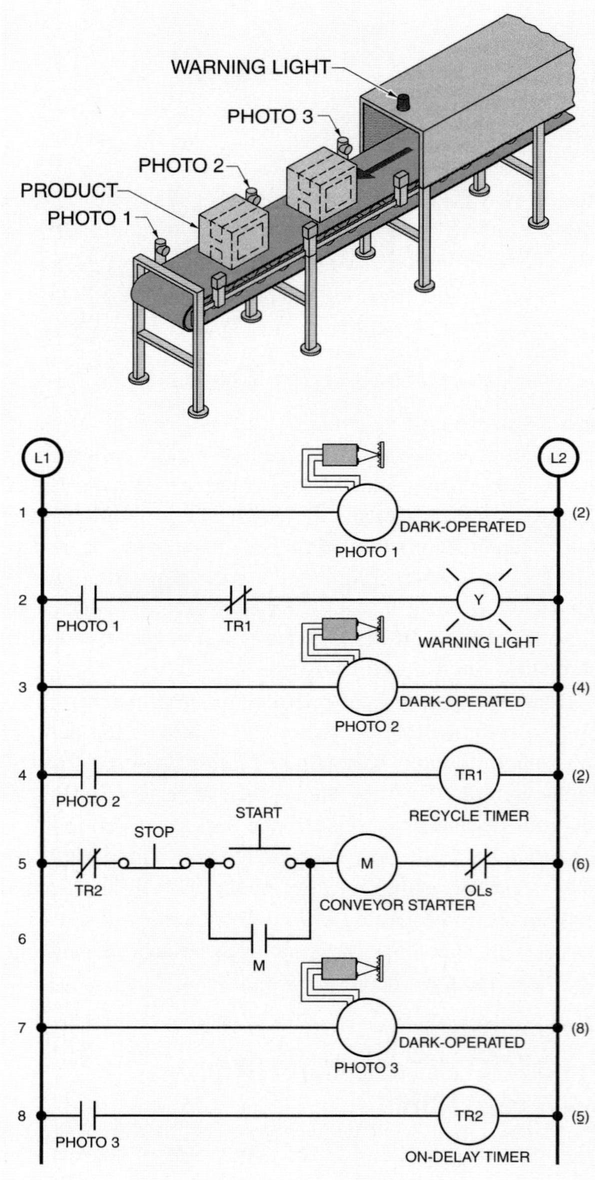

Figure 15-16. Photoelectric controls may be used to detect a backup of a product on a conveyor line.

At this time, the operator should unload the conveyor. If the conveyor is not unloaded, and product backs up to photoelectric control 3 (photo 3), an ON-delay timer is activated, which, after a few seconds, stops the conveyor motor, thus preventing a problem.

All upstream conveyors and machines must also be turned OFF to prevent a jam. In this application, retroreflective scan is used for ease of installation and because of vibration of the conveyor line.

ULTRASONIC SENSORS

An *ultrasonic sensor* is a solid-state sensor that can detect the presence of an object by emitting and receiving high-frequency sound waves. Ultrasonic sensors can provide an analog voltage output or a digital voltage output (switched output). The high-frequency sound waves are typically in the 200 kHz range. Ultrasonic sensors are used to detect solid and liquid targets (objects) at a distance of up to approximately 1 m (3.3′).

Ultrasonic sensors can be used to monitor the level in a tank, detect metallic and nonmetallic objects, and detect other objects that easily reflect sound waves. Soft materials such as foam, fabric, and rubber are difficult for ultrasonic sensors to detect and are better detected by photoelectric or proximity sensors.

Ultrasonic sensors are used to detect clear objects (glass and plastic), which are difficult to detect with photoelectric sensors. For this reason, ultrasonic sensors are ideal for applications in the food and beverage industry, or for any application that uses clear glass or plastic containers.

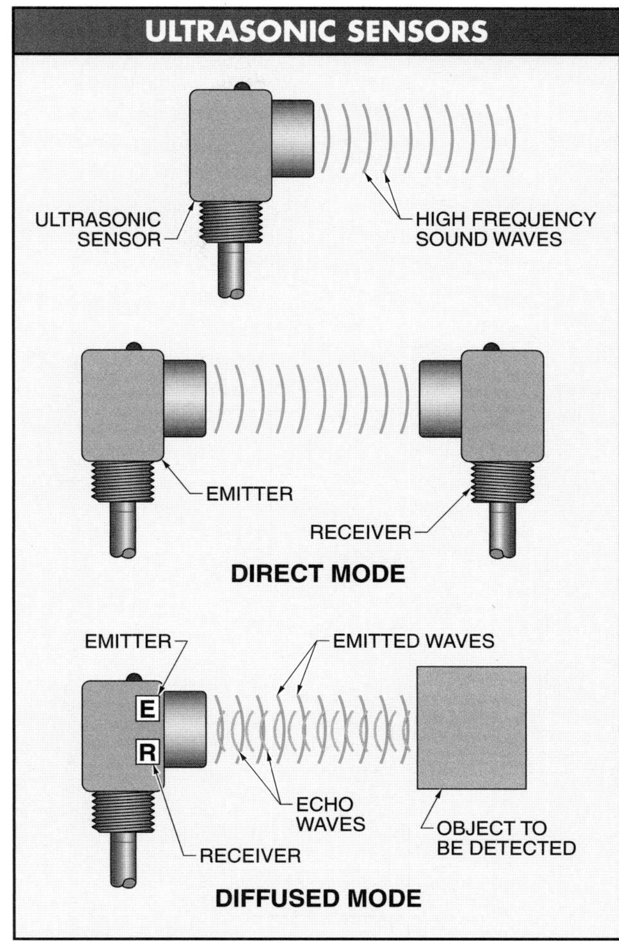

Figure 15-17. Ultrasonic sensors detect objects by bouncing high-frequency sound waves off the object.

Operating Modes

The two basic operating modes of ultrasonic sensors are the direct mode and the diffused mode. In the direct mode, an ultrasonic sensor operates like a direct scan photoelectric sensor. In the diffused mode, an ultrasonic sensor operates like a scan diffuse photoelectric sensor. **See Figure 15-17.**

Direct mode is a method of ultrasonic sensor operation in which the emitter and receiver are placed opposite each other so that the sound waves from the emitter are received directly by the receiver. Ultrasonic sensors used in the direct mode usually include an output that is activated when a target is detected. The output is normally a transistor (PNP or NPN) that can be used to switch a DC circuit. Outputs are available in normally open and normally closed switching modes. Ultrasonic sensors include an adjustment for adjusting (tuning) the sensor sensing distance. Tuning the receiver to the emitter minimizes interference from ambient noise sources that may be present in the area.

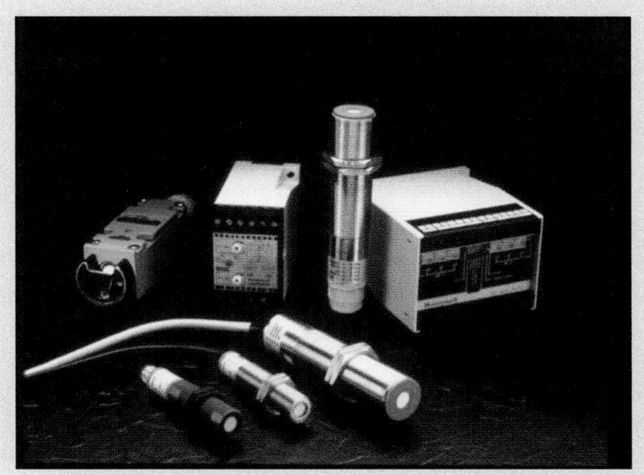

Honeywell

Ultrasonic sensors use sound waves rather than light waves and are ideal for use in harsh environments because of their encapsulated solid-state design.

Diffused mode is a method of ultrasonic sensor operation in which the emitter and receiver are housed in the same enclosure. In the diffused mode, the emitter sends out a sound wave and the receiver listens for the sound wave echo bouncing back off an object. Ultrasonic sensors used in the diffused mode may include a digital output or an analog output. The analog output provides an output voltage that varies linearly with the target's distance from the sensor. **See Figure 15-18.** The sensor typically includes a light emitting diode (LED) that glows with intensity proportional to the strength of the echo. The analog output sensor includes an adjustable background suppression feature that allows the sensor to better detect only the intended target and not background objects.

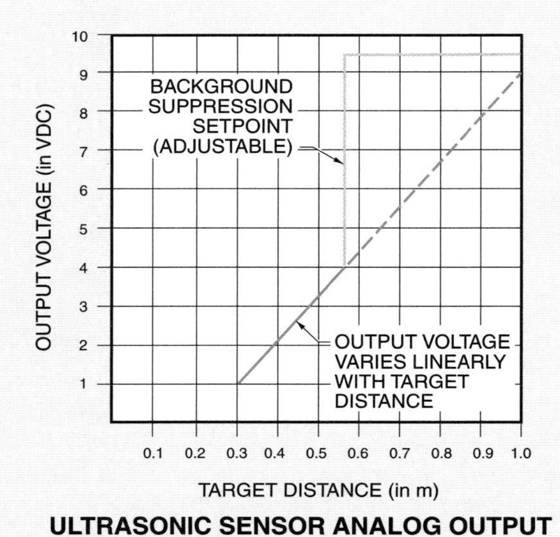

Figure 15-18. An ultrasonic sensor used in the diffused mode can provide an analog output that varies linearly with the target's distance from the sensor.

PROXIMITY SENSORS

A *proximity sensor (proximity switch)* is a solid-state sensor that detects the presence of an object by means of an electronic sensing field. A proximity sensor does not come into physical contact with the object. Proximity sensors can detect the presence or absence of almost any solid or liquid. Proximity sensors are extremely versatile, safe, and reliable, and may be used in applications where limit switches and mechanical level switches cannot be used.

Proximity sensors can detect very small objects, such as microchips, and very large objects, such as automobile bodies. All proximity sensors have encapsulated solid-state circuits that may be used in high-vibration areas, wet locations, and fast-switching applications. To meet as many application requirements as possible, proximity sensors are available in an assortment of sizes and shapes. **See Figure 15-19.**

The two basic proximity sensors are the inductive proximity sensor and the capacitive proximity sensor. An *inductive proximity sensor* is a sensor that detects only conductive substances. Inductive proximity sensors detect only metallic targets. A *capacitive proximity sensor* is a sensor that detects either conductive or nonconductive substances. Capacitive proximity sensors detect solid, fluid, or granulated targets, whether conductive or nonconductive. The proximity sensor used depends on the type and material of the target.

Banner Engineering Corp.

Figure 15-19. Proximity sensors are solid-state sensors that detect the presence of an object by means of an electronic sensing field.

Inductive Proximity Sensors

Inductive proximity sensors operate on the eddy current killed oscillator (ECKO) principle. The ECKO principle states that an oscillator produces an alternating magnetic field that varies in strength depending on whether or not a metallic target is present. The generated alternating field operates at a radio frequency (RF). **See Figure 15-20.**

When a metallic target is in front of an inductive proximity sensor, the RF field causes eddy currents to be set up on the surface of the target material. These eddy currents upset the AC inductance of the sensor oscillator circuit, causing the oscillations to be reduced. When the oscillations are reduced to a certain level, the sensor triggers, which indicates the presence of a metallic object. Inductive proximity sensors detect ferrous materials (containing iron, nickel, or cobalt) more readily than nonferrous materials (all other metals, such as aluminum, brass, etc.).

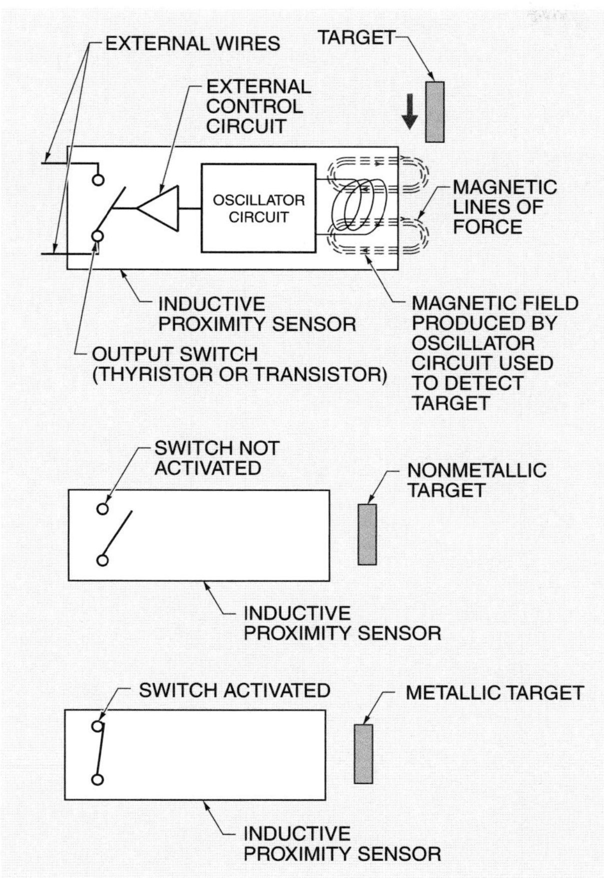

Figure 15-20. Inductive proximity sensors use a magnetic field to detect the presence of a target.

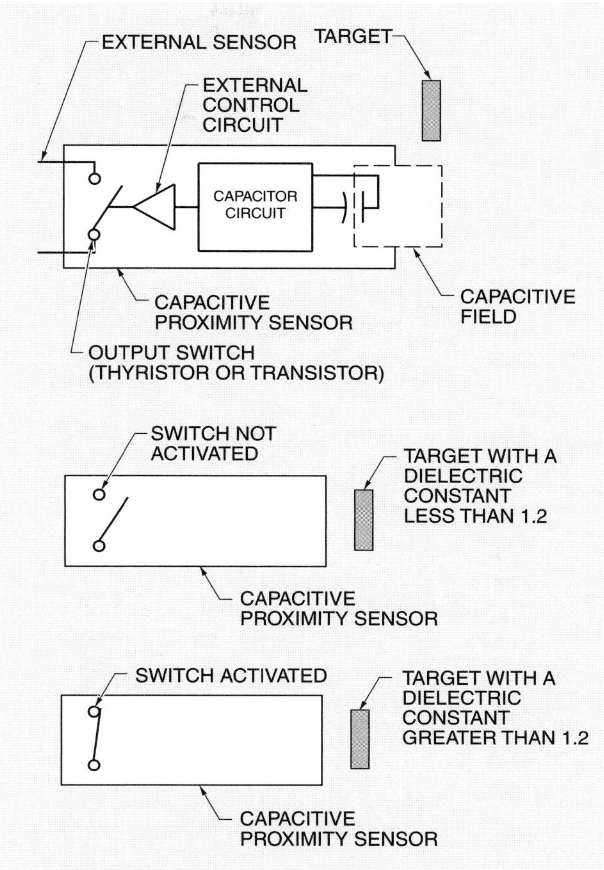

Figure 15-21. Capacitive proximity sensors use a capacitive field to detect the presence of a target.

Nominal sensing distances range from .5 mm to about 40 mm. Sensitivity varies depending on the size of the object and the type of metal. Iron may be sensed at about 40 mm. Aluminum may be sensed at approximately 20 mm. Applications of inductive proximity sensors include positioning of tools and parts, metal detection, drill bit breakage detection, and solid-state replacement of mechanical limit switches.

Capacitive Proximity Sensors

A capacitive proximity sensor measures a change in capacitance which is caused by the approach of an object to the electrical field of a capacitor. A capacitive proximity sensor detects all materials that are good conductors in addition to insulators that have a relatively high dielectric constant. A *dielectric* is a nonconductor of direct electric current. Capacitive sensors can detect materials such as plastic, glass, water, moist wood, etc. **See Figure 15-21.**

Namco Controls Corporation

Inductive proximity sensors typically include a "hard coat," nonstick material that prevents materials from adhering to the sensor.

Two small plates that form a capacitor are located directly behind the front of the sensor. When an object approaches the sensor, the dielectric constant of the capacitor changes, thus changing the oscillator frequency, which activates the sensor output. Nominal sensing distances range from 3 mm to about 15 mm. The maximum sensing distance depends on the physical and electrical characteristics (dielectric) of the object to be detected. The larger the dielectric constant, the easier it is for a capacitive sensor to detect the material. Generally, any material with a dielectric constant greater than 1.2 may be detected. **See Figure 15-22.**

Hall Effect Sensors

A *Hall effect sensor* is a sensor that detects the proximity of a magnetic field. The Hall effect principle was discovered in 1879 by Edward H. Hall at Johns Hopkins University. Hall found that when a magnet is placed in a position where its field is perpendicular to one face of a thin rectangle of gold through which current was flowing, a difference of potential appeared at the opposite edges. He found this voltage is proportional to the current flowing through the conductor, and the magnetic induction is perpendicular to the conductor.

Today, semiconductors are used for the sensing element (Hall generator) in Hall effect sensors. Hall voltages obtained with semiconductors are much higher than those obtained with gold and are also less expensive.

Theory of Operation. A *Hall generator* is a thin strip of semiconductor material through which a constant control current is passed. **See Figure 15-23.** When a magnet is brought near the Hall generator with its field directed at right angles to the face of the semiconductor, a small voltage (Hall voltage) appears at the contacts placed across the narrow dimension of the Hall generator. When the magnet is removed, the Hall voltage drops to zero. The Hall voltage is dependent on the presence of the magnetic field and on the current flowing in the Hall generator. The output of the Hall generator is zero if either the current or the magnetic field is removed. In most Hall effect devices, the control current is held constant and the magnetic induction is changed by movement of a permanent magnet.

Sensor Packaging

To meet many different application requirements, Hall effect sensors are packaged in a number of different configurations. **See Figure 15-24.** Typical configurations include cylinder, proximity, vane, and plunger. Cylinder and proximity Hall effect sensors are used to detect the presence of a magnet. Vane Hall effect sensors include a sensor on one side and a magnet on the other, and are used to detect an object passing through the opening. Plunger Hall effect sensors include a magnet that is moved by an external force acting against a lever.

DIELECTRIC CONSTANT

MATERIAL	NUMBER
Acetone	20.7
Acrylic Resin	2.7 – 4.5
Air	1.000590
Ammonia (Liquid)	15 – 24
Aniline	5.5 – 7.8
Aqueous Solutions	50 – 80
Benzene	2.3
Carbon Dioxide	1.000985
Carbon Tetrachloride	2.2
Cement Powder	5 – 10
Cereal (Dry)	3 – 5
Chlorine Liquid	2.0
Ebonite	2.5 – 2.9
Epoxy Resin	3.3 – 3.7
Ethanol	24
Ethylene Glycol	37
Fly Ash	1.9 – 2.6
Flour	2.5 – 3.0
Freon 12	2.4
Gasoline	2.0
Glass	3.7 – 10
Glycerine	47 – 68
Lime	2.2 – 2.5
Marble	8.5
Melamine Resin	4.7 – 10.9
Mica	7.0
Nylon	3 – 4.4
Paraffin	2.0 – 2.5
Paper (Dry)	2.0
Petroleum Jelly	2.2 – 2.9
Phenol Resin	4.9
Polyacetal	3.6 – 3.7
Polyester Resin	3 – 4
Polypropylene	1.5
Polytetrafluoroethylene Resin	2.0
Polyvinyl Chloride Resin	3.3 – 4.5
Porcelain	6 – 8
Powdered Milk	1.8
Pressboard	2 – 5
Rubber	3.0
Salt	3 – 15
Sand (Dry)	5.0
Shellac	2.0 – 3.8
Silicon Varnish	2.8 – 3.3
Soybean	2.8
Styrene Resin	2.55 – 2.95
Sugar	3.0
Sulfur	1.6 – 1.7
Toulene (Liquid)	2.0 – 2.4
Turpentine	2.2
Urea Resin	6.2 – 9.5
Water	80 – 88
Wood, Dry	2 – 6
Wood, Wet	10 – 30

* Values will vary with changes in temperature

Figure 15-22. Capacitive sensors work based on the dielectric of the material to be sensed.

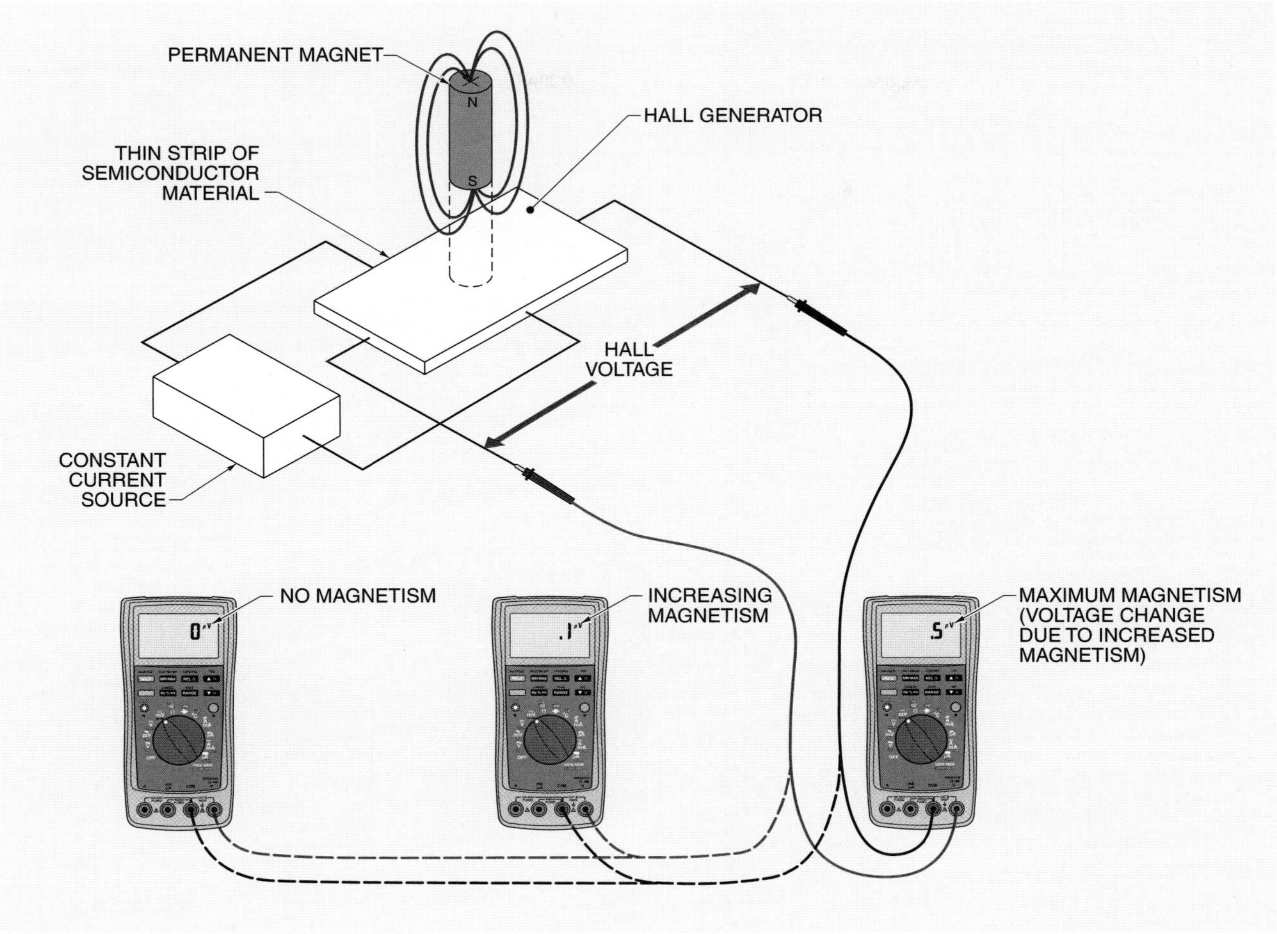

Figure 15-23. A Hall generator is a thin strip of semiconductor material through which a constant control current is passed.

Honeywell

Figure 15-24. Hall effect sensors are available in a variety of packages for different applications.

Hall Effect Sensor Actuation

Hall effect sensors may be activated by head-on, slide-by, pendulum, rotary, vane, ferrous proximity shunt, and electromagnetic actuation. The actuation method depends on the application.

Head-on Actuation. *Head-on actuation* is an active method of sensor activation in which a magnet is oriented perpendicular to the surface of the sensor and is usually centered over the point of maximum sensitivity. **See Figure 15-25.**

The direction of movement is directly toward and away from the Hall effect sensor. The actuator and Hall effect sensor are positioned so the south (S) pole of the magnet approaches the sensitive face of the sensor.

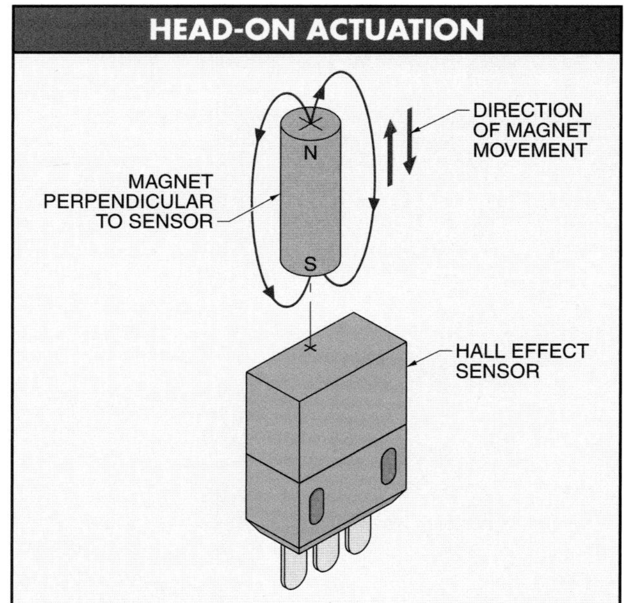

Figure 15-25. In head-on actuation, a magnet is oriented perpendicular to the surface of the sensor and is usually centered over the point of maximum sensitivity.

Slide-by Actuation. *Slide-by actuation* is an active method of sensor activation in which a magnet is moved across the face of a Hall effect sensor at a constant distance (gap). **See Figure 15-26.** The primary advantage of slide-by actuation over head-on actuation is that less actuator travel is needed to produce a signal large enough to cycle the device between operate and release.

Pendulum Actuation. *Pendulum actuation* is a method of sensor activation that is a combination of the head-on and the slide-by actuation methods. **See Figure 15-27.** The two methods of pendulum actuation are single-pole and multiple-pole. Single or multiple signals are generated by one actuator.

Rotary Actuation. *Rotary actuation* is an active method of sensor activation in which a multipolar ring magnet or collection of magnets is used to produce an alternating magnetic pattern. **See Figure 15-28.** The induction pattern (+ and –) is seen by the sensor located in the threaded tubular housing. Ring magnets with up to 60 or more poles are available for multiple output signals.

Vane Actuation. *Vane actuation* is a passive method of sensor activation in which an iron vane shunts or redirects the magnetic field in the air gap away from the Hall effect sensor. **See Figure 15-29.** When the iron vane is moved through the air gap between the Hall effect sensor and the magnet, the sensor is turned ON and OFF sequentially at any speed due to the shunting effect. The same effect is achieved with a rotary-operated vane.

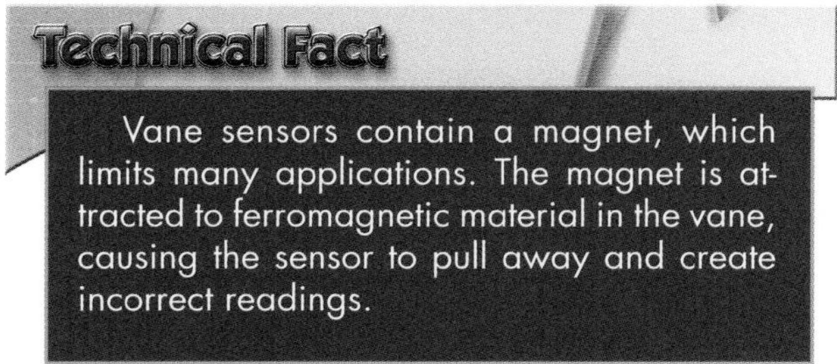

Vane sensors contain a magnet, which limits many applications. The magnet is attracted to ferromagnetic material in the vane, causing the sensor to pull away and create incorrect readings.

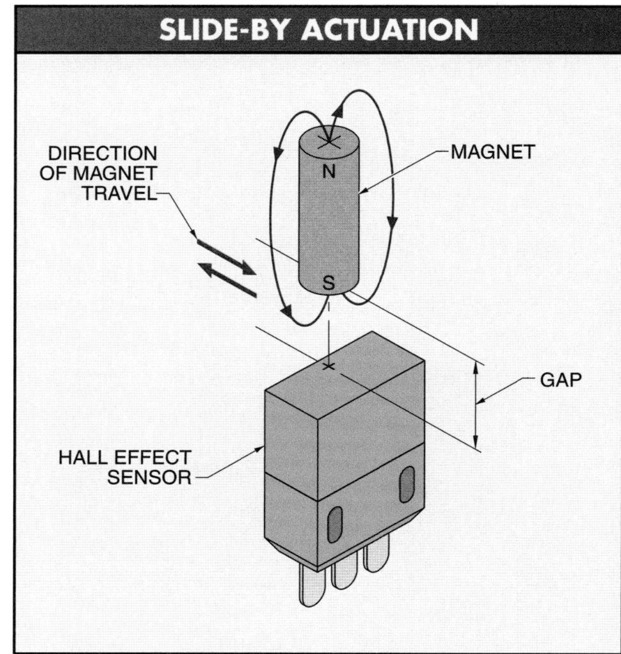

Figure 15-26. In slide-by actuation, a magnet is moved across the face of a Hall effect sensor at a constant distance.

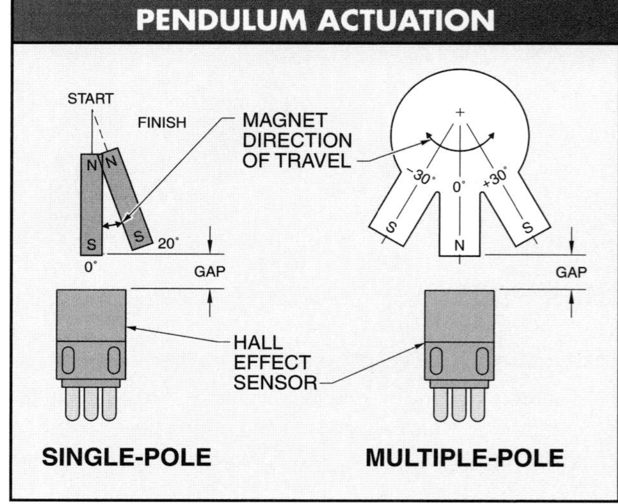

Figure 15-27. Pendulum actuation is a combination of the head-on and the slide-by actuation methods.

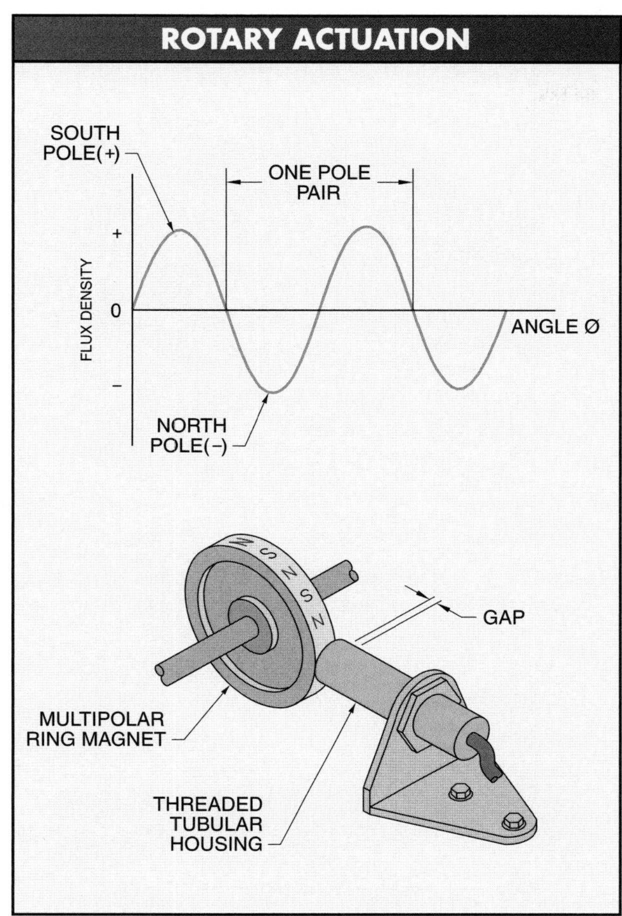

Figure 15-28. Rotary actuation uses a multipolar ring magnet or collection of magnets to produce an alternating magnetic pattern.

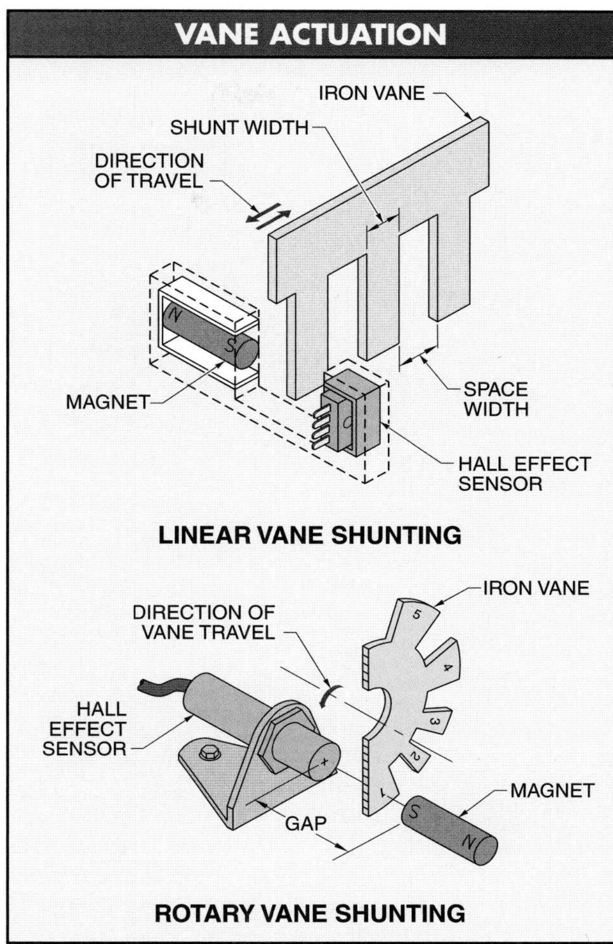

Figure 15-29. In vane actuation, an iron vane shunts or redirects the magnetic field in the air gap away from the Hall effect sensor.

Ferrous Proximity Shunt Actuation. *Ferrous proximity shunt actuation* is a passive method of sensor activation in which the magnetic induction around the Hall effect sensor is shunted with a gear tooth. **See Figure 15-30.**

The gear tooth causes the magnetic induction to be shunted from the sensor when a tooth is present and passes through the sensor when the tooth is absent. This variable magnetic field concentration causes the Hall effect sensor to be activated, producing an output signal.

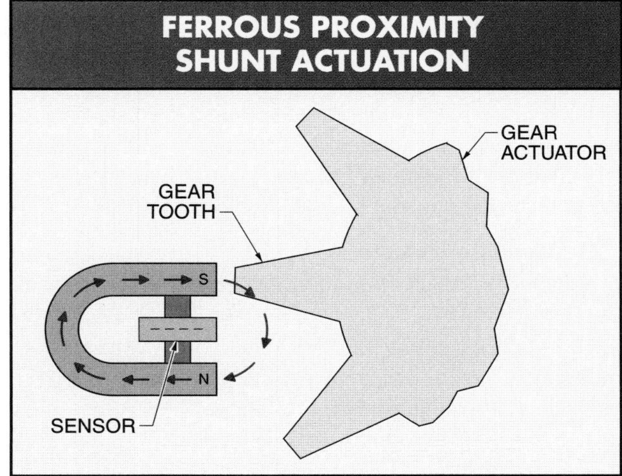

Figure 15-30. In ferrous proximity shunt actuation, the magnetic induction around the Hall effect sensor is shunted with a gear tooth.

Electromagnetic Actuation. *Electromagnetic actuation* is a passive method of sensor activation in which a magnetic field produced by a coil of wire is used to activate a Hall effect sensor. **See Figure 15-31.** The more current and coverage of wire, the stronger the magnetic field. The coil serves the same purpose as a bar magnet. Attention must be given to thermal considerations because electromagnets dissipate heat. Care must be exercised to ensure that the current and temperature limits are not exceeded.

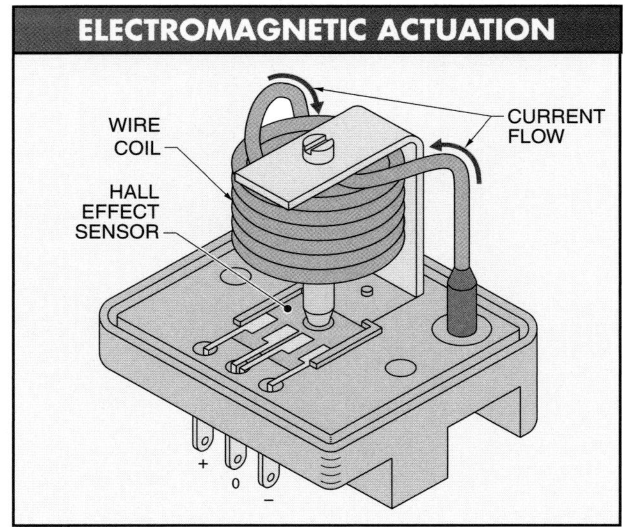

Figure 15-31. In electromagnetic actuation, a magnetic field produced by a coil of wire is used to actuate a Hall effect sensor.

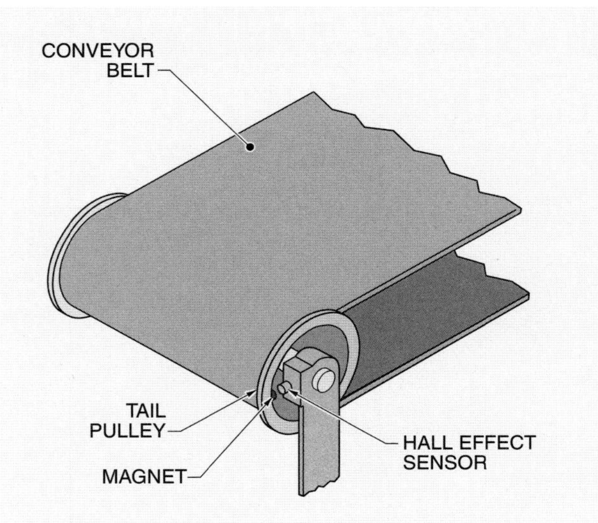

Figure 15-32. A Hall effect sensor may be used for monitoring a remote conveyor operation.

Hall Effect Sensor Applications

Hall effect sensors are used in a wide range of applications requiring the detection of the presence (proximity) of objects. Hall effect sensors are used in slow-moving and fast-moving applications to detect movement and are also used to replace mechanical limit switches in applications that require the detection of an object's position. Hall effect sensors are also used to provide a solid-state output in applications that normally use a mechanical output, such as a standard pushbutton on a level switch.

Conveyor Belt. A Hall effect sensor may be used for monitoring a remote conveyor operation. **See Figure 15-32.** In this application, a cylindrical Hall effect sensor is mounted to the frame of the conveyor. A magnet mounted on the tail pulley revolves past the sensor to cause an intermittent visual or audible signal at a remote location to ensure that the conveyor is running. Any shutdown of the conveyor interferes with the normal signal and alerts the operator. Maintenance is minimal because the sensor makes no physical contact and has no levers or linkages to break.

Current Sensor. A Hall effect current sensor may be used to develop a fast-acting, automatically resetting circuit. **See Figure 15-33.**

An overload signal changes state from low to high, or vice versa, when the current exceeds the circuit design trip point. This signal is used to trigger a warning alarm or to control the current directly by electronic means. Hall effect current sensors are available in sizes ranging from the diameter of a dime to about half the size of a deck of playing cards. They indicate overload currents from a few milliamperes to several amperes.

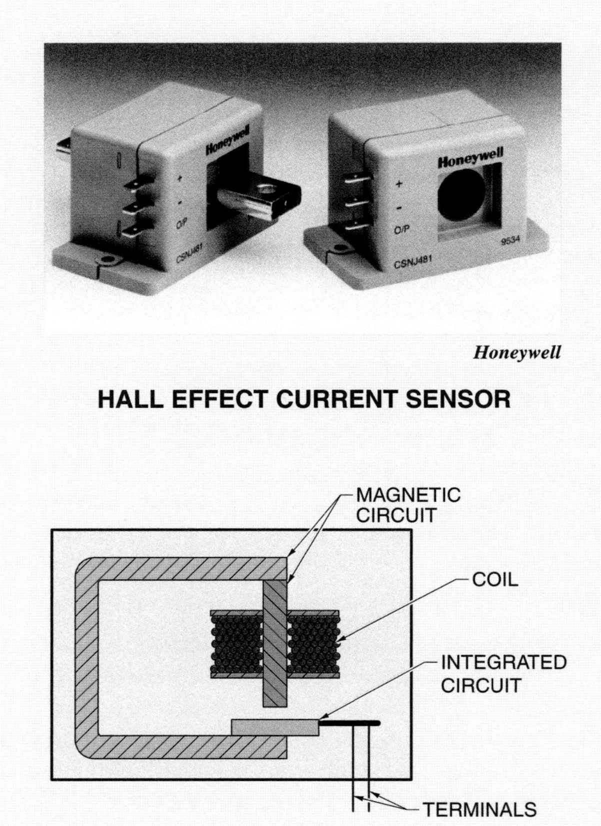

Figure 15-33. A Hall effect current sensor may be used to develop a fast-acting, automatically resetting circuit.

Different reset characteristics can be achieved by tailoring the electromagnetic design to the sensor's requirements. In addition, use of a linear Hall effect sensor in place of a digital Hall effect sensor can provide an output signal proportional to the input current, but electrically isolated from it.

Speed Sensing. A Hall effect sensor may be used with a ring magnet in speed-sensing applications. **See Figure 15-34.** In this circuit, the sensor's output provides an electrical waveform in which repetition frequency varies directly with shaft speed, but in which amplitude does not.

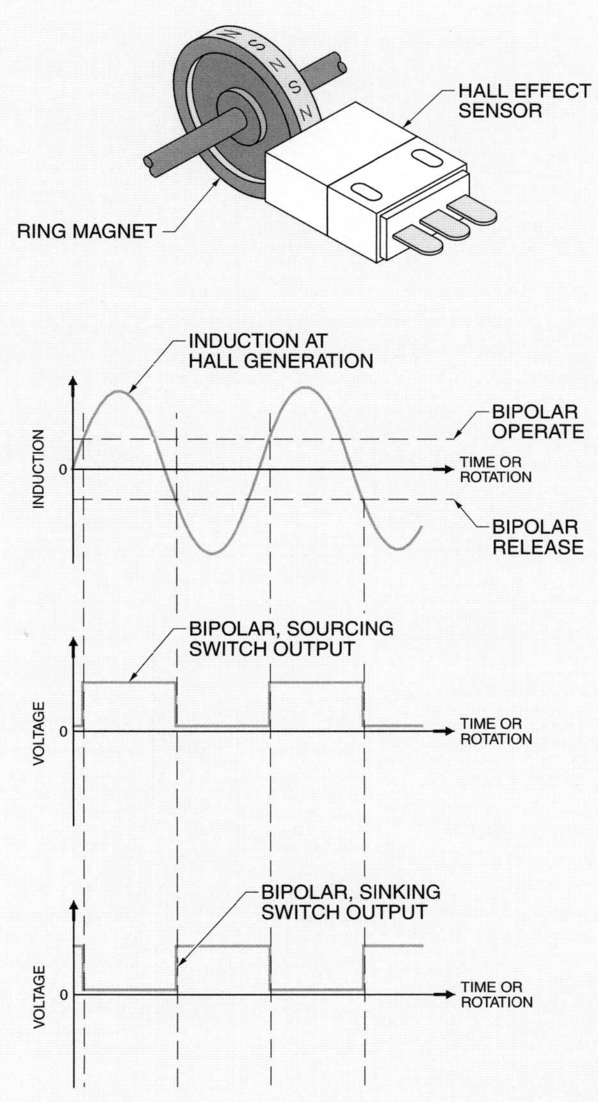

Figure 15-34. A Hall effect sensor may be used with a ring magnet in speed-sensing applications.

The use of a ring magnet with a Hall effect sensor provides a valuable alternative to the coil pickup or variable reluctance speed sensing methods. Techniques that depend on inducing a voltage in a coil, either by passing magnets near it or by changing the air gap reluctance of a fixed magnet, have the disadvantage that both the frequency and amplitude of the voltage waveform change with speed. Hall effect sensors have no minimum speed of operation.

Instrumentation. A Hall effect sensor and ring magnet combination may be used in an instrumentation application. **See Figure 15-35.** In this circuit, the Hall effect sensor is actuated by a ring magnet, which initiates resistance measurements of electrical circuits being life tested.

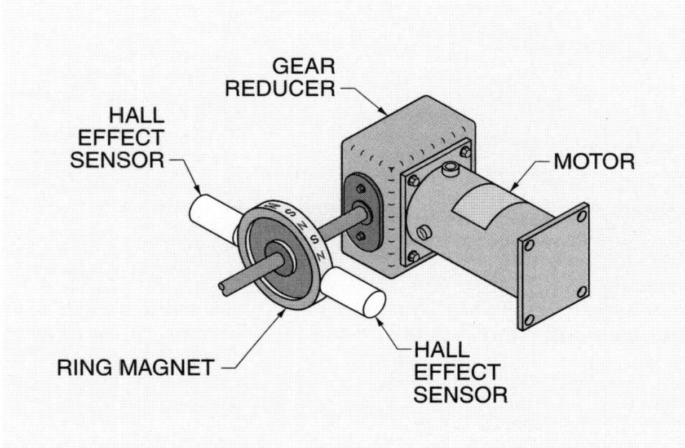

Figure 15-35. A Hall effect sensor and ring magnet combination may be used in an instrumentation application.

A second Hall effect sensor actuated 180° after the first sensor checks that the circuits are open. The no-touch actuation and long life are ideal in instrument and apparatus designs.

Beverage Gun. Hall effect sensors are used in beverage gun applications because of their small size, sealed construction, and reliability. **See Figure 15-36.** The Hall effect sensor's small size allows seven sensors to be installed in a hand-held device. Hall effect sensors cannot be contaminated by syrups, other liquids, or foodstuffs because they are completely enclosed in the beverage gun. The beverage gun is completely submersible in water for easy cleaning and requires low maintenance.

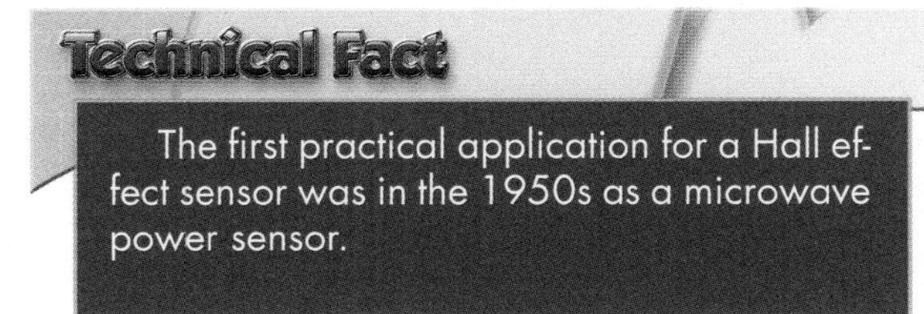

The first practical application for a Hall effect sensor was in the 1950s as a microwave power sensor.

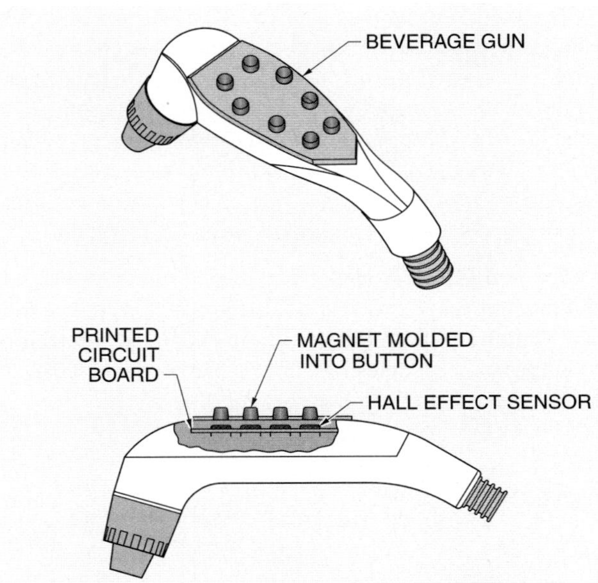

Figure 15-36. Hall effect sensors are used in beverage gun applications because of their small size, sealed construction, and reliability.

Sequencing Operations. Sequencing operations can be controlled by activating Hall effect sensors through the use of metal disks clamped to a common shaft. **See Figure 15-37.** The metal disks are rotated in the gaps of Hall effect vane sensors. A disk rotating in tandem with others can be used to create a binary code, which is used to establish a sequence of operations. Programs can be altered by replacing the disks with others having a different air-to-ferrous cam ratio.

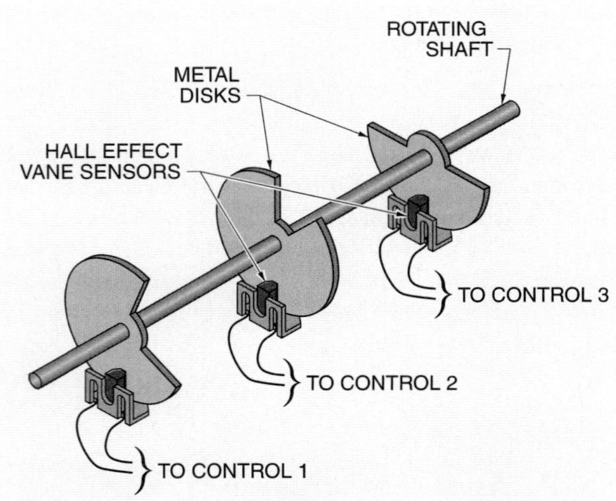

Figure 15-37. Sequencing operations can be controlled by activating Hall effect sensors through the use of metal disks clamped to a common shaft.

Length Measurement. Length measurement can be accomplished by mounting a disk with two notches on the extension of a motor drive shaft. **See Figure 15-38.** In this circuit, a Hall effect vane sensor is mounted so that the disk passes through the gap. Each notch represents a fixed length of material and can be used to measure tape, fabric, wire, rope, thread, aluminum foil, plastic bags, etc.

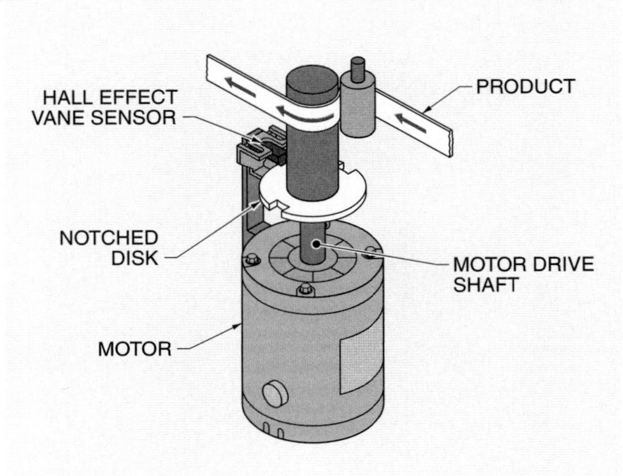

Figure 15-38. Length measurement can be accomplished by mounting a disk with two notches on the extension of a motor drive shaft.

Shaft Encoding. A cylindrical Hall effect sensor can be used in shaft encoding applications. **See Figure 15-39.** A ring magnet is mounted on the motor shaft. Each pair of north (N) and south (S) poles activates the Hall effect sensor. Each pulse represents angular movement.

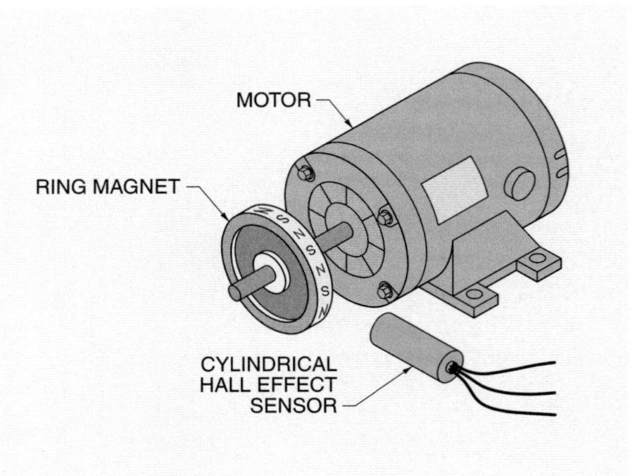

Figure 15-39. A cylindrical Hall effect sensor can be used in shaft encoding applications.

Level/Degree of Tilt. Hall effect sensors may be installed in the base of a machine to indicate the level or degree of tilt. **See Figure 15-40.** Magnets are installed above the Hall effect sensors in a pendulum fashion. The machine is level as long as the magnet remains directly over the sensor. A change in state of output (when a magnet swings away from a sensor) is indication that the machine is not level. The sensor/magnet combination may also be installed in such a manner as to indicate the degree of tilt.

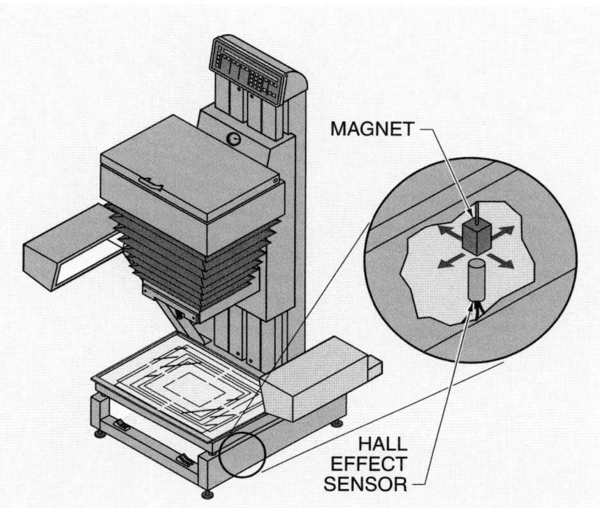

Figure 15-40. Hall effect sensors may be installed in the base of a machine to indicate the level or degree of tilt.

Joystick. Hall effect sensors may be used in a joystick application. **See Figure 15-41.** In this application, the Hall effect sensors inside the joystick housing are actuated by a magnet on the joystick. The proximity of the magnet to the sensor controls activation of different outputs used to control cranes, operators, motor control circuits, wheelchairs, etc. Use of an analog device also achieves degree of movement measurements such as speed.

Paper Detector. A Hall effect plunger sensor can be used in printers to detect paper flow. **See Figure 15-42.** Hall effect sensors have extremely long life and low maintenance costs, have no contacts to become gummy or corroded, and may be directly interfaced with logic circuitry.

Flow Detection Sensor

A *flow detection sensor (solid-state flow sensor)* is a sensor that detects the movement (flow) of a liquid or gas using a solid-state device. Because the flow detector is a solid-state device, there are no moving parts (mechanical parts) that can become damaged due to corrosion or product deposits.

Flow detection sensors are used to monitor lubricant and coolant flow in pumps, generators, welding machines, high-speed machine tools, and other applications in which the loss of flow may be dangerous or critical to the operation of a machine. Flow detection sensors are also used to monitor fan and ventilation flow in applications that require a flow of air to maintain a healthy work environment.

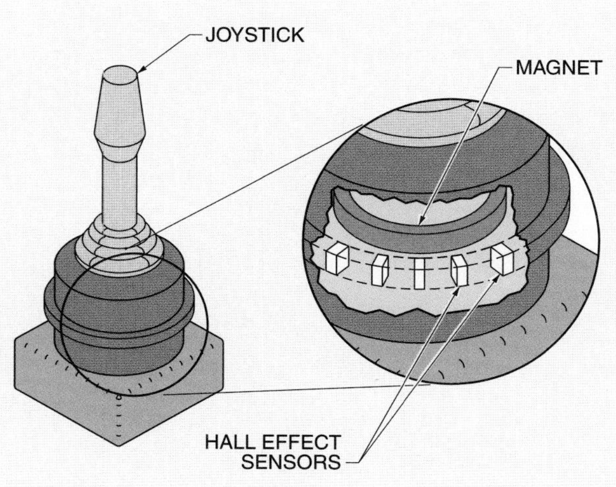

Figure 15-41. Hall effect sensors may be used in a joystick application.

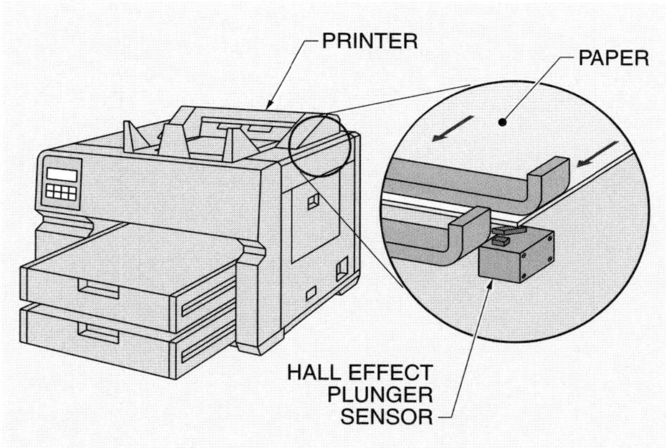

Figure 15-42. A Hall effect plunger sensor can be used in printers to detect paper flow.

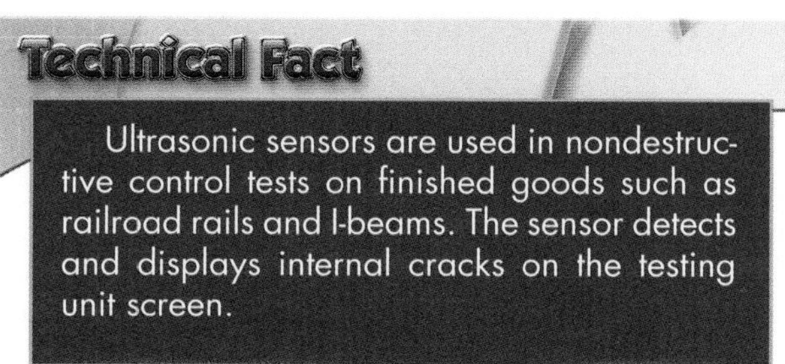

A solid-state flow detection sensor operates on the principle of thermal conductivity. The sensor head, which is in contact with the medium (liquid or air) to be detected, is heated to a temperature that is a few degrees higher than the medium to be detected. When the medium is flowing, the heat produced at the sensor head is conducted away from the sensor, cooling the sensor head. When the medium stops flowing, the heat produced by the sensor head is not conducted away from the sensor head. A thermistor in the sensor head converts the heat not conducted away into a stronger electrical signal than is produced when the heat is conducted away. This electrical signal is used to operate the sensor's output (contacts, transistor, etc.). **See Figure 15-43.**

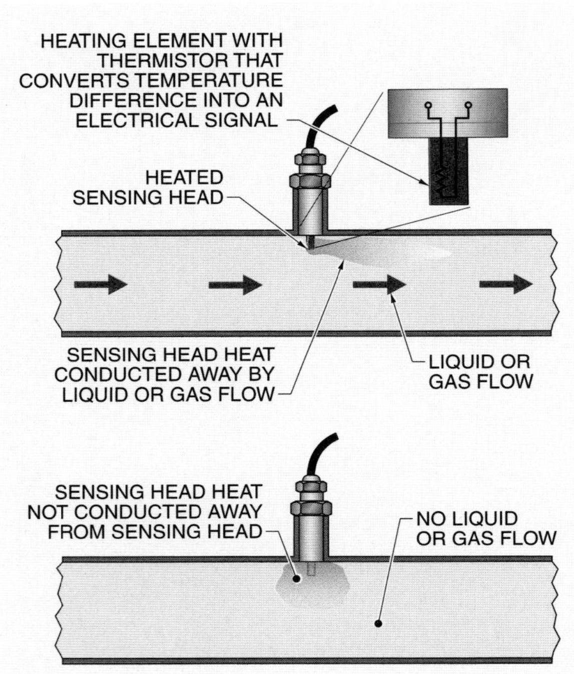

Figure 15-43. A solid-state flow detection sensor operates on the principle of thermal conductivity.

Because there is a small delay (usually less than 30 sec) before the sensor heats enough to signal no flow, the flow sensor acts like a motor starter overload, which allows for a short time delay to pass before signaling a problem. This short time delay helps prevent false alarms. The sensor also includes adjustments for setting the sensor to detect different product types and flow rates (speed past the sensor). Different color LEDs are usually included on the sensor for indicating different operating conditions.

When used to monitor the flow of a liquid, a flow detection sensor is mounted within a pipe in which there should be flow during normal operation. The sensor can be mounted in a vertical or horizontal pipe and the product can flow in either direction. **See Figure 15-44.**

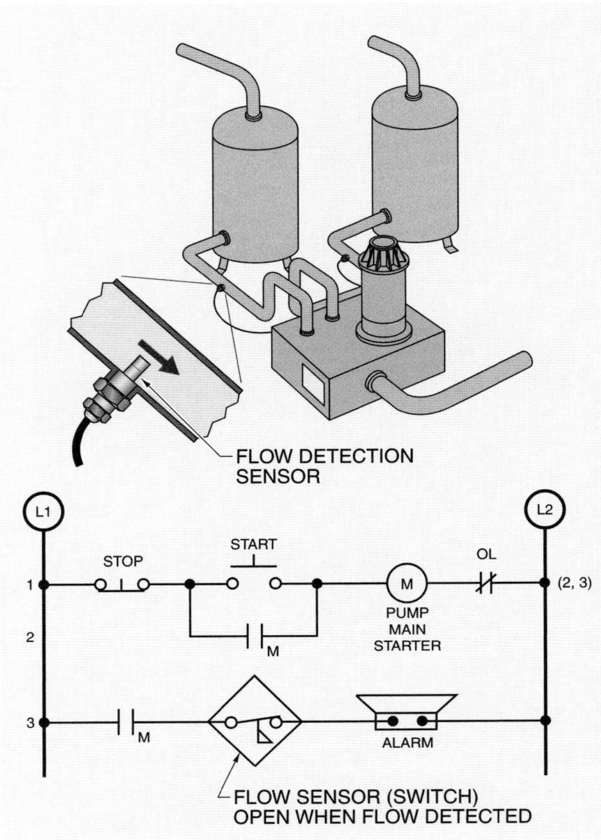

Figure 15-44. A flow detector sensor can be used to monitor product flow in a pipe.

In this application, the motor starter contacts in line 3 close each time the pump motor is operating. If there is product flow, a normally closed (held open during flow) switch is used to prevent the alarm from sounding. The alarm sounds if the flow stops and the motor starter is still ON.

When used to monitor the flow of a gas (usually air), the flow detection sensor can be mounted within a duct in which there should be flow during normal operation. However, if the flow is critical for a safe work environment, several flow sensors can be used to ensure the flow is moving in all parts of the exhaust system. **See Figure 15-45.**

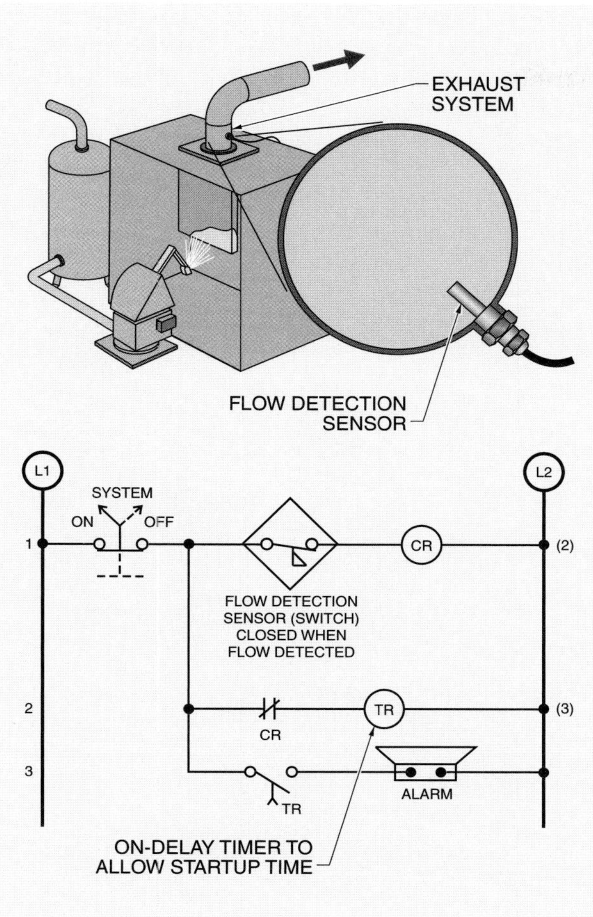

Figure 15-45. A flow detector sensor can be used to monitor airflow in painting or welding exhaust system applications.

In this circuit, a normally open (held closed during flow) switch is used to operate a control relay. The control relay operates an on-delay timer. The on-delay timer sounds an alarm if flow stops for longer than the timer's set time period. The time delay prevents the alarm from sounding during the first few seconds of system startup.

PHOTOELECTRIC AND PROXIMITY OUTPUTS

Photoelectric and proximity sensors use solid-state outputs to control the flow of electric current. The solid-state output of a photoelectric or proximity sensor may be a thyristor, NPN transistor, or PNP transistor. The thyristor output is used for switching AC circuits. The NPN and PNP transistor outputs are used for switching DC circuits. The output selected depends on specific application needs. Considerations that affect the solid-state output include:

- Voltage type to be switched (AC or DC).
- Amount of current to be switched. Most proximity sensors can only switch a maximum of a few hundred milliamperes. An interface is needed if higher current switching is required. The solid-state relay is the most common interface used with photoelectric and proximity sensors.
- Electrical requirements of the device to which the output of the proximity sensor is to be connected. Compatibility with a controller such as a programmable controller may require that a certain type of solid-state output be used as the input to a specific controller.
- Required polarity of the switched DC output. NPN outputs deliver a negative output and PNP outputs deliver a positive output.
- Electrical characteristics such as load current, operating current, and minimum holding current because solid-state outputs are never completely open or closed.

AC Photoelectric and Proximity Sensors

AC photoelectric and proximity sensors switch alternating current circuits. An AC sensor is connected in series with the load that it controls. The sensor is connected between line 1 and the load to be controlled. **See Figure 15-46.**

Because AC sensors are connected in series with the load, special precautions must be taken. The three main factors to be considered when connecting AC sensors include load current, operating (residual) current, and minimum holding current.

Load Current. *Load current* is the amount of current drawn by a load when energized. Because a solid-state sensor is wired in series with the load, the current drawn by the load must pass through the solid-state sensor. For example, if a load draws 5 A, the sensor must be able to safely switch 5 A. Five amperes burns out most solid-state proximity sensors because they are normally rated for a maximum of less than 0.5 A. An electromechanical or solid-state relay must be used as an interface to control the load if a solid-state sensor must switch a load above its rated maximum current. A solid-state relay is the preferred choice for an interface with a sensor. **See Figure 15-47.**

Technical Fact

Diffuse reflective sensors have an operating range of 10″, retroreflective sensors have an operating range of 10′, and through-beam sensors have an operating range of 120′.

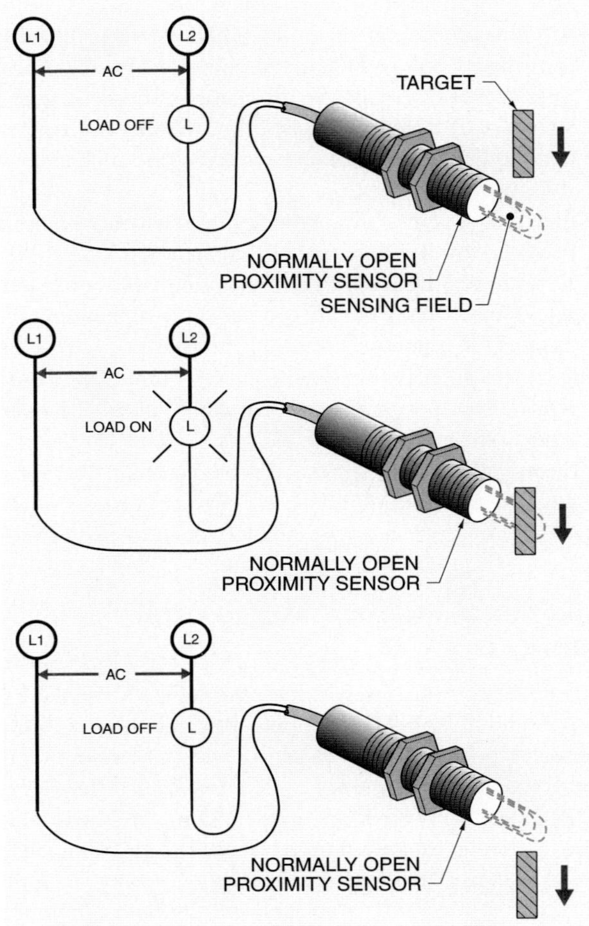

Figure 15-46. AC photoelectric and proximity sensors are connected in series with the load.

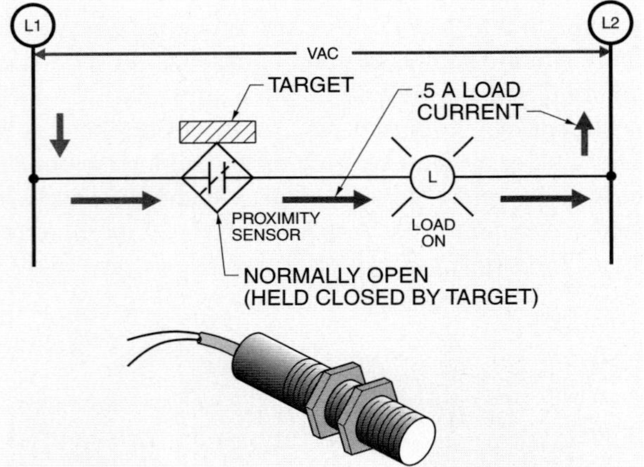

Figure 15-47. Load current is the amount of current drawn by a load when energized, and flows through AC photoelectric and proximity sensors.

Operating Current. *Operating current* (*residual or leakage current*) is the amount of current a sensor draws from the power lines to develop a field that can detect a target. When a sensor is in the OFF condition (target not detected), a small amount of current passes through both the sensor and the load. This operating current is required for the solid-state detection circuitry housed within the sensor. Operating currents are normally in the range of 1.5 mA to 7 mA for most sensors. **See Figure 15-48.**

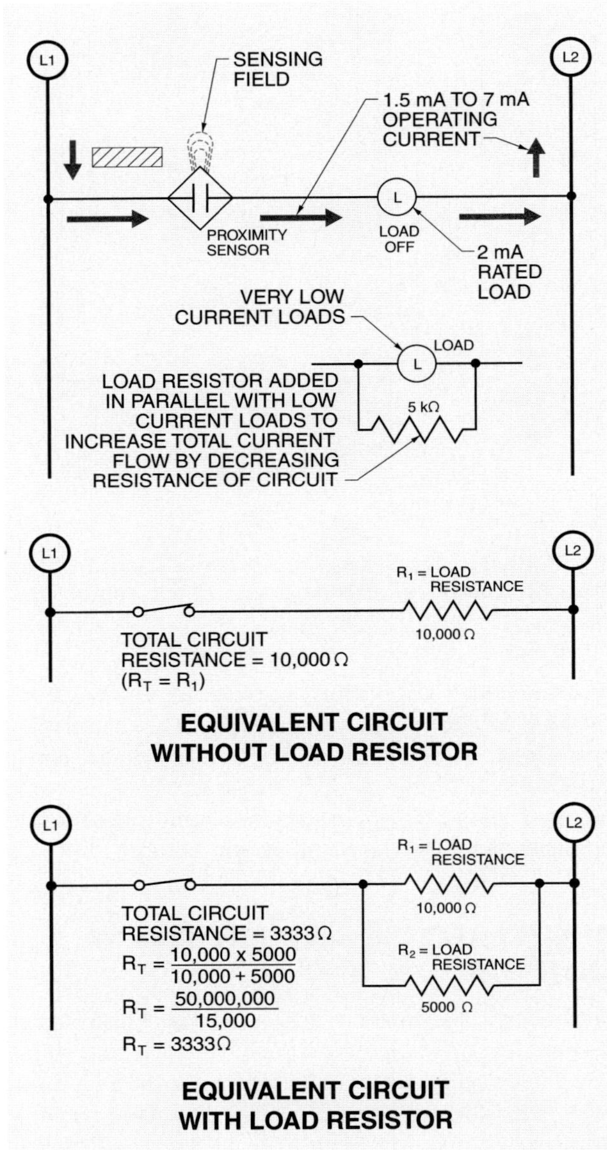

Figure 15-48. Operating current is the amount of current a sensor draws from the power lines to develop a field that can detect the target.

The small operating current normally does not have a negative effect on low-impedance loads or circuits such as mechanical relays, solenoids, and magnetic motor starters. However, the operating current may be enough to activate high-impedance loads such as programmable controllers, electronic timers, and other solid-state devices. In this case, the load is activated regardless of whether a target is present or not. This problem may be corrected by placing a load resistor in parallel with low current loads. The resistance value should be selected to ensure that the effective load impedance (load plus resistor) is reduced to a level that prevents false triggering due to the operating current, and that the minimum current required to operate the load is provided. This resistance value is normally in the range of 4.5 kΩ to 7.5 kΩ. A general rule is to use a 5 kΩ, 5 W resistor for most conditions.

Minimum Holding Current. *Minimum holding current* is the minimum amount of current required to keep a sensor operating. When the sensor has been triggered and is in the ON condition (target detected), the current drawn by the load must be sufficient to keep the sensor operating. Minimum holding currents range from 3 mA to 20 mA for most solid-state sensors. The amount of current a load draws must be correct for the proper operation of a sensor. Excessive current (operating current) burns up the sensor. Low current (minimum holding current) prevents proper operation of the sensor. **See Figure 15-49.**

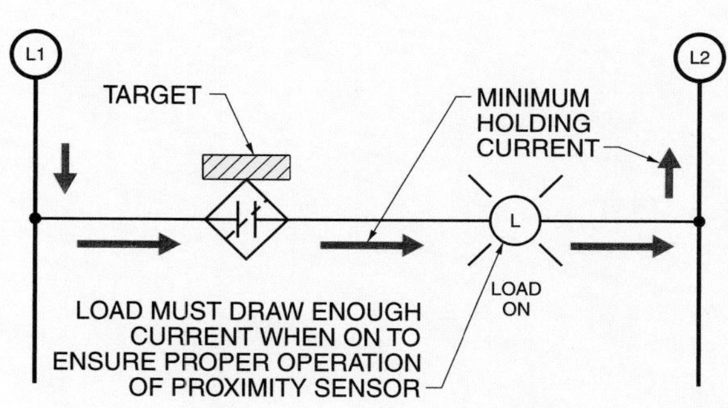

Figure 15-49. Minimum holding current is the minimum amount of current required to keep a sensor operating.

Series/Parallel Connections. All AC, two-wire photoelectric and proximity sensors may be connected in series or parallel to provide both AND and OR control logic. When connected in series (AND logic), all sensors must be activated to energize the load. When connected in parallel (OR logic), any one sensor that is activated energizes the load. **See Figure 15-50.**

As a general rule, a maximum of three sensors may be connected in series to provide AND logic. Factors that limit the number of AC, two-wire sensors that may be wired in series to provide AND logic include:

- AC supply voltage. Generally, the higher the supply voltage, the higher the number of sensors that may be wired in series.
- Voltage drop across the sensor. Voltage drop varies for different sensors. The lower the voltage drop, the higher the number of sensors that may be connected in series.
- Minimum operating load voltage. This varies depending on the load that is controlled. For every proximity sensor added in series with the load, less supply voltage is available across the load.

As a general rule, a maximum of three sensors may be connected in parallel to provide OR logic. Factors that limit the number of AC, two-wire sensors that may be wired in parallel to provide OR logic include:

- Photoelectric and proximity switch operating current. The total operating current flowing through a load is equal to the sum of each sensor's operating current. The total operating current must be less than the minimum current required to energize the load.
- Amount of current a load draws when energized. The total amount of current a load draws must be less than the maximum current rating of the lowest rated sensor. For example, if three sensors rated at 125 mA, 250 mA, and 275 mA are connected in parallel, the maximum rating of the load cannot exceed 125 mA.

DC Photoelectric and Proximity Sensors

Photoelectric and proximity sensors that switch DC circuits normally use transistors as the switching element. The sensors use NPN transistors or PNP transistors. For most applications, the exact transistor used does not matter, as long as the switch is properly connected into the circuit. However, NPN transistor sensors are far more common than PNP transistor sensors.

Technical Fact

NPN transistors are the standard transistors used in Japan, China, and the Pacific Rim while PNP transistors are standard in Europe. NPN transistors are faster and easier to implement, making them more popular than PNP transistors. In PNP transistors, the common is 0 V. In NPN transistors, the common is 24 V.

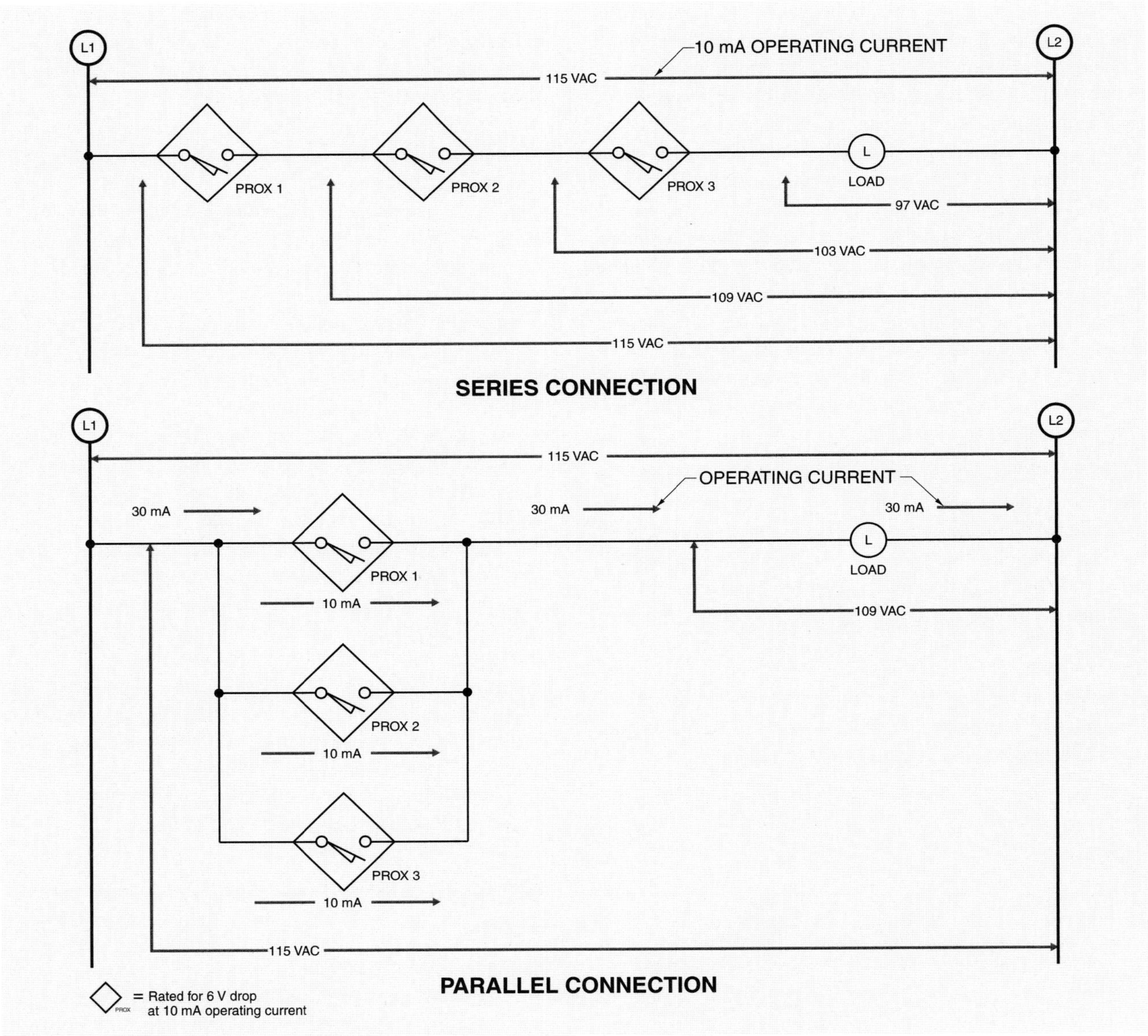

Figure 15-50. AC photoelectric and proximity sensors may be connected in series or parallel.

Technical Fact

Photoelectric sensors measure at distances from 4″ up to 10′.

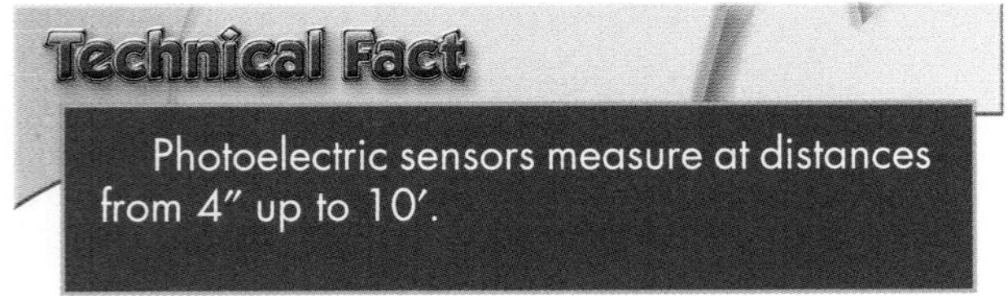

NPN Transistor Switching. When using an NPN transistor, the load is connected between the positive terminal of the supply voltage and the output terminal (collector) of the sensor. When the sensor detects a target, current flows through the transistor and the load is energized. **See Figure 15-51.**

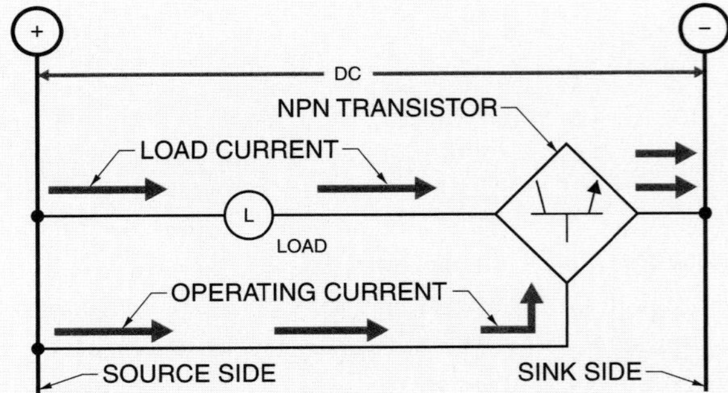

Figure 15-51. When an NPN (current sink) transistor is used, the load is connected between the positive terminal of the supply voltage and the output terminal of the sensor.

Output devices that use an NPN transistor as the switching element are current sink devices. The negative terminal of a DC system is the sink due to conventional current flowing into it. A current sinking switch "sinks" the current from the load.

PNP Transistor Switching. When using a PNP transistor, the load is connected between the negative terminal of the supply voltage and the output terminal (collector) of the sensor. When the sensor (current source) detects a target, current flows through the transistor and the load is energized. **See Figure 15-52.**

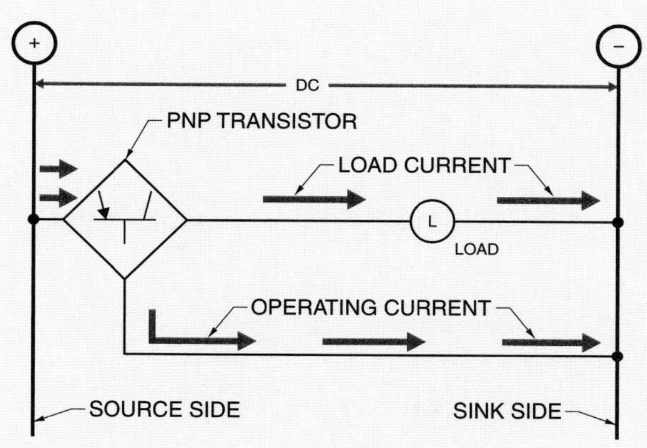

Figure 15-52. When a PNP (current source) transistor is used, the load is connected between the negative terminal of the supply voltage and the output terminal of the sensor.

PROXIMITY SENSOR INSTALLATION

Proximity sensors have a sensing head that produces a radiated sensing field. The sensing field detects the target of the sensor. The sensing field must be kept clear of interference for proper operation. *Interference* is any object other than the object to be detected that is sensed by a sensor. Interference may come from objects close to the sensor or from other sensors. General clearances are required for most proximity sensors.

Flush-Mounted Inductive and Capacitive Proximity Sensors

A distance equal to or greater than twice the diameter of the sensors is required between sensors when flush-mounting inductive and capacitive proximity sensors. The diameter of the largest sensor is used for installation when two sensors of different diameters are used. For example, at least 16 mm is required between sensors if two 8 mm inductive proximity sensors are flush-mounted. **See Figure 15-53.**

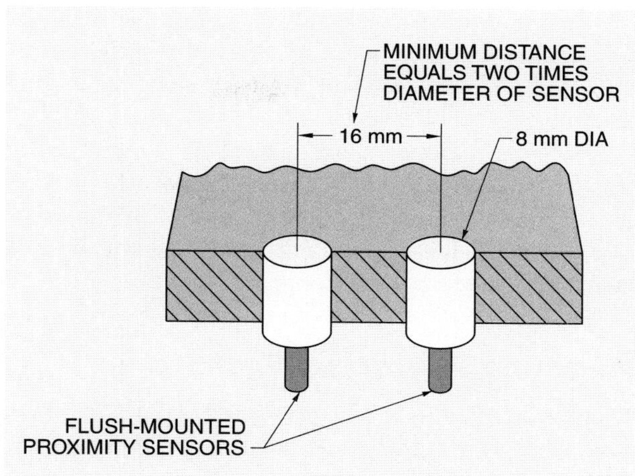

Figure 15-53. A distance equal to or greater than twice the diameter of the sensors is required between sensors when flush mounting inductive and capacitive proximity sensors.

Non-Flush-Mounted Inductive and Capacitive Proximity Sensors

A distance of three times the diameter of the sensor is required within or next to a material that may be detected when using non-flush-mounted inductive and capacitive proximity sensors. For example, at least 48 mm is required between sensors if two 16 mm capacitive proximity sensors are non-flush-mounted. **See Figure 15-54.**

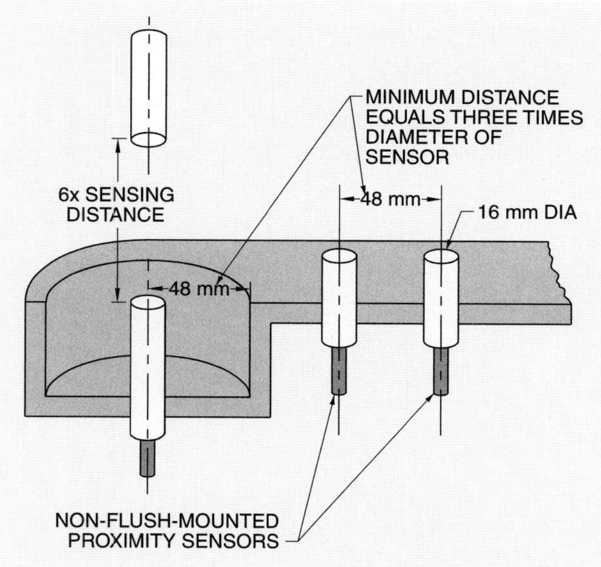

Figure 15-54. A distance of three times the diameter of the sensor is required within or next to a material that may be detected when using non-flush-mounted inductive and capacitive proximity sensors.

Three times the diameter of the largest sensor is required when inductive and capacitive proximity sensors are installed next to each other. Spacing is measured from center to center of the sensors.

Six times the rated sensing distance is required for proper operation when inductive and capacitive proximity sensors are mounted opposite each other. Six times the rated sensing distance is required because the sensing field causes false readings on the opposite sensor.

Mounting Photoelectric Sensors

A photoelectric sensor transmits a light beam. The light beam detects the presence (or absence) of an object. Only part of the light beam is effective when detecting the object. The *effective light beam* is the area of light that travels directly from the transmitter to the receiver. The object is not detected if the object does not completely block the effective light beam.

The receiver is positioned to receive as much light as possible from the transmitter when mounting photoelectric sensors. Greater operating distances are allowed and more power is available for the system to see through dirt and debris in the air and on the transmitter and receiver lenses when more light is available at the receiver. The transmitter is mounted on the clean side of the detection zone because light scattered by debris on the transmitter lens affects the system more than light scattered by debris on the receiver lens. **See Figure 15-55.**

TROUBLESHOOTING PHOTOELECTRIC AND PROXIMITY SENSORS

Photoelectric and proximity sensors typically have solid-state output switches. A solid-state switch has no moving parts (contacts). A solid-state switch uses a triac, SCR, current sink (NPN) transistor, or current source (PNP) transistor output to perform the switching function. The triac output is used for switching AC loads. The SCR output is used for switching high-power DC loads. The current sink and current source outputs are used for switching low-power DC loads. Solid-state switches include normally open, normally closed, or combination switching outputs. **See Figure 15-56.**

OUTPUT SWITCHING DEVICES	
Device	**Use**
TRIAC	SWITCH AC LOADS
SCR	SWITCH HIGH-POWER DC LOADS
NPN TRANSISTOR, PNP TRANSISTOR	SWITCH LOW-POWER DC LOADS

PROXIMITY SWITCH USING A SOLID-STATE SWITCHING DEVICE
OUTPUT SWITCHING DEVICE
TRIGGER CIRCUIT
SENSING FIELD
COIL
OSCILLATOR

SOLID-STATE SWITCHES

Figure 15-56. A solid-state switch uses a triac, SCR, current sink transistor, or current source transistor output to perform the switching function.

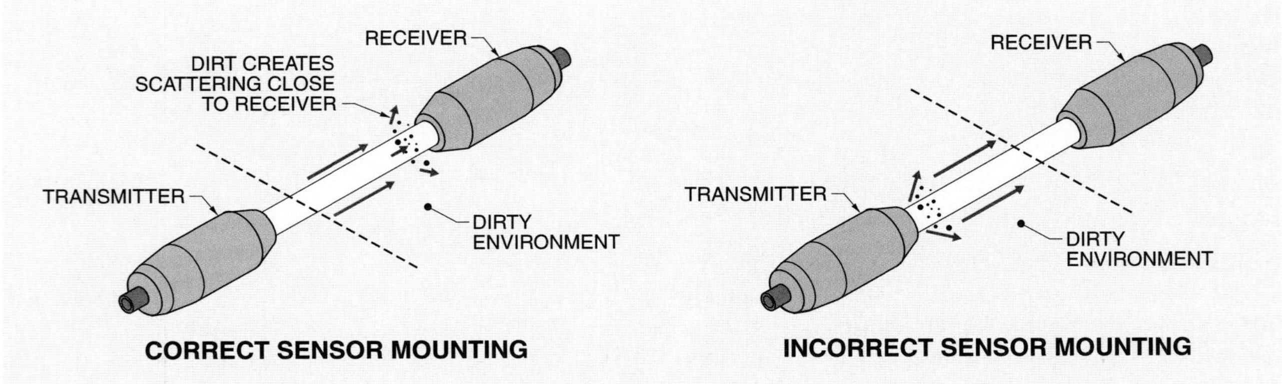

Figure 15-55. A transmitter is mounted on the clean side of the detection zone because light scattered by dirt on the transmitter lens affects the system more than light scattered by dirt on the receiver lens.

Two-Wire Solid-State Switches

A two-wire solid-state switch has two connecting terminals or wires (exclusive of ground). A two-wire switch is connected in series with the controlled load. A two-wire solid-state switch is also referred to as a load-powered switch because it draws operating current through the load. The operating current flows through the load when the switch is not conducting (load OFF). This operating current is inadequate to energize most loads. Operating current is also referred to as residual current or leakage current. Operating current may be measured with a DMM set to measure amperes when the load is OFF. **See Figure 15-57.**

The current in a circuit is a combination of the operating current and load current when a switch is conducting (load ON). A solid-state switching device must be rated high enough to carry the current of the load. Load current is measured with an ammeter when the load is ON.

Banner Engineering Corp.

Conveyor-beam™ sensors are designed to operate on either 24-240 VAC or 12-240 VDC and draw a maximum of only 2 W with a sensing range of 13.1′.

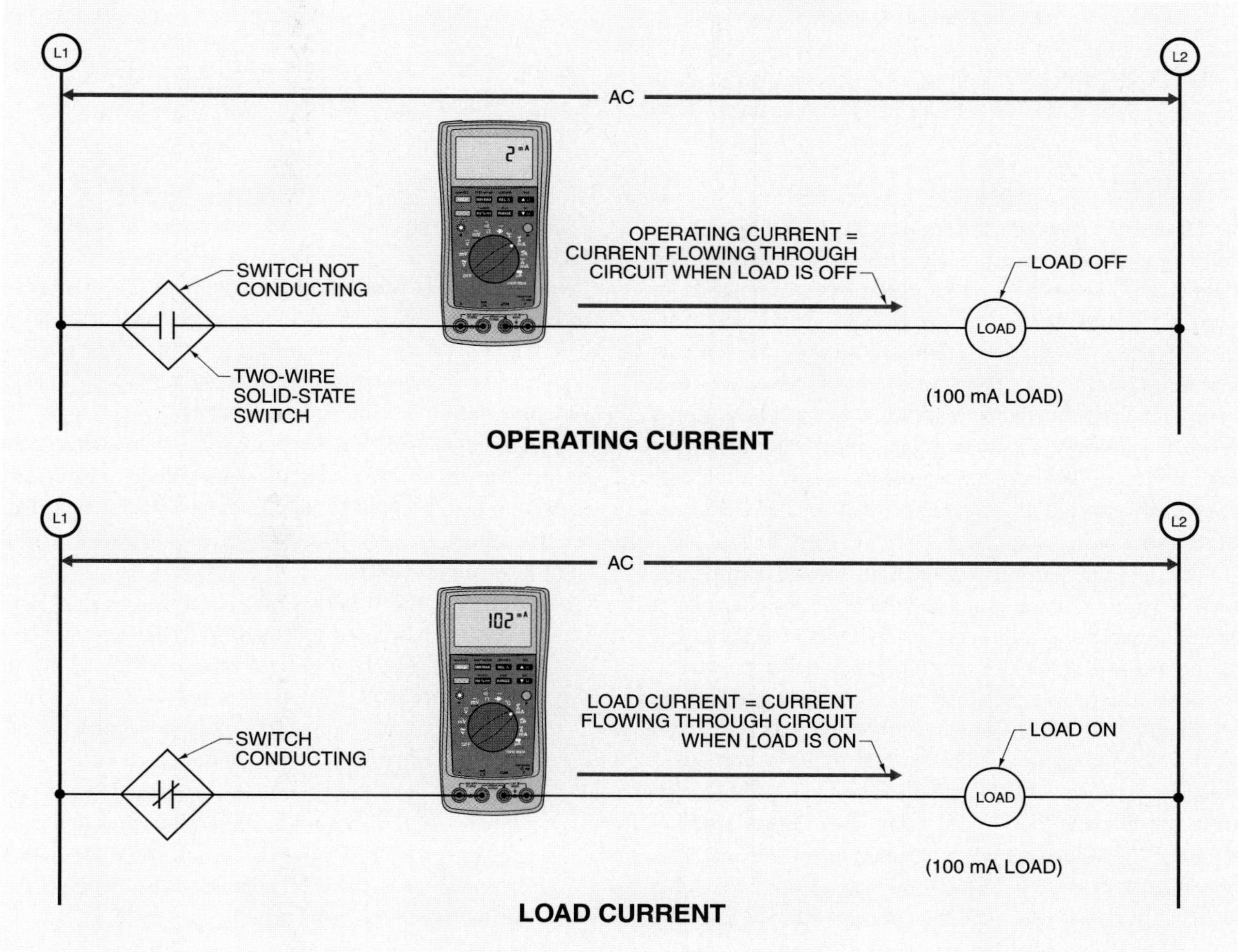

L1
L2
AC
2 mA
SWITCH NOT CONDUCTING
TWO-WIRE SOLID-STATE SWITCH
OPERATING CURRENT = CURRENT FLOWING THROUGH CIRCUIT WHEN LOAD IS OFF
LOAD OFF
LOAD
(100 mA LOAD)
OPERATING CURRENT

L1
L2
AC
102 mA
SWITCH CONDUCTING
LOAD CURRENT = CURRENT FLOWING THROUGH CIRCUIT WHEN LOAD IS ON
LOAD ON
LOAD
(100 mA LOAD)
LOAD CURRENT

Figure 15-57. A two-wire solid-state switch is also referred to as a load-powered switch because it draws operating current through the load.

Advanced Assembly Automation Inc.

Proximity sensors are used in manufacturing and assembly systems to control the positioning of products.

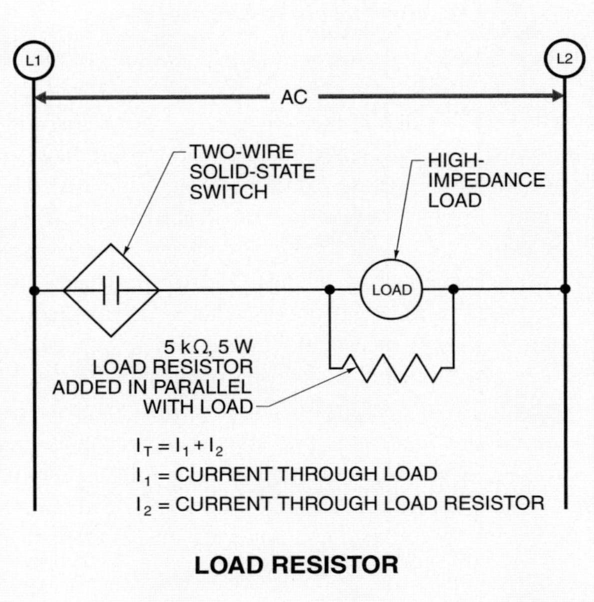

Figure 15-58. A load resistor acts as an additional load which increases the total current in the circuit.

The current draw of a load must be sufficient to keep the solid-state switch operating when the switch is conducting (load ON). Minimum holding current values range from 2 mA to 20 mA.

Operating current and minimum holding current values are normally not a problem when a solid-state switch controls a low-impedance load, such as a motor starter, a relay, or a solenoid. Operating current and minimum holding current values may be a problem when a solid-state switch controls a high-impedance load, such as a PLC or other solid-state device.

The operating current may be high enough to affect the load when the switch is not conducting. For example, a programmable controller may see the operating current as an input signal. A load resistor must be added to the circuit to correct this problem. A load resistor is connected in parallel with the load. The load resistor acts as an additional load which increases the total current in the circuit. Load resistors range in value from 4.5 kΩ to 7 kΩ. A 5 kΩ, 5 W resistor is used in most applications. **See Figure 15-58.**

Two-wire solid-state switches connected in series affect the operation of the load because of the voltage drop across the switches. A two-wire switch drops about 3 V to 8 V. The total voltage drop across the switches equals the sum of the voltage drop across each switch. No more than three solid-state switches should be connected in series. **See Figure 15-59.**

Two-wire solid-state switches connected in parallel affect the operation of the load because each switch has its operating current flowing through the load. The load may turn ON if the current through the load becomes excessive. The total operating current equals the sum of the operating current of each switch. No more than three solid-state switches should be connected in parallel.

A suspected fault in a two-wire solid-state switch may be tested using a DMM set to measure voltage. The DMM is used to test the voltage into and out of the switch. **See Figure 15-60.** To test a two-wire solid-state switch, apply the following procedure:

1. Measure the supply voltage into the switch. The problem is located upstream from the switch when there is no voltage present or the voltage is not at the correct level. The problem may be a blown fuse or open circuit. Voltage to the switch must be re-established before the switch may be tested.
2. Measure the voltage out of the switch. The voltage should equal the supply voltage minus the rated voltage drop (3 V to 8 V) of the switch when the switch is conducting (load ON). Replace the switch if the voltage output is not correct.

Warning: Always ensure power is OFF before changing a control switch. Use a DMM to ensure the power is OFF.

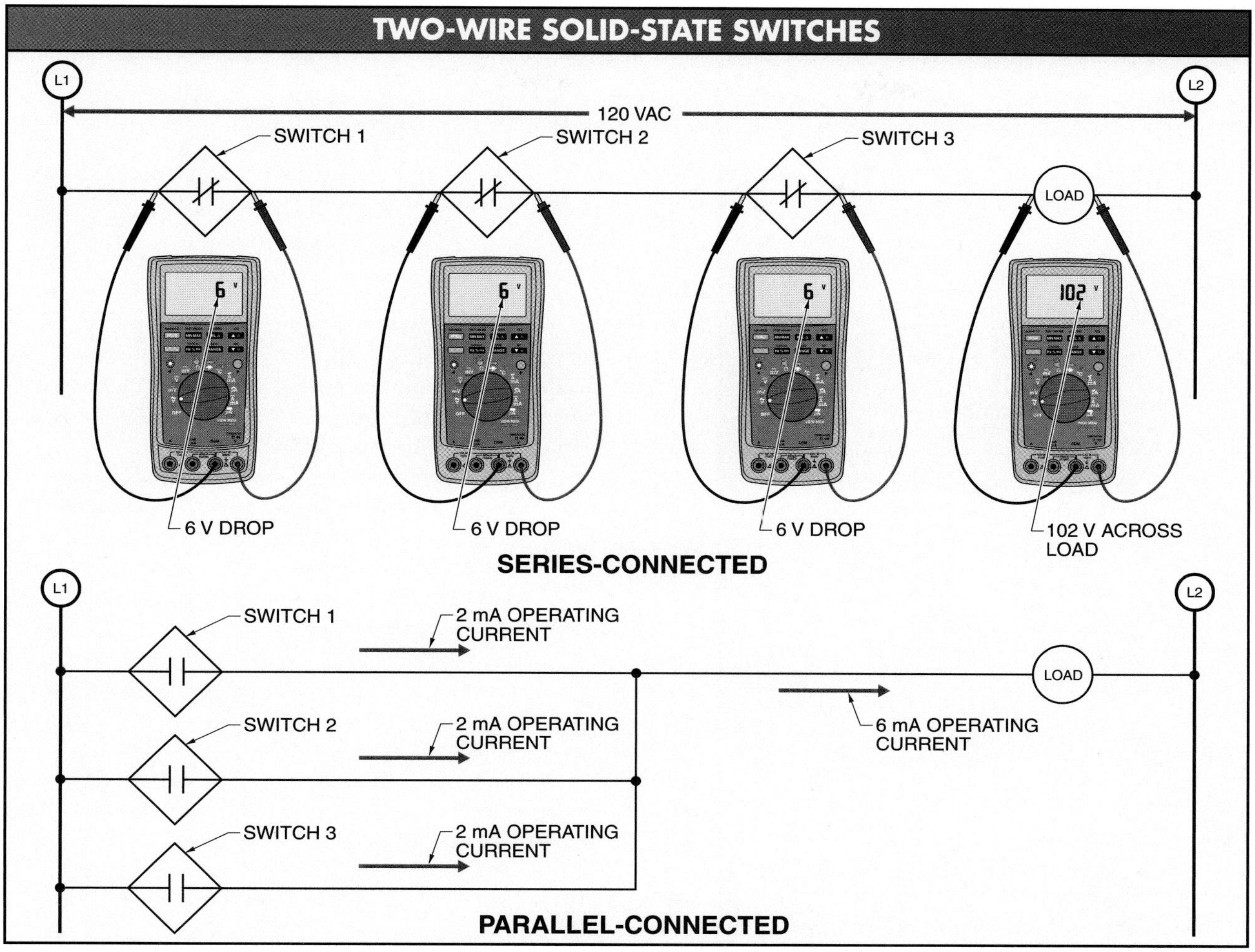

Figure 15-59. A two-wire solid-state switch has two connecting terminals or wires (exclusive of ground).

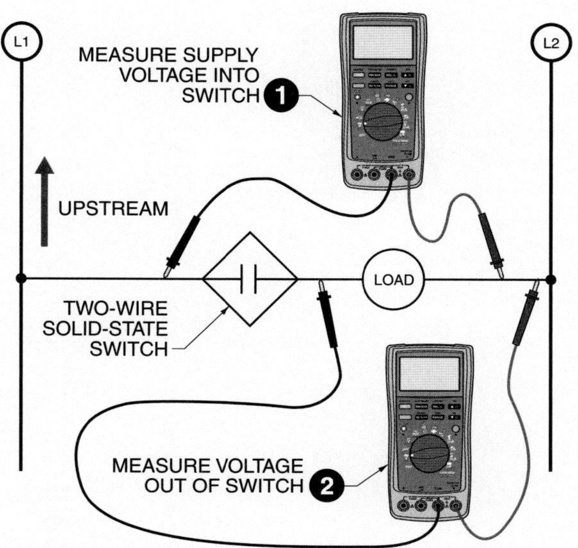

TWO-WIRE SOLID-STATE SWITCH TESTING

Figure 15-60. A suspected fault in a two-wire solid-state switch may be tested using a DMM set to measure voltage.

Three-Wire Solid-State Switches

Three-wire solid-state switches have three connecting terminals or wires (exclusive of ground). A three-wire solid-state switch draws its operating current directly from the power lines. The operating current does not flow through the switch. **See Figure 15-61.**

Three-wire solid-state switches connected in series affect the operation of the load because each switch downstream from the previous switch must carry the load current and the operating current of each switch. A DMM set to measure current may be used to measure operating and load current values. The measured values must not exceed the manufacturer's maximum rating.

Three-wire solid-state switches connected in parallel affect the operation of the load because the nonconducting switch may be damaged due to reverse polarity. A blocking diode should be added to each switch output to prevent reverse polarity on the switch.

THREE-WIRE SOLID-STATE SWITCHES

SWITCH 1 CARRIES LOAD CURRENT AND OPERATING CURRENT OF EACH SWITCH DOWNSTREAM
SWITCH 2
LOAD CURRENT
LOAD
SWITCH 1
SWITCH 2 OPERATING CURRENT
SWITCH 1 OPERATING CURRENT

SERIES-CONNECTED SWITCHES

SWITCH 1
BLOCKING DIODE
LOAD
SWITCH 2
BLOCKING DIODE ADDED TO PROTECT SWITCH

PARALLEL-CONNECTED SWITCHES

Figure 15-61. A three-wire solid-state switch draws its operating current directly from the power lines.

Namco Controls Corporation

Proximity sensors can withstand extreme shock and have 70 mm to 250 mm sensing ranges.

Technical Fact

A light year is the distance traveled by a pulse of light in one year, which is equal to 5.88 trillion miles. A jet traveling at 500 mph would take 1.34 million years to travel 1 light year.

A suspected fault with a three-wire solid-state switch may be tested using a DMM set to measure voltage. The DMM is used to test the voltage into and out of the switch. **See Figure 15-62.** To test a three-wire solid-state switch, apply the following procedure:

1. Measure the voltage into the switch. The problem is located upstream from the switch if there is no voltage present or the voltage is not at the correct level. The problem may be a blown fuse or open circuit. Voltage to the switch must be re-established before the switch may be tested.
2. Measure the voltage out of the switch. The voltage should equal the supply voltage when the switch is conducting (load ON). Replace the switch if the voltage out of the switch is not correct.

Warning: Always ensure power is OFF before changing a control switch. Use a DMM set to measure voltage to ensure the power is OFF.

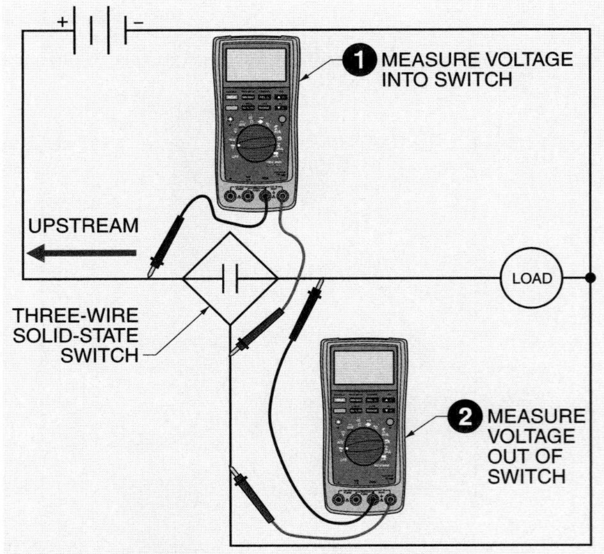

Figure 15-62. A suspected fault with a three-wire solid-state switch is tested by checking the voltage into and out of the switch.

Protecting Switch Contacts

Switches are rated for the amount of current they can switch. The switch rating is usually specified for switching a resistive load. Resistive loads are the least destructive loads to switch. Inductive loads are the most destructive loads to switch. Most loads that are switched are inductive loads such as solenoids, relays, and motors.

A large induced voltage appears across the switch contacts when inductive loads are turned OFF. The induced voltage causes arcing at the switch contacts. Arcing may cause the contacts to burn, stick, or weld together. Contact protection should be added when frequently switching inductive loads to prevent or reduce arcing. **See Figure 15-63.**

A diode is added in parallel with the load to protect contacts that switch DC. The diode does not conduct when the load is ON. The diode conducts when the switch is open, providing a path for the induced voltage in the load to dissipate.

A resistor and capacitor are connected across the switch contacts to protect contacts that switch AC. The capacitor acts as a high-impedance load (resistor) at 60 Hz, but becomes a short circuit at the high frequencies produced by the induced voltage of the load. This allows the induced voltage to dissipate across the resistor when the load is switched OFF.

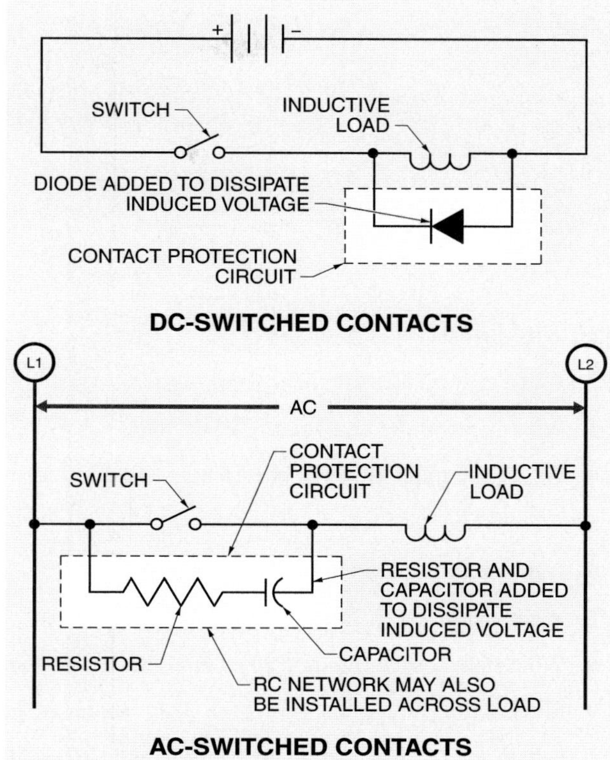

Figure 15-63. Contact protection should be added when frequently switching inductive loads to prevent or reduce arcing.

Review Questions

1. What is direct scan?
2. When may an aperture have to be used?
3. What is retroreflective scan?
4. What scan method works best in high-vibration areas?
5. What is specular scan?
6. What is diffuse scan?
7. What is convergent beam scan?
8. What light source (modulated or unmodulated) should generally be the first choice for a given application?
9. In what type of applications is the rated response time of a photoelectric control most important?
10. What type of cable should be used in applications that require cable flexing?
11. When does a dark-operated photoelectric control energize the output switch?
12. When does a light-operated photoelectric control energize the output switch?
13. What are the two basic operating modes of ultrasonic sensors?
14. What are the two basic types of proximity sensors?
15. What does a Hall effect sensor detect?
16. What type of solid-state output is used to switch AC circuits?
17. What type of solid-state output is used to switch DC circuits?
18. What is load current?
19. What is another name for operating current?
20. What is minimum holding current?
21. What is a flow detection sensor?
22. What is the maximum number of two-wire AC sensors that should generally be connected in series?
23. What type of transistor (NPN or PNP) is used most frequently as an output?
24. What type of proximity sensor is used to detect metallic objects?
25. What type of proximity sensor is used to detect materials that have a relatively high dielectric constant?

PROGRAMMABLE CONTROLLERS

A *programmable controller (PLC)* is a solid-state control device that is programmed and reprogrammed to automatically control an industrial process or machine. PLCs are capable of many industrial functions and applications and are widely used in automated industrial applications.

The automotive industry was the first to recognize the advantages of PLCs. Annual model changes required constant modifications of production equipment controlled by relay circuitry. In some cases, entire control panels had to be scrapped and new ones designed and built with new components. This resulted in increased production costs.

The automotive industry was looking for equipment that could reduce changeover costs required by model changes. In addition, the equipment had to operate in a harsh factory environment of dirty air, vibration, electrical noise, and wide temperature and humidity ranges.

To meet this need, a ruggedly constructed computer-like control was developed. The PLC could easily accommodate constant circuit changes using a keyboard to introduce new operation instructions. In 1968, the first PLC was delivered to General Motors (GM) in Detroit by Modicon.

The first PLCs were large and costly. Their initial use was in large systems with the equivalent of 100 or more relays. Today, PLCs are available in all sizes from micro, which are cost-effective equivalents of as few as 10 relays, to large units with the equivalent of thousands of inputs and outputs.

PLCs are popular because they can be programmed and reprogrammed using ladder (line) diagrams that plant personnel understand. Required machine operation is programmed and read as a line diagram showing open and closed contacts. This is the same approach used to describe relay logic circuits. Programming machine operation as a line diagram allows the use of a computer-like device without learning a computer language.

Advantages of using a PLC include the following:

- Reduced hard wiring and reduced wiring cost
- Reduced space requirements due to small size compared to using standard relays, timers, counters, and other control components
- Flexible control because all operations are programmable
- High reliability using solid-state components
- Storage of large programs and data due to microprocessor-based memory
- Improved on-line monitoring and troubleshooting by monitoring and diagnosing its own failures as well as those of the machines and processes it controls
- Elimination of the need to stop a controlled process to change set parameters
- Provision for analog, digital, and voltage inputs as well as discrete inputs such as pushbuttons and limit switches
- Modular design allows components to be added, substituted, and rearranged as requirements change
- Programming languages used are familiar and follow industrial standards, such as line diagrams

Although the first PLCs were designed to replace relays, today's PLCs are used to achieve factory automation and interfacing with robots, numerical control (NC) equipment, CAD/CAM systems, and general-purpose computers. PLCs are used in almost all segments of industry where automation is required. **See Figure 16-1.**

Giddings & Lewis, Inc.

Figure 16-1. PLCs are used to achieve factory automation and interfacing with robots, numerical control equipment, CAD/CAM systems, and general-purpose computers.

PLC Usage

Some electrical components, such as pushbuttons and fuses, are used in most types of residential, commercial, and industrial electrical systems. Other electrical components, such as PLCs, are used primarily in only one type of electrical system. PLCs are commonly used in industrial electrical systems that are designed to manufacture a product. Industrial electrical systems designed to produce products are commonly divided into discrete parts manufacturing and process manufacturing. In discrete parts and process manufacturing, PLCs have become the standard component used to control the operation from start to finish.

Discrete Parts Manufacturing. The discrete parts manufacturing market represents durable goods such as automobiles, washers, refrigerators, and tractors. Discrete parts manufacturing is done primarily by stand-alone machines that bend, drill, punch, grind, and shear metals. All of these machines can be automated with PLCs. **See Figure 16-2.**

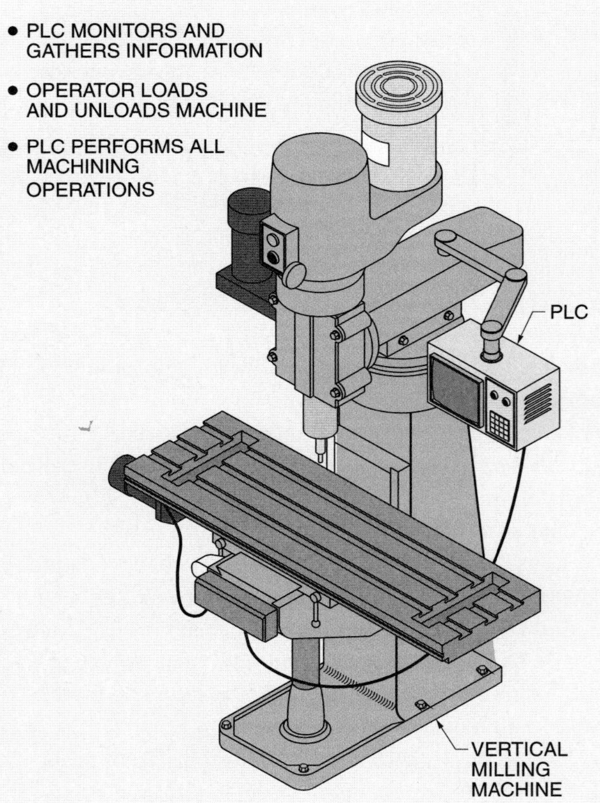

Figure 16-2. A PLC can be used to control all electrical functions on a machine used in discrete parts manufacturing.

The PLC allows each machine to have its own unique capability using standard hardware. The PLC allows easy modification of the controls when the functional requirements of the machine change. Modular replacement of PLCs reduces downtime of the machine. PLC use helps reduce startup and debug time, and allows manufacturers to incorporate additional user requirements for changes in machine operations after startup.

Today, the PLC has become the standard for machine builders. Increased capabilities in a reduced size allow today's PLCs to control one machine or link up to many machines in any network configuration.

In addition to allowing each machine to have its own unique capabilities, a PLC can also be used to interface and control the operation of all or parts of the machines along a production line. PLCs can be used to control the speed of a production line, divert production to other lines when there is a problem, make product changes, and maintain documentation such as inventory and losses. **See Figure 16-3.**

Process Manufacturing. The process manufacturing industry produces consumables such as food, gas, paint, pharmaceuticals, paper, and chemicals. Most of these processes require systems to blend, cook, dry, separate, or mix ingredients. **See Figure 16-4.**

Automation is required for opening and closing valves and controlling motors in the proper sequence and at the correct time. A PLC allows for easy modifications to the system if the time, temperature, or flow requirements of the products change.

Today's PLCs control process manufacturing activities such as the conveying, palletizing, storing, and treatment of the product, and the alarms, interlocks, and preventive maintenance functions for the system. The PLC can also generate reports that are used to determine production efficiency.

PLC manufacturers offer a variety of PLCs from micro to very large units. **See Figure 16-5.** A micro or small PLC is the best choice for machines and processes that have limited capability and little potential for future expansion.

PLCs AND PCs

PLCs have grown in popularity for applications that were once handled exclusively by personal computers (PCs). PCs feature fast number manipulation and powerful text-handling capabilities. PLCs offer several advantages compared to PCs for industrial control applications.

The first difference between a PLC and a PC is that a PLC is designed to communicate directly with inputs from the machine and process and control outputs. The PLC recognizes these inputs and outputs (I/Os) as part of its internally programmed system. Inputs include limit switches, pushbuttons, temperature controls, photoelectric controls, analog signals, American Standard Code for Information Interchange (ASCII), serial data, and other inputs. The outputs include voltage or current levels that drive end devices such as solenoids, motor starters, relays, and lights. Other outputs are analog devices, digital binary coded decimal (BCD) displays, ASCII-compatible devices, and other PLCs and computers.

The second difference between PLCs and PCs is the ease of programming a PLC. A PLC uses simple programming techniques that are easily learned and understood. Simple ladder (line) diagram programming does not require knowledge of computer languages. A PLC can be programmed and reprogrammed on-line while a process is running. Hardware modifications are not required.

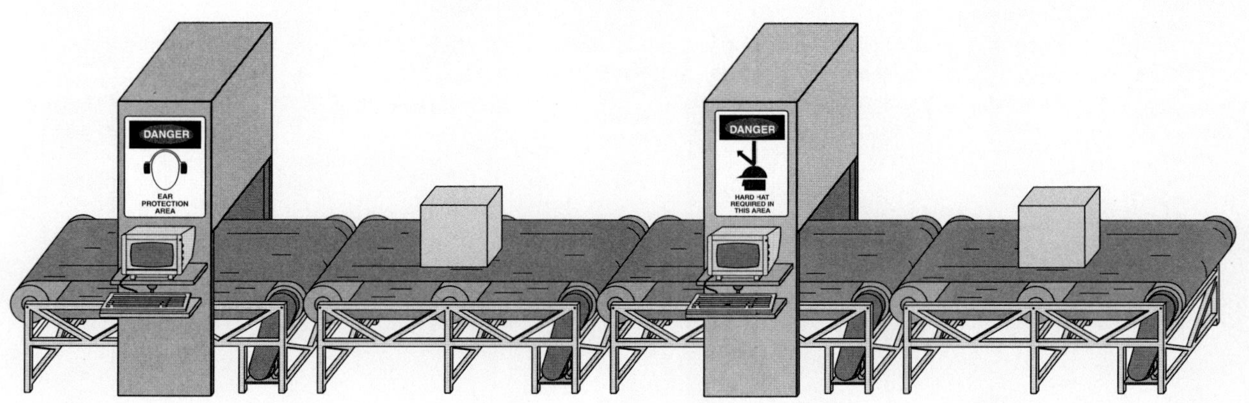

Figure 16-3. PLCs can be used to control the speed of a production line, divert production to other lines when there is a problem, make product changes, and maintain documentation such as inventory and losses.

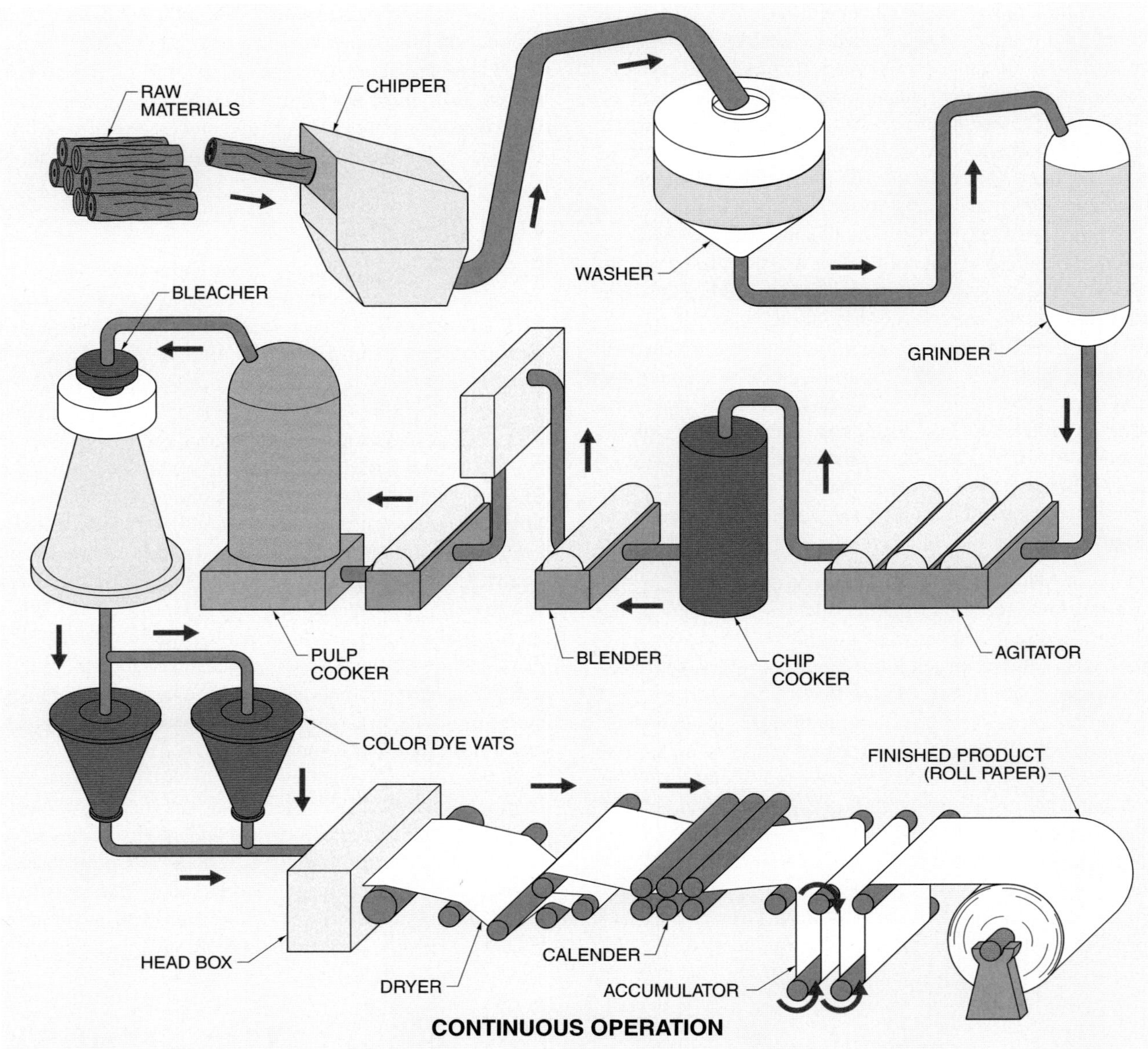

Figure 16-4. The process manufacturing industry produces consumables, such as food, gas, paint, pharmaceuticals, and chemicals, which require systems to blend, cook, dry, separate, or mix ingredients.

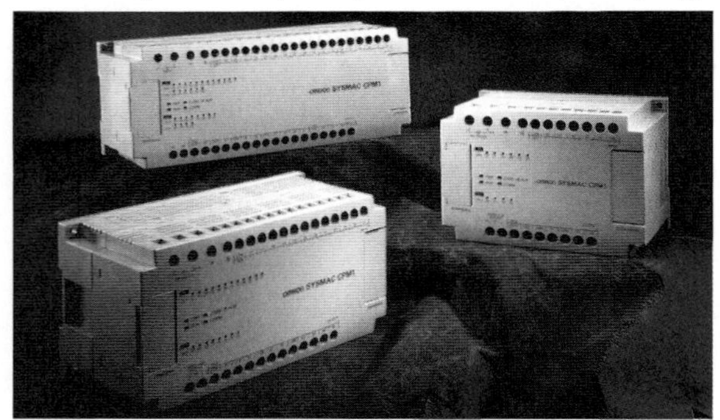

Omron Electronics, Inc.

Figure 16-5. PLC manufacturers offer a variety of PLCs for machines and processes that have limited capability and little potential for future expansion, and for processes that have complex control requirements.

The third difference between PLCs and PCs is that a PLC is designed specifically for use in an industrial environment. **See Figure 16-6.** Variations in levels of noise, vibration, temperature, and humidity do not adversely affect PLC operation. A PC cannot withstand typical industrial environments.

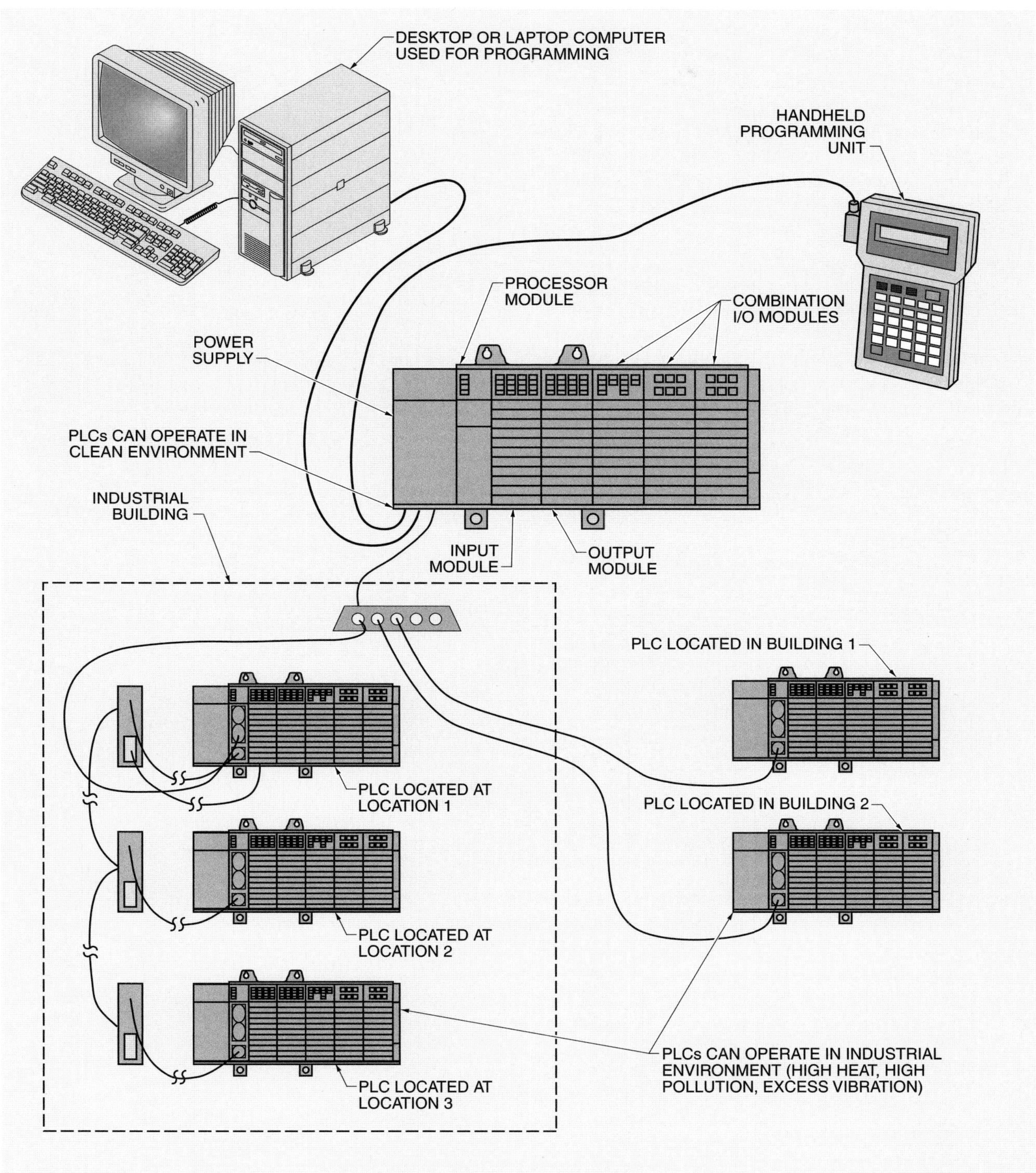

Figure 16-6. PLCs are designed to withstand fluctuations in noise, vibration, temperature, and humidity in the industrial environment.

A PLC can be programmed using either a handheld programming unit or a PC. Most PLCs are programmed using a PC because handheld programming units are specific to a manufacturer's PLC type only. A PC is used to develop, store, and monitor the program. The PC program (normally in line diagram format) is downloaded to the PLC. The PLC and PC are interconnected through their input and output ports. This allows the PLC to be located in a harsh industrial environment and the PC to be located in a less harsh environment. In addition to a standard PC, a laptop PC can also be used to program and/or monitor a PLC operation. The laptop PC can be used as a portable field device.

PLC Configurations

PLCs can be used as stand-alone control devices or configured into a system. A PLC can be configured into a system that uses a PC, a handheld programming unit, an operator interface panel, other PLCs, or other devices that connect into an electrical system. The amount and level of control between the components in a PLC depends on the system operating requirements and cost. The configuration of a PLC system is determined by the type and quantity of required I/O devices, required communication between devices, programming types and requirements, and future needs. **See Figure 16-7.**

PLC PARTS

All PLCs have four basic parts. The four basic parts of a PLC include the power supply, input/output section, processor section, and programming section. **See Figure 16-8.**

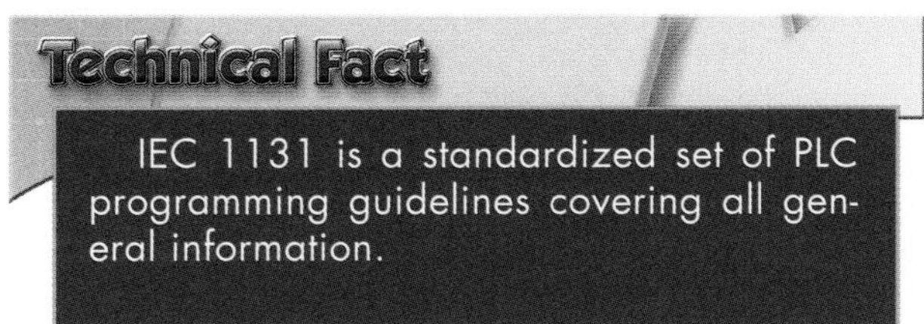

PLC SYSTEM CONFIGURATIONS

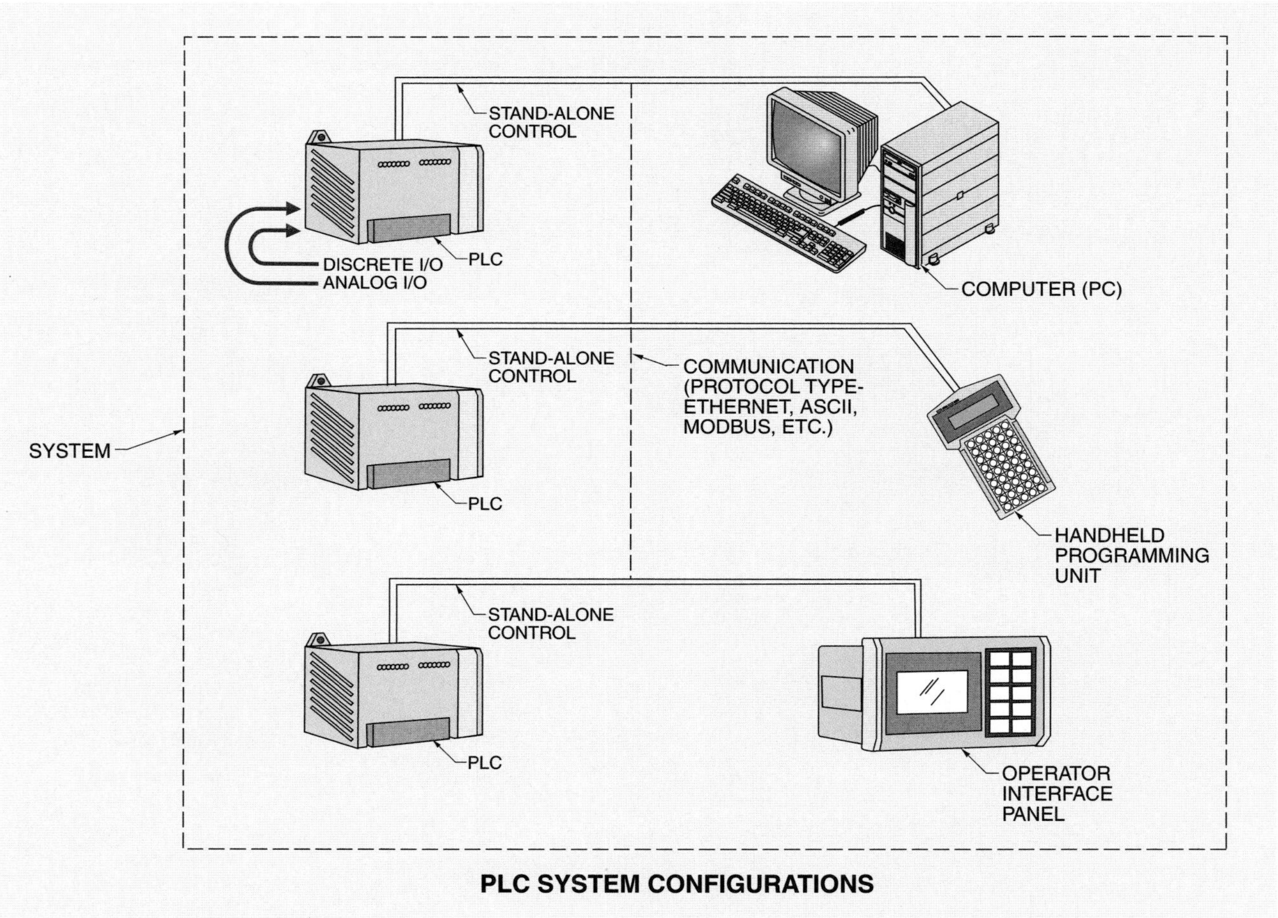

Figure 16-7. The configuration of a PLC system is determined by the type and quantity of required I/O devices, required communication between devices, programming types and requirements, and future needs.

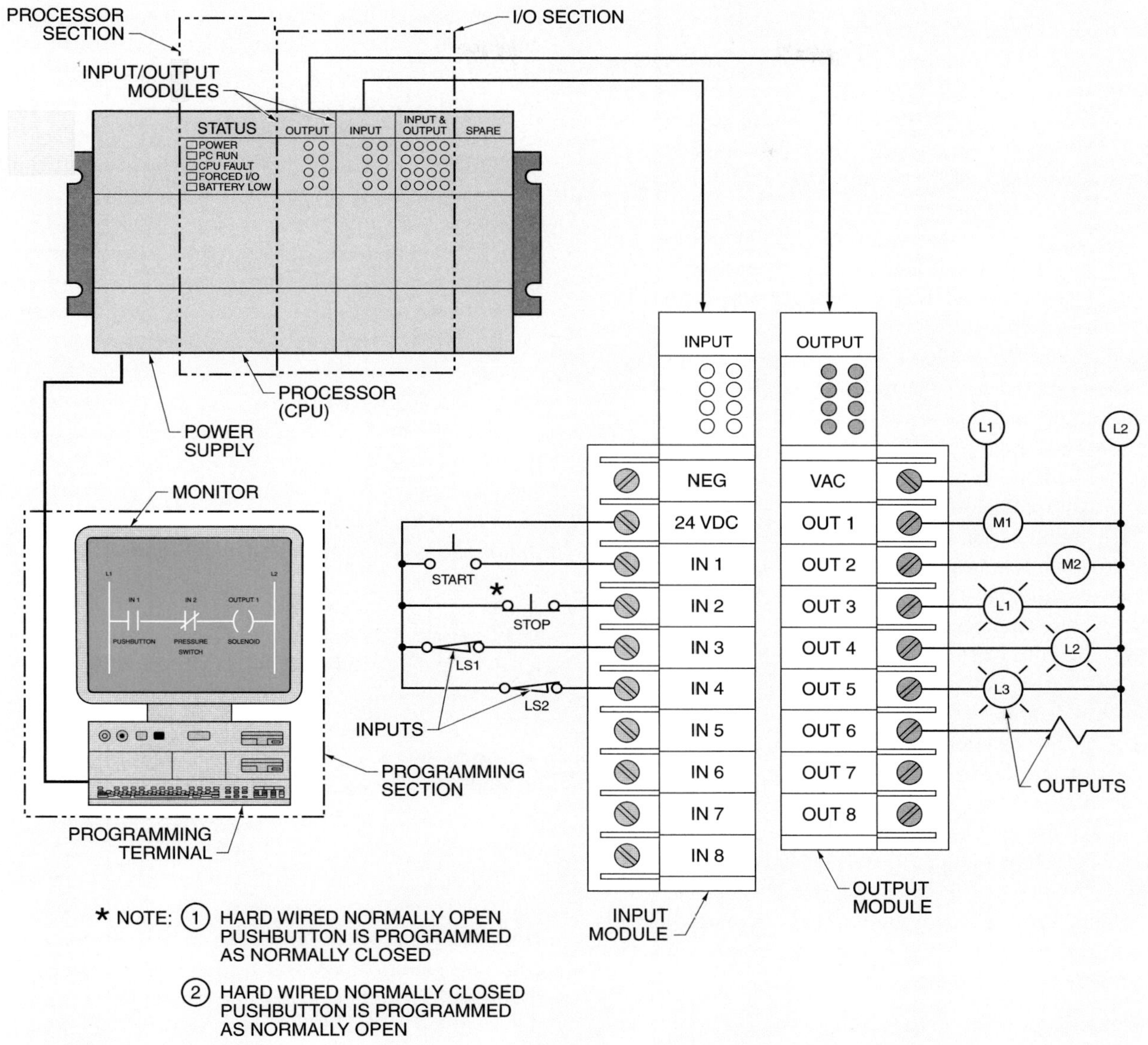

Figure 16-8. The four basic parts of a PLC include the power supply, input/output section, processor section, and programming section.

The programs used in manufacturing parts, equipment, etc., and processing goods and other consumables are stored in and retrieved from memory as required. Sections of the PLC are interconnected and work together to allow the PLC to accept inputs from a variety of sensors, make a logical decision as programmed, and control outputs such as motor starters, solenoids, valves, and drives.

Power Supply

The power supply provides necessary voltage levels required for the internal operations of the PLC. In addition, it may provide power for the input/output modules. The power supply can be a separate unit or built into the processor section. It takes the incoming voltage (normally 120 VAC or 240 VAC) and changes the voltage as required (normally 5 VDC to 32 VDC).

The power supply must provide constant output voltage free of transient voltage spikes and other electrical noise. The power supply also charges an internal battery in the PLC to prevent memory loss when external power is removed. The operating life of lithium batteries is from 3 years to 5 years.

Input/Output Section

The input/output section functions as the eyes, ears, and hands of the PLC. The input section is designed to receive information from pushbuttons, temperature switches, pressure switches, photoelectric and proximity switches, and other sensors. The output section is designed to deliver the output voltage required to control alarms, lights, solenoids, starters, and other loads.

The input section receives incoming signals (normally at a high voltage level) and converts them to low-power digital signals that are sent to the processor section. The processor then registers and compares the incoming signals to the program.

The output section receives low-power digital signals from the processor and converts them into high-power signals. These high-power signals can drive industrial loads that can light, move, grip, rotate, extend, release, heat, and perform other functions.

The input/output section can either be located on the PLC (onboard) or be part of expansion modules. Onboard inputs and outputs are a permanent part of the PLC package. Expansion modules are removable units that include inputs, outputs, or combinations of inputs and outputs.

Onboard inputs and outputs usually include a fixed number of inputs and outputs that define the limits of the PLC. For example, a small PLC may include up to 16 inputs and eight outputs. This means that up to 16 inputs and eight outputs may be connected to the PLC. PLCs that use expansion modules allow the total number of inputs and/or outputs to be changed by changing or adding modules. Onboard PLCs are normally used for individual machines and small systems. Expansion PLCs are normally used for large systems or small systems that require flexible changes.

Discrete I/Os. Discrete I/Os are the most common inputs and outputs. Discrete I/Os use bits, with each bit representing a signal that is separate and distinct, such as ON/OFF, open/closed, or energized/de-energized. The processor reads this as the presence or absence of power.

Examples of discrete inputs are pushbuttons, selector switches, joysticks, relay contacts, starter contacts, temperature switches, pressure switches, level switches, flow switches, limit switches, photoelectric switches, and proximity switches. Discrete outputs include lights, relays, solenoids, starters, alarms, valves, heating elements, and motors.

Data I/Os. In many applications, more complex information is required than the simple discrete I/O is capable of producing. For example, measuring temperature may be required as an input into the PLC and numerical data may be required as an output. Data I/Os are inputs and outputs that produce or receive a variable signal. They may be analog, which allows for monitoring and control of analog voltages and currents, or they may be digital, such as BCD inputs and outputs.

When an analog signal such as voltage or current is input into an analog input card, the signal is converted from analog to digital by an analog-to-digital (A/D) converter. The converted value, which is proportional to the analog signal, is sent to the processor section. After the processor has processed the information according to the program, the processor outputs the information to a digital-to-analog (D/A) converter. The converted signal can provide an analog voltage or current output that can be used or displayed on an instrument in a variety of processes and applications.

Examples of data inputs are potentiometers, rheostats, encoders, bar code readers, and temperature, level, pressure, humidity, and wind speed transducers. Examples of data outputs are analog meters, digital meters, stepper motor (signals), variable voltage outputs, and variable current outputs.

I/O Capacity. The size of a PLC is based on the controller I/O capacity. Common I/O capacities of different size PLCs include:

- Mini/micro – 32 or fewer I/Os, but may have up to 64
- Small – 64 to 128 I/Os, but may have up to 256
- Medium – 256 to 512 I/Os, but may have up to 1023
- Large – 1024 to 2048 I/Os, but may have many thousands more on very large units

The inputs and outputs may be directly connected to the PLC or may be in a remote location. I/Os in a location remote from the processor section can be hard wired to the controller, multiplexed over a pair of wires, or sent by a fiber-optic cable. In any case, the remote I/O is still under the control of the central processor section. Common PLCs may have 16, 32, 64, 128, or 256 remote I/Os.

Fiber-optic communication modules route signals from the inputs, to the processor section, and then to the outputs. Fiber-optic communication modules are unaffected by noise interference and are commonly used for process applications in the food industry and petrochemical industry, and in hazardous locations.

Processor Section

The processor section is the brain of the PLC. The *processor section* is the section of a PLC that organizes all control activity by receiving inputs, performing logical decisions according to the program, and controlling the outputs. **See Figure 16-9.**

Rockwell Automation, Allen-Bradley Company, Inc.

Figure 16-9. The processor section organizes all control activity by receiving inputs, performing logical decisions, and controlling the outputs.

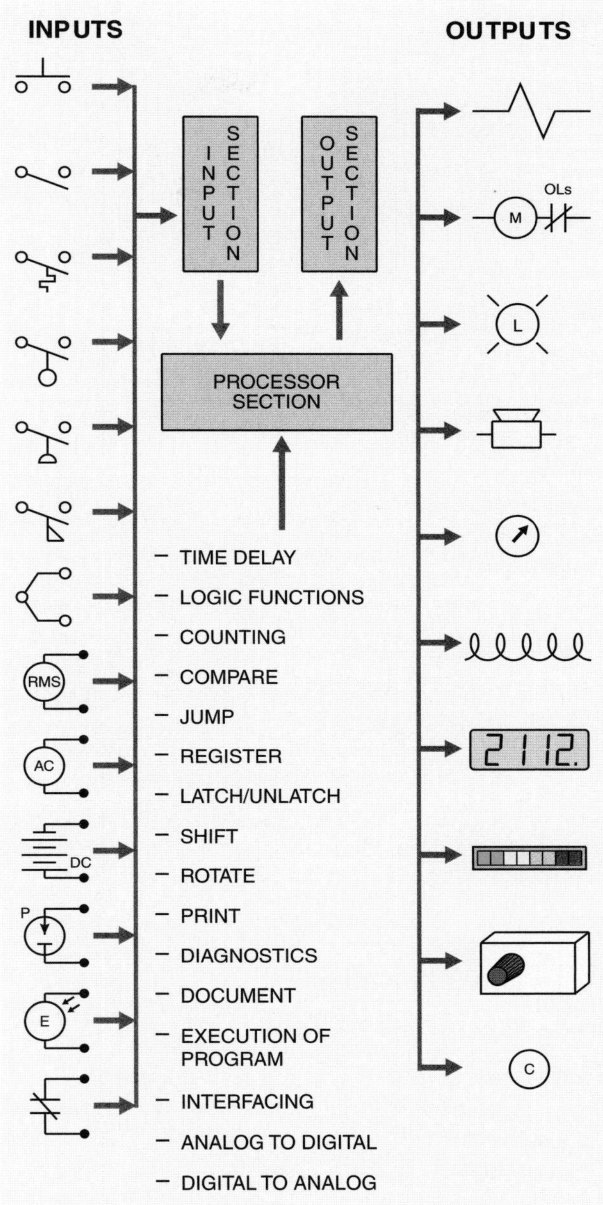

Figure 16-10. The processor continuously examines the status of the inputs and outputs and updates them according to the program.

The processor section evaluates all input signals and levels. This data is compared to the memory in the PLC, which contains the logic of how the inputs are interconnected in the circuit. The interconnections are programmed into the processor by the programming section. The processor section controls the outputs based on the input conditions and the program. The processor continuously examines the status of the inputs and outputs and updates them according to the program. **See Figure 16-10.**

Scan is the process of evaluating the input/output status, executing the program, and updating the system. *Scan time* is the time it takes a PLC to make a sweep of the program. Scan time is normally given as the time per 1 kilobyte of memory and normally is listed in milliseconds (ms). Scanning is a continuous and sequential process of checking the status of inputs, evaluating the logic, and updating the outputs.

The processor section of a PLC has different modes. The different modes allow the PLC to be taken on-line (system running) or off-line (system on standby). Processor modes include the program, run, and test modes. The program mode is used for developing the logic of the control circuit. In the program mode, the circuit is monitored and the program is edited, changed, saved, and transferred.

The run mode is used to execute the program. In the run mode, the circuit may be monitored and the inputs and outputs forced. Program changes cannot normally be made in the run mode. The test mode is used to check the program without energizing output circuits or devices. In the test mode, the circuit is monitored and inputs and outputs are forced (without actually energizing the load connected to the output). **See Figure 16-11.**

Warning: A PLC is switched from the program mode to the run mode by placing the controller in the run mode. The machine or process is started when the controller is placed in the run mode. Extreme care must be taken to ensure that no damage to personnel or equipment occurs when switching the controller to the run mode. Only qualified personnel should change processor modes, and key-operated switches should always be used in any dangerous application.

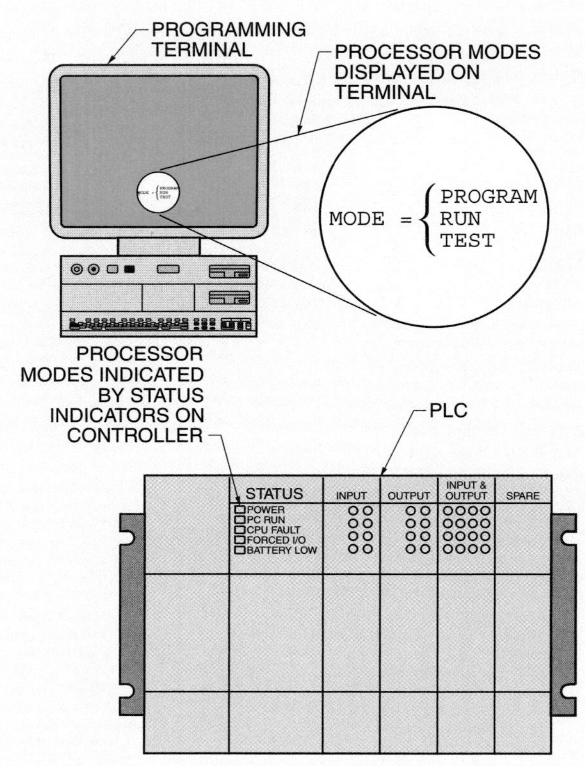

Figure 16-11. The processor section of a PLC has different modes that allow the PLC to be taken on-line (system running) or off-line (system on standby).

Programming Section

The *programming section* of a PLC is the section that allows input into the PLC through a keyboard. The processor must be given exact, step-by-step directions. This includes communicating to the processor such things as load, set, reset, clear, enter in, move, and start timing. Programming a PLC involves the programming device that allows access to the processor and the programming language that allows the operator to communicate with the processor section.

Programming Devices. Programming devices vary in size, capability, and function. Programming devices are available as simple, small text display units or complex color CRTs with monitoring and graphics capabilities. **See Figure 16-12.**

A programming device may be connected permanently to the PLC or connected only while the program is being entered. Once a program is entered, the programming device is no longer needed, except to make changes in the program or for monitoring functions. Some PLCs are designed to use an existing personal computer (PC) for programming. *Off-line programming* is the use of a personal computer to program a PLC that is not in the run mode. This permits the computer to be used for other purposes when not being used with the PLC.

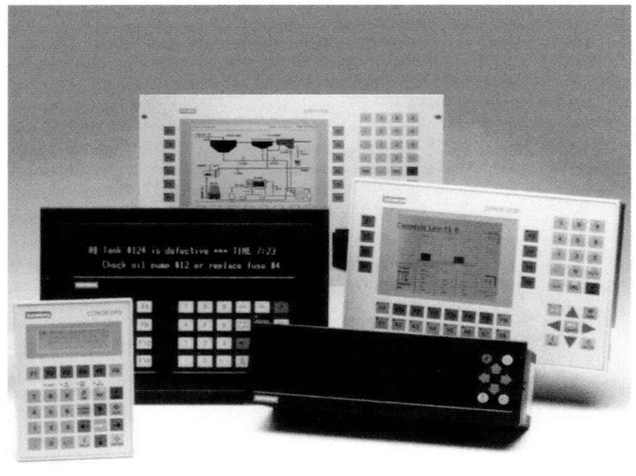

Siemens Corporation

Figure 16-12. Programming devices are available as simple, small text display units or complex color CRTs with monitoring and graphics capabilities.

Programming Symbols

PLC programs are designed using PLC software. PLC software uses different types of symbols, letters, and numbers to designate each component. Components such as inputs, outputs, relays, timers, and counters each have their own symbol and addressing (assigned values and numbers). When designing and programming a circuit using a PLC, component symbols are selected and placed on the screen as the circuit is developed. The symbols are commonly selected from the tool palette that is displayed on the computer screen. **See Figure 16-13.**

Technical Fact

PLC programming software is available in several programming languages (based on the IEC 1131-3 standard) including VersaPro, Cscape, fxControl, Logicmaster90, and State Logic, so that PLC programs can be efficiently applied to specific industrial applications.

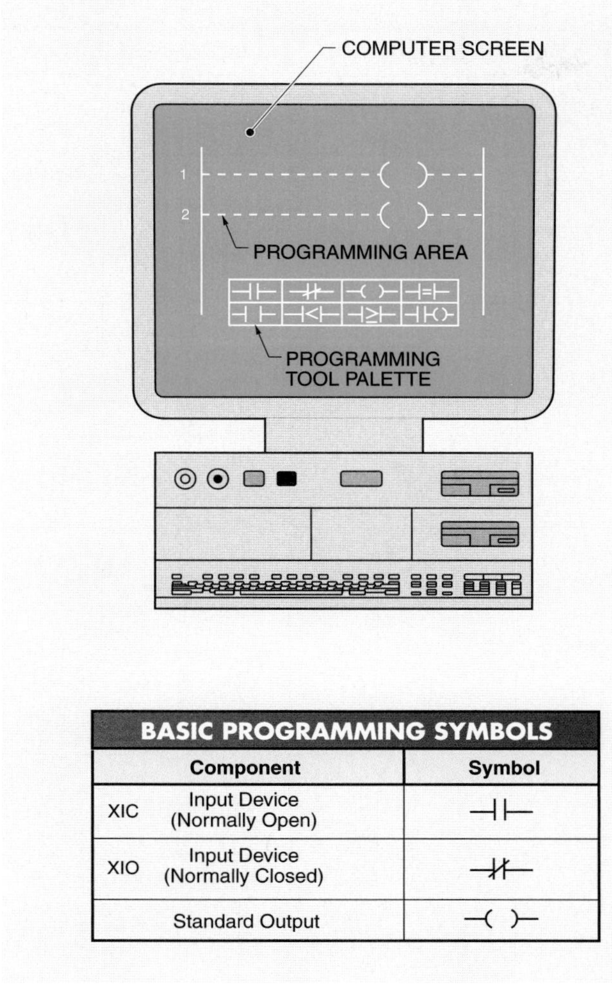

BASIC PROGRAMMING SYMBOLS

	Component	Symbol
XIC	Input Device (Normally Open)	—| |—
XIO	Input Device (Normally Closed)	—|/|—
	Standard Output	—( )—

EXPANDED/SPECIAL PROGRAMMING SYMBOLS

Component	Symbol
Equal-To Contact	—|=|—
Not-Equal-To Contact	—|≠|—
Greater-Than Contact	—|≥|—
Less-Than Contact	—|<|—
Addition	—(+)—
Subtraction	—(-)—
Multiplication	—(×)—
Division	—(÷)—
End	—(END)—
Set or Latch	—(SET)— —(L)—
Reset or Unlatch	—(RSET)— —(U)—
Timer	—(TMR)— or —[TMR]—
Counter	—(CNT)— or —[CNT]—

Figure 16-13. When designing and programming a circuit using a PLC, component symbols are selected and placed on the screen as the circuit is developed.

Standard symbols used in programming a circuit include normally open inputs, normally closed inputs, and standard outputs. Inputs are pushbuttons, limit switches, pressure switches, and other devices that are used to send information to a circuit. Standard outputs include lamps, solenoids, motor starters, alarms, and other devices that are used to perform work or give an indication of circuit operation. All PLC programs also include expanded (special) components that can be programmed into a circuit. Special components include timers, counters, logic functions, and common control functions (latch, unlatch, etc.). Although exact symbols differ slightly with each manufacturer, most have common symbols, shapes, and designations.

PLC Language. The first PLCs used line diagrams as a language for inputting information. Line diagrams are still commonly used as a language for PLCs throughout the world. Other languages used include Boolean, Functional Blocks, and English Statement. Line diagrams and Boolean are basic PLC languages. Functional Blocks and English Statement are higher-level languages required to execute more powerful operations such as data manipulation, diagnostics, and report generation.

The line diagram is drawn as a series of rungs. Each rung contains one or more inputs and the output (or outputs) controlled by the inputs. The rung relates to the machine or process controls, and the programming instructions communicate the desired logic to the processor.

Basic logic functions are used to enter the circuit's logical operation into the processor section. **See Figure 16-14.** The program is entered into the controller through the keyboard.

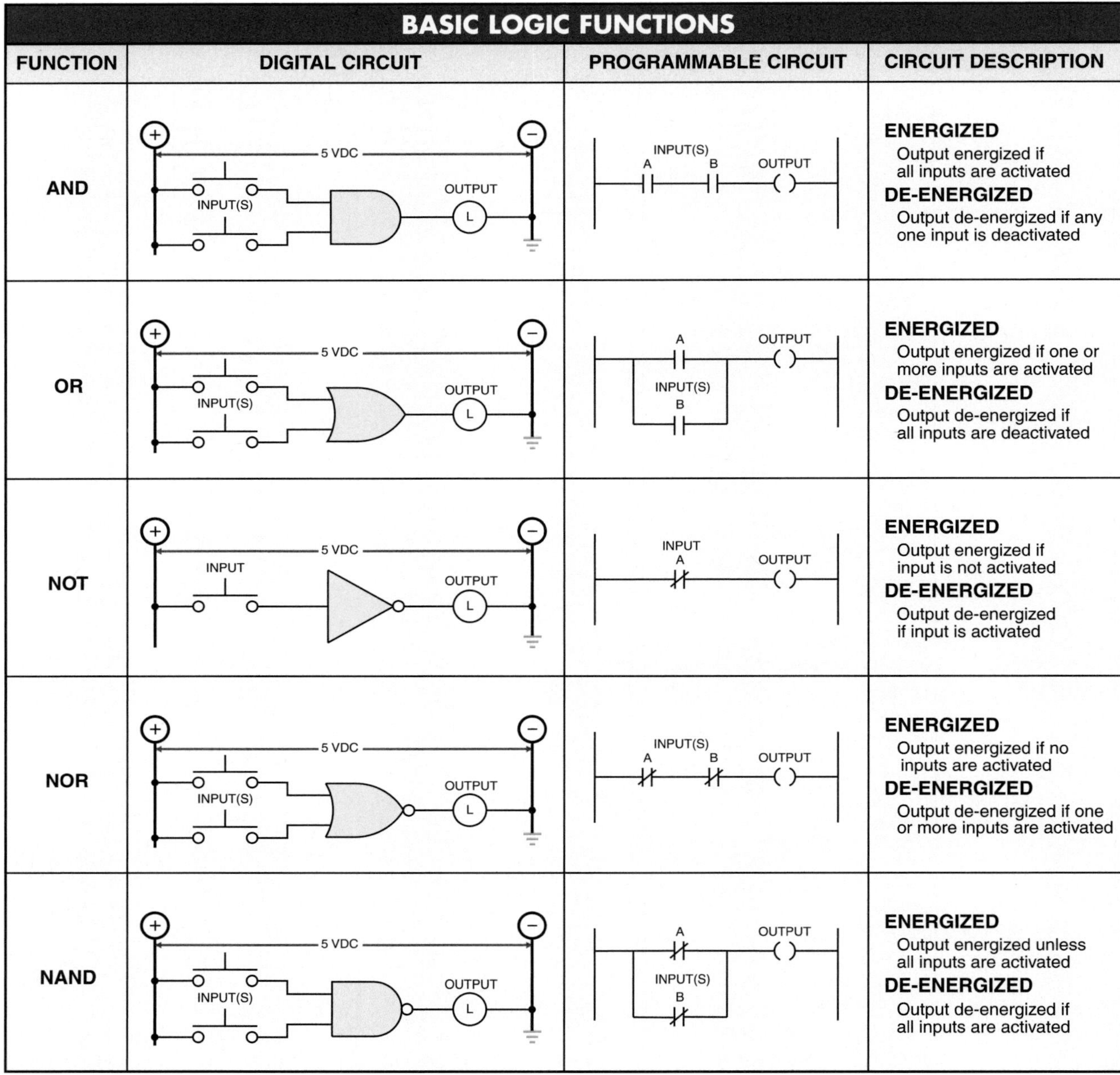

BASIC LOGIC FUNCTIONS			
FUNCTION	**DIGITAL CIRCUIT**	**PROGRAMMABLE CIRCUIT**	**CIRCUIT DESCRIPTION**
AND	+ 5 VDC − INPUT(S) OUTPUT L	INPUT(S) A B OUTPUT	**ENERGIZED** Output energized if all inputs are activated **DE-ENERGIZED** Output de-energized if any one input is deactivated
OR	+ 5 VDC − INPUT(S) OUTPUT L	A OUTPUT INPUT(S) B	**ENERGIZED** Output energized if one or more inputs are activated **DE-ENERGIZED** Output de-energized if all inputs are deactivated
NOT	+ 5 VDC − INPUT OUTPUT L	INPUT A OUTPUT	**ENERGIZED** Output energized if input is not activated **DE-ENERGIZED** Output de-energized if input is activated
NOR	+ 5 VDC − INPUT(S) OUTPUT L	INPUT(S) A B OUTPUT	**ENERGIZED** Output energized if no inputs are activated **DE-ENERGIZED** Output de-energized if one or more inputs are activated
NAND	+ 5 VDC − INPUT(S) OUTPUT L	A OUTPUT INPUT(S) B	**ENERGIZED** Output energized unless all inputs are activated **DE-ENERGIZED** Output de-energized if all inputs are activated

Figure 16-14. Basic logic functions are used to enter the circuit's logical operation into the processor section.

Programming a PLC follows a logical process. Inputs and outputs are entered into the controller in the same manner as if connecting them by hard wiring. The difference in programming is that although a circuit is the same, each manufacturer has a different method of entering that circuit. There are more similarities than differences from manufacturer to manufacturer.

Technical Fact

Every aspect of industry from power generation to automobile manufacturing to food packaging uses PLCs to expand and enhance production.

PLC Line Diagrams. Except for a few differences, PLC line diagrams are similar to standard hard wired line diagrams. PLC line diagrams have two vertical power lines that represent L1 and L2 except that no voltage potential exists between the two lines. Horizontal lines represent the current paths between the vertical power lines. These horizontal lines are referred to as rungs. Each rung may have several input and output devices. **See Figure 16-15.**

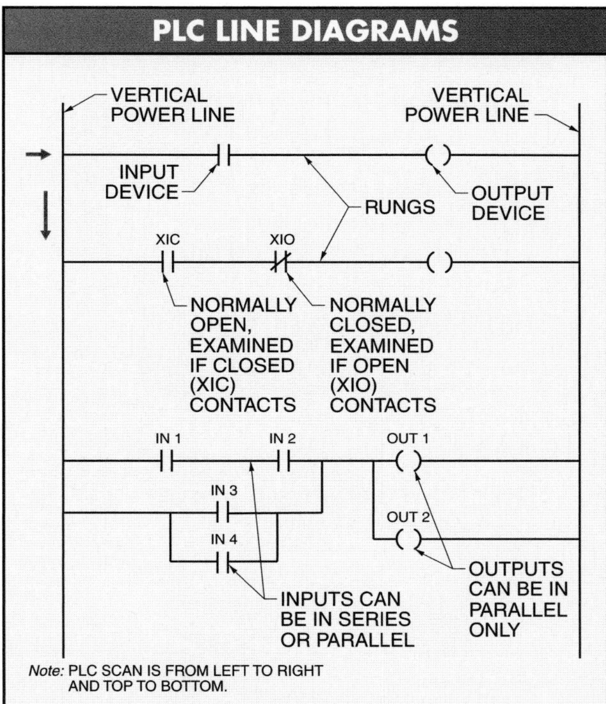

Figure 16-15. PLC diagrams follow the same basic rules as standard hard wired line diagrams.

Input devices are either normally open (NO) or normally closed (NC). The NO devices are referred to as "examined if closed" (XIC) contacts. The NC devices are referred to as "examined if open" (XIO) contacts. A *PLC scan* is one execution cycle of a line diagram. A typical PLC scan starts in the upper-left corner of the line diagram and scans from left to right and top to bottom. During each scan, the NO contacts are examined for being closed (XIC) and the NC contacts are examined for being opened (XIO).

Programming Rules

When programming a PLC circuit, basic rules must be followed if the circuit is to be accepted by the software before downloading. **See Figure 16-16.**

Basic PLC circuit programming rules include:

Rule 1: Inputs (normally open, normally closed, and special) are placed on the left side of the circuit between the left rung and the output. Outputs are placed on the right side of the circuit.

Rule 2: Only one output can be placed on a rung. This means that outputs can be placed in parallel but never in series.

Rule 3: Inputs can be placed in series, parallel, or in series/parallel combinations.

Rule 4: Inputs can be programmed at multiple locations in the circuit. An input (same input) can be programmed as normally open and/or normally closed at multiple locations.

Rule 5: Standard outputs cannot be programmed at multiple locations in the circuit. There is a special output called an "or-output" that allows an output to be placed in more than one location but only if the "or-out" special function is identified when programming the output.

Developing Typical Programs. Several steps must be taken before a program can be entered into a PLC. The first step is to develop the logic required of the circuit into a line diagram. **See Figure 16-17.** In this circuit, pressing any one of the three start pushbuttons energizes the motor starter. Once the motor starter is energized, the start pushbutton may be released. The motor starter remains energized because the M1 contact closes and provides a parallel path for current flow around the start pushbuttons. Pressing one of the stop pushbuttons stops the flow of current through the motor starter and de-energizes it.

Rockwell Automation, Allen-Bradley Company, Inc.

The output modules receive low-power digital signals from the processor and convert them into high-power signals to control the loads connected to the PLC.

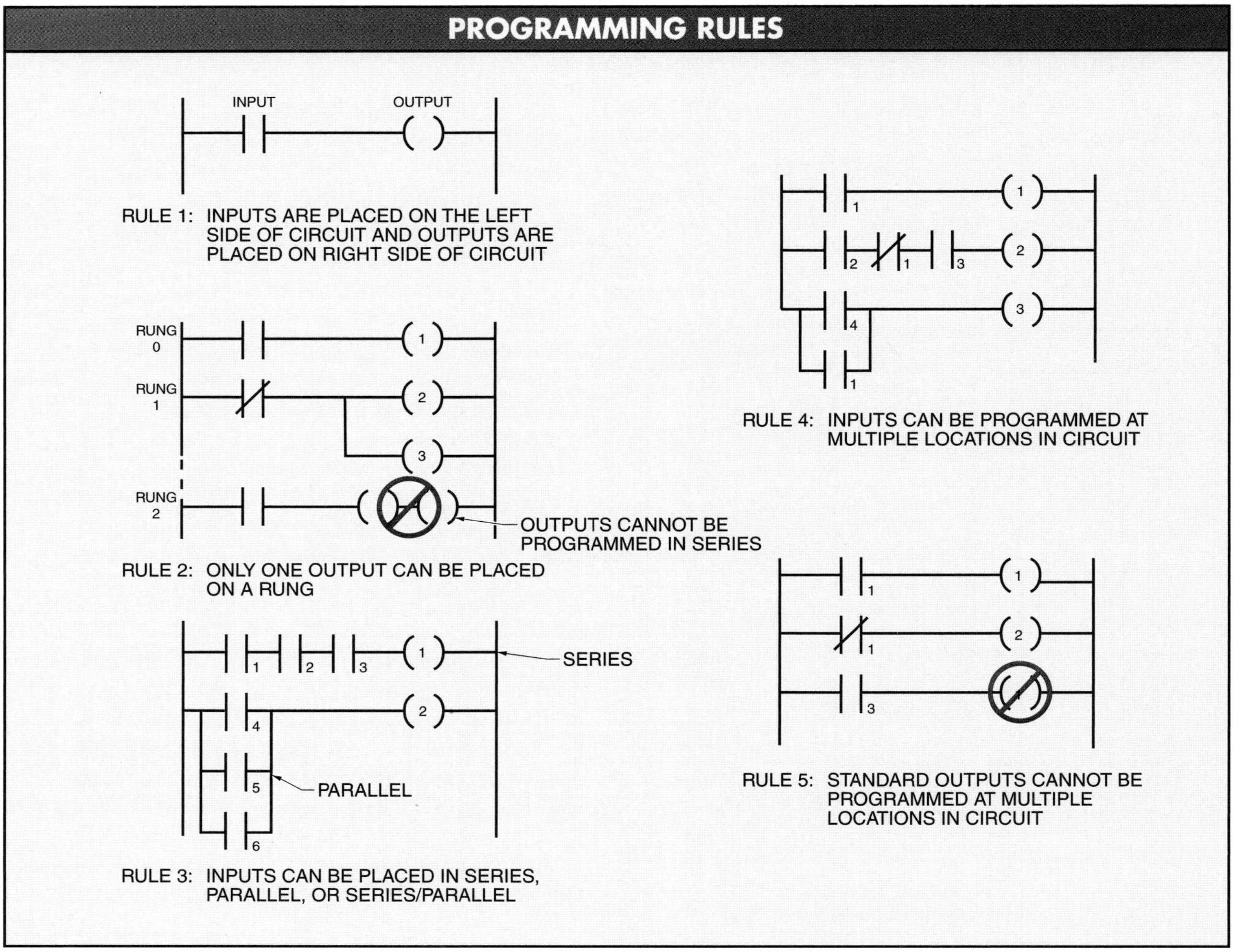

PROGRAMMING RULES
INPUT
OUTPUT
RULE 1: INPUTS ARE PLACED ON THE LEFT SIDE OF CIRCUIT AND OUTPUTS ARE PLACED ON RIGHT SIDE OF CIRCUIT
RUNG 0
RUNG 1
RUNG 2
OUTPUTS CANNOT BE PROGRAMMED IN SERIES
RULE 2: ONLY ONE OUTPUT CAN BE PLACED ON A RUNG
SERIES
PARALLEL
RULE 3: INPUTS CAN BE PLACED IN SERIES, PARALLEL, OR SERIES/PARALLEL
RULE 4: INPUTS CAN BE PROGRAMMED AT MULTIPLE LOCATIONS IN CIRCUIT
RULE 5: STANDARD OUTPUTS CANNOT BE PROGRAMMED AT MULTIPLE LOCATIONS IN CIRCUIT

Figure 16-16. Basic rules must be followed when programming a PLC if the circuit is to be accepted by the software before downloading.

L1
STOP
STOP
STOP
START
MOTOR STARTER
OLs
L2
M1
(4)
START
START
M1
1
2
3
4

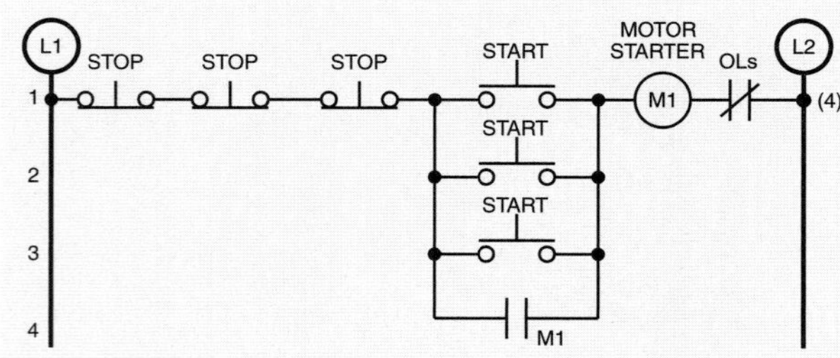

Figure 16-17. Circuit logic must be developed into a line diagram to enter the circuit into a PLC.

The line diagram shows the logic of the circuit but not the actual location of each component. A wiring diagram shows the location of the components in an electrical circuit. **See Figure 16-18.** The wiring diagram of the three start/stop pushbutton stations shows the location of each pushbutton.

The phantom line around each start/stop pushbutton station indicates that the two pushbuttons are located in the same enclosure. Each pushbutton in the wiring diagram is connected in the exact manner as in the line diagram. Any additions (or changes) to this hard wired control circuit require that the circuit be rewired.

The second step is to take the line diagram and convert it into a programming diagram. A *programming diagram* is a line diagram that better matches the PLC's language. Like a standard (hard wired) line diagram, a PLC programming diagram shows the flow of current through the control circuit. The PLC programming diagram does not use distinct symbols for each input/output. Instead, there are two basic symbols for inputs and one basic symbol for outputs. One of the input symbols represents normally open (NO) inputs and the other represents normally closed (NC) inputs. **See Figure 16-19.**

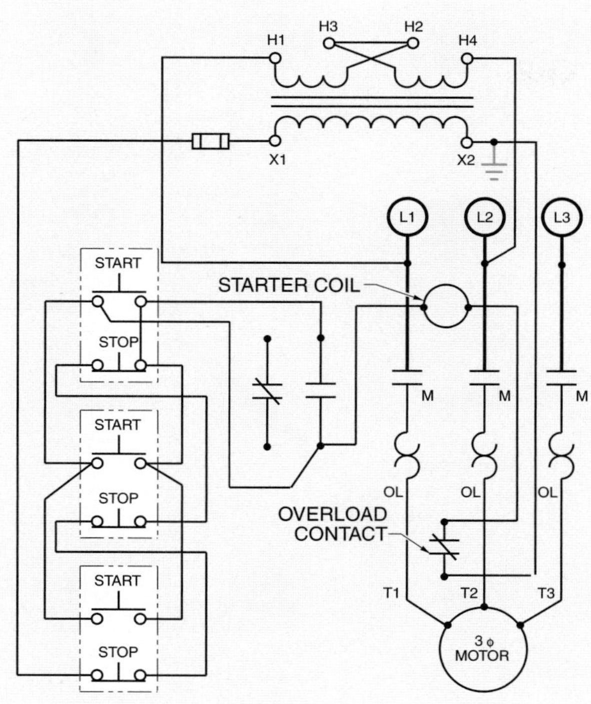

Figure 16-18. A wiring diagram shows the location of the components in an electrical circuit.

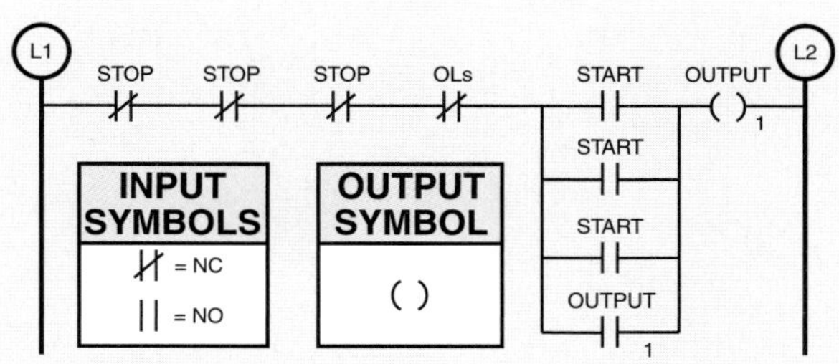

Figure 16-19. A programming diagram is a line diagram that better matches the PLC's language.

In this circuit, pressing any one of the three start pushbuttons energizes the motor starter. Once the motor starter is energized, the start pushbutton may be released and remain energized. This is because the contacts of output 1 close and provide a parallel path for current flow around the start pushbuttons. The motor starter de-energizes when the current flow to output 1 is de-energized by pressing the stop pushbutton.

The PLC wiring diagram is very different from a hard wired wiring diagram. **See Figure 16-20.** In a PLC wiring diagram, each input is wired to a designated input terminal and each output is wired to a designated output terminal. The way the inputs and outputs are connected does not determine the logic of the circuit's operation. The circuit's logic is controlled by the way the circuit is programmed into the PLC. Any changes to the circuit are made by changing the program, not the wiring of the inputs and outputs.

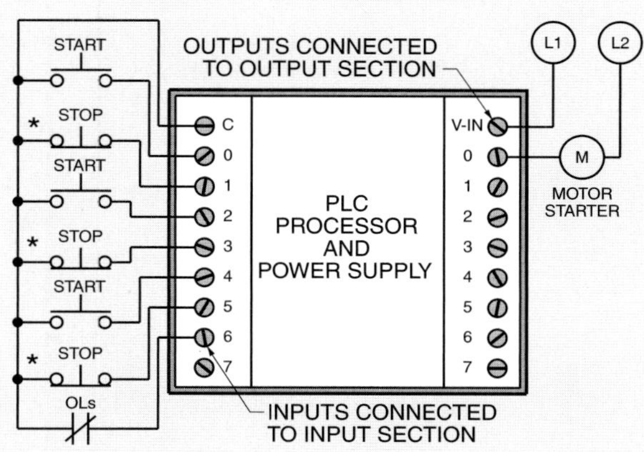

Figure 16-20. In a PLC wiring diagram, each input is wired to a designated input terminal and each output is wired to a designated output terminal.

The third step is to enter the desired logic of the circuit into the controller. Every manufacturer has a slightly different set of steps and functions to enter the program into the PLC. The program is entered in the program mode and then saved for future use.

Storing and Documentation. Once a program has been developed it may be necessary to store the program outside of the controller or document the program by printing it out. **See Figure 16-21.** This allows for a means of storing and retrieving control programs, which makes for fast changes in a process or operation. Storage of a program is commonly achieved using the hard drive, CD-ROMs, or memory cards. For example, one file may have the program for filling 8 oz bottles and a second file may have the program for filling 16 oz bottles.

Technical Fact

The first application of the Boolean algebra logic system was Shannon's 1938 research which analyzed switching circuits.

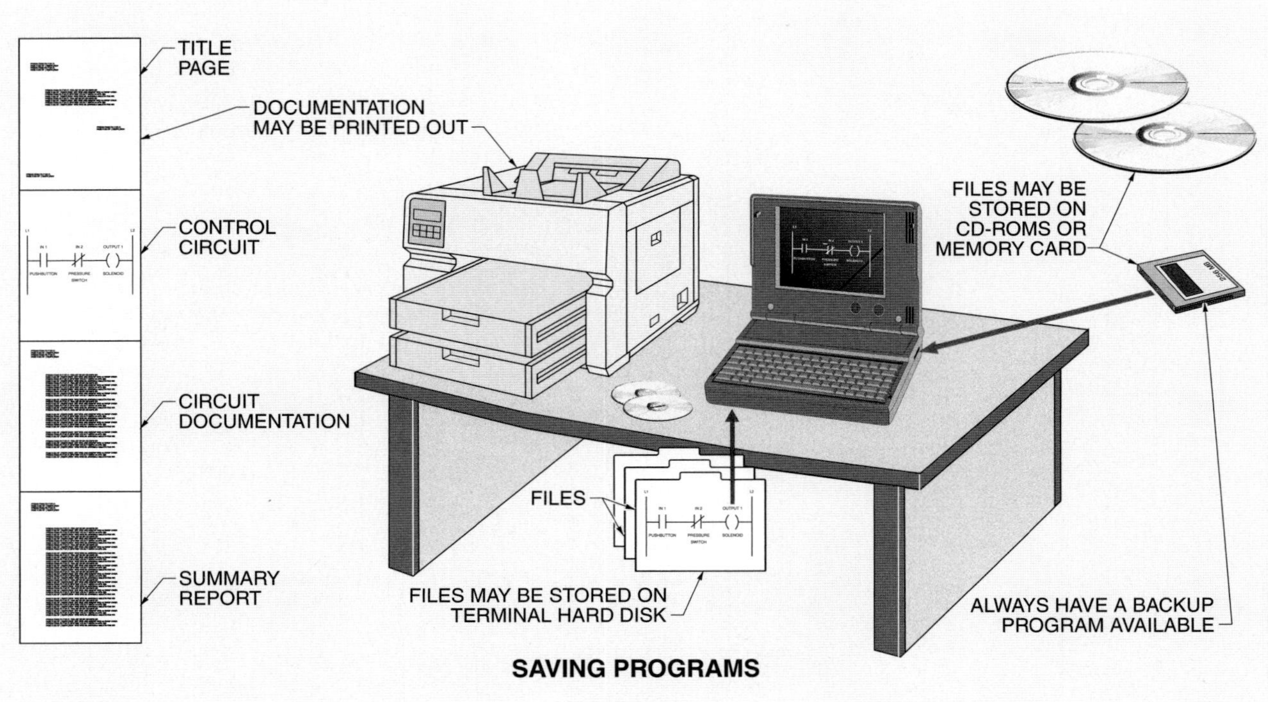

Figure 16-21. Storage of a program is commonly achieved using the hard drive, CD-ROMs, or memory cards.

When a change from one size bottle to another is required, the PLC is loaded with the correct disk or file to start the line with all the proper control settings. Even if the PLC is not likely to ever have its program changed, the program should be stored on a disk or CD-ROM. This ensures the safety of the program in the event of a problem.

Once a program has been entered into the PLC, a copy of the program and other circuit documentation can be made by connecting the controller to a printer. The printout can be used as a hard copy of the program for documentation and future reference.

PLC Status and Fault Indicators

All PLCs include indicator lamps (LEDs) that show the condition of the components in the PLC. PLCs include indicator lamps that show the condition of the inputs and outputs and indicator lamps that show the operating conditions of the PLC. **See Figure 16-22.**

An input/output (I/O) indicator lamp shows the status of the input and output devices. The input indicator lamps are energized (ON) when an electrical signal is received at an input terminal. This occurs when an input contact is closed or a signal is present. The output indicator lamps are energized (ON) when an output device is energized. Each input and output on the PLC has its own indicator lamp.

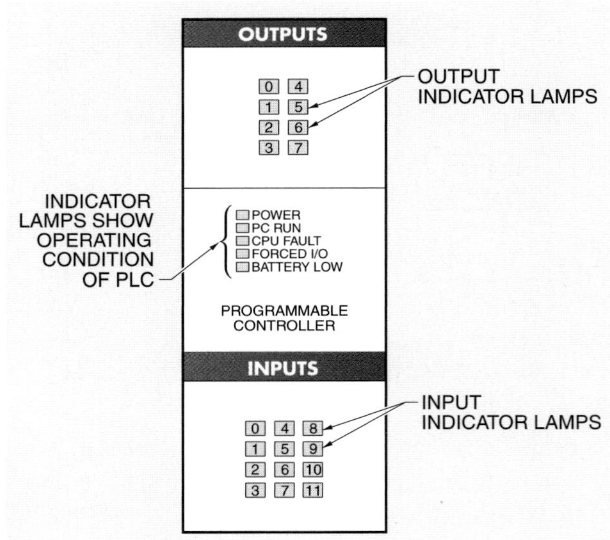

Figure 16-22. PLC indicator lamps show the operating condition of the inputs, outputs, and PLC.

Most PLCs have several status indicator lamps that show the operating condition of the PLC. The status indicator lamps commonly include power, PC run, CPU fault, forced I/O, and battery low lamps. **See Figure 16-23.**

STATUS INDICATOR CONDITION			
Status Indicator	**Problem**	**Possible Cause**	**Corrective Action**
■ POWER ■ PC RUN □ CPU FAULT □ FORCED I/O □ BATTERY LOW	Normal situation.		None.
□ POWER □ PC RUN □ CPU FAULT □ FORCED I/O □ BATTERY LOW	No or low system power.	Blown fuse, tripped CB, or open circuit.	Test line voltage at power supply. Line voltage must be within 10% of the controller s rated voltage. Check for proper power supply jumper connections when voltage is correct. Replace the power supply module when module has power coming into it but is not delivering correct power.
■ POWER □ PC RUN □ CPU FAULT □ FORCED I/O □ BATTERY LOW	Programmable controller not in run mode.	Improper mode selection. Faulty memory module, memory loss, or memory error caused by a high voltage surge, short circuit, or improper grounding.	Check line voltage and use an ohmmeter to check system ground.
■ POWER □ PC RUN ■ CPU FAULT □ FORCED I/O □ BATTERY LOW	Fault in controller.	Faulty memory module, memory loss, or memory error caused by a high voltage surge, short circuit, or improper grounding.	Turn power OFF and restart system. Remove power and replace the memory module when fault indicator is still ON.
■ POWER □ PC RUN □ CPU FAULT □ FORCED I/O ▨ BATTERY LOW	Fault in controller due to inadequate or no power.	Loss of memory when power OFF and battery charge was inadequate to maintain memory.	Replace battery and reload program.
■ POWER ■ PC RUN □ CPU FAULT ■ FORCED I/O □ BATTERY LOW	System does not operate as programmed.	Input or output device(s) in forced condition.	Monitor program and determine forced input and output device(s). Disable forced input or output device(s) and test system.
■ POWER ■ PC RUN □ CPU FAULT □ FORCED I/O □ BATTERY LOW	System does not operate.	Defective input device, input/output device, output module, or program.	Monitor program and check condition of status indicators on the input and output modules. Reload program when there is a program error.

Figure 16-23. Status indicator lamps show the operating condition of the PLC.

The power lamp indicates when power is applied to the PLC and the processor is energized. The power lamp should normally be energized. The PC run lamp indicates when the processor is in the run mode. The PC run lamp may be energized at all times in run mode or may strobe ON and OFF as the PLC is running and processing input and output data. The CPU fault lamp indicates that there is an error in the PLC system. PLCs are normally designed to de-energize (turn OFF) all outputs when the CPU fault lamp is energized. On most PLCs, an error message that indicates the error is also displayed.

The forced I/O lamp indicates when the PLC is in the forced operating mode. In the forced operating mode, the inputs and/or outputs are being forced ON or OFF through the software. Extreme caution must be used when forcing inputs and outputs, and any time the forced I/O lamp is energized. The battery low lamp indicates a low battery charge problem or that it is time to replace the battery. The battery is used to maintain processor memory during a power failure. The PLC battery should be replaced as recommended by the manufacturer or every five years.

Force and Disable

The force command opens or closes an input device or turns ON or OFF an output device. The force command is designed for use when troubleshooting the system. Forcing an input or output device allows checking the circuit using software. **See Figure 16-24.**

An input device may be forced to test the circuit operation. Forcing an input device may also be used when service is required on a defective input device. The defective input device may be forced ON until the device may be serviced if the input device is not critical to production. The force command is removed after the device is fixed.

An output device turns ON regardless of the programmed circuit's logic when the force ON command is used. The output device remains ON until the force OFF command is used. Care must be taken when using the force command because it overrides all safety features designed for the program.

The disable command prevents an output device from operating. The disable command is the opposite of the force command. The disable command is used to prevent one or all of the output devices from operating. Ensure that all force and disable commands are removed before returning a system to normal operation.

PLC Communication Networks

PLCs may be used as stand-alone control devices that control their outputs (solenoids, motor starters, etc.) as their inputs (limit switches, etc.) send signals to the PLC. PLCs may also be connected through communication ports to other devices such as human machine interfaces (HMIs), PCs, variable speed drives, and other PLCs. **See Figure 16-25.**

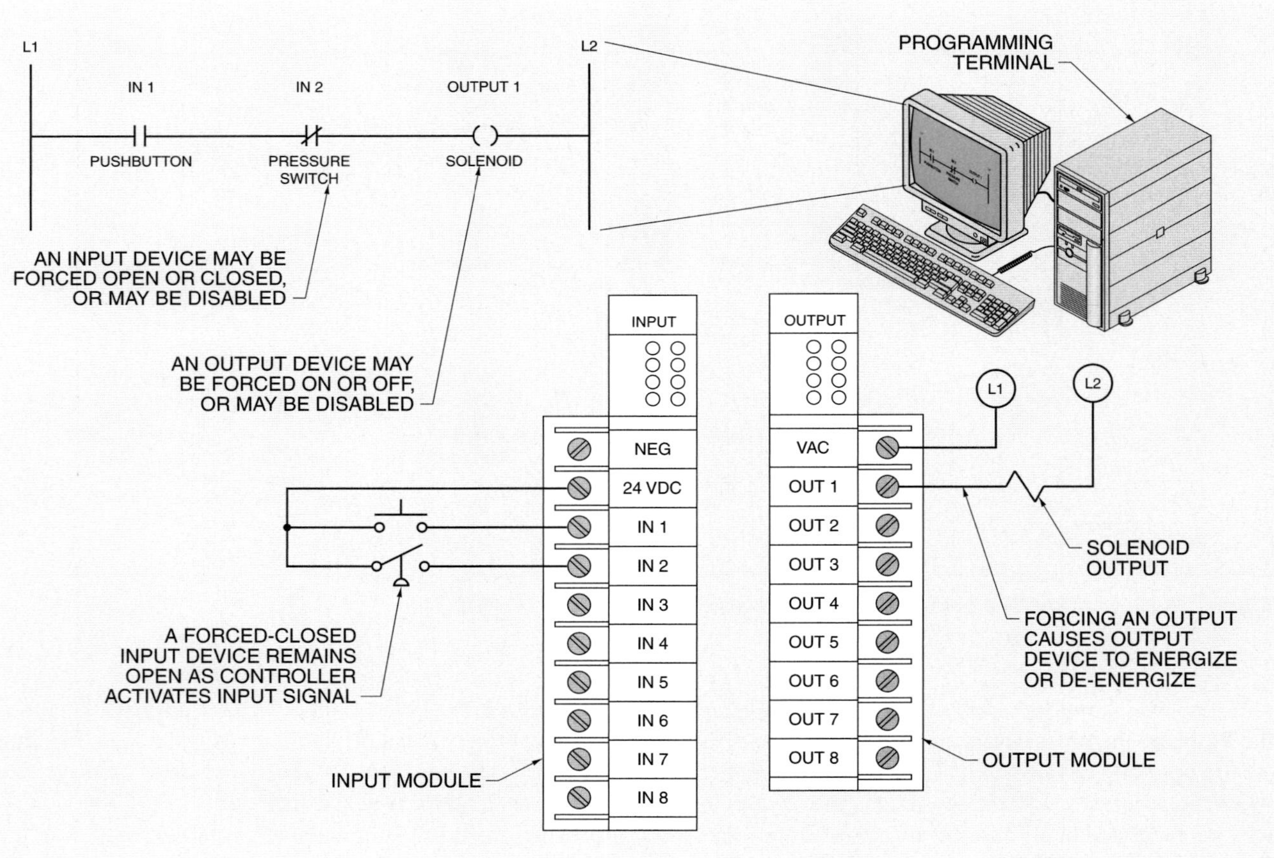

Figure 16-24. Forcing an input or output device allows checking the circuit using software.

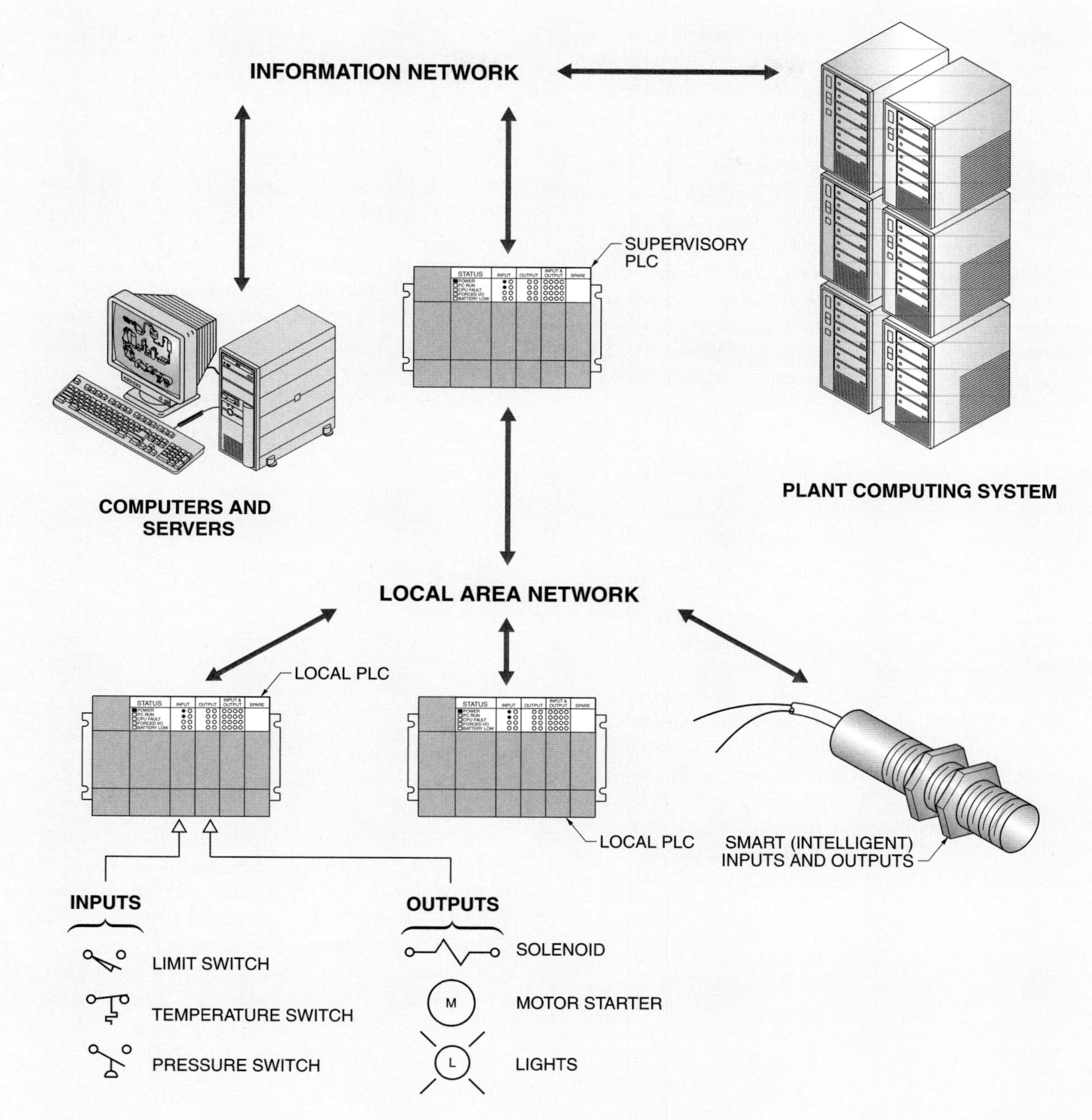

Figure 16-25. PLCs may be connected into a network for system control.

In a typical network system, the PLCs become part of a large control system. Field input devices (limit switches, photoelectric switches, etc.) and output devices (solenoids, motor starters, etc.) are connected to the PLCs. The PLC is then connected to the local area network (LAN) system. Smart (intelligent) input and output devices can be directly connected to the LAN system.

The LAN system is a collection of data and power lines that are used to communicate information among individual devices and to supply power to individual devices connected to the system. Depending on the size of the process, there can be any number of LANs. Local area networks can be connected to information networks through other PLCs. The information network system is also a collection of data and power lines that are used to communicate information among individual devices and to supply power to individual devices connected to the system. The information network devices are used to control the LAN devices and allow the flow of information between the central computing system and each local input and output device.

Each network system monitors and controls variables such as time, temperature, speed, weight, voltage, current, power, flow rate, level, volume, density, color, brightness, and pressure. These variables can be controlled, measured, displayed, and recorded. This system allows for a closed loop operation. A *closed loop operation* is an operation that has feedback from the output to the input. Monitoring the outputs and sending information to the inputs controls the system so that the inputs are automatically adjusted to meet the needs of the outputs.

INTERFACING SOLID-STATE CONTROLS

PLCs can have many types of inputs including pushbuttons, level switches, temperature controls, and photoelectric controls. Inputs such as pushbuttons and temperature controls are normally easy to input. However, more complex solid-state control inputs such as proximity and photoelectric inputs require special consideration because of their function.

Solid-state proximity and photoelectric controls are used in many automated systems. **See Figure 16-26.** These controls normally have a solid-state output and are ideal for inputting to PLCs. Photoelectric controls can be input into PLCs for detection, inspection, monitoring, counting, and documentation. Available outputs include two- and three-wire types with thyristor and transistor outputs that can be connected individually or in series/parallel combinations.

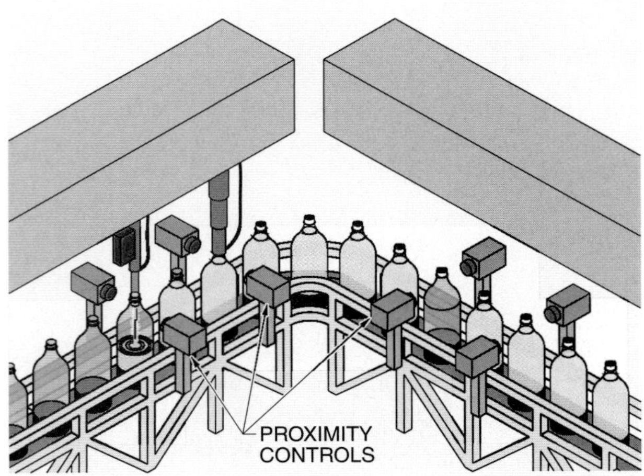

Figure 16-26. Solid-state proximity controls normally have a solid-state output and are ideal for inputting to PLCs.

Two-Wire Thyristor Output Sensors

Two-wire thyristor output sensors are available in a supply voltage range of 20 VAC to 270 VAC at about 180 mA to 500 mA range in either NO or NC versions. Two-wire thyristor output sensors have only two wires and are wired in series with the load like a mechanical switch. **See Figure 16-27.** The power to operate these sensors is received through the load when the load is not being operated. As with any thyristor output device, some consideration must be given to off-state leakage current and minimum load current. Unlike a mechanical switch, there is current consumed by the proximity sensor in the inactivated mode. The current is small enough that most industrial loads are not affected. This leakage current may be enough to activate the load on some high-impedance loads and PLCs.

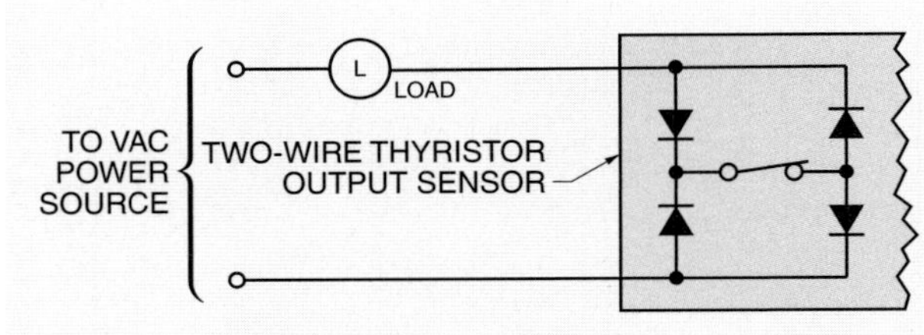

Figure 16-27. Two-wire thyristor and output sensors have only two wires and are wired in series with the load like a mechanical switch.

This problem can be corrected by placing a load resistor across the input device. **See Figure 16-28.** The resistor value should be chosen to ensure that minimum load current is exceeded and the effective load impedance is reduced. This prevents off-state leakage current turn-on. This resistance value is normally in the range of 4.5 kΩ to 7.5 kΩ. A general rule is to use a 5 kΩ, 5 W resistor for most applications.

Electrical Noise Suppression

Electrical noise is unwanted signals that are present on a power line. Electrical noise enters through input devices, output devices, and power supply lines. Unwanted noise pickup may be reduced by placing the controller away from noise-generating equipment such as motors, motor starters, welders, and drives.

Noise suppression should be included in every PLC installation because it is impossible to eliminate noise in an industrial environment. Certain sensitive input devices (analog, digital, and thermocouple) require a shielded cable to reduce electrical noise.

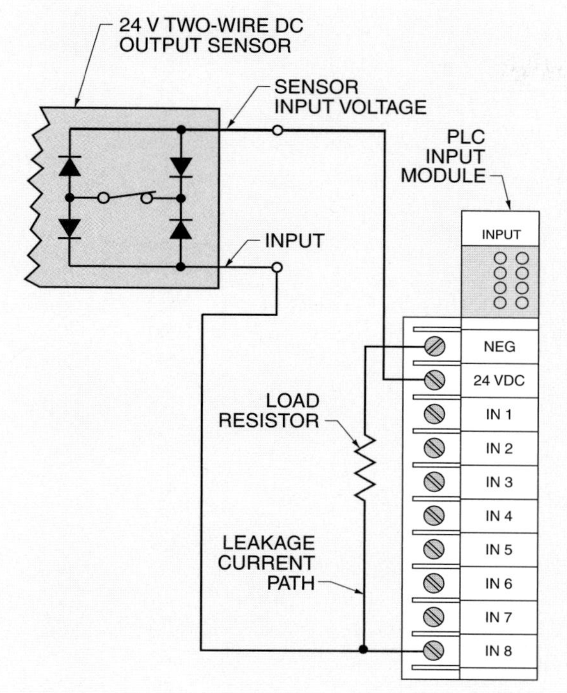

Figure 16-28. A load resistor may be required when connecting a sensor to the PLC to prevent leakage current of the sensor from inputting into the controller.

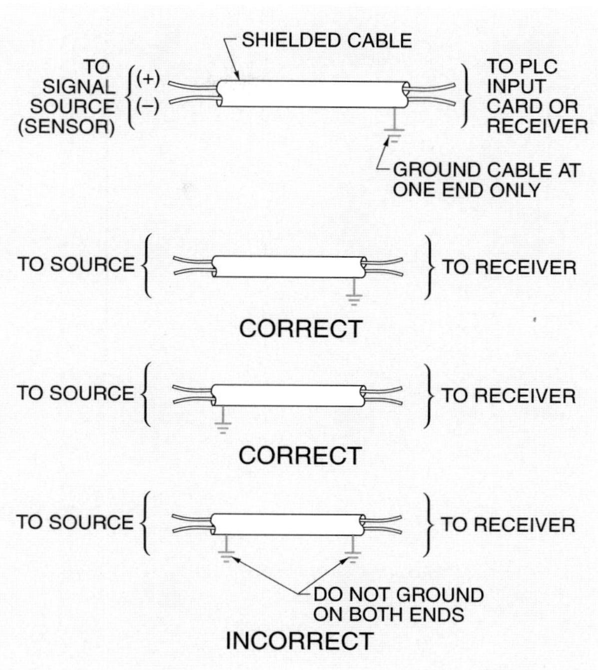

Figure 16-29. A shielded cable uses an outer conductive jacket (shield) which blocks magnetic interference from the two inner, signal-carrying conductors.

A shielded cable uses an outer conductive jacket (shield) to surround the two inner signal-carrying conductors. The shield blocks electromagnetic interference. The shield must be properly grounded to be effective. Proper grounding includes grounding the shield at only one point. A shield grounded at two points tends to conduct current between the two grounds. Do not route low-voltage DC signals near high-voltage (120 V) AC signals. If the signals must cross, ensure the cables cross at 90° to minimize interference. **See Figure 16-29.**

A high-voltage spike is produced when inductive loads such as motors, solenoids, and coils are turned OFF. These spikes may cause problems in a PLC. High-voltage spikes should be suppressed to prevent problems. A snubber circuit is used to suppress a voltage spike. Typical snubber circuits use an RC (resistor/capacitor), MOV (metal oxide varistor), or diode, depending on the load. **See Figure 16-30.**

Three-Wire Transistor Output Sensors

Three-wire transistor output sensors are available in a supply range of 10 VDC to 40 VDC at about 200 mA. These sensors are easily interfaced with other electronic circuitry and PLCs. Output sensor types consist of either an open collector NPN or PNP transistor. Both NO or NC versions are available.

These sensors receive their power to operate through two of the leads (positive and negative, respectively) from the power source. The third lead is used to switch power to the load either using the same source of power as the proximity switch or an independent source of power. **See Figure 16-31.** When an independent source is used, one lead of that source is common with one lead of the source used to power the sensor. The voltage level must be within the specifications of the sensor used when using an independent power source for the load.

Technical Fact

Electromagnetic interference (EMI) is unwanted electrical noise on a power line. This noise, also known as a magnetic field, may leak from power lines and affect equipment that is not connected to the power line. These noise signals have an adverse affect on electronic equipment and cause intermittent data problems.

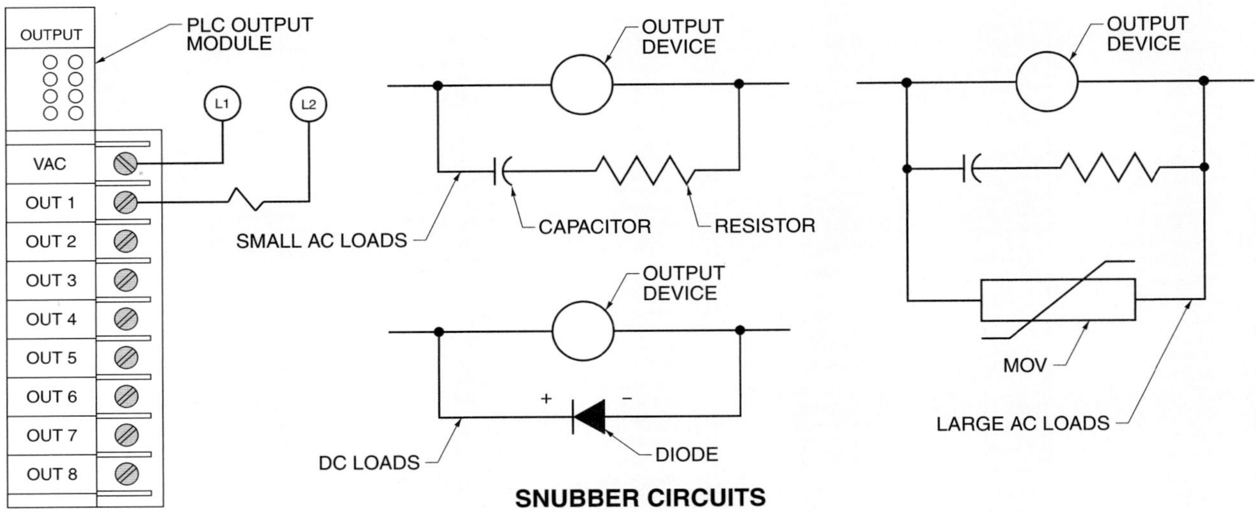

Figure 16-30. Snubber circuits are used to suppress voltage spikes in PLCs.

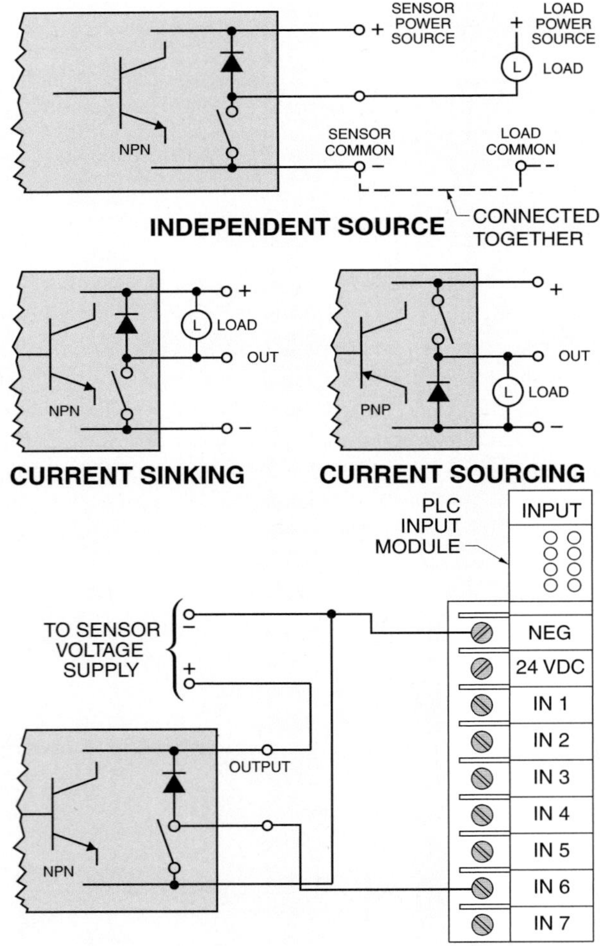

Figure 16-31. Three-wire transistor output sensors use either NPN or PNP transistors to control the load.

PROGRAMMING PROGRAMMABLE TIMERS

All PLCs include internal programmable timers (timing functions). The programmable timers are programmed, not wired, into the control circuit. When programming a programmable timer, the type of timer (ON-delay or OFF-delay), a preset time value (15 sec, 2 min, etc.), and a time base (second, hundredth of a second, etc.) are selected. Preset time is the length of time for which a timer is set. Programmable timers must be identified and set as would be done with a stand-alone timer. **See Figure 16-32.**

The method of timer identification in the circuit and program varies with different PLC manufacturers. The timer instructions include a timer reference number and a preset timer value. The timer reference number specifies the timer number as used in the control circuit and program. The preset timer value is the required time delay value for the application in which the timer is used. Once programmed, the on-screen timer display normally shows the timer type, number, time base, preset time setting, and accumulated time (since the timer started). Accumulated time is the length of time that has passed since the timer started.

Status Bits

Programmable timers use status bits to control the timer. A status bit is an indicator used to show an ON or OFF state. The status bit can be changed from high to low, ON to OFF, or true to false. Status bits include enable and done bits. An enable bit is a controlling contact or input that sends power to the timer (or counter) and starts the timer timing or the counter counting. A done bit is the controlled contact (or output) that changes from open or closed (OFF to ON) after the preset time (or number of counts) has been reached.

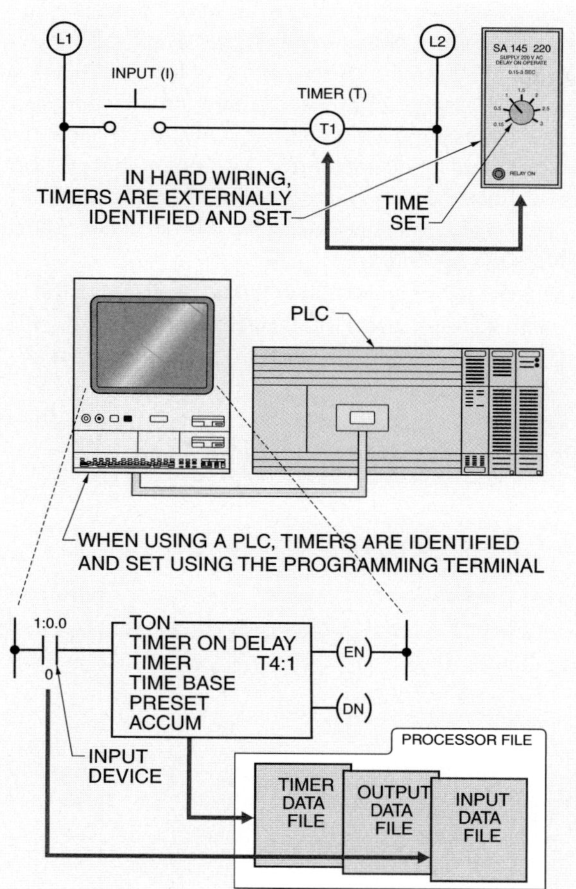

Figure 16-32. Programmable timers must be identified and set as would be done with a stand-alone timer.

The enable bit is ON (true) any time there is an input signal applied to the timer. The enable bit is OFF (false) or reset at all other times. The timer contacts are activated and the timer no longer times when the accumulated value reaches the preset value. The done bit is turned ON when the timer contacts are activated. The done bit is OFF when the timer contacts are not activated.

In an ON-delay timer, the done bit is ON (true) when the accumulated value equals the preset value. The done bit is OFF (false) when the enable bit is OFF. In an OFF-delay timer, the done bit is ON (true) when the enable bit is ON. The done bit is OFF (false) only after the enable bit is OFF and the preset time delay passes. **See Figure 16-33.**

PLC APPLICATIONS

PLCs are useful in increasing production and improving overall plant efficiency. PLCs can control individual machines and link the machines together into a system. The flexibility provided by a PLC has resulted in many applications in manufacturing and process control.

Process control has gone through many changes. In the past, process control was mostly accomplished by manual control. Flow, temperature, level, pressure, and other control functions were monitored and controlled at each stage by production workers.

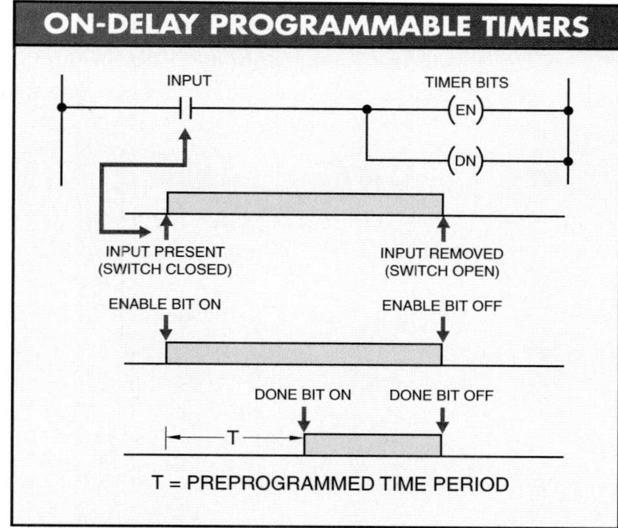

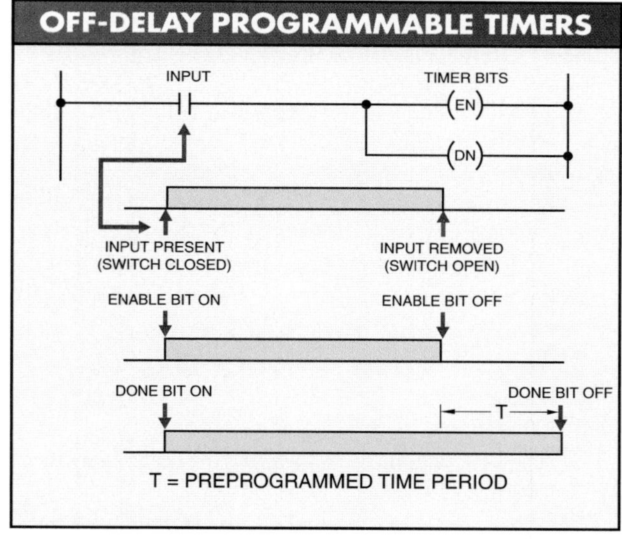

Figure 16-33. In an ON-delay programmable timer, the done bit is ON (true) when the accumulated value equals the preset value. In an OFF-delay timer, the done bit is ON (true) when the enable bit is ON.

Today, using PLCs, an entire process can be automatically monitored and controlled with few or no workers involved. Process applications in which PLCs are used include the following:

- Grain operations involving storage, handling, and bagging
- Syrup refining involving product storage tanks, pumping, filtration, clarification, evaporators, and all fluid distribution systems
- Fats and oils processing involving filtration units, cookers, separators, and all charging and discharging functions
- Dairy plant operations involving all process control from raw milk delivered to finished dairy products
- Oil and gas production and refining from the well pumps in the fields to finished product delivered to the customer
- Bakery applications from raw material to finished product
- Beer and wine processing, including the required quality control and documentation procedures

PLC Timer Applications

A stoplight circuit is a circuit that uses timers to sequence the turning ON and OFF of the red, yellow, and green lamps. A stoplight timing circuit can use stand-alone timers or programmable timers to control the timing sequence.

The circuit is drawn in standard line diagram format when stand-alone timers are used. Four ON-delay timers are used for a basic stoplight timing sequence application. **See Figure 16-34.** To simplify the circuit, the stoplight circuit shows the basic operation of sequencing three lamps, and the time values have been reduced to seconds. Once the circuit start pushbutton is pressed and released, the lamp sequence is:

1. The red lamp turns ON for 30 sec.
2. The red lamp turns OFF and the yellow lamp turns ON for 5 sec.
3. The yellow lamp turns OFF and the green lamp turns ON for 40 sec.
4. The green lamp turns OFF and the yellow lamp turns ON for 5 sec.
5. The yellow lamp turns OFF and the red lamp turns ON for 30 sec.
6. Timer 4 resets the circuit and the sequence starts over.

The circuit is drawn in PLC line diagram format when a programmable timer is used. The PLC line diagram is similar to the standard line diagram format. The difference is that a standard line diagram is drawn by hand or by using a computer program. The programmable timer line diagram is automatically drawn on the computer screen as the circuit is programmed. **See Figure 16-35.**

In this diagram, each line of the control circuit is referred to as a rung. A number, starting with the number 0, identifies the rungs. Input, output, relay, and timer identification varies by manufacturer. A common manufacturer identification method is:

- The inputs are addressed as I:0.0-0 (stop button) and I:0.0-1 (start button).
- The outputs are addressed as 0:0.0-0 (red lamp), 0:0.0-1 (yellow lamp), and 0:0.0-2 (green lamp).
- The relays are addressed as B3-0 (first relay), and B3-1 (second relay).
- The timers are addressed as T4:1 (timer 1), T4:2 (timer 2), T4:3 (timer 3), and T4:4 (timer 4).

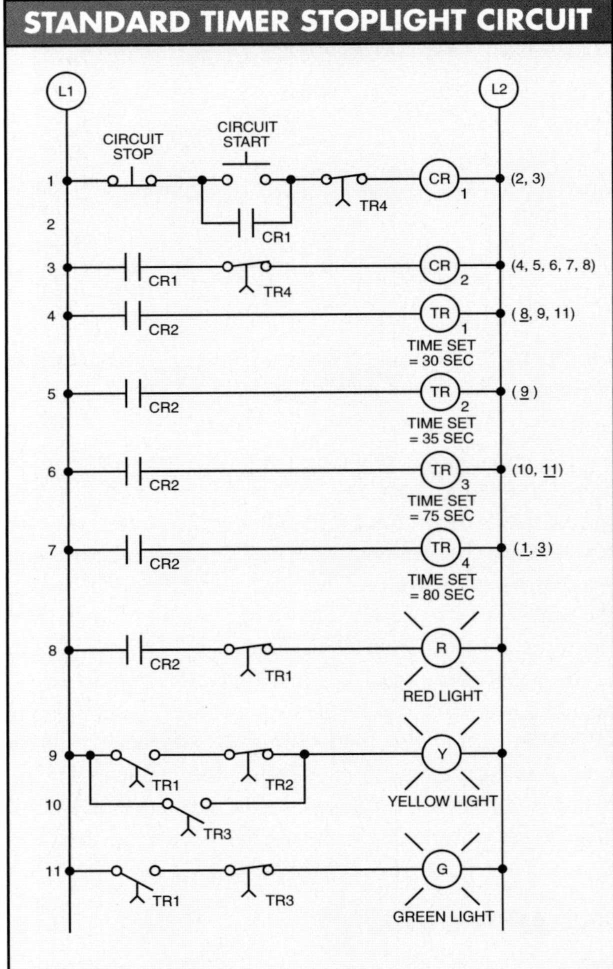

Figure 16-34. Four ON-delay stand-alone timers are used for a basic stoplight timing sequence application.

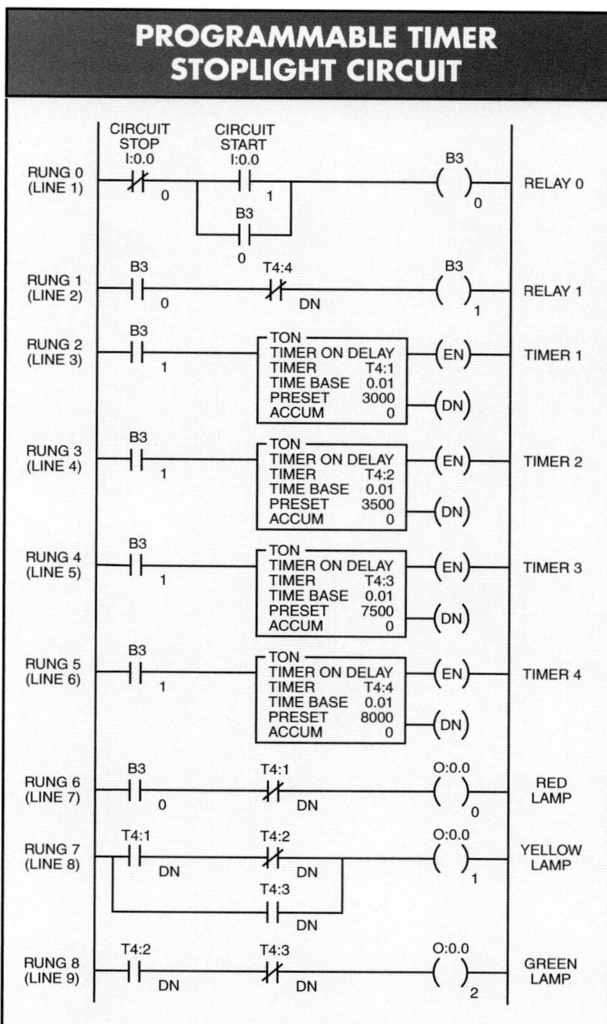

Figure 16-35. The programmable timer line diagram is automatically drawn on the computer screen as the circuit is programmed.

Welding

In manufacturing of discrete parts, welding is a major part of the system. PLCs may be used to control and automate industrial welding processes. **See Figure 16-36.** In this application, the PLC can control the length of the weld and the power required to produce the correct weld. The PLC is programmed to allow the weld to occur only if all inputs and conditions are correct. These include the following:

- Presence and correct position of all the parts
- The correct weld cycle speed and power setting
- The correct rate of speed on the line for the given application
- Proper functioning of all interlocks and safety features

FANUC Robotics North America

Figure 16-36. PLCs can be used to control and automate industrial welding processes.

In addition, the PLC can be used to determine if parts are running low and be set to automatically turn the line ON and OFF as required. Documentation of production efficiency can be generated for quality control and inventory requirements.

A PLC may be used to control and interlock many welders. Welders at one station may require more power than is available if all the welders are ON simultaneously. In this case, a large power draw can cause poor-quality welds. A requirement in a system using many welders is to limit the amount of power being consumed at any one time. This is accomplished by time-sharing the power feed to each welder.

A PLC may be programmed for a maximum power draw. The controller can determine if power is available when a welder requires power. The weld takes place if the correct power level is available. If not, the controller remembers the request and when power is available, it permits the welder to proceed with the weld cycle. The PLC can also be programmed to determine which welder has priority.

Machine Control

Controls must be synchronized when machines are linked together to form an automated system. **See Figure 16-37.** In this application, each machine may be controlled by a PLC, with another controller synchronizing the operation. This is likely if the machines are purchased from different manufacturers. In this case, each machine may include a PLC to control all the functions on that machine only. If the machines are purchased from one manufacturer or designed in-plant, it is possible to use one large PLC to control each machine and synchronize the process.

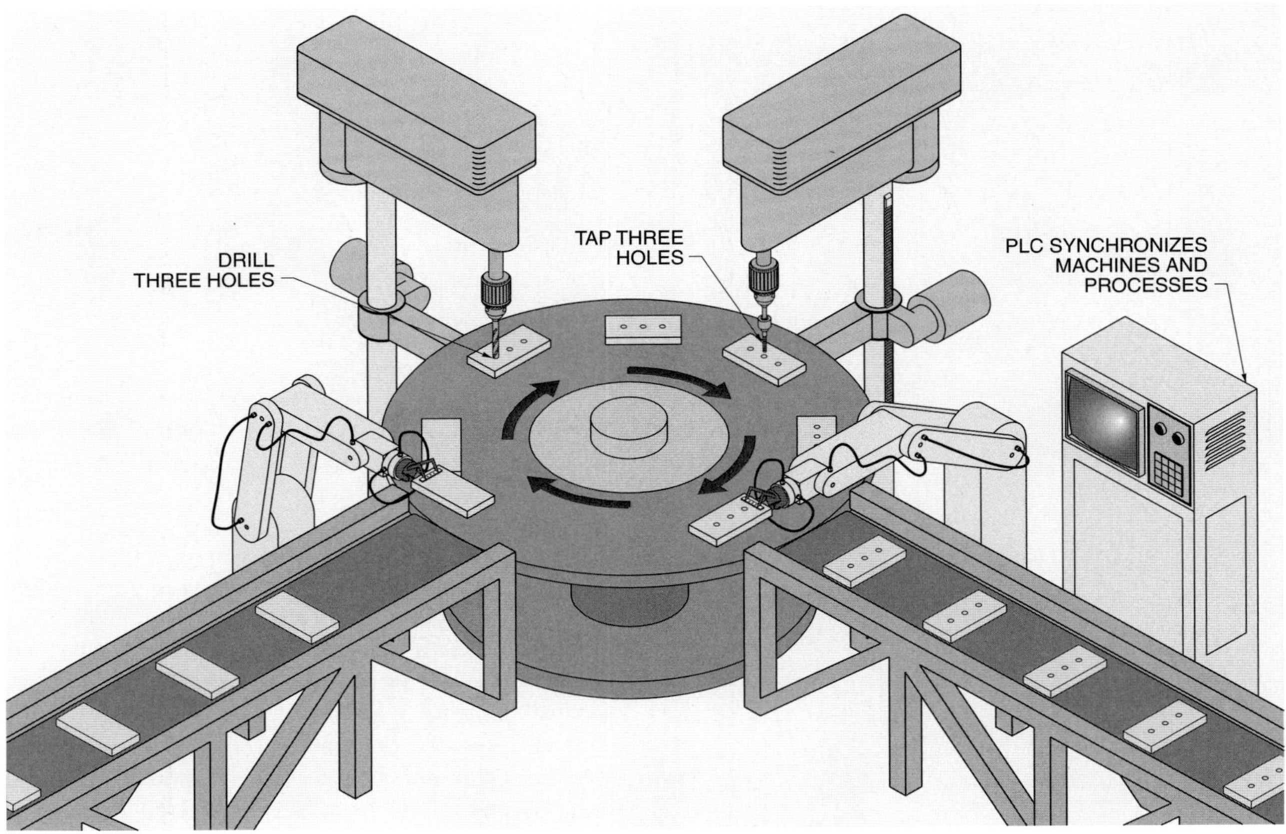

Figure 16-37. PLCs are used to control and synchronize individual machine operations with other machines.

Industrial Robot Control

PLCs are ideal devices for controlling any industrial robot. **See Figure 16-38.** The PLC can be used to control all operations such as rotate, grip, withdraw, extend, and lift. A PLC is recommended because most robots operate in an industrial environment.

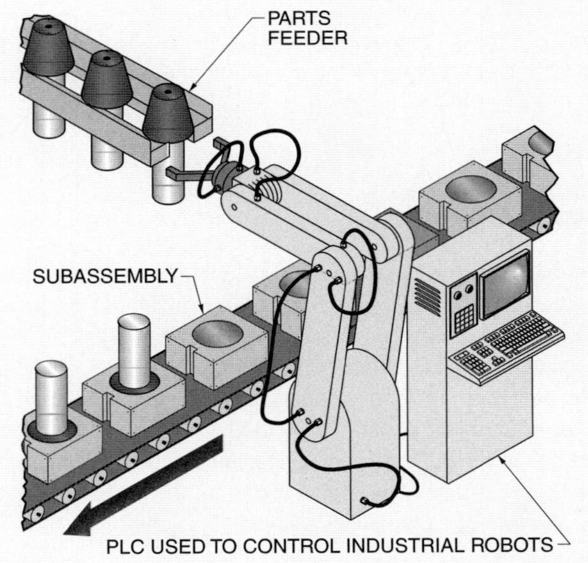

Figure 16-38. PLCs can be used to control the operations of an industrial robot.

Fluid Power Control

Fluid power cylinders are normally chosen when a linear movement is required in an automated application. Pneumatic cylinders are common because they are easy to install and most plants have compressed air. Pneumatics work well for most robot grippers, drives, and positioning cylinders; machine loading and unloading; and tool-working applications. Hydraulic cylinders are used when a manufacturing process requires high forces. Hydraulic systems of several thousand psi are often used to punch, bend, form, and move components.

PLCs may be used to control linear and rotary actuators in an industrial fluid power circuit. **See Figure 16-39.** This system, as is any fluid power circuit, is ideal for control by a PLC. The controller's output module is connected to control the four solenoids. Solenoid A moves the cylinder in, solenoid B moves the cylinder out, solenoid C rotates the rotary actuator in the forward direction, and solenoid D rotates the rotary actuator in the reverse direction.

The PLC is used to control the energizing or de-energizing of the solenoids. Solenoids control the directional control valves, which control the actuators.

Industrial Drive Control

Motors have normally been connected directly to the power lines and operated at a set speed. As systems become more automated, variable motor speed is required. Adjustable speed controls are available to control the speed of AC and DC motors. These controls are normally manually set for the desired speed, but many allow for automatic control of the set speed. A PLC may be used to control AC drives. **See Figure 16-40.** The drives can accept frequency and direction commands in a BCD format that the PLC can provide with a BCD output module.

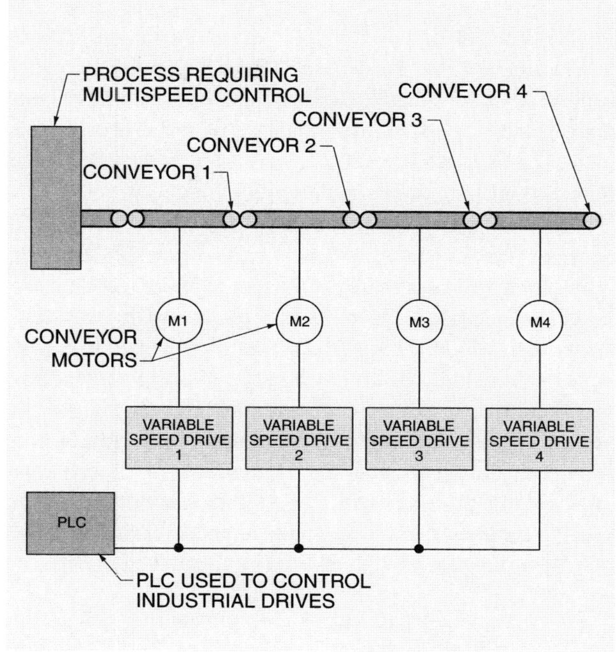

Figure 16-40. PLCs can be used to control and synchronize the speed of conveyors on an assembly line.

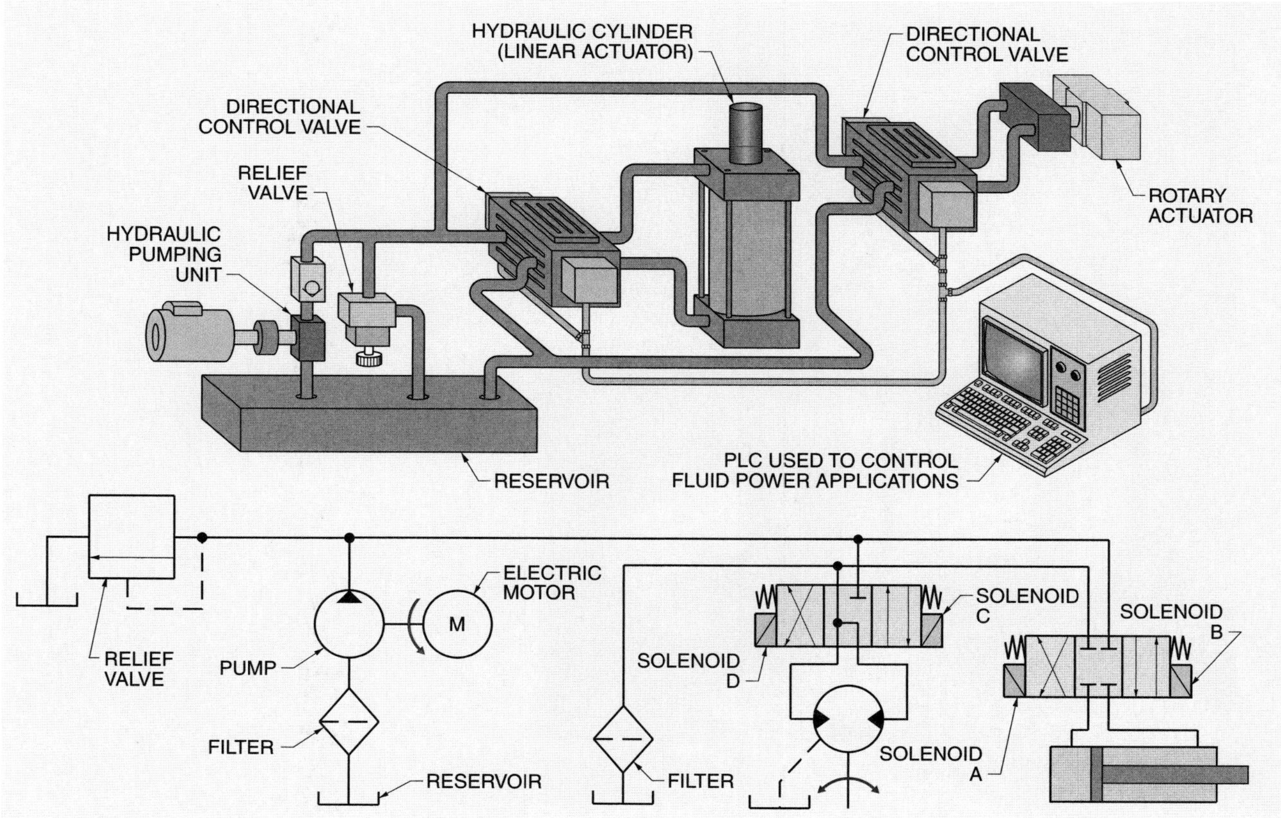

Figure 16-39. PLCs can be used to control linear and rotary actuators in an industrial fluid power circuit.

Pulp and Paper Industries. Pulp and paper industries use PLCs to control each production process and diagnose problems in the system. **See Figure 16-41.** The pulp and paper production process can involve equipment that covers a large area. The control of pulp and paper production processes is ideal for a PLC because most control logic is start/stop, time delay, count sequential, and interlock functions.

The PLC allows for the required I/O which, when multiplexed, can transmit multiple signals over a single pair of wires. The basic operation of a paper mill is to receive raw material such as logs, pulpwood, or chips, and process, size, store, and deliver the material. This includes a large conveyor system that has diverter gates, overtravel switches, speed control, and interlocking. A break in any part of the system can shut down the entire system. Finding a fault is time-consuming because the system covers a large area. To solve this problem, a PLC with fault diagnostics can be used to analyze the system and give an alarm and printout of where the problem exists with suggested solutions.

Batch Process Control Systems. Batch processing blends sequential, step-by-step functions with continuous closed-loop control. Process control is systems control, and systems are made up of many parts. Individual PLCs can be used to control each part and step of the process, with additional PLCs and computers supervising the total operation.

In a batch process control system, an operator interface is used for instrumentation or other monitoring functions. An operator interface is added as part of the system. This may be in the form of an instrumentation and process control station, a human machine interface (HMI), or any other type of interface. To aid in interfacing and monitoring a programmable-based system, a serial port is used for monitoring and programming a system using a computer. Thus, the individual solenoids, motor starters, heating elements, etc. at each process step are directly controlled by the local PLCs with the host computer supervising all of the PLCs. **See Figure 16-42.**

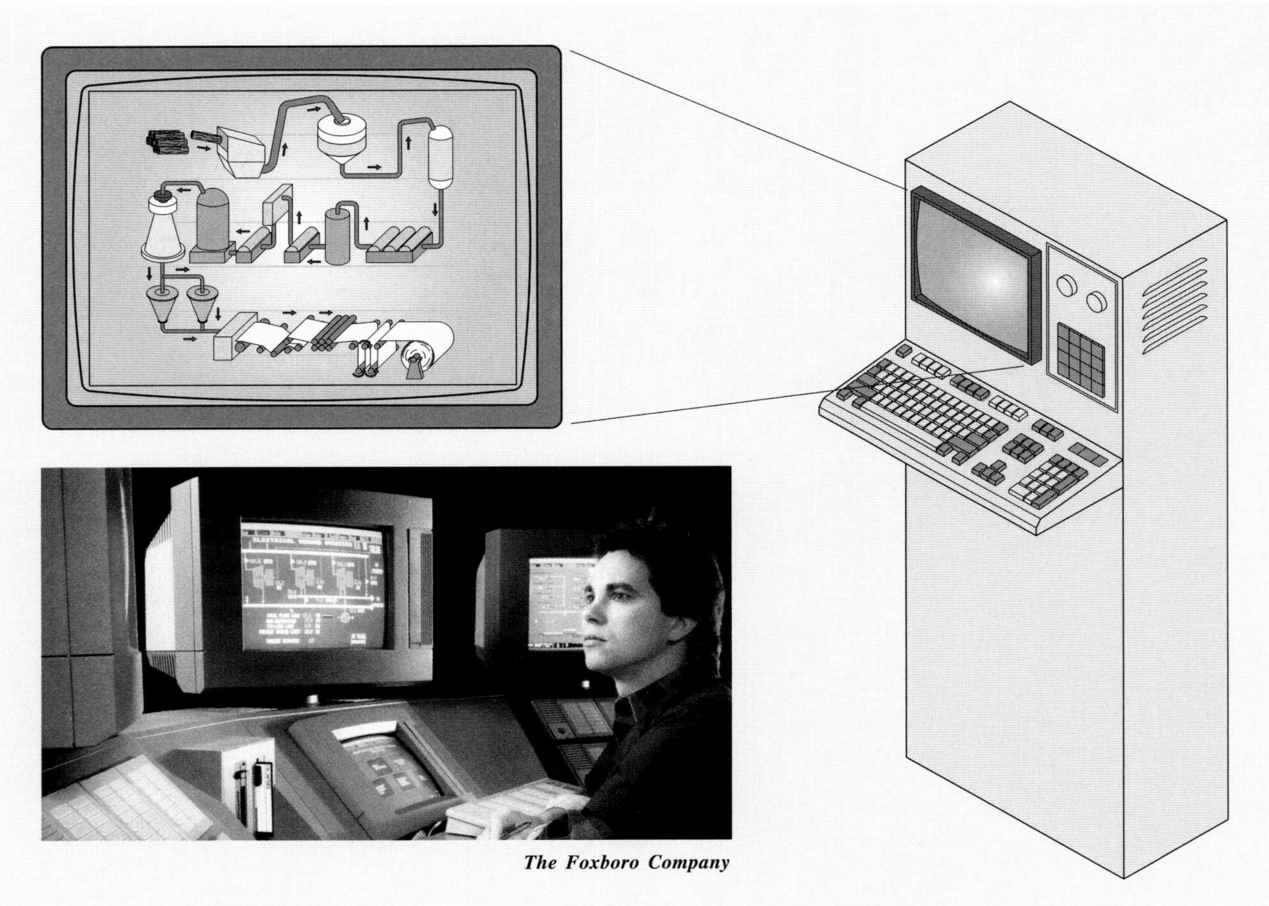
The Foxboro Company

Figure 16-41. In a paper mill, PLCs control each process and diagnose problems in the system.

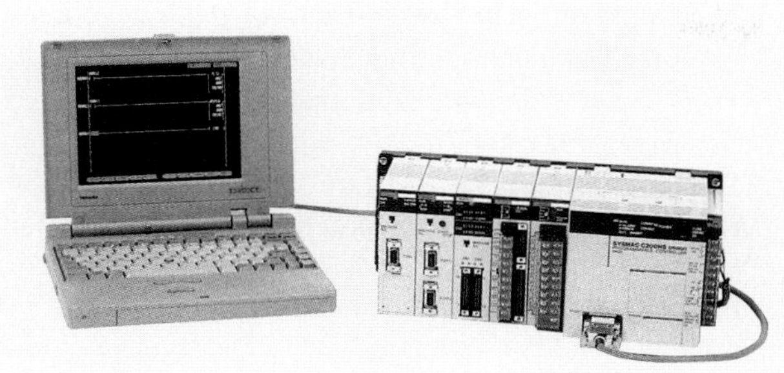

Omron Electronics, Inc.

Figure 16-42. A serial port is used for monitoring and programming a system using a computer.

PLC Circuits

Control circuits that do not use PLCs for control functions have been used for over 100 years. These control circuits do not allow for much flexibility or change. Modern electrical circuits are usually designed with change in mind. Changes include the way the circuit operates or additional safety features. With a PLC, changes in an electrical circuit can easily be made by changing the program.

A basic forward/reversing circuit is an example of a circuit that may require changes. In a basic forward/reversing circuit, very little circuit logic is required. **See Figure 16-43.**

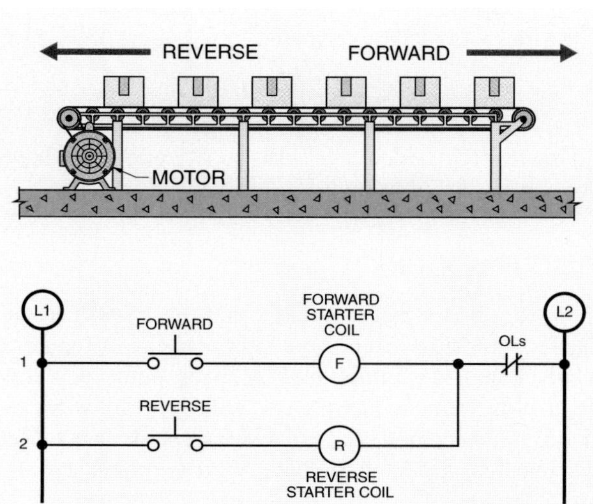

Figure 16-43. In a basic forward/reversing circuit, very little circuit logic is required.

In this circuit, the forward pushbutton operates the forward starter coil, and the reverse pushbutton operates the reverse starter coil. This circuit operates satisfactorily if no operator error occurs. However, if an operator presses both pushbuttons at the same time, both starter coils energize. This causes a short circuit in the power circuit. Interlocking is added to solve this problem. Interlocking prevents the operator from energizing both starter coils at the same time. **See Figure 16-44.**

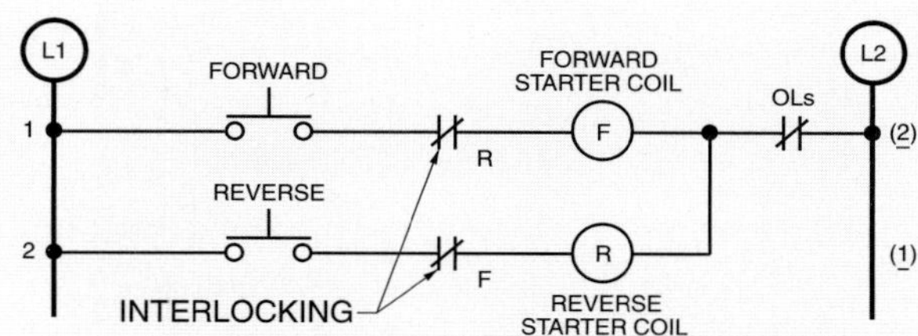

Figure 16-44. Interlocking is added to a circuit to prevent the operator from energizing both starter coils at the same time.

In a hard wired circuit, auxiliary contacts must be added and wired to interlock the circuit. A PLC allows interlocking of the circuit with a simple change in the program and no additional components. **See Figure 16-45.**

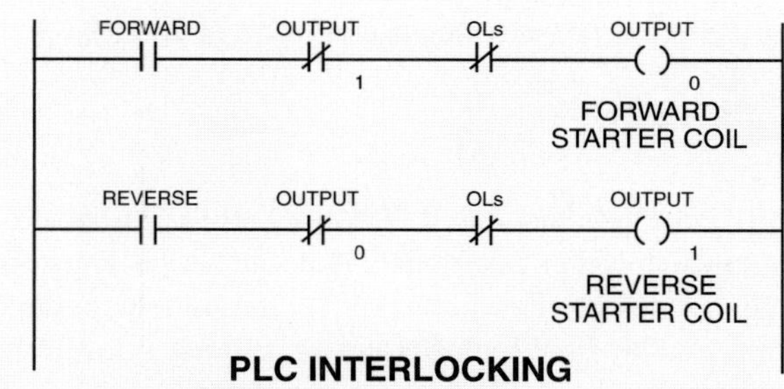

Figure 16-45. A PLC allows interlocking of the circuit with a simple change in the program and no additional components.

Another change that might be required is adding memory to the circuit. In a hard wired circuit, auxiliary contacts are required. In a PLC circuit, the program is changed. **See Figure 16-46.**

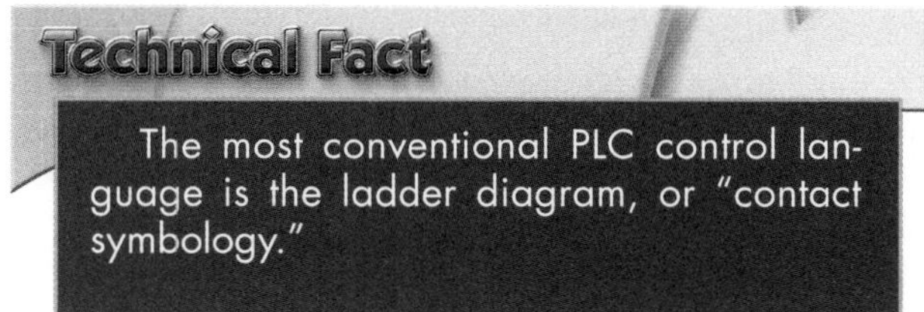

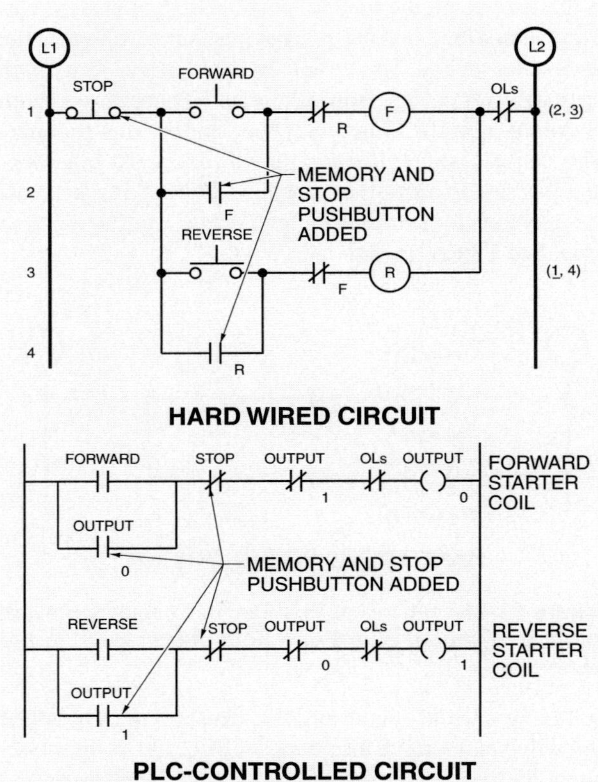

Figure 16-46. Adding memory to a circuit requires adding components and wiring if the circuit is hard wired, or changing the program if the circuit is controlled by a PLC.

In a PLC circuit, many circuit changes can be programmed. However, additional inputs and outputs must be wired to the controller. For example, if a light is required to indicate the direction of motor (or product) travel, the light must be physically wired to the PLC. This is one of the similarities between hard wired and PLC circuits. **See Figure 16-47.**

Input and Output Address Identification

Every time an input or output device on a PLC is programmed, the device must be assigned an address (instruction). The address links external inputs and outputs to data files and processor files within the PLC. Although each manufacturer assigns their own addresses to inputs and outputs, there are more similarities than differences among most manufacturers. For example, a typical manufacturer addressing numbering system uses number/letter assignments. **See Figure 16-48.**

The numbering/letter assignments are made as follows:

I = Input (pushbutton, limit switch, etc.)

O = Output (solenoid, lamp, motor starter, etc.)

T = Timer (internal PLC timer)

C = Counter

: = Slot number (physical slot number [1, 2, 3, etc.] of the I/O module)

For example, I:0/1 identifies an input in slot number 0 at terminal 1 and 0:0/4 identifies an output in slot number 0 at terminal 4.

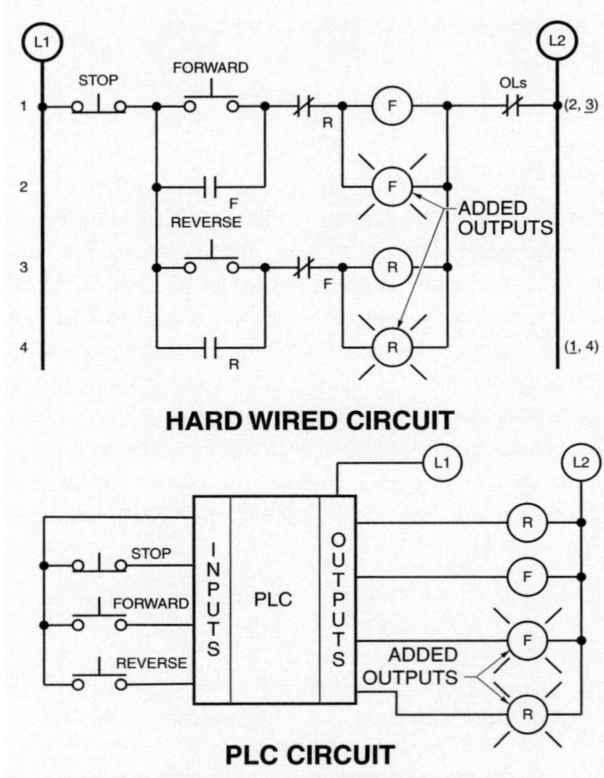

Figure 16-47. In a PLC circuit, additional inputs and outputs must be wired to the controller.

MULTIPLEXING

Multiplexing is a method of transmitting more than one signal over a single transmission system. As the distance between any transmitting and receiving point increases, the cost of multiconductor cable with separate wires for each signal becomes very expensive through installation, maintenance, and replacements. With multiplexing, two wires can serve multiple transmitters and receivers. A multiplexing system (two-wire system) is ideal when used with a PLC, as all inputs and outputs can be connected with just two wires.

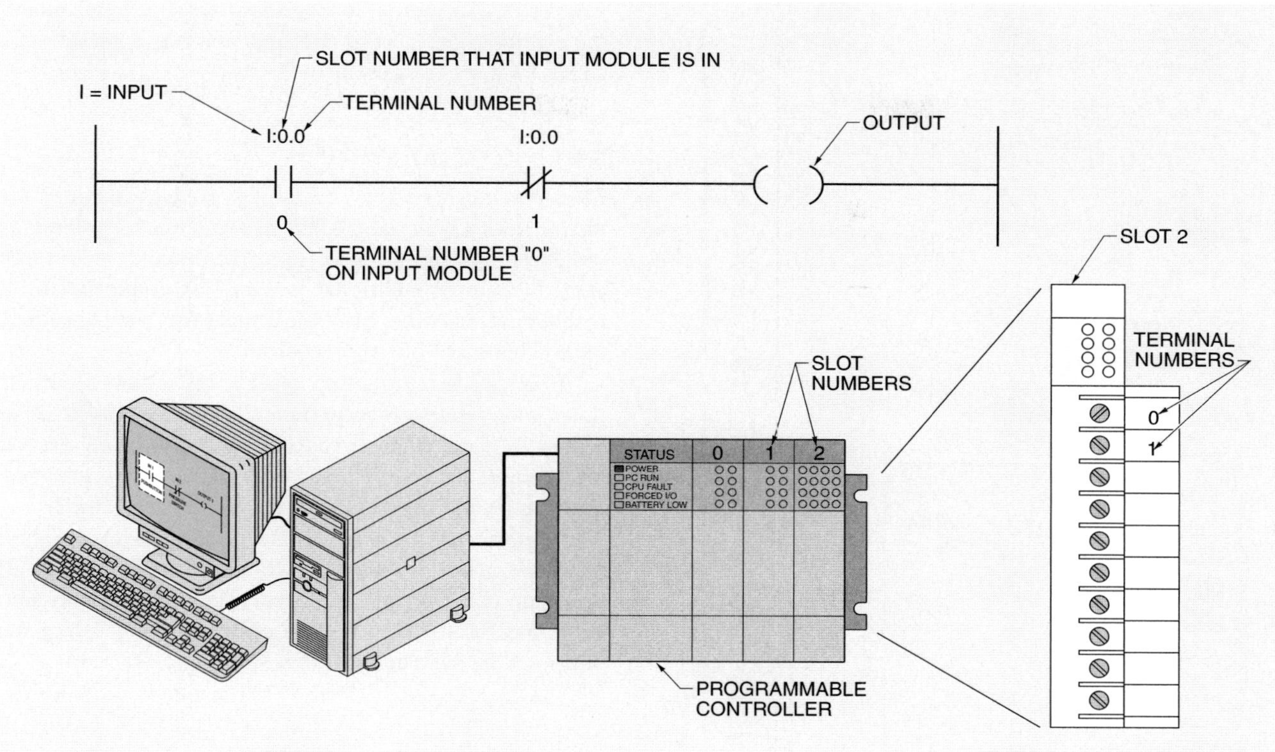

Figure 16-48. Numbers and letters are used to assign (address) inputs, outputs, timers, and other internal and external components.

One of the advantages in using a multiplexing system for control is the elimination of costly hard wiring. **See Figure 16-49.** In this circuit, eight control switches are hard wired to control eight loads. Two wires are required for each control switch. This means that time and money is wasted for even the shortest distance. As the distance between the control switches and loads increases, the cost of time and materials for the hard wired circuit increases. The disadvantages of hard wiring include the following:

- Point-to-point wiring is required between each switch and load.
- Dozens of wires must be pulled for even small applications.
- The number and size of wires required is large.
- A larger conduit size is required.
- Conduit, wire, and labor costs are high.
- Cost increases as distance between switches and loads increases.

This same circuit can be connected using a multiplexing system. Only two wires are required between the eight control switches and eight loads. Additional control switches to be added require no additional transmission wires. Additional transmitters, receivers, displays, or PLCs can all be connected to the same two wires. The advantages of multiplexing include the following:

- No point-to-point wiring is required.
- No conduit or multiple wires are required.
- Can easily expand with no additional transmission wires required.
- Digital and analog signals can be transmitted on the same two-wire system.

Wiring a control circuit becomes more difficult as the circuit increases in size and function. A multiplexing system can send back a signal to indicate that the load is energized. A multiplexing system is much simpler than hard wiring and can be expanded to almost any number of inputs and outputs, all controlled by a PLC. **See Figure 16-50.** The PLC controls all inputs and outputs and makes timing, counting, sequencing, and any other required logic decisions.

A multiplexing system can be used to transmit both analog and digital signals on the same two-wire system. This makes the system ideal for any instrumentation application including the transmission and control of temperatures, BCD signals, speed (rpm), voltage, and current levels, and counts.

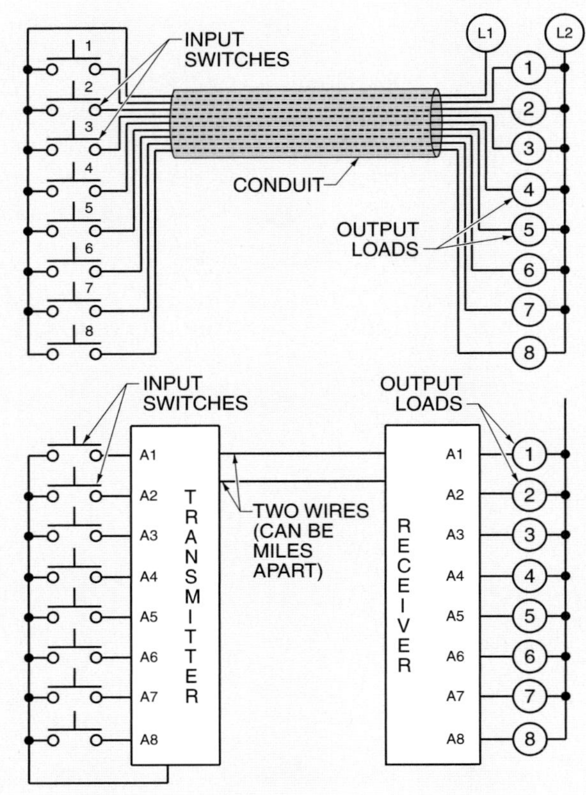

Figure 16-49. Multiplexing eliminates the need for costly hard wiring in a system.

In addition, a 24-hour clock and printer can be added to the system for documentation. This makes it possible to print out the time of day when a certain event has taken place on the multiplexing system.

Security Systems

A multiplexing system can be used in a security system for a plant or building. Each door and window can be connected to the two-wire system. A display is located in a central control location to monitor the total system. **See Figure 16-51.** A clock and printer can be added to record the time each door or window is opened and closed. A PLC may be added to control all required circuit logic.

The PLC's controlling functions on the multiplexing system can be expanded as necessary. For example, if a security guard is to patrol a building, the controller can be programmed to monitor the guard as well as the building. As the guard moves through the building, the controller monitors the movement by recording when a door is opened and/or when the guard activates an assigned switch. The controller knows how long it should take the guard to move from station to station. If something happens to the guard, the controller detects this and takes corrective action, such as alarming a central control station.

Conveyor Systems

Conveyor systems are commonly used in industry for movement of materials. Additional control is required as industrial systems become more automated. Additional control requires additional wires to be connected from machine to machine. Multiplexing can be used to reduce the total number of wires required.

As in any assembly line application, a fault or breakdown at one station requires that all upstream machines be turned OFF to prevent a product jam. Multiplexing may be used to link the system together because this system may cover miles in many applications. **See Figure 16-52.**

A sensor may be used to detect a fault at one location and send a signal over the two-wire system to stop all upstream machines and conveyors. This system may also be connected for total control of all functions using the multiplexing system and a PLC.

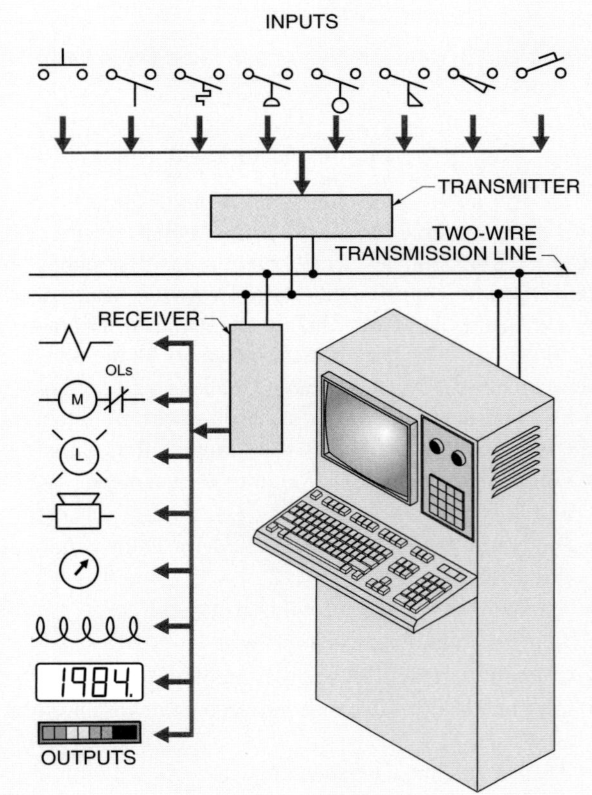

Figure 16-50. A multiplexing system is simpler than hard wiring and can be expanded to almost any number of inputs and outputs, all controlled by a PLC.

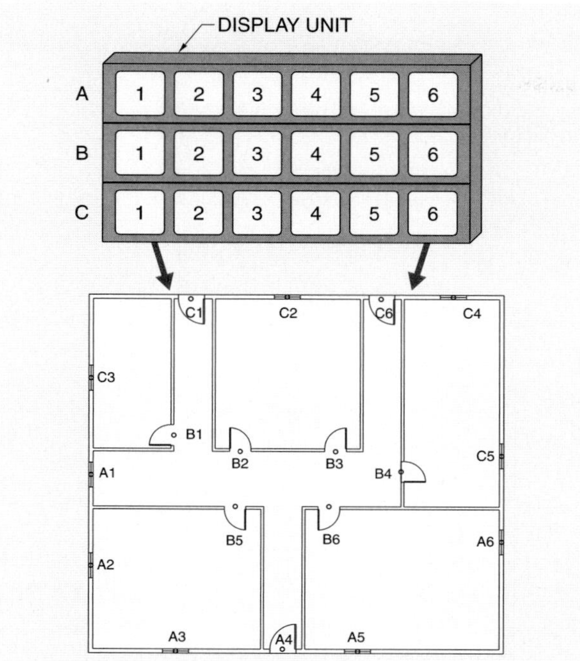

Figure 16-51. A multiplexing system can be used in a security system for a plant or building.

One main advantage of using a PLC is that the PLC eliminates the need for additional external contacts on motor starters, external relays, external timers, and counters. For example, an application requires that three conveyors be sequenced. **See Figure 16-53.** In this circuit, when the start pushbutton is pressed and released, conveyor 1 starts. After timer 1 times out, conveyor 2 starts. After timer 2 times out, conveyor 3 starts. An overload on any conveyor automatically stops all three conveyors.

In the standard line diagram, two external timers are required, in addition to a memory contact M1 on motor starter 1. When this circuit is programmed, the PLC eliminates the need for the external timers because the PLC includes internal timers. Likewise, the memory contact is programmed (not hard wired) using the PLC software. Also note that in following the basic rules of programming a PLC circuit, the three overload contacts must be moved to the left side of the output (conveyor one).

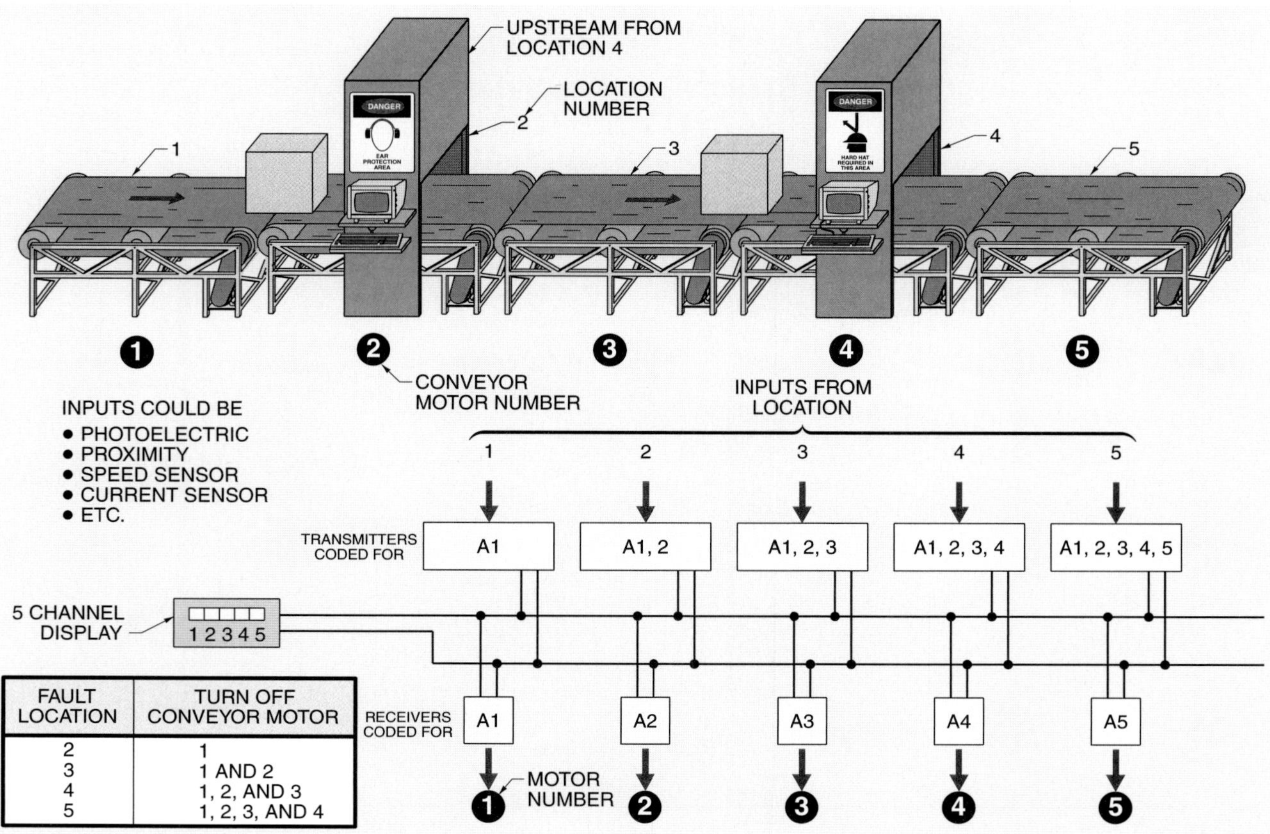

FAULT LOCATION	TURN OFF CONVEYOR MOTOR
2	1
3	1 AND 2
4	1, 2, AND 3
5	1, 2, 3, AND 4

Figure 16-52. An assembly line using several conveyors can be controlled by a two-wire multiplexing system.

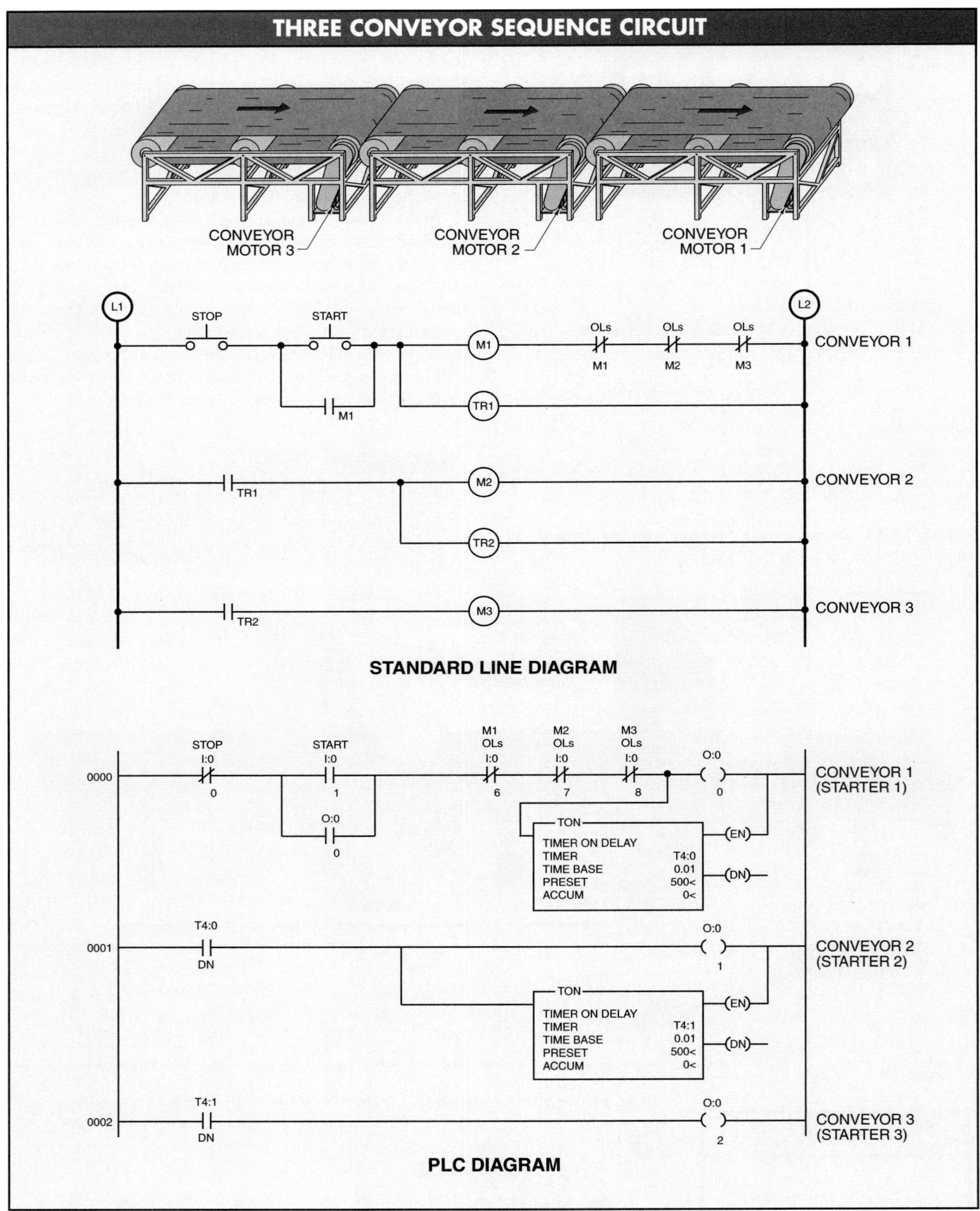

Figure 16-53. One main advantage of using a PLC is that the PLC eliminates the need for additional external contacts on motor starters, external relays, external timers, and counters.

TROUBLESHOOTING PLCs

Troubleshooting PLCs normally involves finding a problem in the hardware or software. Most hardware problems are found in the input and output sections of the PLC and can usually be found using standard DMMs. Software problems require a knowledge of the specific program used and type of manufacturer equipment used.

Troubleshooting Input Modules

A vibrating voltage tester (wiggie) must never be used to measure voltage levels on a PLC. Vibrating voltage testers contain a solenoid. When the test leads of a vibrating voltage tester are removed, the collapsing field of the solenoid can damage the solid-state components of PLC I/O modules.

Signals and information are sent to a PLC using input devices such as pushbuttons, limit switches, level switches, and pressure switches. The input devices are connected to the input module of the PLC. Input devices are connected to terminal screws at the back of the input module. The controller does not receive the proper information if the input device or input module is not operating correctly. **See Figure 16-54.** To troubleshoot an input module of a PLC, apply the following procedure:

1. Measure the supply voltage at the input module to ensure that there is power supplied to the input device(s). Test the main power supply of the controller when there is no power.
2. Measure the voltage from the control switch. Connect the DMM directly to the same terminal screw to which the input device is connected. The DMM should read the supply voltage when the control switch is closed. The DMM should read the full supply voltage when the control device uses mechanical contacts. The DMM should read nearly the full supply voltage when the control device is solid-state. Full supply voltage is not read because .5 V to 6 V is dropped across the solid-state control device. The DMM should read zero or little voltage when the control switch is open.
3. Monitor the status indicators on the input module. The status indicators should illuminate when the DMM indicates the presence of supply voltage.
4. Monitor the input device symbol on the programming terminal monitor. The symbol should be highlighted when the DMM indicates the presence of supply voltage. Replace the control device if the control device does not deliver the proper voltage. Replace the input module if the control device delivers the correct voltage but the status indicator does not illuminate.

Troubleshooting Input Devices

Input devices such as pushbuttons, limit switches, pressure switches, and temperature switches are connected to the input module(s) of a PLC. Input devices send information and data concerning circuit and process conditions to the controller. The processor receives the information from the input devices and executes the program. All input devices must operate correctly for the circuit to operate properly. **See Figure 16-55.** To troubleshoot an input device of a PLC, apply the following procedure:

1. Place the controller in the test or program mode. This step prevents the output devices from turning ON. Output devices are turned ON when the controller is placed in the run mode.
2. Monitor the input devices using the input status indicators (located on each input module), the programming terminal monitor, or the data file. A *data file* is a group of data values (inputs, timers, counters, and outputs) that are displayed as a group and whose status may be monitored.
3. Manually operate each input starting with the first input. Never reach into a machine when manually operating an input. Always use a wooden stick or other nonconductive device.

The input status indicator located on the input module should illuminate and the input symbol should be highlighted in the control circuit on the monitor screen when a normally open input device is closed. The bit status on the programming terminal monitor screen should be set to 1 indicating a high or presence of voltage.

Heidelberg Harris, Inc.

Monitoring the I/O status indicators helps locate any problem(s) when troubleshooting large systems.

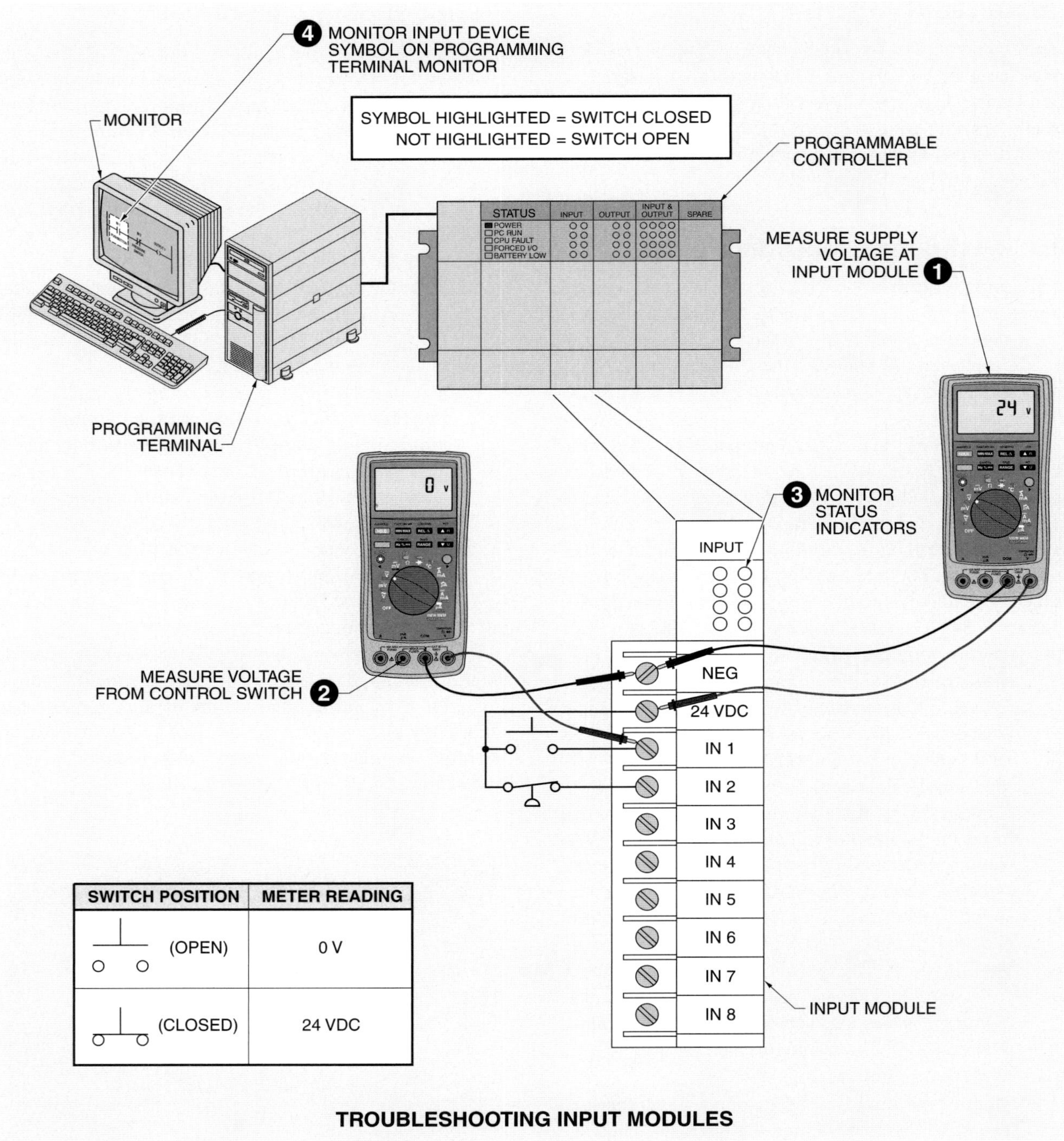

SWITCH POSITION	METER READING
(OPEN)	0 V
(CLOSED)	24 VDC

Figure 16-54. The controller does not receive the proper information if the input device or input module is not operating correctly.

The input status indicator located on the input module should turn OFF and the input symbol should no longer be highlighted in the control circuit on the monitor screen when a normally closed input device is open. The bit status on the programming terminal monitor screen should be set to 0 indicating a low or absence of voltage.

Select the next input device and test it when the status indicator and associated bit status match. Continue testing each input device until all inputs have been tested. Troubleshoot the input device and output device when the status indicator and associated bit status do not match.

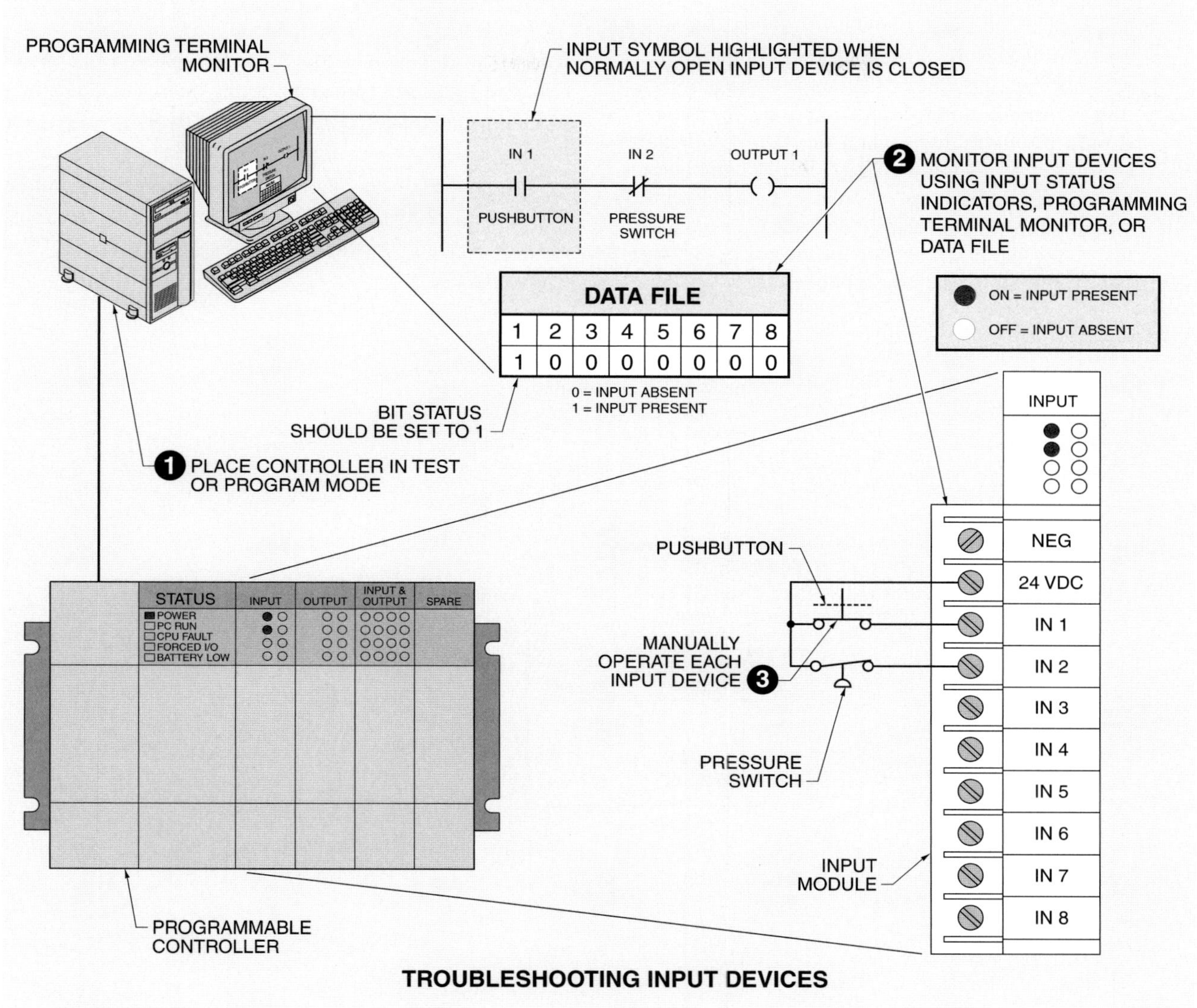

Figure 16-55. All input devices must operate correctly for the circuit to operate properly.

Troubleshooting Output Modules

A PLC turns ON and OFF the output devices (loads) in the circuit according to the program. The output devices are connected to the output module of the PLC. No work is produced in the circuit when the output module or the output devices are not operating correctly. The problem may lie in the output module, output device, or controller when an output device does not operate. **See Figure 16-56.** To troubleshoot an output module of a PLC, apply the following procedure:

1. Measure the supply voltage at the output module to ensure that there is power supplied to the output devices. Test the main power supply of the controller when there is no power.
2. Measure the voltage delivered from the output module. Connect the DMM directly to the same terminal screw to which the output device is connected. The DMM should read the supply voltage when the program energizes the output device. The DMM should read full supply voltage when the output module uses mechanical contacts. The DMM should read almost full supply voltage when the output module uses a solid-state switch. Full voltage is not read because .5 V to 6 V is dropped across the solid-state switch. The DMM should read zero or little voltage when the program de-energizes the output device.

3. Monitor the status indicators on the output module. The status indicators should be energized when the DMM indicates the presence of supply voltage.
4. Monitor the output device symbol on the programming terminal monitor. The output device symbol should be highlighted when the DMM indicates the presence of supply voltage. Replace the output module when the output module does not deliver the proper voltage. Troubleshoot the output device when the output module does deliver the correct voltage but the output device does not operate.

Troubleshooting Output Devices

Output devices such as motor starters, solenoids, contactors, and lights are connected to the output modules of a PLC. An output device performs the work required for the application.

The processor energizes and de-energizes the output devices according to the program. All output devices must operate correctly for the circuit to operate properly. **See Figure 16-57.**

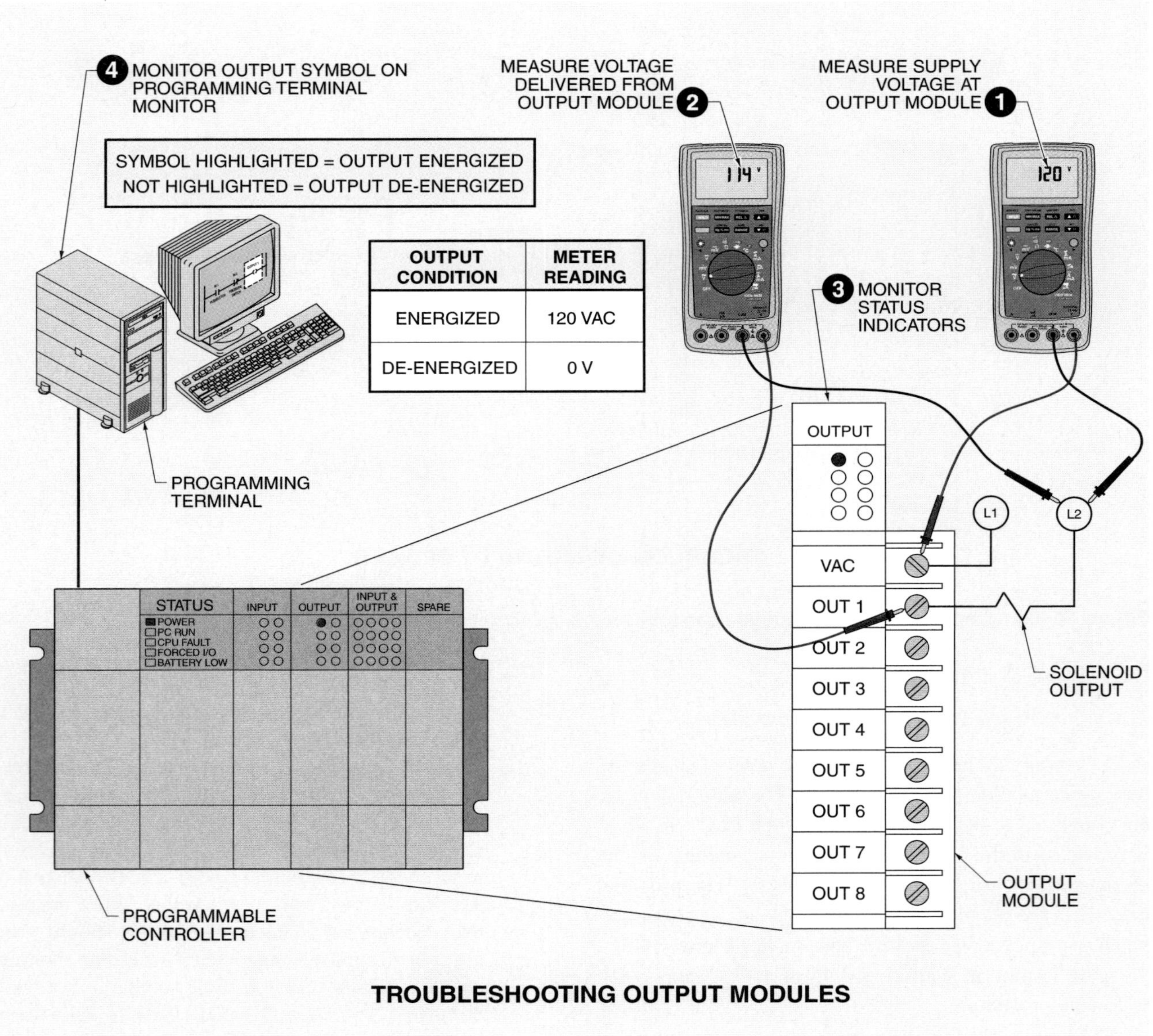

Figure 16-56. No work is produced in the circuit when the output module or output devices are not operating correctly.

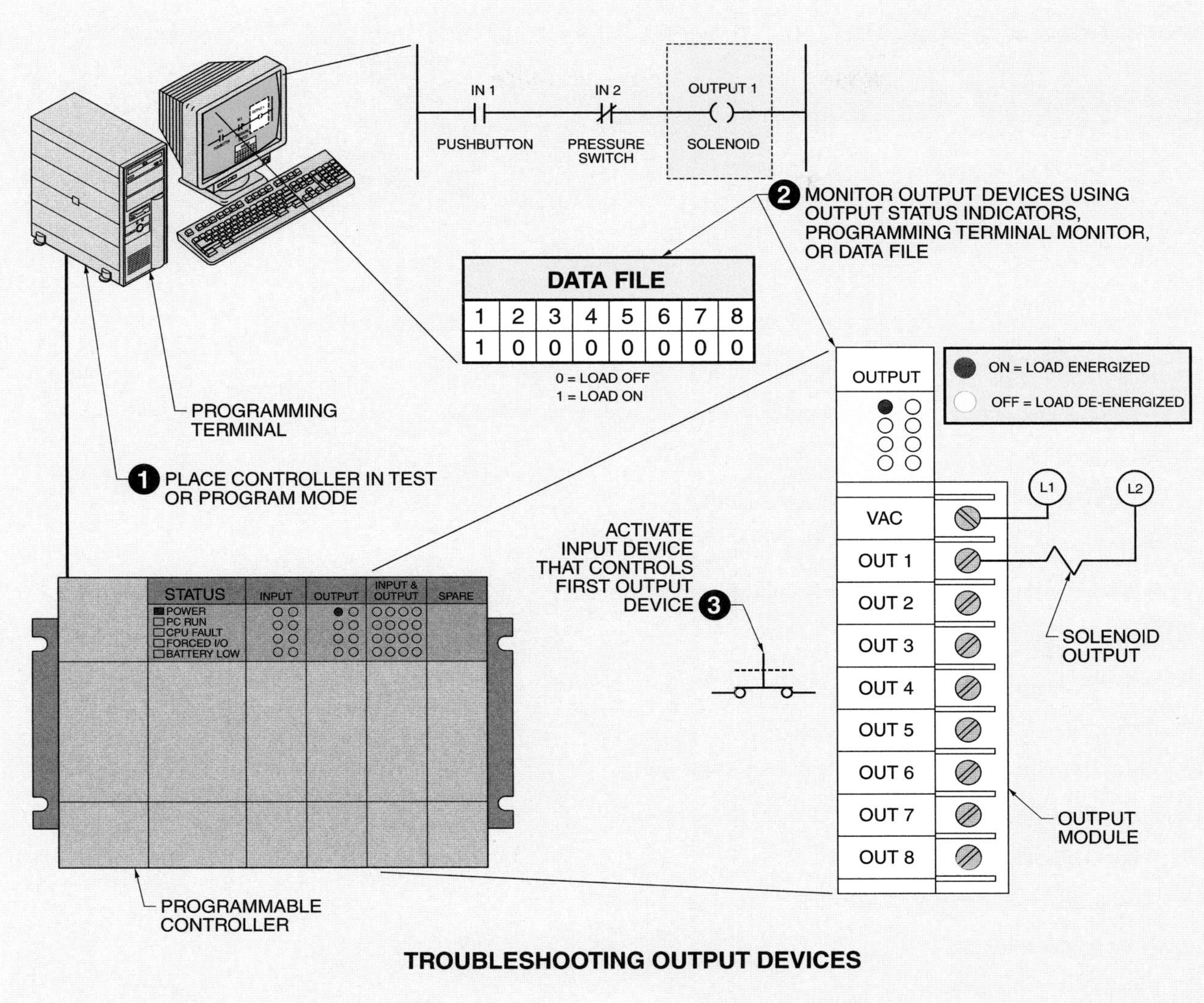

Figure 16-57. All output devices must operate correctly for the circuit to operate properly.

To troubleshoot an output device of a PLC, apply the following procedure:

1. Place the controller in the test or program mode. Placing the controller in the test or program mode prevents the output devices from turning ON. Output devices turn ON when the controller is placed in the run mode.
2. Monitor the output devices using the output status indicators (located on each output module), the programming terminal monitor, or the data file.
3. Activate the input that controls the first output device. Check the program displayed on the monitor screen to determine which input activates which output device. Never reach into a machine to activate an input.

Select the next output device and test it when the status indicator and associated bit status match. Continue testing each output device until all output devices have been tested. Troubleshoot the input device and output device when the status indicator and associated bit status do not match.

Review Questions

1. What are the two major categories of electrical systems designed to produce products?
2. Which manufacturing area represents durable goods type products?
3. Which manufacturing area represents consumable goods type products?
4. What are I/Os?
5. What type of diagram is used to program a PLC?
6. What are the four basic parts of a PLC?
7. What are some examples of discrete inputs?
8. What are some examples of data inputs?
9. What is scan time?
10. What is a programming diagram?
11. Why does a PLC include a battery as part of its system?
12. What command is the opposite of the force command?
13. What is electrical noise?
14. How is interlocking added into a PLC controlled circuit?
15. What is multiplexing?
16. What type of input device is XIC used to describe?
17. What type of input device is XIO used to describe?
18. What type of operation has feedback from the output back to the input?
19. What are the basic programming rules that must be followed if a circuit is to be accepted by software before downloading?
20. What is one main advantage of using a PLC?

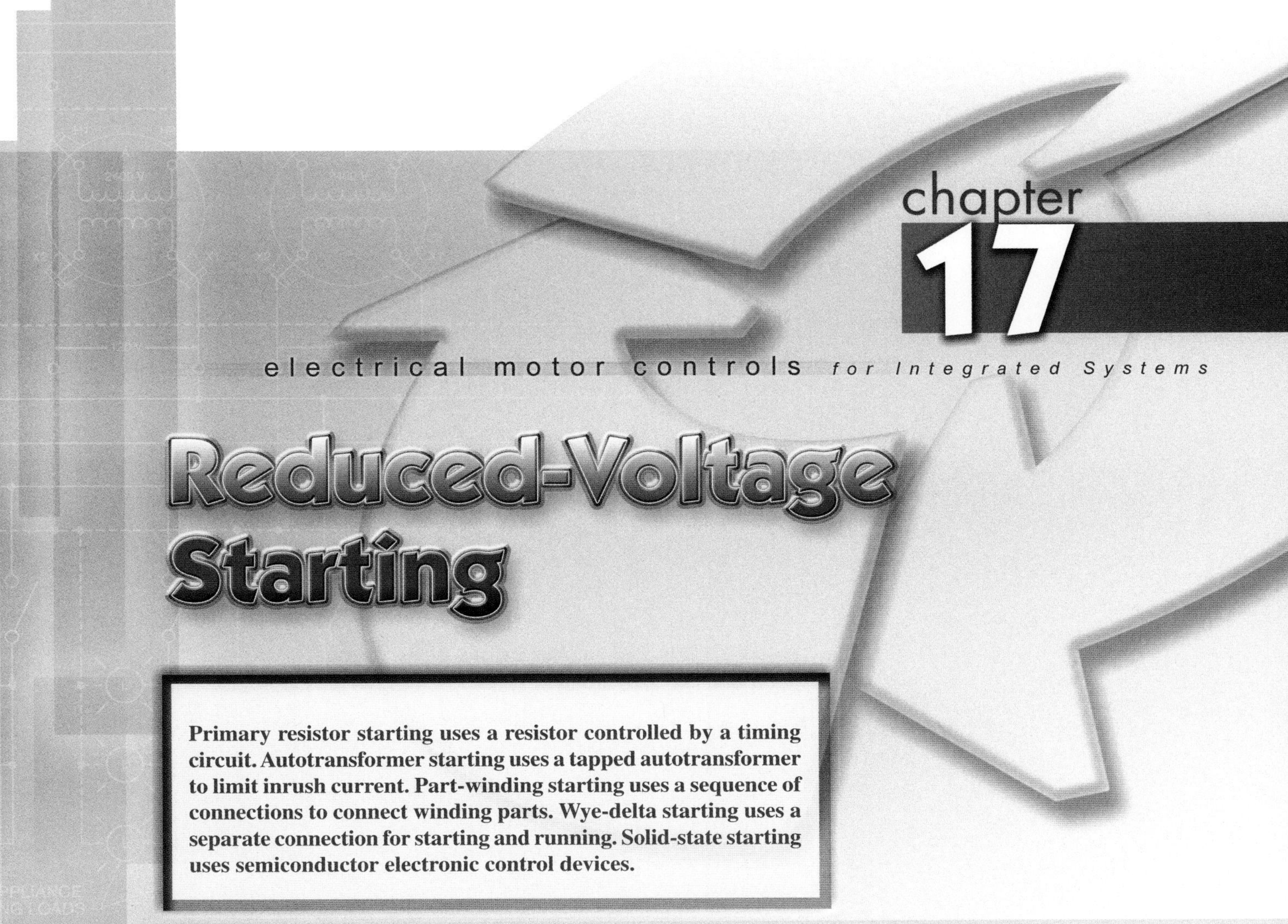

REDUCED-VOLTAGE STARTING

Full-voltage starting is the least expensive and most efficient means of starting a motor for applications involving small horsepower motors. Many applications involve large horsepower DC and AC motors that require reduced-voltage starting because full-voltage starting may create interference with other systems. Reduced-voltage starting reduces interference in the power source, the load, and the electrical environment surrounding the motor.

Power Source

Reduced-voltage starting is used to reduce the large current drawn from the power company lines by the across-the-line start of a large motor. An induction motor acts much like a short circuit in the secondary of a transformer when it is started. The current drawn by a motor at starting is typically about two to six times the current rating found on the motor nameplate. This sudden demand for a large amount of current can reflect back into the power lines and create problems. Reduced-voltage starting reduces the amount of starting current a motor draws when starting. **See Figure 17-1.**

Electric utilities normally limit the inrush current drawn from their lines to a maximum amount for a specified period of time. Such limitations are necessary for smooth, steady power regulation and for eliminating objectionable voltage disturbances such as annoying light flicker.

In these cases, the utility company is not limiting the total maximum amount of current that can be drawn, but rather dividing the amount of current into steps. This permits an incremental start that allows the utility voltage regulators sufficient time to compensate for the large current draw. *Increment current* is the maximum current permitted by the utility in any one step of an increment start. This increment current may be determined by checking with the local utility company. Reduced-voltage starting provides incremental current draw over a longer period of time.

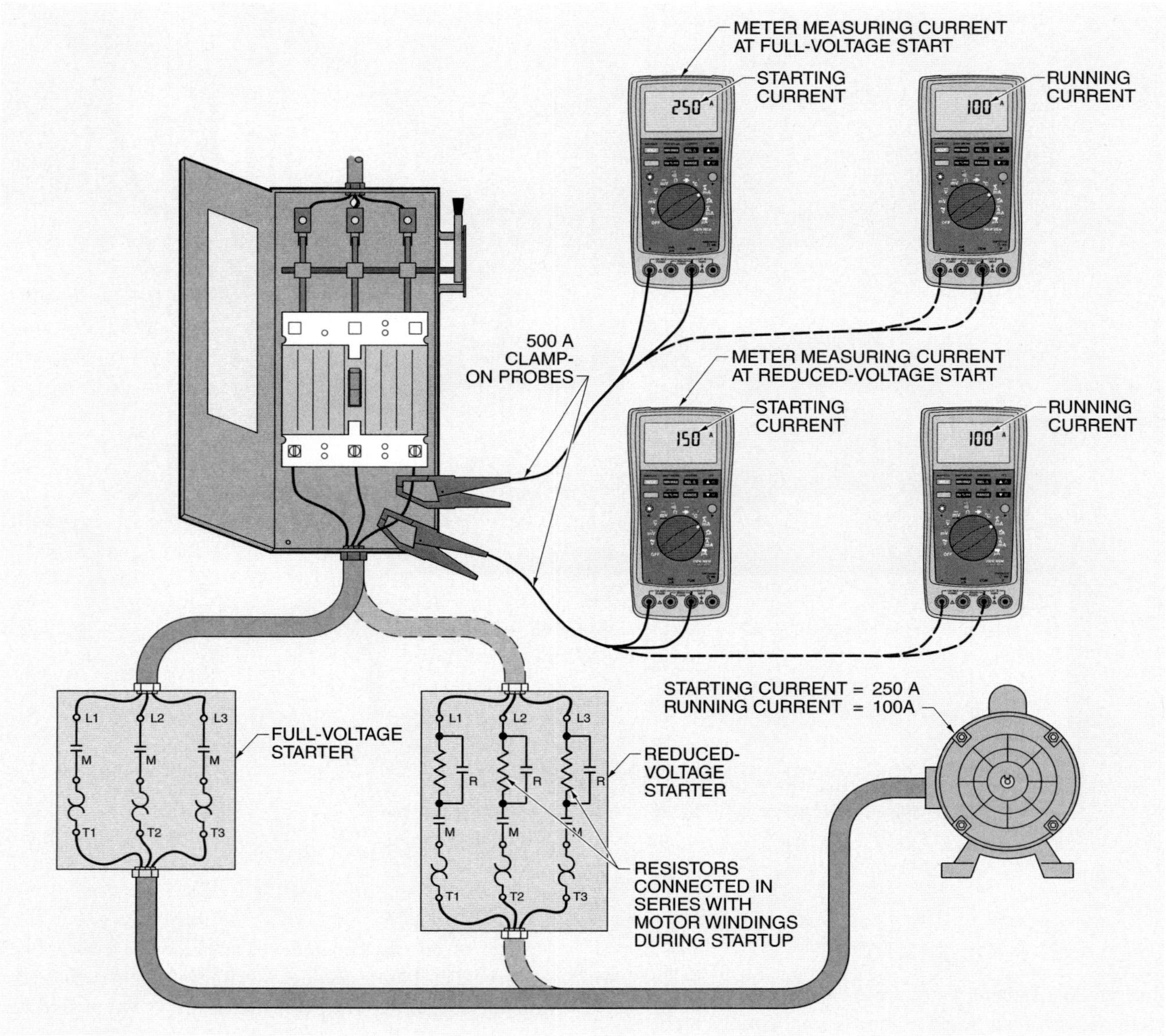

Figure 17-1. Reduced-voltage starting reduces the amount of current a motor draws when starting.

Load Torque and Starting Requirements

In several industries, especially those dealing with paper and other delicate fabrics, care must be exercised to avoid sudden high starting torque (turning force). Such torque could stretch or tear the product.

To prevent product damage or damage to gears, belts, and chain drives, it is necessary to limit starting torque surges. Reduced-voltage starting is used to overcome excessive starting torque by providing a gentle start and smooth acceleration of a motor. **See Figure 17-2.**

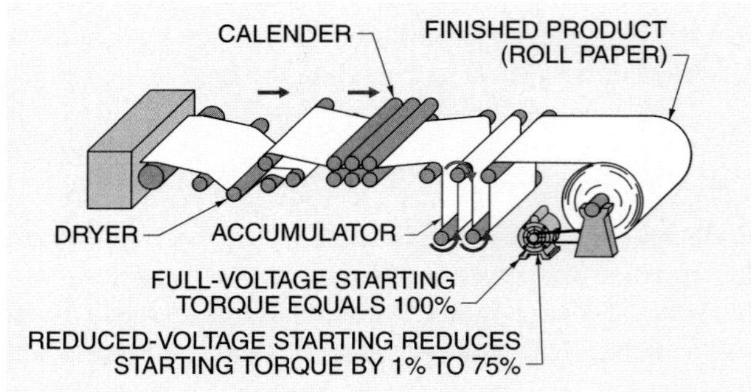

Figure 17-2. Reduced-voltage starting reduces the amount of starting torque produced on a load.

Based on the formula $I = E \div R$, as voltage is reduced, current is reduced; and as current is reduced, torque is reduced because motor torque is proportional to current. A reduction in voltage reduces current, which reduces torque to produce a gentle start. As voltage is increased, current and torque are increased, providing smooth acceleration. Reduced-voltage starting is not speed control. Reduced-voltage starting acts as a buffer or shock absorber to the load when it is starting.

Reduced-voltage starting should not be considered for use on loads that are difficult to start. A load that is difficult to start at full voltage does not start at a reduced voltage.

Electrical Environment

A new electrical system should not create problems for systems that are already installed and working properly. Electric current surges can cause disruptions. For example, a current surge may cause timers to reset or relays and starters to drop out due to the voltage sag caused by the high current draw. A *voltage sag (voltage dip)* is a drop in voltage of more than 10% (but not to 0 V) below the normal rated line voltage lasting from .5 cycles up to 1 minute. In buildings that are totally air conditioned, compressor motors have caused major computers and microcomputers to malfunction due to current surges. Reduced-voltage starting may be used to solve current surge problems even when not required by the utility company.

DC MOTOR REDUCED-VOLTAGE STARTING

A DC motor is used to convert electrical energy into a rotating mechanical force. Although AC and DC motors operate on the same fundamental principles of magnetism, they differ in the way the conversion of the electrical power to mechanical power is accomplished. This difference gives each motor its own operating characteristics. The two fundamental operating characteristics of DC motors that make them the preferred choice for some applications are high torque output and good speed control.

Another factor in using DC motors is the available source of power. For applications such as an automobile starter, a DC motor is compatible with the power source (battery), which delivers only DC. DC motors run by batteries are also used in industrial applications using portable power equipment such as forklift trucks, dollies, and small locomotives used to move materials and supplies. In applications where the motor is to be connected to a power source other than a battery, the available power source may be either AC or DC.

All DC motors are supplied with current directly connected to the armature and field windings. During startup, current is limited by the resistance of the wire in the armature and the field windings when current is connected directly to the motor. The larger the motor, the less the resistance and the larger the current. In DC motors, this starting current may be so high that it damages the motor. To prevent motor damage, reduced-voltage starting must be applied to DC motors larger than 1 HP.

Reduced-voltage starting of DC motors reduces the amount of current during starting. As the motor accelerates, the reduced voltage may be removed because the current in the motor decreases with an increase in motor speed. This decrease in current results from the motor generating a voltage that is opposite to the applied voltage as it accelerates.

This opposing voltage, known as counter electromotive force (counter EMF), depends on the speed of the motor. Counter EMF is zero at standstill and increases with motor speed. Ohm's law is used to calculate motor starting current. To calculate motor starting current, apply the following formula:

$$I = \frac{E}{R}$$

where

I = starting current (in A)

E = applied voltage (in V)

R = resistance (in Ω)

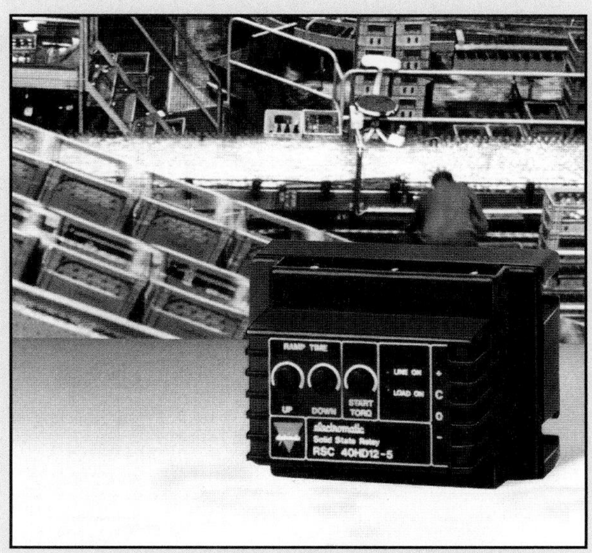

Carlo Gavazzi Inc. Electromatic Business Unit

Microprocessor-based control modules are used for soft starting and soft stopping of 3ϕ induction motors.

Example: Calculating Motor Starting Current

What is the starting current of a DC motor with an armature resistance of 1 Ω that is connected to a 200 V supply?

$$I = \frac{E}{R}$$

$$I = \frac{200}{1}$$

I **= 200 A**

As the motor accelerates, a counter EMF is generated. The counter EMF reduces the current in the motor. To calculate the current drawn by a motor during starting, apply the following formula:

$$I = \frac{E - C_{EMF}}{R}$$

where

I = starting current (in A)

C_{EMF} = generated counter electromotive force (in V)

E = applied voltage (in V)

R = resistance (in Ω)

Rofin Sinar

Many industrial processes require electrical motors that must be designed to operate in hot and severe environments.

Example: Calculating Current during Starting

What is the current during starting of a DC motor with an armature resistance of 1 Ω that is connected to a 200 V supply and is generating 100 V of counter EMF?

$$I = \frac{E - C_{EMF}}{R}$$

$$I = \frac{200 - 100}{1}$$

$$I = \frac{100}{1}$$

I **= 100 A**

A motor at full speed generates an even higher counter EMF. The higher counter EMF further reduces the current in the motor. It is the 200 A (or starting current of any large DC motor) that the motor must be protected from to prevent damage.

Example: Calculating Motor Running Current

What is the running current of a DC motor with an armature resistance of 1 Ω that is connected to a 200 V supply and is generating 180 V of counter EMF?

$$I = \frac{E - C_{EMF}}{R}$$

$$I = \frac{200 - 180}{1}$$

$$I = \frac{20}{1}$$

I **= 20 A**

A starting rheostat or a solid-state circuit is used when reduced voltage is applied to DC motors. The starting rheostat is connected in series with the incoming power line (typically the positive DC line) and the motor. The rheostat reduces the voltage applied to the motor during starting by placing a high resistance in series with the motor. The resistance is decreased as the rheostat is moved to the run position. **See Figure 17-3.**

The starting rheostat is controlled manually, which means that the operator determines the exact starting time. Although a starting rheostat can also be used to control motor speed (the speed of a DC motor varies with the applied voltage), the purpose of a starting rheostat is to reduce the voltage (and thus current and torque) during starting. After the motor is started, a different circuit can be used to control motor speed.

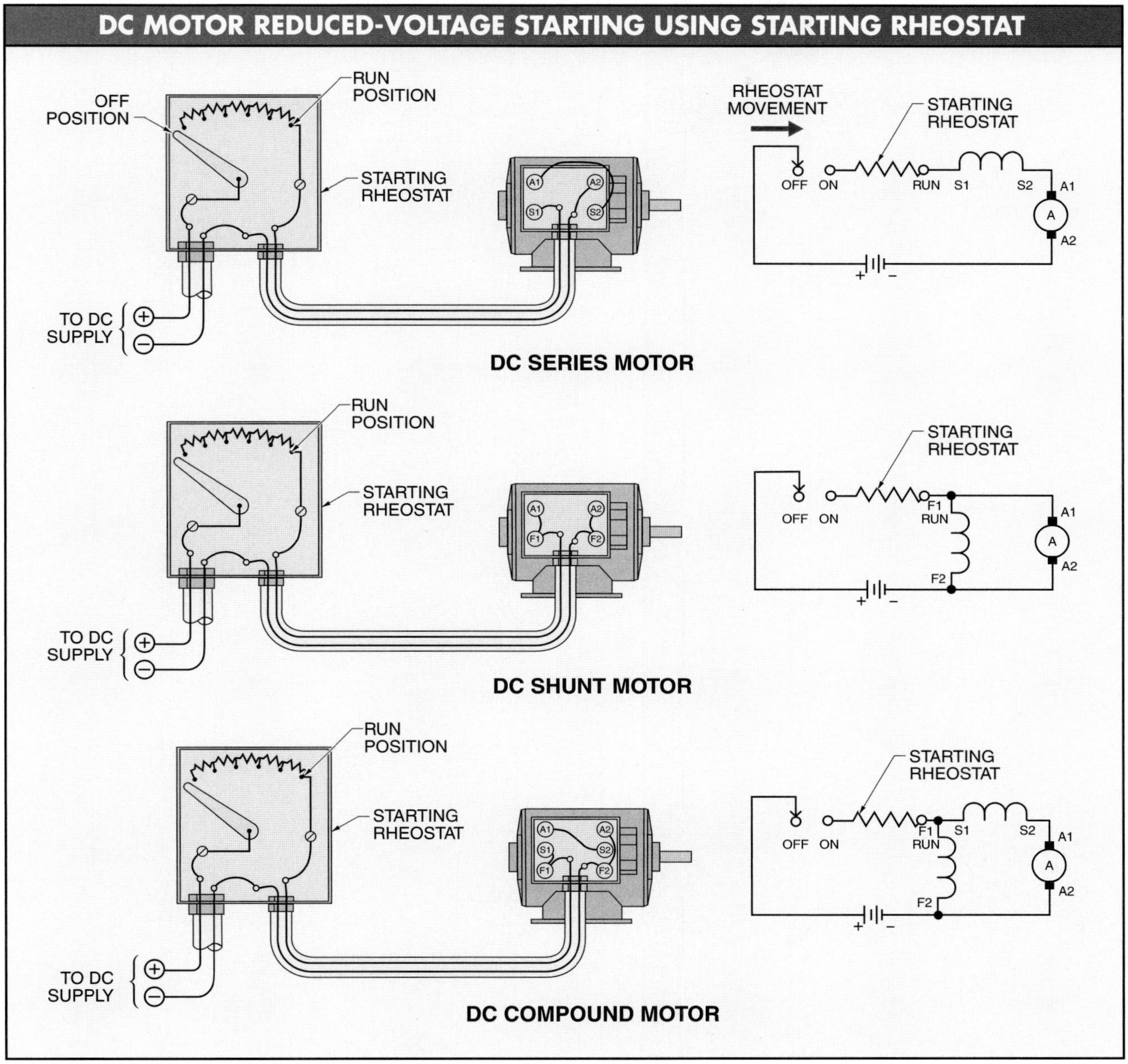

DC MOTOR REDUCED-VOLTAGE STARTING USING STARTING RHEOSTAT

OFF POSITION
RUN POSITION
STARTING RHEOSTAT
TO DC SUPPLY
RHEOSTAT MOVEMENT
OFF ON RUN S1 S2 A1 A2

DC SERIES MOTOR

RUN POSITION
STARTING RHEOSTAT
TO DC SUPPLY
OFF ON F1 RUN F2 A1 A2

DC SHUNT MOTOR

RUN POSITION
STARTING RHEOSTAT
TO DC SUPPLY
OFF ON F1 RUN S1 S2 F2 A1 A2

DC COMPOUND MOTOR

Figure 17-3. A rheostat reduces the voltage applied to a motor during starting by placing a high resistance in series with the motor.

The purpose of a rheostat is to reduce the voltage applied to a motor. A solid-state circuit can also be used to reduce the voltage applied to a motor. A silicon-controlled rectifier (SCR) is typically used to control the applied voltage. The voltage is adjusted through a control circuit that controls the voltage applied to the gate of the SCR. The higher the applied gate voltage, the higher the output voltage applied to the motor. **See Figure 17-4.**

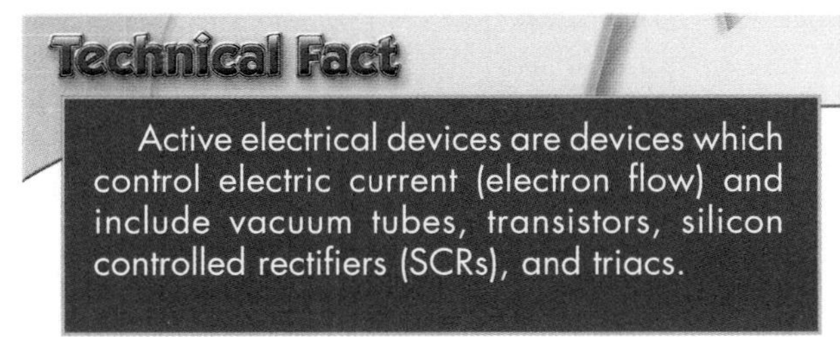

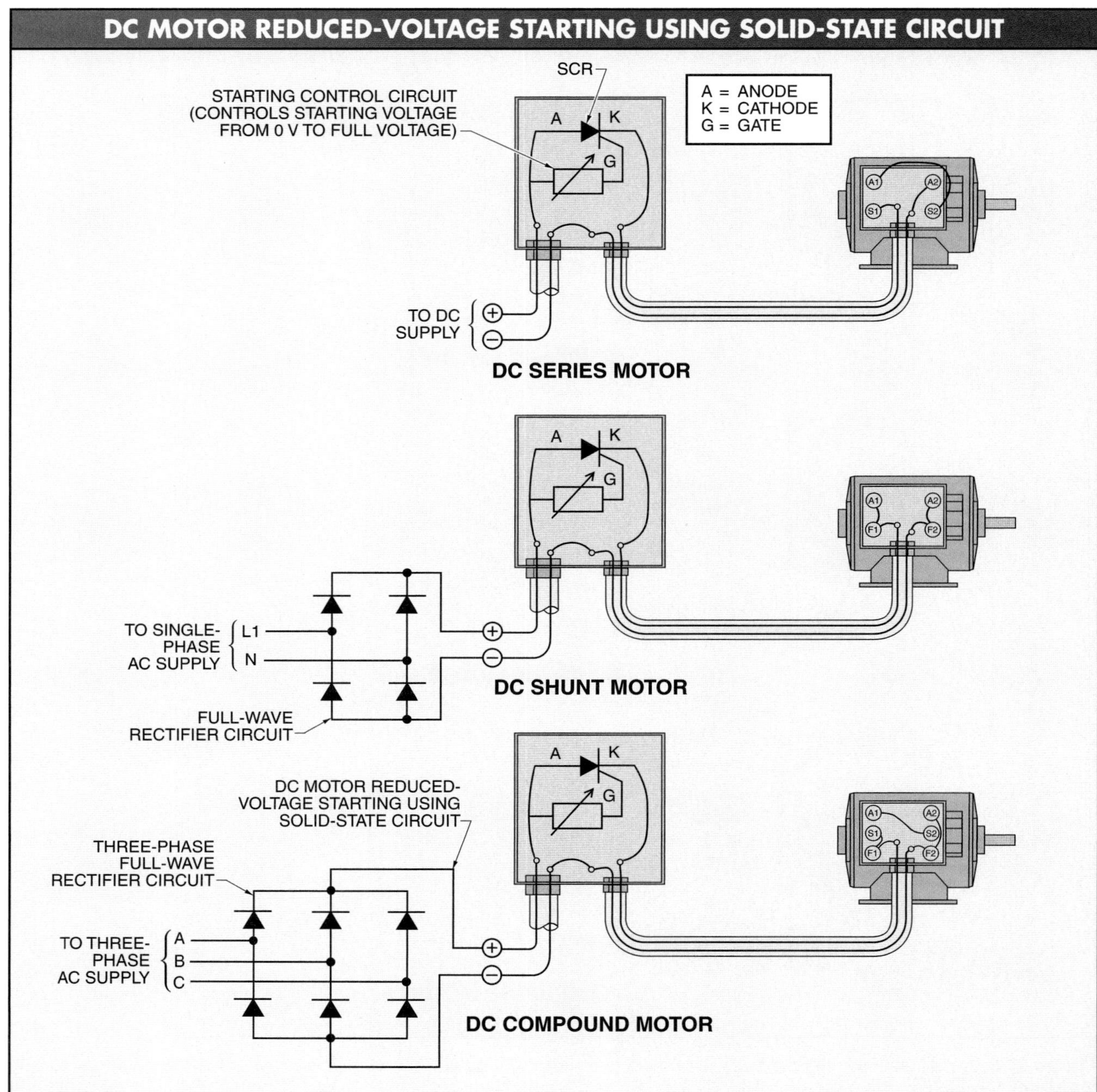

Figure 17-4. A solid-state circuit can be used to reduce the voltage to a motor.

REDUCED-VOLTAGE STARTING FOR THREE-PHASE INDUCTION MOTORS

The majority of industrial applications normally use 3ϕ induction motors. Three-phase induction motors are normally chosen over other types of motors because of their simplicity, ruggedness, and reliability. Both 1ϕ and 3ϕ induction motors have become the standard for AC, all-purpose, constant-speed motor applications. Reduced voltage starting is applied to 3ϕ motors because 1ϕ motors are small (typically 5 HP and less).

AC Motor Reduced-Voltage Starting

A heavy current is drawn from the power lines when an induction motor is started. This sudden demand for large current can reflect back into the power lines and create problems such as voltage sags (temporary low voltage) on the power lines.

The revolving field of the stator induces a large current in the short-circuited rotor bars. The current is highest when the rotor is at a standstill and decreases as the motor speed increases. The current drawn by a motor when starting is excessive because of a lack of counter EMF at the instant of starting. Once rotation begins, counter EMF is built up in proportion to speed, and the current decreases. **See Figure 17-5.**

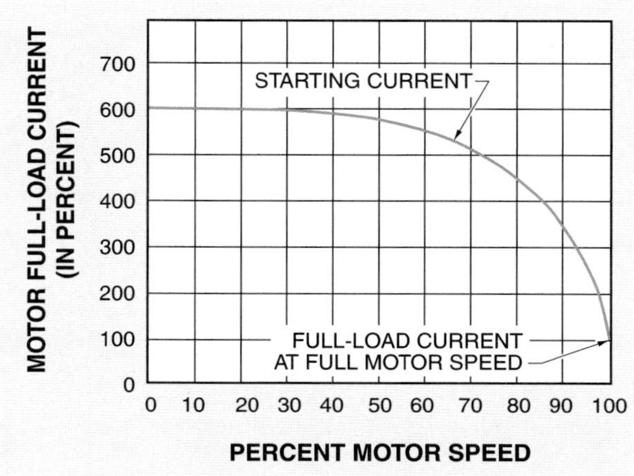

Figure 17-5. The current drawn by a motor when starting is excessive because of a lack of counter EMF at the instant of starting.

The percent of full-load current is marked on the vertical scale and the percent of motor speed is marked on the horizontal scale. The starting current is quite high compared to the running current. The starting current remains fairly constant at this high value as the speed of the motor increases, but then drops sharply during the last few percentages up to 100%.

This illustrates that the heating rate is quite high during acceleration because the heating rate is a function of current. A motor may be considered to be in the locked condition during nearly all of the accelerating period.

Locked rotor current (LRC) is the steady-state current taken from the power line with the rotor locked (stopped) and with the voltage applied. Locked rotor current and the resulting torque produced in the motor shaft (in addition to load requirements) determine whether the motor can be connected across the line or whether the current has to be reduced through a reduced-voltage starter.

Full-load current (FLC) is the current required by a motor to produce full-load torque at the motor's rated speed. Full-load current is the current given on the motor nameplate. The load current is less than what is given on the nameplate if the motor is not required to deliver full torque. This information is required when testing a motor (running a motor without a load). The only torque a motor must produce when not connected to a load is the torque that is needed to overcome its own internal friction and winding losses. For this reason, the current is less than the value given on the motor nameplate.

Open and Closed Circuit Transition

Motors that are started at reduced voltage must be switched to line voltage before reaching full speed. The two methods used to switch motors from starting voltage to line voltage include open circuit transition and closed circuit transition. In open circuit transition, a motor is temporarily disconnected from the voltage source when switching from a reduced starting voltage level to a running voltage level, before reaching full motor speed. In closed circuit transition, a motor remains connected to the voltage source when switching from a reduced starting voltage level to a running voltage level, before reaching full motor speed.

Closed circuit transition is preferable to open circuit transition because closed circuit transition does not cause a high-current transition surge. However, closed circuit transition is the more expensive circuit transition method. Open circuit transition produces a higher current surge than closed circuit transition at the transition point because the motor is momentarily disconnected from the voltage source. **See Figure 17-6.**

The high current surge during open circuit transition is based on the motor speed at the time of transition. Transfer from the low starting voltage to the high line voltage should occur as close to full motor speed as possible. If the transition occurs when the motor is at a low speed, a surge current even higher than the starting current can occur. **See Figure 17-7.**

Baldor Electric Co.

Reduced-voltage starting is used with large industrial motors to reduce the damaging effect of a large starting current.

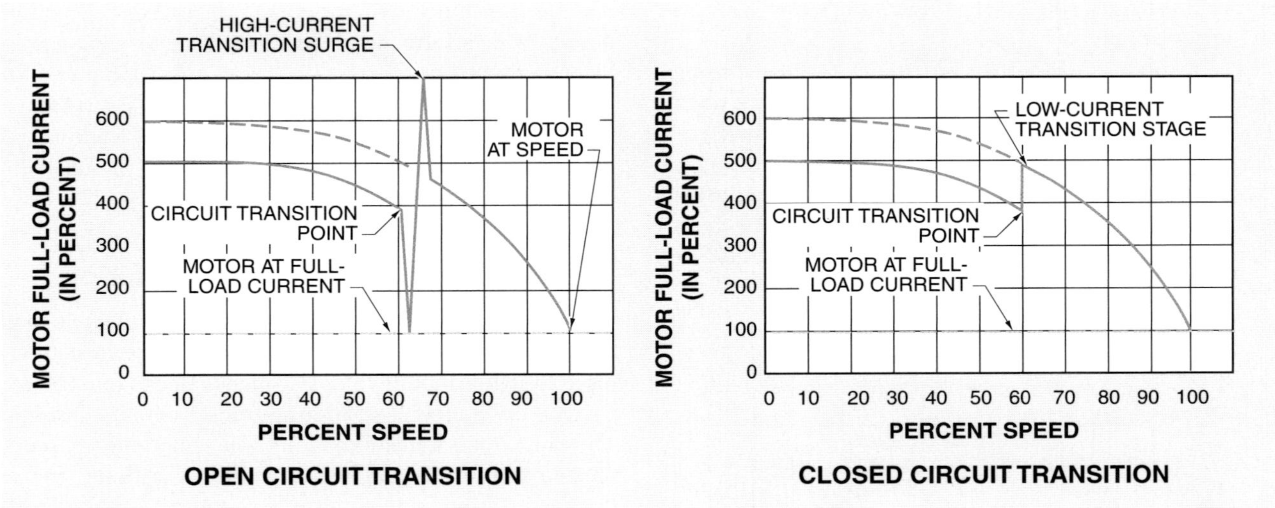

Figure 17-6. Open circuit transition produces a higher current surge than closed circuit transition because in open circuit transition the motor is momentarily disconnected from the voltage source.

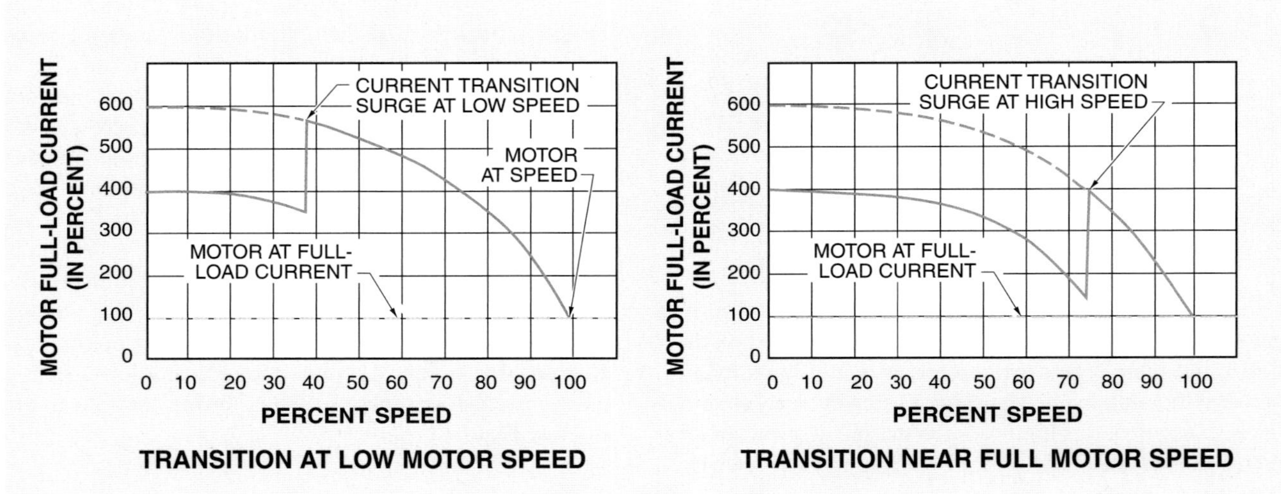

Figure 17-7. Circuit transition should occur at or near full motor speed to minimize current surges.

PRIMARY RESISTOR STARTING

Primary resistor starting is a reduced-voltage starting method that uses a resistor connected in each motor line (in one line in a 1ϕ starter) to produce a voltage drop. This reduces the motor starting current as it passes through the resistor. **See Figure 17-8.** A timer is provided in the control circuit to short the resistors after the motor accelerates to a specified point. The motor is started at reduced voltage but operates at full line voltage.

Primary resistor starters provide extremely smooth starting due to increasing voltage across the motor terminals as the motor accelerates. Standard primary resistor starters provide two-point acceleration (one step of resistance) with approximately 70% of line voltage at the motor terminals at the instant of motor starting. Multiple-step starting is possible by using additional contacts and resistors when extra smooth starting and acceleration are needed. This multiple-step starting may be required in paper or fabric applications where even a small jolt in starting may tear the paper or snap the fabric.

Furnas Electric Co.

Figure 17-8. Primary resistor starting uses a resistor connected in each motor line and a timer to short out the resistors after the motor accelerates to a specified point.

Primary Resistor Starting Circuits

In a primary resistor starting circuit, external resistance is added to and taken away from the motor circuit. **See Figure 17-9.** The control circuit consists of the motor starter coil M, ON-delay timer TR1, and contactor coil C. Coil M controls the motor starter, which energizes the motor and provides overload protection. The timer provides a delay from the point where coil M energizes until contacts C close, shorting resistors R1, R2, and R3. Coil C energizes the contactor, which provides a short circuit across the resistors.

Pressing start pushbutton PB2 energizes motor starter coil M and the ON-delay timer coil TR1. Motor starter coil M closes contacts M to create memory. ON-delay timer coil TR1 causes contacts TR1 to remain open during reset, stay open during timing, and close after timing out. Once timed out, the contactor coil C energizes, causing contacts C to close and the resistors to short.

This circuit is a common reduced-voltage starting circuit. Changes are often made in the values of resistance and wattage to accommodate motors of different horsepower ratings.

AUTOTRANSFORMER STARTING

Autotransformer starting uses a tapped 3ϕ autotransformer to provide reduced-voltage starting. **See Figure 17-10.** Autotransformer starting is one of the most effective methods of reduced-voltage starting. Autotransformer starting is preferred over primary resistor starting when the starting current drawn from the line must be held to a minimum value, yet the maximum starting torque per line ampere is required.

In autotransformer starting, the motor terminal voltage does not depend on the load current. The current to the motor may change because of the motor's changing characteristics, but the voltage to the motor remains relatively constant.

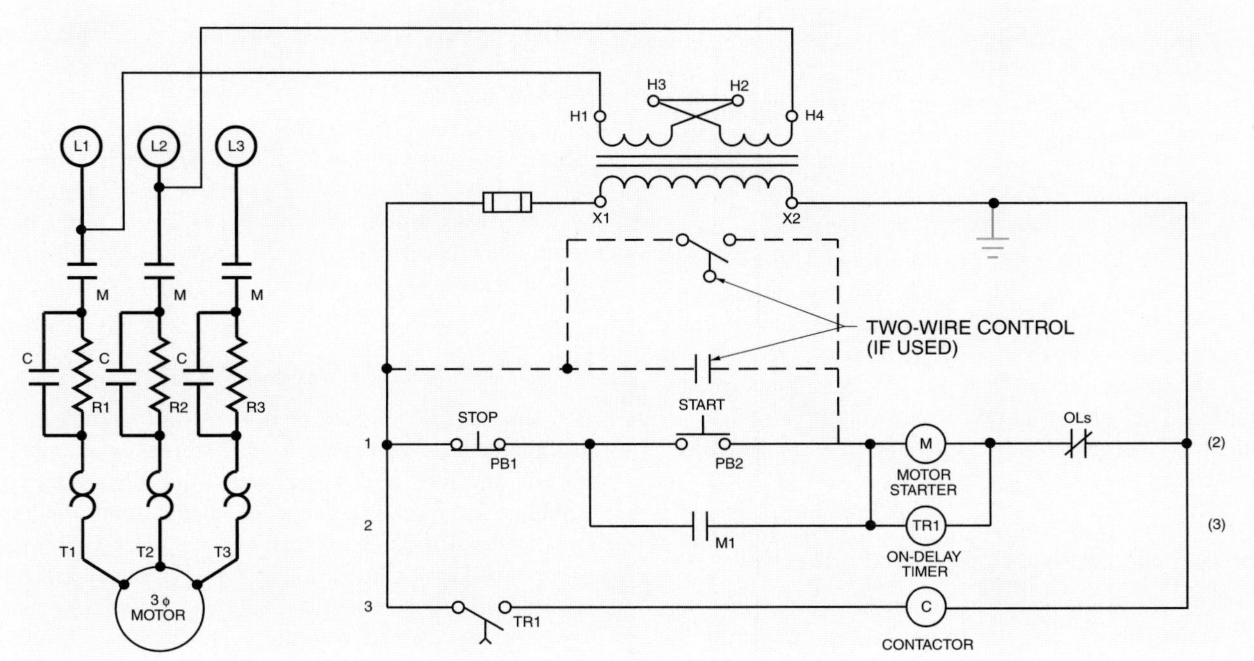

Figure 17-9. In a primary resistor starting circuit, external resistance is added to and taken away from the motor circuit.

Furnas Electric Co.

Figure 17-10. Autotransformer starting uses a tapped 3φ autotransformer to provide reduced-voltage starting.

Autotransformer starting may use its turns ratio advantage to provide more current on the load side of the transformer than on the line side. In autotransformer starting, transformer motor current and line current are not equal as they are in primary resistor starting.

For example, a motor has a full-voltage starting torque of 120% and a full-voltage starting current of 600%. The power company has set a limitation of 400% current draw from the power line. This limitation is set only for the line side of the controller. Because the transformer has a step-down ratio, the motor current on the transformer secondary is larger than the line current as long as the primary of the transformer does not exceed 400%.

In this example, with the line current limited to 400%, 80% voltage can be applied to the motor, generating 80% motor current. The motor draws only 64% line current (0.8 × 80 = 64%) due to the 1 : 0.8 turns ratio of the transformer. The advantage is that the starting torque is 77% (0.8 × 80 of 120%) instead of the 51% obtained in primary resistor starting. This additional percentage may be sufficient accelerating energy to start a load that may be difficult to start otherwise.

Autotransformer Starting Circuits

In an autotransformer starting circuit, the various windings of the transformer are added to and taken away from the motor circuit to provide reduced voltage when starting. **See Figure 17-11.**

The control circuit consists of an ON-delay timer TR1 and contactor coils C1, C2, and C3. Pressing start pushbutton PB2 energizes the timer, causing instantaneous contacts TR1 in line 2 and 3 of the line diagram to close. Closing the normally open (NO) timer contacts in line 2 provides memory for timer TR1, while closing NO timer contacts in line 3 completes an electrical path through line 4, energizing contactor coil C2. The energizing of coil C2 causes NO contacts C2 in line 5 to close, energizing contactor coil C3. The normally closed (NC) contacts in line 3 also provide electrical interlocking for coil C1 so that they may not be energized together. The NO contacts of contactor C2 close, connecting the ends of the autotransformers together when coil C2 energizes. When coil C3 energizes, the NO contacts of contactor C3 close and connect the motor through the transformer taps to the power line, starting the motor at reduced inrush current and starting torque. Memory is also provided to coil C3 by contacts C3 in line 6.

After a predetermined time, the ON-delay timer times out and the NC timer contacts TR1 open in line 4, de-energizing contactor coil C2, and NO timer contacts TR1 close in line 3, energizing coil C1. In addition, NC contacts C1 provide electrical interlock in line 4, and NC contacts C2 in line 3 return to their NC position. The net result of de-energizing C2 and energizing C1 is the connecting of the motor to full line voltage.

Note that during the transition from starting to full line voltage, the motor was not disconnected from the circuit, indicating closed circuit transition. As long as the motor is running in the full-voltage condition, timer TR1 and contactor C1 remain energized. Only an overload or pressing the stop pushbutton stops the motor and resets the circuit. Overload protection is provided by a separate overload block.

In this circuit, pushbuttons are used to control the motor. However, any NO and/or NC device may be used to control the motor. Thus, in an air conditioning system, the pushbuttons would be replaced with a temperature switch, and the circuit would be connected for two-wire control.

PART-WINDING STARTING

Part-winding starting is a method of starting a motor by first applying power to part of the motor coil windings for starting and then applying power to the remaining coil windings for normal running. The motor stator windings must be divided into two or more equal parts for a motor to be started using part-winding starting. Each equal part must also have its terminal available for external connection to power. In most applications, a wye-connected motor is used, but a delta-connected motor can also be started using part-winding starting.

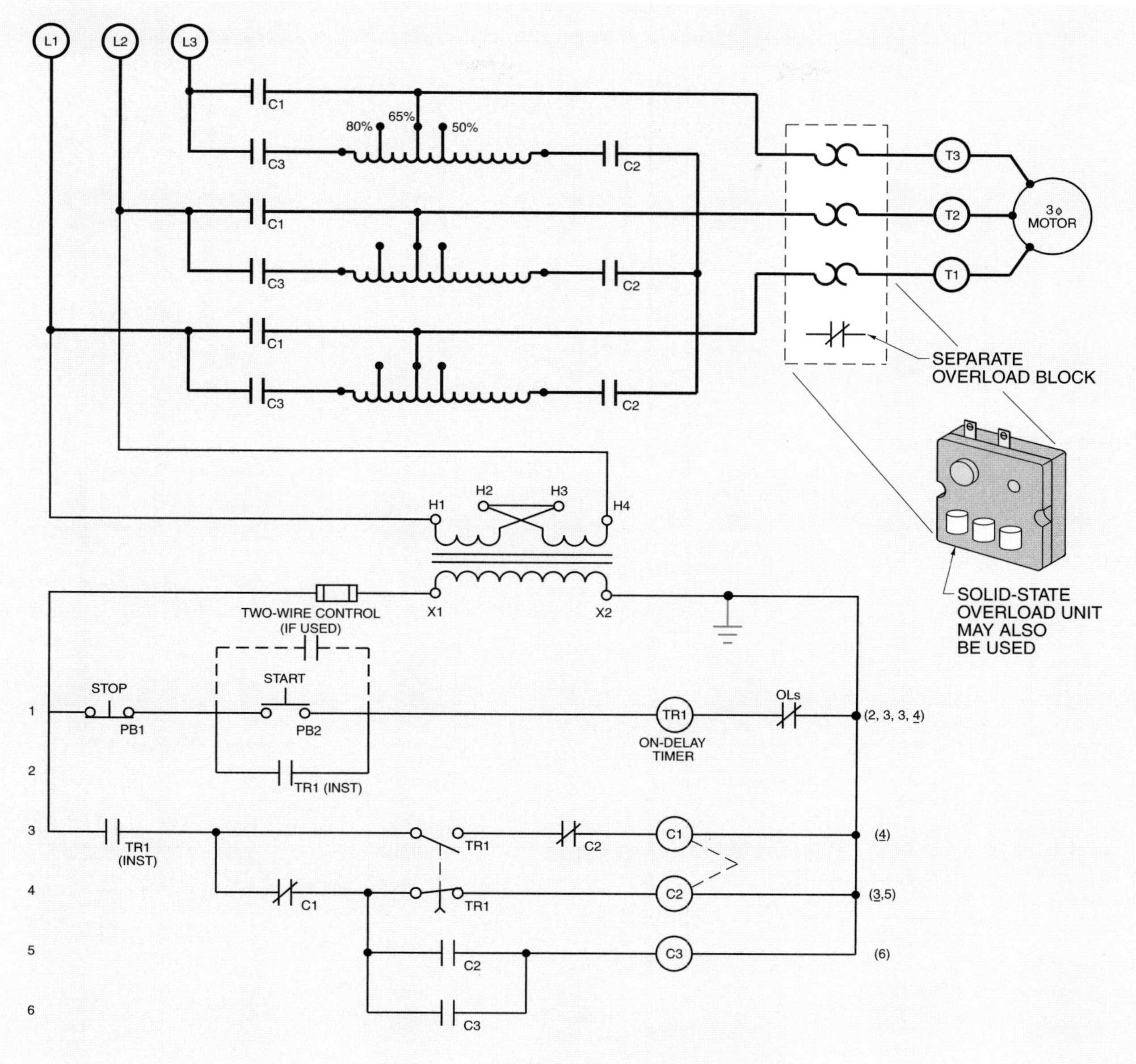

Figure 17-11. In an autotransformer starting circuit, the various windings of the transformer are added to and taken away from the motor circuit to provide reduced voltage when starting.

Wye-Connected Motors

Part-winding starting requires the use of a part-winding motor. A part-winding motor has two sets of identical windings, which are intended to be used in parallel. These windings produce reduced-starting current and reduced-starting torque when energized in sequence. Most dual-voltage 230/460 V motors are suitable for part-winding starting at 230 V. **See Figure 17-12.**

Part-winding starters are available in either two- or three-step construction. The more common two-step starter is designed so that when the control circuit is energized, one winding of the motor is connected directly to the line. This winding draws about 65% of normal locked rotor current and develops approximately 45% of normal motor torque. After about one second, the second winding is connected in parallel with the first winding in such a way that the motor is electrically complete across the line and develops its normal torque.

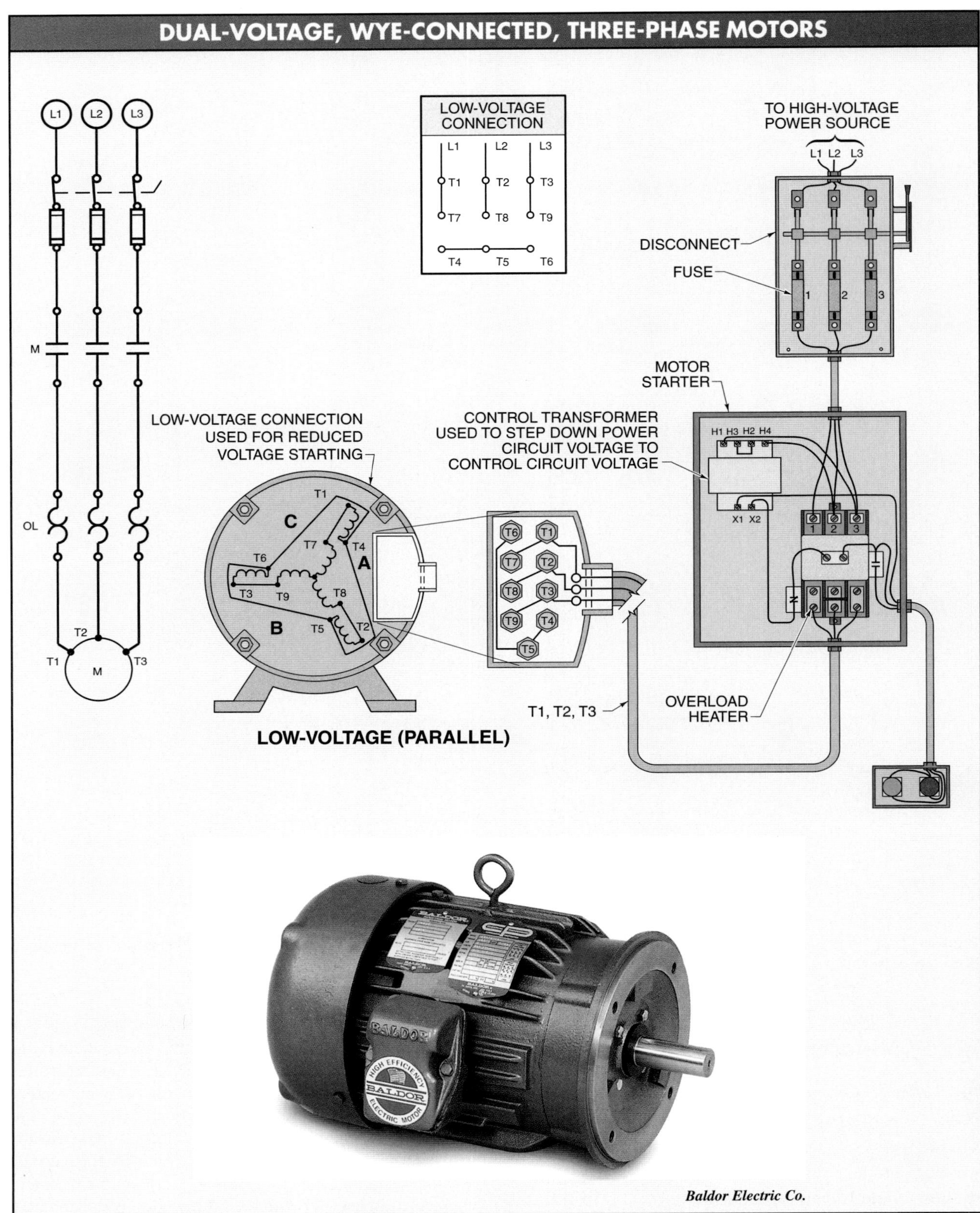

Figure 17-12. A part-winding motor has two sets of identical windings which are intended to be used in parallel.

Part-winding starting is not truly a reduced-voltage starting method. Part-winding starting is usually classified as reduced-voltage starting because of the resulting reduced current and torque.

Delta-Connected Motors

When a dual-voltage, delta-connected motor is operated at 230 V from a part-winding starter having a three-pole starting and a three-pole running contactor, an unequal current division occurs during normal operation, resulting in overloading the starting contactor. To overcome this problem, some part-winding starters are furnished with a four-pole starting contactor and a two-pole running contactor. This arrangement eliminates the unequal current division obtained with a delta-wound motor and enables wye-connected part-winding motors to be given either a one-half or two-thirds part-winding start.

Advantages and Disadvantages of Part-Winding Starting

Part-winding starting is less expensive than most other starting methods because it requires no voltage-reducing components such as transformers or resistors, and it uses only two one-half size contactors. Also, its transition is inherently closed circuit.

Part-winding starting has poor starting torque because the starting torque is fixed. In addition, the starter is almost always an increment start device. Not all motors should be part-winding started. Consult the manufacturer specifications before applying part-winding starting to a motor. Some motors are wound sectionally with part-winding starting in mind. Indiscriminate application to any dual-voltage motor can lead to excessive noise and vibration during starting, overheating, and extremely high transient currents on switching.

The fuses in a part-winding starter must be sized to protect the small contactors and overload devices allowed because of the low-current requirements in part-winding starters. Dual-element fuses are normally required.

Part-Winding Starter Circuits

Part-winding reduced-voltage starting is less expensive than other starting methods and produces less starting torque. **See Figure 17-13.**

The control circuit consists of motor starter M1, ON-delay timer TR1, and motor starter M2. Pressing start pushbutton PB2 energizes starter M1 and timer TR1. M1 energizes the motor, and closes contacts M1 in line 2 to provide memory. With the motor starter M1 energized, L1 is connected to T1, L2 to T2, and L3 to T3, starting the motor at reduced current and torque through one-half of the wye windings.

The ON-delay NO contacts of ON-delay timer TR1 in line 2 remain open during timing and close after timing out, energizing coil M2. When M2 energizes, L1 is connected to T7, L2 to T8, and L3 to T9, applying voltage to the second set of wye windings. The motor now has both sets of windings connected to the supply voltage for full current and torque. The motor may normally be stopped by pressing stop pushbutton PB1 or by an overload in any line. Each magnetic motor starter need be only half-size because each one controls only one-half of the winding. Overloads must be sized accordingly.

WYE-DELTA STARTING

Wye-delta starting accomplishes reduced-voltage starting by first connecting the motor leads in a wye configuration for starting. A motor started in the wye configuration receives approximately 58% of the normal voltage and develops approximately 33% of the normal torque.

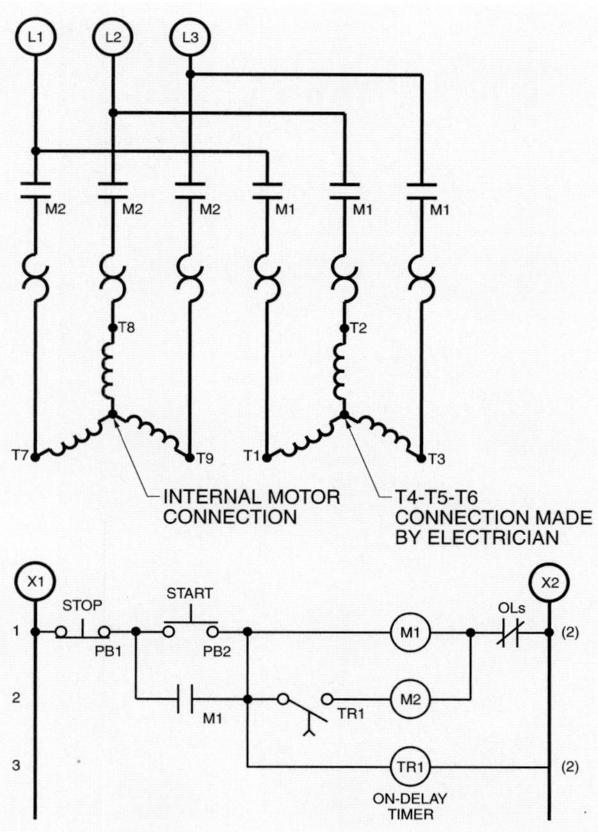

Figure 17-13. Part-winding reduced-voltage starting is less expensive than other starting methods and produces less starting torque.

Wye-delta motors are specially wound with six leads extending from the motor to enable the windings to be connected in either a wye or delta configuration. When a wye-delta starter is energized, two contactors close, with one contactor connecting the windings in a wye configuration, and the second contactor connecting the motor to line voltage. After a time delay, the wye contactor opens (momentarily de-energizing the motor), and the third contactor closes to reconnect the motor to the power lines with the windings connected in a delta configuration. A wye-delta starter is inherently an open transition system because the leads of the motor are disconnected and then reconnected to the power supply.

This starting method does not require any accessory voltage-reducing equipment such as resistors and transformers. Wye-delta starting gives a higher starting torque per line ampere than part-winding starting, with considerably less noise and vibration.

Wye-delta starters have the disadvantage of being open transition. Closed transition versions are available at additional cost. In closed transition wye-delta starters, the motor windings are kept energized for the few cycles required to transfer the motor windings from wye to delta. Such starters are provided with one additional contactor plus a resistor bank.

Wye-Delta Motors

Windings of a wye-delta motor may be joined to form a wye or delta configuration. **See Figure 17-14.** There are no internal connections on this motor as there are on standard wye and delta motors. This allows the electrician to connect the motor leads into a wye-connected motor or into a delta-connected motor.

Technical Fact

Starting current for motors is 2 to 6 times the full load current. Disconnects, fuses, wires, and contactors must be selected accordingly.

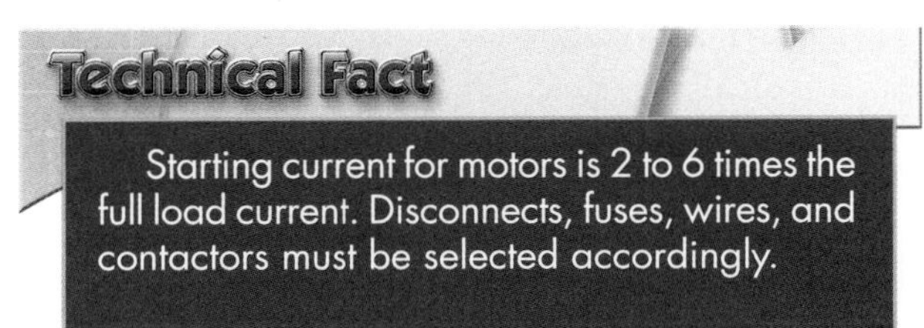

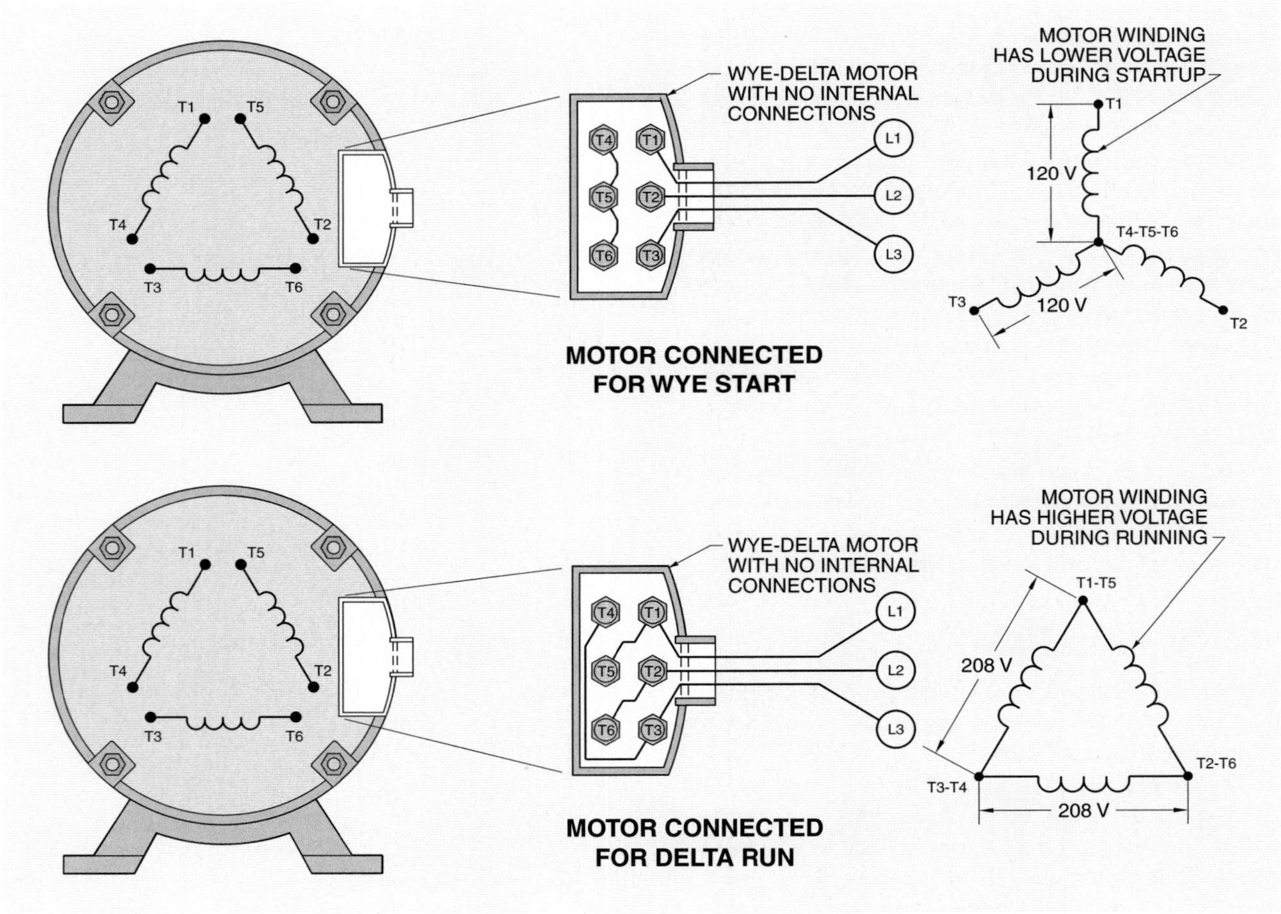

Figure 17-14. Windings on a wye-delta motor may be joined to form a wye or delta configuration.

Technical Fact

Overcurrent protection devices (OCPDs) are installed in the combination starter, safety switch, or fuse panel. The OCPD is part of the circuit but located before the motor starter. The OCPD automatically trips if there is a problem when the motor is started. This turns the motor OFF. The motor is restarted after changing the fuses or resetting the circuit breakers.

Each coil winding in the motor receives 208 V if a delta-connected motor is connected across a 208 V, 3ϕ power line. This is because each coil winding in the motor is connected directly across two power lines. **See Figure 17-15.**

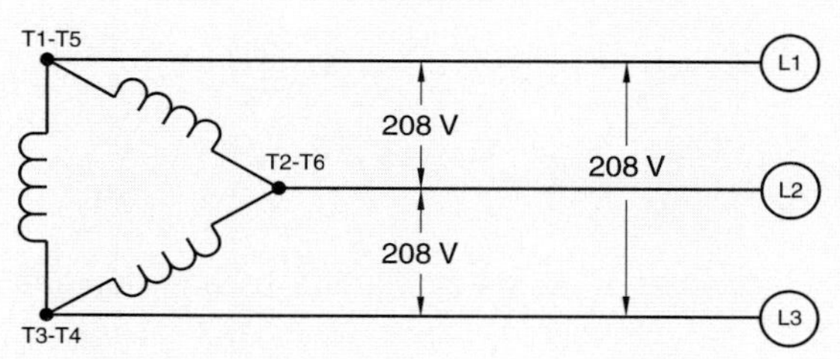

Figure 17-15. A delta-connected motor has each coil winding directly connected across two power lines so each winding receives the entire source voltage of 208 V.

Each coil winding in a wye-connected motor receives 120 V if it is connected across a 208 V, 3ϕ power line. This is because there are two coils connected in series across any pair of power lines. **See Figure 17-16.**

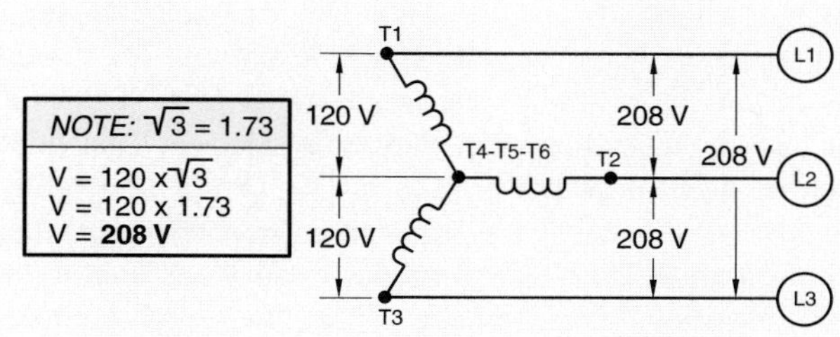

Figure 17-16. A wye-connected motor has two power lines connected across two sets of windings.

When calculating the voltage in the coil for a wye-connected circuit, the voltage is equal to the line voltage divided by the square root of 3 (1.73). The coil voltage is equal to 120 V ($^{208}/_{1.73}$) because the line voltage is equal to 208 V. A wye-delta motor connected to a line voltage of 208 V starts with 120 V (wye) and runs with 208 V (delta) across the motor windings, thus reducing starting voltage.

Wye-Delta Starting Circuits

The control circuit of a typical wye-delta starting circuit consists of motor starter coils M1 and M2, contactor C1, and ON-delay timer TR1. **See Figure 17-17.** Pressing start pushbutton PB2 energizes coil M1, which provides memory in line 2 and connects the power lines L1 to T1, L2 to T2, and L3 to T3. Contactor coil C1 in line 3 is energized, providing electrical interlock in line 2 and connecting motor terminals T4 and T5 to T6 so the motor starts in a wye configuration. TR1 in line 3 is also energized, and after a preset time the ON-delay timer times out, causing the NO TR1 contacts in line 2 to close and the NC TR1 contacts in line 3 to open. The opening of the NC contacts in line 3 disconnects contactor C1, and an instant later the NO contacts in line 2 energize the second motor starter through coil M2.

The short time delay between M2 and C1 is necessary to prevent a short circuit in the power lines and is provided through the NC auxiliary contacts of C1 in line 2. With contactor C1 de-energized, terminals T5, T6, and T4 are connected to power lines T1, T2, and T3 because L1, L2, and L3 are still connected to run in a delta configuration. The circuit can normally be stopped only by an overload in any line or by pressing the stop pushbutton PB1.

Rockwell Automation, Allen-Bradley Company, Inc.

Reduced-voltage starting circuits require two or more interlocked motor starters.

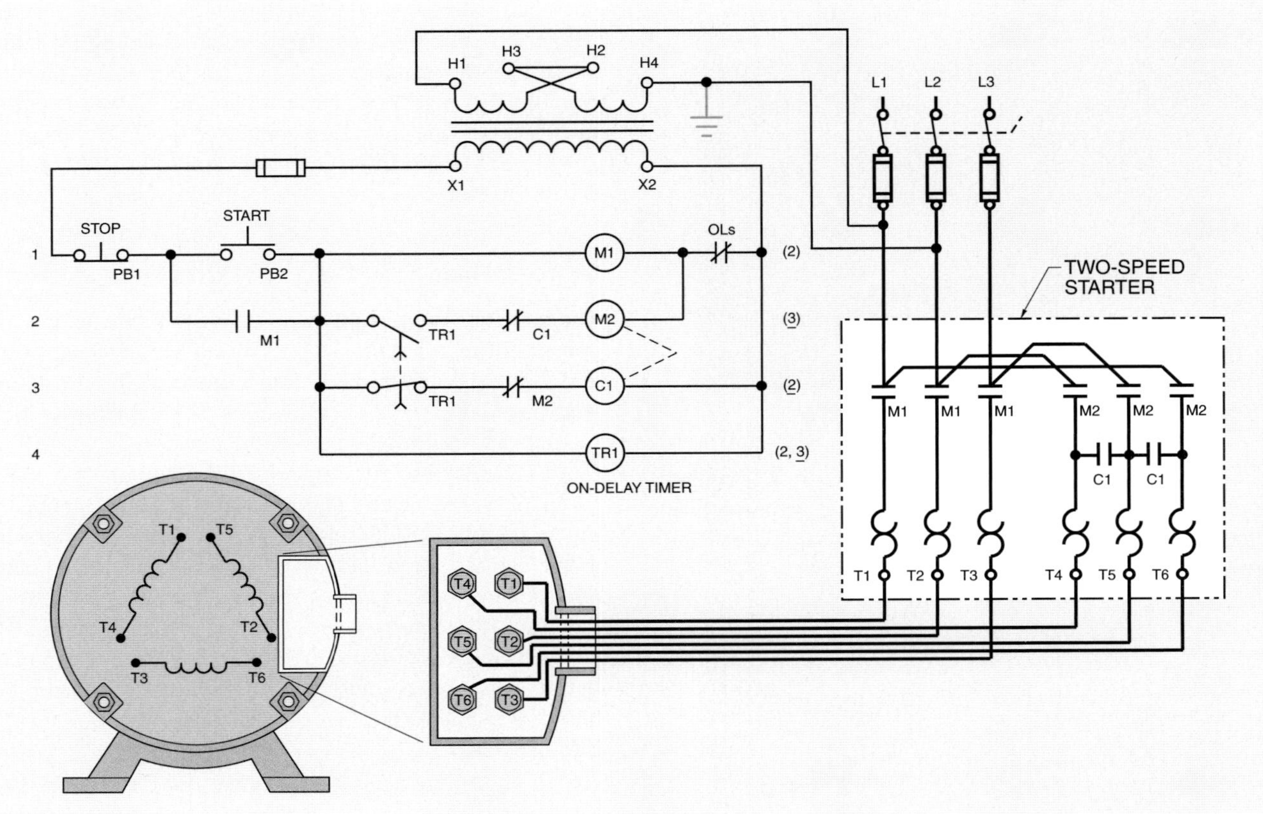

Figure 17-17. The control circuit of a typical wye-delta starting circuit consists of two motor starters, a contactor, and an ON-delay timer.

SOLID-STATE SWITCHES

Solid-state switches are electronic devices that have no moving parts (contacts). Solid-state switches can be used in most motor control applications. Advantages of solid-state switches include fast switching, no moving parts, long life, and the ability to be interfaced with electronic circuits (PLCs and PCs). However, solid-state switches must be properly selected and applied to prevent potential problems. Solid-state switches include transistors, silicon-controlled rectifiers (SCRs), triacs, and alternistors. **See Figure 17-18.**

Transistors

A *transistor* is a three-terminal device that controls current through the device depending on the amount of voltage applied to the base. Transistors may be NPN or PNP transistors. Transistors can be switched ON and OFF quickly. Transistors have a very high resistance when open and a very low resistance when closed. Transistors are used to switch low-level DC only. When transistors are used as switches, a diode can be mounted across the transistor to avoid damage from high-voltage spikes (transients).

Silicon-Controlled Rectifiers

A *silicon-controlled rectifier (SCR)* is a solid-state rectifier with the ability to rapidly switch heavy currents. SCRs are used as solid-state low- and high-level DC switches. An SCR is either ON or OFF. The SCR is turned ON when voltage is applied to its gate. The SCR remains ON as long as current flows through the anode and cathode. The SCR is turned OFF when current flow is stopped. One SCR can be used to switch high-level DC. When controlling high-level AC, two SCRs can be mounted in an antiparallel configuration. Each SCR is used to control one-half of the AC sine wave. The advantage of using two separate SCRs (and not one triac) is greater heat dissipation. One SCR in a diode bridge can be used when low-level current switching is required (often on printed circuit boards).

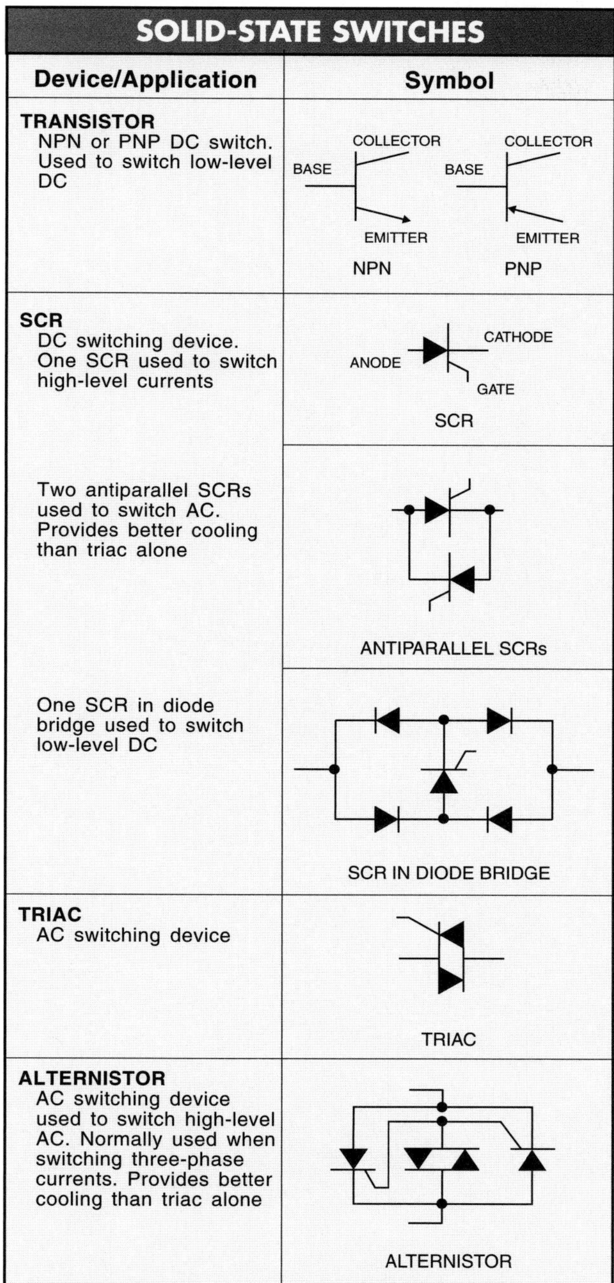

Figure 17-18. Solid-state switches include transistors, silicon-controlled rectifiers (SCRs), triacs, and alternistors.

Triacs

A *triac* is a three-terminal semiconductor thyristor that is triggered into conduction in either direction by a small current to its gate. Triacs are used as solid-state AC switches. Like an SCR, a triac is either ON or OFF. A triac is turned ON when voltage is applied to its gate. Once ON, the triac allows current to flow in both directions (AC). The triac is turned OFF when the gate voltage is removed. A triac that is used as a switch may have a snubber mounted across it to avoid damage from high-current spikes (transients).

Alternistors

An *alternistor* is two antiparallel thyristors and a triac mounted on the same chip. The alternistor was developed specially for industrial AC high-current switching applications. A combination of three alternistors is normally used in three-phase switching applications. An alternistor requires less space than antiparallel SCRs. The components are separated for increased heat dissipation.

SOLID-STATE STARTING

A solid-state starter is a method of reduced-voltage starting for standard motors. The heart of the solid-state starter is the SCR, which controls motor voltage, current, and torque during acceleration.

A solid-state reduced-voltage starter ramps up motor voltage as the motor accelerates, instead of applying full voltage instantaneously (as do across-the-line starters). A solid-state starter reduces inrush current compared to the high inrush current produced by across-the-line starters. Solid-state starters also minimize starting torque (which can damage some loads connected to the motor) and smooth motor acceleration. **See Figure 17-19.**

Solid-state starting provides a smooth, stepless acceleration in applications such as starting conveyors, compressors, pumps, and a wide range of other industrial applications because of its unique switching capability.

The advantage of SCRs is that they are small in size, are rugged, and have no contacts. Unlimited life can be expected when SCRs operate within specifications. The major disadvantage of solid-state starting is its relatively high cost in relation to other systems.

Electronic Control Circuitry

A solid-state controller determines to what degree the SCRs should be triggered ON to control the voltage, current, and torque applied to a motor. A solid-state controller also includes current-limiting fuses and current transformers for protection of the unit. The current-limiting fuses are used to protect the SCRs from excess current. The current transformers are used to feed information back to the controller. Heat sinks and thermostat switches are also used to protect the SCRs from high temperatures.

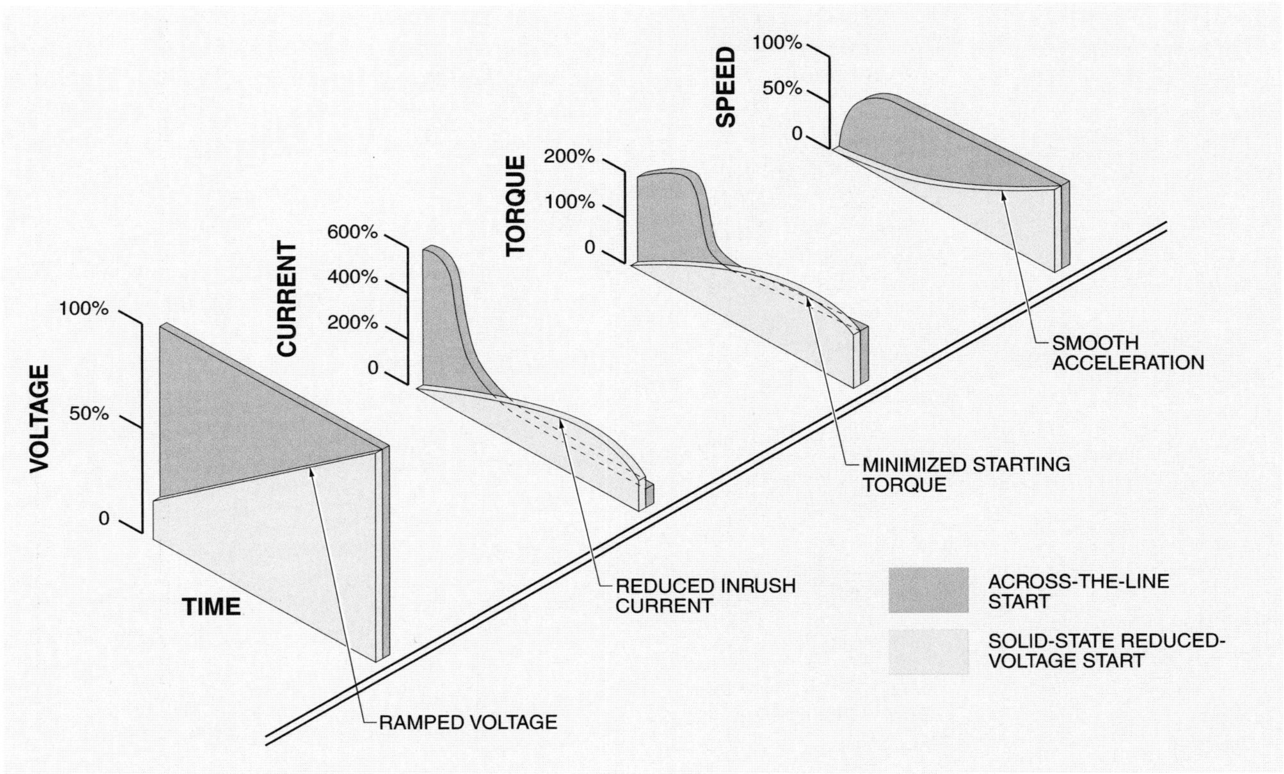

Figure 17-19. A solid-state starter ramps up voltage, reduces inrush current, minimizes starting torque, and smoothes acceleration.

The controller also provides the sequential logic necessary for interfacing other control functions of the starter, such as line loss detection during acceleration. The controller is turned OFF if any voltage is lost or too low on any one line. This may happen if one line opens or a fuse blows.

Silicon-Controlled Rectifier Operation

An SCR includes an anode, a cathode, and a gate. **See Figure 17-20.** The anode and cathode of an SCR are similar to the anode and cathode of a diode. The gate gives the SCR added control that is not possible with an ordinary diode.

With the gate, an SCR can be made to operate as an OFF/ON switch controlled by a voltage signal to the gate. Unlike an ordinary diode, an SCR does not pass current from cathode to anode unless an appropriate signal is applied to the gate.

When the signal is applied to the gate, the SCR is triggered ON, and the anode resistance decreases sharply, such that the resulting current flow through the SCR is only limited by the resistance of the load. The advantage of this device is its ability to turn ON at any point in the half-cycle. **See Figure 17-21.**

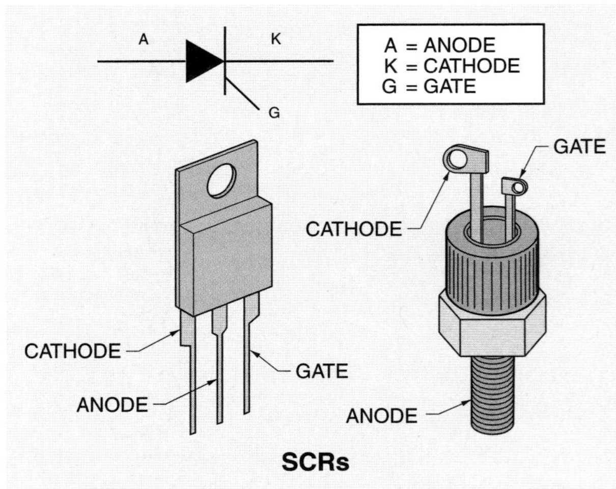

Figure 17-20. An SCR includes an anode, a cathode, and a gate.

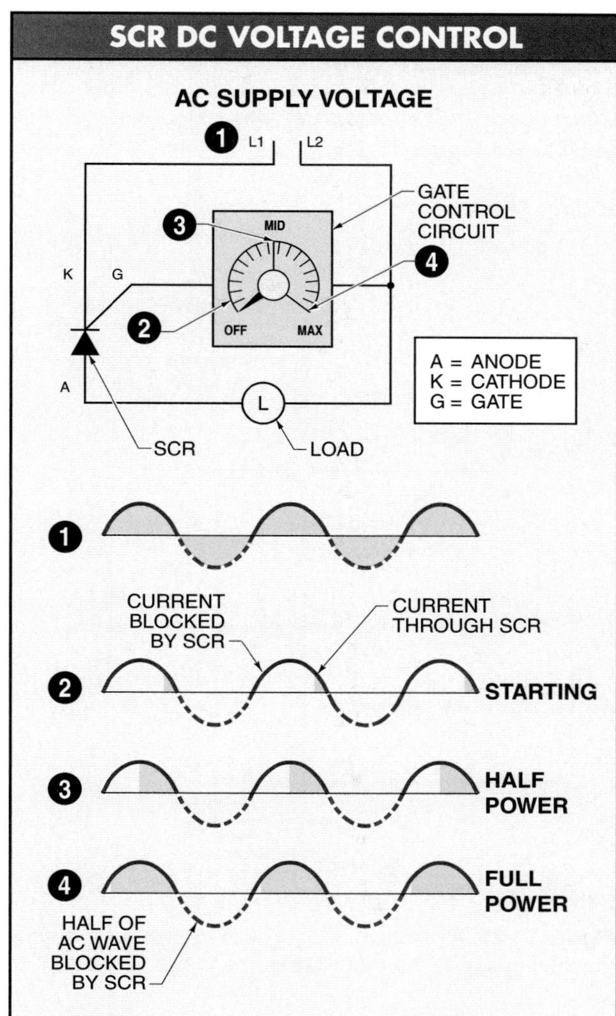

Figure 17-21. When the signal is applied to the gate, the SCR is triggered ON and the anode resistance decreases sharply, such that the resulting current flow through the SCR is only limited by the resistance of the load.

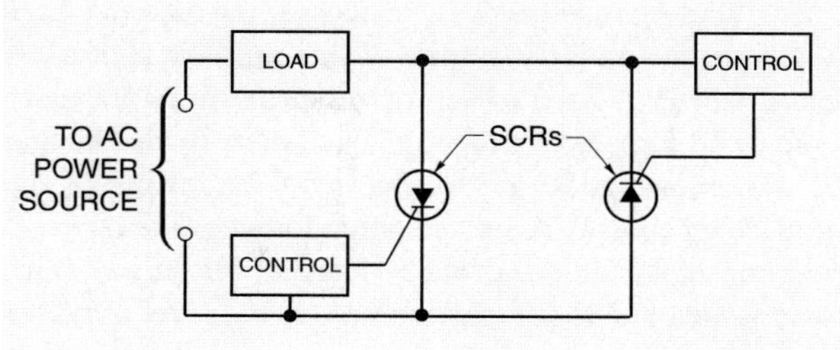

Figure 17-22. SCRs may be used alone in a circuit to provide one-way current control, or may be wired in reverse-parallel circuits to control AC line current in both directions.

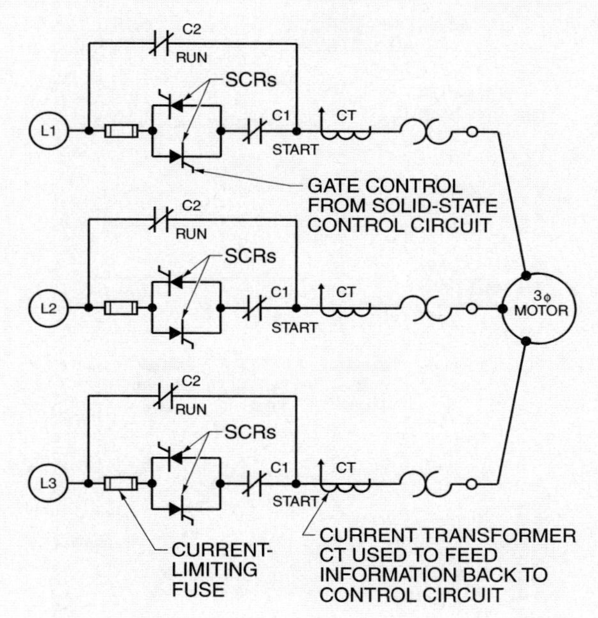

Figure 17-23. An SCR circuit with reverse-parallel wiring of SCRs provides maximum control of an AC load.

The average amount of voltage and current can be reduced or increased by the triggering of the SCRs because the amount of conduction can be varied. SCRs may be used alone in a circuit to provide one-way current control or may be wired in reverse-parallel circuits to control AC line current in both directions. **See Figure 17-22.**

Solid-State Starting Circuits

A typical solid-state starting circuit consists of both start and run contactors connected in the circuit. The start contactor contacts C1 are in series with the SCRs and the run contactor contacts C2 are in parallel with the SCRs. **See Figure 17-23.**

The start contacts C1 close and the acceleration of the motor is controlled by triggering ON the SCRs when the starter is energized. The SCRs control the motor until it approaches full speed, at which time the run contacts C2 close, connecting the motor directly across the power line. At this point, the SCRs are turned OFF, and the motor runs with full power applied to the motor terminals.

Soft Starters

A *soft starter* is a device that provides a gradual voltage increase (ramp up) during AC motor starting. Most soft starters also provide soft stopping (ramp down) capabilities. Soft starters are used to control single-phase and three-phase motors.

Soft starting is achieved by increasing the motor voltage in accordance with the setting of the ramp-up control. A potentiometer is used to set the ramp-up time (normally 1 sec to 20 sec). Soft stopping is achieved by decreasing the motor voltage in accordance with the setting of the ramp-down control. A second potentiometer is used to set the ramp-down time (normally 1 sec to 20 sec). A third potentiometer is used to adjust the starting level of motor voltage to a value at which the motor starts to rotate immediately when soft starting is applied. **See Figure 17-24.**

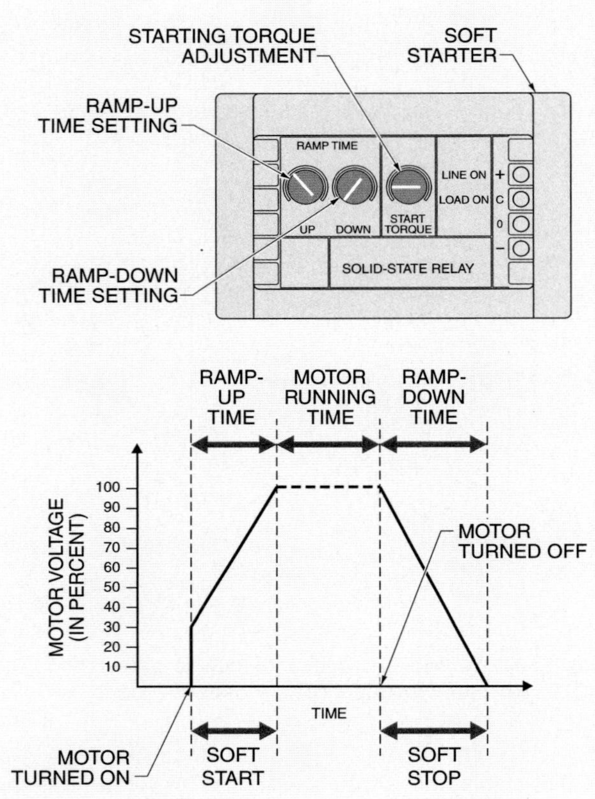

Figure 17-24. A soft starter is a device that provides a gradual voltage increase (ramp up) during motor starting and a gradual voltage decrease (ramp down) during motor stopping.

Like any solid-state switch, a soft starter produces heat that must be dissipated for proper operation. The heat dissipation requires large heat sinks when high-current loads (motors) are controlled. For this reason, a contactor is often added in parallel with a soft starter. The soft starter is used to control the motor when the motor is starting or stopping. The contactor is used to short out the soft starter when the motor is running. This allows for soft starting and soft stopping without the need for large heat sinks during motor running. The soft starter includes an output signal which is used to control the time when the contactor is ON or OFF. **See Figure 17-25.**

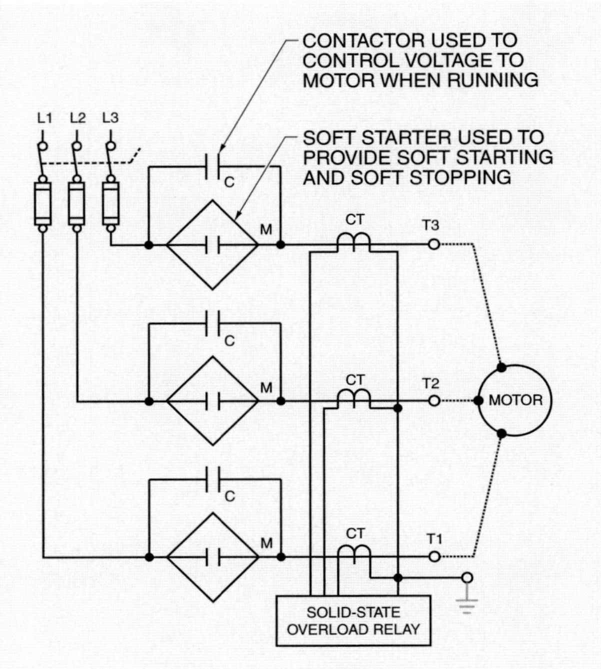

Figure 17-25. A contactor is used with a soft starter to control the voltage to the motor when the motor is running.

STARTING METHOD COMPARISON

Several starting methods are available when an industrial application calls for using reduced-voltage starting. The amount of reduced current, the amount of reduced torque, and the cost of each starting method must be considered when selecting the appropriate starting method.

The selection is not simply a matter of selecting the starting method that reduces the current the most. The motor does not start and the motor overloads trip if the starting torque is reduced too much.

A general comparison can be made of the amount of reduced current for each type of starting method compared to across-the-line starting. **See Figure 17-26.** The amount of reduced current is adjustable when using solid-state or autotransformer starting. Autotransformer starting uses taps so the amount of reduced current is somewhat adjustable. Solid-state starting is adjustable throughout its range. Some primary resistor starters are adjustable, others are not. Part-winding and wye-delta starting are not adjustable.

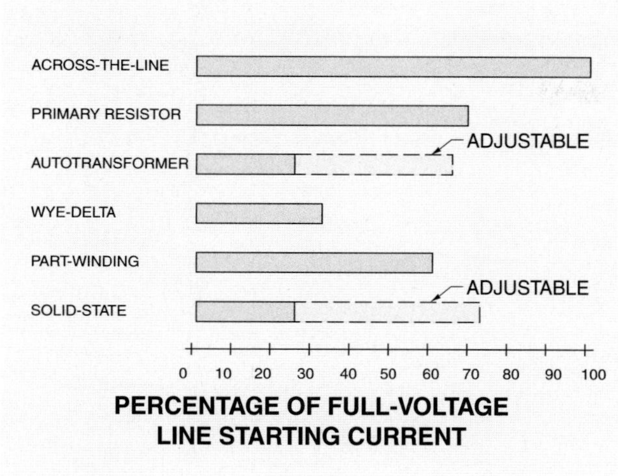

Figure 17-26. The different methods of reduced-voltage starting produce different percentages of full-voltage current.

A general comparison can be made of the amount of reduced torque for each type of starting method compared to across-the-line starting. **See Figure 17-27.** The amount of reduced torque is adjustable when using the solid-state or autotransformer starting method. The autotransformer starting method has taps, so the amount of reduced torque is somewhat adjustable. Solid-state starting is adjustable throughout its range. The motor overloads trip if the load requires more torque than the motor can deliver. The torque requirements of the load must be taken into consideration when selecting a starting method.

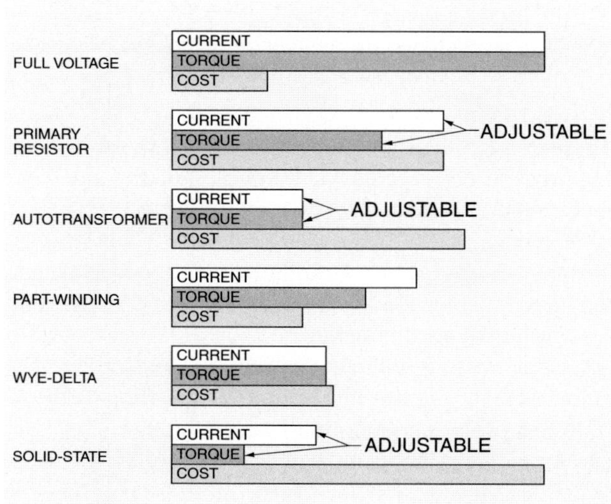

Figure 17-27. The different methods of reduced-voltage starting produce different amounts of reduced torque.

A general comparison can also be made of the costs for each type of starting method compared to across-the-line starting. Although reducing the amount of starting current or starting torque in comparison to the load requirements is the primary consideration for selecting a starting method, cost may also have to be considered. The costs vary for each starting method. **See Figure 17-28.**

The primary resistor starting method is used when it is necessary to restrict inrush current to predetermined increments. Primary resistors can be built to meet almost any current inrush limitation. Primary resistors also provide smooth acceleration and can be used where it is necessary to control starting torque. Primary resistor starting may be used with any motor.

The autotransformer starting method provides the highest possible starting torque per ampere of line current and is the most effective means of motor starting for applications where the inrush current must be reduced with a minimum sacrifice of starting torque. Three taps are provided on the transformers, making it field adjustable. Cost must be considered because the autotransformer is the most expensive type of transformer. Autotransformer starting can be used with any motor.

The part-winding starting method is simple in construction and economical in cost. Part-winding starting provides a simple method of accelerating fans, blowers, and other loads involving low starting torque. The part-winding starting method requires a nine-lead wye motor. The cost is less because no external resistors or transformers are required.

The wye-delta starting method is particularly suitable for applications involving long accelerating times or frequent starts. Wye-delta starting is commonly used for high-inertia loads such as centrifugal air conditioning units, although it can be used in applications where low starting torque is necessary or where low starting current and low starting torque are permissible. The wye-delta starting method requires a special six-lead motor.

The solid-state starting method provides smooth, stepless acceleration in applications such as starting conveyors, compressors, and pumps. Solid-state starting uses a solid-state controller, which uses SCRs to control motor voltage, current, and torque during acceleration. Although the solid-state starting method offers the most control over a wide range, it is also the most expensive.

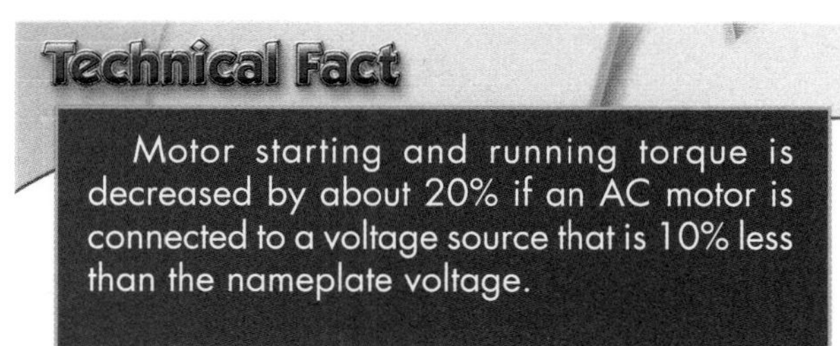

MOTOR STARTING METHOD COMPARISON										
Starter Type	Starting Characteristics			Standard Motor	Transition	Extra Acceleration Steps Available	Installation Cost	Advantages	Disadvantages	Applications
	Volts at Motor	Line Current	Starting Torque							
Across-the-Line	100%	100%	100%	Yes	None	None	Lowest	Inexpensive, readily available, simple to maintain, maximum starting torque	High inrush, high starting torque	Many and various
Primary Resistor	65%	65%	42%	Yes	Closed	Yes	High	Smooth acceleration, high power factor during start, less expensive than autotransformer starter in low HPs, available with as many as 5 accelerating points	Low torque efficiency, resistors give off heat, starting time in excess of 5 sec, requires expensive resistors, difficult to change starting torque under varying conditions	Belt and gear drives, conveyors, textile machines
Autotransformer	80% 65% 50%	64% 42% 25%	64% 42% 25%	Yes	Closed	No	High	Provides highest torque per ampere of line current, 3 different starting torques available through autotransformer taps, suitable for relatively long starting periods, motor current is greater than line current during starting	Is most expensive design in lower HP ratings, low power factor, large physical size	Blowers, pumps, compressors, conveyors
Part-Winding	100%	65%	48%	*	Closed	Yes†	Low	Least expensive reduced-voltage starter, most dual-voltage motors can be started part-winding on lower voltage, small physical size	Unsuited for high-inertia, long-starting loads, requires special motor design for voltage higher than 230 V, motor does not start when torque demanded by load exceeds that developed by motor when first half of motor is energized, first step of acceleration must not exceed 5 sec or motor overheats	Reciprocating compressors, pumps, blowers, fans
Wye-Delta	100%	33%	33%	No	Open‡	No	Medium	Suitable for high-inertia, long-acceleration loads, high torque efficiency, ideal for especially stringent inrush restrictions, ideal for frequent starts	Requires special motor, low starting torque, momentary inrush occurs during open transition when delta contractor is closed	Centrifugal compressors, centrifuges
Solid-State	Adjust	Adjust	Adjust	Yes	Closed	Adjust	Highest	Energy-saving features available, voltage gradually applied during starting for a soft start condition, adjustable acceleration time, usually self-calibrating, adjustable built-in braking features included	High cost, requires specialized maintenance and installation, electrical transients can damage unit, requires good ventilation	Machine tools, hoists, packaging equipment, conveyor systems

* Standard dual-voltage 230/460 V motor can be used on 230 V systems
† Very uncommon
‡ Closed transition available for average of 30% more cost

Figure 17-28. Factors considered when selecting a reduced-voltage starting method include voltage at the motor, line current, starting torque, and installation costs.

TROUBLESHOOTING REDUCED-VOLTAGE STARTING CIRCUITS

As with all motor circuits, the two main sections that must be considered when troubleshooting reduced-voltage starting circuits are the power circuit and control circuit. The power circuit connects the motor to the main power supply.

In addition to including the main switching contacts and overload detection device (which can be heaters or solid-state), the power circuit also includes the power resistors (in the case of primary resistor starting) or autotransformer (in the case of autotransformer starting).

The control circuit determines when and how the motor starts. The control circuit includes the motor starter (mechanical or solid-state), overload contacts, and timing circuit. To troubleshoot the control circuit, the same troubleshooting procedure is used as when troubleshooting any other motor control circuit.

Voltage and current readings are taken when troubleshooting power circuits. Current readings can be taken at the incoming power leads or the motor leads. The current reading during starting should be less than the current reading when starting without reduced-voltage starting. The amount of starting current varies by the starting method.

When troubleshooting the power circuit, voltage and current readings are taken. Current readings can be taken at the incoming power leads or the motor leads, since the current draw is the same at either point. With each starting method, there should be a reduction in starting current, as compared to a full-voltage start. **See Figure 17-29.**

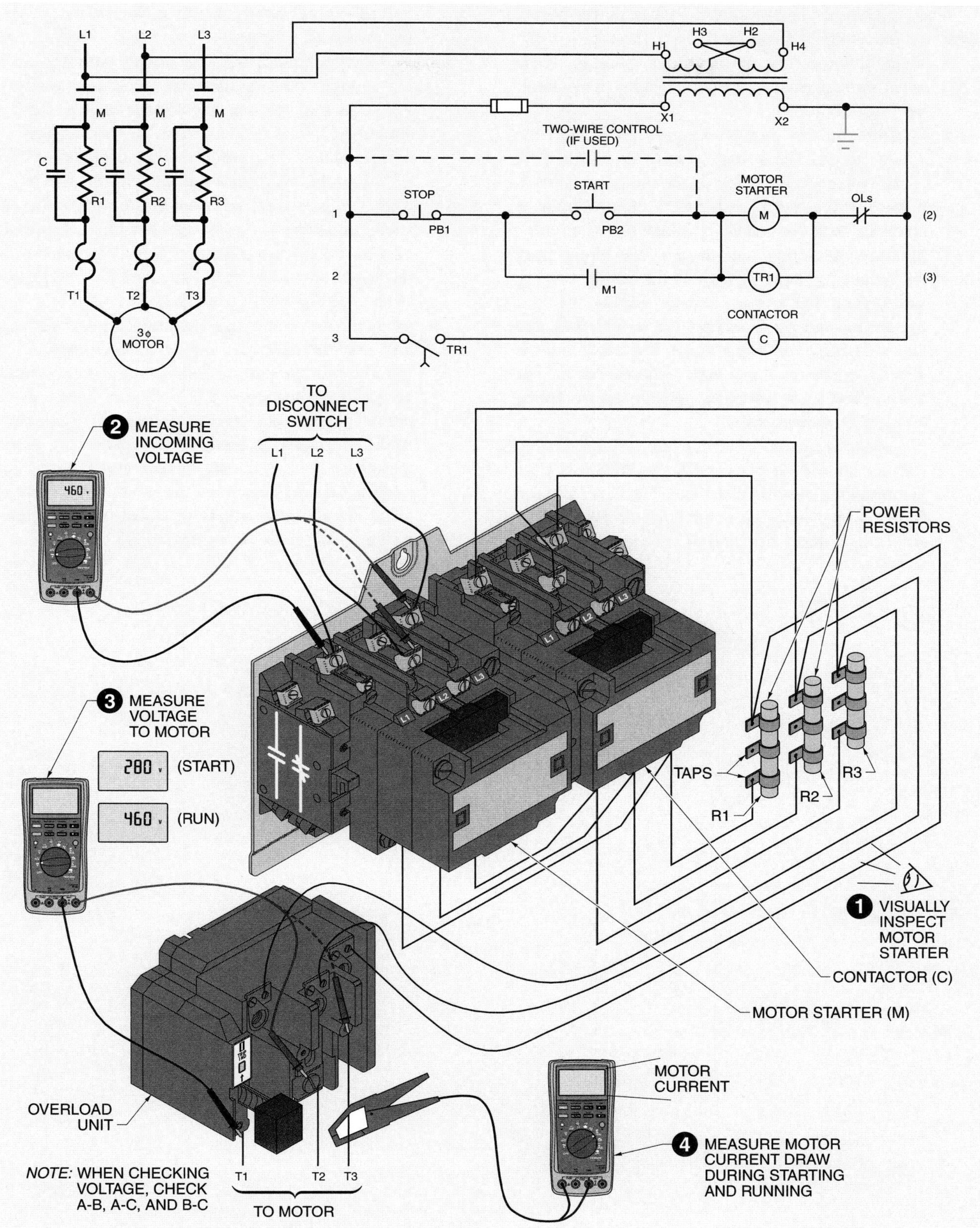

Figure 17-29. Voltage and current readings are taken when troubleshooting reduced-voltage power circuits.

When troubleshooting a reduced-voltage power circuit, apply the following procedure:

1. Visually inspect the motor starter. Look for loose wires, damaged components, and signs of overheating (discoloration).
2. Measure the incoming voltage coming into the power circuit. The voltage should be within 10% of the voltage rating listed on the motor nameplate. If the voltage is not within 10%, the problem is upstream from the reduced-voltage power circuit.
3. Measure the voltage delivered to the motor from the reduced-voltage power circuit during starting and running. For primary resistor starting, the voltage during starting should be 10% to 50% less than the incoming measured voltage. The exact amount depends on the resistance added into the circuit. The resistance is set by using the resistor taps or adding resistors in series/parallel.

   For autotransformer starting, the voltage during starting should be 50%, 65%, or 80% less than the incoming measured voltage. The exact amount depends on which tap connection is used on the autotransformer. For part-winding starting, the voltage during starting should be equal to the incoming measured voltage.

   For wye-delta starting, the voltage during starting should equal the incoming measured voltage. For solid-state starting, the voltage during starting should be 15% to 50% less than the incoming measured voltage. The exact amount depends on the setting of the solid-state starting control switch.

   The voltage measured after the motor is started should equal the incoming voltage with each method of reduced-voltage starting. There is a problem in the power circuit or control circuit if the voltage out of the starting circuit is not correct.
4. Measure the motor current draw during starting and after the motor is running. In each method of reduced-voltage starting, the starting current should be less than the current that the motor draws when connected for full-voltage starting. The current should normally be about 40% to 80% less. After the motor is running, the current should equal the normal running current of the motor. This current value should be less than or equal to the current rating listed on the motor nameplate.

Review Questions

1. What are three reasons for using reduced-voltage starting?
2. What is the difference in starting current between full-voltage starting and reduced-voltage starting?
3. What is the difference in starting torque between full-voltage starting and reduced-voltage starting?
4. What is locked-rotor current?
5. What is full-load current?
6. How can an additional reduction in current be accomplished in primary resistor starting?
7. What is open circuit transition?
8. What is closed circuit transition?
9. What is the advantage of part-winding starting?
10. What is the disadvantage of part-winding starting?
11. How many external leads does a wye-delta motor have?
12. Is wye-delta starting an open transition starting method or a closed transition starting method?
13. In a solid-state reduced-voltage starter, what device is used to reduce the voltage applied to the motor?
14. Is a solid-state reduced-voltage starter limited to a preset number of steps?
15. What types of solid-state device are used to switch low-level DC?
16. What type of solid-state device is used to switch high-level DC?
17. What type of current is a triac used to switch?
18. What is a soft starter used for?
19. What is voltage sag?
20. What is the purpose of a rheostat?
21. What are the benefits of using a solid-state reduced-voltage starter?

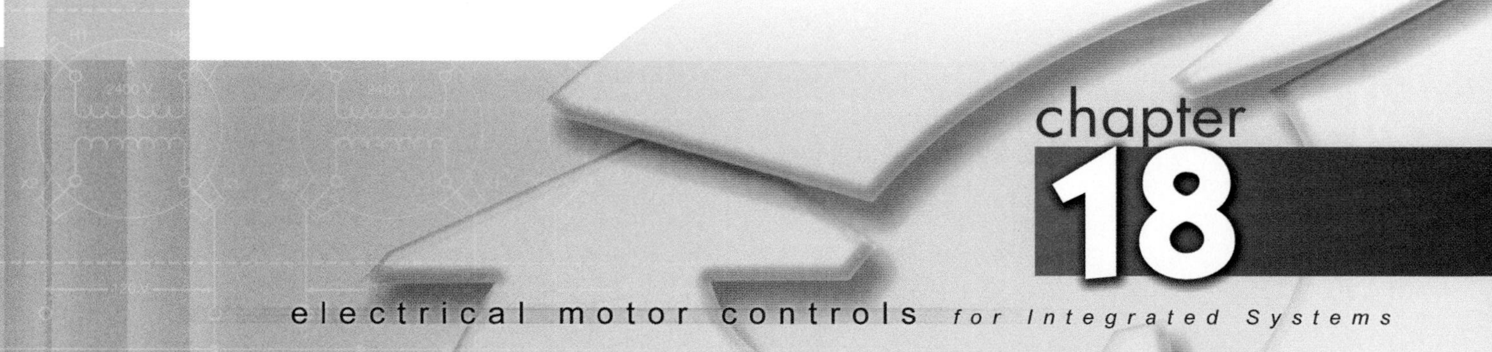

Accelerating and Decelerating Methods

Braking is used when a motor must be stopped more quickly than coasting allows. The speed of a DC motor is controlled by silicon-controlled rectifiers. The speed of an AC motor is controlled by changing the frequency applied to the motor or using a motor with windings that may be reconnected to form different numbers of poles.

BRAKING

A motor coasts to a stop when disconnected from the power supply. The time taken by the motor to come to rest depends on the inertia of the moving parts (motor and motor load) and friction. Braking is used when it is necessary to stop a motor more quickly than coasting allows.

Braking is accomplished by different methods, each of which has advantages and disadvantages. The braking method used depends on the application, available power, circuit requirements, cost, and desired results.

Braking applications vary greatly. For example, braking may be applied to a motor every time the motor is stopped, or it may be applied to a motor only in an emergency. In the first application, the braking action requires a method that is reliable with repeated use. In the second application, the method of stopping the motor may give little or no consideration to the damage braking may do to the motor or motor load.

Hazard braking may be required to protect an operator (hand in equipment, etc.) even if braking is not part of the normal stopping method.

Friction Brakes

Friction brakes normally consist of two friction surfaces (shoes or pads) that come in contact with a wheel mounted on the motor shaft. Spring tension holds the shoes on the wheel and braking occurs as a result of the friction between the shoes and the wheel. Friction brakes (magnetic or mechanical) are the oldest motor stopping method. Friction brakes are similar to the brakes on automobiles. **See Figure 18-1.**

Solenoid Operation. Friction brakes are normally controlled by a solenoid, which activates the brake shoes. The solenoid is energized when the motor is running. This keeps the brake shoes from touching the drum mounted on the motor shaft. The solenoid is de-energized and the brake shoes are applied through spring tension when the motor is turned OFF.

Heidelberg Harris, Inc.

Figure 18-1. Friction brakes normally consist of two friction surfaces that come in contact with a wheel mounted on the motor shaft.

Two methods are used to connect the solenoid into the circuit so that it activates the brake whenever the motor is turned ON and OFF. **See Figure 18-2.**

The first circuit is used if the solenoid has a voltage rating equal to the motor voltage rating. The second circuit is used if the solenoid has a voltage rating equal to the voltage between L1 and the neutral. Always connect the brake solenoid directly into the motor circuit, not into the control circuit. This eliminates improper activation of the brake.

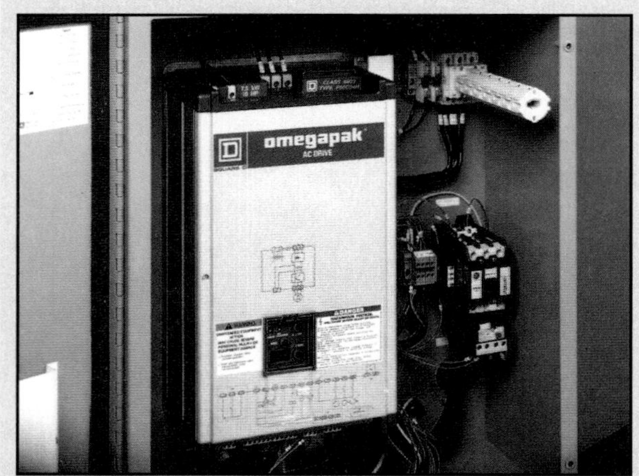

DoALL Company

Cut-off saws from DoALL include an AC inverter which controls the speed of the bank motor and provides infinitely variable band speeds from 40 fpm to 360 fpm.

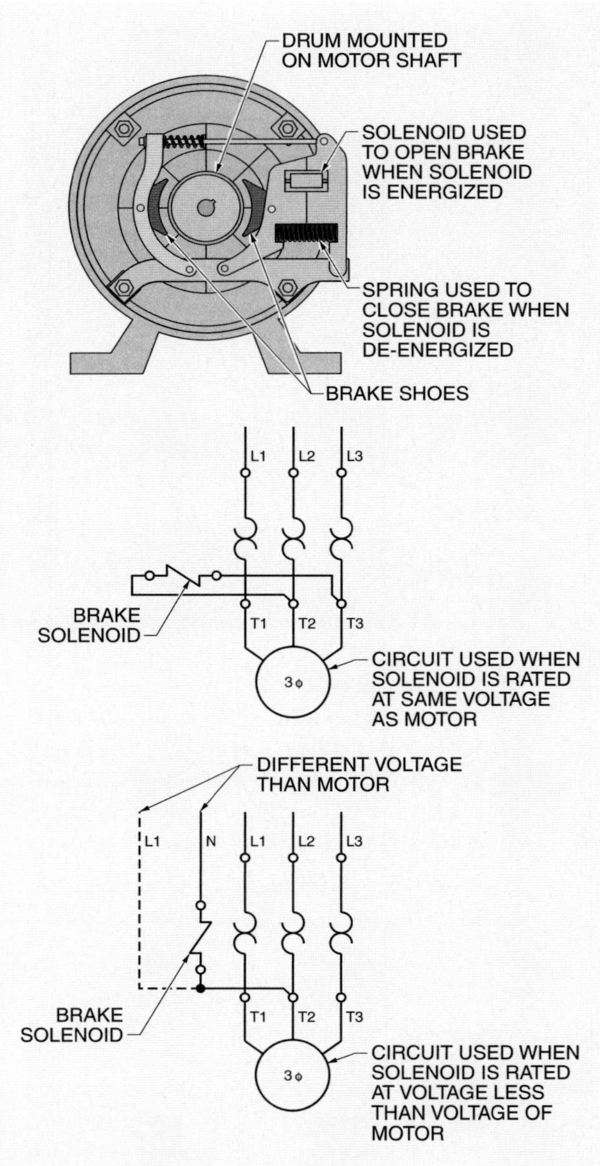

Figure 18-2. A friction brake may be connected to full-line voltage equal to that produced between L1 and the neutral.

Brake Shoes. In friction braking, the braking action is applied to a wheel mounted on the shaft of the motor rather than directly to the shaft. The wheel provides a much larger braking surface than could be obtained from the shaft alone. This permits the use of large brake shoe linings and low shoe pressure. Low shoe pressure, equally distributed over a large area, results in even wear and braking torque. The braking torque developed is directly proportional to surface area and spring pressure. The spring pressure is adjustable on nearly all friction brakes.

Determining Braking Torque. Full-load motor torque is calculated to determine the required braking torque of a motor. To calculate braking torque, apply the following formula:

$$T = \frac{HP \times 5252}{rpm}$$

where

T = full-load motor torque (in lb-ft)

5252 = constant ($\frac{33{,}000}{\pi \times 2} = 5252$)

HP = motor horsepower

rpm = speed of motor shaft

Example: Calculating Braking Torque

What is the braking torque of a 60 HP, 240 V motor rotating at 1725 rpm?

$$T = \frac{HP \times 5252}{rpm}$$

$$T = \frac{60 \times 5252}{1725}$$

$$T = \frac{315{,}120}{1725}$$

T = **182.7 lb-ft**

The torque rating of the brake selected should be equal to or greater than the full-load motor torque. Manufacturers of electric brakes list the torque ratings (in lb-ft) for their brakes.

Braking torque may also be determined using a horsepower-to-torque conversion chart. **See Figure 18-3.** A line is drawn from the horsepower to the rpm of the motor. The point at which the line crosses the torque values is the full-load and braking torque of the motor. For example, a 50 HP motor that rotates at 900 rpm requires a braking torque of 300 lb-ft.

Advantages and Disadvantages of Friction Brakes. The advantages of using friction brakes are lower initial cost and simplified maintenance. Friction brakes are less expensive to install than other braking methods because fewer expensive electrical components are required. Maintenance is simplified because it is easy to see whether the shoes are worn and if the brake is working. Friction brakes are available in both AC and DC designs to meet almost any application. The disadvantage of friction brakes is that they require more maintenance than other braking methods. Maintenance consists of replacing the shoes. Shoe replacement depends on the number of times the motor is stopped. A motor that is stopped often needs more maintenance than a motor that is almost never stopped. Friction brake applications include printing presses, cranes, overhead doors, hoisting equipment, and machine tool control.

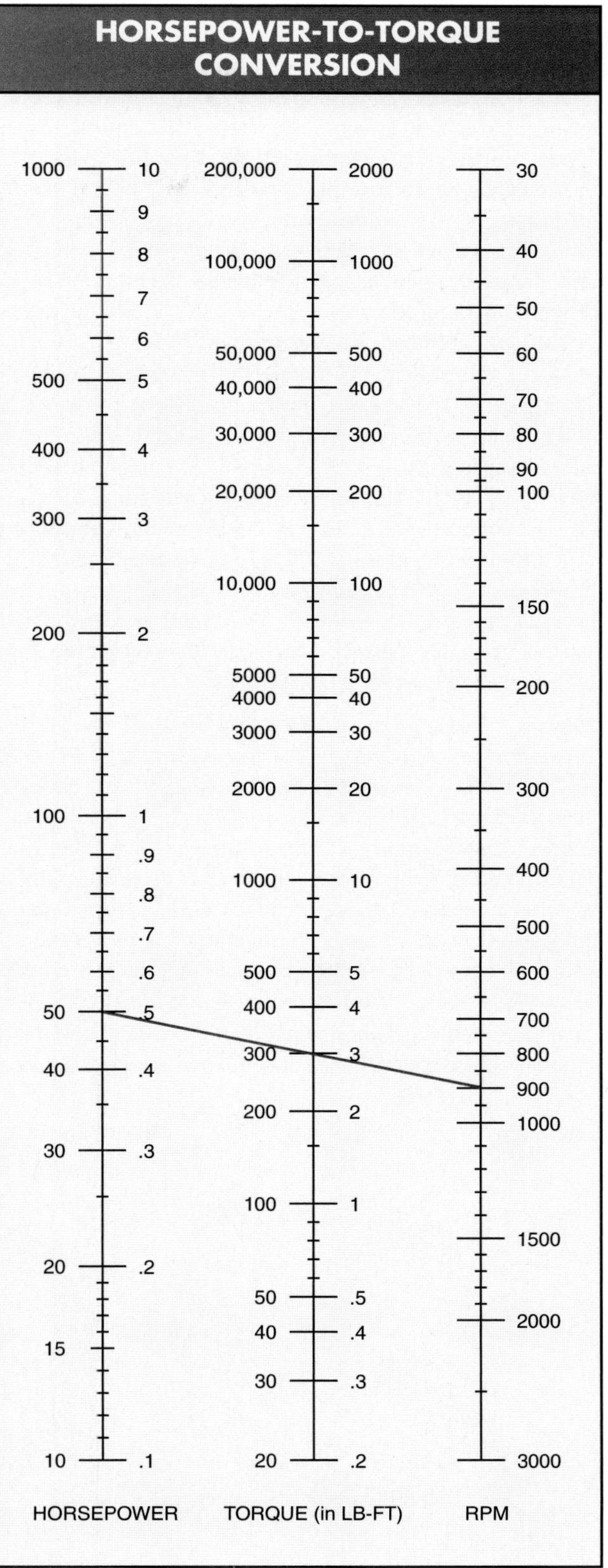

Figure 18-3. Braking torque may be determined by using a horsepower-to-torque conversion chart.

Plugging

Plugging is a method of motor braking in which the motor connections are reversed so that the motor develops a countertorque that acts as a braking force. The countertorque is accomplished by reversing the motor at full speed with the reversed motor torque opposing the forward inertia torque of the motor and its mechanical load. Plugging a motor allows for very rapid stopping. Although manual and electromechanical controls can be used to reverse the direction of a motor, a plugging switch is normally used in plugging applications. **See Figure 18-4.**

A plugging switch is connected mechanically to the shaft of the motor or driven machinery. The rotating motion of the motor is transmitted to the plugging switch contacts either by a centrifugal mechanism or by a magnetic induction arrangement. The contacts on the plugging switch are NO, NC, or both, and actuate at a given speed. The primary function of a plugging switch is to prevent the reversal of the load once the countertorque action of plugging has brought the load to a standstill. The motor and load would start to run in the opposite direction without stopping if the plugging switch were not present.

Plugging Switch Operation. Plugging switches are designed to open and close sets of contacts as the shaft speed on the switch varies. As the shaft speed increases, the contacts are set to change at a given rpm. As the shaft speed decreases, the contacts return to their normal condition. As the shaft speed increases, the contact setpoint (point at which the contacts operate) reaches a higher rpm than the point at which the contacts reset (return to their normal position) on decreasing speed. The difference in these contact operating values is the differential speed or rpm.

In plugging, the continuous running speed must be many times the speed at which the contacts are required to operate. This provides high contact holding force and reduces possible contact chatter or false operation of the switch.

Continuous Plugging. A plugging switch may be used to plug a motor to a stop each time the motor is stopped. **See Figure 18-5.**

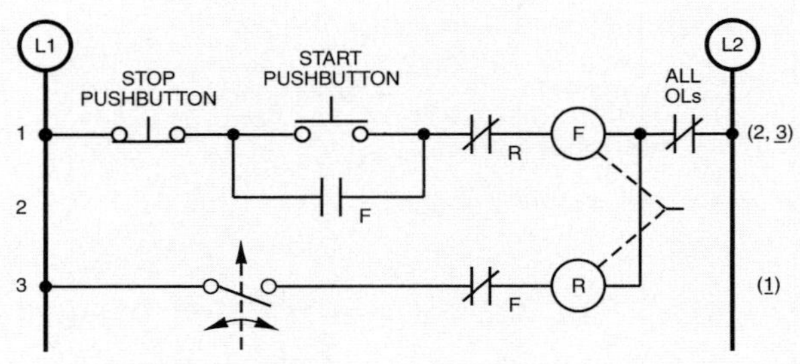

Figure 18-5. A plugging switch may be used to plug a motor to stop each time the motor is stopped.

In this circuit, the NO contacts of the plugging switch are connected to the reversing starter through an interlock contact. Pushing the start pushbutton energizes the forward starter, starting the motor in forward and adding memory to the control circuit. As the motor accelerates, the NO plugging contacts close. The closing of the NO plugging contacts does not energize the reversing starter because of the interlocks. Pushing the stop pushbutton drops out the forward starter and interlocks. This allows the reverse starter to immediately energize through the plugging switch and the NC forward interlock. The motor is reversed and the motor brakes to a stop. After the motor is stopped, the plugging switch opens to disconnect the reversing starter before the motor is actually reversed.

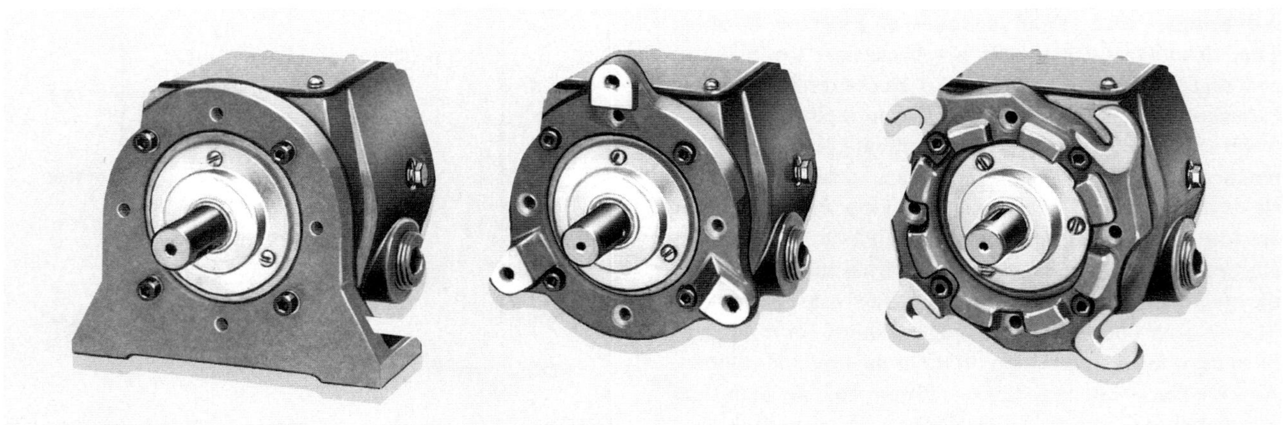

Rockwell Automation, Allen-Bradley Company, Inc.

Figure 18-4. Plugging switches prevent the reversal of the controlled load after the load has stopped.

Plugging for Emergency Stops. A plugging switch may be used in a circuit where plugging is required only in an emergency. **See Figure 18-6.** In this circuit, the motor is started in the forward direction by pushing the run pushbutton. This starts the motor and adds memory to the control circuit. As the motor accelerates, the NO plugging contacts close. Pushing the stop pushbutton de-energizes the forward starter but does not energize the reverse starter. This is because there is no path for the L1 power to reach the reverse starter, so the motor coasts to a stop.

Pushing the emergency stop pushbutton de-energizes the forward starter and simultaneously energizes the reversing starter. Energizing the reversing starter adds memory in the control circuit and plugs the motor to a stop. When the motor is stopped, the plugging switch opens to disconnect the reversing starter before the motor is actually reversed. The de-energizing of the reversing starter also removes the memory from the circuit.

Limitations of Plugging

Plugging may not be applied to all motors and/or applications. Braking a motor to a stop using plugging requires that the motor be a reversible motor and that the motor can be reversed at full speed. Even if the motor can be reversed at full speed, the damage that plugging may do may outweigh its advantages.

Reversing. A motor cannot be used for plugging if it cannot be reversed at full speed. For example, a 1ϕ shaded-pole motor cannot be reversed at any speed. Thus, a 1ϕ shaded-pole motor cannot be used in a plugging circuit. Likewise, most 1ϕ split-phase and capacitor-start motors cannot be plugged because their centrifugal switches remove the starting windings when the motor accelerates. Without the starting winding in the circuit, the motor cannot be reversed.

Heat. All 3ϕ motors, and most 1ϕ and DC motors, can be used for plugging. However, high current and heat result from plugging a motor to a stop. A motor is connected in reverse at full speed when plugging a motor. The current may be three or more times higher during plugging than during normal starting. For this reason, a motor designated for plugging or a motor with a high service factor should be used in all cases except emergency stops. The service factor (SF) should be 1.35 or more for plugging applications.

Plugging Using Timing Relays

Plugging can also be accomplished by using a timing relay. The advantage of using a timing relay is normally lower cost since a timer is inexpensive and does not have to be connected mechanically to the motor shaft or driven machine. The disadvantage is that, unlike a plugging switch, the timer does not compensate for a change in the load condition (which affects stopping time) once the timer is preset.

An OFF-delay timer may be used in applications where the time needed to decelerate the motor is constant and known. **See Figure 18-7.** In this circuit, the NO contacts of the timer are connected into the circuit in the same manner as a plugging switch. The coil of the timer is connected in parallel with the forward starter.

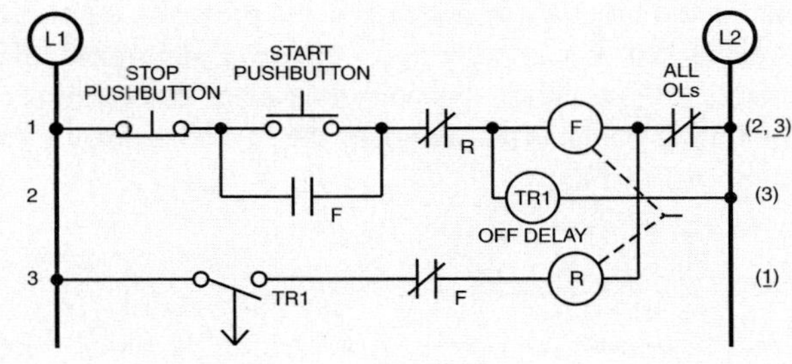

Figure 18-7. An OFF-delay timer may be used in applications where the time needed to decelerate the motor is constant.

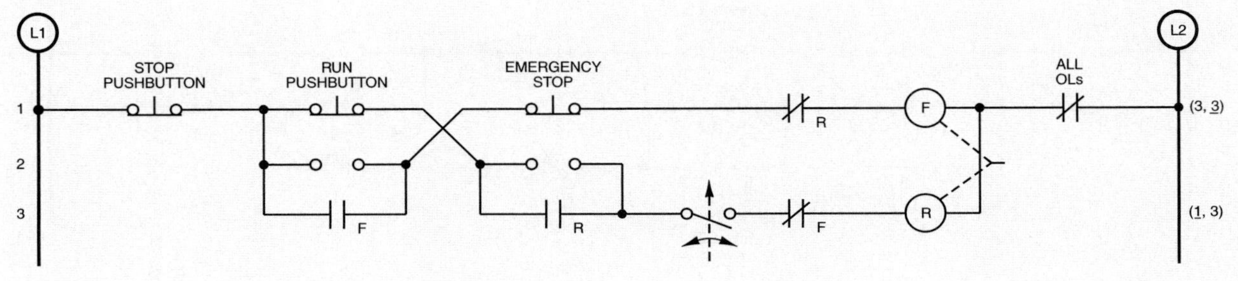

Figure 18-6. A plugging switch may be used in a circuit where plugging is required only in an emergency.

The motor is started and memory is added to the circuit when the start pushbutton is pressed. In addition to energizing the forward starter, the OFF-delay timer is also energized. The energizing of the OFF-delay timer immediately closes the NO timer contacts. The closing of these contacts does not energize the reverse contacts because of the interlocks.

The forward starter and timer coil are de-energized when the stop pushbutton is pressed. The NO timing contact remains held closed for the setting of the timer. The holding closed of the timing contact energizes the reversing starter for the period of time set on the timer. This plugs the motor to a stop. The timer's contact must reopen before the motor is actually reversed. The motor reverses direction if the time setting is too long.

An OFF-delay timer may also be used for plugging a motor to a stop during emergency stops. **See Figure 18-8.** In this circuit, the timer's contacts are connected in the same manner as the plugging switch. The motor is started and memory is added to the circuit when the start pushbutton is pressed. The forward starter and timer are de-energized if the stop pushbutton is pressed. Although the timer's NO contacts are held closed for the time period set on the timer, the reversing starter is not energized. This is because no power is applied to the reversing starter from L1.

If the emergency stop pushbutton is pressed, the forward starter and timer are de-energized and the reversing starter is energized. The energizing of the reversing starter adds memory to the circuit and stops the motor. The opening of the timing contacts de-energizes the reversing starter and removes the memory.

Electric Braking

Electric braking is a method of braking in which a DC voltage is applied to the stationary windings of a motor after the AC voltage is removed. Electric braking is also known as DC injection braking. **See Figure 18-9.** Electric braking is an efficient and effective method of braking most AC motors. Electric braking provides a quick and smooth braking action on all types of loads including high-speed and high-inertia loads. Maintenance is minimal because there are no parts that come in physical contact during braking.

Electric Braking Operating Principles. The principle that unlike magnetic poles attract each other and like magnetic poles repel each other explains why a motor shaft rotates. The method in which the magnetic fields are created changes from one type of motor to another.

In AC induction motors, the opposing magnetic fields are induced from the stator windings into the rotor windings by transformer action. The motor continues to rotate as long as the AC voltage is applied. The motor coasts to a standstill over a period of time when the AC voltage is removed because there is no induced field to keep it rotating.

Electric braking can be used to provide an immediate stop if the coasting time is unacceptable, particularly in an emergency situation. Electric braking is accomplished by applying a DC voltage to the stationary windings once the AC is removed. The DC voltage creates a magnetic field in the stator that does not change polarity.

The constant magnetic field in the stator creates a magnetic field in the rotor. Because the magnetic field of the stator does not change in polarity, it attempts to stop the rotor when the magnetic fields are aligned (N to S and S to N). **See Figure 18-10.** The only force that can keep the rotor from stopping with the first alignment is the rotational inertia of the load connected to the motor shaft. However, because the braking action of the stator is present at all times, the motor brakes quickly and smoothly to a standstill.

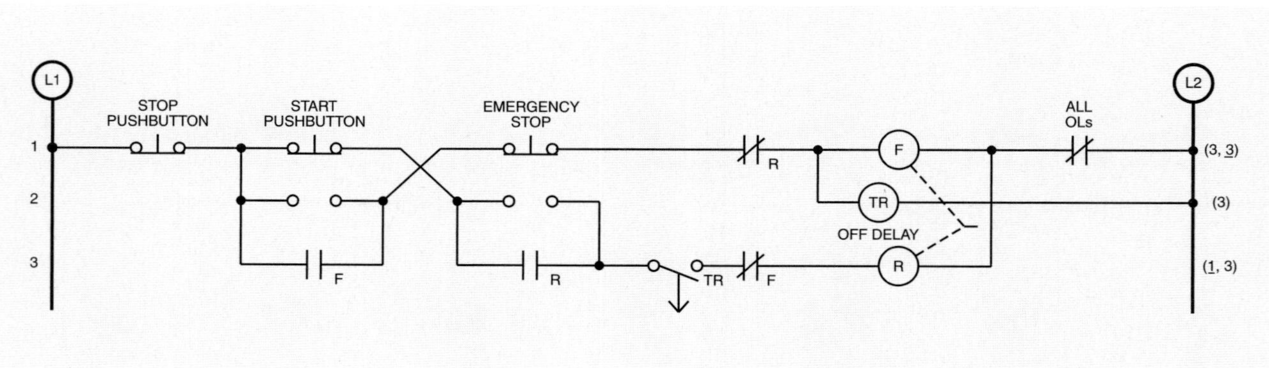

Figure 18-8. An OFF-delay timer may also be used for plugging a motor to stop during emergency stops.

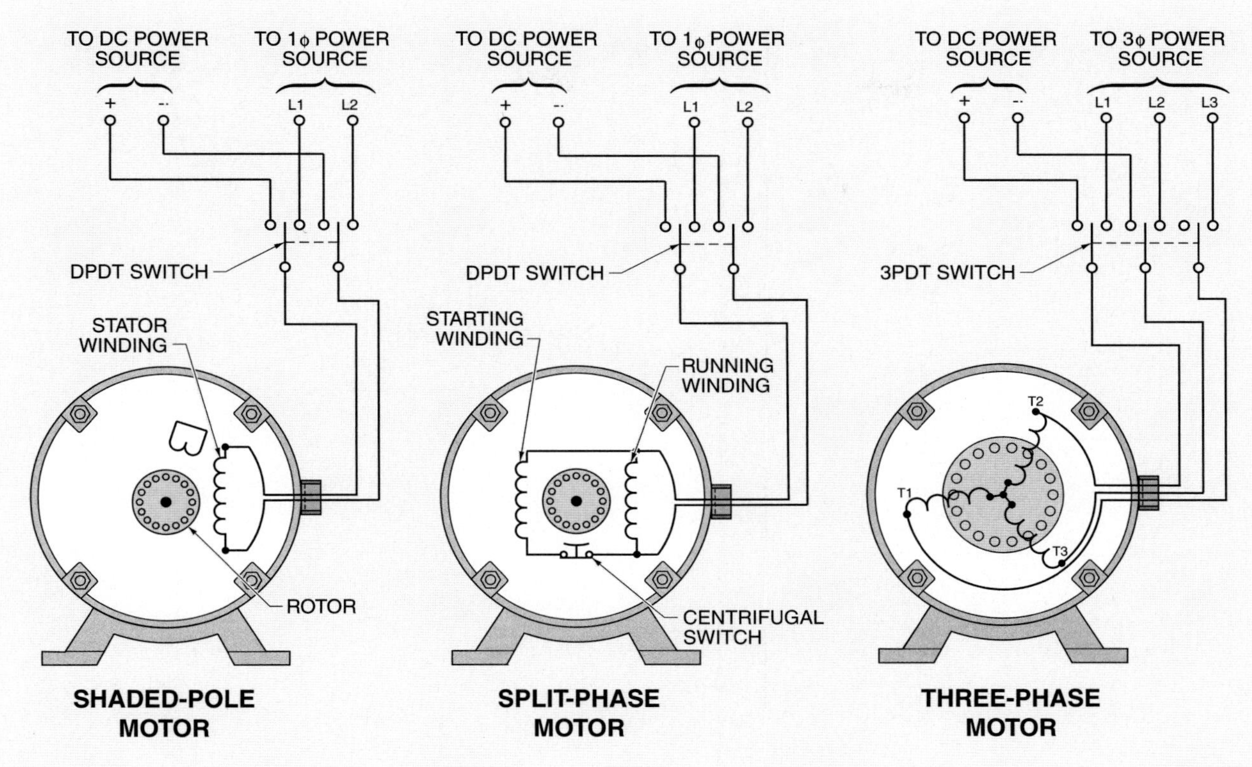

Figure 18-9. Electric braking is achieved by applying DC voltage to the stationary windings once the AC is removed.

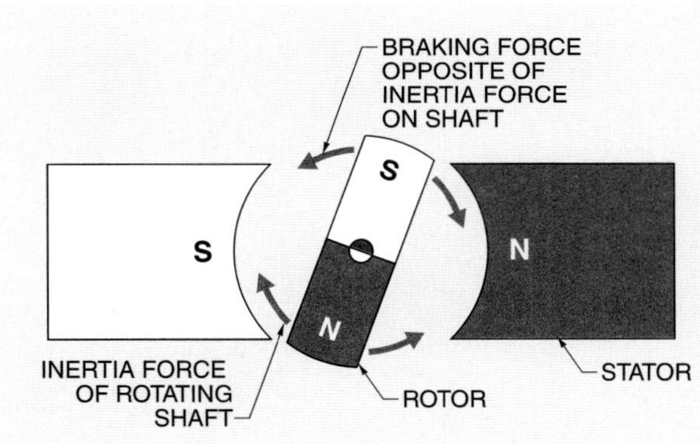

Figure 18-10. The DC voltage applied during electric braking creates a magnetic field in the stator that does not change polarity.

DC Electric Braking Circuits. DC is applied after the AC is removed to bring the motor to a stop quickly. **See Figure 18-11.** This circuit, like most DC braking circuits, uses a bridge rectifier circuit to change the AC into DC. In this circuit, a 3ϕ AC motor is connected to three-phase power by a magnetic motor starter.

The magnetic motor starter is controlled by a standard stop/start pushbutton station with memory. An OFF-delay timer is connected in parallel with the magnetic motor starter. The OFF-delay timer controls a NO contact that is used to apply power to the braking contactor for a short period of time after the stop pushbutton is pressed. The timing contact is adjusted to remain closed until the motor comes to a stop.

The braking contactor connects two motor leads to the DC supply. A transformer with tapped windings is used to adjust the amount of braking torque applied to the motor. Current-limiting resistors could be used for the same purpose. This allows for a low- or high-braking action, depending on the application. The larger the applied DC voltage, the greater the braking force.

The interlock system in the control circuit prevents the motor starter and braking contactor from being energized at the same time. This is required because the AC and DC power supplies must never be connected to the motor simultaneously. Total interlocking should always be used on electrical braking circuits. Total interlocking is the use of mechanical, electrical, and pushbutton interlocking. A standard forward and reversing motor starter can be used in this circuit, as it can with most electric braking circuits.

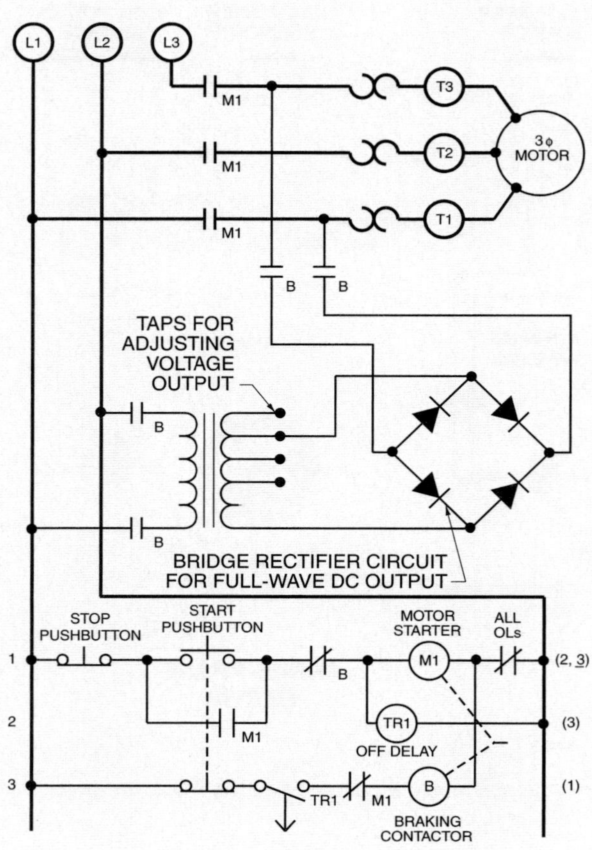

Figure 18-11. DC is applied after AC is removed to bring the motor to a stop quickly.

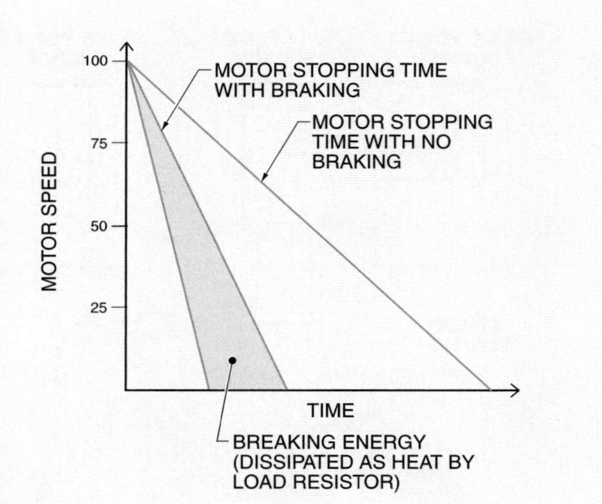

Figure 18-12. Dynamic braking is a method of motor braking in which a motor is reconnected to act as a generator immediately after it is turned OFF.

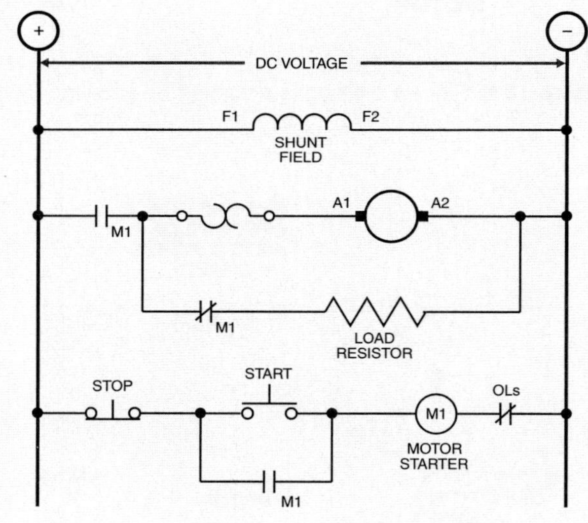

Figure 18-13. Dynamic braking is normally applied to DC motors because there must be access to the rotor windings to reconnect the motor to act as a generator.

Dynamic Braking DC Motors

Dynamic braking is a method of motor braking in which a motor is reconnected to act as a generator immediately after it is turned OFF. Connecting the motor in this way makes the motor act as a loaded generator that develops a retarding torque, which rapidly stops the motor. The generator action converts the mechanical energy of rotation to electrical energy that can be dissipated as heat in a resistor. **See Figure 18-12.**

Dynamic braking is normally applied to DC motors because there must be access to the rotor windings to reconnect the motor to act as a generator. Access is accomplished through the brushes on DC motors. **See Figure 18-13.** Dynamic braking of a DC motor may be needed because DC motors are often used for lifting and moving heavy loads that may be difficult to stop.

In this circuit, the armature terminals of the DC motor are disconnected from the power supply and immediately connected across a resistor that acts as a load. The smaller the resistance of the resistor, the greater the rate of energy dissipation and the faster the motor comes to rest. The field windings of the DC motor are left connected to the power supply. The armature generates a counter electromotive force (CEMF). The CEMF causes current to flow through the resistor and armature.

The current causes heat to be dissipated in the resistor in the form of electrical watts. This removes energy from the system and slows the motor rotation.

The generated CEMF decreases as the speed of the motor decreases. As the motor speed approaches 0 rpm, the generated voltage also approaches 0 V. The braking action lessens as the speed of the motor decreases. As a result, a motor cannot be braked to a complete stop using dynamic braking. Dynamic braking also cannot hold a load once it is stopped because there is no braking action.

Electromechanical friction brakes are often used along with dynamic braking in applications that require the load to be held. A combination of dynamic braking and friction braking can also be used in applications where a large, heavy load is to be stopped. In these applications, the force of the load wears the friction brake shoes excessively; so dynamic braking can be used to slow the load before the friction brakes are applied. This is similar to using a parachute to slow a racecar before applying the brakes.

Dynamic Braking Electric Motor Drives

A motor drive can control the time it takes for a motor to stop. The stopping time is programmed by setting the deceleration parameter. The deceleration parameter can be set for 1 sec (or less) to several minutes. However, for fast stops (especially with high inertia loads), a braking resistor is added. **See Figure 18-14.**

The braking resistor also helps control motor torque during repeated ON/OFF motor cycling. The size (wattage rating) and resistance (in ohms) of the braking resistor are typically selected using the drive manufacturer braking resistor chart. The resistor wattage must be high enough to dissipate the heat delivered at the resistor. The resistor resistance value must be low enough to allow current flow at a level high enough to convert the current to heat, but not so low as to cause excessive current flow. The braking resistor(s) can be housed in an enclosure such as a NEMA 1 enclosure.

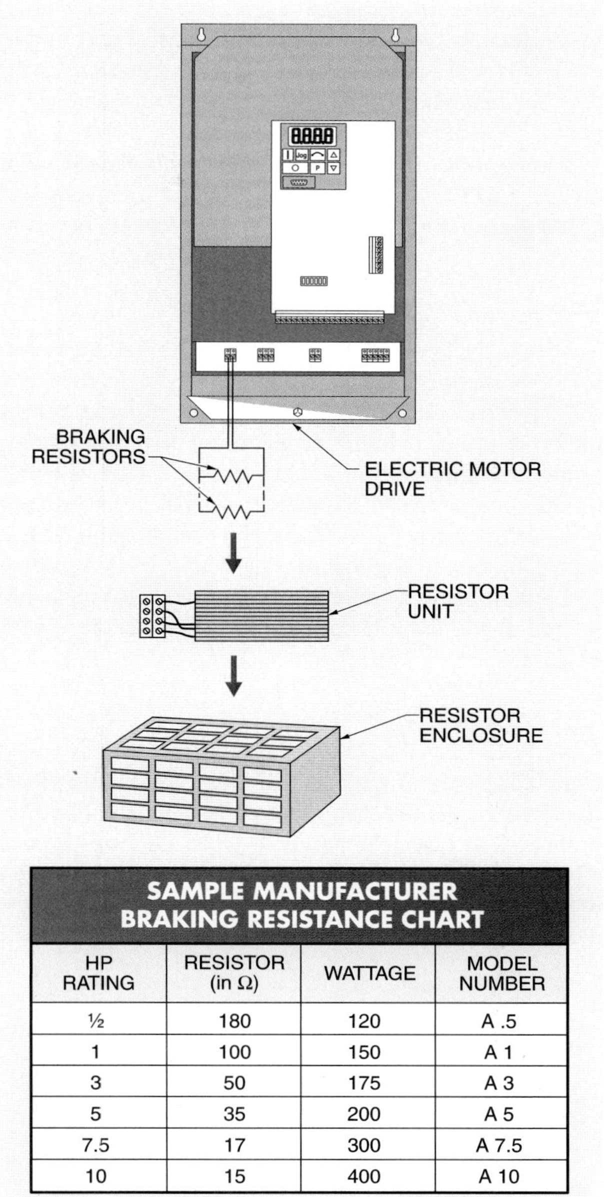

SAMPLE MANUFACTURER BRAKING RESISTANCE CHART

HP RATING	RESISTOR (in Ω)	WATTAGE	MODEL NUMBER
½	180	120	A .5
1	100	150	A 1
3	50	175	A 3
5	35	200	A 5
7.5	17	300	A 7.5
10	15	400	A 10

Figure 18-14. A braking resistor can be added to a motor drive to control fast stops on high-inertia loads.

SPEED CONTROL

Speed control is essential in many residential, commercial, and industrial applications. For example, motors found in washing machines, commercial furnaces and air conditioners, as well as electrical appliances such as mixers and blenders, are required to turn the load at different speeds.

Industrial applications of speed control include mining machines, printing presses, cranes and hoists, elevators, assembly line conveyors, food processing equipment, and metalworking or woodworking lathes.

AC and DC Motors

The motor is the main consideration when choosing speed control for an application. Some motors offer excellent speed control through a total range of speeds, while others may offer only two or three different speeds. Other motors offer only one speed that cannot be changed except by external means such as gears, pulley drives, or changes of power source frequency.

The two basic motors used in speed control applications are AC and DC motors. Each motor has different ranges of speed control, applications, and cost effectiveness.

Load Requirements

The loads that are connected to and controlled by motors vary considerably. Each motor has its own ability to control different loads at different speeds. For example, certain motors are rated at high starting torque with low running torque and others are rated at low starting torque with high running torque. Load requirements must be determined to select the correct motor for a given application. This is especially true in applications that require speed control. The requirements a motor must meet in controlling a load include work, force, torque, and horsepower in relation to speed.

Work

Work is applying a force over a distance. **See Figure 18-15.** *Force* is any cause that changes the position, motion, direction, or shape of an object. Work is done when a force overcomes a resistance. *Resistance* is any force that tends to hinder the movement of an object. If an applied force does not cause motion, no work is produced.

SPM Instrument, Inc.

Overhead cranes are an example of equipment that uses work and force to overcome resistance.

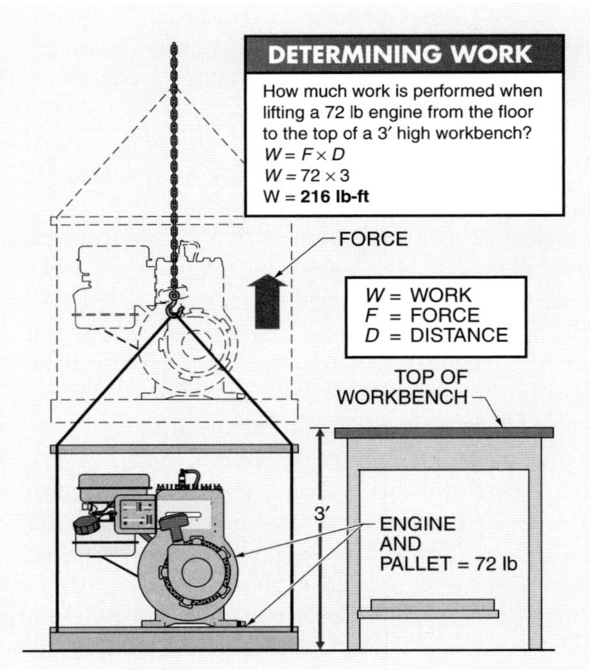

Figure 18-15. Work equals force times distance.

The amount of work (W) produced is determined by multiplying the force (F) that must be overcome by the distance (D) through which it acts. Thus, work is measured in pound-feet (lb-ft). To calculate the amount of work produced, apply the following formula:

$W = F \times D$

where

W = work (in lb-ft)

F = force (in lb)

D = distance (in ft)

Example: Calculating Work

How much work is required to carry a 25 lb bag of groceries vertically from street level to the fourth floor of a building 30′ above street level?

$W = F \times D$

$W = 25 \times 30$

$W =$ **750 lb-ft**

Resistance must be overcome to perform work. More work would be required if the groceries were heavier, the distance longer, or a combination of the two. For example, 1000 lb-ft of work is required to carry a 25 lb bag of groceries to a floor 40′ above the street (25 × 40 = 1000 lb-ft). Less work would be required if the groceries were lighter, the distance were shorter, or a combination of the two. For example, 180 lb-ft of work is required to carry a 12 lb bag of groceries vertically from street level to the second floor of a building 15′ above street level (12 × 15 = 180 lb-ft).

Torque

Torque is the force that produces rotation. Torque causes an object to rotate. Torque (T) consists of a force (F) acting on a radius (r). **See Figure 18-16.** Torque, like work, is measured in pound-feet (lb-ft). However, torque, unlike work, may exist even though no movement occurs. To calculate torque, apply the following formula:

$T = F \times r$

where

T = torque (in lb-ft)

F = force (in lb)

r = radius (in ft)

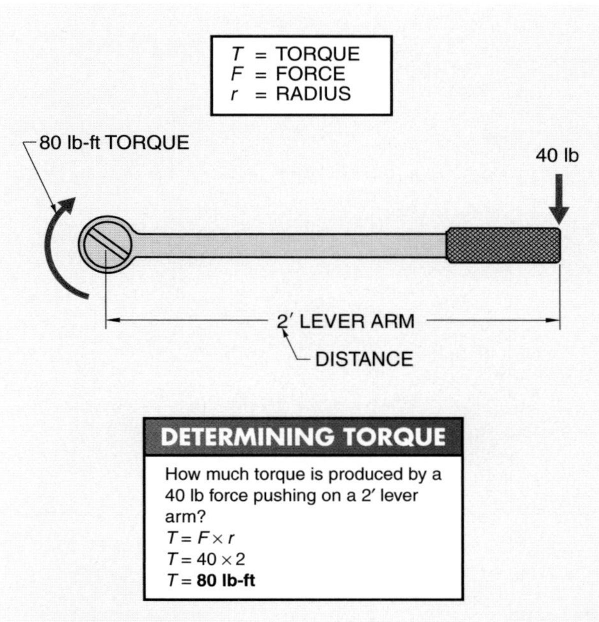

Figure 18-16. Torque is the force that produces or tends to produce rotation.

Example: Calculating Torque

What is the torque produced by a 60 lb force pushing on a 3′ lever arm?

$T = F \times r$

$T = 60 \times 3$

$T =$ **180 lb-ft**

Work is done if the amount of torque produced is large enough to cause movement. More torque would be produced if the force were larger, the lever arm longer, or a combination of the two. Less torque would be produced if the force were smaller, the lever arm shorter, or a combination of the two. For example, 105 lb-ft of torque is produced by a 70 lb force pushing on a 1½′ lever arm (70 × 1½ = 105 lb-ft).

Motor Torque

Motor torque is the force that produces or tends to produce rotation in a motor. A motor must produce enough torque to start the load and keep it moving for the motor to operate the load connected to it. A motor connected to a load produces four types of torque. The four types of torque are locked rotor torque (LRT), pull-up torque (PUT), breakdown torque (BDT), and full-load torque (FLT). **See Figure 18-17.**

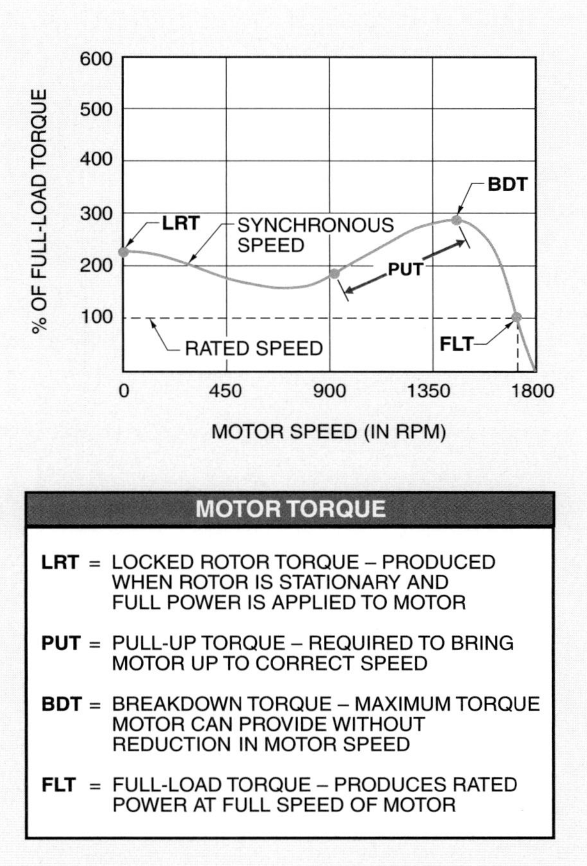

Figure 18-17. Motor torque is the force that produces or tends to produce rotation in a motor.

Locked Rotor Torque. *Locked rotor torque (LRT)* is the torque a motor produces when its rotor is stationary and full power is applied to the motor. **See Figure 18-18.** All motors can safely produce a higher torque output than the rated full-load torque for short periods of time. Since many loads require a higher torque to start them moving than to keep them moving, a motor must produce a higher torque when starting the load. Locked rotor torque is also referred to as breakaway or starting torque. Starting torque is the torque required to start a motor. Starting torque is normally expressed as a percentage of full-load torque.

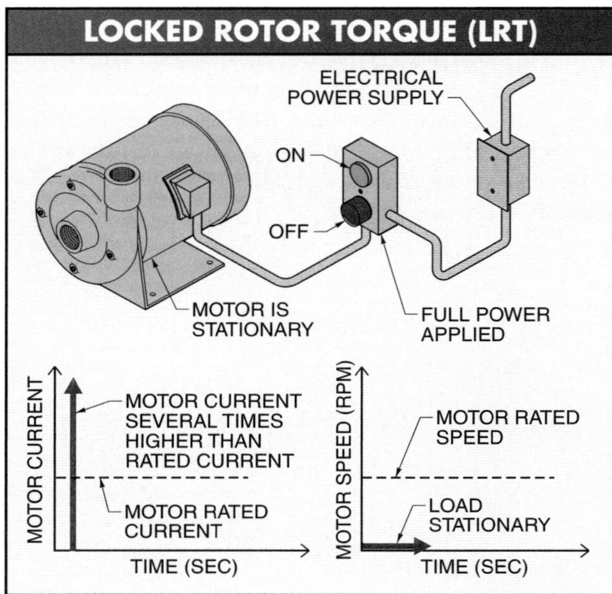

Figure 18-18. Locked rotor torque is the torque a motor produces when its rotor is stationary and full power is applied to the motor.

Cincinnati Milacron

Torque requirements on many machine processes vary with the type of material being machined.

Pull-Up Torque. *Pull-up torque (PUT)* is the torque required to bring a load up to its rated speed. **See Figure 18-19.** If a motor is properly sized to the load, pull-up torque is brief. If a motor does not have sufficient pull-up torque, the locked rotor torque may start the load turning but the pull-up torque cannot bring it up to rated speed. Once the motor is up to rated speed, full-load torque keeps the load turning. Pull-up torque is also referred to as accelerating torque.

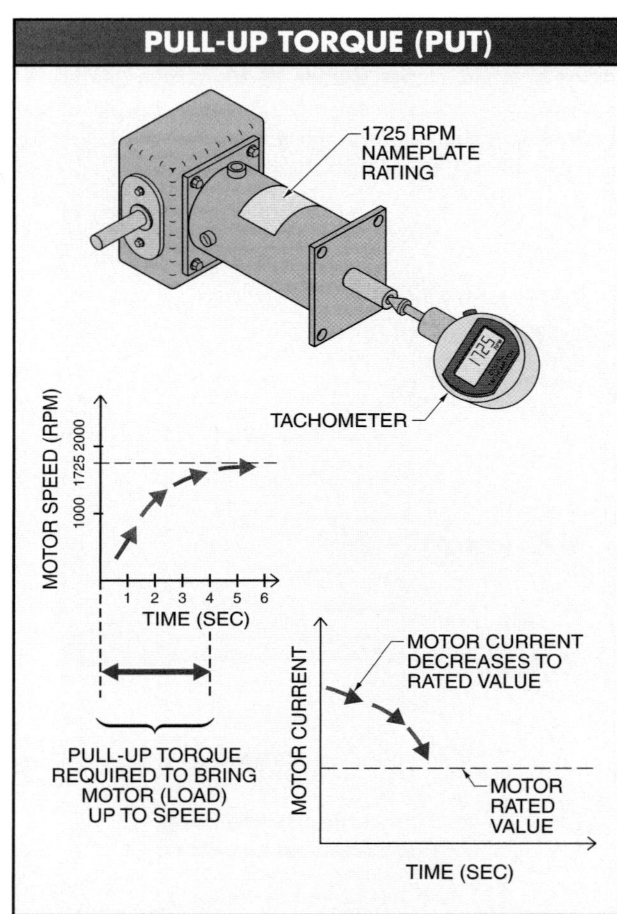

Figure 18-19. Pull-up torque is the torque required to bring a load up to its rated speed.

Technical Fact

All motors produce torque but not all motors produce the same torque characteristics. Motors are classified by the National Electrical Manufacturers Association (NEMA) according to their electrical characteristics. Motor torque characteristics vary with the classification of the motor. Motors are classified as Class A through Class F. Classes B, C, and D are the most common classifications.

Breakdown Torque. *Breakdown torque (BDT)* is the maximum torque a motor can provide without an abrupt reduction in motor speed. **See Figure 18-20.** As the load on a motor shaft increases, the motor produces more torque. As the load continues to increase, the point at which the motor stalls is reached. This point is the breakdown torque.

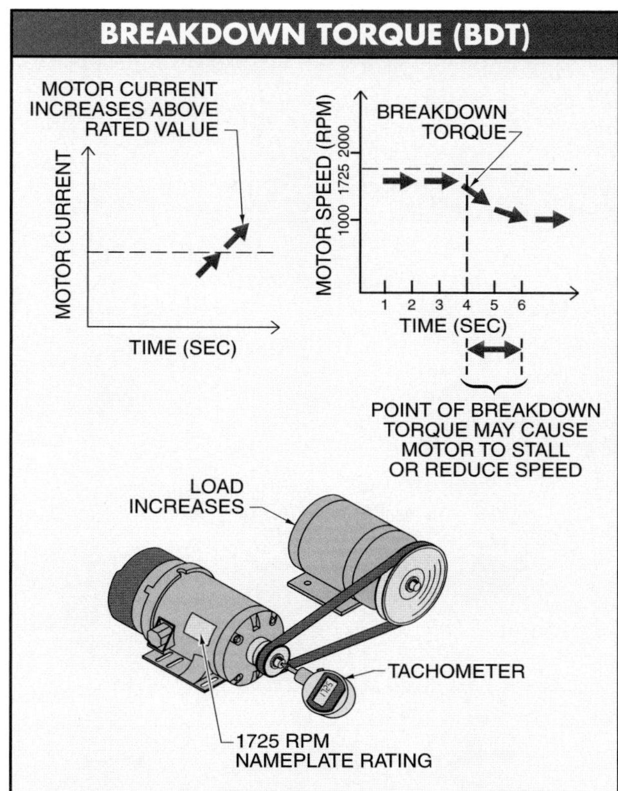

Figure 18-20. Breakdown torque is the maximum torque a motor can provide without an abrupt reduction in motor speed.

Full-Load Torque. *Full-load torque (FLT)* is the torque required to produce the rated power at the full speed of the motor. **See Figure 18-21.** The amount of torque a motor produces at rated power and full speed (full-load torque) can be found by using a horsepower-to-torque conversion chart. To calculate motor full-load torque, apply the following formula:

$$T = \frac{HP \times 5252}{rpm}$$

where

T = torque (in lb-ft)

HP = horsepower

5252 = constant ($\frac{33,000}{\pi \times 2}$ = 5252)

rpm = revolutions per minute

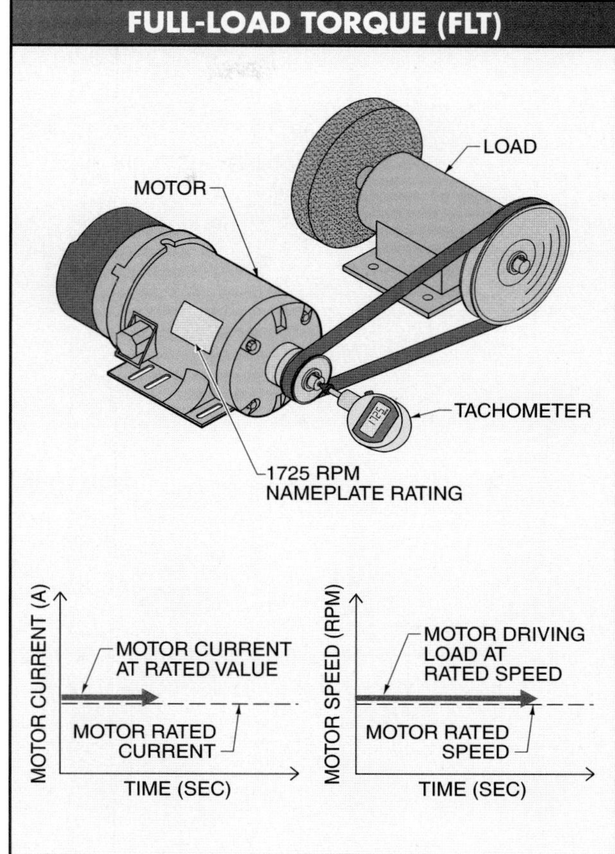

Figure 18-21. Full-load torque is the torque required to produce the rated power at full speed of the motor.

Example: Calculating Full-Load Torque

What is the full-load torque of a 30 HP motor operating at 1725 rpm?

$$T = \frac{HP \times 5252}{rpm}$$

$$T = \frac{30 \times 5252}{1725}$$

$$T = \frac{157,560}{1725}$$

T = **91.34 lb-ft**

If a motor is fully loaded, it produces full-load torque. If a motor is underloaded, it produces less than full-load torque. If a motor is overloaded, it must produce more than full-load torque to keep the load operating at the motor's rated speed. **See Figure 18-22.**

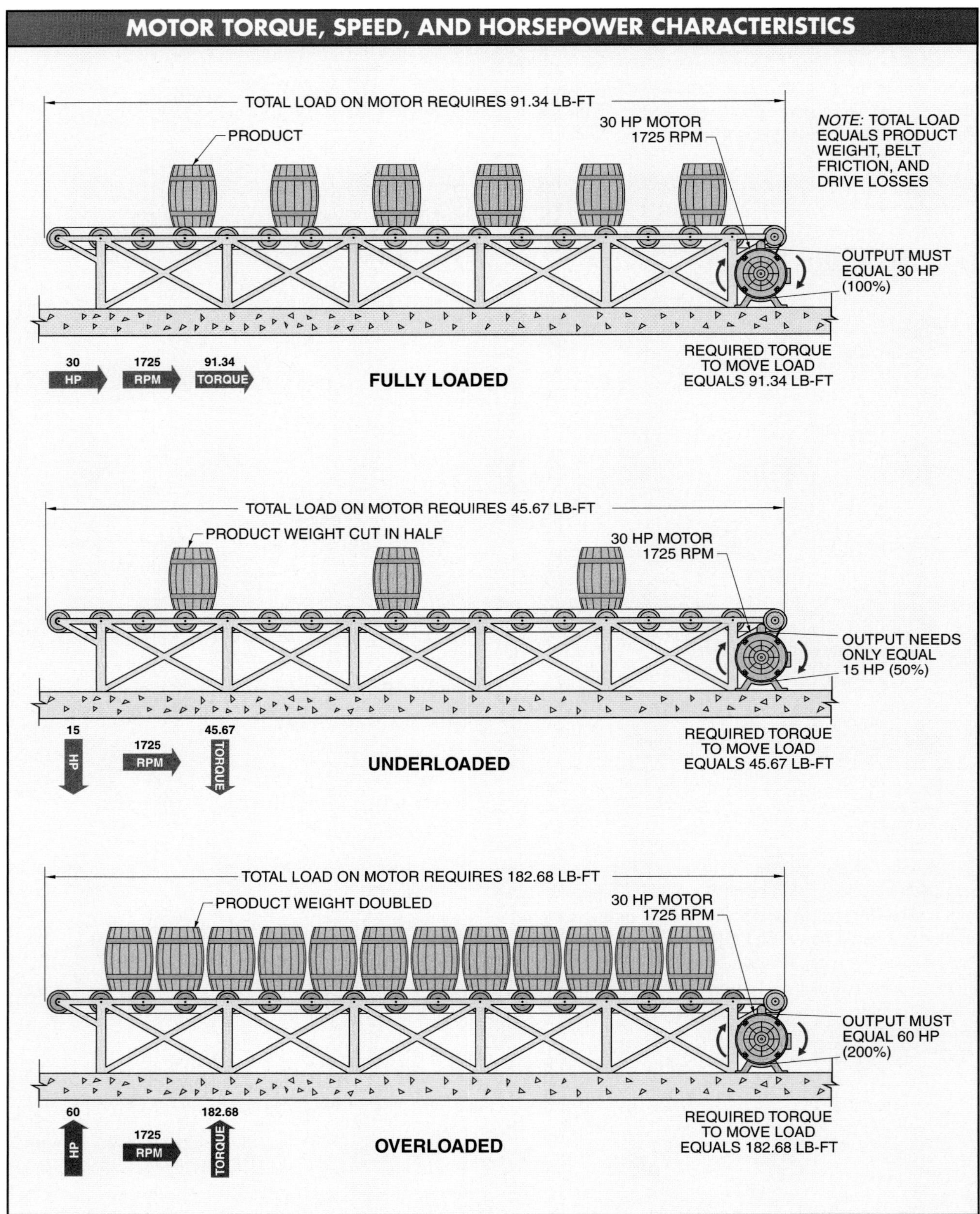

Figure 18-22. A motor may be fully loaded, underloaded, or overloaded.

For example, a 30 HP motor operating at 1725 rpm can develop 91.34 lb-ft of torque at full speed. If the load requires 91.34 lb-ft at 1725 rpm, the 30 HP motor produces an output of 30 HP. However, if the load to which the motor is connected requires only half as much torque (45.67 lb-ft) at 1725 rpm, the 30 HP motor produces an output of 15 HP. The 30 HP motor draws less current (and power) from the power lines and operates at a lower temperature when producing 15 HP.

However, if the 30 HP motor is connected to a load that requires twice as much torque (182.68 lb-ft) at 1725 rpm, the motor must produce an output of 60 HP. The 30 HP motor draws more current (and power) from the power lines and operates at a higher temperature. If the overload protection device is sized correctly, the 30 HP motor automatically disconnects from the power line before any permanent damage is done to the motor.

Technical Fact

Motors are classified by power rating as subfractional, fractional, or integral horsepower.

Horsepower

Electrical power is rated in horsepower or watts. A *horsepower (HP)* is a unit of power equal to 746 W or 33,000 lb-ft per minute (550 lb-ft per sec). A *watt (W)* is a unit of measure equal to the power produced by a current of 1 A across a potential difference of 1 V. A watt is $\frac{1}{746}$ of 1 HP. The watt is the base unit of electrical power. Motor power is rated in horsepower and watts. **See Figure 18-23.**

Horsepower is used to measure the energy produced by an electric motor while doing work. To calculate the horsepower of a motor when current, efficiency, and voltage are known, apply the following formula:

$$HP = \frac{I \times E \times E_{ff}}{746}$$

where

HP = horsepower

I = current (in A)

E = voltage (in V)

E_{ff} = efficiency

746 = constant

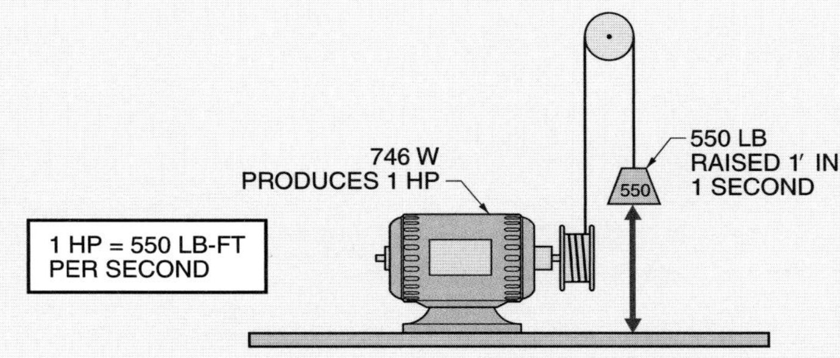

MECHANICAL ENERGY	ELECTRICAL ENERGY
$\frac{1}{2}$ HP	373 W
1 HP	746 W
2 HP	1492 W
5 HP	3730 W
100 HP	74,600 W

DETERMINING HP: VOLTAGE AND CURRENT GIVEN

What is the horsepower of a 120 V motor pulling 7.2 A and having 88% efficiency?

$$HP = \frac{I \times E \times E_{ff}}{746}$$

$$HP = \frac{7.2 \times 120 \times .88}{746}$$

$$HP = \frac{760.32}{746}$$

HP = **1 HP**

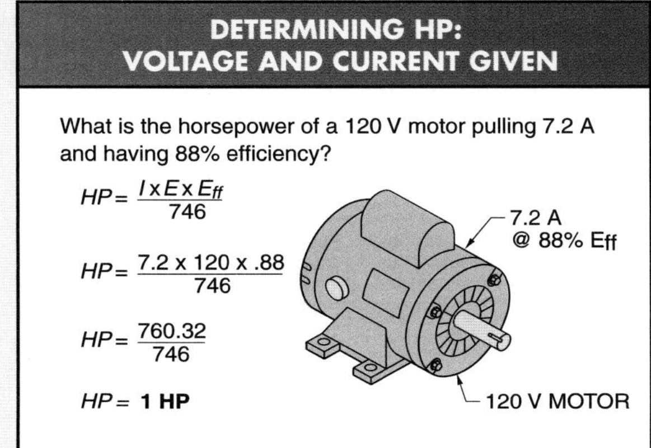

DETERMINING HP: SPEED AND TORQUE GIVEN

What is the horsepower of a 1180 rpm motor with an FLT of 2.25 lb-ft?

$$HP = \frac{rpm \times T}{5252}$$

$$HP = \frac{1180 \times 2.25}{5252}$$

$$HP = \frac{2655}{5252}$$

HP = $\frac{1}{2}$ **HP**

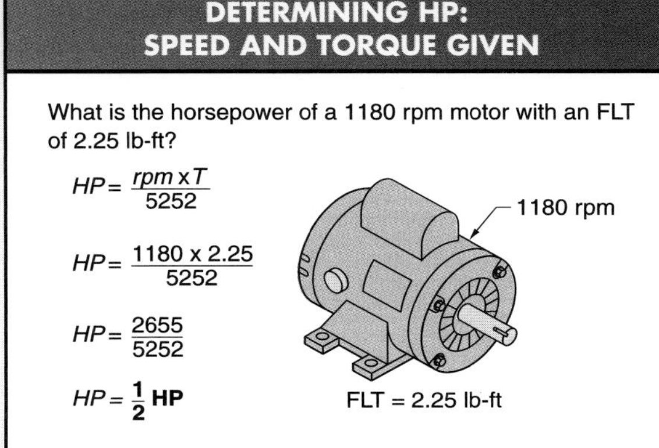

Figure 18-23. Motor power is rated in horsepower and watts.

Rockwell Automation, Allen-Bradley Company, Inc.

Motor controllers are available with ratings from 24 A to 1000 A and options that include pump control, motor braking, slow speed with braking, preset slow speed, and soft stop.

Example: Calculating Horsepower Using Voltage, Current, and Efficiency

What is the horsepower of a 230 V motor pulling 4 A and having 82% efficiency?

$$HP = \frac{I \times E \times Eff}{746}$$

$$HP = \frac{4 \times 230 \times .82}{746}$$

$$HP = \frac{754.4}{746}$$

HP = **1 HP**

The horsepower of a motor determines the size of load a motor can operate and the speed at which the load turns. To calculate the horsepower of a motor when the speed and torque are known, apply the following formula:

$$HP = \frac{rpm \times T}{5252}$$

where

HP = horsepower

rpm = revolutions per minute

T = torque (lb-ft)

5252 = constant ($\frac{33,000}{\pi \times 2} = 5252$)

Example: Calculating Horsepower Using Speed and Torque

What is the horsepower of a 1725 rpm motor with a full-load torque of 3.1 lb-ft?

$$HP = \frac{rpm \times T}{5252}$$

$$HP = \frac{1725 \times 3.1}{5252}$$

$$HP = \frac{5347.5}{5252}$$

HP = **1 HP**

Formulas for determining torque and horsepower are for theoretical values. When applied to specific applications, an additional 15% to 40% capability may be required to start a given load. Loads that are harder to start require the higher rating. To increase the rating, multiply the calculated theoretical value by 1.15 (115%) to 1.4 (140%). For example, what is the horsepower of a 1725 rpm motor with a full-load torque of 3.1 lb-ft with an added 25% output capability?

$$HP = \frac{rpm \times T}{5252} \times \%$$

$$HP = \frac{5347.5}{5252} \times 1.25$$

$$HP = 1 \times 1.25$$

HP = **1.25 HP**

Relationship between Speed, Torque, and Horsepower

A motor's operating speed, torque, and horsepower rating determine the work the motor can produce. These three factors are interrelated when applied to driving a load. **See Figure 18-24.**

If the torque remains constant, speed and horsepower are proportional. (A) If the speed increases, the horsepower must increase to maintain a constant torque. (B) If the speed decreases, the horsepower must decrease to maintain a constant torque.

If speed remains constant, torque and horsepower are proportional. (C) If the torque increases, the horsepower must increase to maintain a constant speed. (D) If the torque decreases, the horsepower must decrease to maintain a constant speed.

If torque and speed vary simultaneously but in opposite directions, the horsepower remains constant. (E) If the torque is increased and the speed is reduced, the horsepower remains constant. (F) If the torque is reduced and the speed is increased, the horsepower remains constant.

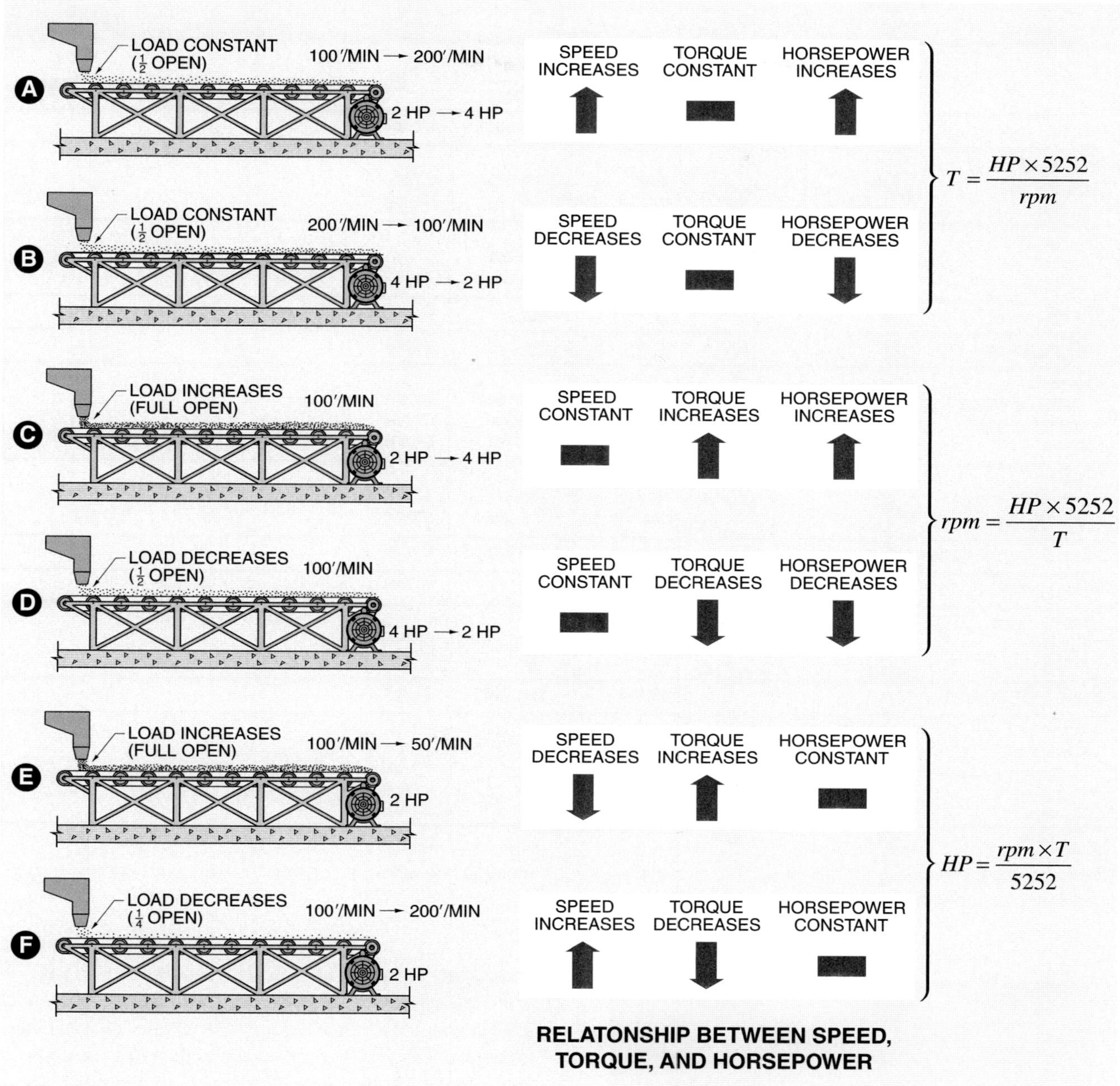

Figure 18-24. Operating speed, torque, and horsepower rating determine the work a motor can produce.

Motor Loads

Motor loads may require constant torque, variable torque, constant horsepower, or variable horsepower when operating at different speeds. Each motor type has its own ability to control different loads at different speeds.

The best type of motor to use for a given application depends on the type of load the motor must drive. Loads are generally classified as constant torque/variable horsepower (CT/VH), constant horsepower/variable torque (CH/VT), or variable torque/variable horsepower (VT/VH). **See Figure 18-25.**

> **Technical Fact**
>
> Single-phase motors are used if a fractional horsepower motor drive is required or if no 3φ power is available.

MOTOR LOADS				
Load	Motor Torque*		Classification	NEMA Motor Design
	LRT	PUT		
Ball Mill (mining)	125-150	175-200	CT/VH	C-D
Band Saws				
Production	50-80	175-225	CT/VH	C
Small	40	150	CT/VH	B
Car Pullers				
Automobile	150	200-225	CH/VT	C
Railroad	175	250-300	CH/VT	D
Chipper	60	225	CT/VH	B
Compressor (air)	60	150	VT/VH	B
Conveyors				
Unloaded at start	50	125-150	CT/VH	B
Loaded at start	125-175	200-250	CT/VH	C
Screw	100-125	50-175	CT/VH	C
Crushers				
Unloaded at start	75-100	150-175	CT/VH	B
With flywheel	125-150	175-200	CT/VH	D
Dryer, Industrial (loaded rotary drum)	150-175	175-225	CT/VH	D
Fan and Blower	40	150	VT/VH	B
Machine Tools				
Drilling	40	150	CT/VH	B
Lathe	75	150	CT/VH	B
Press (with flywheel)	50-100	250-350	CH/VT	D
Pumps				
Centrifugal	50	150	VT/VH	B
Positive displacement	60	175	CT/VH	B
Propeller	50	150	VT/VH	B
Vacuum	60	150	CT/VH	B-C

* in % of FLT

Figure 18-25. Loads are generally classified as constant torque/variable horsepower, constant horsepower/variable torque, or variable torque/variable horsepower.

Ruud Lighting, Inc.

A conveyor is a constant torque/variable horsepower load because the torque remains constant and any change in operating speed requires a change in horsepower.

Constant Torque/Variable Horsepower (CT/VH). A *constant torque/variable horsepower (CT/VH) load* is a load in which the torque requirement remains constant. Any change in operating speed requires a change in horsepower. **See Figure 18-26.** Constant torque/variable horsepower loads include loads that produce friction. Examples include conveyors, gear pumps and machines, metal-cutting tools, load-lifting equipment, and other loads that operate at different speeds.

Although the operating speed may change, a constant torque/variable horsepower load requires the same torque at low speeds as at high speeds. Since the torque requirement remains constant, an increase in speed requires an increase in horsepower.

Constant Horsepower/Variable Torque (CH/VT). A *constant horsepower/variable torque (CH/VT) load* is a load that requires high torque at low speeds and low torque at high speeds. Since the torque requirement decreases as speed increases, the horsepower remains constant. Speed and torque are inversely proportional in constant horsepower/variable torque loads. **See Figure 18-27.**

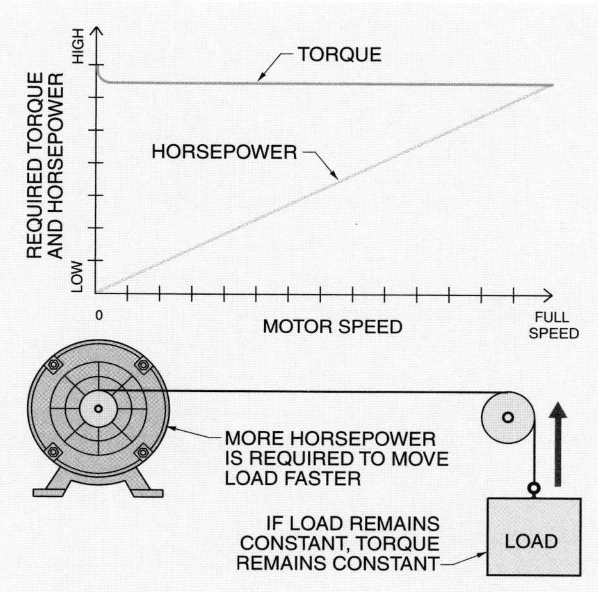

Figure 18-26. Constant torque/variable horsepower loads are loads in which the torque requirement remains constant.

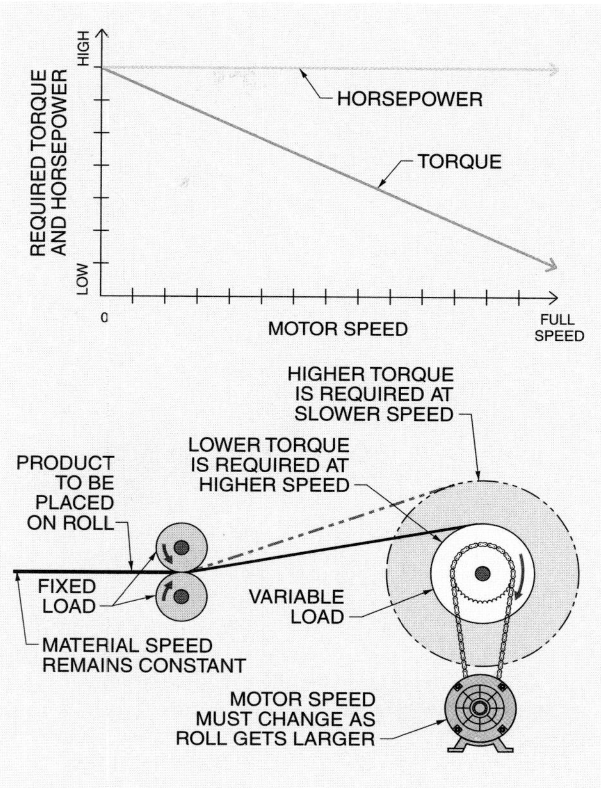

Figure 18-27. Constant horsepower/variable torque loads are loads that require high torque at low speeds and low torque at high speeds.

An example of a constant horsepower/variable torque load is a center-driven winder used to roll and unroll material such as paper or metal. Since the work is done on a varying diameter with tension and linear speed of the material constant, horsepower must also be constant. Although the speed of the material is kept constant, the motor speed is not. The diameter of the material on the roll that is driven by the motor is constantly changing as material is added. At the start, the motor must run at high speed to maintain the correct material speed while torque is kept at a minimum. As material is added to the roll, the motor must deliver more torque at a lower speed. As the material is rolled, both torque and speed are constantly changing while the motor horsepower remains the same.

Variable Torque/Variable Horsepower (VT/VH). A *variable torque/variable horsepower (VT/VH) load* is a load that requires a varying torque and horsepower at different speeds. **See Figure 18-28.** With this type of load, a motor must work harder to deliver more output at a faster speed. Both torque and horsepower are increased with increased speed. Examples of variable torque/variable horsepower loads include fans, blowers, centrifugal pumps, mixers, and agitators.

Heidelberg Harris, Inc.

Printing presses require an increase in torque and horsepower when the press speed is increased.

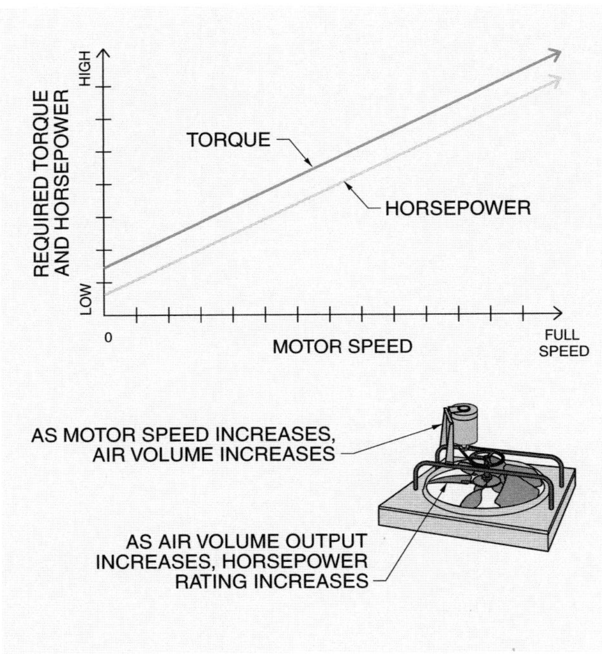

Figure 18-28. Variable torque/variable horsepower loads are loads that require different torque and horsepower at different speeds.

NEMA Design

Different motors are more suited for particular applications because each motor has its own characteristics of horsepower, torque, and speed. The basic characteristics of each motor are determined by the design of the motor and the supply voltage used. These designs are classified and given a letter designation, which can be found on the nameplate of some motors listed as NEMA design. **See Figure 18-29.**

MULTISPEED MOTORS

A motor's torque or horsepower characteristics change with a change in speed when motors are required to run at different speeds. The motor chosen depends on the application in which the motor is used. Once this selection is made, the motor is connected into the circuit. Common motor connection arrangements conforming to NEMA standards are used when wiring motors in a circuit. **See Figure 18-30.**

DC MOTOR SPEED CONTROL

DC motors are used in industrial applications that require variable speed control and/or high torque. DC motors are used in many acceleration and deceleration applications because the speed of most DC motors can be controlled smoothly and easily from 0 rpm to full speed.

In addition to having excellent speed control, DC motors are ideal in applications that require momentarily high torque outputs. This is because a DC motor can deliver three to five times its rated torque for short periods of time. Most AC motors stall with a load that requires twice their rated torque. Good speed control and high torque are the reasons DC motors are used in cranes, hoists, and large machine tools found in the mining industry.

Technical Fact

NEMA standards are of two classes: 1. *The NEMA Standard,* for current products subject to repetitive manufacture, and 2. *Suggested Standard for Future Design,* for future product developments.

MOTOR DESIGN CHARACTERISTICS

NEMA Design	Starting Torque	Starting Current	Breakdown Torque	Full-Load Slip	Typical Applications
A	Normal	Normal	High	Low	Machine tools, fans, centrifugal pumps
B	Normal	Low	High	Low	Machine tools, fans, centrifugal pumps
C	High	Low	Normal	Low	Loaded compressors, loaded conveyors
D	Very High	Low	—	High	Punch presses

Figure 18-29. Motor design characteristics are classified and given a letter designation, which can be found on the nameplate of some motors listed as NEMA design.

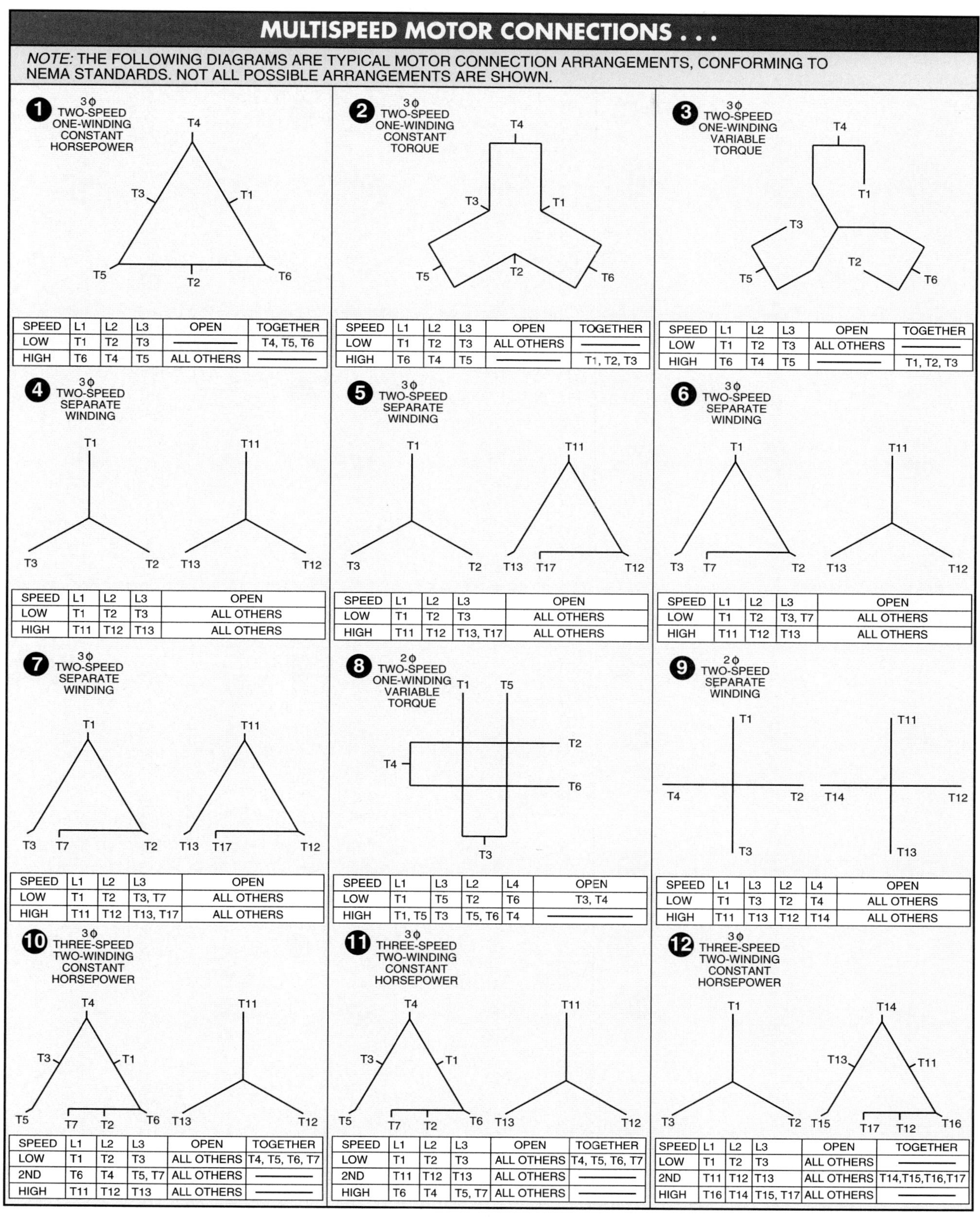

SPEED	L1	L2	L3	OPEN	TOGETHER
LOW	T1	T2	T3	———	T4, T5, T6
HIGH	T6	T4	T5	ALL OTHERS	———

SPEED	L1	L2	L3	OPEN	TOGETHER
LOW	T1	T2	T3	ALL OTHERS	———
HIGH	T6	T4	T5	———	T1, T2, T3

SPEED	L1	L2	L3	OPEN	TOGETHER
LOW	T1	T2	T3	ALL OTHERS	———
HIGH	T6	T4	T5	———	T1, T2, T3

SPEED	L1	L2	L3	OPEN
LOW	T1	T2	T3	ALL OTHERS
HIGH	T11	T12	T13	ALL OTHERS

SPEED	L1	L2	L3	OPEN
LOW	T1	T2	T3	ALL OTHERS
HIGH	T11	T12	T13, T17	ALL OTHERS

SPEED	L1	L2	L3	OPEN
LOW	T1	T2	T3, T7	ALL OTHERS
HIGH	T11	T12	T13	ALL OTHERS

SPEED	L1	L2	L3	OPEN
LOW	T1	T2	T3, T7	ALL OTHERS
HIGH	T11	T12	T13, T17	ALL OTHERS

SPEED	L1	L3	L2	L4	OPEN
LOW	T1	T5	T2	T6	T3, T4
HIGH	T1, T5	T3	T5, T6	T4	———

SPEED	L1	L3	L2	L4	OPEN
LOW	T1	T3	T2	T4	ALL OTHERS
HIGH	T11	T13	T12	T14	ALL OTHERS

SPEED	L1	L2	L3	OPEN	TOGETHER
LOW	T1	T2	T3	ALL OTHERS	T4, T5, T6, T7
2ND	T6	T4	T5, T7	ALL OTHERS	———
HIGH	T11	T12	T13	ALL OTHERS	———

SPEED	L1	L2	L3	OPEN	TOGETHER
LOW	T1	T2	T3	ALL OTHERS	T4, T5, T6, T7
2ND	T11	T12	T13	ALL OTHERS	———
HIGH	T6	T4	T5, T7	ALL OTHERS	———

SPEED	L1	L2	L3	OPEN	TOGETHER
LOW	T1	T2	T3	ALL OTHERS	———
2ND	T11	T12	T13	ALL OTHERS	T14,T15,T16,T17
HIGH	T16	T14	T15, T17	ALL OTHERS	———

Figure 18-30 continued...

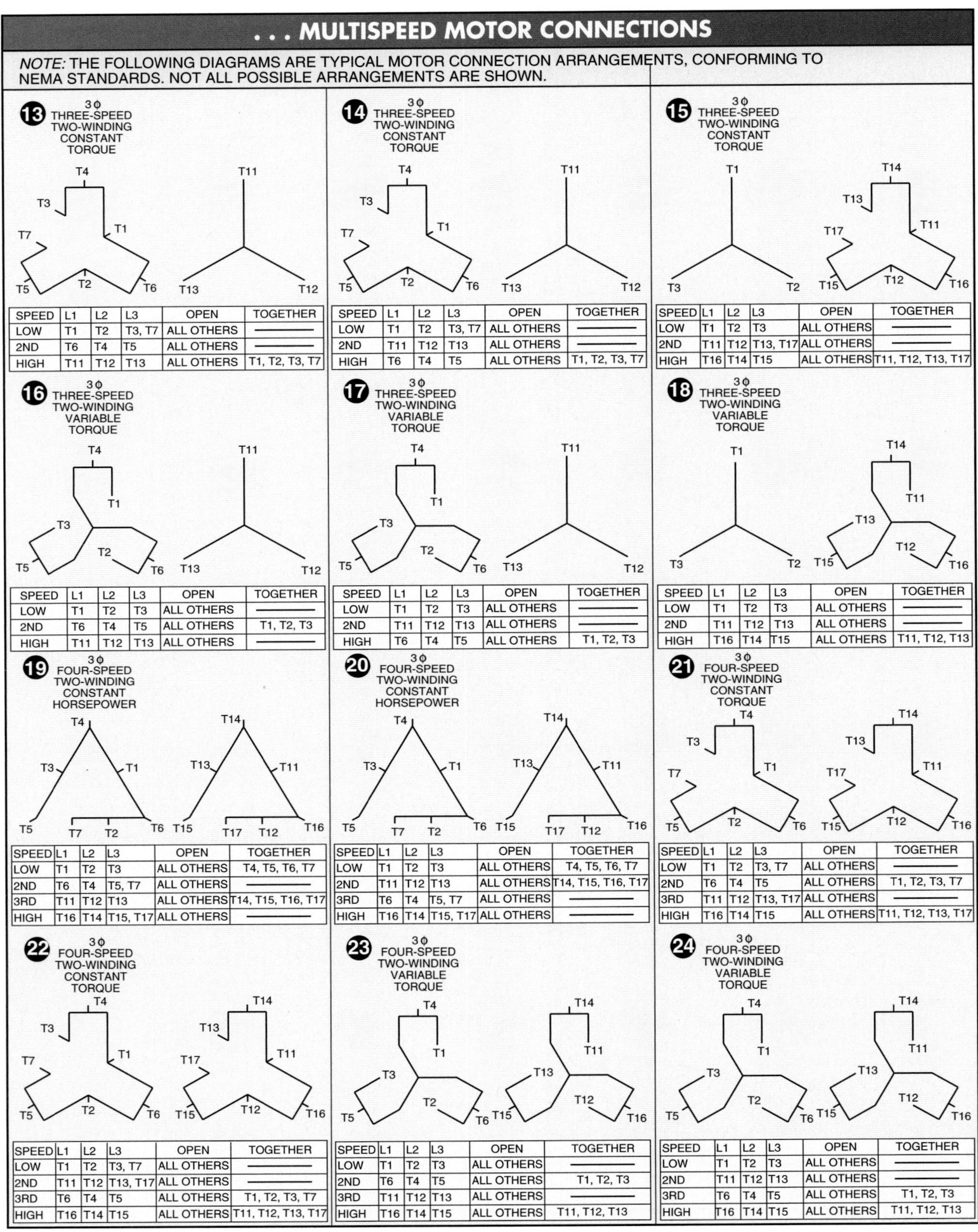

13

SPEED	L1	L2	L3	OPEN	TOGETHER
LOW	T1	T2	T3, T7	ALL OTHERS	—
2ND	T6	T4	T5	ALL OTHERS	—
HIGH	T11	T12	T13	ALL OTHERS	T1, T2, T3, T7

14

SPEED	L1	L2	L3	OPEN	TOGETHER
LOW	T1	T2	T3, T7	ALL OTHERS	—
2ND	T11	T12	T13	ALL OTHERS	—
HIGH	T6	T4	T5	ALL OTHERS	T1, T2, T3, T7

15

SPEED	L1	L2	L3	OPEN	TOGETHER
LOW	T1	T2	T3	ALL OTHERS	—
2ND	T11	T12	T13, T17	ALL OTHERS	—
HIGH	T16	T14	T15	ALL OTHERS	T11, T12, T13, T17

16

SPEED	L1	L2	L3	OPEN	TOGETHER
LOW	T1	T2	T3	ALL OTHERS	—
2ND	T6	T4	T5	ALL OTHERS	T1, T2, T3
HIGH	T11	T12	T13	ALL OTHERS	—

17

SPEED	L1	L2	L3	OPEN	TOGETHER
LOW	T1	T2	T3	ALL OTHERS	—
2ND	T11	T12	T13	ALL OTHERS	—
HIGH	T6	T4	T5	ALL OTHERS	T1, T2, T3

18

SPEED	L1	L2	L3	OPEN	TOGETHER
LOW	T1	T2	T3	ALL OTHERS	—
2ND	T11	T12	T13	ALL OTHERS	—
HIGH	T16	T14	T15	ALL OTHERS	T11, T12, T13

19

SPEED	L1	L2	L3	OPEN	TOGETHER
LOW	T1	T2	T3	ALL OTHERS	T4, T5, T6, T7
2ND	T6	T4	T5, T7	ALL OTHERS	—
3RD	T11	T12	T13	ALL OTHERS	T14, T15, T16, T17
HIGH	T16	T14	T15, T17	ALL OTHERS	—

20

SPEED	L1	L2	L3	OPEN	TOGETHER
LOW	T1	T2	T3	ALL OTHERS	T4, T5, T6, T7
2ND	T11	T12	T13	ALL OTHERS	T14, T15, T16, T17
3RD	T6	T4	T5, T7	ALL OTHERS	—
HIGH	T16	T14	T15, T17	ALL OTHERS	—

21

SPEED	L1	L2	L3	OPEN	TOGETHER
LOW	T1	T2	T3, T7	ALL OTHERS	—
2ND	T6	T4	T5	ALL OTHERS	T1, T2, T3, T7
3RD	T11	T12	T13, T17	ALL OTHERS	—
HIGH	T16	T14	T15	ALL OTHERS	T11, T12, T13, T17

22

SPEED	L1	L2	L3	OPEN	TOGETHER
LOW	T1	T2	T3, T7	ALL OTHERS	—
2ND	T11	T12	T13, T17	ALL OTHERS	—
3RD	T6	T4	T5	ALL OTHERS	T1, T2, T3, T7
HIGH	T16	T14	T15	ALL OTHERS	T11, T12, T13, T17

23

SPEED	L1	L2	L3	OPEN	TOGETHER
LOW	T1	T2	T3	ALL OTHERS	—
2ND	T6	T4	T5	ALL OTHERS	T1, T2, T3
3RD	T11	T12	T13	ALL OTHERS	—
HIGH	T16	T14	T15	ALL OTHERS	T11, T12, T13

24

SPEED	L1	L2	L3	OPEN	TOGETHER
LOW	T1	T2	T3	ALL OTHERS	—
2ND	T11	T12	T13	ALL OTHERS	—
3RD	T6	T4	T5	ALL OTHERS	T1, T2, T3
HIGH	T16	T14	T15	ALL OTHERS	T11, T12, T13

Figure 18-30. Common motor connection arrangements conforming to NEMA standards are used when wiring motors in a circuit.

DC Series Motors

A DC series motor produces high starting torque. **See Figure 18-31.** The field coil (series field) of the motor is connected in series with the armature. Although speed control is poor, a DC series motor produces very high starting torque and is ideal for applications in which the starting load is large. Applications include cranes, hoists, electric buses, streetcars, railroads, and other heavy-traction applications.

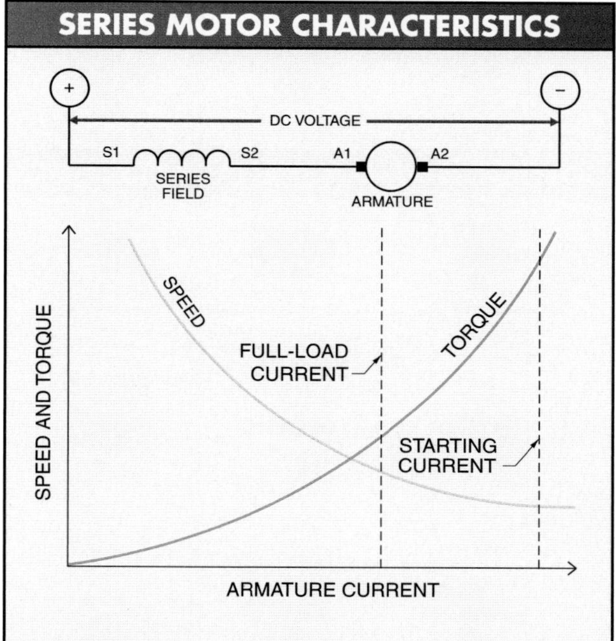

Figure 18-31. A DC series motor produces high starting torque.

The torque that is produced by a motor depends on the strength of the magnetic field in the motor. The strength of the magnetic field depends on the amount of current that flows through the series field. The amount of current that flows through a motor depends on the size of the load. The larger the load, the greater the current flow. Any increase in load increases current in both the armature and series field because the armature and field are connected in series. This increased current flow is what gives a DC series motor a high torque output.

In DC series motors, speed changes rapidly when torque changes. When torque is high, speed is low, and when speed is high, torque is low. This occurs because, as the increased current (created by the load) flows through the series field, there is a large flux increase. This increased flux produces a large counter electromotive force, which greatly decreases the speed of the motor. As the load is removed, the motor rapidly increases speed. Without a load, the motor would gain speed uncontrollably. In certain cases, the speed may become great enough to damage the motor. For this reason, a DC series motor should always be connected directly to the load, and not through belts, chains, etc.

The speed of a DC series motor is controlled by varying the applied voltage. Although the speed control of a series motor is not as good as the speed control of a shunt motor, not all applications require good speed regulation. The advantage of a high torque output outweighs good speed control in certain applications, such as the starter motor in automobiles.

DC Shunt Motors

In a DC shunt motor, the field coil (shunt field) is connected in parallel (shunt) with the armature. **See Figure 18-32.** DC shunt motors have good speed control and are used in applications that require a constant speed at any control setting, even with a changing load. A DC shunt motor is considered a constant speed motor for all reasonable loads.

Speed is determined by the voltage across the armature and the strength of the shunt field.

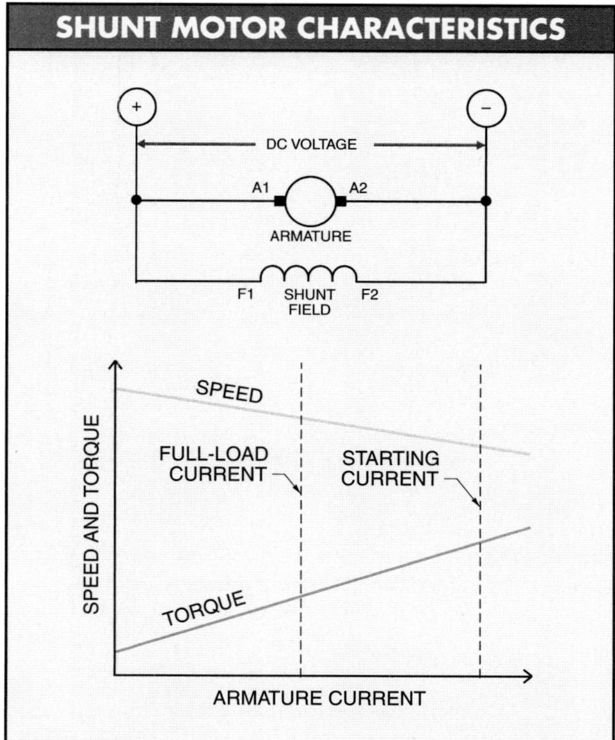

Figure 18-32. In a DC shunt motor, the field coil is connected in parallel with the armature.

In a DC shunt motor, if the voltage to the armature is reduced, the speed is also reduced. If the strength of the magnetic field is reduced, the motor speeds up. DC shunt motors speed up with a reduction in shunt field strength because with less field strength, there is less counter electromotive force developed in the armature. When the counter electromotive force is lowered, the armature current increases, producing increased torque and speed. To control the speed of a DC shunt motor, the voltage to the armature is varied or the shunt field current is varied.

A field rheostat or armature rheostat is used to adjust the speed of a DC shunt motor. **See Figure 18-33.** The rheostat is used to increase or decrease the strength of the field or armature. Once the strength of the field is set, it remains constant regardless of changes in armature current. As the load is increased on the armature, the armature current and torque of the motor increase. This slows the armature, but the reduction of counter electromotive force simultaneously allows a further increase in armature current and thus returns the motor to the set speed. The motor runs at a fairly constant speed at any control setting.

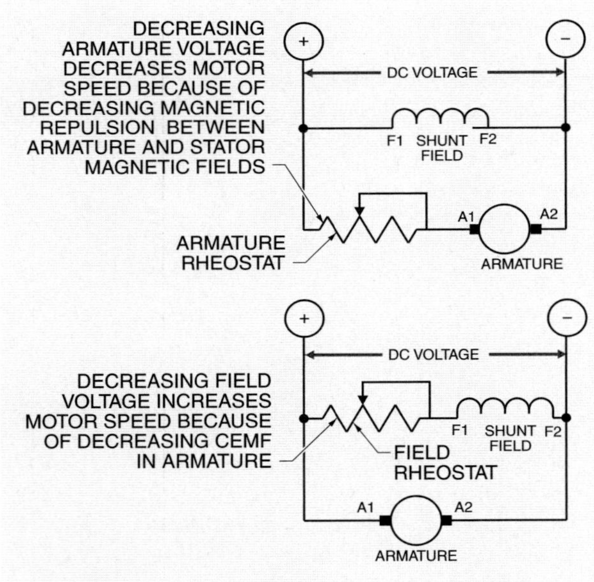

Figure 18-33. A field rheostat or armature rheostat is used to adjust the speed of a DC shunt motor.

A DC shunt motor has relatively high torque at any speed. The motor torque is directly proportional to the armature current. As armature current is increased, so is motor torque, with only a slight drop in motor speed.

DC Compound Motors

A DC compound motor has both a series coil (series field) and shunt coil (shunt field) connected in relationship to the armature. **See Figure 18-34.** A DC compound motor combines the operating characteristics of series and shunt motors. This produces the high starting torque characteristics of a series motor and good speed control characteristics of a shunt motor. However, a DC compound motor does not have as high starting torque as a series motor or as good speed regulation as a shunt motor. DC compound motors are used in applications where the load changes and precise speed control is not required. Applications include elevators, hoists, cranes, and conveyors.

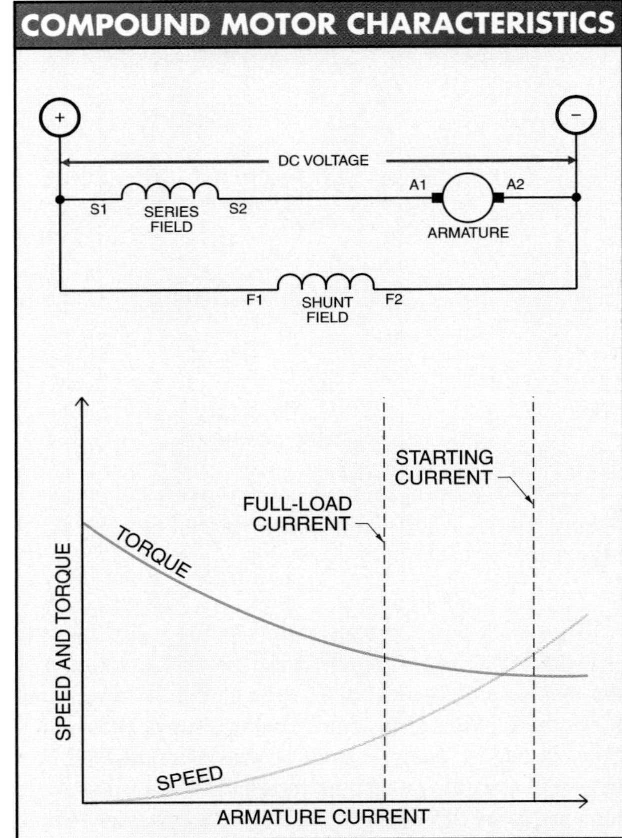

Figure 18-34. A DC compound motor combines the operating characteristics of series and shunt motors.

Speed control is obtained in a DC compound motor by changing the shunt field current strength or changing the voltage applied to the armature. This is accomplished by using a controller that uses resistors to reduce the applied voltage or by using a variable voltage supply.

DC Motor Solid-State Speed Control

Most DC motor speed controls produced today use silicon-controlled rectifiers (SCRs) instead of rheostats for control of DC motor speed. An SCR can perform the same function as resistance in controlling the voltage applied to a load. An SCR is similar to a rheostat (variable resistor) because it can be adjusted throughout its range. Advantages of SCRs over rheostats are that an SCR is smaller in size for the same rating, energy-efficient in not wasting power within itself, and less expensive than a rheostat.

Controlling DC Motor Base Speed

DC motor control is one of the best industrial applications of SCRs. In DC motor control applications, an SCR can be used to control the speed of a DC motor below the base speed by changing the amount of current that flows through the armature circuit.

Speed below the base speed is controlled by changing the armature voltage. In a DC motor, *base speed* is the speed (in rpm) at which the motor runs with full-line voltage applied to the armature and field. Base speed is listed on the motor nameplate.

The speed of a DC motor is controlled by varying the applied voltage across the armature and/or the field. When armature voltage is controlled, the motor delivers a constant torque characteristic. When field voltage is controlled, the motor delivers a constant horsepower characteristic. **See Figure 18-35.**

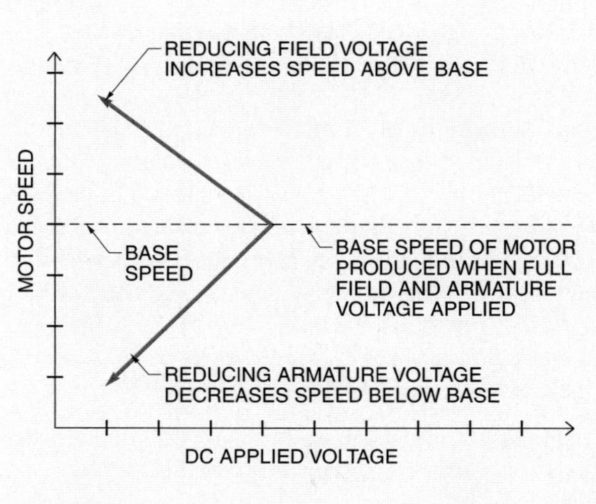

Figure 18-35. The speed of a DC motor is controlled by varying the applied voltage across the armature and/or the field.

An SCR is used to control the speed of a DC motor. **See Figure 18-36.** In this circuit, the speed is controlled from 0 rpm to the base speed using an SCR. The SCR is controlled by the setting of the gate trigger circuit, which varies the ON time of the SCR per cycle. This varies the amount of average current flow to the armature.

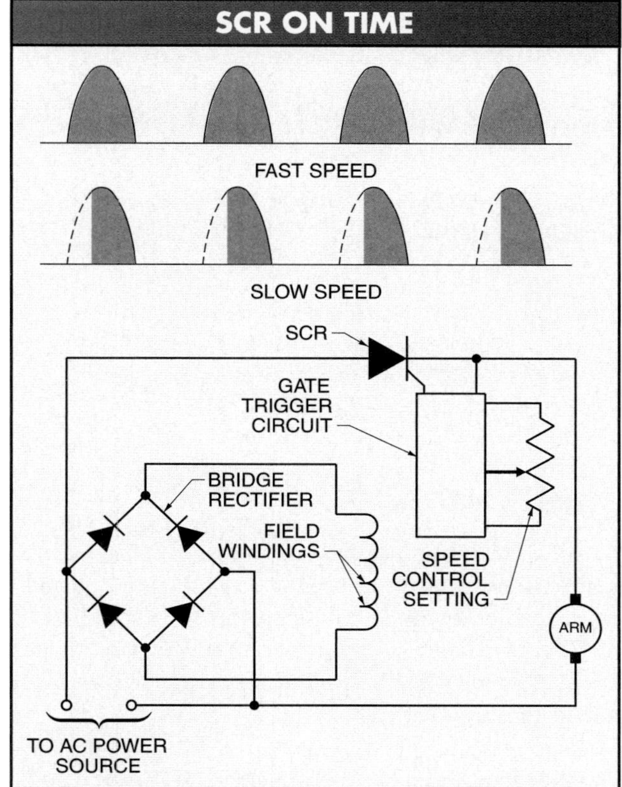

Figure 18-36. An SCR is used to control the speed of a DC motor.

The voltage applied to the SCR is AC because the SCR rectifies (as well as controls) AC voltage. A rectifier circuit is required for the field circuit because the field circuit must be supplied with DC. If speed control above the base is required, the rectifier circuit in the field can also be changed to an SCR control.

AC MOTOR SPEED CONTROL

AC motors are considered constant speed motors. This is because the synchronous speed of an induction motor is based on the supply frequency and the number of poles in the motor winding. Motors designed for 60 Hz use have synchronous speeds of 3600, 1800, 1200, 900, 720, 600, 514, and 450 rpm. To calculate synchronous speed of an induction motor, apply the following formula:

$$rpm_{syn} = \frac{120 \times f}{N_p}$$

where

rpm_{syn} = synchronous speed (in rpm)

f = supply frequency (in cycles/sec)

N_p = number of motor poles

Example: Calculating Synchronous Speed

What is the synchronous speed of a four-pole motor operating at 50 Hz?

$$rpm_{syn} = \frac{120 \times f}{N_p}$$

$$rpm_{syn} = \frac{120 \times 50}{4}$$

$$rpm_{syn} = \frac{6000}{4}$$

rpm_{syn} = **1500 rpm**

All induction motors have a full-load speed somewhat below their synchronous speed. *Percent slip* is the percentage reduction in speed below synchronous speed. Most motors run from 2% to 10% slower than their synchronous speed at no load. As the load increases, the percentage of slip increases.

Supply frequency and number of poles are the only variables that determine the speed of an AC motor. Unlike the speed of a DC motor, the speed of an AC motor should not be changed by varying the applied voltage. Damage may occur to an AC motor if the supply voltage is varied more than 10% above or below the rated nameplate voltage. This is because in an induction motor, the starting torque and breakdown torque vary as the square of the applied voltage. For example, with 90% of rated voltage, the torque is 81% ($.9^2$ = .81 or 81%) of its rated torque.

Voltage drop compensation is required in applications where a motor is located at the end of a long power line. The line voltage drop at the motor may be great enough to keep the motor from starting the load or developing the required torque to operate satisfactorily. The two methods of speed control available for AC motors are changing the frequency applied to the motor or using a motor with windings that may be reconnected to form different numbers of poles.

Multispeed AC motors, designed to be operated at a constant frequency, are provided with stator windings that can be reconnected to provide a change in the number of poles and thus a change in the motor speed. These multispeed motors are available in two or more fixed speeds, which are determined by the connections made to the motor. Two-speed motors normally have one winding that may be connected to provide two speeds, one of which is half the length of the other. Motors with more than two speeds normally include many windings that are connected and reconnected to provide different speeds by changing the number of poles.

In multispeed motors, the different speeds are determined by connecting the external winding leads to a multispeed starter. Although a multispeed starter may be a manual starter, a magnetic motor starter is the most common means used for AC motor speed control. One starter is required for each speed of the motor. Each starter must be interlocked (using mechanical, auxiliary contact, or pushbutton interlocking) to prevent more than one starter from being ON at the same time. The motor can run at only one speed at a time. For two-speed motors, a standard forward/reverse starter is normally used because it provides mechanical interlocking.

Basic Speed Control Circuits

Several control circuits can be developed to control a multispeed motor, depending on the requirements of the circuit. In a two-speed motor control circuit, the motor can be started in the low or high speed. **See Figure 18-37.** In this circuit, the operator may start the motor from rest at either speed. The stop pushbutton must be pressed before changing from low to high speed or from high to low speed.

In a modified control circuit, the motor can be changed from low speed to high speed without first stopping the motor. **See Figure 18-38.** In this circuit, the operator can start the motor from rest at either speed or change from low speed to high speed without pressing the stop pushbutton. The stop pushbutton must be pressed before it is possible to change from high to low speed. This high-to-low arrangement prevents excessive line current and shock to the motor. In addition, the machinery driven by the motor is protected from shock that could result from connecting a motor at high speed to low speed.

Compelling Circuit Logic. In many speed control applications, a motor must always be started at low speed before it can be changed to high speed. *Compelling circuit logic* is a control function that requires the operator to start and operate a motor in a predetermined order. **See Figure 18-39.**

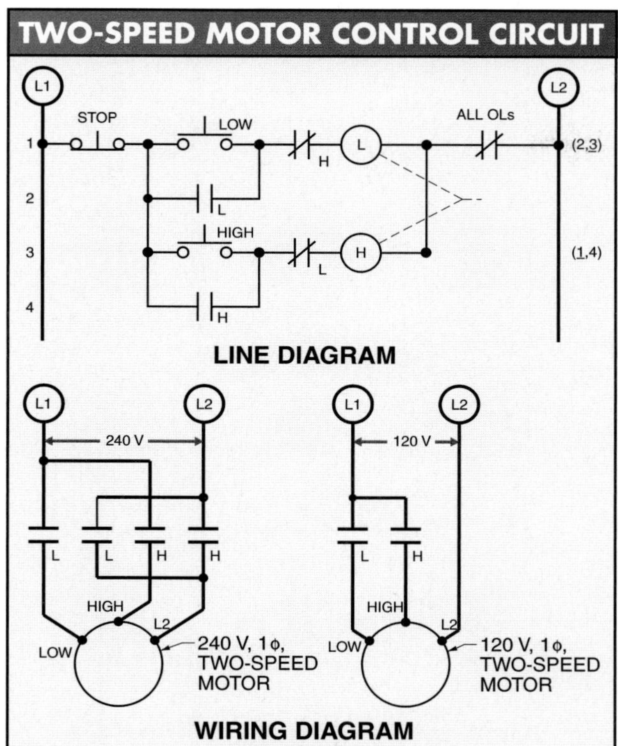

Figure 18-37. In a two-speed motor control circuit, the motor can be started in the low or high speed.

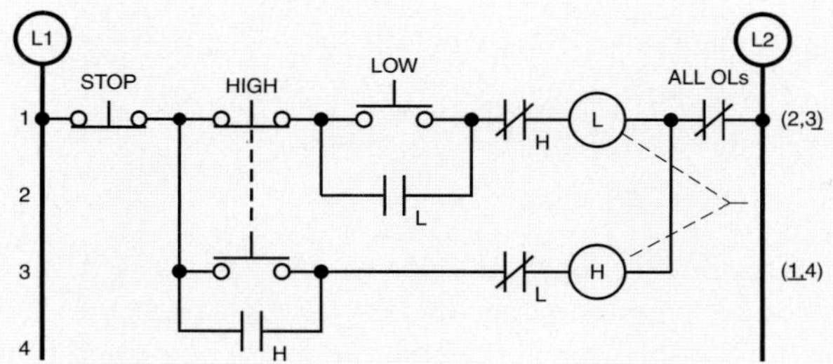

Figure 18-38. In a modified control circuit, the motor can be changed from low speed to high speed without first stopping the motor.

This circuit does not allow the operator to start the motor at high speed. The circuit compels the operator to first start the motor at low speed before changing to high speed. This arrangement prevents the motor and driven machinery from starting at high speed. The motor and driven machinery are allowed to accelerate to low speed before accelerating to high speed. The circuit also compels the operator to press the stop pushbutton before changing speed from high to low.

Accelerating Circuit Logic. In many applications, a motor must be automatically accelerated from low to high speed even if the high pushbutton is pressed first. *Accelerating circuit logic* is a control function that permits the operator to select a high motor speed and the control circuit automatically accelerates the motor to that speed. **See Figure 18-40.**

A circuit with accelerating circuit logic allows the operator to select the desired speed by pressing either the low or high pushbutton. If the operator presses the low pushbutton, the motor starts and runs at low speed. If the operator presses the high pushbutton, the motor starts at low speed and runs at high speed only after the predetermined time set on the timer in the control circuit. This arrangement gives the motor and driven machinery a definite time period to accelerate from low to high speed. The circuit also requires that the operator press the stop pushbutton before changing speed from high to low. In any control circuit, the pushbuttons may be replaced with any control device such as a pressure switch, photoelectric switch, etc., without changing the circuit logic.

Decelerating Circuit Logic. In some applications, a motor or load cannot take the stress of changing from a high to a low speed without damage. In these applications, the motor must be allowed to decelerate by coasting or braking before being changed to a low speed. *Decelerating circuit logic* is a control function that permits the operator to select a low motor speed and the control circuit automatically decelerates the motor to that speed. **See Figure 18-41.**

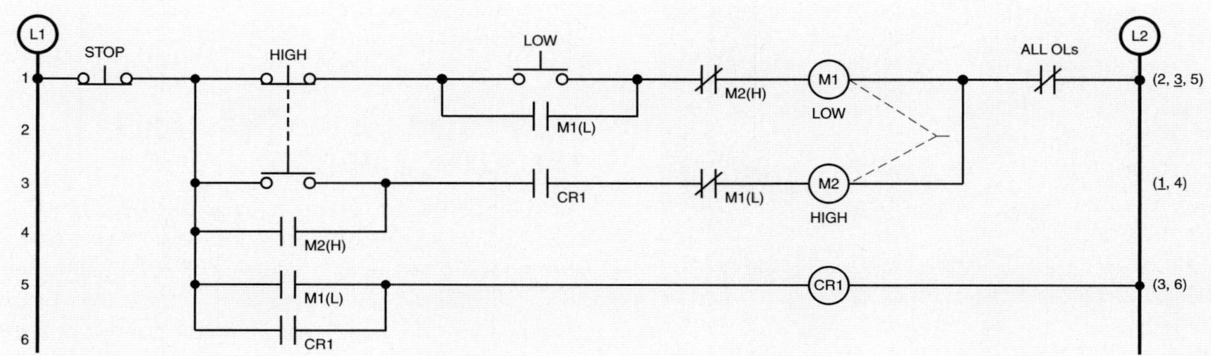

Figure 18-39. Compelling circuit logic is used where a motor must always be started at low speed before it can be changed to high speed.

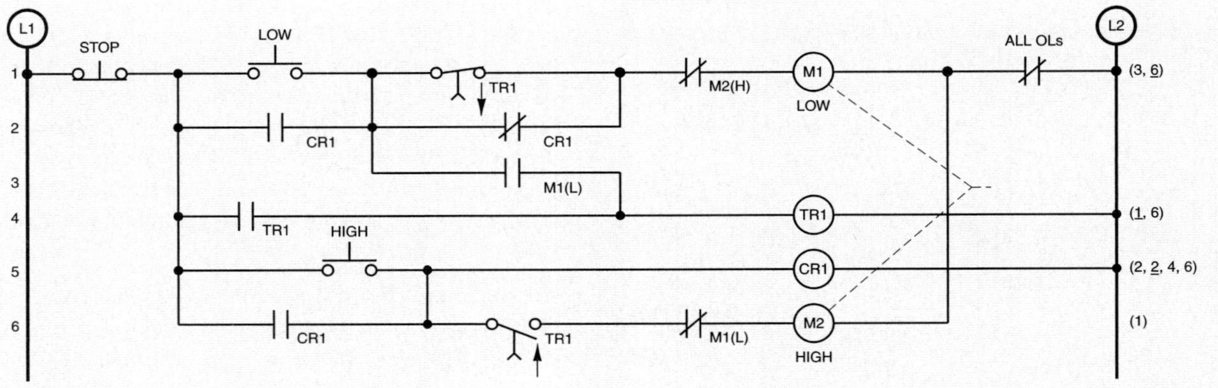

Figure 18-40. In accelerating circuit logic, a motor is automatically accelerated from low speed to high speed even if the high pushbutton is pressed first.

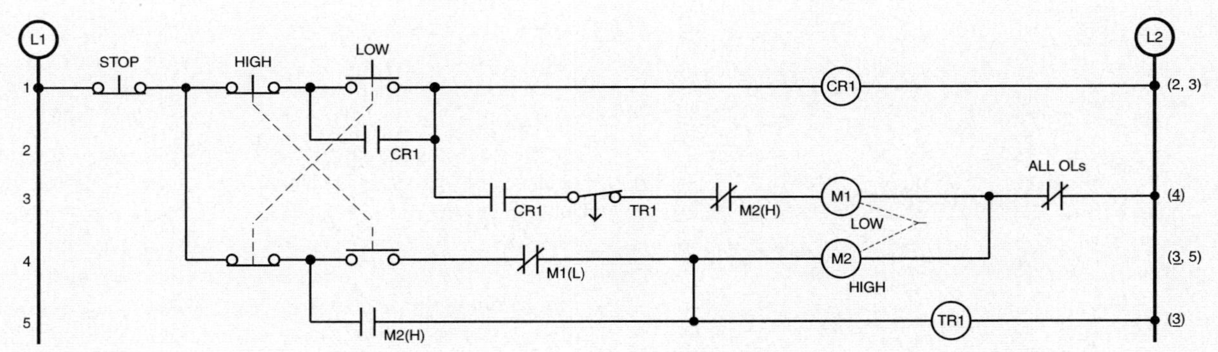

Figure 18-41. In decelerating circuit logic, a motor decelerates before being changed to a low speed.

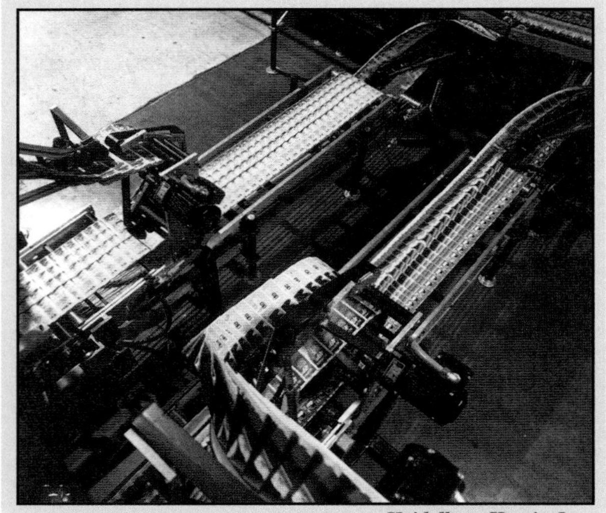
Heidelberg Harris, Inc.

Material handling systems may require controlled stopping and starting methods.

This circuit allows the operator to select the desired speed by pressing either the low or high pushbutton. If the low pushbutton is pressed, the motor starts and runs at low speed. If the high pushbutton is pressed, the motor starts and runs at high speed. If the operator changes from high to low speed, the motor changes to low speed only after a predetermined time. This gives the motor and driven machinery a time period to decelerate from a high speed to a low speed.

AC MOTOR DRIVES (VARIABLE FREQUENCY DRIVES)

AC motor drives are used to change the speed of AC motors by changing the frequency of the voltage applied to the motor. Changing the frequency of the voltage applied to an AC motor changes the speed of the motor. In addition to controlling motor speed, motor drives can control motor acceleration time, deceleration time, motor torque, and motor braking.

A motor drive changes the frequency of the voltage applied to a motor by taking the incoming AC voltage, converting it to a DC voltage, and inverting it back to an AC voltage in which the frequency can be changed. An *inverter* is an electronic device that changes DC voltage into AC voltage. A *converter* is an electronic device that changes AC voltage to DC voltage. **See Figure 18-42.**

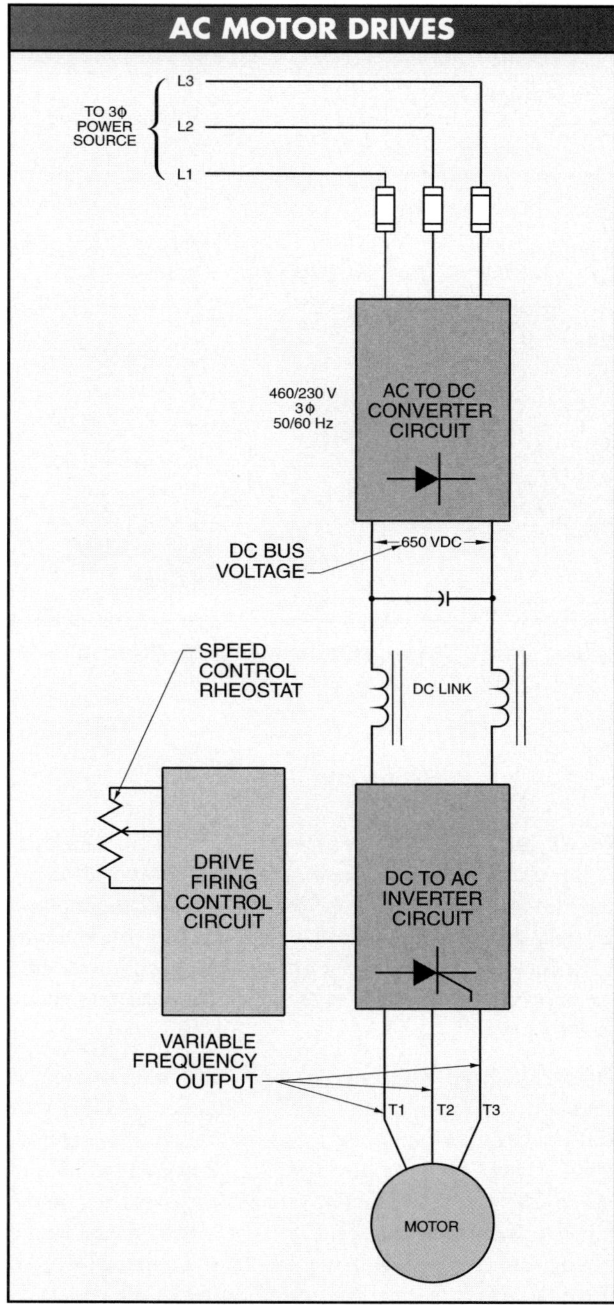

Figure 18-42. An inverter circuit changes DC power to a variable frequency AC output that controls the speed of a motor.

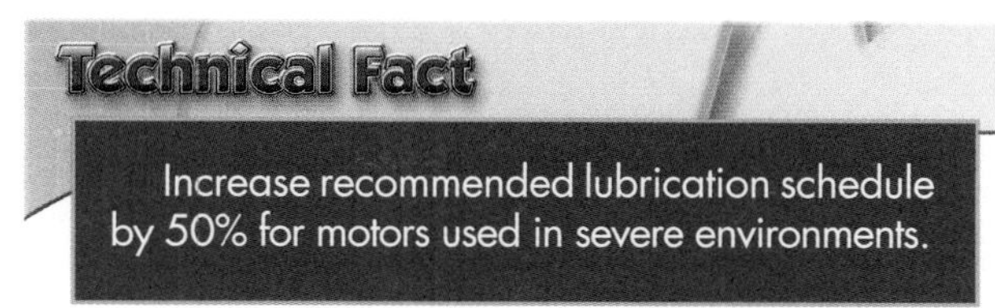

Motor Stopping Methods

A motor coasts to a stop when disconnected from the power supply. The length of time required for a motor to come to rest depends on the inertia of the moving parts (motor and motor load) and friction. *Inertia* is the property of matter by which a mass persists in its state of rest or motion until acted upon by an external force. *Friction* is the resistance to motion that occurs when two surfaces slide against each other.

Motor braking is used when a motor must be stopped more quickly than coasting allows. The braking method used depends on the application. For example, applications that use motors for driving fans, blowers, pumps, and mixers normally do not require braking. However, applications that use motors for material handling (conveyors, packaging, etc.), monorail systems, indexing machines, grinders, lathes, and assembly machines, normally require some degree of braking force.

An AC motor drive decelerates a motor at a controlled rate by placing an electric load on the motor. The advantage in using a motor drive to apply a braking force is that maintenance is kept to a minimum because there are no parts that come in contact during braking. Applications that require a braking force to hold a load for a period of time after the motor has stopped may use the motor drive to stop the motor and a friction brake to hold the motor shaft/load. Friction brakes may also be used with motor drive braking in applications that require an emergency stop function.

AC motor drives can be programmed for different braking (stopping) methods. Common drive stopping methods include ramp stop, coast stop, DC brake stop, and soft stop (S-curve) stopping methods. **See Figure 18-43.**

Ramp Stop. *Ramp stop* is a stopping method in which the level of voltage applied to a motor is reduced as the motor decelerates. Ramp stop is normally the factory (default) setting for controlling the stopping of a motor. When the motor drive receives a stop command, the drive maintains control of the motor speed by controlling the voltage on the motor stator. This allows for a smooth stop from any speed. The length of time the motor drive takes to stop the motor is controlled by the deceleration time parameter on the motor drive. The default setting for the deceleration is normally 10 sec, but can be set from a few seconds (or less) to several minutes. **See Figure 18-44.**

MOTOR DRIVE STOPPING METHODS

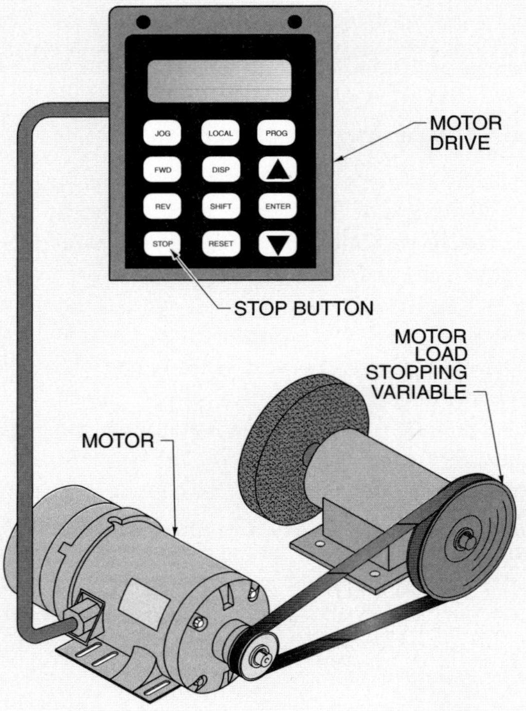

STOPPING METHOD	LOAD DETERMINES STOPPING TIME	DRIVE DETERMINES STOPPING TIME
RAMP		X
COAST	X	
DC BRAKE		X
S-CURVE		X

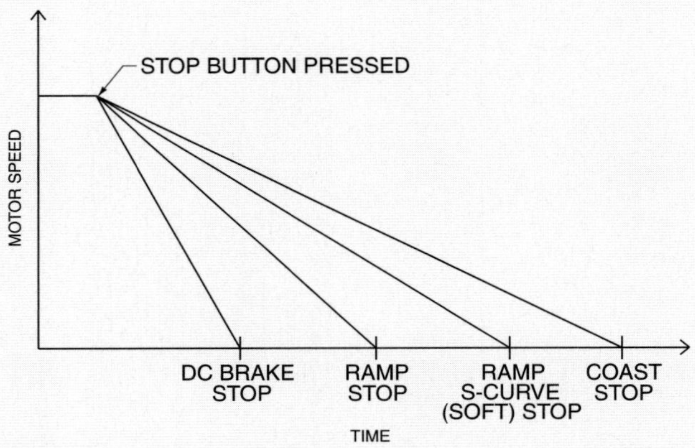

Figure 18-43. Motor drive stopping methods include ramp stop, coast stop, DC brake stop, and soft stop (S-curve) stopping methods.

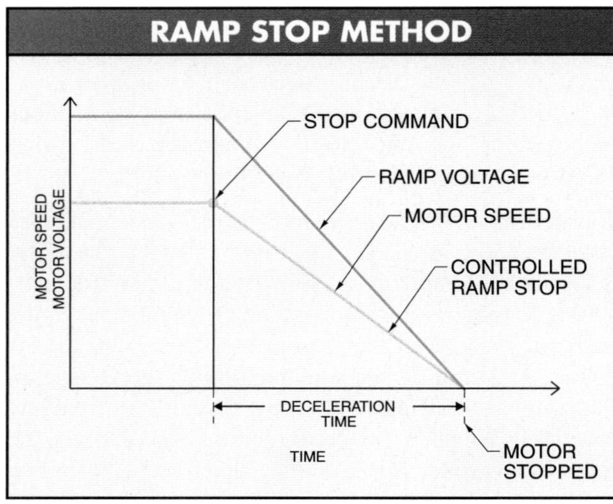

Figure 18-44. A ramp stop provides a smooth, controlled deceleration by reducing the voltage applied to a motor.

Coast Stop. *Coast stop* is a stopping method in which the motor drive shuts OFF the voltage to a motor, allowing the motor to coast to a stop. When the coast stop method is used, the drive does not have any control of the motor after the stop command is entered. The length of time the motor takes to stop depends on the load connected to the motor.

DC Brake Stop. *DC brake stop (DC injection braking)* is a stopping method in which a DC voltage is applied to the stator winding of a motor after a stop command is entered. Unlike ramp stopping, the applied DC voltage is held at the level entered into the DC hold volts parameter on the motor drive. The applied DC voltage is maintained on the motor stator for the length of time entered into the DC hold time parameter on the drive. **See Figure 18-45.**

The DC brake stop method provides a fast stop and can apply a braking force to hold the motor after the motor is stopped. However, the DC brake stop method cannot take the place of a friction brake in applications that require the motor shaft to be held for a long period of time after stopping. This is because the DC hold time parameter normally has a maximum setting of approximately 15 sec. A longer time period produces excessive heat in the motor windings.

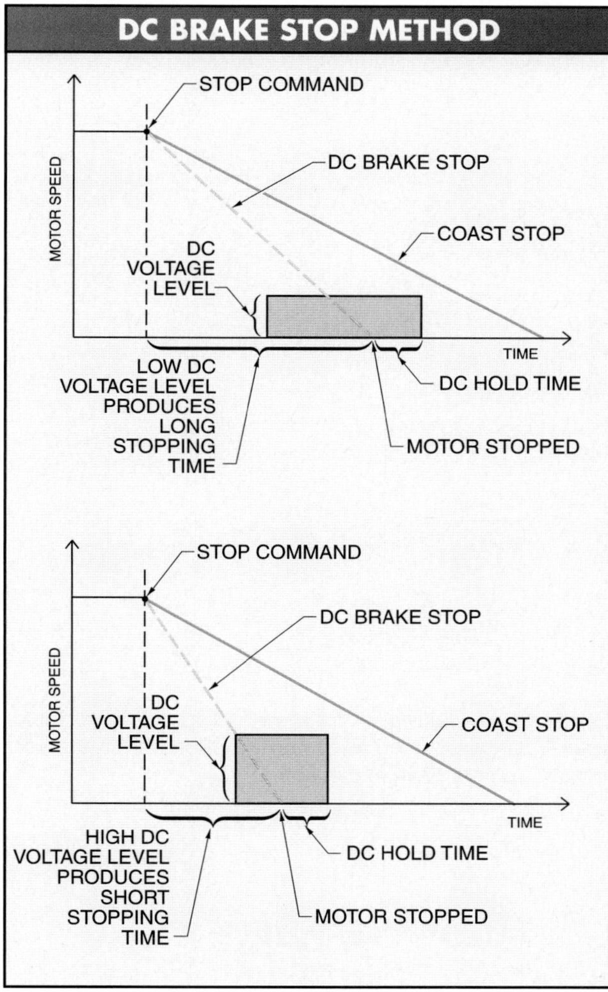

Figure 18-45. In the DC brake stop method, the DC hold level (amount of applied voltage) determines the motor stopping time.

Soft Stop (S-Curve). *Soft stop (S-curve)* is a stopping method in which the programmed deceleration time is doubled and the stop function is changed from a ramp slope to an S-curve slope. The soft stop method provides a very soft stopping operation. A very soft stop may be required to prevent light loads (such as empty cans or bottles) from tipping. In such applications, the motor drive can be reprogrammed for a soft stop (S-curve) stopping method. **See Figure 18-46.**

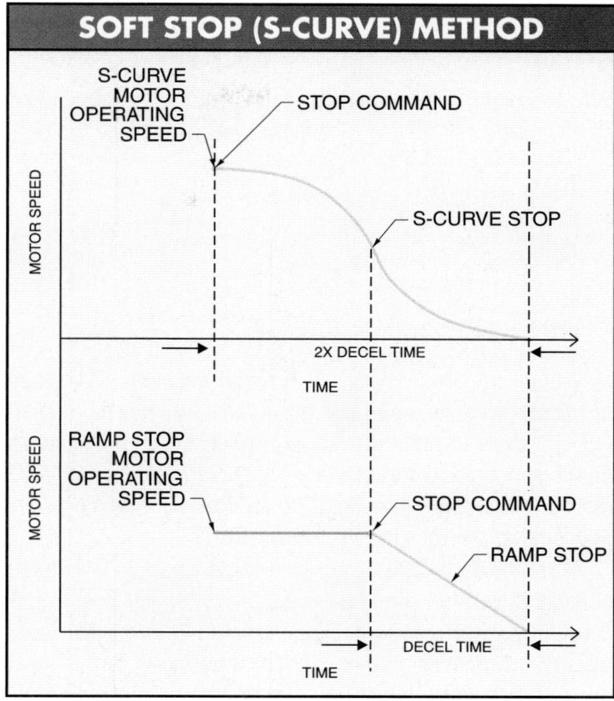

Figure 18-46. The soft stop (S-curve) stopping method has a doubled stopping time and an S-curve slope reduction in voltage.

Voltage and Frequency

The voltage applied to the stator of an AC motor must be decreased by the same amount as the frequency. The motor heats excessively and damage occurs to the windings if the voltage is not reduced when frequency is reduced. The motor does not produce its rated torque if the voltage is reduced more than required. The ratio between the voltage applied to the stator and the frequency of the voltage applied to the stator must be constant. This ratio is referred to as the volts-per-hertz (V/Hz) ratio (constant volts/hertz characteristic). *Volts per hertz (V/Hz)* is the relationship between voltage and frequency that exists in a motor. The motor develops rated torque if this relationship is kept constant (linear).

The volts-per-hertz ratio for an induction motor is found by dividing the rated nameplate voltage by the rated nameplate frequency. To find the volts-per-hertz ratio for an AC induction motor, apply the following formula:

$$V/Hz = \frac{V}{Hz}$$

where

V/Hz = volts-per-hertz ratio

V = rated nameplate voltage (in V)

Hz = rated nameplate frequency (in Hz)

Example: Calculating Volts-per-Hertz Ratio

What is the volts-per-hertz ratio if a motor nameplate rates an AC motor for 230 VAC, 60 Hz operation?

$$V/Hz = \frac{V}{Hz}$$

$$V/Hz = \frac{230}{60}$$

$V/Hz =$ **3.83**

Above approximately 15 Hz, the amount of voltage needed to keep the volts-per-hertz ratio linear is a constant value. Below 15 Hz, the voltage applied to the motor stator may be boosted to compensate for the large power loss AC motors have at low speed. The amount of voltage boost depends on the motor. **See Figure 18-47.**

A motor drive can be programmed to apply a voltage boost at low motor speeds to compensate for the power loss at low speeds. The voltage boost gives the motor additional rotor torque at very low speeds. The amount of torque boost depends on the voltage boost programmed into the motor drive. The higher the voltage boost, the greater the motor torque. **See Figure 18-48.**

Motor drives can also be programmed to change the standard linear volts-per-hertz ratio to a nonlinear ratio. A nonlinear ratio produces a customized motor torque pattern that is required by the load operating characteristics. For example, a motor drive can be programmed for two nonlinear ratios that can be applied to fan or pump motors. Fans and pumps are normally classified as variable torque/variable horsepower loads. Variable torque/variable horsepower loads require varying torque and horsepower at different speeds.

Acceleration and Deceleration Times

One of the main advantages of a motor drive is that it can control the exact speed of a motor. Controlling the speed of a motor also controls the speed of the load connected to the motor. To control motor speed, the speed must be controlled during acceleration, running, and deceleration.

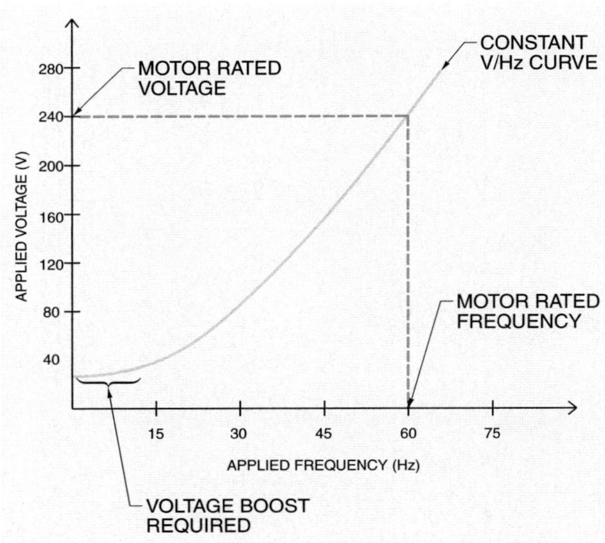

Figure 18-47. Below 15 Hz, the voltage applied to the motor stator may be boosted to compensate for the large power loss AC motors have at low speed.

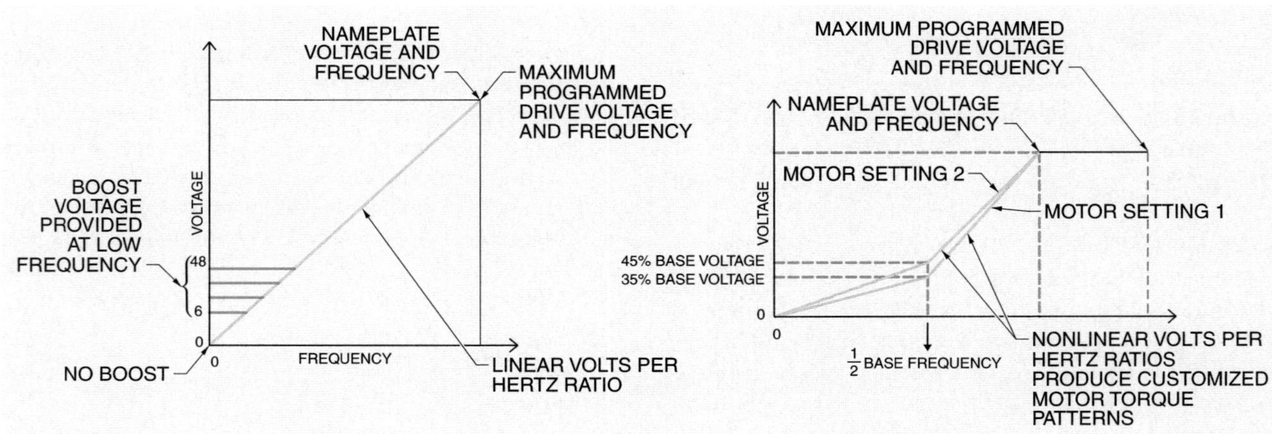

Figure 18-48. Motor drives can be programmed to apply a voltage boost at low motor speeds and to change the standard linear volts-per-hertz ratio to a nonlinear ratio.

In many applications, motor acceleration and deceleration times must be considerable. For example, monorail systems, storage and retrieval systems, assembly conveyors, material handling systems, lathes, and other tooling machines require a considerably long and smooth acceleration and deceleration time. Motor drives allow motor acceleration and deceleration speed to be programmed to follow a standard curve, S-curve, or other customized curve. **See Figure 18-49.**

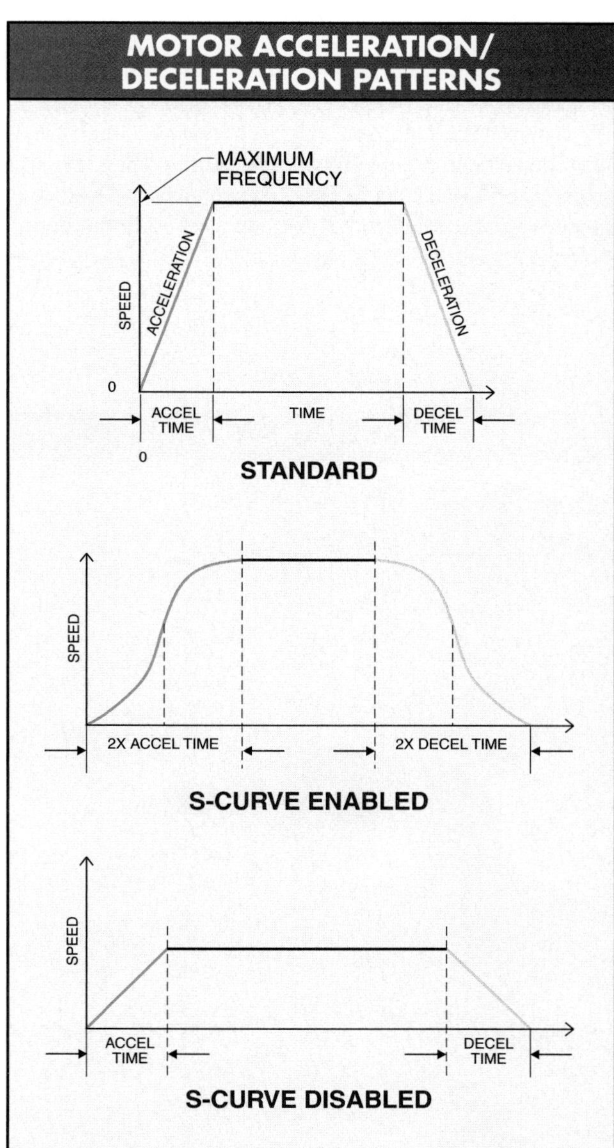

Figure 18-49. Motor drives allow motor acceleration and deceleration speed to be programmed to follow a standard curve, S-curve, or other customized curve.

Frequency Source Selection

Motor drives can normally be controlled by two or three different methods (locations). Methods by which a motor drive can be controlled include local control, remote control, and PLC/PC/HMI control. Once one of the three methods is selected, the other two are not operable. **See Figure 18-50.**

Local Control. In the local control method, motor frequency (speed) is controlled from the motor drive keypad. The arrow up/down keys on the motor drive are used to control motor speed. The local control method is automatically set when power is first applied to a motor drive. To change motor speed selection from local control, the drive must be reprogrammed for remote control.

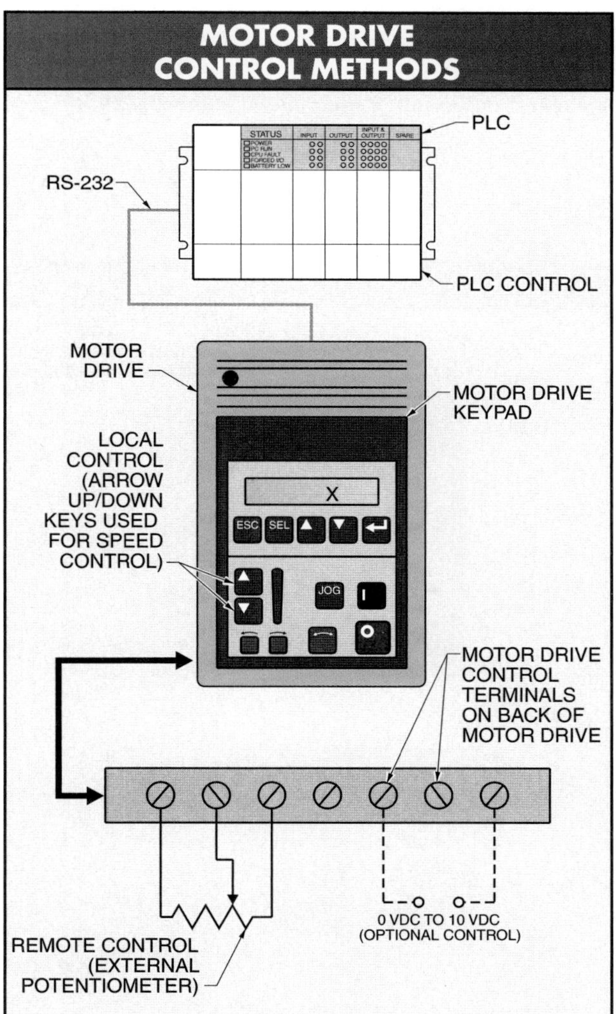

Figure 18-50. Motor drive control methods include local control, remote control, and PLC/PC/HMI control.

Remote Control. In the remote control method, motor frequency (speed) is controlled by an external potentiometer connected at the motor drive control terminals. However, motor drives can be set to change motor speed based on several different variables such as 0 V to 10 V, 4 mA to 20 mA, or a variable resistance (thermistor, photoresistor, etc.). Most motor drives also allow a set number of external switches to be used to preselect motor speeds.

A motor drive can control the speed of a motor over a wide range of operating speeds. Some applications require motor speed to be operable over a wide range. In other applications, only a few predetermined motor speeds are required. External control switches are used to determine which of the different speeds are operable. The external control switches are normally referred to as SW1, SW2, etc. These switches are connected to the drive input section. **See Figure 18-51.**

The external control switches that can be used to select a programmed speed are selector switches, key-operated switches, or any other maintained contact switch. Limit switches, timer contacts, counter contacts, and any other contacts that can be held closed can also be used. Momentary contacts are not normally used because the switch contacts must be held closed to set the motor speed to the preprogrammed speed setting.

Preselected motor speeds simplify the operation of the system. For example, SW1, SW2, and SW3 can be labeled to state the operating function of each switch. For example, switches could be labeled as slow speed, fast speed, normal operation, system setup speed, test speed, etc. In addition to simplification for the operator, labeling the switches better defines system operation.

PLC/PC/HMI Control. Some motor drives allow a programmable controller (PLC), personal computer (PC), or human-machine interface (HMI) to control the speed of a motor. Any one of these devices can be connected to the SW1, SW2, etc., terminal connections. Other drives provide a separate port that allows infinite speed control from the external control unit. On such drives, frequency (speed) is controlled through a standard communication port such as an RS-232C port.

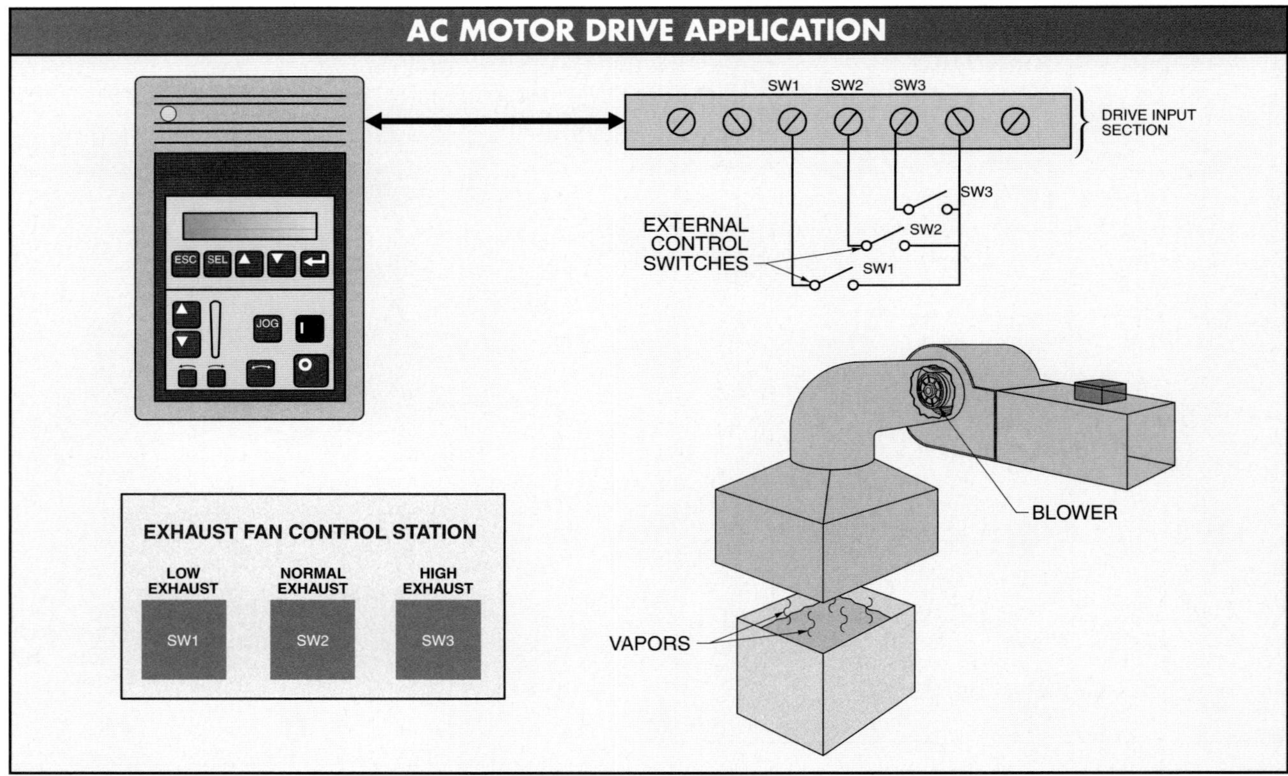

Figure 18-51. A motor drive can control the speed of a motor over a wide range of operating speeds through external control switches connected to the drive input section.

Inverter Duty Motors

AC drives can control the speed of a motor and the speed of the load the motor is driving. An *inverter duty motor* is an electric motor specifically designed to work with AC motor drives. AC motor drives create insulation problems for standard electric motors because motor drives produce voltage spikes. The voltage spikes create heat that damages the insulation and shortens the life expectancy of standard electric motors. Inverter duty motors are designed to withstand the heat created by voltage spikes.

Most motors are self-cooled as air flows over the motor windings. The air is forced over the windings by the motor rotation. Motors can be adequately cooled, even when operated at low speed. However, a motor can overheat if it is operating at maximum torque (or close to it) and the motor cooling is reduced at low speeds. Overheating destroys the motor insulation.

Even if the motor is properly cooled, the high switching frequencies and fast voltage rise times produced by a motor drive can destroy standard motor insulation. Motor manufacturers produce motors designed specifically for use with motor drives. These motors are made with special voltage-spike-resistant wire. Although a drive can be used to control a standard motor, it is best to use motors specifically designed for motor drive use.

Changing Applied Voltage

The speed of standard AC squirrel-cage induction motors is normally varied by changing the number of poles or the applied frequency. Although the speed of an AC motor is determined by the number of poles and applied frequency, it is possible in some applications to control the speed of a load by varying the voltage applied to the motor. This method is not a standard method of speed control and caution must be taken in applying it. By varying the voltage applied to the motor, the torque that the motor can deliver to the load is varied. The torque of a squirrel-cage induction motor varies with the square of the applied voltage.

The greater the torque, the faster the acceleration time. If the torque of a motor is reduced, the speed at which the motor performs the work is also reduced. Although it is possible to reduce the speed of a large motor by reducing the applied voltage, this method could damage the motor.

This damage may come from excess heat buildup in the motor. Most AC motors are not designed to have their voltage varied more than 10% from the nameplate rating. However, manufacturers that know the load requirements and the motor type and size in advance may install this type of speed control. The advantage of less cost for control with a large motor is the determining factor. This type of speed control is limited to applications of soft-start light loads. Fan motors are sometimes controlled this way. Shaded-pole or permanent-magnet motors are normally used in these applications. Except in applications that are specifically designed for this type of speed control, it should not be considered a standard method.

Most variable voltage control circuits use a full-wave triac output to vary the voltage. **See Figure 18-52.** The triac varies the voltage by adjusting the point on the AC sine wave at which the triac is turned ON.

The triac in this application is similar to the SCR used to control the speed of a DC motor. A triac consists of two reverse-parallel-connected SCRs connected to allow the AC sine wave to pass in both directions at a controlled level. The triac output is controlled by varying a potentiometer in the gate triggering circuit. This same basic triac circuit is also used to control the heat or light output of heating elements and incandescent lamps.

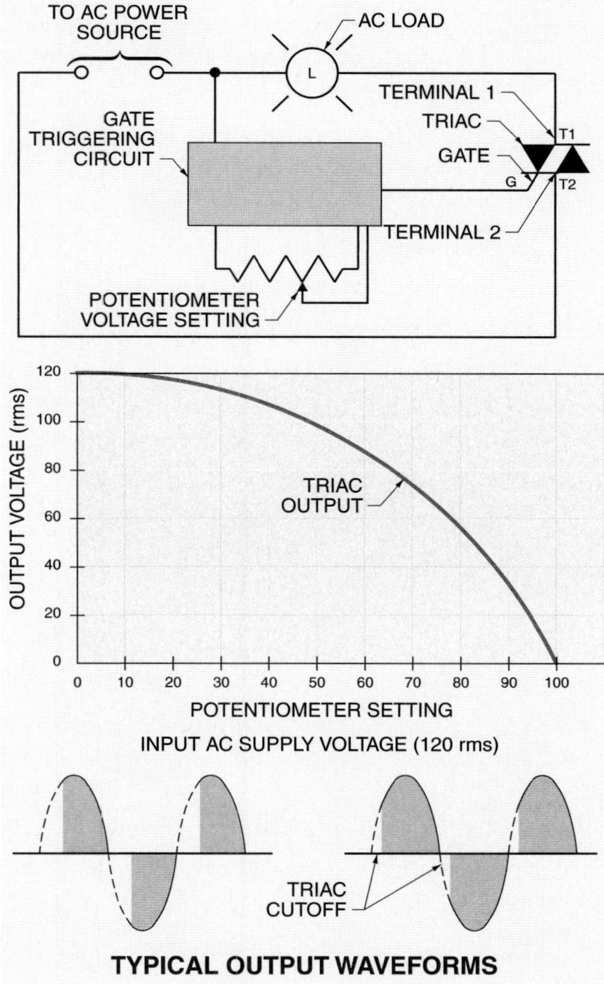

Figure 18-52. Most variable voltage control circuits use a full-wave triac output to vary the voltage.

Motor Mechanical Drives

A standard induction motor runs at a constant speed for a given frequency and number of poles. The most common and economical running speed of an induction motor is about 1800 rpm. Lower speeds require the addition of poles to reduce the speed. Because there are many applications that require some speed other than 1800 rpm, but not a variable speed control, some means must be provided to match the motor output speed to the lower or higher speed required by the load without changing the running speed of the motor. This is accomplished with belts, chains, or gear drives. These belts, chains, or gear drives are used for smooth speed changes between the motor drive and machine drive.

A pulley can be used to change the output speed of a motor, provided the manufacturer limits are not exceeded. To determine the pulley size needed, the speed of the motor (drive rpm), the speed of the machine that is driven (driven rpm), and the diameters of the pulleys on both the drive motor and driven machine must be considered. **See Figure 18-53.** To calculate the driven machine pulley diameter, apply the following formula:

$$PD_m = \frac{PD_d \times N_d}{N_m}$$

where

PD_m = driven machine pulley diameter (in in.)

PD_d = drive pulley diameter (in in.)

N_d = motor drive speed (in rpm)

N_m = driven machine speed (in rpm)

Note: This formula may be rewritten to solve for any unknown value if the other three values are known.

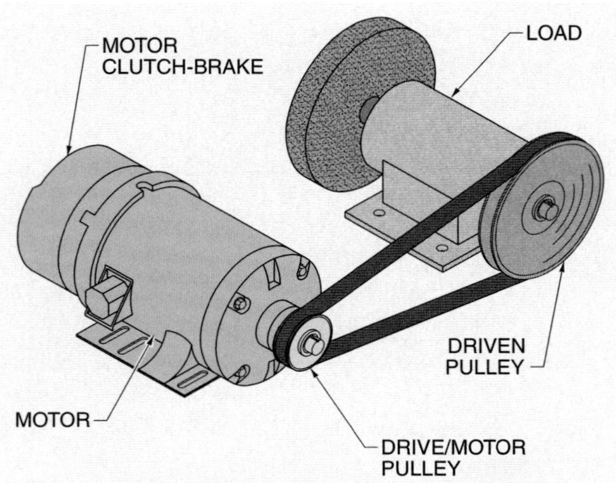

Figure 18-53. The pulley diameter for a driven machine is obtained from the correct motor rpm, driven rpm, and motor pulley diameter.

Example: Calculating Pulley Diameter

What is the required driven machine pulley diameter if a motor running at 1800 rpm has a 6″ pulley, and the driven machine is run at 900 rpm?

$$PD_m = \frac{PD_d \times N_d}{N_m}$$

$$PD_m = \frac{6 \times 1800}{900}$$

$$PD_m = \frac{10,800}{900}$$

$$PD_m = \mathbf{12''}$$

If the drive motor or driven machine does not have a pulley, a common pulley size can be selected for one or the other and the equation used to solve for the unknown size.

For very low speeds that are required in some applications, a gearmotor or motor connected to a gear drive is used. Gearmotors are designed with output speeds as low as 1 rpm. Gearmotors and geardrives work on the same gear reduction principle as most clocks and watches.

TROUBLESHOOTING DRIVE AND MOTOR CIRCUITS

Troubleshooting is the systematic elimination of the various parts of a system or process to locate a malfunctioning part. When a motor circuit is not operating properly, voltage and current measurements are taken to help determine or isolate the problem. Voltage measurements are taken to establish that the voltage is present and at the correct level. Voltage measurements may help determine circuit problems such as blown fuses, improper grounding, contacts not closing, etc. However, voltage measurements alone do not indicate the true condition of a motor because the voltage may be correct at the motor terminals but the motor may be faulty. *Note:* Always wear the required protection and safety equipment, and apply proper and safe procedures when taking voltage measurements. Voltage is measured using standard procedures. **See Figure 18-54.**

1. Measure the voltage at the disconnect. With the power ON, measure the voltage between each power line (L1-L2, L2-L3, and L1-L3). The voltage between the power lines should be within 2% (2 V per 100 V). If the power lines are not within 2%, there is a power supply problem.
2. Check the fuses or circuit breakers. Replace any blown fuses (or reset tripped circuit breakers). *Note:* Ensure the disconnect is in the OFF position and the motor circuit is in the OFF condition before replacing fuses or resetting circuit breakers.

3. With the power ON, measure the voltage into the drive (L1-L2, L2-L3, and L1-L3). The voltage into the drive (or motor starter) should be within 2% (2 V per 100 V). If the voltage into the drive is not within 2%, there is a problem between the disconnect and the drive.
4. With power ON, measure the voltage out of the motor drive (T1-T2, T2-T3, and T1-T3). The voltage out of the drive should be within 2% (2 V per 100 V) when the motor is at full speed. If the voltage out of the motor drive is not within 2%, there is a problem with the motor drive (or control circuit).

Turn the power OFF and apply a lockout/tagout before making any repairs.

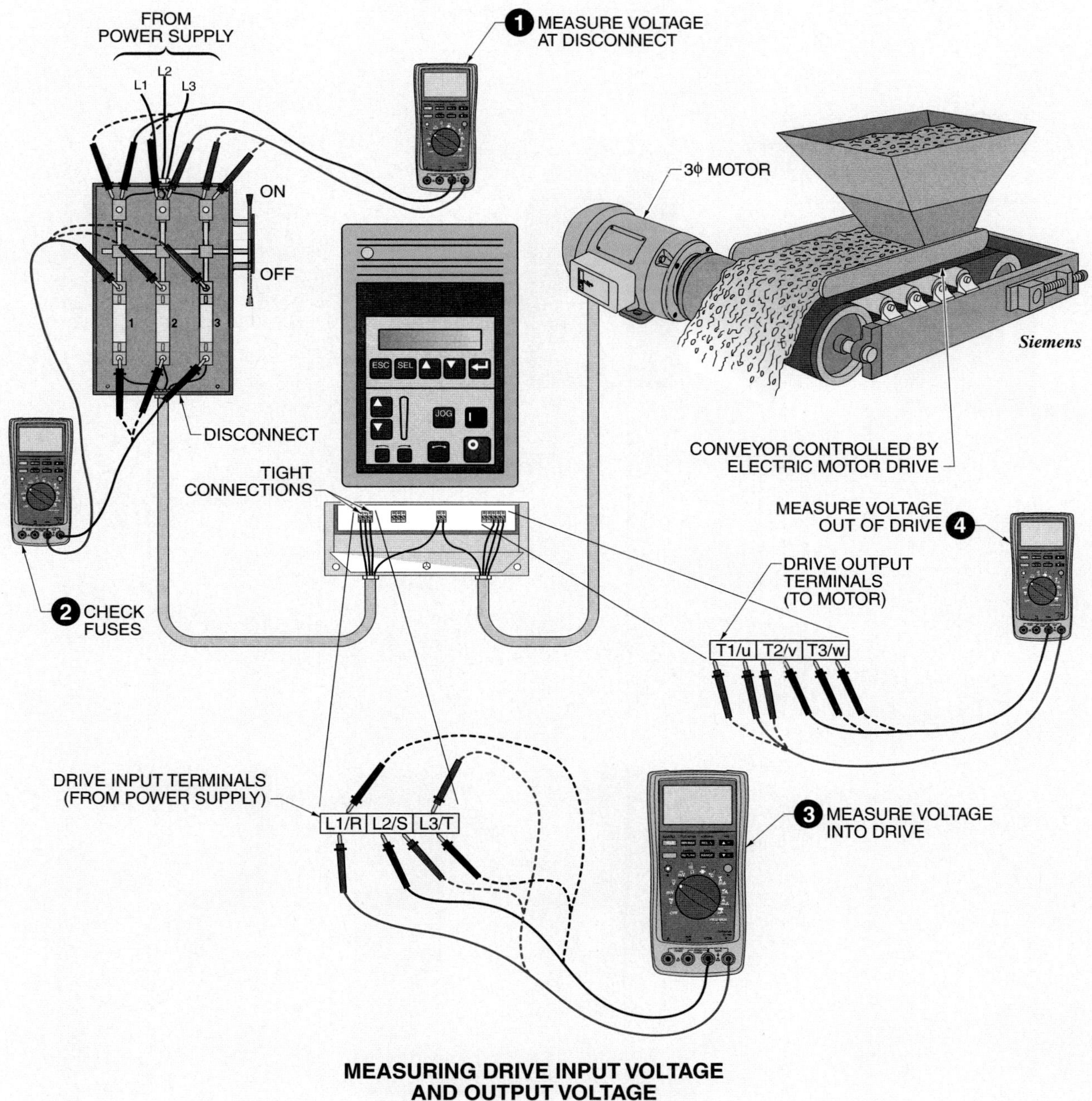

Figure 18-54. Voltage measurements are taken to establish that the voltage is present and at the correct level.

Current measurements are taken to determine the condition of a motor. Current measurements indicate if a motor is underloaded (receiving less than nameplate rated current), fully loaded (receiving nameplate current), or overloaded (receiving higher than nameplate rated current). *Note:* Always wear the required protection and safety equipment, and apply proper and safe procedures when taking current measurements. Current is measured using standard procedures. **See Figure 18-55.**

1. With power ON, measure the current in the lines leading to the motor (T1, T2, and T3). The current on each line should be within 10% of the other lines and less than the motor nameplate rated current. If the current is equal to the nameplate rated current, the motor is fully loaded and may have a problem, or the motor is undersized for the application. If the current is higher than the nameplate rated current, there is a problem with the motor.

Turn the power OFF and apply a lockout/tagout before making any repairs.

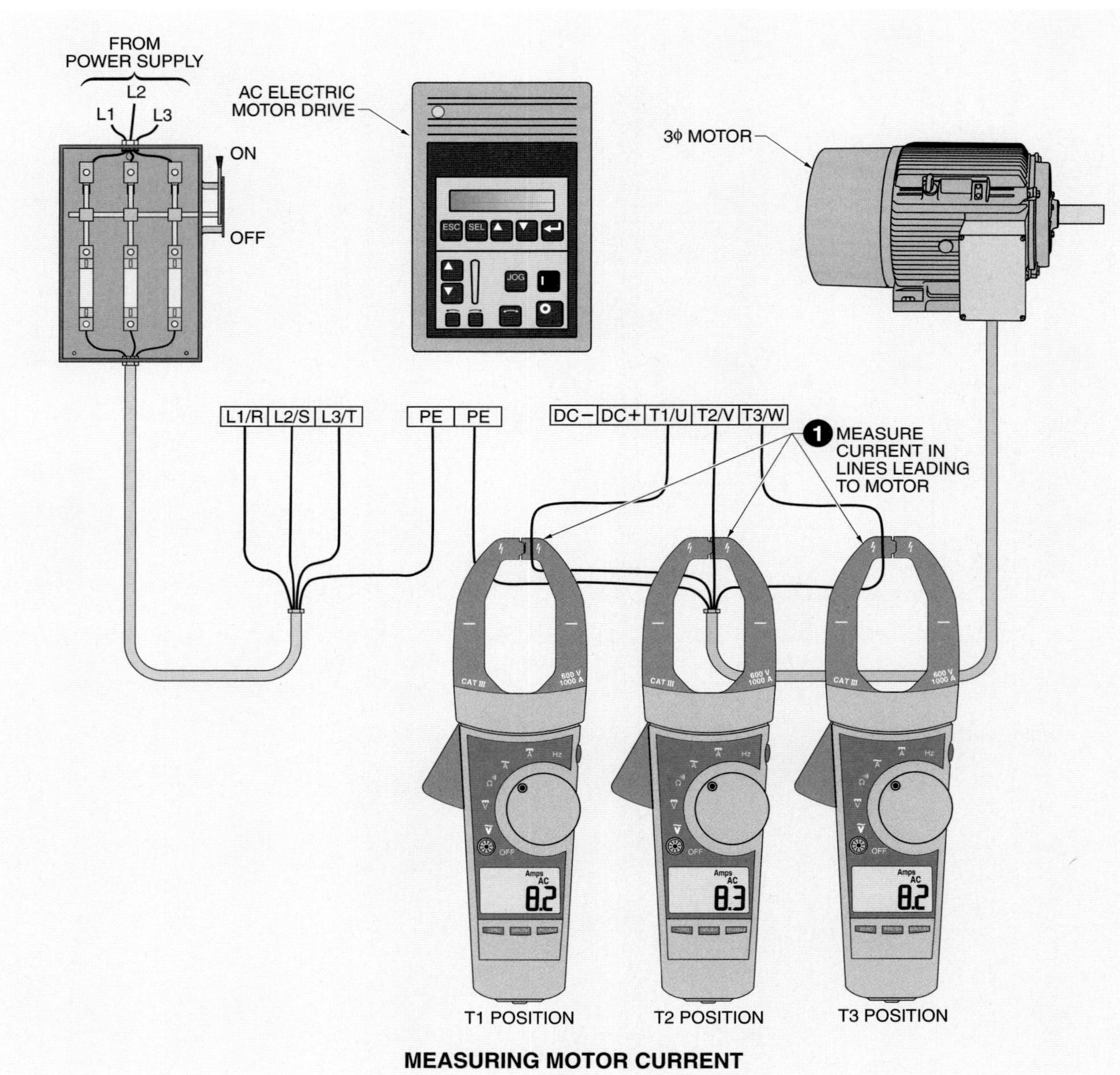

Figure 18-55. Current measurements indicate whether a motor is underloaded, fully loaded, or overloaded.

Review Questions

1. What is the basic operating principle of friction brakes?
2. Why are large brake shoe linings required in friction brakes?
3. What is the formula used to determine braking torque?
4. What is plugging?
5. When do the contacts of a plugging switch change?
6. What are two reasons some motors cannot be plugged?
7. What type of timer is used when a timing relay is used for plugging?
8. What is the basic operating principle of electric braking?
9. Why is interlocking important in electrical braking?
10. What is the basic operating principle of dynamic braking?
11. What is work?
12. What is torque?
13. What is full-load torque?
14. What is locked-rotor torque?
15. What is pull-up torque?
16. What is breakdown torque?
17. What does a NEMA design letter indicate?
18. What is the relationship between speed and torque in a DC series motor?
19. What is the relationship between speed and torque in a DC shunt motor?
20. What is the relationship between speed and torque in a DC compound motor?
21. How is the speed of a DC motor changed?
22. What determines the speed of an AC motor?
23. What is the synchronous speed of an AC motor with eight poles?
24. What is compelling circuit logic?
25. What is accelerating circuit logic?
26. What is decelerating circuit logic?
27. What type of motor drive stopping method allows a motor to come to a stop without applying any force from the drive?
28. What type of motor drive stopping method reduces the voltage applied to a motor as the motor decelerates?

chapter 19

electrical motor controls *for Integrated Systems*

Preventive Maintenance and Troubleshooting

Preventive maintenance consists of inspecting, cleaning, and testing components before failure. Troubleshooting consists of inspecting, cleaning, testing, and replacing components after failure. Troubleshooting normally involves the cost of downtime and loss of production.

PREVENTIVE MAINTENANCE

Today's industrial plants and assembly lines turn out products faster and more economically than at any time in the past. Shutdowns of even short periods of time are extremely costly. Without preventive maintenance, profit and productivity are lower.

Preventive maintenance is maintenance performed to keep machines, assembly lines, production operations, and plant operations running with little or no downtime. In the past, the job of the maintenance department was almost always to repair broken equipment and install new equipment.

Today, preventive maintenance does not take the time and personnel it once did. This is due to the introduction of a number of different inexpensive monitors that can detect and react to almost any problem before it becomes a major problem.

For example, inexpensive monitors are available that can monitor an electrical system for voltage or phase unbalances, voltage losses, phase reversals, over- or undervoltages, currents, and temperatures, and other conditions that may be signs of a major problem.

These monitors are easy to install and operate, can be installed on new or old equipment, and are designed to signal the maintenance department of trouble and take preventive measures until the maintenance personnel arrive. The monitors take measurements 24 hours a day.

Technical Fact

Predictive maintenance is the monitoring of wear conditions and equipment characteristics against a predetermined tolerance to detect possible malfunctions and failures. Troubleshooting skills are used to identify the cause of malfunction or failure.

A preventive maintenance program includes inspection, cleaning, tightening, adjusting and lubricating, keeping equipment dry, and electronically monitoring power circuits. The purpose of a preventive maintenance program is to do the following:

- Maintain equipment in such condition as to ensure uninterrupted operation for as long as possible.
- Maintain equipment in such condition that it always operates at the highest possible efficiency.
- Protect equipment from dirt, dust, moisture, corrosion, and electrical and mechanical overloads.
- Maintain good records of all maintenance work to establish maintenance needs and priorities.

Inspection

Inspection of all equipment normally uncovers evidence of a problem before it causes downtime. In most cases, time can be saved if problems are corrected before they lead to major breakdowns. Inspection consists of observation for signs of overheating, dirt, loose parts, noise, and any other signs of abnormalities.

Fluke Corporation

Scopemeters are used during preventive maintenance and troubleshooting to check for phase unbalance, voltage unbalance, single phasing, improper sequence, voltage variations, and frequency variations.

Cleaning

Keeping equipment clean helps eliminate overheating, high-voltage leakage, and breakdowns. Most equipment can be cleaned by blowing the dust and dirt away with a low-pressure, dry air stream. Care must be taken to remove the power if possible before cleaning. Cleaning should always be done on any equipment that is serviced for any reason.

Tightening

Vibration results in loose connections that eventually cause problems. All connections should be tightened firmly, but not beyond the pressure for which the connection is intended. Always use the correct tools of the proper size.

Adjusting and Lubricating

Routine maintenance such as adjusting and lubricating equipment is part of a good preventive maintenance program. Lubrication of bearings in motors and other rotating equipment helps to eliminate wear and heat. An adjustment in equipment which has been in operation for some time ensures that the equipment operates properly. Always follow the manufacturer recommendations when making adjustments and adding lubrication.

Keeping Equipment Dry

Electrical equipment operates best in a dry atmosphere. Moisture on copper and other metal surfaces used in electrical equipment can cause corrosion and rust, which lead to high resistance and heat. The moisture often causes leakage current or short circuits in the equipment. Always use the correct enclosure for the application.

Technical Fact

A facilities maintenance technician is a maintenance technician who operates, maintains, and repairs systems and equipment in hotels, schools, office buildings, and hospitals. An industrial maintenance technician is a maintenance technician who operates, maintains, and repairs production systems and equipment in industrial settings.

Terminal Connections

A loose terminal connection can cause a load to burn out from undervoltage. A loose connection causes an increase in resistance. A voltage drop occurs at the resistance point and heat develops when current passes through resistance of any kind.

The heat at a terminal can be carried by the wire to the thermal overload inside a circuit breaker if the loose terminal is on a circuit breaker. The heat from the loose contact, added to the current in the overload, may cause the circuit breaker to trip on a current far below its rating. This may lead an electrician to incorrectly suspect an overloaded circuit or faulty breaker.

Loads may burn out, coils (in solenoids, starters, etc.) may drop out, and timers and counters may reset if a loose terminal develops a high enough voltage drop across it. The heat developed at a loose terminal may also destroy the insulation around the terminal, leaving the possibility of a short circuit. This heat may also destroy any device that is connected to or near the loose connection.

To avoid loose connections, ensure that lugs clamp wires tightly. This is especially true with aluminum wire because aluminum is softer than copper and does not hold its shape as well. Aluminum also expands and contracts more than copper, which may cause a loose connection. Ensure that both wires fit tightly if two wires are used in the same lug. Always check possible problem areas.

TROUBLESHOOTING

Electric motors are reliable and require little maintenance. It is not uncommon for motors to run satisfactorily for 20 or more years without repair. For example, refrigerators, air conditioners, and heating systems often last more than 20 years without any work to the motors. Any repair that is required almost always involves the related components.

Today, it is more common to replace a motor that has failed than it is to repair it. Small motors generally cost more to repair than to replace. Most large motors can be replaced with a more energy-efficient motor that justifies the extra expenditure.

The cost involved with a motor that has failed is almost always in downtime of the operation and maintenance time involved. For this reason, motor problems must be located as quickly as possible and the reason the motor failed must be determined to eliminate the cause and prevent the problem from returning.

Troubleshooting is the systematic elimination of the various parts of a system, circuit, or process to locate a malfunctioning part. In most cases, troubleshooting is straightforward. This is because in most cases only one problem exists. Proper test tools, instruments, and equipment are essential to help troubleshoot problems quickly. The basic rules followed when using test instruments include the following:

- Always read the manufacturer instructions. Save the instructions in a file or safe place for future reference.
- Always start with the highest scale available on a test instrument to prevent overloading due to unknown values.
- Always remove the component to be tested or disconnect the line voltage from the circuit before making any resistance measurements.
- Never try to use a test instrument beyond its rated capacity.
- Always close clamp-on instrument jaws tightly. All clamp-on instruments are designed to attach to one conductor. Attaching to two conductors neutralizes the fields and no reading can be taken.
- Ensure all leads are insulated.
- Ensure all connections are tight for accurate readings.
- Always check to ensure that any instrument fuses or batteries are in working condition. A new battery is needed if the needle on an analog DMM cannot be zeroed on the ohm scale. The needle on an analog clamp-on instrument should read in the upper half of the scale for greatest accuracy.
- Always apply basic rules of electrical safety when testing any circuit.

Technical Fact

Always select the proper tool for each job and consult the operator's manual for proper use.

Electronically Monitoring Power Circuits

Prevention of major motor failure and downtime can be accomplished by detecting problems before they can cause any damage. Problems such as phase unbalance, voltage unbalance, single phasing, improper phase sequence, voltage surges, frequency variations, overcycling, improper ventilation, and improper motor mounting can all cause motor failure.

Electronic power monitors are available and can be easily installed to monitor phase unbalance, voltage unbalance, single phasing, improper phase sequence, voltage surges, voltage variations, and frequency variations. Surge protectors can be used to protect against voltage surges. Properly designed control circuits can prevent overcycling, and proper motor installation and maintenance can prevent improper ventilation and mounting problems.

Phase Unbalance

Phase unbalance is the unbalance that occurs when power lines are out of phase. Phase unbalance of a three-phase power system occurs when 1ϕ loads are applied, causing one or two of the lines to carry more or less of the load. An electrician balances the load of a three-phase power system during the installation process. An unbalance begins to occur as additional 1ϕ loads are added to the system. This unbalance causes the three-phase lines to move out of phase so the lines are no longer 120 electrical degrees apart. **See Figure 19-1.**

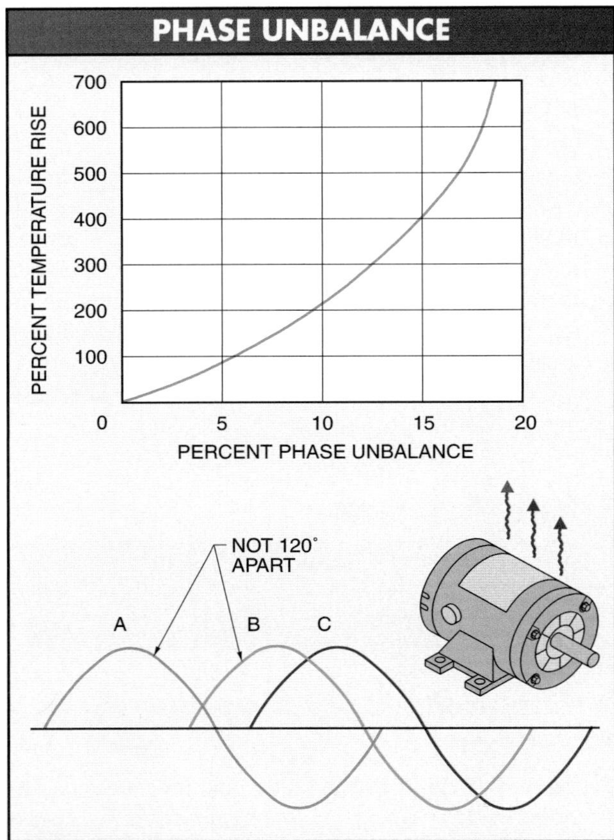

Figure 19-1. Phase unbalance is the unbalance that occurs when power lines are out of phase.

Phase unbalance causes 3ϕ motors to run at temperatures higher than their listed ratings. The greater the phase unbalance, the greater the temperature rise. High temperatures produce insulation breakdown and other related problems.

A 3ϕ motor operating in an unbalanced circuit cannot deliver its rated horsepower. For example, a phase unbalance of 3% causes a motor to work at 90% of its rated power. This requires the motor to be derated. **See Figure 19-2.**

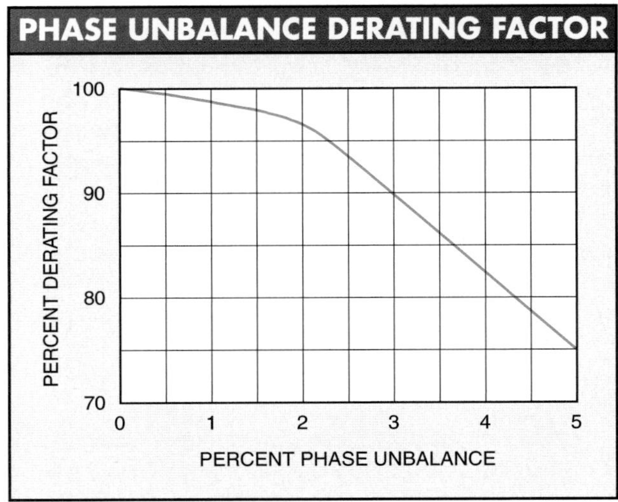

Figure 19-2. A motor operating on a circuit that has phase unbalance must be derated.

Voltage Unbalance

Voltage unbalance is the unbalance that occurs when the voltages at different motor terminals are not equal. One winding overheats, causing thermal deterioration of the winding if voltage is not balanced. Voltage unbalance results in a current unbalance. Line voltage should be checked for voltage unbalance periodically and during all service calls. Whenever more than 2% voltage unbalance is measured, the following steps should be taken:

- Check the surrounding power system for excessive loads connected to one line.
- Adjust the load or motor rating by reducing the load on the motor or oversizing the motor if the voltage unbalance cannot be corrected.
- Notify the power company.

To find voltage unbalance, apply the following procedure:

1. Measure the voltage between each incoming power line. The readings are taken from L1 to L2, L1 to L3, and L2 to L3.
2. Add the voltages.
3. Find the voltage average by dividing by 3.

Technical Fact

According to the National Safety Council, over 1000 people are killed each year in the United States from electrical shock and over 65,000 injuries occur due to failure to properly control hazardous energy sources during maintenance.

4. Find the voltage deviation by subtracting the voltage average from the voltage with the largest deviation.
5. To find voltage unbalance, apply the following formula:

$$V_u = \frac{V_d}{V_a} \times 100$$

where
V_u = voltage unbalance (in %)
V_d = voltage deviation (in V)
V_a = voltage average (in V)
100 = constant

Example: Calculating Voltage Unbalance

Calculate the voltage unbalance of a feeder system with the following voltage readings: L1 to L2 = 442 V; L1 to L3 = 474 V; L2 to L3 = 456 V. **See Figure 19-3.**

1. Measure incoming voltage. Incoming voltage is 442 V, 474 V, and 456 V.
2. Add voltages. 442 V + 474 V + 456 V = 1372 V

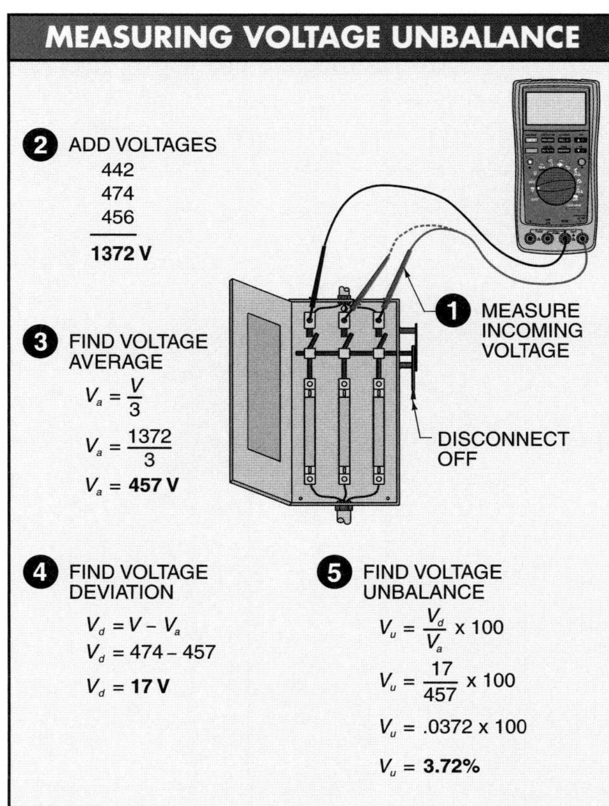

Figure 19-3. Voltage unbalance is the unbalance that occurs when the voltages at different motor terminals are not equal.

3. Find voltage average.

$$V_a = \frac{V}{3}$$

$$V_a = \frac{1372}{3}$$

$$V_a = \mathbf{457\ V}$$

4. Find voltage deviation.

$$V_d = V - V_a$$

$$V_d = 474 - 457$$

$$V_d = \mathbf{17\ V}$$

5. Find voltage unbalance.

$$V_u = \frac{V_d}{V_a} \times 100$$

$$V_u = \frac{17}{457} \times 100$$

$$V_u = .0372 \times 100$$

$$V_u = \mathbf{3.72\%}$$

An electrician can observe the blackening of one delta stator winding or two wye stator windings which occurs when a motor has failed due to voltage unbalance. The winding with the largest voltage unbalance is the darkest.

Single Phasing

Single phasing is the operation of a motor that is designed to operate on three phases but is only operating on two phases because one phase is lost. Single phasing occurs when one of the three-phase lines leading to a 3ϕ motor does not deliver voltage to the motor. Single phasing is the maximum condition of voltage unbalance.

Single phasing occurs when one phase opens on either the primary or secondary power distribution system. This occurs when one fuse blows, when there is a mechanical failure within the switching equipment, or when lightning takes out one of the lines.

Single phasing can go undetected on some systems because a 3ϕ motor running on two phases can run in low torque applications. When single phasing, the motor draws all its current from two lines.

Measuring the voltage at a motor does not normally detect a single phasing condition. The open winding in the motor generates a voltage almost equal to the phase voltage that is lost. In this case, the open winding acts as the secondary of a transformer, and the two windings connected to power act as the primary.

Single phasing is reduced by using the proper size dual-element fuse and by using the correct heater sizes. In motor circuits, or other types of circuits in which a single phasing condition cannot be allowed to exist for even a short period of time, an electronic phase-loss monitor is used to detect phase loss. The monitor activates a set of contacts to drop out the starter coil when a phase loss is detected.

An electrician can observe the severe blackening of one delta winding or two wye windings of the three 3ϕ windings which occurs when a motor has failed due to single phasing. The coil or coils that experienced the voltage loss indicate obvious and fast damage, which includes the blowing out of the insulation. **See Figure 19-4.**

Single phasing is distinguished from voltage unbalance by the severity of the damage. Voltage unbalance causes less blackening (but normally over more windings) and little or no distortion. Single phasing causes burning and distortion to one winding.

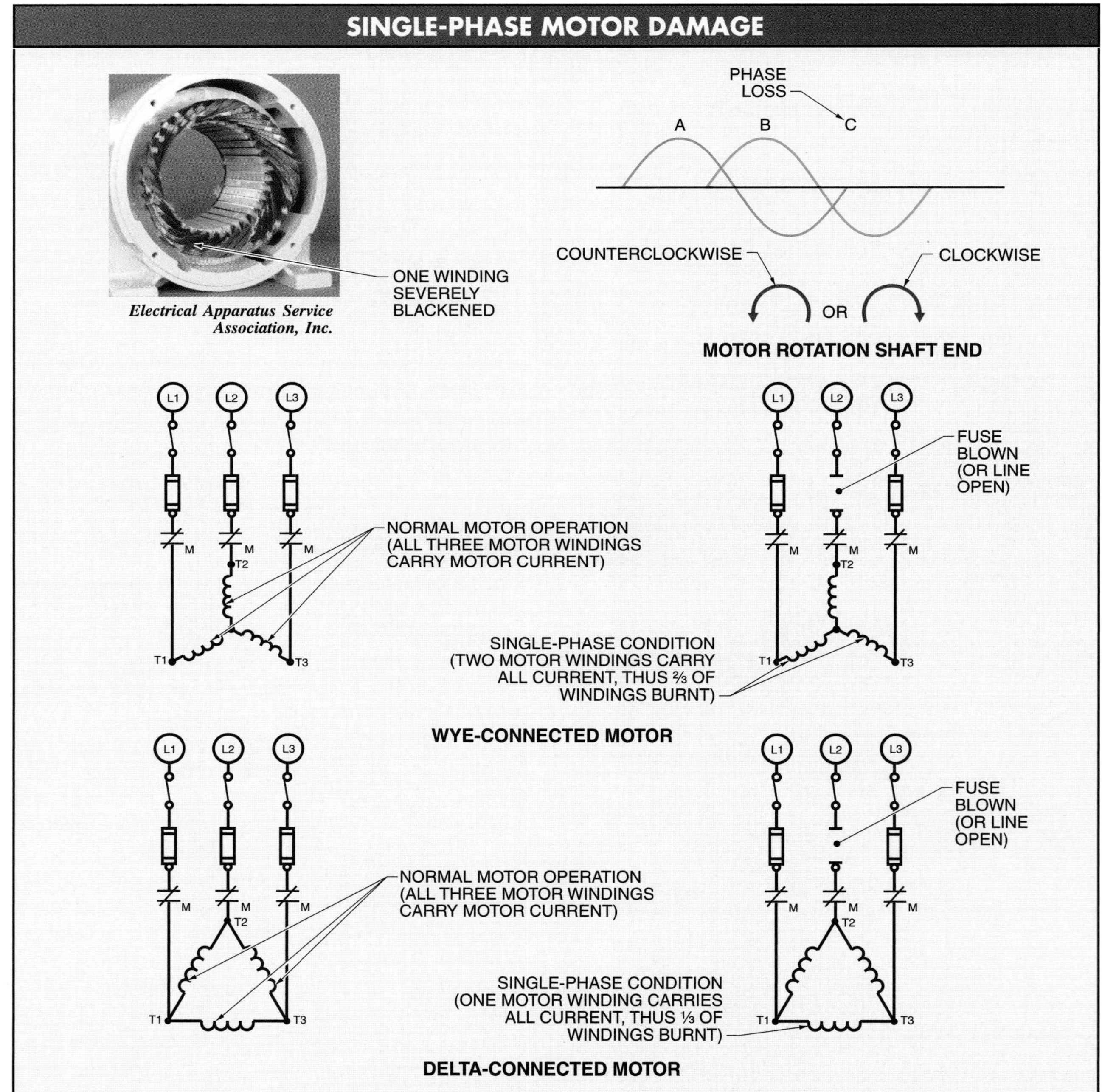

Figure 19-4. Single-phasing causes severe burning and distortion to one or two windings.

Improper Phase Sequence

Improper phase sequence is the changing of the sequence of any two phases (phase reversal) in a 3ϕ motor circuit. Improper phase sequence reverses the motor rotation. Reversing motor rotation can damage driven machinery or injure personnel. Phase reversal can occur when modifications are made to a power distribution system or when maintenance is performed on electrical conductors or switching equipment. The NEC® requires phase reversal protection on all personnel transportation equipment such as moving walkways, escalators, and ski lifts. **See Figure 19-5.**

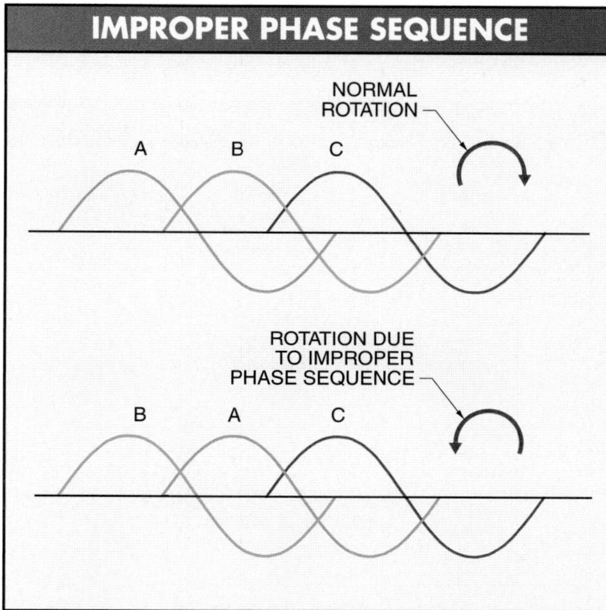

Figure 19-5. Improper phase sequence is the changing of the sequence of any two phases (phase reversal) in a 3ϕ motor circuit.

Voltage Surges

A *voltage surge* is a higher-than-normal voltage that temporarily exists on one or more power lines. Lightning is a major cause of large voltage surges. A lightning surge on a power line comes from a direct lightning hit or induced voltage. The lightning energy moves in both directions on the power lines, much like a rapidly moving wave.

Technical Fact

To help prevent electrical problems, neutral conductors should be the same size as, or larger than, hot conductors.

This traveling surge causes a large voltage rise in a short period of time. The large voltage is impressed on the first few turns of the motor windings, destroying the insulation and burning out the motor.

An electrician can observe the burning and opening of the first few turns of the windings which occur when a motor has failed due to a voltage surge. The rest of the windings appear normal, with little or no damage. **See Figure 19-6.**

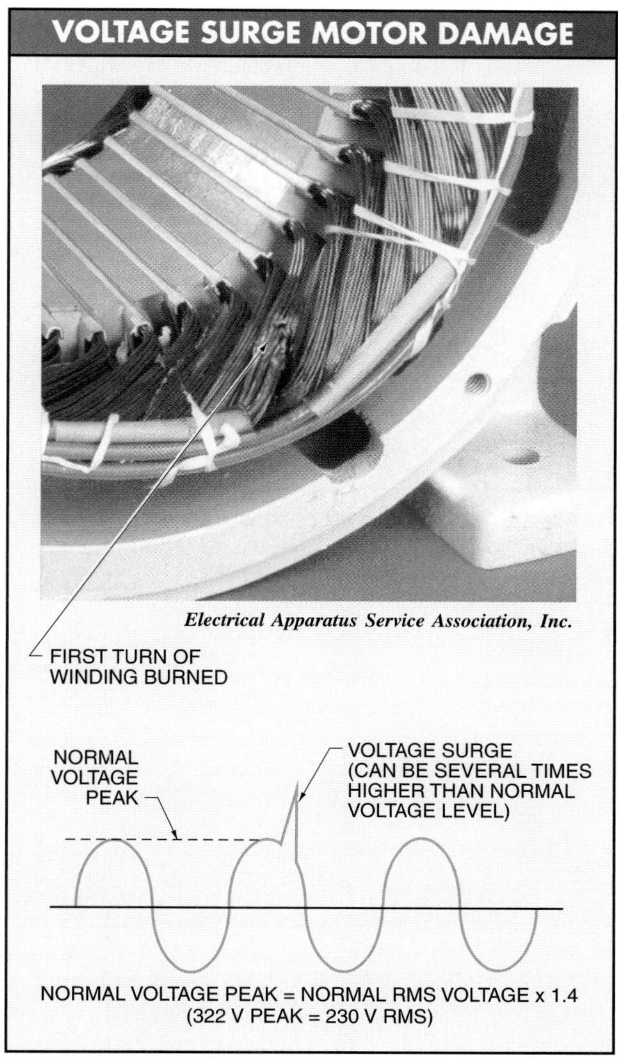

Figure 19-6. A voltage surge causes burning and opening of the first few turns of the windings.

Lightning arresters with the proper voltage rating and connection to an excellent ground assure maximum voltage surge protection. Surge protectors are also available. Surge protectors are placed on equipment or throughout the distribution system.

Voltage surges can also occur from normal switching of high-power circuits. Voltage surges occurring from switching high-power circuits are of lesser magnitude than lightning strikes and normally do not cause motor problems. A surge protector should be used on computer equipment circuits to protect sensitive electronic components.

AC Voltage Variations

Motors are rated for operation at specific voltages. Motor performance is affected when the supply voltage varies from a motor's rated voltage. A motor operates satisfactorily with a voltage variation of ±10% from the voltage rating listed on the motor nameplate. **See Figure 19-7.**

VOLTAGE VARIATION CHARACTERISTICS

Performance Characteristics	10% above Rated Voltage	10% below Rated Voltage
Starting current	+10% to +12%	−10% to −12%
Full-load current	−7%	+11%
Motor torque	+20% to +25%	−20% to −25%
Motor efficiency	Little change	Little change
Speed	+1%	−1.5%
Temperature rise	+3°C to −4°C	+6°C to +7°C

Figure 19-7. A motor operates satisfactorily with a voltage variation of ±10% from the voltage rating listed on the motor nameplate.

AC Frequency Variations

Motors are rated for operation at specific frequencies. Motor performance is affected when the frequency varies from a motor's rated frequency. A motor operates satisfactorily with a frequency variation of ±5% from the frequency rating listed on the motor nameplate. **See Figure 19-8.**

FREQUENCY VARIATION CHARACTERISTICS

Performance Characteristics	5% above Rated Frequency	5% below Rated Frequency
Starting current	−5% to −6%	+5% to +6%
Full-load current	−1%	+1%
Motor torque	−10%	+11%
Motor efficiency	Slight increase	Slight increase
Speed	+5%	−5%
Temperature rise	Slight decrease	Slight decrease

Figure 19-8. A motor operates satisfactorily with a frequency variation of ±5% from the frequency rating listed on the motor nameplate.

DC Voltage Variations

DC motors should be operated on pure DC power. *Pure DC power* is power obtained from a battery or DC generator. DC power is also obtained from rectified AC power. Most industrial DC motors obtain power from a rectified AC power supply. DC power obtained from a rectified AC power supply varies from almost pure DC power to half-wave DC power.

Half-wave rectified power is obtained by placing a diode in one of the AC power lines. Full-wave rectified power is obtained by placing a bridge rectifier (four diodes) in an AC power line. Rectified DC power is filtered by connecting a capacitor in parallel with the output of the rectifier circuit. **See Figure 19-9.**

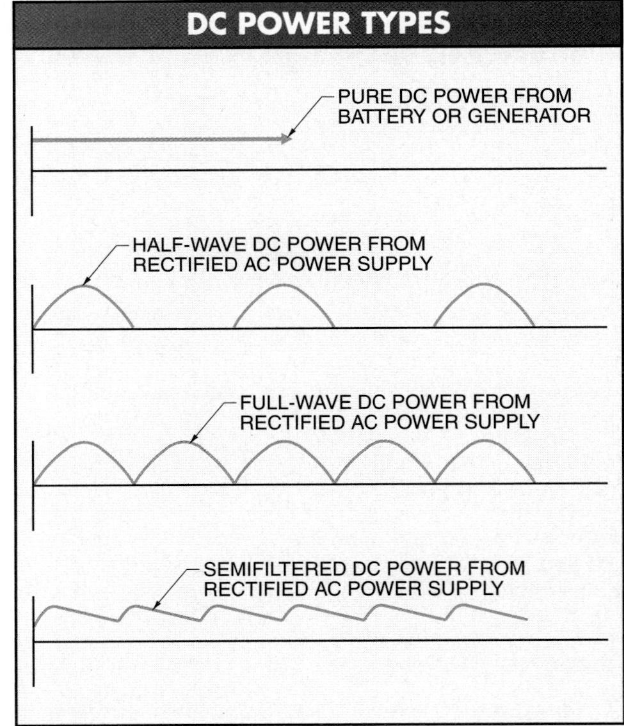

Figure 19-9. DC power obtained from a rectified AC power supply varies from almost pure DC power to half-wave DC power.

DC motor operation is affected by a change in voltage. The change may be intentional as in a speed-control application, or the change may be caused by variations in the power supply. The power supply voltage normally should not vary by more than 10% of a motor's rated voltage. Motor speed, current, torque, and temperature are affected if the DC voltage varies from the motor rating. **See Figure 19-10.**

DC MOTOR PERFORMANCE CHARACTERISTICS				
Performance Characteristics	Voltage 10% below Rated Voltage		Voltage 10% above Rated Voltage	
	Shunt	Compound	Shunt	Compound
Starting torque	−15%	−15%	+15%	+15%
Speed	−5%	−6%	+5%	+6%
Current	+12%	+12%	−8%	−8%
Field temperature	Decreases	Decreases	Increases	Increases
Armature temperature	Increases	Increases	Decreases	Decreases
Commutator temperature	Increases	Increases	Decreases	Decreases

Figure 19-10. Motor speed, current, torque, and temperature are affected if the DC voltage varies from the motor rating.

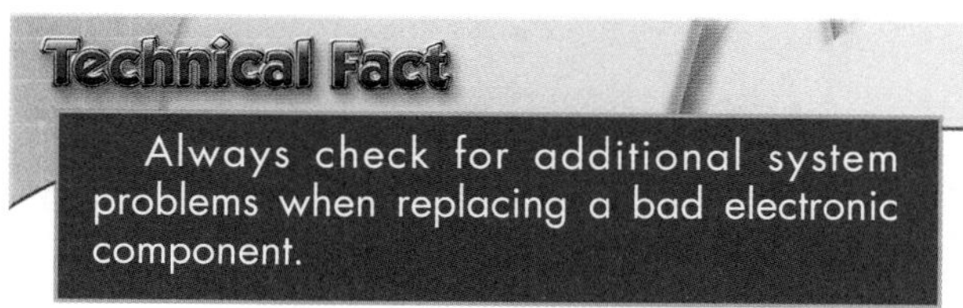

Allowable Motor Starting Time

A motor must accelerate to its rated speed within a limited time period. The longer a motor takes to accelerate, the higher the temperature rise in the motor. The larger the load, the longer the acceleration time. The maximum recommended acceleration time depends on the motor frame size. Large motor frames dissipate heat faster than small motor frames. **See Figure 19-11.**

MAXIMUM ACCELERATION TIME	
Frame Number	Maximum Acceleration Time (in sec)
48 and 56	8
143 – 286	10
324 – 326	12
364 – 505	15

Figure 19-11. A motor must accelerate to its rated speed within a limited time period.

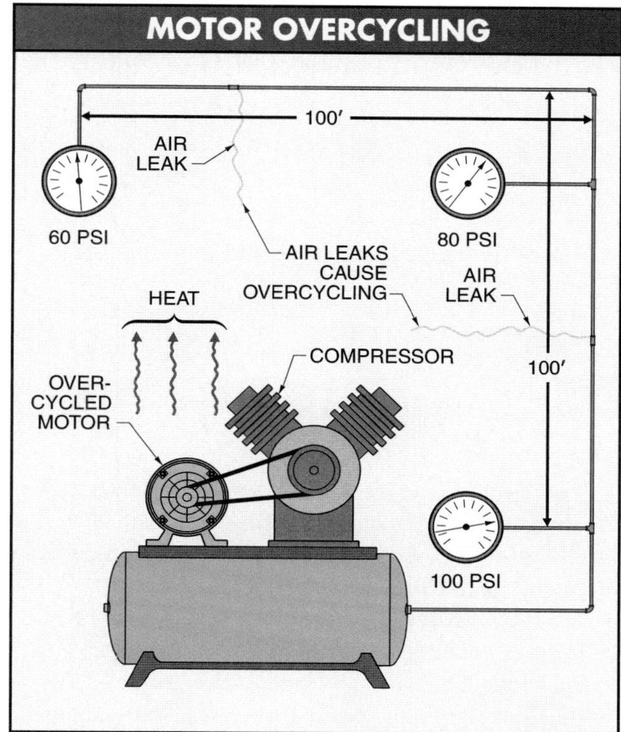

Figure 19-12. Overcycling is the process of turning a motor ON and OFF repeatedly, increasing the temperature of the motor and destroying the motor insulation.

Overcycling

Overcycling is the process of turning a motor ON and OFF repeatedly. Motor starting current is usually six to eight times the full-load running current of a motor. Most motors are not designed to start more than 10 times per hour. Overcycling occurs when a motor is at its operating temperature and still cycles ON and OFF. This further increases the temperature of the motor, destroying the motor insulation. **See Figure 19-12.**

Totally enclosed motors withstand overcycling better than open motors because they are designed to dissipate heat faster without damaging the motor. When a motor application requires a motor to be cycled often, the following steps should be taken:

- Use a motor with a 50°C rise instead of the standard 40°C.
- Use a motor with a 1.25 or 1.35 service factor instead of a 1.00 or 1.15 service factor.
- Provide additional cooling by forcing air over the motor.

Heat Problems

Excessive heat is a major cause of motor failure and other motor problems. Heat destroys motor insulation, which short-circuits the windings. The motor is not functional when motor insulation is destroyed.

The life of motor insulation is shortened as the heat in a motor increases beyond the temperature rating of the insulation. The higher the temperature, the sooner the insulation fails. The temperature rating of motor insulation is listed as the insulation class. **See Figure 19-13.**

MOTOR INSULATION CLASS

Class	°C	°F
A	105	221
B	130	266
F	155	311
H	180	356

°C 0° 100° 200°
°F 0° 32° 212° 392°

Figure 19-13. The temperature rating of motor insulation is listed as the insulation class.

The insulation class is given in Celsius (°C) and/or Fahrenheit (°F). A motor nameplate normally lists the insulation class of the motor. Heat buildup in a motor can be caused by the following conditions:

- Incorrect motor type or size for the application
- Improper cooling, normally from dirt buildup
- Excessive load, normally from improper use
- Excessive friction, normally from misalignment or vibration
- Electrical problems, normally voltage unbalance, phase loss, or a voltage surge

Improper Ventilation. All motors produce heat as they convert electrical energy to mechanical energy. This heat must be removed to prevent destruction of motor insulation. Motors are designed with air passages that permit a free flow of air over and through the motor. Air flow removes heat from a motor. Anything that restricts air flow through a motor causes the motor to operate at higher than design temperature. Air flow through a motor may be restricted by the accumulation of dirt, dust, lint, grass, pests, rust, etc. Air flow is restricted much faster if a motor becomes coated with oil from leaking seals or from overlubrication. **See Figure 19-14.**

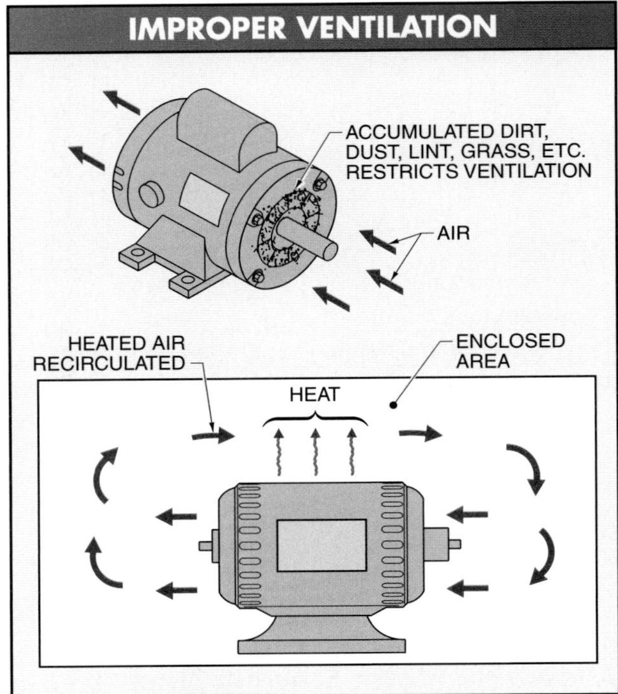

Figure 19-14. All motors produce heat which must be removed to prevent destruction of motor insulation.

Overheating can also occur if a motor is placed in an enclosed area. A motor overheats due to the recirculation of heated air when a motor is installed in a location that does not permit the heated air to escape. Vents added at the top and bottom of the enclosed area allow a natural flow of heated air.

Overloads. An *overload* is the application of excessive load to a motor. Motors attempt to drive the connected load when the power is ON. The larger the load, the more power required. All motors have a limit to the load they can drive. For example, a 5 HP, 460 V, 3ϕ motor should draw no more than 7.6 A. See NEC® Table 430-150.

Technical Fact

Never assume that conductors are properly marked or color coded. All ungrounded (hot) conductors will read a voltage between a conductor and a ground point. Use a DMM set to measure voltage to take voltage measurements to ensure that the voltage is OFF before working on a circuit.

Overloads should not harm a properly protected motor. Any overload present longer than the delay time built into the protection device is detected and removed. Properly sized heaters in the motor starter assure that an overload is removed before any damage is done. **See Figure 19-15.**

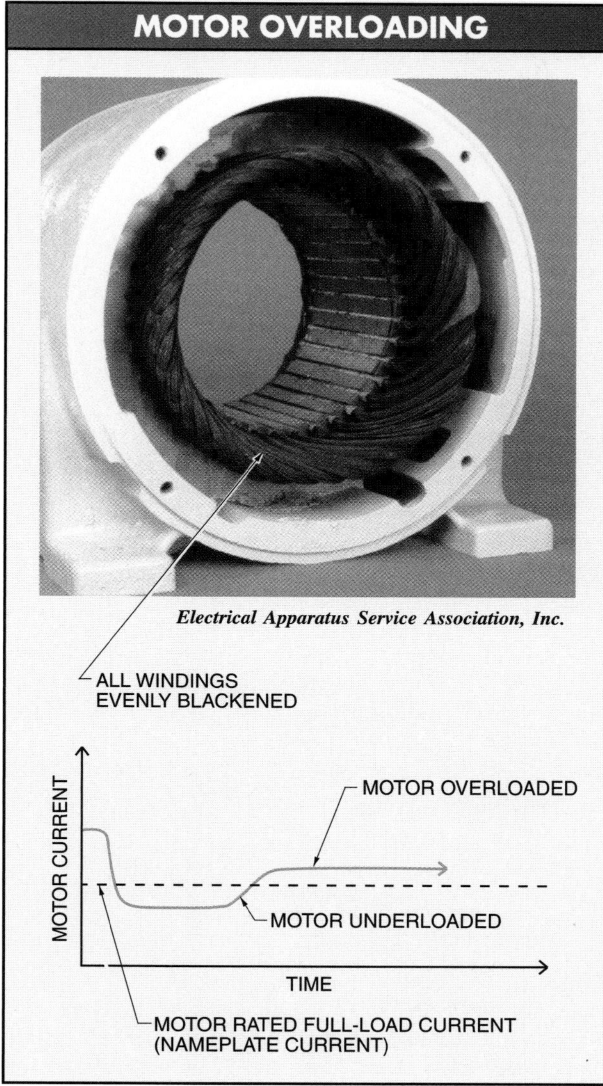

Figure 19-15. Overloading causes an even blackening of all motor windings.

An electrician can observe the even blackening of all motor windings which occurs when a motor has failed due to overloading. The even blackening is caused by the motor's slow destruction over a long period of time. No obvious damage or isolated areas of damage to the insulation are visible.

Current readings are taken at a motor to determine an overload problem. A motor is working to its maximum if it is drawing rated current. A motor is overloaded if it is drawing more than rated current. The motor size may be increased or the load on the motor decreased if overloads are a problem. **See Figure 19-16.**

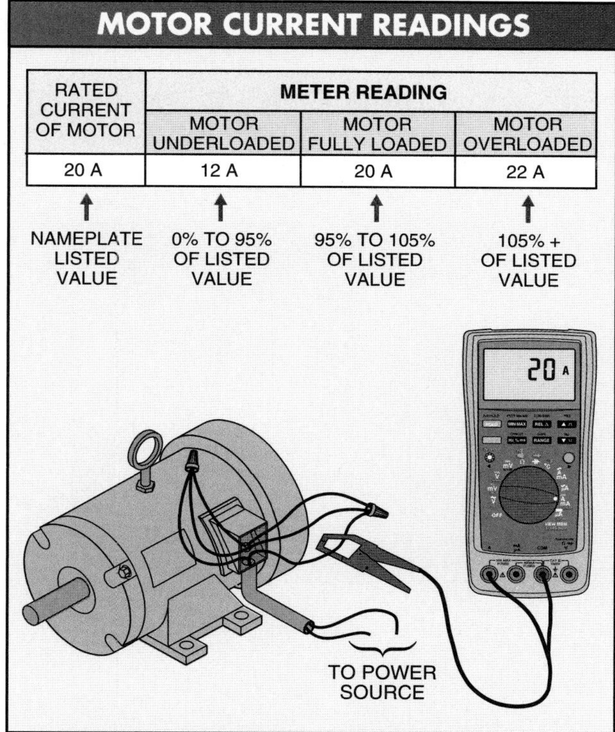

RATED CURRENT OF MOTOR	METER READING		
	MOTOR UNDERLOADED	MOTOR FULLY LOADED	MOTOR OVERLOADED
20 A	12 A	20 A	22 A
NAMEPLATE LISTED VALUE	0% TO 95% OF LISTED VALUE	95% TO 105% OF LISTED VALUE	105% + OF LISTED VALUE

Figure 19-16. Current readings are taken at a motor to determine an overload problem.

Altitude Correction

Temperature rise of motors is based on motor operation at altitudes of 3300′ or less. A motor with a service factor of 1.0 is derated when it operates at altitudes above 3300′. A motor with a service factor above 1.0 is derated based on the altitude and service factor. **See Figure 19-17.**

MOTOR ALTITUDE DERATINGS

Altitude Range (in ft)	Service Factor			
	1.0	1.15	1.25	1.35
3300 – 9000	93%	100%	100%	100%
9000 – 9900	91%	98%	100%	100%
9900 – 13,200	86%	92%	98%	100%
13,200 – 16,500	79%	85%	91%	94%
Over 16,500	Consult manufacturer			

Figure 19-17. A motor with a service factor of 1.0 is derated when it operates at altitudes above 3300′.

Motor Mounting and Positioning

Motors that are not mounted properly are more likely to fail from mechanical problems. A motor must be mounted on a flat, stable base. This helps reduce vibration and misalignment problems. An adjustable motor base aids in proper mounting and alignment.

To ensure a long life span, a motor should be mounted so that it is kept as clean as possible. To reduce the chance of damaging material reaching a motor, a belt cleaner should be used in any application in which the belts are likely to bring damaging material to a motor.

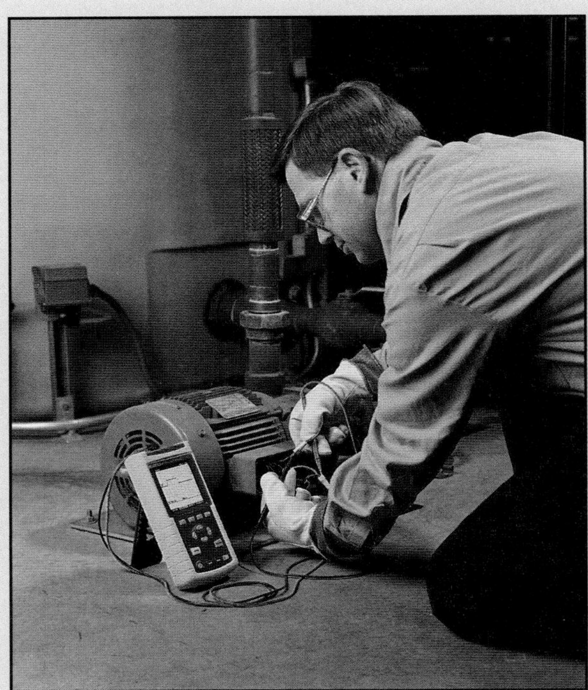

Fluke Corporation

Multimeters are used to check for single phasing because a 3ϕ motor running on two phases may continue to run with no physical signs of trouble, except for excessive noise and continuously burning out.

Motor Short Circuits

A short circuit occurs any time current takes a shortcut around the normal path of current flow. In a motor, a short circuit occurs because:

- The insulation on the motor winding breaks down due to overheating, which occurs when the motor winding must carry higher currents, or the point of the short is the weakest point of the winding insulation.
- The insulation is nicked (or removed) because of a foreign object entering the motor housing (file shaving, etc.).
- There is a manufacturer fault that occurred when the insulation was placed on the winding and the fault only showed up after the motor was operated (subjected to vibration, heat, etc.).

If motor insulation breaks down between two windings, there is a phase-to-phase short circuit. If motor insulation breaks down between a winding and ground, there is a phase-to-ground short. **See Figure 19-18.**

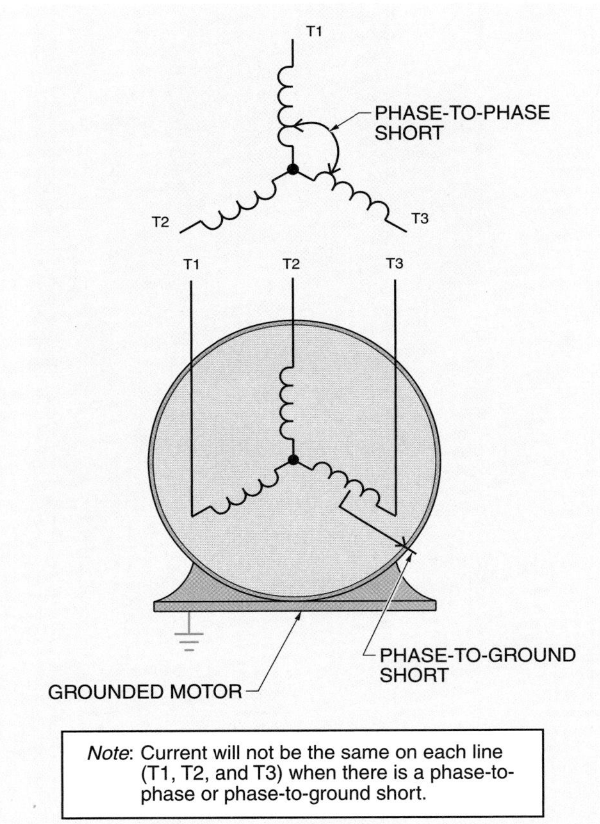

Figure 19-18. A phase-to-phase short circuit occurs when motor insulation breaks down between electrical windings. A phase-to-ground short circuit occurs when motor insulation breaks down between an electrical winding and the ground wire.

Adjustable Motor Bases

An adjustable motor base makes the installation, tensioning, maintenance, and replacement of belts easier. An *adjustable motor base* is a mounting base that allows a motor to be easily moved over a short distance. **See Figure 19-19.** An adjustable motor base simplifies the installation of a motor and the tightening of belts and chains.

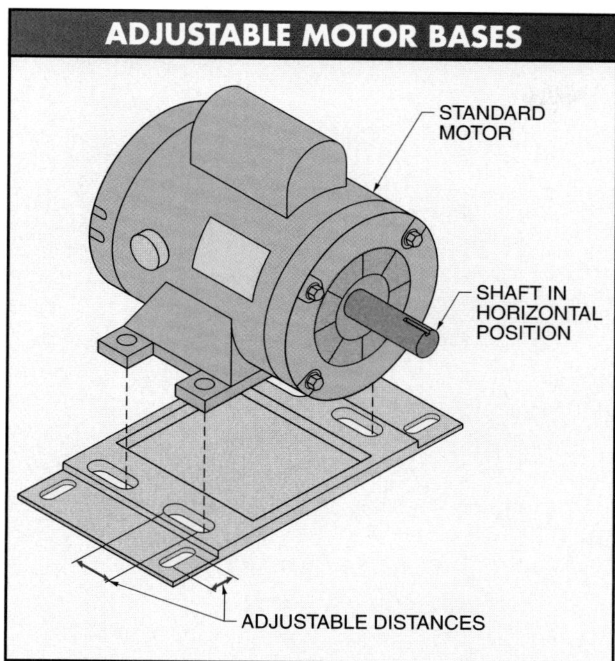

Figure 19-19. An adjustable motor base is a mounting base that allows a motor to be easily moved over a short distance.

Mounting Direction

The position of the driven machine normally determines whether a motor is installed horizontally or vertically. Standard motors are designed to be mounted with the shaft horizontal. The horizontal position is the best operating position for motor bearings. A specially designed motor is used for vertical mounting. Motors designed to operate vertically are more expensive and require more preventive maintenance.

Technical Fact

NEMA classifies motor frames. A standard number is used to indicate the motor's mounting dimensions. Motors with the same frame number can be interchanged. For example, a motor with a number 48 frame from one manufacturer will fit a 48 frame from another manufacturer.

Motor Belt Tension and Mounting

Belt drives provide quiet, compact, and durable power transmission and are widely used in industrial applications. A belt must be tight enough not to slip, but not so tight as to overload motor bearings.

Belt tension is normally checked by placing a straightedge from pulley to pulley and measuring the amount of deflection at the midpoint, or by using a tension tester. As a rule of thumb, belt deflection should equal 1⁄64″ per inch of span. For example, if the span between the center of a drive pulley and the center of a driven pulley is 16″, the belt deflection is 1⁄4″ (16 × 1⁄64 = 1⁄4″). Belt tension is normally adjusted by moving the drive component away from or closer to the driven component. **See Figure 19-20.**

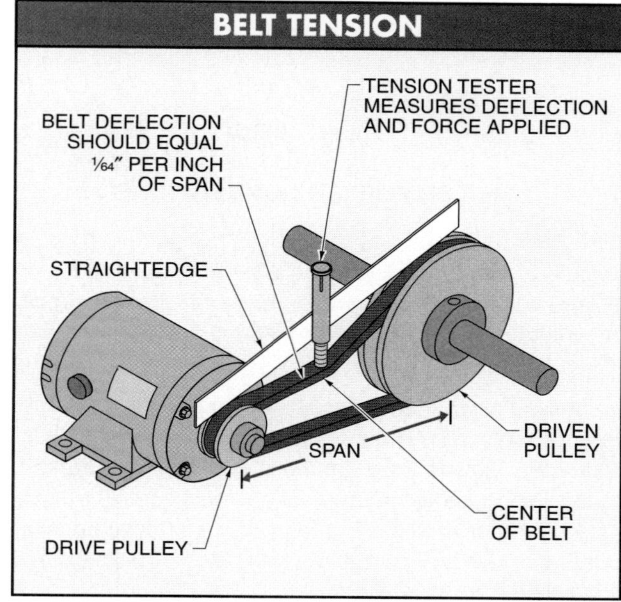

Figure 19-20. A belt must be tight enough not to slip, but not so tight as to overload motor bearings.

Megohmmeter Tests

A *megohmmeter* is a device that detects insulation deterioration by measuring high resistance values under high test voltage conditions. A megohmmeter may be powered by a hand crank, batteries, or 120 VAC. A megohmmeter detects motor insulation deterioration before a motor fails. A megohmmeter is an ohmmeter capable of measuring very high resistances by using high voltages. Megohmmeter test voltages range from 50 V to 5000 V. A megohmmeter is used to perform motor insulation tests to test motor insulation for failure which is caused by moisture, dirt, heat, cold, corrosive vapors or solids, vibration, and aging. A megohmmeter measures the resistance of different windings or the resistance from a winding to ground. An ohmmeter measures the resistance of common windings and components in a motor circuit. **See Figure 19-21.** *Note:* Weather can affect megohmmeter readings.

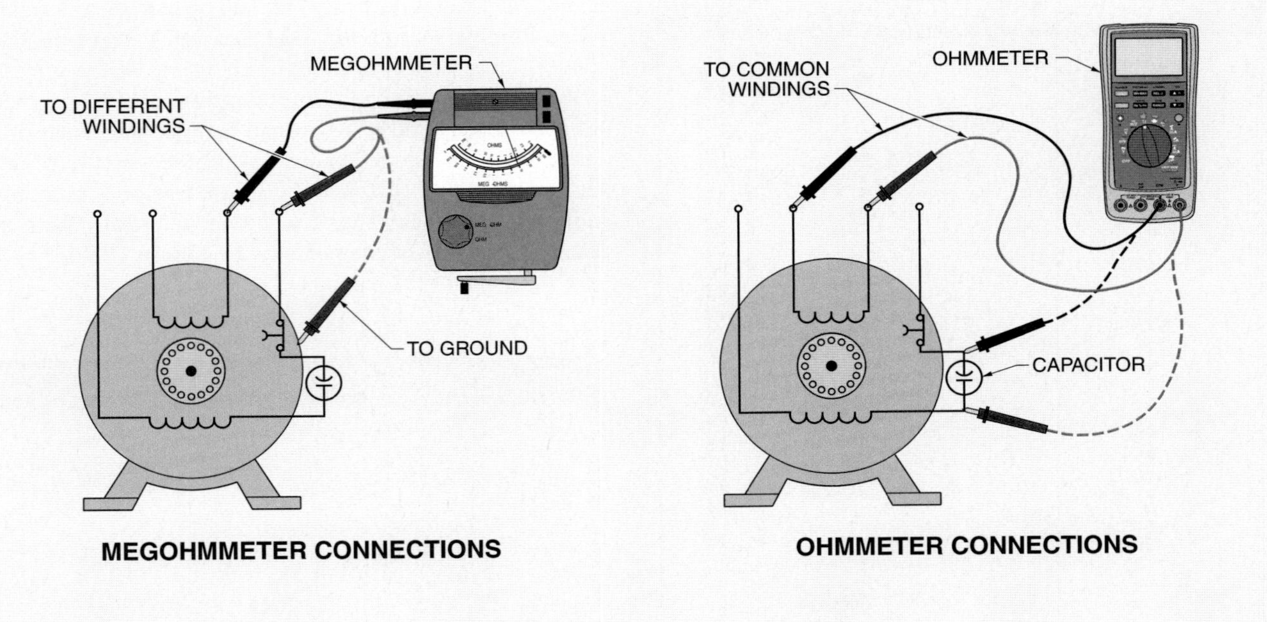

Figure 19-21. A megohmmeter measures the resistance of different windings or the resistance from a winding to ground. An ohmmeter measures the resistance of common windings and components in a motor circuit.

Several megohmmeter readings should be taken over a long period of time because the resistance of good insulation varies greatly. Megohmmeter readings are normally taken when a motor is installed and semiannually thereafter. A motor is in need of service if the megohmmeter reading is below the minimum acceptable resistance. **See Figure 19-22.**

RECOMMENDED MINIMUM RESISTANCE*	
Motor Voltage Rating (from nameplate)	**Minimum Acceptable Resistance**
Less than 208	100,000 Ω
208 – 240	200,000 Ω
240 – 600	300,000 Ω
600 – 1000	1 MΩ
1000 – 2400	2 MΩ
2400 – 5000	3 MΩ

* values for motor windings at 40°C

Figure 19-22. A motor is in need of service if the megohmmeter reading is below the minimum acceptable resistance.

Note: A motor with good insulation may have readings of 10 to 100 times the minimum acceptable resistance. The motor should be serviced if the resistance reading is less than the minimum value.

Technical Fact

Although insulating materials in practical use can be solids, liquids, or gases, resistance measurements generally refer to solid insulation. Insulating properties of liquids are stated in ohm-centimeters.

Caution: A megohmmeter uses high voltage for testing. A megohmmeter ground lead is connected to the motor frame during testing. Follow all manufacturer recommended procedures and safety rules. After performing insulation tests with a megohmmeter, connect the motor windings to ground through a 5 kΩ, 5 W resistor. The winding should be connected for 10 times the motor testing time to discharge energy stored in the insulation.

Insulation Spot Tests. An *insulation spot test* is a test that checks motor insulation over the life of a motor. An insulation spot test is taken when the motor is placed in service and every six months thereafter. The test should also be taken after a motor is serviced. **See Figure 19-23.** To perform an insulation spot test, apply the following procedure:

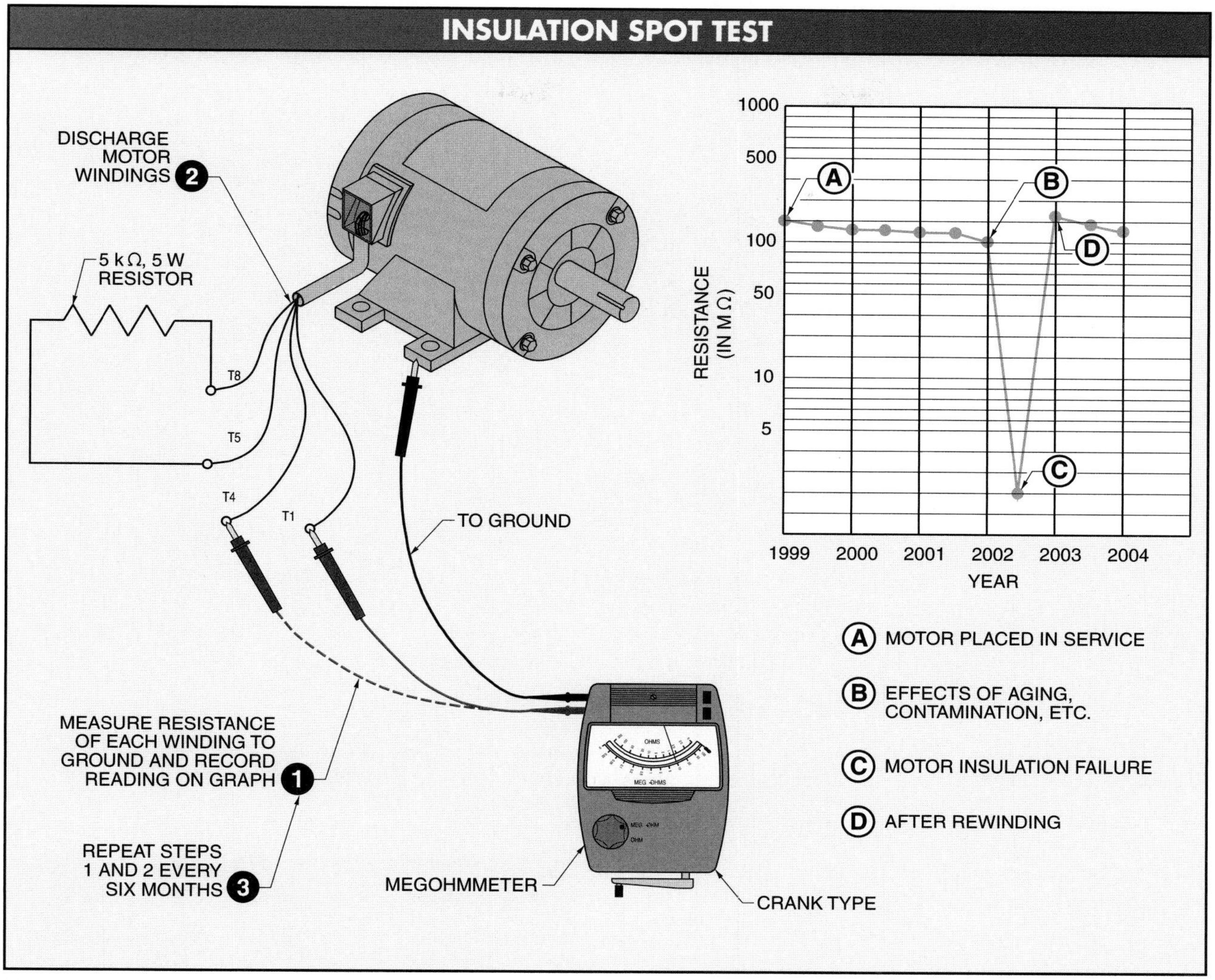

Figure 19-23. An insulation spot test checks motor insulation over the life of a motor.

1. Connect a megohmmeter to measure the resistance of each winding lead to ground. Record the readings after 60 sec. Service the motor if a reading does not meet the minimum acceptable resistance. Record the lowest meter reading on an insulation spot test graph if all readings are above the minimum acceptable resistance. The lowest reading is used because a motor is only as good as its weakest point.
2. Discharge the motor windings.
3. Repeat steps 1 and 2 every six months.

Interpret the results of the test to determine the condition of the insulation. Point A represents the motor insulation condition when the motor was placed in service. Point B represents effects of aging, contamination, etc., on the motor insulation. Point C represents motor insulation failure. Point D represents motor insulation condition after being rewound.

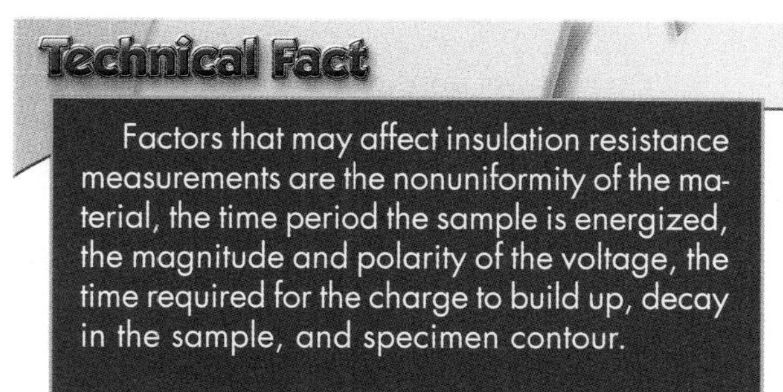

Technical Fact

Factors that may affect insulation resistance measurements are the nonuniformity of the material, the time period the sample is energized, the magnitude and polarity of the voltage, the time required for the charge to build up, decay in the sample, and specimen contour.

Dielectric Absorption Tests. A *dielectric absorption test* is a test that checks the absorption characteristics of humid or contaminated insulation. The test is performed over a 10-min period. **See Figure 19-24.** To perform a dielectric absorption test, apply the following procedure:

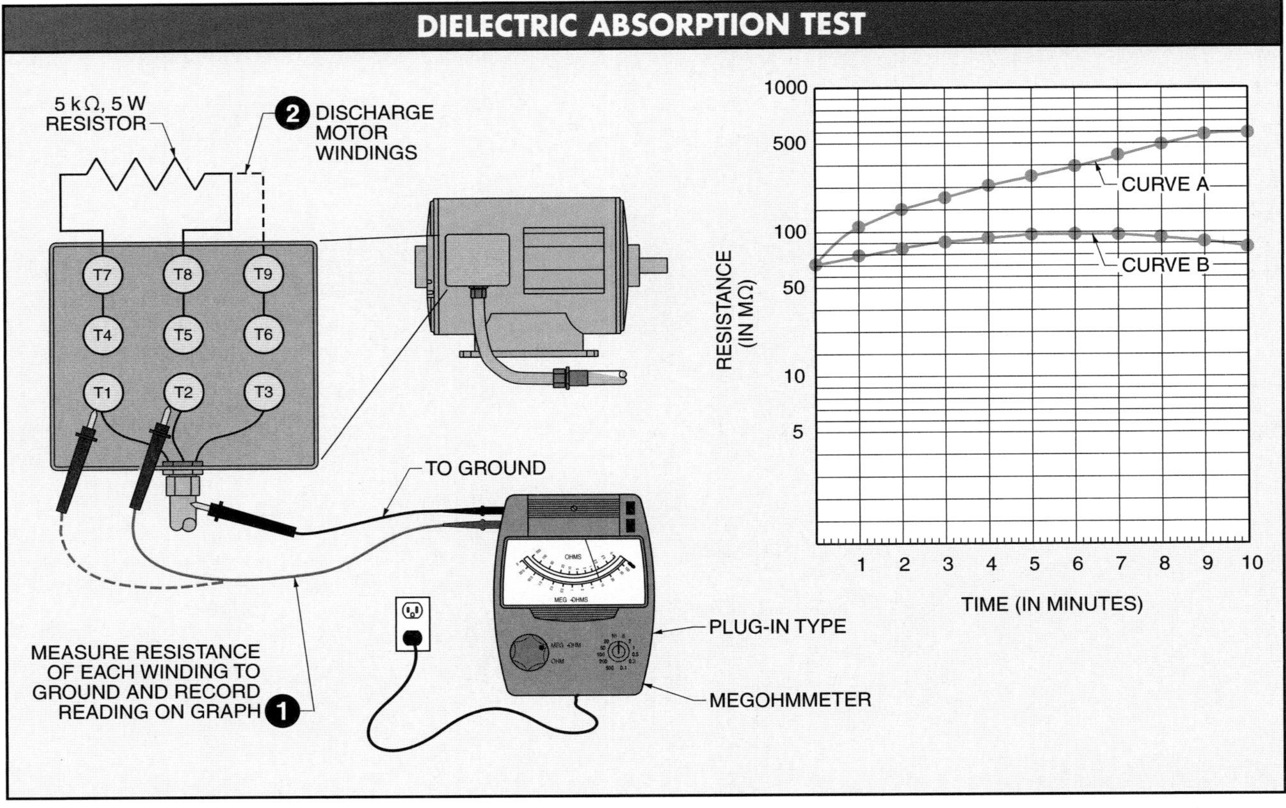

Figure 19-24. A dielectric absorption test checks the absorption characteristics of humid or contaminated insulation.

1. Connect a megohmmeter to measure the resistance of each winding lead to ground.
   Service the motor if a reading does not meet the minimum acceptable resistance. Record the lowest meter reading on a dielectric absorption test graph if all readings are above the minimum acceptable resistance. Record the readings every 10 sec for the first minute and every minute thereafter for 10 min.
2. Discharge the motor windings.
   Interpret the results of the test to determine the condition of the insulation. The slope of the curve shows the condition of the insulation. Good insulation (Curve A) shows a continual increase in resistance. Moist or cracked insulation (Curve B) shows a relatively constant resistance.

A polarization index is obtained by dividing the value of the 10-min reading by the value of the 1-min reading. The polarization index is an indication of the condition of the insulation. A low polarization index indicates excessive moisture or contamination. **See Figure 19-25.**

MINIMUM ACCEPTABLE POLARIZATION INDEX VALUES

Insulation	Value
Class A	1.5
Class B	2.0
Class F	2.0

Figure 19-25. The polarization index is an indication of the condition of the insulation. A low polarization index indicates excessive moisture or contamination.

For example, if the 1-min reading of Class B insulation is 80 MΩ and the 10-min reading is 90 MΩ, the polarization index is 1.125 (90/80 = 1.125). The insulation contains excessive moisture or contamination.

Insulation Step Voltage Tests. An *insulation step voltage test* is a test that creates electrical stress on internal insulation cracks to reveal aging or damage not found during other motor insulation tests. An insulation step voltage test is performed only after an insulation spot test. **See Figure 19-26.** To perform an insulation step voltage test, apply the following procedure:

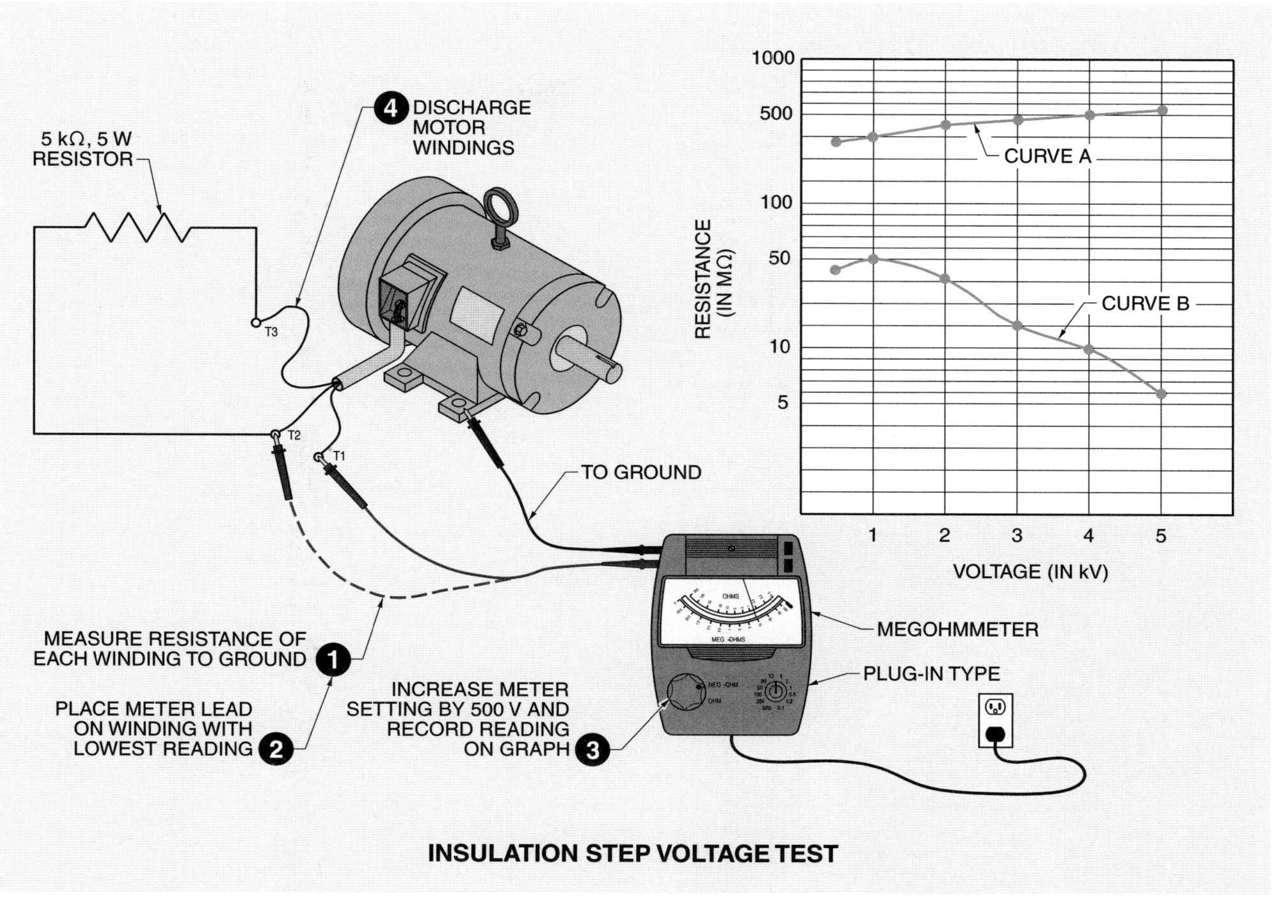

Figure 19-26. An insulation step voltage test is a test that creates electrical stress on internal insulation cracks to reveal aging or damage not found during other motor insulation tests.

1. Set the megohmmeter to 500 V and connect it to measure the resistance of each winding lead to ground. Take each resistance reading after 60 sec. Record the lowest reading.
2. Place the meter leads on the winding that has the lowest reading.
3. Set the megohmmeter on increments of 500 V starting at 1000 V and ending at 5000 V. Record each reading after 60 sec.
4. Discharge the motor windings.

Technical Fact

A megohmmeter is used to measure insulation resistance of rotating machinery, line insulators, bushings, and other field or factory equipment.

Interpret the results of the test to determine the condition of the insulation. The resistance of good insulation that is thoroughly dry (Curve A) remains approximately the same at different voltage levels. The resistance of deteriorated insulation (Curve B) decreases substantially at different voltage levels.

Re-marking Three-Phase Induction Motor Connections

Three-phase induction motors are the most common motors used in industrial applications. Three-phase induction motors operate for many years with little or no required maintenance. It is not uncommon to find 3ϕ induction motors that have been in operation for 10 to 20 years in certain applications. The length of time a motor is in operation may cause the markings of the external leads to become defaced. This may also happen to a new or rebuilt motor that has been in the maintenance shop for some time. To ensure proper operation, each motor lead must be re-marked before troubleshooting and reconnecting the motor to a power source.

The two most common 3ϕ motors are the single-voltage, 3ϕ, three-lead motor and the dual-voltage, 3ϕ, nine-lead motor. Both may be internally connected in a wye or delta configuration.

The three leads of a single-voltage, 3ϕ, three-lead motor can be marked T1, T2, and T3 in any order. The motor can be connected to the rated voltage and allowed to run. Any two leads may be interchanged if the rotation is in the wrong direction. The industry standard is to interchange T1 and T3.

Fluke Corporation

Current clamp accessories allow the checking of circuits for overloads without opening the circuit.

Wye or Delta Connection Determination

A standard dual-voltage motor has nine leads extending from it and may be internally connected as a wye or delta motor. The internal connections must be determined when re-marking the motor leads. A DMM is used to measure resistance or a continuity tester is used to determine whether a dual-voltage motor is internally connected in a wye or delta configuration.

Technical Fact

In a motor, practically all the electromechanical energy conversion occurs in the air gap.

A dual-voltage, wye-connected motor has four separate circuits. A dual-voltage, delta-connected motor has three separate circuits. **See Figure 19-27.** A wye-connected motor has three circuits of two leads each (T1-T4, T2-T5, and T3-T6) and one circuit of three leads (T7-T8-T9). A delta-connected motor has three circuits of three leads each (T1-T4-T9, T2-T5-T7, and T3-T6-T8).

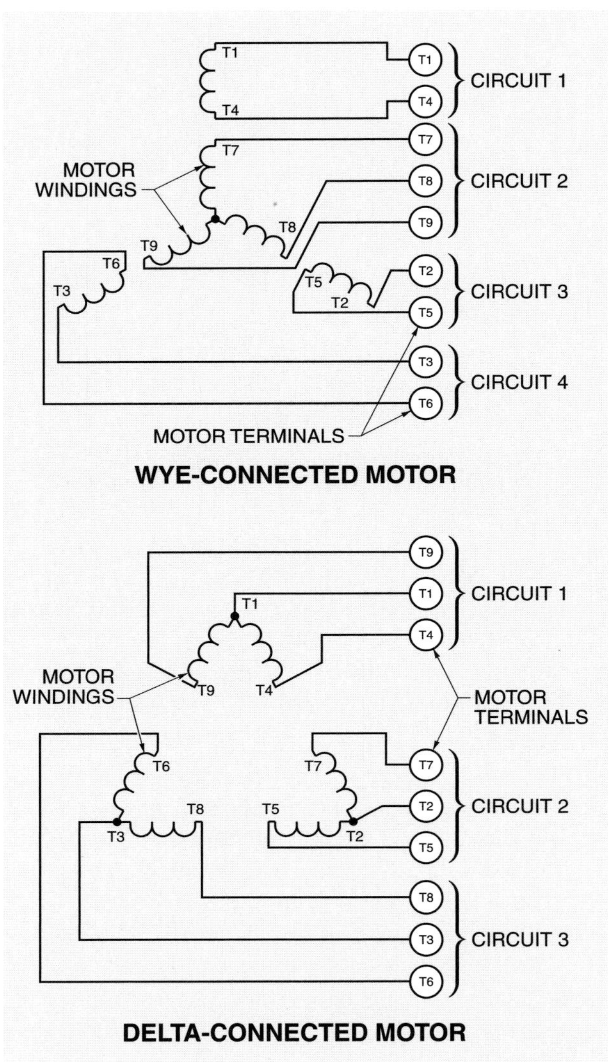

Figure 19-27. The internal connections of a motor must be determined when re-marking the motor leads.

A DMM is used to determine the winding circuits (T1-T4, T2-T5, etc.) on an unmarked motor by connecting one meter lead to any motor lead and temporarily connecting the other meter lead to each remaining motor lead. **See Figure 19-28.** *Note:* Ensure that the motor is disconnected from the power supply. A resistance reading other than infinity indicates a complete circuit.

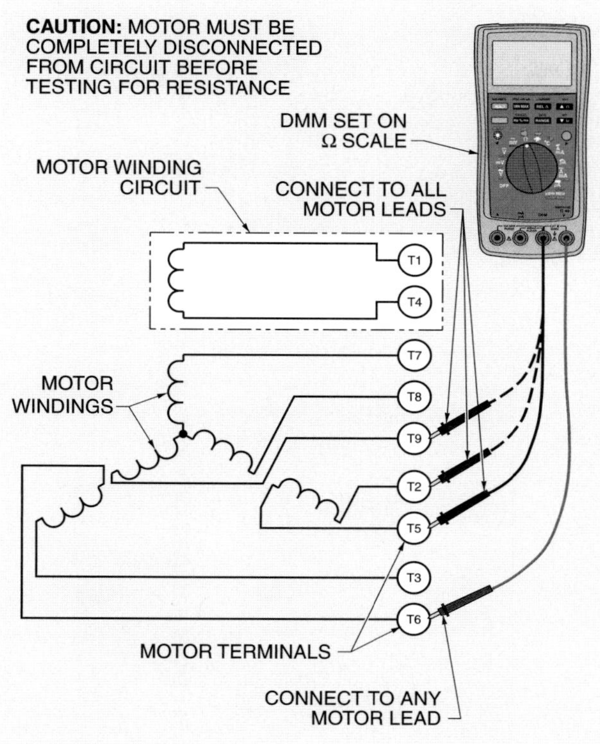

Figure 19-28. A DMM is used to determine a winding circuit on an unmarked motor by connecting one meter lead to any motor lead and temporarily connecting the other meter lead to each of the remaining motor leads.

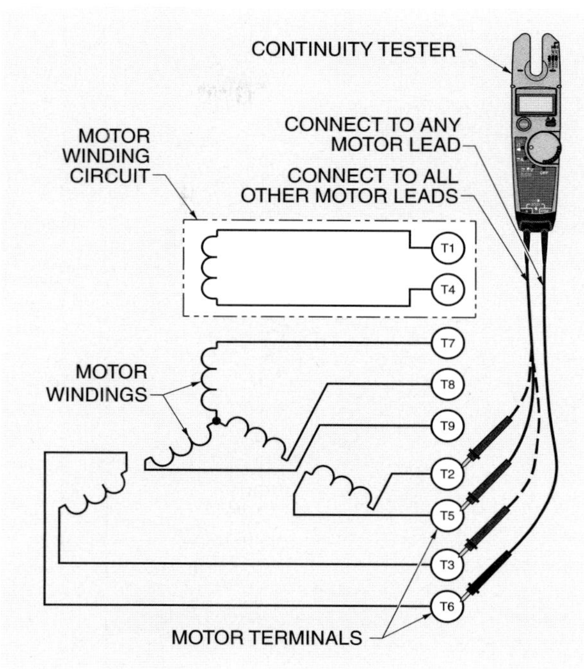

Figure 19-29. A continuity tester is used to determine a winding circuit on an unmarked motor by connecting one test lead to any motor lead and temporarily connecting the other test lead to each of the remaining motor leads.

A continuity tester may also be used to determine the winding circuits on an unmarked motor by connecting one test lead to any motor lead and temporarily connecting the other test lead to each remaining motor lead. **See Figure 19-29.**

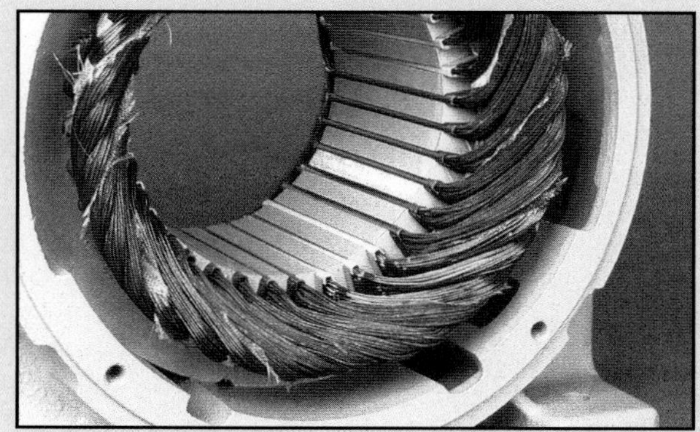

Electrical Apparatus Service Association, Inc.

A single-phased winding failure occurs when one phase of the power supply to the motor is open.

The continuity tester indicates a complete circuit by an audible beep. Mark each connection that indicates a complete circuit by taping or pairing the leads together. Check all pairs of leads with all the remaining motor leads to determine if the circuit is a two- or three-lead circuit. The motor is a wye-connected motor if three circuits of two leads and one circuit of three leads are found. The motor is a delta-connected motor if three circuits of three leads are found.

Re-marking Dual-Voltage, Wye-Connected Motors

To re-mark a dual-voltage, wye-connected motor with no power or load conductors connected, apply the following procedure:

1. Determine the winding circuits using a DMM or continuity tester. **See Figure 19-30.**
2. Mark the leads of the one three-lead circuit T7, T8, and T9 in any order. Separate the other motor leads into pairs, making sure none of the wires touch.

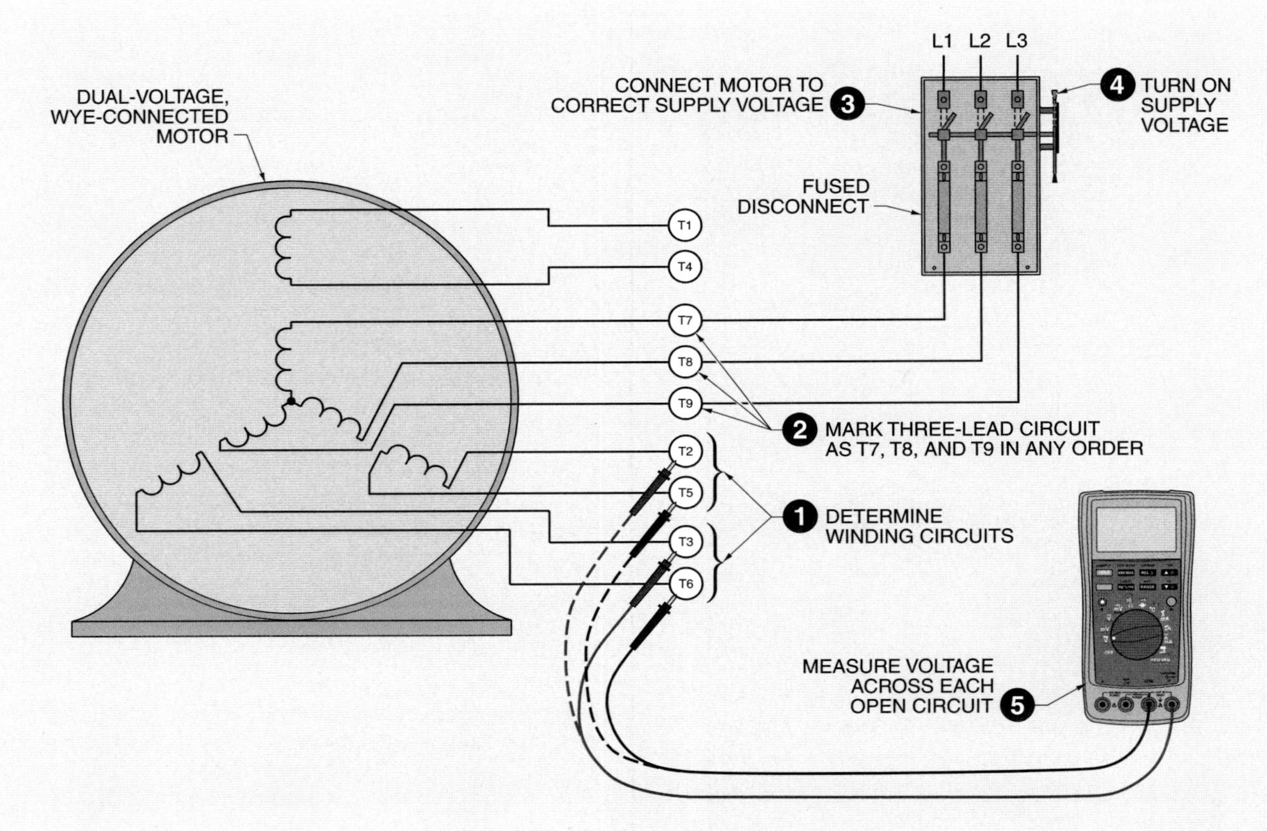

Figure 19-30. When re-marking a dual-voltage, wye-connected motor, the three-lead circuit is connected to the correct supply voltage and the voltage across each of the three open circuits is measured.

3. Connect the motor to the correct supply voltage. Connect T7 to L1, T8 to L2, and T9 to L3. The correct supply voltage is the lowest voltage rating of the dual-voltage rating given on the motor nameplate. The low voltage is normally 220 V because the standard dual-voltage motor operates on 220/440 V. For any other voltage, all test voltages should be changed in proportion to the motor rating.
4. Turn ON the supply voltage and let the motor run. The motor should run with no apparent noise or problems. The starting voltage should be reduced through a reduced-voltage starter if the motor is too large to be started by connecting it directly to the supply voltage.
5. Measure the voltage across each of the three open circuits while the motor is running, using a DMM set on at least the 440 VAC scale. Care must be taken when measuring the high voltage of a running motor. Insulated test leads must be used. Connect only one test lead at a time. The voltage measured should be about 127 V or slightly less, and should be the same on all three circuits.

The voltage is read on all circuits even though the two-wire circuits are not connected to the power lines because the voltage applied to the three-lead circuit induces a voltage in the two-wire circuits.

Draw the wiring diagram for the dual-voltage, wye-connected motor and mark the voltage readings on the wiring diagram. **See Figure 19-31.** Connect one lead of any two-wire circuit to T7 and connect the other lead of the circuit to one side of a DMM. Temporarily mark the lead connected to T7 as T4 and the lead connected to the DMM as T1. Connect the other lead of the DMM to T8 and then to T9. Mark T1 and T4 permanently if the two voltages are the same and are approximately 335 V. Perform the same procedure on another two-wire circuit if the voltages are unequal. Mark the new terminals T1 and T4 if the new circuit gives the correct voltage (335 V). T1, T7, and T4 are found by this first test.

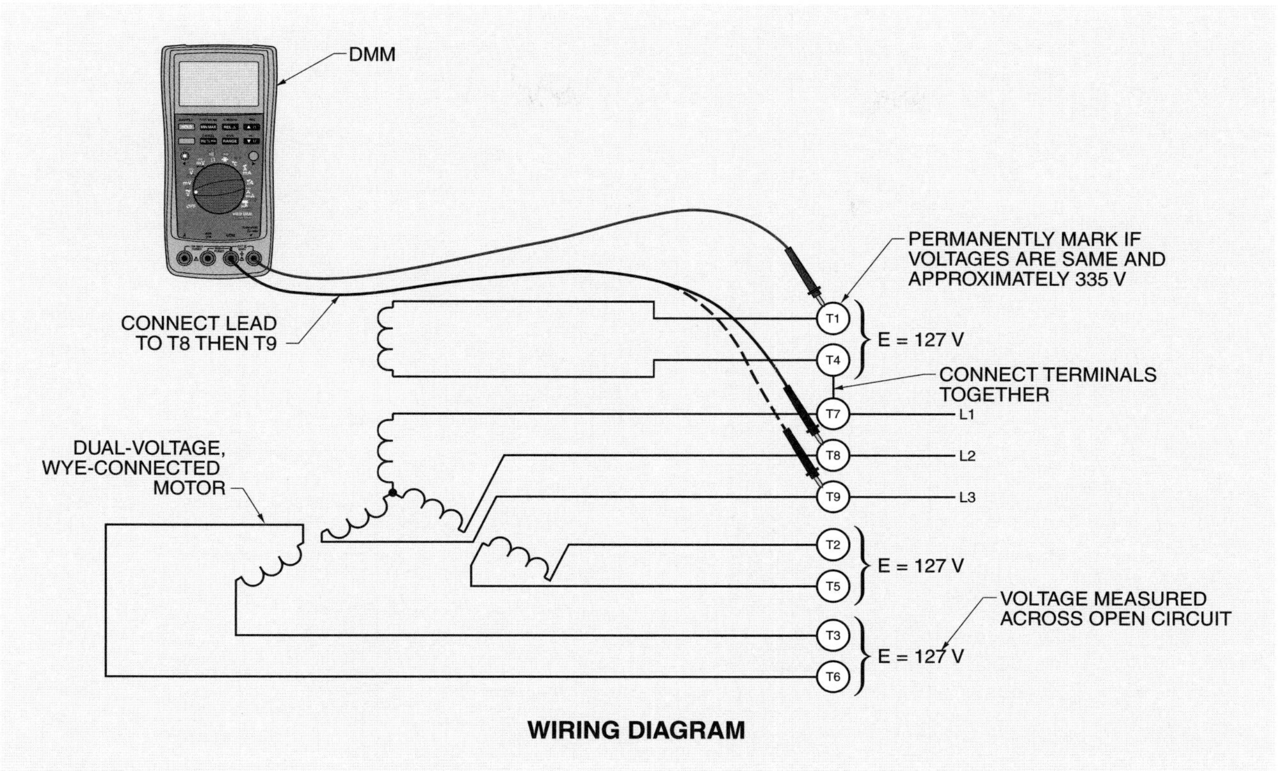

Figure 19-31. A wiring diagram is drawn when re-marking a dual-voltage, wye-connected motor to clarify the internal winding circuits.

Connect one lead of the two remaining unmarked two-wire circuits to T8 and the other lead to one side of the DMM. Temporarily mark the lead connected to T8 as T5 and the lead connected to the DMM as T2. Connect the other side of the DMM to T7 and T9 and measure the voltage. Measurements and changes should be made until a position is found at which both voltages are the same and approximately 335 V. T2, T5, and T8 are found by this second test.

Check the third circuit in the same way until a position is found at which both voltages are the same and approximately 335 V. T3, T6, and T9 are found by this third test.

After each motor lead is found and marked, turn OFF the motor and connect L1 to T1 and T7, L2 to T2 and T8, L3 to T3 and T9, and connect T4, T5, and T6 together. Start the motor and let it run. Check the current on each power line with a clamp-on ammeter. The markings are correct and may be marked permanently if the current is approximately equal on all of the three power lines.

Technical Fact

AC armature windings may be single (polyphase), full (fractional pitch), wye-connected, or delta connected. Practically all AC armature windings, except those in small-capacity AC equipment, have three-phase windings, with coils distributed around the entire armature perimeter for maximum utilization of space and components in the machine.

Re-marking Dual-Voltage, Delta-Connected Motors

A dual-voltage, delta-connected motor has nine leads grouped into three separate circuits. Each circuit has three motor leads connected, which make the circuits T1-T4-T9, T2-T5-T7, and T3-T6-T8. To re-mark a dual-voltage, delta-connected motor with no load, apply the following procedure:

1. Determine the winding circuits using a DMM or continuity tester. **See Figure 19-32.**

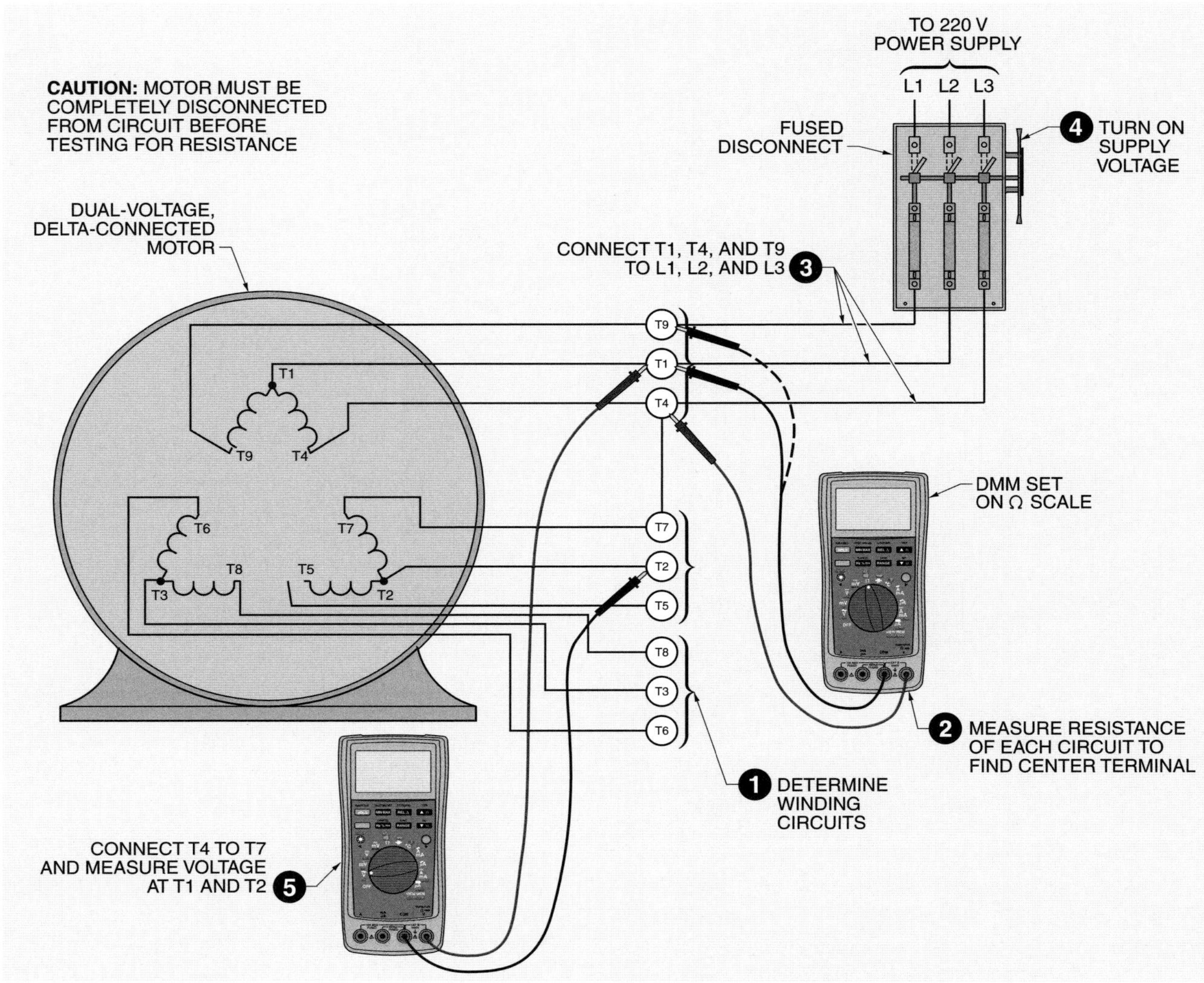

Figure 19-32. A dual-voltage, delta-connected motor has nine leads grouped into three separate circuits.

2. Measure the resistance of each circuit to find the center terminal. The resistance from the center terminal to the other two terminals is one-half the resistance between the other two terminals. Separate the three circuits and mark the center terminal for each circuit as T1, T2, and T3. Temporarily mark the two leads in the T1 group as T4 and T9, the two leads in the T2 group as T5 and T7, and the two leads in the T3 group as T6 and T8. Disconnect the DMM.
3. Connect the group marked T1, T4, and T9 to L1, L2, and L3 of a 220 V power supply. This should be the low-voltage rating on the nameplate of the motor. The other six leads should be left disconnected and must not touch because a voltage is induced in these leads even though these leads are not connected to power.
4. Turn the motor ON and let it run with the power applied to T1, T4, and T9.
5. Connect T4 (which is also connected to L2) to T7 and measure the voltage between T1 and T2. Set the DMM on at least a 460 VAC range. Use insulated test leads and connect one meter lead at a time. The lead markings for T4 and T9, and T7 and T5, are correct if the measured voltage is approximately 440 V. Interchange T5 with T7 or T4 with T9 if the measured voltage is approximately 380 V. Interchange both T5 with T7, and T4 with T9 if the new measured voltage is approximately 220 V. T4, T9, T7, and T5 may be permanently marked if the voltage is approximately 440 V.

Technical Fact

Motor windings consist of coils of insulated copper wire, insulated aluminum wire, or heavy, rigid insulated conductors. Coils are placed around pole pieces called salient poles.

To correctly identify T6 and T8, connect T6 and T8 and measure the voltage from T1 and T3. The measured voltage should be approximately 440 V. Interchange leads T6 and T8 if the voltage does not equal 440 V. T6 and T8 may be permanently marked if the voltage is approximately 440 V.

Turn OFF the motor and reconnect the motor to a second set of motor leads. Connect L1 to T2, L2 to T5, and L3 to T7. Restart the motor and observe the direction of rotation. The motor should rotate in the same direction as with the previous connection. Turn OFF the motor and reconnect the motor to the third set of motor leads (L1 to T3, L2 to T6, and L3 to T8) after the motor has run and the direction is determined.

Restart the motor and observe the direction of rotation. The motor should rotate in the same direction as the first two connections. Start over carefully, re-marking each lead if the motor does not rotate in the same direction for any set of leads.

Turn OFF the motor and reconnect the motor for the low-voltage connection. Connect L1 to T1-T6-T7, L2 to T2-T4-T8, and L3 to T3-T5-T9. Restart the motor and take current readings on L1, L2, and L3 with a clamp-on ammeter. The markings are correct if the motor current is approximately equal on each line.

Fluke Corporation

Multimeters can be used to measure the kW, kVA, and power factor of a circuit.

Re-marking DC Motor Connections

The three basic types of DC motors are the series, shunt, and compound motor. **See Figure 19-33.** All three types may have the same armature and frame but differ in the way the field coil and armature are connected. For all DC motors, terminal markings A1 and A2 always indicate the armature leads. Terminal markings S1 and S2 always indicate the series field leads. Terminal markings F1 and F2 always indicate the shunt field leads.

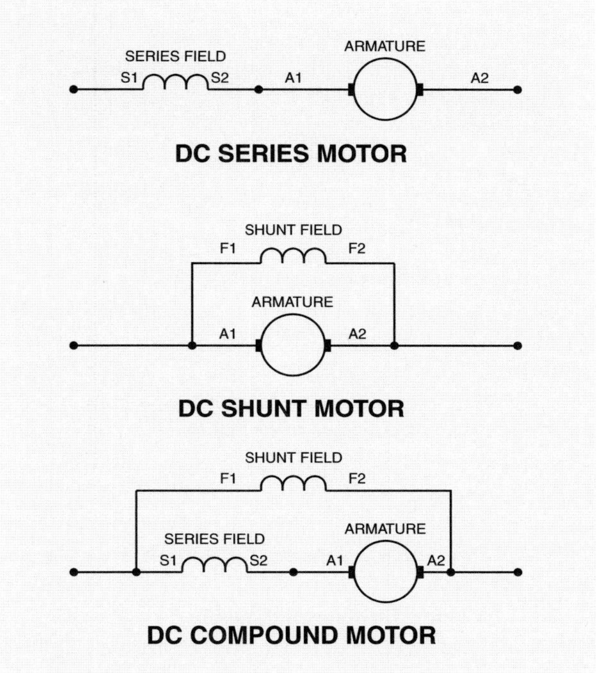

Figure 19-33. The three basic types of DC motors may have the same armature and frame, but differ in the way the field coil and armature are connected.

DC motor terminals can be re-marked using a DMM by measuring the resistance of each pair of wires. A pair of wires must have a resistance reading or they are not a pair.

The field reading can be compared to the armature reading because each DC motor must have an armature. The series field normally has a reading less than the armature. The shunt field has a reading considerably larger than the armature. The armature can be easily identified by rotating the shaft of the motor when taking the readings. The armature varies the DMM reading as it makes and breaks different windings. One final check can be made by lifting one of the brushes or placing a piece of paper under the brush. The DMM moves to the infinity reading.

From this information, a motor is either a DC series or DC shunt motor if it has two pairs of leads (four wires) coming out. A coil is the series field if the reading of the coil is less than the armature coil resistance. A coil is the shunt field if the reading is considerably larger than the armature resistance.

Technical Fact

Kirchoff's voltage law states, "At each instant of time, the algebraic sum of the voltage rise is equal to the algebraic sum of the voltage drops, both being taken in the same direction around the closed loop."

MAINTENANCE AND TROUBLESHOOTING RESOURCES

Maintenance and troubleshooting involves using information and communication technology including operator manuals, service bulletins, electronic monitoring systems, computers, telephones, information from machine operators, and advice from other technicians. Most manufacturers supply maintenance and troubleshooting recommendations and symptom diagnostic guides with their equipment. These materials should be organized for easy access. It is impossible to operate, maintain, and troubleshoot modern industrial equipment without using manufacturer manuals and other resources.

Manufacturer information is traditionally found in the operations and maintenance (operator) manual. Updates to the manual may be distributed as service bulletins. The same information that is in the manuals and service bulletins is often distributed on computer disk for use on-screen or to be printed as needed. Some companies operate online services that allow access to maintenance and troubleshooting information using a computer equipped with a modem.

In addition, equipment sales personnel can be excellent sources of maintenance assistance. Equipment sales personnel may also have contact with manufacturer representatives or company engineers who have useful suggestions.

Trade journals and magazines are often excellent sources of maintenance information if specific maintenance suggestions cannot be found. The magazine's advertisements often contain manufacturer contact numbers and may include manufacturer contact cards that allow a maintenance technician to obtain information from a variety of manufacturers. Often, specific maintenance procedures are discussed in articles or case studies from industry.

General preventive maintenance suggestions can be obtained from manuals for similar equipment. In addition, some manufacturers supply general maintenance suggestions. For example, a maintenance technician may study general material produced by electrical equipment manufacturers. **See Figure 19-34.**

RECOMMENDED MAINTENANCE PROCEDURES

Hermetic Motors

Annually:

- Take insulation resistance test of stator windings. Values below 50 MΩ at an ambient temperature of 85°F or less may indicate moisture in the winding insulation.
- Inspect the contacts in the magnetic motor starter for signs of deterioration.
- Check all line and load side terminals for loose connections.
- Test control relays for proper timing sequence.
- Measure line voltage and current load for proper balance.
- Test motors which have been tripped by any protective devices. Do not restart them until the windings have been tested and the motor starter circuits have been examined to determine the reason for tripping.

Reciprocating Compressors

Annually:

- Sample oil for analysis. The results will indicate any need for a special service or maintenance activity.
- Check the crankcase heater circuit for operation.
- Test the low oil pressure cutoff switch, which should be within the time delay rating and at the pressure differential specified by the compressor manufacturer. Replace it if it fails to function properly.

Figure 19-34. Some manufacturers supply general maintenance suggestions which can be used for equipment troubleshooting.

Codes and standards may be used by state and local authorities when dictating preventive maintenance requirements. The National Electrical Manufacturers Association (NEMA) produces recommendations for establishing a preventive maintenance system for industrial equipment in NEMA ICS 1.3, *Preventive Maintenance of Industrial Control and Systems Equipment*. Electrical maintenance guides are produced by the National Fire Protection Association (NFPA), publication NFPA 70B. Preventive maintenance requirements are also given in the ASME International boiler operating code, Section VI, *Recommended Rules for the Care and Operation of Heating Boilers*, and Section VII, *Recommended Rules for the Care of Power Boilers*.

The American Society of Heating, Refrigeration, and Air-Conditioning Engineers (ASHRAE) codes cover topics such as refrigeration, indoor air quality, ventilation standards, building operation and maintenance, and energy efficiency. Some regional and local building and mechanical codes also specify maintenance activities. Local and national codes are usually located in libraries and state or local offices, or can be purchased along with guides to using and applying the codes. Some insurance companies specify the maintenance that must be completed on equipment that they insure.

Operator Manuals

Maintenance and troubleshooting advice in operator manuals is probably the most commonly used resource for ensuring minimal system downtime. For example, hydraulic system components will be damaged if a technician uses the incorrect flushing fluid. Following the manufacturer instructions is essential to avoiding problems.

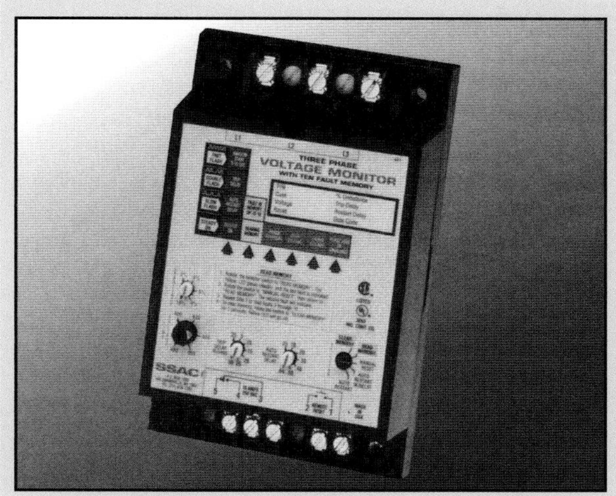

SSAC Inc.

A 3ϕ voltage monitor provides protection against premature equipment failure caused by voltage faults on the 3ϕ lines.

Operator manuals may contain troubleshooting information printed in chart form. The chart lists symptoms, possible causes, and suggestions for repairing the problem. **See Figure 19-35.** For example, a symptom of incorrect motor rotation may be caused by the wiring, and the solution is to interchange two of the load conductors.

Troubleshooting information may also be presented as a flow chart. **See Figure 19-36.** The flow chart is read by beginning at the start ellipse and following the arrows and answering yes or no to the questions in the diamonds. For example, if the answer to the question in the first diamond is no, the action is to turn power OFF then check to ensure the power is OFF. If the answer to the question is yes, the action is to check to ensure the power is OFF. The arrows are followed and the questions are answered by replying yes or no and following the respective paths that lead to the problem. The chart takes lengthy word descriptions and condenses them into a flow chart for quick problem solving.

Technical Fact

The most important function of a facility's maintenance group is implementation of a preventive maintenance program. Preventive maintenance such as cleaning, adjusting, exchanging, and lubricating on a regular basis eliminates and reduces plant shutdowns due to machine failure. Other functions include inspection, repair, overhaul, reconstruction, waste disposal, and plant protection.

Technical Service Bulletins and Troubleshooting Reports

Troubleshooting reports are patterned after the technical service bulletin system operated by the auto industry. After a new car model is introduced, information is gathered on its reliability and operational problems. All breakdowns or problems encountered in the manufacturer repair shops are recorded on a report form. Each report details one problem, its symptom(s), cause(s), and repair procedure(s). As patterns of breakdowns or problems develop, technical service bulletins are distributed to all manufacturer-operated repair facilities where they are used by mechanics. If a mechanic has already diagnosed and solved a problem, another mechanic can save time by using this information to repair the same problem in another vehicle.

The technical service bulletin system works well in a highly organized industry based on similar, widely distributed products such as automobiles. Because these conditions do not exist in most industrial settings, there are few service bulletin systems being operated by manufacturers of industrial equipment. The technical service bulletin system can be adapted for use in a plant by incorporating it into the plant preventive maintenance program as a troubleshooting report. A *troubleshooting report* is a record of a specific problem that occurs in a particular piece of equipment.

ELECTRIC MOTOR DRIVE TROUBLESHOOTING

ELECTRIC MOTOR DRIVE TROUBLESHOOTING MATRIX			
FAULTS			
SYMPTOM/FAULT CODE	**PROBLEM**	**CAUSE**	**SOLUTION**
ELECTRIC MOTOR DRIVE OVERVOLTAGE FAULT	ELECTRIC MOTOR DRIVE OVERVOLTAGE	DECELERATION TIME IS TOO SHORT	INCREASE DECELERATION TIME
		HIGH INPUT VOLTAGE (VOLTAGE SWELL)	*SEE INCOMING POWER TROUBLESHOOTING MATRIX*
		LOAD IS OVERHAULING MOTOR	ADD DYNAMIC BRAKING RESISTOR AND/OR INCREASE DECELERATION TIME
COMPONENT FAILURES			
ELECTRIC MOTOR DRIVE DOES NOT TURN ON. BLOWN FUSE OR TRIPPED BREAKER	DEFECTIVE CONVERTER SECTION (RECTIFIER SEMICONDUCTOR)	HIGH INPUT VOLTAGE (VOLTAGE SWELL)	REPLACE CONVERTER SECTION SEMICONDUCTOR OR REPLACE ELECTRIC MOTOR DRIVE *SEE ALSO INCOMING POWER MATRIX*
		ELECTRIC MOTOR DRIVE COOLING FAN IS DEFECTIVE	REPLACE CONVERTER SECTION SEMICONDUCTOR AND COOLING FAN OR REPLACE ELECTRIC MOTOR DRIVE
PARAMETER PROBLEMS			
UNUSUAL NOISES OR VIBRATIONS WHEN ELECTRIC MOTOR DRIVE POWERING MOTOR	PARAMETERS INCORRECT	PARAMETER(S) INCORRECTLY PROGRAMMED	ADJUST SKIP FREQUENCY PARAMETER
	PROBLEM WITH MOTOR AND/OR LOAD	PROBLEM WITH MOTOR AND/OR LOAD	*SEE MOTOR AND LOAD TROUBLESHOOTING MATRIX*
INPUT AND OUTPUT PROBLEMS			
ELECTRIC MOTOR DRIVE DOES NOT OPERATE CORRECTLY WHEN INPUT MODE IS OTHER THAN KEYPAD, MOTOR AND LOAD ARE CONNECTED, AND DRIVE IS OPERATED AS DESIGNED	EXTERNALLY CONNECTED INPUTS AND OUTPUTS INCORRECT	INPUT(S) AND/OR OUTPUT(S) INCORRECTLY WIRED. INPUT OR OUTPUT DEVICES NOT FUNCTIONAL	TIGHTEN LOOSE WIRES AND/OR REPLACE NON-FUNCTIONAL OR INCORRECT DEVICES FOR APPLICATION
		PROBLEM WITH INPUTS THAT SUPPLY START, STOP, REFERENCE, OR FEEDBACK SIGNALS	CHECK INPUT SYSTEM FOR PROPER INPUT
	PARAMETERS INCORRECT	PARAMETERS INCORRECTLY PROGRAMMED	*SEE ELECTRIC MOTOR DRIVE PARAMETER PROBLEMS*
OPERATIONAL PROBLEMS			
MOTOR ROTATION INCORRECT WHEN POWERED BY ELECTRIC MOTOR DRIVE	INCORRECT PHASING	WIRING	INTERCHANGE TWO OF THE LOAD CONDUCTORS AT THE ELECTRIC MOTOR DRIVE LOAD TERMINAL STRIP
MOTOR ROTATION INCORRECT WHEN IN BYPASS MODE	WIRING	WIRING	INTERCHANGE TWO LINE CONDUCTORS AT DISCONNECT *NOTE: ASSUMES ELECTRIC MOTOR DRIVE AND BYPASS SHARE COMMON FEED*

Figure 19-35. Troubleshooting information may be printed in chart form.

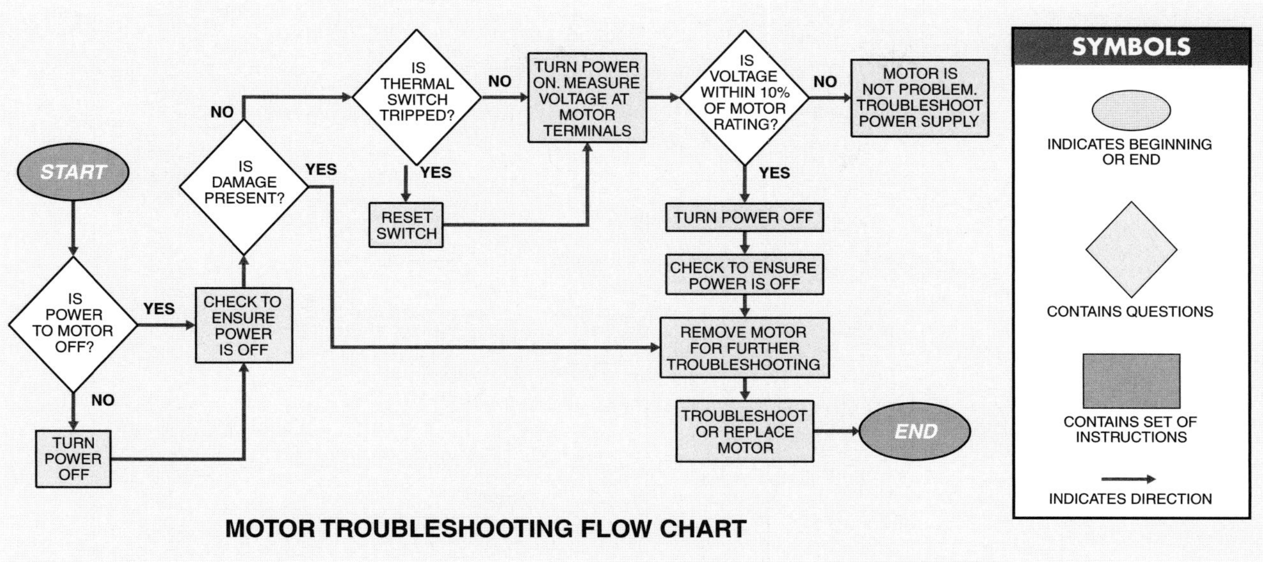

Figure 19-36. Troubleshooting information is often presented as a flow chart.

Maintenance technicians can evaluate troubleshooting reports to improve their own troubleshooting abilities. Troubleshooting reports that are incorporated into a plant preventive maintenance system become part of the equipment history for each machine. If the cause of the problem is discovered, modifications to the machine or adjustments to the machine preventive maintenance work are made. Over time, these modifications reflect the needs of each plant or equipment situation.

A troubleshooting report is filled out for each breakdown or equipment problem immediately after the problem is solved. This information is filed manually or entered into a computer for future reference. The next time a particular machine requires troubleshooting, the technician accesses the machine troubleshooting report to learn if the symptoms of the current problem have occurred previously. If so, the technician uses the information on the troubleshooting report to begin troubleshooting. Such information can result in a tremendous saving of time. Each troubleshooting report should include standard information such as the individual(s) who worked on the problem, the department, the equipment identification number, the problem, symptoms, cause(s), repair procedures, and preventive maintenance action. **See Figure 19-37.**

Over time, most equipment develops tendencies or problems that repeat. The use of troubleshooting reports enables maintenance technicians to develop plant procedures to deal with these repetitive problems. For example, the troubleshooting and replacement of a defective PLC module can be written as a procedure. The maintenance technician follows the procedure when replacing the module. **See Figure 19-38.**

Electronic Monitoring Systems

Many industrial systems are equipped with fault monitoring systems that display error codes. When a problem occurs in the system, numbers or words are displayed on a digital readout attached to the equipment or in a control room. The numbers or words are error codes that indicate the cause of the problem. The explanation of the code is found in the equipment manual and the technician follows instructions in the manual or tests for problems in the area indicated.

Some electronic fault monitoring systems are computerized. When a problem occurs, the monitoring system is accessed using a computer. The computer may be handheld or a desktop model connected to the monitoring system. The error code is displayed on the computer screen followed by instructions for testing components and making repairs.

Technical Fact

Maintenance technicians should hold monthly safety meetings to address problems and accidents that have occurred. Discuss safety regulations or changes in equipment and methods to be adopted for safety reasons. As the size of the organization increases, the extent to which meeting planning can be formalized and the amount of time that should be spent on meeting also increases.

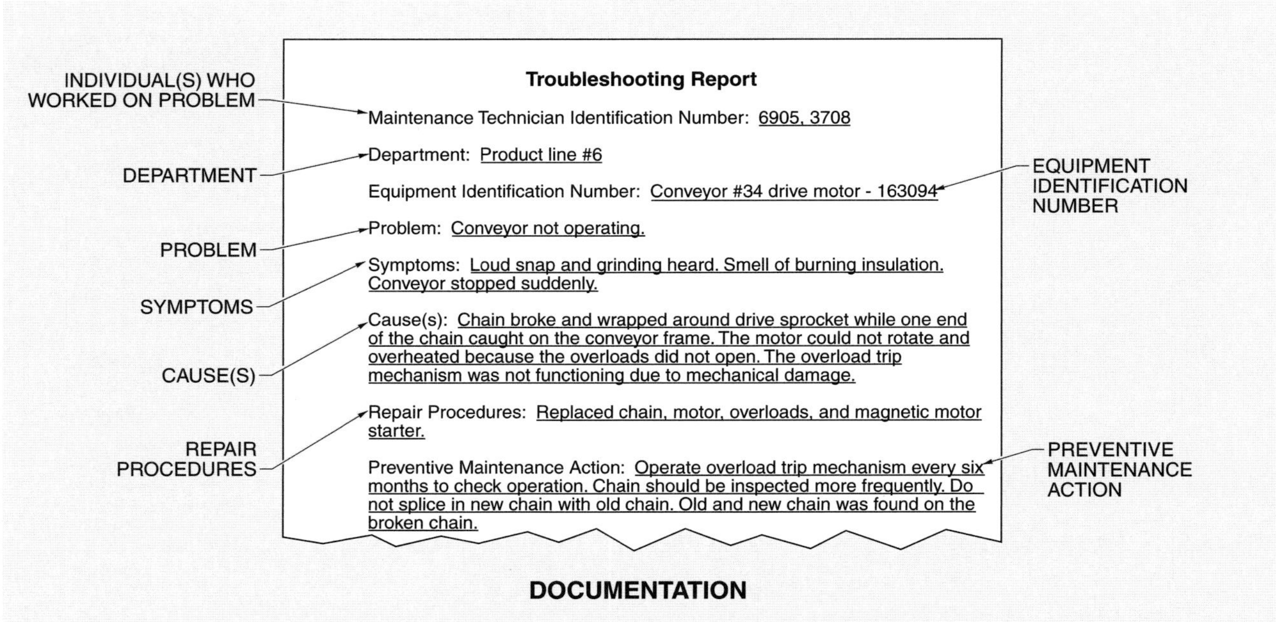

Figure 19-37. A troubleshooting report is a record of a specific problem that occurs in a particular piece of equipment.

Programmable Controller Troubleshooting and Replacement

1. If there is no indication of power (status lights OFF) on the programmable controller, measure the voltage at the incoming terminals on the power supply module.
2. If voltage is present and correct, replace power supply on programmable controller.
3. Remove power from the programmable controller.
4. Disconnect the power lines from the power supply terminals.
5. Disconnect the processor power cable from the power supply output terminal.
6. Remove the four mounting screws on power supply to free power supply from main panel.
7. Grasp the power supply firmly and pull out.
8. Press the replacement power supply into the main panel.
9. Replace and tighten the four mounting screws.
10. Connect the processor power cable.
11. Connect the power lines.
12. Turn power ON.

Figure 19-38. The use of troubleshooting reports enables maintenance technicians to develop plant procedures to deal with repetitive problems.

Computerized Maintenance Management Systems

Preventive maintenance systems use either a computerized maintenance management system (CMMS) or a paper-based system. Both can be purchased as a package or developed in-house. Computerized maintenance management system programs offer flexibility in the development and monitoring of the preventive maintenance system and quick access to maintenance information. Paper-based systems consist of binders, file folders, or other commercially produced organizers. Paper-based systems are less expensive, but do not provide the capabilities of a computerized maintenance management system. For example, in paper-based systems, cross-referencing files requires more time because several files must be searched by hand.

The preventive maintenance system selected depends on the operating budget, facility size, personnel considerations, and the objective of the system. Large facilities require a computerized maintenance management system. Small facilities can be managed from paper-based systems or a basic computerized maintenance management system. Integrating a new preventive maintenance system is best accomplished with an individual production line or facility where cost savings can be realized quickly.

Many computer maintenance management systems include general preventive maintenance tasks. These suggestions are available as part of the computer program and can be used as general suggestions for preventive maintenance tasks. If a computerized maintenance management system is used, a master schedule can be created with work orders projected automatically by day, week, month, or year. **See Figure 19-39.**

Out-of-Plant Services

Outside advice is often necessary when decisions must be made or work completed that requires expertise not found in a maintenance crew. In addition, installing new equipment or making major renovations may require companies with specialized tools and expertise. Outside companies can install new equipment or make major renovations more efficiently than a plant maintenance crew, who must maintain other plant operations. In addition, outside companies provide warranties for their work should problems occur during the equipment break-in period.

Performance of regular, operational inspections is necessary for implementation of troubleshooting procedures.

WORK ORDER PROJECTION - TASK BY WEEK

RF Industries

Week 2

Date	Task No.	WO No.	Equipment No.	Cost Center	Expense Class	Hours
7/12	BEARING-REPLACE	-	MOTOR-EXTRUD1-LN1	7001	MECH	2.0
7/12	BEARING-REPLACE	-	MOTOR-EXTRUD1-LN1	7001	MECH	2.0
Total						4.0

Week 3

Date	Task No.	WO No.	Equipment No.	Cost Center	Expense Class	Hours
7/19	BEARING-REPLACE	-	MOTOR-EXTRUD1-LN1	7001	MECH	2.0
7/19	BEARING-REPLACE	-	MOTOR-EXTRUD1-LN1	7001	MECH	2.0
7/19	EXTRD-MTR-BELT-3M	-	MOTOR-EXTRD2-LN2	7001	MECH	2.0
7/19	EXTRDSCREW-BRNG-6M	-	SCREW-EXTRD1-LN2	7001	MECH	4.0
7/19	EXTRDSCREW-BRNG-6M	-	SCREW-EXTRD1-LN2	7001	MECH	4.0
Total						14.0

Week 4

Date	Task No.	WO No.	Equipment No.	Cost Center	Expense Class	Hours
7/26	BEARING-REPLACE	-	MOTOR-EXTRUD1-LN1	7001	MECH	2.0
7/26	FKLFT-PM-1M	-	-MULTITASK-		COMB	6.0
7/26	BEARING-REPLACE	-	MOTOR-EXTRUD1-LN1	7001	MECH	2.0
7/26	DIE-CLEAN	-	DIE-LN2	7001	MECH	2.0
Total						12.0

Datastream Systems, Inc.

Figure 19-39. A computerized maintenance management system can be used to project future maintenance tasks.

Outside experts, manufacturers, and distributors can be consulted by telephone, fax, or computer (via the Internet or e-mail). Many manufacturers have toll-free numbers that provide access to service personnel. Sometimes, technical support services are part of a service contract and can be costly. Therefore, these resources must be used efficiently. When making calls to any outside support services for troubleshooting help, all symptoms should be written down, manuals and other service material should be nearby, and any computers should be ready to use. Portable phones allow for such calls to be made from the equipment location. All tools and test equipment should be within reach and ready to use because the service technician may need the results of tests and inspections.

In some cases, service technicians can access the equipment fault monitoring system using their own computer and run diagnostic tests from many thousands of miles away. Fax machines are commonly used to send parts lists, questions, and other written material or illustrations of equipment. Such faxed material should be easy to view and written in large, easy-to-read type to avoid communication problems.

Some companies run computer-based information services where information is posted for those using the company's equipment. The service is accessed using a computer through which computer programs or service information can be transferred or questions asked of the service technicians.

Inventory Control

Inventory control is the organization and management of commonly used parts, vendors and suppliers, and purchasing records in the preventive maintenance system. **See Figure 19-40.** Accurate inventory records are necessary for any preventive maintenance program. Without the proper items in inventory, a maintenance or production department will have unnecessary downtime, causing lost production and lost revenue. In some systems, replacement parts are scanned with a barcode reader and recorded in the computer. A *barcode* is a code consisting of a group of variously patterned bars and spaces designed to be scanned and read into computer memory as identification for an object. A *barcode reader* is a device that identifies an object and manipulates information about the object that is recorded in a computer system based on the barcode. In an inventory control system, a part is assigned a number that is printed in computer code on a stick-on label. The barcode reader is passed over the label when a part is removed. The inventory control computer records that one part has been removed and subtracts one part from the total number to keep an up-to-date record of parts available.

The computer issues a purchase order for the number of parts required when supplies are close to depletion. Parts and supplies not used are identified and eliminated from the reorder list. This helps to maintain an adequate inventory to support operations without excessive parts and material. A computerized maintenance management system without a barcode reader requires replacement part information to be manually input into the computer. Inventory amounts are then automatically calculated to reflect inventory added or removed.

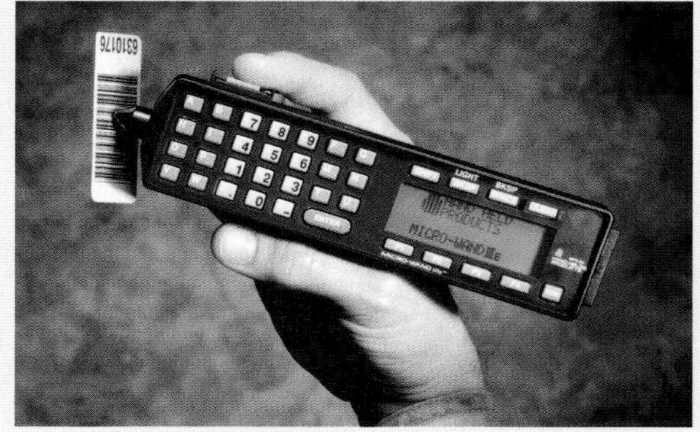

Hand Held Products

Figure 19-40. Inventory control requires organization and management of parts commonly used for maintenance tasks.

Review Questions

1. What is the purpose of preventive maintenance in industrial plants?
2. How many electrical degrees apart is each power line in a balanced three-phase system?
3. As little as a 3% phase imbalance can cause how much of a power loss in a motor?
4. Why does measuring the voltage at a motor that is single phasing give a voltage measurement on the power line that is open?
5. What is a voltage surge?
6. A motor should operate satisfactorily as long as the voltage is within what percent of the motor's voltage rating?
7. What is overcycling?
8. How does improper phase sequence affect a 3ϕ motor?
9. What is a megohmmeter?
10. Why should extra caution be taken when using a megohmmeter?
11. What is an insulation spot test?
12. What is a dielectric absorption test?
13. What is an insulation step voltage test?
14. How many external leads does a standard dual-voltage wye motor have?
15. How many external leads does a standard dual-voltage delta motor have?
16. How many separate circuits does a dual-voltage wye motor have?
17. How many separate circuits does a dual-voltage delta motor have?
18. What do terminal markings A1 and A2 indicate on a DC motor?
19. What do terminal markings S1 and S2 indicate on a DC motor?
20. What do terminal markings F1 and F2 indicate on a DC motor?
21. What is a troubleshooting report?
22. What information should always be included on a troubleshooting report?
23. What is an electronic fault monitoring system?
24. In what ways are computers utilized with electronic fault monitoring systems?
25. What is a computerized maintenance management system?
26. What advantages does a computerized maintenance management system have as opposed to a paper-based system?

Review Question Answer Key

1—ELECTRICAL QUANTITIES AND CIRCUITS

1. Potential energy and kinetic energy
3. Transformer
5. Amperes (A)
7. Watts (W)
9. True power
11. Apparent power
13. Power factor
15. Resistance
17. Impedance
19. Decrease
21. Decrease

2—ELECTRICAL TOOLS AND TEST INSTRUMENTS

1. Pegboards, tool pouches, toolboxes, chests, and cabinets
3. The ground ensures that any short circuit trips the circuit breaker or blows the fuse.
5. No, when using wrenches, never use a pipe extension or other form of "cheater" to increase the leverage of the wrench.
7. Diagonal-cutting pliers and side-cutting pliers
9. Fish tape
11. Fuse puller
13. An electronic device that displays meter readings as numerical values
15. By a light-emitting diode (LED) or a liquid crystal display (LCD)
17. A voltage that appears on a meter that is not connected to a circuit
19. Receptacle tester
21. Voltmeter or multimeter

3—ELECTRICAL SAFETY

1. Grounding is the connection of all exposed non-current-carrying metal parts to the earth.
3. When power is not required to be ON to a piece of equipment to perform a task, when machine guards or other safety devices are removed or bypassed, when the possibility exists of being injured or caught in moving machinery, when jammed equipment is being cleared, or when the danger exists of being injured if equipment power is turned ON
5. Ordinary combustibles
7. Electrical equipment
9. Once a month
11. When protection from flying objects is required
13. Severe muscular contractions, paralysis of breathing, heart convulsions
15. 10′ plus 4″ for every 10 kV over 50 kV

4—ELECTRICAL SYMBOLS AND DIAGRAMS

1. To show the logic of an electrical circuit or system using standard symbol
3. Between the motor and L2
5. By a zigzag symbol
7. Contacts which may be added to a contactor
9. An electrically operated switch (contactor) that includes motor overload protection
11. Any circuit that requires a person to initiate an action for it to operate
13. Identical except for the overloads attached to them
15. From left to right
17. A wiring diagram is used to show as closely as possible the actual location of each component in a circuit.
19. The switch is normally closed.
21. Load, source of electricity, conductors, switch, and protection device
23. A letter or combination of letters that represents a word
25. An assembly of conductors and electrical devices through which current flows

5—LOGIC APPLIED TO LINE DIAGRAMS

1. 1
3. L2
5. By numbering
7. Top left to bottom right
9. Signal(s), decision(s), and action
11. Section of circuit which determines what work is to be done and in what order the work is to occur
13. Load is ON only if all control signal contacts are closed.
15. Load is ON if control signal is OFF.
17. Two or more normally closed contacts are connected in parallel. Only opening all of the control signals' contacts will turn the load OFF.
19. In series (NOR logic)
21. Frequent starting and stopping of a motor for short periods of time
23. A short circuit that opens a circuit as soon as the circuit is energized or when the section of the circuit containing the short is energized

6—SOLENOIDS, DC GENERATORS, AND DC MOTORS

1. Permanent and temporary
3. By increasing the current to increase the voltage, increasing the number of coils, and inserting an iron core through the coils
5. To reduce eddy currents produced by transformer action in the metal
7. A shading coil sets up an auxiliary magnetic field which is out of phase with the main coil magnetic field to help hold in the armature as the main coil magnetic field drops to zero.
9. The coil draws more than its rated current and may overheat.
11.
- Obtain full data on load requirements.
- Allow for possible low-voltage conditions of the power supply.
- Use shortest possible stroke.
- Never use an oversized solenoid.

13. Coil burnout or mechanical damage
15. Amount of voltage applied to the coil and amount of current allowed to pass through the coil
17. Copper or aluminum
19. DC generators consist of field windings, an armature, a commutator, and brushes.
21. Many coils of wire

23. Brush sparking, chattering, or a rough commutator indicates service is required.
25. When they have worn down to about half of their original size
27. A continuity tester can give results quickly when there is a problem.

7—AC GENERATORS, TRANSFORMERS, AND AC MOTORS

1. Field windings produce the magnetic field in a generator.
3. A connection that has each coil end connected end-to-end to form a closed loop
5. +5% to −10%
7. A transformer with more turns on the secondary than on the primary
9. Because of transformer losses such as resistive loss, eddy current loss, and hysteresis loss
11. A split-phase motor is a single-phase AC motor that includes a running winding (main winding) and a starting winding (auxiliary winding).
13. Select a transformer that safely and efficiently provides the maximum current that can be drawn by a load.
15. Capacitors and shaded poles
17. The capacitor may have a short circuit or an open circuit, or may deteriorate to the point where it must be replaced.
19. Shaded-pole motors

8—CONTACTORS AND MOTOR STARTERS

1. Exposed (live) parts; speed of opening and closing contacts is determined solely by operator; soft copper knife switches require replacement after repeated arcing; heat generation; mechanical fatigue
3. To protect the gap between the set of fixed contacts as the contacts make or break the circuit
5. A mechanical interlock ensures that both sets of contacts cannot be closed at the same time.
7. The NEC® requires that the control device shall not only turn a motor ON and OFF, but shall also protect the motor from destroying itself under an overload situation, such as a locked rotor.
9. To not open the circuit while the motor is starting, and to open the circuit if the motor gets overloaded and the fuses do not blow
11. AC contactor assemblies may have several sets of contacts and DC contactor assemblies have only one set.
13. Because the continuous DC supply causes current to flow constantly and with great stability across a much wider gap than does an AC supply of equal voltage
15. A contactor that includes overload protection
17. The percentage of extra demand that can be placed on a motor for short intervals without damaging the motor
19. An overload device located directly on or in a motor to provide overload protection
21. By changing the voltage applied to the motor

9—CONTROL DEVICES

1. One normally open and one normally closed contact
3. Mushroom
5. To select or determine one of several different circuit conditions
7. A joystick is an operator that selects one to eight different circuit conditions by shifting the joystick from the center position into one of the other positions.
9. A circle with a dot and a truth table
11. The part of a limit switch that transfers the mechanical force of the moving part to the electrical contacts
13. A switch that automatically turns lamps ON at dusk and OFF at dawn
15. Normally open
17. The contacts may chatter ON and OFF.
19. A switch that responds to the intensity of heat
21. A switch used to detect the movement of a fluid
23. No
25. Two

10—REVERSING MOTOR CIRCUITS

1. By interchanging any two of the three main power lines
3. By reversing the connections to the starting or running windings
5. To keep the reversing circuit from closing
7. One NO and one NC
9. Normally closed
11. The part of an electrical circuit that connects the loads to the main power lines
13. Advantages: It is the oldest, most straightforward motor control wiring method used. It is wired point-to-point. It will operate properly for a period of time. Disadvantages: Troubleshooting and circuit modifications are difficult and time consuming.

11—POWER DISTRIBUTION SYSTEMS

1. One-line diagram
3. Phase-to-neutral, phase-to-phase, and phase-to-phase-to-phase
5. 240 V
7. X1 and X2
9. If one transformer is damaged or removed from service, the other two transformers can be connected in an open-delta connection.
11. Switchboard
13. A wall-mounted distribution cabinet containing a group of overcurrent devices for lighting, appliance, or power distribution branch circuits
15. To receive incoming power and deliver it to the control circuit and motor loads

12—SOLID-STATE DEVICES AND SYSTEM INTEGRATION

1. Valence electrons
3. An insulating material such as fiberglass or phenolic with conducting paths secured to one or both sides
5. A circuit containing a diode which permits only the positive half-cycles of the AC sine wave to pass
7. A thermally sensitive resistor whose resistance changes with a change in temperature
9. A sensor that produces a voltage depending on the strength of the magnetic field applied to the sensor
11. A diode that produces light when current flows through it
13. By its transistor outline (TO) number
15. A ratio of the amplitude of the output signal to the amplitude of the input signal
17. The voltage required to switch an SCR into a conductive state
19. A triac operates much like a pair of SCRs connected in a reverse-parallel arrangement. The triac conducts if the appropriate signal is applied to the gate.

21. Thousands of semiconductors providing a complete circuit function in one small semiconductor package
23. A very high gain, directly coupled amplifier that uses external feedback to control response characteristics
25. AND, OR, NAND, and NOR

13—TIMERS AND COUNTERS

1. Dashpot, synchronous clock, solid-state, and programmable
3. By the speed at which a synchronous clock motor operates clock hands
5. A device that has a preset time period that must pass after the timer has been energized before any action occurs on the timer contacts
7. A device that does not start its timing function until the power is removed from the timer
9. A device in which the contacts change position immediately and remain changed for a set period of time after the timer has received power
11. A device in which the contacts cycle open and closed repeatedly once the timer has received power
13. The control switch can be wired using low-voltage wiring and switch contacts can be rated at a lower current level.
15. A sensor-controlled timer
17. With an X
19. The relay ON periods
21. A retentive timer maintains its current accumulated time value when its control signal is interrupted.

14—RELAYS AND SOLID-STATE STARTERS

1. EMR and SSR
3. No
5. Yes
7. Single-pole, double-throw
9. 2 (single-break or double-break)
11. 1
13. A higher mechanical life rating
15. Zero switching
17. Voltage at the load starts at a low level and is increased over a period of time.
19. Triacs
21. It increases.
23. The amount of current that leaks through an SSR when the switch is turned OFF
25. When a relay fails to turn OFF because the current and voltage in the circuit reach zero at different times

15—SENSING DEVICES AND CONTROLS

1. A method of scanning in which the transmitter and receiver are placed opposite each other so the light beam from the transmitter shines directly at the receiver
3. A method of scanning in which the transmitter and receiver are housed in the same enclosure and the transmitted light beam is reflected back to the receiver from a reflector
5. A method of scanning in which the transmitter and receiver are placed at equal angles to a highly reflective surface
7. A method of scanning that simultaneously focuses and converges a light beam to a fixed focal point in front of the photoreceiver
9. When the object to be detected moves at a high speed or the object to be detected is not much bigger than the effective beam of the controller
11. When the target is missing
13. Direct mode and diffused mode
15. A magnetic field
17. A transistor or an SCR
19. Residual or leakage current
21. A sensor that detects the movement (flow) of a liquid or gas using a solid-state device.
23. NPN
25. Capacitive

16—PROGRAMMABLE CONTROLLERS

1. Discrete parts manufacturing and process manufacturing
3. Process manufacturing
5. Ladder (line) diagrams
7. Pushbuttons, temperature switches, and limit switches
9. The time it takes a PLC to make a sweep of the program
11. To provide back-up power for the processor memory in case of an external power failure
13. Unwanted signals that are present on a power line
15. A method of transmitting more than one signal over a single transmission system
17. NC
19. **Rule 1:** Inputs (normally open, normally closed, and special) are placed on the left side of the circuit between the left power rung and the output. Outputs are placed on the right side of the circuit.

    Rule 2: Only one output can be placed on a rung. This means that outputs can be placed in parallel but never in series.

    Rule 3: Inputs can be placed in series, parallel, or in series/parallel combinations.

    Rule 4: Inputs can be programmed at multiple locations in the circuit. An input (the same input) can be programmed as normally open and/or normally closed at multiple locations.

    Rule 5: Standard outputs cannot be programmed at multiple locations in the circuit. There is a special output called an "or-output" that allows an output to be placed in more than one location but only if the "or-out" special function is identified when programming the output.

17—REDUCED-VOLTAGE STARTING

1. To reduce interference in the power source, load, and electrical environment surrounding the motor
3. Torque is reduced.
5. The amount of current required by the motor to produce full-load torque at the motor's rated speed
7. When the motor is temporarily disconnected when moving from one incremental voltage to another
9. Less expensive
11. 6
13. SCRs
15. Transistors
17. AC
19. A drop in voltage of more than 10% (but not to 0 V) below the normal rated line voltage lasting from .5 cycles up to 1 minute
21. Solid-state reduced-voltage starters ramp up motor voltage as the motor accelerates instead of applying full voltage instantaneously. Solid-state starters also reduce inrush current and reduce starting torque.

18—ACCELERATING AND DECELERATING METHODS

1. A friction surface comes in contact with a wheel mounted onto the motor shaft

3. $T = \frac{5252 \times HP}{rpm}$
5. At a given rpm
7. OFF-delay
9. To prevent both the AC and DC from being connected at the same time
11. Applying a force over a distance
13. The torque required to produce the rated power at full speed of a motor
15. The torque required to bring a load up to its rated speed
17. The motor's characteristics of HP, torque, and speed
19. The DC shunt motor has fairly high torque at any speed.
21. By changing the voltage across the armature or field
23. 900 rpm
25. A control function that permits the operator to select a high motor speed and the control circuit to automatically accelerate the motor to that speed
27. Coast stop method

19—PREVENTIVE MAINTENANCE AND TROUBLESHOOTING

1. To keep machines, assembly lines, production operations, and plant operations running with little or no downtime
3. A 10% loss
5. Any higher than normal voltage that temporarily exists on one or more of the power lines
7. The process of turning a motor ON and OFF repeatedly
9. An instrument that measures high resistance values under high test voltage conditions
11. A test that checks motor insulation over the life of the motor
13. A test that creates electrical stress on internal insulation cracks to reveal aging or damage
15. 9
17. 3
19. Series field leads
21. A detailed record of a specific problem, its symptom(s), causes(s), and repair procedure(s)
23. A computerized system that displays error codes followed by instructions for testing components and making repairs
25. A preventive maintenance system that utilizes computerized data (rather than paper copy)

Appendix

METRIC PREFIXES			
Multiples and Submultiples	**Prefixes**	**Symbols**	**Meaning**
1,000,000,000,000 = 10^{12}	tera	T	trillion
1,000,000,000 = 10^{9}	giga	G	billion
1,000,000 = 10^{6}	mega	M	million
1000 = 10^{3}	kilo	k	thousand
100 = 10^{2}	hecto	h	hundred
10 = 10^{1}	deka	d	ten
Unit 1 = 10^{0}			
.1 = 10^{-1}	deci	d	tenth
.01 = 10^{-2}	centi	c	hundredth
.001 = 10^{-3}	milli	m	thousandth
.000001 = 10^{-6}	micro	μ	millionth
.000000001 = 10^{-9}	nano	n	billionth
.000000000001 = 10^{-12}	pico	p	trillionth

METRIC CONVERSIONS												
Initial Units	**Final Units**											
	giga	**mega**	**kilo**	**hecto**	**deka**	**base unit**	**deci**	**centi**	**milli**	**micro**	**nano**	**pico**
giga		3R	6R	7R	8R	9R	10R	11R	12R	15R	18R	21R
mega	3L		3R	4R	5R	6R	7R	8R	9R	12R	15R	18R
kilo	6L	3L		1R	2R	3R	4R	5R	6R	9R	12R	15R
hecto	7L	4L	1L		1R	2R	3R	4R	5R	8R	11R	14R
deka	8L	5L	2L	1L		1R	2R	3R	4R	7R	10R	13R
base unit	9L	6L	3L	2L	1L		1R	2R	3R	6R	9R	12R
deci	10L	7L	4L	3L	2L	1L		1R	2R	5R	8R	11R
centi	11L	8L	5L	4L	3L	2L	1L		1R	4R	7R	10R
milli	12L	9L	6L	5L	4L	3L	2L	1L		3R	6R	9R
micro	15L	12L	9L	8L	7L	6L	5L	4L	3L		3R	6R
nano	18L	15L	12L	11L	10L	9L	8L	7L	6L	3L		3R
pico	21L	18L	15L	14L	13L	12L	11L	10L	9L	6L	3L	

COMMON PREFIXES		
Symbol	**Prefix**	**Equivalent**
G	giga	1,000,000,000
M	mega	1,000,000
k	kilo	1000
base unit	—	1
m	milli	0.001
μ	micro	0.000001
n	nano	0.000000001
p	pico	0.000000000001
Z	impedance	ohms —

THREE-PHASE VOLTAGE VALUES
For 208 V × 1.732, use 360
For 230 V × 1.732, use 398
For 240 V × 1.732, use 416
For 440 V × 1.732, use 762
For 460 V × 1.732, use 797
For 480 V × 1.732, use 831
For 2400 V × 1.732, use 4157
For 4160 V × 1.732, use 7205

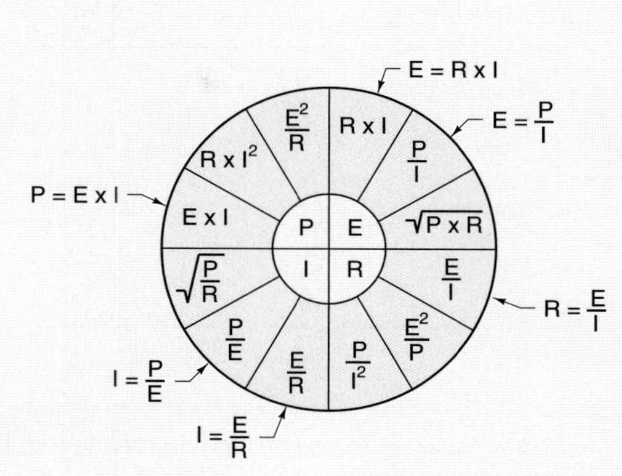

OHM'S LAW AND POWER FORMULA

POWER FORMULA ABBREVIATIONS AND SYMBOLS	
P = Watts	V = Volts
I = Amps	VA = Volt Amps
A = Amps	φ = Phase
R = Ohms	√ = Square Root
E = Volts	

POWER FORMULAS — 1φ, 3φ					
Phase	To Find	Use Formula	Example		
			Given	Find	Solution
1φ	I	$I = \frac{VA}{V}$	32,000 VA, 240 V	I	$I = \frac{VA}{V}$ $I = \frac{32,000\text{ VA}}{240\text{ V}}$ $I = \mathbf{133\text{ A}}$
1φ	VA	$VA = I \times V$	100 A, 240 V	VA	$VA = I \times V$ $VA = 100\text{ A} \times 240\text{ V}$ $VA = \mathbf{24,000\text{ VA}}$
1φ	V	$V = \frac{VA}{I}$	42,000 VA, 350 A	V	$V = \frac{VA}{I}$ $V = \frac{42,000\text{ VA}}{350\text{ A}}$ $V = \mathbf{120\text{ V}}$
3φ	I	$I = \frac{VA}{V \times \sqrt{3}}$	72,000 VA, 208 V	I	$I = \frac{VA}{V \times \sqrt{3}}$ $I = \frac{72,000\text{ VA}}{360\text{ V}}$ $I = \mathbf{200\text{ A}}$
3φ	VA	$VA = I \times V \times \sqrt{3}$	2 A, 240 V	VA	$VA = I \times V \times \sqrt{3}$ $VA = 2 \times 416$ $VA = \mathbf{832\text{ VA}}$

CAPACITORS

Connected in Series		Connected in Parallel	Connected in Series/Parallel
Two Capacitors	Three or More Capacitors		
$C_T = \frac{C_1 \times C_2}{C_1 + C_2}$ where C_T = total capacitance (in µF) C_1 = capacitance of capacitor 1 (in µF) C_2 = capacitance of capacitor 2 (in µF)	$\frac{1}{C_T} = \frac{1}{C_1} + \frac{1}{C_2} + \ldots$	$C_T = C_1 + C_2 + \ldots$	1. Calculate the capacitance of the parallel branch. 2. Calculate the capacitance of the series combination. $C_T = \frac{C_1 \times C_2}{C_1 + C_2}$

TEMPERATURE CONVERSIONS

Convert °C to °F	Convert °F to °C
$°F = (1.8 \times °C) + 32$	$°C = \frac{°F - 32}{1.8}$

UNITS OF POWER

Power	W	ft lb/s	HP	kW
Watt	1	0.7376	$.341 \times 10^{-3}$	0.001
Foot-pound/sec	1.356	1	$.818 \times 10^{-3}$	1.356×10^{-3}
Horsepower	745.7	550	1	0.7457
Kilowatt	1000	736.6	1.341	1

BRANCH CIRCUIT VOLTAGE DROP

$$\%V_D = \frac{V_{NL} - V_{FL}}{V_{FL}} \times 100$$

where

$\%V_D$ = percent voltage drop (in volts)

V_{NL} = no-load voltage drop (in volts)

V_{FL} = full-load voltage drop (in volts)

100 = constant

STANDARD SIZES OF FUSES AND CBs

NEC® 240-6(a) lists standard ampere ratings of fuses and fixed-trip CBs as follows:

15, 20, 25, 30, 35, 40, 45,
50, 60, 70, 80, 90, 100, 110,
125, 150, 175, 200, 225,
250, 300, 350, 400, 450,
500, 600, 700, 800,
1000, 1200, 1600,
2000, 2500, 3000, 4000, 5000, 6000

VOLTAGE CONVERSIONS

To Convert	To	Multiply By
rms	Average	.9
rms	Peak	1.414
Average	rms	1.111
Average	Peak	1.567
Peak	rms	.707
Peak	Average	.637
Peak	Peak-to-peak	2

ELECTRICAL/ELECTRONIC ABBREVIATIONS/ACRONYMS

Abbr/ Acronym	Meaning	Abbr/ Acronym	Meaning	Abbr/ Acronym	Meaning
A	Ammeter; Ampere; Anode; Armature	FU	Fuse	PNP	Positive-Negative-Positive
AC	Alternating Current	FWD	Forward	POS	Positive
AC/DC	Alternating Current; Direct Current	G	Gate; Giga; Green; Conductance	POT.	Potentiometer
A/D	Analog to Digital	GEN	Generator	P-P	Peak-to-Peak
AF	Audio Frequency	GRD	Ground	PRI	Primary Switch
AFC	Automatic Frequency Control	GY	Gray	PS	Pressure Switch
Ag	Silver	H	Henry; High Side of Transformer; Magnetic Flux	PSI	Pounds Per Square Inch
ALM	Alarm			PUT	Pull-Up Torque
AM	Ammeter; Amplitude Modulation	HF	High Frequency	Q	Transistor
AM/FM	Amplitude Modulation; Frequency Modulation	HP	Horsepower	R	Radius; Red; Resistance; Reverse
		Hz	Hertz	RAM	Random-Access Memory
ARM.	Armature	I	Current	RC	Resistance-Capacitiance
Au	Gold	IC	Integrated Circuit	RCL	Resistance-Inductance-Capacitance
AU	Automatic	INT	Intermediate; Interrupt	REC	Rectifier
AVC	Automatic Volume Control	INTLK	Interlock	RES	Resistor
AWG	American Wire Gauge	IOL	Instantaneous Overload	REV	Reverse
BAT.	Battery (electric)	IR	Infrared	RF	Radio Frequency
BCD	Binary Coded Decimal	ITB	Inverse Time Breaker	RH	Rheostat
BJT	Bipolar Junction Transistor	ITCB	Instantaneous Trip Circuit Breaker	rms	Root Mean Square
BK	Black	JB	Junction Box	ROM	Read-Only Memory
BL	Blue	JFET	Junction Field-Effect Transistor	rpm	Revolutions Per Minute
BR	Brake Relay; Brown	K	Kilo; Cathode	RPS	Revolutions Per Second
C	Celsius; Capacitiance; Capacitor	L	Line; Load; Coil; Inductance	S	Series; Slow; South; Switch
CAP.	Capacitor	LB-FT	Pounds Per Foot	SCR	Silicon Controlled Rectifier
CB	Circuit Breaker; Citizen's Band	LB-IN.	Pounds Per Inch	SEC	Secondary
CC	Common-Collector Configuration	LC	Inductance-Capacitance	SF	Service Factor
CCW	Counterclockwise	LCD	Liquid Crystal Display	1 PH; 1ϕ	Single-Phase
CE	Common-Emitter Configuration	LCR	Inductance-Capacitance-Resistance	SOC	Socket
CEMF	Counter Electromotive Force	LED	Light Emitting Diode	SOL	Solenoid
CKT	Circuit	LRC	Locked Rotor Current	SP	Single-Pole
CONT	Continuous; Control	LS	Limit Switch	SPDT	Single-Pole, Double-Throw
CPS	Cycles Per Second	LT	Lamp	SPST	Single-Pole, Single-Throw
CPU	Central Processing Unit	M	Motor; Motor Starter; Motor Starter Contacts	SS	Selector Switch
CR	Control Relay			SSW	Safety Switch
CRM	Control Relay Master	MAX.	Maximum	SW	Switch
CT	Current Transformer	MB	Magnetic Brake	T	Tera; Terminal; Torque; Transformer
CW	Clockwise	MCS	Motor Circuit Switch	TB	Terminal Board
D	Diameter; Diode; Down	MEM	Memory	3 PH; 3ϕ	Three-Phase
D/A	Digital to Analog	MED	Medium	TD	Time Delay
DB	Dynamic Braking Contactor; Relay	MIN	Minimum	TDF	Time Delay Fuse
DC	Direct Current	MN	Manual	TEMP	Temperature
DIO	Diode	MOS	Metal-Oxide Semiconductor	THS	Thermostat Switch
DISC.	Disconnect Switch	MOSFET	Metal-Oxide Semiconductor Field-Effect Transistor	TR	Time Delay Relay
DMM	Digital Multimeter			TTL	Transistor-Transistor Logic
DP	Double-Pole	MTR	Motor	U	Up
DPDT	Double-Pole, Double-Throw	N; NEG	North; Negative	UCL	Unclamp
DPST	Double-Pole, Single-Throw	NC	Normally Closed	UHF	Ultrahigh Frequency
DS	Drum Switch	NEUT	Neutral	UJT	Unijunction Transistor
DT	Double-Throw	NO	Normally Open	UV	Ultraviolet; Undervoltage
DVM	Digital Voltmeter	NPN	Negative-Positive-Negative	V	Violet; Volt
EMF	Electromotive Force	NTDF	Nontime-Delay Fuse	VA	Volt Amp
F	Fahrenheit; Fast; Field; Forward; Fuse	O	Orange	VAC	Volts Alternating Current
FET	Field-Effect Transistor	OCPD	Overcurrent Protection Device	VDC	Volts Direct Current
FF	Flip-Flop	OHM	Ohmmeter	VHF	Very High Frequency
FLC	Full-Load Current	OL	Overload Relay	VLF	Very Low Frequency
FLS	Flow Switch	OZ/IN.	Ounces Per Inch	VOM	Volt-Ohm-Milliammeter
FLT	Full-Load Torque	P	Peak; Positive; Power; Power Consumed	W	Watt; White
FM	Fequency Modulation	PB	Pushbutton	w/	With
FREQ	Frequency	PCB	Printed Circuit Board	X	Low Side of Transformer
FS	Float Switch	PH;	Phase	Y	Yellow
FTS	Foot Switch	PLS	Plugging Switch	Z	Impedance

AC MOTOR CHARACTERISTICS

Motor Type	Typical Voltage	Starting Ability (Torque)	Size (HP)	Speed Range (rpm)	Cost*	Typical Uses
1ϕ						
Shaded-pole	115 V, 230 V	Very low 50% to 100% of full load	Fractional ½ HP to ⅓ HP	Fixed 900, 1200, 1800, 3600	Very low 75% to 85%	Light-duty applications such as small fans, hair dryers, blowers, and computers
Split-phase	115 V, 230 V	Low 75% to 200% of full load	Fractional ⅓ HP or less	Fixed 900, 1200, 1800, 3600	Low 85% to 95%	Low-torque applications such as pumps, blowers, fans, and machine tools
Capacitor-start	115 V, 230 V	High 200% to 350% of full load	Fractional to 3 HP	Fixed 900, 1200, 1800	Low 90% to 110%	Hard-to-start loads such as refrigerators, air compressors, and power tools
Capacitor-run	115 V, 230 V	Very low 50% to 100% of full load	Fractional to 5 HP	Fixed 900, 1200, 1800	Low 90% to 110%	Applications that require a high running torque such as pumps and conveyors
Capacitor-start-and-run	115 V, 230 V	Very high 350% to 450% of full load	Fractional to 10 HP	Fixed 900, 1200, 1800	Low 100% to 115%	Applications that require both a high starting and running torque such as loaded conveyors
3ϕ						
Induction	230 V, 460 V	Low 100% to 175% of full load	Fractional to over 500 HP	Fixed 900, 1200, 3600	Low 100%	Most industrial applications
Wound rotor	230 V, 460 V	High 200% to 300% of full load	½ HP to 200 HP	Varies by changing resistance in rotor	Very high 250% to 350%	Applications that require high torque at different speeds such as cranes and elevators
Synchronous	230 V, 460 V	Very low 40% to 100% of full load	Fractional to 250 HP	Exact constant speed	High 200% to 250%	Applications that require very slow speeds and correct power factors

* based on standard 3ϕ induction motor

DC AND UNIVERSAL MOTOR CHARACTERISTICS

Motor Type	Typical Voltage	Starting Ability (Torque)	Size (HP)	Speed Range (rpm)	Cost*	Typical Uses
DC Series	12 V, 90 V, 120 V, 180 V	Very high 400% to 450% of full load	Fractional to 100 HP	Varies 0 to full speed	High 175% to 225%	Applications that require very high torque such as hoists and bridges
Shunt	12 V, 90 V, 120 V, 180 V	Low 125% to 250% of full load	Fractional to 100 HP	Fixed or adjustable below full speed	High 175% to 225%	Applications that require better speed control than a series motor such as woodworking machines
Compound	12 V, 90 V, 120 V, 180 V	High 300% to 400% of full load	Fractional to 100 HP	Fixed or adjustable	High 175% to 225%	Applications that require high torque and speed control such as printing presses, conveyors, and hoists
Permanent-magnet	12 V, 24 V, 36 V, 120 V	Low 100% to 200% of full load	Fractional	Varies from 0 to full speed	High 150% to 200%	Applications that require small DC-operated equipment such as automobile power windows, seats, and sun roofs
Stepping	5 V, 12 V, 24 V	Very low** .5 to 5000 oz/in.	Size rating is given as holding torque and number of steps	Rated in number of steps per sec (maximum)	Varies based on number of steps and rated torque	Applications that require low torque and precise control such as indexing tables and printers
AC / DC Universal	115 VAC, 230 VAC, 12 VDC, 24 VDC, 36 VDC, 120 VDC	High 300% to 400% of full load	Fractional	Varies 0 to full speed	High 175% to 225%	Most portable tools such as drills, routers, mixers, and vacuum cleaners

* based on standard 3ϕ induction motor

** torque is rated as holding torque

OVERCURRENT PROTECTION DEVICES

Motor Type	Code Letter	FLC (%)				
		Motor Size	TDF	NTDF	ITB	ITCB
AC*	—	—	175	300	150	700
AC*	A	—	150	150	150	700
AC*	B–E	—	175	250	200	700
AC*	F–V	—	175	300	250	700
DC	—	⅛ to 50 HP	150	150	150	150
DC	—	Over 50 HP	150	150	150	175

* full-voltage and resistor starting

FULL-LOAD CURRENTS — DC MOTORS		
Motor rating (HP)	Current (A)	
	120 V	240 V
1/4	3.1	1.6
1/3	4.1	2.0
1/2	5.4	2.7
3/4	7.6	3.8
1	9.5	4.7
1 1/2	13.2	6.6
2	17	8.5
3	25	12.2
5	40	20
7 1/2	48	29
10	76	38

FULL-LOAD CURRENTS — 1φ, AC MOTORS		
Motor rating (HP)	Current (A)	
	115 V	230 V
1/6	4.4	2.2
1/4	5.8	2.9
1/3	7.2	3.6
1/2	9.8	4.9
3/4	13.8	6.9
1	16	8
1 1/2	20	10
2	24	12
3	34	17
5	56	28
7 1/2	80	40

FULL-LOAD CURRENTS — 3φ, AC INDUCTION MOTORS				
Motor rating (HP)	Current (A)			
	208 V	230 V	460 V	575 V
1/4	1.11	.96	.48	.38
1/3	1.34	1.18	.59	.47
1/2	2.2	2.0	1.0	.8
3/4	3.1	2.8	1.4	1.1
1	4.0	3.6	1.8	1.4
1 1/2	5.7	5.2	2.6	2.1
2	7.5	6.8	3.4	2.7
3	10.6	9.6	4.8	3.9
5	16.7	15.2	7.6	6.1
7 1/2	24.0	22.0	11.0	9.0
10	31.0	28.0	14.0	11.0
15	46.0	42.0	21.0	17.0
20	59	54	27	22
25	75	68	34	27
30	88	80	40	32
40	114	104	52	41
50	143	130	65	52
60	169	154	77	62
75	211	192	96	77
100	273	248	124	99
125	343	312	156	125
150	396	360	180	144
200	—	480	240	192
250	—	602	301	242
300	—	—	362	288
350	—	—	413	337
400	—	—	477	382
500	—	—	590	472

TYPICAL MOTOR EFFICIENCIES					
HP	Standard Motor (%)	Energy-Effiecient Motor (%)	HP	Standard Motor (%)	Energy-Efficient Motor (%)
1	76.5	84.0	30	88.1	93.1
1.5	78.5	85.5	40	89.3	93.6
2	79.9	86.5	50	90.4	93.7
3	80.8	88.5	75	90.8	95.0
5	83.1	88.6	100	91.6	95.4
7.5	83.8	90.2	125	91.8	95.8
10	85.0	90.3	150	92.3	96.0
15	86.5	91.7	200	93.3	96.1
20	87.5	92.4	250	93.6	96.2
25	88.0	93.0	300	93.8	96.5

MOTOR FRAME DIMENSIONS											
Frame No.	Shaft		Key			Dimensions – Inches					
	U	V	W	T	L	A	B	D	E	F	BA
48	1/2	1 1/2*	flat	3/64	—	5 5/8*	3 1/2*	3	2 1/8	1 3/8	2 1/2
56	5/8	1 7/8*	3/16	3/16	1 3/8	6 1/2*	4 1/4*	3 1/2	2 7/16	1 1/2	2 3/4
143T	7/8	2	3/16	3/16	1 3/8	7	6	3 1/2	2 3/4	2	2 1/4
145T	7/8	2	3/16	3/16	1 3/8	7	6	3 1/2	2 3/4	2 1/2	2 1/4
182	7/8	2	3/16	3/16	1 3/8	9	6 1/2	4 1/2	3 3/4	2 1/4	2 3/4
182T	1 1/8	2 1/2	1/4	1/4	1 3/4	9	6 1/2	4 1/2	3 3/4	2 1/4	2 3/4
184	7/8	2	3/16	3/16	1 3/8	9	7 1/2	4 1/2	3 3/4	2 3/4	2 3/4
184T	1 1/8	2 1/2	1/4	1/4	1 3/4	9	7 1/2	4 1/2	3 3/4	2 3/4	2 3/4
203	3/4	2	3/16	3/16	1 3/8	10	7 1/2	5	4	2 3/4	3 1/8
204	3/4	2	3/16	3/16	1 3/8	10	8 1/2	5	4	3 1/4	3 1/8
213	1 1/8	2 3/4	1/4	1/4	2	10 1/2	7 1/2	5 1/4	4 1/4	2 3/4	3 1/2
213T	1 3/8	3 1/8	5/16	5/16	2 3/8	10 1/2	7 1/2	5 1/4	4 1/4	2 3/4	3 1/2
215	1 1/8	2 3/4	1/4	1/4	2	10 1/2	9	5 1/4	4 1/4	3 1/2	3 1/2
215T	1 3/8	3 1/8	5/16	5/16	2 3/8	10 1/2	9	5 1/4	4 1/4	3 1/2	3 1/2
224	1	2 3/4	1/4	1/4	2	11	8 3/4	5 1/2	4 1/2	3 3/8	3 1/2
225	1	2 3/4	1/4	1/4	2	11	9 1/2	5 1/2	4 1/2	3 3/4	3 1/2
254	1 1/8	3 1/8	1/4	1/4	2 3/8	12 1/2	10 3/4	6 1/4	5	4 1/8	4 1/4
254U	1 3/8	3 1/2	5/16	5/16	2 3/4	12 1/2	10 3/4	6 1/4	5	4 1/8	4 1/4
254T	1 5/8	3 3/4	3/8	3/8	2 7/8	12 1/2	10 3/4	6 1/4	5	4 1/8	4 1/4
256U	1 3/8	3 1/2	5/16	5/16	2 3/4	12 1/2	12 1/2	6 1/4	5	5	4 1/4
256T	1 5/8	3 3/4	3/8	3/8	2 7/8	12 1/2	12 1/2	6 1/4	5	5	4 1/4
284	1 1/4	3 1/2	1/4	1/4	2 3/4	14	12 1/2	7	5 1/2	4 3/4	4 3/4
284U	1 5/8	4 5/8	3/8	3/8	3 3/4	14	12 1/2	7	5 1/2	4 3/4	4 3/4
284T	1 7/8	4 3/8	1/2	1/2	3 1/4	14	12 1/2	7	5 1/2	4 3/4	4 3/4
284TS	1 5/8	3	3/8	3/8	1 7/8	14	12 1/2	7	5 1/2	4 3/4	4 3/4
286U	1 5/8	4 5/8	3/8	3/8	3 3/4	14	14	7	5 1/2	5 1/2	4 3/4
286T	1 7/8	4 3/8	1/2	1/2	3 1/4	14	14	7	5 1/2	5 1/2	4 3/4
286TS	1 5/8	3	3/8	3/8	1 7/8	14	14	7	5 1/2	5 1/2	4 3/4
324	1 5/8	4 5/8	3/8	3/8	3 3/4	16	14	8	6 1/4	5 1/4	5 1/4
324U	1 7/8	5 3/8	1/2	1/2	4 1/4	16	14	8	6 1/4	5 1/4	5 1/4
324S	1 5/8	3	3/8	3/8	1 7/8	16	14	8	6 1/4	5 1/4	5 1/4
324T	2 1/8	5	1/2	1/2	3 7/8	16	14	8	6 1/4	5 1/4	5 1/4
324TS	1 7/8	3 1/2	1/2	1/2	2	16	14	8	6 1/4	5 1/4	5 1/4
326	1 5/8	4 5/8	3/8	3/8	3 3/4	16	15 1/2	8	6 1/4	6	5 1/4
326U	1 7/8	5 3/8	1/2	1/2	4 1/4	16	15 1/2	8	6 1/4	6	5 1/4
326S	1 5/8	3	3/8	3/8	1 7/8	16	15 1/2	8	6 1/4	6	5 1/4
326T	2 1/8	5	1/2	1/2	3 7/8	16	15 1/2	8	6 1/4	6	5 1/4
326TS	1 7/8	3 1/2	1/2	1/2	2	16	15 1/2	8	6 1/4	6	5 1/4
364	1 7/8	5 3/8	1/2	1/2	4 1/4	18	15 1/4	9	7	5 5/8	5 7/8
364S	1 5/8	3	3/8	3/8	1 7/8	18	15 1/4	9	7	5 5/8	5 7/8
364U	2 1/8	6 1/8	1/2	1/2	5	18	15 1/4	9	7	5 5/8	5 7/8

* not NEMA standard dimensions

MOTOR FRAME LETTERS	
LETTER	**DESIGNATION**
G	Gasoline pump motor
K	Sump pump motor
M and N	Oil burner motor
S	Standard short shaft for direct connection
T	Standard dimensions established
U	Previously used as frame designation for which standard dimensions are established
Y	Special mounting dimensions required from manufacturer
Z	Standard mounting dimensions except shaft extension

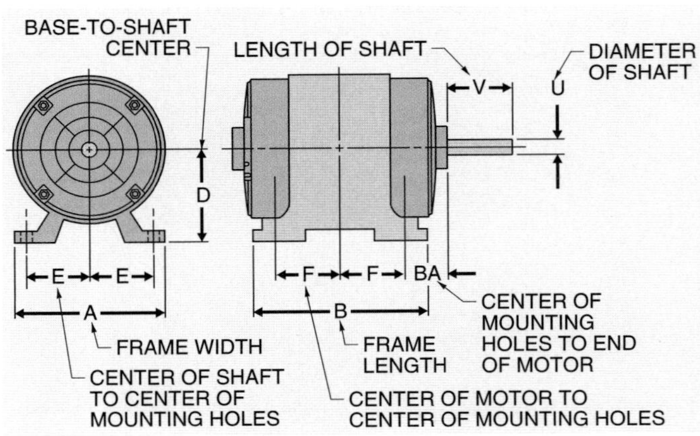

3ϕ, 230 V MOTORS AND CIRCUITS – 240 V SYSTEM

1		2		3	4	5				6	
Size of motor		Motor overload protection Low-peak or Fusetron®				Controller termination temperature rating				Minimum size of copper wire and trade conduit	
						60°C		75°C			
HP	Amp	Motor less than 40°C or greater than 1.15 SF (Max fuse 125%)	All other motors (Max fuse 115%)	Switch 115% minimum or HP rated or fuse holder size	Minimum size of starter	TW	THW	TW	THW	Wire size (AWG or kcmil)	Conduit (inches)
½	2	2½	2¼	30	00	•	•	•	•	14	½
¾	2.8	3½	3 2/10	30	00	•	•	•	•	14	½
1	3.6	4½	4	30	00	•	•	•	•	14	½
1½	5.2	6¼	5 6/10	30	00	•	•	•	•	14	½
2	6.8	8	7½	30	0	•	•	•	•	14	½
3	9.6	12	10	30	0	•	•	•	•	14	½
5	15.2	17½	17½	30	1	•	•	•	•	14	½
7½	22	25	25	30	1					10	½
10	28	35	30	60	2	•	•	•		8	¾
									•	10	½
15	42	50	45	60	2	•	•	•	•	6	1
										6	¾
20	54	60	60	100	3	•	•	•	•	4	1
25	68	80	75	100	3	•	•			3	1¼
								•		3	1
									•	4	1
30	80	100	90	100	3	•	•	•		1	1¼
									•	3	1¼
40	104	125	110	200	4	•	•	•		2/0	1½
									•	1	1¼
50	130	150	150	200	4	•	•	•		3/0	2
									•	2/0	1½
75	192	225	200	400	5	•	•	•		300	2½
									•	250	2½
100	248	300	250	400	5	•	•	•		500	3
									•	350	2½
150	360	450	400	600	6	•	•	•		300-2/ϕ*	2-2½*
									•	4/0-2/ϕ*	2-2*

* two sets of multiple conductors and two runs of conduit required

3φ, 460 V MOTORS AND CIRCUITS — 480 V SYSTEM											
1		2		3	4	5				6	
Size of motor		Motor overload protection — Low-peak or Fusetron®				Controller termination temperature rating — 60°C		75°C		Minimum size of copper wire and trade conduit	
HP	Amp	Motor less than 40°C or greater than 1.15 SF (Max fuse 125%)	All other motors (Max fuse 115%)	Switch 115% minimum or HP rated or fuse holder size	Minimum size of starter	TW	THW	TW	THW	Wire size (AWG or kcmil)	Conduit (inches)
1/2	1	1 1/4	1 1/8	30	00	•	•	•	•	14	1/2
3/4	1.4	1 6/10	1 6/10	30	00	•	•	•	•	14	1/2
1	1.8	2 1/4	2	30	00	•	•	•	•	14	1/2
1 1/2	2.6	3 2/10	2 6/10	30	00	•	•	•	•	14	1/2
2	3.4	4	3 1/2	30	00	•	•	•	•	14	1/2
3	4.8	5 6/10	5	30	0	•	•	•	•	14	1/2
5	7.6	9	8	30	0	•	•	•	•	14	1/2
7 1/2	11	12	12	30	1	•	•	•	•	14	1/2
10	14	17 1/2	15	30	1	•	•	•	•	14	1/2
15	21	25	20	30	2	•	•	•	•	10	1/2
20	27	30	30	60	2	•	•	•		8	3/4
									•	10	1/2
25	34	40	35	60	2	•	•	•		6	1
									•	8	3/4
30	40	50	45	60	3	•	•	•		6	1
									•	8	3/4
40	52	60	60	100	3	•	•	•		4	1
									•	6	1
50	65	80	70	100	3	•	•	•		3	1 1/4
									•	4	1
60	77	90	80	100	4	•	•	•		1	1 1/4
									•	3	1 1/4
75	96	110	110	200	4	•	•	•		1/0	1 1/2
									•	1	1 1/4
100	124	150	125	200	4	•	•	•		3/0	2
									•	2/0	1 1/2
125	156	175	175	200	5	•	•	•		4/0	2
									•	3/0	2
150	180	225	200	400	5	•	•	•		300	2 1/2
									•	4/0	2
200	240	300	250	400	5	•	•	•		500	3
									•	350	2 1/2
250	302	350	325	400	6	•	•	•		4/0-2/φ*	2-2*
									•	3/0-2/φ*	2-2*
300	361	450	400	600	6	•	•	•		300-2/φ*	2-1 1/2 *
									•	4/0-2/φ*	2-2*

* two sets of multiple conductors and two runs of conduit required

ENCLOSURES					
Type	**Use**	**Service Conditions**	**Tests**	**Comments**	**Type**
1	Indoor	No unusual	Rod entry, rust resistance		1
3	Outdoor	Windblown dust, rain, sleet, and ice on enclosure	Rain, external icing, dust, and rust resistance	Do not provide protection against internal condensation or internal icing	
3R	Outdoor	Falling rain and ice on enclosure	Rod entry, rain, external icing, and rust resistance	Do not provide protection against dust, internal condensation, or internal icing	
4	Indoor/outdoor	Windblown dust and rain, splashing water, hose-directed water, and ice on enclosure	Hosedown, external icing, and rust resistance	Do not provide protection against internal condensation or internal icing	4
4X	Indoor/outdoor	Corrosion, windblown dust and rain, splashing water, hose-directed water, and ice on enclosure	Hosedown, external icing, and corrosion resistance	Do not provide protection against internal condensation or internal icing	4X
6	Indoor/outdoor	Occasional temporary submersion at a limited depth			
6P	Indoor/outdoor	Prolonged submersion at a limited depth			
7	Indoor locations classified as Class I, Groups A, B, C, or D, as defined in the NEC®	Withstand and contain an internal explosion of specified gases, contain an explosion sufficiently so an explosive gas-air mixture in the atmosphere is not ignited	Explosion, hydrostatic, and temperature	Enclosed heat-generating devices shall not cause external surfaces to reach temperatures capable of igniting explosive gas-air mixtures in the atmosphere	7
9	Indoor locations classified as Class II, Groups E or G, as defined in the NEC®	Dust	Dust penetration, temperature, and gasket aging	Enclosed heat-generating devices shall not cause external surfaces to reach temperatures capable of igniting explosive gas-air mixtures in the atmosphere	9
12	Indoor	Dust, falling dirt, and dripping noncorrosive liquids	Drip, dust, and rust resistance	Do not provide protection against internal condensation	12
13	Indoor	Dust, spraying water, oil, and noncorrosive coolant	Oil explosion and rust resistance	Do not provide protection against internal condensation	

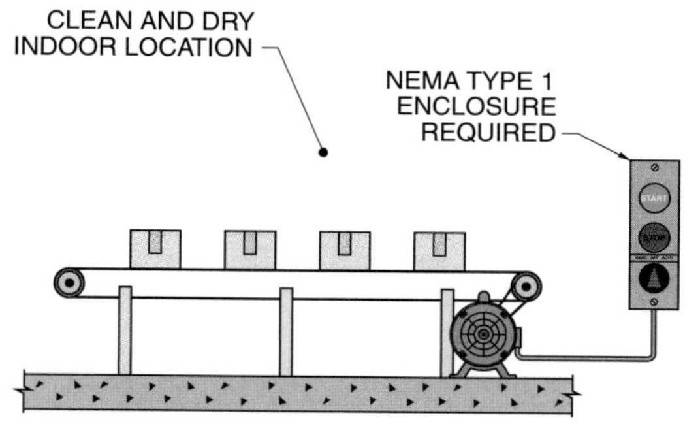

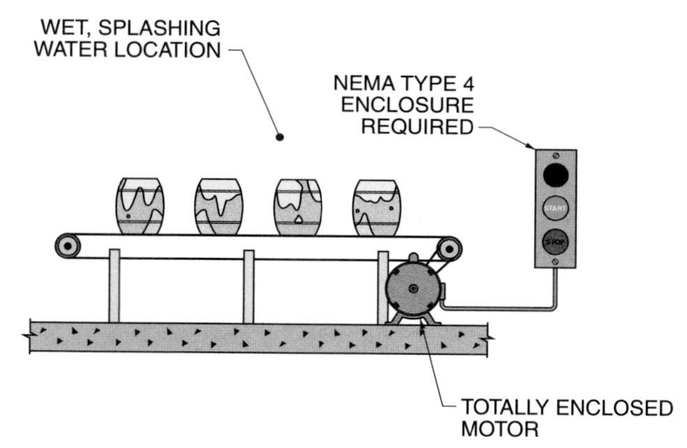

MOTOR RATINGS		
Classification	**Rating**	**Size**
Milli	W	1, 1.5, 2, 3, 5, 7.5, 10, 15, 25, 35
Fractional	HP	1/20, 1/12, 1/8, 1/6, 1/4, 1/3, 1/2, 3/4
Full	HP	1, 1½, 2, 3, 5, 7½, 10, 15, 20, 25, 30, 40, 50, 60, 75, 100, 125, 150, 200, 250, 300
Full-Special Order	HP	350, 400, 450, 500, 600, 700, 800, 900, 1000, 1250, 1500, 1750, 2000, 2250, 2500, 3000, 3500, 4000, 4500, 5000, 5500, 6000, 7000, 8000, 9000, 10,000, 11,000, 12,000, 13,000, 14,000, 15,000, 16,000, 17,000, 18,000, 19,000, 20,000, 22,500, 30,000, 32,500, 35,000, 37,500, 40,000, 45,000, 50,000

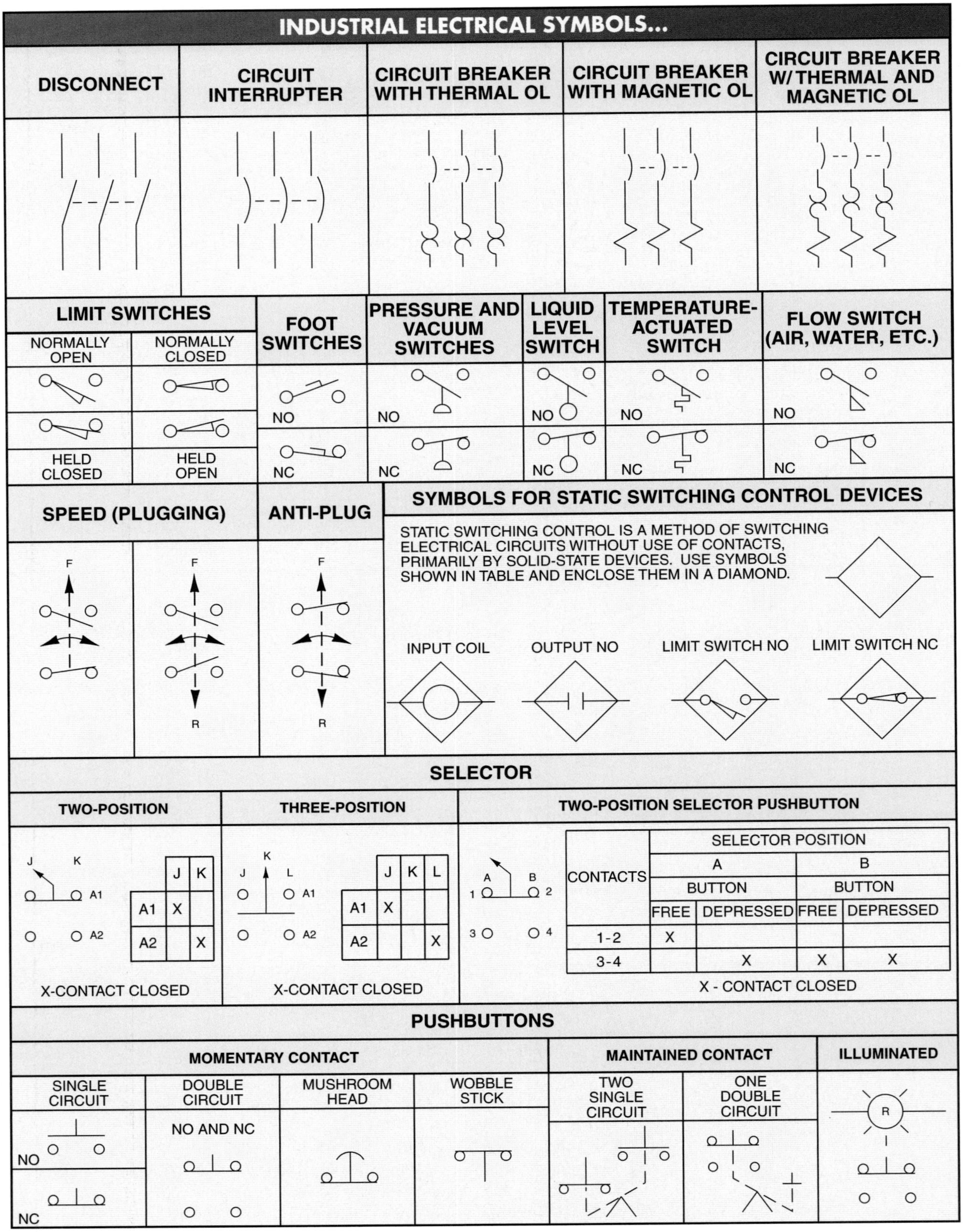

INDUSTRIAL ELECTRICAL SYMBOLS...

DISCONNECT	CIRCUIT INTERRUPTER	CIRCUIT BREAKER WITH THERMAL OL	CIRCUIT BREAKER WITH MAGNETIC OL	CIRCUIT BREAKER W/ THERMAL AND MAGNETIC OL

LIMIT SWITCHES

NORMALLY OPEN	NORMALLY CLOSED
HELD CLOSED	HELD OPEN

FOOT SWITCHES: NO, NC

PRESSURE AND VACUUM SWITCHES: NO, NC

LIQUID LEVEL SWITCH: NO, NC

TEMPERATURE-ACTUATED SWITCH: NO, NC

FLOW SWITCH (AIR, WATER, ETC.): NO, NC

SPEED (PLUGGING): F, R

ANTI-PLUG: F, R

SYMBOLS FOR STATIC SWITCHING CONTROL DEVICES

STATIC SWITCHING CONTROL IS A METHOD OF SWITCHING ELECTRICAL CIRCUITS WITHOUT USE OF CONTACTS, PRIMARILY BY SOLID-STATE DEVICES. USE SYMBOLS SHOWN IN TABLE AND ENCLOSE THEM IN A DIAMOND.

INPUT COIL — OUTPUT NO — LIMIT SWITCH NO — LIMIT SWITCH NC

SELECTOR

TWO-POSITION

J K; A1; A2

	J	K
A1	X	
A2		X

X-CONTACT CLOSED

THREE-POSITION

J K L; A1; A2

	J	K	L
A1	X		
A2			X

X-CONTACT CLOSED

TWO-POSITION SELECTOR PUSHBUTTON

A B; 1 2; 3 4

CONTACTS	SELECTOR POSITION			
	A		B	
	BUTTON		BUTTON	
	FREE	DEPRESSED	FREE	DEPRESSED
1-2	X			
3-4		X	X	X

X - CONTACT CLOSED

PUSHBUTTONS

MOMENTARY CONTACT

SINGLE CIRCUIT	DOUBLE CIRCUIT	MUSHROOM HEAD	WOBBLE STICK
NO / NC	NO AND NC		

MAINTAINED CONTACT

TWO SINGLE CIRCUIT	ONE DOUBLE CIRCUIT

ILLUMINATED

R

...INDUSTRIAL ELECTRICAL SYMBOLS...

CONTACTS

INSTANT OPERATING				TIMED CONTACTS - CONTACT ACTION RETARDED AFTER COIL IS:			
WITH BLOWOUT		WITHOUT BLOWOUT		ENERGIZED		DE-ENERGIZED	
NO	NC	NO	NC	NOTC	NCTO	NOTO	NCTC

OVERLOAD RELAYS

THERMAL	MAGNETIC

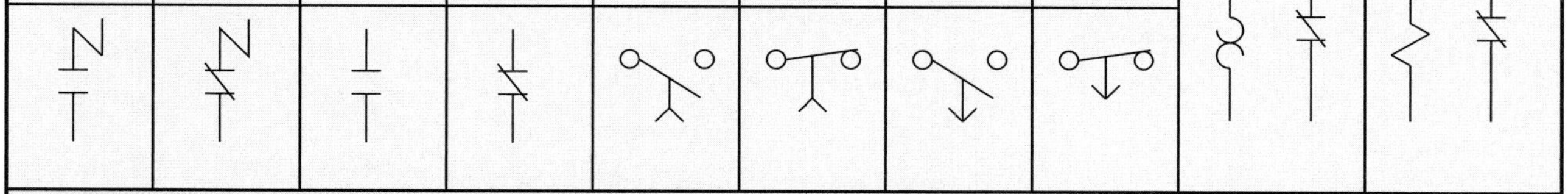

SUPPLEMENTARY CONTACT SYMBOLS

SPST NO		SPST NC		SPDT	
SINGLE BREAK	DOUBLE BREAK	SINGLE BREAK	DOUBLE BREAK	SINGLE BREAK	DOUBLE BREAK

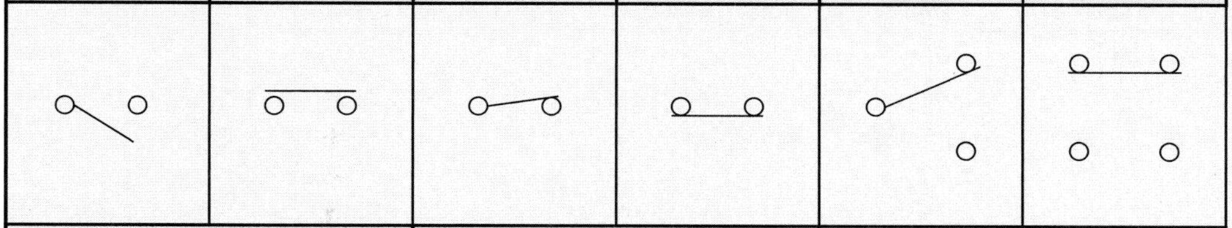

DPST, 2NO		DPST, 2NC		DPDT	
SINGLE BREAK	DOUBLE BREAK	SINGLE BREAK	DOUBLE BREAK	SINGLE BREAK	DOUBLE BREAK

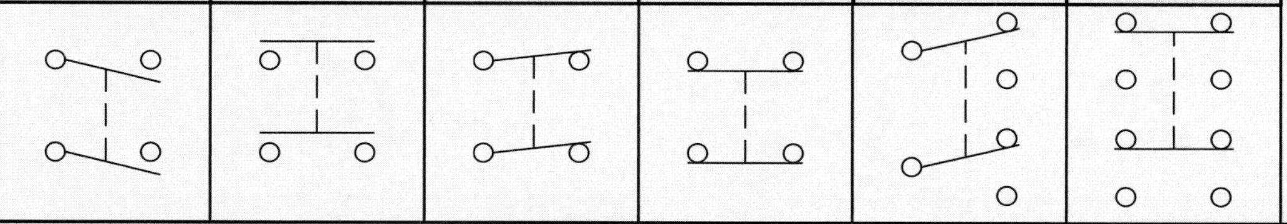

TERMS

SPST
SINGLE-POLE, SINGLE-THROW

SPDT
SINGLE-POLE, DOUBLE-THROW

DPST
DOUBLE-POLE, SINGLE-THROW

DPDT
DOUBLE-POLE, DOUBLE-THROW

NO
NORMALLY OPEN

NC
NORMALLY CLOSED

METER (INSTRUMENT)

INDICATE TYPE BY LETTER

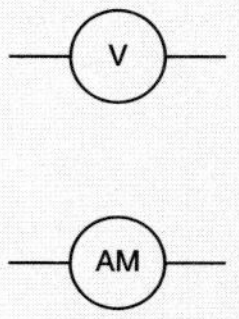

V

AM

TO INDICATE FUNCTION OF METER OR INSTRUMENT, PLACE SPECIFIED LETTER OR LETTERS WITHIN SYMBOL.

AM or A	AMMETER	VA	VOLTMETER
AH	AMPERE HOUR	VAR	VARMETER
µA	MICROAMMETER	VARH	VARHOUR METER
mA	MILLAMMETER	W	WATTMETER
PF	POWER FACTOR	WH	WATTHOUR METER
V	VOLTMETER		

PILOT LIGHTS

INDICATE COLOR BY LETTER

NON PUSH-TO-TEST	PUSH-TO-TEST
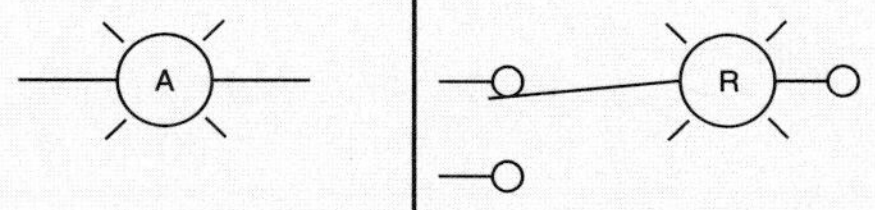 A	R

INDUCTORS

IRON CORE

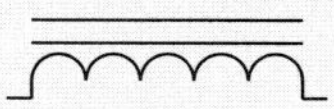

AIR CORE

COILS

DUAL-VOLTAGE MAGNET COILS

HIGH-VOLTAGE	LOW-VOLTAGE
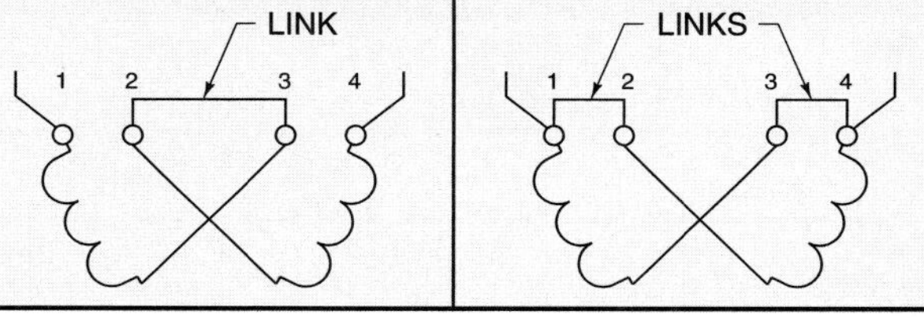 LINK 1 2 3 4	LINKS 1 2 3 4

BLOWOUT COIL

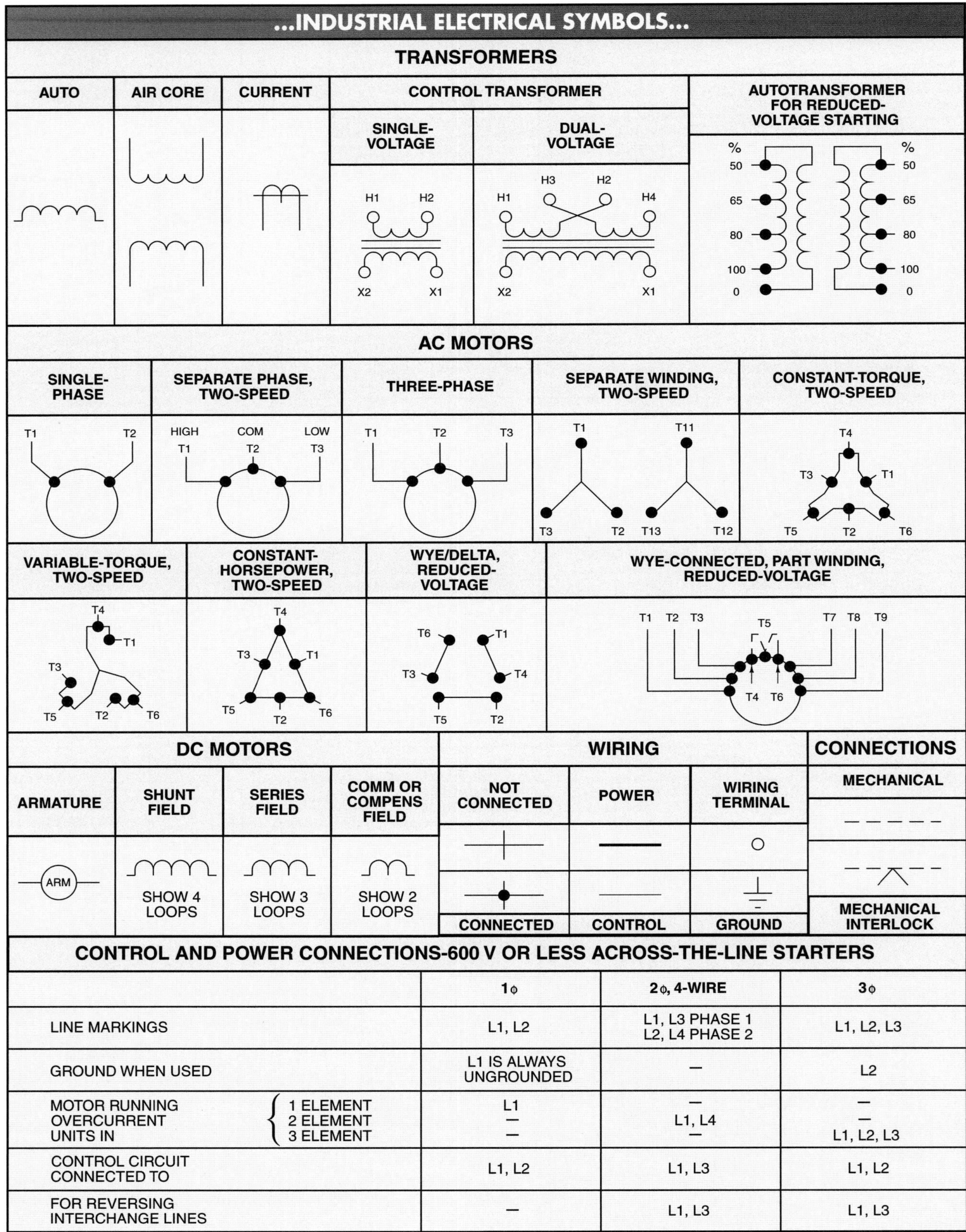

CONTROL AND POWER CONNECTIONS-600 V OR LESS ACROSS-THE-LINE STARTERS

		1ϕ	2ϕ, 4-WIRE	3ϕ
LINE MARKINGS		L1, L2	L1, L3 PHASE 1 L2, L4 PHASE 2	L1, L2, L3
GROUND WHEN USED		L1 IS ALWAYS UNGROUNDED	–	L2
MOTOR RUNNING OVERCURRENT UNITS IN	1 ELEMENT	L1	–	–
	2 ELEMENT	–	L1, L4	–
	3 ELEMENT	–	–	L1, L2, L3
CONTROL CIRCUIT CONNECTED TO		L1, L2	L1, L3	L1, L2
FOR REVERSING INTERCHANGE LINES		–	L1, L3	L1, L3

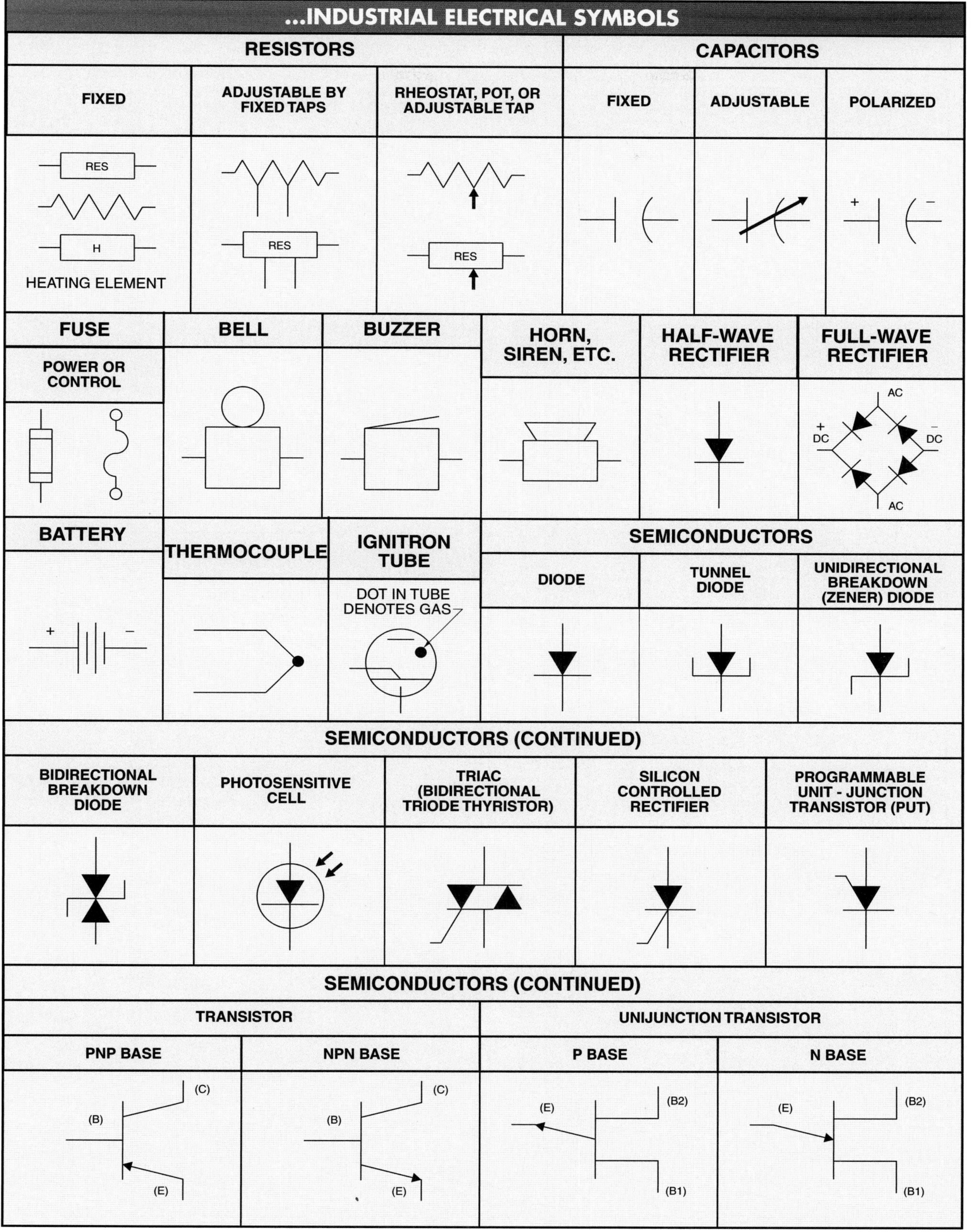
...INDUSTRIAL ELECTRICAL SYMBOLS
RESISTORS
FIXED
RES
H
HEATING ELEMENT
ADJUSTABLE BY FIXED TAPS
RES
RHEOSTAT, POT, OR ADJUSTABLE TAP
RES
CAPACITORS
FIXED
ADJUSTABLE
POLARIZED
+
−
FUSE
POWER OR CONTROL
BELL
BUZZER
HORN, SIREN, ETC.
HALF-WAVE RECTIFIER
FULL-WAVE RECTIFIER
AC
+ DC
− DC
AC
BATTERY
+
−
THERMOCOUPLE
IGNITRON TUBE
DOT IN TUBE DENOTES GAS
SEMICONDUCTORS
DIODE
TUNNEL DIODE
UNIDIRECTIONAL BREAKDOWN (ZENER) DIODE
SEMICONDUCTORS (CONTINUED)
BIDIRECTIONAL BREAKDOWN DIODE
PHOTOSENSITIVE CELL
TRIAC (BIDIRECTIONAL TRIODE THYRISTOR)
SILICON CONTROLLED RECTIFIER
PROGRAMMABLE UNIT - JUNCTION TRANSISTOR (PUT)
SEMICONDUCTORS (CONTINUED)
TRANSISTOR
PNP BASE
(C)
(B)
(E)
NPN BASE
(C)
(B)
(E)
UNIJUNCTION TRANSISTOR
P BASE
(E)
(B2)
(B1)
N BASE
(E)
(B2)
(B1)

NON-LOCKING WIRING DEVICES

2-POLE, 3-WIRE

WIRING DIAGRAM	NEMA ANSI	RECEPTACLE CONFIGURATION	RATING
	5-15 C73.11		15 A 125 V
	5-20 C73.12		20 A 125 V
	5-30 C73.45		30 A 125 V
	5-50 C73.46		50 A 125 V
	6-15 C73.20		15 A 250 V
	6-20 C73.51		20 A 250 V
	6-30 C73.52		30 A 250 V
	6-50 C73.53		50 A 250 V
	7-15 C73.28		15 A 277 V
	7-20 C73.63		20 A 277 V
	7-30 C73.64		30 A 277 V
	7-50 C73.65		50 A 277 V

4-POLE, 4-WIRE

WIRING DIAGRAM	NEMA ANSI	RECEPTACLE CONFIGURATION	RATING
	18-15 C73.15		15 A 3ϕY 120/208 V
	18-20 C73.26		20 A 3ϕY 120/208 V
	18-30 C73.47		30 A 3ϕY 120/208 V
	18-50 C73.48		50 A 3ϕY 120/208 V
	18-60 C73.27		60 A 3ϕY 120/208 V

3-POLE, 3-WIRE

WIRING DIAGRAM	NEMA ANSI	RECEPTACLE CONFIGURATION	RATING
	10-20 C73.23		20 A 125/250 V
	10-30 C73.24		30 A 125/250 V
	10-50 C73.25		50 A 125/250 V
	11-15 C73.54		15 A 3ϕ 250 V
	11-20 C73.55		20 A 3ϕ 250 V
	11-30 C73.56		30 A 3ϕ 250 V
	11-50 C73.57		50 A 3ϕ 250 V

3-POLE, 4-WIRE

WIRING DIAGRAM	NEMA ANSI	RECEPTACLE CONFIGURATION	RATING
	14-15 C73.49		15 A 125/250 V
	14-20 C73.50		20 A 125/250 V
	14-30 C73.16		30 A 125/250 V
	14-50 C73.17		50 A 125/250 V
	14-60 C73.18		60 A 125/250 V
	15-15 C73.58		15 A 3ϕ 250 V
	15-20 C73.59		20 A 3ϕ 250 V
	15-30 C73.60		30 A 3ϕ 250 V
	15-50 C73.61		50 A 3ϕ 250 V
	15-60 C73.62		60 A 3ϕ 250 V

Glossary

A

abbreviation: A letter or combination of letters that represents a word.

accelerating circuit logic: A control function that permits the operator to select a high motor speed and the control circuit automatically accelerates the motor to that speed.

AC sine wave: A symmetrical waveform that contains 360 electrical degrees.

actuator: The part of a limit switch that transfers the mechanical force of the moving part to the electrical contacts.

AC voltage: The voltage that reverses its direction of flow at regular intervals.

adjustable motor base: A mounting base that allows a motor to be easily moved over a short distance.

alternating current (AC): The current that reverses its direction of flow at regular intervals.

alternating current (AC) motor: A motor that uses alternating current to produce rotation.

alternation: The half of a cycle.

alternistor: Two antiparallel thyristors and a triac mounted on the same chip.

ambient temperature: The temperature of the air surrounding a motor.

ampere: The number of electrons passing a given point in one second.

amplification: The process of taking a small signal and making it larger.

analog display: An electromechanical device that indicates a value by the position of a pointer on a scale.

analog multimeter: A meter that can measure two or more electrical properties and that displays the measured properties along calibrated scales using a pointer.

analog switching relay: An SSR that has an infinite number of possible output voltages within the relay's rated range.

AND gate: A device with an output that is high only when both of its inputs are high.

apparent power (P_A): The product of the voltage and current in a circuit calculated without considering the phase shift that may be present between the voltage and the current in a circuit.

arc blast: An explosion that occurs when the surrounding air becomes ionized and conductive.

arc chute: A device that confines, divides, and extinguishes arcs drawn between contacts opened under load.

arcing: The discharge of an electric current across a gap, such as when an electric switch is opened.

arc suppressor: A device that dissipates the energy present across opening contacts.

armature: **1.** The movable part of a solenoid. **2.** The movable coil of wire in a generator that rotates through the magnetic field. **3.** The rotating part of a DC motor.

asymmetrical recycle timer: A timer which has independent adjustments for the ON and OFF time periods.

atom: The smallest particle that an element can be reduced to and still keep the properties of that element.

avalanche current: Current passed when a diode breaks down.

B

barcode: A code consisting of a group of variously patterned bars and spaces designed to be scanned and read into computer memory as identification for an object.

barcode reader: A device that identifies an object and manipulates information about the object that is recorded in a computer system based on the barcode.

bar graph: A graph on a digital display composed of segments that function as an analog pointer.

base speed: The speed (in rpm) at which the motor runs with full-line voltage applied to the armature and field.

base unit: A number that does not include a metric prefix.

battery-powered cable cutter: A tool designed to cut various diameters of electrical and fiber-optic cables.

bellows: A cylindrical device with several deep folds that expand or contract when pressure is applied.

bench testing: Testing performed when equipment under test is brought to a designated service area.

bimetallic overload relay: An overload relay that resets automatically.

bimetallic sensor: A sensor that bends or curls when the temperature changes.

bipolar device: A device in which both holes and electrons are used as internal carriers for maintaining current flow.

branch circuit: A portion of a distribution system between the final overcurrent protection device and the outlet or load connected to it.

break: The number of separate places on a contact that open or close an electrical circuit.

breakdown torque (BDT): The maximum torque a motor can provide without an abrupt reduction in motor speed.

brush: The sliding contact that rides against the commutator segments; used to connect the armature to the external circuit.

busway: A metal-enclosed distribution system of busbars available in prefabricated sections.

C

cable tie gun: A handheld device that is used to hold and tighten several plastic or steel cable ties.

capacitance (C): The ability of a component or circuit to store energy in the form of an electrical charge.

capacitive circuit: A circuit in which current leads voltage (voltage lags current).

capacitive level switch: A level switch that detects the dielectric variation when the product is in contact (proximity) with the probe and when the product is not in contact with the probe.

capacitive proximity sensor: A sensor that detects either conductive or non conductive substances.

capacitive reactance (X_C): The opposition to current flow by a capacitor.

capacitor: An electric device that stores electrical energy by means of an electrostatic field.

capacitor motor: A 1ϕ AC motor that includes a capacitor in addition to the running and starting windings.

capillary tube sensor: A sensor that changes internal pressure with a change in temperature.

caution signal word: A word used to indicate a potentially hazardous situation which, if not avoided, may result in minor or moderate injury.

circuit analysis method: A method of SSR replacement in which a logical sequence is used to determine the reason for the failure.

circuit breaker: An overcurrent protection device with a mechanical mechanism that may manually or automatically open the circuit when an overload condition or short circuit occurs.

cladding: The first layer of protection for the glass or plastic core of the optical fiber cable.

clamp-on ammeter: A meter that measures the current in a circuit by measuring the strength of the magnetic field around a single conductor.

closed-loop operation: An operation that has feedback from the output to the input.

coast stop: A stopping method in which the motor drive shuts OFF the voltage to a motor, allowing the motor to coast to a stop.

cold junction (reference junction): The end of a thermocouple that is kept at a constant temperature in order to provide a reference point.

cold trip: The trip point from the time the motor starts until the first time the overloads trip (motor operating below nameplate rated current).

communication: The transmission of information from one point to another by means of electromagnetic waves.

commutator: A ring made of segments that are insulated from one another.

compelling circuit logic: A control function that requires the operator to start and operate a motor in a predetermined order.

complementary metal-oxide semiconductor (CMOS) ICs: A group of ICs that employ MOS transistors.

compound-wound generator: A generator that includes series and shunt field windings.

conductive probe level switch: A level switch that uses liquid to complete the electrical path between two conductive probes.

conductor: A material that has very little resistance to current flow and permits electrons to move through it easily.

conduit bender: A device used to radius (bend) electrical metallic tubing, intermediate metallic conduit, and rigid steel and aluminum conduit in sizes ranging from 1/2″ to 1 1/2″ diameter.

confined space: A space large enough and so configured that an employee can physically enter and perform assigned work, that has limited or restricted means for entry and exit, and is not designed for continuous employee occupancy.

constant horsepower/variable torque (CH/VT) load: A load that requires high torque at low speeds and low torque at high speeds.

constant torque/variable horsepower (CT/VH) load: A load in which the torque requirement remains constant.

contact: The conducting part of a switch that operates with another conducting part of the switch to make or break a circuit.

contact block: The part of the pushbutton that is activated when the operator is pressed.

contact-controlled timer: A timer that does not require the control switch to be connected in line with the timer coil.

contact life: The number of times a relay's contacts switch the load controlled by the relay before malfunctioning.

contact protection circuit: A circuit that protects contacts by providing a nondestructive path for generated voltage as a switch is opened.

contactor: A control device that uses a small control current to energize or de-energize the load connected to it.

continuity tester: A test instrument that tests for a complete path for current to flow.

control circuit: The part of the circuit that determines when the output component is energized or de-energized.

control switch: A switch that controls the flow of current in a circuit.

control transformer: A transformer that is used to step down the voltage to the control circuit of a system or machine.

conventional current flow: The current flow from positive to negative.

convergent beam scan: A method of scanning that simultaneously focuses and converges a light beam to a fixed focal point in front of the photoreceiver.

converter: An electronic device that changes AC voltage into DC voltage.

core: The actual path for light in a fiber-optic cable.

counter: A counting device that accounts for the total number of inputs entering the counter and can provide an output (mechanical or solid-state contacts) at predetermined counts in addition to displaying the counted value.

current (I): The amount of electrons flowing through an electrical circuit.

cutoff region: The point at which the transistor is turned OFF and no current flows.

cycle: One complete positive and negative alternation of a wave form.

D

danger signal word: A word used to indicate an imminently hazardous situation which, if not avoided, results in death or serious injury.

dark-operated photoelectric control: A photoelectric control that energizes the output switch when a target is present (breaks the beam).

dashpot timer: A timer that provides time delay by controlling how rapidly air or liquid is allowed to pass into or out of a container through an orifice (opening) that is either fixed in diameter or variable.

daylight switch: A switch that automatically turns lamps ON at dusk and OFF at dawn.

DC brake stop (DC injection braking): A stopping method in which a DC voltage is applied to the stator winding of a motor after a stop command is entered.

DC compound motor: A DC motor with the field connected in both series and shunt with the armature.

DC permanent-magnet motor: A motor that uses magnets, not a coil of wire, for the field windings.

DC series motor: A DC motor that has the series field coils connected in series with the armature.

DC shunt motor: A DC motor that has the field connected in shunt (parallel) with the armature.

DC voltage: The voltage that flows in one direction only.

deadband (differential): The amount of pressure that must be removed before the switch contacts reset for another cycle after the setpoint has been reached and the switch has been actuated.

dead short: A short circuit that opens the circuit as soon as the circuit is energized or when the section of the circuit containing the short is energized.

decelerating circuit logic: A control function that permits the operator to select a low motor speed and the control circuit automatically decelerates the motor to that speed.

decibel (dB): A unit of measure used to express the relative intensity of sound.

delta connection: A connection that has each coil end connected end-to-end to form a closed loop.

diac: A three-layer bidirectional device used primarily as a triggering device.

diaphragm: A deflecting mechanism that moves when force (pressure) is applied.

dielectric: A nonconductor of direct electric current.

dielectric absorption test: A test that checks the absorption characteristics of humid or contaminated insulation.

dielectric material: A medium in which an electric field is maintained with little or no outside energy supply.

dielectric variation: The range at which a material can sustain an electric field with a minimum dissipation of power.

diffused mode: A method of ultrasonic sensor operation in which the emitter and receiver are housed in the same enclosure.

diffuse scan (proximity scan): A method of scanning in which the transmitter and receiver are housed in the same enclosure and a small percentage of the transmitted light beam is reflected back to the receiver from the target.

digital display: An electronic device that displays readings on a meter as numerical values.

digital logic probe: A special DC voltmeter that detects the presence or absence of a signal.

digital multimeter (DMM): A meter that can measure two or more electrical properties and that displays the measured properties as numerical values.

diode: An electronic component that allows current to pass through it in only one direction.

direct current (DC): The current that flows in only one direction.

direct current (DC) motor: A motor that uses direct current connected to the field and armature to produce shaft rotation.

directional control valve: A valve that is used to direct the flow of fluid throughout a fluid power system.

direct mode: A method of ultrasonic sensor operation in which the emitter and receiver are placed opposite each other so that the sound waves from the emitter are received directly by the receiver.

direct scan (transmitted beam, thru-beam, opposed scan): A method of scanning in which the transmitter and receiver are placed opposite each other so that the light beam from the transmitter shines directly at the receiver.

disconnect: A device used only periodically to remove electrical circuits from their supply source.

double-break contacts: Contacts that break an electrical circuit in two places.

doping: The addition of impurities to the crystal structure of a semiconductor.

drop-out voltage: The voltage that exists when voltage is reduced sufficiently to allow a solenoid to open.

drum switch: A manual switch made up of moving contacts mounted on an insulated rotating shaft.

dynamic breaking: A method of motor breaking in which a motor is reconnected to act as a generator immediately after it is turned OFF.

E

earmuff: An ear protection device worn over the ears.

earplug: An ear protection device made of moldable rubber, foam, or plastic and inserted into the ear canal.

eddy current: Unwanted current induced in the metal structure of a device due to the rate of change in the induced magnetic field.

edge card: A PC board with multiple terminations (terminal contacts) on one end.

edge card connector: A connector that allows the edge card to be connected to the system's circuitry with the least amount of hardware.

effective light beam: The area of light that travels directly from the transmitter to the receiver.

electric breaking: A method of breaking in which a DC voltage is applied to the stationary windings of a motor after the AC voltage is removed.

electrical circuit: An assemblage of conductors and electrical devices through which current flows.

electrical noise: Unwanted signals that are present on a power line.

electrical shock: A shock that results any time a body becomes part of an electrical circuit.

electrical warning signal word: A word used to indicate a high-voltage location and conditions that could result in death or serious personal injury from an electrical shock if proper precautions are not taken.

electrolyte: A conduction medium in which the current flow occurs by ion migration.

electromagnet: A magnet whose magnetic energy is produced by the flow of electric current.

electromagnetic actuation: A passive method of sensor activation in which a magnetic field produced by a coil of wire is used to activate a Hall effect sensor.

electromagnetism: The magnetism produced when electric current passes through a conductor.

electromechanical relay (EMR): A switching device that has sets of contacts that are closed by a magnetic field.

electron: A negatively charged particle that orbits the nucleus of an atom at great speeds in shells.

electron current flow: The current flow from negative to positive.

electronic overload: A device that has built-in circuitry to sense changes in current and temperature.

energy: The capacity to do work.

equipment grounding conductor (EGC): An electrical conductor that provides a low-impedance ground path between electrical equipment and enclosures within the distribution system.

eutectic alloy: A metal that has a fixed temperature at which it changes directly from a solid to a liquid state.

exact replacement method: A method of SSR replacement in which a bad relay is replaced with a relay of the same type and size.

explosion warning signal word: A word used to indicate locations and conditions where exploding parts may cause death or serious personal injury if proper precautions and procedures are not followed.

extended button operator: A pushbutton that has the button extended beyond the guard.

F

face shield: An eye and face protection device that covers the entire face with a plastic shield, and is used for protection from flying objects.

ferrous proximity shunt actuation: A passive method of sensor activation in which the magnetic induction around the Hall effect sensor is shunted with a gear tooth.

fiber optics: A technology that uses a thin flexible glass or plastic optical fiber (POF) to transmit light.

field windings: **1.** Magnets used to produce the magnetic field in a generator. **2.** The stationary winding or magnets of a DC motor.

fish tape: A retractable tape, usually of a rectangular cross section, that is pushed through an inaccessible space such as a run of conduit or a partition in order to draw in wires.

555 timer: An integrated circuit designed to output timing pulses for control of certain types of circuits.

flip-flop: An electronic circuit having two stable states or conditions normally designated set and reset.

flow: The travel of fluid in response to a force caused by pressure or gravity.

flow detection sensor (solid-state flow sensor): A sensor that detects the movement (flow) of a liquid or gas using a solid-state device.

flow switch: A control switch that detects the movement of a fluid.

flush button operator: A pushbutton with a guard ring surrounding the button that prevents accidental operation.

foot protection: Shoes worn to prevent foot injuries that are typically caused by objects falling less than 4′ and having an average weight of less than 65 lb.

foot switch: A control switch that is operated by a person's foot.

force: Any cause that changes the position, motion, direction, or shape of an object.

fork lever actuator: An actuator operated by either one of two roller arms.

forward-bias voltage: The application of the proper polarity to a diode.

forward breakover voltage: The voltage required to switch an SCR into a conductive state.

friction: The resistance to motion that occurs when two surfaces slide against each other.

full-load current (FLC): The current required by a motor to produce full-load torque at the motor's rated speed.

full-load torque (FLT): The torque required to produce the rated power at the full speed of the motor.

fuse: An overcurrent protection device (OCPD) with a fusible link that melts and opens the circuit on an overcurrent condition.

fuse puller: A device that is used for the safe removal of fuses from electrical boxes and cabinets.

G

gain: A ratio of the amplitude of an output signal to the amplitude of an input signal.

gas switch: A switch that detects a set amount of a specified gas and activates a set of electrical contacts.

general-purpose relay: A mechanical switch operated by a magnetic coil.

generator (alternator): A machine that converts mechanical energy into electrical energy by means of electromagnetic induction.

ghost voltage: A voltage that appears on a meter not connected to a circuit.

goggles: An eye protection device with a flexible frame that is secured on the face with an elastic headband.

graph: A diagram that shows a variable in comparison to other variables.

grounded circuit: A circuit in which current leaves its normal path and travels to the frame of the motor.

grounded conductor: A conductor that has been intentionally grounded.

ground electrode: A conductor embedded in the earth to provide a good ground.

ground fault circuit interrupter (GFCI): A device that protects against electrical shock by detecting an imbalance of current in the normal conductor pathways and opening the circuit.

grounding: The connection of all exposed non-current-carrying metal parts to the earth.

grounding electrode conductor (GEC): A conductor that connects grounded parts of a power distribution system (equipment grounding conductors, grounded conductors, and all metal parts) to the NEC®-approved earth grounding system.

ground resistance tester: A device used to measure ground connection resistance of electrical installations such as power plants, industrial plants, high-tension towers, and lightning arrestors.

H

half-shrouded button operator: A pushbutton with a guard ring that extends over the top half of the button.

half-wave rectifier: A circuit containing a diode which permits only the positive half-cycles to the AC sine wave to pass.

half-waving: A phenomenon that occurs when a relay fails to turn OFF because the current and voltage in the circuit reach zero at different times.

Hall effect sensor: A sensor that detects the proximity of a magnetic field.

Hall generator: A thin strip of semiconductor material through which a constant control current is passed.

hammer: A striking or splitting tool with a hardened head fastened perpendicular to a handle.

hammer drill: A drill that rotates and drives simultaneously.

hasp: A multiple lockout/tagout device.

head-on actuation: An active method of sensor activation in which a magnet is oriented perpendicular to the surface of the sensor and is usually centered over the point of maximum sensitivity.

heater coil: A sensing device used to monitor the heat generated by excessive current and the heat created through ambient temperature rise.

holding current: The minimum current necessary for an SCR to continue conducting.

holes: The missing electrons in a crystal structure.

horsepower (HP): A unit of power equal to 746 W or 33,000 lb-ft per minute (550 lb-ft per second).

hot junction (measuring junction): The joined end of a thermocouple that is exposed to the process where the temperature measurement is desired.

hot trip: The trip point after the overloads have tripped and have been reset (motor operating near or over nameplate rated current).

human interface module (HIM): A manually operated input control unit that includes programming keys, system operating keys, and normally a status display.

hybrid relay: A combination of electromechanical and solid-state technology used to overcome unique problems that cannot be solved by one device or the other alone.

I

improper phase sequence: The changing of the sequence of any two phases (phase reversal) in a 3ϕ motor circuit.

increment current: The maximum current permitted by the utility in any one step of an increment start.

index pin: A metal extension from the transistor case.

inductance (L): The property of a circuit that causes it to oppose a change in current due to energy stored in a magnetic field.

induction motor: A motor that has no physical electrical connection to the rotor.

inductive circuit: A circuit in which current lags voltage.

inductive proximity sensor: A sensor that detects only conductive substances.

inductive reactance (X_L): The opposition of an inductor to alternating current.

inertia: The property of matter by which a mass persists in its state of rest or motion until acted upon by an external force.

infinity: An unlimited number or amount.

infrared light: Light that is not visible to the human eye.

inherent motor protector: An overload device located directly on or in a motor to provide overload protection.

in-phase: The state in which voltage and current in an AC circuit reach their maximum amplitude and zero level simultaneously.

input circuit: The part of the circuit to which the control component is connected.

instant-ON switching relay: An SSR that turns ON the load immediately when the control voltage is present.

insulation spot test: A test that checks motor insulation over the life of a motor.

insulation step voltage test: A test that creates electrical stress on internal insulation cracks to reveal aging or damage not found during other motor insulation tests.

integrated circuit (IC): A circuit composed of thousands of semiconductor devices, providing a complete circuit function in one small semiconductor package.

interference: Any object other than the object to be detected that is sensed by a sensor.

International Electrotechnical Commission (IEC): An organization that develops international safety standards for electrical equipment.

inventory control: The organization and management of commonly used parts, vendors and suppliers, and purchasing records in the preventive maintenance system.

inverter: An electronic device that changes DC voltage into AC voltage.

inverter duty motor: An electric motor specifically designed to work with AC motor drives.

J

jogging: The frequent starting and stopping of a motor for short periods of time.

joystick: An operator that selects one to eight different circuit conditions by shifting the joystick from the center position into one of the other positions.

jumbo mushroom button operator: A pushbutton that has a large curved operator extending beyond the guard.

K

kinetic energy: The energy of motion.

knee pad: A rubber, leather, or plastic pad strapped onto the knees for protection.

L

laser diode: A diode similar to an LED but with an optical cavity, which is required for lasing (emitting coherent light) production.

laser distance estimator: A device that uses sonar to take measurements.

lateral service: Electrical service in which service-entrance conductors are run underground from the utility service to a dwelling.

leakage current: Current that flows through insulation.

leather protectors: Gloves worn over rubber insulating gloves to prevent penetration of the rubber insulating gloves and provide added protection against electrical shock.

left-hand generator rule: The relationship between the current in a conductor and the magnetic field existing around the conductor.

legend plate: The part of a switch that includes the written description of the switch's operation.

level switch: A switch that detects the height of a liquid or solid (gases cannot be detected by level switches) inside a tank.

lever actuator: An actuator operated by means of a lever that is attached to the shaft of the limit switch.

light-activated SCR (LASCR): An SCR that is activated by light.

light emitting diode (LED): A semiconductor diode that produces light when current flows through it.

lightning arrester: A device that protects transformers and other electrical equipment from voltage surges caused by lightning.

light-operated photoelectric control: A photoelectric control that energizes the output switch when the target is missing (removed from the beam).

limit switch: A mechanical input that requires physical contact of the object with the switch actuator.

linear scale: A scale that is divided into equally spaced segments.

line (ladder) diagram: A diagram that shows the logic of an electrical circuit or system using standard symbols.

liquid crystal display (LCD): A display device consisting of a liquid crystal hermetically sealed between two glass plates.

load: Any device that converts electrical energy to motion, heat, light, or sound.

load current: The amount of current drawn by a load when energized.

locked rotor: A condition when a motor is loaded so heavily that the motor shaft cannot turn.

locked rotor current (LRC): The steady-state current taken from the power line with the rotor locked (stopped) and with the voltage applied.

locked rotor torque (LRT): The torque a motor produces when its rotor is stationary and full power is applied to the motor.

lockout: The process of removing the source of electrical power and installing a lock which prevents the power from being turned ON.

M

machine control relay: An EMR that includes several sets (usually two to eight) of NO and NC replaceable contacts (typically rated at 10 A to 20 A) that are activated by a coil.

main bonding jumper (MBJ): A connection at the service equipment that connects the equipment grounding conductor, the grounding electrode conductor, and the grounded conductor (neutral conductor).

magnet: A substance that produces a magnetic field and attracts iron.

magnetic level switch: A switch that contains a float, a moving magnet, and a magnetically operated reed switch to detect the level of a liquid.

magnetic motor starter: An electrically operated switch (contactor) that includes motor overload protection.

manual contactor: A control device that uses pushbuttons to energize or de-energize the load connected to it.

manual control circuit: Any circuit that requires a person to initiate an action for the circuit to operate.

manual starter: A contactor with an added overload protection device.

mechanical interlock: The arrangement of contacts in such a way that both sets of contacts cannot be closed at the same time.

mechanical level switches: Level switches that use a float that moves up and down with the level of the liquid and activates electrical contacts at a set height.

mechanical life: The number of times a relay's mechanical parts operate before malfunctioning.

mechanical switch: Any switch that uses silver contacts to start and stop the flow of current in a circuit.

megohmmeter: A device that detects insulation deterioration by measuring high resistance values under high test voltage conditions.

minimum holding current: The minimum amount of current required to keep a sensor operating.

molecular theory of magnetism: The theory that states that all substances are made up of an infinite number of molecular magnets that can be arranged in either an organized or disorganized manner.

momentary power interruption: A decrease to 0 V on one or more power lines lasting from .5 cycles up to 3 sec.

motor: A machine that converts electrical energy into mechanical energy by means of electromagnetic induction.

motor drive: An electronic unit designed to control the speed of a motor using solid-state components.

motor torque: The force that produces or tends to produce rotation in a motor.

multimeter: A meter that is capable of measuring two or more electrical quantities.

multiplexing: A method of transmitting more than one signal over a single transmission system.

N

NAND gate: A device that provides a low output when both inputs are high.

neutron: A particle contained in the nucleus of an atom that has no electrical charge.

nonlinear scale: A scale that is divided into unequally spaced segments.

non-permit confined space: A confined space that does not contain or, with respect to atmospheric hazards, have the potential to contain any hazards capable of causing death or serious physical harm.

nonretentive timer: A timer that does not maintain its current accumulated time value when its control input signal is interrupted or power to the timer is removed.

NOR gate: A device that provides a low output when either or both inputs are high.

negative resistance characteristic: The characteristic that current decreases with an increase in applied voltage.

N-type material: Material created by doping a region of a crystal with atoms of an element that has more electrons in its outer shell than the crystal.

nucleus: The heavy, dense center of an atom; has a positive electrical charge.

O

OFF-delay (delay on release) timer: A device that does not start its timing function until the power is removed from the timer.

off-line programming: The use of a personal computer to program a PLC that is not in the run mode.

Ohm's law: The relationship between voltage, current, and resistance in a circuit.

ON-delay (delay on operate) timer: A device that has a preset time period that must pass after the timer has been energized before any action occurs on the timer contacts.

one-line diagram: A diagram that uses single lines and graphic symbols to indicate the path and components of an electrical circuit.

one-shot (interval) timer: A device in which the contacts change position immediately and remain changed for the set period of time after the timer has received power.

op-amp: A very high gain, directly coupled amplifier that uses external feedback to control response characteristics.

open circuit: An electrical circuit that has an incomplete path that prevents current flow.

open circuit transition switching: A process in which power is momentarily disconnected when switching a circuit from one voltage supply (or level) to another.

operating current (residual or leakage current): The amount of current a sensor draws from the power lines to develop a field that can detect a target.

operator: The device that is pressed, pulled, or rotated by the individual operating the circuit.

optical level switches: Level switches that use a photoelectric beam to sense the liquid.

optocoupler: A device that consists of an IRED as the input stage and a silicon NPN phototransistor as the output stage.

oscilloscope: An instrument that displays an instantaneous voltage.

output (load-switching) circuit: The load switched by the SSR.

OR gate: A device with an output that is high when either or both inputs are high.

overcurrent protection device (OCPD): A disconnect switch with circuit breakers (CBs) or fuses added to provide overcurrent protection for the switched circuit.

overcycling: The process of turning a motor ON and OFF repeatedly.

overhead power lines: Electrical conductors designed to deliver electrical power; located in an above-ground aerial position.

overhead service: Electrical service in which service-entrance conductors are run from a utility pole through the air and to a dwelling.

overload: The application of excessive load to a motor.

P

pads: Small round conductors to which component leads are soldered.

panelboard: A wall-mounted distribution cabinet containing a group of overcurrent and short-circuit protection devices for lighting, appliance, or power distribution branch circuits.

parallel connection: A connection that has two or more components connected so there is more than one path for current flow.

part-winding starting: A method of starting a motor by first applying power to part of the motor coil windings for starting and then applying power to the remaining coil windings for normal running.

peak inverse voltage (PIV): The maximum reverse-bias voltage that a diode can withstand.

peak switching relay: An SSR that turns ON the load when the control voltage is present and the voltage at the load is at its peak.

pendulum actuation: A method of sensor activation that is a combination of the head-on and the slide-by actuation methods.

percent slip: The percentage reduction in speed below synchronous speed.

permanent magnet: A magnet that can retain its magnetism after the magnetizing force has been removed.

permeability: The ability of a material to carry magnetic lines of force.

permit-required confined space: A confined space that has specific health and safety hazards associated with it.

personal protective equipment (PPE): Clothing and/or equipment worn by a technician to reduce the possibility of injury in the work area.

phase sequence indicator: A device used to determine phase sequence and open phases.

phase shift: The state in which voltage and current in an AC circuit do not reach their maximum amplitude and zero level simultaneously.

phase unbalance: The unbalance that occurs when power lines are out of phase.

photoconductive cell (photocell): A device which conducts current when energized by light.

photoconductive diode (photodiode): A diode which is switched ON and OFF by light.

photoelectric sensor (photoelectric switch): A solid-state sensor that can detect the presence of an object without touching the object.

phototransistor: A device that combines the effect of a photodiode and the switching capability of a transistor.

phototriac: A triac that is activated by light.

photovoltaic cell (solar cell): A device that converts solar energy to electrical energy.

pick-up voltage: The minimum voltage that causes an armature to start to move.

pictorial drawing: A drawing that shows the length, height, and depth of an object in one view.

pigtail: An extended, flexible connection or a braided copper conductor.

PIN photodiode: A diode with a large intrinsic region sandwiched between P-type and N-type regions.

piston: A cylinder that is moved back and forth in a tight fitting chamber by the pressure applied in the chamber.

PLC scan: One execution cycle of a line diagram.

pliers: A hand tool with opposing jaws for gripping and/or cutting.

plugging: A method of motor braking in which the motor connections are reversed so that the motor develops a countertorque that acts as a braking force.

PN junction: The area on a semiconductor material between the P-type and N-type material.

point-to-point wiring: Wiring in which each component in a circuit is connected (wired) directly to the next component as specified on the wiring and line diagrams.

polarity: The positive (+) or negative (-) state of an object.

polarized scan: A method of scanning in which the receiver responds only to the depolarized reflected light from corner cube reflectors or polarized sensitive reflective tape.

pole: The number of completely isolated circuits that a relay can switch.

position: The number of locations within a valve in which a spool is placed to direct fluid through the valve.

potential energy: The stored energy a body has due to its position, chemical state, or condition.

power: The rate of doing work or using energy.

power cable puller: A device used to pull large cables and wires into place.

power circuit: The part of an electrical circuit that connects the loads to the main power lines.

power distribution: The process of delivering electrical power to where it is needed.

power factor: (PF): The ratio of true power used in an AC circuit to apparent power delivered to the circuit.

power formula: The relationship between power (P), voltage (E), and current (I) in an electrical circuit.

power source: A device that converts various forms of energy into electricity.

pressure: The force exerted over a surface divided by its area.

pressure sensor: A transducer that changes resistance with a corresponding change in pressure.

pressure switch: A switch that detects a set amount of force and activates electrical contacts when the set amount of force is reached.

preventive maintenance: Maintenance performed to keep machines, assembly lines, production operations, and plant operations running with little or no downtime.

primary division: A division on an analog scale with a listed value.

primary resistor starting: A reduced-voltage starting method that uses a resistor connected in each motor line (in one line in a 1ϕ) to produce a voltage drop.

primary winding: The coil of a transformer that draws power from the source.

printed circuit (PC) board: An insulating material such as fiberglass or phenolic with conducting paths laminated to one or both sides of the board.

processing: An activity or systematic sequence of operations that produces a specified result.

processor section: The section of a PLC that organizes all control activity by receiving inputs, performing logical decisions according to the program, and controlling the outputs.

programmable controller (PLC): A solid-state control device that is programmed and reprogrammed to automatically control an industrial process or machine.

programmable timer: A timer (timing function) included in electrical control devices such as PLCs.

programming diagram: A line diagram that better matches the PLC's language.

programming section: The section that allows input into the PLC through a keyboard.

protective clothing: Clothing that provides protection from contact with sharp objects, hot equipment, and harmful materials.

protective helmet: A hard hat that is used in the workplace to prevent injury from the impact of falling and flying objects, and from electrical shock.

proton: A particle contained in the nucleus of an atom that has a positive electrical charge.

proximity sensor (proximity switch): A solid-state sensor that detects the presence of an object by means of an electronic sensing field.

P-type material: Material with empty spaces (holes) in its crystal structure.

pull-up torque (PUT): The torque required to bring a load up to its rated speed.

pure DC power: Power obtained from a battery or DC generator:

pushbutton station: An enclosure that protects the pushbutton, contact block, and wiring from dust, dirt, water, and corrosive fluids.

push-roller actuator: An actuator operated by direct forward movement into the limit switch.

PVC cutter: A handheld tool designed to cut up to 2″ diameter PVC pipe, polyethylene pipe, and hose quickly and accurately without the use of a vise.

Q

qualified person: A person who is trained and has specific knowledge of the construction and operation of electrical equipment or a specific task, and is trained to recognize and avoid electrical hazards that might be present with respect to the equipment or specific task.

R

ramp stop: A stopping method in which the level of voltage applied to a motor is reduced as the motor decelerates.

RC circuit: A circuit in which resistance (R) and capacitance (C) are used to help filter the power in a circuit.

reactance: The opposition to the flow of alternating current in a circuit due to inductance.

reactive power (VAR): The power supplied to a reactive load (capacitor or coil).

receptacle tester: A device that is plugged into a standard receptacle to determine if the receptacle is properly wired and energized.

reciprocating saw: A multipurpose cutting tool in which the blade reciprocates (quickly moves back and forth) to create the cutting action.

rectification: The changing of AC to DC.

rectifier: A device that converts AC voltage to DC voltage by allowing the voltage and current to flow in only one direction.

recycle timer: A device in which the contacts cycle open and closed repeatedly once the timer has received power.

reed relay: A fast-operating, single-pole, single-throw switch with normally open (NO) contacts hermetically sealed in a glass envelope.

relay: A device that controls one electrical circuit by opening and closing contacts in another circuit.

resistance: **1.** The opposition to the flow of electrons. **2.** Any force that tends to hinder the movement of an object.

resistive circuit: A circuit that contains only resistance, such as heating elements and incandescent lamps.

response time: The number of pulses (objects) per second a controller can detect.

retentive timer: A timer that maintains its current accumulated time value when its control input signal is interrupted or power to the timer is removed.

retroreflective scan (retro scan): A method of scanning in which the transmitter and receiver are housed in the same enclosure and the transmitted light beam is reflected back to the receiver from a reflector.

reverse-bias voltage: The application of the opposite polarity to a diode.

rotor: The rotating part of an AC motor.

rotary actuation: An active method of sensor activation in which a multipolar ring magnet or collection of magnets is used to produce an alternating magnetic pattern.

rubber insulating gloves: Gloves made of latex rubber and are used to provide maximum insulation from electrical shock.

rubber insulation matting: A floor covering that provides technicians protection from electrical shock when working on live electrical circuits.

S

safety glasses: An eye protection device with special impact-resistant glass or plastic lenses, reinforced frames, and side shields.

safety label: A label that indicates areas or tasks that can pose a hazard to personnel and/or equipment.

saturation region: The maximum current that can flow in the transistor circuit.

scaffold: A temporary or movable platform and structure for workers to stand on when working at a height above the floor.

scan: The process of evaluating the input/output status, executing the program, and updating the system.

scanning: The process of using the light source and photosensor together to measure a change in light intensity when a target is present in, or absent from, the transmitted light beam.

scan time: The time it takes a PLC to make a sweep of the program.

schematic diagram: A diagram that shows the electrical connections and functions of a specific circuit arrangement with graphic symbols.

screwdriver: A hand tool with a tip designed to fit into a screw head for fastening operations.

seal-in voltage: The minimum control voltage required to cause an armature to seal against the pole faces of a magnet.

secondary division: A division on an analog scale that divides primary divisions in halves, thirds, fourths, fifths, etc.

secondary winding: The coil of a transformer that delivers the energy at the transformed or changed voltage to the load.

selector switch: A switch with an operator that is rotated (instead of pushed) to activate the electrical contacts.

self-excited shunt field: A shunt field connected to the same power supply as the armature.

semiconductor devices: Devices that have electrical conductivity between that of a conductor (high conductivity) and that of an insulator (low conductivity).

sensor-controlled timer: A timer controlled by an external sensor in which the timer supplies the power required to operate the sensor.

separately excited shunt field: A shunt field connected to a different power supply than the armature.

series connection: A connection that has two or more components connected so there is only one path for current flow.

series/parallel connection: A combination of series- and parallel- connected components.

series-wound generator: A generator that has its field winding connected in series with the armature and the external circuit (load).

service factor (SF): A number designation that represents the percentage of extra demand that can be placed on a motor for short intervals without damaging the motor.

shaded-pole motor: A 1ϕ AC motor that uses a shaded stator pole for starting.

shading coil: A single turn of conducting material (normally copper or aluminum) mounted on the face of a magnetic laminate assembly or armature.

short circuit: A circuit in which current takes a shortcut around the normal path of current flow.

shunt-wound generator: A generator that has its field windings connected in parallel (shunt) with the armature and the external circuit (load).

silicon-controlled rectifier (SCR): A solid-state rectifier with the ability to rapidly switch heavy currents.

single phasing: The operation of a motor that is designed to operate on three phases but is only operating on two phases because one phase is lost.

single-voltage motor: A motor that operates at only one voltage level.

slide-by actuation: An active method of sensor activation in which a magnet is moved across the face of a Hall effect sensor at a constant distance (gap).

slip rings: Metallic rings connected to the ends of the armature; used to connect the induced voltage to the brushes.

smart (intelligence) device: A device that includes an electronic circuit (chip) that provides communication and diagnostic capabilities to the device.

smoke switch: A switch that detects a set amount of smoke caused by smoldering or burning material and activates a set of electrical contacts.

snubber circuit: A circuit that suppresses noise and high voltage on power lines.

soft starter: A device that provides a gradual voltage increase (ramp up) during AC motor starting.

soft stop (S-curve): A stopping method in which the programmed deceleration time is doubled and the stop function is changed from a ramp slope to an S-curve slope.

solenoid: An electric output device that converts electrical energy into linear mechanical force.

solid-state power source: A semiconductor device that controls electrons, electric fields, and magnetic fields in a solid material and typically has no moving parts.

solid-state relay (SSR): A switching device that has no contacts and switches entirely by electronic means.

solid-state switch: A switch that uses a triac, SCR, current sink (NPN) transistor, or current source (PNP) transistor to perform the switching function.

solid-state timer: A timer whose time delay is provided by solid-state electronic devices enclosed within the timing device.

specular scan: A method of scanning in which the transmitter and receiver are placed at equal angles from a highly reflective surface.

split-phase motor: A 1ϕ motor that includes a running winding (main winding) and a starting winding (auxiliary winding).

stator: The stationary part of an AC motor.

status indicator: A light that shows the condition of the components in a system.

subdivision: A division on an analog scale that divides secondary divisions in halves, thirds, fourths, fifths, etc.

supply voltage timer: A timer that requires the control switch to be connected so that it controls power to the timer coil.

surge suppressor: An electrical device that provides protection from high-level transients by limiting the level of voltage allowed downstream from the surge suppressor.

sustained power interruption: A decrease to 0 V on all power lines for a period of more than 1 min.

switchboard: A piece of equipment into which a large block of electric power is delivered from a substation and broken down into smaller blocks for distribution throughout a building.

symbol: A graphic element that represents a quantity or unit.

symmetrical recycle timer: A timer that operates with equal ON and OFF timer periods.

synchronous clock timer: A timer that opens and closes a circuit depending on the position of the hands of the clock.

T

tagout: The process of placing a danger tag on the source of electrical power, which indicates that the equipment may not be operated until the danger tag is removed.

tap: A connection brought out of a winding at a point between its endpoints to allow changing the voltage or current ratio.

temperature switches: Control devices that react to heat intensity.

temporary magnet: A magnet that retains trace amounts of magnetism after the magnetizing force has been removed.

temporary power interruption: A decrease to 0 V on one or more power lines lasting for more than 3 sec up to 1 min.

thermal resistance (R_{TH}): The ability of a device to impede the flow of heat.

thermistor: A temperature-sensitive resistor whose resistance changes with a change in temperature.

thermocouple: A temperature sensor that consists of two dissimilar metals joined at the end where heat is to be measured (hot junction), and that produces a voltage output at the other end (cold junction) proportional to the measured temperature.

throw: The number of closed contact positions per pole.

tie-down troubleshooting method: A testing method in which one DMM probe is connected to either the L2 (neutral) or L1 (hot) side of a circuit and the other DMM probe is moved along a section of the circuit to be tested.

torque: The force that produces rotation.

totalizer: A counting device that keeps track of the total number of inputs and displays the counted value.

traces (foils): Conducting paths used to connect components on a PC board.

transducer: A device used to convert physical parameters, such as temperature, pressure, and weight, into electrical signals.

transformer: An electric device that uses electromagnetism to change voltage from one level to another or to isolate one voltage from another.

transient voltages: A temporary, unwanted voltage in an electrical circuit.

transistor: A three-terminal device that controls current through the device depending on the amount of voltage applied to the base.

transistor-controlled timer: A timer that is controlled by an external transistor from a separately powered electronic circuit.

transistor-transistor logic (TTL) ICs: A broad family of ICs that employ a two-transistor arrangement.

triac: A three-terminal semiconductor thyristor that is triggered into conduction in either direction by a small current to its gate.

trip class setting: The length of time it takes for an overload relay to trip and remove power from the motor.

troubleshooting: The systematic elimination of the various parts of a system, circuit, or process to locate a malfunctioning part.

troubleshooting report: A record of a specific problem that occurs in a particular piece of equipment.

true power (P_T): The actual power used in an electrical circuit.

U

ultrasonic sensor: A solid-state sensor that can detect the presence of an object by emitting and receiving high frequency sound waves.

unijunction transistor (UJT): A transistor consisting of N-type material with a region of P-type material doped within the N-type material.

uninterruptible power system (UPS): A power supply that provides constant on-line power when the primary power supply is interrupted.

up counter: A device used to count inputs and provide an output (contacts) after the preset count value is reached.

up/down counter: A device used to count inputs from two different inputs, one that adds a count and the other that subtracts a count.

V

valence electrons: Electrons in the outermost shell of an atom.

vane actuation: A passive method of sensor activation in which an iron vane shunts or redirects the magnetic field in the air gap away from the Hall effect sensor.

vapor: A gas that can be liquefied by compression without lowering the temperature.

variable torque/variable horsepower (VT/VH) load: A load that requires a varying torque and horsepower at different speeds.

varistor: A resistor whose resistance is inversely proportional to the voltage applied to it.

vise: A portable or stationary clamping device used to firmly hold work in place.

voltage: The amount of electrical pressure in a circuit.

voltage sag (voltage dip): A drop in voltage of more than 10% (but not to 0 V) below the normal rated line voltage lasting from .5 cycles up to 1 minute.

voltage surge: A higher-than-normal voltage that temporarily exists on one or more power lines.

voltage tester: A device that indicates approximate voltage level and type (AC or DC) by the movement and vibration of a pointer on a scale.

voltage unbalance: The unbalance that occurs when the voltages at different motor terminal are not equal.

volts per hertz (V/Hz): The relationship between voltage and frequency that exists in a motor.

W

warning signal word: A word used to indicate a potentially hazardous situation which, if not avoided, could result in death or serious injury.

watt (W): A unit of measure equal to the power produced by a current of 1 A across a potential difference of 1 V. A watt is $\frac{1}{746}$ of 1 HP.

way: A flow path through a valve.

wire stripper/crimper/cutter: A device used for the removal of insulation from small-diameter wire.

wiring diagram: A diagram that shows the connection of all components in a piece of equipment.

wobble-stick actuator: An actuator operated by means of any movement into the switch, except a direct pull.

work: Applying a force over a distance.

wraparound bar graph: A bar graph that displays a fraction of the full range on the graph at one time.

wrench: A hand tool with jaws at one or both ends that are designed to turn bolts, nuts, or pipes.

wye connection: A connection that has one end of each coil connected together and the other end of each coil left open for external connections.

Y

yawing: A side-to-side movement.

Z

zener diode: A silicon PN junction that differs from a rectifier diode in that it operates in the reverse breakdown region.

zero switching relay: An SSR that turns ON the load when the control voltage is applied and the voltage at the load crosses zero (or within a few volts of zero).

Index

C

D

E

M

N

O

P

Q

R

S

T

U

V

W

Y

Z

USING THE *ELECTRICAL MOTOR CONTROLS FOR INTEGRATED SYSTMS* CD-ROM

Before removing the CD-ROM from the protective sleeve, please note that the book cannot be returned for refund or credit if the CD-ROM sleeve seal is broken.

System Requirements

The *Electrical Motor Controls for Integrated Systems* CD-ROM is designed to work best on a computer meeting the following hardware/software requirements:

- Intel® Pentium® processor or equivalent
- Microsoft® Windows® 95, 98, 98 SE, Me, NT®, 2000, or XP® operating system
- 64 MB of free available system RAM (128 MB recommended)
- 90 MB of available disk space
- 800 × 600 16-bit (thousands of colors) color display or better
- Sound output capability and speakers
- CD-ROM drive

Opening Files

Insert the CD-ROM into the computer CD-ROM drive. Within a few seconds, the home screen will be displayed allowing access to all features of the CD-ROM. Information about the usage of the CD-ROM can be accessed by clicking on USING THIS CD-ROM. The Chapter Quick Quizzes™, Illustrated Glossary, Media Clips, and Reference Material can be accessed by clicking on the appropriate button on the home screen. Clicking on the American Tech web site button (www.go2atp.com) accesses information on related educational products. Unauthorized reproduction of the material on this CD-ROM is strictly prohibited.

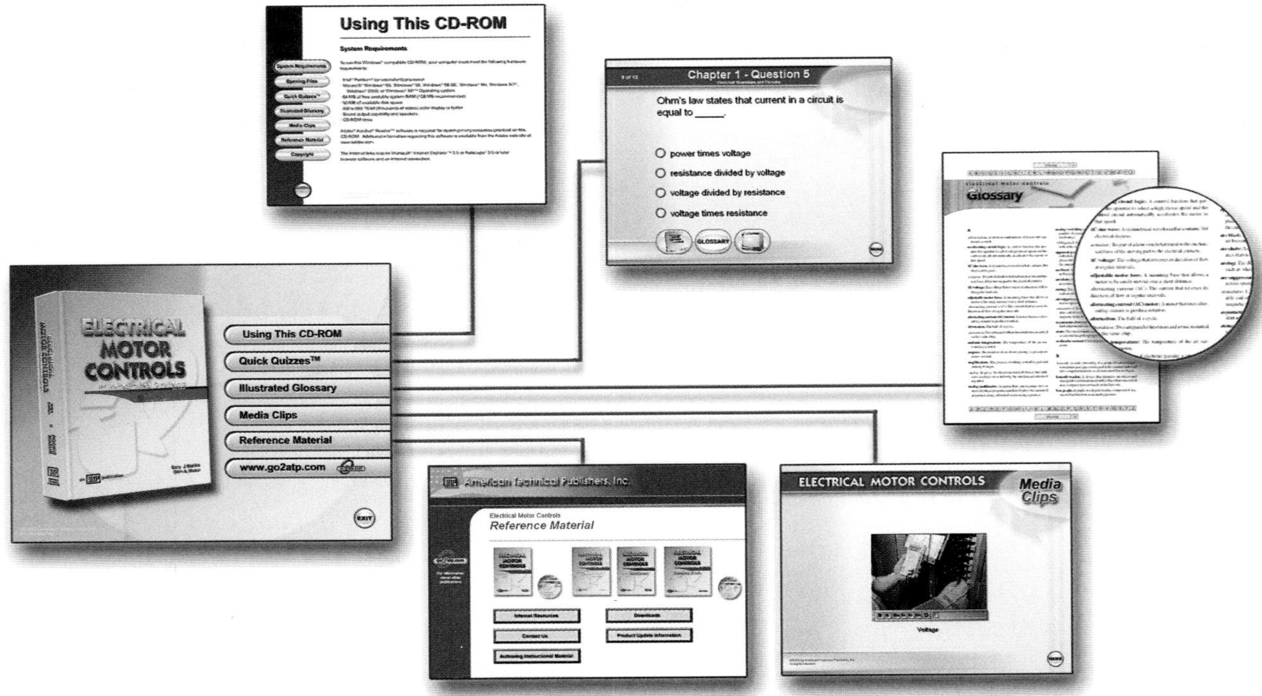